Exploring the Earth's Crust—
History and Results of Controlled-Source Seismology

By

Claus Prodehl
Geophysical Institute
University of Karlsruhe
Karlsruhe Institute of Technology
Hertzstr. 16
76187 Karlsruhe
Germany

and

Walter D. Mooney
U.S. Geological Survey
345 Middlefield Road, MS 977
Menlo Park, California 94025
USA

THE
GEOLOGICAL
SOCIETY
OF AMERICA®

Memoir 208

3300 Penrose Place, P.O. Box 9140 ▪ Boulder, Colorado 80301-9140, USA

2012

Published by The Geological Society of America, Inc.
3300 Penrose Place, P.O. Box 9140, Boulder, Colorado 80301-9140, USA
www.geosociety.org

Printed in U.S.A.

GSA Books Science Editor: Kent Condie and F. Edwin Harvey

Library of Congress Cataloging-in-Publication Data

Prodehl, C., 1936–
  Exploring the Earth's crust : history and results of controlled-source seismology / by Claus Prodehl, Walter D. Mooney.
       p. cm. — (Memoir ; 208)
   Includes bibliographical references.
   ISBN 978-0-8137-1208-6 (cloth)
     1. Seismic reflection method. 2. Seismic waves—Speed. 3. Seismology—Methodology. 4. Earth—Crust. 5. Earth—Mantle. I. Mooney, Walter D. II. Title.
   TN269.84.P76 2011
   551.1′3—dc22

                                                                2011007063

Cover: Topographic map showing data points and crustal thickness around the world as derived from global seismic studies, based on the Global Seismic Profiles Catalog. See Figure 10.8-04, Chapter 10.

15553422

mh

Contents

Appendix on Accompanying DVD

Acronyms

AB	Alberta Basement (LITHOPROBE)
ACCRETE	(U.S. experiment in the Canadian Cordillera, 1994)
ACEX	Arctic Coring Expedition (2004)
ACORP	Australian Continental Reflection Profiling Program
ACRUP	Antarctica Crustal Profile (1993–1994)
ADC	ADCOH—Appalachian Ultradeep Core Hole (1985)
ADK	Adirondack reflection survey (COCORP 1980–1981)
AFTAC	Air Force Technical Applications Center
AG	Abitibi—Grenville (LITHOPROBE)
AGSO	Australian Geological Survey Organization
ALP75	Alpine Longitudinal Profile (1975)
AMC reflector	Axial Magma Chamber reflector
AMOCO	(Oil company name, USA)
ANCORP 96	Andean Continental Research Project (1996)
ANU	Australian National University
ARKTIS VV/2	Polish expedition to the Antarktis (1999)
ASFA	Arbeitsgruppe für Seismische Feldmessungen und Auswertung (Working Group for Seismic Fieldwork and Interpretation)
ASPER	Airgun and Sonobuoy Precision Echo Recorder
ASTRA	Anisotropy experiment Voronesh Massif, Russia (1991)
AWI	Alfred-Wegener-Institut, Bremerhaven, Germany
BABEL	Baltic and Bothnian Echoes from the Lithosphere (1989)
BALTIC	Finnish BALTIC shield project (1982)
BASIN	North German BASIN reflection project (1996)
BASIX	Bay Area Seismic Imaging Experiment (1991)
BELCORP	Belgian Continental Reflection Program
BGR	Bundesanstalt fuer Geowissenschaften und Rohstoffe, Hannover, Germany
BIRPS	British Institutions Reflection Profiling Syndicate
BMR	Bureau of Mineral Resources, Canberra, Australia
BOHEMA	Bohemian Massif passive seismic tomography project (2001–2003)
BOLIVAR	Broadband Ocean Land Investigation of Venezuela and the Antilles arc Region (2004)
BPI	Bernhard Price Institute (University of Witwatersrand at Johannesburg, South Africa)
BRIDGE	British Mid-Ocean Ridge Initiative
BS	Bachelor of Science
BUMP	Bass Strait Upper Mantle Project (1966)
CALCRUST	California Consortium for Crustal Studies
CARDEX	Caldera and Rift Deep seismic Experiment (1981)
CCSS	Commission on Controlled Source Seismology
CDP	Common depth point
CD-ROM	Continental Dynamics–Rocky Mountains
CEL	CELEBRATION profile

CELEBRATION 2000   Central European Lithospheric Experiment Based on Refraction (2000)
CESAR   Canadian Expedition to Study the Alpha Ridge (1983)
CHP   Charleston reflection survey (COCORP, 1978, and USGS, 1981)
CINCA 95   Crustal Investigations offshore and onshore Nazca plate/Central Andes 1995
CJES   Canadian Journal of Earth Sciences
CMP   Common midpoint
CMRE   Continental Margin Refraction Experiment (1969)
CNIPSE   Linear broadband array (New Zealand 2001–2002)
CNR   Italian National Research Council
COCORP   Consortium for Continental Reflection Profiling
COCRUST   Consortium of Canadian University and Government Crustal Seismologists
CONDOR   Chilean Offshore Natural Disaster and Ocean Environmental Research
COOLE   Caledonian Onshore Offshore Lithospheric Experiment (1985 ff)
COP   Constant Offset Profile
COTCOR   Comparative Transections in Costa Rica project (1996)
CRC   Collaborative Research Center (German Research Association) = SFB
                       (SonderForschungsBereich)
CROP   Crosta Profonda (Italian continental reflection seismic program)
CRUMP   Carpentaria Region Upper Mantle Project (1966)
CSB   China Seismological Bureau
CSSP   Caledonian Suture Seismic Project (1982)
DACIA PLAN   Danube and Carpathian Integrated Action on Process in the Lithosphere and
                       Neotectonics (2001)
DEKORP   Deutsches Kontinentales Reflexionsseismisches Programm (German continental
                       reflection seismic program)
DEPAS   Deutscher Pool für Aktive Seismologie
DESERT   Dead Sea Rift Transect (2000)
DESIRE   Dead Sea Integrated Research project (2006)
DFG   Deutsche Forschungs-Gemeinschaft (German Research Association)
DIAS   Dublin Institute for Advanced Studies
DICE THROW   (Code name, Rio Grande rift, 1976)
DOBRE   Donbas Refraction and Reflection (1999, 2005–2006)
DOBS   Digital ocean-bottom seismometer
DRACULA   Deep Reflection Acquisition Constraining Unusual Lithospheric Activity (2004)
DRUM   Deep reflections from the upper mantle (BIRPS, 1984)
DSDP   Deep Sea Drilling Project
DSS   Deep seismic sounding
EAGE   European Association of Geoscientists & Engineers
EAGLE   Ethiopia Afar Geoscientific Lithospheric Experiment (2001)
ECCO   Estudio Cortical de la Cuenca Oriental (2001)
ECOGUAY   Estudios de la Estructura Cortical del Escudo de Guayana (1998)
ECOOE   East Coast Onshore-Offshore Experiment (1965)
ECORS   Etude de la Croûte Continentale et Océanique par Réflexion et Réfraction
                       Sismique
ECRIS   European Cenozoic Rift System
ECSOOT   Eastern Canadian Shield Onshore-Offshore Transect (LITHOPROBE)
EDA   (Company name, Canada)
EDGE   (Pacific Gas & Electric experiment off Morro Bay, California, 1986)
EDGE   (Mid-Atlantic seismic experiment off Virginia, 1990)
EDL   Earth Data Loggers
EDZOE   West-Canadian Upper Mantle Seismic Program (1969)
EEC   East European Craton
EGT   European GeoTraverse
ENEL   (Company name, Italy)

ENI-AGIP	(Company name, Italy)
EPLAC	(Brasilian Continental Margin Project, 1992)
ESC	European Seismological Commission
ESCI	Estructura Sismica de la Corteza Iberica (1990s)
ESCIN	Estructura Sismica de la Corteza Iberica Norte (1990s)
ESF	European Science Foundation
ESP	Expanding spread profile
ESTRID	Explosion Seismic Transects around a Rift in Denmark (2004)
ESRU	EUROPROBE Seismic Reflection profiling in the Urals (1993–2003)
ETH	Eidgenössische Technische Hochschule
EUGEMI	European Geotraverse middle part
EUGENO-S	European GeoTraverse North—Southern extension
EUROBRIDGE	(East European craton seismic traverse, 1995, 1996, 1997)
EUROPROBE	European research program, 1990 ff
EXCO	Exchange between Crust and Ocean (1995)
FAMOUS	Franco-American Mid-Ocean Undersea Study
FAST	Geo-Prakla profile across Faeroe-Shetland Trough (1990s)
FENNOLORA	Fennoscandian Long-Range project
FIRE	Faeroe-Iceland Ridge Experiment (1994)
FIRE	Finland deep seismic reflection survey (2001–2003)
FKPE	Forschungs-Kollegium für die Physik des Erdkörpers (Research Consortium for the Physics of the Earth)
FLARE	Geo-Prakla profile across Faeroe-Shetland Trough (1990s)
FSH Hispanoil	(Company name, Spain)
FRUMP	Fremantle Region Upper Mantle Project (1967)
GAA + GAB + GAC	Southern Appalachians reflection (COCORP, 1978, 1979; USGS, 1981)
GEDEPTH	German Deep Profiling of Tibet and the Himalaya (1994)
GEOS	(Company name, USA)
GFT	Georgia-Florida Transect (COCORP, 1983–1985)
GFZ	GeoForschungsZentrum (GeoScienceCenter), Potsdam, Germany
GLIMPCE experiment	Great Lakes International Multidisciplinary Program on Crustal Evolution (1986)
GMM	Gulf of Maine marine reflection (USGS, 1985)
GMP	Grandfather Mountain Profile (USGS, 1979)
G-PRIME	Galapagos marine survey (2000)
GRANIT	Russian Platform seismic transect (1991, Ukraine to Siberia)
GRANU-95	Saxonian Granulite Mountain metamorphic core complex project (1995)
GRAP-88	Ontario–New York refraction experiment (1988)
GRC	Geophysical Research Corporation (U.S. company founded in 1925)
GRID	GRID lines north of Scotland (BIRPS, 1986–1987)
GSC	Geological Survey of Canada
GSC	Global Seismic profiles catalog (computer database/master catalog of 1-D global crustal and mantle structures profiles)
GSI	Geophysical Service Incorporated (U.S. company founded in 1930)
GSN	Global Seismic Network (USA)
GTT	Goettingen seismic station (Germany)
GURALP	Trade name, digital seismic instrumentation
GVR	Great Valley wide angle reflection survey (Princeton, 1982–1984)
HADES	Hatton Deep Seismic project (2002)
HIMPROBE	Himalaya Project from India to Tibet (1990s)
HRC	Hardeman Basin (COCORP, 1975)
HVO	Hawaiian Volcano Observatory
Hz	Cycles per second
IAM	Iberian Atlantic Margins European project (1993)
IASPEI	International Association for Seismology and Physics of the Earth

IBERSEIS	Iberian Massif seismic reflection transect (2001)
ICDP	International Continental Scientific Drilling Program
ICSSP	Irish Caledonian Suture Seismic Project (1982)
IFREMER	Institut Français de Recherche et d'Exploitation de la Mer
IGES	Imperial Geophysical Experimental Survey (Australia, firm name)
IGS	Institute of Geological Sciences, U.K.
IGY	International Geophysical Year (1957)
ILIHA	Iberian Lithosphere Heterogeneity and Anisotropy (1989)
ILP	International Lithosphere Program
INAG	Institut National d'Astronomie et de Géophysique, Paris
INDEPTH	International Deep Profiling of Tibet and the Himalaya (1992 ff)
IODP	Integrated Ocean Drilling Project
IPOD	International Phase of Oceanic Drilling
IRIS	Incorporated Research Institutions for Seismology
ISLE	Irish Seismological Lithospheric Experiment (2002–2003)
ISSA	Integrated Seismological experiment in the Southern Andes (2000)
ISZ	Iapetus Suture Zone
IUGG.	International Union for Geodesy and Geophysics
JAMSTEC	Japan Marine Science and Technology Center
JEDS	Japanese Deep Sea Expedition
JOIDES	Joint Oceanographic Institutions for Deep Earth Sampling
JRC	James River Corridor, Virginia (USGS, 1980–1981)
JTEX	Jemez Tomography Experiment (1993 and 1995)
KAN	Kansas reflection survey (COCORP, 1981)
KIMBA	Kimberley Array, NW Australia (1993–1996)
KMFZ	Killarney-Mallow Fault Zone, Ireland
KRISP	Kenya Rift International Seismic Project (1985, 1990, 1994)
KTB	Kontinentale Tief-Bohrung (Continental Deep Drillhole)
LARSE	Los Angeles Region Seismic Experiment (1994, 1999)
LASA	Large Aperture Seismic Array (1966, 1968)
LASE	Large Aperture Seismic Experiment (1981)
LE	LITHOPROBE East (1988 and 1991)
LEGS	Leinster Granite Seismics (1999)
LIP	Long Island Platform marine survey (USGS, 1979)
LISA	Ligurian Sardinian margins project (1995)
LISPB	Lithospheric Seismic Profile through Britain (1974)
LITHOPROBE	Probing the Earth's Lithosphere (Canadian Research Program)
LOBS	Land-ocean-bottom seismometers
LUST	Lewisian Units Seismic Traverse (NW Scotland, 1976)
MAGMA	East Pacific Rise seismic refraction (1982)
MAMBA	Geophysical Measurements across the Continental Margin of Namibia experiment (1995)
MAMUT	Makran-Murray Traverse (1997)
MARBE	Mid-Atlantic Ridge Bull's Eye (1996)
MARS-66	Magnetband-Apparatur (= Magnetic Tape Recorder) for Refraction-Seismics)
MAVIS	Midland Valley Investigation by Seismology (central Scotland, 1984)
MBA	Michigan reflection survey (COCORP, 1978)
MCS	Multichannel seismic
MELT	Mantle Electromagnetic and Tomography experiment (1996)
MEMSAC	Mega Earth Mobile Seismic Array Consortium
MIN	Minnesota reflection survey (COCORP, 1979)
MIS	Mississippi embayment reflection survey (USGS, 1978–1979)
MOBIL	Measurements over Basins to Image Lithosphere (BIRPS, 1987)
MOIST lines	Moine and Outer Isles Seismic Traverse (BIRPS, 1981)

MONA LISA	Marine and Onshore North Sea Acquisition for Lithospheric Seismic Analysis (1993)
MOOSE	Marlborough Onshore Offshore Seismic Experiment (1994, New Zealand)
MRP	Maine refraction survey (USGS, 1985)
MVE-90	Muenchberg-Vogtland-Erzgebirge seismic reflection experiment (1990, East Germany)
NARS	Network of Autonomously Recording Stations (European GeoTraverse)
NASP	North Atlantic Seismic Project (1972)
NBT	Newark Basin rift Tectonics (USGS, 1984)
NCSN	Northern California Seismic Network
NEC	New Hampshire reflection survey (COCORP, 1980–1981)
NEC	North East Coast line (BIRPS, 1985)
NERC	National Environmental Research Council (Great Britain)
NFP 20	Swiss National-Fond Project (1986 and 1988)
NIGHT	North Island Geophysical Transect (New Zealand, 2001)
NORSAR	Norwegian Seismic Array
NPE	Department of Energy's Nuclear Non-Proliferation Experiment (1993)
NSDP	North Sea Deep Profile (BIRPS, 1984)
NSF	U.S. National Science Foundation
NTS	Nevada Test Site
NVI	Near-vertical incidence seismic reflections
OBH	Ocean-bottom hydrophone
OBS	Ocean-bottom seismometer
OBVSA	Ocean-bottom vertical seismometer array
ODP	Ocean Drilling Program
OFR	Open-File Report
OGS	Osservatorio Geofisico Sperimentale
OMEGA	(Trade name)
OSE	Oblique Seismic Experiment (1977, western Pacific)
OU	University of Oklahoma
OUC + OUP	Ouachita seismic transect (COCORP, 1981; PASSCAL, 1986)
PACE	Pacific to Arizona Crustal Experiment (1985, 1987, 1989)
PAGANINI	Panama basin and Galapagos plume—New Investigations of Intraplate Magmatism (1999)
PANCARDI	Pannonian basin–Carpathians–Dinarides
PASSCAL	Program for Array Seismic Studies of the Continental Lithosphere
PCM	Pulse Code Modulation
PDAS	Trade name of digital instrument
PETROBAS	(Company name, Brasilia)
PISCO 94	Proyecto de Investigaciones Sismologicas de la Cordilllera Occidental (1994)
PLP	P-wave reflected from lower lithosphere
PMP, PmP, or P^M	P-wave reflected from Moho
P_n	P-wave refracted in upper mantle
PNE	Peaceful Nuclear Explosions
POLONAISE	Polish Lithosphere Onsets—An International Seismic Experiment (1997)
PRAKLA	(Company name, Germany)
PRECORP	Pre-Continental Research Project (Andes, 1995)
PRS-1	Portable Refraction Seismograph PRS-1
PSINE	Parkfield Seismic Imaging Ninety Eight (1998)
PUSSes	Pull-up Shallow-water Seismometers
PVF	Potrillo Volcanic Field (2003 experiment)
QMT	Quebec–Maine reflection Transect (USGS, 1984–1985)
RAMESSES	Reykjanes Axial Melt Experiment: Structural Synthesis from Electromagnetics and Seismics (1993)

RAPIDS	Rockall and Porcupine Irish Deep Seismic project (1988, 1990, 1999, 2002)
RCEG	Research Center of Exploration Geophysics, China
RefTek	(Company name, USA)
RGVE	Romanian Group for Vrancea strong Earthquakes
ROSE	Rivera Ocean Seismic Experiment (1979)
RRISP	Reykjanes Ridge Seismic Project (1977)
SAF	San Andreas Fault
SAFOD	San Andreas Fault Observatory at Depth
SALIERI	South American Lithosphere transects across volcanic Ridges (2001)
SALT	Zechstein SALT profile (BIRPS, 1983)
SALT95	Southern Alberta Lithospheric Transect (Canada, 1995)
SAPSE	Southern Alps Passive Seismic Experiment (1995–1996)
SAREX	South Alberta seismic Refraction Experiment (1995)
SCAS	Southwest Center for Advanced Studies, Dallas, Texas
SCoRE	Southern Cordillera Refraction Experiment (1989–1990)
SCORD9	Southern Cordillera refraction experiment
SCR	Seismic Cassette Recorders
SCREECH	Study of Continental Rifting and Extension on the Eastern Canadian Shelf (2000)
SEISGRECE	French-Greek seismic project (Ionian Sea, 1997)
SEISMARMARA	Seismic marine project in MARMARA Sea (2001)
SEISMOS	(Company name, Germany)
SERIS	Seismic Experiment Ross Ice Shelf (1990–1991)
SFB	SonderForschungsBereich (CRC—Collaborative Research Center of German Research Association)
SGR	Seismic Group Recorders
SHET	Shetland survey (BIRPS, 1984)
SHIPS	Seismic Hazards Investigation of Puget Sound (1998, 1999, 2000, 2002)
SIE	(Company name, USA)
SIGHT	South Island Geophysical Transect (1995–1996)
SIGMA	Seismic Investigation of the Greenland Margin (1996)
SISSE	South Irish Sea Seismic Experiment (1977)
SKIPPY	12-station mobile array in Australia (1993–1996)
SLAVE	Synthetic Large Aperture Velocity Experiment (BIRPS, 1987)
SNORCLE	Slave Northern Cordillera Lithospheric Evolution (LITHOPROBE, 1997)
SNST	Southern North Sea Tie (BIRPS, 1983)
SOHMA	Servicio de Oceanografie, Hidrografia y Meteorologia da la Armada, Uruguay
SPOC	Subduction Processes off Chile (2001)
SRGE	Special Regional Geophysical Expedition, Russia
SRI	Stanford Research Institute
SSCD	Southern Sierra Continental Dynamics (1993)
STREAMERS	French marine field project between Italy and Greece (1993)
SUDETES	International seismic project centered on the Sudetes Mountains, Czech Republic-Hungary-Poland (2003)
SVEKA	Finnish field project (1981)
SVEKALAPKO	Svecofennian-Karelian-Lapland-Kola Transect project (1998–1999)
SWAT profiles	Southwest Approaches Traverse (BIRPS and ECORS, 1983)
SWESE	SW England Seismic Experiment (1979)
TACT	Trans-Alaska Crustal Transect project (1984–1990)
TAICRUST	marine seismic project off Taiwan (1995)
TASS	Trans-Australia Seismic Survey (1972)
TASYO	Spanish marine seismic project Gulf of Cadiz (2001)
TCXO	Temperature-compensated oscillator
TECALB	Spanish marine seismic project Alboran Sea (2004)
TESZ	Trans-European Suture Zone

TEXAN	1-component digital instrument
THORE	Trans-Hudson Orogen seismic-refraction experiment (1993)
THOT	Trans-Hudson Orogen Transect (LITHOPROBE, 1991–1994)
TICOSECT	Trans Isthmus Costa Rica Scientific Exploration of a Crustal Transect project (1995)
TIPTEQ	The incoming plate to mega-thrust earthquake processes (2004–2005)
TOR	Teleseismic Tomography Tornquist project (1996–1997)
TRANSALP	Trans-Alpine seismic traverse through Eastern Alps (1998–2001)
TTZ	Tornquist-Teysseire Zone
TXCO	Temperature-compensated internal oscillator
TWT	Two-way traveltime
TVZ	Taupo Volcanic Zone, New Zealand
UNST	Geo-Prakla profile across Faeroe-Shetland Trough (1990s)
URSEIS	Urals Seismic project (1995)
USGS	U.S. Geological Survey
UTEP	University of Texas at El Paso
UWARS	Urals Wide Angle Reflection Seismic experiment (1990s)
VALSIS	Valencia trough marine seismic profiles (early 1990s)
VARNET	Variscan Front Network (1996)
VAULT	Vibroseis augmented listen time (Canada, 1995)
VCO	Voltage-controlled oscillator
VHF	Very high frequency
VOLTAIRE	Spanish marine seismic project Gulf of Cadiz (2002)
VRANCEA	Seismic profiles through seismogenic Vrancea area, Romania (1999–2001)
WAM line	Western Approaches Margin (BIRPS and ECORS, 1985)
WHOI	Woods Hole Oceanographic Institution, USA
WHUMP	Wellington-Hikurangi Upper Mantle Profile (1979)
WINCH lines	Western Isles and North Channel traverse (BIRPS, 1982)
WIRELINES	West of Ireland LINES (BIRPS, 1987)
WISE	Western Isles Seismic Experiment (western Scotland, 1979 and 1981)
WISPA	Weardale Integrated S- and P-wave Analysis (BIRPS, 1988)
WMT	Wichita Mountains reflection survey (COCORP, 1979–1980)
WRAMP	Warramunga Seismic Array Mantle Projects (1966)
WS	Western Superior (LITHOPROBE, 1996)
WWII	World War II
WWV	(National Bureau of Standards Radio Station's time signal, USA)

The Geological Society of America
Memoir 208
2012

Exploring the Earth's Crust—
History and Results of Controlled-Source Seismology

Claus Prodehl*
Geophysical Institute, University of Karlsruhe, Karlsruhe Institute of Technology, Hertzstr. 16, 76187 Karlsruhe, Germany
Walter D. Mooney*
U.S. Geological Survey, 345 Middlefield Road, MS 977, Menlo Park, California 94025, USA

ABSTRACT

This volume presents a comprehensive, worldwide history of seismological studies of the Earth's crust using controlled sources from 1850 to 2005. Essentially all major seismic projects on land and the most important oceanic projects are presented. The time period of 1850 to 1939 is presented as a general synthesis, and from 1940 onward the history and results are subdivided into a separate chapter for each decade, with the material ordered by geographical region. Each chapter highlights the major advances achieved during that decade in terms of data acquisition, processing technology, and interpretation methods. For all major seismic projects, we provide specific details regarding the field observations, interpreted crustal cross section, and key references. We conclude with global and continental scale maps of all field measurements and interpreted Moho contours. An accompanying DVD contains important out-of-print publications and an extensive collection of controlled-source data, location maps, and crustal cross sections.

*claus.prodehl@gmx.net; mooney@usgs.gov

Prodehl, C., and Mooney, W.D., 2012, Exploring the Earth's Crust—History and Results of Controlled-Source Seismology: Geological Society of America Memoir 208, 764 p., doi:10.1130/2012.2208. For permission to copy, contact editing@geosociety.org. © 2012 The Geological Society of America. All rights reserved.

∼ CHAPTER 1 ∽

Introduction and Acknowledgments

This book deals with the exploration of the Earth's crust and uppermost mantle using the method of controlled-source seismology. It presents the historical development of controlled-source seismology from its very beginning at around 1850 until 2005. We summarize the main scientific and technical achievements that have led to an increase in our knowledge of the structure of the Earth's crust and uppermost mantle, as obtained by seismic controlled-source methods.

It is the intention of the book to present a comprehensive history of controlled-source seismic studies of the Earth. As far as they were accessible in literature, we have compiled the results from all major field projects, on land and at sea, that have used man-made seismic energy sources with the aim to explore the Earth's crust, the crust-mantle boundary, and details of the uppermost mantle to depths of ~100 km.

We have not reinterpreted the seismic data to obtain new scientific results. Rather, we summarize published interpretations and show how controlled-source seismology developed gradually over the years, in particular how the amount and quality of data grew hand-in-hand with the development of better technology, as well as enhanced seismic theory and computational methods.

We have subdivided the history and results of controlled-source seismology into decades. For each decade we discuss the outstanding achievements of that time period, including which major field experiments were carried out and their main scientific results. For each decade from 1940 onwards, the reader will find short summaries of all the continental and marine seismic reflection and refraction campaigns that we could access. Within each decade, we have followed consistent geographic order.

Marine projects in continental shelf areas are discussed in the "Oceans" chapters, as long as they are pure offshore marine surveys. As soon as a major onshore component is involved, we have placed the project under the corresponding continent and country. Also, all British BIRPS and French ECORS marine seismic reflection surveys are discussed as continental projects, as well as all projects located in the North and Baltic Seas, in the Mediterranean, and in the Red Sea.

This book has been prepared with several purposes in mind. First, for each major project, we present location information, some technical details of the data acquisition, and then the scientific results, in many cases as velocity-depth cross sections and/or contour maps of crustal thickness. In seismic reflection surveys, authors tend to give depths in s TWT (two-way traveltime). We have not converted such depths into km, if the authors did not do it themselves. We provide a list of basic references for the reader who seeks further information. In addition, for many major

projects we have compiled copies of the original seismic data in the form of record sections and present these in the Appendix on a DVD. We do not, however, attempt to summarize all details of the seismic interpretations, nor the implications in terms of geodynamic or tectonic interpretations. These implications can be found in the original papers.

A second purpose of this book is to provide some background information regarding scientists whose efforts drove major projects to be realized, or on experimental details, many of which have changed over the decades. The intention is to document the special efforts needed to carry out the major projects reported here, and to highlight the very individual and personal input of scientists, students, technicians and others. In particular, we mention many of those people who have promoted controlled-source seismic experiments in theory and practice. Often these personalities contributed greatly to the success of the most logistically challenging and adventurous projects, e.g., those located in such places as arid deserts, in high mountains, at sea, or on ice. Our observation is that, in each country with major projects in controlled-source seismology, there have been strong personalities acting as driving forces behind the scientific vitality.

A third purpose is to present a project that is as complete as possible. We have tried to mention nearly all major projects, together with experimental details and the most important results. With a composite work such as the present one, it is, however, impossible to ensure a precise balance between the various projects. Some individual projects have been described in greater detail, whereby we often use the original phrasing of the corresponding authors, whereas in other cases the published sources have not been detailed enough to describe a major project with full details.

A fourth purpose is to present a summary of the whole earth that is as complete as possible. But, according to our own professional experience, we are most familiar with results from North America and central Europe, but have less familiarity with the former USSR and, in general, marine projects. Considering the large quantity of controlled-source seismic experiments carried out around the world since World War II and the enormous number of relevant publications being widely spread throughout the international literature, we are aware that we may have missed the one or other larger project and we apologize for such omissions.

A fifth purpose is to highlight the most important developments within each decade, be it interpretational procedures, experimental innovations, or landmark publications that summarized the state-of-the art at that point in time. Any use of trade product or firm names is for descriptive purpose only and does not imply endorsement by the German or U.S. governments.

We have prepared the text such that the individual field projects are described in the decade when the experiment was actually carried out, and not when results were published. While in most cases experiments and subsequent interpretations described in the individual chapters were accomplished in the same decade, in some cases the corresponding publications followed many years later. Examples are the special publications of Giese et al. (1976) or Prodehl (1979), published in the mid-to-late 1970s, but describing work essentially accomplished in the 1960s, or the PNE (Peaceful Nuclear Explosions) experiments in Russia, carried out in the 1970s and early 1980s but published only after 1990.

Some large and/or colored figures reprinted in this historic overview of explosion seismology may be difficult to read because we have avoided using foldouts. To enable the reader to view the figures in full, we have reproduced all figures of the printed version as jpg files in Appendix A1-1. In the figure captions, we have furthermore added the corresponding book or journal details where the original figure was published.

The data collected were usually compiled as record sections aligned either in linear profiles or fans. In our text, the description of the individual projects will not always contain data examples. But, as far as it was available to us, a significant amount of seismic data can be viewed on the accompanying DVD in Appendices A2–A10, arranged according to the relevant decade. The DVD contains collections of record sections in a reduced and compressed scale for various projects, with or without correlations, that were easily available from personal collections and open-file reports or from publications. For several projects of the U.S. Geological Survey in the 1980s and early 1990s, data reports were available in PDF that we have included in this data collection part. Most data presented in the Appendix on the DVD come from projects in which one of the authors or his institution was actively involved. It would be beyond the scope of this book, however, to reproduce complete sets of digital data for any project which people might be able to reprocess at later times. In the Appendix, we have also included PDF files of a few publications of historical interest that cannot easily be found in libraries.

This compilation would not have been possible without the input of many colleagues with whom the authors have cooperated in numerous field projects, meetings, conferences and personal discussions. The authors are also indebted to the numerous colleagues who helped enlarge the personal databases for further use without formalities by exchanging the data from the many joint national and international cooperative research projects in which they had been involved.

In particular we want to acknowledge Jim Mechie (Geo-ForschungsZentrum Potsdam, Germany), who has critically read the entire manuscript, and Jörg Ansorge (ETH Zürich, Switzerland) and Peter Giese (FU Berlin) who have read first versions of the early history of explosion seismology and supplied additional material. Strong support came from Karl Fuchs and Friedemann Wenzel (both University of Karlsruhe), Randy Keller (University of Oklahoma), and Rolf Meissner (University of Kiel). Joachim Ritter (University of Karlsruhe) provided figures and details from his unpublished report on the early history of the University of Göttingen. R. Giese (GeoForschungsZentrum Potsdam, Germany), Charlotte Krawczyk (University of Hannover), Ewald Lüschen (Leibniz-Institut für Angewandte Geophysik, Hannover, Germany), and Peter Wigger (Free University Berlin) helped with digital figures and manuscripts and/or reprints of projects of their institutions in which they were personally involved. Edward Perchuc (Academy of Sciences, Warsaw, Poland) had invited one of the authors for a visit at Warsaw to discuss the Polish history of explosion seismology in detail. On this occasion, Marek Grad (University of Warsaw) and Alexander Guterch (Academy of Sciences, Warsaw, Poland) supported the authors' project by providing numerous materials in digital form and reprints. Victor Raileanu (National Institute for Earth Physics, Bucharest) provided manifold informations and literature on crustal research in Romania, Alfred Hirn (University of Paris) provided reprints and information on French projects, Peter Maguire (Leicester University, UK) added information to the early British exploration history, and Clive Collins and Douglas Finlayson (both at the then Bureau of Mineral Resources—now Geoscience Australia—Canberra, Australia) critically edited the early history of explosion seismology in Australia; in particular, Douglas Finlayson provided basic material to be included in the Appendix. Tom Brocher and Janice Murphy (both of the U.S. Geological Survey) had prepared a CD containing all major Open-File Reports of the U.S. Geological Survey in PDF which dealt with seismic experiments from the late 1970s to the early 1990s and which could be included in the Appendix of this explosion-seismic history. Reprints or PDF files were furthermore provided by Brian O'Reilly for Ireland, by Josep Gallart and Ramon Carbonell (both CSIC, Barcelona) for the Iberian Peninsula, by Wilfried Jokat for Greenland, and by Gary Fuis and David P. Hill for the United States. Adrian Egger (formerly at ETH Zurich) provided his Ph.D. thesis to be included in the Appendix, and the support of Alan Green and Edi Kissling (both of ETH Zurich) is greatly acknowledged.

The personal and technical support of Shane Detweiler (U.S. Geological Survey) is particularly acknowledged. Scientists and students working part-time at the U.S. Geological Survey have helped in improving and scanning figures. In particular the help of Nihal Okaya for the final version of the majority of figures is greatly acknowledged. Jillian McLaughin, Roxanne Renedo, Alex Ferguson, Max Gardner, Kealani Kitaura, and Jennifer Lee helped with the production of figures and revising tables. In general, we acknowledge the technical support provided at any time by the Geophysical Institute of the University of Karlsruhe (Germany) and by the Earthquake Science Center of the U.S. Geological Survey at Menlo Park, California.

For review, the manuscript was divided so that some reviews dealt with general aspects while others specialized in particular regions. The constructive criticism and manifold suggestions were extremely valuable and of great importance in improving the manuscript. All reviewers added valuable information and additional references. We want to thank in particular Ron Clowes

(Vancouver, B.C., Canada), Fred J. Davey (Lower Hutt, New Zealand), Doug M. Finlayson (Canberra, Australia), Robert D. Hatcher (Knoxville, Tennessee, USA), David P. Hill (Menlo Park, California, USA), Takaya Iwasaki (Tokyo, Japan), G. Randy Keller (Norman, Oklahoma, USA), Peter K. H. Maguire (Leicester, UK), Jim Mechie (Potsdam, Germany), Rolf Meissner (Kiel, Germany), Edward Perchuc (Warsaw, Poland), David W. Scholl (Menlo Park, California, USA), and Ralph Stephen (Woods Hole Oceanographic Institution, USA).

Last but not least, we want to thank the GSA Science Editor Dr. Pat Bickford and his assistant Joanne Ranz at Syracuse University, New York, USA, for their continuing support from submission of our manuscript through the review process towards final acceptance. Also we want to express our sincere thanks to the GSA Publications staff for their efforts to complete the editing and publication process.

Most of the data from western Europe (Scandinavia, British Isles, France, Germany, Alpine countries, and Mediterranean) and from the United States shown as record sections on jpg-files in the Appendix are from the personal data archives of either one of the authors. In most cases they had been collected prior to publication of the corresponding projects. Due to the complexity and costs of large-scale seismic projects, a major number of research institutions always cooperated, both on national and international basis. However, it was impossible to include in the authors' lists of those publications the members of all institutions which had directly or indirectly been involved in the realization of the fieldworks. Therefore, in practice, mostly only a few actively interpreting scientists appeared as authors. In Germany, for example, the organization of manpower and instrumentation (widely distributed in small numbers amongst all seismically active institutions) for seismic-refraction experiments was enabled by ASFA (working group for seismic field measurements and interpretation), in which all university and state institutions with seismic instrumentation cooperated and which was headed by one of the authors from 1970 to 2001. Similar organizations existed in Great Britain, Italy, and Spain. It was by mutual agreement that scientists of the participating institutions were permitted to reuse freely the data for reinterpretations and republications without further formalities, after the first report on the corresponding project had been published.

The history and results of controlled-source seismology would not be complete without the inclusion of many of the original figures. Numerous publishing companies and institutions have provided permission for their figures to be reprinted here. We greatly acknowledge the friendly cooperation of all. These are in particular:

American Association of Petroleum Geologists (*AAPG Bulletin*)
American Association for the Advancement of Science, Washington, D.C., USA (*Science*)
American Geophysical Union (*Eos, Journal of Geophysical Research, Geophysical Research Letters*, American Geophysical Union Monographs)

Annual Reviews, Palo Alto, California, USA (*Annual Reviews*)
Australian Geological Survey Organisation (AGSO) (*BMR Journal of Australian Geology and Geophysics*)
Australian National University (ANU, Research School of Earth Sciences)
Bergakademie Freiberg, Germany (*Freiberger Forschungshefte*)
Birkhäuser Verlag AG, Basel, Switzerland (*PAGEOPH*)
Cambridge University Press, New York, USA (books)
Canadian Society of Petroleum Geologists (*Bulletin of Canadian Petroleum Geology*)
Carnegie Institution for Science, Washington, D.C. (a.k.a. Carnegie Institution of Washington)
CROP Atlas, Italy, Prof. D. Scrocca
Czech Academy of Sciences, Prague, Czech Republic (*Studia in Geophysica et Geodaetica*)
Deutsche Geophysikalische Gesellschaft, Potsdam, Germany (*DGG-Mitteilungen*)
Earthquake Research Institute, Tokyo, Japan (*Bulletin of the Earthquake Research Institute*, University of Tokyo)
Edition Technip (Paris, France)
Elsevier, Amsterdam, Netherlands (books, *Earth and Planetary Science Letters, Journal of South American Earth Science, Physics of the Earth and Planetary Interiors, Tectonophysics*, books—Academic Press)
Eötvös Loránd Geophysical Institute of Hungary, Budapest (*Geophysical Transactions*, special editions)
ETH Zürich, Institute of Geophysics, Department of Earth Sciences, Switzerland
European Association of Geoscientists and Engineers (EAGE) (*First Break*)
European Science Foundation, Strasbourg, France
Freie Universität Berlin, Germany (*Berliner Geowissenschaftliche Abhandlungen*)
Geological Society of America (*Geology, Geological Society of America Bulletin*, Special Papers, Memoirs)
Geological Society, London, UK (Special Publications, Memoirs, *Petroleum Geology, Journal of the Geological Society*)
Geological Society of Australia (*Australian Journal of Earth Sciences*)
Geological Society of India (Memoirs)
Geologische Vereinigung (*Geologische Rundschau*)
Instituto Geográfico Nacional, Madrid, Spain (Monografia)
International Union of Geological Sciences (*Episodes*)
Interscience Publishers, Inc.
Istituto Nazionale di Oceanografia e di Geofisica Sperimentale, Trieste, Italy (*Bollettino di Geofisica Teorica e Applicata*)
John Wiley and Sons (*Geophysical Journal of the Royal Astronomical Society, Geophysical Journal International, First Break, Terra Nova*)
Komitet Badań Polarnych (*Polish Polar Research*)
LITHOPROBE, Canada
National Research Council, Canada
Nature Publishing Group
NRC Research Press (*Canadian Journal of Earth Science*)

Polish Academy of Sciences (Publications of the Institute of Geophysics, Polish Academy of Sciences, *Acta Geophysica Polonica*)

Royal Astronomical Society, London, UK

Russian Academy of Sciences, Schmidt United Institute of Physics of the Earth, Russian Academy of Sciences (books, Nauka Moscow)

Schweizerbart'sche Verlagsbuchhandlung OHG (Naegele u. Obermiller, Stuttgart, Germany) (books)

Schweizerische Geologische Gesellschaft (*Eclogae Geologicae Helvetiae*)

Seismological Society of America (*Bulletin of the Seismological Society of America*)

Seismological Society of Japan (*Earth, Planets and Space*, a continuation of *Journal of Physics of the Earth*)

Servicio Nacional de Geologia y Mineria, Chile (*Revista de Chile/Andean Geology*)

Society of Exploration Geophysicists, Tulsa, Oklahoma, USA (*Geophysics*)

Springer-Verlag, Berlin-Heidelberg-New York (books, *Zeitschrift für Geophysik,* continued as *Journal of Geophysics, Geologische Rundschau, Die Naturwissenschaften*)

Swedish Research Council

TOTAL-CSTJF (Centre Scientifique et Technique Jean Feger), Pau, France (*Bulletin Centre Recherche Pau–SNPA*)

U.S. Geological Survey

University of Bergen (Norway, Prof. Sellevoll)

University of Wyoming, Laramie, Wyoming, USA (*Rocky Mountain Geology*)

∾ CHAPTER 2 ∾

History of Controlled-Source Seismology—A Brief Summary

2.1. INTRODUCTION AND WORLDWIDE REVIEWS

Controlled-source seismology (also called "explosion seismology" or "deep-seismic sounding") is a special method to explore the velocity-depth structure of the Earth's crust and uppermost mantle, approximately to a depth range of about 50–100 km, by the investigation of the propagation of seismic body waves. The particular advantage of controlled-source seismology is that it uses man-made seismic sources, such as quarry blasts, borehole or underwater explosions, vibrators on land and in water (airguns), etc., where time and origin of the seismic source are precisely known. This has the particular advantage that the instruments can be switched on and off at any pre-established time and can be arranged and properly oriented on profiles or fans.

The beginning of controlled-source seismology can be defined in several ways. In 2009, controlled-source seismology could have celebrated its 160th birthday, because in 1849, Robert Mallet used dynamite explosions to measure the speed of elastic waves in surface rocks. In 2006, controlled-source seismology could have celebrated its 100th birthday, because in 1906, Emil Wiechert in Goettingen developed the first mobile seismograph with which profiles could be laid out. In 2008, it was 100 years since Ludger Mintrop made the first experiments to investigate the uppermost sedimentary layers using a weight drop and recording with Wiechert's portable seismographs. The stepwise improvement and development of this method led to applied geophysics and to its first commercial successes in the 1930s, when gravity and seismic prospecting guided geologists to locate oil resources in sedimentary basins.

The scientific exploration of the Earth's crust using controlled-source seismology as we know it today, however, effectively only started after World War II, when simultaneously the systematic use of large explosions and of quarry blasts, the development of proper instrumentation, and theoretical work on wave propagation became a joint and worldwide effort to study the physical properties of the Earth's crust and uppermost mantle. With time, borehole explosions on land and the development of non-explosive sources for oceanic research as well as the development of recording devices, both on land and in water, became determining factors for large-scale experiments, hand in hand with the development of theory and corresponding fast computer programs to handle the steadily increasing amount of data.

Controlled-source seismology involves both seismic-refraction and seismic-reflection investigations. In general, reflections (with much higher frequencies than used in refraction work) deliver completely different information than do refraction-wide angle studies. Often, reflection sections show structures and tectonics directly and with relatively high resolution, while refraction-wide angle studies reveal major interfaces and variations in velocities that are important for petrologic interpretations.

Since the beginning of explosion seismology, text books on seismic theory and interpretation methods (e.g., Nettleton, 1940) as well as reviews on crustal and upper-mantle research, achieved by seismic and seismological methods, were published from time to time, summarizing the results for the whole world. Table 2.1 gives an overview of summary publications on worldwide seismic crustal and uppermost mantle studies, sorted by publication date.

2.2. REVIEWS AT REGIONAL SCALES

Many publications exist in which the results of seismic research projects for particular regions were reviewed. There are, for example, crustal study summaries for individual countries or groups of countries such as Scandinavia, or summaries on larger regions such as the western United States. Other summaries concentrate on tectonic units such as the Alps, or the Afro-Arabian rift system. Furthermore, crustal thickness (depth to the Mohorovičić discontinuity, or Moho) compilations exist for Europe, the former USSR, the United States, and Australia, for example. Those summaries were updated from time to time by the same or by other authors, depending on the state of the art at the time. In Table 2.2, publications were collected which review results on regional scales covering countries, continents and/or large tectonic units, sorted by publication date.

2.3. THE FIRST 100 YEARS (1845–1945)

Since the beginning of the twentieth century, seismic waves have been used to study Earth's interior. Such studies involved both investigations of the whole Earth by distant earthquakes and of the Earth's crust by local natural and artificial events. Thus, the earliest studies of the Earth's crust employed whatever data were available, either earthquake traveltimes or controlled sources. Mohorovičić's classic study that defined the crust was based on earthquake traveltimes rather than on controlled sources. Nevertheless, Mohorovičić is still considered as the father of seismic studies of the Earth's crust. The rapid development of this special branch of seismology would not have been possible without the earlier technical developments of seismographs and sensitive recording devices during the nineteenth century. An early historic review was published by Mintrop as early as 1947, describing the history of the first 100 years of earthquake research and explosion

TABLE 2.1. REVIEWS OF WORLDWIDE SEISMIC CRUSTAL AND UPPERMOST MANTLE STUDIES

Year	Author or editor	Contents	
1947	Mintrop	100 years of explosion seismology	A
1951	Gutenberg	Crustal layers of the continents and oceans	A
1951	Macelwane	Tables of seismic investigations until 1950	A
1954	Reinhardt	Quarry blasts and explosion seismology history	B
1955	Poldervaart	Crust of the Earth	E
1955	Gutenberg	Wave velocities in the Earth's crust	A
1961	Closs and Behnke	Velocity–cross section around the world	A
1961	Steinhart and Meyer	Explosion studies of continental structure	E
1966	James and Steinhart	Critical review of explosion studies 1960–1965	A
1969	Hart	The Earth's crust and upper mantle	E
1971	Heacock	Structure and physical properties of the Earth's crust	E
1973	Mueller	Structure of the Earth's crust based on seismic data	E
1975	Ansorge	Uppermost mantle under Europe and North America	B
1975	Christensen and Salisbury	Structure and constitution of lower oceanic crust	A
1977	Heacock	The Earth's crust—Its nature and properties	E
1978	Klemperer et al.	Seismic probing of continents and their margins	E
1980	Spudich and Orcutt	Seismic velocity structure of the oceanic crust	A
1981	Soller et al.	A global crustal thickness map	B
1984	Prodehl	Tables of Earth's crust and uppermost mantle structure	A
1986	Meissner	The continental crust	A
1986	Barazangi and Brown	Reflection seismology on continents—Global perspective	E
1987	Fuchs et al.	Mantle heterogeneities: High-resolution experiments	A
1987	Matthews and Smith	Seismic reflection profiling of continental lithosphere	E
1987	Mooney and Brocher	Coincident seismic-reflection/refraction studies	A
1987	Orcutt	Oceans	A
1988	Leven et al.	Seismic probing of continents and their margins	E
1991	Meissner et al.	Continental lithosphere: deep seismic reflections	E
1992	Mooney and Meissner	Review of seismic-reflection profiling of lower crust	A
1992	Holbrook et al.	Seismic velocity structure of deep continental crust	A
1992a,b,c	Ziegler	Geodynamics of rifting	E
1994	Clowes and Green	Seismic reflection profiling of continents and margins	E
1995	Christensen and Mooney	Seismic velocity structure and composition of continents	A
1995	Olsen	Continental rifts: Evolution, structure, tectonics	E
1996	White et al.	Seismic reflection profiling of continents and margins	E
1997	Jacob et al.	Lithospheric structure and evolution in continental rifts	E
1998	Mooney et al.	CRUST5.1: A global model at 5° x 5°	A
1998	Klemperer and Mooney	Deep seismic probing of the continents	E
1999	Jones	Oceans	A
2000	Carbonell et al.	Deep seismic profiling of continents and margins	E
2000	Jacob et al.	Active and passive seismic techniques reviewed	E
2002	Mooney	Continental crust	A
2002	Mooney et al.	Seismic velocity structure from controlled-source data	A
2002	Minshull	Seismic structure of the oceanic crust	A
2002	Thybo	Deep seismic probing of continents and margins	E
2004	Davey and Jones	Continental lithosphere	E
2005	Fowler	The solid Earth—Global geophysics	A
2006	Snyder et al.	Seismic probing of continents and their margins	E
2007	Romanowicz and Dziewonski	Seismology and structure of the earth	E
2012	Prodehl and Mooney (this volume)	Controlled-source seismology history and global model	B

Note: A—article in a book or journal, B—single-author book, E—editors of a book with several seismic crustal study articles.

seismology (Mintrop, 1947; see Appendix A3-1). While the average structure of the Earth's crust could be detected by the detailed study of local earthquake records, the accuracy of seismological studies was limited for more refined studies of the Earth's crust because of too many unknown parameters, i.e., the exact time and the exact location of natural earthquakes.

In the middle of the nineteenth century, the first studies with controlled events were started. Explosion seismology was born in 1849 when Robert Mallet used dynamite explosions to measure the speed of elastic waves in surface rocks (Mintrop, 1947; Dewey and Byerly, 1969; Jacob et al., 2000).

The final instrumental progression to record artificial earthquakes was in 1906, when Emil Wiechert in Göttingen constructed a transportable seismograph which amplified the horizontal component of the ground movements by 50,000. In 1908, active seismological experiments started in Goettingen, when Ludger Mintrop used a weight drop as a controlled source and recorded it with Wiechert's portable seismographs. Thus he obtained the first seismograms, which included the fine details of precursor waves, now called P and S body waves (Mintrop, 1947).

For academic crustal structure research, however, it was not until 1923 that "artificial earthquakes" were introduced by the use of large explosions (Angenheister, 1927, 1928; Wiechert, 1923, 1926, 1929). Not only had large explosions been recorded in Germany. For example, in France, large surface explosions in

TABLE 2.2. REVIEWS OF REGIONAL SEISMIC CRUSTAL AND UPPERMOST MANTLE STUDIES

Year	Author or editor	Region	
1963	Closs and Labrouste	Western Alps	A
1964	Pakiser and Steinhart	Western Hemisphere	A
1966	Steinhart and Smith	North America	E
1967	Morelli et al.	Europe	A
1969	Kosminskaya et al.	USSR	A
1969	Healy and Warren	United States	A
1970	Maxwell	Oceans	E
1970	Shor et al.	Pacific basin	A
1971	Vogel	Northern Europe	E
1972	Sollogub et al.	Eastern and southeastern Europe	E
1973	Willmore et al.	British Isles	A
1973	Sellevoll	Northern Europe	A
1973	Belyaevsky et al.	USSR	A
1973	Cleary	Australia	A
1973	Warren and Healy	United States	A
1973	Berry	Canada	A
1973	Massé	North America	A
1973	Furumoto et al.	Hawaii and Central Pacific Basin	A
1975	Woollard	Pacific Ocean	A
1976a	Giese et al.	Central Europe	E
1977	Makris	Eastern Mediterranean and Hellenides	B
1980	Morelli and Nicolich	Western Mediterranean	B
1980	Zverev and Kosminskaya	USSR	B
1980	Brewer and Oliver	USA COCORP	A
1980	Jacoby et al.	Iceland	E
1987	Meissner et al.	Europe	A
1987	Orcutt	Oceans	A
1988	Dooley and Moss, Moss and Dooley	Australia	A
1989	Yan and Mechie	Alps	A
1989	Pakiser and Mooney	United States	E
1989	Braile et al.	North America	A
1989	Smithson and Johnson	Western United States	A
1989	Phinney and Roy-Chowdury	Eastern United States	A
1989	Trehu et al.	Continental margin, North America	A
1989	Mereu et al.	Canada COCRUST	A
1990	Meissner and Bortfeld	Germany DEKORP Atlas	B
1991	Bois and ECORS Scienific Parties	France ECORS review	A
1991	Heitzmann et al.	Swiss Alps	A
1991	Fuis et al.	TACT Southern Alsaska	A
1991	Collins	Australia	A
1991	Drummond	Australia	E
1992	Blundell et al.	Europe EGT	E
1992	Freeman and Mueller	Europe EGT maps	E
1992	Clowes et al.	Canada LITHOPROBE	A
1992	Klemperer and Hobbs	British Isles BIRPS Atlas	B
1993	Mechie et al.	USSR PNE data	A
1994	Mahadevan	India	B
1994	Ludden	Canada LITHOPROBE Abitibi-Grenville I	E
1994	Percival	Canada LITHOPROBE Kapuskasing transect	E
1994	Prodehl et al.	Kenya Rift	E
1995	Prodehl et al.	Central European Rift	A
1995	Ludden	Canada LITHOPROBE Abitibi-Grenville II	E
1995	Cook	Canada LITHPOROBE S Canadian Cordillera	E
1995	Braile et al.	East African Rift	A
1996	Pavlenkova	USSR	A
1997	Fuis et al.	TACT Northern Alaska	A
1997	Fuchs et al.	Afro-Arabian Rift	E
1997	Prodehl et al.	Afro-Arabian Rift	A
1998	Quinlan	Canada LITHOPROBE Newfoundland Appalachians	E
1998	Li and Mooney	China	A
1999	Reutter	Chile: Central Andean Deformation	E
2000a	Ludden and Hynes	Canada LITHOPROBE Abitibi-Grenville III	E
2000	Ross	Canada LITHOPROBE Alberta basement I	E
2000	Clitheroe et al.	Australia	A
2002	Iwasaki et al.	Japan	A
2002	Ross	Canada LITHOPROBE Alberta basement II	E
2002a	Wardle and Hall	LITHOPROBE eastern Canadian Shield on-offshore	E
2002	Chulick and Mooney	North America and adjacent oceans	A

(continued)

TABLE 2.2. REVIEWS OF REGIONAL SEISMIC CRUSTAL AND UPPERMOST MANTLE STUDIES (*continued*)

Year	Author or editor	Region	
2003	Collins et al.	Australia	A
2005	Landes et al.	Ireland and Irish seas	A
2005a	Hajnal et al.	Canada LITHOPROBE Trans-Hudson Orogen	E
2005	Cook and Erdmer	Canada LITHOPROBE northwestern Canada	E
2006	Gee and Stephenson	Europe EUROPROBE	E
2006	Percival and Helmstaedt	Canada LITHOPROBE West Superior Province	E
2006	Li et al.	China	A
2006	Gebrande et al.	Alps—TRANSALP	E
2007	Guterch et al.	Central and eastern Europe—long-range profiles	A
2009	Díaz and Gallart	Iberian peninsula	A
2009	Grad et al.	Moho map of European plate	A
2010	Finlayson	Australia deep seismic profiling chronicle	B

Note: A—article in a book or journal; B—single-author book; E—editors of a book or journal with several seismic crustal study articles.

1924 near La Courtine were recorded and reported. In California and in the eastern United States, seismic investigations of quarry blasts had been undertaken, but successful interpretations were not published until 1935. Seismic refraction and reflection investigations in water-covered areas started as early as 1927 (Rosaire and Lester, 1932) and were continued in the 1930s at the Atlantic coastal shelf (e.g., Ewing et al., 1937).

An overview on seismic velocities obtained from explosions and blasts in France, Germany, Italy, Switzerland and California up to 1939 was published by J.B. Macelwane in table 41 of the first edition of Volume VII of *Physics of the Earth* (edited by B. Gutenberg) on the internal constitution of the Earth (see second edition: Macelwane, 1951, table 41, reproduced in Table 3.3-02 in Chapter 3). A historical review of early explosion seismology work until the early 1950s was also summarized by Reinhardt (1954; see Appendix A4-1). Table 2.3 summarizes outstanding historical events on the early development of controlled-source seismology.

2.4. THE 1940s (1940–1950)

Controlled-source seismology investigations of the Earth's crust and the first international cooperation started effectively after 1945. The large explosions on Heligoland in 1947 and in the Black Forest in 1948 (Schulze, 1947; Reich et al., 1948) had the greatest impacts on crustal studies at this time. The explosion

on the island of Heligoland is the factual beginning of controlled-source seismology for science in Germany (Schulze, 1974).

Early crustal studies in the 1940s had also been undertaken in other parts of the world, in part during World War II. In 1950, Twaltwadzse reported on a series of very large explosions which had occurred between 1941 and 1945 in the Soviet Republic of Georgia. In 1949 and 1950, underwater explosions in the lakes Issyk-Kul and Kara-Kul served to investigate the crustal structure under the northern Tien-Shan region (Gamburtsev, 1952). In Canada, rock bursts occurring between 1938 and 1945 in the mining area near Kirkland Lake, Ontario, were used for the first time for crustal structure studies (e.g., Hodgson, 1947) and these were continued in 1947–1951. In the United States, a large explosion of ammunition occurred in 1944 near Port Chicago in California (Byerly, 1946). The first nuclear tests were also recorded in North America (Gutenberg and Richter, 1946). In the Appalachian Mountains, large quarry blasts were recorded by optical-mechanical-electrical mobile field stations at distances up to 350 km (Tuve et al., 1948).

At sea during World War II, the techniques of seismic measurement were further developed so that after 1945 the experiments could be extended from shallow coastal waters into the deep ocean basins using hydrophones. The data obtained in the late 1930s were supplemented by new expeditions, in particular in the northwestern Atlantic Ocean in 1948 and 1949 (e.g., Drake et al., 1952; Ewing et al., 1950), but offshore investigations were

TABLE 2.3. MAJOR STEPS TOWARDS CONTROLLED-SOURCE SEISMIC INVESTIGATIONS OF
THE CONTINENTAL CRUST IN THE BEGINNING OF MODERN SEISMOLOGY

Chapter	Year	Project	Location	Reference
3.3	1849	First controlled-source seismic experiment	Ireland	Mallet (1852)
3.3	1906	Wiechert's first portable seismometer	Germany	Wiechert (1923)
3.3	1908	Mintrop's iron ball experiment	Germany	Mintrop (1947)
3.3	1923	Start of using "artificial earthquakes" (blasts)	Central Europe	Wiechert (1926)
3.3	1926–1929	Quarry blast recording in southern California	California	Wood and Richter (1931)
3.3	1927	Observations at detonations	Germany	Angenheister (1927)
3.3	1927	Coastal lakes Louisiana seismic-refraction tests	United States	Rosaire and Lester (1932)
3.3	1931	Quarry blast recording in southern California	California	Wood and Richter (1933)
3.3	1935	Richmond quarry blasts	California	Byerly and Wilson (1935b)
3.3	1935	Atlantic coastal plain seismic-refraction surveys	Atlantic shelf	Ewing et al. (1937)
3.3	1936–1938	New England quarry blasts	Eastern United States	Leet (1938)
3.3	1938	Bermuda seismic-refraction survey	Atlantic	Woollard and Ewing (1939)
3.3	1939	Shelf east of Britain	North Sea	Bullard et al. (1940)
3.3	1940	Continental slope off Britain	North Atlantic	Bullard and Gaskell (1941)

also undertaken in the eastern Atlantic Ocean in 1949 (Hill and Swallow, 1950) and in the Pacific Ocean off California in 1948 (Raitt, 1949).

An overview of the early seismic projects undertaken in the 1940s is mentioned in Chapter 4 and can be seen in Table 2.4.

By the early 1950s, the overall picture of the crust had been established. Gutenberg (1951b, 1955) summarized worldwide results; his crustal columns (Fig. 2.4-01) provided a rough picture of the seismic structure of the Earth's crust around the world (Gutenberg, 1955). Reinhardt (1954) wrote a fundamental study on the use of quarry blasts. In his review of crustal investigations up to 1954, he plotted the basic scheme (see Fig. 4.1-01) of seismic refraction work, compiled a world map (Fig. 2.4-02) showing the locations of crustal studies around the world which

had used explosive sources, and also summarized the results in crustal columns. Beneath a layer of sediments, the Earth was divided into an upper crust of granitic composition and a lower crust consisting of gabbroic rocks, underlain by the Mohorovičić discontinuity and a peridotitic mantle layer (Gutenberg, 1951b, 1955; Reinhardt, 1954).

2.5. THE 1950s (1950–1960)

Since the beginning of the 1950s, commercial quarry blasts have been increasingly used, since they proved to be a powerful and low-cost energy source for a systematic investigation of the detailed structure of the Earth's crust (Reinhardt, 1954). Systematic seismic crustal research started in central Europe in

TABLE 2.4. MAJOR CONTROLLED-SOURCE SEISMIC INVESTIGATIONS OF THE CRUST IN THE 1940S

Chapter	Year	Project	Location	Reference
4.1.0	1947	Heligoland	Germany	Reich et al. (1951)
4.1.0	1948	Haslach	Southern Germany	Reich et al. (1948)
4.2.1	1941ff	Large explosions	Central Asia, USSR	Twaltwadze (1950)
4.2.1	1949–1950	Tien-Shan	Central Asia, USSR	Gamburtsev (1952)
4.2.2	1938ff	Rock bursts	Canada	Hodgson (1953)
4.2.2	1945ff	Appalachian quarry blasts	Eastern United States	Tuve et al. (1948)
4.2.2	1945–1946	First nuclear tests	New Mexico, United States; Pacific	Gutenberg (1946)
4.2.2	End of 1940s	Big Horn County, Montana, United States	Western United States	Junger (1951)
4.2.3	1948–1949	Whitwatersrand tremors	South Africa	Willmore et al. (1952)
4.2.4	1948	Gulf of Maine	Northwest Atlantic	Drake et al. (1952)
4.2.4	1948	Pacific off Southern California, United States	Eastern Pacific	Raitt (1949)
4.2.4	1948–1949	North American Basin	Northwest Atlantic	Ewing et al. (1952)
4.2.4	1948–1949	North American Basin	Northwest Atlantic	Hersey et al. (1952)
4.2.4	1949	Gulf of Maine	Northwest Atlantic	Katz et al. (1953)
4.2.4	1949	West of British Isles	Northeast Atlantic	Hill and Swallow (1950)

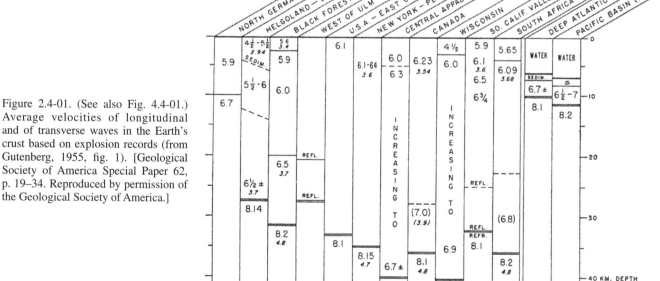

Figure 2.4-01. (See also Fig. 4.4-01.) Average velocities of longitudinal and of transverse waves in the Earth's crust based on explosion records (from Gutenberg, 1955, fig. 1). [Geological Society of America Special Paper 62, p. 19–34. Reproduced by permission of the Geological Society of America.]

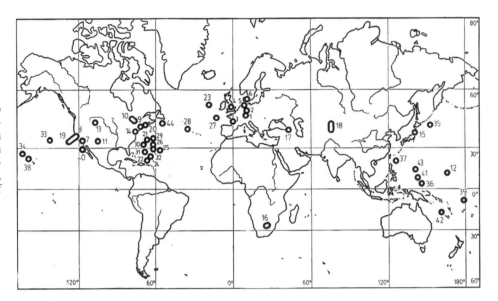

Figure 2.4-02. (See also Fig. 4.4-02.) Location map of large explosion-seismic studies of the Earth's crust (from Reinhardt, 1954, fig. 3). For explanation of numbers, see Table 4.4-02 in Chapter 4. [Freiberger Forschungshefte, C15, p. 9–91. Reproduced by permission of TU Freiberg, Germany.]

1954, when the Alps became a special target of crustal research by the foundation of the Subcommission of Alpine Explosions under the umbrella of the International Union of Geodesy and Geophysics (IUGG), initiating the first major inter-European fieldwork in 1954 (Closs and Labrouste, 1963). In 1957 in Germany, a priority program funded by the German Research Society, "Geophysical Investigation of Crustal Structure in Central Europe," was initiated. This program comprised both reflection and refraction seismic experiments and involved all geophysical university institutes of Germany and geophysical departments of the German state geological surveys. It also prompted cooperation with geophysical institutions in neighboring countries. The priority program included the development of new recording systems (Closs and Behnke, 1961; Closs, 1969; Giese et al., 1976a). It was strongly supported by commercial geophysical exploration companies in Germany which showed particular interest in the scientific deep-seismic sounding programs and helped these efforts by recording up to 12 seconds two-way traveltime (e.g., Dohr, 1959; Liebscher, 1964).

In North America and Canada, the studies of the Earth's crustal structure using blasts and rockbursts were continued more systematically from the late 1940s. Tatel and Tuve (1955) and Katz (1955) carried out seismic-refraction experiments in various geologic provinces and regions of the United States. Steinhart and Meyer (1961) edited a special volume describing the experimental work of several projects in detail, the interpretation method used at this time and a critical review of worldwide results (see Appendix A5-1).

In the USSR, the first period of deep-seismic sounding experiments started at the end of the 1940s. The first deep seismic research was conducted in 1948–1954 under the leadership of G.A. Gamburtzev, E. Galperin, and I.P. Kosminskaya in central Asia and in the southern Caspian area. Since the middle of the 1950s, deep-seismic sounding profiles were recorded on the Russian platform as well as on the Russian part of the Baltic Shield,

in central Asia, in the Caucasus, and in the Urals covering thousands of kilometers (results published in Russian, for references see Pavlenkova, 1996). A major project covered the transition zone from the Asian continent to the Pacific Ocean (Galperin and Kosminskaya, 1964).

In Japan, controlled-source seismology started in 1950, when the construction of the Isibuti dam in north-central Honshu required a large explosion of 57 tons "carlit" to be detonated simultaneously. Within the short time span of only ten days, an active working group, the Research Group for Explosion Seismology, was established and became very active in the following years to provide instrumentation and to organize shotpoints for deep-seismic sounding investigations (Research Group for Explosion Seismology, Tokyo, 1951).

On continental Australia, the first definitive measurement of Moho depth was interpreted from recordings of nuclear explosions at Maralinga (South Australia) westwards across the Nullarbor Plain (Bolt et al., 1958; Finlayson, 2010; Appendix 2-2).

At the same time, when the continental crust was studied by the first systematic experiments, the seismic refraction method for work at sea was already well established. The experiments of Ewing and coworkers in the 1950s were concentrated mainly on the Northwestern Pacific, but the Gulf of Mexico and the Caribbean Sea were also investigated in great detail. Worldwide sea expeditions led U.S. and British researchers into the eastern Atlantic, into the Pacific, and into the Mediterranean Sea, and the Indian Ocean became the focus of several seismic refraction investigations. A detailed overview and summary on oceanic crustal structure studies was published by Raitt (1963), based on experiments obtained by the application of the seismic-refraction method in the 1950s.

By the end of the 1950s, a basic knowledge on crustal structure around the Earth had been established, based on a considerable number of seismic refraction investigations recording man-made explosions out to distances of several hundred kilometers, as shown in reviews such as Steinhart and Meyer (1961; see Table 5.7-01) and of Closs and Behnke (1961; Fig. 2.5-01). Steinhart and Meyer (1961; see Appendix A5-1) as well as Ewing (1963a) also gave detailed and critical reviews on the state of the art in the methodology and its limitations, including the present state of instrumentation and major field problems. An overview of the major seismic projects undertaken in the 1950s and mentioned in Chapter 5 can be seen in Table 2.5.

2.6. THE 1960s (1960–1970)

The first experimental phase of the 1950s to develop new types of instruments continued into the 1960s and finally led to the production of powerful instruments for wide-angle seismic profiling in large numbers, in particular in western Europe, North America, and the USSR. With this instrumentation, a major breakthrough in the study of the Earth's crust was achieved. In central Europe, in the western United States, and in the southern part of the USSR, major seismic-refraction fieldwork was undertaken. In central Europe, quarry blasts were the main source for seismic profiling; in the western United States and southwestern USSR, seismic energy was provided mainly by borehole explosions. Thus, at the end of the 1960s, major networks of seismic profiles existed in all three areas.

During the 1960s, experimental reflection profiling surveys were also undertaken in many parts of the world, in particular in Canada (e.g., Kanasewich and Cumming, 1965; Clowes et al., 1968), Germany (e.g., Liebscher, 1964; Dohr and Fuchs, 1967) and Russia (e.g., Beloussov et al., 1962; Kosminskaya and Riznichenko, 1964). They demonstrated that near-vertical profiling methods used in oil and gas exploration could be used to image geological structures within basement rocks and that at long recording times, reflections from the deep crust and the Moho could be obtained. This paved the way for COCORP and subsequent deep reflection programs around the world in the 1970s and 1980s.

In parallel with the fieldwork, the art of interpretation was pushed forward by new developments in the theory of seismic wave propagation and by applying the results using the rapidly developing new computer technology. The known methods were made more efficient, and at the same time new methods were developed to interpret the increasing number of recently observed data. In the 1960s, the interpretations were almost exclusively based on the correlation of waves by travel times, read from picked arrival times, plotted in time-distance graphs, and correlated by straight or curved lines, but gradually seismic phase correlation using record sections became common.

The use of record sections created a major breakthrough in understanding the character of seismic waves and their relation to the structure of the Earth's crust and mantle which would guide the interpretation of seismic refraction observations through the following decades. In an internal report (Prodehl, 1998) a major collection of record sections was compiled which had been produced from observed data and had been published from the early 1960s to the end of the 1990s from major controlled-source seismic projects around the world. The report included an abstract and location map from the most relevant publication of the corresponding project. This report is reproduced in Appendix A2-1 and the data shown there will be referred to in the following chapters. Furthermore, for the Australian continent and its margins a comprehensive chronicle on the history of deep-seismic profiling was prepared showing maps and many data (Finlayson, 2010; Appendix A2-2). Additional data and abstracts can be found in Appendices A3 to A10 accompanying the corresponding decades. A more complete compilation of seismic projects carried out in the 1960s using controlled seismic sources and mentioned in Chapter 6 is compiled in Table 2.6.

While most of the seismic-refraction and -reflection experiments in the 1960s were carried out on a national basis, major international projects were pushed forward also. These efforts were strongly supported by the formulation of major international research programs, which were supported and financed by national research foundations and thus allowed the systematic study of specific tectonic regions of the Earth. In Europe, the European Seismological Commission (ESC) had been founded with meetings every two years, alternatively held in eastern or in

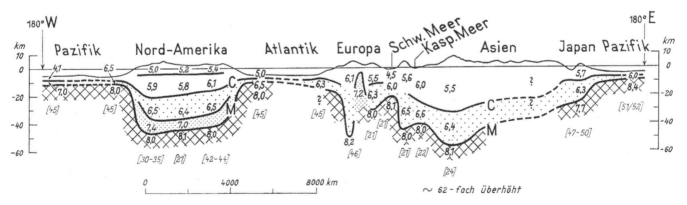

Figure 2.5-01. (See also Fig. 5.7-01.) West-east profile across the Earth at 45°N, compiled after individual results (for detailed references see Closs and Behnke, 1961). [Geologische Rundschau, v. 51, p. 315–330. Reproduced with kind permission of Springer Science+Business Media.]

TABLE 2.5 MAJOR CONTROLLED-SOURCE SEISMIC INVESTIGATIONS OF THE CRUST IN THE 1950S

Chapter	Year	Project	Location	Reference
5.1	1952	Blaubeuren deep reflections	Germany	Reich (1953)
5.1	1955ff	Deep reflection recording	Germany	Dohr (1959)
5.1	1956–1958	Lac Nègre and other lake shots	Western Alps	Closs and Labrouste (1963)
5.1	1957ff	Quarry blast profiling	Germany	Giese et al. (1976)
5.1	1957ff	Seismic reflection work, southern Germany	Southern Germany	Dohr (1959)
5.1	1960	Ivrea zone borehole shot profiles	Western Alps	Closs and Labrouste (1963)
5.2	1950ff	Quarry blast and test explosions profiling	United States	Tatel and Tuve (1955)
5.2	1950–1953	New York–Pennsylvania quarry blasts	Eastern United States	Katz (1955)
5.2	1955	Underwater explosions	Alaska	Tatel and Tuve (1956)
5.2	1956–1959	Nevada Test Site and quarry blast profiles, Nevada-Utah	Western United States	Berg et al. (1960)
5.2	1957	Central Plateau, Mexico	Mexico	Meyer et al. (1961a)
5.2	1958	Arkansas-Missouri	Central United States	Steinhart et al. (1961a)
5.2	1958–1959	Wisconsin–Upper Michigan	Central United States	Steinhart et al. (1961b)
5.2	1959	Rocky Mountains, Montana	Western United States	Meyer et al. (1961b)
5.3	1948–1954	Central Asia, southern Caspian area	USSR	Pavlenkova (1996)
5.3	1955ff	Deep seismic sounding profiling	USSR	Pavlenkova (1996)
5.3	1957–1958	Deep seismic sounding profiling	Sea of Ochotsk	Galperin and Kosminskaya (1964)
5.4	1950	First Japanese profiles on Honshu	Japan	Matuzawa et al. (1959)
5.4	1953–1956	Nuclear explosions	Australia	Bolt et al. (1958)
5.4	1956–1957	Eaglehawk quarry blasts	East Australia	Doyle et al. (1959)
5.4	1959	Prospect blue metal quarry blasts	East Australia	Bolt (1962)
5.5.1	1952	Wellington refraction line	New Zealand	Officer (1955)
5.5.3	1950s	Andes	Peru and Chile	Tatel and Tuve (1958)
5.5.3	1950s–1960s	Arctic interior upper crust	Antarctica	Bentley (1973)
5.5.3	1959–1960	Belgian Antarctic expedition	Antarctica	Dieterle and Peterschmitt (1964)
5.6.1	1950–1967	Seismic reflection profiles, Atlantic and Pacific	Atlantic and Pacific	Ewing and Ewing (1970)
5.6.1	1950	Bermuda area	Northwest Atlantic	Officer et al. (1952)
5.6.1	1950	North Atlantic Ocean	Atlantic	Gaskell and Swallow (1951)
5.6.1	1950+	Continental slope south of Nova Scotia	Northwest Atlantic	Officer and Ewing (1954)
5.6.1	1950+	Grand Banks	Northwest Atlantic	Press and Beckmann (1954)
5.6.1	1950+	Continental slope south of Grand Banks	Northwest Atlantic	Bentley and Worzel (1956)
5.6.1	1950–1951	English Channel	English Channel	Hill and King (1953)
5.6.1	1950–1952	Bermuda—North American shelf	Northwest Atlantic	Katz and Ewing (1956)
5.6.1	1951	North American Basin	Northwest Atlantic	Ewing et al. (1954)
5.6.1	1952	Gulf of Maine	Northwest Atlantic	Drake et al. (1952)
5.6.1	1953	Deep-sea reflection profile (Bermuda)	Northwest Atlantic	Officer (1955a)
5.6.1	1954–1956	US Atlantic continental margin	Northwest Atlantic	Hersey et al. (1959)
5.6.1	1957	Caribbean Sea—Lesser Antilles	Caribbean Sea	Edgar et al. (1971)
5.6.2	1950	Mid Pacific expedition	Central Pacific	Raitt (1956)
5.6.2	1951	North Pacific Ocean	Pacific	Gaskell and Swallow (1952)
5.6.2	1952–1953	Capricorn expedition	Central Pacific	Raitt (1956)
5.6.2	1952	Indian Ocean and Mediterranean Sea	Middle-East	Gaskell and Swallow (1953)
5.6.2	1957	Peru-Chile Trench	South Pacific	Raitt (1964)
5.6.2	1957	East Pacific Rise	South Pacific	Raitt (1964)
5.6.2	1959	First Japanese deep-sea expedition	Japan trench	Ludwig et al. (1966)

western Europe. These meetings served as most effective communication centers between scientists in eastern and western Europe and secured personal contacts of scientists also at times of difficult political situations. A special subcommission of the ESC dealt in particular with experiments and results of explosion seismology investigations. In 1964 under the umbrella of IASPEI (International Association of Seismology and Physics of the Earth's Interior), the Upper Mantle Project was started as an international program of geophysical, geochemical, and geological studies concerning the "upper mantle and its influence on the development of the Earth's crust."

In central and western Europe, national research programs in Germany, Italy, France, and Britain were extended into collaborative investigations and triggered joint studies. The cooperative research in the Western Alps of the 1950s was continued in the 1960s. The Eastern Alps and the Apennines became the focus of international experiments mainly based on the fruitful cooperation of Italian, Austrian and German institutions. The Scandinavian countries undertook major efforts to investigate the Baltic Shield in close cooperation with German and British institutions. The Moho map in Figure 2.6-01, compiled by Morelli et al. (1967), is primarily the result of international cooperation.

The East European countries developed a separate methodology of deep-seismic sounding. They extended their national research projects into a cooperative effort by establishing a network of international profiles, which covered the entire southeast of Europe and included all Eastern Block countries. The resulting network of deep-seismic sounding profiles reached from the East European Craton of the USSR into Poland as well

TABLE 2.6. MAJOR CONTROLLED-SOURCE SEISMIC INVESTIGATIONS OF THE CRUST IN THE 1960S

Chapter	Year	Project	Location	Reference
6.2.2	1960ff	Quarry blast profiling	Germany	Giese et al. (1976a)
6.2.2	1964	Common-depth-point Bavarian Molasse Basin	Germany	Meissner (1966)
6.2.2	1968	Common-depth-point profile Rhenish Massif	Germany	Meissner et al. (1976a)
6.2.3	1960	Continental shelf	West of Britain	Bunce et al. (1964)
6.2.3	1962ff	Underwater explosions	British Isles and seas	Bamford (1971)
6.2.3	1965	NORSAR deep seismic sounding profiles	Southern Norway	Sellevoll (1973)
6.2.3	1967	North Sea traverse	Norway–Scotland	Sornes (1968)
6.2.3	1969	Continental margin project	Atlantic off Ireland	Bamford (1971)
6.2.3	1969	Rockall Plateau off Ireland	Atlantic off Ireland	Scrutton (1972)
6.2.3	1969	Trans-Scandinavian deep seismic sounding project	Scandinavia	Vogel and Lund (1971)
6.2.4	1961, 1962	Lago Lagorai profiles	Eastern Alps	Prodehl (1965)
6.2.4	1961–1967	French Massif Central	France	Perrier and Ruegg (1973)
6.2.4	1964	Lago Bianco profiles	Central Alps	Giese and Prodehl (1976)
6.2.4	1965	Alpine quarry and construction blasts	Western Alps	Giese et al. (1967)
6.2.4	1965–1968	Puglia–Sicily	South Italy	Giese et al. (1973)
6.2.4	1966	Lac Negre underwater shots	Alps and Appenines	Röwer et al. (1977)
6.3	1963ff	Deep seismic international profiles	Southeastern and eastern Europe	Sollogub et al. (1972)
6.3.2	1960–1965	Deep seismic sounding profiles	Poland	Guterch et al. (1967)
6.3.2	1965ff	Continuous profiling Tornquist-Teysseire Zone and Sudetes	Poland	Toporkiewicz (1986)
6.3.2	1969	International Profile V and VII	Poland	Uchman (1972)
6.3.2	1969	International Profile V and VI	Czechoslovakia	Beránek et al. (1972)
6.3.2	1969	International Profile VI	E Germany	Knothe and Schröder (1972)
6.3.2	1969	International Profile III, IV and V	Hungary	Mituch and Posgay (1972)
6.3.2	1969	International Profile III	Yugoslavia	Prosen et al. (1972)
6.3.2	1969	International Profile II and X	Bulgaria	Dachev et al. (1972)
6.3.2	1969	International Profile II and XI	Romania	Constantinescu (1972)
6.4	1960ff	Deep seismic sounding profiles	USSR Europe and Asia	Kosminskaya et al. (1969)
6.4	1963ff	Deep seismic sounding profiles	Southwest USSR	Sollogub and Chekunov (1972)
6.5.1	1961	USGS profiles in Colorado	Western United States	Jackson et al. (1963)
6.5.1	1962–1963	USGS California-Nevada profiles	Western United States	Prodehl (1979)
6.5.1	1964	Tonto Forest Array project, Arizona	Southwest United States	Warren (1969)
6.5.1	1965	USGS Rocky Mountain profiles	Western United States	Prodehl and Pakiser (1980)
6.5.1	1965	Kansas refraction line	Central United States	Steeples and Miller (1989)
6.5.1	1966	LASA array deep seismic sounding project, Montana	Northwest United States	Warren et al. (1973)
6.5.1	1967	Coast Ranges deep seismic sounding, California	Western United States	Walter and Mooney (1982)
6.5.1	1969	Wind River reflection, Wyoming	Western United States	Perkins and Phinney (1969)
6.5.2	1962ff	Great Plains borehole shots profiles	Central United States	Healy and Warren (1969)
6.5.2	1963, 1964	Lake Superior shots	Canada and United States	Mansfield and Evernden (1966)
6.5.2	1965	Cumberland Obs, Appalachians	Eastern United States	Prodehl et al. (1984)
6.5.2	1965	East Coast Offshore-Onshore Experiment	Eastern United States	Hales et al. (1968)
6.5.4	1964	Near-vertical incidence seismic reflection studies	Central Canada	Kanasewich and Cumming (1965)
6.5.4	1965	Hudson Bay experiment	Northeast Canada	Hobson (1967)
6.5.4	1966	Yellowknife Seismic Array project	Northwest Canada	Weichert and Whitham (1969)
6.5.4	1966	Bear, Slave, Churchill Precambrian	Northwest Canada	Berry (1973)
6.5.4	1967–1968	Alaska Range—Tanana Basin profile	Central Alaska	Hanson et al. (1968)
6.5.4	1968	Grenville Front seismic experiment	Eastern Canada	Berry and Fuchs (1973)
6.5.4	1969	Yellowknife Seismic Array project	Northwest Canada	Clee et al. (1974)
6.5.5	1965	EARLY RISE Lake Superior 5-t shots	North America	Iyer et al. (1969); Massé (1973)
6.5.5	1969	EDZOE Upper mantle seismic project	Western North America	Berry and Forsyth (1975)
6.6	1963–1964	Deep seismic sounding Kurayoshi-Hanasuba profile	Honshu, southwest Japan	Research Group for Explosion Seismology (1966b)
6.6	1966–1967	Deep seismic sounding Atumi-Noto profile	Honshu, central Japan	Aoki et al. (1972)
6.6	1968–1969	Deep seismic sounding Shakotan-Erimo profile	Hokkaido, northeast Japan	Okada et al. (1973)
6.6	1970	Deep seismic sounding Kurayoshi-Hanasuba profile	Honshu, southwest Japan	Yoshii et al. (1974)
6.7.1	1959–1964	Quarry blasts, SW Western Australia	Australia	Everingham (1965)
6.7.1	1960	Marine shots, Perth Basin	Australia	Hawkins et al. (1965)
6.7.1	1962	Marine shots off Perth	Australia	Hawkins et al. (1965)
6.7.1	1963	Marine shots off Perth	Australia	Hawkins et al. (1965)
6.7.1	1963–1967	Quarry blasts, South Australia	Australia	White (1969)
6.7.1	1960ff	Continuous deep reflection profiles	Australia	Moss and Dooley (1988)

(continued)

TABLE 2.6. MAJOR CONTROLLED-SOURCE SEISMIC INVESTIGATIONS OF THE CRUST IN THE 1960S (*continued*)

Chapter	Year	Project	Location	Reference
6.7.1	1965	Offshore shots off New South Wales	Eastern Australia	Cleary (1973)
6.7.1	1966	CRUMP Offshore shots	North Australia	Cleary (1973)
6.7.1	1966	WRAMP Inland borehole shots	North Australia	Cleary (1973)
6.7.1	1966	BUMP Bass Street experiment	Southeast Australia	Cleary (1973)
6.7.1	1967	FRUMP Western Australia	Southwest Australia	Finlayson (2010)
6.7.1	1968–1969	Deep reflection experiments	Southwest Australia	Moss and Dooley (1988)
6.7.1	1969	Western Australia Geotraverse	Southwest Australia	Mathur (1974)
6.7.1	1970	ORD RIVER experiment	North Australia	Denham et al. (1972)
6.7.2	1969	First Kenya Rift Seismic Project	Kenya	Griffiths et al. (1971)
6.7.3	1968	Peru-Bolivia Altiplano project	West South America	Ocola and Meyer (1972)
6.7.4	1969	14th Soviet Antarctica Expedition	Antarctica	Bentley (1973)
6.8.2	1950–1967	Seismic reflection profiles, Atlantic and Pacific	Atlantic and Pacific	Ewing and Ewing (1970)
6.8.2	1962	"*Arlis II*" ice island drift Arctic	Arctic Ocean	Kutschale (1966)
6.8.2	1966	"*Meteor* (1964)" expedition M04	Reykjanes Ridge	Sarnthein et al. (2008)
6.8.2	1967	Seamounts "Atlantic Kuppelfahrten"	Northeast Atlantic	Closs et al. (1968)
6.8.2	1968	Mid-Atlantic Ridge at 45°N	Atlantic	Keen and Tramontini (1970)
6.8.2	1969	Gulf of Mexico upper mantle experiment	Gulf of Mexico	Hales (1973)
6.8.2	1969	"*Meteor* (1964)" expedition M17	Mediterranean Sea	Closs et al. (1972)
6.8.2	1969	"*Meteor* (1964)" expedition M17	Arabian Sea	Closs et al. (1969a)
6.8.2	1969	"*Meteor* (1964)" expedition M18	Norwegian Sea	Closs (1972)
6.8.2	1969	Iceland–Faeroe Ridge	North Atlantic	Bott et al. (1971)
6.8.3	1959, 1961	Japanese Deep Sea Expedition	Japan trench	Ludwig et al. (1966)
6.8.3	1960ff	Oceanic seismic refraction lines	Indian and Pacific Ocean	Shor and Raitt (1969)
6.8.3	1963ff	Kuril Islands–South Kamchatka	Western Pacific	Kosminskaya et al. (1973)
6.8.3	1964, 1966	Hawaii Big Island projects	Hawaii	Hill (1969)
6.8.3	1965	Juan de Fuca and Gorda Ridges	Off Oregon-California	Shor et al. (1968)
6.8.3	1965	Cook Islands	Pacific east of Fiji	Hochstein (1968)
6.8.3	1965	Philippine Sea	West Pacific	Murauchi et al. (1968)
6.8.3	1964ff	Upper mantle anisotropy experiment	East Pacific	Raitt et al. (1971)
6.8.3	1966	Northwest Pacific Basin	West Pacific	Den et al. (1969)
6.8.3	1967	Nova expedition Melanesian borderland	West Pacific	Shor et al. (1971)
6.8.3	1967	Coral Sea—Queensland Basin	West Pacific	Ewing et al. (1970)
6.8.3	1967, 1969	Bismarck archipelago project	Northern Melansia	Finlayson et al. (1972)
6.8.4	1962	Central Indian Ridge	Northwest Indian Ocean	Francis and Shor (1966)
6.8.4	1962	Ninetyeast Ridge	Southern Indian Ocean	Francis and Raitt (1967)
6.8.4	1962	Seychelles Bank	Northwest Indian Ocean	Shor and Pollard (1963)
6.8.4	1962	Broken and Naturaliste Plateaus	Southeast Indian Ocean	Francis and Raitt (1967)
6.8.4	1962	Agulhas Bank and Transkei Basin	Southeast Indian Ocean	Green and Hales (1966)
6.8.4	1962	Southeast African continental margin	Southwest Indian Ocean	Ludwig et al. (1968)
6.8.4	1968	Agulhas Bank and Plateau	Southern Indian Ocean	Hales and Nation (1973)

as into the Bohemian Massif, the Carpathians, and the Dinarides (Sollogub et al., 1972).

In the territory of the USSR, not only the southwestern part was intensively investigated by deep-seismic sounding profiles, but many other projects dealt with detailed crustal structure investigations covering also the Asian part of the USSR, which culminated in some 215 crustal sections along deep-seismic sounding profiles of over 50,000 km length (Belyaevsky et al., 1973), the results of which could be compiled into a Moho map covering the whole USSR (Fig. 2.6-02).

In North America a major seismic refraction survey was started by the U.S. Geological Survey (Pakiser, 1963), but also several university projects dealt with seismic crustal research. The main target was the investigation of the crustal structure of the Basin and Range province in the west, which included the adjacent Sierra Nevada and Coast Ranges in the west and the Rocky Mountains and adjacent Great Plains in the east. Some areas in the central and the eastern United States were also investigated. On the basis of the numerous new data, gathered in the 1960s, Warren and Healy (1973) created fence diagrams of crustal cross

sections throughout the United States and compiled a Moho map (Fig. 2.6-03).

Another focus became the Lake Superior region (Steinhart and Smith, 1966). The Lake Superior experiments which involved all major North American research institutions did not only comprise detailed crustal studies, but in particular the EARLY RISE project of 1965 opened another dimension by recording man-made events to distances of several thousand kilometers thus demonstrating that parts of the uppermost mantle could be systematically studied with controlled-source seismology. Another large cooperative project, EDZOE, covered much of Canada and aimed to study the structure of the North American Great Plains and the Rocky Mountains.

In Southeast Asia, from 1964 onwards, considerable progress was made in Japan in instrumentation and seismic fieldwork for crustal studies (Research Group for Explosion Seismology, 1966a). Some of the experiments were carried out under the Upper Mantle Project, providing first cross sections of Honshu Island (Aoki et al., 1972; Yoshii and Asano, 1972; Okada et al., 1973; Yoshii et al., 1974). From these data, it could be concluded

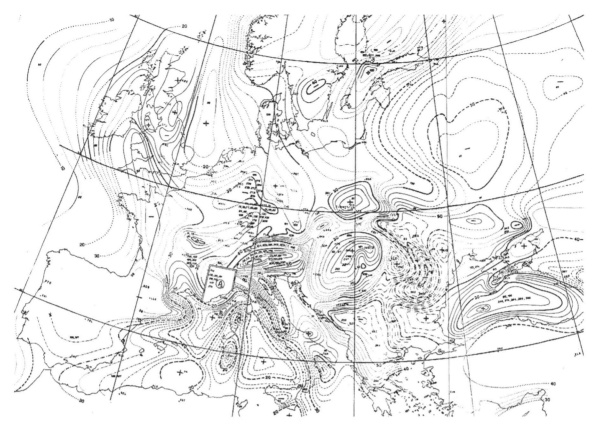

Figure 2.6-01. Moho contour map of Europe (from Morelli et al., 1967, fig. 2). [Bolletino di Geofisica Teorica Applicata, v. 9, p. 142–157. Reproduced by permission of Bolletino di Geofisica, Trieste, Italy]

Figure 2.6-02. Moho contour map of the territory of the USSR (from Belyaevsky et al., 1973, fig. 1). [Tectonophysics, v. 20, p. 35–45. Copyright Elsevier.]

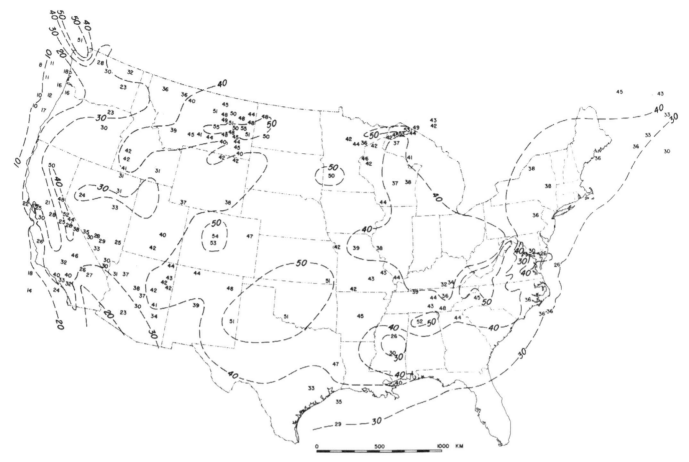

Figure 2.6-03. Moho contour map of the United States of America (from Warren and Healy, 1973, fig. 8). [Tectonophysics, v. 20, p. 203–213. Copyright Elsevier.]

that the northern part of the NE Japan arc is characterized by a low (~7.5 km/s) uppermost mantle velocity (Yoshii and Asano, 1972; Okada et al., 1973).

In Australia, several large-scale experiments provided first results on crustal thickness for southeastern and northern Australia (Denham et al., 1972; Cleary, 1973). Also, reflection profiling in basins in central and southeastern Australia provided strong reflections from the Moho (Dooley and Moss, 1988; Moss and Dooley, 1988; Finlayson, 2010; Appendix 2-2). In Africa, the first seismic investigation of the East African rift system was initiated by British scientists in Kenya (Griffiths et al., 1971). The very first explosion seismic investigation in South America concentrated on the Andes with a reconnaissance survey of the Peru-Bolivia Altiplano involving recording distances of 320–400 km (Ocola and Meyer, 1972).

At sea, all instruments in use in the 1950s and 1960s were essentially echo sounders utilizing pulses of low frequency sonic energy and a graphic recording system that displayed the data in the form of cross sections. The analysis methods, available also during the 1960s (Ewing, 1963a), used least-squares slope-intersect solutions for picked first arrivals, but did not allow for velocity gradients. Techniques and equipment for continuous

seismic profiling were developed early in the 1960s, enabling studies of local variations in sediment thickness and details of its stratification. To eventually overcome the work with explosives, continuous research tried to develop efficient non-explosive energy sources (Ewing and Zaunere, 1964). On the basis of many observations in the Atlantic Ocean, a detailed overall picture of the oceanic crust was established but significant deviations from this average were found on the flanks of the Mid-Atlantic Ridge. Numerous new data were also gathered in the Indian and Pacific Ocean (Shor and Raitt, 1969). Detailed surveys involved, for example, the surroundings of the islands of Hawaii and archipelagos to the northeast of Australia (Furumoto et al., 1973). For the first time, the existence of seismic anisotropy of the uppermost mantle was discussed and investigated by a series of special anisotropy experiments in various areas of the Pacific Ocean (e.g., Raitt et al., 1971).

The 1960s accumulated a wealth of data, based on the integration of gradual improvement in seismic-refraction and -reflection techniques, instrumental development and the art of interpretation. In many experiments seismic energy produced by effective explosive sources was recorded out to large distances of several 100 km. Moho contour maps published for Europe (e.g., Morelli et al., 1967; Fig. 2.6-01), the USSR (e.g., Belyaevsky et al., 1973; Fig. 2.6-02) and the United States (Warren and Healy, 1973; Fig. 2.6-03) were based on a large number of published models, available by the end of the 1960s. They not only resulted in a relatively complete picture of the gross velocity-depth structure of the Earth's crust underneath the northern hemisphere around the world, but also detected specific properties of different tectonic areas such as shields, platforms, orogens, basins, rift zones, etc. However, the long-range data had not yet been interpreted in terms of subcrustal fine structure. In order to understand the seismic wave field, various groups had worked on the theoretical background, so that by the end of the 1960s the art of computing synthetic seismograms was ready for application. Other groups had concentrated on the character and basic features of seismic phases so that gradually a fine structure of the hitherto homogeneously layered crust was detected. In particular the character of the crust-mantle boundary was attracting increased interest.

2.7. THE 1970s (1970–1980)

In the 1960s, the personnel situation in the geophysical departments of universities and other research institutions of western Europe had gained by public support of science, leading to an increase of existing facilities as well as to the foundation of new geophysical departments. By the beginning of the 1970s many young scientists had become professors and heads of departments.

With the powerful instrumentation developed in the 1960s and acquired in major numbers in the 1970s, in particular in western Europe, North America, and the USSR, controlled-source seismology approached new frontiers. The 1970s can be characterized by several highlights.

Making use of the rapidly developing new computing facilities, the traveltime routines of the 1960s enabled many different computer programs to be developed. They now allowed a faster interpretation of the large quantity of new data. Furthermore, the consequent application of the reflectivity method published by Fuchs and Müller (1971) enabled the calculation of synthetic seismograms for the full wave field in laterally homogeneous structures. By comparing observational and synthetic data, the models calculated by traveltime routines could now be improved and/or verified by including the amplitude information from the correlated phases. Braile and Smith (1975) published a whole series of synthetic record sections calculated for typical crustal models. The further development and application of the time-term method widely used by British scientists to the data of dense networks of seismic-refraction profiles led unexpectedly to the detection of uppermost mantle velocity anisotropy under continents (Bamford, 1973; Bamford et al., 1979).

Underwater shots in the Lake Superior experiment of 1965 had shown that controlled seismic sources could efficiently be recorded to at least 2000 km distance if the recording conditions were favorable. In particular, it was recognized that the hitherto uniform P_n phase at distances of several hundred kilometers in

reality consisted of a number of different phases reflected from various depth levels in the uppermost mantle. This enabled the seismic investigation of hitherto unknown depth ranges below the Moho and led to the detection of fine structures in the lower lithosphere down to 80–100 km depth. A considerable number of long-range profiles was subsequently organized and recorded throughout Europe (Fig. 2.7-01). One-thousand-km-long lines through France (Hirn et al., 1973, 1975), Britain (Faber and Bamford, 1979) and along the axis of the Alps (Alpine Explosion Seismology Group, 1976), and a 2000-km-long line through Scandinavia (Guggisberg and Berthelsen, 1987), for example, became well-known experiments (see also Fuchs et al., 1987). In the USSR, super-long profiles were recorded using nuclear devices as sources (Fig. 2.7-01). However, the data from these profiles did not become generally available until the early 1990s (Pavlenkova, 1996).

Japanese research activities in the 1970s were characterized by two programs. Big offshore shot experiments were carried out in 1974–1976 in the Pacific Ocean and the Sea of Japan and onshore seismic refraction surveys started in 1979 under the framework of the national project of earthquake prediction. The onshore activities were continued up to 2003, providing various scale crustal heterogeneities existing within NE and SW Japan arcs (e.g., Asano et al., 1981; Yoshii, 1994). In Australia, long-term recording devices had been developed (Finlayson, 2010; Appendix 2-2). A long-range seismic profile through the center of Australia recorded earthquakes from South East Asia and studied the upper mantle (Hales et al., 1980). Crustal profiles based mainly on quarry blasts investigated the crustal structure of Precambrian Shield in western Australia (Drummond 1979, 1981) and under the Lachlan Fold Belt in southeastern Australia (Finlayson et al., 1979). In southwest Africa, a large-scale seismic refraction experiment investigated the Damara orogen and the adjacent Kalahari craton, using the German MARS-66 equipment and long-term recording devices developed at the University of Johannesburg (Baier et al., 1983).

In the oceans, long-range profiles were recorded that penetrated well below Moho. In the Pacific Ocean, e.g., a 600-km-long line was laid out in the northeastern Pacific between the Clarion and Molokai fracture zones (Orcutt and Dorman, 1977). In the western Pacific basin, explosions were recorded by ocean-bottom seismometers up to distances of 1300 km during two Longshot experiments in 1973 in the East Mariana basin (Asada and Shimamura, 1976). From 1974 to 1980, intensive long-range experiments were undertaken in the northwestern Pacific, which detected large scale lithospheric structure and anisotropy in the upper-mantle (Asada and Shimamura, 1979; Asada et al., 1983; Shimamura et al., 1983). In the Atlantic Ocean, from 40° to 43°N a long-range profile was recorded near the Azores (Steinmetz et al., 1977), dealing also with the lower lithosphere. Another long-range seismic refraction experiment in 1977 established an 800 km long line along the southeastern flank of the Reykjanes Ridge which aimed to resolve both crust and upper mantle to greater depths than previously possible (RRISP Working Group,

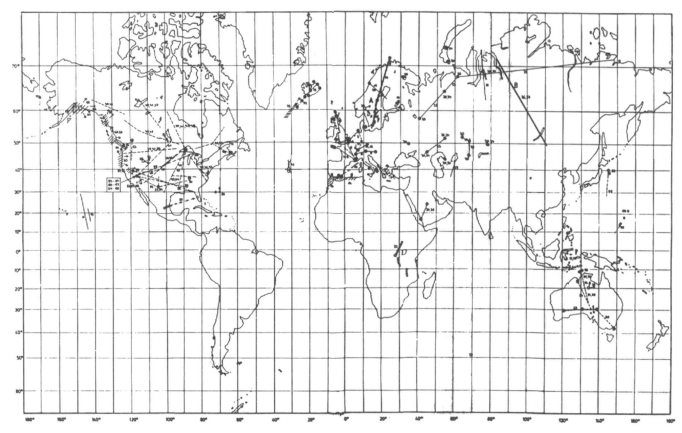

Figure 2.7-01. World map showing long-range profiles until 1980 (from Fuchs et al., 1987, fig. 1). [*In* Fuchs, K., and Froidevaux, C., eds., Composition, structure and dynamics of the lithosphere-asthenosphere system: American Geophysical Union Geodynamics Series, v. 16, p. 137–154. Reproduced by permission of American Geophysical Union.]

1980). Figure 2.7-01 gives an overview, where in the world controlled-source seismic experiments were performed up to 1980 and which penetrated well below Moho (Prodehl, 1984).

Another highlight of the 1970s was an advanced understanding of rifting. Detailed studies were carried out in Europe in the central European Rift System through Germany and France (Prodehl et al., 1995; Fig. 2.7-02), along the Tornquist-Teysseire Zone in Poland (Guterch et al., 1983), in the Afro-Arabian rift system, in the western Jordan–Dead Sea transform, in the Afar triangle of Ethiopia (Ginzburg et al., 1979a; Berckhemer et al., 1975, see Fig. 2.9-02), and in North America in the Mississippi embayment (Mooney et al., 1983), the Rio Grande rift (Olsen et al., 1979) and the Yellowstone–Snake River Plain area, a suggested hot spot (Smith et al., 1982).

In several countries surrounding the Mediterranean, crustal studies were started in the 1970s such as in Portugal, Spain, Morocco, Greece, and Israel. In Italy, both the northern Apennines and adjacent Mediterranean Sea including the island of Corsica and southernmost Italy were the focus of several seismic land and sea investigations (e.g., Morelli et al., 1977).

Seismic near-vertical incidence reflection experiments were organized on a large scale for the first time in the 1970s. They covered distances of 100 km or more which brought new insight

into details of crustal structure and composition of the Earth's crust down to Moho. On the continents sedimentary basins and at sea the ocean basins had already been studied in much detail, the former by petroleum seekers since the 1930s and the latter by marine scientists largely during the decades following World War II. For the study of sedimentary basins, the seismic reflection profiling method had been highly developed over the years by the petroleum industry. Consequently, academic researchers now took advantage of the expertise, techniques, and equipment of the industry to study the whole crystalline crust and obtain not only geophysical models with velocity, density and attenuation, but also determine the extent and the configuration of reflecting horizons (Oliver, 1986). So, in the 1970s the first national large-scale seismic-reflection programs were initiated. COCORP in the United States (e.g., Oliver et al., 1976), a purely seismic reflection program, was soon followed by the Canadian equivalent COCRUST (Mereu et al., 1989), which, however, comprised a combination of crustal reflection and refraction studies. Similar large-scale seismic near-vertical incidence reflection profiles, accompanied by wide-angle reflection observations, were recorded in Germany (Bartelsen et al., 1982; Meissner et al., 1980).

Besides special upper-mantle studies, detailed crustal research continued in the oceans. For the Pacific Ocean, for example,

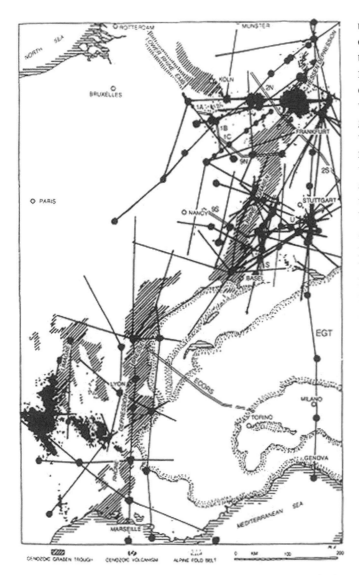

Figure 2.7-02. Location map of explosion seismology surveys in the European Cenozoic rift system (ECRIS) until 1989 (from Prodehl et al., 1995, fig. 1). Double lines—seismic-reflection observations (numbers and letters denote DEKORP and other lines); single lines—seismic-refraction profiles (dots = shotpoints, EGT = European Geotraverse line of 1986). [*In* Olsen, K.H., ed., Continental rifts: evolution, structure, tectonics: Amsterdam, Elsevier, p. 133–212. Copyright Elsevier.]

Woollard (1975) reviewed the available data and published several cross sections with details of crustal structure. In the 1970s, the main interest of oceanic crust and upper mantle research had shifted from the investigation of crustal and upper mantle structure under ocean basins to special structures with anomalous properties such as mid-ocean ridges, hotspots, ocean islands and ocean-continent transition zones. The 1970s were also the decade when ocean-bottom seismometers (OBS) and ocean-bottom hydrophones (OBH) were developed in various institutions and successfully tested in experiments around the world. From the numerous new investigations carried out in the 1970s,

using airguns and expendable sonobuoys, a more detailed picture of the crust under the ocean basins was obtained. The application of the reflectivity method to calculate synthetic seismograms for deep-ocean data proved to be most powerful in unraveling details of the basement structure (Spudich and Orcutt, 1980). It was shown that the structure was better represented by velocity gradient zones than by discrete constant-velocity layers. There was also research work going on to determine if seismic anisotropy could be observed in the crust. Bibee and Shor (1976) investigated a large number of standard marine refraction studies and concluded that anisotropy in the crust was insignificant, but that mantle velocities, however, exhibited a high correlation with both age and azimuth, indicating an increase of velocity with age and ~5% anisotropy with the highest velocity in the direction perpendicular to the local magnetic anomalies. In OBS refraction experiments in the 1970s on the northern East Pacific Rise, a fast-spreading ridge, a localized low-velocity zone was detected for the first time (Orcutt et al., 1976; Minshull, 2002) which was interpreted as resulting from the presence of a magma chamber containing partially molten rocks. Other seismic studies in the 1970s were concentrated on the Mid-Atlantic Ridge, a slow-spreading ridge. Here, early OBS refraction experiments found neither a low-velocity zone corresponding to magma chamber nor a strong velocity contrast at Moho depths. Instead, the crust-mantle boundary was defined by a gradual increase in velocities (Fowler, 1976, 1978; Minshull, 2002).

Numerous other crustal studies were performed during the 1970s, leading to a large amount of crustal data available by 1980 (Fig. 2.7-03; for details, see Appendix 7-1). The main features of crustal and uppermost mantle studies available by the end the 1970s were compiled as tables giving details such as thicknesses of the whole crust, upper and lower crust and corresponding average velocities (Prodehl, 1984). An overview of the large number of major seismic projects undertaken in the 1970s around the world and mentioned in Chapter 7 can be seen in Table 2.7.

Fundamental for future decades were new steps undertaken both in theory and instrumental development. During the second half of the 1970s, the ray method was developed (Červený et al., 1977), which led to the development of several ray tracing methods which could handle data in complex tectonic structures and which would become the standard tool for interpreting seismic data in the 1980s and following decades up to the present day. Also other groups were writing synthetic seismogram routines on the basis of ray theory which were widely applied, for example, the program by McMechan and Mooney (1980) in the United States or the program by Spence et al. (1984) in Canada and elsewhere.

At the same time, J.H. Healy started to raise new interest in explosion seismology in the United States by developing a new type of equipment, later called the cassette recorders, which was ready for its first field test in Saudi Arabia in 1978. The Saudi Arabian long-range profile set new standards and laid the foundation for new activities in the following 1980s (Healy et al., 1982). This new system was small, was timed by a sophisticated clock

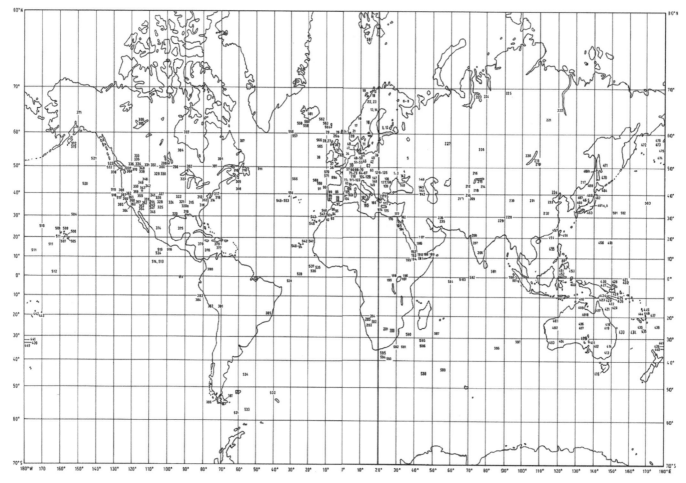

Figure 2.7-03. World map showing points where seismic data on crustal structure were available by 1980–1981 (from Prodehl, 1984, fig. 10). [*In* Hellwege, K.-H., editor in chief, Landolt Börnstein New Series: Numerical data and functional relationships in science and technology. Group V, Volume 2a: Fuchs, K., and Soffel, H., eds., Physical properties of the interior of the earth, the moon and the planets: Berlin-Heidelberg, Springer, p. 97–206. Reproduced with kind permission of Springer Science+Business Media.]

of high accuracy, was built in large quantities, could easily be operated by untrained personnel, and was able to record mostly automatically and run unattended for several days.

2.8. THE 1980s (1980–1990)

In the 1980s, the success of controlled-source seismology continued, but with increased interest into details of the crust. Computer programs for ray tracing, which had been developed in the 1970s and early 1980s, were now ready for applications. In contrast to the reflectivity method, ray theory enabled the interpretation of complicated structures in two or three dimensions, and therefore the interpretation of the fast-growing amount of data in crustal and upper mantle research work was carried out more and more by applying ray theory (Červený et al., 1977).

With these techniques, the requirement to obtain a denser data coverage, particularly in tectonically complicated regions, grew rapidly. The station spacing of 5 km or more for detailed crustal structure studies which had been regarded as sufficient

up to the 1970s no longer suited the accuracy aimed for. In continental Europe, the MARS-66 analogue system, of which by the early 1980s ~200 units were available, experienced a late peak of magnetic-tape recording systems. Thus it became possible to decrease the station spacing to 2 km in most seismic-refraction experiments.

In North America, the 1980s saw a transformation in the development of new instrumentation. Here too it had also become clear that a station spacing of 5 km or more for detailed crustal structure studies was no longer sufficient. Smaller shotpoint and station spacings were required to achieve the scientific results needed. The development of the cassette recorders (Healy et al., 1982; Murphy, 1988; Appendix A7-5.6) at the U.S. Geological Survey at the end of the 1970s was the first attempt to fill this gap. This system provided a large number of easily maintained instruments with fast play-back facilities and led to an increasing number of fine-tuned experiments in North America. The increasing demand for more recording equipment sped up the developments of a new generation of recording units. It was in Canada where

TABLE 2.7. MAJOR CONTROLLED-SOURCE SEISMIC INVESTIGATIONS OF THE CRUST IN THE 1970S

Chapter	Year	Project	Location	Reference
7.2.1	1971	Crustal project France	France	Sapin and Prodehl (1973)
7.2.1	1971	Lower Lithosphere Project 1971	France	Hirn et al. (1973)
7.2.1	1972	Asthenosphere Project 1972	W-Europe	Steinmetz et al. (1974)
7.2.1	1973	Lower Lithosphere Verification 1971	France	Hirn et al. (1975)
7.2.2	1972	Upper Rhinegraben project	Central Europe	Edel et al. (1975)
7.2.2	1972	Rhônegraben project	South France	Sapin and Hirn (1974)
7.2.3	1970–1973	Southern and Western Portugal	Portugal	Moreira et al. (1977)
7.2.3	1970	ANNA Western Mediterranean project	Mediterranean Sea	Leenhardt et al. (1972)
7.2.3	1971	Southern Italy Puglia and Calabria	Italy	Colombi et al. (1973)
7.2.3	1971	"*Meteor* (1964)" M22 Ionian Sea	Greece	Weigel (1974)
7.2.3	1971	Meteor offshore-onshore Peleponnese	Greece	Weigel (1974)
7.2.3	1973	Island of Crete project	Greece	Makris and Vees (1977)
7.2.3	1973–1974	Central Greece–Aegean Sea	Greece	Makris (1977)
7.2.3	1974	"*Meteor* (1964)" expedition M33	Cretan Sea	Makris et al. (1977)
7.2.3	1974	Northern Appenines to Corsica	Italy-France	Morelli et al. (1977)
7.2.3	1974	Long range project Ligurian Sea	Mediterranean Sea	Hirn et al. (1977)
7.2.3	1974–1975	Betic Cordillera long range	Southern Spain	Banda and Ansorge (1980)
7.2.3	1975	Moroccan Meseta and Atlas	Morocco	Makris et al. (1985)
7.2.3	1976	Balearic islands land-sea project	Spain	Banda et al. (1980)
7.2.3	1976, 1981	Eastern Spain	Spain	Zeyen et al. (1985)
7.2.3	1977, 1979	Canary islands land-sea project	Off North Africa	Banda et al. (1981a)
7.2.3	1977, 1980	Betic Cordillera crust	South Spain	Barranco et al. (1990)
7.2.3	1978	Pyrenees seismic refraction lines	Spain-France	Gallart et al. (1980)
7.2.3	1978	"*Meteor* (1964)" expedition M50	Ionian Sea	Avedik et al. (1981)
7.2.3	1979	Long-range project Thyrrenian Sea	Mediterranean Sea	Steinmetz et al. (1983)
7.2.4	1974	Lithospheric Seismic Profile through Britain, Crustal Seismic Project Britain	British Isles	Bamford et al. (1976, 1978)
7.2.4	1974	Lithospheric Seismic Profile through Britain, Lower Lithosphere Britain	British Isles	Faber and Bamford (1979)
7.2.4	1976	Lewisian Units Seismic Traverse	British Isles	Hall (1978)
7.2.4	1977	South Irish Sea Seismic Experiment Wales	British Isles	Ransome (1979)
7.2.4	1979	Western Isles Seismic Experiment offshore W Scotland	British Isles	Summers et al. (1982)
7.2.4	1975	ALP75Alpine Longitudinal Profile	Alps	Miller et al. (1978)
7.2.4	1977	SÜDALP77	Northern Italy	Ansorge et al. (1979b)
7.2.4	1979	Institute of Geological Sciences, Wiltshire Survey	British Isles	Kenolty et al. (1981)
7.2.5	1973, 1975	Hunsrück seism refl surveys	West Germany	Meissner et al. (1980)
7.2.5	1974	North German Basin hydrocarbon survey	North Germany	Yoon et al. (2008)
7.2.5	1975–1976	North German Plain	North Germany	Reichert (1993)
7.2.5	1976	Northern Rhônegraben	France	Ansorge and Mueller (1979)
7.2.5	1978	Hunsrück-Eifel refraction line	West Germany	Meissner et al. (1983)
7.2.5	1978	Urach refraction seismics	Southwest Germany	Jentsch et al. (1982)
7.2.5	1979	Urach reflection seismics	Southwest Germany	Bartelsen et al. (1982)
7.2.6	1972	Blue Road profile	Norway-Sweden	Hirschleber et al. (1975)
7.2.6	1979	Rhenish Massif long range project	Germany-France	Mechie et al. (1983)
7.2.6	1979	FENNOLORA Scandinavia crust	Scandinavia	Guggisberg et al. (1991)
7.2.6	1979	FENNOLORA Scandinavia asthenosphere	Scandinavia	Guggisberg and Berthelsen (1987)
7.2.7	1970ff	M1-M13 Fore-Sudetic seismics	Poland	Toporkiewicz (1986)
7.2.7	1970ff	LT2-LT5Tornquist-Teisseye Zone	Poland	Guterch (1977)
7.2.7	1970ff	Deep seismic international profiles	Eastern Europe	Guterch et al. (1991)
7.2.7	1970ff	Deep seismic international profiles	Southeastern Europe	Radulescu et al. (1976)
7.3	1970ff	Deep seismic sounding profiles	Southwest USSR	Sollogub et al. (1973)
7.3	1972ff	Deep seismic sounding Voronezh Shield	Southwest USSR	Tarkov and Basula (1983)
7.3	1970ff	Deep seismic sounding crustal profiles USSR	Europe and Asia	Zverev and Kosminskaya (1980)
7.3	1970ff	Peaceful nuclear explosions asthenosphere USSR	Europe and Asia	Egorkin and Pavlenkova (1981)
7.3	1977–1979	Peaceful nuclear explosions athenosphere profiles	USSR	Mechie et al. (1993)
7.4.1	1971–1972	Bingham, Utah–Wasatch Front	Western USA	Braile et al. (1974)
7.4.1	1973	Long Valley Caldera seismics	Western USA	Hill (1976)
7.4.1	1976	DICE THROW Rio Gande Rift	Western USA	Olsen et al. (1979)
7.4.1	1978	Arizona Basin and Range province	Western USA	Sinno et al. (1981)
7.4.1	1978, 1980	Yellowstone–Snake River Plain	Western USA	Smith et al. (1982)
7.4.2	1978–1979	USGS seismic reflection, New Madrid	Eastern USA	Hamilton (1986)
7.4.2	1979	Imperial Valley, California	Western USA	Fuis et al. (1984)
7.4.2	1979–1980	Mississippi embayment reflections survey	Central USA	Mooney et al. (1983)
7.4.3	1975ff	COCORP	USA	Brewer and Oliver (1980)
7.4.3	1975	COCORP Hardeman County	Central USA	Oliver et al. (1976)
7.4.3	1975–1976	COCORP Rio Grande Rift	Western USA	Brown et al. (1979)
7.4.3	1976	COCORP San Andreas Fault project	Western USA	Long et al. (19
7.4.3	1977	COCORP Wind River Mountains	Western USA	Smithson et al. (1979)

(continued)

TABLE 2.7. MAJOR CONTROLLED-SOURCE SEISMIC INVESTIGATIONS OF THE CRUST IN THE 1970S (*continued*)

Chapter	Year	Project	Location	Reference
7.4.3	1978	COCORP Michigan basin Michigan reflection survey	Central USA	Jensen et al. (1979)
7.4.3	1978–1979	COCORP S Appalachians GAA and GAB	Eastern USA	Cook et al. (1979)
7.4.3	1978–1979	COCORP Charleston–Georgia GAC/CHP	Eastern USA	Cook et al. (1979)
7.4.3	1979	COCORP Wichita Uplift Oklahoma Wichita Mountains reflection survey	Central USA	Brewer and Oliver (1980)
7.4.3	1979	COCORP Minnesota Archean Minnesota reflection survey	Central USA	Gibbs et al. (1984)
7.4.3	1979	USGS Grandfather Mountain Profile	Eastern USA	Harris et al. (1981)
7.4.4	1970	Baffin Bay	Canada	Keen et al. (1972)
7.4.4	1972–1973	Cordillera seismic reflection profiles	Canada	Berry and Mayr (1977)
7.4.4	1977ff	COCRUST Williston Basin	Canada	Green et al. (1986)
7.5.1	1971	Djibouti crustal survey	Djibouti	Ruegg (1975)
7.5.1	1972	Afar crustal survey	Ethiopia	Berckhemer et al. (1975)
7.5.2	1977	Jordan–Dead Sea transform	Israel	Ginzburg et al. (1979)
7.5.2	1978	Cyprus-Israel	Israel	Makris et al. (1983a)
7.5.2	1978	Saudi Arabia lithosphere	Saudi Arabia	Mooney et al. (1985)
7.5.2	1978	Northern Red Sea–Egypt	Egypt	Rihm et al. (1991)
7.5.2	1978	Zagros Belt reconnaissance	Iran	Giese et al. (1984)
7.6.1	1972–1975	Deep seismic sounding S-India, Indian Shield	South India	Kaila et al. (1979)
7.6.1	1972ff	Deep seismic sounding N-India, Kashmir	North India	Kaila et al. (1978)
7.6.1	1970s	Cambay basin	India	Kaila et al. (1981a)
7.6.1	1970s	Koyna Dam	India	Kaila et al. (1981b)
7.6.2	1975–1977	Tibetan plateau	China	Teng (1987)
7.6.2	1978ff	Qaidam basin, Yongping eplosions	China	Teng (1979); Yuan et al. (1986)
7.6.3	1975	Shizuoka district, central Japan	Japan	Ikami (1978)
7.6.3	1976	Western Kanto district	Japan	Kaneda et al. (1979)
7.6.3	1979	Mishima-Shimoda profile, Izu peninsula	Central Japan	Asano et al. (1982)
7.6.3	1979	Seismic intrusion detector onshore line Shikoku Island	Southwest Japan	Ikami et al. (1982)
7.7.1	1970–1971	Deep reflection experiments	Australia	Moss and Dooley (1988)
7.7.1	1972	Trans-Australia seismic survey	Australia	Finlayson et al. (1974)
7.7.1	1973	Bowen Basin, Queensland	Eastern Australia	Collins (1978)
7.7.1	1974–1975	Fiordland South Island NZ	New Zealand	Davey and Broadbent (1980)
7.7.1	1975	Long range earthquake profile	Central Australia	Hales et al. (1980)
7.7.1	1976–1978	Lachlan Foldbelt, NewSouthWales	Eastern Australia	Finlayson et al. (1979)
7.7.1	1976–1978	Deep seismic reflection experiments	Northeast Australia	Mathur (1983)
7.7.1	1977, 1979	Pilbara-Yilgarn craton projects	Northwest Australia	Drummond (1979)
7.7.1	1979	Mount Isa–Tennant Creek profile	Central Australia	Finlayson (1982)
7.7.2	1975	Damara orogen	Namibia	Baier et al. (1983)
7.7.2	1975	Damara orogen	Namibia	Baier et al. (1983)
7.7.3	1973	Narino III Pacific Ocean–Northern Andes	Colombia	Meyer et al. (1976)
7.7.3	1976	Cordilleras Northern Andes	Colombia	Mooney et al. (1979)
7.7.3	1978	Cordilleras Northern Andes	Colombia	Flueh et al. (1981)
7.7.4	1979–1980	Polish Antarctica expedition	West Antarctica	Guterch et al. (1985)
7.8.2	1970	Northeast Pacific Ocean	Canada	Keen and Barrett (1971)
7.8.2	1970–1971	"*Vitiaz*" cruise no 49	Pacific	Kosminskaya et al. (1973)
7.8.2	1972–1973	Peru-Chile trench	Pacific-Peru	Hussong et al. (1976)
7.8.2	1973	East Papua crustal survey	Papua New Guinea	Finlayson et al. (1977)
7.8.2	1973–1974	Cocos Plate east of East Pacific Rise	Pacific	Lewis and Snydsman (1979)
7.8.2	1973–1974	Longshot Experiment off Japan	Pacific	Asada and Shimamura (1976)
7.8.2	1973, 1976	Mariana long range experiments	Japan	Nagumo et al. (1981)
7.8.2	1974	East Pacific Rise at Siqueiros Fracture Zone	Pacific	Orcutt et al. (1976)
7.8.2	1974	Offshore large shot experiment, Pacific	Off Japan	Okada et al. (1979)
7.8.2	1976	Offshore large shot experiment, Sea of Japan	Off Japan	Okada et al. (1978)
7.8.2	1976	600 km long-range profile	Northeast Pacific	Orcutt (1977)
7.8.2	1976	Banda Sea	North off Australia	Jacobson et al. (1979)
7.8.2	1976	Timor Sea	North off Australia	Rynn and Reid (1983)
7.8.2	1976	Great Australian Bight	South off Australia	Talwani et al. (1979)
7.8.2	1976	Explorer Ridge W Vancouver	Canada	Cheung and Clowes (1981)
7.8.2	1976	East Pacific Rise crest	Eastern Pacific	Herron et al. (1978)
7.8.2	1976, 1978	Hawaii Big Island projects	Hawaii	Zucca et al. (1982)
7.8.2	1977–1979	Shirshov Institute of Oceanology world cruise	Oceans	Neprochnov (1989)
7.8.2	1979	Rivera Ocean Seismic Experiment (ROSE)	East Pacific Rise	Ewing and Meyer (1982)
7.8.3	1975	Offshore Sumatra	Indian Ocean	Kiekhefer et al. (1980)
7.8.3	1978	Agulhas Plateau	Southern Indian Ocean	Tucholke et al. (1981)
7.8.3	1978	Madagascar Ridge, Crozet Plateau	Indian Ocean	Goslin et al. (1981)
7.8.4	1972	NASP Iceland to Scotland	North Atlantic	Bott and Gunnarson (1980)
7.8.4	1973	Mid-Atlantic Ridge at 37°N	Southeast of Azores	Whitmarsh (1975)
7.8.4	1974–1975	Mid-Atlantic Ridge long range	North of Azores	Steinmetz et al. (1977)

(*continued*)

TABLE 2.7. MAJOR CONTROLLED-SOURCE SEISMIC INVESTIGATIONS OF THE CRUST IN THE 1970S (*continued*)

Chapter	Year	Project	Location	Reference
7.8.4	1974	"*Meteor* (1964)" expedition M33	Off West Africa	Sarnthein et al. (2008)
7.8.4	1975	"*Meteor* (1964)" expedition M39	Off West Africa	Sarnthein et al. (2008)
7.8.4	1975	Mid-Atlantic Ridge at 45°N	Atlantic	Fowler (1978)
7.8.4	1975	Hebridean Margin Seismic Project	West of Scotland	Bott et al. (1979)
7.8.4	1975–1976	OBS tests in Bay of Biscay	West of France	Avedik et al. (1978)
7.8.4	1976–1977	OBS tests in Norwegian Sea–Blue Norma	West of Norway	Avedik et al. (1978)
7.8.4	1976, 1978	Spitsbergen expeditions	Spitsbergen	Sellevoll et al. (1982)
7.8.4	1977	Kane fracture zone, Mid-Atlantic Ridge	Atlantic	Detrick and Purdy (1980)
7.8.4	1977	"*Meteor* (1964)" expedition M46	Off West Africa	Sarnthein et al. (2008)
7.8.4	1977	Reykjanes Ridge at 60°30′W	Atlantic	Bunch (1980)
7.8.4	1977	"*Meteor* (1964)" M45 Reykjanes Ridge	Atlantic	Reykjanes Ridge Seismic Project Working Group (1980)
7.8.4	1977	Reykjanes Ridge–Iceland Seismic Project	Iceland	Reykjanes Ridge Seismic Project Working Group (1980)
7.8.4	1977–1978	Eastern Canada margin	Canada	Keen and Barrett (1981)
7.8.4	1978	Azores–Biscay rise	Atlantic Ocean	Whitmarsh et al. (1982)
7.8.4	1978	Mid-Atlantic Ridge 45°N	Atlantic	White and Whitmarsh (1984)
7.8.4	1978	"*Meteor* (1964)" expedition M48	Iceland–Faeroe	Sarnthein et al. (2008)
7.8.4	1979	Long Island Platform marine multichannel seismic survey	Eastern USA	Hutchinson et al. (1986)
7.8.4	1979ff	Onshore-offshore reflection work	Eastern USA	Behrendt (1986)

the first digital equipment, the Portable Refraction Seismograph (PRS1), was being developed by the Geological Survey of Canada and afterwards built by EDA Instruments Ltd. (Asudeh et al., 1992). By the end of the 1980s both in Canada and in the United States, not only had a large number of instruments become available, but also the age of digital recorders had started, pushed forward in particular by the foundation of LITHOPROBE in Canada and of IRIS/PASSCAL in the United States.

In Europe, three different approaches to detailed crustal studies were undertaken. First, crustal research in Europe received a new impulse from reflection seismology. Second, the European Geotraverse involved a large-scale seismic-refraction traverse. Third, continental deep drilling was accompanied by detailed reflection-refraction surveys.

Following the establishment of COCORP in the United States and COCRUST in Canada in the late 1970s, in Europe national groups such as BIRPS in Britain (Klemperer and Hobbs, 1991; see Fig. 8.3.1-01), DEKORP in Germany (Meissner et al., 1991a; see Fig. 8.3.1-07), or ECORS in France (Bois et al., 1986; see Fig. 8.3.1-03) were rapidly formed in the early 1980s. In Switzerland, Czechoslovakia, and Italy (CROP), similar research activities were initiated, and finally the multinational program BABEL investigating the Baltic Sea was born. Since 1985, the Australian Geological Survey Organisation (AGSO) has run a large program of deep seismic reflection surveys offshore and onshore Australia (Finlayson, 2010; Appendix 2-2). They all followed the ideas developed in the late 1970s by COCORP in North America to investigate the Earth's crust in great detail by applying vertical-incidence reflection work in a big style. In contrast to most of the early near-vertical incidence seismic reflection work in the 1950s and 1960s in Canada, Germany, the USSR, United States, and in Australia, the initiators of COCORP had hired a commercial reflection company for the field work and also applied their processing and interpretation

techniques (Oliver et al., 1976) on the new data involving the whole crust. The same philosophy was now also applied by the initiators of the national European large-scale seismic reflection programs. All seismic reflection surveys accomplished by 1988 were compiled in a location map by Sadowiak et al. (1991) which is reproduced in Figure 2.8-01.

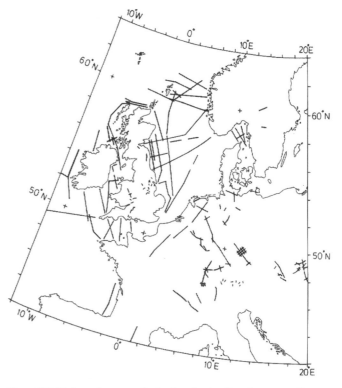

Figure 2.8-01. Location map of seismic reflection lines in western Europe observed until 1988 (from Sadowiak et al., 1991, fig. 1). [Geophysical Monograph 105, p. 45–54. Copyright John Wiley & Sons Ltd.]

In 1984, a series of international symposia on the seismic probing of the continents and their margins was started which in the beginning had the main focus to discuss large-scale seismic-reflection surveys and their impetus on our knowledge of the Earth's crustal structure (Barazangi and Brown, 1986a, 1986b; Matthews and Smith, 1987; Leven et al., 1990; Meissner et al., 1991a).

In the frame of the large-scale interdisciplinary project "European Geotraverse" (Fig. 2.8-02), covering a corridor through all major tectonic units of Europe from the North Cape to Tunisia, North Africa, a large-scale seismic-refraction survey was initiated (Ansorge et al., 1992). The seismic-refraction investigation was split into northern, central, and southern sections and was carried out over a period of several years. The Swiss national seismic reflection profile NFP20 through the central Alps was integrated into the EGT refraction observations (Valasek et al., 1991) and detected the European subduction under the Alps. Finally the Iberian Lithosphere Heterogeneity and Anisotropy (ILIHA) Project was designed as project no. 11 of the European Geotraverse, because only the Iberian peninsula had dimensions where large sea shots could be recorded up to 600–800 km distance on reversed long-range profiles on more or less homogeneous Hercynian crust.

The increasing interest in deep continental drilling and the search for suitable super-deep drill sites led to a variety of large-scale seismic-refraction and reflection surveys, particularly in Russia around the Kola super-deep drillhole and in central

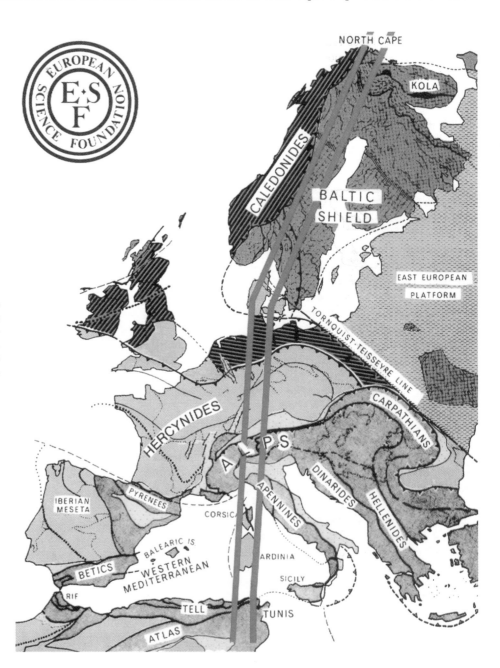

Figure 2.8-02. Location map of the European GeoTraverse (from European Science Foundation, 1990, cover). [European Science Foundation, Strasbourg, 67 p. Reproduced by kind permission of the European Science Foundation, Strasbourg, France.]

Europe at two proposed drill sites in Germany, which partly co-incided with the national reflection programs (Emmermann and Wohlenberg, 1989).

Besides BIRPS, other British projects in the 1980s investigated the North Sea, northern Britain, and the Irish Sea. The North Sea project involved both refraction and reflection work (Barton, 1986). The Caledonian Suture Seismic Project profile traversed northern Britain with shots in the North and Irish Seas (Bott et al., 1983) and stimulated the first seismic crustal profile through Ireland (Jacob et al., 1985). In addition to the seismic reflection profiling of BIRPS, a major seismic investigation followed using both land and sea profiling in and around Ireland (Landes et al., 2005).

In the USSR, the third period of Russian deep-seismic sounding investigations, which had started at the end of the 1970s, continued through the 1980s (Pavlenkova, 1996). The network of seismic profiles covered almost the whole territory of the USSR (Fig. 2.8-03). This seismic research included three-component magnetic recordings of shots of varying sizes recorded by up to 300 stations on profiles with 2500–3000 km length, involving both large conventional explosives and peaceful nuclear explosions (PNE).

In North America, COCORP continued its seismic-reflection program and many new areas were systematically covered (Brown et al., 1986; see Fig. 8.5.3-01). Furthermore, many seismic-refraction experiments were carried out. The new seismic-refraction equipment of the U.S. Geological Survey led to considerably increased activity, parallel to the efforts of COCORP. For example, a large-scale seismic-refraction experiment traversed the Basin and Range province, more or less parallel to the 40° COCORP survey (Catchings and PASSCAL Working Group, 1988). The goal of the PACE (Pacific to Arizona Crustal Experiment) project was to study the evolutionary history of metamorphic core complexes of the western Cordillera (McCarthy et al., 1991). New crustal data were collected in the southern Rio Grande rift area. In northeastern United States and adjacent Canada a seismic refraction/wide-angle reflection experiment crossed the northern Appalachians and the Adirondack Massif and ended in the Grenville province of the North American craton (Hughes and Luetgert, 1991). Braile et al. (1989) and Mooney and Braile (1989) summarized all seismic-refraction profiles, recorded in North America until ca. 1988 (Fig. 2.8-04).

The Canadian geophysical institutions subsequently formulated the joint programs COCRUST (Mereu et al., 1989; see

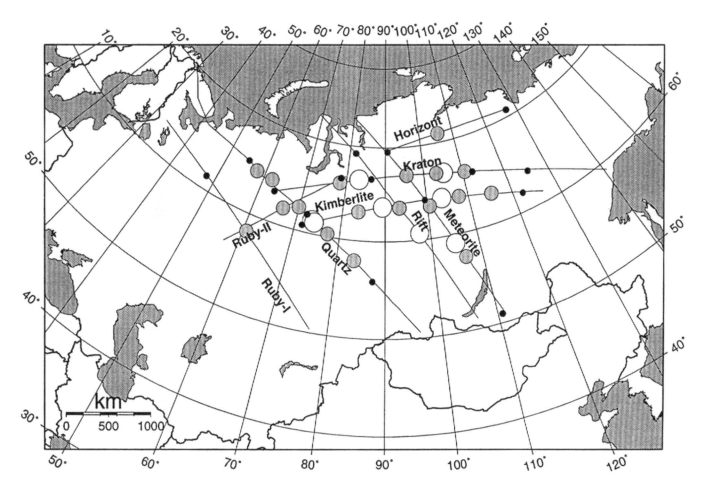

Figure 2.8-03. Location map of deep seismic sounding profiles in the USSR and locations of the peaceful nuclear explosions (PNE) (from Ryberg et al., 1998, fig. 1). [Journal of Geophysical Research, v. 103, p. 811–822. Reproduced by permission of American Geophysical Union.]

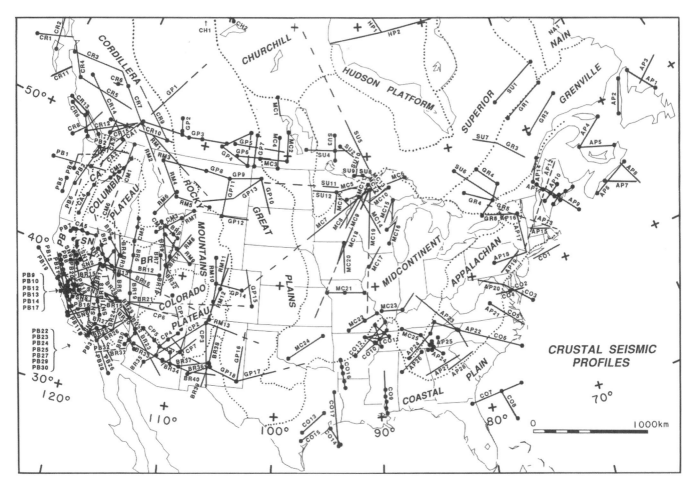

Figure 2.8-04. Location of seismic refraction surveys in Canada and the United States until 1988 (from Braile et al., 1989, fig. 1). For additional seismic profiles in Alaska, northern Canada and Mexico see Mooney and Braile (1989, fig. 1). [*In* Pakiser, L.C., and Mooney, W.D., eds., Geophysical framework of the continental United States: Geological Society of America Memoir 172, p. 655–680. Reproduced by permission of the Geological Society of America.]

also Fig. 8.5.2-01) and LITHOPROBE (e.g., Clowes, 1993) for a systematic crustal and uppermost-mantle investigation of the Canadian territory by a joint application of seismic-refraction and -reflection methodology. Most of the Canadian profiles are included in Figure 2.8-04. Furthermore, a seismic transect through Alaska, the Trans-Alaska Crustal Transect project (TACT; see Fig. 8.5.4-01), was undertaken in several steps from 1984 to 1990 (Plafker and Mooney, 1997).

The Afro-Arabian rift saw major activities of crustal research. Following the 1977 crustal investigation of the Jordan–Dead Sea transform in Israel (Ginzburg et al., 1979a), in 1984 a second survey explored the eastern part in Jordan (El-Isa et al., 1987a). The East African Rift in Kenya became the target of the first KRISP operation in 1985 (Henry et al., 1990). Most surveys, however, concentrated on the Red Sea area (for locations, see Fig. 2.9-02). Several symposia were held to report on the state of the art (e.g., Le Pichon and Cochran, 1988; Makris et al., 1991).

In India a total of 6000 km of long-range refraction and wide-angle reflection observations had been obtained by the end

of the 1980s by deep-seismic sounding surveys throughout India along 20 profiles (Mahadevan, 1994; see Fig. 8.7.1-01). The investigations aimed in particular on the deep structure of basins and rift systems, adopting continuous profiling over major portions of the deep-sounding profiles using geophone spacings of 200 m and shotpoint intervals of 20–40 km. Recordings with useful energy were obtained up to distances of 400 km.

In China, a major cooperative program between Chinese institutions and European and North American institutions was started. The first explosion seismology operation was a joint Sino-French study, which was carried out in Tibet investigating the Himalayan border and the adjacent Lhasa block to the north. A 500-km-long east-west line consisting of a system of reversed and overlapping profiles was recorded. For detailed crustal studies of whole of China, some 250 standardized instruments for deep-seismic sounding, recording on two-channel magnetic tape cassettes, were distributed among various research groups within China belonging to the State Seismological Bureau. Thus, in the 1980s a major activity of seismic research started (see Fig. 8.7.2-06) and has continued since then (Li and Mooney, 1998).

In Japan, onshore seismic crustal research in the 1980s was mainly carried out in the frame of the national Earthquake Prediction Program and focused on the upper crustal structure, searching particularly for major fault zones and tectonic lines (e.g., Ikami et al., 1986; Matsu'ura et al., 1991).

In Australia, several major seismic reflection surveys were undertaken (Finlayson, 2010; Appendix 2-2; see Fig. 8.8.1-03). For example, during 1980–1982 deep reflection data were obtained in the central Eromanga basin in southwestern Queensland. It resulted in 1400 km of traversing in a regional grid on continuous profiles up to 270 km long (Moss and Mathur, 1986). Following these successful studies, an Australian Continental Reflection Profiling Program (ACORP) was initiated and subsequently, in 1985, two major north-south–oriented deep seismic reflection surveys were conducted in central Australia acquiring 486 line kilometers across the Arunta block and the Amadeus basin (Wright et al., 1990).

Major activities of crustal seismic research started in South America, when in 1982 at the Free University of Berlin, Germany, a geoscientific interdisciplinary research group "Mobility of active continental margins" was established. Its main aim was the investigation of the Andes of Chile. Its funding enabled the recording of a number of onshore seismic-refraction lines up to distances of 260 km (see Fig. 8.8.3-01) in northern Chile and adjacent Bolivia and Argentina (Wigger et al., 1994). Energy was primarily obtained by using large quarry blasts of various copper mines, but also some self-organized borehole shots were added. Underwater shots fired by the Chilean navy close to the Chilean coast in the Pacific Ocean were also arranged.

With improved technology, extremely hostile climatic areas such as Antarctica became the focus of extensive seismic research projects. The Institute of Geophysics of the Polish Academy of Sciences undertook several expeditions to explore the structure underneath West Antarctica (Janik, 1997) beginning in 1979 and 1980 and continuing in 1984–1985 and 1987–1988 (see Fig. 8.8.4-01). McMurdo Sound at the southern end of the Ross Sea was the focus of the U.S. Louisiana State University to investigate the east-west boundary of the McMurdo Sound (McGinnis et al., 1985) with seismic refraction profiles in 1980 and 1981.

With advanced techniques, the number of marine seismic experiments carried out in the past three decades literally exploded. From the early 1980s onwards, large dynamic ranges and dense spatial sampling were targeted for the investigation of the crustal structure. For this reason experiments were designed for obtaining large offsets and large-aperture seismic refraction/wide-angle reflection data as well as near-vertical incidence reflections. In the 1980s, the development of non-explosive sources had become effective enough to be successfully recorded over long distance ranges of several 100 km at sea and also, in onshore-offshore experiments, on land. Of greatest importance for the advance of marine deep-seismic sounding was the fact that gradually over the years more and more ocean bottom seismographs came into use, giving much better signal-to-noise ratios than was possible to obtain with strings of hydrophones which suffered also from the noise produced by the moving towing ship.

A worldwide study of the lower crust and upper mantle using ocean bottom seismographs (OBS) and big airguns in the oceans was performed by the Shirshov Institute of Oceanology (Neprochnov, 1989) in the period from 1977 to 1984 (see Fig. 8.9.2-01). The North Atlantic Transect (NAT Study Group, 1985) provided a major improvement on the knowledge of the structure of the oceanic crust and its variability on a large regional scale. Other spectacular images of the internal structure of the oceanic crust were published by White et al. (1990). They showed widespread occurrence of intracrustal reflectivity in the western Central Atlantic Ocean The research project RAPIDS (Rockall and Porcupine Irish Deep Seismic) project of the Dublin Institute for Advanced Studies and partners consisted of two orthogonal wide-angle seismic profiles totaling 1600 km. The individual lines were typically 200–250 km long and produced a 1000-km-long east-west seismic profile from Ireland to the Iceland Basin crossing various troughs, basins and intervening banks (e.g., Hauser et al., 1995).

Various projects dealt with the ocean-continent transition at continental margins. For example, seismic surveys were carried out in the vicinity of the northern Japan trench (Suyehiro and Nishizawa, 1994) or across the east Oman continental margin north of the Masirah Island ophiolite (Barton et al., 1990). An extended offshore marine survey targeted the crustal structure off Norway on the Voering plateau (Zehnder et al., 1990) and along the Lofoten margin (Mjelde et al., 1992). Detailed seismic investigations aimed for the deep structure in the transition zones west and north of Spitsbergen (Czuba et al., 1999) as well as eastern Greenland and its margin on the western side of the northernmost Atlantic Ocean (e.g., Mandler and Jokat, 1998). The 1981 Large Aperture Seismic Experiment (LASE) was one of the first experiments along the Atlantic continental margin of North America (e.g., Trehu et al., 1989b).

Only a few examples of research projects could be mentioned in this overview. A summary of all projects mentioned in Chapter 8 is compiled in Table 2.8.

A particular aspect of the new era of continental crustal research in the 1980s was the compatibility of results obtained either from the very detailed seismic-reflection projects or from the less dense seismic-refraction observations. In the beginning of COCORP, accompanying wide-angle piggyback experiments were rare. The different techniques and different frequency ranges of the seismic signals of near-incidence reflection research work and of wide-angle reflection profiling led to quite different presentations of crustal structure, and it took a while before the different philosophies were jointly discussed.

However, in central Europe as well as in Canada and Australia, close cooperation between the "reflection" and "refraction" groups started. The Canadian research programs COCRUST and LITHOPROBE involved the simultaneous use of both methods.

TABLE 2.8. MAJOR CONTROLLED-SOURCE SEISMIC INVESTIGATIONS OF THE CRUST IN THE 1980S

Chapter	Year	Project	Location	Reference
8.3.1.1	1980–1982	IGS land reflection surveys	British Isles	Whittaker and Chadwick (1983)
8.3.1.1	1981	BIRPS MOIST line	Northwest British Isles	Matthews and Cheadle (1986)
8.3.1.1	1982	SHELL UK82-101	North Sea	Klemperer and Hurich (1990)
8.3.1.1	1982	BIRPS WINCH lines	Northwest British Isles	Matthews and Cheadle (1986)
8.3.1.1	1983	BIRPS-ECORS Southwest AT lines	English Channel	Matthews and Cheadle (1986)
8.3.1.1	1983	NOPEC SNST83-07	South North Sea	Holliger and Klemperer (1990)
8.3.1.1	1983	North Sea BIRPS along SALT	North Sea	Barton (1986)
8.3.1.1	1984	BIRPS SHET (Sheltland survey)	North Sea	McGeary (1987)
8.3.1.1	1984–1985	BIRPS North Sea Deep Profile	North Sea	Matthews and Cheadle (1986)
8.3.1.1	1985	SAP-5 and N-11 (Rockall Trough)	Northwest of Ireland	Joppen and White (1990)
8.3.1.1	1985	BIRPS NEC (North East Coast line)	North Sea	Freeman et al. (1988)
8.3.1.1	1985	BIRPS WAM	Southwest of Channel	Peddy et al. (1989)
8.3.1.1	1986–1987	BIRPS GRID lines	North of Scotland	Blundell and Docherty (1987)
8.3.1.1	1987	BIRPS SLAVE	ESP experiment	Blundell and Docherty (1987)
8.3.1.1	1987	BIRPS WIRE	West of Ireland	Blundell and Docherty (1987)
8.3.1.1	1987	BIRPS MOBIL	South North Sea	Blundell and Docherty (1987)
8.3.1.1	1988	BIRPS WISPA	Onshore United Kingdom	Ward and Warner (1991)
8.3.1.1	1990	BIRPS WESTLINE	West of Ireland	England (1995)
8.3.1.2	1983–1984	ECORS Paris basin	Northern France	Bois et al. (1986)
8.3.1.2	1984	ECORS Aquitaine Basin N-S line	France	Marillier et al. (1988)
8.3.1.2	1985ff	ECORS Aquitaine Basin–Pyrenees	France	ECORS Pyrenees team (1988)
8.3.1.2	1985ff	ECORS Gulf of Biscay	France	Pinet et al. (1987)
8.3.1.2	1986–1987	ECORS Bresse-Jura	France	Guellec et al. (1990)
8.3.1.3	1986	CROP-ECORS Torino-Geneva	Italy-France	Scrocca et al. (2003)
8.3.1.3	1988	CROP-NFP20 South-Central Alps	Italy-Switzerland	Scrocca et al. (2003)
8.3.1.3	1988	CROP-ECORS Gulf of Lions–Sardinia	Western Mediterranean	Scrocca et al. (2003)
8.3.1.4	1982	VibroSeis Swiss unfolded Jura	Northern Switzerland	Finck et al. (1986)
8.3.1.4	1986, 1988	NFP20 Swiss reflection line	Switzerland	Valasek et al. (19
8.3.1.5	1984	DEKORP-2 South Main-Danube line	Southern Germany	DEKORP Research Group (1985)
8.3.1.5	1984	Black Forest reflection line	Southwest Germany	Lüschen et al. (1987)
8.3.1.5	1985	DEKORP-4 KTB–Eastern Bavaria	Southeastern Germany	DEKORP Research Group (1988)
8.3.1.5	1986	DEKORP-2 North Rhenish Massif	Western Germany	Franke et al. (1990)
8.3.1.5	1987, 1988	BELCORP-DEKORP-1 Rhen Massif	Western Germany	DEKORP Research Group (1991)
8.3.1.5	1988	DEKORP-9 N Northern Rhinegraben	Southwest Germany	Wenzel et al. (1991)
8.3.1.5	1988	ECORS-DEKORP-9S South Rhinegr	Southwest Germany	Brun et al. (1991)
8.3.1.5	1989	DEKORP 3-D survey KTB drill site	Southeastern Germany	Dürbaum et al. (1992)
8.3.1.5	1990	DEKORP-3 and DEKORP-MVE	Eastern Germany	DEKORP Research Group (1994)
8.3.1.6	1980s	West Carpathians deep reflection lines	Czechoslovakia	Tomek et al. (1987)
8.3.1.7	1989	BABEL Baltic Shield	Baltic Sea	BABEL WG (1991)
8.3.1.7	1989	BALTIC SEA profile	Baltic Sea	Ostrovsky (1993)
8.3.2	1980–1981	North Sea wide-angle SALT	North Sea	Barton (1986)
8.3.2	1981	WISE offshore W Scotland	British Isles	Summers et al. (1982)
8.3.2	1982	CSSP Caledonian Suture	Northern England	Bott et al. (1983)
8.3.2	1982	ICSSP Caledonian Suture	Ireland	Jacob et al. (1985)
8.3.2	1983	"Meteor (1964)" expedition M66	Skagerak	Behrens et al. (1986)
8.3.2	1984	MAVIS Midland Valley Scotland	British Isles	Dentith and Hall (1990)
8.3.2	1985	COOLE Caledonian onshore	Ireland	Lowe and Jacob (1989)
8.3.2	1987	BB87 North Sea–Ireland	Ireland	Bean and Jacob (1990)
8.3.3.1	1982	Wildflecken fan profiles	Southern Germany	Zeis et al. (1990)
8.3.3.1	1984	Black Forest wide-angle seismic	Southwest Germany	Gajewski and Prodehl (1987)
8.3.3.1	1984	Black-Zollern-Forest wide-angle seism	Southwest Germany	Gajewski et al. (1987)
8.3.3.2	1985	KTB wide angle seismics	Southeastern Germany	Gebrande et al. (1989)
8.3.4.1	1981	EGT SVEKA wide-angle profile	Central Finland	Luosto et al. (1984)
8.3.4.1	1981–1983	Kola peninsula seismic profiles	Northwest Russia	Azbel et al. (1989)
8.3.4.1	1982	BALTIC wide-angle profile	Southeastern Finland	Luosto and Korhonen (1986)
8.3.4.1	1983	Skagerrak offshore lines	Denmark-Norway	Behrens et al. (1986)
8.3.4.1	1984	Tornquist-Teysseire Zone	Denmark-Sweden	EUGENO-S WG (1988)
8.3.4.1	1985	POLAR wide-angle profile	Finland-Norway	Luosto et al. (1989)
8.3.4.2	1981, 1984	Northern German Basin hydrocarbon survey	Northern Germany	Yoon et al. (2008)
8.3.4.2	1983	EGT N-Appenines crustal profiles	Northern Italy	Biella et al. (1987)
8.3.4.2	1986	EGT central segment	Germany-Italy	Aichroth et al. (1992)
8.3.4.2	1987	Alpine wide-angle profiles	Switzerland	Ye (1991)
8.3.4.3	1983	EGT southern section Corsica-Sardinia	Western Italy	Egger et al. (1988)

(*continued*)

TABLE 2.8. MAJOR CONTROLLED-SOURCE SEISMIC INVESTIGATIONS OF THE CRUST IN THE 1980S (*continued*)

Chapter	Year	Project	Location	Reference
8.3.4.3	1985	EGT southern segment Tunisia	Tunisia	Research Group for Lithospheric Structure in Tunisia (1992)
8.3.4.3	1983, 1986	Atlas and Anti Atlas crust	Morocco	Wigger et al. (1992)
8.3.4.4	1980 ff	Northwest corner of Iberia	Spain	Cordoba et al. (1987)
8.3.4.4	1988	Valencia Trough	Spain	Pascal et al. (1992); Torné et al. (1992)
8.3.4.4	1989	Valencia Trough	Spain	Danobeitia et al. (1992)
8.3.4.4	1989	EGT ILIHA long range lines	Spain and Portugal	ILIHA DSS Group (1993)
8.3.5	1983–1985	West Carpathians reflection lines	Czechoslovakia	Tomek et al. (1987)
8.3.5	1986	LT-7 Tornquist-Teisseyre Zone	Northwest Poland	Guterch et al. (1991b)
8.3.5	1987	GB2 near-vertical recording	Northwest Poland	Guterch et al. (1991a)
8.4	1981–1988	Crustal data on PNE lines	USSR	Egorkin et al. (1991)
8.4	1981–1988	PNE athenosphere profiles	USSR	Mechie et al. (1993)
8.4	1981, 1983	Mirnyi kimberlite field in Siberia	USSR	Suvorov et al. (2006)
8.5.2	1980	COCRUST Vancouver Island project	Western Canada	Green et al. (1986)
8.5.2	1981	COCRUST Williston Basin	Central Canada	Hajnal et al. (1984)
8.5.2	1982	COCRUST Bonnechere-Grenville	Southeastern Canada	Mereu et al. (1986)
8.5.2	1984	COCRUST Kapuskasing	South-central Canada	Mereu et al. (1989)
8.5.2	1984	LITHOPROBE Vancouver Island	Western Canada	Green et al. (1990a)
8.5.2	1984–1985	LITHOPROBE LE off Newfoundland	Southeastern Canada	Marillier et al. (1994)
8.5.2	1984–1988	Canadian Arctic ice shelf project	Northern Canada	Forsyth et al. (1990)
8.5.2	1985	LITHOPROBE Vancouver on-offshore	Western Canada	Clowes et al. (1987b)
8.5.2	1985	COCRUST Peace River Arch	Northwest Canada	Stephenson et al. (1989)
8.5.2	1985	LITHOPROBE Cordillera reflection	Western Canada	Cook et al. (1987)
8.5.2	1986	LITHOPROBE GLIMPCE GreatLakes	South-central Canada	Green et al. (1989)
8.5.2	1988	GRAP-88 Ontario–New York	Southeastern Canada	Marillier et al. (1994)
8.5.2	1989	LITHOPROBE off Newfoundland	Southeastern Canada	Van der Velden (2004)
8.5.2	1989–1990	LITHOPROBE SCoRE Cordillera	Western Canada	Clowes et al. (1995)
8.5.3.1	1980–1981	COCORP Adirondack–New England	Northeastern United States	Brown et al. (1983a)
8.5.3.1	1980–1981	James River corridor, Virginia	Eastern United States	Pratt et al. (1987)
8.5.3.1	1981	COCORP Kansas reflection survey	Central United States	Brown et al. (1983b)
8.5.3.1	1981	COCORP Ouachita Arkansas OUC	South-central United States	Nelson et al. (1982)
8.5.3.1	1981	USGS Carolina-Georgia GAC-CHP	Southeastern United States	Behrendt (1986)
8.5.3.1	1981–1985	USGS Northeast California	Western United States	Zucca et al. (1986)
8.5.3.1	1982	COCORP west-central Utah	Western United States	Allmendinger et al. (1986)
8.5.3.1	1982	USGS Central California reflection survey	Western United States	Hamilton (1986)
8.5.3.1	1982–1983	Mono Craters–Long Valley refraction lines	Western United States	Hill et al. (1985)
8.5.3.1	1982–1984	COCORP 40°N Transect	Western United States	Allmendinger et al. (1987)
8.5.3.1	1983	USGS Maine–N Appalachians reflection survey	Northeastern United States	Hamilton (1986)
8.5.3.1	1983–1985	COCORP Georgia-Florida Transect	Southeastern United States	Nelson et al. (1985)
8.5.3.1	1984	COCORP Northwest Cordillera	Northwest United States	Potter et al. (1986)
8.5.3.1	1984	USGS Newark Basin rift tectonics	Eastern United States	Ratcliffe et al. (1986)
8.5.3.1	1984	USGS Alaska pipeline reflect survey	Alaska	Hamilton (1986)
8.5.3.1	1985	COCORP Death Valley	Western United States	de Voogd et al. (1988)
8.5.3.1	1985	USGS Southern California–Arizona reflection survey	Western United States	Hamilton (1986)
8.5.3.1	1985	ADCOH site study ADC	Eastern United States	Coruh et al. (1987)
8.5.3.1	1986	COCORP Arizona PACE line	Western United States	Hauser et al. (1987a)
8.5.3.1	1987	COCORP Montana plains	Northwest United States	Latham et al. (1988)
8.5.3.2	1980–1981	USGS Western Mojave desert	Western United States	Harris et al. (1988)
8.5.3.2	1980–1982	USGS Livermore–Santa Cruz Mountains	Western United States	Williams et al. (1999)
8.5.3.2	1982	USGS Great Valley Californa	Western United States	Walter and Mooney (1987)
8.5.3.2	1982–1983	USGS Morro Bay, California	Western United States	Murphy and Walter (1984)
8.5.3.2	1982–1983	USGS Long Valley Caldera	Western United States	Meador et al. (1983, 1985)
8.5.3.2	1983	USGS Coalinga, California, refraction lines	Western United States	Walter (1990)
8.5.3.2	1984	USGS Columbia Plateau, Oregon	Western United States	Catchings and Mooney (1988a)
8.5.3.2	1984	USGS Newberry Volcano, Oregon	Western United States	Catchings and Mooney (1988a)
8.5.3.2	1986	USGS San Luis Obispo, California	Western United States	Sharpless and Walter (1988)
8.5.3.2	1986	PG&E EDGE Central California margin	Western United States	Howie et al. (1993)
8.5.3.2	1989	Cascade Mountains off-shore project	Western United States	Trehu and Nakamura (1993)
8.5.3.3	1980–1982	Yucca Mountain, Nevada	Western United States	Hoffman and Mooney (1983)

(*continued*)

32 *Chapter 2*

TABLE 2.8. MAJOR CONTROLLED-SOURCE SEISMIC INVESTIGATIONS OF THE CRUST IN THE 1980S (*continued*)

Chapter	Year	Project	Location	Reference
8.5.3.3	1988	Yucca Mountain, Nevada	Western United States	Brocher et al. (1990)
8.5.3.3	1980–1983	Southern Rio Grande Rift	Western United States	Sinno et al. (1986)
8.5.3.3	1981	CARDEX Valles Caldera, New Mexico	Western United States	Ankeny et al. (1986)
8.5.3.3	1985, 1987	CALCRUST PACE, California–Arizona	Western United States	McCarthy et al. (1991)
8.5.3.3	1986	PASSCAL Basin & Range province, Nevada	Western United States	Whitman and Catchings (1987)
8.5.3.3	1989	PACE Colorado Plateau	Western United States	McCarthy et al. (1994)
8.5.3.4	1983	Canadian part Quebec-Maine reflection survey	Southeastern Canada	Stewart et al. (1986)
8.5.3.4	1984	USGS Maine refraction survey QMT	Eastern United States	Spencer et al. (1989)
8.5.3.4	1985	USGS Maine refraction survey MRP	Eastern United States	Luetgert et al. (1987)
8.5.3.4	1988	Adirondack-Grenville refraction line	Northeastern United States	1990 Luetgert et al. (1990)
8.5.4	1984–195	Southern Alaska	Alaska	Fuis et al. (1991)
8.5.4	1987	Alaska Range–Yukon River	Alaska	Brocher et al. (2004a)
8.5.4	1988	Prince William Sound–Gulf of Alaska	Alaska	Brocher et al. (1991a)
8.5.4	1990	Brooks Range, northern Alaska	Alaska	Fuis et al. (1997)
8.6.1	1981	Northern Red Sea–Egypt	Egypt	Rihm et al. (1991)
8.6.1	1984	Jordan–Dead Sea transform refraction	Jordan	El Isa et al. (1987a)
8.6.1	1984	Jordan–Dead Sea transform reflection	Israel	Rotstein et al. (1987)
8.6.1	1986	Gulf of Suez–Northern Red Sea	Red Sea	Gaulier et al. (1988)
8.6.1	1988	Red Sea–Sudan and Red Sea–Yemen	Red Sea	Egloff et al. (1991)
8.6.2	1985	KRISP85 East African Rift	Kenya	Henry et al. (1990)
8.7.1	1980–1988	Multichannel reflection lines	India	Mahadevan (1994)
8.7.2	1981–1982	Sino-French exploration Tibet	China	Hirn and Sapin (1984)
8.7.2	1981–1982	Tibetan plateau	China	Teng (1987)
8.7.2	1980–1986	Seismic refraction surveys	Eastern China	Li and Mooney (1998)
8.7.2	1982	Yunnan Province, Southwest China	Southwest China	Kan et al. (1986)
8.7.2	1988	Altai Mountains–Altyn Tagh Fault	Northwest China	Wang et al. (2003)
8.7.3	1980	Ito-Matsuzaki Profile, Izu Peninsula	Central Japan	Yoshii et al. (1985)
8.7.3	1981	DSS in and around Nagano prefecture	Central Japan	Ikami et al. (1986)
8.7.3	1982	DSS Nagano and Yanamashi prefectures	Central Japan	Sasatani et al. (1990)
8.7.3	1984	Niikappu-Samani line, Southwest Hokkaido	Northern Japan	Iwasaki et al. (1998)
8.7.3	1985	Haruno-Tsukude profile, Honshu	Central Japan	Matsu'ura et al. (1991)
8.7.3	1988	DSS Kii peninsula	Southwest Japan	Research Group for Explosion Seismology (1992)
8.7.3	1989	Fujihashi–Kamigori, W Honshu	Japan	Research Group for Explosion Seismology (1995)
8.7.3	1990	Kitakami Massif refraction line	Japan	Iwasaki et al. (1994)
8.8.1	1980–1982	Eromanga Basin, Queensland	Australia	Finlayson et al. (1989)
8.8.1	1982–1983	Central Volcanic Region North Island	New Zealand	Stern (1985)
8.8.1	1983	Yilgarn craton projects	Northwest Australia	Drummond (1988)
8.8.1	1983	Central South Island	New Zealand	Smith et al. (1995)
8.8.1	1984	Eromanga–East Coast, Queensland	Australia	Finlayson et al. (1989)
8.8.1	1984	New England batholith traverse	East Australia	Finlayson and Collins (1993)
8.8.1	1984	Hikurangi zone N Island–Pacific	New Zealand	Davey et al. (1986)
8.8.1	1985	AGSO reflection central Australia	Australia	Wright et al. (1990)
8.8.1	1985	Hikurangi zone south of N Island	New Zealand	Davey (1987)
8.8.1	1987–1989	Lachlan orogen	East Australia	Finlayson (2010)
8.8.1	1989	Bowen Basin–New England Orogen	East Australia	Korsch et al. (1992)
8.8.1	1989	Hikurangi zone Southwest of N Island	New Zealand	Davey and Stern (1990)
8.8.2	1982	Whitwatersrand refraction lines	South Africa	Durrheim (1986)
8.8.2	1988	Whitwatersrand 112 km reflection line	South Africa	Durrheim et al. (1991)
8.8.2	1987–1988	Reconnaissance seismic survey	Botswana	Wright and Hall (1990)
8.8.3	1982–1984	Chuquicamata mine profiles Andes	Chile-Bolivia	Wigger (1986)
8.8.3	1987–1989	Mine blasts and ocean shots Andean lines	Chile-Bolivia	Wigger et al. (1994)
8.8.4	1981–1982	McMurdo Sound, Ross Sea	Antarctica	McGinnis et al. (1985)
8.8.4	1984–1985	Bransfield Strait Polish Expedition	Western Antarctica	Janik (1997)
8.8.4	1987–1988	Bransfield Strait Polish Expedition	Western Antarctica	Janik (1997)
8.9.2	1981	Juan de Fuca Ridge	Northeast Pacific	Morton et al. (1987)
8.9.2	1981–1982	Tohoku subduction zone	East off Japan	Suyehiro and Nishizawa (1994)
8.9.2	1979ff	Refraction around ODP hole 504B	Costa Rica rift	Detrick et al. (1994)

(*continued*)

TABLE 2.8. MAJOR CONTROLLED-SOURCE SEISMIC INVESTIGATIONS OF THE CRUST IN THE 1980S (*continued*)

Chapter	Year	Project	Location	Reference
8.9.2	1982	Shirshov Institute of Oceanology world cruise	Northern Pacific Ocean	Neprochnov (1989)
8.9.2	1982	Hawaiian–Emperor Seamount chain	Central Pacific	Watts et al. (1985)
8.9.2	1982	MAGMA East Pacific Rise	East Pacific	Orcutt et al. (1984)
8.9.2	1982, 1984	Valu Fa Ridge, Lau Basin	Southwest Pacific	Morton and Sleep (1985)
8.9.2	1983	Ngendei expedition south Pacific	South Pacific	Shearer and Orcutt (1985)
8.9.2	1985	East Pacific Rise 8°50 N to 13°30 N	East Pacific	Detrick et al. (1987)
8.9.2	Mid-1980s	Juan de Fuca Ridge	East Pacific	Rohr et al. (1988)
8.9.2	1985–1989	Australian margins marine deep seismics	Southern and eastern Australia	Finlayson (2010)
8.9.2	1989	Tasmania margin and Southern Ocean	Off western Tasmania	Finlayson (2010)
8.9.2	1988	East Pacific Rise at 9°30	East Pacific	Toomey et al. (1990)
8.9.2	1989	Society Island Hotspot Chain	Pacific Ocean	Grevemeyer et al. (2001b)
8.9.2	1990	Juan de Fuca Ridge	Northeast Pacific	Mcdonald et al. (1994)
8.9.3	1984	Shirshov Institute of Oceanology world cruise	Indian Ocean	Neprochnov (1989)
8.9.3	1985	Agulhas Bank off South Africa	Indian Ocean	Durrheim (1987)
8.9.3	1985	Kerguelen Plateau	Indian Ocean	Ramsay et al. (1986)
8.9.3	1986	Owen Basin off Oman	Indian Ocean	Barton et al. (1990)
8.9.3	1986	Exmouth Plateau	Off Northwest Australia	Mutter et al. (1989)
8.9.3	1986	North Perth Basin	Off Western Australia	Finlayson (2010)
8.9.3	1988	South Perth Basin	Off Western Australia	Finlayson (2010)
8.9.4	1980s	North Atlantic Transect near 23°N	North Atlantic	NAT StudyGroup (1985)
8.9.4	1981	LASE Baltimore Cyn	North American Atlantic margin	LASE StudyGroup (1986)
8.9.4	1982	Kane fracture zone, Mid-Atlantic Ridge	Atlantic	Cormier et al. (1984)
8.9.4	1982	Charlie-Gibbs Fracture Zone	North Atlantic	Whitmarsh and Calvert (1986)
8.9.4	1982	Tydeman Fracture Zone	North Atlantic	Calvert and Potts (1985)
8.9.4	1983	Shirshov Institute of Oceanology world cruise	Southern Atlantic Ocean	Neprochnov (1989)
8.9.4	1983	Alpha Ridge Arctic Ocean	Arctic Ocean	Forsyth et al. (1986)
8.9.4	1983–1984	Southwest Newfoundland transform margin	Eastern Canada	Todd et al. (1988)
8.9.4	1984	"*Meteor* (1964)" expedition M67	Off Northwest Africa	Sarnthein et al. (2008)
8.9.4	1984–1985	USGS Gulf of Maine	Northeastern United States	Hutchinson et al. (1987)
8.9.4	1984–1987	Newfoundland continental margin	Eastern Canada	Keen et al. (1990)
8.9.4	Mid-1980s	Labrador Sea	Northwest Atlantic	Osler and Louden (1992)
8.9.4	1985	WAM Western Approaches Margin	Southwest of Britain	Peddy et al. (1989)
8.9.4	1985	COOLE Caledonian offshore	South and southwest off Ireland	O'Reilly et al. (1991)
8.9.4	1985	Hatton bank volcanic margin	West off Ireland	Morgan and Barton (1990)
8.9.4	1985	Svalbard Polish expedition	W and N Spitsbergen	Czuba et al. (1999)
8.9.4	1985	LASE Carolina Trough	North American Atlantic margin	Trehu et al. (1989b)
8.9.4	1987	Canary basin	West off Canary Islands	Banda et al. (1992)
8.9.4	1987	Goban Spur continental margin	Southwest of Britain	Horsefield et al. (1994)
8.9.4	1987	Aegir rift	East of Iceland	Grevemeyer et al. (1997)
8.9.4	1987–1989	Soviet ice-station North Pole-28 (NP-28)	Arctic Ocean	Langinen et al. (2009)
8.9.4	1988	Madeira-Tore Rise, Josephine seamount	West of northwest Africa	Peirce and Barton (1991)
8.9.4	1988	Lofoten-Voring margin	Off Norway	Mjelde et al. (1992)
8.9.4	1988, 1989	Scoresby Sud on-offshore	Eastern Greenland	Weigel et al. (1995)
8.9.4	1988	LASE Southeastern Georgia embayment	North American Atlantic margin	Oh et al. (1991)
8.9.4	1988, 1989	Jameson Land onshore	Eastern Greenland	Mandler and Jokat (1998)
8.9.4	1988, 1990	RAPIDS Rockall and Porcupine basins	West off Ireland	Shannon et al. (1994)
8.9.4	End-1980s	PROBE survey off equatorial Africa	South Atlantic	Rosendahl et al. (1991)
8.9.4	1989	On-offshore wide-angle seismics	Western Greenland	Clement et al. (1994)
8.9.4	1989	Makarov Basin, Arctic Ocean	Arctic Ocean	Sorokin et al. (1999)
8.9.4	1990	Scoresby Sud on-offshore	Eastern Greenland	Mandler and Jokat (1998)
8.9.4	1990	EDGE Mid-Atlantic MSC experiment	North American Atlantic shelf	Holbrook et al. (1992b)
9.7.1	1987ff	AGSO Australian offshore lines	Australia	Goleby et al. (1994)
9.7.1	1989–1992	AGSO on-offshore program	Australia	Goleby et al. (1994)

In Europe, ECORS profiling, for example, was always accompanied by simultaneous wide-angle operations as were the first long reflection profiles accompanying the search for deep drill sites in southern Germany. Also the Alpine part of the EGT refraction line was covered by the Swiss reflection seismic program. Mooney and Brocher (1987) compiled a global review of coincident seismic reflection/refraction studies of the continental lithosphere up until the mid-1980s. An important result was that the Moho was identical, if observed by refraction–wide-angle observations or by near-vertical incidence reflection profiles. At almost all seismic-reflection profiles of BIRPS, ECORS, and DEKORP, as far as they were recorded over Caledonian and Variscan basement, the Moho was clearly identified as the lowest boundary of the laminated lower crust. The same observation was made on seismic-reflection profiles recorded in the Basin and Range province and other extensional areas of the mobile western United States.

Comprehensive reviews on the seismic velocity structure of the deep continental lower crust which represent the state of the art by the end of the 1980s have been published, e.g., by Holbrook et al. (1992a), including a table of lower-crust velocities and corresponding references typical for different tectonic environments. Mooney and Meissner (1992) investigated the multi-genetic origin of crustal reflectivity by reviewing deep seismic reflection profiles around the world. Other compilations contain summaries of crustal and upper mantle structure based on controlled-source seismology observations, both seismic reflection and refraction, for different continents or large parts of those continents, e.g., by Meissner et al. (1987b) for Europe (Fig. 2.8-05), Pavlenkova (1996) for the territories of the former Soviet Union, Braile et al. (1989) and Mooney and Braile (1989a) for North America (Fig. 2.8-06), or Mechie and Prodehl (1988) for the Afro-Arabian rift. Based on a series of discussions at meetings of experts (Olsen, 1995; Ziegler, 1992a, 1992b, 1992c), different authors compiled reviews for continental rifts around the world.

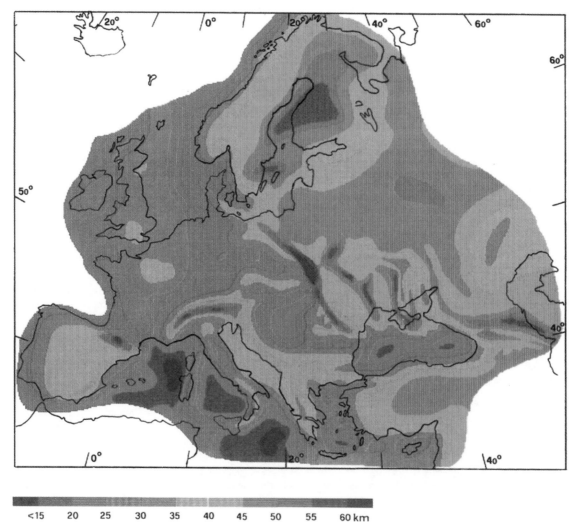

<15 20 25 30 35 40 45 50 55 60 km

Figure 2.8-05. Contour map of crustal thickness across Europe (from Meissner et al., 1987b, fig. 3). [Annales Geophysicae, 5B, p. 357–364. Reproduced by permission of the author.]

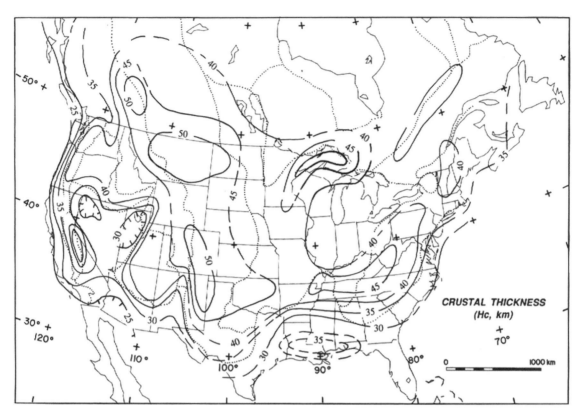

Figure 2.8-06. Contour map of crustal thickness for North America (from Braile et al., 1989, fig. 3). [*In* Pakiser, L.C., and Mooney, W.D., eds., Geophysical framework of the continental United States: Geological Society of America Memoir 172, p. 655–680. Reproduced by permission of the Geological Society of America.]

2.9. THE 1990s AND EARLY 2000s (1990–2005)

With advanced technology, seismic projects had become more and more expensive, using multiple energy sources and a large number of sophisticated recording devices. In the early 1990s, the change from analog to digital recording caused a major breakthrough toward modern recording and interpretation techniques. Most projects were neither projects of individual researchers nor of individual research institutions, but became mostly imbedded in large-scale research programs which involved a multitude of cooperating institutions and an interdisciplinary cooperation of scientists from various geoscientific fields. Starting in the 1980s, and continuing in the 1990s, this was reflected in the formulation of large reflection programs on national scales such as COCORP and BIRPS, which were followed by the foundation of IRIS/PASSCAL in the United States and LITHOPROBE in Canada and by international and interdisciplinary geoscientific programs in Europe, dealing with large-scale tectonic topics.

An important prerequisite for the successful worldwide cooperation in seismic projects was the development of new digital equipment, which had started by the end of the 1980s and continued into the 1990s both in North America and in Europe. In particular the North American equipment not only stimulated and enabled new large-scale seismic-refraction/reflection experiments in Canada and in the United States, but it would also enable many large-scale projects in Europe and Africa. The U.S. Geological Survey–Stanford instrumentation SGR led to the powerful RefTek generation and was fundamental for successful joint European–United States projects in Kenya, France, and Spain in the early 1990s. The IRIS/PASSCAL and LITHOPROBE instrument pools were combined for many large-scale projects in North America, and the IRIS/PASSCAL instrument pool was vital for the large-scale projects in eastern Europe and in Ethiopia at the end of the 1990s and beginning of the 2000s. European research groups also supported several seismic projects in North America.

In 1992, supported by the European Science Foundation, EUROPROBE, a lithosphere dynamics program concerned with the origin and evolution of the continents (Gee and Zeyen, 1996; Gee and Stephenson, 2006) was founded in Europe. EUROPROBE's focus was mainly on, but not restricted to, Europe and was particularly driven to emphasize and encourage East-Central-West European collaboration. It was dedicated to enable the realization of major projects, which aimed to investigate the whole lithosphere and which required the close multinational cooperation of geologists, geophysicists, and geochemists.

Particular EUROPROBE projects (Fig. 2.9-01) that involved major controlled-source seismic experiments were the TESZ project, investigating the Trans-European fault zone TTZ in Scandinavia and Poland; the Uralides, with major seismic campaigns in the Urals; Georift, emphasizing the Dniepr-Donets basin; Eurobridge, establishing a seismic crustal traverse from Lithuania to the Ukraine; and PANCARDI, an interdisciplinary geoscientific frame for large-scale investigations in the Carpathian area, in particular focusing on the deep-earthquake region of Vrancea in the Romanian Carpathians.

Other large-scale interdisciplinary projects in Europe were a seismic-reflection traverse through the Eastern Alps (Gebrande et al., 2006), a German priority program focusing on the tectonics of the Central European Variscides (Franke et al., 2000), and

VARNET, a major research program studying the Variscan front in Ireland (Landes et al., 2003).

In North America during the LITHOPROBE project (see Fig. 9.4.1-01), a large number of seismic projects systematically investigated crust and uppermost mantle of Canada (e.g., Clowes et al., 1999). The results were published in the context of interdisciplinary summary volumes for the different LITHOPROBE transects (e.g., Ludden, 1994, 1995; Wardle and Hall, 2002a; Hajnal et al., 2005a). Numerous detailed seismic studies were undertaken in the western United States along the coast, of which the projects SHIPS around Seattle (Snelson, 2001), BASIX (Brocher et al., 1991b) around San Francisco, and LARSE (Fuis et al., 1996, 2001b) around Los Angeles investigated the crustal structure in earthquake-prone regions in much detail. A similar project, JTEX, targeted the crustal structure beneath a large

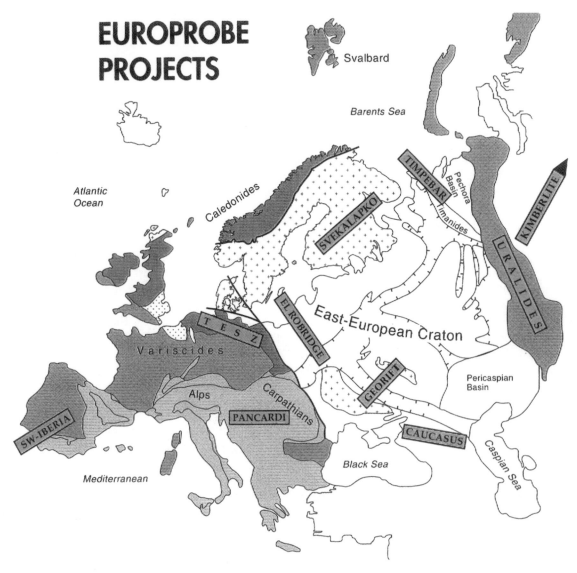

Figure 2.9-01. Map of EUROPROBE projects (from Gee and Zeyen, 1996, p. 12). [EUROPROBE Secretariat, Upssala University, 138 p., Copyright EUROPROBE.]

continental silicic magmatic system in the Jemez Mountains of New Mexico (Baldridge et al., 1997). The "Deep Probe" experiment of 1995 (Snelson et al., 1998; Gorman et al., 2002) followed approximately the 110th meridian and spanned a distance of 3000 km from the southern Northwest Territories to southern New Mexico. It targeted the velocity structure from the base of the crust to depths as great as the mantle transition zone near 400 km depth. Also in 1995, 14 institutions of the United States created the interdisciplinary geoscience project CD-ROM (Continental Dynamics–Rocky Mountain) with the aim to realize a detailed crustal and upper mantle interdisciplinary investigation of the Southern Rocky Mountains from central Wyoming to central New Mexico, following approximately the 101st meridian and involving tectonics, structural geology, regional geophysics, geochemistry, geochronology, xenolith studies, and seismic studies (Karlstrom and Keller, 2005). The seismic component involved seismic reflection and seismic refraction experiments in 1999 as well as teleseismic studies along a 1000-km-long north-south–directed traverse along the southern Rocky Mountains.

Major efforts were undertaken to unravel the details of crustal structure beneath the Afro-Arabian rift system, based on the experiences collected by the earlier expeditions. In 1990 and 1994, two major international campaigns explored the crust and upper mantle under the East African rift of Kenya combining seismic refraction and teleseismic tomography investigations (Prodehl et al., 1994a, 1997a; Fuchs et al., 1997). Figure 2.9-02 summarizes all projects undertaken within the Afro-Arabian Rift System from 1969 until 1995. In 2001, the northern end of the East African Rift system in Ethiopia was the target of a major seismic experiment (Maguire et al., 2003), and the Dead Sea transform saw the beginning of a major international seismic campaign in 2000, covering both sides of the rift in Jordan and Israel (DESERT Group, 2004).

Following the large-scale seismic reflection programs in North America and Europe, the National Geophysical Research Institute at Hyderabad undertook large-scale COCORP-equivalent studies in India. Special targets were the structure and tectonics of the Aravalli-Delhi fold belt in northwestern India (Rajendra Prasat et al., 1998) and in southern India. Other investigations in the Himalayas and southern India used seismological observations (Rai et al., 2006; Krishna et al., 1999).

The seismic investigations of the crust and upper mantle of China with explosion seismology of the 1980s were continued even more intensively with many deep-seismic sounding profiles in mainland China in the 1990s (e.g., Li et al., 2006). It was, however, Tibet in particular that attracted scientists from all over the world. One of the largest cooperative seismic projects was INDEPTH in southern Tibet, involving both near-incident and wide-angle seismic reflection methodology as well as teleseismic tomography, which was accomplished with major U.S. and German participation in several phases, starting in 1992 (e.g., Nelson et al., 1996).

Controlled-source seismology in Japan was activated in the 1990s as part of the Japanese Earthquake Prediction Program.

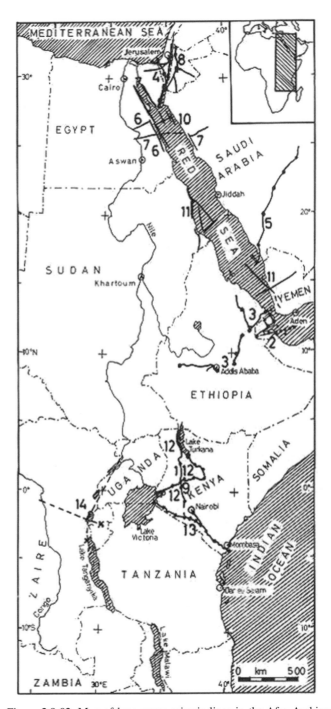

Figure 2.9-02. Map of long-range seismic lines in the Afro-Arabian rift system. Continuous lines: seismic-refraction surveys; full circles mark shotpoints where locations were published. Dashed lines: approximate lines through epicenters of local earthquakes. 1—Kenya rift 1968; 2—Djibouti 1971; 3—Afar depression 1972; 4—Jordan–Dead Sea–Gulf of Aqaba transform system 1977; 5—Arabian Shield 1978; 6—Northern Red Sea 1978; 7—Northern Red Sea 1981; 8—Jordan 1984; 9—Kenya rift 1985; 10—Northern Red Sea; 11—Southern Red Sea; 12—Kenya rift 1990; 13—southern Kenya 1994; 14—Western Rift: local earthquakes recorded at two stations UVI and BTR (marked by crosses) of the IRSAC network near Bukavu, evaluated as seismic profiles UVI-N and BTR-WNW. For references, see Prodehl et al. (1997a, fig. 1). [Tectonophysics, v. 278, p. 1–13. Copyright Elsevier.]

Following the destructive Kobe earthquake of 17 January 1995, seismic refraction (Research Group on Underground Structure in the Kobe-Hanshin Area, 1997; Research Group for Explosion Seismology, 1997; Ohmura et al., 2001) and reflection (Sato et al., 1998) measurements were carried out in the surroundings of Kobe. Other experiments in the first half of the 1990s in Japan focused on the investigation on the deeper crustal structure including lower crust and Moho beneath the SW and NE Japan arcs (e.g., Iwasaki et al., 1994, 1998). In 1994, for the first time in Japan, a seismic reflection survey for deep structural studies was undertaken through the Hidaka Collision zone, Hokkaido (Arita et al., 1998). The subsequent reflection surveys in 1996–1997 provided a clear image of a delamination structure and a reflective lower crust in the collision zone (Tsumura et al., 1999; Ito, 2000, 2002). Since then, seismic surveys using both of reflection and refraction methods became more common in Japan and detailed crustal sections were presented for NE Japan and Hokkaido (e.g., Iwasaki et al., 2001a, 2004; Sato et al. 2002).

Since 1985, the Australian Geological Survey Organisation (AGSO) ran a large program of deep seismic reflection surveys offshore and onshore Australia (Drummond et al., 1998; Finlayson et al., 1996, 1998; Finlayson, 2010, Appendix 2-2; Glen et al., 2002; Goleby et al., 2002; Korsch et al., 1992, 1997, 2002; Petkovic et al., 2000). Several reviews (e.g., Collins et al., 2003; Finlayson, 2010; Goleby et al., 1994) have summarized the results of crust and upper-mantle studies for Australia. Collins et al. (2003) published a crustal thickness map based primarily on the results of controlled-source seismology investigations. In general, within Archean regions of western Australia, the Moho appeared to be relatively shallow with a large velocity contrast, while the Moho is significantly deeper under the Proterozoic North Australian platform, under central Australia and under Phanerozoic southeastern Australia.

To understand the processes involved in continental collision, New Zealand, being deformed by the oblique collision of several plates, was the goal of a joint U.S.–New Zealand geophysical project SIGHT (South Island Geophysical Transect), undertaken in 1995 and 1996. The project involved both active source and passive seismology. The experiment had two main components. The first was a wide-angle reflection-refraction experiment along two land transects across the central South Island. The second experiment consisted of three offshore-onshore transects, two along the two land profiles of the first phase and a third along southeastern South Island. This third transect was a tie line across the eastern part of the two main transects (Davey et al., 1998). Another project in 2001 and 2002 investigated the Central Volcanic Region or Taupo Volcanic Zone (TVZ) occupying the northern half of the North Island of New Zealand (Harrison and White, 2006).

In South America, several seismic profiling projects investigated the area of the deep drilling project in the Chicxulub impact crater at the coast of the Yucatan peninsula, Mexico (e.g., Snyder et al., 1999; Morgan et al., 2005), northern Venezuela (Schmitz et al., 2002, 2005), and central Brazil (Berrocal et al., 2004).

The majority of seismic investigations in South America, however, concentrated on the Andean region of Chile and the adjacent Pacific Ocean (Fig. 2.9-03). For example, a Collaborative Research Center 267 (CRC 267) program "Deformation Processes in the Andes" at the Free University of Berlin, the GeoScience Center of Potsdam, and the University of Potsdam (Germany) was funded by the German Research Society for 15 years (Giese et al., 1999).

The new investigations enabled extended seismic research, but also involved other geoscientific disciplines available at the three research institutions and included strong support by various South American research facilities. In the years 1994–1996, three major seismic projects—PISCO 94, CINCA 95, and ANCORP 96—were conducted in northern Chile and adjacent parts of Bolivia and Argentina (Fig. 2.9-03, northern box; see also Fig. 9.7.3-10). Besides northern Chile, southern Chile became the target of the same research groups, who had cooperated in 1995 in the CINCA project. Also in 1995, within the multidisciplinary CONDOR (Chilean Offshore Natural Disaster and Ocean Environmental Research) project, a marine operation investigated the Valparaiso Basin offshore from Valparaiso, central Chile, along two marine-seismic reflection and refraction profiles north and south of the latitude 33°S (Flueh et al., 1998a). In 2000, the Collaborative Research Center CRC 267 at Berlin and Potsdam, Germany, started a detailed research project in southern Chile (Fig. 2.9-03, southern box; see also Fig. 9.7.3-24). The first approach was the project ISSA 2000 (Sick et al., 2006) and consisted of a temporary seismological network and a seismic-refraction profile. This project was followed by the project SPOC in 2001 (Krawczyk et al., 2006), which involved a shipborne geophysical experiment and two predominantly land-based onshore-offshore experiments. All seismic arrays also recorded teleseismic, regional, and local events. Furthermore broadband stations were deployed along some of the lines.

During the 1990s, digital equipment for marine seismic research also became available, with both digital streamers for details of sedimentary structure beneath the ocean bottom as well as a new generation of ocean-bottom seismometers, which were built in large numbers. Underwater explosions as energy sources had been banned almost completely; instead powerful airgun arrays became available, providing sufficient energy to be recorded over hundreds of kilometers.

Seismic research projects in the 1990s and ongoing in the 2000s were largely devoted to the investigation of details of the mid-ocean rises, such as the East Pacific Rise. Here, for example, projects around the Galapagos hot spot and the Cocos-Nazca Spreading Center (e.g., Sallares et al., 2003) and the Garrett Fracture Zone (Grevemeyer et al., 1998) were undertaken. A large number of Japanese marine surveys, some of which also had land components, targeted the trench system in the Philippine Sea and northwestern Pacific Ocean around Japan. In particular, the Nankei Trough and the Japan trench south and east of Honshu was intensively investigated (e.g., Kodaira et al., 2000; Miura et al., 2003, 2005; Takahashi et al., 2004; Tsuru et al., 2002).

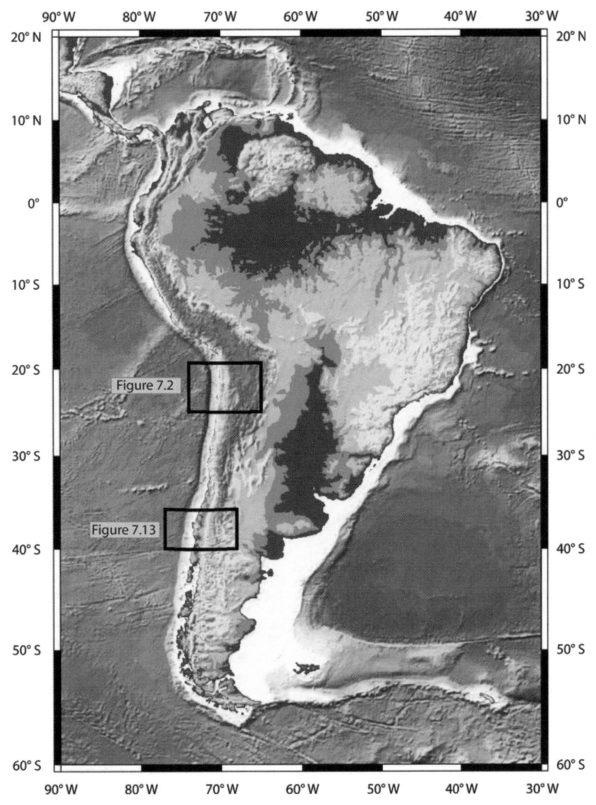

Figure 2.9-03. Topographic map of South America showing the seismic investigation areas in Chile (from Sick et al., 2006, fig. 7-1). Northern box labeled "Figure 7.2" indicates the location of the research projects PISCO94 CINCA95, ANCORP96, and PRECORP; the southern box labeled "Figure 7.13" indicates the location of the research projects ISSA 2000 and SPOC 2001. [Oncken et al., 2006, The Andes: Berlin-Heidelberg-New York, Springer, p. 147–169. Reproduced with kind permission of Springer Science+Business Media.]

A large number of seismic reflection surveys studied the margins of Australia (Finlayson, 2010; Appendix 2-2). In the Indian Ocean the Ninetyeast Ridge was investigated (Grevemeyer et al., 2001a). In the Atlantic the Mid-Atlantic Ridge (e.g., Hooft et al., 2000), but also the continent-ocean transitions were major targets. In particular in the northern Atlantic Ocean projects targeted the microcontinents and intervening basins off Ireland (e.g., Mackenzie et al., 2002), the Norwegian margin (e.g., Raum et al., 2002), the Faeroe–Iceland Ridge (e.g., Smallwood et al., 1999), the northwestern Barents Sea southeast of Spitsbergen (Breivik et al., 2002) and eastern Greenland (e.g., Korenaga et al., 2000, Schmidt-Aursch and Jokat, 2005). The Arctic-2000 transect in the Arctic Ocean between 164°W and 165°E (Lebedeva-Ivanova et al., 2006) and the investigation of the North American margin off Canada (e.g., Funck et al., 2003) were other large marine projects.

In this overview of projects conducted during the various decades, only a limited selection of major research projects could be mentioned. For the 1990s, a summary of all projects mentioned in Chapter 9 is presented in Table 2.9.

Beginning in the 1990s, large-scale teleseismic tomography projects (so-called passive studies, i.e., long-term recording of earthquakes) were carried out in association with large-scale seismic refraction programs (so-called active source studies, i.e., recording controlled sources, such as quarry blasts, borehole and underwater explosions, vibrators, or airguns), thus extending crustal and uppermost mantle research to greater depth ranges. Examples, to name a few, are the KRISP and EAGLE investigations of the East African Rift System in Kenya (Prodehl et al., 1994a; Fuchs et al., 1997) and Ethiopia (Maguire et al., 2003); the onshore-offshore investigations of the Andean region in South America (e.g., ANCORP Working Group, 2003; Flueh et al., 1998a; Giese et al., 1999; Krawczyk et al., 2003, 2006; Rietbrock et al., 2005); the INDEPTH expeditions to Tibet (e.g., Brown et al., 1996; Nelson et al., 1996; Zhao et al., 1993, 1997, 2001); the SKIPPY experiments across Australia (Van der Hilst et al., 1994, 1998; Finlayson, Appendix 2-2); and the CD-ROM project in the Southern Rocky Mountains (Karlstrom and Keller, 2005). The interpretation of these data required special inversion methods used originally in passive experiments, but which in the course of time proved to be practical also for controlled-source seismology data. In particular, the method for studies of crust and upper mantle with P-wave receiver functions, which was being developed since the early 1980s, and later S-wave receiver function studies in the late 1990s, proved to be a very efficient and relatively cheap interpretation tool of the large-scale teleseismic data.

An overview of interpretation methods used in the 1990s to interpret combined active and passive studies was given at one of the Commission on Controlled Source Seismology workshop meetings, held in Dublin in 1999 (Jacob et al., 2000). Introduced in the 1970s (Červený et al., 1977), the ray-tracing method has remained an almost universal method for data interpretation. The most commonly used programs include ray-theoretical and

Gaussian-beam synthetic seismograms (Červený, 1985). Based on finite-difference methods for calculating the full wavefield in horizontally inhomogeneous media, it became possible to make calculations for refraction/wide-angle reflection data on a crustal/lithosphere scale over several hundreds of kilometers and at realistic frequencies. Other groups also wrote ray synthetic seismogram routines, e.g., McMechan and Mooney (1980) in the United States or Spence et al. (1984) in Canada. Ray-tracing using a finite-difference approximation of the eikonal equation was improved in the 1990s and extended in a way that reflected arrivals and second arrival refractions from prograde traveltime branches could be calculated (Hole and Zelt, 1995). In addition, commonly used methods in the processing of near-vertical incidence seismic reflection data, such as normal moveout correction and migration, were applied to refraction/wide-angle reflection data (e.g., Lafond and Levander, 1995).

In the 1990s, theory and associated computer programs on traveltime tomography developed by Colin Zelt and others became popular and evolved to a method widely applied as a first approach to model large amounts of data as well as a last check of the validity of a model (Zelt and Smith, 1992; Zelt, 1998, 1999). Nowadays, there is hardly a publication on crustal and upper mantle interpretation which does not first apply a tomographic approach to the seismic data, before refined raytracing modeling is applied (for example, interpretations of the Polonaise and Celebration 2000 data, see, e.g., Guterch et al., 2003a, 2003b).

2.10. OUTLOOK

Textbooks on seismic theory, published in large numbers over the decades, have accompanied and guided the development of interpretation methods to deal with the proper interpretation of active source seismic data obtained on land and at sea (e.g., Nettleton, 1940; Bullen, 1947; Worzel and Ewing, 1947; Grant and West, 1965; Musgrave, 1967; Maxwell, 1970; Kosminskaya, 1971; Officer, 1974; Červený et al., 1977; Kennett, 1983; Bullen and Bolt, 1985; Yılmaz, 1987; Lay and Wallace, 1995; Aki and Richards, 1980, 2002; Jones, 1999; Kennett, 2001; Chapman, 2004; Borcherdt, 2009). There are in addition numerous articles that address various aspects concerning the theory for interpretations of controlled-source seismology data (e.g., Steinhart et al., 1961c; Ewing, 1963a; Ludwig et al., 1970; Fuchs and Mueller, 1971; Bessonova et al., 1974; Braile and Smith, 1975; Giese, 1976; McMechan and Mooney, 1980; Červený and Horn, 1980; Spence et al., 1984; Červený, 1979, 1985; Zelt and Smith, 1992; Hole, 1992; Hole and Zelt, 1995; Zelt, 1999). Similarly, compilations of interpretation methods commonly used were also summarized from time to time (e.g., James and Steinhart, 1966; Mooney, 1989; Braile et al., 1995). Levander et al. (2007) provided a convenient review of theory and application of controlled-source seismic data.

The large number of digital recording devices available by the year 2000, as well as the ability to record continuously over long time periods, has opened a new dimension in crustal investigations.

TABLE 2.9. MAJOR CONTROLLED-SOURCE SEISMIC INVESTIGATIONS OF THE CRUST IN THE 1990s

Chapter	Year	Project	Location	Reference
9.2.1	1996	VARNET96 Variscan crust	SW Ireland	Landes et al. (2003)
9.2.1	1999	LEGS Leinster Granite Seismics	SE Ireland	Hodgson et al. (2000)
9.2.2	1990	DEKORP 3/MVE	Central eastern Germany	DEKORP Research Group (1994)
9.2.2	1990–1991	Rhinegraben D-90/DJ-91	SW Germany	Mayer et al. (1997)
9.2.2	1991–1992	French Massif Central	South-central France	Zeyen et al. (1997)
9.2.2	1994	Rhinegraben RU-94	SW Germany	Mayer et al. (1997)
9.2.2	1995	GRANU95 Saxon Granulite Mountain	Central eastern Germany	Enderle et al. (1998)
9.2.2	1995	DEKORP 95 Saxon Granulite Mountain	Eastern Germany	Krawczyk et al. (2000)
9.2.2	1996	DEKORP-BASIN	Northern Germany	Krawczyk et al. (1999)
9.2.2	1996	BASIN'96 Bornholm	Baltic Sea	Bleibinhaus et al. (1999)
9.2.2	1998–2001	TRANSALP Eastern Alps	Germany-Italy	Lueschen et al. (2004)
9.2.3	1990	ESCI-Catalan–Mallorca	NE Spain	Gallart et al. (1994)
9.2.3	1990	CROP 03 southern Apennines	Southern Italy	Scrocca et al. (2003)
9.2.3	1990	Southern Calabria Hercynian crust	Southern Italy	Lueschen et al. (1992)
9.2.3	1991	CROP MARE 1	NW, SW and S off Italy	Scrocca et al. (2003)
9.2.3	1991, 1993	ESCIN Northern Iberian Pensinsula	Northern Spain/Bay Biscay	Alvarez-Marron (1996)
9.2.3	1992	Cantabrian Mountains	Northern Spain	Fernandez-Viejo (2000)
9.2.3	1992	STREAMERS Ionian Sea	Italy-Greece	Hirn et al. (1996)
9.2.3	1992–1993	CROP 04 northern Apennines	Central Italy	Scrocca et al. (2003)
9.2.3	1993–1994	CROP MARE 2	SW and S and E off Italy	Scrocca et al. (2003)
9.2.3	1995	CROP 18 northern Apennines	Central Italy	Scrocca et al. (2003)
9.2.3	1996–1999	CROP 11 central Apennines	Central Italy	Scrocca et al. (2003)
9.2.3	1998–1999	CROP 1A Transalp	Northern Italy	Scrocca et al. (2003)
9.2.4	1992	Pannonian Geotraverse	Hungary	Hajnal et al. (1996)
9.2.3	1993	IAM Iberian Atlantic Margins	Iberia-Atlantic	Banda et al. (1995)
9.2.3	1995	LISA Ligurian Sea	Western Mediterranean	Nercessian et al. (2001)
9.2.3	1993	ESCI-Betics reflection lines	Southern Spain	Carbonell et al. (1998a)
9.2.3	1997	SEISGRECE Ionian islands	W Greece	Clement et al. (2000)
9.2.4	1997	POLONAISE Polish lithosphere	Poland	Guterch et al. (1999)
9.2.4	1999	VRANCEA 1999 SE Carpathians	Romania	Hauser et al. (2001)
9.2.4	2000	CELEBRATION E and SE Europe	Poland-Hungary	Guterch et al. (2003b)
9.2.4	2001	VRANCEA 2001 SE Carpathians	Romania	Hauser et al. (2007a)
9.2.5	1992	BIRPS reflection/refraction project	North Sea	Singh et al. (1998a)
9.2.5	1993	MONA LISA	North Sea	MONA LISA WG (1997)
9.2.5	1995	COAST Profile Baltic Sea	Sweden	Lund et al. (2001)
9.3	1991	GRANIT anisotropy ASTRA	Russia	Lueschen (1992)
9.3	1991	GRANIT Middle Urals	Russia	Juhlin et al. (1996)
9.3	1992	Kola borehole seismics	NW Russia	Ganchin et al. (1998)
9.3	1993–2003	ESRU Middle Urals	Russia	Kashubin et al. (2006)
9.3	1994–1996	EUROBRIDGE'95 and '96	Lithuania-Ukraine	EUROBRIDGE WG (1999)
9.3	1995	Kola borehole–Franz-Josef land	NW Russia	Sakoulina et al. (2000)
9.3	1995	URSEIS'95 reflection seismics Urals	Russia	Knapp et al. (1998)
9.3	1995	URSEIS'95 refraction seismics Urals	Russia	Carbonell et al. (2000b)
9.3	1995–2001	Barents Sea–Novaya Zemlya-	Russia	Roslov et al. (2009)
9.3	1997	EUROBRIDGE'97	Ukraine	Thybo et al. (2003)
9.3	1997–00	Kola borehole CMP lines to N and W	NW Russia	Berzin et al. (2002)
9.3	1999	DOBRE Donbas Foldbelt	Ukraine-Russia	DOBRE WG (2003)
9.3	2000–02	Mezen Basin–Timan Range	NW Russia	Kostyuchenko et al. (2006)
9.4.1.1	1991	LITHOPROBE LE91 on-offshore	SE Canada	Mariller et al. (1994)
9.4.1.2	1991, 1996	LITHOPROBE ESCOOT	SE Canada	Wardle and Hall (2002)
9.4.1.3	1991–1992	Abitibi-Grenville Transect	Canada	Clowes et al. (1992)
9.4.1.4	1996	Western Superior Transect	Canada	Musacchio et al. (2004)
9.4.1.5	1991–1994	THOT Trans-Hudson Orogen Transect	Canada	Hajnal et al. (2005b)
9.4.1.5	1993	THORE Trans-Hudson Long Refraction	Canada	Nemeth et al. (2005)
9.4.1.6	1992	CAT92 3D seismic reflection experiment	W Canada	Kanasewich et al. (1995)
9.4.1.6	1994, 1995	PRAISE94/SALT95/VAULT	W Canada	Mandler and Clowes (1998)
9.4.1.6	1995	SAREX/Deep Probe	W Canada	Clowes et al. (2002)
9.4.1.7	1994	ACCRETE marine shots	W Canada	Hammer et al. (2000)
9.4.1.7	1997	SNORCLE transect refraction	W Canada	Clowes et al. (2005)
9.4.2.1	1990	Brooks Range, Northern Alaska	N Alaska	Fuis et al. (1997)
9.4.2.1	1994	EDGE Prince William Sound	Southern Alaska	Ye et al. (1997)
9.4.2.1	1994	Aleutian volcanic arc	Southern Alaska	Fliedner and Klemperer (1999)
9.4.2.1	1994	Continental shelf	Alaska-Siberia	Brocher et al. (1995a)
9.4.2.2	1991	Pacific NW Experiment	Washington, USA	Trehu et al. (1993)
9.4.2.2	1991, 1995	Western Washington on-offshore	Washington, USA	Parsons et al. (1999)
9.4.2.2	1994	Cape Blanco, southern Oregon	Oregon, USA	Brocher et al. (1995b)
9.4.2.2	1996	Oregon continental margin	Oregon, USA	Flueh et al. (1997)
9.4.2.2	1998	Wet SHIPS, Puget Sound	Washington, USA	Brocher et al. (1999)
9.4.2.2	1999	Dry SHIPS, Seattle basin	Washington, USA	Brocher et al. (2000)
9.4.2.2	2000	Kingdome SHIPS	Washington, USA	Snelson (2001)

(continued)

TABLE 2.9. MAJOR CONTROLLED-SOURCE SEISMIC INVESTIGATIONS OF THE CRUST IN THE 1990s *(continued)*

Chapter	Year	Project	Location	Reference
9.4.2.3.1	1993–1994	Mendocino Triple Junction	California, USA	Godfrey et al. (1998)
9.4.2.3.2	1990	Loma Prieta refraction line	California, USA	Brocher et al. (1992)
9.4.2.3.2	1991	BASIX San Francisco Bay Area	California, USA	Holbrook et al. (1996)
9.4.2.3.2	1991, 1993	USGS refraction San Francisco Bay area	California, USA	Kohler and Catchings (1994)
9.4.2.3.2	1995	San Francisco Bay inline and fan lines	California, USA	Parsons (1998)
9.4.2.3.3	1995	San Andreas fault project	California, USA	Thurber et al. (1996)
9.4.2.3.4	1994	LARSE I Los Angeles area	California, USA	Fuis et al. (1996)
9.4.2.3.4	1999	LARSE II Los Angeles area	California, USA	Fuis et al. (2001)
9.4.2.4	1992–1993	Ruby Mountains seismic surveys	Nevada, USA	Satarugsa and Johnson (1998)
9.4.2.4	1993	South Sierra Nevada	California, USA	Ruppert et al. (1998)
9.4.2.4	1993	Delta Force experiment, Basin and Range	California-Nevada-Arizona, USA	Hicks (2001)
9.4.2.4	1994	Yucca Mountain project	Nevada, USA	Brocher et al. (1996)
9.4.2.5	1993, 1995	JTEX Valles Caldera	New Mexico, USA	Baldridge et al. (1997)
9.4.2.5	1995	Deep Probe 110°W long range	Canada–New Mexico	Snelson et al. (1997)
9.4.2.5	1999	CD-ROM Southern Rocky Mountains	Wyoming–New Mexico	Karlstrom and Keller (2005)
9.4.2.6	1991	Atlantic coastal plain profile	South Carolina, USA	Luetgert et al. (1994)
9.4.2.6	1996–1997	E Tennessee Seismic Zone	Eastern Tennessee, USA	Hawman et al. (2001)
9.5.1	1990	KRISP90 Kenya Rift seismics	Kenya	Prodehl et al. (1994a)
9.5.1	1994	KRISP94 southern Kenya Rift	Kenya	Fuchs et al. (1997)
9.5.3	2000	Jordan–Dead Sea Transform	Israel-Jordan	DESERT Group (2004)
9.6.1	1993	Southern India aftershock lines	India	Krishna et al. (1996)
9.6.1	1996	Nagaur-Kunjer deep reflection project	India	Rajendra Prasat (1998)
9.6.1	1998	Active source Merapi	Java, Indonesia	Wegler and Lühr (2001)
9.6.2	1992	INDEPTH I Tibet	China	Zhao et al. (1993)
9.6.2	1994–1995	INDEPTH II Tibet	China	Nelson et al. (1996)
9.6.2	1994, 1997	Dabie Shan Orogen	China	Wang et al. (2000)
9.6.2	1998	INDEPTH III Tibet	China	Zhao et al. (2001)
9.6.2	1999*	Sino-French network NE Tibet	China	Galve et al. (2002)
9.6.2	2000*	Altyn Tagh Range	China	Zhao et al. (2006)
9.6.2	2000*	1000 km NE-Tibet to Ordos basin	China	Liu et al. (2006)
9.6.3	1991	180 km line central Honshu	Japan	Research Group for Explosion Seismology (1994)
9.6.3	1992	Central Hokkaido	Japan	Iwasaki et al. (1998)
9.6.3	1993	Deep seismic sounding in Kanto and Tohoku districts	Japan	Research Group for Explosion Seismology (1996)
9.6.3	1994–1997	Hokkaido deep reflection survey	Japan	Arita et al. (1998)
9.6.3	1995	Kobe eq region profiles	Japan	Research Group for Explosion Seismology (1997)
9.6.3	1997	N Honshu on-offshore	Japan	Research Group for Explosion Seismology (1997)
9.6.3	1998	Backbone range in northern Honshu	Japan	Research Group for Explosion Seismology (2008)
9.6.3	1998–1999	Hokkaido reflection lines	Japan	Iwasaki et al. (2004)
9.6.3	1999–2000	Hokkaido refraction lines	Japan	Iwasaki et al. (2004)
9.6.3	2000	On-offshore Kuril Arc, Hokkaido	Japan	Nakanishi et al. (2009)
9.6.3	2000	NE-Hokkaido coast	Japan	Taira et al. (2002)
9.7.1.1	1987ff	AGSO Australian offshore lines	Australia	Goleby et al. (1994)
9.7.1.1	1989–1992	AGSO on-offshore program	Australia	Goleby et al. (1994)
9.7.1.1	1991	Bowen B–New England Orogen	Eastern Australia	Korsch et al. (1997)
9.7.1.1	1991	Kalgoorlie traverse EGF01	Western Australia	Drummond et al. (2000a)
9.7.1.1	1992	Otway basin onshore lines	Southern Australia	Finlayson et al. (1996)
9.7.1.1	1993	550 km reflection line Musgrave block	Central Australia	Korsch et al. (1998)
9.7.1.1	1994	Mount Isa inlier line	NE Australia	Drummond et al. (1998)
9.7.1.1	1994–1995	Otway basin offshore lines	Southern Australia	Finlayson et al. (1998)
9.7.1.1	1995	Tasmania offshore refl lnes	Southern Australia	Drummond et al. (2000b)
9.7.1.1	1996–1999	Broken Hill reflection lines	Eastern Australia	Finlayson (2010)
9.7.1.1	1996, 1998	OBS-Transects off N Australia	N Australia	Petkovic et al. (2000)
9.7.1.1	1997, 1999	Lachlan Fold Belt reflection lines	Eastern Australia	Finlayson (2010)
9.7.1.1	1997, 1999	Goldfield Province	W Australia	Goleby et al. (2002)
9.7.1.1	2001	AGSNY Goldfield Province	Western Australia	Goleby et al. (2004)
9.7.1.2	1990	Bay of Plenty, N-Island marine	New Zealand	Davey and Lodolo (1995)
9.7.1.2	1994	MOOSE, S-Island marine	New Zealand	Henrys et al. (1995)
9.7.1.2	1994	Central South Island land survey	New Zealand	Kleffman et al. (1998)
9.7.1.2	1994	Stewart Island southern S Island	New Zealand	Davey (2005)
9.7.1.2	1995–1996	SIGHT Southern Alps	Southern New Zealand	Davey et al. (1998)
9.7.1.2	1996	Stewart Island–Puysegur Bank	Southern New Zealand	Melhuish et al. (1999)
9.7.2	1995	MAMBA Continental margin Namibia	Namibia	Bauer et al. (2000)
9.7.2	1999	ORYX Erongo Mountains	Namibia	www.gfz-potsdam.de/GIPP
9.7.3.1	1995	TICOSECT Nicoya Peninsula	Off Costa Rica	Christeson et al. (1999)

(continued)

TABLE 2.9. MAJOR CONTROLLED-SOURCE SEISMIC INVESTIGATIONS OF THE CRUST IN THE 1990s (*continued*)

Chapter	Year	Project	Location	Reference
9.7.3.1	1996	COTCOR Nicoya Peninsula	Costa Rica	Sallares et al. (1999)
9.7.3.1	1996	Chicxulub BIRPS reflection lines	Mexico	Snyder et al. (1999)
9.7.3.2	1997	Toncantins Province	Brazil	Berrocal et al. (2004)
9.7.3.2	1998	ECOGUAY Guayana Shield	Venezuela	Schmitz et al. (1999)
9.7.3.2	2001	ECCO Oriental basin	Venezuela	Schmitz et al. (2005)
9.7.3.3	1994	PISCO W Cordillera	Northern Chile-Bolivia	Schmitz et al. (1999)
9.7.3.3	1995	CINCA off-onshore Andes	Northern Chile	Patzwahl et al. (1999)
9.7.3.3	1995	PRECORP reflection test profile Andes	Northern Chile	Yoon et al. (2003)
9.7.3.3	1995	CONDOR off-onshore refl lines	Central Chile	Flueh et al. (1998a)
9.7.3.3	1996	ANCORP reflection line Andes	Northern Chile	ANCORP Working Group (2003)
9.7.3.3	2000	ISSA southern Andes	Southern Chile	Lueth et al. (2003)
9.7.3.3	2001	SPOC Subduction Offshore Chile	Chile	Krawczyk et al. (2003)
9.7.4	1990–1991	Antarctic Peninsula Polish expedition	Antarctica	Grad et al. (1997)
9.7.4	1990–1991	SERIS Ross Ice shelf	Antarctica	ten Brink et al. (1993)
9.7.4	1993–1994	ACRUP southern Ross Sea	Antarctica	Trey et al. (1999)
9.8.2	1989	Society Island Hotspot Chain	Pacific Ocean	Grevenmeyer et al. (2001b)
9.8.2	1991	TERA East Pacific Rise	Pacific Ocean	Detrick et al. (1993)
9.8.2	1992	Izu-Ogasawara island arc	South off Japan	Suyehiro et al. (1996)
9.8.2	1993	East Pacific Rise at 9°–10°N	East Pacific	Christeson et al. (1997)
9.8.2	1993	Hess Deep rift valley	Pacific Ocean	Wiggins et al. (1996)
9.8.2	1994	Macquarie Ridge	SW off New Zealand	Finlayson (2010)
9.8.2	1994	Nankai Trough ERI94 line	SW Japan	Kodaira et al. (2000)
9.8.2	1995	Nankai Trough ERI95 line	SW Japan	Kodaira et al. (2000)
9.8.2	1995	EXCO S East Pacific Rise	Pacific Ocean	Grevemeyer et al. (1998)
9.8.2	1995	TAICRUST off Taiwan	Pacific Ocean	Schnuerle et al. (1998)
9.8.2	1996	MELT, East Pacific Rise 15°–18°S	E Pacific	Canales et al. (1998)
9.8.2	1997	East Pacific Rise at 9°05′N	E Pacific	Singh et al. (2006)
9.8.2	1996–2001	MCS Japan trench off northern Honshu	Off N Japan	Tsuru et al. (2007)
9.8.2	1997	Nankai Trough MO104 line	SW Japan	Kodaira et al. (2000)
9.8.2	1997	Sanriku region, off northern Honshu	Off NE Japan	Takahashi et al. (2004)
9.8.2	1998	Japan trench cross lines	Off N Japan	Miura et al. (2003)
9.8.2	1999	On-offshore in Shikoku	SW Japan	Kodaira et al. (2002)
9.8.2	1999	Japan trench off Myagi, Honshu	Off northern Japan	Miura et al. (2005)
9.8.2	1999	PAGANINI Galapagos Hot Spot	Pacific Ocean	Sallares et al. (2003)
9.8.2	2000	GEOPECO off Peru	Peru-Pacific	Krabbenhoeft et al. (2004)
9.8.2	ca. 2000	Nankai Trough MCS lines	S off Japan	Park et al. (2002)
9.8.2	2000	G-PRIME Galapagos Hot Spot	Pacific Ocean	Sallares et al. (2005)
9.8.2	2001	SALIERI Galapagos Hot Spot	Pacific Ocean	Sallares et al. (2005)
9.8.3	1990–1993	AGSO profiling off NW Australia	Australia margin	Finlayson (2010)
9.8.3	1991	Kerguelen Plateau	Indian Ocean	Operto and Charvis (1996)
9.8.3	1990s	SW Indian Ridge	Indian Ocean	Muller et al. (1999)
9.8.3	1992	Banda Sea	Indian Ocean	Finlayson (2010)
9.8.3	1997	Makran subduction off Pakistan	Indian Ocean	Kopp et al. (2000)
9.8.3	1998	Ninetyeast Ridge	Indian Ocean	Grevemeyer et al. (2001a)
9.8.4.1	1989–1991	Continental margin basins off NE Brazil	Southern Atlantic Ocean	Mohriak et al. (1998)
9.8.4.1	1991	Ghana transform margin	South of Ghana	Edwards and Whitmarsh (1997)
9.8.4.1	1992	LEPLAC project off NE Brazil	Southern Atlantic Ocean	Gomes et al. (2000)
9.8.4.1	1995	MAMBA continental margin Namibia	Southern Atlantic Ocean	Bauer et al. (2000)
9.8.4.1	1998	Colorado Basin off Argentina	Southern Atlantic Ocean	Franke et al. (2006)
9.8.4.2	1990	Great Meteor and other seamounts	NE Atlantic	Weigel and Grevemeyer (1999)
9.8.4.2	1993	RAMESSES, Reykjanes Ridge	North Atlantic	Navin et al. (1998)
9.8.4.2	1995	ICEMELT	Central Iceland	Darbyshire et al. (1998)
9.8.4.2	1996	B96	North-central Iceland	Menke et al. (1998)
9.8.4.2	1996	MARBE Mid-Atlantic Ridge	North Atlantic Ocean	Canales et al. (2000)
9.8.4.2	1996?	BRIDGE Reykjanes Ridge	North Atlantic Ocean	Sinha et al. (1999)
9.8.4.2	2000	*Meteor* (1986)" M47 5°S Fracture	Off NW Africa	Borus (2001)
9.8.4.3	1990	Scoresby Sud on-offshore	Eastern Greenland	Mandler and Jokat (1998)
9.8.4.3	1992	Voering Margin off Lofoten	Atlantic off Norway	Mjelde et al. (1997)
9.8.4.3	1994	FIRE Faeroe-Iceland Ridge Exp	North Atlantic Ocean	Smallwood et al. (1999)
9.8.4.3	1990s	FAST-UNST-FLARE	Faroe– Shetland Isl	England et al. (2005)
9.8.4.3	1994	Scoresby Sud on-offshore	Eastern Greenland	Schmidt-Aursch and Jokat (2005)
9.8.4.3	1994	SE Greenland coast	SE Greenland	Dahl-Jensen et al. (1998)
9.8.4.3	1996	Voering Basin off Lofoten	Atlantic off Norway	Mjelde et al. (1998)
9.8.4.3	1996	SIGMA Greenland Margin	Eastern Greenland	Korenaga et al. (2000)
9.8.4.3	1997	Porcupine Basin	West off Ireland	Reston et al. (2004)
9.8.4.3	1998	NW Barents Sea SE Spitsbergen	North Atlantic Ocean	Breivik et al. (2002)
9.8.4.3	1999	RAPIDS 3 Rockall Project	North Atlantic off Ireland	Mackenzie et al. (2002)
9.8.4.3	1999	ARKTIS VV/2 NW Spitsbergen	North Atlantic Ocean	Czuba et al. (2004)
9.8.4.3	2000	SCREECH Newfoundland basin	Canada	Funck et al. (2003)
9.8.4.3	2000	Mendelejev Ridge	Arctic Ocean	Lebedeva-Ivanova et al. (2006)

*Date of experiment assumed, not indicated by authors.

For example, by the year 2000, the PASSCAL and UTEP instrument pools in the United States had a total of 840 of the 1-component Texan instruments. In Europe, various groups had purchased 200 of the same instrument (Guterch et al., 2003a). Jones (1999) reports in detail on the equipment used by the end of the 1990s in marine seismic surveys.

The new decade starting in 2000 brought continuing advances in recording techniques. Digital technology, which had taken over recording devices for controlled-source seismology in the 1990s, was further improved, and, after many successful deployments, an updated design (RefTek-125A) for the Texan instrument was finalized in 2004. PASSCAL and the EarthScope program began to purchase these models immediately, and the UTEP (University of Texas at El Paso, USA) group focused on an upgrade path for the existing instruments. Via grants from several sources, UTEP obtained the funds needed to upgrade almost 400 of its units, and upgrades were under way by 2005 (Keller et al., 2005b). In a similar manner, the Australian equipment was steadily improved (Finlayson, 2010; Appendix 2-2).

In Britain, Leeds University purchased the Orion digital recording system and in 2000 the British seismic community (the NERC Geophysical Equipment Pool, Leicester, Cambridge, Leeds, and Royal Holloway) was awarded a grant to acquire large numbers of Guralp 6TD and 40T seismometers, together with a number of the Guralp 3Ts, bringing the British scientists to the forefront of observational broadband seismology (Maguire and SEIS-UK, 2002). In Germany during 2000, the GeoScience Center Potsdam started to gradually renew its instrumental pool for short-period and broadband seismology and replaced the worn-out RefTek and PDAS data loggers with a new system, Earth Data PR6-24. Up to mid-2007 almost 240 of these new data loggers have been acquired and made available to the geophysical community. In addition, for special purposes 10 broadband GURALP units were bought. A major German cooperation between land and sea investigations promises a new instrumental pool, named DEPAS (Deutscher Pool für Aktive Seismologie). In early 2006, it consisted of 30 OBS and 65 GURALP seismometers plus Earth Data Loggers (EDL). This was the result of a joint venture between the German institutions GFZ (GeoForschungsZentrum Potsdam) and AWI (Alfred-Wegener-Institut Bremerhaven) for marine seismology (Schulze and Weber, 2006; Schmidt-Aursch et al., 2006). When completed, 100 Earth Data Loggers PR6-24 and 100 GURALP CMG-3ESP Compact units will be available for the land part of DEPAS. For the offshore part, 68 standard and 12 deep-sea OBS (Guralp CMG-40T seismometers plus hydrophones) and 5 OBH (hydrophones only) will be available.

With the large number of recording devices available, projects can now be planned which extend seismic surveys into three dimensions. Large-scale research programs which involved a multitude of cooperating institutions and interdisciplinary cooperation of scientists from various geoscientific fields continued to dominate the scene in the early 2000s. However, with the increasing number of recording devices and the capacity to deploy instruments over large areas, tomographic methodologies such as teleseismic tomography became viable. Many earth scientists started to prefer long-term deployments of instruments using natural events as energy sources instead of short-term projects using expensive controlled sources.

IRIS/PASSCAL in the United States and LITHOPROBE in Canada continue to support large seismic (active) and seismological (passive) projects, occasionally combined with international and interdisciplinary geoscientific priority programs in Europe and Africa, dealing with large-scale tectonic topics. EUROPROBE, supported by the European Science Foundation, continues to support programs concerned with the origin and evolution of the continents (Gee and Stephenson, 2006) emphasizing East-Central-West European collaboration and close multinational cooperation of geologists, geophysicists and geochemists. Examples include the multinational projects ALP 2002 and SUDETES 2003. Another example of trans-Atlantic partnership is the EAGLE project in Ethiopia.

Special sessions on large national and international seismic programs became important parts of the annual meetings of the various national and international geoscientific organizations, such as, e.g., the American Geophysical Union or the European Geophysical Union. Special meetings of earth scientists at regular intervals also continue such as the meetings of the special subcommission of the ESC (European Seismological Commission) "Structure of the Earth's Interior," and the series "International Symposia on Deep Seismic Profiling of the Continental Lithosphere," which continues biannually into the twenty-first century with meetings at Ulvik, Norway, in 2000 (Thybo, 2002), in Taupo, New Zealand, in 2003 (Davey and Jones, 2004), in Mont-Tremblant, Quebec, Canada, in 2004 (Snyder et al., 2006), in Hayama, Japan, in 2006 (Ito et al., 2009a), in Saariselkä, Finland, in 2008, and in Cairns, Queensland, Australia, in 2010.

In a similar way the seismic investigation of the oceanic lithosphere depended greatly on the development of instrumentation, logistics and theory and needed close cooperation with other marine sciences. The seismic structure of the oceanic crust and passive margins was reviewed by Minshull (2002) and published in Part A of the International Handbook of Earthquake and Engineering Seismology, edited by Lee et al. (2002). The structure of the oceanic crust, as known by 2000, is shown in chapters on P- and S-wave velocity structure, anisotropy and attenuation, and variations of crustal structure with spreading rate and with age. Other chapters deal with the seismic structure of the Moho and the uppermost oceanic mantle, of mid-ocean ridges, of oceanic fracture zone and segment boundaries, and hotspots, ocean islands, aseismic ridges and oceanic plateaus. In a textbook *Marine Geophysics*, Jones (1999) has described in much detail the instrumentation and methodologies, as used in marine seismic exploration until the end of the 1990s. Another summary of our knowledge of the structure of the lithosphere under the oceans, as obtained until 2000, was published in the *Encyclopedia of Ocean Sciences* (Steele et al., 2001), with individual contributions on instrumentation and on results on the seismic structure of the ocean in general and of special features as, e.g., mid-ocean ridges. For example, to concentrate the efforts of mid-oceanic ridge research worldwide, "InterRidge" was founded in 1993 to promote interdisciplinary, international studies of oceanic spreading centers through scientific exchange and the sharing of new technologies and facilities among international partners. Its members are research institutions dealing with marine research around the world. InterRidge is dedicated to sharing knowledge amongst the public, scientists, and governments and to provide a unified voice for ocean ridge researchers worldwide (www.interridge.org).

We have compiled a small number of major seismic projects, undertaken from 2001 to 2005 and mentioned in Chapter 10, in Table 2.10. For many projects undertaken since 2001, only limited information has become available by the end of 2005. We therefore emphasize that Table 2.10 is far from complete.

We have terminated our historical review of controlled-source seismic experiments with projects planned and carried out through 2005. The large number of recording devices available nowadays and their ability to record continuously over long time periods has enabled seismic surveys to be extended into three dimensions in a tendency to plan for seismic tomography surveys with teleseismic and/or local events as energy sources. Neverthe-

less, the interest within the geological and exploration communities for detailed crustal and upper mantle structure has remained until the present day. Many new controlled-source seismic projects have been performed just recently or are in the planning stage. In many situations, details of crustal structure can only be achieved by controlled-source seismic near-vertical incidence reflection seismics and/or seismic refraction/wide-angle reflection experiments with densely spaced controlled sources and a multitude of seismic recorders.

Based on the incredible wealth of data gathered during the past 50 years on seismic crustal structure studies, Walter Mooney started to build up a database, which we describe in some detail in Chapter 10. Its preliminary versions have so far enabled the construction of worldwide syntheses of crustal parameters (e.g., Christensen and Mooney, 1995; Mooney et al., 1998; Mooney, 2002). Another example is a synthesis of the seismic structure of the crust and uppermost mantle of North America and adjacent oceanic basins (Chulick and Mooney, 2002). Based on the database, crustal thickness maps were plotted for the world and for the individual continents, accompanied by maps showing the data points available in the database until ca. 2008.

TABLE 2.10. MAJOR CONTROLLED-SOURCE SEISMIC INVESTIGATIONS OF THE CRUST IN 2001–2005

Chapter	Year	Project	Location	Reference
10.2.1	2001	IBERSEIS southwest Iberia Vibroseis survey	Spain	Carbonell et al. (2004)
10.2.1	2001	IBERSEIS southwest Iberia wide-angle lines	Spain	Palomeras et al. (2009)
10.2.1	2000–2002	Baltic Shield Kola peninsula lines	Russia	Kostyuchenko et al. (2006)
10.2.1	2001–2003	FIRE deep seismic reflection survey	Finland	FIRE consortium (2006)
10.2.1	2002–2003	ISLE teleseismic project	Ireland	Landes et al. (2005)
10.2.1	2004–2005	ESTRID Denish basin	Denmark	Thybo et al. (2006)
10.2.2	2001	VRANCEA 2001 southeast Carpathians	Romania	Hauser et al. (2007a)
10.2.2	2001	DACIA PLAN Focsani basin	Romania	Panea et al. (2005)
10.2.2	2001	SEISMARMARA	Turkey	Laigle et al. (2007)
10.2.2	2002	ALP 2002 Alps and eastern plains	Poland-Hungary	Brueckl et al. (2003)
10.2.2	2002–2003	ESRU Europrobe Urals	Russia	Kashubin et al. (2006)
10.2.2	2003	SUDETES Bohemian Massif	Czech Republic and Poland	Guterch et al. (2003a)
10.2.2	2003	GRUNDY 2003	Poland	Malinowsky et al. (2007)
10.2.2	2004	DRACULA southeast Carpathians	Romania	Enciu et al. (2009)
10.2.2	2005	Black Sea basin	Turkey	Scott et al. (2006)
10.2.2	2005–2006	DOBRE 2 Donbas fold belt	Ukraine	Starostenko et al. (2006)
10.3.1	2002	SHIPS Georgia Street	Northwestern USA–Canada	Brocher et al. (2003)
10.3.1	2004	SAFOD central California refraction line	California, USA	Hole et al. (2006)
10.3.2	2002	Walker Lane refraction survey	California-Nevada, USA	Louie et al. (2004)
10.3.2	2004	Northwest Basin and Range	California-Nevada, USA	Lerch et al. (2007)
10.3.2	2003	Rio Grande rift (Potrillo Volcanic Field)	Southwest New Mexico, USA	Averill (2007)
10.4.1	2001	EAGLE East African Rift	Ethiopia	Maguire et al. (2003)
10.4.1	2002	*Meteor* (1986)" M52 offshore Israel	Eastern Mediterranean	Hübscher et al. (2003)
10.4.2	2004	Jordan–Dead Sea Transform	Israel-Jordan	ten Brink et al. (2006)
10.4.2	2006	DESIRE southern Dead Sea	Israel-Jordan	Mechie et al. (2009)
10.5.1	2004	Naga thrust and fold belt	NE India	Jaiswal et al. (2008)
10.5.2	2000*	Altyn Tagh Range	China	Zhao et al. (2006)
10.5.2	2000*	1000 km northeast Tibet to Ordos basin	China	Liu et al. (2006)
10.5.3	2001	South-central Honshu	Japan	Iidaka et al. (2004)
10.5.3	2000s	Nankai Trough–central Japan on-offshore	Central Japan	Kodaira et al. (2004)
10.5.3	2002–2003	Reflection lines Tokyo area	Central Japan	Sato et al. (2006)
10.5.3	2002	WNW-ESE line south Korean peninsula	South Korea	Kim et al. (2007)
10.5.3	2004	Reflection lines Kinki area	Southwestern Japan	Sato et al. (2006)
10.5.3	2004	NNW-SSE line south Korean peninsula	South Korea	Kim et al. (2007)
10.6.1	2001	AGSNY Goldfield Province	Western Australia	Goleby et al. (2004)
10.6.1	2003	Gawler Craton South Australia	Southern Australia	Drummond et al. (2006)
10.6.1	2003–2004	Curnamona region southern Australia	Southern Australia	Finlayson (2010)

(continued)

TABLE 2.10. MAJOR CONTROLLED-SOURCE SEISMIC INVESTIGATIONS OF THE CRUST IN 2001–2005 (*continued*)

Chapter	Year	Project	Location	Reference
10.6.1	2005	Yilgarn Orogen on-offshore experiment	Western Australia	Finlayson (2010)
10.6.1	2005	Thomson-Lachlan orogens traverse	Eastern Australia	Finlayson (2010)
10.6.1	2005	Tanami region Northern Territory	Central Australia	Finlayson (2010)
10.6.1	2001	NIGHT Taupo Volcanic Zone	Northern New Zealand	Henrys et al. (2003)
10.6.1	2001	Chatham Rise, offshore New Zealand	Off New Zealand	Davy et al. (2008)
10.6.1	2003	Bounty Trough east of South Island	Off New Zealand	Grobys et al. (2007)
10.6.1	2003	Great South Basin east of South Island	Off New Zealand	Grobys et al. (2009)
10.6.1	2005	Marine reflection experiment off east coast North Island	Off New Zealand	Barker et al. (2009)
10.6.2	2003	Onshore-offshore SW coast of South Africa	South Africa	Hirsch et al. (2009)
10.6.2	2005	On-offshore Cape Fold Belt–Karoo Basin	South Africa	Stankiewicz et al. (2007)
10.6.2	2005	Seismic reflection survey Karoo Basin	South Africa	Lindeque et al. (2007)
10.6.3	2001	ECCO Oriental basin	Venezuela	Schmitz et al. (2005)
10.6.3	2004	BOLIVAR off Venezuela	Venezuela-Caribbean	Magnani et al. (2009)
10.6.3	2001	SPOC Subduction processes off Chile	Chile	Krawczyk et al. (2003)
10.6.3	2004–2005	TIPTEQ seismic array project	Chile	Rietbrock et al. (2005)
10.7.2	2000–2004	Australian margins	East and south off Australia	Finlayson (2010)
10.7.2	2001	SALIERI Galapagos volcanic	East Pacific	Sallares et al. (2003)
10.7.2	2001	Tonga Ridge	Southwest Pacific	Crawford et al. (2003)
10.7.2	2004	New Caledonia	Southwest Pacific	Lafoy et al. (2005)
10.7.2	2004?	Nankai Trough 3 parallel lines	South off Japan	Kodaira et al. (2006)
10.7.2	2004	Izu arc wide-angle project	South off Japan	Kodaira et al. (2007a)
10.7.2	2005	Bonin arc wide-angle project	South off Japan	Kodaira et al. (2007b)
10.7.2	2005	Refr survey of Nicaragua	Off Central America	Ivandic et al. (2010)
10.7.3	2000–2004	Australian margins	West and south off Australia	Finlayson (2010)
10.7.3	2003	Seychelles-Laxmi Ridge	Indian Ocean	Collier et al. (2004)
10.7.4	2002	RAPIDS-4 Rockall basin off Ireland	North Atlantic	O'Reilly et al. (2006)
10.7.4	2002	HADES Hatton Deep	North Atlantic	Chabert et al. (2006)
10.7.4	2003	Onshore-offshore southwestern South Africa	South Atlantic	Hirsch et al. (2009)
10.7.4	2003	Continental margin off French Guiana	South Atlantic	Greenroyd et al. (2006)
10.7.4	2003	EUROMARGINS eastern Greenland	North Atlantic	Schmidt-Aursch and Jokat (2005)
10.7.4	2004	Continental margin off Uruguay	South Atlantic	Temmler et al. (2006)
10.7.4	2004	"*Meteor* (1986)" M62-3 Cape Verde Island	South Atlantic	Grevemeyer et al. (2009)
10.7.4	2004	"*Meteor* (1986)" M62-4 Mid-Atlantic Ridge 7°S	South Atlantic	Reston et al. (2009)
10.7.4	2004	"*Meteor* (1986)" M61-2 off Ireland	North Atlantic	Reston et al. (2006)
10.7.4	2004	Porcupine Basin off Ireland	North Atlantic	Hauser et al. (2007b)
10.7.4	2005	Onshore-offshore southern coast South Africa	South Atlantic	Parsiegla et al. (2007)

*Date of experiment assumed; not indicated by authors.

The First 100 Years (1845–1945)

3.1. THE BEGINNING OF SEISMOLOGY

Since the beginning of the twentieth century, seismic waves have been used to study the Earth's interior. This concerns both the study of the whole Earth by distant earthquakes and of Earth's crust by local natural and artificial events. The rapid development of this special branch of seismology would not have been possible without the early technical developments of seismographs and sensitive recording devices of the foregoing century. An early historic review was published by Mintrop as early as 1947, describing the history of the first 100 years of earthquake research and explosion seismology (Mintrop, 1947; see also Appendix 3-1).

The first seismoscope is believed to have been constructed by the Chinese philosopher Chang Hêng in 132 A.D. Europeans wrote about earthquake-detecting instruments from the early eighteenth century. Earthquakes in Naples in 1731, the earthquake of Lisbon in 1755, earthquakes in Calabria in 1783, the New Madrid earthquakes of 1811 and 1812 in America, a series of small earthquakes near Comrie in Perthshire, Scotland, in 1839, and others triggered the construction of various seismoscopes. The general interest in recording earthquakes grew systematically.

Physical earthquake research started in 1848 with a publication by Hopkins (1848), who for the first time applied the refraction and reflection laws of optics to the propagation of earthquake waves and who introduced the expressions apparent surface velocity v, volume velocity V and the relation v/V = i, giving the angle at which elastic waves enter the Earth's surface (Fig. 3.1-01; Mintrop, 1947).

The years between 1850 and 1870 saw several significant contributions to seismological instrumentation. These included Palmieri's seismoscope for recording the time of an earthquake, and a suggestion by Zöllner that the horizontal pendulum might be used in a seismometer. Palmieri's *sismografo*, for example, seems to have been an effective earthquake detector for its time and was used by Palmieri on Mount Vesuvius. Later, Palmieri's *sismografo* was used by seismologists in Japan; where, from 1875 to 1885, 565 earthquakes were detected in Tokyo. In 1889, while investigating tidal signals with a horizontal pendulum at the Telegrafenberg in Potsdam, Germany, Ernst von Rebeur-Paschwitz identified, for the very first time, a signal that was caused by seismic waves from an earthquake near Tokyo.

The foregoing work set the stage for the late 1800s and early 1900s, when many fundamental advances in seismology were made. In Japan, three English professors, John Milne, James Ewing, and Thomas Gray, working at the Imperial College of Tokyo, invented the first seismic instruments sensitive enough to be used in the scientific study of earthquakes. They

Figure 3.1-01. Wave fronts and rays radiating from an earthquake source according to Hopkins (1848) (from Mintrop, 1947, fig. 1). [Die Naturwissenschaften, v. 34, p. 257–262, 289–295. Published by permission of Springer-Alerts.]

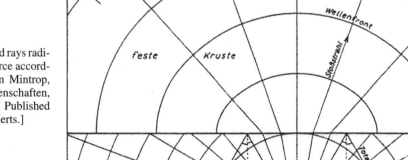

obtained the first known records of ground motion as a function of time, and they learned what such records could reveal about the nature of earthquake motion. They used their instruments to study the propagation of seismic waves, and they used them to study, for engineering purposes, the behavior of the ground in earthquakes. Under the influence of J. Milne, the Seismological Society of Japan was founded in the spring of l880 after a larger earthquake in Yokahama (Milne, 1885). After 1885, routine earthquake recording in Tokyo was started by new seismographs just developed in Japan. Nearly two decades later, Milne was largely responsible for having similar seismographs set up at stations throughout the world, in order to collect data which could be evaluated at a central observatory. In 1882, A.B. Briggs constructed his own seismograph in Launceston, Tasmania, Australia, and made recordings of Tasmania's earthquakes from 1883 to 1885 (D.M. Finlayson, 2010, personal commun.). Ewing's "duplex-pendulum" seismometer, invented in Japan, is of particular interest because in 1887 and 1888, this type of seismometer was placed at ten sites in Northern California and Nevada. In Italy, earthquake research begun by Cavalleri and Palmieri was continued by Italian seismologists in the 1870s. In 1874 the first journal devoted to solid-earth geophysics, the *Bulletino del Vulcanismo Italiano*, was founded and edited by M. De Rossi. In 1869, Zöllner described a horizontal pendulum with the suspension which has since been associated with his name. Horizontal pendulums were to be widely used in seismographs after 1880, because they could be given long periods and could still be compact. The Zöllner suspension was used, e.g., in the Galitzin horizontal seismograph, constructed in 1910 in Russia. A detailed summary of the early history of seismometry has been published by Dewey and Byerly (1969).

3.2. EARLY CRUSTAL STRUCTURE INVESTIGATIONS FROM SEISMOLOGICAL OBSERVATIONS SINCE 1898

During the following decades, earthquake recording stations were established in many countries of the world, in particular in Europe, Japan, and North America, where a high technological standard was available. As an example, the early development of seismology in Germany may be described in some detail.

In Germany, geophysics became a recognized field of science when the University of Goettingen established the Institute for Geophysics in 1898 (for details see Appendix A3-2). Here, Emil Wiechert (1861–1928) was appointed as its director and thus became the first professor of geophysics in Germany. He was one of the first scientists in Germany to study Earth's interior. Wiechert established a working group for seismology which was leading science in this type of research at the beginning of the twentieth century. The most famous co-workers of E. Wiechert were G.H. Angenheister, L. Geiger, B. Gutenberg, G. Herglotz, L. Mintrop, W. Schlüter, G.v.d. Borne, and K. Zoeppritz (Ritter et al., 2000; Ritter, 2001, Appendix A3-2).

The first seismological measurements in Goettingen were made in 1898. In the following years, the theory for seismological instruments was improved and more precise measurements were developed at a new institute on the Hainberg, outside the city to avoid cultural noise (Wiechert, 1903). Modern earthquake recording began in January 1903 (Figure 3.2-01).

Since that time seismicity has been continuously observed in Goettingen, using the famous automatically recording Wiechert-seismographs (Figure 3.2-02) which recorded on smoked paper. From July 1903 onwards, seismic bulletins were published (Linke and Wiechert, 1903). In 1905, this station (GTT) was officially appointed as the main seismic station in Prussia. Additional stations were soon added in the Hartz Mountains (Clausthal) and on the island of Heligoland. The disastrous 1906 earthquake of San Francisco was well recorded in Goettingen (Figure 3.2-03). Worldwide observations were initiated by establishing seismic stations at Tsingtau, China, and on the Samoan archipelago (Apia) in the western Pacific Ocean. In 1909, Father Edward Pigot established the Riverview Observatory, Sydney, Australia, using a system derived from the work of Galitzin in Russia and the seismometers designed by Wiechert (D.M. Finlayson, 2010, personal commun.).

Up to 1914, numerous basic discoveries were made at Goettingen (e.g., Wiechert, 1907), for example, the law for amplitude ratios of reflected, transmitted, and converted waves at discontinuities (Zoeppritz equations; Wiechert and Zoeppritz, 1907), contributions to the existence of Earth's core (Gutenberg, 1914), and an inversion algorithm for determining the depth-dependent distribution of seismic velocity from arrival times (Herglotz-Wiechert method; Herglotz, 1907). During World War I, seismic waves from large guns were detected at arrays of seismic stations and used to pinpoint gun emplacement sites. This basic research laid the foundations for the first experiments with manmade seismic events in the early 1920s after the First World War.

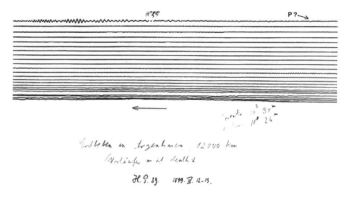

Figure 3.2-01. Example of an early recording on photo-sensitive paper with a horizontal pendulum (from Ritter et al., 2000). The time increases from right to left; one line corresponds to nearly one hour (1 min = 6 mm). On 12 April 1899, a far-distant earthquake from Argentina was recorded. [History of Seismology in Göttingen ("Overview" by Joachim Ritter, fig. 1), unpubl., reproduced in Appendix A3-2. Published by permission of J. Ritter.]

Figure 3.2-02. Drawing of the 17 t horizontal pendulum built by Wiechert in 1904 (from Ritter et al., 2000). [History of Seismology in Göttingen ("Emil Wiechert 1861–1928" by Sebastian Rost, fig. 4), unpubl., reproduced in Appendix A3-2. Published by permission of J. Ritter.]

The very first detection of Earth's crust as a separate unit, however, was detected by the study of local earthquakes in southeastern Europe (Mohorovičić, 1910). Many of the numerous studies of early earthquake records aimed to learn about the structure and physical properties of the Earth. So, in 1909, A. Mohorovičić at Zagreb, during his study of seismograms of a strong local earthquake, constructed a travel-time–distance plot (Figure 3.2-04). This event had occurred on 8 October 1909 in the nearby Kulpa Valley (~40 km south of the observatory) and had many aftershocks recorded throughout central Europe. Mohorovičić noticed that exclusively for distances between 300 km and 720 km, an additional P-wave and a corresponding S-wave could be identified from which he deduced a discontinuity with a velocity jump from 5.68 to 7.75 km/s at a depth which he calculated to be 54 km. He stated, "Since the $\bar{P}$-wave can only reach down to a depth of 50 km, this depth marks the limit of the upper layer of the Earth's crust. At this surface, there must be a sudden change of the material which makes up the interior of the Earth, because there a step in the velocity of the seismic wave must exist" (Mohorovičić, 1910). This boundary, based on the phase $\bar{P}$, later labeled P_n, was shortly thereafter defined as the crust-mantle boundary and was named the Mohorovičić discontinuity (subsequently shortened to "Moho") separating the crust with average velocity of 6.0–6.5 km/s and the uppermost mantle with velocities around 8 km/s.

It took another 15 years before the first fine structure of the Earth's crust was detected. In 1925, when investigating the records of the Tauern earthquake of 20 November 1923, Conrad detected a phase P* and inferred from it an intracrustal

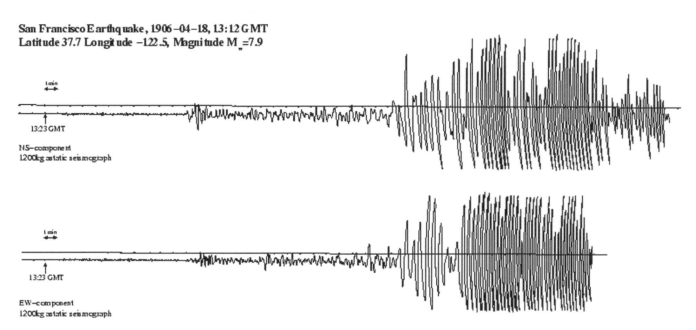

San Francisco Earthquake, 1906–04–18, 13:12 GMT
Latitude 37.7 Longitude –122.5, Magnitude M_w =7.9

Figure 3.2-03. Recording of the 1906 San Francisco earthquake at Göttingen, (digitized by Elmar Rothert, from Ritter et al., 2000). [History of Seismology in Göttingen (seismogram digitized from original paper record), unpubl., reproduced in Appendix A3-2. Published by permission of J. Ritter.]

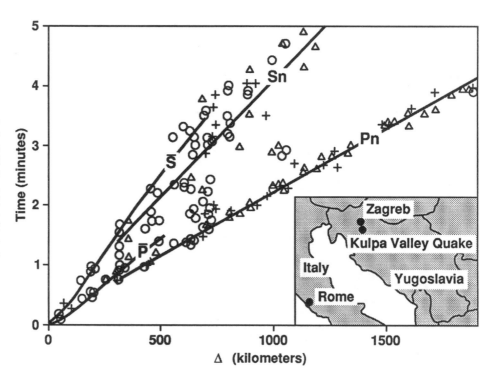

Figure 3.2-04. Mohorovičić's (1910) traveltime versus distance plot for the 8 October 1909 Kulpa Valley earthquake and its aftershocks. After Mohorovičić (1910) and Skoko and Mokrović (1982) and redrawn from Jarchow (1991, ch. 2, fig. 2). [Ph.D. thesis, Department of Geophysics, Stanford University, Stanford, California, ch. 2: The nature of the Mohorovičić discontinuity, p. 2.01–2.53. Published by permission of the author.]

discontinuity (Conrad, 1925). He could establish its existence, but with varying depths, when he investigated a 1927 earthquake (Conrad, 1928). Subsequently, many other investigators worldwide confirmed this discontinuity and it was finally named the Conrad-discontinuity.

In his book *The Earth*, Jeffreys (1929) discussed in much detail the detection of the subdivision of the crust based on near-earthquake observations in continental regions. In his summary on the upper layers of the Earth, he concludes that three layers are concerned: an upper layer, 10 km thick, with P-velocities 5.4–5.6 km/s; an intermediate layer, 20 km thick, with 6.2–6.3 km/s; and a lower layer with 7.8 km/s. Comparing the velocities with laboratory measurements on the compressibility of rocks, he suggested that the three layers are probably composed of granite, tachylyte (glassy basalt) and dunite, and that there is probably no layer of crystalline basalt. Though the conditions below the oceans had been "less thoroughly studied," he saw evidence that the granitic layer there was thin or absent.

Suggestions that the structure of the Earth's crust was even more complicated came from Gutenberg in 1934. He had already used the variation of amplitudes in local earthquake recordings to deduce the existence of a weak low-velocity zone for P-waves in the upper mantle at a depth between 70 and 80 km (Gutenberg, 1932a, vol. 4, p. 213). Now, in 1934, he identified an additional low-velocity zone in the upper crust of southern California just above the Conrad discontinuity (Gutenberg, 1934), which was confirmed by showing that the arrivals of P-waves from explosions were earlier than those from earthquakes (Macelwane, 1951). Gutenberg also speculated on the possibility that there also existed low-velocity zones in the lower crust above the

Mohorovičić discontinuity. Also, for oceanic areas, early estimations of crustal thickness were made. For example, Hayes (1936) and Bullen (1939) used records from earthquakes to determine the crustal thickness around New Zealand.

Gutenberg (1924) also acknowledged another fundamental property of the Earth's crust, namely a fundamental difference between continental and oceanic crust. He confirmed observations, which Tams, Angenheister, and Macelwane had made in 1921 and 1922, that the velocities of propagation of surface waves were faster across the oceanic than across the continental portions of the Earth's surface (Gutenberg, 1924; see also tables 54–65 of Macelwane, 1951). He proposed a method of inversion for the dispersion recognized in surface waves to determine upper mantle structure that was similar to the method ultimately applied in the late 1950s. His inversion for crustal thickness gave a thick crust under the continents and a thinner crust under the oceans, with a crustal thickness of only 5 km under the Pacific. From these results, Gutenberg became convinced that there were large structural differences between continents and oceans in the outermost parts of the Earth, a view that was to play a significant part in his model of continental drift (Gutenberg, 1936).

In Volume VII of *Physics of the Earth*, edited by B. Gutenberg, on the internal constitution of the Earth, first published in 1939 and re-published with revisions in 1951, Macelwane (1951) summarized velocity measurements for the entire earth in 38 tables (tables 36–73), based on 244 references. Table 43 of Macelwane (1951) shows the varying velocities of the phase P* defining the Conrad-discontinuity published until 1939, and a summary of P_n-velocities published until 1939 is given by Macelwane (1951) in table 44.

3.3. THE FIRST STEPS INTO CONTROLLED-SOURCE SEISMOLOGY—1851–1945

While the basic structure of the Earth's crust was detected by the detailed study of earthquake records, the accuracy of seismological studies was limited for more refined studies of the Earth's crust because of too many unknown parameters, i.e., the exact time and the exact location of natural earthquakes. In the middle of the nineteenth century, the first studies with controlled events were started. Explosion seismology was born in 1849, when Robert Mallet used dynamite explosions to measure the speed of elastic waves in surface rocks (Mallet, 1852; Mintrop, 1947; Dewey and Byerly, 1969; Jacob et al., 2000). He wished to obtain approximate values for the velocities with which earthquake waves were likely to travel. However, he obtained only 500 m/s in granite, i.e., the energy was not sufficient to observe the first arrivals at his 2 km distant seismoscope even using 5500 kg of black powder. It was Henry Larcon Abbot who, in 1876, for the first time obtained a realistic velocity in gneissic rocks of 6.24 km/s recording several explosions with charges between 11,000 and 2000 kg dynamite with similar instruments to Mallet's, but with much larger amplification (Abbot, 1878; Mintrop, 1947).

The further development of explosion seismology depended largely on the construction of suitable instrumentation. In 1881, with his new 3-component seismic instrument invented in Tokyo, J. Milne described a trial to record an explosion of 1 kg of dynamite at 65 m distance on smoked paper (Milne, 1885). The problem, however, was to detect the short-period P-waves preceding the long-period surface waves. Also Hecker's experiment with 1500 kg of explosives recorded up to 350 m distance gave only weak indications of these preceding short-period arrivals (Hecker, 1900). Slightly more successful were Fouqué and Lévy (1889) experimenting in a mine to obtain the precursors using, for the first time, a photographic recording device when recording charges of 4 kg black powder at 145 m distance. Mintrop (1947) described in detail the first steps of controlled-source seismology.

Many of the explosion seismic experiments had hinted at practical uses of the method. In the late nineteenth century, various publications appeared on the theory of wave propagations. Schmidt (1888) proposed the study of time-distance records of artificial disturbances to determine the variation of the speed of seismic waves with depth (Nettleton, 1940). Knott (1919) published a paper on the propagation of seismic waves and their refraction and reflection at elastic discontinuities. Weatherby (1940) cites the objectives as expressed by Belar in 1901 that "the modern, sensitive instruments may easily be used to special advantage wherever we wish to learn beforehand the composition of the Earth's crust to carry out to advantage, e.g., the construction of a tunnel. A series of tests carried out along the tunnel line on the surface would be sufficient to enable us to form a reliable judgment on the elasticity conditions, or, say, on the rigidity of the ground, of an earth stratum which would not be accessible by

other means." A few years later, Wiechert and Zoeppritz (1907) had worked out the theory of seismic wave transmission through the Earth and gave solutions to the problems of seismic wave propagation, refraction and reflection, which placed the theoretical solution of the problem far ahead of the experimental solution (Weatherby, 1940). Finally, Wiechert (1910) expressed what would be the subsequent practice of refraction profiling, that the farther away from the source seismic waves are observed, the deeper they penetrate into the Earth and the more of the assumed layers they will have traversed. Correspondingly, one can compute the paths of all rays through these layers.

The final instrumental progression to record artificial earthquakes was when, in 1906, E. Wiechert in Goettingen constructed a transportable seismograph which amplified the horizontal component of the ground movements by 50,000. In 1908, active seismological experiments started in Goettingen in the garden of the institute where Ludger Mintrop made the first experiments to investigate the uppermost sedimentary layers using a weight drop and recording with Wiechert's portable seismographs (Figure 3.3-01).

To examine the effect of a falling weight, he built a mechanism to drop a 4000 kg iron ball from a height of 14 m. In this way, Mintrop recorded the first seismograms including the fine details of precursor waves (now called P and S body waves) from a controlled source (Figure 3.3-02).

As mentioned above, during World War I, seismic waves were recorded to compute the position of large guns. The British

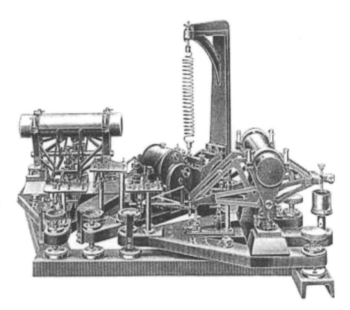

Figure 3.3-01. Drawing of Mintrop/Wiechert mobile seismometers built to observe higher frequencies in all three components (after Galitzin, 1914) (from Schweitzer and Lee, 2003). [*In* Lee, W., Kanamori, H., Jennings, P.C., and Kisslinger, C., eds., International Handbook of Earthquake and Engineering Seismology, Part B: Amsterdam, Academic Press. Copyright Elsevier.]

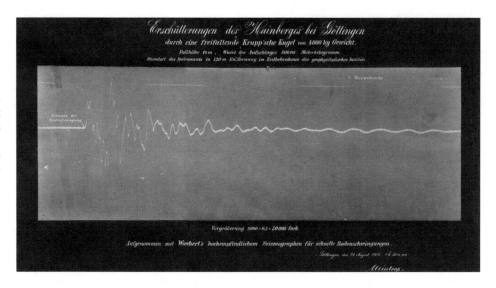

Figure 3.3-02. Recording of the wavelet generated by the drop of the 4000 kg ironball by the Wiechert-seismograph on 21.08.1908 (from Ritter et al., 2000). [History of Seismology in Göttingen ('L. Mintrop' by Andreas Barth, fig. 1), unpubl., reproduced in Appendix A3-2. Published by permission of J. Ritter.]

army used a 6-channel Eindhoven string galvanometer system, tuning fork timing and an array of carbon microphones to record the seismic energy (D.M. Finlayson, 2010, personal commun.).

The Canadian inventor Reginald Fessenden was the first to conceive of using reflected seismic waves to infer geology. In 1913, he had worked in the United States and had conducted various experiments. On the basis this research, he obtained "patents (Fessenden, 1917) on a method of exploration that involved the principle of the seismic method" (Nettleton, 1940). Due to World War I, however, he was unable to follow up on the idea, but worked on methods of detecting submarines.

In the United States, John Clarence Karcher discovered seismic reflections independently while working for the U.S. Bureau of Standards (now the National Institute of Standards and Technology). John Clarence Karcher had entered the University of Oklahoma (OU) in the autumn of 1912 to study electrical engineering, but later changed his major to physics. Contacts he made in both departments become an important part of the story. Karcher graduated from OU in 1916 with a B.S. in physics and started graduate work at the University of Pennsylvania. Unfortunately, World War I began and Karcher went to the U.S. Bureau of Standards. William P. Haseman, who was the head of the Physics Department at OU responsible for getting Karcher into the University of Pennsylvania, took a temporary leave and also went to work during the war at the U.S. Bureau of Standards, where scientists were engaged in the development of sound-ranging equipment to be used in locating the enemy artillery. During this work, seismic energy as well as the air waves were studied. When Karcher tried to observe artillery via ground waves, he noticed some extra waves that seemed to be reflections from layers of rock inside the Earth. It occurred to him that by means of reflected energy, geological structures could be mapped. When he showed these reflections to Haseman, it was Haseman who suggested they form a company to use these reflected waves to find oil and gas. They agreed to form this company after the war. Karcher went back to the University of Pennsylvania to finish

his education before starting on his big venture with Haseman. In 1919, they started experimental work. After numerous tests in many areas, a two-trace record was obtained that positively indicated a shallow reflection. In 1920, they finally organized the Geological Engineering Company and began field operations in Oklahoma (Musgrave, 1967). The first field tests were conducted near Oklahoma City, Oklahoma, in 1921.

The company soon folded due to a drop in the price of oil. In 1925, oil prices had rebounded, and Karcher helped to form Geophysical Research Corporation (GRC) as part of the oil company Amerada. In 1930, Karcher left GRC and helped to found Geophysical Service Incorporated (GSI). GSI was one of the most successful seismic contracting companies for the following 50 years and was the parent of an even more successful company, Texas Instruments (see Chapter 10.8). Early GSI employee Henry Salvatori left that company in 1933 to found another major seismic contractor, Western Geophysical.

In Germany, Mintrop had developed a method to estimate the distance to cannons by using seismic and sonic signals. He also developed a portable vertical seismograph with a photographic device. His interpretation of travel-time curves of seismic waves was based on the methods developed by Wiechert and Zoeppritz for global seismology. In particular, he made use of the seismic head waves (Mintrop, 1930, 1947). In 1919, he finally applied for a patent for a "method to investigate geological structures" (Figure 3.3-03).

In the period up to 1921 Mintrop (1930, 1947, 1949a) had developed seismic prospecting methods, the refraction and reflection methods, for which, in 1921, he obtained the exclusive rights (German Patent no. 371,973, 7 December 1919) without indicating how the method worked. In the same year, 1921, L. Mintrop founded an exploration company in Germany, SEISMOS, which started with successful seismic measurements on oil, coal, ore, and quartzite. In 1923, a team was sent overseas to Mexico and later on to Oklahoma and Texas where the application of the seismic-refraction method led to good results. Nettleton (1940)

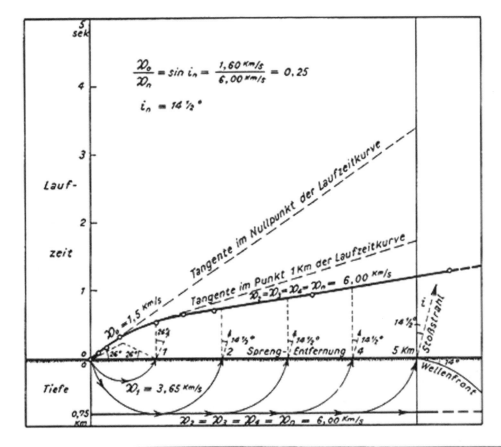

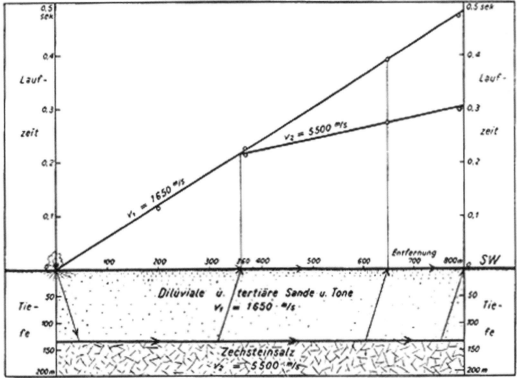

Figure 3.3-03. Two traveltime curves, which were used by Mintrop in 1919 and 1920 (from Ritter et al., 2000). [History of Seismology in Göttingen ('L. Mintrop' by Andreas Barth, fig. 2), unpubl., reproduced in Appendix A3-2. Published by permission of J. Ritter.]

and Weatherby (1940) have described in detail the further development of applied seismic refraction work by various companies in the Gulf Coast area in the 1920s.

The development of near-vertical reflection seismics in Germany was described by Köhler (1974). In 1925, the first electronic amplifier was constructed at the Geophysical Research Corporation which allowed the passing of relatively low-frequency reflected waves. This enabled the development of a central recording unit, i.e., the recording of several traces side by side on one seismogram, and thus allowed recognition of impulses by eye occurring on all traces at the same time and correlating them as reflections.

These achievements enabled new scientific research on shallow geological structures. As suggested by Wiechert in 1926, Mothes (1929) carried out seismic-reflection measurements by detonating dynamite charges in ice to study a glacier in the Alps. A few years later similar measurements were carried out on other European glaciers and in Greenland to measure the thickness of the ice cap (Brockamp, 1931a).

In ca. 1930, the seismic refraction method started to find natural limitations in the Gulf Coast. At the same time, the successful application of the seismic-reflection method started in Oklahoma. Two years later, the reflection method had almost completely replaced the refraction method and was applied in all parts of the world where extensive geologic exploration for oil was conducted (Barton, 1929; Nettleton, 1940). Weatherby (1940) described this development in some more detail.

In the period 1929–1931 the Imperial Geophysical Experimental Survey (IGES) in Australia again used the system that had been used by the British army in World War I to record shots

buried at 4 m depth. The whole system was housed in a portable hut that could be moved in about one hour from site to site. Work was conducted at Gulgong and Tallong ~150 km north of Canberra (D.M. Finlayson, 2010, personal commun.).

In Germany, the first seismic reflections were recorded in 1933 with mechanical seismographs. The clear recognition of reflections, however, was only enabled by introducing central recording, for which electronic amplifiers, electrical seismographs, and recording with galvanometers were essential. In 1933, the first German seismic-reflection instrument was built and in 1934 was successfully used for the first time. The first useful field seismometer was built in 1936 in the lab of SEISMOS (Fig. 3.3-04). As energy sources, explosions in boreholes were used exclusively. The number of traces of a reflection-seismic instrumentation increased with time. In the beginning, only 4–6 channels were available, increasing to 12–14 channels during the years. Also, in the beginning, in order to increase the distance to the individual seismograph from the shot hole when working with a fixed pre-set amplification, the charge had to be increased and the shot repeated. Only in the early 1940s did the techniques advance such as to allow regulation of the amplification per individual trace (Köhler, 1974). This regulation of arriving seismic amplitudes varying as much as 10^5 was a precondition for later deep crustal studies (Meissner, 2010, personal commun.).

This technical achievement enabled the successful application of the seismic-reflection method in Germany. Mintrop (1947) discussed, with the aid of various figures and corresponding references, how the refraction and reflection method gradually developed and how it was used for commercial applications both in Germany and in the United States from 1920 on. The

Figure 3.3-04. Electrical seismometer developed by Trappe and Zettel in 1936 (from Mintrop, 1947, fig. 33). [Die Naturwissenschaften, v. 34, p. 257–262, 289–295. Published by permission of Springer-Alerts.]

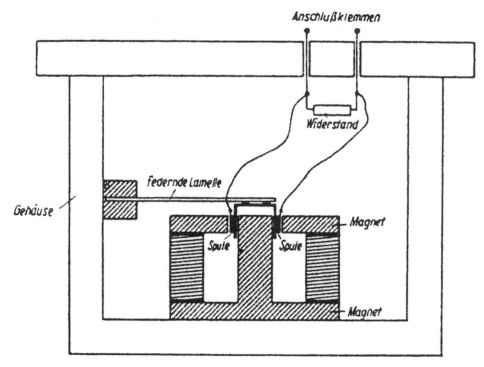

technical development of the seismic method (Barsch and Reich, 1930) was also stimulated by the fact that from 1934 on, the whole area of Germany was systematically being investigated for useful oil and mineral resources by refraction-seismic (mainly fan-shooting), magnetic and gravimetric measurements (Reich, 1937; Closs, 1974). This so-called "Reichsvermessung" was a unique cooperation of geologists and geophysicists from universities and industry. However, mainly data were collected but not simultaneously interpreted at that time.

In the United States, contrary to Europe, with improved instrumentation, the success of the reflection method was so immediate and pronounced that it gave rise to a phenomenal growth in seismic activity. By 1937, there were between 225 and 250 crews working in the country compared to four in 1929. In the 1930s, substantial improvements in instruments and techniques had been made. Automatically controlled amplifiers became general; their sensitivity was either a function of time or of amplitude and they were adjusted so that the early part of the record was compressed and made readable. Narrower band pass filters were used to eliminate extraneous energy from the record. Also, some overlap was already applied, feeding the output of each detector into two or more recording channels. Because of the better control of amplitudes, more seismic traces could be recorded on a given width paper. Near the end of the 1930s, as many as 10 traces could be recorded on the same paper. Some companies even made trials with up to 20 traces on wider paper. More attention was also paid to the depth of shot holes. For example, it was recognized that better energy was provided by placing the charges below the groundwater table (Weatherby, 1940).

The early worldwide state-of-the-art techniques for seismograph prospecting for oil were first summarized in a symposium of the American Institute of Mining and Metallurgical Engineers in Los Angeles in 1938 with contributions by W.H. Tracy (theory of seismic reflection prospecting), A. Nomann (instruments for reflection seismograph prospecting), F. Ittner (seismograph field operations), and P.C. Kelly (determining geologic structure from seismograph records), edited by English (1939). In his textbook *Seismograph Prospecting for Oil*, Nettleton (1940) describes in great detail instrumentation, data collection, theory, and interpretation methods available at the end of the 1930s.

The worldwide history of exploration geophysics, however, will not be the subject of this publication. Rather, the reader is referred to Lawyer et al. (2001), who have written a personalized history on the commercial application of controlled-source seismology starting at the time of World War I. This work was first published under a similar title in 1982, but was thoroughly revised and updated, carrying the story into the new millennium.

For academic crustal structure research, it was not until 1923 that "artificial earthquakes" were introduced by the use of large explosions. Reinhardt (1954) researched the use of quarry blasts for crustal studies up to 1954 and has published a historical review of early explosion seismology work up to the early 1950s.

The artificial event which started detailed investigations using blasts was a large catastrophic explosion near Oppau, Germany, in 1921, which was recorded by earthquake stations up to 307 km distance (e.g., Hecker, 1922). Wiechert (1923) noticed that elastic waves produced by large quarry blasts could be detected by stationary vertical seismographs with 2-million-fold amplification at distances of several hundred kilometers. This fact caused him to name such events "artificial earthquakes" and to treat them as such. He organized the construction of portable recording units and recorded the vibrations from these explosions along profiles. In addition, equipment development enabled the exact explosion time to be transmitted and recorded by the recording stations. Compared to natural earthquakes, the artificial ones have the advantage that both time and location were known exactly. The so-called Goettingen travel-time curves for longitudinal and transversal precursors from far explosions by Wiechert (1929) and their depth interpretation by Brockamp (1931b) are shown in Figure 3.3-05. However, in spite of recording distances up to 230 km, signals from the Mohorovičić discontinuity could not be detected, most probably due to low energy at the source and to insufficient sensitivity of the recording devices. Despite many publications (Wiechert 1923, 1926, 1929) and oral reports, Wiechert tried without success to obtain support from other geophysical institutes in Germany (see also Schulze, 1974).

After Wiechert's death in 1928, G.H. Angenheister continued the observation of quarry blasts and sound transmission, as well as the study of the Earth's interior and the application of the seismic method for the geophysical exploration of the uppermost layer of the Earth's crust, drawing attention to their practical importance (e.g., Angenheister, 1927, 1928, 1935, 1942; Brockamp 1931b). For example, he developed the theory to determine boundaries by using traveltimes of seismic waves and incidence angles. The portable seismic recording instruments were improved as well as the recording of time signals via radio broadcasting. By 1929, recordings were obtained at distances beyond 50 km. Angenheister plotted all arrival times of seismic longitudinal waves from explosions in a diagram, using the same procedure as used for local earthquakes, but the values scattered considerably. For the time being, the only conclusion deduced was that the Eurasian continent does not possess flat discontinuities with interlayering of homogeneous rock sequences (Angenheister, 1950). These investigations were fundamental for subsequent seismic-refraction investigations in Western Germany after World War II (Engelhard, 1998).

At the beginning of the 1930s, the principle of seismic head wave propagation was still not totally accepted in Germany. In his textbook on applied geophysics, Haalk (1934) mentioned that there are three competing models of head wave propagation. The differences occur in the spreading wavefront between the surface and the boundary layer: (a) a totally vertical ray path, (b) a diagonal ray path according to Fermat's law of shortest traveltimes, and (c) a curved ray path similar to version b, but with a sharp refracted wave instead of a real head wave (Figure 3.3-06). Indeed the author preferred the second (correct) model, but obviously the opinions varied. Using seismic measurements on the Rhône glacier in 1931 and 1932, Gerecke (1933) could prove that, according

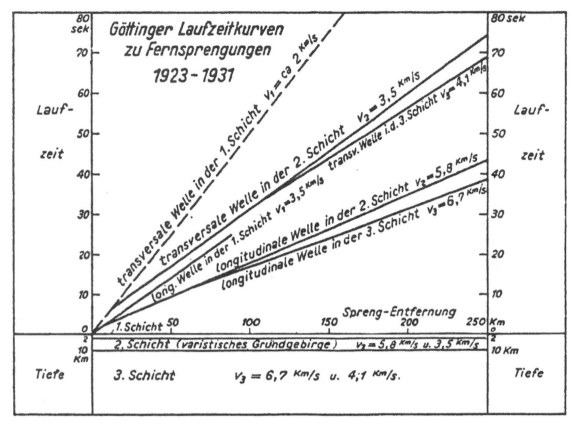

Figure 3.3-05. Traveltime curves of Göttingen of longitudinal and transversal precursors from far explosions and their depth interpretation by Wiechert and Brockamp (from Mintrop, 1947, fig. 38). [Die Naturwissenschaften, v. 34, p. 257–262, 289–295. Published by permission of Springer-Alerts.]

to Fermat's principle, the seismic waves followed the inclined ray path based on the calculated traveltimes of refracted waves.

Not only had large explosions been recorded in Germany. In France, results of the investigation of large surface explosions in 1924 near La Courtine had been reported upon (Maurain et al., 1925; Rothé et al., 1924). Here only the direct waves with an average velocity of 5.5 km/s had been recorded and later arrivals with 2.8 km/s had been interpreted as "long waves."

In California, seismic investigations using quarry blasts had been undertaken (e.g., Wood and Richter, 1931, 1933). A successful interpretation failed, however, due to the fact that the time of the explosion was not exactly known. More detailed velocity-depth models were given by Byerly and co-workers who investigated quarry blasts at Richmond (e.g., Byerly and Wilson, 1935b) and suggested a 3-layer crust of 31 km thickness beneath northern California (Table 3.3-01). Based on near-earthquake recordings, Gutenberg (1932b) derived a 39-km-thick 4-layer crust for southern California (Table 3.3-01).

For the purpose of oil prospecting in the 1930s, the principles of close station spacing were employed at the same time in west Texas and elsewhere in the oil industry (Ewing et al., 1939).

In the eastern United States, several quarry blast studies were reported. In Pennsylvania, Ewing et al. (1934) recorded surface velocities of 6.4 km/s and interpreted them as wave propagation in limestones. The first blast measurement of crustal thickness seems to have been that of Leet (1936) from a study of quarry blast recordings on station seismographs in New England. His tentative depth was 23 km (Steinhart, 1961). Numerous quarry blasts were recorded in the following years by the Harvard observatory (Leet, 1936, 1938) from which a 14.5-km-thick granitic upper crustal layer was deduced overlying an intracrustal layer with 6.77 km/s velocity. Later, Leet (1941) added observations of near-earthquakes to his data set from various permanent stations in northeastern America up to 10° distance and obtained another intracrustal layer with 7.17 km/s velocity, which was underlain by the Moho at 35–36 km depth with an upper-mantle velocity of 8.43 km/s.

In reviewing the studies of the crust by explosions before World War II, Steinhart (1961; see Appendix A5-1) concluded that a number of seismologists recognized the potential value of explosion studies, but instrumentation was inadequate, the theoretical basis of the refraction methods used in geophysical prospecting was suspect, and financial support for large field experiments was not available. The interpretations of seismic data concerning the structure of the crust followed the ideas of Daly (1914), who postulated a "granitic" layer overlying a "basaltic"

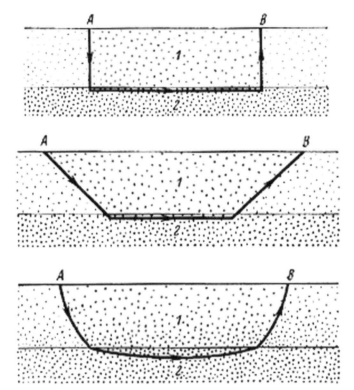

Figure 3.3-06. Ray paths after Haalk, 1934 (from Ritter et al., 2000). [History of Seismology in Göttingen ('L. Mintrop' by Andreas Barth, fig. 3), unpubl., reproduced in Appendix A3-2. Published by permission of J. Ritter.]

TABLE 3.3-01. DEPTH-VELOCITY STRUCTURE DEDUCED FOR CALIFORNIA BY GUTENBERG (1932B) AND BYERLY AND WILSON (1935A)

Southern California (Gutenberg, 1932b)		Northern and central California (Byerly and Wilson, 1935a)	
Depth of layer (km)	P-velocity (km/s)	Depth of layer (km)	P-velocity (km/s)
00–14	5.55	01–13	5.6
14–26	6.05	13–25	6.6
26–30	6.83	25–31	7.3
30–39	7.60	>31	8.0
>39	7.94		

substratum that was of no great strength, and Jeffreys (1926) who designated P_1 as P_g and referred to P_b as a wave traveling in the basaltic layer (Steinhart, 1961) and explained the recorded waves as head waves from the top of the two layers superimposed on the mantle.

Early investigations using explosion data were also made in the former USSR just before World War II. Koridalin, from commercial blasts in 1939, divided the crust near the Ural Mountains into nine layers. At the Institute of Terrestrial Physics in Moscow, experiments were carried out from 1938 to 1940 by Gamburtsev and co-workers for the improvement of refraction prospecting methods. The principle suggested close spacings of seismometers, so that wave groups and individual phases may be traced for as long a distance as possible. Experiments to test these methods were performed on the European platform of the USSR from 1938 to 1941 and on the Aspheron peninsula in 1944 (for references see Steinhart, 1961, Appendix A5-1).

Seismic-refraction techniques were also developed to investigate shallow sea and coastal areas. Reflection and refraction measurements in water-covered areas were made as early as 1927 for the purpose of locating oil-baring structures (Rosaire and Lester, 1932). For various areas the thickness of sediments and the velocity at the top of the underlying basement was determined both on land and at sea (e.g., Ewing et al., 1937, 1939,

1940). The first tests were made in 1935 in the vicinity of a deep borehole; the next tests were made on board of a ship. Both the charge and the geophones were placed on the ocean floor; otherwise, the method was similar as used on land. Both reflection and refraction measurements were successfully applied, but Ewing et al. (1937) stated that the refraction method gave more important results.

At that stage of development, the method was not applicable for water depths exceeding 100 fathoms, i.e., beyond the edge of the continental shelf, though successful tests were made to place both charges and recording system on the bottom of the ocean. Already in 1937 and 1938, tests were made in which seismographs were placed on the ocean floor at depths exceeding 2000 m in several attempts. In these experiments, an automatic oscillograph, four geophones, and four bombs, all distributed along an electric cable ~1 km long, had been lowered to the sea floor, laid down to form a straight line, and left undisturbed for 15 minutes while the profile was shot and recorded automatically. In 1939 and 1940, a new wireless system was developed and a small amount of data successfully recorded at two positions, one at 2600 m near Cape Cod, Massachusetts, and one at 4500 m water depth 550 km NW of Bermuda (Ewing and Vine, 1938; Ewing et al., 1946). These instruments were allowed to drop to the bottom under ballast which was detached automatically to allow them float to the surface after the tests were completed (Ewing and Ewing, 1961).

In general, however, the work was restricted to shallow waters, as hydrophones were not yet invented. Ewing et al. (1937) have described in much detail the experiment and the interpretation methodology both for land and for sea stations. As a result of their onshore and offshore experiments in Massachusetts and in Virginia, Ewing et al. (1937) could determine the configuration of the surface of the crystalline basement from the foot of the Piedmont Plateau to the edge of the continental shelf. The thickness of the unconsolidated and semi-consolidated material (P-wave velocity ~2.4 km/s) near the edge of the continental shelf was ~3600 m. In 1938, Woollard and Ewing (1939) applied the refraction method to the investigation of the Bermudas. They were probably the first to investigate the structure of an atoll using seismic methods. They found that on Bermuda calcareous sediments were underlain by presumably volcanic rocks with a relatively low velocity of ~4.9 km/s. Later measurements by Officer et al. (1952) showed that the velocity of the volcanic cone of Bermuda is 5.05–5.35 km/s without changing the simple model proposed by Woollard and Ewing (1939). Similar work at sea around Britain at the end of the 1930s was also reported upon by British scientists (Bullard and Gaskell, 1941; Bullard et al., 1940). Bullard et al. (1940) found velocities close to 5 km/s for rocks of the Paleozoic floor in southeastern England.

In tables 36–50, Macelwane (1951) has summarized velocities for the outer surface layers of the earth, i.e., for sediments and crustal crystalline rocks, obtained from quarry blast and near earthquake recordings between 1900 (e.g., Hecker, 1900) and 1939 with a few additions of results from 1940 to 1949. Tables 42–47 concentrate on P- and S-velocities in the crust for different depth ranges: P_g, P*, P_n, and S_g, S*, and S_n respectively, and in table 48, velocities of other seismic phases from near earthquakes are cited. Table 41 of Macelwane (1951) (Table 3.3-02) concentrates on P and S velocities of waves caused by explosions and blasts measured in France (Maurain et al., 1925; Rothé et al., 1924), Germany (Müller, 1934; Wrinch and Jeffreys, 1923; Hecker, 1922), Italy (Agamemnone, 1937; de Quervain, 1931), Switzerland (de Quervain, 1931), California (Gutenberg et al., 1932; Byerly and Wilson, 1935b; Wood and Richter, 1931, 1933) and New England (Leet, 1938). With a few exceptions, explosion seismology recordings up until 1939 only resulted in velocities being determined for sedimentary layers and the uppermost crystalline crust, the so-called granitic layer. In Macelwane's table 41 (Table 3.3-02) only one value is given for a P_2-phase of 6.77 km/s, based on quarry blast observations by Leet (1938). Also, the interpretation of Brockamp, 1931b, Figure 3.3-05) shows a third layer with P-velocity of 6.7 km/s, which, however, is not listed in Macelwane's table 41.

TABLE 3.3-02. VELOCITIES OF WAVES RECORDED BETWEEN 1921 AND 1938 CAUSED BY EXPLOSIONS AND BLASTS

Location	Wave type	Velocity (km/s)	Authority
Europe			
France:			
La Courtine	P	4.9	Maurain et al. (1925)
	P	5.3	Maurain et al. (1925)
	P	5.5	Maurain et al. (1925)
	P	5.6	Maurain et al. (1925)
	P	6.2	Maurain et al. (1925)
	P	5.5	Rothe et al. (1924)
Germany:			
Hainberg	P	3.36	Müller (1934)
Oppau	P_g	5.53	Jeffreys (1937)
	P	5.4	Wrinch and Jeffreys (1923)
	P	5.73	Hecker (1922)
	P	5.4–5.6	Gutenberg (1926)
Italy:			
Carrara	P	4.6	Agamemnone (1937)
	S	3.0	Agamemnone (1937)
Falconara	P	6.2–6.4	de Quervain (1931)
	S	3.64	de Quervain (1931)
Switzerland:			
Alpnach	P	4.7	de Quervain (1931)
Grenchen	P	5.1–5.25	de Quervain (1931)
North America			
United States			
California:			
Los Angeles basin	P	2.9–3.5	Gutenberg et al. (1932)
Richmond	P_1	4.3	Byerly and Wilson (1935b)
	P_2	5.4	Byerly and Wilson (1935b)
	S_1	2.4	Byerly and Wilson (1935b)
	S_2	3.1	Byerly and Wilson (1935b)
	S_3	3.8	Byerly and Wilson (1935b)
San Gabriel	P	5.5	Wood and Richter (1931)
Southern California	P	4.1	Wood and Richter (1931)
	P	5.0	Wood and Richter (1931)
	P	5.4	Wood and Richter (1931)
	P	5.55	Wood and Richter (1931)
	P	5.9	Wood and Richter (1931)
	P	6.0	Wood and Richter (1931)
	S	2.7	Wood and Richter (1931)
	S	3.0	Wood and Richter (1931)
	S	3.15	Wood and Richter (1931)
	S	3.21	Wood and Richter (1931)
	S	3.25	Wood and Richter (1931)
	S	3.4	Wood and Richter (1931)
	S	3.5	Wood and Richter (1931)
Ventura basin	P	2.9–3.5	Gutenberg et al. (1932)
Victorville	P	5.5	Wood and Richter (1931)
New England	P	6.0	Leet (1936)
	P	8.0	Leet (1936)
	S	3.5	Leet (1936)
	S	4.6	Leet (1936)
	P_1	6.01	Leet (1938)
	P_2	6.77	Leet (1938)
	S_1	3.45	Leet (1938)
	S_2	3.93	Leet (1938)

Note: From Macelwane (1951, table 41).

CHAPTER 4

The 1940s (1940–1950)

4.1. THE LATE 1940s IN CENTRAL EUROPE

Controlled-source seismology investigations of the Earth's crust and the first international cooperations started effectively after 1945. In the first edition of Volume VII of *Physics of the Earth* (edited by B. Gutenberg) on the internal constitution of the Earth in 1939, J.B. Macelwane summarized essentially the seismic studies up to 1939 on velocities for the outer surface layers of the Earth, i.e., for sediments and crustal crystalline rocks, obtained from quarry blast and near earthquake recordings between 1900 and 1939. In the second edition of this volume in 1951 (Macelwane, 1951), he added some results obtained between 1940 and 1949, in particular results from the seismic profiles using the military explosions in Germany, Heligoland in 1947 (Schulze, 1947; British National Committee for Geodesy and Geophysics, 1948; Reich et al., 1951; Mintrop, 1949b), and Haslach (Black Forest, Southwest Germany) in 1948 (Reich et al., 1948; Rothé et al., 1948; Förtsch, 1951) and from some early explosion studies of Tuve et al. (1948) in the United States.

By the early 1950s, the overall picture of the crust had already been established. Reinhardt (1954), in his review of crustal investigations up to 1954, plotted the basic scheme (Fig. 4.1-01). Beneath a layer of sediments, the Earth was divided into an upper crust of granitic composition and a lower crust consisting of gabbroic rocks, underlain by a peridotitic mantle layer, the top of which was determined by the Mohorovičić discontinuity.

Based on the pre-war investigations by Wiechert, Angenheister, and others, such investigations would eventually be continued, after the Second World War had ended. Former friendships between scientists in Britain, France, and Germany could be renewed and so, only one year after the war, the first international scientific cooperations were organized. The very first crustal investigations in central Europe were enabled by using large explosions undertaken by the British and French armies, particularly in Germany, to destroy great quantities of ammunitions, fortifications, and other military objects.

A large explosion of ammunition in 1944 near Burton-on-Trent in England was recorded by many earthquake stations in Europe (Jeffreys, 1947). In 1946, surface detonations of large ammunition reserves were carried out near Soltau in the North German Plain and recorded up to 50 km away (Schulze and Förtsch, 1950; Willmore, 1949). The recorded traveltimes resulted in upper crustal velocities being determined between 5.8 and 6.6 km/s. Of particular interest were the azimuth-dependent variations, evidently caused by salt domes. Other ammunition

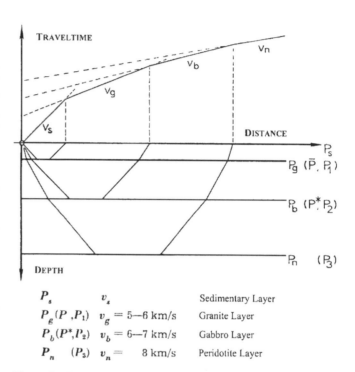

P_s	v_s		Sedimentary Layer
$P_g(P,P_1)$	v_g	$= 5—6$ km/s	Granite Layer
$P_b(P^*,P_2)$	v_b	$= 6—7$ km/s	Gabbro Layer
P_n (P_3)	v_n	$= 8$ km/s	Peridotite Layer

Figure 4.1-01. Schematic division of the Earth's crust due to seismic measurements (from Reinhardt, 1954, fig. 1). [Freiberger Forschungshefte, C15, p. 9–91. Reproduced by permission of TU Freiberg, Germany.]

reserves were detonated in 1946 in Danish waters (Lehmann, 1948). Near Kahla in Thuringia, an underground factory was demolished and the explosions were recorded up to 438 km away (Sponheuer and Gerecke, 1949). The interpretation of the seismograms resulted in a layer with 6.35 km/s velocity being interpreted at 9 km depth. P_n arrivals, however, were not recorded.

The very first nuclear explosions in North America, which were recorded by earthquake stations in North America, also have to be mentioned in this context. However, as their origin time had not been announced in advance, the recordings did not supply any new information on the Earth's crustal and upper mantle structure.

The greatest impact on crustal studies at this time was created by the large explosions on Heligoland in 1947 and in the Black Forest in 1948.

In 1947, 4000 metric tons of explosives were detonated instantaneously in fortifications on the island of Heligoland in the North Sea off northern Germany (Charlier, 1947; Willmore, 1949; Engelhard, 1998). This explosion was located 50 m underground.

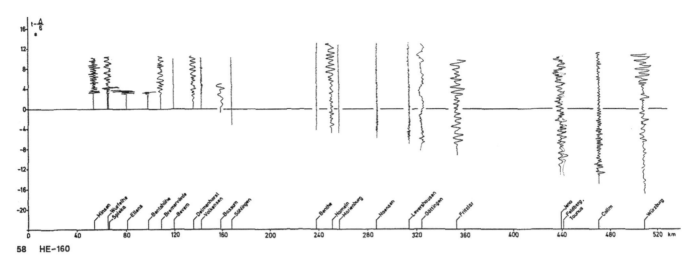

58   HE-160

Figure 4.1-02. Seismograms of the Heligoland-explosions of 1947 recorded on a N-S line through Germany (HE-l60), compiled by P. Friese and J. Ansorge (from Giese et al., 1976, map 2, fig. 58, reproduced after Ansorge et al., 1979a, fig. 6). [*In* Giese, P., Prodehl, C., and Stein, A., eds., Explosion seismology in central Europe—data and results: Berlin-Heidelberg-New York, Springer, 429 p. Reproduced with kind permission of Springer Science+Business Media.]

The explosion on the island of Heligoland (Fig. 4.1-02) is the factual beginning of controlled-source seismology in Germany (Schulze, 1974). The use of the explosion on Heligoland for science was enabled by British geophysicists with good army contacts. Engelhard (1998) reviewed this event and its consequences for explosion seismology in Germany. Twenty-four recording stations were arranged along three profiles throughout northern Germany, the recordings being organized by the Geophysical Institute of Goettingen and the State Geological Survey at Hannover. The longest profile extended for more than 300 km and ran through the troop training area of Soltau and farther toward Goettingen. Another profile ran into Schleswig-Holstein across the magnetic high of Husum.

In southwestern Germany in 1948, an underground ammunition factory near Haslach in the Black Forest was destroyed by several very large explosions. The pre-war friendship of E. Peterschmitt (Strasbourg) and W. Hiller (Stuttgart) enabled the use of these subsurface explosions for science and recording along two lines, one toward the WNW across the Rhine valley toward Strasbourg and one toward the ESE into Bavaria (Fig. 4.1-03).

Both the Heligoland and the Haslach explosions were recorded mainly by mechanical seismometers with mechanical-optical and mechanical-electrical-optical recording systems on photographic paper. The mechanical-optical system (system of Schulze and Förtsch), for example, had seismometers (Fig. 4.1-04) with an eigenperiod of 1/6 s, an amplification of 1:25,000 for 1 m light path, air damping (could be modified), 1 kg mass, and a total weight of 4.5 kg (Reinhardt, 1954). Effectively a 40,000-fold amplification was reached (Reich et al., 1948). The mechanical-electrical-optical instruments operated with induction and worked usually with electronic amplification which increased the sensitivity but caused complications when amplitudes and/or frequencies were evaluated.

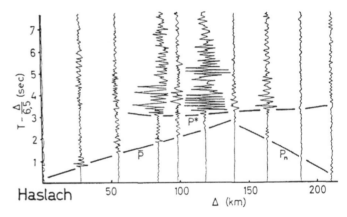

Figure 4.1-03. Seismograms of the Haslach explosion of 1948 (HA-120), recorded on an ESE profile (from Giese et al., 1976, Map 3 fig. 60). [*In* Giese, P., Prodehl, C., and Stein, A., eds., Explosion seismology in central Europe—data and results: Berlin-Heidelberg-New York, Springer, 429 p. Reproduced with kind permission of Springer Science+Business Media.]

The interpretation of the Heligoland and the Haslach explosions resulted in representative models of crustal structure for Central Europe. From the Heligoland observations (Reich et al., 1951; Schulze and Förtsch, 1950; Willmore, 1949) a depth of 26–30 km was derived for the Mohorovičić discontinuity under northern Germany. The internal crustal structure, however, was interpreted differently by the various authors. Reich et al. (1951) and Schulze and Förtsch (1950) favored two crustal layers underneath the sedimentary cover: a granitic upper crust with 5.34 km/s velocity and an essentially thicker gabbroic lower crust, the velocity of which varied from model to model between 6.19 and 6.60 km/s. Willmore (1949) favored a one-layer crustal model, but also discussed the possibility that the crust may con-

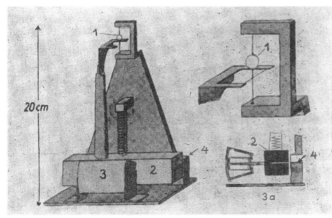

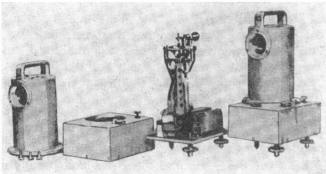

Figure 4.1-04. Mechanical seismometer of Goettingen (from Reinhardt, 1954, fig. 7) for long-range recording after Schulze and Förtsch (1950). Top: Scheme of seismometer: 1—mirror, 2—mass, 3—damping (3a—cross section of damping), 4—plate spring. Bottom: Seismometer closed (right) and opened (left). [Freiberger Forschungshefte, C15, p. 9–91. Reproduced by permission of TU Freiberg, Germany.]

sist of two layers. Mintrop (1949b) considered later arrivals on stations close to Heligoland as possible deep reflections from layers in the upper mantle.

The interpretation of the observations of the Haslach-explosion gave better results than the Heligoland measurements (Reich et al., 1948; Rothé et al., 1948; Rothé and Peterschmitt, 1950; Förtsch, 1951). A 16–21-km-thick granitic upper crust

with velocities of 5.9–6.0 km/s is underlain by a 10–12-km-thick gabbroic lower crust with 6.55 km/s velocity, and the Moho is between 30 and 33 km deep. The P_n phase resulted in uppermost mantle velocities of 8.15–8.34 km/s. The nine seismograms, recorded on the profile from Haslach toward the east-southeast (Haslach-120, Fig. 4.1-03), clearly showed the P_MP-reflection and are among the first of many worldwide examples demonstrating that the Mohorovičić-discontinuity, the Moho, is a first-order boundary between Earth's crust and mantle under many regions of the Earth. At that time, however, this phase was not recognized as a wide-angle reflection, but was interpreted as a refracted wave from which the depth to the Conrad-discontinuity was derived. The model of Rothé and Peterschmitt (1950) showed the division of the crust (Fig. 4.1-05) into a granitic layer reaching to a depth of 15–20 km and a basaltic layer of ~10 km thickness, and a crustal thinning from 31 to 32 km under most of SW Germany to 28 km under the Black Forest and 25 km under the Rhinegraben. The Haslach profile into Bavaria furthermore led to the detection of a new discontinuity which later in the literature was named the "Förtsch-discontinuity," a boundary between upper and middle crust at ~10 km depth (Förtsch, 1951). It was also identified later in a statistical evaluation of commercial deep-seismic reflection seismograms recorded throughout Southern Germany (Liebscher, 1962, 1964), but was not regularly observed elsewhere. It was later identified as the top of a low-velocity zone in the upper crust (Giese, 1968a; Landisman and Mueller, 1966; Mueller and Landisman, 1966).

4.2. CRUSTAL STUDY ACTIVITIES IN THE 1940s OUTSIDE CENTRAL EUROPE

4.2.1. Soviet Union

Early crustal studies in the 1940s had also been undertaken in other parts of the world, partly even during World War II. In 1950 Twaltwadzse reported on a series of very large explosions amounting up to 220,000 kg. They had occurred between 1941 and 1945 in the Soviet Republic of Georgia. Twaltwadzse (1950) assumed that the 20-km-thick granitic upper layer with

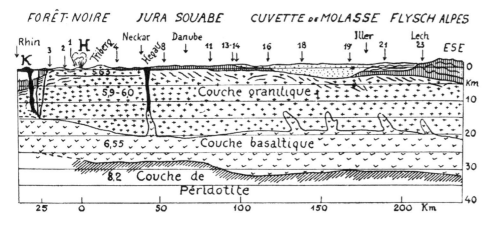

Figure 4.1-05. Seismic-refraction cross section for the ESE-profile from Haslach (after Rothé and Peterschmitt, 1950, fig. 14) (reproduced in Giese et al., 1976, p. 315: fig. 2 of Prodehl et al, 1976a). [*In* Giese, P., Prodehl, C., and Stein, A., eds., Explosion seismology in central Europe—data and results: Berlin-Heidelberg-New York, Springer, 429 p. Reproduced with kind permission of Springer Science+Business Media.]

a velocity of 5.6 km included solidified sediments in its upper part, the physical properties of which are close to those of granite. The existence of a basaltic intermediate layer was also reported, but its lower boundary could not be determined from seismic-refraction data. However, deep reflections indicated a depth to Moho of 48 km. A similar crustal structure was also assumed for other parts of the Caucasus.

Based on the experiments by Gamburtsev and co-workers in 1938–1941 on the European platform of the former USSR in 1938–1941 and on the Aspheron peninsula in 1944, mentioned in Chapter 3, and theoretical work by Riznichenko and Gamburtsev and co-workers during the 1940s, the method of "deep seismic sounding" was developed (for details and references see Steinhart, 1961, Appendix A5-1). A specialty of this method was the use of continuous profile sections of 10–20 km with close geophone spacing, alternating with gaps of 10–30 km.

The deep seismic sounding method was applied in 1949–1950 to investigate the crustal structure under the northern Tien-Shan region (Gamburtsev, 1952). Underwater explosions in the lakes Issyk-Kul and Kara-Kul served as energy sources. As seismic reflection measurements had not given satisfying results, modified refraction instrumentation of the prospecting industry with 0.1 s eigenperiod and increased sensitivity were used to record seismic energy at offsets up to 400 km. Along the profile, specially selected 20–25-km-long stretches were very densely occupied with seismometers. The measurements resulted in the interpretation of a 10–15-km-thick granitic layer which was underlain by a 30–40-km-thick basaltic layer. It was noted in particular that both discontinuities deepened considerably toward the south in the direction of the Kirgisian mountains—an indication of a crustal root?

4.2.2. North America

In Canada, rock bursts occurring between 1938 and 1945 in the mining area near Kirkland Lake, Ontario, were used for the first time for crustal structure studies. Some events were recorded by observatories up to 1000 km distance. The timing of the events was successfully achieved by a geophone set up inside the mine. The interpretation (Hodgson, 1942, 1947) resulted in the interpretation of a depth to Moho of 36 km.

The Canadian investigations of rock bursts in the early 1940s were continued in 1947–1951. In addition, some large explosions in this area were included. Two recording stations were moved regularly to new positions so that finally a complete refraction profile of 174 km length resulted. The data were interpreted by a one-layer 36-km-thick crust with a mean velocity of 6.1 km/s, overlying a mantle with 8.18 km/s velocity. The observed wide-angle reflections indicated that the Moho was a first-order discontinuity. Some later arrivals were recorded from some of the explosions and resulted in a velocity of 7.1 km/s giving hints on the possible existence of an intracrustal discontinuity (Hodgson, 1953).

In the United States, a large explosion of ammunition occurred in 1944 near Port Chicago in California. From records in North America, a velocity of 7.72 km/s was derived, but a depth of only 14 km was deduced (Byerly, 1946). Also recorded in North America were the first nuclear tests: on 16 July 1945 the first test was undertaken in New Mexico (Gutenberg, 1946), a second test followed on the Bikini Island in the Pacific on 24 July 1946 (Gutenberg and Richter, 1946). As the origin times of the detonations were not known, new knowledge on crustal structure was not achieved, however.

Immediately after World War II the Carnegie Institution of Washington undertook a major field program to study the Earth's crust with controlled explosions. In the first years of 1946–1949 optical-mechanical-electrical mobile field stations were developed which were capable of detecting ground motions of the order of 10^{-8} at 4–10 Hz. The detectors used had a resonant frequency of 2 Hz (Steinhart, 1961, see Appendix A5-1).

In the Appalachians, large quarry blasts and shots fired in water in Chesapeake Bay were recorded up to 350 km away (Tuve et al., 1948, Tuve and Tatel, 1950a). From the time-distance curves of the refraction observations Tuve et al. (1948) derived the following model for the region around Washington, D.C.: 0–10 km—6.0–6.17 km/s, 10–24 km—6.7 km/s, 24–42 km—7.05 km/s, 42 km and below—8.15 km/s. When interpreting the data, strong secondary arrivals were detected in the ranges predicted for critical reflections from the Moho and from depths of 90–120 km. The results were consistent with the refraction model, but it also was noted that several models could be deduced that were consistent with the data (Steinhart, 1961). Junger (1951) reported on deep seismic reflections from the basement obtained during commercial seismic prospecting work in Big Horn County, Montana. During experimental seismic work by Shell Company, a number of deep reflections from 7.0 to 8.5 seconds were observed. The reflecting surfaces were flat and were interpreted to come from inside the basement at depths of 18–21 km.

At subsequent recordings of several large shots in Tennessee, seismograms were obtained at distances out to 1200 km at several azimuths, but the arrivals disagreed in different areas by as much as 1 second, indicating some real difference in crustal structure along the various paths. As more measurements were made, Tuve and Tatel (1950a) began to doubt the interpretation of a simply layered crust. They also were concerned about the explanation of "reverberations" on seismic records and the lack of coherence in waves over distances of a few kilometers, leading to new experiments in the 1950s (Steinhart, 1961).

A discussion on the nature of the P_g wave had already started in the early 1940s. In two publications of 1943, Gutenberg (1943a, 1943b) had compared observations of near-earthquakes and quarry blasts in southern California and had concluded that the velocities of 5.5–5.6 km for P_1-waves from earthquakes should not be compared with the 6 km/s-velocity from quarry blasts, because earthquake waves which were generated at greater depths were not influenced by relatively thin sediments on top of the basement. Leet (1946), however, concluded that arrivals with 6 km/s would be strongly limited in their distance range, as the sediments were not very thick and their properties would substantially change locally. Rather he suggested that another phase with 5.9–6.1 km/s determined by Gutenberg and named P_g would better fit with the quarry blast observations. In 1949, the Carnegie Institution of Washington conducted experiments in southern California in an attempt to resolve the apparent discrepancies in the velocities reported for P_1 from earthquakes and from blasts. It could be shown that in fact the traveltimes coincided within the experimental error (Tuve and Tatel, 1950b, Steinhart, 1961).

4.2.3. South Africa

In South Africa, very early seismic investigations were carried out in the mid to late 1940s (Gane et al., 1946; Willmore et al., 1952). These experiments were enabled by the fact that frequent earth tremors occur in the Witwatersrand gold mining area. The first investigation by Gane et al. (1946) had made it clear that many of the tremors were large enough and occurred often enough (about two tremors per working day) to permit observations at relatively large distances (up to 400 km) to be made in a relatively short time using sensitive mobile equipment. This experience was used by Willmore et al. (1952) for more extended seismic field surveys in 1948 and 1949 with recording distances up to 500 km. The data received amounted to ~200 seismograms. Two simple models of crustal structure were deduced assuming flat layering. The first model contained two layers: underneath the Witwatersrand system with 5.65 km/s velocity and 4.5 km thickness the so-called crustal layer 1 of 6.09 km/s P-velocity overlies the mantle with 8.27 km/s P_n velocity, the Moho depth being ~36 km. The second model took into account that a phase corresponding to an intermediate layer is evident and thus contained two crustal layers 1 and 2 underlying the Witwatersrand system giving a Moho depth of ~39 km.

4.3. SEISMIC-REFRACTION WORK AT SEA

During World War II, the techniques of seismic measurements at sea were further developed so that after 1945 the experiments could be extended from shallow coastal waters into the deep ocean basins.

As the method to place geophones on the sea bed, which had been common in the 1930s, restricted the investigations to less than 100 fathoms water depth, after 1945 the hydrophone came into use as sensor for seismic waves. Some of the earliest seismic investigations in the deep ocean were carried out with a single hydrophone suspended from a stationary or slowly drifting vessel (Hersey, 1963; Hersey and Ewing, 1949). Small TNT charges were fired at depths of 1–100 m and a hydrophone suspended 50–300 m below the ship detected echoes from the seabed and reflectors below (Fig. 4.3-01). After amplification and filtering, the hydrophone signals were photographed on an oscilloscope.

The seismic data obtained in the late 1930s were now, after WWII, supplemented by new expeditions, in particular in the

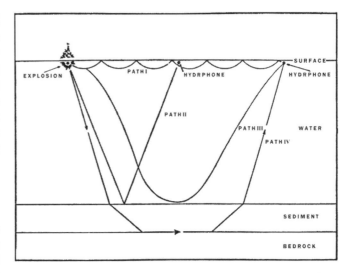

Figure 4.3-01. Sound paths between explosion and hydrophone (from Hersey et al., 1952, fig. 5). [Bulletin of the Geological Society of America, v. 42, p. 291–306. Reproduced by permission of the Geological Society of America.]

northern Atlantic Ocean in 1948 and 1949 (e.g., Drake et al., 1952; Ewing et al., 1950, 1954; Gaskell and Swallow, 1951; Hersey et al., 1952; Hill, 1952; Katz et al., 1953). In general, the seismic-refraction method was applied. Results of these investigations were published since 1949.

Reinhardt (1954, Appendix A4-1) has summarized coordinates, dates, references, some technical details, and main results of selected early offshore experiments in Table VII.

For the work at sea, the seismic refraction method was used at that time. In contrast to land seismics, at sea the shotpoint moved and the recording station remained at a fixed position. Officer et al. (1952) and Hersey et al. (1952) have described the techniques in some detail. Usually two ships were used. One ship would heave to and put the hydrophones over the side. The other ship would proceed away from the first ship and fire 6–20 shots to make up half of the reversed profile. The size of the shots varied from 0.5 to 5 kg for the close shots, fired at 5–15 min intervals, to 25–150 kg for the more distant ones, fired at 30–60 min intervals. The ships were equipped with hydrophones, which were lowered some tens of meters below the effective depth of ocean swell, but usually not exceeding 45 m (Ewing et al., 1950; Galperin and Kosminskaya, 1958; Shor, 1963). The hydrophones had a low-frequency response down to 5 Hz. They detected the seismic signal and fed it into a five-channel amplifier, each channel of which could be filtered in any required manner. Each channel was provided with two output stages at different dynamic levels for recording weak and strong signals on the same record. The amplifier outputs fed a bank of galvanometers, the deflections of which were recorded on photographic paper. One channel was used to record the direct water wave and the bottom-reflected waves. Another galvanometer was connected to a radio receiver to record the shot instant markers from the other ship and one galvanometer finally was connected to a break circuit chronometer to give the time scale.

Profile lengths were between 25 and 100 km and were usually reversed. Some profiles also had intermittent shotpoints or intermittent stations. On a typical record, three types of arrivals were received: the refraction waves from the basement and sediments, the direct wave traveling through the surface sound, and the bottom-reflected waves (Fig. 4.3-02). Before the final travel-time curves could be drawn, numerous corrections had to be applied, for example to account for the speed of the shooting ship, for the time interval between dropping the charge overboard and shooting, or for the depth of the shot and of the hydrophone. For the interpretation the arrivals were plotted on a time-distance graph and connected by straight lines to obtain velocities and then the seismic refraction and reflection formulas were applied for depth determinations. For a geological identification of the layers, most authors compared the obtained seismic velocities with experimental results published, e.g., by Leet (1946) or Birch et al. (1942).

One of the earliest marine surveys in the East Atlantic was undertaken in 1949 in the ocean weather ship *Weather Explorer* by the Department of Geodesy and Geophysics of the University of Cambridge (Hill and Swallow, 1950). The experiments were primarily to test if the method of seismic refraction shooting at sea, which had been developed since 1946, was applicable to deep water. Surface sonobuoys containing radio transmitters for sending hydrophone signals to a recording ship were pioneered by Hill (1952) during these experiments. The sonobuoys were free-floating and held hydrophones on a compliant suspension at depths of 50–150 m to keep ambient noise levels at an acceptable low level.

The area in which the work was undertaken lay within 30 miles of the meteorological position at latitude 53°50′N, 18°40′W, where the water depth was ~1300 fathoms. Depth charges were detonated 30 m deep as sources of the shock waves. They were detected by quartz hydrophones suspended ~45 m below sono-radio buoys which transmitted the information to the recording instruments in the ship. The charges were fired at distances up to 30 km from the buoys, four of which were in use simultaneously spread over a line ~5 km long. The evaluation showed the presence of two interfaces below the sea bed at depths of ~300 m and 500 m. In the surface layer, the velocity was between 1.9 and 2.2 km/s, in the intermediate layer 4.9–5.2 km/s were measured, and the deepest layer showed 6.2 and 6.4 km/s at two different positions.

The majority of investigations in the late 1940s concentrated on the western half of the Atlantic Ocean. Hersey et al. (1952), for example, described an expedition to carry out a series of underwater sound transmission experiments to study the ocean floor north of the Brownson Deep by means of the seismic refraction method. The general area was ~250 km north of Puerto Rico and 100 km north of the axis of the Brownson Deep, at ~21°N, 65°W (Fig. 4.3-03). In this case, two recording vessels and a shooting vessel were used. Shots of 150 kg were fired at a depth of 90 m. Several reversed profiles of up to 50 km distance were recorded. Figure 4.3-02 shows an example of a travel time plot from this experiment.

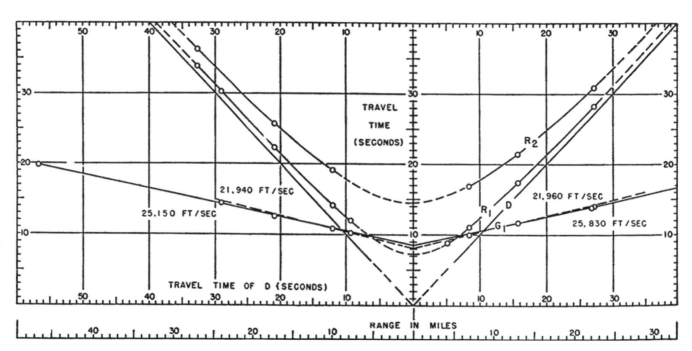

Figure 4.3-02. Traveltime versus distance graph for shots by the moving vessel *Atlantis* received on the stationary vessel *Caryn* (from Hersey et al., 1952, fig. 6). Straight line starting at zero distance and time = travel time curve of direct wave travelling through the surface sound, ellipsoidal curves = bottom-reflected waves, straight line starting at + 9 seconds = refraction wave from the first interface below the sea bed. [Bulletin of the Geological Society of America, v. 42, p. 291–306. Reproduced by permission of the Geological Society of America.]

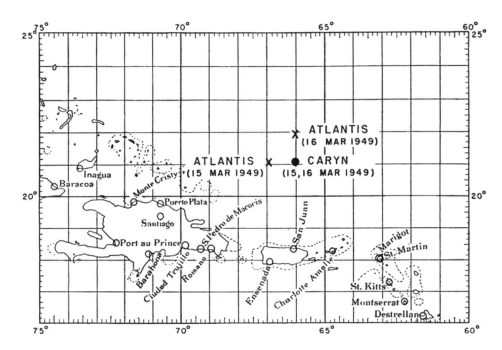

Figure 4.3-03. Location of seismic-refraction profiles shot in 1949 in the area of the Brownson Deep north of Jamaica (from Hersey et al., 1952, fig. 1). [Bulletin of the Geological Society of America, v. 42, p. 291–306. Reproduced by permission of the Geological Society of America.]

Also in the North American basin, northwest and south of Bermuda, Ewing et al. (1952) recorded six reversed seismic refraction profiles of 60–100 km length. Five of these profiles were long enough to record P_n arrivals. Ewing et al. regarded these profiles as typical for the western North Atlantic. For the sedimentary layer, an average thickness of 1.2 km and a velocity of 1.4 km/s were determined; for an intermediate basement layer, a velocity of 6.4 km/s and a thickness of 4.4 km was found, and a velocity of 7.9 km/s started at a depth of 10.3 km below sea level.

Drake et al. (1952) reported on seismic refraction measurements in the Gulf of Maine in 1948 and 1952 as part of a program to investigate the continental margin. During these voyages, they observed sections along the coast of Maine, but they also recorded seismic data along a section from Cape Ann, Massachusetts, to Yarmouth, Nova Scotia. Basement rocks with velocities of 4.95–5.4 km/s pinched out rapidly away from the shore on all sections, and sub-basement rocks with velocities of 5.7–6.3 km/s were found to underlie the entire Gulf of Maine.

In 1949, another marine seismic research project was carried out by Katz et al. (1953) in the Gulf of Maine. A partially reversed refraction profile was shot across the northern Gulf from a point east of Portland eastward beyond Matinicus rock (Fig. 4.3-04). Three portable seismographs were set up at Falmouth near Portland, Maine, on Mount Desert Island, Maine, and at Crowell, Nova Scotia, which recorded shots from 30 positions. The most distant station—Crowell, Nova Scotia—was too far to the east to obtain clear ground arrivals from most of the shots, and an engine failure of the ship stopped the shooting operation beyond shot 30. The distance ranges successfully covered were 36–131 km for Falmouth, 63–158 km for Mount Desert Island, and 116–158 km for Crowell. Interpretation using the standard method, slightly

modified for interpreting reversed profiles, found a surface layer at 5.1 km depth east of Falmouth with an unreversed velocity of 5.3 km/s thinning toward east over a distance of 50 km. The underlying material had a velocity of 6.25 km/s (partly reversed), with the interface between the two layers sloping upwards to the east at ~3°.

Moho was not always detected by these marine experiments, but in some areas, velocities between 7 and 7.6 km/s were found, interpreted as a transition zone between basaltic lower crust and peridotitic upper mantle. In the area around the Bermuda islands, a depth range of 8–12 km was reported for the Moho (Ewing et al., 1952; Hersey et al., 1952).

Offshore investigations were also undertaken in the Pacific off California since 1948 (Raitt, 1949). Since 1948, seismic-refraction studies of the Pacific Ocean were made in a region extending from San Diego to the Marshall Islands and south almost to the Tropic of Capricorn (Raitt, 1956). The investigations included three different regions: the deep Pacific basin proper, atolls and islands, and the continental margin of North America and continued into the early 1950s.

Along the continental shelf, the velocity in the basement increased from 4.5 to 6.3 km/s with increasing distance from the coast. In the ocean basin proper, west of San Diego, underneath thin sediments the thickness of the basement was determined to be 5 km and the velocity 6.5 km/s. At the crust-mantle boundary, the Moho, at 9 km depth below sea level, a sudden velocity increase to 8 km/s was determined.

The experiments also included seismic-reflection shooting with 5 kg TNT charges. The charges were fired a few feet (~1–2 m) below the surface of the sea and the low-frequency reflections from the bottom were received on a hydrophone hung below the ship. Among many confusing sub-bottom echoes was

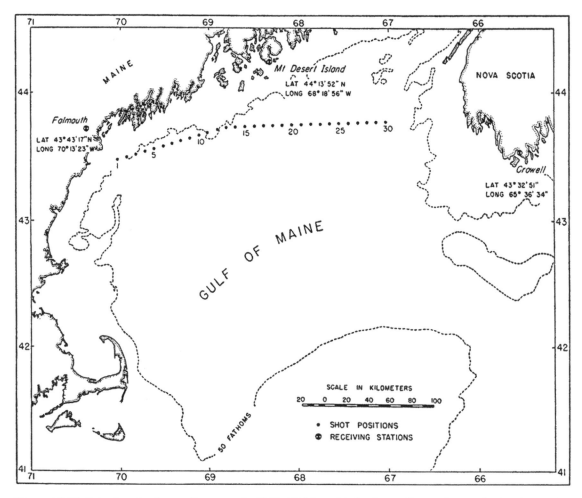

Figure 4.3-04. Seismic-refraction profile across the Gulf of Maine with fixed recording stations on land and a moving shotpoint at sea (from Katz et al., 1953, fig. 1). [Bulletin of the Geological Society of America, v. 64, p. 249–251. Reproduced by permission of the Geological Society of America.]

one persistent reflection from a horizon at a depth of ~600 m below the sea bed (Hill and Swallow, 1950).

The seismic-reflection techniques proved difficult in the beginning. Many hundreds of single-hydrophone observations revealed reflecting horizons at depths greater than 1 km below the ocean floor. However, the echoes between widely spaced stations could not unambiguously be correlated.

4.4. STATE OF THE ART AT THE BEGINNING OF THE 1950s

The philosophy on how to obtain the structure of the crust is based on the assumption that the Earth is divided into more or less flat layers of rocks of different physical properties and that the propagation of seismic waves follows the laws known from optics. This implies that waves are reflected and that beyond the critical distance, head waves develop which follow the discontinuities and produce arrivals aligning on straight lines in a time-distance graph. Thus the methodology to interpret seismograms

in the late 1940s to the mid 1960s was to pick arrivals from the seismograms, to plot them into a time-distance graph, and to correlate those arrivals which were identified and assumed to belong to the same phase. Those arrivals were then connected by a number of straight lines with gradients interpreted as the velocity of the corresponding head waves (see, e.g., Fig. 5.2-02).

For a geological identification of the layers, most authors compared the obtained seismic velocities with experimental results published, e.g., by Leet (1946) or in a handbook edited by Birch et al. (1942). In this handbook, various authors had compiled tables summarizing laboratory measurements of densities, velocities, and other physical properties of minerals and rock samples from all over the world under varying pressure and temperature conditions. Amongst others, the information collected by Leet and Birch (1942) on seismic velocities and summarized in tables were of particular importance.

Gutenberg (1951b) summarized what was known on the crustal structure by the end of the 1940s. He concluded that the relatively large phase $\bar{P}$ in records of nearby earthquakes, which

TABLE 4.4-01. VELOCITIES OF LONGITUDINAL WAVES IN CONTINENTAL AREAS
FROM ARTIFICIAL EXPLOSIONS (FROM GUTENBERG, 1951B, TABLE 1)

Region	Reference	Depth (km)	Velocity (km/s)	Depth (km)	Velocity (km/s)	Depth (km)	Velocity (km/s)
Northern Germany	Wiechert (1926)	0–?	5.98				
Northern Germany	Brockamp (1931b)	0–8	5.9	8–?	6.7		
Helgoland	Willmore (1949)						
	A	7–27	5.95			27	8.18
	B	6–14	5.57	14–30	6.5	30	8.18
Helgoland	Schulze and Förtsch (1950)	6–11	5.34	11–27	6.2–6.6	27	8.19
Black Forest	Reich et al. (1948)	0–21	5.9–6.0	21–31	6.55	31	8.2
Black Forest	Rothé and Peterschmitt (1950)	2–20	5.97	20–30	6.54	30	8.15
Canadian Shield	Hodgson (1947)	0–17	6.15	17–36	6.45	36	8.20
New England	Leet (1936)	0–23	6.0			23	8
Washington, D.C.	Tuve (1948)	0–10	6.0–6.7	10–42	6.7–7.1	42	8.15
Southern California	Wood and Richter (1931)	0–?	6.0				
Corona, California	Gutenberg (1951a)	0–6	5.7–6.0	10–?	6.5–7	40±	8.1–8.2
Corona, California	Tuve and Tatel (1950b)	3–14	6.3	14–?	6.8		

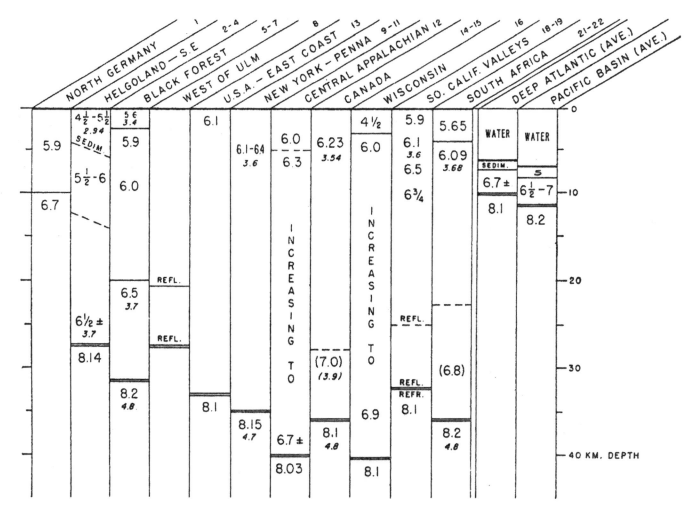

Figure 4.4-01. Average velocities of longitudinal and of transverse waves in the Earth's crust based on explosion records (from Gutenberg, 1955, fig. 1). [Geological Society of America Special Paper 62, p. 19–34. Reproduced by permission of the Geological Society of America.]

TABLE 4.4-02. LOCATIONS OF LARGE
EXPLOSION-SEISMIC STUDIES OF THE EARTH'S
CRUST AS SHOWN IN FIGURES 4.4-02 AND 4.4-03

No.	Location
1	Oppau, Haslach, Southern Germany, Blaubeuren
2	Göttingen, Kahla (central Germany)
3	La Courtine (France)
4	Burton-on-Trent (England)
5	Soltau, Heligoland (Northern Germany)
6	Denmark
7	Southern California
8	Richmond, Port Chicago (Northern California)
9	New England
10	Kirkland Lake (Eastern Canada)
11	New Mexico
12	Bikini
13	Big Horn County, Montana
14	Pennsylvania, Appalachians
15	Northeastern Japan
16	Western Transvaal (South Africa)
17	Soviet Republic of Georgia
18	Tien-Shan (USSR)
19	Californian coast (East Pacific)
20	Gulf of Maine
21	NW Bermuda, North American Basin (western Atlantic)
22	N Brownson Deep (western Atlantic)
23	Eastern Atlantic
24–26	Atlantic (Martinique, Bermuda), (western Atlantic)
27–31	Northern Atlantic (western Atlantic)
32	Bermuda (western Atlantic)
33–43	Pacific (central Pacific, western Pacific)
44	Newfoundland

hitherto was assumed to be the direct longitudinal wave and corresponded with its velocity of 5.6 km/s to the granitic layer, could not be reconciled with observations of the longitudinal waves from blasts. He stated that the velocity of longitudinal waves below the sediments was likely to be 6 km/s and increases to 6.5 km/s at depths of ~10 km. At greater depths between 10 and 15 km, this velocity might decrease, which was to be expected in rocks with considerable amounts of quartz, based on laboratory experiments (Birch et al., 1942). At the bottom of a deeper layer with higher velocity (7.0–7.5 km/s), he expected the Mohorovičić discontinuity at a depth of 30–40 km, but deeper under some mountain chains, forming the boundary between the simatic lower crust and the ultrabasic material below with P-wave velocities of 8.2 km/s. In a table (Table 4.4-01) he summarized the velocities for P-waves in continental areas from artificial explosions, which corresponds to his results published in 1955 and shown in Figure 4.4-01.

Gutenberg (1951b) also pointed out that there seemed to be a greater difference between the structure of the Pacific Ocean and the surrounding continental areas than between the bottom of the Atlantic or Indian Oceans and their surrounding shelves and continents. In the Pacific, he assumed a crustal thickness of only a few kilometers, while under the Atlantic Ocean, and probably also under the Indian Ocean, he saw a more gradual transition from the continent to the basins and the Mohorovičić

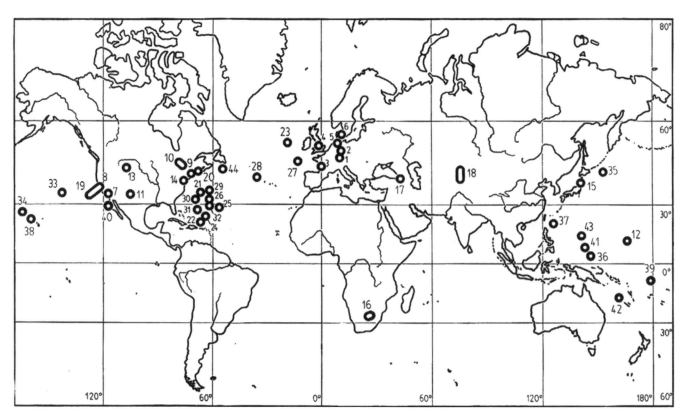

Figure 4.4-02. Location map of large explosion-seismic studies of the Earth's crust (from Reinhardt, 1954, fig. 3). [Freiberger Forschungshefte, C15, p. 9–91. Reproduced by permission of TU Freiberg, Germany.]

discontinuity at greater depths than under the Pacific but much shallower than under continents.

In 1954, a symposium was held at Columbia University ("The Crust of the Earth") where results of earthquake and explosion-seismic studies achieved until 1954 were presented (Poldervaart, 1955). Gutenberg (1955) summarized the velocity values published from earthquake observations from 1948 to 1954 and the corresponding depths for the Conrad and Mohorovičić discontinuities. In addition, he summarized separately apparent wave velocities and depth values obtained from blasts and rock bursts, including some early results from deep ocean basins (Fig. 4.4-01).

Ewing and Press (1955) pointed to the essential difference between oceanic and continental crust, with the silicic crust under ocean basins being no more than one fifth of that under continents. They also saw a remarkable unity of results reported for the oceanic crust: It is a layer of mafic rocks somewhat less than 5 km thick with P-wave velocity of 6.4–6.9 km/s. The Moho is at 10–11 km beneath the sea surface and is underlain by ultramafic rocks with 7.9–8.2 km/s P-wave velocities. They also pointed out that the crust under the Gulf of Mexico, the Caribbean, and the Mediterranean resembles the oceanic type rather than the continental type.

At the same time in the early 1950s, Otto Meisser, head of the Geophysical Institute of the Technical University of Freiberg, East Germany, initiated a diploma thesis to summarize the available knowledge on crustal structure obtained by experimental seismic work based on man-made events with emphasis on quarry blasts. In this thesis, which was published as a monograph in 1954, Reinhardt (1954) compiled all experiments and their main results as far as published until 1952 (Appendix A4-1).

The experiments were summarized in a map and in crustal structure columns, which are reproduced in Table 4.4-02 and Figures 4.4-02 and 4.4-03. The key data (coordinates, dates, references, details of recording equipment, and charges and key results) were compiled by Reinhardt (1954) in a series of tables.

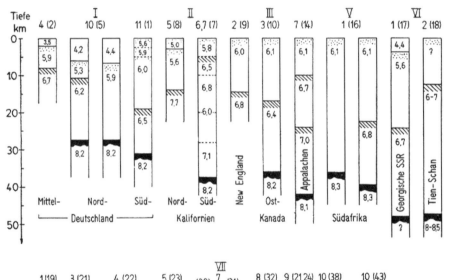

Figure 4.4-03. Crustal structure derived from large explosion-seismic studies (modified from Reinhardt, 1954, fig. 4). Se—sediments, Gr—upper crust (granite), Gb—lower crust (gabbro), Pe—uppermost mantle (peridotite). [Freiberger Forschungshefte, C15, p. 9–91. Reproduced by permission of TU Freiberg, Germany.]

CHAPTER 5

The 1950s (1950–1960)

5.1. SEISMIC-REFRACTION INVESTIGATIONS IN WESTERN EUROPE IN THE 1950s

Since the beginning of the 1950s, commercial quarry blasts have been increasingly used since they proved to be a powerful and low-cost energy source for a systematic investigation of the detailed structure of the Earth's crust (Reinhardt, 1954). Also, it was recognized early on, that underwater explosions are much more effective than other blasts due to better damming of the charge by water (Förtsch and Schulze, 1948; Schulze, 1974). The crustal investigations involved both the seismic-refraction and reflection methods. In 1950, Reich (1953) observed for the first time clear reflections from the Mohorovičić-discontinuity in southern Germany using a quarry blast.

Reinhardt (1954) also gave a detailed description how seismic measurements for crustal studies had to be organized and which technical requirements had to be fulfilled at that time, for example, the recording of the zero-time of an explosion, the organization of a central time signal, and the spacing of recording stations as well as minimum distances depending on the goal of an experiment. He also discussed the various seismometers available at that time.

The leading generation of post-war geophysicists in Germany had a very clear primary focus: to generate a climate of cooperation between the many small Earth science institutes or departments in Germany and to return to international cooperation. Heinz Menzel at Clausthal and Otto Rosenbach at Mainz belonged to the young driving forces in this cooperation of Earth sciences, including geophysics institutions in Germany. They had both studied mathematics and physics at Königsberg University before and during the war. After the war, they had begun their professional career with PRAKLA at Hannover in the exploration industry. In seismology and deep-seismic sounding, the other driving forces were located in Stuttgart with Wilhelm Hiller and the even younger generation with international contacts to Lamont: Hans Berckhemer and Stephan Mueller. In Munich, Hermann Reich attracted in particular scientists such as Otto Förtsch, a pioneer in the early deep-seismic sounding studies in Germany, and the young Peter Giese with a strong geological background.

In 1957, systematic seismic crustal research started in Germany when a priority program ("Geophysical Investigation of Crustal Structure in Central Europe") was initiated, funded by the German Research Society and involving all geophysical university institutes and geophysical departments of the German state geological surveys. At the beginning of these investigations

of crustal structure, however, it was necessary to develop new recording systems with higher magnification than the mechanical-optical ones used to record the Heligoland and Haslach explosions in the late 1940s. Also, more flexibility in field operations was required. In Germany, two systems were constructed and applied, each in connection with highly sensitive electromagnetic and electrodynamic seismographs with a natural frequency of 1–2 cps. The one system was based on low-sensitive galvanometers operating with electronic amplifiers, of which various types were built in the laboratories of the individual institutions. The other system used only high-sensitive galvanometers of the Kipp and Zonen type. Both systems reached a maximum amplification of 10^6 in the frequency range of 5 to 20 cps (Giese et al., 1976a).

Geophysical companies in Germany showed particular interest in the scientific deep-seismic sounding programs and supported these efforts by recording up to 12 s two-way traveltime (TWT) (e.g., Dohr, 1957, 1959; Schulz, 1957; Liebscher, 1962, 1964). Due to their much higher frequency contents, these reflection seismic data offered completely new information than did the seismic-refraction surveys. In 1952, reflections from great depths were reported by Reich (1953) on recordings of quarry blasts in southern Germany by PRAKLA with commercial reflection instrumentation.

Schulz (1957) interpreted deep reflections at 4.0 and 5.4 s TWT as being caused by reflectors at 10.4 and 13.4 km depth, respectively. Liebscher (1962, 1964) prepared histograms, i.e., the number of reflections per time interval of 0.2 seconds, to derive the depth of the main crustal boundaries (e.g., the Förtsch, the Conrad, and the Mohorovičić-discontinuities; see Fig. 6.2.2-04) and published contour maps for the C- and M-boundaries in southern Germany (see Fig. 6.2.2-03).

In 1954, the Alps became a special target of crustal research following the founding of the Subcommission of Alpine Explosions under the umbrella of the International Union of Geodesy and Geophysics (IUGG). Under the leadership of Y. Labrouste (University of Paris) and H. Closs (BGR Hannover), a series of larger explosions was arranged in two lakes of the French Alps: in Lac Rond des Rochilles between Briançon and Modane in 1956, and in Lac Nègre north of Monaco in 1958 (Closs and Labrouste, 1963). Numerous British, French, German, Italian, and Swiss institutions cooperated in the observation of these explosions and set out recording stations on several profiles and fans throughout the Western Alps (Groupe d'Etudes des Explosions Alpines, 1963) which, due to topography, resulted in a more or less areal coverage by seismic stations (Fig. 5.1-01, see also Fig. 6.2.4-01). With the specific goal of studying the zone of Ivrea, which is

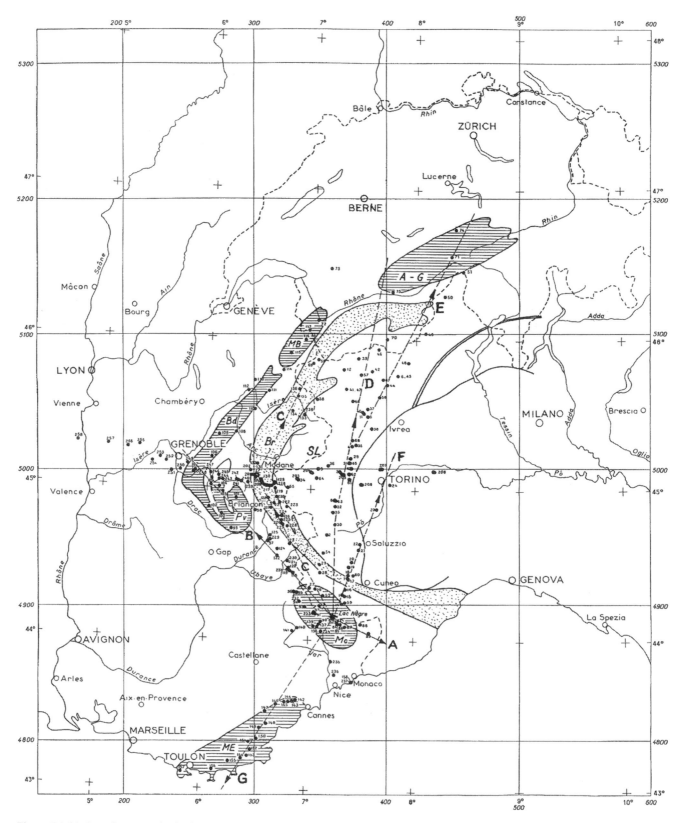

Figure 5.1-01. Location map of seismic observations in the Western Alps in 1956 and 1958. Hatched areas: crystalline massifs of the Western Alps. *A* to *G*: originally planned profiles (from Fuchs et al., 1963a, fig. 1). [*In* Closs, H., and Labrouste, Y., Séismologie: Recherches séismologiques dans les Alpes occidentals au moyen de grandes explosions en 1956, 1958 et 1960. Mémoir Collectif, Année Géophysique Internationale, Centre National de la Recherche Scientifique, Série XII, Fasc. 2, p. 118–176. Reproduced by permission of the author.]

characterized by an anomalous gravity high, some smaller explosions near Levone (SW of Ivrea) and at Monte Bavarione (near Lago Maggiore) were organized in 1960 by L. Solaini (Istituto di Geofisica Applicata del Politecnco, Milano) and observed by the same international community.

The correlation of observed phases was extremely difficult due to the topography, the areal distribution of the recording sites, and differing types of instruments. E. Peterschmitt used a mechanical affinograph which allowed tracing the seismograms in a normalized scale (unique time scale and normalized amplitudes) on Plexiglass. These Plexiglass seismograms were then placed at their proper location on a large topographic map which allowed a three-dimensional view of all phases. This procedure finally allowed Fuchs et al. (1963a) to recognize and correlate six wave types and to determine their velocities and intercept times. An example of seismograms and their correlation by Fuchs et al. (1963a) is shown in Figure 5.1-02.

While the phases i and g were interpreted as P-waves, the phase m was regarded as a converted phase sPp by Fuchs et al. (1963a), arguing that on 40% of the seismograms weak pPp arrivals could be recognized 4 seconds earlier which could be cor-

related as P_n given the Moho-depth calculated from phase m as sPp phase. Using this assumption, under the Western Alps the Moho depth increased from ~30 km under the outer crystalline massifs in the west to a maximum of 53 km in the east under the zone of Ivrea. Figure 5.1-03 shows a W-E cross section through the Western Alps. If the phase m had been interpreted as a P-wave, a maximum Moho depth of 70 km would have resulted, which seemed unlikely to the authors at that time.

In contrast to Fuchs et al. (1963a, 1963b), Labrouste et al. (1963) derived a quite different model assuming that only one crustal layer exists which drops from near 10 km depth under the Ivrea gravity high to 40 km depth under the western crystalline massifs, thus assuming that the surface of the zone of Ivrea is part of the Moho. The discrepancy was later solved and discussed during a symposium in 1968, which is discussed in Chapter 6.

At about the same time, when the Western Alps were investigated in great detail, the first seismic investigations of the Eastern Alps started (Reich, 1958, 1960). Quarry blasts and organized drillhole explosions at the northern margin of the Eastern Alps enabled the determination of the thickness of the Molasse sediments. Because of the high velocities in the Triassic limestones

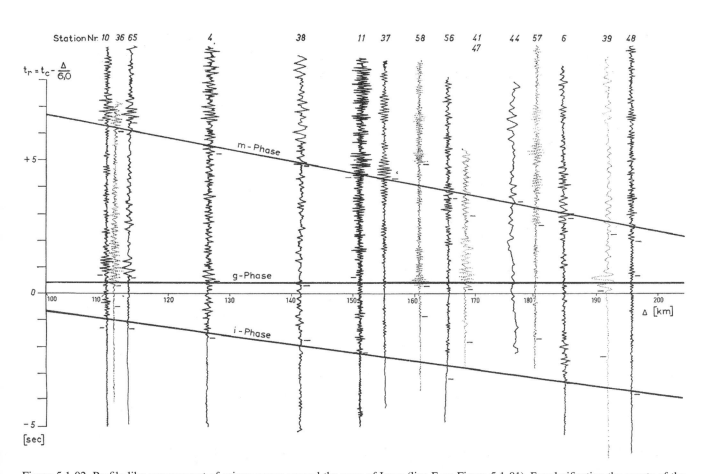

Figure 5.1-02. Profile-like arrangement of seismograms around the zone of Ivrea (line E on Figure 5.1-01). For clarification the onsets of the phases *i*, *g*, and *m* are connected by straight lines (Fuchs et al., 1963a, fig. 4). [*In* Closs, H., and Labrouste, Y., Séismologie: Recherches séismologiques dans les Alpes occidentals au moyen de grandes explosions een 1956, 1958 et 1960. Mémoir Collectif, Année Géophysique Internationale, Centre National de la Recherche Scientifique, Série XII, Fasc. 2, p. 118–176. Reproduced by permission of the author.]

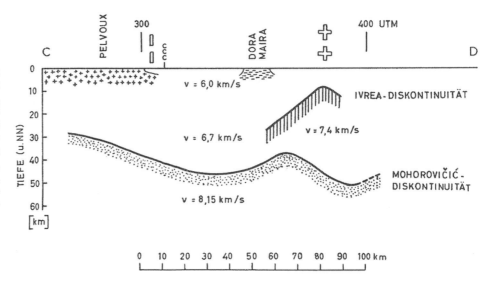

Figure 5.1-03. West-east cross section through the Western Alps from south of Grenoble to Torino (Fuchs et al., 1963a, fig. 8 below). [*In* Closs, H., and Labrouste, Y., Séismologie: Recherches séismologiques dans les Alpes occidentals au moyen de grandes explosions en 1956, 1958 et 1960. Mémoir Collectif, Année Géophysique Internationale, Centre National de la Recherche Scientifique, Série XII, Fasc. 2, p. 118–176. Reproduced by permission of the author.]

of the Northern Alps, however, it was problematic to observe the depth to the crystalline basement. Therefore the investigations there concentrated on the uppermost 10 km of the crust and touched only the northern margin of the Eastern Alps (see Fig. 6.2.4-01 for shotpoint locations).

5.2. THE BEGINNING OF SEISMIC-REFRACTION INVESTIGATIONS IN NORTH AMERICA

Since the late 1940s, studies of the Earth's crustal structure using blasts and rockbursts were also undertaken in North America (Gutenberg, 1952; Tuve et al., 1948, 1954) and Canada (Hodgson, 1953). Tatel and Tuve (1955) and Katz (1955) carried out seismic-refraction experiments in various regions of the United States. Tatel and Tuve (1955) have described in some detail the techniques of their time and the difficulties of choosing useful observation sites with a sufficiently low ground noise to detect the weak impulses of sometimes only a few Angstrom units in displacement amplitude. The portable seismometers used were electrodynamic: their output was electronically amplified and recorded on a pen-and-ink oscillograph. The earlier equipment had an extended frequency range of 0.3–400 cps, but experience showed that more effective records could be produced with a spread of 3–30 cps. This part of the spectrum was then employed in the newer equipment. Timing was obtained by recording second impulses of a chronometer simultaneously with the time signals from the National Bureau of Standards Radio Station WWV, obtaining times reliable to 0.03 seconds or better.

Some data examples are shown in Figure 5.2-01. Distances from shot to receiver positions were located on the best maps available and errors were estimated to be well below 0.5 km. The resulting travel times of correlated arrivals were plotted to 0.1 second accuracy, and true fluctuations were observed around an average line of approximately ± 0.5 seconds.

Tatel and Tuve (1955) concluded that the variations in their measurements were due to the Earth's characteristics and not to the measurements themselves. The data obtained were plotted on charts of time versus distance. To obtain better accuracy, reduced times were plotted; e.g., the difference between the actual time and the time required for a wave to traverse the distance at the mean compressional velocity of the waves. This mean velocity varied from 5.5 km/s in Utah to 6.2 km/s in the Shenandoah Valley (Fig. 5.2-02).

The experiments covered various geologic provinces. From 1948 to 1953, underwater shots by the U.S. Navy in the Patuxent River and in the Chesapeake Bay of 600–2400 lbs enabled the investigation of the Atlantic coastal region in Maryland and Virginia. Three profiles were recorded in 1950 in the Appalachians, along strike up to 1100 km toward NNE and 150 km toward SSW, and up to 1500 km perpendicular to the strike, enabled by several blasts of up to 650 tons during the construction of a dam in northern Tennessee. Open-pit mine blasts were recorded in 1951 on the southern Canadian Shield in Minnesota.

In California, a SW-trending profile was recorded from a blast at Corona, and a second set of shots was recorded from explosions set of off the California coast near Santa Barbara. In 1951, depth charges were exploded in the Puget Sound in Washington and recorded on two lines. Critical reflections, named PP, could not be observed, and the P_n-velocity (named P_2 by the authors) could not be accurately determined in either in California or in Washington. Recordings of explosions in the Bingham mine near Salt Lake City, Utah, caused some technical problems. Also here critical reflections were not observed, but the crossover distance of the wave P_2 (today's P_n) showed that the velocity transition was small and its depth of 29 km was similar to that found in Arizona and New Mexico in 1954. There, 9 explosions could be arranged and 30 good seismograms were obtained. Tatel and Tuve were surprised to "obtain a close-in set of second arrivals and cross-over distances" of P_n with a shallow depth of 30 km for

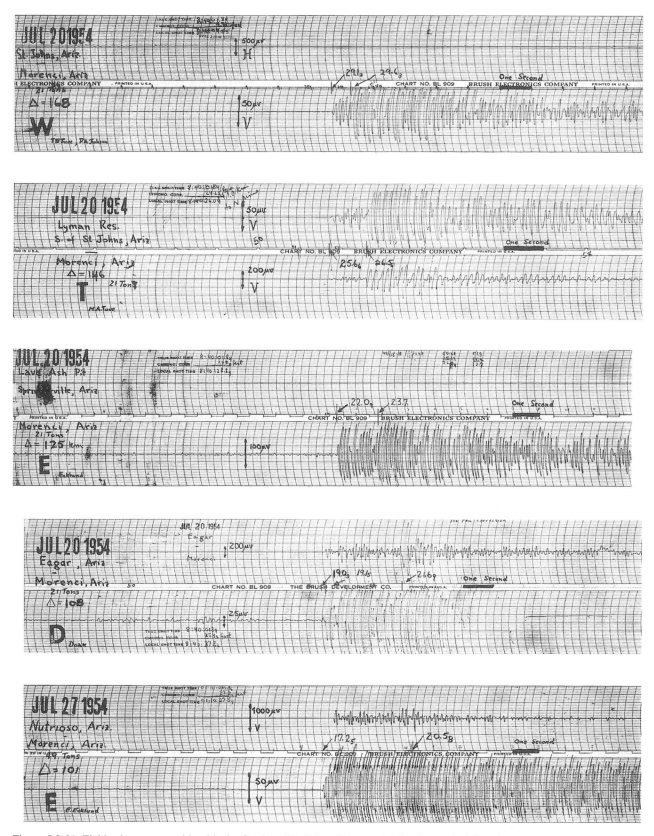

Figure 5.2-01. Field seismograms with critical reflections PP (3.2 to 0.5 seconds after first arrival from bottom to top seismogram) obtained in the Colorado Plateau in 1954 (from Tatel and Tuve, 1955, plate 5). [Geological Society of America Special Paper 62, p. 35–50. Reproduced by permission of the Geological Society of America.]

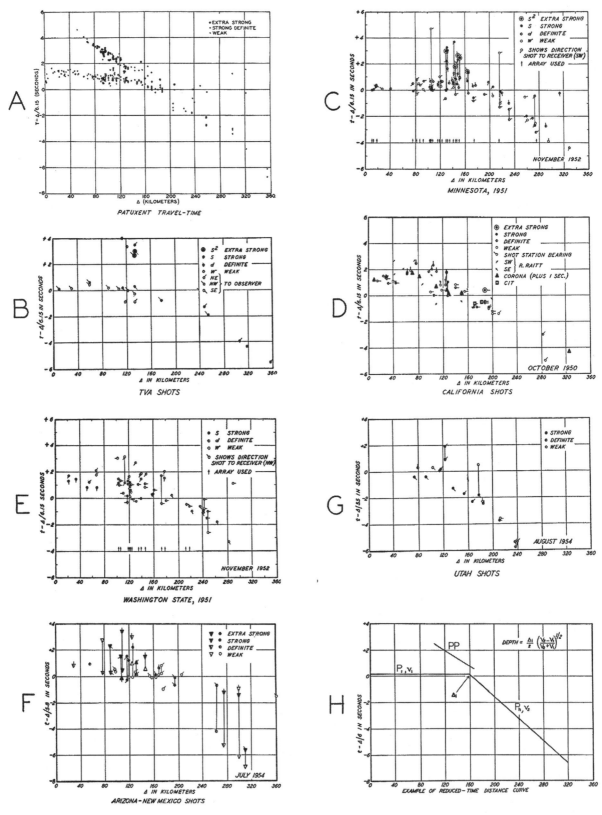

Figure 5.2-02. Time-distance plots for (A) Maryland-Virginia (coastal plain); (B) Tennessee-Virginia (Appalachians); (C) Minnesota (southern Canadian Shield); (D) southern California; (E) Washington (Puget Sound); (F) Arizona–New Mexico (southern Colorado Plateau); (G) Utah (central Rocky Mountains); (H) sketch of the prominent features (from Tatel and Tuve, 1955, fig. 4). [Geological Society of America Special Paper 62, p. 35–50. Reproduced by permission of the Geological Society of America.]

the velocity discontinuity. The results of their investigations were summarized in a table (Fig. 5.2-03).

Tatel and Tuve (1956) also attempted to determine the crustal structure in central Alaska along lines from College Fjord to Dawson and Fairbanks; these data were later (in the 1960s) reinterpreted together with other data (see, e.g., Berg, 1973).

Using similar equipment as Tatel and Tuve and seismometers built at the Lamont Geological Observatory, Katz (1955) reported on quarry blasts in northern New York and central Pennsylvania recorded from 1950 to 1953 to a distance of 309 km (Fig. 5.2-04). Two sample seismograms recorded at 181.4 km distance around the P_n crossover distance and at 220.5 km with P_n as the first arrival are shown in Figure 5.2-05 and a typical travel time plot in Figure 5.2-06. He derived an essentially homogeneous crust with an average thickness of 34.4 km for both regions, "with elastic wave velocities possibly increasing with depth" observing P-velocities of 6.31–6.39 km/s for New York and 6.04 km/s for Pennsylvania. His P_n velocity was 8.14 km/s. He did not see any vertical reflections from the Mohorovičić discontinuity. Katz mentioned some abnormally small first arrivals between 80 and 140 km distance as well as a large arrival at 80–95 km distance following the P_1 phase one second later on all three profiles; since he could not interpret it at the time, he speculated on the possible existence of a low-velocity layer.

In 1958, researchers at the University of Wisconsin observed a non-reversed seismic-refraction profile in Arkansas and Missouri, using quarry blasts of 2000–24,000-lb charges near Little Rock (Steinhart et al., 1961a; Appendix A5-1). In 1958 and 1959, crustal profiles were recorded in Wisconsin and Upper Michigan using underwater shots in Lake Superior and a quarry blast to reverse one of the profiles (Steinhart et al., 1961b; Appendix A5-1).

The University of Wisconsin researchers also did some early studies of the Rocky Mountain area (Steinhart and Meyer, 1961).

In 1959, several profiles were recorded in Montana (Meyer et al., 1961b; Appendix A5-1). Most of the observations in Montana were reversed profiles. The preliminary results were used to construct a fence diagram of the area investigated (McCamy and Meyer, 1964; Figure 5.2-07), but the authors pointed out that at several crossing points the results were inconsistent and would require further analysis.

In 1957, the University of Wisconsin undertook an expedition to measure the thickness of the Earth's crust under the Central Plateau of Mexico (Meyer et al., 1958, 1961a; Appendix A5-1). A series of shots, fired in connection with open pit mining operations near Durango, Mexico, was recorded to ~320 km distance. To obtain observations near the main shotpoint, a reverse profile of 30 km length was also arranged. Several models ranging from a one-layer crust to a three-layer crust resulted in Moho depths between 40 and 45 km.

West of the Rocky Mountains, from 1956 to 1959, Berg et al. (1960) recorded quarry blasts at Promontory and Lakeside in Utah with shot sizes from 50,000 to 2,138,000 lbs and two nuclear explosions near Mercury, Nevada, equivalent to 1.7 and 23 ktons, using 17 temporary and 15 permanent stations. Most stations were located around the quarries Promontory and Lakeside out to ~150 km distance with station spacings between 10 and 30 km, but single stations were placed, e.g., at Gold Hill, Utah, and at Wells, Elko, and Eureka, Nevada, up to distances of 355 km. Also, a few stations outside the Basin and Range province, e.g., in Wyoming and Montana, recorded these events (Fig. 5.2-08). A three-layer model was proposed for the eastern Basin and Range province assuming horizontal layering (Fig. 5.2-09): The upper crust with velocity of 5.73 km/s is 9 km thick; the lower crust with 6.33 km/s extends to 25 km depth, and is underlain by a layer with 7.59 km/s reaching to 72 km depth where a velocity of 7.97 km/s is obtained. The authors hesitated to name the 25-km-deep discontinuity the

Region	v_1, P_1 km/sec	v_2, P_2 km/sec	Critical reflection	Δ_i, km	h (km)	Approximate surface elevation (ft)	Topography
Maryland-Virginia	6.1	8.1	Yes	145–160	26–29	100	Flat
Minnesota	6.1	8.1	Yes	200 ± 10	37	1400	Flat
Tennessee-Virginia	6.0	8.1	Yes	210 ± 20	39	2500	Mountains—Deep Valleys
California							
Coast				120	~23	200	
	6	8	No				Coast—Mountains
Corona				120–170	23–32	1500	
Washington							
East–South				~160	~30	1200	
	6	8	No				Coast—Mountains
West				~100	~19	700	
Arizona-New Mexico							
North				175	34	7000	Colorado Plateau
	5.8	8.1	Yes				—Mountains
South				145	28	4500	
Utah	5.5	8	No	135	29	~6500	Plains—Mountains

$h = \frac{1}{2}\Delta_i \left(\frac{v_2 - v_1}{v_2 + v_1} \right)^{\frac{1}{2}}$ —*assumes no increase of velocity with depth*—*for purposes of comparison only*

Figure 5.2-03. Regional comparison of velocities of waves P_1 (crust) and P_2 (today's P_n), "being recorded," an estimate of the cross-over distance Δ_i, depth of discontinuity (Moho), approximate surface elevation (feet), and topography (from Tatel and Tuve, 1955, table 1). [Geological Society of America Special Paper 62, p. 35–50. Reproduced by permission of the Geological Society of America.]

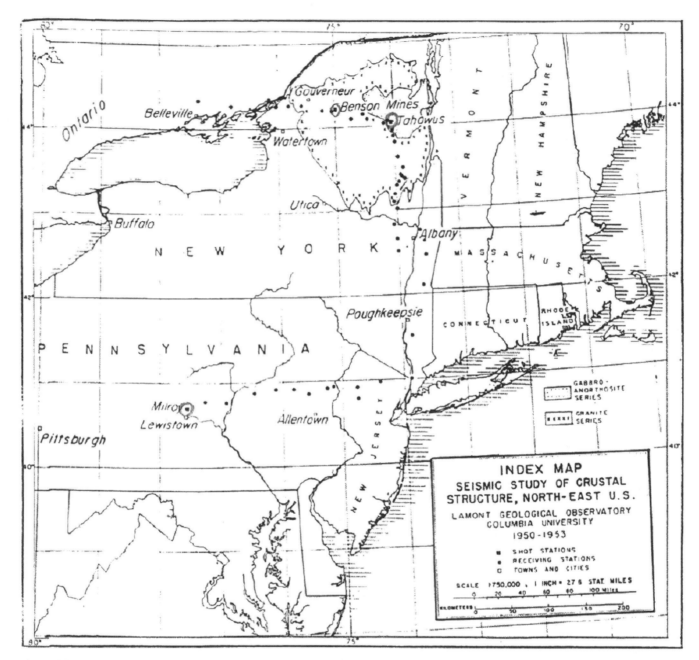

Figure 5.2-04. Index map of seismic-refraction lines in 1950–1953 in the northeastern United States (from Katz, 1955, fig. 1). [Bulletin of the Seismological Society of America, v. 45, p. 303–325. Reproduced by permission of the Seismological Society of America.]

Moho because of the low velocity of only 7.5 km/s underneath, but nevertheless they inferred that this material is "intimately associated with the mantle" and suggested that this mantle has undergone expansion and, referring to Kennedy (1959), discussed a possible phase change caused by increased heat.

Press (1960) has discussed a non-ideal but sufficient two-way coverage of data recorded in southern California from a nuclear blast in Nevada in the north and from quarry blasts near Corona and Hectorville, California, in the south, concluding that no significant crustal variations occur within the region. He ob-

tained a P_2 velocity of 7.66 km/s and a P_n velocity of 8.11 km/s and discussed his favorite model of a 50-km-thick double-layered crust with an intermediate layer at a depth of 24 km with an unusual high velocity (Fig. 5.2-10).

Starting in ca. 1952, the U.S. Geological Survey conducted a modest program of crustal studies using gravity and seismic-refraction methods. The gravity studies of crustal structures were made in conjunction with the Geological Survey's regular program of regional gravity surveying. The immediate purpose of most of these regional gravity surveys was the elucidation of

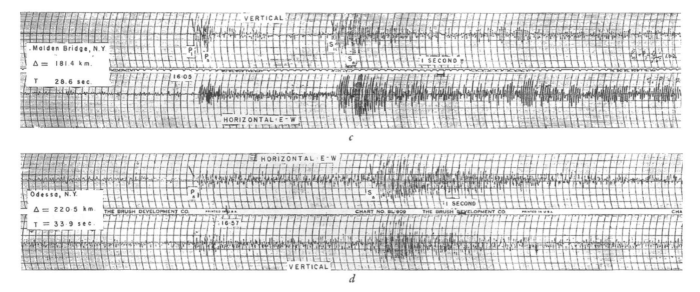

Figure 5.2-05. Seismograms recorded from quarry blasts in New York and Pennsylvania (from Katz, 1955, fig. 5c+d). [Bulletin of the Seismological Society of America, v. 45, p. 303–325. Reproduced by permission of the Seismological Society of America.]

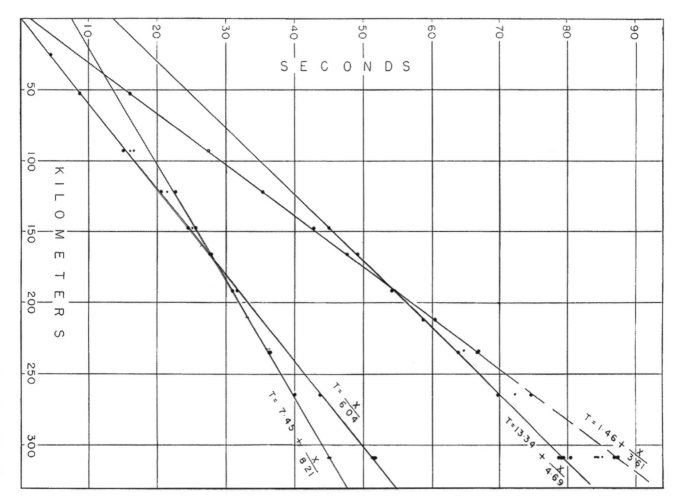

Figure 5.2-06. Travel-time distance correlation of seismic observations in 1950–1953 (from Katz, 1955, fig. 4). [Bulletin of the Seismological Society of America, v. 45, p. 303–325. Reproduced by permission of the Seismological Society of America.]

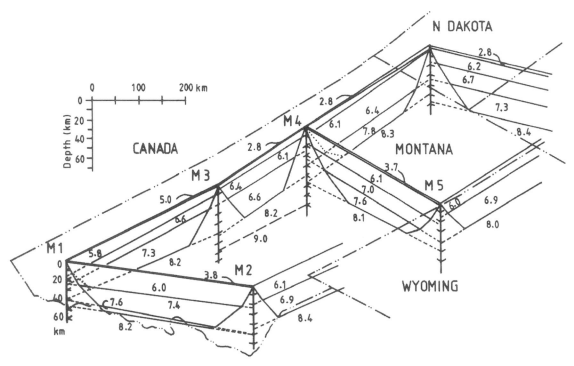

Figure 5.2-07. Fence diagram showing the crustal structure in Montana (McCamy and Meyer 1964, fig. 1, reprinted from Prodehl and Lipman, 1989, fig. 2). [*In* Pakiser, L.C., and Mooney, W.D., eds., Geophysical framework of the continental United States. Geological Society of America Memoir 172, p. 249–284. Reproduced by permission of the Geological Society of America.]

large-scale geologic features of interest in the Geological Survey's regular geologic mapping program. The gravity coverage, however, was sufficiently broad to permit interpretation of the gravity data in terms of major crustal features.

Regional gravity work had been done, for example, in the Sierra Nevada and adjoining areas of California (Pakiser et al., 1960; Pakiser, 1961; Kane and Pakiser, 1961; Oliver et al., 1961). Broad gravity coverage had also been obtained over a large part of the Basin and Range province in California and Nevada and the related Mojave block in southern California (Mabey, 1960a, 1960b) as well as in other parts of the western United States, resulting in new knowledge of the nature of crustal structure and isostatic compensation in the Sierra Nevada region as well as in the Basin and Range province.

In conjunction with the U.S. Geological Survey's geophysical investigations at the Nevada Test Site during the test programs of 1957 and 1958, a number of recordings of seismic waves generated by nuclear explosions were made along a line from the Nevada Test Site to the vicinity of Kingman, Arizona, using seismic instruments available to the Geological Survey at that time (Diment et al., 1961). As a result of these recordings and related work by Press (1960) and Berg et al. (1960), new information on the thickness and structure of the Earth's crust was obtained in the region surrounding the Nevada Test Site.

Alaska and adjacent northwestern Canada also became a target of seismic research in the 1950s. During an expedition of a small field party from the Department of Terrestrial Magnetism

of the Carnegie Institution of Washington in 1955, a series of near-shore 1-ton underwater explosions in College Fjord, Alaska, was recorded on two 300-km-long lines toward the SW along the Kenai peninsula and north toward Fairbanks and on a line toward Dawson, Canada, with stations at distances between 200 and 450 km (Fig. 5.2-11). Also in 1955, during the same expedition, another series of near-shore 1-ton underwater explosions near Skagway, Alaska, was recorded on several lines up to 300 km distance (Fig. 5.2-12) ranging in a northerly direction into the Yukon Territory (Hales and Asada, 1966; Berg, 1973). One of the main results was the observation of clear differences in travel times for various azimuths in many of the areas studied. From the College Fjord shots to the northeast the crustal thickness appeared to be 48–53 km, while to the southwest in the region of the Kenai Peninsula it appeared 10–15 km thinner. Northwest of Skagway, the Moho was found at 36–42 km depth, but north of Skagway it was only 35 km.

5.3. SEISMIC-REFRACTION INVESTIGATIONS IN EASTERN EUROPE IN THE 1950s

The history of deep-seismic sounding experiments in the former USSR is described in detail by Pavlenkova (1996). The first period started at the end of the 1940s. The first deep-seismic research was conducted in 1948–1954 under the leadership of G.A. Gamburtzev, E. Galperin, and I.P. Kosmininskaya in central Asia and in the southern Caspian area. Since the middle of the

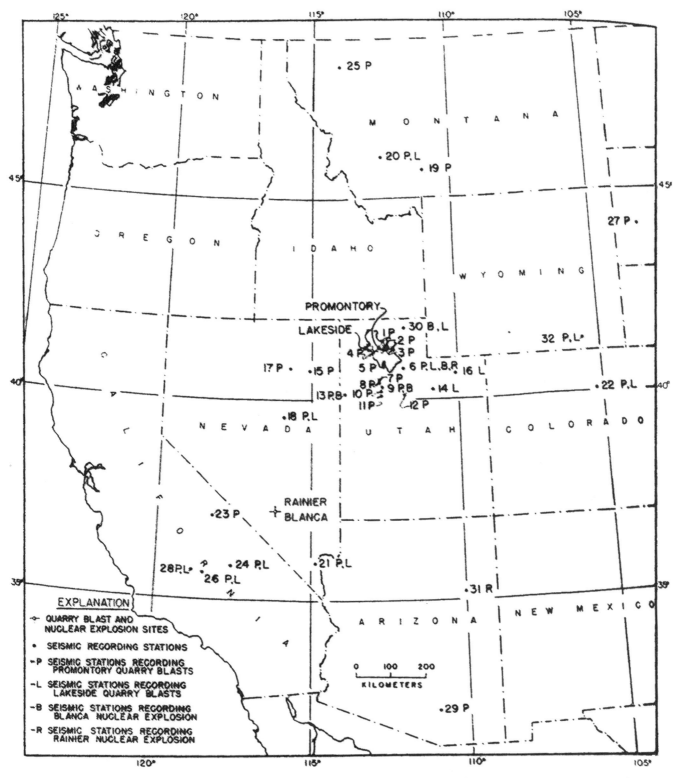

Figure 5.2-08. Index map of quarry blasts in Utah and nuclear blasts in Nevada and recording stations in the western United States (from Berg et al., 1960, fig. 1). [Bulletin of the Seismological Society of America, v. 50, p. 511–535. Reproduced by permission of the Seismological Society of America.]

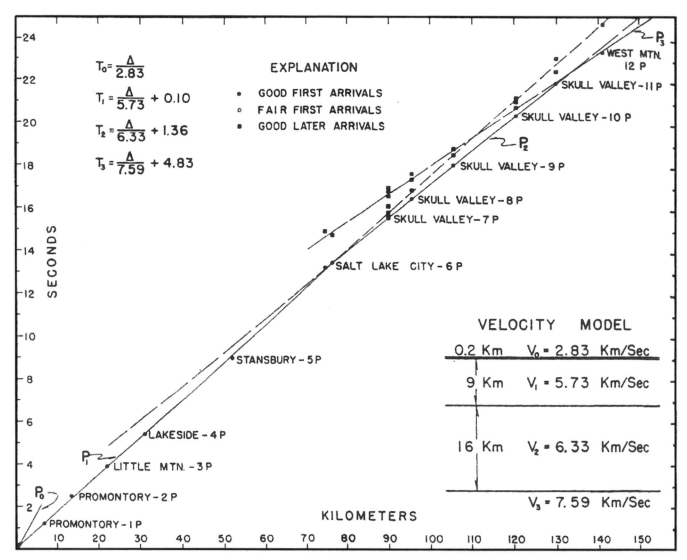

Figure 5.2-09. Travel-time distance correlation of seismic observations in Utah and a model for the eastern Basin and Range province (from Berg et al., 1960, fig. 5). [Bulletin of the Seismological Society of America, v. 50, p. 511–535. Reproduced by permission of the Seismological Society of America.]

Figure 5.2-10. Experimental data from seismic refraction work in southern California (from Press, 1960, fig. 5). [Journal of Geophysical Research, v. 65, p. 1039–1051. Reproduced by permission of American Geophysical Union.]

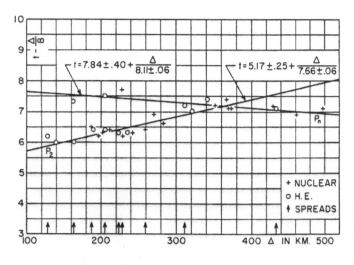

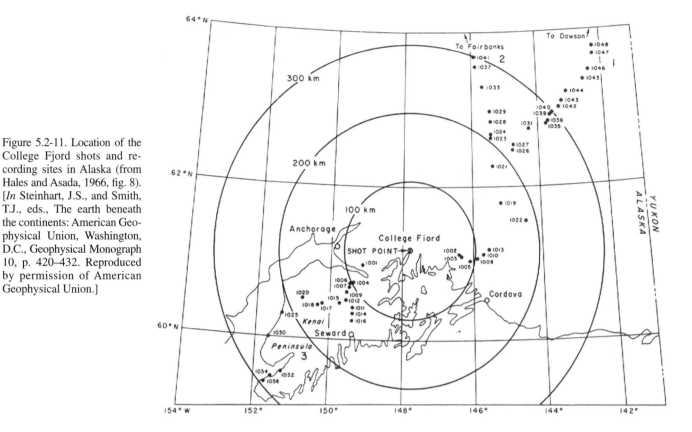

Figure 5.2-11. Location of the College Fjord shots and recording sites in Alaska (from Hales and Asada, 1966, fig. 8). [*In* Steinhart, J.S., and Smith, T.J., eds., The earth beneath the continents: American Geophysical Union, Washington, D.C., Geophysical Monograph 10, p. 420–432. Reproduced by permission of American Geophysical Union.]

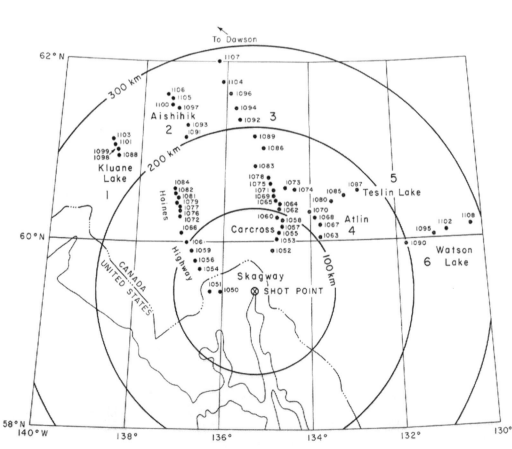

Figure 5.2-12. Location of the Skagway shots and recording sites in Yukon Territory, Canada (from Hales and Asada, 1966, fig. 1). [*In* Steinhart, J.S., and Smith, T.J., eds., The earth beneath the continents: American Geophysical Union, Washington, D.C., Geophysical Monograph 10, p. 420–432. Reproduced by permission of American Geophysical Union.]

1950s, deep-seismic sounding profiles were recorded on the Russian platform as well as on the Russian part of the Baltic Shield, in central Asia, in the Caucasus, and in the Urals covering thousands of kilometers (results published in Russian; for references see Pavlenkova, 1996).

In 1957 and 1958, a major research project was carried out in the transition zone from the Asian continent to the Pacific Ocean (Galperin and Kosminskaya, 1964). Nearly 30 profiles of several 100 km in length were laid out across the Sea of Ochotsk between Kamchatka and the islands of Sakhalin and Hokkaido and across the southern Kuril islands, mainly perpendicular to the strike of the continental margin (Fig. 5.3-01). Some of the data were reinterpreted 50 years later (Pavlenkova et al., 2009).

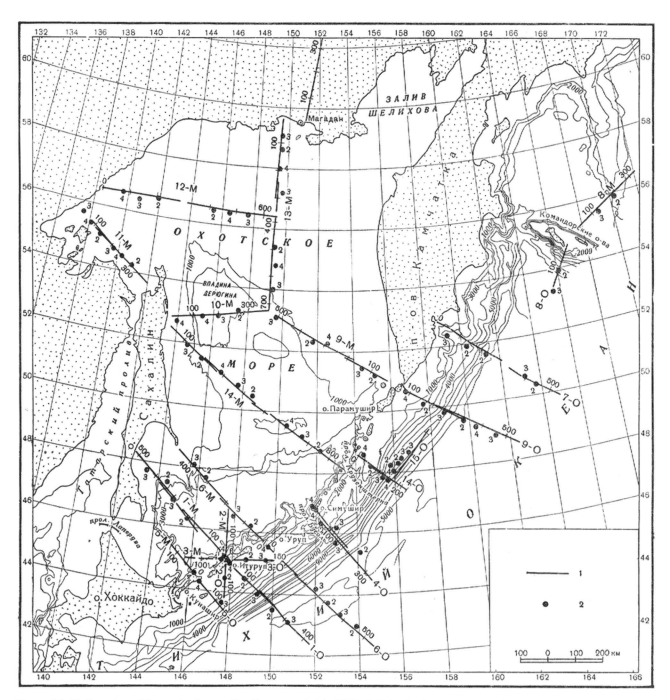

Figure 5.3-01. Location of deep-seismic sounding lines recorded in 1957–1958 in the Sea of Ochotsk (from Galperin and Kosminskaya, 1964, fig. 1.1). [Structure of the earth's crust in the transition zone between Asia and the Pacific: Moscow, Nauka (in Russian). Reproduced by permission of the Schmdt Institute of Physics of the Earth RAS.]

Data examples and a system of correlated travel-time curves for one of the profiles (1-M) are shown in Figure 5.3-02 and Figure 5.3-03. The resulting crustal thickness (Fig. 5.3-04) decreases from 30 km to less than 25 km in the center of the Sea of Ochotsk between Sachalin and Kamchatka, but also shows transitions to oceanic crust in the south near Hokkaido as shown, e.g., along line 1-M (Fig. 5.3-03 and Figure 5.3-04). Oceanic crust of 10 km thickness or less is encountered at the outermost southeastern end of the lines traversing the continental margin (Galperin and Kosminskaya, 1964).

Figure 5.3-02. Seismograms of deep-seismic sounding observations in the Sea of Ochotsk from station 3 along profile 1-M (from Galperin and Kosminskaya, 1964, fig. 8.7). [Structure of the earth's crust in the transition zone between Asia and the Pacific: Moscow, Nauka (in Russian). Reproduced by permission of the Schmdt Institute of Physics of the Earth RAS.]

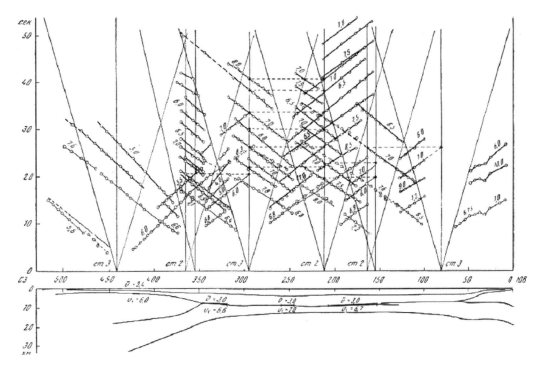

Figure 5.3-03. Travel-time diagram of deep-seismic sounding observations in the Sea of Ochotsk along profile 1-M (from Galperin and Kosminskaya, 1964, fig. 3.4). [Structure of the earth's crust in the transition zone between Asia and the Pacific: Moscow, Nauka (in Russian). Reproduced by permission of the Schmdt Institute of Physics of the Earth RAS.]

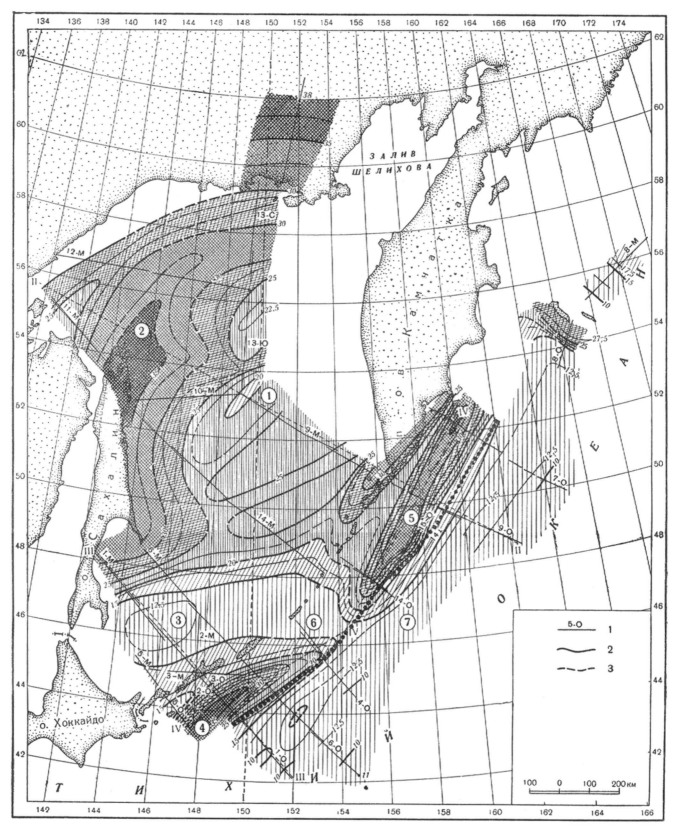

Figure 5.3-04. Crustal thickness map and seismic lines based of deep-seismic sounding observations in the Sea of Ochotsk (from Galperin and Kosminskaya, 1964, fig. 12.5). [Structure of the earth's crust in the transition zone between Asia and the Pacific: Moscow, Nauka (in Russian). Reproduced by permission of the Schmdt Institute of Physics of the Earth RAS.]

5.4. EARLY EXPLOSION-SEISMIC STUDIES IN JAPAN

In Japan, controlled-source seismology started on 25 October 1950, when the construction of the dam Isibuti in north-central Honshu required a large explosion of 57 tons of "carlit" to be detonated simultaneously (Fig. 5.4-01).

Within ten days, the Research Group for Explosion Seismology was established and eight temporary stations were placed at ~20 km intervals along the Tohoku railway line to a maximum distance of 120 km. The firing time was fixed such that the radio time signal JJY could be recorded. The recording equipment used was electromagnetic seismographs with 1, 3, and 10 cps, but also mechanical and mechanical-optical seismographs of 1 cps were used in combination with high-gain amplifiers and electromagnetic oscillographs. The speed of the recording paper or film varied between 2 and 40 mm/s. Sample records are shown in Figure 5.4-02. Two phases with velocities of 5.26 and 6.13 km/s were correlated and a depth of 1.3 km was determined for the second layer: this interpretation caused difficulties for the authors (Research Group for Explosion Seismology, 1951).

In the following years (1951 and 1952), two additional, but weaker explosions from the same source could be recorded, both with 18 stations on two profiles. Finally, in December 1952, a fourth explosion could be recorded from another source near Kamaisi near the coast in northeastern Honshu (Fig. 5.4-01). Unfortunately, the energy was not as strong as expected due to delayed firing of the 29.7 tons of dynamite (7 steps with 25 ms delay) (Research Group for Explosion Seismology, 1952, 1953). Nevertheless the shot was recorded up to 510 km distance on the southern profile. On this occasion three velocities were reported (Research Group for Explosion Seismology, 1954): 6.19, 7.37, and 8.20 km/s for which the authors determined layer thicknesses of 27.2 and 5.1 km, i.e., a total thickness of 32.3 km. However, the authors had doubts about the existence of layer 2 with a 7.37 km/s velocity.

The next series of explosions was fired in 1954 and 1955 at dam constructions near Lake Nozori, ~300 km farther south, of which two with 3.7 and 1.55 ton charges were recorded in a north-westerly direction up to 250 km distance. In addition, a reversing shot was organized by the Research Group for Explosion Seismology near Hokoda in 1956 and 1957, 180 km NE from Nozori,

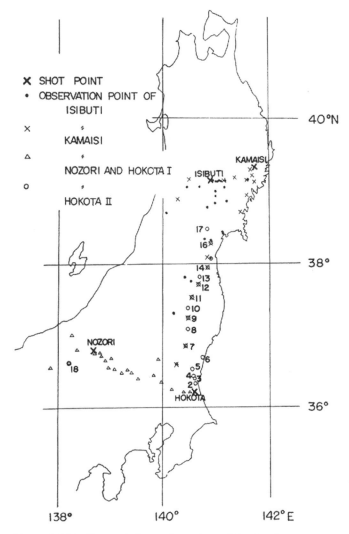

Figure 5.4-01. Shot and observation points of the first Japanese controlled-source seismology experiments (from Research Group for Explosion Seismology, 1959, fig. 1). [Bulletin of the Earthquake Research Institute, Tokyo, v. 37, p. 495–508. Published by permission of Earthquake Research Institute, University of Tokyo.]

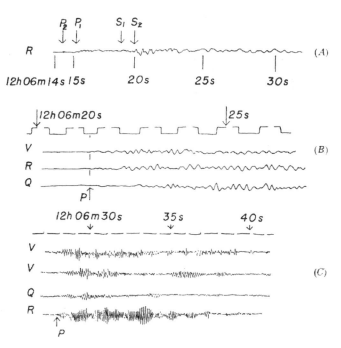

Figure 5.4-02. Seismograms of the first Japanese controlled-source seismology experiment. Distance from shotpoint: (A) 38.42 km, (B) 81.37 km, (C) 121.91 km (from Research Group for Exploration Seismology, 1951, fig. 7). [Bulletin of the Earthquake Research Institute, Tokyo, v. 29, p. 97–105. Published by permission of Earthquake Research Institute, University of Tokyo.]

where 1 ton of dynamite was distributed in 6 boreholes of 60–70 m depth and blasted simultaneously. In total, 33 temporary observation stations, with paper running speeds of 3 cm/s and more, recorded the shots (Fig. 5.4-01). The station separation varied from 5 to 25 km (Research Group for Explosion Seismology, 1958). Velocities of 5.5, 6.1, and 7.7 km/s were derived from the travel-time plots (Usami et al., 1958) and five differing models were discussed. The favorite model shows varying existence of layer 1 (5.5 km/s) along the profile, up to 5 km thick where it exists and an average depth of 25 km of the 6.1 km/s layer. The authors point out that the Moho is not 50 km deep as used before in earthquake studies as the standard model for the crust, but is between only 20 and 30 km depth (Usami et al., 1958). Finally, a second explosion was fired in August 1957 by the Research Group for Explosion Seismology near Hokoda to record a profile to the north and thus to reverse the S-profile from Kamaisi observed in 1952 (Research Group for Explosion Seismology, 1959). The resulting model by Matuzawa et al. (1959) shows the Moho dipping from 20 km in the north to 27.5 km in the south, the average velocity of the crust being 6.2 km/s and the P_n velocity 7.7 km/s.

In 1959, the first Japanese deep sea expedition (JEDS) was undertaken in close cooperation with U.S. institutions investigating the Japan Trench (Ludwig et al., 1966). Results were interpreted together with a second expedition undertaken in 1961 (see Chapter 6).

5.5. EARLY EXPLOSION-SEISMIC STUDIES IN THE SOUTHERN HEMISPHERE

Since the mid 1950s, studies of the Earth's crustal structure using blasts, rock bursts, and atomic explosions were also undertaken in Australia (Bolt et al., 1958), South Africa (Willmore et al., 1952), and South America (Tatel and Tuve, 1958; Steinhart and Meyer, 1961).

5.5.1. Australia and New Zealand

It was nuclear explosions which started seismic crustal studies in Australia (Bolt et al., 1958; Cleary, 1967; Doyle, 1957; Finlayson 2010, Appendix 2-2). British atomic explosions at Emu in 1953 and at Maralinga in 1956, both in South Australia, were recorded along the trans-Australian railway from South Australia to Perth up to 700 km distance and enabled the interpretation of P and S phases. From the recordings of the Maralinga explosion westwards across the Nullarbor Plain the first definitive measurement of Moho depth on continental Australia was interpreted (Bolt et al., 1958). Assuming a one-layer crust, because intermediate arrivals could not be recognized, the interpretation of the P data resulted in a crustal thickness of 32 ± 3 km and 39 ± 3 km being interpreted from P_n and S_n data respectively. Upper mantle P_n and S_n velocities of 8.21 km/s and 4.75 km/s were interpreted. Furthermore, large explosions for dam construction at Eagle-hawk quarry in the Snowy Mountains in New South Wales with charges between 50 and 100 tons were recorded up to 375 km

distance toward Sydney and Melbourne (Doyle et al., 1959). The locations of these early shotpoints are included in Figure 6.7.1-01 (Cleary, 1973) in Chapter 6.7. Systematic recording of seven large blasts of the Prospect blue metal quarry near Sydney in 1959 ranging in size from 2000 to 6260 kg at permanent seismographs resulted in P-wave velocities of 6.16 km/s for crustal rocks of the Southern Highlands of New South Wales and 5.88 km/s within the Sydney Basin (Bolt, 1962).

It is worthwhile noting that, similar to the observations in Germany (e.g., Dohr, 1959), since 1957 the Bureau of Mineral Resources undertook efforts to obtain near-incident vertical seismic reflections from the lower crust and upper mantle in at least one location on most seismic surveys by running near-vertical reflection records up to 38 s or by using large explosions at the end of surveys. Techniques for recording deep reflections improved with experience, but it was not until 1960 that the first definite result was obtained (Moss and Dooley, 1988).

In New Zealand, first preliminary explosion-seismic studies were performed near Auckland in 1951 and Clutha River, South Island, where surplus military explosives had been detonated (Davey, 2010, personal commun.). In 1952, the first crustal seismic-refraction profile was recorded in New Zealand in the Wellington area (for location, see Figure 8.8.1-05). The profile was ~175 km long and unreversed with shots fired at its southern end in Wellington Harbour. Strong reflections at 20 km depth were interpreted by Officer (1955b) as crustal thickness, but a subsequent reinterpretation gave 36 km crustal thickness and no refraction horizon associated with the strong reflection (Eiby, 1955, 1957; Garrick, 1968; Stern et al., 1986).

5.5.2. South Africa

The investigations in South Africa in the late forties have been discussed in Chapter 4.2.3. Further attempts of deep crustal structure in the 1950s are not known to the authors.

5.5.3. South America and Antarctica

In the 1950s, the Carnegie Institution, Washington, D.C., USA, undertook a seismic expedition into the Andean region of Peru and Chile (Tatel and Tuve, 1958). They found 65–70 km depths to Moho under the Altiplano of Peru and Chile and 51–56 km depths at the flanks of the plateau. Location and results were also included in the overview of Steinhart and Meyer (1961; Fig. 10.5 and Table 10.1; see Appendix A5-1-10).

In the 1950s, expeditions were also undertaken into the Antarctica to study the Earth's crust. In 1959 and 1960, a Belgian Antarctic expedition was organized to perform seismic-refraction and other geophysical measurements near the Belgian station in the Breidvika Bay (~70°S, 24°E). A 12-channel seismic recording device was positioned at 14 points on the ice shelf located along a 125-km-long line running southwards from the Soer-Rondane Mountains and shots were detonated around each station up to 20 km distance. In total, 68 shots were fired. The expedition achieved ice thickness results, but unfortunately, due to bad weather conditions, shots at larger distances to achieve a long-distance refraction profile could not be realized (Dieterle and Peterschmitt, 1964).

As part of the U.S. Antarctic research program, during the 1950s and continuing into the 1960s, some three dozen seismic-refraction profiles were completed during reconnaissance exploration of the Antarctic interior. Most of them were long enough to provide information about seismic velocities below the ice, but did not penetrate further than into the topmost upper crust (Bentley, 1973).

Not with active seismics, but with surface wave dispersion studies, the crustal thickness underneath Antarctica was determined by Evison et al. (1959).

5.6. SEISMIC-REFRACTION OCEANIC EXPLORATIONS OF THE CRUST IN THE 1950s

5.6.1. Introduction

Oceanic seismic-refraction experiments, which had started immediately after the end of World War II (see Chapter 4), were continued both in the Atlantic (Ewing et al., 1954) and in the Pacific and Indian Oceans (e.g., Raitt, 1956; Gaskell et al., 1958). The initial investigations concentrated particularly on the ocean basins. As described in some detail in Chapter 4, usually two ships were used for firing and recording where hydrophones floating in the sea water at ~50 feet depth and ~100 feet behind the ship recorded the vibrations of explosive charges of 4–300 lbs. The recording system had several channels to record both high-frequency water waves and lower-frequency ground waves (Ewing et al., 1950, 1954). The shot-receiver distance was calculated from the velocity of the water wave. More information on the techniques of seismic work at sea in the 1940s and 1950s has been given by Officer et al. (1952) and Shor (1963). There were also expeditions with one ship only. In this case, the ship was used for recording and a 20-foot whaleboat for shooting. At the same time, Russian scientists also undertook great efforts to investigate the deep structure under the oceans (e.g., Galperin and Kosminskaya, 1958, 1964; Neprochnov, 1960).

Ewing and Ewing (1959) described the experimental details of two-ship operations briefly as follows.

Throughout the work, the seismic detectors were crystal hydrophones suspended at a depth of 30–50 m in the water. The output of each hydrophone was amplified, filtered, and recorded by a photographic oscillograph on the receiving ship [*see example in Figure 5.6.3-03*]. During the operations, the receiving ship remained hove to while the shooting ship fired a line of shots out to the required distance. The functions of the ships were then reversed, and the former receiving ship fired shots on a line up to the new receiving position. This procedure was referred to as "reversing the station." ... All shots were fired with burning fuse, and the shot instant or "time break" was transmitted to the receiving ship by radio. In addition, events on both ships were related to absolute time by chronometers which were periodically checked against the radio station WWV.

There was no time to shoot short reverse profiles to determine the upper layers in addition to long profiles to determine the deeper layers. Additional deviation from an ideal reversed station resulted from the drift of the receiving ship.

Although the instrumentation of various research groups working in marine seismic experiments varied, all instruments were essentially echo sounders utilizing pulses of low frequency sonic energy and a graphic recording system that displayed the data in the form of cross sections. The sources were most commonly high-voltage sparkers, boomers, gas exploders, magneto-strictive transducers and explosives (Ewing and Zaunere, 1964). However, the majority of the sources was restricted to small offsets and could therefore only be applied for sedimentary structures. Only explosive charges were powerful enough to penetrate into the crystalline crust to Moho and therefore were used exclusively until the mid-1960s for scientific research.

To detect water-borne seismic signals, the hydrophone became the exclusively used receiver after WWII. They are generally made from a piezoelectrical material such as barium titanite, lead zirconate, or lead metaniobate. Small currents are generated in the transducer circuit by pressure changes in the water caused by a seismic source. These are amplified and transmitted by conventional conductor cable or radio link to display units or a data storage device. As hydrophones are small, they are essentially omnidirectional. In the 1940s and 1950s, single hydrophones were used and their signals directed to various channels, as described in detail by Hersey et al. (1952, see Chapter 4.3). Later hydrophones were grouped into array, which have important noise-reducing properties, including a directional response to seismic energy (Jones, 1999).

Already in 1937–1940, tests had been made to place seismographs on the ocean floor at depths exceeding 2000 m in several attempts and connected by an electric cable or wireless to a recording system, as was described in some detail in Chapter 3 (Ewing and Vine, 1938; Ewing et al., 1946; Ewing and Ewing, 1961). These tests were interrupted by World War II. The renewal of this effort was started in 1951 but proceeded slowly, primarily due to the lack of funding. Successful tests with the telemetering ocean-bottom seismograph could not be made before 1959 and were then continued into 1961. By modulation of a supersonic signal sent out from the ocean-bottom seismograph, the information was telemetered to a surface ship. The unit used on the ocean-bottom seismograph was a short-to-medium–period seismometer and an amplifier whose signals modulated the frequency of a 12 Hz acoustical source that sent signals in a broad beam toward the sea surface. The battery was adequate to operate the unit for 7–8 days. Successful tests were made at four positions at water depths between 3700 and 5600 m, shooting at various distances ranging from 15 to 55 km (Ewing and Ewing, 1961).

For the interpretation, first arrivals were picked and plotted into a time-distance graph. Arrivals were then connected by a straight line using a least-squares fit. Examples of travel-time plots have been reproduced for some experiments described in the following subchapters (see Figs. 5.6.3-02 and 5.6.3-05).

The reverse slope of the line would give the apparent velocity, and its backward prolongation to distance 0 would give the intercept time of the refracted wave, from which the depth of the "refracting" layer could be calculated. Ewing et al. (1954) and Ewing (1963a) have described in detail the theory for multilayer cases including inclined interfaces. Figure 5.6.1-01 shows the basic ray diagram, a sketch of the assumed structure.

Nafe and Drake (1957, 1963) have studied the dependence of the P-wave velocity in marine sediments from the thickness of the overburden and have derived corresponding velocity-depth functions using the results from the seismic-refraction measurements from the coastal shelf of eastern North America and the adjacent ocean basins. Having plotted average velocity-depth curves for the sediments in both environments, they found a pronounced difference in the dependence of the P-wave velocity in marine sediments from the thickness of the overburden; i.e., the velocity-depth relationship in shelf sediments differed distinctly from the one in deep-water sediments. They have also shown that, at the same depth of overburden, porosity is much greater in deep-water sediments than in shallow.

During the 1940s and 1950s, oceanographic research had made considerable progress, and publications had appeared in healthy numbers throughout the journals of the world. Therefore, by the end of the 1950s, it was suggested by scientists of the Scripps Institution of Oceanography to sum up the present knowledge in a comprehensive work. In 1963, three volumes of *The Sea* were published, all edited by M.N. Hill. Volume 1 dealt with "new thoughts and ideas on Physical Oceanography," volume 2 dealt with the "composition of sea-water and comparative and descriptive oceanography," and volume 3 contained "the Earth beneath the sea and history." The achievements of marine controlled-source seismic work were described in much detail in the first seven chapters of section 1, "Geophysical exploration," of volume 3 (p. 3–109): "Elementary theory of seismic refraction and reflection measurements" (Chapter 1 by J.I. Ewing); "Refraction and reflection techniques and procedure" (Chapter 2 by G.G. Shor Jr.), "Single ship seismic refraction shooting" (Chapter 3 by M.N. Hill), "Continuous reflection profiling" (Chapter 4 by J.B. Hersey), "The unconsolidated sediments" (Chapter 5 by J.I. Ewing and J.E. Nafe), "The crustal rocks" (Chapter 6 by R.W. Raitt), and "The mantle rocks" (Chapter 7 by J.I. Ewing). In 1970, a second comprehensive publication followed, published as volume 4 of *The Sea* and consisting of two voluminous parts, both edited by A.E. Maxwell: *The Sea, New Concepts of Sea Floor Evolution.*

In the following subchapters, we will discuss various marine expeditions in the Atlantic and Pacific Oceans, carried out in the 1950s. The term *station* was used in the corresponding publications as stationary position of the receiving ship, where it remained hove, while a second ship sailed off and fired a line of shots.

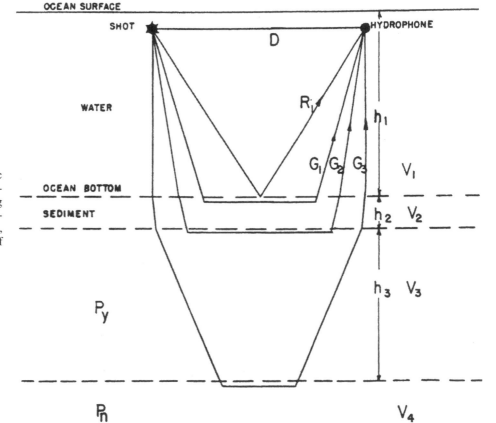

Figure 5.6.1-01. Ray diagram and basic structure derived from the seismic refraction measurements (from Ewing et al., 1954, fig. 2). [Bulletin of the Seismological Society of America, v. 44, p. 21–38. Reproduced by permission of the Seismological Society of America.]

5.6.2. Atlantic Ocean

Already at the end of the 1930s and again since the second half of the 1940s and in the following 1950s, seismic-refraction investigations of the Atlantic Ocean were undertaken in great numbers. Numbered publications were published in the *Bulletins* of the Geological and Seismological Societies of America, entitled "Seismic refraction measurements in the Atlantic Ocean" (e.g., Ewing et al., 1950, Part I; Tolstoy et al., 1953, Part III; Ewing et al., 1952, Part IV; Ewing et al., 1954, Part VI; Katz and Ewing, 1956, Part VII), "Geophysical investigations in the emerged and submerged Atlantic coastal plain" (e.g., Ewing et al., 1939, Part III; Ewing et al., 1940, Part IV; Officer and Ewing, 1954, Part VII; Press and Beckmann, 1954, Part VIII; Bentley and Worzel, 1956, Part X), or "Crustal structure and surface-wave dispersion" (e.g., Ewing and Press, 1952, Part II). British scientists were also heavily involved in marine projects to investigate the oceanic crustal structure applying the seismic-refraction method. Only a few of these many projects can be discussed in the following.

Officer et al. (1952) described an extensive two-ship refraction operation made in 1950. A total of 70 refraction profiles were recorded on tracks from Bermuda to Charleston, South Carolina; Norfolk, Virginia, to Bermuda; Bermuda to the Nares Deep, the deep basin south of the Bermuda rise, extending to the ridge just north of Puerto Rico trough; Bermuda to Halifax, Nova Scotia; Halifax to Sable Island; and Halifax to Woods Hole, Massachusetts. During 1951, another cruise obtained 33 profiles covering the Caribbean, the Puerto Rico trough, and tracks from Puerto Rico to Bermuda, and Bermuda to Woods Hole. The data were plotted on travel time graphs showing direct water-wave travel time in seconds as distance axis versus recorded travel time in seconds. On the profiles near the Nares Basin, four seismic layers were determined, unconsolidated sediments with 1.7 km/s, consolidated volcanics or sediments with 4.51 km/s, a basement with 6.63 km/s, and "a second basement" with 8.03 km/s average velocities, the Moho being at an average depth of 10 km. Over the Bermuda Rise only one high-velocity basement could be determined. The thickness of the volcanics decreased away from Bermuda from 4.5 km to 0.5 km.

The interpretation of six profiles in the North American Basin (Fig. 5.6.2-01), gathered by Ewing et al. (1954) in 1951, resulted in 3 layers: the unconsolidated sediments of 0.5–2 km thickness at the bottom of the ocean, the intermediate layer of 3–5 km thickness with an average velocity of 6.43 km/s and the mantle with an average velocity of 7.89 km/s. The Moho is ~10 km deep below water level (Ewing et al., 1954).

Katz and Ewing (1956) described the results of 25 seismic-refraction stations in the Atlantic Ocean west of Bermuda. They were occupied in 1950 and 1952 and their recordings were incorporated with other available data into four seismic cross sections through the oceanic crust and continental margins along tracts from Bermuda (65°W, 32°N) to the North American continent. From land measurements, Katz and Ewing (1956) inferred, that under the northeastern American continent, the velocity of

near-surface rocks was close to 6 km/s and might increase with depth to near 7 km/s at ~35 km depth, with a mantle velocity of 7.7 km/s below. The upper continental crust (6.0–6.3 km/s) was inferred to thin seawards, disappearing near the foot of the continental slope. For the deep stations of their investigation (mean depth 4.9 km), the mean thickness of sediments was 1.3 km. The underlying crustal rocks had a mean thickness of 5.1 km, with velocities up to 7.1 km/s, but significantly lower near Bermuda. Arrivals from the mantle were observed on five of the longer stations, giving velocities of 7.7–8.5 km/s at 9.4–13.4 km below sea level. Subtracting the water column, the oceanic crust turned out to be less than one-fifth as thick as the continental crust. Toward Bermuda, the mean crustal velocity decreased, observed values ranged from 5.6 to 6.5 km/s. This decrease was associated with the structure of the Bermuda volcanoes.

Also in the early 1950s, during 1951, 1952, and 1953, seismic-refraction measurements were made on several cruises in the Atlantic Ocean as well as in the Mediterranean and Norwegian Seas (Fig. 5.6.2-02). They provided the first opportunity to make measurements in many parts of the ocean, and the data allowed a comparison of the crustal structure in these yet unexplored areas with better known regions. Ewing and Ewing (1959) reported on the results from 71 stations, of which 24 were in the western basins of the Atlantic, 6 in the eastern basins, 5 in the Mediterranean Sea, 19 on the Mid-Atlantic Ridge, 5 in the Norwegian and Greenland seas, 8 on the continental shelves of Britain and Norway and 4 were on the continental margin of eastern North America. As a general remark, the authors stated that both in western and the eastern ocean basins, they might have missed a masked 4.5–5.5 km/s layer. Figure 5.6.2-03 shows a typical travel time plot

Ewing and Ewing (1959) have published their results in a table showing for each station coordinates for the two reversing positions, water depth, and velocities and thicknesses for one to three sedimentary layers as well as for one or two layers of "high-velocity rocks" and the velocity for the Moho. For the western Atlantic Ocean basins, their average result was 5.02 km of water, 1.07 km of sediments with velocities ranging from 1.7 to 4 km/s, 6.15 km depth to and 5.16 km thickness of the oceanic layer with a mean velocity of 6.71 km/s, and 11.25 km depth to the mantle with 8.08 km/s mean velocity. Data example records from a station in the western Atlantic Ocean are shown in Figure 5.6.2-04.

In the eastern North Atlantic Ocean Basin, it was difficult to obtain a good average crustal section because of the scarcity of long reversed stations. Ewing and Ewing (1959) have therefore combined their results with those obtained in earlier operations (Hill, 1952; Hill and Laughton, 1954) and thus came to the following average result: 4.56 km of water, 1.18 km of sediments, 4.87 km thickness of oceanic layer with a mean velocity of 6.52 km/s, and a mantle velocity of 7.81 km/s. The only difference Ewing and Ewing (1959) observed versus the larger basins in the west was that the mantle velocity of 7.81 km/s was less than elsewhere.

The seismic measurements over the Mid-Atlantic Ridge were difficult on account of the great topographic relief, but definite

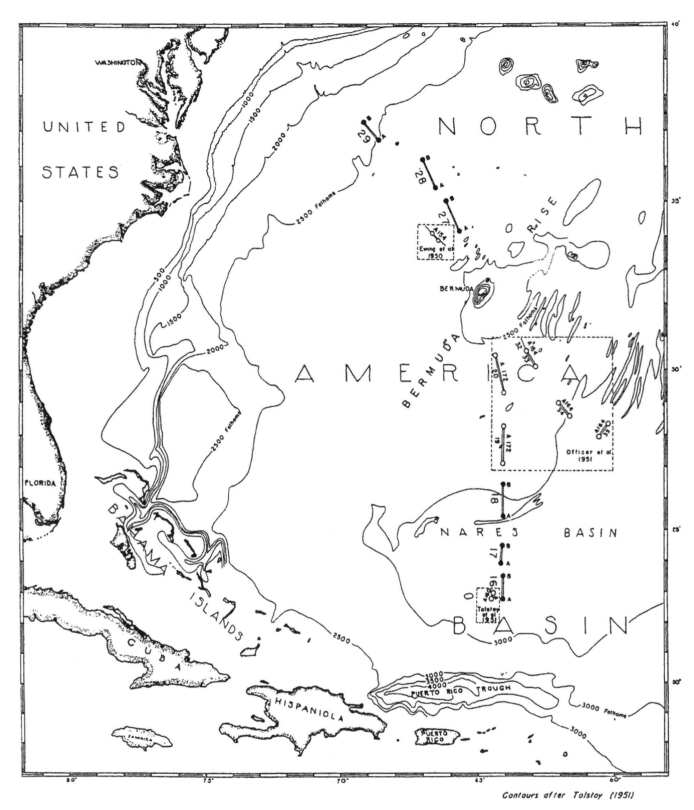

Figure 5.6.2-01. Geographic location of seismic profiles recorded in 1951 in the North American Basin (from Ewing et al., 1954, fig. 1). [Bulletin of the Seismological Society of America, v. 44, p. 21–38. Reproduced by permission of the Seismological Society of America.]

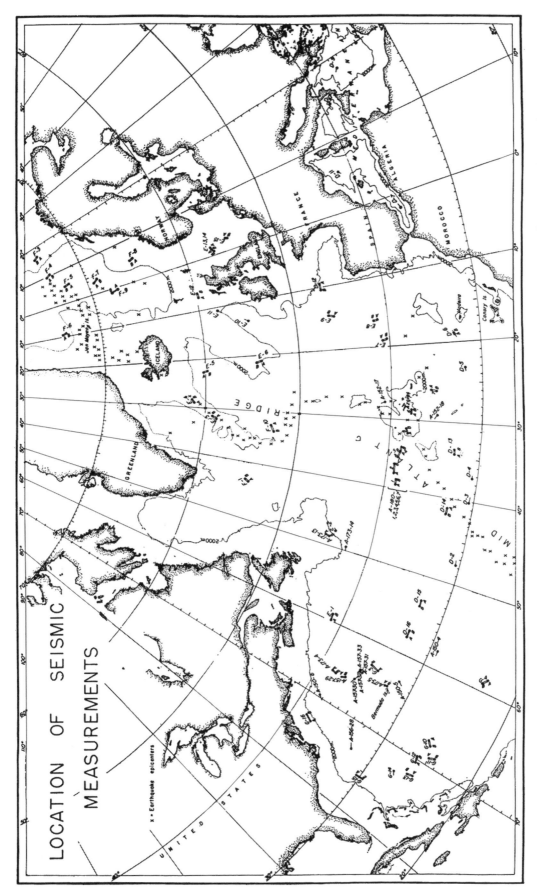

Figure 5.6.2-02. Geographic location of seismic profiles recorded from 1951 to 1953 in the North Atlantic Ocean, Mediterranean and Norwegian Sea (from Ewing and Ewing, 1959, fig. 1). [Bulletin of the Geological Society of America, v. 70, p. 291–318. Reproduced by permission of the Geological Society of America.]

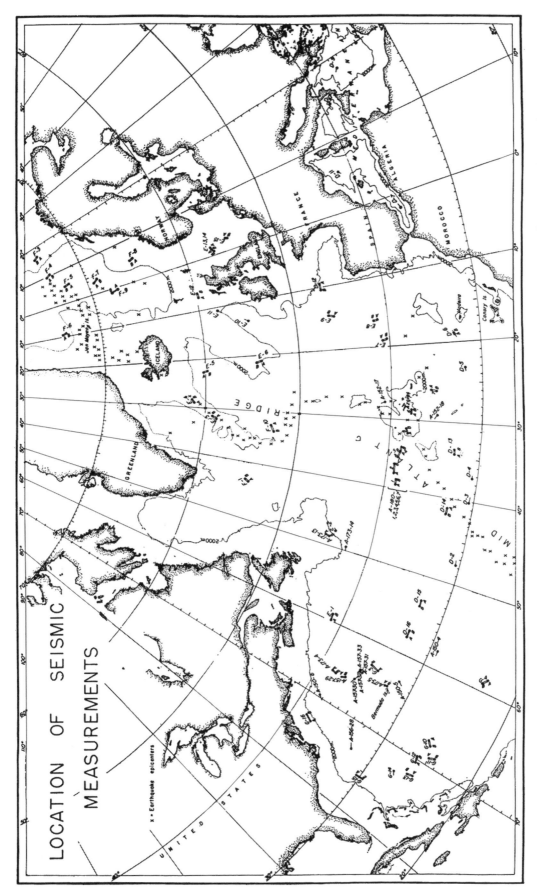

93

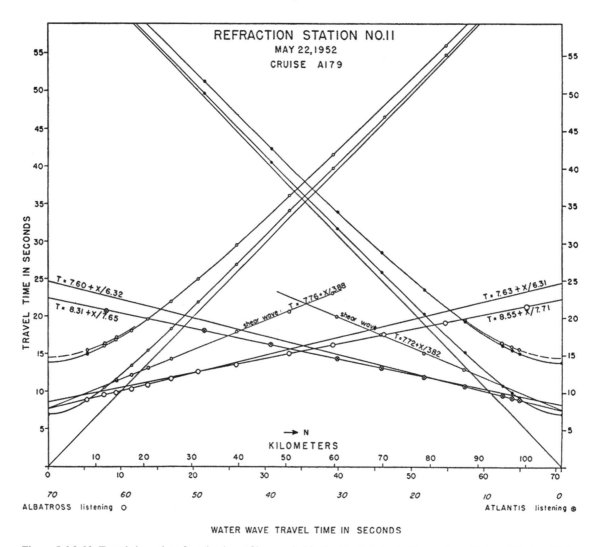

Figure 5.6.2-03. Travel-time plot of a seismic profile recorded in the North Atlantic Ocean (from Katz and Ewing, 1958, fig. 20). The distance is shown as water wave traveltime in seconds. [Bulletin of the Geological Society of America, v. 67, p. 475–510. Reproduced by permission of the Geological Society of America.]

results could nevertheless be obtained. The crustal layering turned out to differ significantly from the ocean basins (Figs. 5.6.2-05 to Fig. 5.6.2-07). The upper layer of the ridge with a velocity of 5.15 km/s was identified as basaltic volcanic rock, whose value agreed well with values given by Birch et al. (1942) and which also had been measured on Bermuda (Officer et al., 1952, see above). The deeper layer with average velocity of 7.21 km/s was tentatively identified as mixture of mantle and oceanic layer. Ewing and Ewing (1959) speculated that the formation of the ridge had required the addition of a great quantity of basalt magma which again raised the question of its source. As already described by Hess (1954), this was attributed to a rising convection current in the mantle which caused extensional forces to the crust to produce the axial rift.

Eight stations could be occupied in the Mediterranean Sea: two near the coast of Cyprus, two near Malta, three in deep water in the eastern basin, and one in shallow water southwest of Malta.

At the deep-water stations, only one refracting layer (4.42–4.57 km/s) was found underlying low-velocity sediments. The shallow station showed a wider range of refracting velocities from 2.96 to 6.70 km/s. The other stations corresponded to the values found in this area by Gaskell and Swallow (1953) in their cruise in 1952.

In the Norwegian and Greenland Seas, only five deep-water and two shallow-water stations were occupied. In spite of the small number of experiments, the results were reasonably consistent and therefore appeared justified. The Norwegian Sea (Fig. 5.6.2-08) differed from that found usually under deeper oceanic areas. The upper high-velocity layer was fairly consistent in velocity (4.96–5.37 km/s) and thickness (2.5–3.0 km). The velocity in the deepest layer varied more (6.97–8.04 km/s). The depth of this layer below sea level was quite consistently 7 km. On the continental shelf and slope of Britain and Scandinavia, five stations were occupied with another three in the North Sea. Previous seismic work on the continental shelf and slope of the British

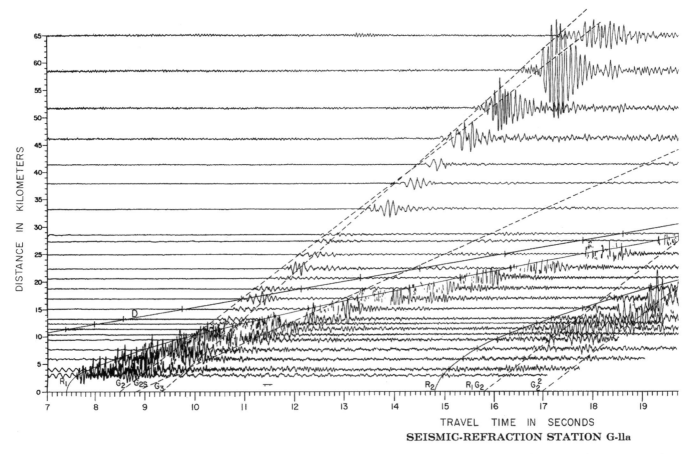

Figure 5.6.2-04. Record section of a seismic profile recorded in the North Atlantic Ocean (from Ewing and Ewing, 1959, pl. 1) [Bulletin of the Geological Society of America, v. 70, p. 291–318. Reproduced by permission of the Geological Society of America.]

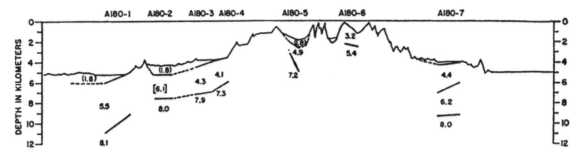

Figure 5.6.2-05. Topographic profile and seismic section across the Mid-Atlantic Ridge south of the Azores (from Ewing and Ewing, 1959, fig. 2). [Bulletin of the Geological Society of America, v. 70, p. 291–318. Reproduced by permission of the Geological Society of America.]

Isles had been reported in earlier years by Bullard et al. (1940), Bullard and Gaskell (1941), Hill and King (1953), and Hill and Laughton (1954). Ewing and Ewing (1959) reported that the velocities in unconsolidated sediments could be reasonably well determined, averaging 1.7 km/s. In all stations, they also found a semi-consolidated and a consolidated layer with velocities of 2.2 and 3.3 km/s, respectively. For the basement, they found an average of 5.11 km/s. The velocities in the central North Sea were

about the same as at the shelf stations, but basement with depths greater than 4 km was not reached.

Finally Ewing and Ewing (1959) made measurements on the continental slope of North America, southeast of Cape May, New Jersey (Fig. 5.6.2-02), and near the southeastern tip of the Grand Banks. Their results were in general agreement with those of other investigations in this area. Their table again distinguishes unconsolidated, semi-consolidated and consolidated sediments

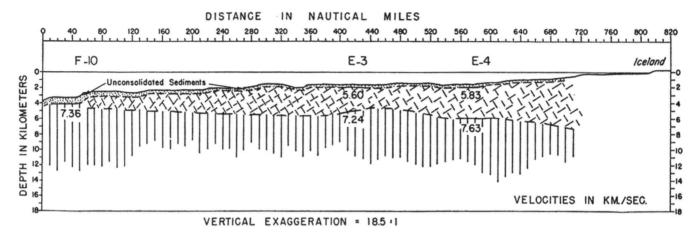

Figure 5.6.2-06. Structure section along the Mid-Atlantic Ridge south of Iceland (from Ewing and Ewing, 1959, fig. 4). [Bulletin of the Geological Society of America, v. 70, p. 291–318. Reproduced by permission of the Geological Society of America.]

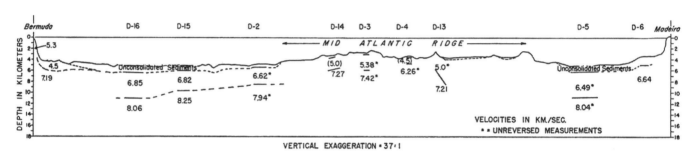

Figure 5.6.2-07. Structure section between Bermuda and Madeira (from Ewing and Ewing, 1959, fig. 5). [Bulletin of the Geological Society of America, v. 70, p. 291–318. Reproduced by permission of the Geological Society of America.]

with average velocities of 1.7, 2.4, and 3.8 km/s and thicknesses of 0.6, 1.9, and 2.1 km. For the basement, they determined a velocity of 5.9 km/s.

Using data from 22 seismic-refraction stations over the northern Mid-Atlantic Ridge and combining them with earlier data, Le Pichon et al. (1965) deduced a refined model of the crustal structure of the Mid-Atlantic Ridge in the North Atlantic Ocean. In the axial zone, they found a 5.8 km/s layer overlying a 7.3 km/s layer. They concluded that the seismic crustal structure is the same at equal water depths and does not thicken at the flanks of the northern Mid-Atlantic Ridge when water depth decreases, but is an uplifted ocean basin crust. In order for gravity and seismic results to agree, Talwani et al. (1965) computed fitting models postulating that the anomalous upper mantle under the ridge extends to the flanks below the normal mantle. They compared three different areas of the North Atlantic Ocean with sections from the East Pacific Rise and concluded that the northern Mid-Atlantic Ridge could be assumed to be in a later stage of evolution, the thickness and velocity of the basement layers being considered the best indicators of the degree of maturation of the mid-ocean ridges.

The first experiments of Ewing and coworkers were followed by many others. From the large amount of investigations

in the Atlantic Ocean (e.g., Katz and Ewing, 1956; Ewing and Ewing, 1959; Officer et al., 1959), an overall structure of the consolidated oceanic crust could be derived. The crust under the ocean basin consisted on average of a 1.7-km-thick layer 2 with velocity 5.07 km/s and a 4.9-km-thick layer 3 with velocity 6.7 km/s, underlain by mantle rocks with an average velocity of 8.13 km/s at the top. A detailed overview and summary of the results has been given by Raitt (1963) for the crust obtained by the application of the seismic-refraction method in the 1950s, listing the individual experiments in tables.

In 1953, Officer (1955a) reported on a series of seismic-reflection measurements on the abyssal plain in the NW Atlantic Ocean north of Bermuda, made on a cruise from Woods Hole, Massachusetts, to Bermuda and return. Similar experiments were also reported by Ewing and Nafe (1963). The method of continuous reflection profiling was discussed by Hersey (1963) in detail and compared with the seismic-refraction method, but at that time, the reflection method did not result in data reaching into the crystalline crust below the sedimentary cover.

From 1954 to 1956, the eastern continental margin of the United States, the intervening Blake plateau and adjacent deep-water areas of the Atlantic Ocean was investigated by 40 seismic-refraction profiles (Hersey et al., 1959), the sur-

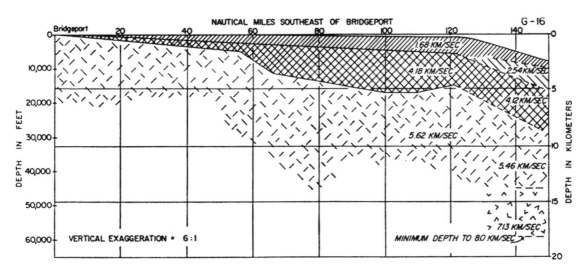

Figure 5.6.2-08. Structure section in the Norwegian Sea (from Ewing and Ewing, 1959, fig. 6). [Bulletin of the Geological Society of America, v. 70, p. 291–318. Reproduced by permission of the Geological Society of America.]

veys extending from 29°39′ to 36°30′N latitude and 73°30′ to 81°10′W longitude (Fig. 5.6.2-09). The results on the continental shelf could be correlated with adjacent continental geology. The deepest horizon traced along the shelf was interpreted as granitic basement with P-wave velocities of 5.82–6.1 km/s. At the southernmost profile near Jacksonville the greatest depth of 6 km to basement was found, toward north its surface rose to as low as 0.9 km near Cape Fear and deepened again north of Cape Hatteras to 3 km depth. North of Charleston, South Carolina, the measured depth could well be correlated with the granitic basement in a coastal borehole. On the Blake Plateau, a smooth and nearly flat area extending north from the Bahamas (located south of point D′ in Figure 5.6.2-09) to Cape Hatteras, several sedimentary layers with velocities from 1.83 to 4.5 km/s were found to overlie the crystalline crust with a 5.5 km/s and 6.2 km/s layer. Higher velocities, 8.0 km/s, as well as 7.3 km/s, were found at markedly different depths. The deep-water area is a continental slope and rise modified by the Blake Plateau and by a ridge trending southeastward from Cape Fear and deepening from ~2700 m to more than 3700 m (1500–2000 fathoms contours in Figure 5.6.2-09).

The ridge was found to be underlain by thick low-velocity layers, interpreted as sediments, and higher-velocity layers which formed a linear structure and had the same trend as the ridge. South of the ridge, the profiles were similar to those in ocean basins (Hersey et al., 1959). The cross section E–E′ shown in Figure 5.6.2-10 starts near Jacksonville, Florida, and crosses continental shelf and Blake Plateau and ends in the deep-water area near 77°W. The cross section H–H′ of Figure 5.6.2-11 follows the approximate continuation of cross section E–E′ to the northeast into the deep-water area and runs about parallel to the continental slope.

Geophysical measurements were also made in the western Caribbean Sea and in the Gulf of Mexico (Fig. 5.6.2-12; Ewing et al., 1960; Antoine and Ewing, 1963). For the center of the Gulf of Mexico, a typical oceanic crust was obtained with Moho depths between 16 and 19 km below sea level. Toward the western Caribbean Sea, Moho depths varied considerably, indicating an interfering of oceanic and continental crust (Fig. 5.6.2-13).

Another detailed survey to determine the crustal structure of the Caribbean Sea was undertaken in 1957 (Edgar et al., 1971). In total, 30 seismic-refraction profiles were measured. Some of them were partly aligned to longer lines, one across the ridge between the Colombia Basin to the west and the Venezuela Basin to the east, while another line was laid out parallel to coast of Venezuela and various scattered profiles were measured in the Venezuela Basin and around the Lesser Antilles islands. The results were summarized in five cross sections. Particular attention was given to areas with complex structures such as Tobago Island, the Curacao Ridge, the Smaller Antilles (called "Netherlands West Indies" by the authors) and others. The main result of the detailed survey was to map the velocity structure of the sedimentary layers and the depth to basement.

A cross section drawn from Cuba to Curacao shows, underneath several sedimentary layers, an ~2–5-km-thick basement at depths varying between 5 and 10 km, a rather thick lower crust with velocities of 6.5–6.8 km/s and a Moho depth near 20 km. The measured upper-mantle velocities varied from 7.9 to 8.5 km/s. On other profiles the crust-mantle boundary was rarely seen. Along the Venezuelan coast only the uppermost basement was reached. Other short sections in complex areas show complex structures. For example under the Curacao Ridge a deeper discontinuity was found near 15 km separating a 4.0–4.4 km/s basement from 6.7 to 7.0 km/s lower crust.

Based on gravity and seismic results, Talwani et al. (1959) published a cross section across the Puerto Rico Trench. The section started in the Nares Basin, 450 km north of San Juan, to the Venezuela Basin, 250 km south of San Juan. The depth to

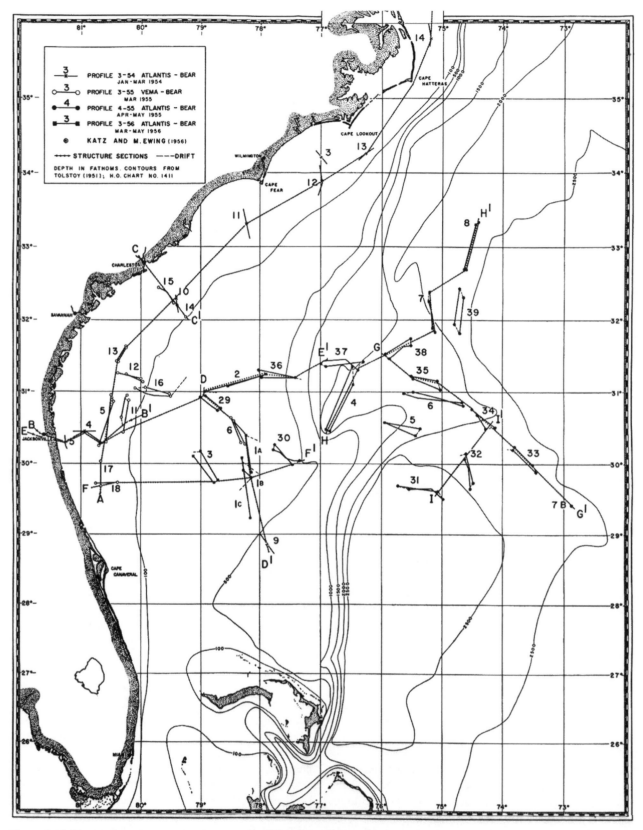

Figure 5.6.2-09. Location of seismic profiles recorded from 1954 to 1956 in the western Atlantic Ocean to the southeast of the United States (from Hersey et al., 1959, pl. 1). [Bulletin of the Geological Society of America, v. 70, p. 437–466. Reproduced by permission of the Geological Society of America.]

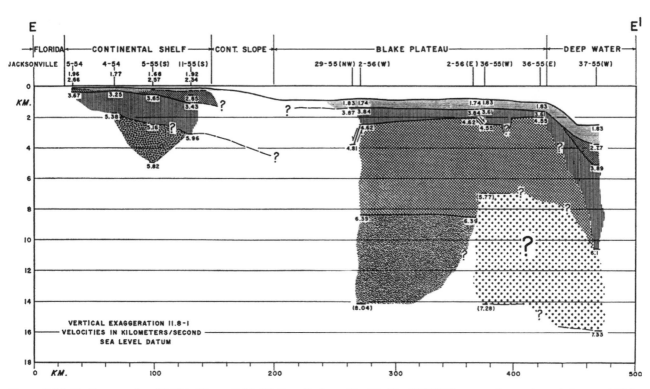

Figure 5.6.2-10. Cross section (E–E′ in Figure 5.6.2-09) from Jacksonville, Florida (81°30′W), across the continental slope and Blake Plateau into the deep-water area near 77°W (from Hersey et al., 1959, fig. 7). [Bulletin of the Geological Society of America, v. 70, p. 437–466. Reproduced by permission of the Geological Society of America.]

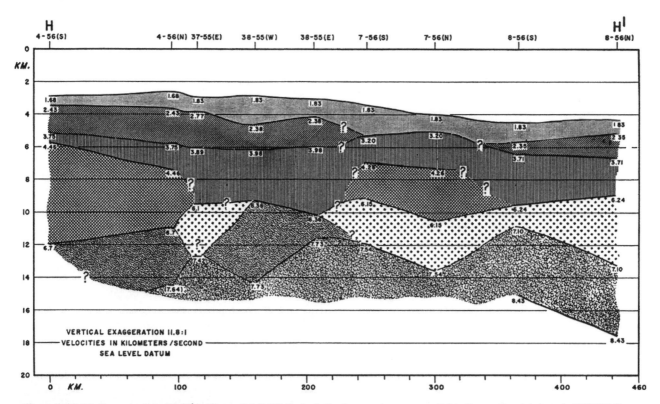

Figure 5.6.2-11. Cross section (H–H′ in Figure 5.6.2-09) through the deep-water area parallel to the continental slope in SW-NE direction (from Hersey et al., 1959, fig. 10). [Bulletin of the Geological Society of America, v. 70, p. 437–466. Reproduced by permission of the Geological Society of America.]

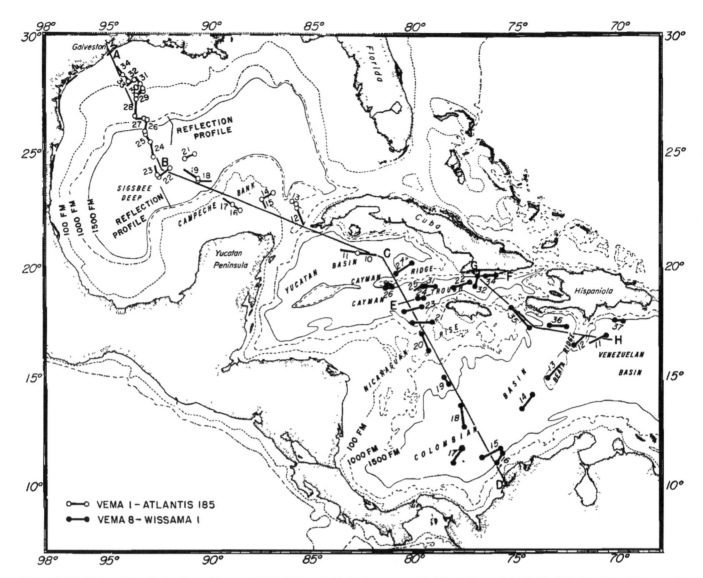

Figure 5.6.2-12. Location of seismic profiles recorded in the late 1950s in the western Caribbean Sea and the Gulf of Mexico (from Ewing et al., 1960, fig. 1) [Journal of Geophysical Research, v. 65, p. 4087–4126. Reproduced by permission of American Geophysical Union.]

Moho was ~20 km under the trench and decreased sharply to both sides. Northwards, Moho reached a minimum depth of 10 km under the Outer Ridge and then deepened gradually to ~13 km under the Nares Basin. To the south, Moho reached a depth of 17 km under the Puerto Rico shelf and then increased to 30 km beneath Puerto Rico. South of Puerto Rico under the Venezuela Basin, a depth of 14 km was obtained.

The first seismic-refraction investigation around the Mid-Atlantic Ridge in Iceland was undertaken by the Seismological Laboratory of Uppsala, Sweden, in cooperation with Icelandic earth scientists (Båth, 1960). The instrument used was a complete 12-channel refraction apparatus of Swedish construction with electromagnetic seismometers of 6 cps. The seismic signals were recorded via amplifier and oscillograph on paper of 150 mm width. The time marks were produced by a synchronous motor driven by an oscillator which was controlled by a temperature-compensated tuning fork. Fourteen galvanometers recorded in parallel. A series of explosions was set off in a lake with 45 m water depth in the southeast of Iceland. The break off of a tone when firing the shot was transmitted by telephone to the radio station of Reykjavik and from there broadcasted in the radio between or after their programs.

Two lines were recorded: a NNE line of 253 km length and a NE line of 337 km length, the distance between stations increasing with increasing distance from the shotpoint. In total, 19 shots were fired resulting in a total of 19 spreads. Due to strong absorption of energy in central Iceland, however, 150 km was the maximum distance on the NE line with recordable energy. A three-layer model resulted from these observations (Fig. 5.6.2-14) with a total crustal thickness of 27.8 km.

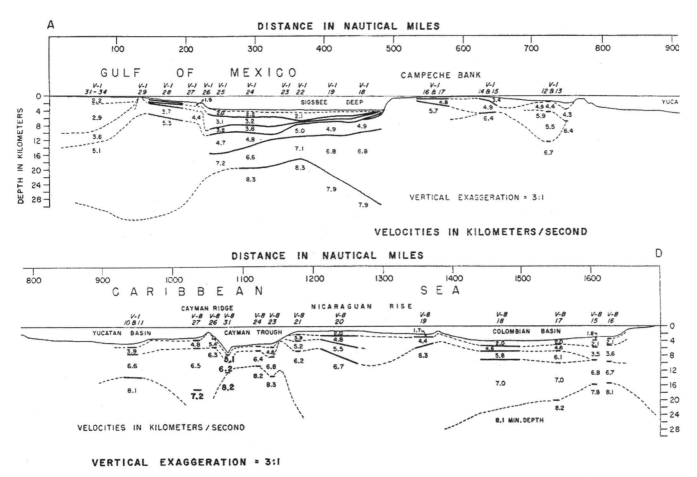

Figure 5.6.2-13. Crustal cross section through the Gulf of Mexico and western Caribbean Sea along line A–D (see Figure 5.6.2-12) recorded in 1951 (from Ewing et al., 1960, fig. 1). [Journal of Geophysical Research, v. 65, p. 4087–4126. Reproduced by permission of American Geophysical Union.]

Figure 5.6.2-14. Traveltime curves and crustal model of western Iceland (from Båth, 1960, fig. 6). [Journal of Geophysical Research, v. 65, p. 1793–1807. Reproduced by permission of American Geophysical Union.]

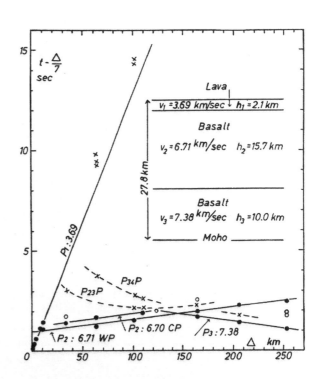

In the 1950s, expeditions into the Arctic were also undertaken to study the Earth's crust. The French Polar expeditions of 1949 and 1951 (Joset and Holtzscherer, 1953) had the purpose of investigating the inland ice of Greenland. Of particular interest were the dependence of the velocity in ice on the temperature and the velocity increase within the uppermost ice layers.

Katz and Ewing (1956) have made a statistical evaluation of velocities measured in the Atlantic Ocean crust at water depths exceeding 300 m (Fig. 5.6.2-15). Their frequency diagram clearly separated sediments with velocities around 2.4 km/s and transitional rocks with 4.4 km/s and showed peaks for crustal rocks at 5.5 and 6.3–6.8 km/s. Mantle rocks were defined from velocities of 7.4 km/s upward with an average of 8.1 km/s.

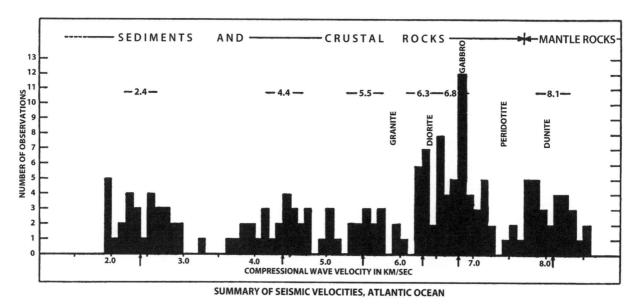

Figure 5.6.2-15. Frequency diagram of seismic velocities in the Atlantic Ocean, exceeding water depths of 300 m (from Katz and Ewing, 1956, fig. 25). [Bulletin of the Geological Society of America, v. 67, p. 475–510. Reproduced by permission of the Geological Society of America.]

5.6.3. Pacific and Indian Oceans

Other marine expeditions investigated the Pacific and Indian Oceans (e.g., Raitt, 1956; Gaskell et al., 1958). In the early 1950s, two major expeditions were undertaken into the Central Equatorial Pacific (Fig. 5.6.3-01): in 1950 the Mid-Pacific expedition explored the Pacific north of the equator, while in 1952 and 1953, the Capricorn expedition aimed for the Pacific between 10° and 20°S. At all stations two ships were used, following the methodology described by Officer et al. (1952) and briefly repeated in Chapter 4. Examples of a time-distance plot and of an oscillogram are shown in Figures 5.6.3-02 and 5.6.3-03.

In total, seismic-refraction observations were made at 42 positions, scattered widely, extending from latitudes 22°S to 28°N and longitudes 162°E to 112°W. At 29 of these stations, velocities of ~8 km/s or more were reached at the depth of greatest penetration of the refracted waves. The mean velocity was determined to 8.24 km/s. The crustal thickness was defined as the depth below the seafloor at which 8 km/s velocity was reached. It ranged from 4.8 to 13.0 km. The group of stations which was regarded as typical of the deep Pacific Basin had an average thickness of 6.31 km with a standard deviation of 1.01 km (Raitt, 1956).

One of the special goals of the Capricorn expedition of 1952–1953 was two weeks of research work in the Tonga Trench and archipelago (Raitt et al., 1955), making echo-sounding, coring, seismic-refraction and towed-magnetometer studies. Seismic-refraction profiles were laid out parallel to the axis of the trench (Fig. 5.6.3-04). The interpretation of the data (Fig. 5.6.3-05) indicated that the basement beneath the sediments had P-wave velocities of 5.2 km/s throughout the area. The sediments thickness was ~2 km in the Tofua trough to the west, thinner than 200 m on average in the inner gorge of the trench, and ~400 m

thick on the east flank, 50 km away from the trench axis. Beneath the 5.2 km/s basement, the crustal velocity was 7 km/s beneath the Tofua Trench and 6.5 km/s at the trench axis and on the east flank. The Moho, characterized by a velocity of 8.1 km/s, was estimated to be at 20 km depth below sea level beneath the trench axis, and 12 km below the eastern flank. Under the Tofua Trench, the highest observed velocity was 7.6 km/s, estimated to occur at 12 km depth below sea level.

At the same time, another series of experiments, lasting from 1950 to 1952, explored various sites in the Pacific Ocean (Gaskell and Swallow, 1952). Sixteen stations were occupied during this time and explored special types of geological structures at very different geographic locations. Four stations explored deep ocean basins: one was placed in the center of the northern Pacific at ~35°N, 143°W, one was located to the northwest of Hawaii, a third one was near the Kuril arch half-way between Kamchatka and Japan, and the fourth one was to the north of New Guinea. All stations showed a thin sedimentary layer overlying material with high P-wave velocity near 6.15 km/s. Other positions of seismic-refraction recordings were in deep ocean waters near islands. One station near Hawaii and another one near Funafuti showed thin layers of sediments followed by 2–3-km-thick material with P-velocities of 3.3–4.5 km/s, overlying a thin high-velocity layer (6.6–6.9 km/s apparent velocities, not reversed). In the western Pacific, two stations were placed in deep water southwest of Formosa (Taiwan). They showed a smaller velocity for the main layer under 600 m of sediments. Continental-type observations finally were made at stations off the North American coast (Southern California), but there were no reversing shots. Nevertheless, two distinct layers were indicated at each station, and the velocity in the deeper layer was substantially lower than at the deep-ocean stations. Some stations were placed near deep

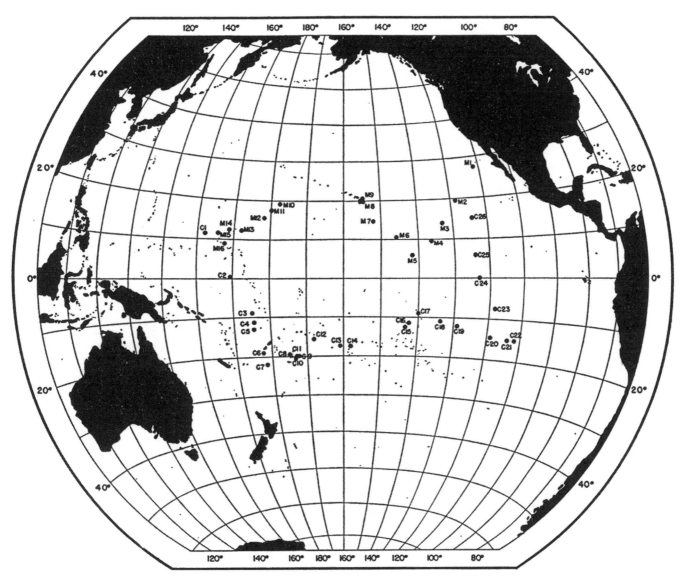

Figure 5.6.3-01. Seismic refraction stations of the Mid-Pacific and Capricorn expeditions (from Raitt, 1956, fig. 1). [Bulletin of the Geological Society of America, v. 67, p. 1623–1640. Reproduced by permission of the Geological Society of America.]

trenches, two, for example, southwest of Guam, but the results were not reliable.

Gaskell and Swallow (1953) made a third voyage into the Indian Ocean, passing through the Red Sea (reported upon by Drake and Girdler, 1964) and the Mediterranean Sea where they also did some measurements. On the basis of their worldwide study, they defined different structural types. In the ocean basins, the basement velocity was found to be higher than 6 km/s, sometimes as high as 7 km/s, and they assumed the presence of basic rocks of at least 2.4 km thickness. At two stations they could determine the peridotitic mantle layer at 8–9 km depth below sea level. Near island arcs layers with velocities of 3.0–4.5 km/s were found above the crystalline basement which Gaskell and Swallow interpreted as volcanics or limestones. Near the continents, they determined sediments with velocities of 2.5–3.3 km/s to be

consolidated, and they concluded that the basement velocities between 5 and 6 km/s infer that the basement is of continental granitic character.

In 1957, the IGY (International Geophysical Year) made it possible to send two ships on the Expedition Downwind into the large unexplored region east of the Capricorn Expedition (Raitt, 1964). Exploration of two major features was planned: the broad East Pacific Rise and the great Peru-Chile Trench forming the western boundary of most of South America. On the travel time plots from 90-km-long seismic profiles across the Peru-Chile Trench, three significant velocities could be identified: 4.5 km/s, 7.0 km/s, and 8.4 km/s. Two cross sections were constructed by Raitt (1964), one across the continental margin of Chile and one of Peru. On the Chile section, the results of four profiles, at 260 km, 170 km, 80 km (above the trench axis), and 20 km west

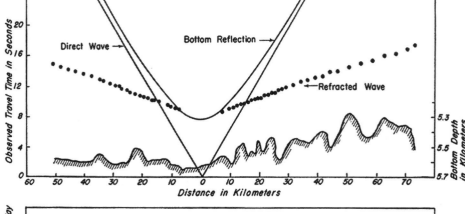

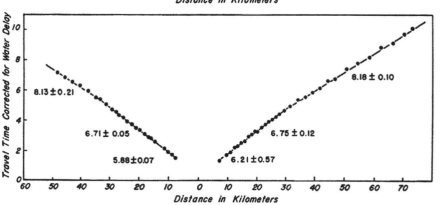

Figure 5.6.3-02. Example of travel-time plots, here for station M7 of the Mid-Pacific expedition (from Raitt, 1956, fig. 7). [Bulletin of the Geological Society of America, v. 67, p. 1623–1640. Reproduced by permission of the Geological Society of America.]

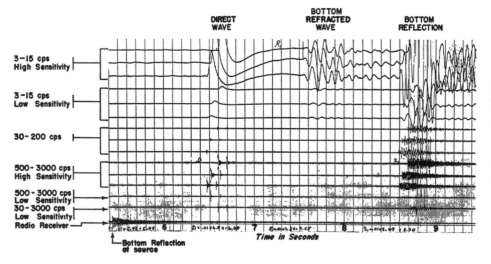

Figure 5.6.3-03. Sample of a complete oscillogram of seismic waves recorded at short range on the Mid-Pacific expedition in 1950 (from Raitt, 1956, fig. 4). [Bulletin of the Geological Society of America, v. 67, p. 1623–1640. Reproduced by permission of the Geological Society of America.]

of the coast, were compiled, showing a downwarp of Moho from 13 km depth at the 260-km station to ~23 km at the 20-km station. Both the 6.6–7.0 km/s refractor and the Moho were about parallel to the sea floor from the trench axis seawards. At the Peru continental margin, only two of the four measured stations gave a result for Moho: at 310 km off the coast, Moho depth was near 11 km, and under the trench axis, Moho depth was near 17 km. The 6.8–6.9 km/s refraction could be followed to 70 km off the coast, showing a generally increasing depth toward the coast. In summary, the crust west of the Chile Trench was 2 km thicker than west of the Peru Trench. Under both trench axes, the same crustal thickness of 10.5 km was found, though the sea floor under the Chile Trench is 2 km deeper than under the Peru Trench (Raitt, 1964).

A summarizing cross section across the East Pacific Rise by Raitt (1964) showed a slight decrease of the total crustal thickness under the axis of the Rise and a decrease of mantle velocities to 7.5 km/s. However, the mantle velocity decrease appeared to be concentrated on a narrow section across the axis of the Rise only, associated with high heat flow values. Outside

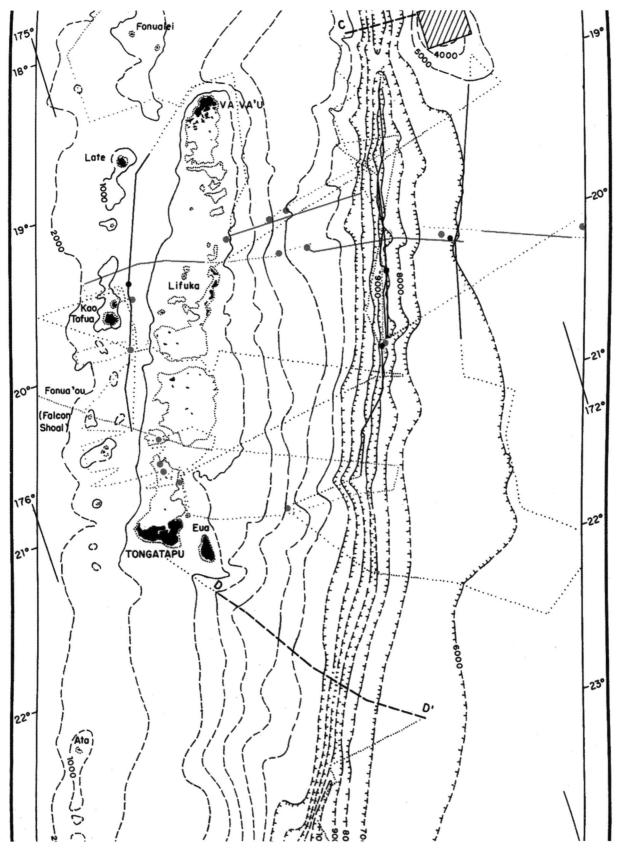

Figure 5.6.3-04. Location of seismic-refraction profiles in the Tonga area (N-S running straight lines) (from Raitt et al., 1955, pl. 1). [Geological Society of America Special Paper 62, p. 237–254. Reproduced by permission of the Geological Society of America.]

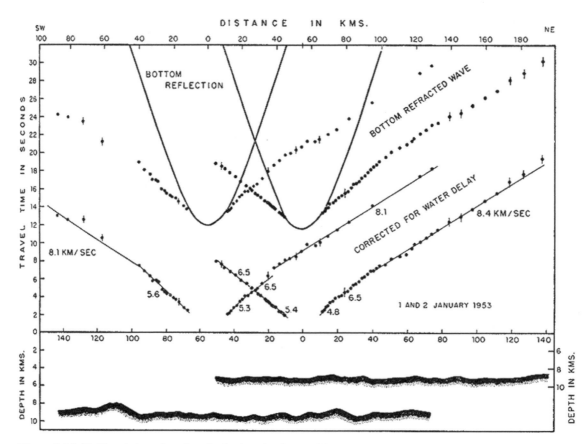

Figure 5.6.3-05. Travel-time plots the of seismic refraction profiles along the axis and the eastern flank of the Tonga trough (from Raitt et al., 1955, fig. 2). Axial profile to the left. Note that the depth axis of the flank profile is shifted upwards to avoid overlap. [Geological Society of America Special Paper 62, p. 237–254. Reproduced by permission of the Geological Society of America.]

the narrow central region, the mantle velocities were normal: equal or above 8 km/s.

Other information on the general crustal thickness in the Pacific area came from the study of surface waves (Officer, 1955b), in particular on the southwestern part including New Zealand and the Antarctica.

5.6.4. Summary of Seismic Observations in the Oceans in the 1950s

An overview of seismic measurements in the oceans was compiled from the database assembled by one of the authors (W.D.M.) and his seismic-refraction group at the U.S. Geological Survey in Menlo Park, California (Fig. 5.6.4-01). It shows all surveys that were published until 1959 in generally accessible journals and books. The map shows the numerous observations in the northwestern Atlantic Ocean, mainly organized through the initiative of William Maurice Ewing and his colleagues and performed by the Woods Hole Oceanographic Institution in Massachusetts and the Lamont Geological Observatory of the Columbia University at New York. The Mid-Atlantic Ridge and the northeastern Atlantic Ocean saw a considerable number of marine expeditions. Also in the Pacific Ocean, a substantial number of seismic

measurements were already achieved in the 1950s. Few measurements were reported from the Indian Ocean. Ewing and Ewing (1970) have compiled a map showing all seismic profiler traverses accomplished by Lamont ships around the world until 1967.

The many seismic-refraction profiles which had been shot in the 1950s, confirmed that the crustal structure of ocean basins is quite different from that of continents. But, as Raitt (1963) had concluded, the results revealed that the seismic structure was remarkably similar from ocean to ocean. The approximately linear segments on the travel time plots (see, e.g., Fig. 5.6.3-02) showed that the oceanic basement is covered by a sedimentary blanket over much of the sea floor and that the basement is composed of two main units: layer 2 with P-velocities of 4–6 km/s and layer 3 with P-velocities of ~6.7 km/s (Jones, 1999). The mantle structure under the oceans as derived from the marine seismic profiles could only be revealed for the uppermost kilometers, because the length of the profiles did not exceed 200 km, but had generally less than 100 km length, therefore the velocity measurements could not be extended into any appreciable depth (Ewing, 1963b). Ewing (1963b) plotted ~180 recorded seismic P-velocities greater than 7 km/s from profiles in the Atlantic, Pacific, and Indian Oceans, the Caribbean Sea, and the Gulf of Mexico. He selected only those profiles where an overlying layer

1950-1959: worldwide oceanic crust with bathymetry < −250 m

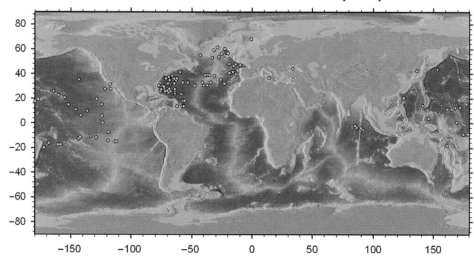

Figure 5.6.4-01. Seismic refraction measurements in the Atlantic, Pacific and Indian Oceans performed between 1950 and 1959 (data points from papers published until 1959 in easily accessible journals and books).

of high-velocity crustal rock (6–7 km/s) was present. This eliminated profiles from the Mid-Atlantic Ridge and some others on continental margins. The peak in the frequency was undoubtedly at 8.1–8.2 km/s, indicating that this is the average value of compressional waves in the upper mantle in deep-ocean areas.

5.7. RESULTS AND ACHIEVEMENTS

The progress in research was due to the efforts of many individual scientists who tried to organize sophisticated fieldwork under usually commonly difficult conditions. Young scientists, who later would become heads of university departments or research centers and whose names are authors in the reference list, started their career in the late 1950s and pushed their experiments with much vigor, thus setting the starting point for centers of excellence, which in the coming decades would achieve worldwide fame. In North America, the early crustal studies were pushed forward mainly by scientists of the Seismological Laboratory of the California Institute of Technology at Pasadena, California, the Dominium Observatory in Ottawa, Canada, the Department of Terrestrial Magnetism of the Carnegie Institution at Washington, D.C., the Lamont Geological Observatory at Columbia University, New York, the U.S. Geological Survey at Denver, Colorado, the University of Wisconsin at Madison, and the University of Utah at Salt Lake City. In Germany in particular, the State Geological Survey of Lower Saxony and the geophysical departments of the Universities of Clausthal, Freiberg, Goettingen, Kiel, Munich, and Stuttgart were combining their efforts on crustal research, supported also by German exploration companies, and partly in cooperation with the French Universities of Paris and Strasbourg. In Russia, crustal research was mainly concentrated in Moscow, Kiev, and Novosibirsk. Crustal research was also started in Sweden by the Seismological Laboratory of Uppsala and in Japan by the Earthquake Research Institute at Tokyo, and British scientists supported ongoing research in Australia and South Africa.

By the end of the 1950s, a basic knowledge on crustal structure around the Earth had been established, based on a considerable number of seismic-refraction investigations recording man-made explosions out to several hundred kilometers on land and recording a vast amount of marine seismic profiles. A comprehensive state of the art summary was compiled by Steinhart and Meyer (1961; Appendix A5-1). In Chapter 2, J.S. Steinhart presents a very detailed review on the development of crustal structure research of the continents, with particular emphasis of the achievements in the United States. In Chapter 10, J.S. Steinhart and G.P. Woollard show detailed location maps of seismic-refraction profiles around the world together with the corresponding references and a table (Table 5.7-01) summarizing the main results of the individual investigations (see also Appendix A5-1-10). An enormous effort had in particular been undertaken to explore the deep ocean basins. Detailed descriptions of the numerous expeditions in various parts of the oceans around the world in the 1950s were published in a comprehensive volume, *The Sea*, volume 3, edited by Hill (1963a). By the end of the 1950s, the basic knowledge of the structure of the oceanic crust was obtained, which would be modified only a little in the following decades (Minshull, 2002).

Not only scientific results, but also instrumentation, management of fieldwork, data presentation, and interpretation of seismic-refraction work on land in the1950s was presented in detail by Steinhart and Meyer (1961). The individual chapters include discussion of the methods in the 1950s for interpreting seismic-refraction measurements and their uncertainties as well as the discussion of instrumentation and field problems. The formulas for a multi-layered medium assuming horizontal flat layering and also inclined layers published by Steinhart et al. (1961c; see Appendix A5-1) became the basis for refined computer programs when appropriate computers became available in the 1960s. For the seismic-refraction work at sea, Ewing (1963a) published a similar detailed overview on the elementary theory

TABLE 5.7-01. CONTINENTAL CRUSTAL STRUCTURE, A REVIEW OF THE STATE OF THE ART AT THE END OF THE 1950s (FROM STEINHART AND MEYER, 1961, TABLE 10.1, SEE ALSO APPENDIX A5-1-10).

Profile no. & note	V_1 Km/Sec	h_1 Thickness Km	V_2 Km/Sec	h_2 Thickness Km	V_3 Km/Sec	h_3 Thickness Km	V_4 Km/Sec	h_4 Thickness Km	V_n Velocity below M	D Depth to M	$\bar{V}$ Mean Velocity above M, Km/Sec	B Bouguer Anomalies, mgals	I_1 Isostatic Anomalies, local mgals	I_2 Isostatic Anomalies, 1° squares (Tanni) mgals	R Isostatic Anomalies, 5° squares	E Elevation, meters	Location
Europe																	
1	3.6	6.0	5.4	7.5	6.5	13.9			8.2	27.4	5.564	−5	+5	+6	+10	50	Northern Germany
2	5.63	2.4	5.97	17.7	6.54	10.1			8.15	30.2	6.134	−25	+25	+20	+10	610	Haslach, Germany
3	5.62	10.9	6.35	19.9					8.15	30.8	6.092	−20	+10	+11	+13	305	Prague, Czechoslovakia
4	2.4	2.0	5.90	17.1	6.65	4.6			8.10	23.7	5.750	+10	+20	+27	+15	90	Hungary
Central Asia																	
5	4.5	10.0	[6.0]	13.0	6.5	[24.0]			[8.1]	23.0	5.348	+50?	0?	0	+15	−500?	Black Sea
6	4.3	7.0	5.6	17.0	6.6	15.0			[8.0]	[48.0]	5.860	−30?		+40	+25	400	Caucasus
7	4.8	15.0	6.0	10.0	5.5	9.0			8.0	40.0	5.775	−50?		−5	−20	−150	Caspian Sea
8	3.5	8.3	5.0	9.9	6.3	16.0			8.0	46.3	5.362	−75		−30	−20	0	Transcaspian Depression
9	3.5	3.9	5.5	13.6	6.3	19.0	6.3	19.0	8.0	28.8	5.865	−10		−30	−20	300	Great Balkhan Range
10	4.0	6.4	5.5	20.0					8.0	39.0	5.645	−30		−15	−20	450	Margin of Kopet Dag Mts.
11	5.8	30.0	6.4	18.0					7.9	50.0	6.040	−110		−35	−25	1400	Stalinabad
12	5.5	33.0	6.4	18.0					8.1	51.0	5.818	−300		−80	−50	3000	38°30'–70°20'
13	5.5	43.0	6.4	29.0					8.1	72.0	5.863	−410		−90	−50	5000	Academy Sci. Mts.
14	5.5	32.0	6.4	25.0					8.1	57.0	5.895	−390		−65	−50	4000	Zaalaisk Mts.
15	5.5	20.0	6.4	24.0					8.1	44.0	5.991	−190		−35	−20	1150	Osh
16	5.5	10.5	6.4	41.0					8.1	51.5	6.217	−210		−40	−35	1830	Lake Issk Kul, Tien Shan
17	5.5	15.0	6.4	23.0					8.1	38.0	6.045	−160		−50	−35	3300	Zailisk Mts.
18	5.5	11.0	6.4	36.0					8.1	47.0	6.189	−80		−30	−15	600	East Lake Balkhash
19	5.5	16.0	6.4	27.0					8.1	43.0	6.065	−30		−50	−15	400	West Lake Balkhash
20	(9 layers, increasing from 2.1–7.6 km/sec)								8.0	41.2	6.411	−75?		+20	0	500	Ural Mts.
Southern Africa																	
21	5.65	4.5	6.09	22.7	6.83	15.5			8.27	38.2	6.312	−130	−10		+20	1600	South Africa
22	5.4	1.3	6.20	36.6					8.21	37.9	6.173	−70	0		+20	1400	South Africa
23	6.03	28.2	7.19	8.4					7.96	36.6	6.296	−130	+20		+20	1500	South Africa
24	5.4	1.3	6.09	34.7					8.42	36.0	6.065	−140	+10		+20	1300	South Africa
Australia–New Zealand																	
25	6.03	37.4	6.04	34.8					8.32	37.4	6.03	−45	−30		−35	300	Western Australia
26	[4.5]	1.5	5.5	3.1					8.03	37.0	5.863	−40	−15		−20	625	Eastern Australia
27	3.5	0.6			6.07	3.3	6.22	10.6	8.02	18.1	5.807					100	Wellington NE, New Zealand
South America																	
28	5.3	4.1	6.2	21.2	6.7	39.6			8.0	64.9	6.448				+10	4300	Peru, Altiplano
29	5.3	4.1	6.2	21.2	6.7	26.4			8.0	51.7	6.384				+30	1520	Peru, (flank of plateau)
30	5.5	6.0	6.35	28.4	7.0	35.9			8.0	70.3	6.609					4500	Altiplano, Chile
31	5.5	6.0	6.35	28.4	7.0	22.2			8.0	56.5	6.526					1520	Chile (flank of plateau)
Japan																	
32	1.74	1.0	5.5	4.3	6.2	21.7			7.7	27	5.923	+125		+15	+35	25	Sakhalin Is.
33	2.51	0.53	5.8	9.0	6.15	15.5			7.75	25	5.951	+50		+50	0	200	
34	5.8	1.5	6.15	22.5					7.75	24	6.128	+125		−47	0	360	
35	5.5	12.0	6.3–6.7	23.0					8.0	35.0	6.157	+140	0		−20	0	
36	1.5	3.0	6.3–6.7	17.0					8.0	20.0	5.750	+170	+30		0	−3000	Sea of Okhotsk
37	1.5	5.0	6.3–6.7	9.5					8.0	14.5	4.776	+410	+50		0	−5000	Pacific Margin

(continued)

TABLE 5.7-01. CONTINENTAL CRUSTAL STRUCTURE, A REVIEW OF THE STATE OF THE ART AT THE END OF THE 1950s (FROM STEINHART AND MEYER, 1961, TABLE 10.1, SEE ALSO APPENDIX A5-1-10). (*continued*)

	P_1		P_2		P_3		P_4		V_n	D	$\overline{V}$	B	I_1	I_2	R	E	
Profile no. & note	V_1, Km/Sec	h_1, Thickness, Km	V_2, Km/Sec	h_2, Thickness, Km	V_3, Km/Sec	h_3, Thickness, Km	V_4, Km/Sec	h_4, Thickness, Km	Velocity below M	Depth to M	Mean Velocity above M, Km/Sec	Bouguer Anomalies, mgals	Isostatic Anomalies, local mgals	Isostatic Anomalies, 1° squares (Tanni) mgals	Isostatic Anomalies, 5° squares	Elevation, meters	Location
										Iceland							
38	3.69	2.12	6.71	15.70	7.38	10.02			[8.0]	27.8	6.731	−10			+15	150	Western Iceland
										North America							
39	5.7	9.1	6.6	15.0	7.3	22.2			8.3	46.1	6.756	+10	+25		+20	0	Alaska: Prince William Sd.
40	5.7	6.1	6.6	10.6	7.3	32.2			8.3	48.9	6.948	−70	0		+20	850	Southern Alaska
41	5.5	4.0	6.1	10.0	6.7	24.0			[8.1]	38.0	6.416	−75	0		0	1000	Alaska: Skagway area
42	3.6	2.0	6.1	1.0	6.2	26	7.2	14	8.2	43.0	6.402	−115	−12		0	900	Alberta, Canada
43	6.0	increasing to 7.0 km/sec at base of crust							8.0	30.0	6.5	−45	−18		−15	300	Washington: Puget Sound
44	5.0	3.72	5.95	26.16	7.44	5.48			7.94	35.36	6.081	−160	+10		+10	1500	Western Montana North
45	5.0	2.29	5.95	19.89	7.44	23.91			7.94	46.09	6.676	−200	+3		+10	1850	Western Montana South
46	[3.6]	.7	5.63	14.77	6.70	13.51	7.24	25.72	7.87	54.7	6.691	−135	+5		+10	1500	Northern Montana
47	[3.6]	2.78	6.08	12.93	6.88	24.61			8.15	40.32	6.397	−170	+15		+20	1700	Southern Montana
48	3.50	2.09	6.08	15.13	6.97	16.91	7.58	22.82	8.07	56.92	6.854	−105	+8		+20	900	East Montana North
49	3.58	3.38	6.08	19.62	6.97	16.86	7.58	10.09	8.07	49.97	6.512	−105	+20		+20	950	East Montana South
50	[3.6]	3.06	6.14	13.62	6.87	11.02	7.25	18.32	7.97	46.02	6.588	−100	+20		+25	800	Northeast Wyomlng
51	5.6	1.0	6.0	increasing to 7.0 km/sec					8.1	42.5	6.5	−8	+27		+10	430	Central Minn.
52	5.40	3.67	6.11	12.17	6.51	26.50			8.03	42.34	6.300	−53	−15		+10	300	N.W. Wisconsin
53	4.75	1.85	6.44	13.84	6.67	22.57			8.15	38.27	6.492	+3	+10		0	200	Keweenaw
54	5.40	.98	6.11	11.25	6.51	25.17			8.03	37.40	6.361	−63	−23		−10	300	Central Wisconsin
55	4.16	2.7	6.03	increasing at (1+.0038Z) to			6.9	37.95	8.17	40.65	6.317	−45	−15		0	200	Central Wisconsin
56	4.58	1.44	5.74	6.05	6.22	30.0			8.17	37.5	6.078	−45	−26		0	195	Central Wisconsin
57	4.5	.63	5.94	increasing at (1+.0049Z) to			7.09	42.95	[8.17]	43.6	6.488	−37	−11		−10	170	Central Wisconsin
58	6.23	28.1	7.09	12.1					8.17	40.2	6.489	−45	−18		−20	305	Ontario
59	5.9	?	6.08	32.5					[8.17]	32.5					−5	−40	St. Lawrence River Basin
60	6.3	(increasing by 2.5 × 10^{-4} km/sec/km to 7.2 krn/sec)							8.07	36.0	6.750	−40	−10		−10	315	Adirondacks
61	5.6	1.4	6.01	31.3					8.21	32.7	5.992	−50	−20		−10	270	Pennsylvania
62	5.6	0.7	6.15	increasing to 7.0 at base of crust					8.1	32.5	6.5	−5	−10		010	30	Chesapeake Bay
63	6.01	5.3	6.33	8.42	6.73	31.58			8.06	45.3	6.572	−75	−25		−15	760	East Tennessee
64	4.64	2.03	5.18	8.19	6.64	30.98			8.16	41.2	6.251	+13	+22		0	87	Arkansas
65	4.8	4.3	6.1	21.7	7.36	22.1	7.34	8.8	8.15	48.1	6.563	−210	−20		−10	2000	Arizona–New Mexico
66	5.2	5.9	5.8	11.7	6.26	21.3	7.34	8.8	8.18	47.7	6.215	−193	0		+10	1000	Utah
67	3.8	2.5	6.15	33.5					8.15	36.0	5.987	−130	−20		−10	1830	Nevada
68	6.11	23.0	7.66	26.0					8.11	49.0	6.932	−105	−25		−20	920	Southern Calif. (inland)
69	5.68	11.3	7.18	32.6					8.10	43.9	6.794	−50	−55		−20	450	Southern Calif.
70	5.9	1.0	6.1	5.0	6.5	10.0	6.85	24.0	8.20	40.0	6.645	−50	−55		−20	200	Southern Calif.
71	5.0	0.5	5.9	5.5	6.07	20.0	7.0	6.0	8.2	32.0	6.198	−20	−60		−20	100	Southern Calif. (near coast)
72	5.5	10.0	6.5	21.5					7.98	31.5	6.183	−3	−27		−15	200	Central Calif.
73	3.0	.80	4.95	3.40	6.01	28.46	7.63	10.69	8.38	43.36	6.269	−210	−20		−20	2200	Mexico

of seismic-refraction and seismic-reflection measurements in the introductory chapter of the comprehensive volume *The Sea*, volume 3 (Hill (1963).

Closs and Behnke (1961, 1963) reviewed the publications on crustal structure available by the end of the 1950s and compiled useful summarizing crustal cross sections of results obtained until 1960, representing the state of the art at that time. The profound difference between continental and oceanic crustal structure is demonstrated in Figure 5.7-01, showing a simplified east-west profile around the northern hemisphere. Some more details are shown in Figures 5.7-02 to 5.7-04. A summarizing cross section for North America is shown in Figure 5.7-02. For the southern hemisphere, some selected crustal columns are shown for the individual continents (Fig. 5.7-03), and finally for Europe and

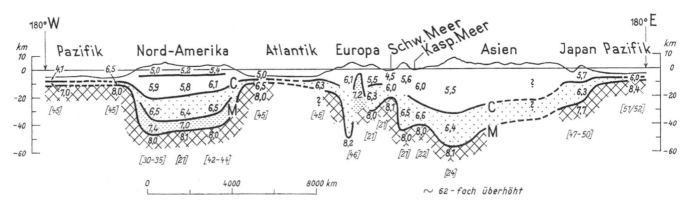

Figure 5.7-01. West-east profile across the earth at 45°N, compiled after individual results (for detailed references see Closs and Behnke, 1961). [Geologische Rundschau, v. 51, p. 315–330. Reproduced with kind permission of Springer Science+Business Media.]

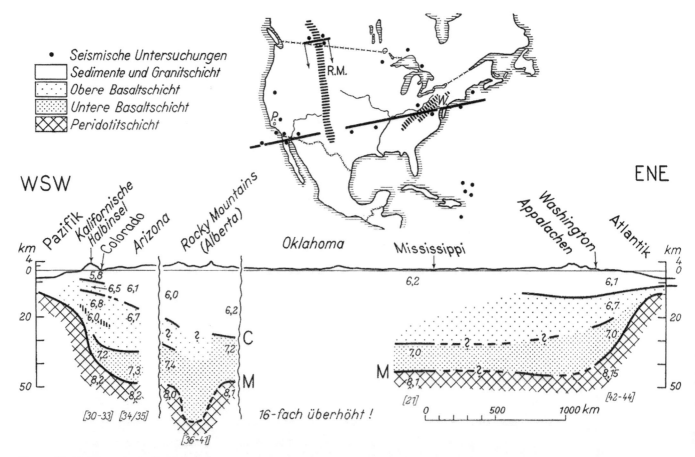

Figure 5.7-02. West-east profile through North America, compiled after individual results (for detailed references see Closs and Behnke, 1961). [Geologische Rundschau, v. 51, p. 315–330. Reproduced with kind permission of Springer Science+Business Media.]

Asia, a cross section was compiled reaching from central Europe into central Asia (Fig. 5.7-04).

The great interest in obtaining more detailed information on the Earth's crust became evident in general recommendations of the two geoscientific international unions: At its meeting in 1960 the International Union for Geodesy and Geophysics (IUGG) recommended exploring the Earth's crust more intensely rather than to bring it up-to-date. At about the same time, the International Congress for Geology established a commission which should recommend and advise major geophysical projects.

Table 5.7-01 shows the main results of continental crustal structure, as known by the end of the 1950s and summarized by Steinhart and Meyer (1961, Chapter 10), subdivided into continents. The table also includes for comparison average gravity anomalies and elevation of the areas investigated.

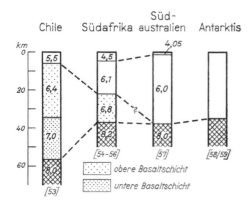

Figure 5.7-03. Velocity-depth columns for singular projects on the southern hemisphere, compiled after individual results (for detailed references see Closs and Behnke, 1961). [Geologische Rundschau, v. 51, p. 315–330. Reproduced with kind permission of Springer Science+Business Media.]

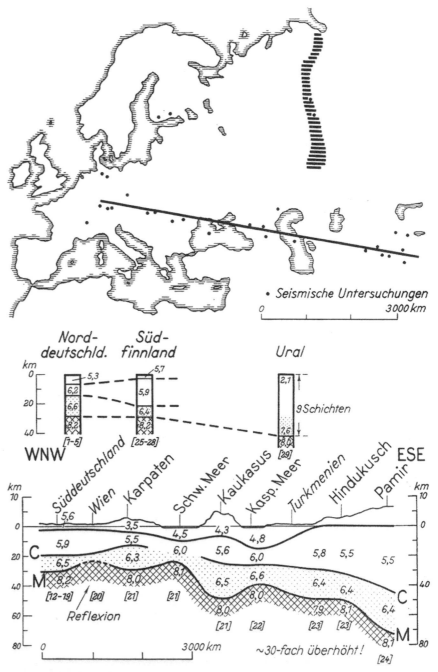

Figure 5.7-04. West-east profile from southern Germany to Pamir, compiled after individual results (for detailed references see Closs and Behnke, 1961). [Geologische Rundschau, v. 51, p. 315–330. Reproduced with kind permission of Springer Science+Business Media.]

The 1960s (1960–1970)

6.1. CRUSTAL STRUCTURE RESEARCH BY QUARRY BLAST OBSERVATIONS AND BOREHOLE SHOOTING

The 1960s saw a major breakthrough in the study of the Earth's crust. During the first experimental phase of crustal studies in the 1950s, new types of instruments had been developed and tested. This experimental stage continued at the beginning of the 1960s and finally led to the production of powerful instrumentation in major numbers, in particular in western Europe, North America, and the former USSR. In parallel to the fieldwork, the art of interpretation was pushed forward when scientists at different places in the world started to make use of rapidly developing computer technology and searched both for improvements of the known methods and at the same time developed new methods to interpret the increasing number of recently observed data.

The majority of the seismic-refraction experiments in the 1960s were carried out on a national basis, but there was also major international cooperation involved. The investigation of the Alps involved scientists from all bordering nations as well as British scientists. Under the heading of the USSR, a series of international profiles was recorded throughout southeastern Europe, and the Lake Superior experiments involved all major North American research institutions.

For the Atlantic Ocean, a detailed overall picture of the oceanic crust was established, but significant deviations from this average were found on the flanks of the Mid-Atlantic Ridge. Numerous new data were gathered in the Indian and Pacific Oceans. For the first time, the existence of seismic anisotropy of the uppermost mantle was investigated by special anisotropy experiments in the Pacific Ocean (e.g., Raitt et al., 1971).

These efforts were strongly supported by the formulation of major research programs, which were financed by national research foundations and thus allowed the systematic study of specific tectonic regions of the Earth. In Europe, the European Seismological Commission (ESC) had been founded with meetings every two years held alternately in eastern or in western Europe that served as a most effective communication center between scientists in eastern and western Europe. A special subcommission of the ESC dealt in particular with experiments and results of explosion seismology investigations. In 1964, under the umbrella of the IASPEI (International Association of Seismology and Physics of the Earth's Interior) the Upper Mantle Project was started as an international program of geophysical, geochemical, and geological studies concerning the "upper mantle and its influence on the development of the earth's crust."

In 1968, a conference of experts on explosion seismology was organized in Leningrad, where leading scientists such as A.S. Alekseyev (Novosibirsk), M.J. Berry (Ottawa), P. Giese (München), A.L. Hales (Dallas, Texas), J.H. Healy (Menlo Park, California), I.W. Litvinenko (Leningrad), I.P. Kosminskaya (Moscow), R. Meissner (Frankfurt), R.P. Meyer (Madison, Wisconsin), P.N.S. O'Brien (Sunbury-on-Thames), N.I. Pavlenkova (Kiev), N.N. Puzyrev (Novosibirsk), V.Z. Ryaboy (Kiev), V.B. Sollogub (Kiev), A. Stein (Hannover), I.S. Volvovsky (Moscow), and S.M. Zverev (Moscow) exchanged their ideas and experiences and discussed in detail the current problems of explosion seismology.

By the end of the decade, a series of publications on explosion seismology, introduced by Kosminskaya (1969), was jointly published (Hart, 1969) presenting major results obtained within the Upper Mantle Project.

6.2. SEISMIC-REFRACTION INVESTIGATIONS IN WESTERN EUROPE IN THE 1960s

6.2.1. Instrumentation

In Germany, the priority program "Geophysical Investigation of Crustal Structure in Central Europe," funded by the German Research Society, was continued until 1964 and involved university institutions, state geological surveys, and exploration companies. Following the start of the Upper Mantle Project initiated in 1960 by the International Union of Geodesy and Geophysics, a national priority program with the same name was started and funded by the German Research Society in 1965.

The development of portable seismic equipment was vigorously continued, and the number of individual instruments was gradually increased. For example, at Munich, up to 15 field stations could be mobilized by 1965 which, due to their highly sensitive galvanometers, could be easily arranged as one, three, or multi-component stations and were at the same time very easy to handle (Figs. 6.2.1-01 and 6.2.1-02). However, due to the individual development at the various research institutions, the diversity of instrumentation was large, and the demand for a unique recording system became stronger and stronger.

The first step was accomplished in 1960 when a new portable highly sensitive electromagnetic seismograph (2 Hz) constructed by H. Berckhemer at Stuttgart (Fig. 6.2.1-03) and manufactured by the company Stroppe (Schwenningen, Germany) became available. This seismograph, named FS-60, could be used either as a vertical or as a horizontal sensor.

Figure 6.2.1-01. Mobile field station of the Institute of Applied Geophysics, University of Munich (photograph by C. Prodehl). The grey box below is the central recording unit with galvanometer and a photographic film device driven by a windshield wiper motor, top left: a radio receiver for time signal reception, top right: FS-60 seismometer.

Figure 6.2.1-03. Electromagnetic seismograph constructed by H. Berckhemer, University of Frankfurt, Germany (photograph by C. Prodehl).

Figure 6.2.1-02. Mobile three-component field station of the Institute of Applied Geophysics, University of Munich (photograph by C. Prodehl). The larger wooden box below is the central recording unit with three galvanometers and a photographic film device driven by a windshield wiper motor. Left and right of the central unit: two time signal receivers (left, a commercial radio; right, a special army radio for short-wave reception); top left: small box for carrying an FS-60 seismometer; in foreground left and to the right: batteries.

The second step toward obtaining unique equipment was the development of a new recording device. In close cooperation with H. Berckhemer and under the supervision of E. Wielandt, the company Lennartz at Tübingen finally developed new portable magnetic-tape recording equipment, which has been described in detail by Berckhemer (1970). In this system, frequency-multiplex modulation recording techniques on 1/4″ magnetic tape were applied (Fig. 6.2.1-04).

Each recording unit consisted of three 2-Hz field seismometers, with the corresponding amplifiers and modulators in one unit, a tape recorder, and a time-signal receiver (Figs. 6.2.1-05 and 6.2.1-06). The signal frequency ranged between 0.3 and 100 cps. The highly dynamic range of 60 dB was obtained by a very effective flutter and wow compensation. Special playback centers were established at selected institutions. By 1966, the MARS-66 system (Magnetic Apparatus for Recording Seismic sounding, established in 1966) was ready for use.

By a special grant from the Volkswagenstiftung issued to the FKPE (Forschungs-Kollegium für die Physik des Erdkörpers = Research Consortium for the Physics of the Earth, an organization in which members were the directors of all geophysical research institutions in Western Germany), it was possible to buy 40 calibrated instruments at once which were distributed evenly amongst the state and university research institutions. By the aid of further individual grants, the German

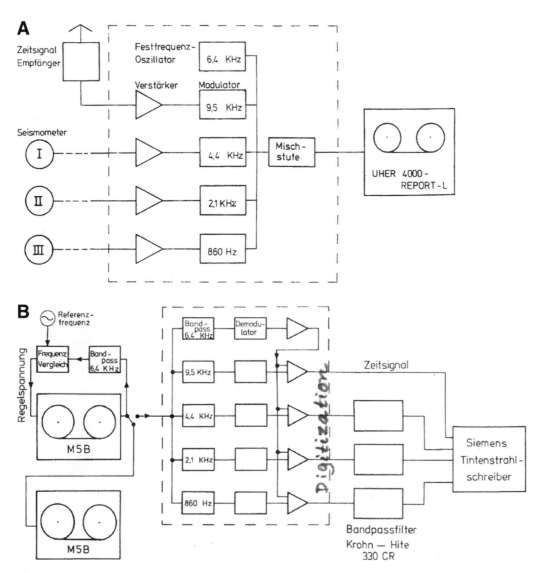

Figure 6.2.1-04. (A) block diagram of the recording unit MARS 66. (B) Block diagram of the playback unit (from Berckhemer, 1970, figs. 4 and 5). [Zeitschrift für Geophysik, v. 36, p. 501–518. Reproduced with kind permission of Springer Science+Business Media.]

scientific community could finally increase the number to over 70 instruments.

This new equipment not only enabled the recording of seismic profiles with a unique and calibrated system, but also allowed digital data preparation. The first digitizing system was developed and installed by W. Kaminski at Hamburg and later successfully reestablished at Karlsruhe. In the 1970s, MARS-66 equipment became widely distributed throughout Europe, in particular in France and Italy, so that in the last field project with this equipment in 1986, more than 200 MARS-66 units could be assembled for the seismic fieldwork along the central part of the European Geotraverse (EUGEMI Working Group, 1990).

A major problem in the early to mid-1960s, before MARS-66 became available, had been the timing of the seismic observa-

tions. Only in a few cases could commercial radio stations be convinced to broadcast a special time signal for a limited time. For example, the broadcast station Munich would, on request, interrupt its normal radio program and transmit a time signal for 2 minutes when quarry blasts at locations in Bavaria were scheduled. This middle-frequency radio station could be received by ordinary portable commercial radio receivers. Special short-wave stations like Pontoise in France had a very special schedule and required a very sophisticated planning of shooting, but through trial and error during the years, some other continuously operating short-wave transmitters were located, such as, e.g., Potsdam at 4525 kHz or Moscow at 10,000 kHz. It required, however, a special skill to receive these time signals with special army radios, which were difficult to find and expensive to buy

Figure 6.2.1-05. Recording unit MARS 66 (photograph by C. Prodehl). Left: modulator, lower right: tape recorder.

Figure 6.2.1-06. Recording unit MARS 66 equipped with a mechanical clock (upper right) (photograph by C. Prodehl). Upper left: T75A time-signal receiver; middle left: modulator; lower left: tape recorder. FS-60 seismometers in the background to the right.

in large quantities (see, e.g., Fig. 6.2.1-02). In the early 1960s, a time signal receiver was offered by Lennartz which was equipped with quartzes for special short-wave frequencies like those of Pontoise, Moscow, or Potsdam. A continuous time signal such as, e.g., WWV in North America, described in the next section, only became available in the mid-1960s when a continuous time signal broadcasted at 77.5 kHz was installed by the Federal Technical Office at Braunschweig which could be received by small specially designed receivers T75A almost everywhere in Europe (see, e.g., Fig. 6.2.1-06). This time signal receiver was incorporated into the MARS-66 recording unit. Also useful for projects in central and southern Europe since the mid-1960s were the Swiss time signal HBG, broadcasted continuously at 75 kHz, and in western and northern Europe, the British time signal MSF at 66 kHz could be well received.

6.2.2. Seismic-Refraction Experiments in Central and Western Europe

In the above-mentioned priority program "Geophysical Investigation of Crustal Structure in Central Europe," a systematic use of major quarry blasts had already been initiated in the late 1950s, offering the most economic seismic energy source. This program was continued in the 1960s, with quarry blasts serving as the major seismic energy source for crustal research work in Germany. Under the leadership of Prof. H. Closs, the state geological survey of Lower Saxony (Niedersächsisches Landesamt für Bodenforschung) overtook the task of contacting all major quarry companies in western Germany and surrounding countries and to inform all research institutions that had suitable equipment a few days before a major blast was scheduled. Thus a system of reversed and unreversed seismic-refraction lines of several hundred kilometers in length was gradually established (Fig. 6.2.2-01).

For the majority of these profiles, record sections were constructed and published on three maps in a special volume *Explosion Seismology in Central Europe* (Giese et al., 1976a). All record sections are reproduced in Appendix A6-1. Details of the shotpoints were published in Chapter 3.1 (Stein and Schröder, 1976) and details of the profiles in Chapter 3.3 (Giese et al., 1976b) of this special volume and are reprinted in Appendix A6-2.

In subsequent years, the organization of seismic fieldwork in Germany and adjacent countries was handed over to a special working group ASFA (Arbeitsgruppe für Seismische Feldmessungen und Auswertung = Working Group for Seismic Fieldwork and Interpretation) of the above mentioned FKPE, being responsible for a centralized organization of seismic fieldwork and interpretation which has remained active until 2006.

The exclusive use of quarry blasts in the 1960s had the disadvantage of a rather unequal distribution of available shotpoints inadequate to tectonic research goals. Most of the usable quarries were located in the Rhenish Massif and Hessisches Bergland, where Paleozoic rocks and Tertiary basalts are outcropping. Here, the location of several quarries allowed an overlapping arrangement of reversed profiles. Due to the uneven spacing and the large distances between such quarries, however, the term *reversed profile* had to be regarded with caution. Other shotpoints were concentrated in the Vosges, the Black Forest, and the Bohemian Massif. A very productive shotpoint with large explosions was a quarry near Eschenlohe (01 in Fig. 6.2.2-01), located near the northern margin of the Eastern Alps, which allowed successful recording up to distances of more than 200 km and also offered the chance for extending the observations into the Alpine region. Figure 6.2.2-02 shows a good example of data from a quarry blast near Hilders (Rhön), compiled into a record section.

First results of this cooperative research program were reported in 1964 (German Research Group for Explosion Seismol-

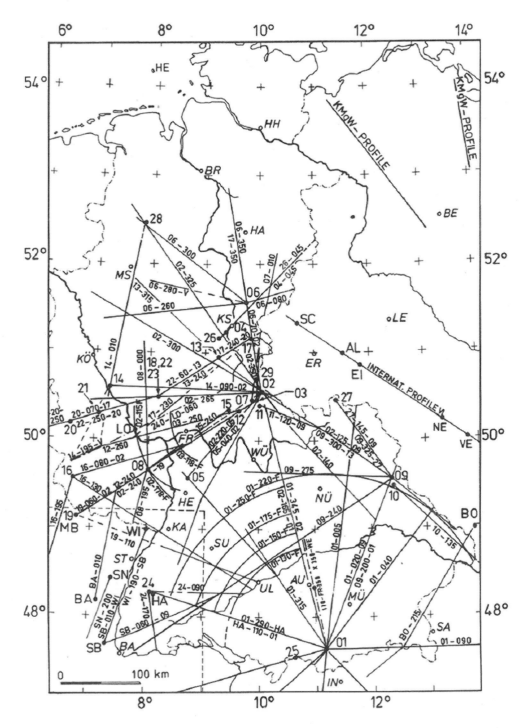

Figure 6.2.2-01. Location map of seismic-refraction observations in central Europe (from Giese et al., 1976a, fig. 1). Numbers are quarry blast locations active in the 1950s and 1960s, letters are centers of special surveys (FR, LO) and quarry and borehole locations outside Germany. Also included are the Heligoland (HE) and Haslach (HA) locations of 1947–1948. Italic letters designate major cities. Code of observations: 16-080-02—reversed profiles (shotpoint code–azimuth–reversing shotpoint code); 01-040—unreversed profiles (shotpoint code–azimuth); 14-195-V—non-completed lines (V = test); 01-150-F—fan observations (shotpoint code–average distance–fan). [*In* Giese, P., Prodehl, C., and Stein, A., eds., Explosion seismology in central Europe—data and results: Berlin-Heidelberg-New York, Springer, p. 73-112. Reproduced with kind permission of Springer Science+Business Media.]

ogy, 1964) presenting first contour maps for central Europe, such as the surface of the crystalline basement and the Conrad and the Moho discontinuities (Fig. 6.2.2-03).

These maps included an evaluation of deep reflections (Fig. 6.2.2-04) recorded during routine seismic exploration surveys of various oil companies. Already in the 1950s, Dohr (1959) had detected that commercial seismic prospecting offered the chance to record deep reflections from the Moho if the recording time was prolonged. So he had set up a research program to obtain

seismograms recorded at least up to 12 seconds that could be evaluated statistically.

Liebscher (1962, 1964) applied this method to a number of seismic prospecting areas in southern Germany and was able to compile histograms, cross sections, and first contour maps for the Conrad and the Moho discontinuities in southern Germany.

Following the early successful recordings of near-vertical reflections from the Moho near Blaubeuren in 1952 (Reich, 1953),

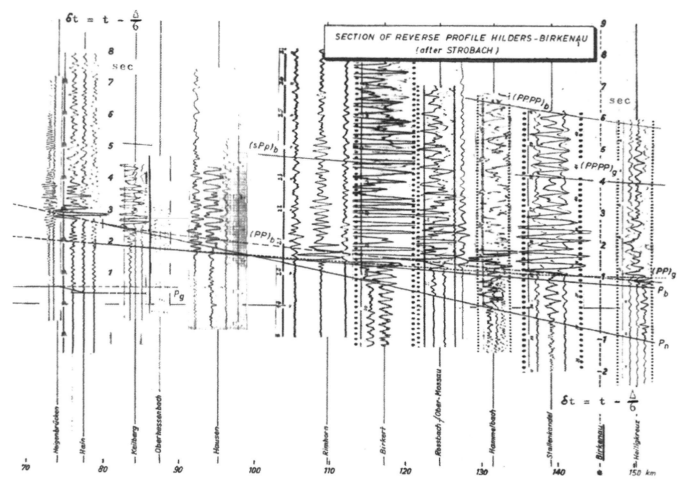

Figure 6.2.2-02. Part of a record section compiled from recordings of a blast in a basalt quarry near Hilders/Rhön (from German Research Group for Explosion Seismology, 1964, fig. 4). [Zeitschrift für Geophysik, v. 30, p. 209–234. Reproduced with kind permission of Springer Science+Business Media.]

other trials were made by university groups that had inherited older reflection instrumentation from exploration companies to record deep near-vertical reflections near multiple shots of large-scale refraction experiments, such as the Lago Lagorai experiments in 1961 and 1962. These attempts were not very successful, however.

It was R. Meissner, based in Frankfurt at that time, who first had the idea to apply the common depth point techniques of exploration seismics to the investigation of the whole crust. He designed and carried out an experiment in 1964 in the Bavarian Molasse basin (FR in Fig. 6.2.2-01) where shotpoints and a relatively small number of mobile recording units would systematically move apart from the central common depth point FR and record in the wide-angle distance range for the corresponding depth range (Meissner, 1966).

In 1968, R. Meissner arranged a second experiment of similar design in the Rhenish Massif (LO in Fig. 6.2.2-01) across its southern ranges Hunsrück and Westerwald. Both near-vertical (Glocke and Meissner, 1976) and wide-angle (Meissner et al.,

1976a) reflections could successfully be recorded and evaluated (Fig. 6.2.2-05).

These special projects where wide-angle reflection profiles using noncommercial equipment were recorded using especially arranged borehole shots (Meissner, 1966; Meissner et al., 1976a) added to the success of the German crustal research program. Furthermore, in northern Germany, a commercial seismic-refraction program was made available (Dohr, 1983), and on a profile in southeast Bavaria, borehole shots (BO in Fig. 6.2.2-01) were recorded that were fired along the International Profile VII traversing Poland and the Bohemian Massif in Czechoslovakia.

Special attention was given to the Rhinegraben, where at a symposium in 1966 a special research program was initiated which started with quarry blast observations both in Germany and in France (shotpoints BA and MB in Fig. 6.2.2-01). It finally led to a special program in 1972 with borehole shots in Alsace (shotpoints SB and WI in Fig. 6.2.2-01), which were recorded on various lines through the Rhinegraben, Vosges, and Black Forest

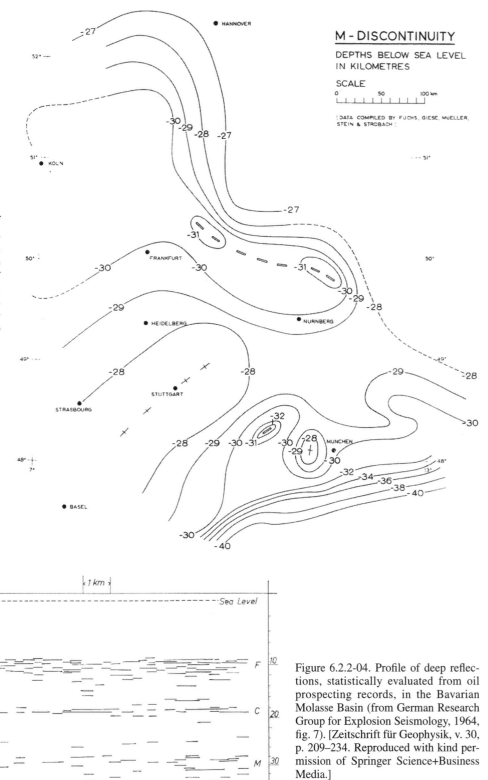

Figure 6.2.2-03. Depth contour map of the Mohorovičić discontinuity compiled until August 1963 by K. Fuchs, P. Giese, St. Mueller, A. Stein, and K. Strobach (from German Research Group for Explosion Seismology, 1964, fig. 10). [Zeitschrift für Geophysik, v. 30, p. 209–234. Reproduced with kind permission of Springer Science+Business Media.]

Figure 6.2.2-04. Profile of deep reflections, statistically evaluated from oil prospecting records, in the Bavarian Molasse Basin (from German Research Group for Explosion Seismology, 1964, fig. 7). [Zeitschrift für Geophysik, v. 30, p. 209–234. Reproduced with kind permission of Springer Science+Business Media.]

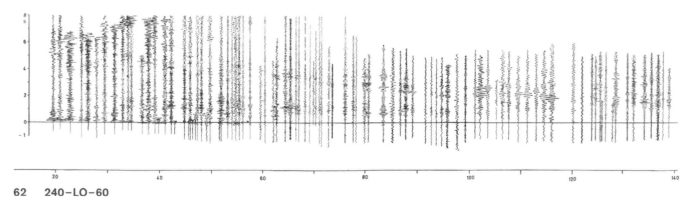

62   240-LO-60

Figure 6.2.2-05. Record section of profile 240-LO-60 (from Giese et al., 1976, map 3, fig. 62, reproduced from Meissner et al., 1976) [*In* Giese, P., Prodehl, C., and Stein, A., eds., Explosion seismology in central Europe—data and results: Berlin-Heidelberg-New York, Springer, Map 3. Reproduced with kind permission of Springer Science+Business Media.]

(Prodehl et al., 1976a). This program will be discussed in more detail in Chapter 7.

For southern Germany, Prodehl (1964, 1965) compiled all available seismic-refraction data (Fig. 6.2.2-06). With increasing numbers of observations, more details could be obtained and more accurate maps could be produced (e.g., Prodehl, 1965). The large amount of data collected in central Europe by the systematic observation of numerous quarry blasts led P. Giese to a more generalized interpretation of seismic data by mapping characteristic parameters for phases which could be regionally correlated, such as, e.g., the critical distance for P_MP (crust-mantle boundary) reflections, the crossover distance of P_n and P_MP, average velocities for certain depth ranges, the P_n velocity, etc. This allowed already for the drawing of some conclusions on general features of the crustal structure before applying any depth computations (Giese and Stein, 1971; Giese, 1976a, 1976c).

6.2.3. Seismic Research in Britain and Scandinavia

The continental region of the British Isles adjoins the Northeast Atlantic margin and includes the onshore regions of the British Isles as well as the stretched continental crust beneath the seas to the south and west of Ireland and northern Norway and beneath the North Sea.

The first crustal research was undertaken on the continental shelf to the west of Britain in 1960 and reported by Bunce et al. (1964). They reported that the deepest horizon sampled may have been "an intermediate layer rather than the mantle."

Another offshore study involved a major crustal profile across the North Sea in 1967, linking Scandinavia and the British Isles (Sornes, 1968). However, Sornes (1968) deferred presenting profiles showing the crustal structure "until more data about the upper layers became available." Nevertheless it was the first such "trans–North Sea" work, which in time was followed by the major Barton and Wood (1984) work (see Section 8.3.2).

Onshore crustal research in Britain started in ca. 1962. From the very beginning, British scientists made use of the favorable position of the British Isles surrounded by water. Most activities involved marine work, with offshore underwater shots being recorded on land. For the interpretation of most surveys, the time-term method, developed by Willmore and Bancroft (1960) was applied. A comprehensive review of early crustal studies in the British Isles is given by Blundell and Parks (1969) and was republished by Willmore (1973) with additions of post-1969 work.

The first crustal investigation in the British Isles reaching the Moho was undertaken to calibrate the Eskdalemuir seismological array (Agger and Carpenter, 1965) with depth charges fired in the North Sea and the Irish Sea in July 1962, resulting in a mean Moho depth of ~25 km. However, Blundell and Parks (1969) noted that the "geological significance of the variations in crustal thickness in their model is very doubtful," considering that there was "no measure of any near surface structural complexities" available.

In 1965, 25 shots were fired again in the Irish Sea and recorded by permanent land stations in Ireland, Wales, and Scotland, including the Eskdalemuir seismological array (Blundell and Parks, 1969). This and the following experiments were especially designed to permit the time-term approach (Willmore and Bancroft, 1960; lines and models 2 and 3 in Fig. 6.2.3-01).

Another sea-to-land seismic-refraction experiment followed in 1966. Eleven seismic recording stations (including three arrays) in Cornwall, northwestern France (Brittany) near Brest, and southeast Ireland (Waterford) and on the Scilly isles recorded two lines of shots at sea, one extending from northwestern France near Brest across the westernmost tip of Cornwall to the Irish coast near Waterford, and the other line running from Cornwall (lands end) along the seaward extension of the granite batholith of southwest England (Holder and Bott, 1971; unmarked lines and model 1 in Fig. 6.2.3-01). In total, 45 × 136 kg depth charges were fired. The seismic recordings on a cross array, two linear arrays, and a long profile of stations allowed the use of phase correlation to obtain apparent velocities. A major development of the Holder and Bott (1971) survey was

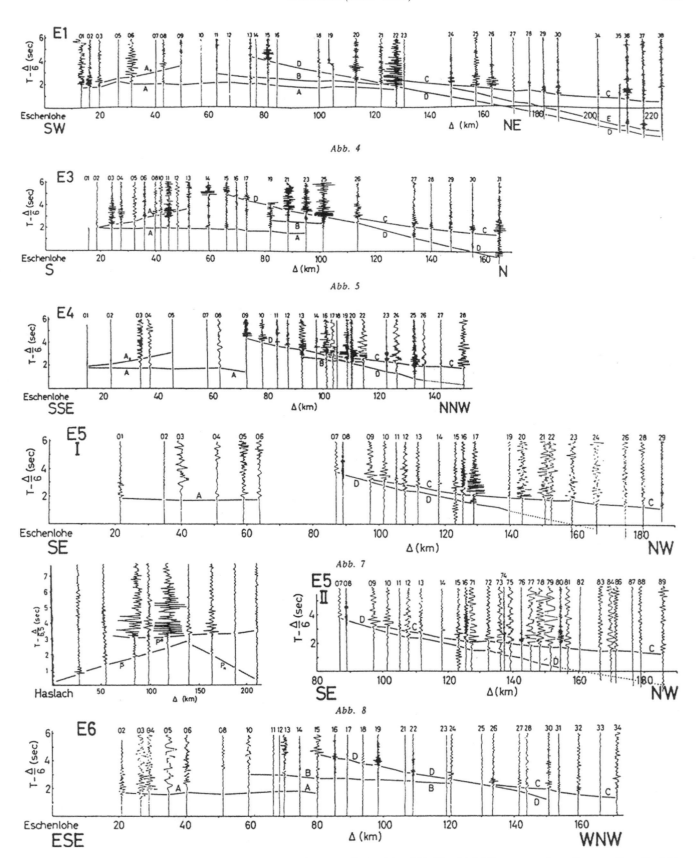

Figure 6.2.2-06. Record sections of data recorded from quarry blasts near Eschenlohe (01 in Fig. 6.2.2-01) across southern Germany (from Prodehl, 1965, figs. 4–9) [Bollettino di Geofisica Teorica e Applicata, v. 7, p. 35–88. Reproduced by permission of Bollettino di Geofisica, Trieste, Italy.]

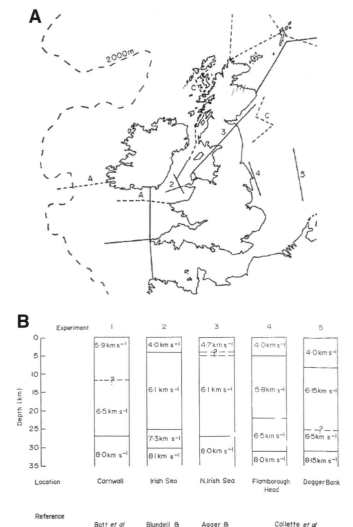

Figure 6.2.3-01. (A) Seismic-refraction profiles in the British Isles and surrounding sea areas (from Bamford, 1971, fig. 1). 1—Bott et al. (1970); 2—Blundell and Parks (1969); 3—Agger and Carpenter (1965); 4–5—Colette et al. (1970); dashed lines A, C and others—unpublished. (B) Associated crustal structure (after Bamford, 1971, fig. 2). Model 1 corresponds to the two unmarked perpendicular lines south of Ireland and west of Cornwall. [Geophysical Journal of the Royal Astronomical Society, v. 24, p. 213–229. Copyright John Wiley & Sons Ltd.]

the use of supercritical reflections to constrain the Moho depth. This was a significant introduction to the analysis of crustal refraction data.

Finally a Continental Margin Refraction Experiment (CMRE) was carried out in 1969 with 68 depth charges fired along lines south and southwest of Ireland and recorded by 11 temporary stations in Ireland, Wales, and England, and 5 sonobuoy stations at sea (A in Fig. 6.2.3-01). The seismic arrivals were recorded at distances up to 400 km. The CMRE shot-station network was designed for time-term analysis, although some of

the data was also suited for a classical interpretation as a reversed profile (Bamford, 1971).

Further investigations also included two seismic-refraction profiles on the Dogger Bank in the North Sea (Collette et al., 1970; lines and models 4 and 5 in Fig. 6.2.3-01). The Moho depths of all these surveys were fairly uniform, lying between 27 and 31 km (Bamford, 1971, 1972).

By the end of the 1960s, these marine surveys were extended into the Atlantic to the west of Ireland. In 1969, data were collected across the Rockall Plateau, northwest of the British Isles between 15–20°W and 54–59°N. The velocities of 6.4 and 7.1 km/s for the main crustal layers and the Moho depths varying from 21 to 30 km were regarded as typical for a microcontinent (Scrutton, 1972).

Marine shots were also the prime source for the early crustal investigations of Scandinavia, resulting in a number of profiles recorded in Norway, Finland, and Denmark (Fig. 6.2.3-02).

In 1965, two lines were recorded across the Caledonian mountain range in southern Norway to determine the crustal structure underneath the seismological array NORSAR. These resulted in Moho depths of 36–38 km under the central mountains and 28–30 km under the coast of southwest Norway (Sellevoll and Warrick, 1971). Preliminary results based on various progress reports were summarized in a map for Scandinavia compiled and published in a review of crustal structure in western and northern Europe by Closs (1969).

More details were published by various authors in the proceedings of a colloquium held in Uppsala in December 1969 (Vogel, 1971) dealing in particular with preliminary results along a Trans-Scandinavian Deep Seismic Sounding project carried out in summer 1969. Nine explosions of 1000–4000 kg TNT at five shotpoints were recorded along five lines through Norway, Sweden, and Finland, resulting in 150 recordings (data examples in Figs. 6.2.3-03 to 6.2.3-05). The average station spacing on all profiles on land except for the Kola profile was between 7 and 15 km. P-wave velocities and Moho depths (Fig. 6.2.3-02) increasing from 30 km at the coast of Norway to more than 40 km under central Sweden, the northern Baltic Sea, and Finland were later published by Sellevoll (1973).

When by 1971 these results were published, clear seismic sections were being presented (as shown in the text) providing a major advance, the reader being able to assess the phase arrivals, rather than merely being presented with the time-term data as common for work undertaken in the 1960s.

6.2.4. Crustal Research in the Alps, Southern France, and Italy

Following the international cooperative investigation of the Western Alps of 1956, 1958, and 1960 described in the former chapter, the Central and Eastern Alps also became major targets of the international community in the 1960s (Fig. 6.2.4-01).

In 1961 and 1962, a series of shots was organized by C. Morelli in Lago Lagorai located in the Dolomites in northern

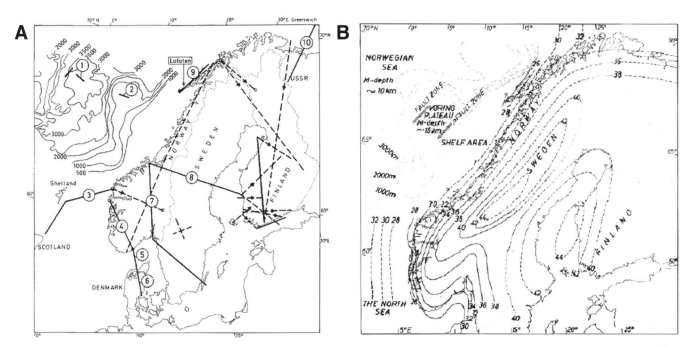

Figure 6.2.3-02. Seismic-refraction profiles (A) and crustal thickness (B, in km) in Fennoscandia (from Sellevoll, 1973, figs. 2 and 3). Full lines = reversed profiles (references in Sellevoll, 1973): 1–3—North Sea profiles; 4—southern Norway; 5–6—Northern Jutland–Skagerrak profiles; 7–8—Trans-Scandinavian project; 9—Lofoten; 10—Kola-Barents Sea. Dashed lines with arrows = unreversed profiles; short cross lines in south-central Sweden = Hagfors observatory array. [Tectonophysics, v. 20, p. 359–366. Copyright Elsevier.]

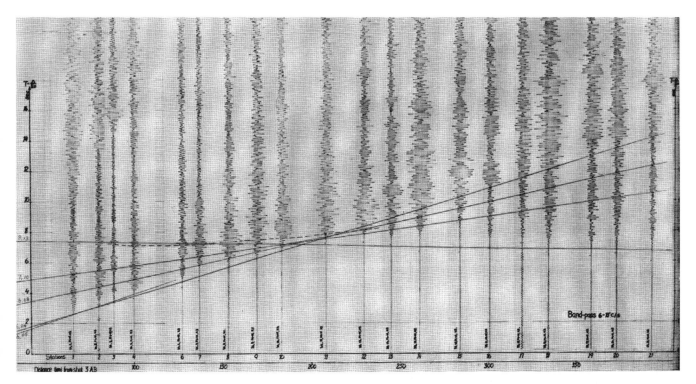

Figure 6.2.3-03. Seismogram section for line 8 (Fig. 6.2.3-02), shotpoint 3 across Norway and Sweden, band-pass filter 6-15 Hz, overlain by linear approximations correlating direct and refracted waves (solid lines) and Moho reflections (dashed lines) of a theoretical model (from Vogel and Lund, 1971, fig. 7). [*In* Vogel, A., ed., 1971, Deep seismic sounding in Northern Europe: Swedish Natural Science Research Council (NFR), Stockholm, 9 p. Permission granted by Swedish Research Council, Stockholm.]

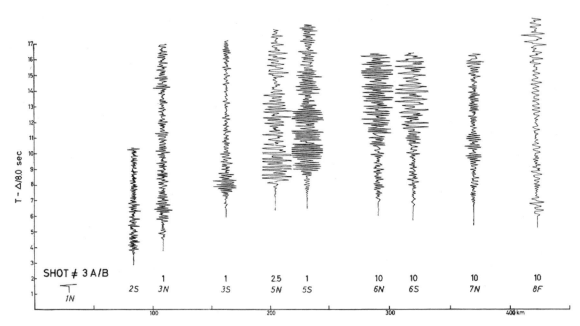

Figure 6.2.3-04. Seismogram section for line 7 (Fig. 6.2.3-02), northern shotpoint, along southeastern Norway, (from Kanestrom and Haugland, 1971, fig. 2). [*In* Vogel, A., ed., 1971, Deep seismic sounding in Northern Europe: Swedish Natural Science Research Council (NFR), Stockholm, 98 p. Permission granted by Swedish Research Council, Stockholm.]

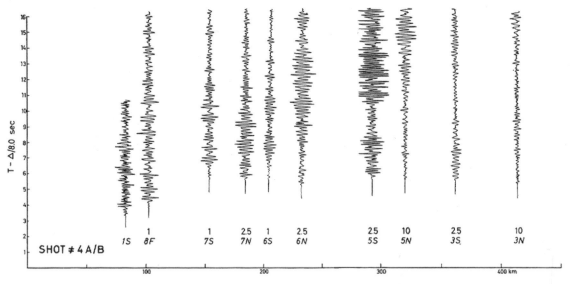

Figure 6.2.3-05. Seismogram section for line 7 (Fig. 6.2.3-02), southern shotpoint, along southeastern Norway, (from Kanestrom and Haugland, 1971, fig. 3). [*In* Vogel, A., ed., 1971, Deep seismic sounding in Northern Europe: Swedish Natural Science Research Council (NFR), Stockholm, 98 p. Permission granted by Swedish Research Council, Stockholm.]

Italy, which were recorded on several lines toward the east, west, south, and north, the northern line reversing the quarry blast shots of Eschenlohe (Prodehl, 1965; data examples in Fig. 6.2.4-02). In 1963 and 1964, similar experiments were organized by A.E. Süsstrunk, centered on a series of shots in Lago Bianco near the northern end of the anomalous zone of Ivrea (Behnke, 1971; Behnke and Giese, 1970; data examples in Fig. 6.2.4-03). All

record sections published by Giese and Prodehl (1976) are reproduced in Appendix A6-3 with more details on shotpoints and record sections in Appendix A6-4.

The first models for crustal structure of the Eastern Alps showing the Conrad and Moho discontinuities were published by Behnke et al. (1962) and Peterschmitt et al. (1965) using the observations of the Lago Lagorai experiments. From the data of the

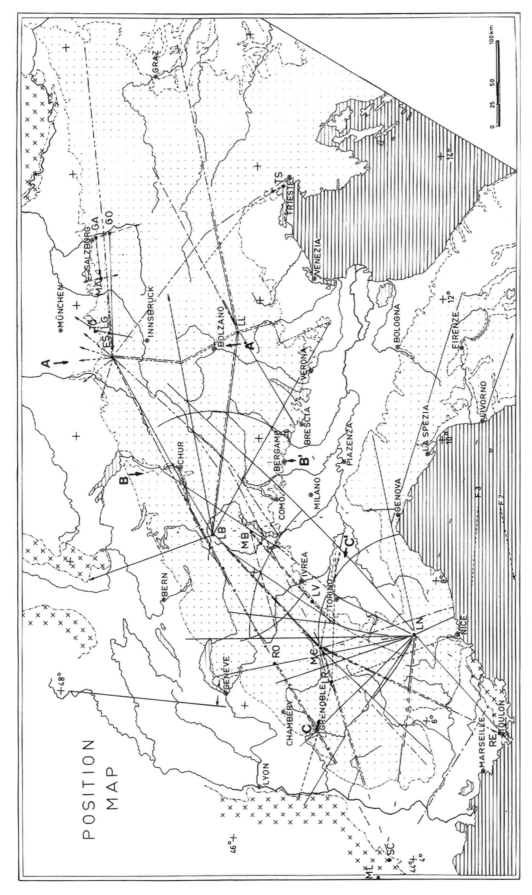

Figure 6.2.4-01. Location of shotpoints and seismic-refraction profiles in the Alps. Shotpoints: ES—Eschenlohe, GA—Gartenau, GO—Golling, LB—Lago Bianco, LG—Lenggries, LL—Lago Lagorai, LN—Lac Nègre, LR—Lac Rond, LV—Levone, MA—Marquartstein, MB—Monte Bavarione, MC—Mont Cenis, ML—Mont Lozère, RE—Le Revest, RO—Roselend, SC—Ste. Cécile d'Andorge, TÖ—Tölz, TS—Trieste (from Giese and Prodehl, 1976, fig. 1). [In Giese, P., Prodehl, C., and Stein, A., eds., Explosion seismology in central Europe—data and results: Berlin-Heidelberg-New York, Springer, p. 347–375. Reproduced with kind permission of Springer Science+Business Media.]

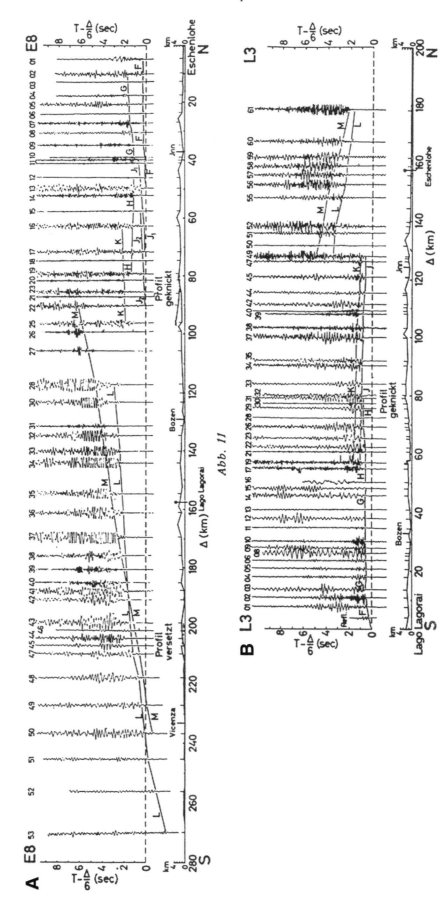

Abb. 11

Figure 6.2.4-02. Record sections of profiles from Eschenlohe–Lago Lago Lagorai. (A) Eschenlohe–S; (B) Lago Lagorai–N (from Prodehl, 1965, figs. 11 and 12). [Bollettino di Geofisica Teorica e Applicata, v. 7, p. 35–88. Reproduced by permission of Bollettino di Geofisica, Trieste, Italy.]

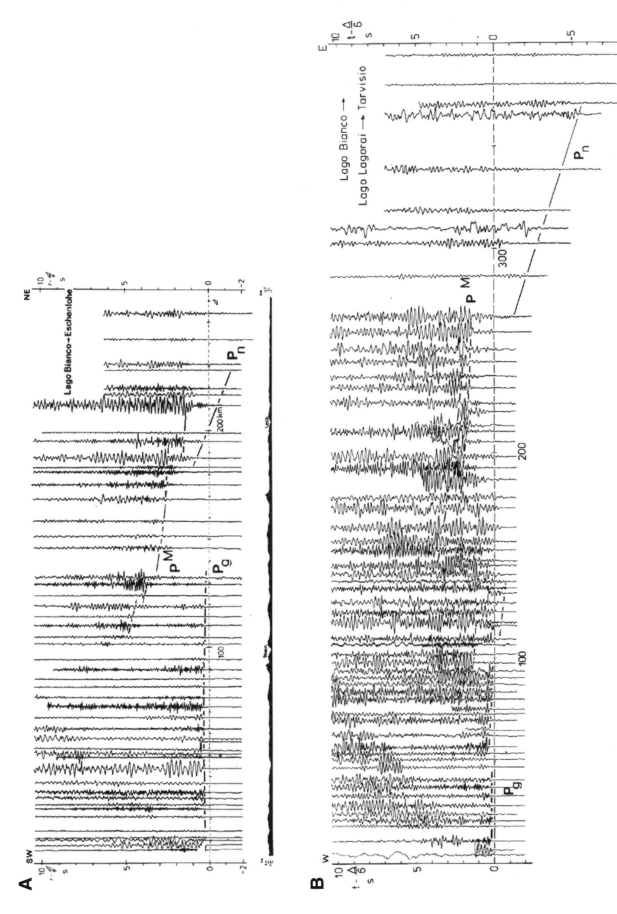

Figure 6.2.4-03. Record sections of profiles from Lago Bianco: top: towards NE, bottom: towards E (from Giese and Prodehl, 1976, figs. 20 and 21). [*In* Giese, P., Prodehl, C., and Stein, A., eds., Explosion seismology in central Europe—data and results: Berlin-Heidelberg-New York, Springer, p. 347–375. Reproduced with kind permission of Springer Science+Business Media.]

reversed profile Eschenlohe–Lago Lagorai, using a wave-front method, Prodehl (1965) constructed a cross section through the Eastern and Southern Alps, introducing a high-velocity layer at the base of the lower crust (7.2–7.4 km/s).

A new concept for Alpine crustal modeling was introduced by Giese (1968a). He interpreted the late appearance of high-amplitude reflections in most of the Alpine profiles as being caused by a strong velocity inversion within the crust. Furthermore, he interpreted the crust-mantle boundary as a transition zone of several kilometers width, where the velocity gradually increases from lower-crustal values to 8 or more km/s.

A special summary of all available data along the Geotraverse Ia through the Eastern Alps was published by Angenheister et al. (1972).

At the same time, from 1961 to 1967, the French seismic-refraction investigations concentrated on the French Massif Central and the Western Alps (Giese et al., 1967, 1973; Labrouste et al., 1968; Perrier and Ruegg, 1973, Appendix A6-4-6), which demonstrated that the crust outside the Alps does not show any anomalous behavior (Fig. 6.2.4-04).

In 1965, Y. Labrouste organized various shotpoints in the Western Alps (Roselend and Mont Cenis, RO and MC in Fig. 6.2.4-01) and in the Massif Central (Ste. Cécile d'Andorge and Mont Lozère, Fig. 6.2.4-04; SC and ML in Fig. 6.2.4-01) (Giese et al., 1967). In 1966, she revived the shotpoint Lac Nègre (LN in Fig. 6.2.4-01) allowing long-range recordings on various lines through the Alps and the Northern Apennines (Röwer et al., 1977).

The peculiarities of the structure of the Ivrea zone caused strong discrepancies in the models of Fuchs et al. (1963a, 1963b) and Labrouste et al. (1963) when the seismic data from 1956, 1958, and 1960 was interpreted. This was discussed in detail and resolved during a symposium in 1968, where several authors (Ansorge, 1968; Berckhemer, 1968; Giese, 1968b) introduced a strong upper crustal velocity inversion underlying the zone of Ivrea and thus explained the phase *m* of Fuchs et al. (1963a, 1963b) as a pure P-wave without excessive Moho depths.

Similar to the earlier interpretation of Labrouste et al. (1963), the anomalous body of Ivrea was now regarded as a mantle wedge gradually detaching from the mantle under the Po plain and intruding the crust under the Western Alps, its surface being at less than 10 km under the eastern edge of the Western Alps and underlain by upper to lower crustal rocks. Reinterpreting all available data, Giese et al. (1971) investigated the areal extension of this anomaly. A cross section through the Western Alps and the anomalous zone of Ivrea as compiled by Giese and Prodehl (1976) is shown in Figure 6.2.4-05.

Choudhury et al. (1971) compiled all data then available and elaborated a structural model for the Alps based on a unique concept of interpretation which was first presented at the General Assembly of the IUGG in 1967. Many record sections, selected cross sections, and contour maps published before 1971 can be viewed in a summary article compiled by Giese and Prodehl (1976) including a contour map of the crust-mantle boundary shown in Figure 6.2.4-06. Under the Western Alps north of Torino, the

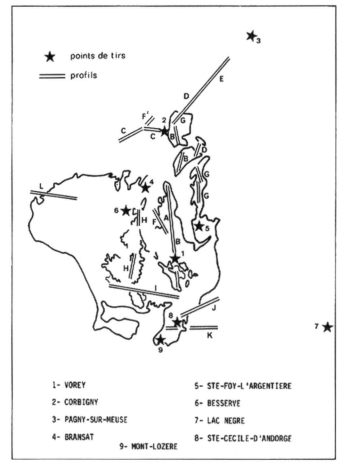

Figure 6.2.4-04. Seismic profiles observed in the Central Massif of France before 1968. 1—Vorey, 2—Corbigny, 3—Pagny-sur-Meuse, 4—Bransat, 5—Ste-Foy-L'Argentiére, 6—Besserve, 7—Lac Nègre, 8—Ste-Cecile-d'Andorge (modified after Perrier and Ruegg, 1973, fig. 1 [Annales de Géophysique, v. 29, p. 435–502.])

crust-mantle boundary of the Po Plain gradually turns over into the Ivrea body, while the deep crust-mantle boundary of the Western Alps under the low-velocity wedge under Ivrea gradually dissolves eastward as shown in detail in Figure 6.2.4-05.

The international cooperation in the Alps, in particular the close cooperation of German and Italian institutions also stimulated a major crustal seismic research program in southern Italy (Giese et al., 1973) in the late 1960s. The first profiles were observed in 1965 and 1966 in Puglia, and two seismic lines were recorded through Sicily in 1968 (Giese et al., 1973). The lines through Sicily were described in much detail by Cassinis et al. (1969). The crustal research was continued in the early 1970s (see Chapter 7, where the location of all projects in southern Italy and Sicily is shown in Fig. 7.2.3-05).

Of particular interest were the long-range profiles across the Apennines of central Italy, recorded in 1966 from the shotpoint Lac Nègre in the Western Alps (see Fig. 6.2.4-01). From the two prominent reflections P^M (A) and P^M (E), correlated between 70

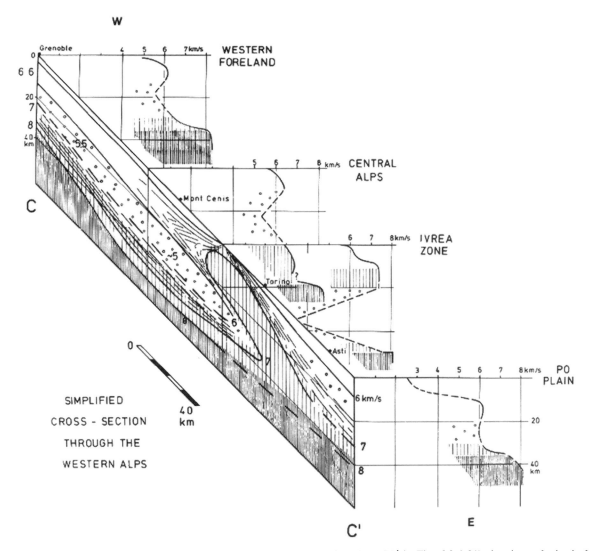

Figure 6.2.4-05. Simplified W-E cross section through the western part of the Alps (*CC'* in Fig. 6.2.4-01) showing velocity-isolines and average velocity-depth functions (from Giese and Prodehl, 1976, fig. 27). [*In* Giese, P., Prodehl, C., and Stein, A., eds., Explosion seismology in central Europe—data and results: Berlin-Heidelberg-New York, Springer, p. 347–375. Reproduced with kind permission of Springer Science+Business Media.]

and 170 km distance (Fig. 6.2.4-07; see also Appendix A7-2-10), the authors derived two overlapping crust-mantle boundaries: the upper crust mantle boundary is the shallow Adriatic crust at 25 km depth in the southeast which is being shifted across the Alpine Moho, and the lower crust-mantle boundary, at 46 km depth in the west and north. Together with crustal structure data of Fahlquist and Hersey (1969) for the adjacent sea areas, Giese and Morelli (1973) were able to construct a Moho map for all of Italy combining the previously published maps for northern and southern Italy.

6.2.5. Advances in Theory and Interpretation

Stimulated by the wealth of data obtained particularly during the 1960s, new methodologies were developed, most of them in close cooperation with or as a consequence of the numerous

high-quality data obtained by deep seismic sounding fieldwork, and which have been described in some detail in the previous subchapters.

In Germany, the priority program "Geophysical Investigation of Crustal Structure in Central Europe" had generated refraction/wide-angle reflection observations of high quality and growing and dense quantities. The young generation of scientists, such as Peter Giese at Munich, Rolf Meissner at Frankfurt, Karl Fuchs at Clausthal, Gerhard Müller at Mainz, and others realized very soon, that the previous methods of interpretation, based primarily on traveltimes, were inadequate compared to the richness of information in the new data.

Mueller and Landisman (1966) demonstrated evidence for a world-wide existing crustal low-velocity zone (Fig. 6.2.5-01): the concept might bring crustal seismic-refraction and seismic-

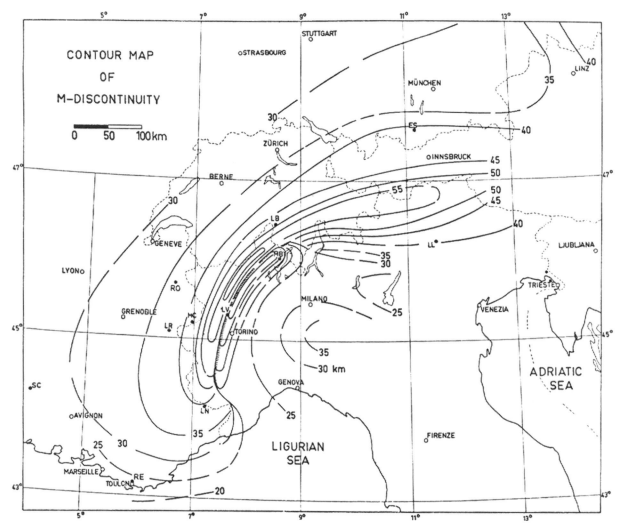

Figure 6.2.4-06. Contour map of the depth of the strongest velocity gradient of the Alpine area in the velocity range 7.5–8.2 km/s, which is regarded as the crust-mantle boundary (from Giese and Prodehl, 1976, fig. 35). [*In* Giese, P., Prodehl, C., and Stein, A., eds., Explosion seismology in central Europe—data and results: Berlin-Heidelberg-New York, Springer, p. 347–375. Reproduced with kind permission of Springer Science+Business Media.]

reflection data into better agreement, as Landisman and Mueller (1966) proved from examples from southern Germany.

Independently, Giese (1968a) investigated in detail the characteristics of reflected waves. In the ideal case of a sharp discontinuity, the reflection appears as a hyperbola of the corresponding traveltime curve in a time-distance diagram. The curvature of the traveltime curve and its position changes when the discontinuity is replaced by a transition zone with a strong, but not infinite, velocity gradient (Fig. 6.2.5-02).

At the same time, P. Giese undertook an approach to explain delays of phases versus the foregoing phases as could be seen in most of the record sections obtained in central Europe and the Alps (Giese, 1968a; Giese and Stein, 1971). He developed a particular approximation method to rapidly calculate the velocity-depth distribution directly from any given traveltime curve system, including options of low-velocity layers and veloc-

ity gradients instead of first-order discontinuities (Giese, 1976b). This method was later applied by Prodehl (1970, 1979) on data from the western United States, where the method is also explained in some detail (see Section 6.5.1).

Based both on wide- and near-angle observations, Meissner (1966, 1967a, 1967b, 1976) discussed the increased reflectivity of the lower crust. In a later, more detailed paper, Meissner (1973) described the "Moho" as a transition zone. Often sparse wide-angle reflection data in many continental areas were used to derive an overall velocity gradient of a few kilometers. From denser near-vertical and subcritical reflection data, Meissner deduced a stepwise character of the gradient zone. From wide-angle P_MP and P_cP arrivals in the Rhenish Massif (LO-profile in Fig. 6.2.2-05), showing a nearly perfect correlation and interpreted by Meissner et al. (1976a), Meissner has plotted vertical reflections versus the velocity-depth function (Fig. 6.2.5-03).

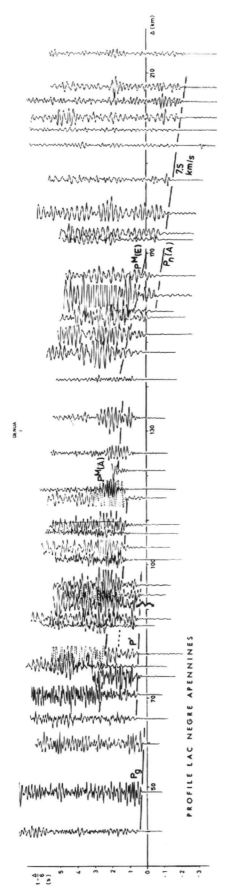

Figure 6.2.4-07. Record section of the profile from Lac Nègre, France, across the Apennines of central Italy (from Morelli et al., 1977, fig. 7.2). [Bollettino di Geofisica Teorica e Applicata, v. 19, p. 199–260. Reproduced by permission of Bollettino di Geofisica, Trieste, Italy.]

The Moho as well as the boundaries CD (Conrad discontinuity) and SCD (sub-Conrad discontinuity) all reflect energy in the near-vertical and the wide-angle range (Fig. 6.2.5-03). The overall picture from different seismic investigations consequently showed the Moho as a laminar transition zone of a few kilometers thick in which the general stepwise increase of the velocity may be interrupted by layers with lower velocities until values around 8 km/s for P-waves are reached in the uppermost part of the upper mantle. As possible petrological explanations, these features may deal with layers of partial melts, crystallization seams, intrusions, and some peeling of mantle matter (Meissner, 1973).

The new computer technology available since the mid-1960s enabled sophisticated traveltime routines by programming the formulas for multi-layered structures, so that models could be checked by recalculating the theoretical traveltimes and comparing them with the observations. Gerhard Müller and Karl Fuchs were not satisfied with traveltime calculations alone, but started to develop methods to calculate synthetic seismograms. A first approach was published by Fuchs (1968).

Initially, K. Fuchs and G. Müller, working separately and independently at different places and having different and common forerunners and academic teachers, developed the basic ideas for the reflectivity method (e.g., Fuchs, 1968, 1969; Müller, 1970), which successfully enabled the calculation of the dynamics of the seismic wave field, i.e., amplitudes, frequency contents and reverberations. Subsequently, they jointly programmed and published the method (Fuchs and Müller, 1971; Müller and Fuchs, 1976), which for the first time allowed the computation of whole synthetic seismograms allowing more severe testing of crustal models (Fig. 6.2.5-04).

A very different approach to interpreting seismic data was the time-term method. It was first suggested by Scheidegger and Willmore (1957) and elaborated further by Willmore and Bancroft (1960) to be applied to seismic-refraction data. This method was described in some detail both by Berry and West (1966) and Smith et al. (1966) and was successfully applied to various data sets in Britain and surroundings and for the interpretation of the first-arrival data of the 1963 Lake Superior experiment, which is described in some detail in Section 6.5.2.

6.3. THE INTERNATIONAL DEEP SEISMIC SOUNDING INVESTIGATION OF SOUTHEASTERN EUROPE

6.3.1. Introduction

In 1963, a large seismic crustal project was initiated in eastern and southeastern Europe (Fig. 6.3.1-01), and 13 international profiles were planned traversing southern Russia, the Ukraine, the Black Sea, Bulgaria, Romania, Yugoslavia, Hungary, Czechoslovakia, Poland, and East Germany (Sollogub, 1969). In 1971, five of the international profiles were more or less completed and ready for a first publication (Sollogub et al., 1972, 1973a, 1973b; Vogel, 1971; Radulescu and Pompilian, 1991).

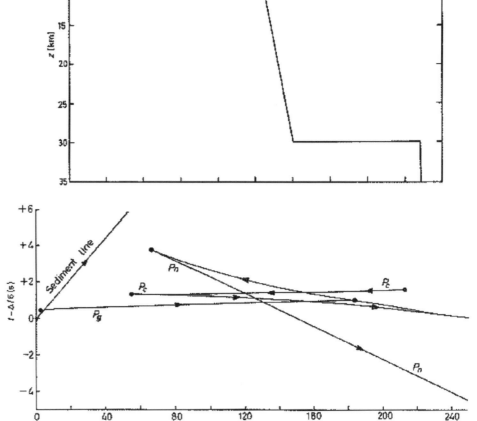

Figure 6.2.5-01. Top: New Crustal Model proposed by Mueller and Landisman (1966) illustrating a well-developed sialic low-velocity zone overlying an abrupt velocity increase. Bottom: Reduced traveltimes for New Crustal Model (from Mueller and Landisman, 1966, fig. 2). [Geophysical Journal of the Royal Astronomical Society, v. 10, p. 525–538. Copyright John Wiley & Sons Ltd.]

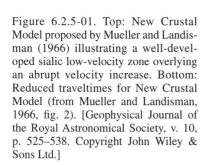

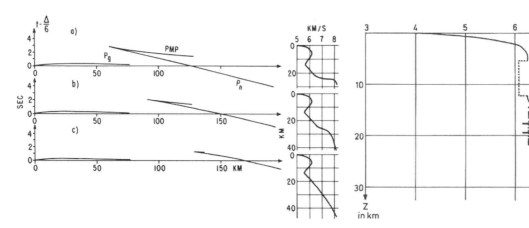

Figure 6.2.5-02. Traveltime diagrams and corresponding models for different structures of the crust-mantle boundary (after Giese, 1968, from Prodehl, 1977, fig. 1). [*In* Heacock, J.G., ed., The earth's crust: American Geophysical Union Geophysical Monograph 20, p. 349–369. Reproduced by permission of American Geophysical Union.]

Figure 6.2.5-03. Near-vertical reflections (horizontal bars) in a velocity-depth curve from the Rhenish Massif (profile LO in Fig. 6.2.2-01) (from Meissner, 1973, fig. 10). Discontinuities: CD = Conrad, SCD = sub-Conrad, MD = Moho. [Geophysical Surveys, v. 1, p. 195–216. Reproduced by kind permission of the author.]

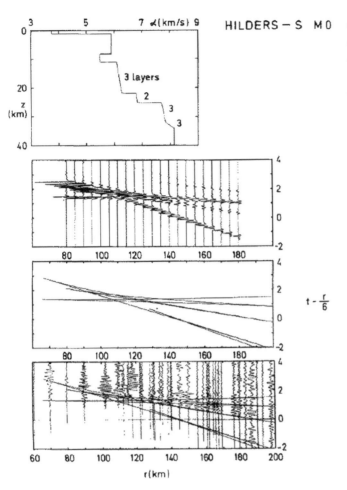

HILDERS – S M O

Figure 6.2.5-04. Velocity-depth function, synthetic seismograms (vertical displacement), traveltime curves and traveltime curves overlain on the observed seismograms shown for the example of the profile from Hilders-S (shotpoint 02 in Fig. 6.2.2-01). Only reflections from the bottom of the layers are included (from Müller and Fuchs, 1976, fig. 3). [*In* Giese, P., Prodehl, C., and Stein, A., eds., Explosion seismology in central Europe—data and results: Berlin-Heidelberg-New York, Springer, p. 178–188. Reproduced with kind permission of Springer Science+Business Media.]

The length of the individual profiles varied from 100–110 km to 160–180 km, with shotpoint separations of 20–25 km for the shorter profiles and 40–50 km for the longer lines (Subbotin et al., 1968; Sollogub, 1969; Sollogub et al., 1972). Figure 6.3.1-02 shows a typical seismogram, recorded at 233 km distance. The phase correlation allowed the deduction of detailed traveltime diagrams (see, e.g., Fig. 6.3.3-02), with P^s identifying sedimentary phases and P_g, with P^c and P^M, identifying intracrustal phases (intracrustal waves) from the Moho.

As can be seen in Figure 6.3.1-01, besides the southeastern USSR, other countries in central-eastern and southeastern Europe were also covered fairly well by these international profiles, performance and interpretation of which was executed by the scientists of those countries and discussed in individual

chapters of a monograph *The Crustal Structure of Central and Southeastern Europe Based on the Results of Explosion Seismology*, edited by V.B. Sollogub, D. Prosen, and H. Militzer, and published in English in 1972 (Sollogub et al., 1972; see Appendix A6-5).

The fieldwork and the interpretation of the data followed in general a unique style, which was proposed by the Soviet scientists, but details of the organization and realization remained the responsibility of the scientists of each country. The equipment, usually national constructions, and the style of shooting also differed from country to country.

6.3.2. The 1960s in Poland

Jan Uchmann was the first to start explosion seismology in Poland. Aided by Alexander Guterch, he planned in the late 1950s a 485 long seismic N-S traverse through Poland, starting at the Baltic Sea in the Gdansk Bay and ending in the West Carpathians near Raciborz (line A in Fig. 6.3.2-01). The project was finally realized in 1960 and 1961 (Guterch et al., 1973). Eighteen prospecting multi-channel seismic stations allowed an average station spacing of 25 km by recording explosions from three shotpoints. In the north, underwater explosions were carried out in the Gdansk Bay, the southern shotpoint was an old water-filled quarry, and a middle shotpoint was placed in the Konin area (Wojtczak-Gadomska et al., 1964).

This first seismic reconnaissance survey was followed in 1963 and 1964 by a second line B, 242 km long, running in an E-W direction in southern Poland (line B in Fig. 6.3.2-01), again with three shotpoints, the western one of which being the same as used for the N-S line. This time, 15 stations were used. On both profiles, the charges ranged from 500 to 3000 kg (Wojtczak-Gadomska et al., 1964; Guterch et al., 1967).

Since 1965, continuous profiling was introduced using 60-channel recording equipment and geophone spacing of 100 m. This method was first applied on the northeastern section of profile A, and later on a third, 136-km-long profile (line C in Fig. 6.3.2-01). This profile was recorded south of Warsaw, applying continuous profiling with shots from the two end points. The shotpoints consisted of multiple boreholes with 60 kg per hole, totaling a 600–800 kg charge size (Betlej et al., 1967).

In the middle of the 1960s, a special investigation of the fore-Sudetic region was started, aiming primarily to study the sedimentary cover for location of possible oil reserves. A dense network of seismic-refraction lines was carried out with the goal of determining the morphology of the basement (Toporkiewicz, 1986). Most of the lines were recorded in a SW-NE direction, but some were also recorded parallel to the strike of the Tornquist-Teisseyre Tectonic Zone (TTZ). From 1967 onward, magnetic tape recording equipment was used, with distances of 100 m between channels. Shotpoints on the individual lines were ~12 km apart. Shots were fired in grouped boreholes at 40–45 m depth, with charge sizes from 30 to 600 kg. The so-called M-lines were the longest of this dense

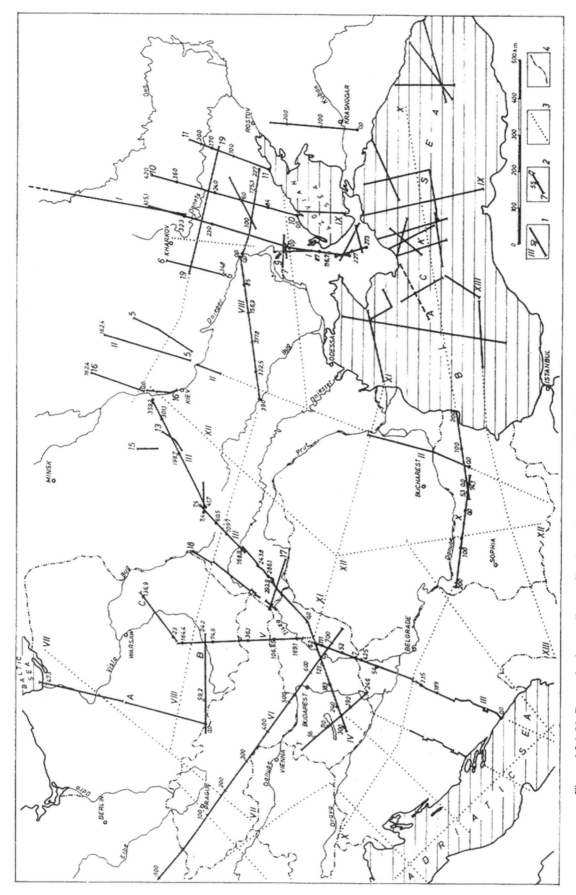

Figure 6.3.1-01. Deep seismic sounding profiles in southeastern Europe (from Sollogub et al., 1972, fig. 1). Solid lines: completed until 1969, dotted lines: planned. [*In* Crustal structure of central and southeastern Europe based on the results of explosion seismology (in Russian). English translation edited by György, S., 1972, Geophysical Transactions, spec. ed., Müszaki Könyvkiadó, Budapest, 172 p. Reproduced by permission of Eötvös Lorand Geophysical Institute of Hungary.]

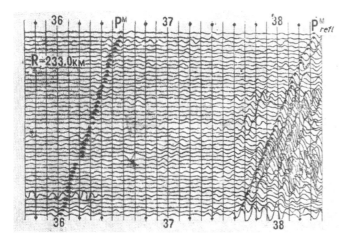

Figure 6.3.1-02. Sample recording showing the refracted P_n (here named P^M) and reflected P_MP phases, observed on the Ukrainian Shield (from Sollogub et al., 1972, fig. 11). [*In* Sollogub, V.B., Prosen, D., and Militzer, H., eds., 1971, Crustal structure of central and southeastern Europe based on the results of explosion seismology (in Russian). English translation edited by György, S., 1972, Geophysical Transactions, spec. ed., Müszaki Könyvkiado, Budapest, 172 p. Reproduced by permission of Eötvös Lorand Geophysical Institute of Hungary.]

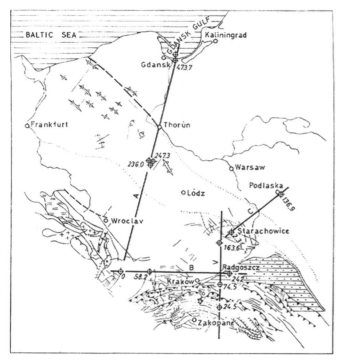

Figure 6.3.2-01. Deep seismic sounding profiles performed in Poland until 1969 (from Sollogub et al., 1972, fig. 29). Stars in circles: shotpoints. The unnamed N-S line in the southeast, leading into the Carpathians, is the International Profile V. [*In* Crustal structure of central and southeastern Europe based on the results of explosion seismology (in Russian). English translation edited by György, S., 1972, Geophysical Transactions, spec. ed., Müszaki Könyvkiado, Budapest, 172 p. Reproduced by permission of Eötvös Lorand Geophysical Institute of Hungary.]

network of lines and served to investigate the fore-Sudetic Monocline, which is part of the Paleozoic platform (Toporkiewicz, 1986; Guterch et al., 1991a). The investigations extended far into the 1970s. This survey therefore will be discussed in more detail in Chapter 7.

Finally, in close cooperation with Russian, Czech, and Hungarian scientists, under the heading of V.B. Sollogub, Kiev, the so-called International Profile V was planned and performed in 1969 (unnamed N-S line in Fig. 6.3.2-01). It started at the East European platform on Polish territory, crossed the western Carpathians, and finally ended in the Hungarian lowlands (Uchman, 1972, 1973). The combined interpretation of this line, as published by Beránek et al. (1972b), is shown in Figure 6.3.3-02.

6.3.3. International Profiles through Czechoslovakia

For Czechoslovakia, three international profiles—V, VI, and VII—had been planned, but in the 1960s, only the two lines V and VI were completed (Fig. 6.3.3-01). Profile VI traversed the southwestern part of the country and continued into eastern Germany. Profile V traversed the Carpathians in the Slovakian part, the interpretation of which is shown in Figure 6.3.3-02 indicating a deep crustal root under the Carpathians at least for southern Poland. Two institutions were responsible for the organization of the fieldwork: the Institute for Applied Geophysics at Brno and the Central Geological Institute at Prague (Beránek et al., 1972a).

6.3.4. Seismic Research in Eastern Germany in the 1960s

One of the Czech lines, profile VI, continued into East Germany, the German Democratic Republic (Knothe and Schröder, 1972). Here, from 1960 to 1967, quarry blasts had been used to study the Earth's crust, mainly along profiles, with scattered observations and different techniques. The work was mainly organized by the Technical University of Freiberg (School of Mines) and the National Geophysical Company at Leipzig. Between 1963 and 1967, this company recorded a few profiles in northeastern Germany, which mainly served the search for hydrocarbons but also obtained some random data from deeper crustal horizons. In 1964, a deep-seismic-reflection survey had also been integrated into the crustal study (Schumann and Oesberg, 1966). The fieldwork along the International Profile VI (Fig. 6.3.1-01) started in 1966 and was completed in 1968. In total, five shotpoints in Germany and in the western part of Czechoslovakia were used to prolong the line by 200 km into eastern Germany. The charges were between 800 and 1200 kg and were either detonated in 40–50 m deep boreholes (1966) or in 4–5 m deep water reservoirs (1968). The system of recording was continuous in-line profiling with 150 m spacing of seismometers, using 8-channel recorders. Responsible for the realization and a first interpretation of the data was H. Knothe and co-workers of the University of Freiberg (Knothe and Walther, 1968).

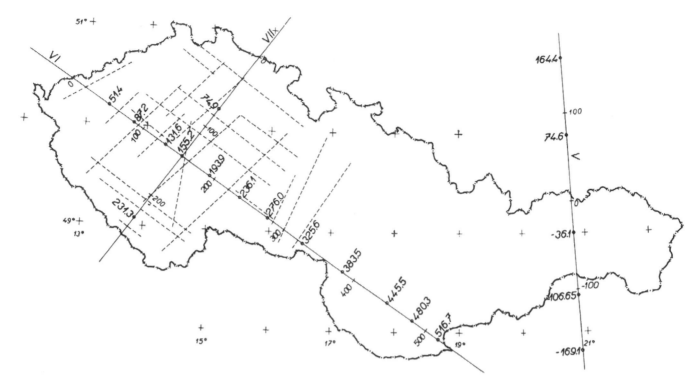

Figure 6.3.3-01. International Profiles recorded on the territory of the CSSR (from Sollogub et al., 1972, fig. 38). [*In* Crustal structure of central and southeastern Europe based on the results of explosion seismology (in Russian). English translation edited by György, S., 1972, Geophysical Transactions, spec. ed., Müszaki Könyvkiado, Budapest, 172 p. Reproduced by permission of Eötvös Lorand Geophysical Institute of Hungary.]

6.3.5. International Profiles through Hungary

In Hungary, four of the International Profiles (III, IV, V, and VI; Fig. 6.3.1-01), crossed each other in order to investigate the anomalous thin crust in much detail (Mituch and Posgay, 1972). As an example of the interpretation of the data, the model along the International Profile V, recorded across the Carpathians from Poland into Hungary, is shown above (Fig. 6.3.3-02).

Already in 1958, a 120-km-long profile had been recorded along the later International Profile VI, but the isolated recordings had not allowed a continuous tracing of phases. Besides the four international profiles, a national NW-SE–running line was also added in western Hungary, traversing Transdanubia from Sopron via Lake Balaton toward the Danube River. Besides in-line continuous and point wise recording, broadside profiling also was applied where two parallel lines, spaced at 60 km, were covered with shotpoints and recording stations. Here, all shots from one line were recorded by four spreads of 2.2 km length each on the other line, resulting in a total length of 8.8 km per shotpoint. The imaginary intermediate profile corresponded to the International Profile. The leading scientist was K. Posgay of ELGI, Budapest.

A preliminary crustal thickness contour map of Hungary constructed from these data showed that the Moho depth ranged between less than 25 km to 28 km, with only one exception north of Lake Balaton in northwestern Hungary (Fig. 6.3.5-01).

6.3.6. International Profile III through the Dinarides of Yugoslavia

The International Profile III (Fig. 6.3.1-01) was continued through the Dinarides of central Yugoslavia to the Adriatic coast (Prosen et al., 1972). Preparations started in 1964 and the fieldwork was realized in 1968 with continuous in-line profiling (150 m geophone spacing) as far as geography allowed, from one single underwater shotpoint only, with charges detonated at 50–100 m depth ranging in size from 100 to 1400 kg, depending on the observation distance.

The project was continued in 1970 with more detailed observations from additional borehole shotpoints. The preliminary interpretation showed a 32-km-thick crust near the Adriatic coast and a crustal root of 52 km under the Outer Dinarides. Further inland, the crustal model showed a gradually thinning to near 30 km in thickness (Fig. 6.3.6-01).

6.3.7. International Profiles through Bulgaria

The territory of Bulgaria was investigated by two crossing lines, the International Profile X running for 550 km distance in an E-W direction from the Black Sea to the Yugoslavian border, and a Profile II, running from the Danube delta in Romania to the south (Fig. 6.3.1-01), which by the end of the 1960s terminated at the intersection with Profile X (Dachev et al., 1972).

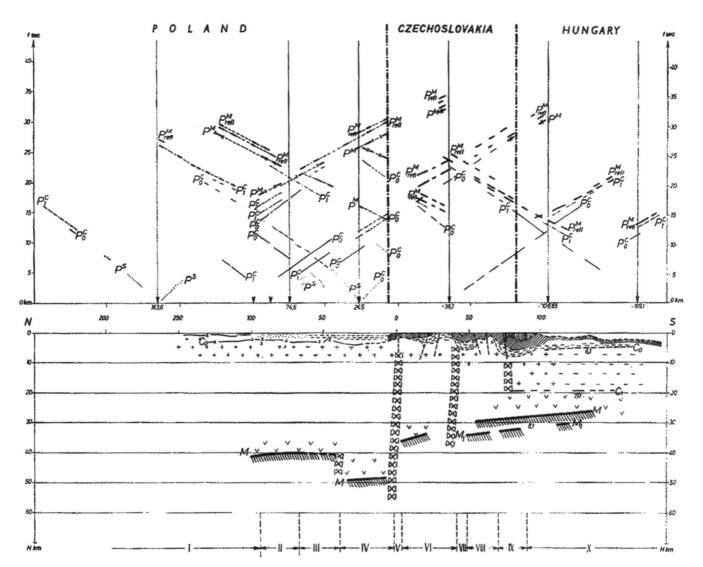

Figure 6.3.3-02. Interpretation of the DSS data of the International Profile V, crossing the Carpathians and reaching from Poland into Hungary (for location see Figs. 6.3.1-01 and 6.3.3-01, from Sollogub et al., 1972, fig. 72). [*In* Sollogub, V.B., Prosen, D., and Militzer, H., eds., 1971, Crustal structure of central and southeastern Europe based on the results of explosion seismology (in Russian). English translation edited by György, S., 1972, Geophysical Transactions, spec. ed., Müszaki Könyvkiado, Budapest, 172 p. Reproduced by permission of Eötvös Lorand Geophysical Institute of Hungary.]

Three institutions were responsible for organizing the project: the Geophysical and Geological Mapping Company, the Physics Department of the University of Sofia, and the Geological Board at Sofia. Small shots spaced between 6.5 and 12 km served to continuously trace the basement, while seven unevenly spaced large shots supplied energy up to 140 km away to obtain seismic waves penetrating the whole crust. The resulting preliminary cross section of Profile X (Fig. 6.3.7-01) showed only scattered pieces of Moho at around 30–37 km depth.

6.3.8. International Profiles through Romania

Following some first crustal structure experiments in 1962 in the central Dobrogea of Romania, fieldwork on the International Profile II in Romania (Fig. 6.3.1-01) was performed in 1967–

1970, in cooperation with Soviet and Bulgarian scientists (Constantinescu and Cornea, 1972). The line was mainly observed with continuous profiling, here and there with isolated (critical) soundings to obtain Conrad and Moho arrivals from shotpoints 40–45 km away. The preliminary results showed a 30–35-km-thick crust in the southern part of southeastern Romania, and a 40–45-km-thick crustal section in the northern part, separated by a deep fault (Fig. 6.3.8-01). First point wise observations were also made on Profile XI at the end of the 1960s, which was planned to traverse Romania in ESE-WNW direction.

6.4. SEISMIC CRUSTAL STUDIES IN THE USSR

In the USSR in 1948–1955, G.A. Gamburtsev had designed the deep seismic sounding method which is based on

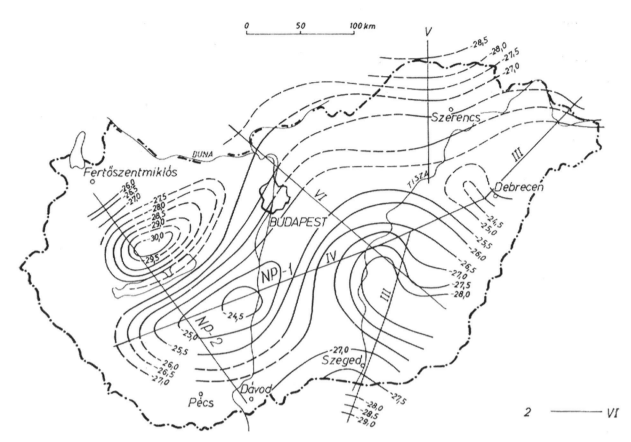

Figure 6.3.5-01. Moho contour map for Hungary and location of seismic DSS profiles, from Sollogub et al., 1972, fig. 70). [*In* Sollogub, V.B., Prosen, D., and Militzer, H., eds., 1971, Crustal structure of central and southeastern Europe based on the results of explosion seismology (in Russian). English translation edited by György, S., 1972, Geophysical Transactions, spec. ed., Müszaki Könyvkiado, Budapest, 172 p. Reproduced by permission of Eötvös Lorand Geophysical Institute of Hungary.]

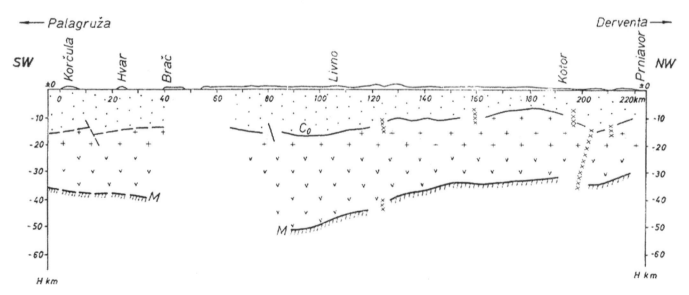

Figure 6.3.6-01. Interpretation of the DSS data of the Yugoslavian part of the International Profile III, across the Dinarides (for location see Fig. 6.3.1-01, from Sollogub et al., 1972, fig. 48). [*In* Sollogub, V.B., Prosen, D., and Militzer, H., eds., 1971, Crustal structure of central and southeastern Europe based on the results of explosion seismology (in Russian). English translation edited by György, S., 1972, Geophysical Transactions, spec. ed., Müszaki Könyvkiado, Budapest, 172 p. Reproduced by permission of Eötvös Lorand Geophysical Institute of Hungary.]

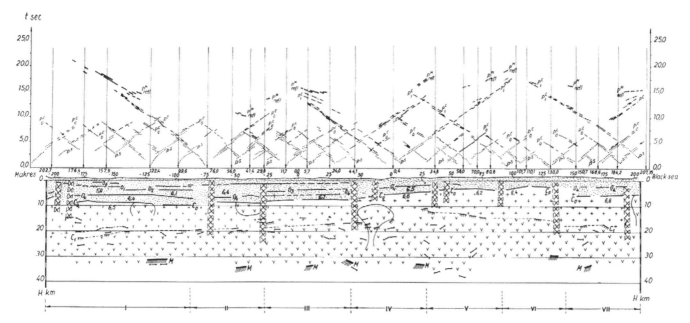

Figure 6.3.7-01. Interpretation of the DSS data of the E-W–directed International Profile X through northern Bulgaria (from Sollogub et al., 1972, fig. 52, for location see Fig. 6.3.1-01). [*In* Sollogub, V.B., Prosen, D., and Militzer, H., eds., 1971, Crustal structure of central and southeastern Europe based on the results of explosion seismology (in Russian). English translation edited by György, S., 1972, Geophysical Transactions, spec. ed., Müszaki Könyvkiado, Budapest, 172 p. Reproduced by permission of Eötvös Lorand Geophysical Institute of Hungary.]

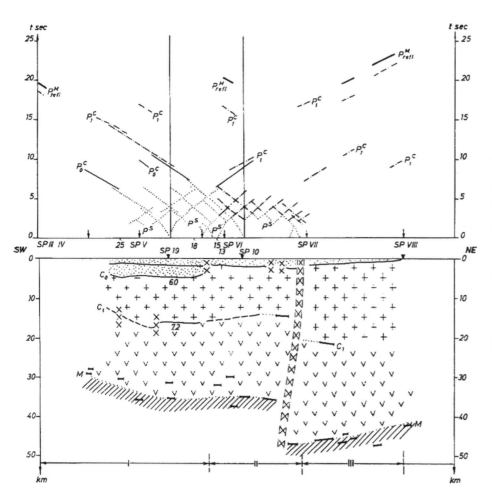

Figure 6.3.8-01. Interpretation of the DSS data of the Romanian part of the International Profile II, across the Dobrogea (for location see Fig. 6.3.1-01; from Sollogub et al., 1972, fig. 56). [*In* Sollogub, V.B., Prosen, D., and Militzer, H., eds., 1971, Crustal structure of central and southeastern Europe based on the results of explosion seismology (in Russian). English translation edited by György, S., 1972, Geophysical Transactions, spec. ed., Müszaki Könyvkiado, Budapest, 172 p. Reproduced by permission of Eötvös Lorand Geophysical Institute of Hungary.]

the correlation of seismic waves generated by small explosions (100 kg charges). The optimal frequency range to produce a depth range of 40–50 km for land investigations that used 1–3 ton charges turned out to be 5–15 Hz. Applying this methodology, seismic profiles of several hundred kilometers in length were recorded across different tectonic zones (Fig. 6.4-01). The explosions were recorded by multichannel seismic prospecting stations with a frequency range of 5–30 Hz using 3–5 Hz seismometers (Kosminskaya et al., 1969). As far as possible, reversed and overlapping observations were obtained, with observation distances ranging between 200 and 300 km and shotpoint intervals of 50–100 km.

As already described in Section 6.3, the crustal profiles in southern Russia and the Ukraine were part of the large international seismic crustal project, which was initiated in 1963 by V.B. Sollogub and co-workers (Fig. 6.3.1-01) and which consisted of 13 international profiles traversing southern Russia, the Ukraine, the Black Sea, Bulgaria, Romania, Yugoslavia, Hungary, Czechoslovakia, Poland, and East Germany (Subbotin et al., 1968; Sollogub et al., 1972; Appendix 6.5), as discussed in some detail in Section 6.3.

As can be seen in Figure 6.3.1-01, the majority of profiles covered the Ukrainian Shield and the Black Sea and delivered a very detailed picture of crustal structure for this part of the USSR (Sollogub and Chekunov, 1972). The recording equipment used in the Ukraine and southern Russia consisted of geophones and low-frequency 60-channel seismic stations with amplifiers allowing the recording of a frequency range

of 6–15 Hz (Sollogub, 1969). A typical data set was shown in Figure 6.3.1-02.

A good example, which also shows the complex crustal structure, as interpreted at this time, is the national profile 10, running parallel to International Profile I to the east (Figs. 6.3.1-01 and 6.4-02), crossing the Ukrainian Shield and the Dniepr-Don aulacogen north of the Azovian Sea. More data examples, to illustrate the recorded phases and their interpretations, were published by Subbotin et al. (1968). The corresponding figures are reproduced in Appendix A6-5-2.

In order to compare the data, usually obtained with a geophone spacing of 100 m, and their interpretation of the Russian deep seismic sounding method with the data and interpretations obtained in western Europe at the same time, but with a much wider sensor spacing of 3–5 km in average, Jentsch (1979) has digitized the data of the Russian national profile 10 and presented as record sections with differing sensor spacing: 3 km, 1 km, and 200 m (Fig. 6.4-02, Appendix A6-5-3).

It was obvious that the key phases (reflections from the Moho, for example) could be detected equally well, but that details such as deep crustal faults recognized in the dense data sets with 100 m spacing could not be resolved by station spacings of 1 km or larger. The main crustal boundaries were the Conrad and the Moho discontinuities, but due to the high observation density along parts of the recording lines, the main seismic interfaces were interpreted by Sollogub and Chekunov (1972) as unconformable, i.e., deep reaching faults were introduced which penetrated the whole crust (Fig. 6.4-03).

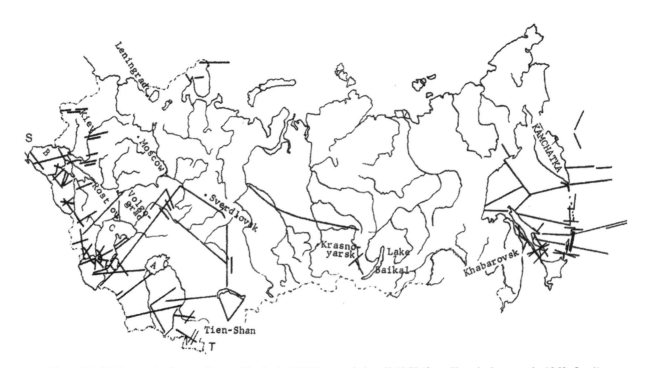

Figure 6.4-01. Deep seismic sounding profiles in the USSR, recorded until 1967 (from Kosminskaya et al., 1969, fig. 1). A—Aral Sea, B—Black Sea, C—Caspian Sea. [*In* Hart, P.J., ed., The Earth's crust and upper mantle: American Geophysical Union, Geophysical Monograph 13, p. 195–208. Reproduced by permission of American Geophysical Union.]

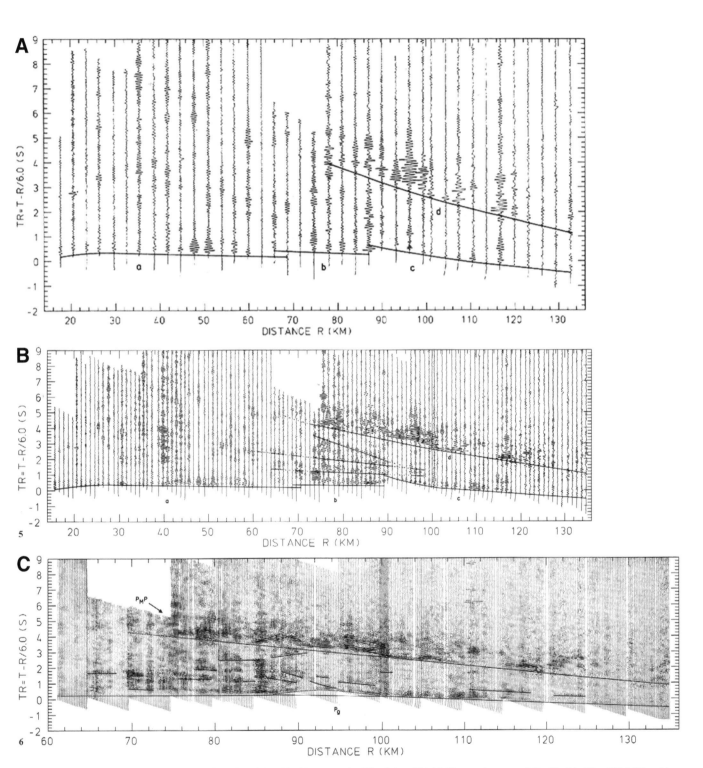

Figure 6.4-02. Record sections of the DSS National Profile 10 across the Ukrainian Shield (line to the east of Profile I in Fig. 6.3.1-01, with a mean spacing of 3 km (A), 1 km (B) and 200 m (C), from Jentsch, 1979, figs. 3, 5, and 6, selected from an original data set of Sollogub et al., 1972). [Journal of Geophysics, v. 45, p. 355–372. Reproduced with kind permission of Springer Science+Business Media.]

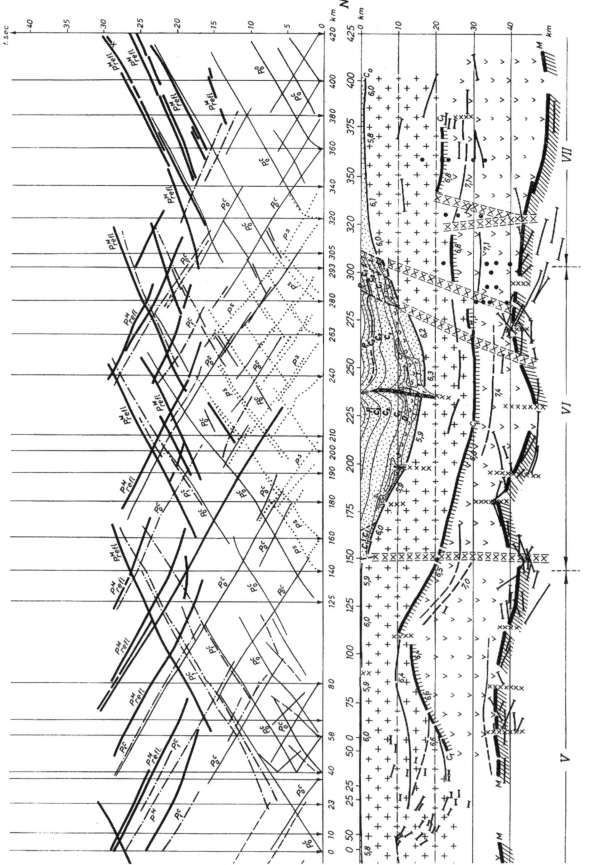

Figure 6.4-03. Interpretation of the DSS data of the part of the National Profile 10 across the Ukrainian Shield (line to the east of Profile I in Fig. 6.3.1-01, from Sollogub et al., 1972, fig. 24). [In Sollogub, V.B., Prosen, D., and Militzer, H., eds., 1971, Crustal structure of central and southeastern Europe based on the results of explosion seismology (in Russian). English translation edited by György, S., 1972, Geophysical Transactions, spec. ed., Müszaki Könyvkiado, Budapest, 172 p. Reproduced by permission of Eötvös Lorand Geophysical Institute of Hungary.]

The Moho contour map (Fig. 6.4-04) of Sollogub and Chekunov (1972) displays best the complex structure of the whole area of southern Russia and the Ukraine and wider surroundings, as resulting in the interpretations of the Soviet scientists until the end of the 1960s.

As far as possible, the method of continuous profiling was also applied in other parts of the USSR. Seismic profiles of several 100 km in length were recorded across different tectonic zones, such as old shields, e.g., the Ukrainian (described in some detail above) and Baltic Shields, old and young platforms, mountain regions, sub-montane basins, graben structures, and continental margin areas. Figure 6.4-05 shows a data example recorded in the Fergana intermontane basin, and Figure 6.4-06 shows the observation system and the corresponding interpreted traveltime curves of one of the profiles recorded in the Kopetdag–Central Kyzulkum area. The corresponding distances and traveltimes are included as a table.

In some regions of the Soviet Union, where a complex relief prohibited continuous profiling, a piece-continuous profiling was applied, i.e., the observations were restricted to discrete intervals. In the almost inaccessible taiga regions of Siberia and in the Far East, a special point sounding method was used (Puzyrev et al., 1973, 1978). Here, one explosion was recorded at one point only at a distance optimal for observing a single or several prominent deep-penetrating waves.

Deep-sea investigations of Neprochnov et al. (1970) dealt with the structure of crust and upper mantle under the Black and Caspian Seas. The cross section from the Black Sea to the Caspian Sea (Fig. 6.4-07A) and its continuation to the Tien Shan (Fig. 6.4-07B), which also crosses the Caucasus mountains, shows a summary of all seismic data obtained for the southwestern USSR until the end of the 1960s.

The overall complex structure of the USSR also becomes apparent in Figures 6.4-08 and 6.4-09, where Figure 6.4-08 shows

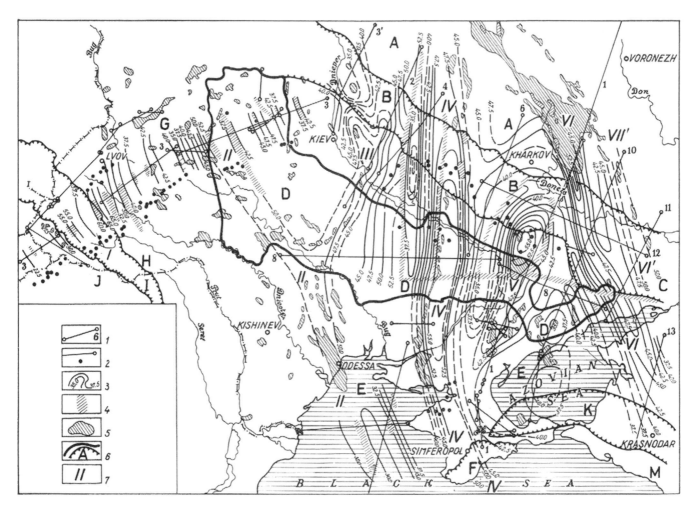

Figure 6.4-04. Crustal thickness map for the Ukraine and the European part of southern Russia and adjacent areas (from Sollogub et al., 1972, fig. 25). [*In* Sollogub, V.B., Prosen, D., and Militzer, H., eds., 1971, Crustal structure of central and southeastern Europe based on the results of explosion seismology (in Russian). English translation edited by György, S., 1972, Geophysical Transactions, spec. ed., Müszaki Könyvkiado, Budapest, 172 p. Reproduced by permission of Eötvös Lorand Geophysical Institute of Hungary.]

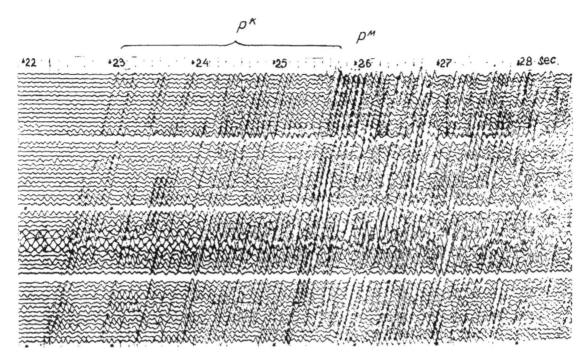

Figure 6.4-05. Seismogram observed in the Fergana intermontane basin at 160 km shotpoint distance (from Kosminskaya et al., 1969, fig. 2). P^k crustal phases, P^M mantle phases. [*In* Hart, P.J., ed., The Earth's crust and upper mantle: American Geophysical Union, Geophysical Monograph 13, p. 195–208. Reproduced by permission of American Geophysical Union.]

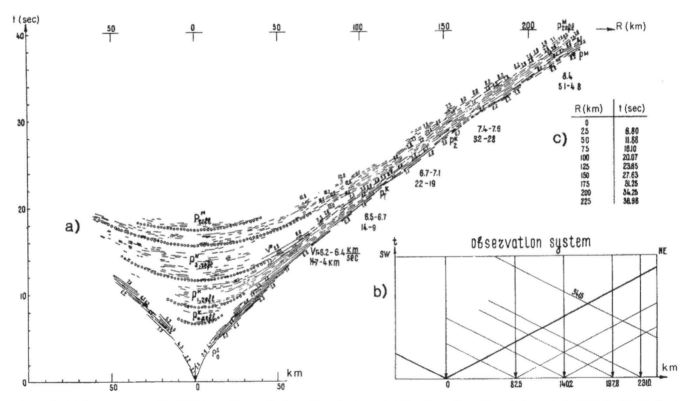

Figure 6.4-06. Interpretation of the DSS data of the profile from Kopetdag to central Kyzylkum (from Kosminskaya et al., 1969, fig. 3). (a) phase travel-time curves of deep waves, showing reflected (index: refl) and refracted P-waves (indices 1, 2 for intracrustal phases k, index M for Pn phase); (b) observation system; (c) average travel times for the first arrivals. [*In* Hart, P.J., ed., The Earth's crust and upper mantle: American Geophysical Union, Geophysical Monograph 13, p. 195–208. Reproduced by permission of American Geophysical Union.]

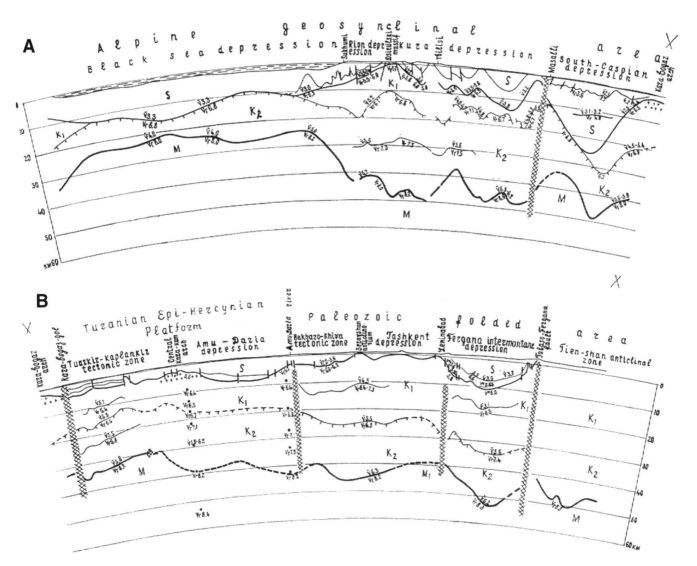

Figure 6.4-07. Section of the Earth's crust from the Black Sea through the Caucasus and the Caspian Sea to the central Kara-Kun-Tien Shan (from Kosminskaya et al., 1969, fig. 8). S—sedimentary layers, K—crustal layers, M—Moho. Velocities in km/s. [*In* Hart, P.J., ed., The Earth's crust and upper mantle: American Geophysical Union, Geophysical Monograph 13, p. 195–208. Reproduced by permission of American Geophysical Union.]

generalized crustal models for different tectonic areas throughout the USSR, and Figure 6.4-09 shows a crustal thickness contour map for the USSR and adjacent areas. Other transcontinental crustal sections throughout the USSR and adjacent areas were also summarized and published, e.g., by Belyaevsky et al. (1968).

6.5. SEISMIC-REFRACTION INVESTIGATIONS IN NORTH AMERICA IN THE 1960s

6.5.1. Western United States

In 1960, under the VELA UNIFORM program of the Advanced Research Projects Agency of the U.S. Department of Defense, the U.S. Geological Survey at Denver, Colorado, re-

ceived major funding to start a large program of crustal studies, headed by L.C. Pakiser.

At first, staff had to be organized, largely by reassignment of available personnel, which involved scientists such as W.H. Diment, J.P. Eaton, J.H. Healy, D.P. Hill, W.H. Jackson, J.C. Roller, A. Ryall, S.W. Stewart, D.J. Stuart, R.W. Warrick, and others. In addition, the gravity studies of major crustal features in the western United States were continued and the organization of a new and more extensive program "Study of Seismic Propagation Paths and Regional Traveltimes in the California-Nevada Region" was planned.

Later on, in 1965–1966, the same group of the U.S. Geological Survey, again under the leadership of L.C. Pakiser, moved from Denver, Colorado, to Menlo Park, California, where the

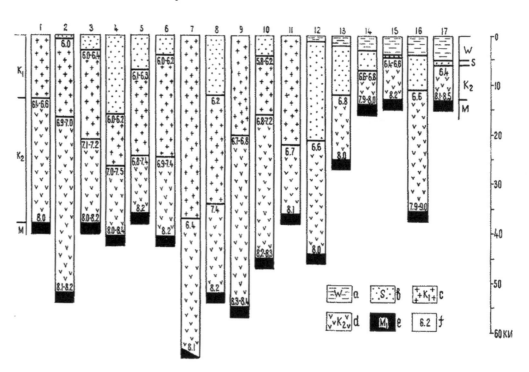

Figure 6.4-08. Generalized crustal models for different tectonic areas in the USSR (from Kosminskaya et al., 1969, fig. 9). (a) water; (b) unconsolidated crust (sediments); (c) consolidated upper crust; (d) consolidated lower crust; (e) upper mantle; (f) boundary velocities in km/s. [*In* Hart, P.J., ed., The Earth's crust and upper mantle: American Geophysical Union, Geophysical Monograph 13, p. 195–208. Reproduced by permission of American Geophysical Union.]

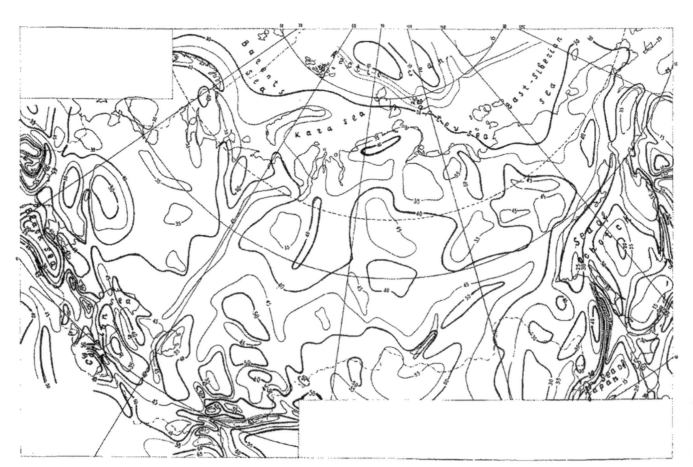

Figure 6.4-09. Crustal thickness map for the territory of the USSR and adjacent areas (from Kosminskaya et al., 1969, fig. 11). [*In* Hart, P.J., ed., The Earth's crust and upper mantle: American Geophysical Union, Geophysical Monograph 13, p. 195–208. Reproduced by permission of American Geophysical Union.]

National Center of Earthquake Research was established to start a major earthquake research program.

Following detailed discussions with scientists and technicians of various laboratories in the United States to design the detailed performance specifications, in 1960 a new seismic-refraction system was developed and constructed by Dresser Electronics, SIE Division (Figs. 6.5.1-01 to 6.5.1-03).

The first seismic system was delivered, tested, and accepted in early 1961, and by the end of 1961, ten units had been placed in operation, each installed in a four-wheel-drive vehicle. In addition, mounted in two aluminum trailers and in a four-wheel-drive master vehicle for use at the shotpoints, communication and timing equipment was developed.

Compatible communications and timing systems were installed in each recording unit. The units (Figs. 6.5.1-02 and 6.5.1-03), recording simultaneously at two levels of amplification on magnetic tape (30 dB separation) and by an oscillograph on photographic paper (15 dB separation), were equipped with 8

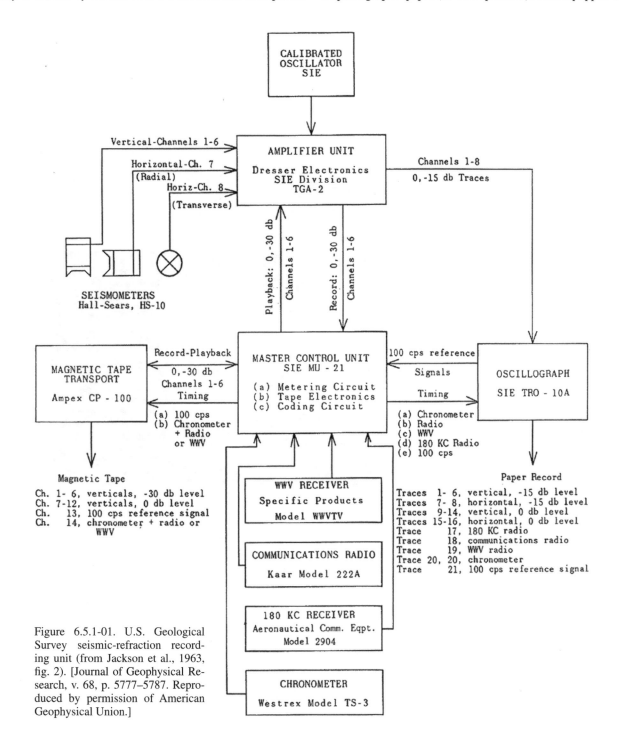

Figure 6.5.1-01. U.S. Geological Survey seismic-refraction recording unit (from Jackson et al., 1963, fig. 2). [Journal of Geophysical Research, v. 68, p. 5777–5787. Reproduced by permission of American Geophysical Union.]

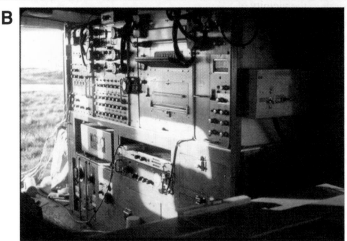

Figure 6.5.1-02. U.S. Geological Survey seismic-refraction equipment installed in a four-wheel drive recording truck. (A: from Warrick et al., 1961, fig. 2). [Geophysics, v. 26, p. 820–824. Permission granted by Society of Exploration Geophysicists, Tulsa, Oklahoma, USA.] (B: Photograph by C. Prodehl.)

seismic and two timing channels, which allowed a 2.5-km-wide array of 6 vertical seismometers (with natural frequencies of 1–2 Hz) with average spacing of 500 m where terrain permitted, while two horizontal seismometers were usually placed in the center. Additional traces were available for recording the time signals of the broadcast station WWV, a calibrated chronometer, and when possible the shot instant. The details of this new seismic system were discussed by Warrick et al. (1961).

The first seismic measurements were made from April to August 1961 along two unreversed profiles in eastern Colorado. Fieldwork was started with one recording unit, and six units were in operation at the end of the Colorado field program. One pro-

file extended north from Nee Granda Reservoir, between Eads and Lamar, Colorado, to the vicinity of Sterling, Colorado. The profile was ~325 km long. Seismic recordings were made at 113 spread locations, and a total of 64 shots, ranging in size from a few to 1500 kg, were exploded in Nee Granda Reservoir and in drill holes along the shore of the reservoir (for location of Lamar, see Fig. 6.5.1-12).

A data example recorded at 227.5 km distance with P_n as first arrivals (P_7) is shown in Figure 6.5.1-04, showing the monitor seismogram (upper two sets) and magnetic-tape playback with filtering (lower two sets) with low (top) and high (bottom) gain traces in each set. Figure 6.5.1-05 shows parts of the result-

Figure 6.5.1-03. U.S. Geological Survey seismic-refraction recording truck, with Sam Stewart laying out an array of 6 seismometers (photograph by C. Prodehl).

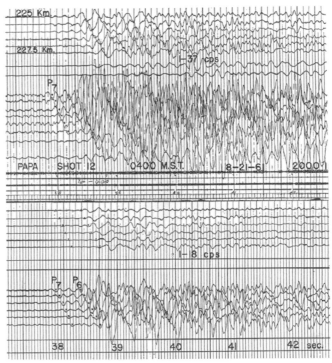

Figure 6.5.1-04. Monitor seismogram (upper two sets of traces) and magnetic-tape playback with filtering (lower two sets) recorded 227.5 km north of Nee Granda Reservoir (Lamar). The upper traces in each set are recorded with low gain, the lower traces with high gain. The 6 upper traces in each set are recordings from vertical seismometers, traces 7 and 8 on monitor only are in-line and transverse horizontal motion (from Jackson et al., 1963, fig. 6). [Journal of Geophysical Research, v. 68, p. 5777–5787. Reproduced by permission of American Geophysical Union.]

ing record section with traveltime curves correlated by Jackson et al. (1963). As seen in Figures 6.5.1-04 and 6.5.1-05, the data achieved by the new equipment at its first field experiment show an extremely good quality.

The second profile in Colorado extended west from Nee Granda Reservoir to the vicinity of Salida, Colorado, with a total length of 290 km. Along this profile, 13 shots in Nee Granda Reservoir were recorded at 54 spreads.

The data of both profiles were later digitized by Dave Warren (unpublished U.S. Geological Survey Open-File Report, data reproduced in Appendix A6-8, together with the 1965 Rocky Mountain profile data; see below).

The seismic-refraction recording of 1961 resulted in 45–50 km crustal thickness under eastern Colorado, as determined from interpretation of the profile from Nee Granda Reservoir to Sterling. However, the fieldwork served primarily to test thoroughly the new seismic-refraction recording, communications, and timing equipment and to train observers from the U.S. Geological Survey and United ElectroDynamics, Inc., in operation of the new seismic system. Furthermore, efficient field procedures had to be developed for the later major program of fieldwork in the California-Nevada region. In the remainder of 1961 and in the following two years, a network of 64 seismic-refraction profiles was recorded by the U.S. Geological Survey in Nevada and California and adjacent areas in Idaho, Wyoming, Utah, and Arizona (Fig. 6.5.1-06). The investigation concentrated mainly on the Great Basin of the Basin and Range Province and the Sierra Nevada. In addition, two recording lines extended into the western Snake River Plain of Idaho and the southern Cascade Range in California. Other profiles were recorded in the Coast Ranges of California, in the Colorado Plateau, and in the middle Rocky Mountains.

Nuclear test sites were also included in the program. From the first nuclear test site in southeastern New Mexico, GNOME, a line had been recorded in the Great Plains northwards, paral-

lel to and east of the front of the southern Rocky Mountains (Stewart and Pakiser, 1962). From SHOAL, Nevada (no. 10 in Fig. 6.5.1-06), a line was recorded eastward through the Basin and Range province and reversed later by chemical shots near Delta, Utah. Finally, the repeated tests at the Nevada Test Site (no. 19 in Fig. 6.5.1-06) allowed the recording of long-range profiles of several 1000 km in length through the western United States in all directions.

Except for the nuclear tests, seismic energy was generated by chemical explosions fired in the Pacific Ocean, in lakes, or in drill holes. Ocean shots ranged from 2000 to 6000 lb and were placed on the sea bottom. The charges in lakes ranged from 2000 to 10,000 lb and were placed on the lake bottoms. At all the other shotpoints, the charges were fired in drill holes and ranged in size from 250 to 20,000 lb. Usually 10 recording units were available, and the average spacing of the recording units was 10 km, so that most of the lines required more than one layout. Only in a few instances were the units placed closer together.

The complete data set was later published in a U.S. Geological Survey Professional Paper in the form of record sections by Prodehl (1979). Data examples are shown for the Sierra Nevada (Fig. 6.5.1-07A), the Basin and Range province and western

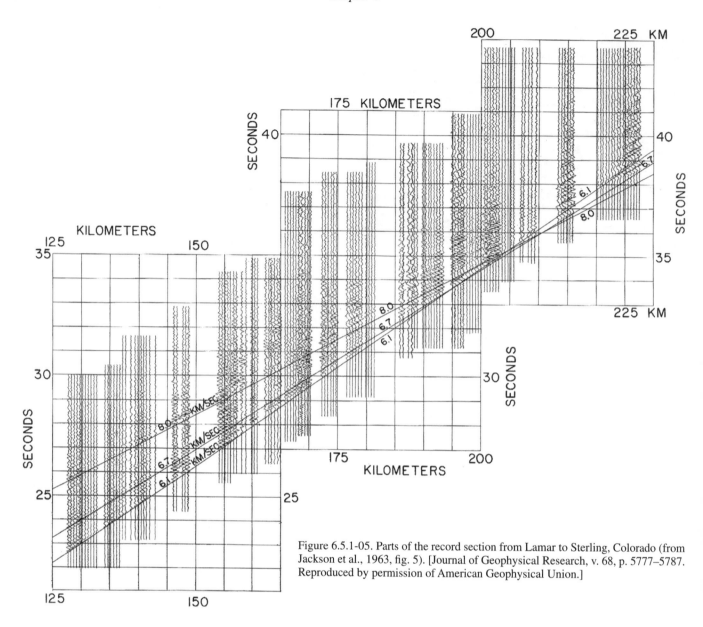

Figure 6.5.1-05. Parts of the record section from Lamar to Sterling, Colorado (from Jackson et al., 1963, fig. 5). [Journal of Geophysical Research, v. 68, p. 5777–5787. Reproduced by permission of American Geophysical Union.]

Snake River Plain (Fig. 6.5.1-07B), the middle Rocky Mountains and the Colorado Plateau (Fig. 6.5.1-07C). All record sections are reproduced in Appendix A6-6 and the corresponding basic data of shotpoint and recording site locations can be found in Appendix A6-7.

First interpretations of parts of this extensive fieldwork of 1961 and 1962, including the earlier line from the Nevada Test Site to Kingman, Arizona (Diment et al., 1961), were presented at a symposium of the American Geophysical Union and subsequently published in a joint volume in 1963 (Journal of Geophysical Research, vol. 68, no. 20, p. 5747–5849), introduced by an overview of Pakiser (1963). Over most of the Basin and Range province, including the Sierra Nevada, P_n velocities of 7.8–7.9 km/s had been obtained (Eaton, 1963; Pakiser and Hill, 1963; Roller and Healy, 1963; Ryall and Stuart, 1963). Along the Cali-

fornia coastline (Healy, 1963), in the Colorado Plateau (Ryall and Stuart, 1963), and in the Great Plains province (Stewart and Pakiser, 1962), P_n velocities ranged between 8.0 and 8.2 km/s.

Two preliminary crustal cross sections in W-E and S-N directions with many question marks had been compiled for this symposium (Fig. 6.5.1-08). These sections showed a fairly thin crust but a questionable existence of an intracrustal boundary for the whole Basin and Range province. Toward the north, a thick intermediate layer underlying a very thin uppermost crust resulted for the Snake River Plain. Toward the west, a crustal root was indicated under the Sierra Nevada, while a thin crust again resulted for most of California.

In the frame of a series of maps published along a transcontinental survey covering the latitudes 35°–39°N (Hart, 1964; Pakiser and Zietz, 1965), Warren (1968a) compiled the most

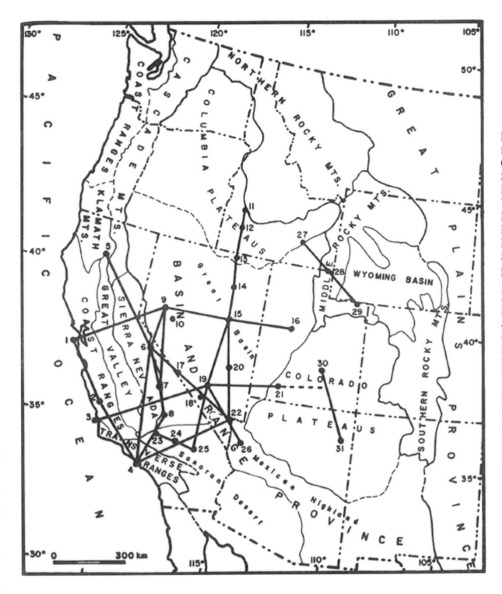

Figure 6.5.1-06. Location of seismic pro-
files recorded in 1961 to 1963 by the U.S.
Geological Survey in the western United
States (from Prodehl, 1970, fig. 1).
MD—Mojave Desert, OV—Owens
Valley, DV—Death Valley, LP—Lassen
Peak National Park. Shotpoints: 1—San
Francisco, 2—Camp Roberts, 3—San
Luis Obispo, 4—Santa Monica Bay, 5—
Shasta Lake, 6—Mono Lake, 7—Inde-
pendence, 8—China Lake, 9—Fallon,
10—SHOAL, 11—Boise, 12—Strike
Reservoir, 13—Mountain City, 14—
Elko, 15—Eureka, 16—Delta, 17—Lida
Junction, 18—Lathrop Wells, 19—NTS
(Nevada Test Site), 20—Hiko, 21—
Navajo Lake, 22—Lake Mead, 23—
Mojave, 24—Barstow, 25—Ludlow,
26—Kingman, 27—American Falls
Reservoir, 28—Bear Lake, 29—Flam-
ing Gorge Reservoir, 30—Hanksville,
31—Chinle. [Geological Society of
America Bulletin, v. 81, p. 2629–2646.
Reproduced by permission of the Geo-
logical Society of America.]

prominent seismic-refraction lines within these latitudes and
constructed preliminary cross sections, which he published in
fence diagrams (see also Healy and Warren, 1969, fig. 1).

For a unified interpretation of the complete set of data recorded by the U.S. Geological Survey in 1961–1963 in the western United States, Prodehl (1970, 1979) used the same methodology as had been proposed and applied by Giese (1968a, 1976b)
for the interpretation of many data in central Europe and the Alps.

Two crustal cross sections along the same lines as those
shown by Pakiser (1963; Fig. 6.5.1-08) are reproduced in Figure 6.5.1-09 and the corresponding record sections for the N-S
line Boise to Lake Mead are shown in Figure 6.5.1-07B. Figure
6.5.1-10 shows a fence diagram for the crustal structure under
California and Nevada and adjacent areas obtained by Prodehl
(1970, 1979) from the reinterpretation of the whole set of data.

In 1964, the U.S. Geological Survey fieldwork was extended to a network of profiles in Arizona across the Tonto Forest

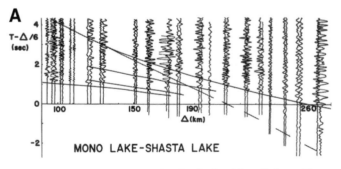

Figure 6.5.1-07. (A) Record section for the line Mono Lake to Shasta
Lake, Sierra Nevada (from Prodehl, 1979, fig. 47). [U.S. Geological
Survey Professional Paper 1034, 74 p. Copyright U.S. Geological Sur-
vey.] (*Continued on following page.*)

B

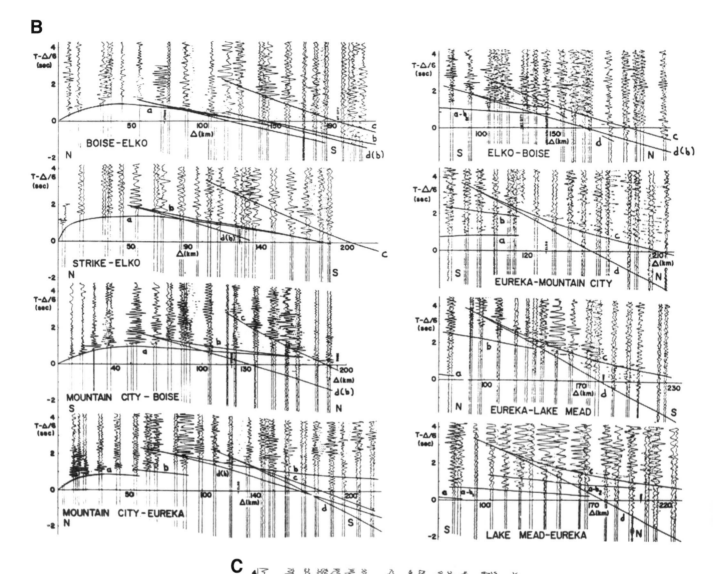

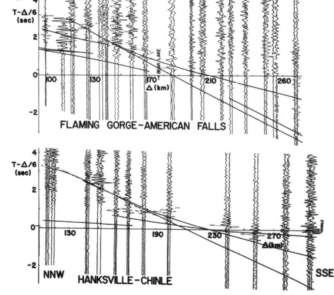

C

Figure 6.5.1-07, *continued*. (B) Record sections for the line Boise, Idaho, Columbia Plateaus, to Lake Mead, Nevada, Basin and Range province (from Prodehl, 1970, figs. 3 and 5). [Geological Society of America Bulletin, v. 81, p. 2629–2646. Reproduced by permission of the Geological Society of America.] (C) Record sections for the lines Flaming Gorge, Utah, to American Falls, Idaho, Middle Rocky Mountains and Hanksville, Utah, to Chinle, Arizona, Colorado Plateau (from Prodehl, 1979, figs. 78 and 74). [U.S. Geological Survey Professional Paper 1034, 74 p. Copyright U.S. Geological Survey.]

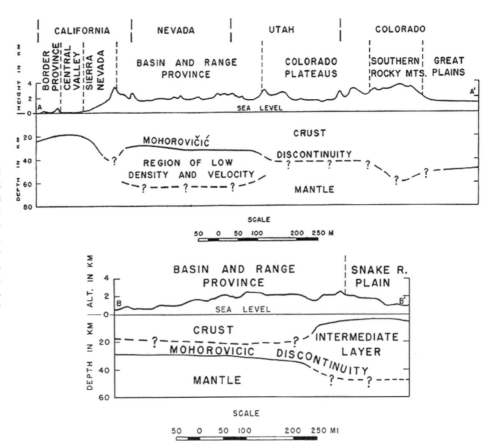

Figure 6.5.1-08. Variations in crustal thickness along two cross sections through the Basin and Range province (from Pakiser, 1963, figs. 3 and 4). Top: W-E section from San Francisco, California (A), to Lamar, Colorado (A′). Bottom: S-N section, including an intermediate crustal layer, from Ludlow, California (B), to Boise, Idaho (B′). [Journal of Geophysical Research, v. 68, p. 5747–5756. Reproduced by permission of American Geophysical Union.]

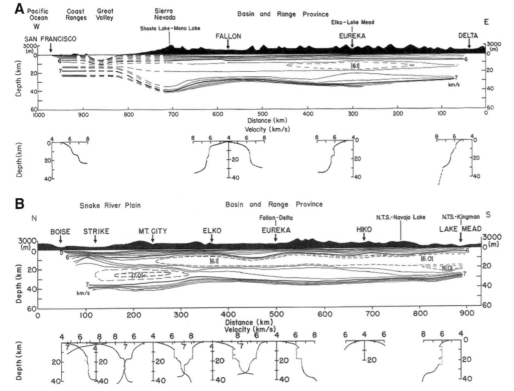

Figure 6.5.1-09. Crustal cross sections through the Basin and Range province and adjacent areas (from Prodehl, 1979, figs. 12 and 25). (A) W-E section from San Francisco, California, to Delta, Utah. (B) N-S section from Boise, Idaho, to Lake Mead, Nevada. [U.S. Geological Survey Professional Paper 1034, 74 p. Copyright U.S. Geological Survey.]

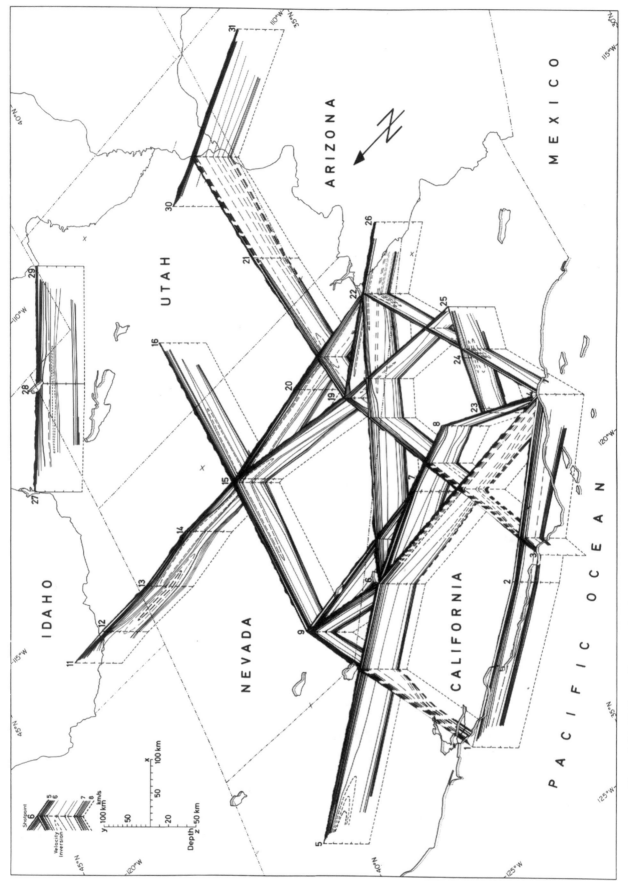

Figure 6.5.1-10. Fence diagram, showing the crustal structure under California and Nevada and adjacent areas obtained from the reinterpretation of the whole set of seismic-refraction data recorded by the U.S. Geological Survey in 1961–1963 (from Prodehl, 1970, fig. 10). [Geological Society of America Bulletin, v. 81, p. 2629–2646. Reproduced by permission of the Geological Society of America.]

Seismological Observatory, located 10 miles south of the Mogollon Rim near Payson in central Arizona (Warren, 1969). Two recording lines 400 km long intersected at the observatory (Fig. 6.5.1-11) and recorded chemical explosions from a total of 14 shotpoints. One line trended approximately parallel to the Mogollon Rim in SE-NW direction; the other one followed a SW-NE direction. The Moho contour map of Warren (1969) shows a rapid increase in depth from less than 25 km under the Basin and Range province in southwestern Arizona to ~40 km under the Colorado Plateau of northern Arizona (Fig. 6.5.1-11).

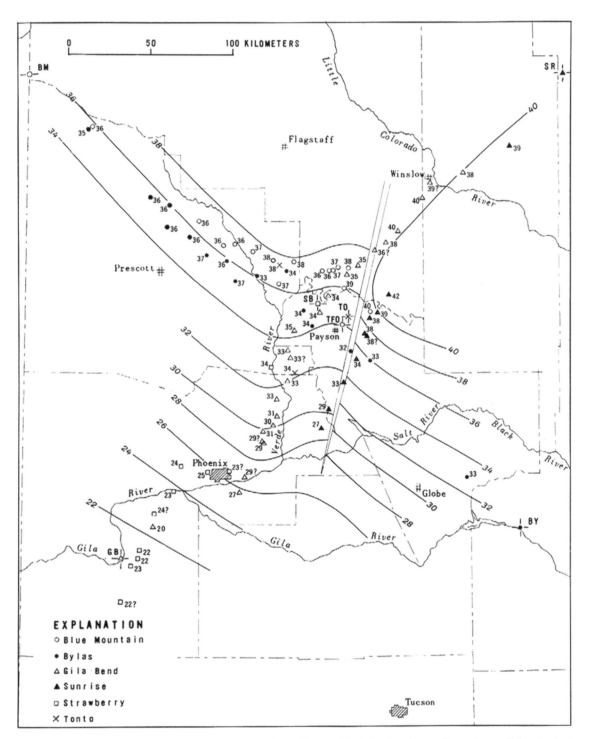

Figure 6.5.1-11. Moho configuration under Arizona (from Warren, 1969, fig. 13). TFO—Tonto Forest Seismological Observatory; stars named by two bold letters are shotpoint locations. [Geological Society of America Bulletin, v. 80, p. 257–282. Reproduced by permission of the Geological Society of America.]

In 1965, a network of profiles in the Southern Rocky Mountains and adjacent areas in the Great Plains of Colorado and Kansas was added. Some of the lines are contained in the compilation of Warren (1968b) along the transcontinental survey covering the latitudes 35°–39°N, showing preliminary crustal cross sections (see also Healy and Warren, 1969, fig. 2). Only one of the lines in the Rocky Mountains was interpreted in more detail at the time (Jackson and Pakiser, 1965). The remainder of the data was reinterpreted later (see, e.g., Prodehl, 1977; Prodehl and Pakiser, 1980; Prodehl and Lipman, 1989). Most of the data, recorded in the region of the Southern Rocky Mountains and adjacent Great Plains in the 1960s, were compiled as record sections and published in various publications (Prodehl and Pakiser, 1980; Prodehl and Lipman, 1989; Prodehl et al., 2005; Steeples and Miller, 1989) and, when the appropriate techniques had become available, were subsequently digitized by David Warren and compiled in internal U.S. Geological Survey Open-File Reports in the seventies. These reports and the corresponding record sections are reproduced in Appendix A6-8.

A more detailed crustal and upper mantle seismic investigation of the area followed only in the late 1990s (see Chapter 9; Karlstrom and Keller, 2005). The map in Figure 6.5.1-12 shows all profiles recorded in the 1960s but also includes more recent seismic-refraction projects Deep Probe of 1995 and CD-ROM (Continental Dynamics of the Rocky Mountain experiment) of 1999 (Prodehl et al., 2005).

The Moho contour map for Wyoming and Colorado (Fig. 6.5.1-13), which Keller et al. (1998) compiled before the CD-ROM data became available, is mainly based on data recorded in the 1950s and 1960s. Only the area of the Rio Grande rift in New Mexico is based on data recorded in the 1970s and 1980s (see Chapters 7 and 8).

In eastern Montana, a crustal study project of LASA (Large Aperture Seismic Array) (Fig. 6.5.1-14), located in the Great Plains between the earlier shotpoints Fort Peck (M4 in Fig. 5.2-07) and Acme Pond (M5 in Fig. 5.2-07), followed in 1966. The crustal structure study of Montana of 1956–1960 by various groups (see Chapter 5, Fig. 5.2-07) had indicated major changes

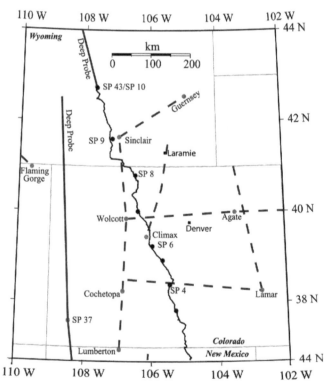

Figure 6.5.1-12. Seismic refraction lines, recorded in the Southern Rocky Mountains and adjacent Great Plains (from Prodehl et al, 2005, part of fig. 1). Dashed lines and grey circles—profiles recorded in the sixties; full crooked line—project CD-ROM of 1999; full straight lines—project Deep Probe of 1995. [The Rocky Mountain Region: An evolving lithosphere—tectonics, geochemistry, and geophysics: American Geophysical Union, Geophysical Monograph 154, p. 201–216. Reproduced by permission of American Geophysical Union.]

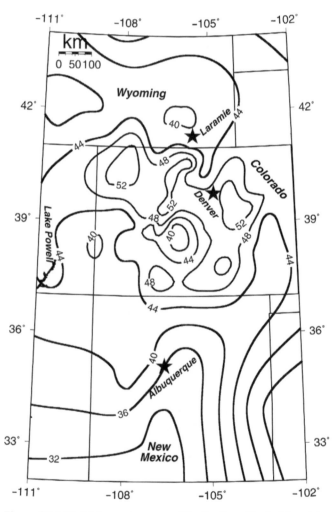

Figure 6.5.1-13. Moho contour map of the Southern Rocky Mountains Region and adjacent areas (from Keller et al, 1998, fig. 2). [Rocky Mountain Geology, v. 33, p. 217–228. Published by permission of Rocky Mountain Geology, Laramie, Wyoming, USA.]

in crustal structure throughout Montana, but had also left marked differences in the internal structure of the crust (McCamy and Meyer, 1964).

Therefore, in 1966, a crustal calibration seismic-refraction line with five shotpoints was recorded by the U.S. Geological Survey in a NE-SW direction across the center of the LASA array (Warren et al., 1972, 1973; Fig. 6.5.1-14), the analysis of which confirmed that marked changes of crustal structure existed within the array, calling for a more detailed study of the crust under LASA.

This was realized in 1968 (Warren et al., 1972, 1973). The USGS recording trucks (Figs. 6.5.1-01 to 6.5.1-03) were used to establish short arrays along profile lines; in addition, 20 portable 1- or 3-component seismic stations and the standard LASA array stations operated continuously for about two months recording both a number of teleseisms as well as the arrivals from the high-explosive shots fired in the calibration program.

Fifteen shots of either 4000 or 20,000 pounds were detonated and recorded by the expanded array of seismic stations (Fig. 6.5.1-15), resulting in a total of 325 seismic traces of the expanded LASA clusters plus 33 seismograms of the recording trucks and temporary seismic stations for each shot. The data were compiled as record sections and published as a U.S. Geological Survey Open-File report (Warren et al., 1972; reproduced in Appendix A6-9). The crustal thickness map resulting from a first analysis of the complex data set (Warren et al., 1972, 1973) shows depth variations from 45 to 55 km over relatively short horizontal distances (Fig. 6.5.1-16).

Finally, in 1967, two sparsely sampled seismic-refraction lines were recorded in the Coast Ranges south of the San Francisco Bay Area on both sides of the San Andreas fault system (Stewart, 1968a), with five shotpoints along each line (Fig. 6.5.1-17). The line in the Diablo Range to the northeast of the San Andreas fault was 200 km long, but only the central part between the shotpoints Cedar Mountain and Mount Stakes of 70 km length was well covered with recording stations. The Gabilan Range line in the Salinian block to the southwest of the San Andreas fault was 235 km long, but only the central

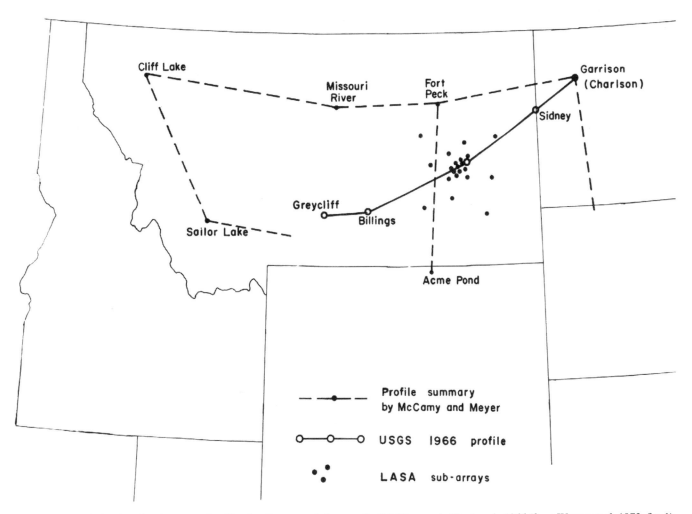

Figure 6.5.1-14. Seismic refraction, crustal calibration line, recorded across the LASA array in Montana in 1966 (from Warren et al, 1972, fig. 1). [U.S. Geological Survey Open-File Report, p. 1–58 and A1–A105.]

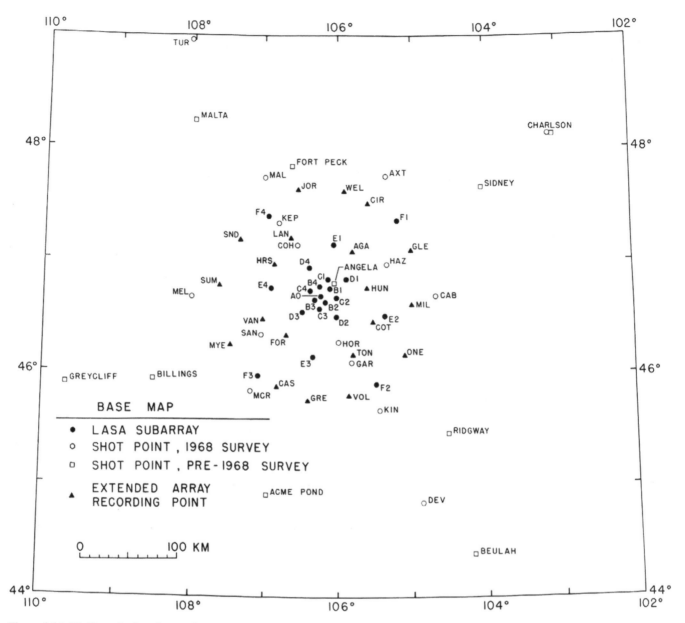

Figure 6.5.1-15. Shotpoint locations and temporary recording locations of the 1968 calibration project and LASA subarrays (from Warren et al, 1973, fig. 3). [Journal of Geophysical Research, v. 78, p. 8721–8734. Reproduced by permission of American Geophysical Union.]

part between shotpoints San Juan and Fernandez of 80 km length was sufficiently covered. The data are reproduced in Appendix A6-10.

Nevertheless, due to the overlapping data from five shotpoints along each line, the complete data could be successfully interpreted (Stewart, 1968a, Walter and Mooney, 1982). Another reinterpretation followed in conjunction with local earthquakes (Blümling and Prodehl, 1983; see also Chapter 7). The resulting overall crustal thicknesses, also including the investigations of other data sets by Steppe and Crosson (1978) for the northwestern side of the San Andreas fault system and of Healy and Peake (1975), and Prodehl (1979) for the Salinian side varied from author to author, as Walter and Mooney (1982) demonstrated, ranging between 26 and 30 km under the Diablo Range and between 24 and 26 km under the Gabilan Range. In earlier publications investigating the Salinian side, Hamilton et al. (1964), using quarry blasts for a profile between San Francisco and Salinas, had estimated 20–24 km, while McEvilly (1966) and Chuaqui and McEvilly (1968), using earthquake data, obtained 20 km total crustal thickness.

In their review of the first years of COCORP in the 1970s, Brewer and Oliver (1980) briefly review earlier seismic-reflection experiments. Besides the numerous large-scale seismic-refraction experiments to explore the Earth's

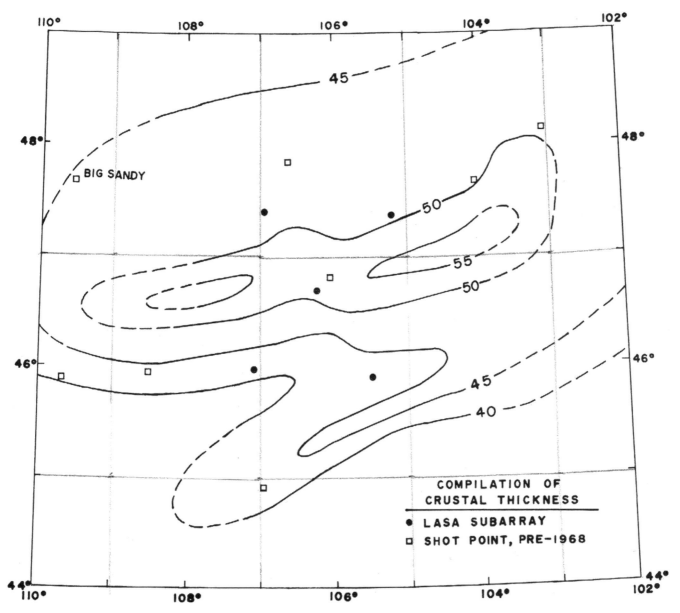

Figure 6.5.1-16. LASA Moho depth map derived from the 1968 calibration project (from Warren et al, 1973, fig. 18; Healy and Warren, 1974, fig. 5). [Journal of Geophysical Research, v. 78, p. 8721–8734; Tectonophysics, v. 20, p. 203–213. Reproduced by permission of American Geophysical Union.]

crust of the western United States, several attempts were undertaken already in the 1960s to study the crust using seismic-reflection techniques. One of the first attempts was a study by Narans et al. (1961) in Utah resulting in two reflecting horizons which roughly correlated with results from refraction profiles, but indicated a yet more complex structure. Another seismic-reflection study was undertaken in eastern Colorado. Hasbrouk (1964) tentatively identified the Moho and noted a complex crust and mantle structure. In an experiment in the Mojave Desert, Dix (1965) documented many reflectors of low dip between 8 and 35 km depth. In Wyoming in 1969, five deep crustal reflection profiles were obtained on

a line crossing the Wind River Mountains. Here, deep crustal reflections at times as great as 14 s TWT (two-way traveltime) were recorded (Perkins and Phinney (1971).

6.5.2. Central and Eastern North America

In the south-central United States, several seismic-refraction studies concentrated on the structure of the Great Plains (Healy and Warren, 1969; Warren, 1968a, 1968b). Stewart and Pakiser (1962) interpreted an unreversed line, recorded northwards along the New Mexico–Texas border from the first nuclear test in southeastern New Mexico, GNOME. The model was republished

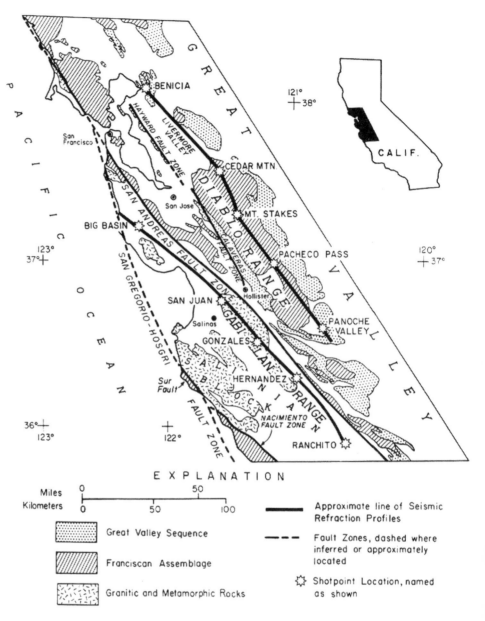

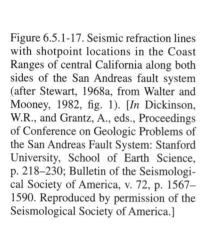

Figure 6.5.1-17. Seismic refraction lines with shotpoint locations in the Coast Ranges of central California along both sides of the San Andreas fault system (after Stewart, 1968a, from Walter and Mooney, 1982, fig. 1). [*In* Dickinson, W.R., and Grantz, A., eds., Proceedings of Conference on Geologic Problems of the San Andreas Fault System: Stanford University, School of Earth Science, p. 218–230; Bulletin of the Seismological Society of America, v. 72, p. 1567–1590. Reproduced by permission of the Seismological Society of America.]

in the compilation of Warren (1968b; see also Healy and Warren, 1969, fig. 2). The data were later reinterpreted by Mitchell and Landisman (1971) in conjunction with the interpretation of a 340-km-long reversed line through Oklahoma (Tryggvason and Qualls, 1967; see Fig. 6.5.2-01) with the shotpoints Manitou in the southwest and Chelsea in the northeast (Mitchell and Landisman (1971, data shown in Appendix A6-11). A >50-km-thick crust was the result of the data interpretations.

Stewart (1968b) interpreted a seismic-refraction network in Missouri, consisting of two 300-km-long lines, which were recorded in 1963 (Fig. 6.5.2-01; data shown in Appendix A6-11). Each line was reversed and contained a central shotpoint. South of the Missouri network, a 420-km-long line was recorded in 1962 from Cape Girardeau, Missouri, toward the southwest to Little Rock, Arkansas (Fig. 6.5.2-01), aiming to investigate the

crustal structure of the northern Mississippi embayment. Five shotpoints were located in a bend of the Mississippi River and spaced such that they would be 1 km apart when projected on the recording line. Shots at each location were repeated when the recording stations had moved to new positions, until the largest distance of 420 km was reached (McCamy and Meyer, 1966). The average crustal thickness for Missouri resulting from these data was around 40 km. All lines, described above, were compiled into the fence diagram in Figure 6.5.2-01, which shows a unified simplified reinterpretation by Warren (1968c).

Finally, a detailed seismic-refraction study, located south of the Transcontinental Geophysical Survey and therefore not shown by Warren (1968c), concentrated on the structure of the southern part of the Mississippi embayment north of New Orleans (Warren et al., 1966). It was an almost 500-km-long line,

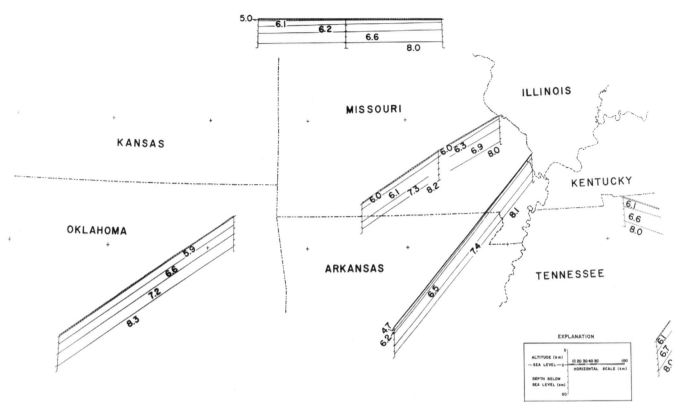

Figure 6.5.2-01. Fence diagram of crustal structure in the central-eastern portion of the Transcontinental Geophysical Survey (after Warren, 1968c, from Healy and Warren, 1969, fig. 3). Numbers are average velocities (km/s) of P-waves. [*In* Hart, P.J., ed., The Earth's crust and upper mantle: American Geophysical Union, Geophysical Monograph 13, p. 208–220. Reproduced by permission of American Geophysical Union.]

trending north from Ansley, Louisiana, to Oxford, Mississippi, where ~200 seismograms were obtained. Along the southern 200 km section between Ansley and Raleigh, Mississippi, 22 shots were fired at 5 shotpoints spaced at intervals of ~50 km, creating an overlapping coverage. The section north of Raleigh was not reversed.

The main result was a crustal thickness of 41 km near Ansley at the Gulf Coast, decreasing toward north to near 29 km at Raleigh (Fig. 6.5.2-02). A location map and some of the data are shown in Appendix A6-11.

In the north-central United States and adjacent Canada, a target of special research became Lake Superior. As part of the

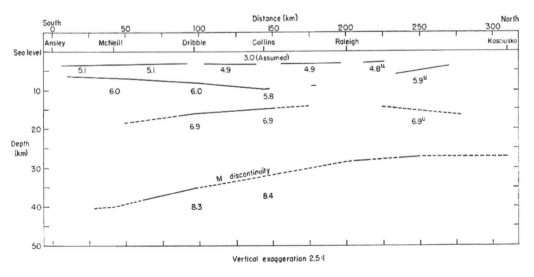

Figure 6.5.2-02. Crustal structure of the Mississippi embayment (after Warren et al., 1966, fig. 16). Numbers are average velocities (km/s) of P-waves. [Journal of Geophysical Research, v. 71, p. 3437–3458. Reproduced by permission of American Geophysical Union.]

VELA UNIFORM program in 1963 and 1964, three series of explosions were organized in the lake. In July 1963, 80 1-ton shots, in July 1964, 40 principally 1-ton shots, and in October 1964, 10 shots of 2–20 tons (Fig. 6.5.2-03) were recorded by 13 institutions from the United States and Canada at various azimuths and up to distances of more than 2500 km (Mansfield and Evernden, 1966).

The crust under Lake Superior was found to be unusual and rapidly varying in comparison with the Canadian Shield surrounding it (Smith et al., 1966). The depth of the Moho, for example, varies from 20 km west of the lake to 55 km and more in the eastern part of the lake. Cohen and Meyer (1966) used the explosions to study the Midcontinent gravity high along two reconnaissance profiles extending to the SSW through Minnesota and Wisconsin and found Moho depths of 46 km under the ridge of high gravity and 42 km under the low-gravity trough to the east.

In the southern Appalachians, a seismic-refraction survey with two 300-km-long crossing lines was carried out by the U.S.

Geological Survey in 1965 across the Cumberland Plateau Seismic Observatory near Minville, Tennessee (Fig. 6.5.2-04). The data were preliminarily interpreted by Warren (1968d) and later reinterpreted (Prodehl, 1977; Prodehl et al., 1984; data shown in Appendix A6-11). A third but less densely recorded reconnaissance line with only 20–50 km spacing of the recording units ran through the eastern coastal plains to the Atlantic Ocean and was connected with the ECOOE (East Coast On-shore Off-shore Experiment) survey (profiles along the Atlantic coast line in Fig. 6.5.2-04; Hales et al., 1967, 1968; Warren, 1968d). For the majority of these lines, a preliminary interpretation was published in fence diagrams (Fig. 6.5.2-04; Warren, 1968d; Healy and Warren, 1969, fig. 4) in the frame of a series of maps published along a transcontinental survey covering the latitudes 35°–39°N (Hart, 1964; Pakiser and Zietz, 1965).

The more detailed interpretation of the network across the Cumberland Plateau Seismic Observatory near Minville, Tennes-

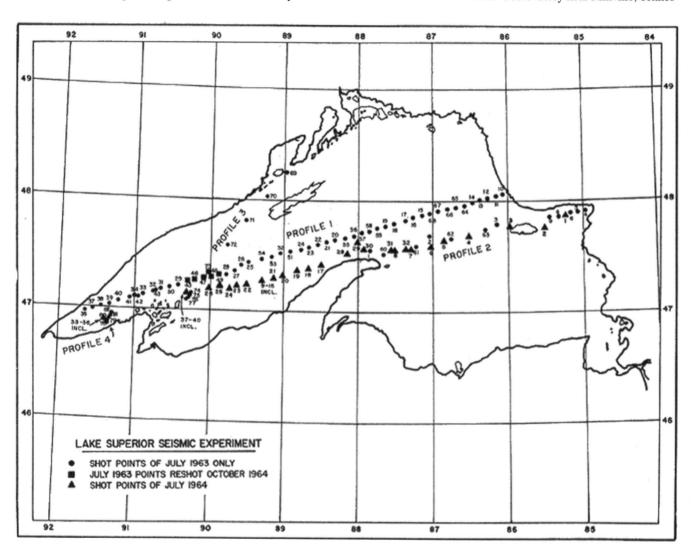

Figure 6.5.2-03. Lake Superior shot point map of 1963–1964 (from Mansfield and Evernden, 1966, fig. 1). [*In* Steinhart, J.S., and Smith, T.J., eds., The Earth beneath the continents: American Geophysical Union, Geophysical Monograph 10, p. 249–269. Reproduced by permission of American Geophysical Union.]

see (Fig. 6.5.2-04), where each profile had two end and two middle shotpoints, revealed that the lower crust was characterized by two distinct layers with velocities in the range of 6.7–6.8 km/s and 7.3–7.4 km/s. However, the boundaries between crustal layers as well as the crust-mantle boundary were not sharp but appeared as velocity transition zones ranging up to 11 km in thickness (Fig. 6.5.2-05). The high-velocity gradients between the various layers instead of distinct boundaries resulted from the fact that in the corresponding record sections clear reflected phases could hardly be detected. Instead, however, the first arrivals between 50 and 250 km shotpoint distance could be correlated as refracted phases resulting in unusual high velocities, and beyond 250 km distance clear P_n arrivals were present in the data (Fig. 6.5.2-06). Also, the total crustal thickness was rather high, between 49 and 55 km (Prodehl, 1977; Prodehl et al., 1984).

6.5.3. Achievements in the United States

The available interpretations on crustal structure in the United States of North America were reviewed from time to time (see, e.g., Pakiser and Steinhart, 1964; Healy and Warren, 1969) and in particular compiled for a transcontinental geophysical survey throughout the United States between 35° and 39°N latitude (Pakiser and Zietz, 1965).

A comprehensive summary on the state of the art of crustal structure interpretation as available in 1968, not including the reinterpretation of Prodehl (1970), can be found in Healy and Warren (1969). As already described above, it shows in maps fence diagrams of crustal cross sections with a unified simplified reinterpretation by D.H. Warren (Warren, 1968a, 1968b, 1968c, 1968d; see, e.g., Figs. 6.5.2-01 and 6.5.2-04), based on preliminary interpretations by various authors as, e.g., cited by Pakiser (1963), for most of the seismic-refraction profiles observed by the U.S. Geological Survey from 1961 to 1965.

The overview was complemented in 1970 by Warren and Healy (1973) with a map of all profiles recorded until 1970 within the United States and a table of references (Fig. 6.5.3-01 and Table 6.5.3-01) and a Moho contour map (Fig. 6.5.3-02).

The most common interpretation methods, which were used for the interpretation of seismic data by the end of the 1960s in North America, were described in much detail by Warren et al. (1972; Appendix A6-9) at the example of the interpretation of the detailed data set obtained during the seismic investigation of the deep structure under the LASA, Montana, array. A quite

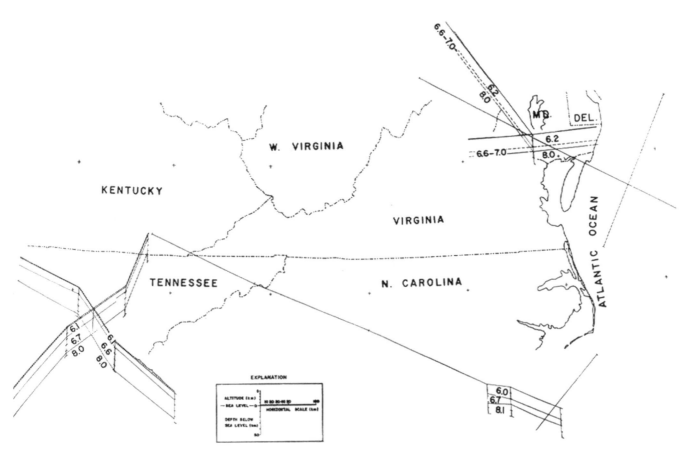

Figure 6.5.2-04. Fence diagram of crustal structure in the central-eastern portion of the Transcontinental Geophysical Survey (after Warren, 1968d, from Healy and Warren, 1969, fig. 4). Numbers are average velocities (km/s) of P-waves. [*In* Hart, P.J., ed., The Earth's crust and upper mantle: American Geophysical Union, Geophysical Monograph 13, p. 208–220. Reproduced by permission of American Geophysical Union.]

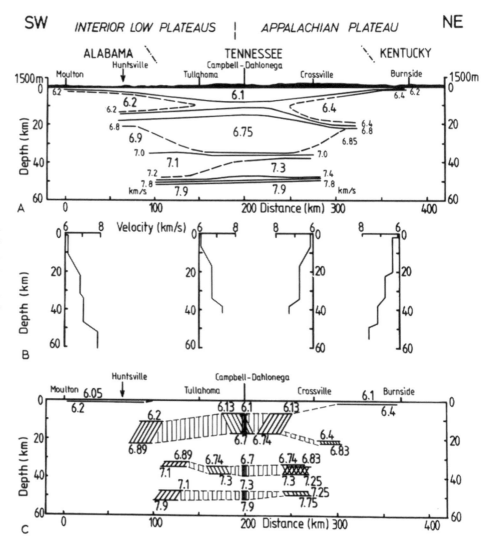

Figure 6.5.2-05. Crustal cross section between Moulton, Alabama and Burnside, Kentucky (from Prodehl et al., 1984, fig. 7). (A) Lines of equal velocity. (B) Individual velocity-depth profiles. (C) Position of velocity transition zones. [Tectonophysics, v. 109, p. 61–76. Copyright Elsevier.]

different approach, proposed by Giese (1968a, 1976b) for a rapid calculation of the velocity-depth distribution from any given travel-time curve system (see section 6.2.5) was described and applied by Prodehl (1979) on the majority of seismic-refraction profiles recorded by the U.S. Geological Survey in the western United States.

Based on a combination of data from deep seismic sounding, underground nuclear explosions and earthquakes, Herrin and Taggart (Herrin, 1969; Fig. 1) have compiled a map showing the estimated P_n velocity. The overall velocity was 8.0 km/s and above, except for the areas west of the Rocky Mountain front to the Pacific coast, where velocities smaller than 8.0 km/s prevailed.

6.5.4. Canada and Alaska

The achievements of crustal and uppermost mantle structure research until the end of the 1960s were summarized by Berry (1973). Figure 6.5.4-01 shows the seismic lines and average values of crustal thickness and upper-mantle P-wave velocities.

The Cordillera of western Canada had been studied by a kind of reconnaissance survey for over a decade (Berry, 1973). As a result, the Rocky Mountain trench appeared to be the western limit of thick crust and also of the North American craton (Berry et al., 1971). Under the central plateaus, between the Rocky and Coast Mountains, a crustal thickness of 30–35 km was obtained (White et al., 1968).

Detailed seismic-refraction experiments in southern Alberta and Saskatchewan revealed a 47-km-thick crust under the prairies of southern Alberta which decreased to the north with upper-mantle velocities of 8.2–8.3 km/s (Kanasewich, 1966).

A pioneering study was undertaken in 1964 by a group of seismologists of the University of Alberta under the leadership of Ernic Kanasewich. For the first time in North America, they used the near-vertical incidence seismic-reflection techniques, as used by the exploration industry, for an investigation of the whole crust and demonstrated that this technique was a viable tool of determining detailed crustal structure and imaging the Moho (Kanasewich and Cumming, 1965; Clowes et al., 1968;

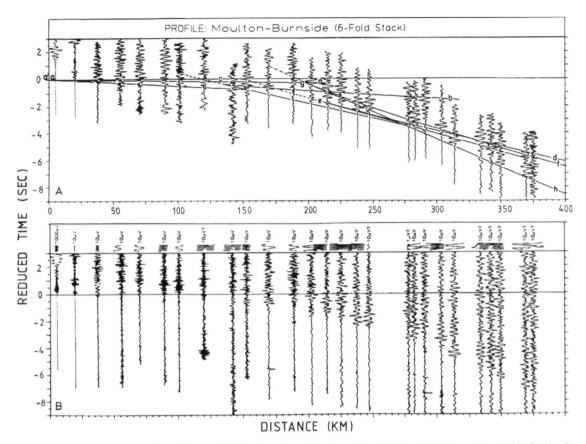

Figure 6.5.2-06. Record section of the profile Moulton–Burnside (from Prodehl et al., 1984, fig. 4). (A) Six-fold optimally stacked traces. (B) "Best" traces for each array. [Tectonophysics, v. 109, p. 61–76. Copyright Elsevier.]

Kanasewich et al., 1969). As outstanding results, they reported on reflected energy from the Moho "M" from 38 to 47 km depth and from the Reil boundary "R" at 26–34 km depth, the layer in between having velocity of 6.5–7.2 km/s. The seismic-reflection data revealed a surprisingly complex structure, including topography of 8 km over a horizontal distance of 25 km and dips of as great as 20° on some reflecting horizons at depth (Clowes et al., 1968).

In the same area, on the basis of combined reflection and refraction data, good evidence was found for major steeply dipping faults in the middle and lower crust which penetrated across Moho. Clowes and Kanasewich (1970) used the seismic near-vertical incidence reflection data and undertook detailed studies of an intra-crustal reflector (Conrad discontinuity), the Moho, and crustal Q. They concluded that there was significant fine structure at the base of the crust and proposed a lamellar model of alternating high and low velocity material at the base of the crust, as had been proposed by Fuchs (1969) for southern Germany. In another study, Clowes and Kanasewich (1972) investigated the special characteristics of the reflections by applying industry digital filtering techniques to enhance crustal reflections and enable a migrated "squiggle" section to be developed. Following the initial success of work in southern Alberta, Ernie

Kanasewich helped to carry out a similar survey in Wyoming in 1969, in cooperation with Bob Phinney and co-workers (Perkins and Phinney, 1971; see subchapter 6.5.1; Clowes, 2010, personal commun.). In conjunction with the earlier work on deep reflections in Germany (Dohr, 1957, 1959; Krey et al., 1961; Liebscher, 1962, 1964; German Research Group for Explosion Seismology, 1964; Dohr and Fuchs, 1967) and Russia (Beloussov et al., 1962; Kosminskaya and Riznichenko, 1964; Zverev, 1967) the studies by the University of Alberta group paved the way for COCORP and subsequent deep reflection programs around the world in the 1970s and 1980s.

A large experiment was carried out in 1966 in northwestern Canada to calibrate the Yellowknife Seismic Array (Fig. 6.5.4-01). For this purpose, 17 explosions of 1–2 ton charges were detonated in lakes at distances ranging from 15 to 480 km. An average Moho depth of 37.5 km resulted (Weichert and Whitham, 1969). Another research project in 1969 investigated the fine crustal structure of the Yellowknife area by a combined reflection-refraction experiment on two lines with 250 m station spacing (Clee et al., 1974).

The Lake Superior experiments in 1963 and 1964 as well as the Early Rise experiment in 1965, which will be discussed in more detail below, also provided data for crustal structure studies

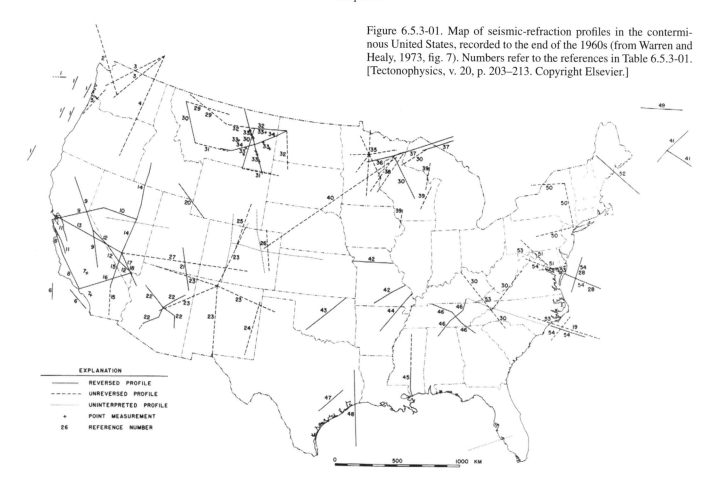

Figure 6.5.3-01. Map of seismic-refraction profiles in the contermi- nous United States, recorded to the end of the 1960s (from Warren and Healy, 1973, fig. 7). Numbers refer to the references in Table 6.5.3-01. [Tectonophysics, v. 20, p. 203–213. Copyright Elsevier.]

EXPLANATION

— REVERSED PROFILE
----- UNREVERSED PROFILE
·········· UNINTERPRETED PROFILE
+ POINT MEASUREMENT
26 REFERENCE NUMBER

TABLE 6.5.3-01. REFERENCES TO THE MAP
OF SEISMIC-REFRACTION PROFILES IN THE
CONTERMINOUS UNITED STATES IN FIG. 6.5.3-01

1. Shor et al. (1968)	28. Merkel and Alexander (1969)
2. White and Savage (1965)	29. Asada et al. (1961)
3. Johnson and Couch (1970)	30. Steinhart and Meyer (1961)
4. Hill (1972)	31. Aldrich et al. (1960)
5. Berg et al. (1966)	32. McCamy and Meyer (1964)
6. Shor and Raitt (1958)	33. Warren et al. (1972)
7. Shor (1955)	34. Gibbs (1972)
8. Healy (1963)	35. Tuve (1953)
9. Eaton (1966)	36. Smith et al. (1966)
10. Eaton (1963)	37. O'Brien (1968)
11. Stewart (1968a)	38. Cohen and Meyer (1966)
12. Johnson (1965)	39. Slichter (1951)
13. Carder et al. (1970)	40. Roller and Jackson (1966)
14. Hill and Pakiser (1966)	41. Barrett et al. (1964)
15. Gibbs and Roller (1966)	42. Stewart (1968b)
16. Roller and Healy (1963)	43. Tryggvason and Qualls (1967)
17. Roller (1964)	44. McCamy and Meyer (1966)
18. Diment et al (1961)	45. Warren et al. (1966)
19. Steinhart et al. (1962)	46. Roller et al (1967)
20. Willden (1965)	47. Cram (1961)
21. Roller (1965)	48. Hales et al. (1970)
22. Warren (1969)	49. Ewing et al. (1966)
23. Warren and Jackson (1968)	50. Katz (1955)
24. Stewart and Pakiser (1962)	51. Hart (1954)
25. Jackson and Pakiser (1965)	52. Steinhart et al. (1963)
26. Jackson et al. (1963)	53. James et al. (1968)
27. Ryall and Stuart (1963)	54. Hales et al. (1968)

Note: From Warren and Healy (1973, table 1).

of the Churchill and Superior Precambrian provinces of Canada. They showed in particular that the crustal thickness under the Canadian Shield varies substantially. After Hodgson (1953) had derived a 35-km-thick crust from the observation of rockbursts in the Kirkland Lake area of northern Ontario, the new data and results of the Lake Superior experiments stimulated a variety of new experiments to study the Shield in more detail.

In the Hudson Bay experiment of 1965 (Hobson, 1967) and the Grenville Front seismic experiment in 1968 (Mereu and Jobidon, 1971; Berry and Fuchs, 1973), several Canadian research insti- tutions cooperated. The Grenville Front seismic experiment, investigating the Grenville Front, was a large-scale reconnaissance refraction experiment with three parallel, SE-NW–trending pro- files investigated the Grenville and Superior provinces of the Cana- dian Shield (Berry and Fuchs, 1973), north of the St. Lawrence River and immediately west of Appalachia (Fig. 6.5.4-01). Here, 26 shots of sizes between 2000 and 6000 lbs were fired by the Dominion Observatory in six different lakes and recorded up to 800 km distance traversing a major part of the Canadian Shield between Hudson Bay and the Gulf of St. Lawrence.

The seismic-refraction studies of the crust of the Appala- chian region which had started as early as 1956 (Willmore and Scheidegger, 1956) were enforced in the early 1960s (Dainty et al., 1966, Ewing et al., 1966, Berry, 1973). At the flanks of the

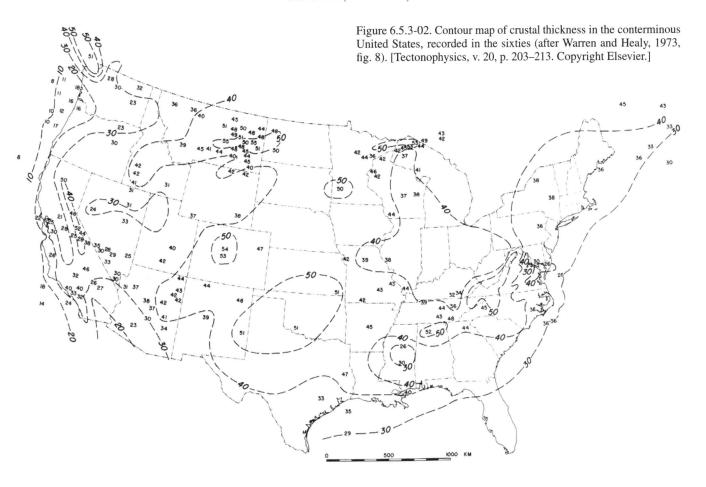

Figure 6.5.3-02. Contour map of crustal thickness in the conterminous United States, recorded in the sixties (after Warren and Healy, 1973, fig. 8). [Tectonophysics, v. 20, p. 203–213. Copyright Elsevier.]

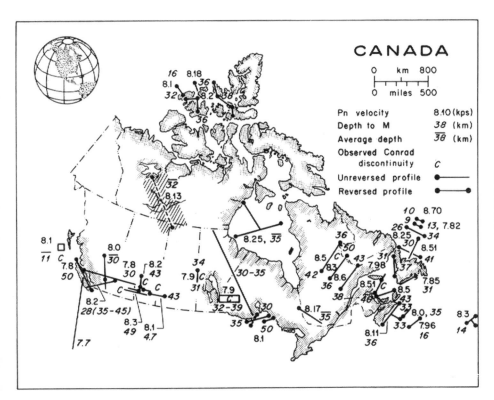

Figure 6.5.4-01. Map of seismic-refraction profiles in Canada, recorded to the end of the sixties (from Berry, 1973, fig. 1). [Tectonophysics, v. 20, p. 183–201 Copyright Elsevier.]

Appalachian system, crustal thicknesses between 30 and 40 km were obtained with an average crustal velocity of less than 6.5 km/s and an uppermost-mantle velocity close to 8.0 km/s. In the center of the system, a thick intermediate layer with 7.3–7.5 km/s velocity was seen causing the crustal thickness to increase to more than 40 km. Also, a much higher uppermost-mantle velocity of 8.5–8.7 km/s resulted (Figs. 6.5.4-01 and 6.5.4-02). Well-defined high uppermost-mantle velocities of 8.7–8.8 km/s at 52 km depth were also observed on data recorded from a number of shots from the north on a profile across the Gaspe Peninsula (Rankin et al., 1969).

In central Alaska during the winter of 1967–1968, an unreversed 217-km-long seismic-refraction profile was recorded from blasts in the Alaska Range toward north through the Tanana Basin, ending ~30 km north of Fairbanks (Fig. 6.5.4-03). Assuming a true P_n velocity of around 8 km/s, the interpretation of the observed 8.8.km/s apparent velocity results at a Moho depth of 31 km under Fairbanks dipping southward under the Alaska Range to ~48 km depth (Hanson et al., 1968; Berg, 1973). Carder et al. (1967) investigated the seismic data obtained from the LONG-SHOT underground nuclear explosion on Amchitka Island. It was recorded at 27 sites in Alaska and 2 sites in Yukon Territory at distances from 2.6° to 26.6°. Upper-mantle traveltimes correlated to distances of 11° resulted in an average velocity of 7.85 km/s and 22–26 km Moho depth beneath the island chain.

6.5.5. Upper Mantle Projects in North America

The possibility of using lakes as shotpoints to record very long seismic-refraction profiles was recognized immediately. One of the first long-range lines was recorded in 1964 by the U.S. Geological Survey from Lake Superior to Denver at ~1400 km distance (Roller and Jackson, 1966). A special volume of the American Geophysical Union, edited by Steinhart and Smith (1966) and dedicated to M.A. Tuve, deals in particular with results from these surveys, but also contains other seismic, seismological and compositional studies related to the Earth's crust and mantle in various parts of North America and other continents, including a critical review of recent explosion studies by James and Steinhart (1966).

Following the experience from the previous Lake Superior experiments, in 1965 the Project Early Rise emerged with

Figure 6.5.4-02. Crustal models through the Canadian Appalachian system (from Berry, 1973, fig. 2, after Ewing et al., 1966, with additions). [Tectonophysics, v. 20, p. 183–201. Copyright Elsevier.]

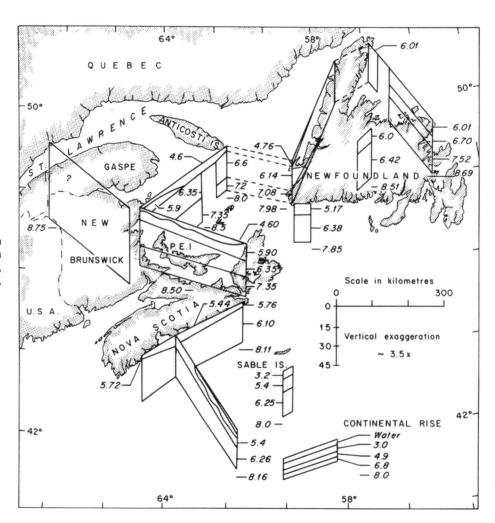

recording distances of 1000 km and longer. Thirty-eight 5-ton shots were detonated in the lake and recorded along lines in many directions from Lake Superior (Fig. 6.5.5-01) by groups from North American universities and government agencies. Due to its success in obtaining explosion seismic energy to large recording distances, this experiment, as with the Nevada Test Site observations, resulted in insight into the uppermost mantle below the Moho, due to the long recording distances.

A data example in Figure 6.5.5-02 shows the data recorded from Lake Superior into Colorado. The data of the Early Rise Project have been interpreted by many authors. Comprehensive interpretations were published, e.g., by Iyer et al. (1969), Mereu and Hunter (1969), Ansorge (1975), and Massé (1973).

A second large-scale crustal and upper-mantle seismic experiment, the Project Edzoe, was undertaken jointly by Canadian and American scientists in 1969 and aimed to study the structure of the North American Great Plains and the Rocky Mountains. Twenty explosions with charges between 5700 and 7700 kg were fired by the Dominion Observatory in Greenbush Lake near Revelstoke, British Columbia, and were recorded on several lines (up to 800–1000 km distance). The observations covered both Canada and the United States. Contrary to the Early Rise project of 1965, however, a joint publication evi-

dently never exceeded an internal report status for project Edzoe of 1969, so a location map of all lines did not become publicly available. The interested reader has to rely on publications of the many individual lines.

Individual interpretations of the long-range profiles recorded in Canada were published. These were profiles recorded across the Canadian Cordillera (Berry and Forsyth, 1975; Forsyth et al., 1974), through the North Cascade Mountains of Washington and British Columbia (Johnson and Couch, 1970), along the Rocky Mountain Front Range (Mereu et al., 1977), through the Great Plains of southern Alberta (Chandra and Cumming, 1972) and southern Saskatchewan and western Manitoba (Bates and Hall, 1975). In the United States, one line extended southwestward through the Columbia Plateau and was interpreted by Hill (1972). Two other lines were essentially north-south–oriented and traversed the central United States (Hales and Nation, 1973a). The first line was directed toward the Big Bend area in western Texas and traversed mainly the Rocky Mountains. The second line extended toward the SE corner of Kansas and covered the Great Plains area. Hales and Nation (1973a) correlated essentially one phase on both lines. On the Rocky Mountain line, they correlated a P_n phase in the first arrivals from 185 to 700 km distance, while on the Great Plains profile they observed P_n up to 1000 km distance. On both lines the average P_n velocity was ~8 km/s. Contrary to this interpretation, Mereu et al. (1977) correlated a high-velocity upper-mantle traveltime branch, beginning at distances beyond 300 km, and attributed it to a high-velocity layer with 8.5–8.8 km/s below 60 km depth, in correspondence to results found on European long-range profiles observed in the 1970s, which will be discussed in detail in Chapter 7.

6.6. EXPLOSION SEISMOLOGY RESEARCH IN SOUTHEAST ASIA IN THE 1960s

Following the start of the Upper Mantle Project, considerable progress was made in instrumentation and seismic fieldwork from 1964 onwards in Japan (Research Group for Explosion Seismology, 1966a). Several new lines on the island of Honshu were added to the early observations of the 1950s, described above (Aoki et al., 1972; Yoshii and Asano, 1972; Okada et al., 1973; Yoshii et al., 1974). For example, in southern Honshu, two reversed 300-km-long profiles were recorded, one along a north-south line at 139°E longitude, where at the same time an earthquake could also be recorded, and the other along an east-west line at ~35.5°N latitude. Furthermore, a series of 10 underwater shots off the eastern coast was recorded along a SE-NW line across northern Honshu and added new information to the very first quarry blast lines observed in the early 1950s.

A comprehensive map (Fig. 6.6-01) shows all seismic-refraction observations from 1950 to 1970. As a result, the one-layer crustal model of the 1950s could be replaced by a two-layer model, based on first arrivals with velocity greater than 6.4 km/s. In conclusion, the depth to the Moho is between 20 and 30 km, but increases by ~5 km in the central mountainous areas

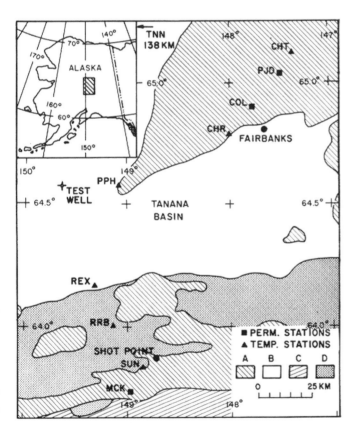

Figure 6.5.4-03. Location of a seismic-refraction line around Fairbanks, central Alaska (from Berg, 1973, fig. 8, after Hanson et al., 1968, fig. 1). [Tectonophysics, v. 20, p. 165–182. Copyright Elsevier.]

Figure 6.5.5-01. Location of recording stations for Project Early Rise of 1965. Shots were fired from the same location in Lake Superior. The codes for groups that recorded are indicated at the corresponding lines: AFTAC—Air Force Technical Applications Center; UA—University of Alberta; UMA—University of Manitoba; UMI—University of Michigan; USGS—U.S. Geological Survey; SCAS—Southwest Center for Advanced Studies; SRI—Stanford Research Institute; UT—University of Toronto; UW—University of Wisconsin; UWO—University of Western Ontario (from Healy and Warren, 1969, fig. 11). [*In* Hart, P.J., ed., The Earth's crust and upper mantle: American Geophysical Union, Geophysical Monograph 13, p. 208–220. Reproduced by permission of American Geophysical Union.]

of Honshu. The existence of an additional intermediate layer with 6.7–6.8 km/s is not unique, but if it is assumed, the Moho depths would increase by 10 km (Aoki et al., 1972; Yoshii and Asano, 1972; Research Group for Explosion Seismology, 1966a, 1966b, 1973; Yoshii et al., 1974). In general, the upper-mantle velocity is ~7.7 km/s under Japan. A crustal cross section through central Japan along line 3 is shown in Figure 6.6-02. A crustal profile was also recorded along the southern coast of Hokkaido (Fig. 6.6-01; Okada et al., 1973; Research Group for Explosion Seismology, 1973). The crustal cross section also includes marine seismic-refraction results (Ludwig et al., 1966).

6.7. EXPLOSION SEISMOLOGY RESEARCH IN THE SOUTHERN HEMISPHERE CONTINENTS IN THE 1960s

6.7.1. Australia

Crustal research in Australia using explosion seismology was already quite extensive in the 1960s due to the existence of a small but very active geophysical community.

Between 1959 and 1964, the structure of crust and upper mantle of southwest Western Australia was the goal of a variety of seismic-refraction investigations. The Mundaring Geophysical Observatory of the Bureau of Mineral Resources, Geology and Geophysics (BMR) acquired seismic data from a variety of quarry blasts and marine shots recorded at stations throughout southwestern Australia. In 1960, the Lamont Geological Observatory (USA) research vessel *Vema* conducted seismic investigations on the offshore Perth Basin. In 1962 the Scripps Oceanographic Institute (USA) vessel *Argo* exploded four charges off Perth. In 1963, the Royal Australian Navy ship HMAS *Diamantina* detonated six charges. The data resulted in crustal models both for southwestern Australia and the Western Australia margin (Everingham, 1965; Hawkins et al., 1965a, 1965b; Finlayson, 2010; Appendix 2-2). The crustal investigations using local earthquakes and quarry blasts were continued during 1963–1967 with a network of widely spaced stations in the southern part of Southern Australia (White, 1969).

Following the seismic observations of the Emu and Maralinga nuclear explosions of 1953 and 1956 and of some large explosions at Eaglehawk quarry in the Snowy Mountains in

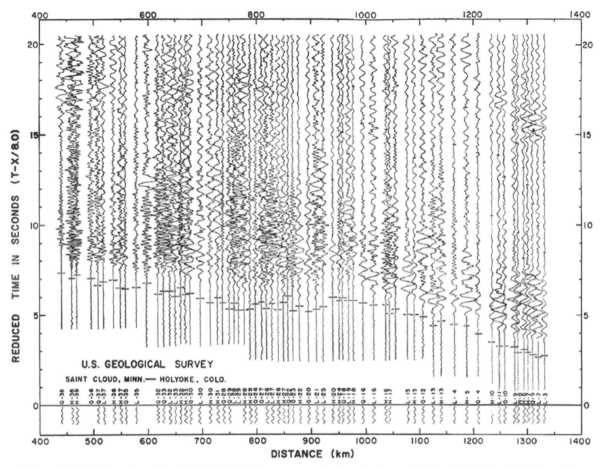

Figure 6.5.5-02. Record section of the Early Rise long-range seismic-refraction profile from Lake Superior, Wisconsin, to Denver, Colorado recorded by the U.S. Geological Survey in 1965 (from Healy and Warren, 1969, fig. 12). [*In* Hart, P.J., ed., The Earth's crust and upper mantle: American Geophysical Union, Geophysical Monograph 13, p. 208–220. Reproduced by permission of American Geophysical Union.]

1956–1957, several seismic experiments followed in the 1960s which are summarized in Figure 6.7.1-01 (Cleary, 1973). The Eaglehawk quarry observations were followed by several series of offshore shots, both off New South Wales in 1965 and those of the BUMP (Bass Strait Upper Mantle Project) experiment in the Bass Street in 1966.

The resulting crustal model (Fig. 6.7.1-02) shows a 40+ km thick crust under southeastern Australia, thickening by ~10 km under the Eastern Highlands. The data of the Maralinga explosions were supplemented in 1969 when a seismic Western Australia Geotraverse was conducted within the Archaean shield. In 1969, deep crustal refractions and reflections were recorded along the Geotraverse in southwestern Australia (Mathur, 1974). The line extended eastward from Perth to Coolgardie and continued in a southeast direction toward Point Culver on the Great Australian Bight. Figure 6.7.1-01 shows the location of the four large shots. In 1967, the Fremantle Region Upper Mantle Project (FRUMP) investigated the crust in the southwestern region of Western Australia (Finlayson, 2010; Appendix 2-2).

In northern Australia, several experiments were carried out: the CRUMP (Carpentaria Region Upper Mantle Project) experiment with offshore shots in 1966, the WRAMP (Warramunga Seismic Array Mantle Projects) experiment with some inland shots in 1966, and the ORD RIVER experiment with a series of shots in 1970–1971. Data recorded in northern and western Australia are relatively scarce but indicated crustal thicknesses of around 40+ km within the Australian continent and a gradual thinning toward the coastal areas. There appeared a tendency for a systematic increase of P_n velocities across Australia from east to west. There was also good evidence for the existence of a Conrad discontinuity at ~20 km depth (Denham et al., 1972; Cleary, 1973).

A historic overview on early seismic-reflection work in Australia was published by Dooley and Moss (1988) and Moss and Dooley (1988). In southeast Australia in November 1960, strong reflections from the Moho were recorded in the Murray Basin during extended recording time experiments. The first attempt to record continuous deep-reflection profiles was in 1960 over the Comet Ridge, Queensland. Records were obtained on 800 m split

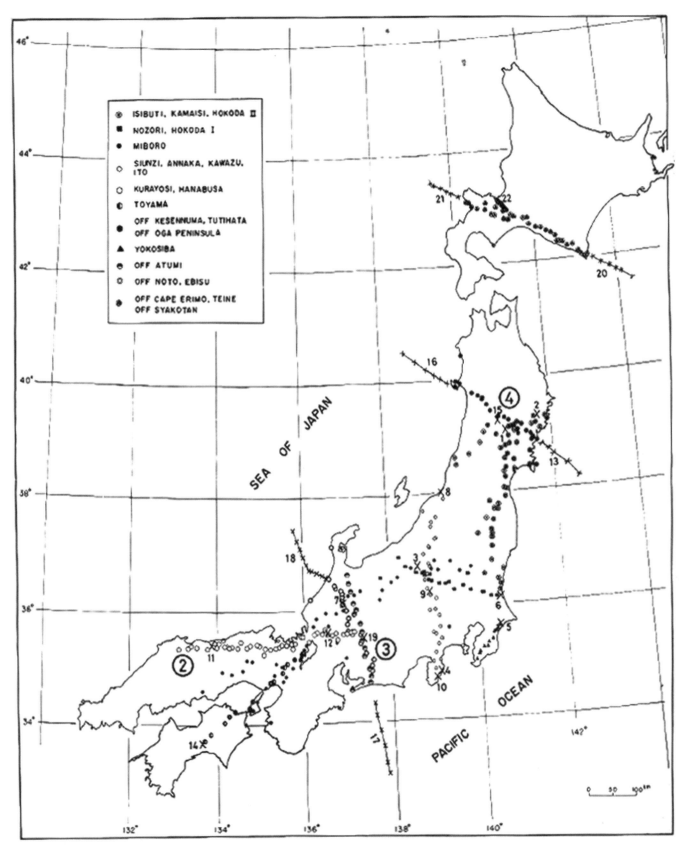

Figure 6.6-01. Shot and observation points in Japan from 1950 to 1970 (from Research Group for Explosion Seismology, 1973, fig. 1). [Tectono-physics, v. 20, p. 129–135. Copyright Elsevier.]

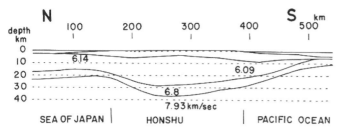

Figure 6.6-02. Crustal structure along the profile Atumi–Noto (line 3 in Fig. 6.6-01), central Japan (from Research Group for Explosion Seismology, 1973, fig. 3). [Tectonophysics, v. 20, p. 129–135. Copyright Elsevier.]

spreads along a 5.2 km traverse. Similar attempts were made at Bushbrook in the Perth Basin, Western Australia, in 1964, where recordings were made on 1100 m spreads along a 6.6 km traverse. Other experimental and offset recordings were made in the Roma shelf area of the Surat Basin, Queensland, in 1967. All data showed events between 5 and 12 s, but no complete continuity. Following this earlier work, seismic profiling trials near Mildura, Victoria, and Broken Hill, New South Wales, in 1968 and 1969 demonstrated convincingly that the Moho could be clearly imaged at ~10.2 s TWT using reflection profiling techniques (Branson et al., 1976). Additional tests were carried out in 1968 and 1969 in hard-rock areas at Tidbinbilla, Australian Capital Territory, and Braidwood, New South Wales, to complement the Mildura and Broken Hill experiments. Both large and small charges gave good results. During 1968, seismic-reflection profiling across the

southern margin of the Ngalia Basin in central Australia imaged clearly the dipping boundary fault and underlying sedimentary sequences beneath overlying Precambrian basement rocks, thus clearly demonstrating the viability of seismic imaging in hard rock terrains. Data of good quality were also obtained in 1969 in the northern part of the Amadeus Basin, Northern Territory, using short cross section, a cross-spread, and a partial expanding spread (Brown, 1970).

6.7.2. Africa and the Afro-Arabian Rift System

While crustal research in Australia using explosion seismology was quite substantial already in the 1960s due to the existence of a small but very active geophysical community, crustal structure investigations in the two other continents of the southern hemisphere, Africa and South America, was mainly limited to earthquake-related research. Active-seismic projects required much more sophisticated logistics and were therefore very difficult to handle with the equipment available at that time.

The Afro-Arabian rift, extending from Lebanon through the Jordan–Dead Sea transform, the Red Sea, the Gulf of Aden, the Afar triangle of Ethiopia into the East African rift system was the goal of first seismic investigations already in the 1960s. Marine studies investigated the Red Sea (Drake and Girdler, 1964; Tramontini and Davies, 1969) and the Gulf of Aden (Laughton and Tramontini, 1970).

Only two seismic-refraction surveys were made in the Red Sea. The first one was a survey in 1958 which was reported upon by Drake and Girdler (1964), and the second survey was done in

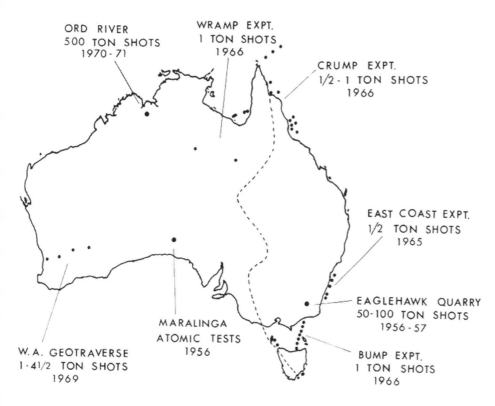

Figure 6.7.1-01. Seismic refraction experiments in Australia (from Cleary, 1973, fig. 1). The dashed line indicates the eastern limit of exposed Precambrian rocks. [Tectonophysics, v. 20, p. 241–248. Copyright Elsevier.]

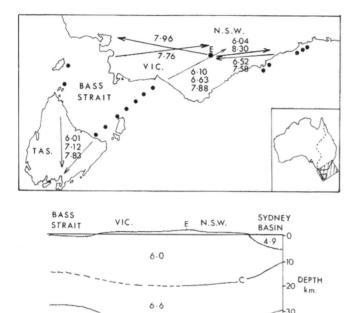

Figure 6.7.1-02. Top: Location of seismic-refraction experiments in Southeast Australia. Bottom: Crustal cross section through Southeast Australia (from Cleary, 1973, fig. 2). E—position of Eaglehawk quarry. [Tectonophysics, v. 20, p. 241–248. Copyright Elsevier.]

the late 1960s (Tramontini and Davies, 1969). Drake and Girdler had classified rock types attributed to velocities: 1.7–3.0 km/s unconsolidated sediments, 3.0–5.0 km/s evaporites (with possible volcanic material), 5.5–6.4 km/s shield rocks, 6.4–7.3 km/s basic intrusive rocks. The latter was referred to as layer 3, although the velocity range was wider than the average values for the world's oceans. Tramontini and Davies (1969) chose to survey the area at the northern limit of the large magnetic anomalies. They shot 20 seismic lines over the axial trough and the eastern flanks. No mantle velocities were detected at any of the stations. No shield rock was detected in the central trough but, without exception, the trough was found to be underlain by layer 3 material (6.4–7.3 km/s). The depth to the surface of layer 3 was ~4 km below sea level, i.e., 2 km below the ocean floor. Under the main trough, i.e., northern part of the Red Sea, where the axial trough has not yet developed, the depth to layer 3 increased to ~7 km below the sea surface. On the western flanks of the main trough shield rocks were detected and layer 3 was absent. There was some evidence that layer 3 was not only deeper on the eastern flank, but had also a lower velocity.

The seismic-refraction data in the Gulf of Aden showed that the whole of the Gulf of Aden between continental margins is oceanic in nature, leading to the conclusion that the Gulf of Aden had formed as a result of the continental separation of Arabia and Africa, driven by the same subcrustal forces that created the mid-ocean ridge systems. The crustal columns derived for the Gulf of Aden are also shown by Laughton et al. (1970, fig. 7) in their

summary paper on the Indian Ocean and were later republished in a review of crustal studies along the Afro-Arabian rift system (Prodehl et al., 1997a). From the marine seismic-refraction and other geophysical studies in the Red Sea (Tramontini and Davies, 1969) the association with the Gulf of Aden was clear, but the opinion was divided on the amount of continental separation involved in the formation of the Red Sea.

In 1969, the University of Birmingham undertook the first seismic-refraction experiment in the East African rift (Griffiths et al., 1971; Long et al., 1973). With shots fired in Lake Turkana and in Lake Baringo recorded by 10 stations, a reversed line of 300 km length was established along the Gregory rift. Due to their limited number, however, the observations were contradictory (Long et al., 1973) and only fully understood when, 16–25 years later, the KRISP projects (Kenya Rift International Seismic Project) of 1985, 1990, and 1994 investigated the rift in much more detail (see Chapter 9; Prodehl et al., 1994a; Fuchs et al., 1997).

6.7.3. South America

The very first explosion seismic investigation in South America concentrated on the Andes (Fig. 6.7.3-01). In 1968, a reconnaissance survey of the Peru-Bolivia Altiplano involved recording distances of 320–400 km (Figs. 6.7.3-02 and 6.7.3-03). However, no P_n-first arrivals were obtained. From weak reflections, a total crustal thickness of more than 70 km but less than 80 km was estimated (Ocola and Meyer, 1972; Appendix A6-12).

6.7.4. Antarctica

Several Antarctica marine surveys were carried out in the 1960s and 1970s by Don Griffiths' group of the Birmingham University (Allen, 1966; Ashcrost, 1970; Harrington et al., 1972) investigating crust of the Scotia Sea and adjacent Antarctica peninsula region. It was either a two ship or a single ship–custom buoys operation using explosives as sources.

As was mentioned in Chapter 5, during a reconnaissance exploration of the Antarctic interior as part of the U.S. Antarctic research program during the 1950s and 1960s, some three dozen seismic-refraction profiles had been recorded which, however, provided only information about seismic velocities below the ice, but did not penetrate further than into the topmost upper crust (Bentley, 1973).

The very first time explosion seismic research on land in Antarctica reached through the ice into deeper levels of the crust, was in 1969, when two deep seismic sounding profiles were completed by members of the 14th Soviet Antarctic Expedition (Bentley, 1973; Kogan, 1972) in the vicinity of Novolazarevskaya near the northern coast of Queen Maud Land (Fig. 6.7.4-01). The interpretation resulted in an average thickness of the crust of 40 km, with a few kilometers thicker near the mountains and shallowing to only 27 km below sea level near the coast. The measured uppermost mantle velocity was 7.9 km/s.

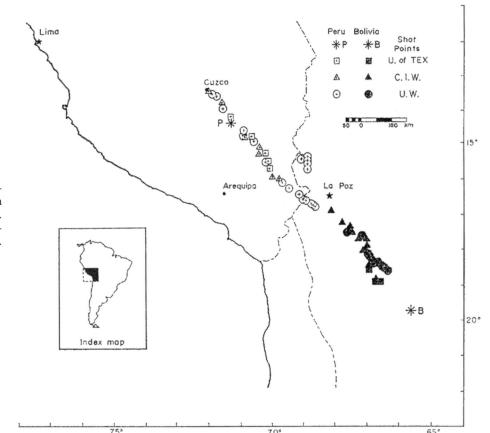

Figure 6.7.3-01. Seismic refraction experiment of 1968 in Peru and Bolivia (from Ocola and Meyer, 1972, fig. 1). [Geophysical Journal of the Royal Astronomical Society, v. 30, p. 199–209. Copyright John Wiley & Sons Ltd.]

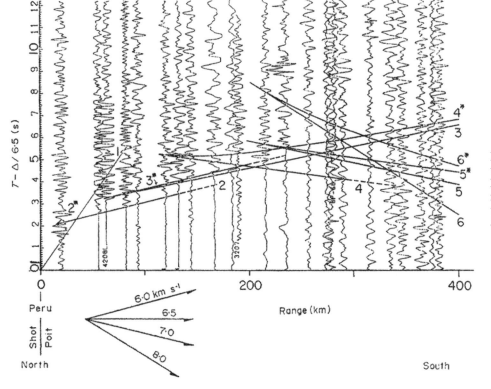

Figure 6.7.3-02. Record section for the 1968 Peru profile (from Ocola and Meyer, 1972, fig. 2). [Geophysical Journal of the Royal Astronomical Society, v. 30, p. 199–209. Copyright John Wiley & Sons Ltd.]

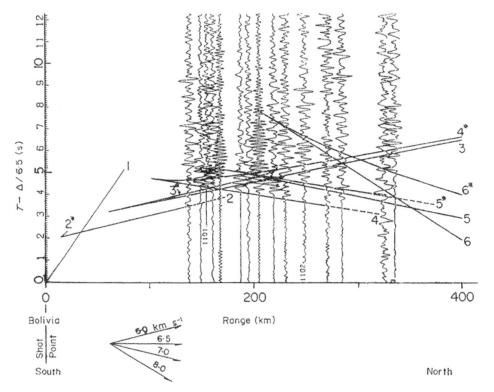

Figure 6.7.3-03. Record section for the 1968 Bolivia profile (from Ocola and Meyer, 1972, fig. 4). [Geophysical Journal of the Royal Astronomical Society, v. 30, p. 199–209. Copyright John Wiley & Sons Ltd.]

6.8. SEISMIC-REFRACTION EXPLORATIONS OF THE CRUST BENEATH THE OCEANS IN THE 1960s

6.8.1. Introduction

Following the reviews of Jones (1999) and Minshull (2002), the basic structure of the oceanic crust had been established and published in the comprehensive Volume 3 of *The Sea* (Hill, 1963a), before plate tectonics was introduced after the discovery of seafloor spreading (Hill, 1963a). In 1970, a second comprehensive publication followed, published as Volume 4 of *The Sea* and consisting of two voluminous parts edited by A.E. Maxwell: *The Sea, New Concepts of Sea Floor Evolution*, which comprised all new results assembled during the 1960s. Until the beginning of the 1960s, the averaging seismic-refraction method was applied which did not allow the measurement of local variations in sediment thickness or details of its stratification. The data were, however, of sufficient quality that they could be reinterpreted in the 1970s when the reflectivity method became available and synthetic seismograms could be computed (Orcutt et al., 1976; Spudich and Orcutt, 1980).

The 1960s were the decade when plate tectonics was born, and knowledge of the properties of the oceanic crust was critical. The proper interpretation of marine magnetic anomalies came at this time (Menard, 1964). The Lamont ships had surveyed a vast amount of profiles in all oceans (Ewing and Ewing, 1970, fig. 1), and the Scripps Institution also had accomplished extensive work. Other nations were active as well, for example, Canada (e.g., Dainty et al., 1966), Japan (e.g., Den et al., 1969; Murauchi

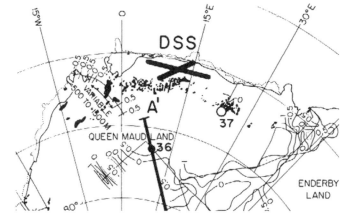

Figure 6.7.4-01. Location of the Soviet Deep Seismic Sounding (DSS) experiment in Antarctica (from Bentley, 1973, fig. 1). The line from A′ southward was part of shallow seismic investigations not discussed here. [Tectonophysics, v. 20, p. 229–240. Copyright Elsevier.]

et al., 1968; Okada et al., 1973), the former USSR (e.g., Galperin and Kosminskaya, 1964; Neprochnov et al., 1967; Neprochnov et al., 1970; Zverev and Tulina, 1973), and Germany. The German research vessel *Meteor* (1964) was the second research vessel carrying the name *Meteor*.

Already between 1924 and 1939 the first *Meteor* (1924) had explored the oceans and had contributed to the detection of the existence of mid-ocean ridges. The second *Meteor* (1964) concentrated its research expeditions on the northeastern half of the Atlantic Ocean, the Mediterranean, and the Red Sea. Between 1964 and

1985, it undertook 73 expeditions and covered 650,000 nautical miles; at least 12 expeditions included major seismic investigations of the oceanic crust (Bröckel, 1994; see also http://www.dfg-ozean.de/berichte/fs_meteor, report M0; Sarnthein et al., 2008).

As was pointed out in Chapter 5, all instruments in use in the 1950s and 1960s were essentially echo sounders utilizing pulses of low-frequency sonic energy and a graphic recording system that displayed the data in the form of cross sections. Techniques and equipment for continuous profiling were developed early in the 1960s and successfully applied in the following years, enabling studies of local variations in sediment thickness and details of its stratification. Ewing and Ewing (1970) have described in detail the instrumentation that was used in all their excursions to obtain detailed information on the sedimentary structure. Their figure 1 shows the seismic profiler traverses made by Lamont ships and personnel through 1967 (*in* Maxwell, A.E., ed., The Sea, Volume 4, New Concepts of Ocean Floor Evolution: New York, Wiley-Interscience, p. 53–84). Work on the application of ocean-bottom seismographs was also an important research aim (e.g., Neprochnov et al., 1968).

The use of internally recording sonobuoys allowed seismic-refraction experiments to conduct from a single ship. Since the late 1960s, low-cost expendable military sonobuoys for launching from ships became available for shooting wide-angle profiles. On entering the water, VHF transmitter aerials and hydrophones were deployed and salt-water batteries actuated (Jones, 1999). One or more hydrophones were suspended from the buoy to detect the seismic head waves which were then transmitted back to the firing ship, timed and recorded. The buoys were designed to sink after ~10 hours.

To overcome eventually the work with explosives, continuous research tried to develop efficient nonexplosive energy sources. High-voltage sparkers, boomers, gas exploders, and magnetostrictive transducers (Ewing and Zaunere, 1964), proved useful for shallow research. Working since 1961 on several types of pneumatic sources, Ewing and Zaunere finally found a pneumatic gun which worked highly satisfactorily in continual operation for several months and produced excellent profiler data both in shallow and deep ocean provinces. In connection with a towed receiving array of eight detectors, the pneumatic gun produced sufficient energy to penetrate deep-sea sediments of over a thousand meters while the ship proceeded with a speed at more than 10 knots. Ewing and Zaunere (1964) predicted that this gun could produce sufficient energy to penetrate the total sedimentary section in all deep ocean areas. Hinz (1969), for example, described in detail his experiences of seismic-reflection work around the Great Seamount in the North Atlantic Ocean using a pneumatic sound source. It would only be in the 1990s, however, that for deep crustal and upper-mantle research at sea, airguns became the almost exclusive powerful and reliable sources and finally replaced explosives almost completely.

The analysis methods, available in the 1950s and also during the 1960s (Ewing, 1963a), used least-squares slope-intersect solutions for picked first arrivals, but did not allow for velocity gradients. The igneous crust had been found to be subdivided into a 2–3-km-thick upper layer 2 and a 3–5-km-thick lower layer 3, while the overlying sediments were labeled as layer 1. The two-layer structure of the ocean basins which had been deduced from the many marine expeditions in the late 1940s and 1950s was basically confirmed in the 1960s.

A first approach for a changed view of crustal structure assuming transitional rather than sharp boundaries between layers was proposed by Helmberger (1968). With an example of marine data obtained in the Bering Sea, he developed the theory to treat discontinuities as layered transition zones and calculated corresponding synthetic seismograms.

Until 1967, the oceans and their side branches, such as the Mediterranean Sea or the Gulf of Mexico, had been covered by cruises of the Lamont ships (Ewing and Ewing, 1970, fig. 1), and an enormous wealth of deep crustal seismic data had been obtained as was summarized in great detail in eight summary volumes of *The Sea*, of which Volumes 3 (Hill, 1963a) and 4 (Maxwell, 1970) specialized in seismic research. Introductions to the results of the 1960s in Volume 4 were given by Ewing and Ewing (1970) for the reflection method and Ludwig et al. (1970) for the refraction method. On all cruises aiming for deep seismic research of the whole oceanic crust, for which the refraction seismics was the only available tool at that time, detailed reflection surveys were always a major component. So, a very detailed picture on the sedimentary structure overlying the basement resulted from these measurements. In their sediment map of the North Pacific, Ewing and Ewing (1970, fig. 5) have shown an impressive summary what was known by 1967 on the thickness of the sediments for example in the northern Pacific Ocean.

A few ocean-floor samples had become available during the 1960s and confirmed the suggestion of Cann (1970) on the possible composition of the oceanic crust. Layer 2 consisted of basaltic lavas and their feeder dikes which were unmetamorphosed in their upper parts. The underlying layer 3 consisted of dense dike swarms of basic composition and massive plutonic gabbros. The two-layer crust was underlain by mantle peridotites. In places, serpentinite diapirs from the upper mantle would cut through the oceanic basement. Such a structure also conformed with land-based evidence from ophiolite sequences which had been interpreted as uplifted sections of oceanic crust. An example of such a structure was the Troodos Massif in Cyprus, which had been studied by Vine and Moores (1972).

6.8.2. The Atlantic Ocean

Searching through the literature for seismic-refraction work in the 1960s, there are not very many publications which report on new marine investigations in the Atlantic Ocean and adjacent seas, such as the Gulf of Mexico and Caribbean Sea. It appears that after the enormous effort of the 1950s, the researchers of the large laboratories in North America, which concentrated on deep sea research, regarded the experimental work in the Atlantic Ocean as being "done." Rather, the experimental activities in the 1960s shifted into the Pacific and Indian Oceans. Based on hundreds of kilometers of traverse recorded in the Atlantic Ocean in the 1950s, Ewing (1969) published a generalized structural section from the North American continental shelf to the Mid-Atlantic Ridge (Fig. 6.8.2-01). Ewing and Ewing (1970) have compiled a map showing all seismic profiler traverses accomplished by Lamont ships around the world until 1967.

Ewing (1969) described the results of numerous individual investigations and summarized them as follows: The main crustal layer, named oceanic layer or layer 3, is very uniform in wave velocity and thickness. It is ~5 km thick and has velocities between 6.5 and 7.1 km/s, while the velocities of the basement layer or layer 2 are between 4.5 and 5.5 km/s. This layer 2 was often not detected when thick sediments accumulated above. Velocity histograms showed that 80% of mantle velocities fell between 7.7 and 8.3 km/s, but no obvious relationship between upper-mantle velocity and depth or thickness of overlying crustal material could be detected. However, an inverse relationship did appear to exist between mantle depth and water depth. On average, the depth from the water surface to the Moho is 12 km, but significant deviations from this average were found on the flanks of the Mid-Atlantic Ridge. In the crestal zone the Moho disappears as a distinct boundary and mantle velocities were found at shallow depths of 9–10 km, and a thick layer of intermediate velocity exists instead of normal oceanic crust and upper mantle.

A brief review on crustal structure research of the Gulf of Mexico until the end of the 1960s was published by Hales (1973) based on profiles observed in the 1950s (see section 5.6; Ewing et al.,1960; Antoine and Ewing, 1963) and a profile recorded by the Texas A&M University in 1965 (Fig. 6.8.2-02).

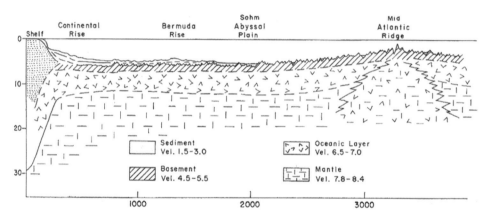

Figure 6.8.2-01. Generalized structural section from the North American continental shelf to the Mid-Atlantic ridge (from Ewing, 1969, fig.1). Horizontal and vertical scales in km, velocities in km/s. [*In* Hart, P.J., ed., The Earth's crust and upper mantle: American Geophysical Union, Geophysical Monograph 13, p. 220–225. Reproduced by permission of American Geophysical Union.]

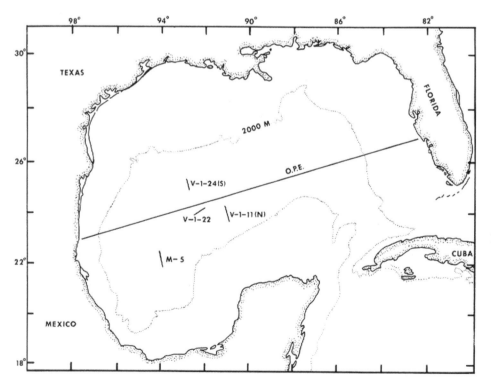

Figure 6.8.2-02. Location of seismic refraction profiles in the Gulf of Mexico (from Hales, 1973, fig. 1). *V-1* lines recorded in the fifties, *M-5* line of 1965, *O.P.E.* large-scale seismic refraction line of 1969. [Tectonophysics, v. 20, p. 217–225. Copyright Elsevier.]

A key result was that the crust underneath the deep part of the Gulf was similar to that of typical ocean basins. Depth to Moho beneath sea level on the crustal profiles shown in Figure 6.8.2-02 was between 16 and 19 km (Hales, 1973). Also from the interpretation of the observations in Mexico from the long line of shooting (O.P.E. in Fig. 6.8.2-02), Hales et al. (1970) concluded that only a typical oceanic crust in both the eastern and western deep basins of the Gulf could satisfy the observations. In addition, the long-range observations of the 1500-km-long O.P.E. line showed a phase beyond 1200 km distance (Fig. 6.8.2-03) which indicated a seismic discontinuity in the mantle at 57 km depth (Hales et al., 1970; Hales, 1973).

Other new activities in the Atlantic Ocean focused less on the general structure of the deep ocean basins than on anoma-lous features. In Germany in 1964, the second research vessel named *Meteor* (1964) had started its service. The expedition M04 of 1966 investigated the Reykjanes Ridge. In 1967, Ger-man researchers undertook a reconnaissance survey (expedi-tion M09) with the German vessel *Meteor* (Aric et al., 1970; Closs et al., 1968, 1969b; Hinz, 1969) to the chain of seamounts (Meteor, Cruiser, Plato, and Atlantis), located at 28°–30°W, 30°–34°N to the west of the triple junction Mid-Atlantic Ridge and the Azores–Gibraltar zone. The area was revisited with the third *Meteor* (1986) in 1990 (Weigel and Grevemeyer, 1999, see Chapter 9.8.2). In 1969, the expedition M17 investigated the Mediterranean and the northern Arabian Sea (Closs et al., 1969a, 1969c). The last expedition of 1969 with a seismic component, M18, was active in the Norwegian Sea (Closs, 1972).

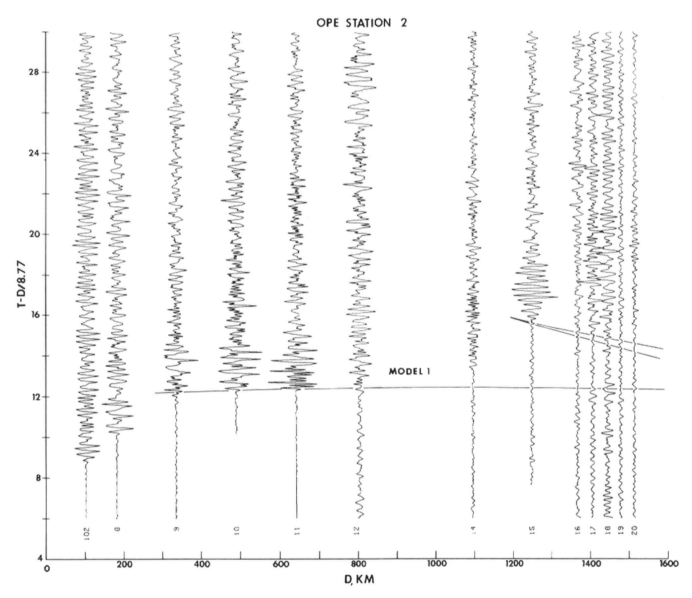

Figure 6.8.2-03. Record section of observations along the *O.P.E.* large-scale seismic refraction line of 1969 in the Gulf of Mexico (from Hales, 1973, fig. 4). [Tectonophysics, v. 20, p. 217–225. Copyright Elsevier.]

In 1968, Keen and Tramontini (1970) shot a seismic-refraction experiment on the western flank of the Mid-Atlantic Ridge between 45° and 46°N. The object was to obtain a reliable determination of the crustal structure near the axis of the ridge. Two ships, one shooting and one receiving, and six sono-buoys, whose positions could be made almost stationary, were used. The seismic arrivals observed by the sonobuoys were transmitted to the recording vessel which also towed a hydrophone array. Charges of up to 150 kg were fired at intervals of 5 or 10 minutes. In total, four profiles were completed The result was a "normal" oceanic crust yielding velocities of 4.6, 6.6, and 8.1 km/s for the first P-wave arrivals within ~30 km of the rift axis, but at 10 km from the western edge of the median valley, the first arrivals resulted in a 7.5 km/s velocity. Moho occurred at a mean depth of 7.5 km.

In 1969, a series of seismic-refraction lines was recorded between Iceland and the Faeroe Islands (Bott et al., 1971). The presence of a 6.8 km/s refractor was seen underneath the Iceland–Faeroe Ridge at a depth of ~7 km and a short unreversed segment with apparent velocity of 7.84 km/s was interpreted as head wave from the Moho at ~16 km depth. The results were later reinterpreted in a different manner (see Chapter 7.8.4; Bott and Gunnarson, 1980).

First surveys also explored the Arctic Ocean (Kutschale, 1966, Hutchins, 1969). Kutschale (1966) describes an experiment of 1962, when an ice island (Arlis II) drifted over a portion of the southern half of the Siberian basin. Amongst other geophysical studies, scientists from the Lamont Geological Observatory of Columbia University also carried out seismic-reflection measurements while aboard Arlis II using special ocean-bottom and ice-surface seismographs. At least 3.5 km of stratified sediments were found to overlie the crust, composed of basement and presumably the "oceanic layer," 6–8 km south of the Alpha ridge and thickening to ~22 km below the ridge.

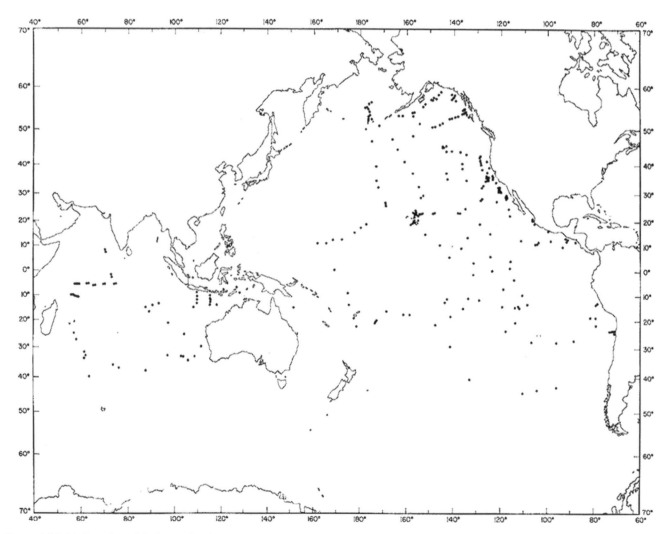

Figure 6.8.3-01. Location of Scripps refraction stations in water depths greater than 2000 m (from Shor and Raitt, 1969, fig. 1). [*In* Hart, P.J., ed., The Earth's crust and upper mantle: American Geophysical Union, Geophysical Monograph 13, p. 225–230. Reproduced by permission of American Geophysical Union.]

6.8.3. The Pacific Ocean

Shor and Raitt (1969) reviewed the results in the Pacific and Indian Oceans. Since the time when Ewing (1963b) and Raitt (1963) published summaries on crustal and upper-mantle velocities obtained from seismic-refraction data in the oceans, considerable additional data had been obtained. To demonstrate the amount of data collected, the surveys which were performed by the Scripps Institution of Oceanography in the Pacific and Indian Oceans are shown in Figure 6.8.3-01.

In a subsequent summary, Shor et al. (1970) summarized the results of more than 200 individual stations recorded in the Pacific Basin in a table, showing coordinates, azimuth, water column, velocities and thicknesses of sediments, transitional and oceanic crustal layers, mantle velocity, and depth to Moho, and they plotted frequency diagrams on velocities and thicknesses for the whole crust and for individual layers.

The many new data confirmed that the averages published by Raitt (1963) for the oceans continued to be valid (Shor and Raitt, 1969; Shor et al., 1970). The velocity of the sediments varied between 1.5 and 3.4 km/s, with their median thickness at 0.34 km. The velocities in layer 2, the transition, varied between 3.4 and 6.0 km/s, with a median of 1.21 km in thickness and a median velocity of 5.15 km/s. The oceanic layer 3 of 4.57 km average thickness showed a median velocity of 6.82 km/s, and the median mantle velocity was 8.15 km/s.

Consistent regional variations were observed for sediments where turbidite flows can be received from continents. Mantle velocities and depths were frequently less than normal along the mid-ocean rift system, and mantle depths appeared to become greater than normal beneath trenches and near the foot of the continental slope where no trenches were visible.

Shor et al. (1968) reported on a series of profiles made in 1965 across the Juan de Fuca Ridge, west of the coast of Oregon. They interpreted these data to mean that the ridge was the surface expression of an upraised oceanic crust and mantle. Near the ridge, the upper-mantle velocity appeared to be below normal and the Moho was found as shallow as 7 km below sea level. Profiles in the Mendocino escarpment in the vicinity of the Gorda Ridge off northern California gave evidence that normal oceanic depths near 11 km of the mantle and high velocities of 8.4–8.5 km/s exist on the south side of the escarpment. They were the same on the north side except near the Gorda Ridge, which seemed to have a similar structure as the Juan de Fuca Ridge.

In 1967, the Hawaii Institute of Geophysics started a program of single-ship seismic wide-angle reflection and refraction measurements aimed primarily at defining sediments and upper crustal layers. A 7000 J sparker was used as the energy source. A year later, in 1968, an airgun source was added to the system. In 1969, with sonobuoys as receivers and a 20-cubic-inch airgun as a repetitive source, a series of ASPER (Airgun and Sonobuoy Precision Echo Recorder) refraction stations were run in waters exceeding 5 km at sites well away from any obvious structural transition or topographic high (Fig. 6.8.3-02).

For the first time, the entire crust could be penetrated and the mantle be reached. In particular, a high-velocity basal crustal layer was detected with velocities between 7.1 and 7.7 km/s at several locations in the Pacific Ocean basin. Sutton et al. (1971)

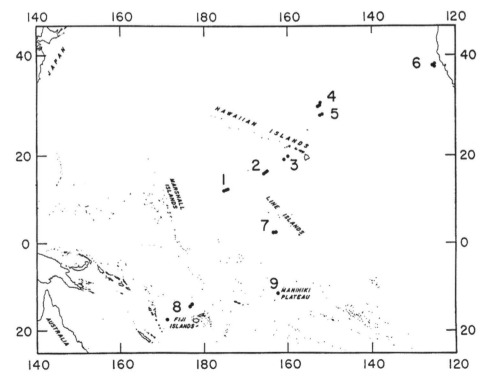

Figure 6.8.3-02. Locations of the ASPER stations conducted in 1969–1970 (from Sutton et al., 1971, fig. 1). The Solomon Islands referred to below and discussed in detail by Furumoto et al. (1973) are located to the NE of no. 8 (Fiji). [*In* Heacock, J.G., ed., 1971, The structure and physical properties of the Earth's crust: American Geophysical Union Geophysical Monograph 14, p. 193–209. Reproduced by permission of American Geophysical Union.]

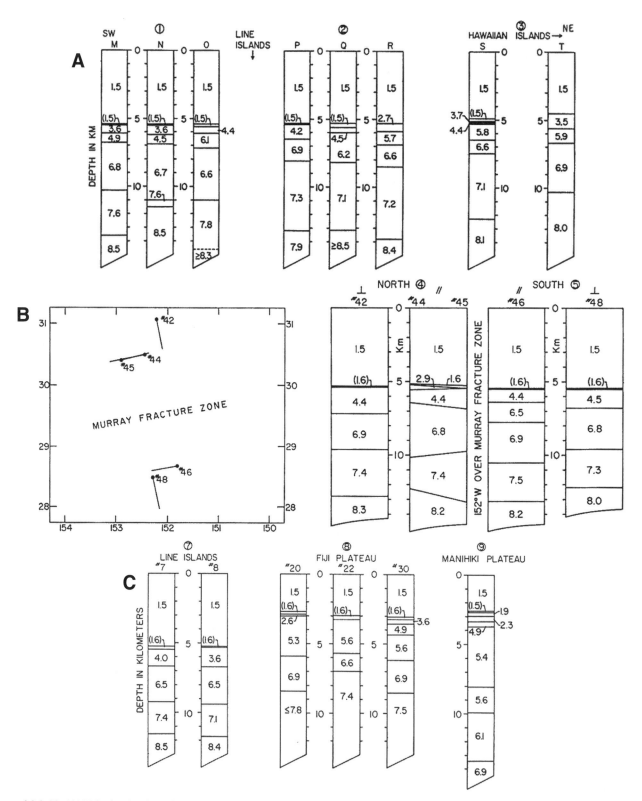

Figure 6.8.3-03. (A) Velocity-depth profiles of the crust in oceanic basin south of Hawaii (from Sutton et al., 1971, fig. 6). Location nos. 1–3 in Fig. 6.8.3-02. (B) Left: Location of ASPER stations on either side of the Murray fracture zone. Right: Velocity-depth profiles of the crust around the Murray fracture zone (from Sutton et al., 1971, figs. 7 and 8). Location nos. 4 and 5 in Fig. 6.8.3-02. (C) Velocity-depth profiles of the crust in oceanic basin south of Hawaii (from Sutton et al., 1971, fig. 10). For location see Fig. 6.8.3-02. [*In* Heacock, J.G., ed., 1971, The structure and physical properties of the Earth's crust: American Geophysical Union Geophysical Monograph 14, p. 193–209. Reproduced by permission of American Geophysical Union.]

could show that this layer, which in other data evidently was masked because it is seen primarily in secondary arrivals, occurred widely in the Pacific Ocean. On all sites, except on the Fiji and Manihiki Plateaus (sites 8 and 9 in Figs. 6.8.3-02 and 6.8.3-03C), the upper mantle velocity underlying the basal crustal layer was near or above 8.0 km/s

In total, Sutton et al. (1971) constructed 21 crustal columns from these measurements. The results of stations 1–3 (Fig. 6.8.3-03A), obtained in 1969, were spaced along a line extending almost over 1000 km over a deep ocean basin between the Marshall Islands in the SW and the Hawaiian Islands in the NE. Other data (Fig. 6.8.3-03B) were obtained, also in 1969, north and south of the Murray Fracture Zone (stations 4 and 5 in Fig. 6.8.3-02). In 1970, more data (Fig. 6.8.3-03C) could be recorded near the Line Islands, on the Fiji Plateau, and on the Manihiki Plateau (stations 7–9 in Fig. 6.8.3-02).

The results of Sutton et al. (1971) were included in the publication of Furumoto et al. (1973), who summarized in particular the many seismic surveys around the Hawaiian archipelago (Figs. 6.8.3-04 and 6.8.3-05), as well as those around northern Melanesia (Figs. 6.8.3-08 and 6.8.3-09). The crustal thickness of the island of Hawaii varies from 12 to 17 km, the crust of Maui and Lanai is ~15 km thick, and at 19–20 km, the crust underneath

Oahu is the thickest (Fig. 6.8.3-04). The crust under Kauai and Nihoa is ~17 km thick, but under the smaller islands to the northwest, crustal thickness is on average only 12 km (Fig. 6.8.3-05).

A special crustal survey of the main island of Hawaii was also performed and published by Hill (1969). In the experiment of 1964, shots with charges between 150 and 200 kg were fired at sea within 2 km of the shore at 10-km intervals along each of the three coasts. The shots fired along a particular coastline were recorded by 5 seismic-refraction units spaced at ~25 km intervals along each coast and by 13 short-period seismograph stations of the permanent HVO (Hawaiian Volcano Observatory) seismic network.

Furthermore, in 1965, three 500-ton charges of high explosives were detonated on the ground surface on Kahoolawe Island south of Maui and recorded by the HVO network stations on Hawaii. For the interpretation, the position of shotpoints and recording stations were projected onto straight lines implying that variations in subsurface structure perpendicular to the "profiles" could be corrected by special treatment. Figure 6.8.3-06 shows an example of the record section of a station located at the northern edge of the island of Hawaii (corresponding to position 5 of Fig. 6.8.3-04) and its recordings of shots along the northeastern coast. The interpretation gave an 11–12-km-thick crust along the

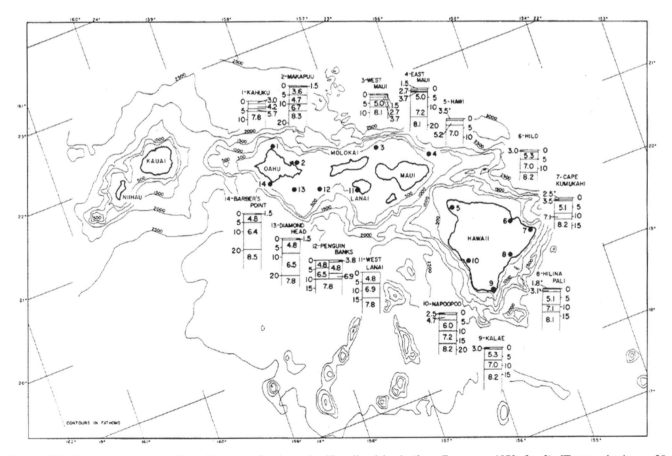

Figure 6.8.3-04. Velocity-depth profiles of the crust for the major Hawaiian islands (from Furumoto, 1973, fig. 2). [Tectonophysics, v. 20, p. 153–164. Copyright Elsevier.]

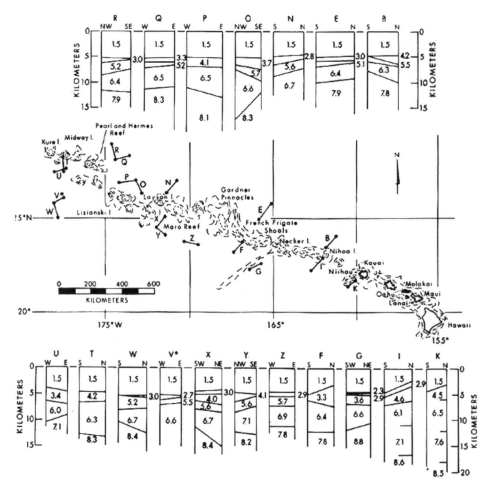

Figure 6.8.3-05. Velocity-depth profiles and locations of the seismic traverses around the Hawaiian archipelago west of the main islands (from Furumoto, 1973, fig. 3). [Tectonophysics, v. 20, p. 153–164. Copyright Elsevier.]

northeast and southwest flanks of the volcano Kilauea, but indicated an increase in thickness from north to south along the west coast from 14 km to ~18 km. The upper-mantle velocity under most of the island was 8.2 km/s.

The Lamont-Doherty Geological Observatory collected a large set of 190 sonobuoy stations seismic data in the northern Pacific, also in 1969 (Houtz et al., 1970), north of latitude 10°S and east of longitude 150°E (Fig. 6.8.3-07). In contrast to those of Sutton et al. (1971), these measurements had not aimed to study the whole crust and reached neither the previously discussed basal layer nor the Moho, but aimed instead for detailed research of the behavior of layer 2, the upper basement layer beneath the sediments only. On average, a layer-2 thickness of 1.11 km resulted with an average velocity of 5.52 km/s, overlying layer 3 with an average velocity of 6.63 km/s.

The Bismarck archipelago is the northwestern part of northern Melanesia and includes the islands of New Britain and New Ireland bordered by the Bismarck Sea to the northwest and the Solomon Sea in the southeast. Sea experiments were carried out here by the Australian Bureau of Mineral Resources, Canberra, in the area of the Bismarck and Solomon Seas near the continental margin north of Australia and New Guinea. In 1967 and 1969, a total of ~100 shots of up to 1 ton of TNT were recorded

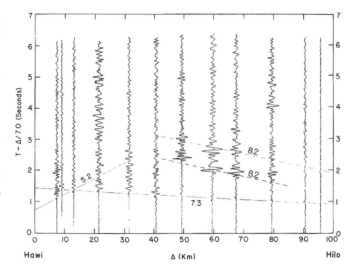

Figure 6.8.3-06. Record section of observations of a station located at the northern edge of the island of Hawaii (corresponding to position 5 of Fig. 6.8.3-04) and its recordings of shots along the northeastern coast of the main island of Hawaii (from Hill, 1969, fig. 5). [Bulletin of the Seismological Society of America, v. 59, p. 101–130. Reproduced by permission of the Seismological Society of America.]

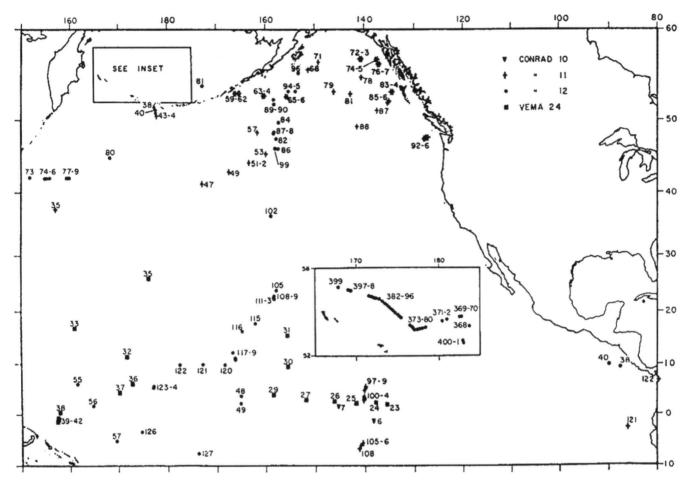

Figure 6.8.3-07. Chart of the North Pacific Ocean showing location of sonobuoy stations recorded in 1969 (from Houtz et al., 1970, fig. 1). [Journal of Geophysical Research, v. 75, no. 26, p. 5093–5111. Reproduced by permission of American Geophysical Union.]

by up to 47 stations. The recording stations were moved during the course of the survey to give wider coverage of New Britain and New Ireland (Fig. 6.8.3-08). Maximum distances were 300 km (Finlayson et al., 1972). A time-term analysis applied to the data by Finlayson and Cull (1973) resulted in a 10-km-thick crust including 2 km of water for the Bismarck Sea and a 12-km-thick crust including 5 km of water for the Solomon Sea to the southeast, with a crustal thickening to 20 km under the intervening New Britain. East of New Ireland, the crust of the Pacific plate appeared to be 35–40 km thick (Finlayson and Cull, 1973). The results have been summarized together with other surveys around the Solomon Islands and the southeastern part of Northern Melanesia by Furumoto et al. (1973) and are shown in Figure 6.8.3-09.

Seismic investigations were also performed in the area of the Cook Islands. In 1965, during the New Zealand Eclipse Expedition, seismic-refraction studies were made to investigate the structure of some of these volcanic islands. Near Aitutaki, two reversed profiles with a total length of 120 km determined a downwarp of the crust-mantle boundary by 1.6 km, the crustal thickness being 6.4 km and the upper-mantle velocity 7.7 km/s

under the islands (Hochstein, 1968). The Lamont group was also active in this area in the 1960s and 1970s (F.J. Davey, 2010, personal commun.).

In 1967, the Scripps Institution investigated most of the Melanesian Borderland (Shor et al., 1971) bounded on the west by Australia and on the east by New Zealand and the Tonga-Kermadec trench (Fig. 6.8.3-10). Two ships were used for the seismic measurements, using the techniques described by Shor (1963) with only a few changes. The electronic equipment had been slightly modified, and reflection profiling was attempted throughout much of the refraction survey work. A shipboard computer enabled rapid computation of preliminary results. Explosives of up to 200 kg per shot continued to be the source of energy. Almost all profiles could be reversed and the profile directions were laid along strike of topographic features as much as possible.

The data analysis followed the conventional method by fitting lines to the traveltime data by least squares with the requirement that reversed times must agree for each layer. An example is shown in Figure 6.8.3-11. The results are shown as velocity-depth cross sections in Figure 6.8.3-12. The Lord Howe Rise

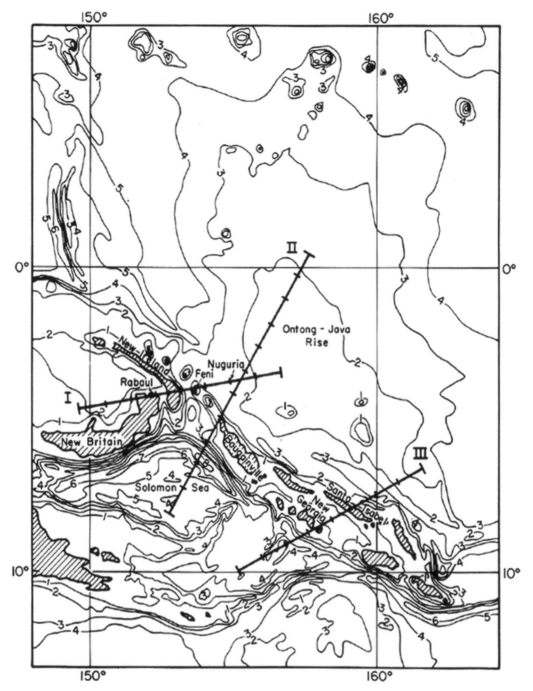

Figure 6.8.3-08. Map of northern Melanesia showing the locations of the crustal cross sections of Fig. 6.8.3-09 (from Furumoto et al., 1973, fig. 5). [Tectonophysics, v. 20, p. 153–164. Copyright Elsevier.]

and the Norfolk Ridge are topped by thick sediments and have deep crustal roots and thick layers with velocities measured on the Australian continent. The Kermadec and Lau ridges, on the other hand, have structure and velocities typical for island arcs. The South Fiji Basin resembles an oceanic area with additional sedimentary fill. The western part of the Fiji Plateau has thin sediments and the structure of a normal deep-ocean basin that has been uplifted by 2 km (Shor et al., 1971).

Also in 1967, the continental margin off northern Queensland was investigated by a two-ship wide-angle seismic profiling program, where 19 profiles were shot predominantly along a traverse from Cairns, Queensland, to Milne Bay, Papua New Guinea, and from Townsville across the Queensland Plateau to the deep Coral Sea Basin and Lousiade Rise (Ewing et al., 1970; Mutter and Karner, 1979; Finlayson 2010, Appendix 2-2).

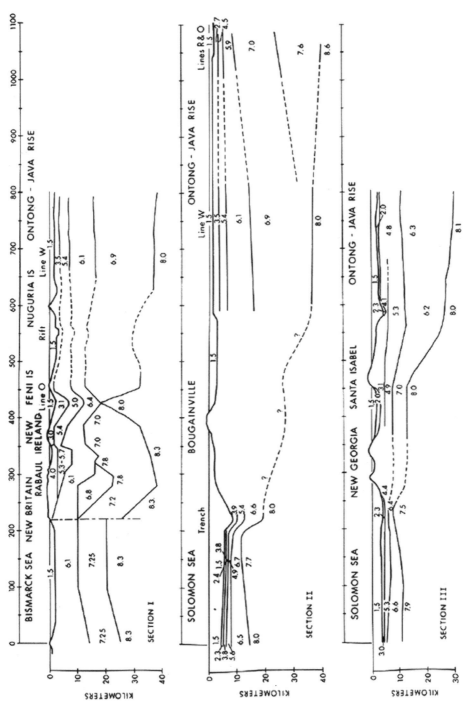

Figure 6.8.3-09. Velocity-depth profiles of the crust of northern Melanesia (from Furumoto et al., 1973, fig. 6). Locations are shown in Fig. 6.8.3 08. [Tectonophysics, v. 20, p. 153–164. Copyright Elsevier.]

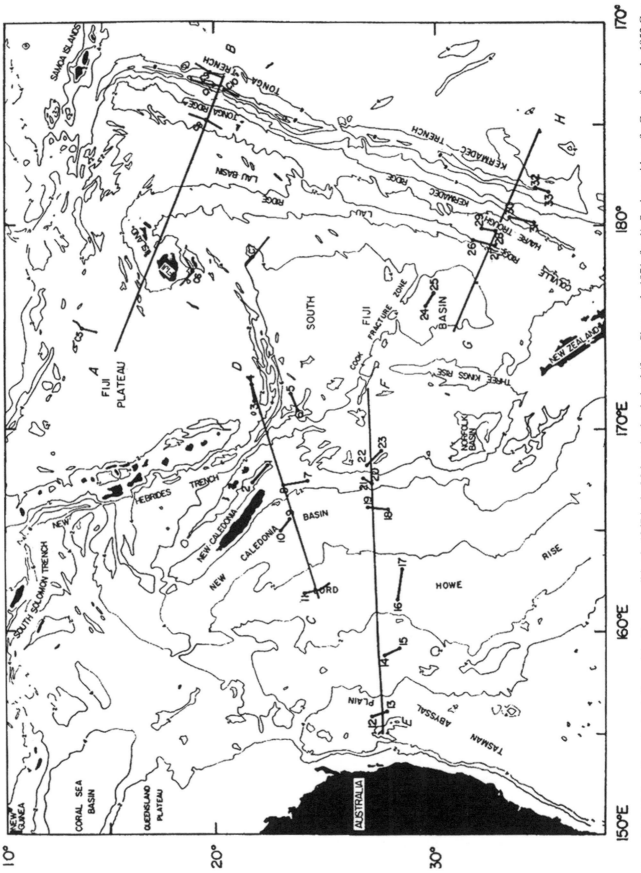

Figure 6.8.3-10. Location of seismic lines of the Nova expedition in 1967 in the Melanesian borderland (from Shor et al., 1971, fig. 1). Stations with prefix C are from the 1952 Capricorn expedition. [Journal of Geophysical Research, v. 76, no. 11, p. 2562–2586. Reproduced by permission of American Geophysical Union.]

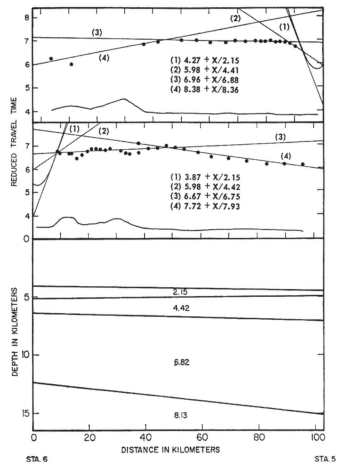

Figure 6.8.3-11. Traveltime plots and layer solutions for two of the Nova expedition reversing seismic refraction stations (from Shor et al., 1971, fig. 4). [Journal of Geophysical Research, v. 76, no. 11, p. 2562–2586. Reproduced by permission of American Geophysical Union.]

From various profiles in the Philippine Sea near latitude 25°N (Fig. 6.8.3-13), a crustal cross section (Fig. 6.8.3-14) was constructed by Murauchi et al. (1968) between the continental shelf of the eastern China Sea and the northwest Pacific Basin showing distinct crustal thickenings under the Nansei Shoto Ridge, the Oki-Daito Ridge, and the Honshu-Mariana Ridge, intervened by thin oceanic crust with Moho depths of 11–14 km below sea level. Only under the Honshu-Mariana Ridge, Moho was found with 8.0 km/s upper mantle velocity at 16 km depth; under the other ridges, the upper mantle was not reached.

Another interesting anomalous feature, found in the 1960s in the Northwest Pacific Basin along latitude 32°N and first published by Den et al. (1969), was the Shatsky Rise (Ludwig et al., 1970, fig. 13). Layer 3 with a velocity near 6.9 km/s at 8–14 km depth underneath the adjacent oceanic crust was found to be subdivided into two layers with average velocities of 6.9 and 7.5 km/s and reaching a total thickness of near 15 km, its lowest depth being near 24 km below sea level, its root being slightly shifted to the west versus the highest topographic eleva-

tion (Fig. 6.8.3-15). For comparison, Den et al. (1969) included to the left of their crustal section two crustal columns which had been obtained by Ludwig et al. (1966) in the westernmost part of the northwest Pacific Basin.

Russian scientists continued their exploration of the transition zone from the Asian continent to the Pacific Ocean with a second phase starting in 1963. The seismic investigations concentrated on the western part of the Pacific Ocean between the southern Kuril Islands and South Kamchatka (Kosminskaya et al., 1973). The experiments focused in particular on the Sakhalin-Hokkaido marginal zone, the Yamato Bank, the southern Kuril Islands and the earthquake focal zone east of Kamchatka. A particularly detailed profile was recorded across Iturup Island and further into the Pacific Ocean (Zverev and Tulina, 1971).

In 1964, Hess published an article on seismic anisotropy of the upper mantle under the oceans (Hess, 1964). To exclude the possibility that the observed directional dependence of seismic velocity was caused by lateral velocity changes, a series of special anisotropy experiments was subsequently carried out in various areas of the Pacific Ocean (Keen and Barrett, 1971; Morris et al., 1969; Raitt et al., 1969, 1971).With a special scheme of shot and receiver positions (Fig. 6.8.3-16), it was ensured that P-waves traversed the same area of the upper mantle in various azimuths.

In the subsequently applied time-term analysis of the P-traveltimes, both topography of the crust-mantle boundary and directional dependence of the velocities were taken into account. In the areas of Hilo and Show near Hawaii (Morris et al., 1969), this dependency became very clear: the anisotropy here was near 8%. In the areas of Flora and Quaret near the Californian coast (Raitt et al., 1969), the resulting anisotropy was 3% and 4%. Fuchs (1975, 1977) published a compilation of all anisotropy values obtained in the Pacific Ocean (Fig. 6.8.3-17). The compilation clearly shows that the directions of maximum velocity are almost perpendicular to the mid-ocean rises and run parallel to the major fault zones. In the Atlantic Ocean, only one experiment had been carried out by the end of the 1960s, resulting in 8% anisotropy (Keen and Tramontini, 1970).

6.8.4. The Indian Ocean

The crustal structure of the Indian Ocean was also investigated in the 1960s by several seismic refraction measurements, distributed to almost all directions. Laughton et al. (1970) have compiled a synthesis of the present knowledge on the Indian Ocean, as available up to January 1969, from seismic, magnetic, geothermal, topographic, earthquake epicenter, paleontological, and other geophysical and geological information, obtained during numerous cruises of research vessels mainly in the 1960s. In this context, we have extracted the information on the deep crustal structure as revealed from seismic data measured in the 1960s.

The Central Indian Ridge striking in N-S direction to the west of the Central Indian Basin was traversed in 1962 by a seismic-refraction profile at 6°S, 68°E (Fig. 6.8.4-01). However,

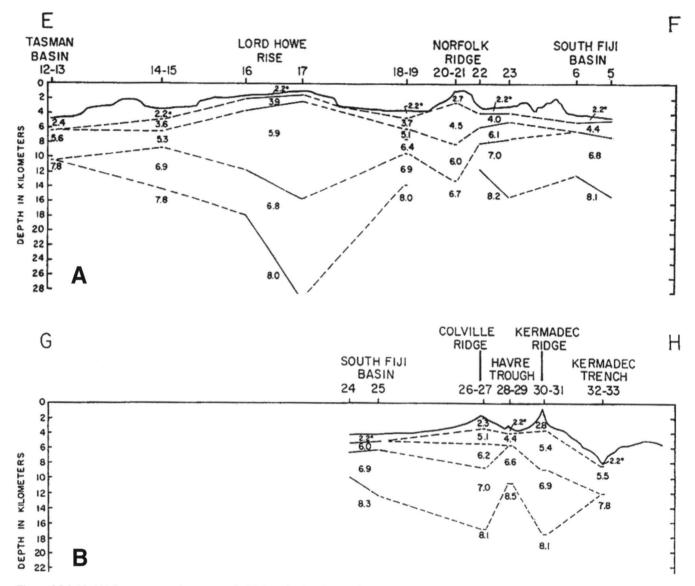

Figure 6.8.3-12. (A) Structures sections across the Melanesian borderland from the Tasman Basin to the Fiji Basin (for location see Fig. 6.8.3-10) (from Shor et al., 1971, figs. 2 and 3). (B) Structures sections across the Melanesian borderland from the South Fiji Basin to the Kermadec Trench (for location see Fig. 6.8.3-10) (from Shor et al., 1971, figs. 2 and 3). [Journal of Geophysical Research, v. 76, no. 11, p. 2562–2586. Reproduced by permission of American Geophysical Union.]

no station was recorded over the central 235 km (Francis and Shor, 1966). On the flanks, normal oceanic crust was found with crustal thicknesses of 4–7 km and subcrustal velocities of 8–8.4 km/s. Russian seismic-refraction data showed in their survey area, which lay in the center of the profile of Francis and Shor, a high-velocity layer with 7–7.5 km/s beneath 2.5 km of layer 2 with velocities of 4.5–5 km/s (Fig. 6.8.4-02). They found, however, no indication of subcrustal velocities (Neprochnov et al., 1967).

The Ninetyeast Ridge striking in a N-S direction to the east of the Central Indian Basin (Fig. 6.8.4-03; for location, see also Fig. 9.8.3-01) was first studied by a seismic survey in 1952 when

it was thought to be an isolated seamount (Gaskell et al., 1958). Further seismic work followed in 1962, and it was then believed that the Ninetyeast Ridge at 10°S, 89°E was a horst-type feature in which oceanic crust had been upthrust (Francis and Raitt, 1967).

The Seychelles Bank at 5°S, 56°E was shown to be underlain by continental crust, based on seismic-refraction and gravity data obtained in 1962, while further south, the structure of the Saya de Malha Bank appeared as a coral capping, 1.5 km thick, which covers an oceanic type crust (Shor and Pollard, 1963).

In the eastern half of the Indian Ocean, three possibly continental-type aseismic plateaus were defined (Laughton et al., 1970). From seismic studies in 1962, Francis and Raitt (1967)

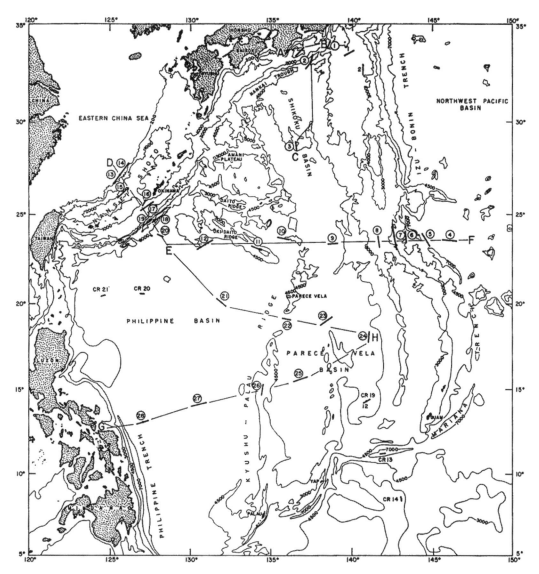

Figure 6.8.3-13. Geographic location of seismic profiles in the Philippine Sea (from Murauchi et al., 1968, fig. 1). [Journal of Geophysical Research, v. 73, no. 10, p. 3143–3171. Reproduced by permission of American Geophysical Union.]

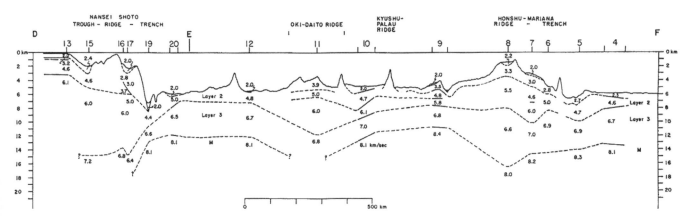

Figure 6.8.3-14. Structure section from the continental shelf of the Eastern China Sea into the Philippine Basin in NW-SE direction and continued into the NW Pacific basin in W-E direction (for location see Fig. 6.8.3-13) (from Murauchi et al., 1968, fig. 3). [Journal of Geophysical Research, v. 73, no. 10, p. 3143–3171. Reproduced by permission of American Geophysical Union.]

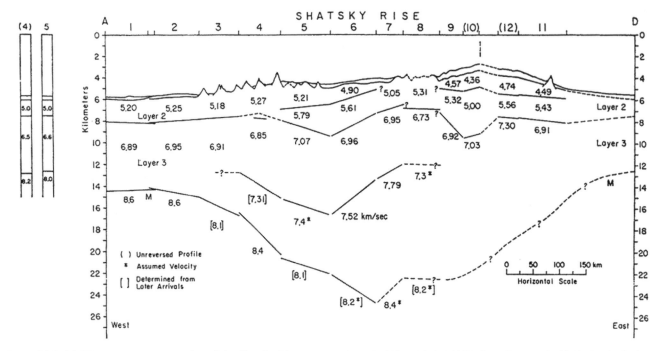

Figure 6.8.3-15. West-east structure section of the Shatsky Rise at approximately 32°N, 151°–160°E (from Den et al., 1969, fig. 3). Left: two crustal columns of Ludwig et al. (1966) from the Northwest Pacific basin. [Journal of Geophysical Research, v. 74, no. 6, p. 1421–1434. Reproduced by permission of American Geophysical Union.]

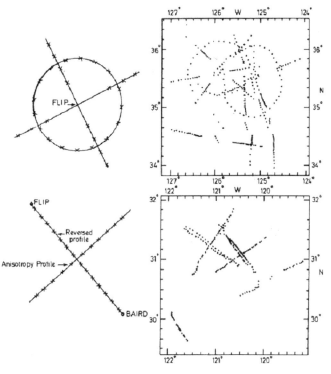

Figure 6.8.3-16. Shot and receiver arrangements in specific areas of the Pacific Ocean (from Fuchs, 1975, fig. 3). [Geologische Rundschau, v. 64, p. 700–716. Reproduced with kind permission of Springer Science+Business Media.]

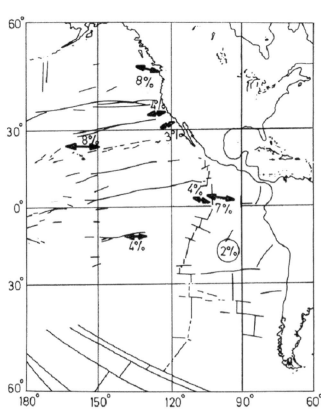

Figure 6.8.3-17. Distribution of anisotropy-values in the Pacific Ocean in direction and percentage (from Fuchs, 1975, fig. 6). [Geologische Rundschau, v. 64, p. 700–716. Reproduced with kind permission of Springer Science+Business Media.]

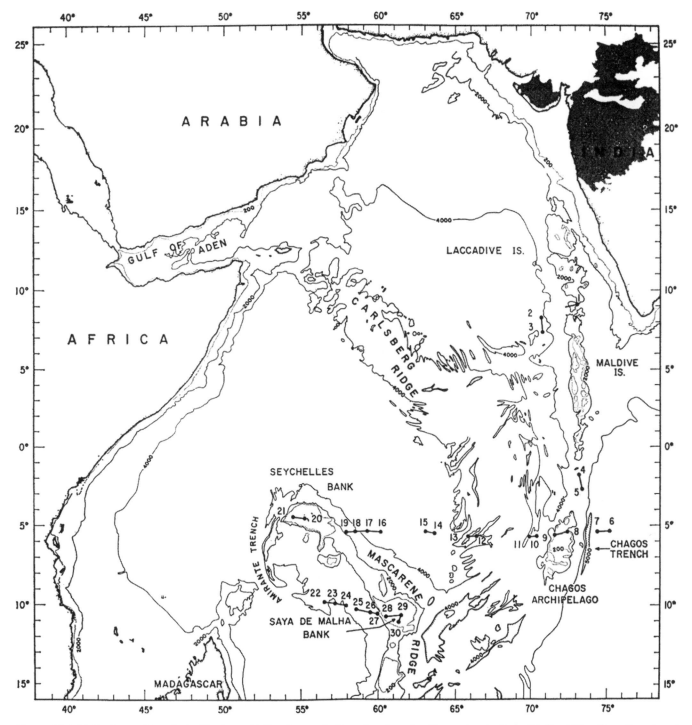

Figure 6.8.4-01. Station positions in the Northwest Indian Ocean, the line of stations 6 to 20 crossing the Central Indian Ridge (from Francis and Shor, 1966, fig. 1). [Journal of Geophysical Research, v. 71, p. 427–449. Reproduced by permission of American Geophysical Union.]

had suggested that the Broken Ridge and the Naturaliste Plateau were once part of the Western Australian Shield and that "the crust has either been thinned tectonically or the mantle has risen at the expense of the crust by some process at the Moho" (Fig. 6.8.4-04). The third Wallaby Plateau was not surveyed. In their

summary, Laughton et al. (1970) concluded that the majority of the aseismic ridges might well be called "microcontinents."

Other seismic investigations in the Indian Ocean in the 1960s concentrated on the continental margins. On the Agulhas Bank, located south of Africa immediately to the southeast of the

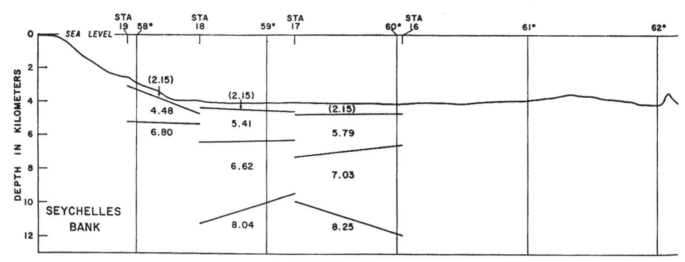

Figure 6.8.4-02 (*on this and following page*). Seismic profile between 57° and 67°E along a line between 5° and 6°S crossing the Central Indian Ridge (from Francis and Shor, 1966, fig. 4).). [Journal of Geophysical Research, v. 71, p. 427–449. Reproduced by permission of American Geophysical Union.]

Cape, the shelf extends 200 km from the coast (Fig. 6.8.4-05). Seismic-refraction studies were made in 1962 by Green and Hales (1966) and others were added later in the same year by Ludwig et al. (1968). A series of stations on the shelf showed continental crustal velocities that could be correlated with known lithologic structures (Fig. 6.8.4-06). An E-W section between the continental shelf off Durban and the Mozambique Basin along latitude 29°30′ S (Fig. 6.8.4-07) by Ludwig et al. (1968) suggested that there is oceanic crust on both sides of the ridge and that between the ridge and the continent is considerable cover. The Transkei Basin further south was established as being oceanic.

Additional seismic-refraction studies were carried out in the southern Indian Ocean in 1968 (Hales and Nation, 1972, 1973b) on the Agulhas Bank and Plateau and in deep water in the Transkei Basin adding to the previous work of Green and Hales (1966) described above. Free-floating recording buoys, built in 1963 and 1964 in the Southwest Center for Advanced Studies at Dallas, Texas, recorded shots which were alternatively floating and sinking or suspended at comparatively shallow depths (5–10 m) to produce bubble-pulse periods of 0.5–1 s. Several large shots of 900 kg, shot on a 190-km-long north-south profile along the Agulhas Bank, were also recorded by two land stations in South Africa. Arrival times beyond 120 km distance showed an anomalously high apparent velocity of 8.9 km/s. Hales and Nation (1972) did not believe that this high velocity was caused only by a dipping Moho.

Apart from the entrance to the Gulf of Aden, where the mid-oceanic Shaba Ridge penetrates the continent (crustal experiments see subchapter 6.7.2), the continental margins of Africa and Arabia as far north as Pakistan showed certain characteristics that distinguished them from the North Atlantic–type margins (Laughton et al., 1970). The question was raised as to whether the true edge of the African continent lay halfway between the coast and the Seychelles. To investigate this problem, a seismic-refraction sec-

tion was laid out between Kenya and the Seychelles (Francis et al., 1966, reproduced by Laughton et al., 1970, fig. 10c). The western half of the section showed an abnormally thick crust (9–19 km) that comprised 4–12 km of sediments lying above a basement which, although having the appropriate velocity, lacked the magnetic properties of an oceanic crust. In contrast, the eastern half of the section showed an abnormally thin crust of 3 km in thickness with typical oceanic magnetic anomalies.

In the northern Arabian Sea, 17 seismic-refraction profiles were shot at the continental margins of Somalia and western India during the expedition M01 of the German research vessel *Meteor* (64) (Bungenstock et al., 1966; Closs et al., 1969a). These lines served to construct structural profiles, one from the Oman Basin across the Murray Ridge and three across the continental margin between Karachi and Bombay. In all the profiles, which extended 220 km seaward from the continental shelf edge, the basement was seen to be covered by 5–8 km of sediments. Basement velocities varied from 6 to 6.9 km/s, which seemed low for the oceanic crustal layer that was expected at a depth of 8 km. No base of this crustal layer was found.

From the various seismic profiles discussed above, several profiles also touched the Indian Ocean basins (Francis et al., 1966; Francis and Shor, 1966; Francis and Raitt, 1967; Neprochnov et al., 1967) and composite sections of the available seismic data were compiled and discussed by Laughton et al. (1970, fig. 11). They indicated that the structure under the Indian Ocean basins did not differ appreciably from that in other basins of the world. Apparent anomalous regions could be explained by extraordinary locations of the corresponding seismic stations close to disturbed structures. For example, huge sediment thicknesses near the Indus and Ganges cones were derived from erosion of the Himalayan mountain chain and extend far into the Central Indian Ocean Basin as far south as the equator with a thickness of greater than 2 km.

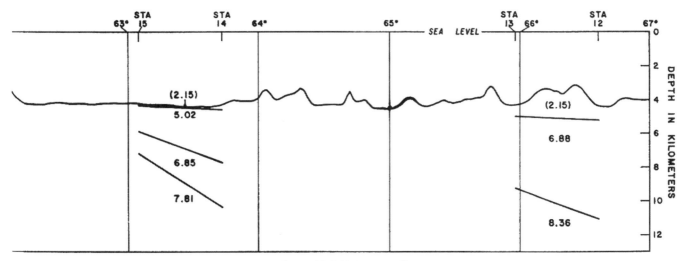

Figure 6.8.4-02 *(continued)*.

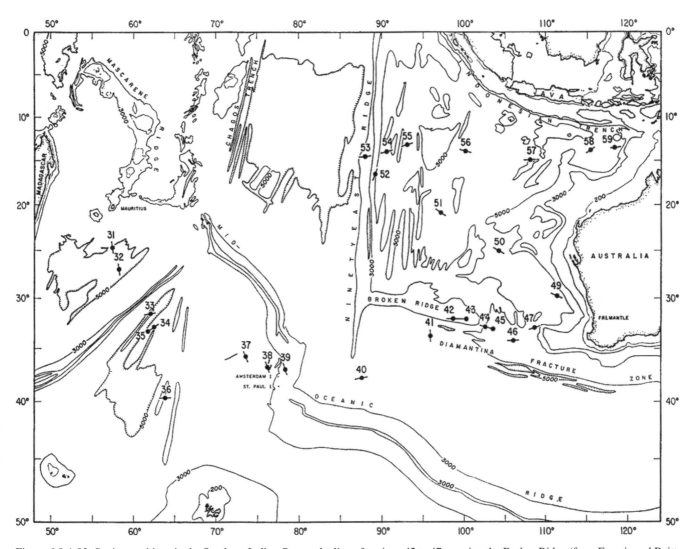

Figure 6.8.4-03. Station positions in the Southern Indian Ocean, the line of stations 42 to 47 crossing the Broken Ridge (from Francis and Raitt, 1967, fig. 1). [Journal of Geophysical Research, v. 72, p. 3015–3041. Reproduced by permission of American Geophysical Union.]

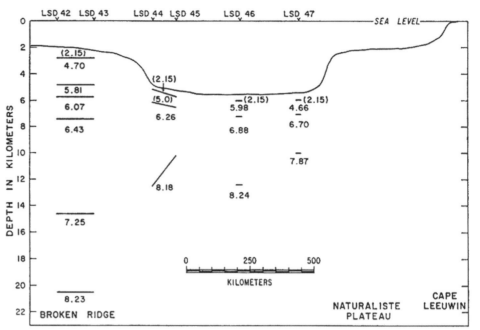

Figure 6.8.4-04. Seismic section on a line between station 42 and the tip of Australia (Cape Leeuwin) (from Francis and Raitt, 1967, fig. 3). [Journal of Geophysical Research, v. 72, p. 3015–3041. Reproduced by permission of American Geophysical Union.]

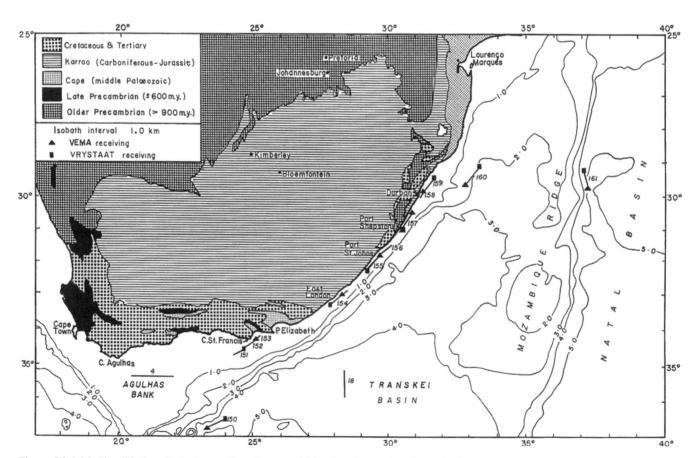

Figure 6.8.4-05. Simplified geological map of southeastern Africa showing the locations of offshore seismic refraction profiles (from Ludwig et al., 1968, fig. 1). [Journal of Geophysical Research, v. 73, p. 3707–3719. Reproduced by permission of American Geophysical Union.]

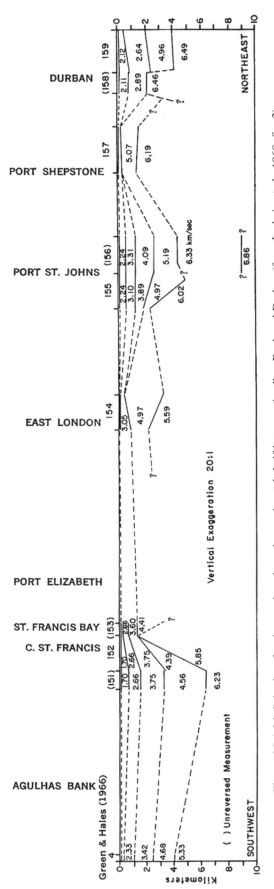

Figure 6.8.4-06. Seismic refraction section along the continental shelf between Agulhas Bank and Durban (from Ludwig et al., 1968, fig. 2). [Journal of Geophysical Research, v. 73, p. 3707–3719. Reproduced by permission of American Geophysical Union.]

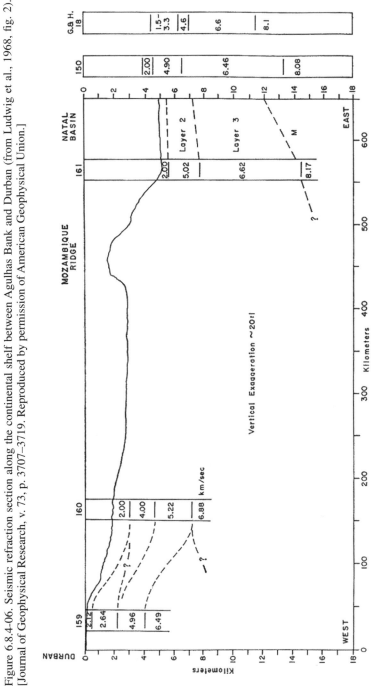

Figure 6.8.4-07. Seismic refraction section between the continental shelf off Durban and the Natal basin along latitude 29°30′ S (from Ludwig et al., 1968, fig. 3). [Journal of Geophysical Research, v. 73, p. 3707–3719. Reproduced by permission of American Geophysical Union.]

6.8.5. Summary of Seismic Observations in the Oceans in the 1960s

An overview of seismic measurements in the oceans, compiled from the database assembled at the U.S. Geological Survey in Menlo Park, California, for the 1960s shows all surveys that were published until 1969 in generally accessible journals and books (Fig. 6.8.5-01). The map shows the numerous observations in continental shelf and adjacent deep ocean areas of North America and eastern Asia. Also in the Red Sea, Gulf of Aden, as well as in the entire Indian Ocean, a large number of observations were made in the 1960s. In the northern Atlantic Ocean the Mid-Atlantic Ridge and in the Pacific Ocean the Hawaii islands were of special interest.

On the basis of many observations in the Atlantic Ocean, a detailed overall picture of the oceanic crust was established, but significant deviations from this average were found on the flanks of the Mid-Atlantic Ridge. Numerous new data were also gathered in the Indian and Pacific Oceans. Detailed surveys involved, for example, the surroundings of the islands of Hawaii and archipelagos to the northeast of Australia (not shown on the map). For the first time, the existence of seismic anisotropy of the uppermost mantle was discussed and investigated by a series of special anisotropy experiments in various areas of the Pacific Ocean (e.g., Raitt et al., 1971).

6.9. THE STATE OF THE ART AT THE END OF THE 1960s

The 1960s accumulated a wealth of data and the art of seismic-refraction techniques was gradually improved in both instrumental development and in the art of interpretation. In many experiments, seismic energy produced by effective explosion sources was recorded at large distances of up to several 100 km. Moho contour maps published for Europe (e.g., Morelli et al., 1967), the USSR (e.g., Belyaevsky et al., 1973), and the United States (Warren and Healy, 1973) gave evidence for the large amount of published models, available by the end of the 1960s. They not only resulted in a quite complete picture of the gross velocity-depth structure of the Earth's crust underneath the northern hemisphere around the world, but also detected specific properties of different tectonic areas such as shields, platforms, orogens, basins, rift zones, etc. However, the long-range data had not yet been interpreted in terms of subcrustal fine structure.

On the other hand, groups in Germany (e.g., Dohr, 1957, 1959; Krey et al., 1961; Liebscher, 1962, 1964; German Research Group for Explosion Seismology, 1964; Dohr and Fuchs, 1967), the USSR (e.g., Beloussov et al., 1962; Kosminskaya and Riznichenko, 1964; Zverev, 1967), and Canada (e.g., Kanasewich and Cumming, 1965; Clowes et al., 1968; Kanasewich et al., 1969) had demonstrated the feasibility of determining fine crustal structure using near-vertical incidence reflection methods, and such techniques have been used extensively in subsequent decades.

In order to understand the seismic wave field, various groups had worked on the theoretical background so that by the end of the 1960s, the art of computing synthetic seismograms was ready for application. Other groups had concentrated on the character and basic features of seismic phases so that gradually a fine structure of the hitherto homogeneously layered crust was detected. In particular the character of the crust-mantle boundary was attracting increased interest.

Furthermore, two volumes on oceanic research edited by Maxwell (1970) summarized all achievements and subsequent concepts on seafloor evolution, based mainly on the numerous marine seismic investigations in the oceans collected since the end of World War II until the end of the 1960s.

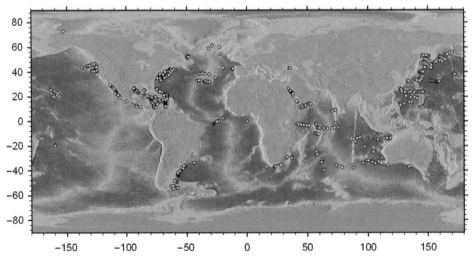

Figure 6.8.5-01. Seismic refraction measurements in the Atlantic, Pacific and Indian Oceans performed between 1960 and 1969 (data points from papers published until 1969 in easily accessible journals and books).

The 1970s (1970–1980)

7.1. THE DECADE OF INTERNATIONAL COOPERATION

While it took almost 100 years from its beginnings in 1851 to lay the foundations of successful controlled-source seismic experiments, the development of this particular branch of seismology grew rapidly in the first 25 years following the end of the Second World War in 1945. In the two foregoing chapters we have tried to present an overview of all major developments and experiments carried out in the 1950s and 1960s. However, by the end of the 1960s, the number of controlled-source seismology experiments had reached a level such that a complete description of all major projects for the following decades would be forbidding. Therefore, we shall restrict ourselves to particular highlights providing major impact on our knowledge of crustal and uppermost-mantle structure.

The personnel situation in the geophysical departments of the universities and other research institutions of Western Europe had improved significantly owing to the favorable development in the 1960s. By the beginning of the 1970s, many young scientists had become professors and heads of departments. In Paris, L. Steinmetz, assisted by A. Hirn, G. Perrier, J.P. Ruegg, and many others, had started the Groupe Grands Profils Sismiques. In Great Britain, M.H.P. Bott and R.E. Long at Durham, D.H. Griffiths and R. King with their assistants and students D. Bamford, D.J. Blundell, M.A. Khan, M. Brooks and K. Nunn at Birmingham, as well as P.L. Willmore and B. Jacob in Edinburgh, pursued explosion seismology research. In Germany, explosion seismology was pushed forward in particular by H. Berckhemer and B. Baier (Frankfurt), J. Dürbaum and K. Hinz (BGR Hannover), K. Fuchs and C. Prodehl (Karlsruhe), P. Giese (Free University of Berlin), H. Gebrande and H. Miller (Munich), J. Makris and W. Weigel (Hamburg), and R. Meissner (Kiel). In Switzerland, it was S. Mueller, who in 1971, together with J. Ansorge, E. Wielandt, and others, moved from Karlsruhe to Zurich, Switzerland, and started a new period of active and passive seismology research at the Eidgenössische Technische Hochschule (ETH). In Italy, it was C. Morelli in particular who, since the early 1960s, had established a strong explosion seismology group, assisted by G. de Visintini and R. Nicolich and supported strongly by the Milan group with R. Cassinis and S. Scarascia. In northern Europe, M.A. Sellevoll at Oslo, Norway, C.-E. Lund and A. Vogel at Uppsala, Sweden, and M. Korhonen and U. Luosto in Finland were actively involved in explosion seismology projects.

In the former USSR, deep seismic sounding projects were continued in particular by I.P. Kosminskaya (Moscow), V.B. Sollogub (Kiev), and N.N. Puzyrev (Novosibirsk). Furthermore, in the USSR, L.P. Vinnik (Moscow), A.S. Alexeyev (Novosibirsk), S.M. Zverev (Moscow), and N.I. Pavlenkova (who had moved from Kiev to Moscow) pursued explosion seismology in active experiments and methodology studies. In Poland, A. Guterch (Warsaw) and in Czechoslovakia, V. Červený (Prag) became well-known scientists to the western explosion seismology community.

In the United States, wide-angle reflection and refraction seismology played a minor role throughout the first half of the 1970s, but in 1975, strong interest in exploring the crust by deep seismic-reflection surveys led to the foundation of COCORP (Consortium for Continental Reflection Profiling), pushed forward by, e.g., J.A. Brewer, L.D. Brown, S. Kaufman, J.E. Oliver (all at Cornell University, Ithaca, New York), R.A. Phinney (Princeton, New Jersey), and S.B. Smithson (Laramie, Wyoming). Seismic-refraction activities were concentrated at a few universities, e.g., in Wisconsin (R.P. Meyer and W.D. Mooney), Salt Lake City (R.B. Smith, assisted by L.W. Braile, G.R. Keller, and others), and Dallas (A. Hales. who soon moved to Australia). At the U.S. Geological Survey, while the main activity of the seismology group concentrated on earthquake research, J.H. Healy started to raise new interest by developing a new type of equipment. With one hundred pieces of the so-called cassette recorder, the U.S. Geological Survey started a new style of fieldwork, first tested in the Saudi Arabian long-range profile in 1978, which laid the foundation for major new activities in the following decade.

In Canada, it was, e.g., M.J. Berry (Ottawa), R.M. Clowes, D.A. Forsyth, Z. Hajnal, R.F. Mereu (London, Ontario), and others who continued controlled-source seismology projects, which finally, at the end of the 1970s, led to the foundation of the COCRUST project. This project was different to other national pure reflection programs, as it included both refraction and reflection surveys.

In the southern hemisphere, it was especially in Australia where an active group of scientists such as J.C. Dooley, D. Denham, D.M. Finlayson, and C.D.N. Collins (Bureau of Mineral Resources, Canberra), as well as K. Muirhead and B.J. Drummond (Australian National University, Canberra) pursued crustal and upper-mantle research by both active and passive source seismology. In South Africa, it was R. Green in the University of Witwatersrand at Johannesburg, South Africa, who pushed crustal and upper mantle research by developing instrumentation that enabled the performance of one-man seismic-refraction surveys.

The 1970s can be characterized by several highlights. The first was the detection of fine structures in the lithosphere below the Moho all over the world, using controlled-source seismic experiments with observation distances beyond 300 km.

The second highlight was advanced research and understanding of continental rifting in Europe, Asia, Africa, and North America. The consequent application of the time-term method widely used by British scientists on data from dense networks of seismic-refraction profiles led unexpectedly to the detection of uppermost mantle velocity anisotropy also under continents.

A third highlight was the increased investigation of the Mediterranean area, and special research efforts concentrated on details of the crustal structure above hot spots such as Hawaii and the Yellowstone–Snake River Plain area.

The 1970s also saw the opening of the first national large-scale seismic-reflection program: COCORP in the United States, which was soon followed by the Canadian equivalent, COCRUST.

The interpretation of the new data was guided by the improvement of lithosphere models using the reflectivity method, which allowed the calculation of amplitudes, frequency contents, and reverberations in laterally homogeneous media.

A similar rapid development was achieved in oceanic research. Numerous marine experiments were carried out, their character changing gradually with development of instrumentation, logistics, and theoretical background. Here, the reader is referred to Chapter 7.8.1.

During the second half of the decade, the ray method was developed, which led to the development of several ray tracing methods that could also handle data in complex tectonic structures and which would become the standard tool for interpreting seismic data in the 1980s and following decades up to the present day.

From the large amount of new data and their interpretations, Soller et al. (1981) have prepared a global thickness map, and Prodehl (1984) has prepared maps of the world showing the location of crustal and upper-mantle surveys available at that time and has summarized the main crustal features in tables (reprinted in Appendix A7-1).

7.2. CONTROLLED-SOURCE SEISMOLOGY IN EUROPE

7.2.1. Fine Structure of the Lower Lithosphere—The First West European Long-Range Profile

Though many seismic-refraction profiles observed in the 1960s reached recording distances of several hundred kilometers, the fine structure of the P_n phase had barely been noticed. A first report on systematic amplitude variations of the P_n phase was published in 1966 by a Soviet scientist (Ryaboi, 1966). Ansorge (1975) took up this idea and started to investigate the long-range observations of the Lake Superior experiments in North America, which at the same time were also more closely investigated by Massé (1973), and compared them with data recorded in the past 25 years in central and northern Europe with distances ranging from 400 to 2200 km (Heligoland-South 1947, Lac de l'Eychauda 1963, Norway 1965, Lac Nègre 1966, Folkestone

1967, Trans-Scandinavian Profile 1969, quarry blast observations at Böhmischbruck, Germany 1960–1971). He found that a similar fine structure described by Ryaboi (1966) for the P_n phase could also be seen in these sparse data, which indicated that a rather complex structure in the uppermost mantle down to 120 km depth was to be expected (Ansorge, 1975; Ansorge and Mueller, 1971).

So, the idea was born to systematically plan for an experiment out to 1000 km in western Europe to be placed within a tectonically unique surrounding. Experience from the Early Rise project and the theoretical investigations of Wielandt (1972, 1975) had shown that underwater shooting of relatively small depth charges (1000–3000 kg) at optimal water depths would produce sufficient energy to be recorded across a profile of possibly up to 1000 km long. A profile planned diagonally through France from the Bretagne across the Massif Central to the Mediterranean coast being located mainly on Variscan rocks (Fig. 7.2.1-01) would almost fulfill this requirement and would furthermore offer the chance to work with sufficiently large offshore shots at both ends.

Following the recommendation of their German colleagues, in 1969, the French geophysical community had bought 30 MARS-66 recording devices from the German company Lennartz. The purchase of the German equipment by the University of Paris continued the close cooperation between French and German geophysicists which had started with the Haslach explosion in 1948 and was continued in the investigation of the Western Alps in the 1950s and 1960s. This equipment was thoroughly tested and used in detailed research across the Massif Central, which will be described in more detail below, where the French colleagues also gained experience with organizing borehole shots on land (Perrier and Ruegg, 1973; Hirn and Perrier, 1974). Also the Swiss ETH and the Portuguese Meteorological Survey bought new MARS (magnetic tape recording system) equipment, thus increasing the total amount of seismic systems to be used in the subsequent international cooperative projects.

The joint effort of L. Steinmetz and A. Hirn at Paris, S. Mueller and J. Ansorge at Zurich, and K. Fuchs and C. Prodehl at Karlsruhe raised the necessary funding and so, with the German and the recently acquired French equipment, in 1971, the first long-range profile in Europe with a total length of 1200 km was realized (Groupe Grands Profils Sismiques and German Research Group for Explosion Seismology, 1972). In addition to the offshore sea shots crustal control became involved by placing borehole shots at 200–300 km intervals along the line (see record sections in Appendix A2-1, p. 36–37, France 1971). The station spacing was from 5 km up to 600 km distance and 10 km for distances between 600 and 1200 km. The shots in the Atlantic Ocean off Brest and the intervening land shots along the line carried energy as expected, even to the island of Corsica at 1400 km distance, but due to unfavorable locations as well as bad-weather conditions the reverse upper-mantle shots off St. Tropez in the Mediterranean caused a considerable decrease in seismic efficiency and could not be recorded along the entire line.

As expected and hoped for, the crustal profiles (Sapin and Prodehl, 1973; Fig. 7.2.1-02) revealed a rather homogeneous

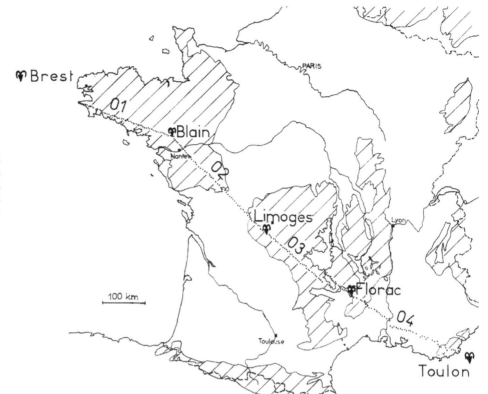

Figure 7.2.1-01. Location of shots and recording sites of the first long-range profile in 1971 through France. Dashed areas are basement outcrops (published also by Groupe Grands Profils Sismiques and German Research Group for Explosion Seismology, 1972, fig. 1). [Annales de Géophysique, v. 28, p. 247–256.]

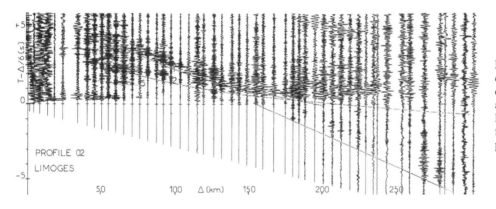

Figure 7.2.1-02. Record section of a crustal profile through France in 1971, data recorded from shotpoint Limoges along profile 02 toward northwest (published also by Sapin and Prodehl, 1973, fig. 4). [Annales de Géophysique, v. 29, p. 127–145.]

structure and a more or less flat Moho at 30–31 km depth, meaning that the mantle data could be interpreted without particular traveltime corrections.

At first glance, the first arrivals of the sea shot recordings beyond 300 km distance showed a rather uniform picture and the first-arrival energy propagated with an average velocity of ~8.1 km/s between 150 and more than 1000 km distance from the shotpoint. However, the close station separation of only 5 km revealed a fine structure as was known from Ansorge's previous investigations (Fig. 7.2.1-03).

The proper P_n phase with a velocity of 8.15 km/s disappeared beyond 210 km, and from 320 km onward, signals with strong amplitudes and a hardly noticeable delay in traveltime replaced the P_n phase.

A first interpretation of these mantle data correlated two mantle phases: P_I with 8.1–8.3 km/s between 320 and 480 km and P_{II} with 8.3–8.6 km/s between 450 and 620 km (Hirn et al., 1973). A reinterpretation of the same data shortly afterwards revealed a third upper-mantle phase P_{III} beyond 650 km distance (Kind, 1974). A similar interpretation as that of Kind (1974) was proposed by Ansorge (1975), but he adjusted the model by introducing first-order discontinuities to enable a direct comparison with the older less detailed long-range observations published earlier by Ansorge and Mueller (1971). On the basis of the 1971 long-range profile, Bottinga et al. (1973) searched the P_n-P_I range data of other long profiles obtained in France in 1970 and 1972. They detected a similar fine structure of the subcrustal lithosphere throughout France and speculated about its petrologic nature.

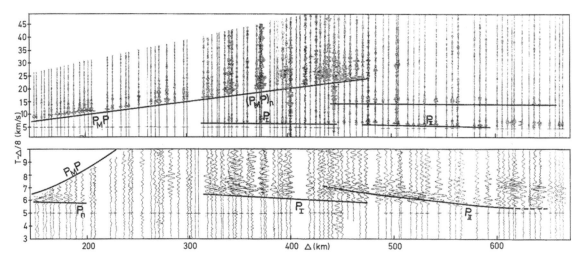

Figure 7.2.1-03. Record sections of long-range profiles through France in 1971 with different time and amplitude scales (from Hirn et al., 1973, fig. 3). P_MP reflection from Moho, P_n refracted head wave guided at Moho, P_I, P_{II} reflections from subcrustal interfaces at 55 km and 80 km depth. The lower section shows a clear separation of P_I and P_{II}, which is not clearly visible in the upper section. (See Fig. 7.2.1-06.) [Zeitschrift für Geophysik, v. 39, p. 363–384. Reproduced with kind permission of Springer Science+Business Media.]

In 1972, the long-range observations were extended to 1500 km distance. During the time of the Rhônegraben experiment jointly carried out by French, German, and Swiss institutions in southern France (see below in section 7.2.2) a large shot of 10 tons was organized off the coast of Scotland by B. Jacob and P.L. Willmore. They had already experimented with another 10-ton shot in the late 1960s and had successfully collected seismograms from permanent European stations which proved its energy efficiency (Jacob and Willmore, 1972). For this 10-ton shot off Scotland, the recording stations operating in France at that time were distributed throughout the French Massif Central over a range of 900 to 1500 km (Steinmetz et al., 1974). Simultaneously another profile was organized in western Germany, where stations recorded on a line into the Bavarian Forest between 800 and 1400 km distance (Bonjer et al., 1974). From the data recorded in France, Steinmetz et al. (1974) deduced a model to the base of the asthenosphere at 200 km depth.

In 1973, the experiment of 1971 was practically repeated (Hirn et al., 1975), but with slight variations. It was now known that the upper-mantle phases correlated in the 1971 data were not caused by near-surface irregularities at particular station locations, but that the traveltimes and amplitudes really varied with shotpoint distance, and thus they were directly dependent on the depth penetration of the corresponding P-waves. In the 1973 experiment, the first (northwestern) half of the 1971 line was re-occupied, i.e., for 650 km, and the shotpoint off Brest was shifted 95 km toward the coast (shotpoint Brest 2 in Fig. 7.2.1-04). In addition, fan profiles were set up in the distance range of 450–650 km where both P_I and P_{II} had been correlated and another 1-ton shot was fired at the same position as in the 1971 experiment (shotpoint Brest 1 in Fig. 7.2.1-04). Finally, a second profile was recorded 90 km to the north and parallel to the main line. The shotpoint Issoudun (Fig. 7.2.1-04) carried sufficient energy along

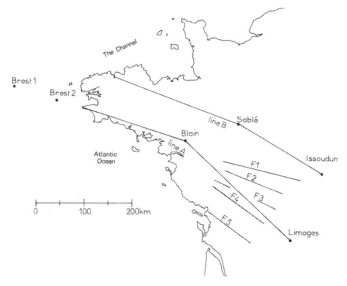

Figure 7.2.1-04. Location of upper-mantle profiles recorded in western France in 1971 and 1973. Shots at location Brest 1 were recorded on line A in 1971 and on fan profiles F1 to F5 in 1973, shot Brest 2 was recorded on line A in 1973. Blain, Limoges: borehole shots for reversed crustal profiles on line A, Issoudun, Sablé: borehole shots for upper mantle and crustal observations along line B in 1973 (published also by Hirn et al., 1975, fig. 1). [Annales de Géophysique, v. 31, p. 517–530.]

the whole line, which was 500 km long. Unfortunately, the corresponding reversing sea shot failed.

In all cases, a similar pattern of upper-mantle phases could be correlated (Figs. 7.2.1-03 and 7.2.1-05), proving that the model obtained for the sub-crustal lithosphere in 1971 is of more general nature. Figure 7.2.1-05 shows the observed fan data, and Figure 7.2.1-06 summarizes the results and also shows corresponding synthetic data calculated with the reflectivity method for the local

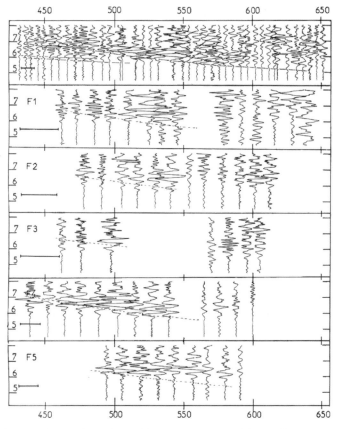

Figure 7.2.1-05. Observed data on the main profile Brest 1 (1971) and fan profiles F1 to F5 obtained in 1973 from a shot at the same position Brest 1 (published also by Hirn et al., 1975, fig. 4). Reduction velocity is 8 km/s. The dotted line on each section indicates the travel time curve correlated for P_{II} on Brest 1, line A (top section). Overall magnifications are uniform for all traces in one section, but note its variation from one section to another (0.5 μ/s amplitude bar in lower left corner). [Annales de Géophysique, v. 31, p. 517–530.]

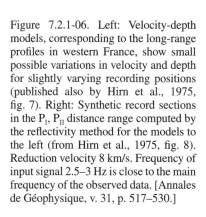

Figure 7.2.1-06. Left: Velocity-depth models, corresponding to the long-range profiles in western France, show small possible variations in velocity and depth for slightly varying recording positions (published also by Hirn et al., 1975, fig. 7). Right: Synthetic record sections in the P_I, P_{II} distance range computed by the reflectivity method for the models to the left (from Hirn et al., 1975, fig. 8). Reduction velocity 8 km/s. Frequency of input signal 2.5–3 Hz is close to the main frequency of the observed data. [Annales de Géophysique, v. 31, p. 517–530.]

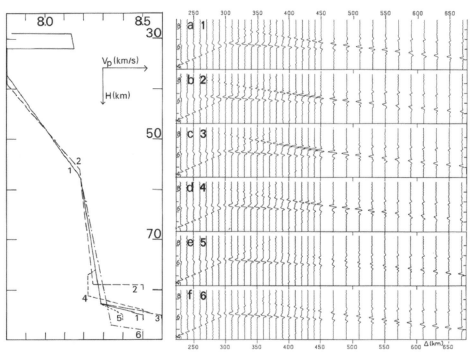

variations of the general model. Below the Moho, there is a distinct velocity inversion where the velocity drops from 8.15 km/s to ~7.8 km/s. Underlying the low-velocity zone, a reflector at 55–60 km depth causes the reflection P_I. At ~80–90 km, there is a second boundary producing phase P_{II}. The variations of some kilometers in depth for the reflectors or some tenths of km/s for the velocities are likely for the region (Hirn et al., 1975).

Compilations of many of these models and comparison with upper-mantle models worldwide down to 150 km depth, including Anderson's general model (Anderson, 1971), were prepared by Prodehl (Prodehl et al., 1976b; Prodehl, 1984). Figure 7.2.1-07 shows a diagram with models for western Europe as published until 1976.

7.2.2. Crustal Research of the Central European Rift System and the Variscan Basement

The first investigations of crustal structure of the Rhinegraben had already started in the mid-1960s but had exclusively been based on quarry blast observations. The results had been discussed and published by Mueller et al. (1967) and Ansorge et al. (1970) in two successive symposium volumes: Rhinegraben Progress Report 1967 (Rothé and Sauer, 1967) and Graben Problems (Illies and Mueller, 1970). The most famous result was the detection of a high-velocity rift cushion in the lowermost crust, located under the axis of the graben proper.

As mentioned above, with the new MARS-66 purchased in 1969, in 1970 the French group started a detailed investigation of the French Massif Central and its intervening rift structures, the Limagnegraben and the Bresse-Rhônegraben system (Perrier and Ruegg, 1973; Hirn and Perrier, 1974). At the same time, the German scientists at Karlsruhe started a major operation in Portugal (see below). Finally, from 1971 onwards, forces were put together, in particular pushed forward on the French side by A. Hirn, G. Perrier, E. Peterschmitt, J.P. Ruegg, and L. Steinmetz as well as on the German side by J. Ansorge, K. Fuchs, St. Mueller, and C. Prodehl.

Following the first highly successful joint effort of French and German scientists to unravel details of the subcrustal lithosphere under France by controlled-source seismic experiments as described above, the same group started a major crustal research of the Central European rift system with specially designed borehole shots. Two major experimental phases were carried out in 1972.

The first project was undertaken in May 1972 to investigate the Upper Rhinegraben in much detail (Rhinegraben Research Group for Explosion Seismology, 1974). For this project, borehole shots were planned along the axis of the graben. Due to logistical problems, these were arranged on the French side only, with the southernmost shotpoint being located near Steinbrunn (SB in Fig. 7.2.2-01), in the border region between the Rhinegraben and the Swiss Jura, and the northernmost shotpoint being close to the French-German border near Wissembourg (WI in Fig. 7.2.2-01).

In addition to the reversed axial profile, a series of fan profiles across the Vosges in the west and the Black Forest in the east was recorded (Rhinegraben Research Group for Explosion Seismology, 1974; solid lines in Fig. 7.2.2-01).

The data (reproduced in Appendix A2-1, p. 11–14) showed profound differences of phases to be correlated under the graben and under the flanks (Fig. 7.2.2-02), as was discussed in detail by Edel et al. (1975). As a result of the extended data set, the idea of a high-velocity lower-crust rift cushion was abandoned and replaced by a transitional crust-mantle boundary centered under the rift. Under the graben flanks the Moho appeared as a first-order discontinuity (Fig. 7.2.2-03).

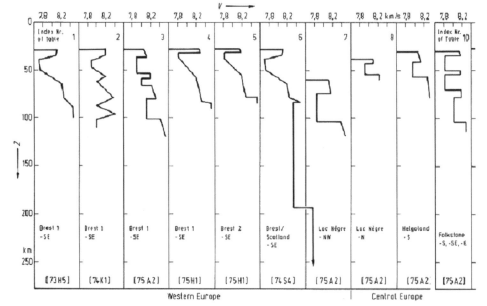

Figure 7.2.1-07. Velocity-depth models of the subcrustal lithosphere in western Europe for profiles recorded until 1973 (from Prodehl, 1984, fig. 15). 73H5—Hirn et al. (1973); 74K1—Kind (1974); 75A2—Ansorge (1975); 75H1—Hirn et al. (1975); 74S4—Steinmetz et al. (1974). [*In* K.-H. Hellwege, editor in chief, Landolt Börnstein New Series: Numerical data and functional relationships in science and technology. Group V, Volume 2a: K. Fuchs and H. Soffel, eds., Physical properties of the interior of the earth, the moon and the planets: Springer, Berlin-Heidelberg, p. 97–206. Reproduced with kind permission of Springer Science+Business Media.]

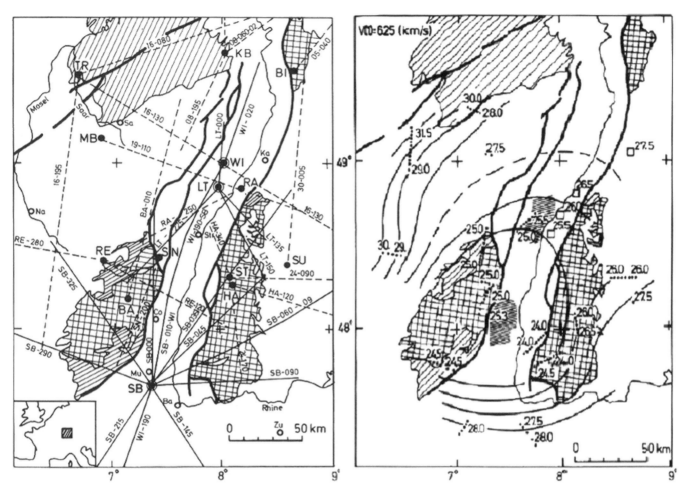

Figure 7.2.2-01. Left: Location map of seismic-refraction profiles in the surroundings of the southern Upper Rhine Graben (from Prodehl et al., 1976, fig. 3). Right: Moho contour map of the southern Upper Rhinegraben (from Prodehl et al., 1976, fig. 12). [*In* P. Giese, C. Prodehl, and A. Stein, eds., Explosion seismology in central Europe—data and results: Springer, Berlin-Heidelberg-New York, p. 313–331. Reproduced with kind permission of Springer Science+Business Media.]

The new data furthermore evidenced a broad warping in an upward direction of the Moho by up to 5 km and a normal P_n velocity of ~8 km/s (Edel et al., 1975; Prodehl et al., 1976a, 1995). The interpretation of Edel et al. (1975) incorporated both old and new data, which were all published in small-scale format.

During the Rhônegraben experiment carried out in July 1972 the international working force of French and German institutions was joined by Swiss and Portuguese participants and their equipment (Sapin and Hirn, 1974; Michel, 1978; Prodehl et al., 1995).

This survey did not only consist of a reversed profile along the rift axis, but also supplied reversed lines along the flanks to the west in the Massif Central and to the east in the sub-Alpine region. Furthermore intermediate shots were added in the middle of all lines, and transverse profiles across the Rhône valley were arranged. The data in the southern Rhône Valley was further complemented by the crustal data of the NW-SE long-range profile of 1971 (Fig. 7.2.2-04). This large amount of data was of extremely high quality and was split into a southern part (Sapin and Hirn, 1974) and a northern part (Michel, 1978). Most of the data were presented as record sections and are reproduced in Appendix A2-1 (p. 38–48).

7.2.3. The Mediterranean Region

Stimulated by German colleagues, the Portuguese scientists became interested in a detailed crustal research of Portugal. By 1970, the Portuguese community had acquired enough funding to buy MARS-66 equipment. Then, in 1970 and 1972, in close cooperation with the University of Karlsruhe, Germany, a concentrated effort resulted in a number of seismic-refraction profiles (Fig. 7.2.3-01) across the Variscan part of southern Portugal (Mueller et al., 1973, 1974; Moreira et al., 1977). In 1971, a line was added north of Cabo da Roca and reversed from Nazaré (Fig. 7.2.3-01). All shotpoints were at sea, which was enabled by the cooperation with a Portuguese navy vessel and a shooting crew of the German navy (Prodehl et al., 1975).

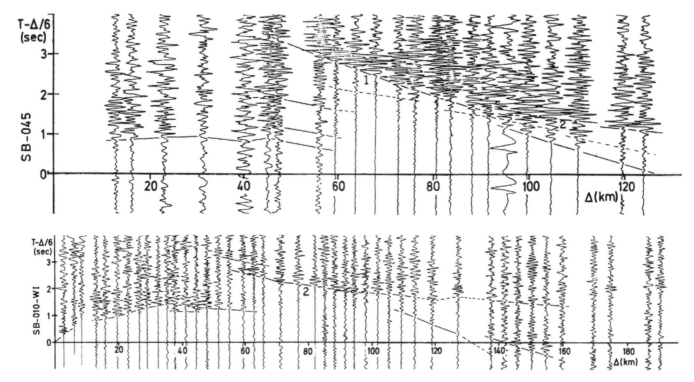

Figure 7.2.2-02. Record sections of the Rhinegraben area of two profiles from Steinbrunn . Top: Profile SB-045 across the flank of the graben, phase 1 = PMP, phase 2 = reflection from mid-crust boundary (from Prodehl et al., 1977, fig. 12). Bottom: Profile SB-010-WI along the graben axis, only phase 2 is visible (from Prodehl et al., 1977, fig. 13). [*In* Heacock, J.G., ed., The Earth's crust: American Geophysical Union Geophysical Monograph 20, p. 349–369. Reproduced by permission of American Geophysical Union.]

The profiles were arranged so that all profiles, except the SW-NE line of 1972, could be reversed (Fig. 7.2.3-01). All data, recorded in Portugal from 1970 to 1972, were published as record sections (reproduced in Appendix A2-1, p. 55).

Two years later, from 1974 to 1975 (Working Group for Deep Seismic Sounding in Spain 1974–1975, 1977; Appendix A7-2-1), and continued later in 1977 and 1980 (Barranco et al., 1990; Appendix A7-2-2), the first deep seismic sounding experiments followed in southern Spain by the cooperation of Spanish, Swiss, Portuguese, French, and German research institutions and the Spanish navy. Profiles crossed the area of the Betic Cordillera with shotpoints both at sea and on land (Banda and Ansorge, 1980). The interpretation resulted in drastically varying crustal thickness, from near 25 km near the southeastern coast at Cartagena and near the southern coast near Adra to as much as 39 km under the central Cordillera. The experiments included a long-range line of 450 km length between Cartagena in the east and Cadiz in the west (Fig. 7.2.3-02). Record sections and detailed maps for these profiles and all following experiments in the Mediterranean area are reproduced in Appendix A7-2.

These first seismic investigations of southern Spain were soon followed by quarry blast profiling on the Meseta (Banda et al., 1981b; Appendix A7-2-3) and by deep seismic sounding experiments in eastern Spain in 1976 and 1981 (Zeyen et al., 1985; Appendix A7-2-4), resulting in 31 km crustal thickness

under the Meseta, which near the southeastern coast gradually decreased to as low as 20 km thickness. Land-sea operations around both the Balearic islands in 1976 (Banda et al., 1980; Fig. 7.2.3-02; Appendix A7-2-5) and the Canary islands in 1977 and 1979 (Banda et al., 1981a; Appendix A7-2-6) gave average crustal thicknesses of 20 km under the Balearic Islands and 12–15 km under the Canary Islands.

In close cooperation with the French group, a detailed investigation of the Pyrenees was performed in 1978 with two E-W lines parallel to the strike of the mountain range (Fig. 7.2.3-02) and some additional lines into the northern and southern forelands (Appendix A7-2-7). Under the Pyrenees, for example, an abrupt thinning by 10 km from the Paleozoic axial zone with a 40–50-km-thick crust toward the North Pyrenean zone with 25–30 km crustal thickness was detected (Explosion Seismology Group Pyrenees, 1980; Gallart et al., 1980; Hirn et al., 1980; Daignières et al., 1982).

On the African side, in Morocco, the structure of the Moroccan Meseta, the High Atlas and the Anti-Atlas was the goal of a major seismic-refraction campaign in 1975 organized by J. Makris of the University of Hamburg, Germany, and A. Demnati of the Geological Survey of Morocco (Fig. 7.2.3-03; see also Fig. 8.3.4-17).

The seismic energy was generated by quarry blasts at several mines and at sea by shots fired from the German research vessel

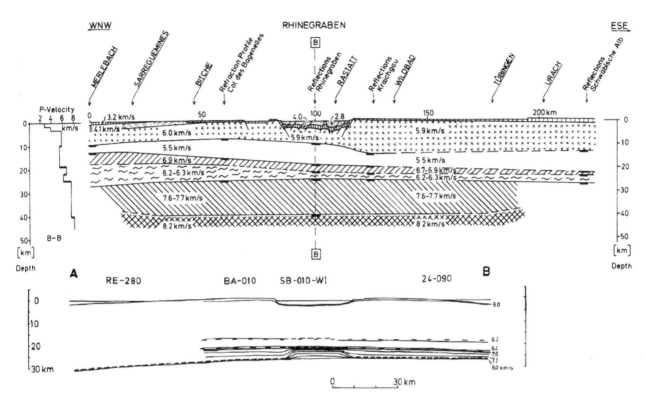

Figure 7.2.2-03. Crustal cross sections through the central part of the Rhinegraben area. Top: The depth range of 25–40 km is interpreted as a rift cushion, i.e. a high-velocity lower crust after Mueller et al., 1969 (from Prodehl et al., 1976a, fig. 6). Bottom: The Moho rises gradually from 30 km west of the graben to about 25 km east of the graben, but becomes a transitional gradient zone under the graben proper, with its top up-warping to less than 25 km depth (from Prodehl et al., 1976a, fig. 14, upper section). Thin lines show equal velocity contours, interval is 0.2 km/s. [*In* P. Giese, C. Prodehl, and A. Stein, eds., Explosion seismology in central Europe—data and results: Springer, Berlin-Heidelberg-New York, p. 313–331. Reproduced with kind permission of Springer Science+Business Media.]

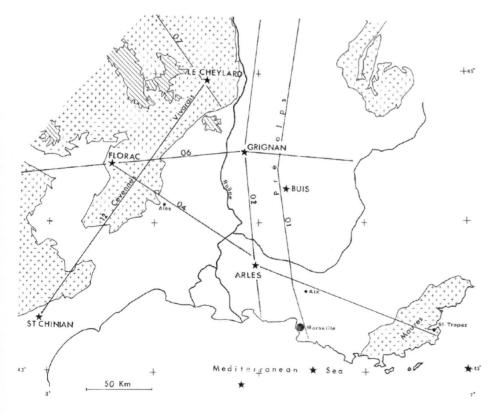

Figure 7.2.2-04. Location map of seismic-refraction profiles in the area of the southern Rhône valley (from Sapin and Hirn, 1974, fig. 1). [Annales de Géophysique, v. 30, p. 181–202. Reproduced by permission of the author.]

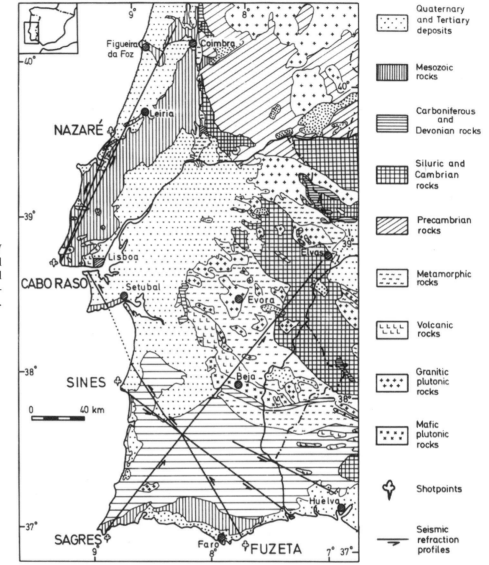

Figure 7.2.3-01. Crustal profiles in SW Portugal in 1970 to 1972 (from Prodehl et al., 1975, fig. 1). Proc. 14th General Assembly of the European Seismological Committee (Trieste, 1974) [Akad. Wiss. DDR, Berlin, p. 235–241.]

Meteor during cruise M39 (Sarnthein et al., 2008) between the Canary Islands and the Moroccan coast (Appendix A7-2-8). The resulting model shows a gradual thickening of the crust from 24 km near the coast, to 30 km under the Meseta and the Anti-Atlas, and to as deep as 38 km under the High Atlas. At the continent-ocean transition, a deep reflector below the Moho was identified at depth ranges between 60 and 80 km (Makris et al., 1985).

Mainly under the heading of the Trieste group, the Western Mediterranean was investigated by several sea-seismic and onshore-offshore campaigns (Hirn et al., 1977; Leenhardt et al., 1972; Morelli et al., 1977; Morelli and Nicolich, 1980; Steinmetz et al., 1983). In 1970, a joint cruise—then called the *Anna* cruise—was undertaken by French and German marine geophysicists, employing both reflection and refraction measurements along three lines located along a north-south–directed traverse from the Rhône delta across the Balearic islands to the African

coast near Algiers (Leenhardt et al., 1972; Appendix A7-2-9). Profile 1, located in the Gulf of Lyon, was only recorded to 25 km distance. Profiles 2 and 3 were north-south lines, located north and south of the island of Mallorca (Fig. 7.2.3-04). They were successfully recorded up to 100 km distances and resulted in a crustal thicknesses of 14 km north (line 2) and 12 km south of Mallorca (Hinz, 1972).

In 1974, a large seismic-refraction program was carried out as a joint project of Italian, German, and French geophysical research institutions (Morelli et al., 1977; Appendix A7-2-10). It covered the Northern Apennines, the Ligurian Sea and Corsica. Only sea shots of 200 kg charges were involved fired at optimum depth at 90–100 m to obtain constructive interferences of the bubble pulse with the surface reflections (Wielandt, 1972, 1975). The shots were arranged on lines in the Corsica–Elba Channel, the Ligurian Sea, and the Provençal Basin and recorded on

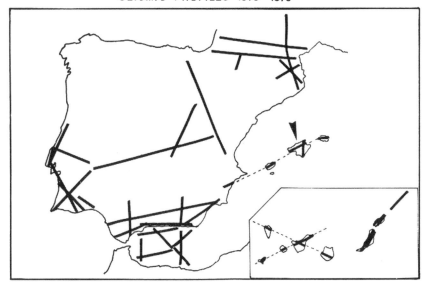

SEISMIC PROFILES 1970 - 1979

Figure 7.2.3-02. Location of deep seismic-sounding experiments on the Iberian Peninsula. Seismic lines in Portugal and Spain from 1970 to 1979 (modified from Cordoba and Banda, fig. 1). [Geofisica Internacional, Mexico, v. 19, p. 285–303.]

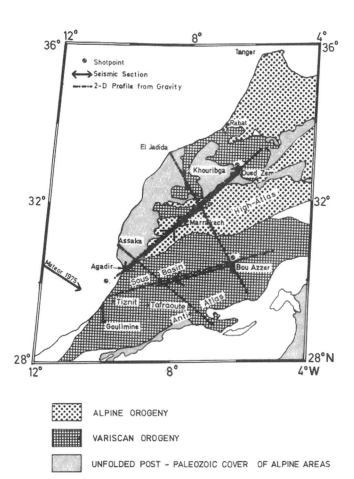

ALPINE OROGENY

VARISCAN OROGENY

UNFOLDED POST - PALEOZOIC COVER OF ALPINE AREAS

Figure 7.2.3-03. Location of the seismic experiment of 1975 in Morocco (from Makris et al., fig. 1). Thick lines: land profiles recording quarry blasts and a line of sea shots fired by the vessel Meteor. [Annales Geophysicae, 3, p. 369–380. Reproduced by permission of the author.]

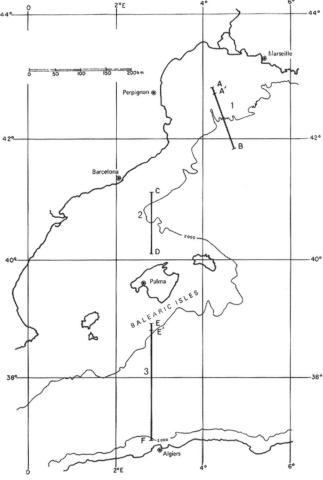

Figure 7.2.3-04. Location of Project Anna 1970 deep seismic-sounding experiments in the Ligurian Sea and surrounding land areas (from Leenhardt et al., 1972, Organization, fig. 1). [Bulletin of the Centre Recherche Pau–SNPA, v. 6, p. 373–381. Reproduced by permission of TOTAL–CSTJF.]

Corsica and along four profiles across the Apennines in northern and central Italy (Fig. 7.2.3-05).

For the interpretation of the data (data example in Fig. 7.2.3-06), various author groups have contributed to the publication of Morelli et al. (1977). The crust underneath the island of Corsica with 30 km thickness proved to be typically continental, updipping rapidly toward the west, thus coinciding with the earlier results of an 11–13-km-thick crust under the Provençal basin (Hirn and Sapin, 1976; Appendix A7-2-11). Toward northeast, the crustal thickness appeared to be at least 20 km under the island of Elba and 15–20 km under the northern Ligurian Sea, thickening to at least 35 km under the crest of the Apennines (Morelli et al., 1977).

This campaign was followed by two long-range experiments performed in the western Mediterranean (Appendix A7-2-12). The first was a 430-km-long line between the Balearic Islands and Corsica. It was recorded in 1974 and consisted of a line of shots spaced ~8–10 km, which were recorded by land stations both in NW Corsica and on the islands of Menorca and Mallorca. Unfortunately, the only ocean-bottom sensor, a proto-

type, failed, and shots started at only 70 km from the nearest land station. The result was a 6-km-thick low-velocity zone with 7.7 km/s beneath a Moho lid (8.25 km/s) at 28 km depth, followed by a 20-km-thick layer with 8.3 km/s, underlain by 8.7 km/s mantle material (Hirn et al., 1977). The second long-range profile was a 550-km-long traverse, recorded in 1979 in the Tyrrhenian Sea between Corsica and southern Italy. It involved recently developed ocean-bottom seismometers and thus enabled the investigation of gross features of the upper mantle (Steinmetz et al., 1983).

In southern Italy, in the early 1970s, profiles had been recorded along Puglia and Calabria with shots recorded from the adjacent Tyrrhenian Sea. In 1971, two profiles were recorded from shotpoint C (Fig. 7.2.3-07). One line ran along Puglia and was reversed from shotpoint B; the other line was unreversed and was recorded toward Napoli (Giese et al., 1973). Colombi et al. (1973) have described in much detail a second experiment of 1971, where recording stations in southern Italy, in Sicily, and on the island of Pantelleria recorded shots in the Tyrrhenian Sea (line E in Fig. 7.2.3-07) and in the Channel of Sicily (line F in

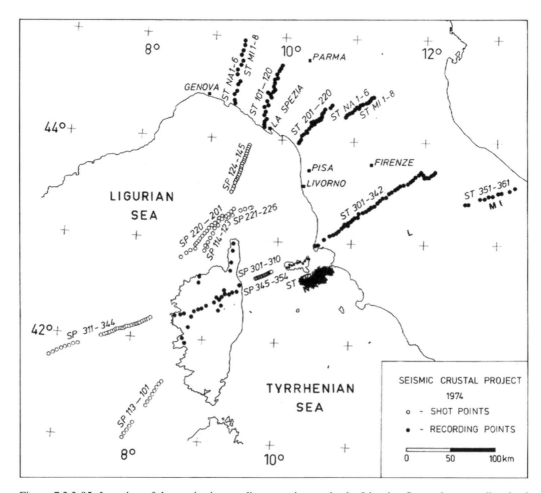

Figure 7.2.3-05. Location of deep seismic-sounding experiments in the Ligurian Sea and surrounding land areas (from Morelli et al., 1977, fig. 1.2). [Bollettino di Geofisica Teorica e Applicata, v. 7, p. 199–260. Reproduced by permission of Bollettino di Geofisica, Trieste, Italy.]

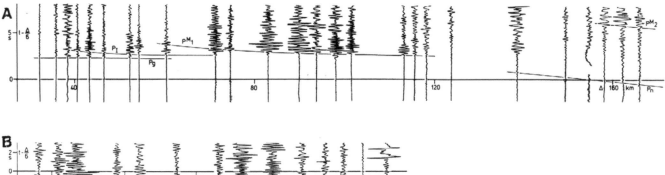

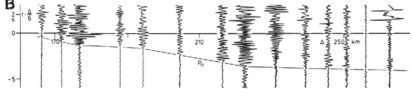

Figure 7.2.3-06. Data example recording along line 3 from the Ligurian Sea into the Appennines (from Morelli et al., 1977, fig. 4.6). [Bollettino di Geofisica Teorica e Applicata, v. 7, p. 199–260. Reproduced by permission of Bollettino di Geofisica, Trieste, Italy.]

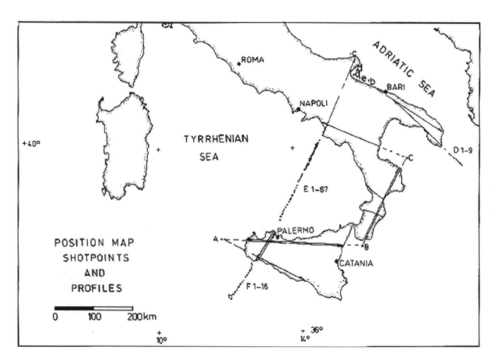

Figure 7.2.3-07. Location map of shotpoints, refraction lines and fans in southern Italy (from Giese et al., 1973, fig. 8). [Tectonophysics, v. 20, p. 367–379. Copyright Elsevier.]

Fig. 7.2.3-07). Record sections, maps, and models are reprinted in Appendix A7-2-13.

Combining all observations of the late 1960s (Cassinis et al., 1969) and early 1970s (Colombi et al., 1973), Giese et al. (1973) compiled a Moho depth map for the area of southern Italy, Sicily, and the adjacent Tyrrhenian Sea (Fig. 7.2.3-08) showing crustal thickness variations from oceanic to continental crust with less than 10 km in the center of the Tyrrhenian Sea to 35 km underneath central Sicily, Calabria, and Puglia.

By the end of the 1970s and beginning of 1980s, Italy was densely covered with seismic-refraction profiles (Morelli, 2003; Fig. 7.2.3-09). Most of the data had been acquired by the Osservatorio Geofisico Sperimentale (OGS) under the leadership of

Prof. Carlo Morelli, by whose influence many international collaborations, e.g., with Prof. Peter Giese, Berlin, or Dr. A. Hirn, Paris, helped to fulfill this goal. In the years from 1956 to 1982 ~21,160 km of seismic profiles had been obtained. Many of the deep seismic sounding profiles were later digitized and reinterpreted by the Istituto di Ricerca del Rischio Sismico CNR, Milano, and were made available to the scientific community (Scarascia et al., 1994). Similar activities of OGS at sea—much thanks to the driving forces of C. Morelli and R. Nicolich—had accumulated a wealth of marine data in the Mediterranean Sea. In the period of 1969–1982, 39,500 km of regional marine seismic-reflection profiles were recorded by the CNR vessel *Marsili*, equipped with a seismic recorder DFS, a neutral buoyant streamer (24 traces,

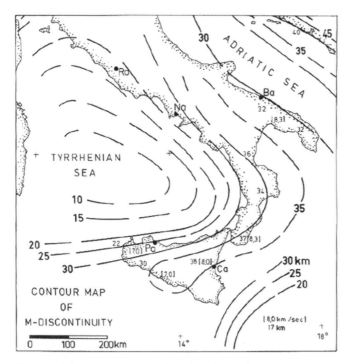

Figure 7.2.3-08. Crust-mantle boundary in southern Italy and surrounding seas, defined as depth of the strongest gradient in the velocity interval from 7.2–8.4 km/s (from Giese et al., 1973, fig. 9). [Tectonophysics, v. 20, p. 367–379. Copyright Elsevier.]

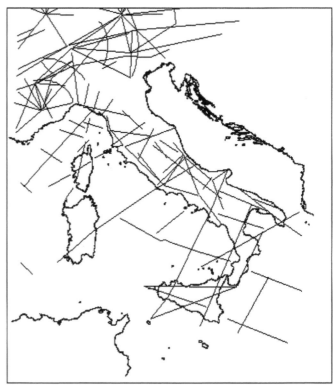

Figure 7.2.3-09. Deep seismic sounding profiles acquired from 1956 to 1982 in Italy by OGS, Trieste (from Morelli, 2003, fig. 1) [*In* Scrocca et al., eds., 2003, CROP Atlas: seismic reflection profiles of the Italian crust: Memorie Descrittive della Carta Geologica d'Italia, Roma, v. 62, p. 194 p. Copyright CROP.]

2400 m long) and two IFP Flexotir guns. Besides detailed sections for the sedimentary strata, also initial information on the crystalline parts of the crust were obtained in all areas (Finetti and Morelli, 1973). Other surveys added to the wealth of data. For example, the deep structure of the Ionian Sea was one of the goals to record a network of seismic-reflection profiles during cruise M50 of the German vessel *Meteor* (64) (Avedik and Hieke, 1981).

The first detailed investigation in Greece and the surrounding maritime areas started in 1969 and 1971 with two onshore-offshore experiments from the Ionian Sea into Greece. In 1969, three seismic land stations near the coast of the Peloponnesus had recorded shots during the *Meteor* cruise M17. A second offshore survey followed in 1971 during the cruise M22 of the German research vessel *Meteor* (Weigel, 1974). This investigation included a series of large shots recorded by four sonobuoy stations and 25 stations on land across the Peloponnesian Peninsula by a number of German institutions headed by H. Berckhemer and J. Makris (Fig. 7.2.3-10). The profile line crossed the Mediterranean Ridge and the Hellenic Trench at sea and ended on land in the east near Aegina on the Peloponnesus (Weigel, 1974). The Moho was not clearly seen along the offshore line, but was inferred from gravity. Under the coast of the Peloponnesus, the Moho reached a depth of ~46 km; farther west, the crust-mantle transition zone, though not clearly marked by a discontinuity, was interpreted to rise to 18 km depth under the western part of the Mediterranean Ridge and the Ionian Abyssal Plain. Under the Malta marginal trough, the Moho was found at 17 km depth (Weigel, 1974).

The island of Crete was the goal of an axial east-west profile in 1973. A NW-SE line recorded in 1973 and 1974 was arranged from central Greece in the northwest across the island of Evia toward a series of smaller islands in the Aegean Sea in the southeast (Makris and Vees, 1977). Another survey of the German research vessel *Meteor* in 1974 concentrated on the substructure of the Cretan Sea north of the island of Crete (Dürbaum et al., 1977). Record sections, maps, and models of all projects in and around Greece are reprinted in Appendix A7-2-14.

Summarizing contributions by Makris (1976, 1985) show the results of these seismic (Fig. 7.2.3-10) and other geophysical surveys and discuss the geodynamic implications for the evolution of the Hellenides. The Moho map for Greece and surroundings (Fig. 7.2.3-11), constructed by Makris (1977, 1985), using seismic and gravity data, demonstrates the complicated crustal structure of the Eastern Mediterranean area.

7.2.4 Fine Structure of the Lower Lithosphere— The Second Stage of Long-Range Profiling

In 1973, a young scientist, David Bamford from Birmingham, received a grant to work for one year at the University of Karlsruhe. This visit was the starting point of a very active cooperation

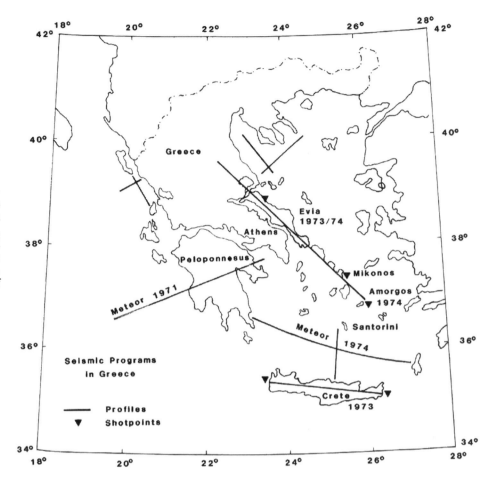

Figure 7.2.3-10. Location of deep seismic-sounding experiments in Greece and surrounding maritime areas (from Makris, 1985, fig. 11.1). [*In* Stanley, J., and Wezel, F-C, eds., Geological evolution of the Mediterranean Basin: Springer, New York-Berlin, p. 231–248. Reproduced with kind permission of Springer Science+Business Media.]

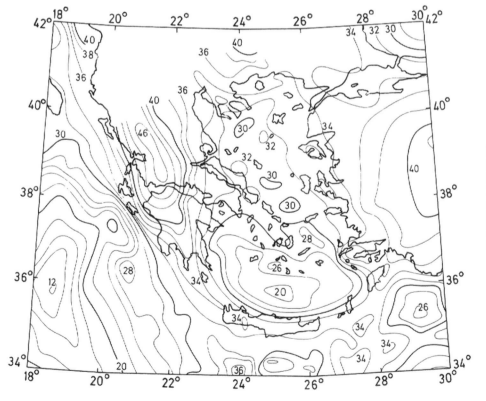

Figure 7.2.3-11. Moho map of Greece and surrounding maritime areas (from Makris, 1985, fig. 11.15). [*In* Stanley, J., and Wezel, F-C, eds., Geological evolution of the Mediterranean Basin: Springer, New York-Berlin, p. 231–248. Reproduced with kind permission of Springer Science+Business Media.]

between British and German scientists for the next 25 years and almost immediately stimulated two exciting projects.

The first project was the research for which the grant had been given: a time term analysis of the P_n data of the dense German network of quarry blast observations including the Rhinegraben (Fig. 7.2.4-01). This project resulted in an unexpected finding, namely the largest continental anisotropy of more than 3% in the uppermost mantle, with the fast velocity in the NNE direction and the slow velocity in the WNW direction (Bamford, 1973; Fig. 7.2.4-01). Bamford's results stimulated K. Fuchs to

discuss the impact of the seismic uppermost-mantle anisotropy with respect to the petrological composition of the subcrustal mantle material (Fuchs, 1975, 1977) and to construct his famous "anvil" model (Fuchs, 1983; Fig. 7.2.4-02).

Some years later, D. Bamford, in cooperation with his Karlsruhe colleagues, extended his research to other continental areas where rather dense networks of seismic profiles existed. In northern England and in the eastern United States, the uppermost mantle appeared to be isotropic within the limits of measurement error. However, a similar result as that for southern Germany

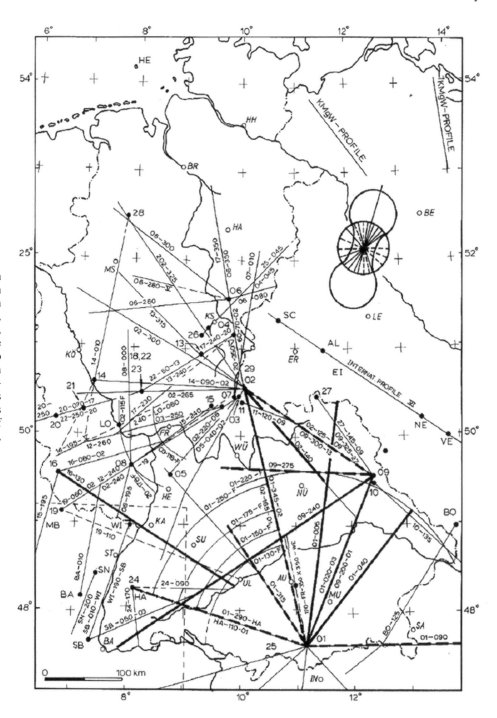

Figure 7.2.4-01. Azimuthal distribution of $P_n/P_M P$ amplitude ratios on a location map of all seismic-refraction profiles in western Germany (from Fuchs, 1983, fig. 12). Thick solid lines: ratios = 1, thick dashed lines, ratios < 1. These amplitude ratios are projected into Bamford's (1973) velocity distribution plotted to the right above. The large-amplitude sector is centered on Bamford's fast direction, the small amplitudes around the slow direction. [Physics of the Earth and Planetary Interiors, v. 31, p. 93–118. Copyright Elsevier.]

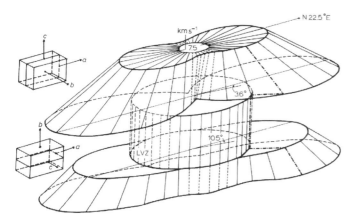

Figure 7.2.4-02. Three-dimensional "anvil" model of Fuchs (1983) showing the continental anisotropic velocity-depth distribution in the uppermost mantle under southern Germany found by Bamford's (1973) time-term analysis (from Fuchs, 1983, fig.13). [Physics of the Earth and Planetary Interiors, v. 31, p. 93–118. Copyright Elsevier.]

emerged from the crustal data of the western United States (Prodehl, 1970, 1979). Here a similarly large anisotropy of ~3% was found, with its high velocity direction 70–80° east of north, parallel to the spreading direction of the Basin and Range province (Bamford et al. (1979).

Both the amount and the direction were very similar to results found for the Pacific Ocean off California (Raitt et al., 1971; Fuchs, 1975, 1977). This coincidence led to the conclusion that this anisotropy is present as a consequence of the subduction of oceanic lithosphere beneath the western United States during Mesozoic times (Fig. 7.2.4-03).

The second project that was stimulated by D. Bamford's visit to Karlsruhe was the planning of the second West European long-range profile. This profile would be a 1000-km-long seismic line along the axis of the British Isles, from northern Scotland to the English Channel (LISPB 1974—Lithospheric Seismic Profile through Britain 1974). This project was to become an excellent collaboration of all major seismically oriented British and German institutions. On the British side the leading scientists were D. Bamford, K. Nunn, R. King, and D. Griffiths of the University of Birmingham, as well as B. Jacob and P. Willmore of the British Geological Survey at Edinburgh, while on the German side the project leaders were K. Fuchs and C. Prodehl.

The involvement of B. Jacob was particularly important for the shooting side. From previous experience, he knew that an optimum depth shot of ~5 tons would be adequate to reach the required distance range of 1000 km (e.g., Iyer et al., 1969). However, this would require a water depth of 180 m, which was not available at the ends of the LISPB line. Also, the main energy of such shots would be near 2 Hz, which was unfavorable for most of the available equipment. To overcome these problems, B. Jacob devised a simple, but effective system that used multiples of much smaller depth charges ("dispersed shots").

He had already experimented in a lake to see how seismic charges could be most efficiently detonated at shallower water

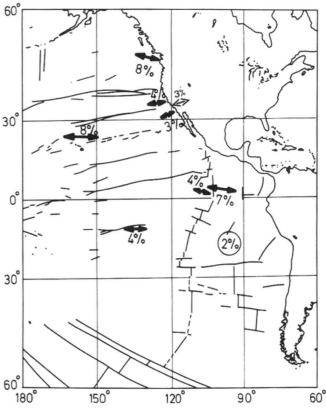

Figure 7.2.4-03. Anisotropy of uppermost-mantle velocity extending from western North America into the Pacific Ocean (from Bamford et al., 1979, fig. 17). [Geophysical Journal of the Royal Astronomical Society, v. 57, p. 397–430. Copyright John Wiley & Sons Ltd.]

depths, making use of the optimum depth which is directly dependent on the charge size. He found that splitting a charge into small units to be detonated simultaneously resulted in a much smaller optimum depth than would be necessary for a single larger charge (Jacob, 1975). As a result, 3–9 units of 200 kg were assigned for the northern offshore shots and 4–6 units for the southern offshore shots, which would be equivalent to unit charges of up to 5000 kg (Fig. 7.2.4-04).

The project was carried out from July to August 1974 (Bamford et al., 1976). The southern portion of the line was shifted

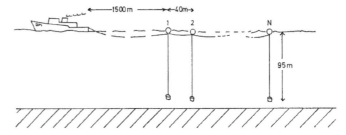

Figure 7.2.4-04. Scheme of sea shots with split charges off northern Scotland and in the English channel for LISPB 1974 (from Bamford et al., 1976, fig. 3). [Geophysical Journal of the Royal Astronomical Society, v. 44, p. 145–160. Copyright John Wiley & Sons Ltd.]

toward the west to run through Wales and Cornwall to avoid a profile line through the industrial centers of central England. The section through Wales was covered by two sea shots and an intermediate land shot for crustal control, while the line through northern England and Scotland contained three land shots and an underwater shotpoint in the Firth of Forth (Fig. 7.2.4-05; Appendix A2-1, p. 56–59).

This particular project was the first major effort to obtain a detailed picture of crustal and uppermost-mantle structure

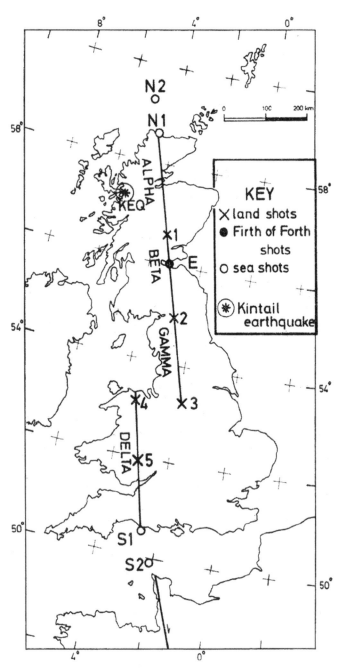

Figure 7.2.4-05. Location map of LISPB 1974 (from Bamford et al., 1976, fig. 1a). [Geophysical Journal of the Royal Astronomical Society, v. 44, p. 145–160. Copyright John Wiley & Sons Ltd.]

under the British Isles proper. It stimulated the participation of all seismological research institutions in Britain and obtained much public attention. A particularly lucky event was the local earthquake with a magnitude of 3 that occurred under western Scotland during the recording window of the northern half of the line. This small Kintail earthquake (Fig. 7.2.4-05) served as an additional "land shotpoint" (Kaminski et al., 1976), even though it was slightly off the main line.

The interpretation, similar to the interpretation of the experiment in France, was divided into a crustal part and an upper-mantle part. The crustal model (Fig. 7.2.4-06) focuses on the northern section of LISPB (Bamford et al., 1978). The southern section DELTA was only subsequently published (Griffiths and Westbrook, 1992; Maguire et al., 2011). While the resulting crustal structure (Bamford et al., 1978) was of particular interest to the British earth scientists, the upper-mantle part was of high interest to the international community.

As in the French long-range profiles, mantle phases could be correlated on several parts of the line from different shotpoints (Fig. 7.2.4-07). The model proposed by Faber (1978) and Faber and Bamford (1979) showed two high-velocity layers with velocities near 8.15 and 8.4 km/s embedded in an upper mantle ranging from 28 to 85 km depth where the background velocity gradually increased from near 7.95 to near 8.25 km/s (Fig. 7.2.4-06, bottom). The model was similar to the previously obtained model in France, except for the depth range around the Moho.

The strong velocity contrast of the French model immediately below the Moho, consisting of a high-velocity lid underlain by a distinct velocity inversion, was missing in Britain. Also, some lateral variations of the sub-crustal structure could be determined which are evidently related to gross geological variations beneath northern Britain (Faber and Bamford, 1981).

The successful completion of LISPB had a large impact on the British earth science community. Jeremy Hall undertook two very important surveys, but concentrating on the upper crust. LUST (the Lewisian Units Seismic Traverse) was recorded in 1976 across the Ben Stack Line separating granulite and amphibolite facies blocks with the Archean basement of NW Scotland (Hall, 1978). The deep crust project SISSE (South Irish Sea Seismic Experiment) in 1977 (Ransome, 1979) concentrated on Wales, and the project WISE (Western Isles Seismic Experiment) was carried out offshore western Scotland in 1979 and 1981 (Summers et al., 1982). Explosive and airgun shots were recorded along two profiles with the aim to map the Lewisian basement from the foreland to the Caledonian mountain belt (Summers et al., 1982). The first profile was shot in 1979 between Barra and Ballantree, and the second one in 1981 between Mull and Kintrye. The shallow seismic-refraction project SWESE (SW England Seismic Experiment) was also carried out in 1979 between Lizard and Stait peninsulas in SW England (Doody and Brooks, 1982).

In 1979 the Deep Geology Unit of the Institute of Geological Sciences (IGS) started to acquire seismic-reflection data to study the continental crust beneath the UK landmass particularly at levels deeper than those examined on a routine basis by explora-

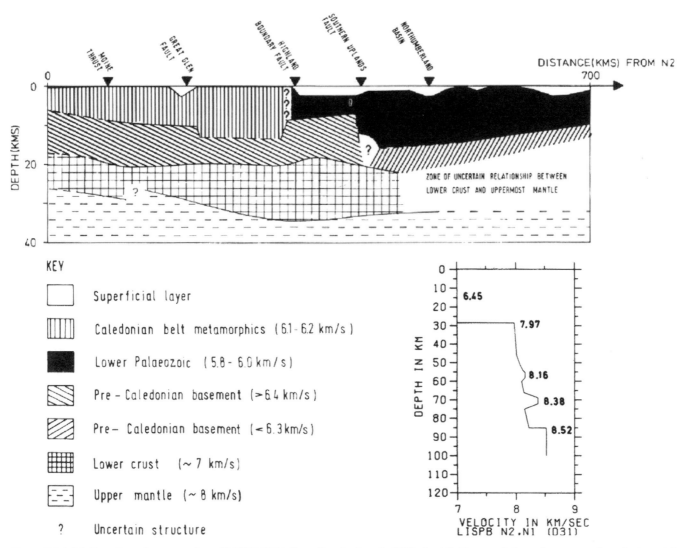

Figure 7.2.4-06. Top: Crustal cross section of LISPB 1974 (from Bamford et al., 1978, fig. 15). Bottom: Schematic cross section of subcrustal reflections (from Faber, 1978, fig. 10.4). [Ph.D. thesis, Universitat Karlsruhe, 132 p. Published by permission of Geophysical Institute, University of Karlsruhe, Germany.]

tion companies (Whittaker and Chadwick, 1983). For these experiments, either commercial companies were hired or lines were shot using IGS in-house equipment but processed commercially, or the record lengths of commercial survey lines were increased obtaining extra grant money. The first project was the IGS 1979 Wiltshire Survey. It comprised two intersecting lines located over the Variscan Front. Higher amplitude events of restricted lateral extent were seen between 8.8 and 9.4 s TWT, possibly reflected from the base of the crust.

Following the successful experiments to explore the lower lithosphere beneath the Variscides of France and the Caledonides of Great Britain, the European international community started to plan for a detailed investigation of the entire lithosphere underneath a young orogen, the Alps.

For this purpose a long-range line was planned that would follow the strike of the main geological units as much as pos-

sible. However, this aim could only be realized for the Central and Eastern Alps, but not for the Western Alps due to the strong curvature of the Alpine arc. A second guideline for the location of the main line was the Bouguer gravity map. It was aimed to position the line as close as possible through the center of the negative Bouguer anomaly (Fig. 7.2.4-08).

For the detailed planning and a first detailed interpretation an international working group was founded consisting of scientists from Austria (Vienna), England (Birmingham), France (Grenoble), Germany (Munich and Karlsruhe), Hungary (Budapest), Italy (Milano and Trieste), and Switzerland (ETH Zurich), headed by J. Ansorge (Zurich) and H. Miller (Munich). The final line was split. The eastern half within Austria followed the Eastern Alps in an E-W direction, and the western half within Switzerland and eastern France was slightly rotated into the ENE-WSW direction. Several side lines served both as fans and lines with

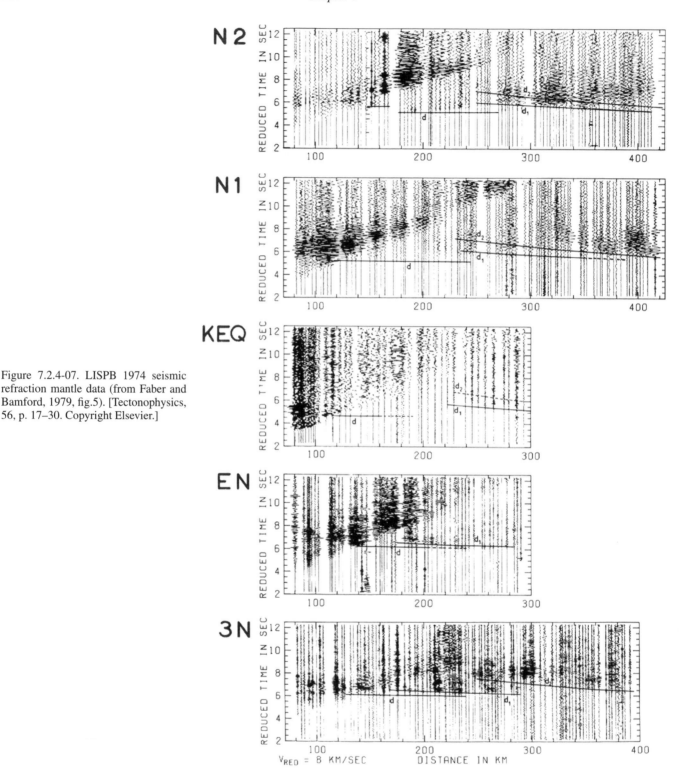

Figure 7.2.4-07. LISPB 1974 seismic refraction mantle data (from Faber and Bamford, 1979, fig.5). [Tectonophysics, 56, p. 17–30. Copyright Elsevier.]

additional shotpoints at their ends and one line between Innsbruck and Vienna tried to continue the direction of the western line toward the ENE in order to catch the ray paths of the deeper penetrating waves as close as possible to the axis of the Central Alps underneath Switzerland (Fig. 7.2.4-08).

In order to obtain a distance range up to 850 km, the line was extended beyond the Alps proper to the east, where colleagues from Hungary managed to organize two large borehole shots of 2 and 4 tons. The other shots within the Alps ranged between 0.6 and 3.3 tons and were mainly borehole shots as well, save for

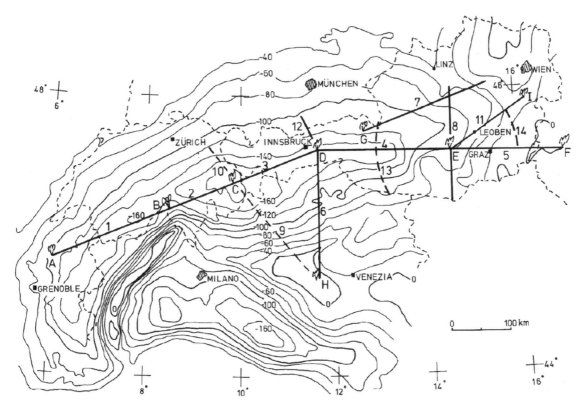

Figure 7.2.4-08. Simplified gravity map of the Alps and the location of ALP75 (from Alpine Explosion Seismology Group, 1975, fig. 2). [PAGEOPH, v. 114, p. 1109–1130. Reprinted by permission from Birkhaeuser Verlag, Basel, Switzerland.]

two locations. Alpine lakes at a high enough elevation were used for underwater shooting, and two shotpoints were quarry blasts. A total of 193 recording stations could be acquired, of which 158 were MARS-66 units. Within Hungary, four 24-trace reflection units were used. The maximum distance range to which energy was recorded was near 620 km.

The crustal data were of high quality (Appendix A2-1, p. 7–10). P_n phases could be observed to ~360 km distance, and two upper-mantle phases at 300–440 km and at 420–620 km distance could be correlated. Due to the complicated crustal structure of the Alps, the first interpretation by the Alpine Explosion Seismology Group (1975) did not go beyond a rough outline of the Moho along ALP75, but many other publications were to follow in the subsequent years as, e.g., two publications written by Miller et al. (1978, 1982).

The most recent interpretation was published by Yan and Mechie (1989; Appendix A2-1, p. 7–10, and Appendix A7-3-1) which confirmed a crustal root under the main part of the line between shotpoint locations B and E as well as several high-velocity lenses in the lower lithosphere beneath the Moho. These lenses, however, are not continuous features but differ in number, as well as lateral extension and depth under the Central and Eastern Alps. Yan and Mechie have also investigated the mantle phases, which could be correlated on several record sections at distances beyond 250 km, following the P_n phase, named d by Yan and Mechie (1989). An example is shown in Figure 7.2.4-09,

other sections and the corresponding 1-D models can be viewed in Appendix A7-3-1.

Shortly after the completion of ALP75, in 1977 a second longitudinal profile was laid out through the Southern Alps, named SUDALP77 (Ansorge et al., 1979b). SUDALP77 was more or less parallel to ALP75, and was complemented in 1978 (Italian Explosion Seismology Group and Institute of Geophysics ETH Zurich, 1981). SUDALP77 involved the cooperation of Italian, Swiss, French, and German institutes, ran from Lago Maggiore to Gemona near the Italian-Yugoslavian border and was based on four shotpoints. The additional line ran southeastward from the Central Alps to the Adriatic Sea near Trieste (Fig. 7.2.4-10; Appendix 7-3-2).

Presenting a lithospheric cross section across the border region between the Central and Southern Alps, Ansorge et al. (1979b) demonstrated the significant difference in structure of the Central and Southern Alps as well as the termination of the anomalous Ivrea zone north of that border region.

Combining all available seismic data at the time being, Mueller et al. (1980) compiled a crustal cross section along the so-called "Swiss Geotraverse." That crustal cross section extended from the Rhinegraben near Basel, Switzerland, to the Po Plain near Como, Italy. It traversed the Swiss folded Jura Mountains, the Molasse basin, the Aar Massif, the Central or Penninic Alps between the Rhine–Rhône Line and the Insubric Line, and the Southern Alps (Fig. 7.2.4-11).

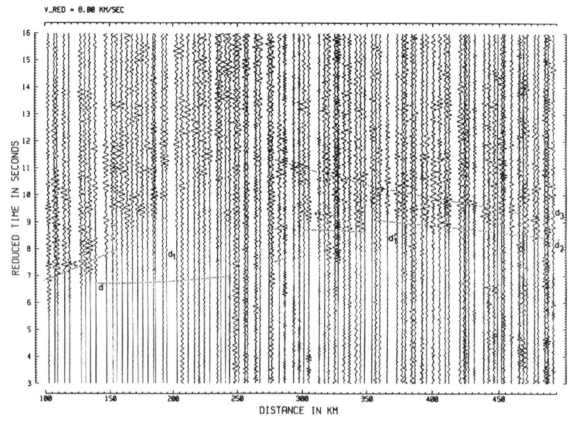

Figure 7.2.4-09. Example of mantle data recorded during ALP75 (shotpoint F to west) (from Yan and Mechie, 1989, fig. 17). Reduction velocity = 8 km/s. [Geophysical Journal International, v. 98, p. 465–488. Copyright John Wiley & Sons Ltd.]

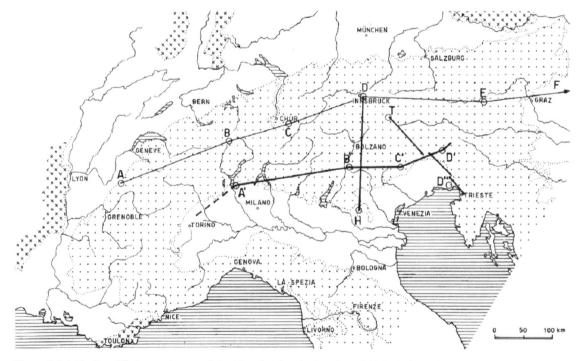

Figure 7.2.4-10. Simplified tectonic map of the Alps showing the location of ALP75 (shotpoints A to H) and SUDALP77 (shotpoints A′ to D′, D′, and T) (from Ansorge et al., 1979, fig. 1). [Bollettino di Geofisica Teorica e Applicata, v. 21, p. 149–157. Reproduced by permission of Bollettino di Geofisica, Trieste, Italy.]

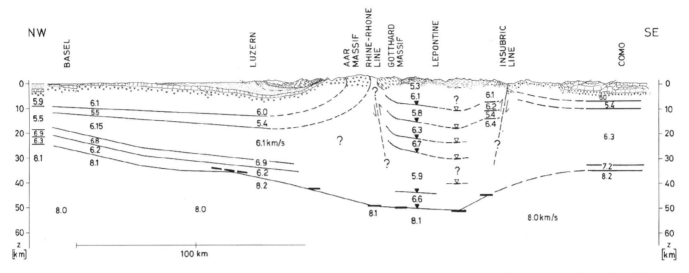

Figure 7.2.4-11. Compilation of seismic crustal data along the Swiss Geotraverse (from Mueller et al., 1980, fig. 7). [Eclogae Geologicae Helvetiae, v. 73, p. 463–483. Reproduced by permission of Swiss Geological Society, Neuchatel, Switzerland.]

7.2.5. Continued Explorations of Anomalous Areas and a Second Stage of Deep-Seismic Reflection Campaigns in Central Europe

In 1975 and 1976, a first attempt was undertaken to unravel the crustal structure of the North German basin by way of a large-scale seismic-refraction experiment from the North Sea north-west of Heligoland to the area near Goettingen (location is shown as line 1 in Fig. 8.3.4-08).

In spite of a high natural noise level, with dense station spacing and 10 shotpoints on land distributed over a distance of ~250 km plus 4 additional sea shots in the North Sea, a rather complex crustal structure model was obtained (Fig. 7.2.5-01). A thick sedimentary sequence of more than 15 km overlies a rather thin crystalline crust, and the depth to the Moho was found to decrease from 33 km in the north and to 27 km in the south. Below the Moho at 37–42 km depth, a velocity inversion was modeled (Reichert, 1993).

In 1974, and continuing into the 1980s (see Chapter 8.3.4.2) the hydrocarbon industry started to acquire an extensive seismic-reflection database in the North German Basin near the Elbe lineament between Hamburg to the south, Kiel to the north, the North Sea to the west, and the Baltic Sea to the east (Dohr et al., 1989; Yoon et al., 2008). The area of the Glückstadt graben, being

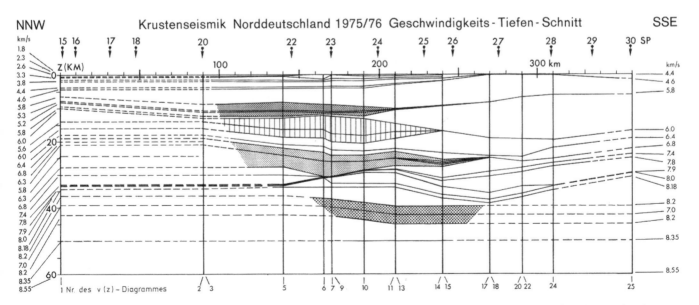

Figure 7.2.5-01. Crustal structure of the North German Basin between Heligoland and Goettingen (from Reichert, 1993, fig. 18). Location shown as line 1 in Fig. 8.3.4-08. [Geol. Jb., Hannover, E, 50, 87 p. Reproduced by permission of Schweizerbart, Stuttgart, Germany (www.schweizertbart.de).]

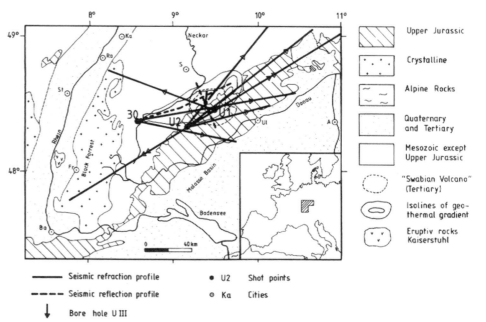

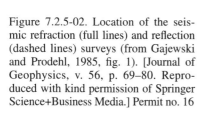

Figure 7.2.5-02. Location of the seismic refraction (full lines) and reflection (dashed lines) surveys (from Gajewski and Prodehl, 1985, fig. 1). [Journal of Geophysics, v. 56, p. 69–80. Reproduced with kind permission of Springer Science+Business Media.] Permit no. 16

the deepest Mesozoic graben structure dominated by strong salt tectonics, was of particular interest as it revealed the deep crustal structure along the transition zone from the Precambrian Fennoscandian Shield to the Caledonian accreted crust in NW Germany. The first N-S directed line 4, located west of Lübeck (see fig. 2 of Yoon et al., 2008) was 66 km long and consisted of 590 shot recordings using 48 receivers. Shotpoint and receiver spacings were 100 m, and the total recording time was 15 s TWT, resulting in an actual common midpoint fold between 15 and 20.

In 1976, the area between the Rhine and Rhône grabens was a renewed goal for French, German, and Swiss scientists. Some of the shotpoints of the 1972 Rhônegraben experiment were revived, and some profiles observed in a northerly direction, both along and across the Saône and Bresse depressions as well as into the Swiss Jura (Michel, 1978; Ansorge and Mueller, 1979). An assumed upward doming of the Moho in the area of the so-called "Burgundische Pforte," the lowland between the Rhine and Rhône valleys, was confirmed.

Another goal of investigation close to the Rhinegraben was the area of the so-called "Swabian Volcano" near Urach, east of Stuttgart, where a geothermal anomaly reaching temperatures of 130–140 °C at 3 km depth had been mapped. Under the auspices of the European Community, a multidisciplinary geoscientific research project of this anomaly had started to look for geothermal resources with the final goal to drill a geothermal hole which, when it was finally realized, reached a depth of 3333 m. To investigate the area of the geothermal anomaly prior to drilling in some detail, in 1978 and 1979, both seismic-refraction and reflection measurements were carried out (Bartelsen et al., 1982; Jentsch et al., 1982; Gajewski and Prodehl, 1985; Fig. 7.2.5-02; Appendix A2-1, p. 25–26). The initial goal was to map the topography and the velocity struc-

ture of the crystalline basement. Two different approaches were undertaken to accomplish this goal.

The wider surroundings were explored by a seismic-refraction survey (Jentsch et al., 1982). Two specially designed borehole shots were arranged along a WSW-ENE line along the axis of the Swabian Jura, (U1 and U2 in Fig. 7.2.5-02) from which, with a recording distance of up to 120 km, a model of the entire crust was to be obtained, while a number of nonreversed additional lines, crossing the area in various directions, were mainly aimed to map the basement (Jentsch et al., 1982). The main line was arranged so that it was located entirely on the limestone ridge of the Swabian Jura and also coincided with the prolongation of one of the fan profiles, which were recorded in 1972 from shotpoint Steinbrunn to the ENE (Edel et al., 1975), and was later interpreted by Gajewski and Prodehl (1985; Appendix A2-1, p. 25–26).

Since the 1950s, it had become evident that reflection seismology, as used for oil exploration studies, was a useful tool also for the investigation of the whole Earth's crust. In general, seismic-reflection surveys, working with much higher frequencies than refraction investigations, deliver completely different information than do refraction–wide-angle studies. Often they show structures and tectonics directly and with high accuracy while

Figure 7.2.5-03. From top to bottom: Stacked reflection section of profile U1 with correlatable events. Record section of wide-angle observations from shots along profile U1 with traveltime curves calculated from velocity model below. Velocity model with ray paths. Final velocity model along profile U1 showing a body of reduced velocity (hatched area). (From Bartelsen et al., 1982, figs. 9, 16, and 13). [*In* Haenel, R., ed., The Urach geothermal project (Swabian Alb, Germany): Schweizerbart, Stuttgart, p. 247–262. Reproduced by permission of Schweizerbart, Stuttgart, Germany (www.schweizertbart.de).]

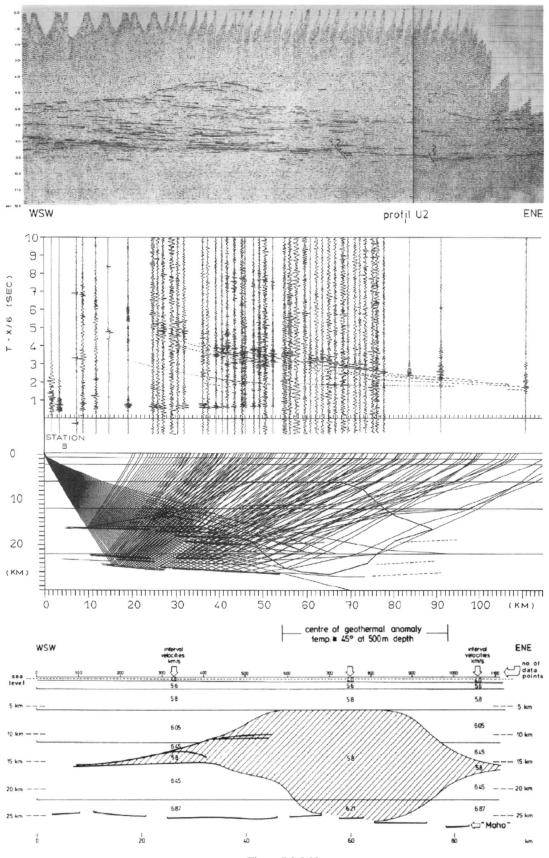

Figure 7.2.5-03.

refraction–wide-angle studies show sequences of velocities, important for petrological structure. It was R. Meissner who organized the first large-scale, purely scientific, seismic-reflection experiments in Germany.

For the detailed investigation of the geothermal anomaly of Urach, R. Meissner proposed a detailed seismic-reflection survey (Bartelsen et al., 1982). Parallel to the main seismic-refraction line but further to the north in the foreland of the Swabian Jura, a seismic-reflection profile of 27 km length was arranged and recorded by a commercial company (dashed lines in Fig. 7.2.5-02). The steep-angle observations were complemented by additional refraction stations recording all shots in the wide-angle distance range up to 100 km. Furthermore, a second shorter steep-angle reflection line was set up, running perpendicular to the first one (Bartelsen et al., 1982) and crossing the center of the anomaly near the town of Urach where the drill hole was finally placed.

This combined seismic reflection-refraction survey with extended spread lengths and an 8-fold coverage provided a high resolution result with significant velocity anomalies. As a result,

a large body with velocity as low as 5.8 km/s could be identified to underlie the area at mid- and lower-crust levels down to the Moho at to 25 km depth, where the geothermal anomaly had been mapped (Fig. 7.2.5-03). Both thermal effects and a large-scale crustal alteration were suggested to explain this result (Bartelsen et al., 1982).

Following the wide-angle profile through the Hunsrück in 1968, as referred to in Chapter 6, in 1973, R. Meissner and H. Murawski started the project Geotraverse Rhenoherzynikum (Meissner et al., 1980).

They followed the idea to systematically fill in seismic-reflection surveys along a SSE-NNW–directed seismic-refraction line between Landau in the Upper Rhinegraben to Aachen at the border between the Rhenish Massif in the south and the Lower Rhine embayment in the north. In 1973 and in 1975, two short reflection seismic surveys were performed. The first one ran across the Hunsrück fault, which separates the Paleozoic Saar-Nahe trough from the Rhenish Massif. The second one was placed near Landau, crossing the western Rhinegraben border zone. Both

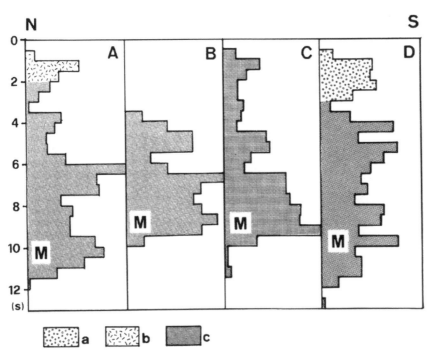

Figure 7.2.5-04. (Top) Comparison of the density of reflections for four areas: A—Hohes Venn near Aachen 1978, B—Hunsrück 1968, C—Hunsrück border zone fault 1973, D—Rhinegraben border fault 1975. M = Moho; (a) sedimentary basin; (b) thrust faults; (c) crystalline crust (from Meissner et al., 1983, fig. 7). (Bottom) Cross section along the Geotraverse Rhenoherzynikum (from Meissner et al., 1983, fig. 8) based on seismic-reflection and refraction studies of Meissner et al. (1980, 1983) and Mechie et al. (1983). [In Fuchs, K., von Gehlen, K., Mälzer, H., Murawski, H., and Semmel, A., eds., Plateau uplift: The Rhenish Massif—a case history: Springer, Berlin-Heidelberg, p. 276–287. Reproduced with kind permission of Springer Science+Business Media.]

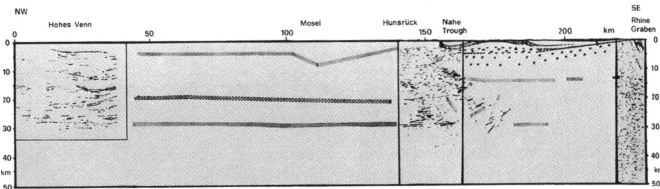

surveys included seismic-refraction stations recording the reflection shots at wide-angle distance ranges.

Later, in 1978, R. Meissner organized another seismic-reflection survey combined with wide-angle observations (Meissner et al., 1980, 1983) in the northernmost part of the Rhenish Massif near its northern margin, close to the city of Aachen (see next section, line L1-L2 in Fig. 7.2.6-01). This time, a reversed seismic-refraction profile of ~100 km in length could be arranged as the wide-angle part with the reflection shots near Aachen in the north, and a number of shots fired at its southern end in the troop training area of Baumholder within the Saar-Nahe trough (shots M1-M2 in Fig. 7.2.6-01; see next section). Here, a few months earlier, the Karlsruhe colleagues, using their Rhinegraben seismic network, had located a strong event which turned out to be an unplanned non-delayed quarry blast within this troop training area.

This detection opened a very fruitful cooperation of German geophysicists with the German army, which in the following years would lead to a number of zero-cost shots for scientific purposes. The interpretation of the reflection survey L1-L2 (Fig. 7.2.6-01) by Meissner et al. (1983) led to the identification of the shape and depth pattern of prominent fault zones, suggesting that a huge thin-skinned nappe forms the northern Variscan deformation front (Fig. 7.2.5-04).

7.2.6. The Third Stage of Long-Range Profiling

In 1979, two major long-range projects were undertaken by the western European geophysical community. The first one in western central Europe targeted the crust and uppermost mantle under the Rhenish Massif and Ardennes, while the second one in Scandinavia targeted the lithosphere and asthenosphere.

The first project took place in May. It consisted of a 600-km-long line through Germany, Luxembourg, and northeastern France and was recorded from the Harz Mountains through the Rhenish Massif and Ardennes into the Paris Basin of northern France and ended south of Reims (Fig. 7.2.6-01; Appendix A2-1, p. 15–18; Appendix A7-4-1). In addition, two side profiles, up

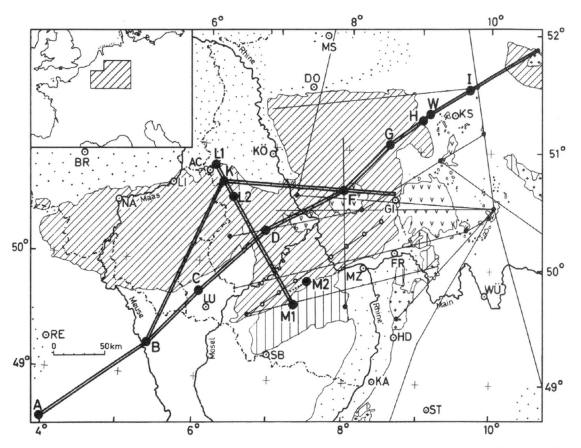

Figure 7.2.6-01. Recording lines through the Rhenish Massif and adjacent areas (from Mechie et al., 1983, fig. 1). Thick double lines: shotpoints and recording lines of 1978–1979. Thin lines: shotpoints and recording lines prior to 1978. Cities: AC—Aachen, BR—Brussels, DO—Dortmund, FR—Frankfurt, GI—Giessen, HD—Heidelberg, KA—Karlsruhe, KÖ—Cologne, KS—Kassel, LI—Liège, LU—Luxembourg, MS—Münster, MZ—Mainz, NA—Namur, RE—Reims, SB—Saarbrücken, ST—Stuttgart, WÜ—Würzburg. [*In* Fuchs, K., von Gehlen, K., Mälzer, H., Murawski, H., and Semmel, A., eds., Plateau uplift: The Rhenish Massif—a case history: Springer, Berlin-Heidelberg, p. 260–275. Reproduced with kind permission of Springer Science+Business Media.]

to 170 km long, were also recorded from Aachen toward the southwest through the Ardennes of eastern Belgium and toward the east into the volcanic area of the Vogelsberg (Mechie et al., 1983). The Geotraverse Rhenoherzynikum of 1978 (Meissner et al., 1983) served as the third side profile toward the SSE (Fig. 7.2.6-01; Appendix A7-4-1).

A reinterpretation of Mooney and Prodehl (1978) had revealed peculiarities of the Moho whenever seismic lines crossed volcanic areas or touched the northern end of the Rhinegraben. Here, the Moho, elsewhere a first-order discontinuity, seemed elevated and disrupted (Fig. 7.2.6-02). The new data showed

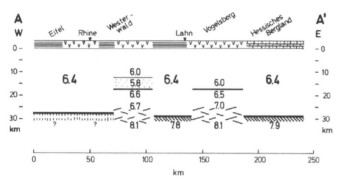

Figure 7.2.6-02. Crustal sketch through the Rhenish Massif, reinterpreted from data prior to 1978 (from Mooney and Prodehl, 1978, fig. 17). [Journal of Geophysics, v. 44, p. 573–601. Reproduced with kind permission of Springer Science+Business Media.]

very few P$_n$ arrivals and, not in all cases, very strong secondary arrival phases. In particular, on the eastern half of the main line, under the Eifel and Westerwald, the secondary arrivals indicated the existence of two phases mixed into each other (phases d and e in Fig. 7.2.6-03).

It was interpreted as a Moho lid, at 30 km depth overlying an ~5–7-km-thick low-velocity zone where the velocity drastically decreases and then gradually increases to 8.4 km/s or more below 35 km depth (Fig. 7.2.6-04). As the long-range line did not cross the volcanic area of the Westerwald, the disrupted Moho of Mooney and Prodehl (1978) did not show up in the new model. West of the Eifel, under the Ardennes, this Moho lid had disappeared, as well as much of the intracrustal structure, modeled elsewhere, and only the lower velocity boundary, at 37 km depth, appeared as a deep-reaching transition between the crust and mantle. A normal crust, with a sharp Moho at 30 km depth, appears again under the Paris Basin (Mechie et al., 1983).

The previous successful experiments with observations up to and beyond 1000 km distance, which had enabled viewing fine structure of the subcrustal lithosphere to 80 km in depth and more, and the successful long-range lines in North America, created the desire to look deeper into the upper mantle beneath Europe. Only Scandinavia offered the chance of planning for a line up to 2000 km in length on relatively homogeneous geological subsurface structure. It was hoped that the homogeneity of crustal terrane would allow a uniform "window" into the underlying mantle.

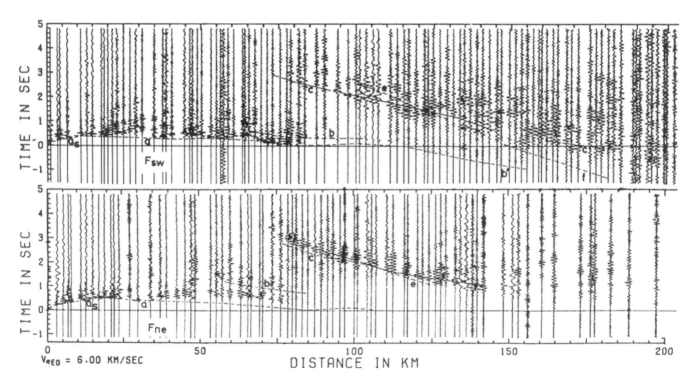

Figure 7.2.6-03. Crustal record sections of the 1979 Rhenish Massif project recorded from shotpoint F towards SW and NE (from Mechie et al., 1983, figs. 2h and 2i). [In Fuchs, K., von Gehlen, K., Mälzer, H., Murawski, H., and Semmel, A., eds., Plateau uplift: The Rhenish Massif—a case history: Springer, Berlin-Heidelberg, p. 260–275. Reproduced with kind permission of Springer Science+Business Media.]

In Scandinavia in 1972, following the 1969 observation of the Trans-Scandinavian profile (TSSP 69 in Fig. 7.2.6-05), another long profile had already been recorded through central Scandinavia along the so-called "Blue Road," starting at about the Arctic circle in Norway and traversing in a SSE direction the Caledonian mountain range and adjacent Baltic Shield of northern Sweden (BR 72 in Fig. 7.2.6-05).

Distance ranges of up to 550 km were reached, but in a first interpretation, the authors (Hirschleber et al., 1975) concluded that no fine structure of the P_n wave could be detected, possibly due to the rather thick crust of ~40 km underneath the whole transect. Later, Lund (1979) reinterpreted the data and found a fine structure of the subcrustal lithosphere similar to that found in France and Britain.

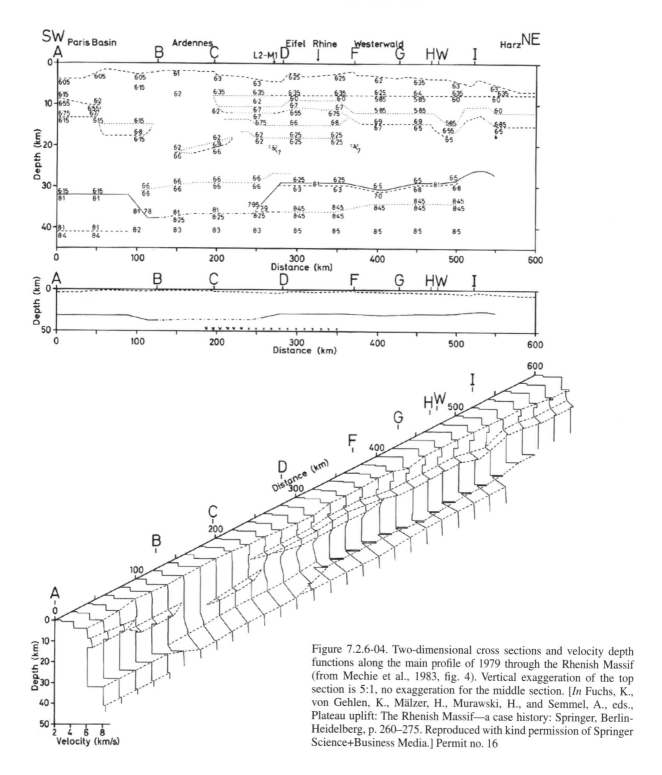

Figure 7.2.6-04. Two-dimensional cross sections and velocity depth functions along the main profile of 1979 through the Rhenish Massif (from Mechie et al., 1983, fig. 4). Vertical exaggeration of the top section is 5:1, no exaggeration for the middle section. [*In* Fuchs, K., von Gehlen, K., Mälzer, H., Murawski, H., and Semmel, A., eds., Plateau uplift: The Rhenish Massif—a case history: Springer, Berlin-Heidelberg, p. 260–275. Reproduced with kind permission of Springer Science+Business Media.] Permit no. 16

Figure 7.2.6-05. Seismic-refraction lines in Scandinavia. Thick lines: FENNOLORA 1979 including side branches through East Germany and Poland–Ukraine. Thin lines: TSSP 69 Trans-Scandinavian Deep Seismic Sounding project of 1969, BR 72 Blue Road project of 1972, SVEKA 81 and BALTIC 82 Finnish projects of 1981 and 1982 (from Guggisberg et al., 1991, fig. 1). [Tectonophysics, v. 195, p. 105–137. Copyright Elsevier.]

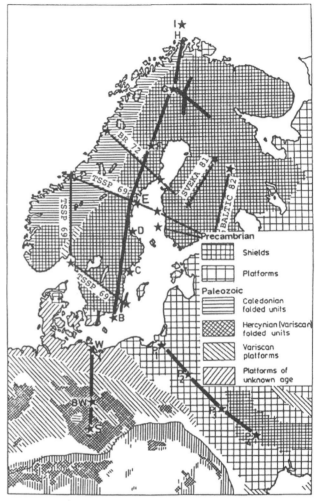

FENNOLORA 1979

In 1979, the long-range project FENNOLORA (Fennoscandian Long Range Project) was launched in Scandinavia (Fig. 7.2.6-05), aiming for a much deeper penetration of the upper mantle than any previous long-range profile.

One of the goals of FENNOLORA was to reach the mantle transition zone at 400 km depth named by seismologists the 20° discontinuity. Therefore observation distances were laid out to almost 2000 km between shotpoints off the North Cape, and off the southern coast of Sweden (Fig. 7.2.6-05). In parallel, a series of intermediate shots, most of them detonated in the Baltic Sea close to the shore of Sweden, served to obtain all necessary details of crustal and uppermost mantle structure (Fig. 7.2.6-06; Appendix A2-1, p. 69–71 and Appendix A7-4-2).

The FENNOLORA shots were also recorded on two lines to the south through eastern Germany and to the southeast on a line through Poland and the Ukraine (Fig. 7.2.6-05). Furthermore, the map shows profiles in Finland, which were recorded in 1981 and 1982, in the framework of the European Geotraverse which will be discussed in detail in Chapter 8.

The interpretation of the crustal profiles (Fig. 7.2.6-06; Guggisberg et al., 1991; Appendix A2-1, p. 69–71; Appendix A7-4-2, p. A3–A58) showed that the crust is not uniform along the whole line. In the southernmost part of Sweden it was found

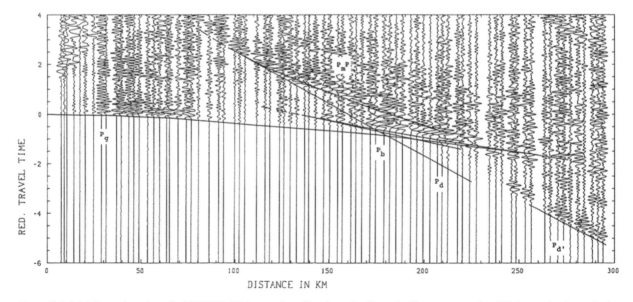

Figure 7.2.6-06. Record section of a FENNOLORA crustal profile, shotpoint B-north, distance range 0 to 300 km, reduction velocity 6 km/s (from Guggisberg et al., 1991, fig. 5). [Tectonophysics, v. 195, p. 105–137. Copyright Elsevier.]

to be less than 40 km thick, but for most of the line, its thickness ranged between 40 and 55 km. There were a few areas with an additional deep lowermost crustal layer with a velocity around 7.5 km/s, and major tectonic peculiarities were mirrored by anomalous features at Moho level (Fig. 7.2.6-07).

The lithosphere-asthenosphere data was interpreted by several authors (e.g., Guggisberg and Berthelsen, 1987; Stangl, 1990). As an appendix to his Ph.D. thesis, Stangl (1990) prepared an Open-File-Report containing all crustal and mantle P- and S-data as record sections (Appendix A7-4-2). Though varying in detail, the overall picture showed a stratified upper mantle where high-velocity areas are intercalated with low-velocity regions with substantial velocity inversions (Fig. 7.2.6-08; Appendix A7-4-2, p. A59–A128).

Of particular interest is a reflection at a distance range of 1650–1900 km reflected from the top of the mantle transition zone (Fig. 7.2.6-09A). Fuchs et al. (1987) produced one of the first velocity-depth models (Fig. 7.2.6-09B) showing a rather complicated velocity-depth structure with the mantle-transition zone at its base, and compared this result with models obtained from other seismological data.

7.2.7. Deep Seismic Sounding Projects in Eastern Europe

With the establishment of the international network of long and detailed seismic-refraction profiles throughout southeastern Europe, initiated in particular by V.B. Sollogub in the 1960s, considerable details of the crustal structure had been obtained. In the 1970s, investigations by deep seismic sounding were particularly continued in Poland due to the engagement of A. Guterch and his co-workers at the Geophysical Institute of the Polish Academy of Sciences in Warszaw. As a continuation and to complement the knowledge obtained, the international profiles VII and VIII were completed in the early 1970s (Fig. 7.2.7-01).

A special study of the Tornquist-Teisseyre tectonic zone was now the goal by observing a series of regional profiles named LT2 to LT5 (e.g., Guterch, 1977; Guterch et al., 1983; Fig. 7.2.7-02). These profiles crossed the Tornquist-Teisseyre tectonic zone perpendicular to its strike, and ran from the fore-Sudetic monocline, or Paleozoic platform, in the southwest to the East European platform in the northeast (Fig. 7.2.7-02). Seismic measurements along these lines were taken using a technique of continuous profiling with distances of 100 or 200 m

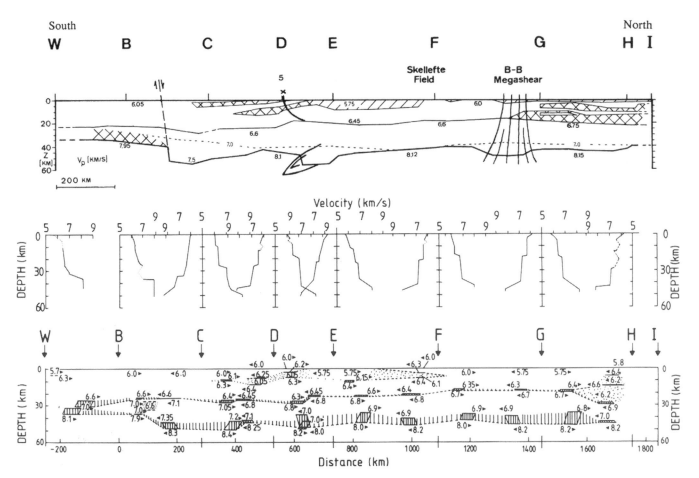

Figure 7.2.6-07. Top: Velocity-depth cross section of the crust underneath FENNOLORA. Center: 1-D velocity-depth functions. Bottom: Depth ranges which generate wide-angle reflections correlated in the data (from Guggisberg et al., 1991, figs. 31 and 18). [Tectonophysics, v. 195, p. 105–137. Copyright Elsevier.]

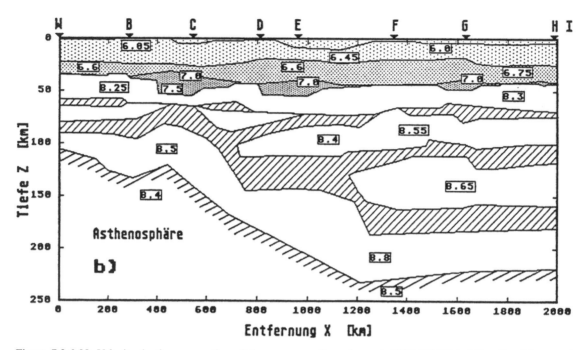

Figure 7.2.6-08. Velocity-depth cross section of the upper mantle underneath FENNOLORA (after Guggisberg and Berthelsen, 1987, fig. 6, from Stangl, 1990, fig. 7–11b). [Ph.D. thesis, University of Karlsruhe, 187 p. Published by permission of Geophysical Institute, University of Karlsruhe, Germany.]

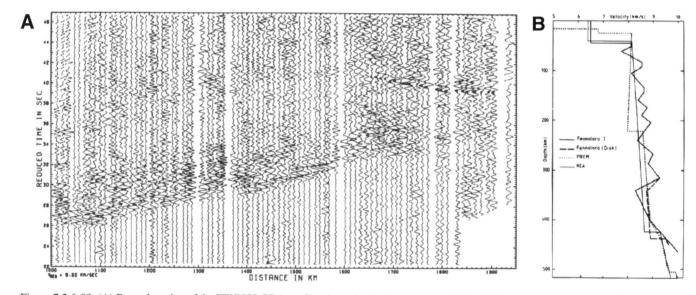

Figure 7.2.6-09. (A) Record section of the FENNOLORA profile, shotpoint I, distance range 1000–1950 km, reduction velocity 9.5 km/s. A later phase near 40 sec reduced traveltime at 1650 and 1900 km is interpreted as reflection from the mantle transition zone (from Fuchs et al., 1987, fig. 5). (B) Preliminary velocity-depth function deduced from the data of shotpoint I, in comparison with other earth models based on seismological data (from Fuchs et al., 1987, fig. 6). [*In* Fuchs, K., and Froidevaux, C., eds., Composition, structure and dynamics of the lithosphere-asthenosphere system: American Geophysical Union Geodynamics Series 16, p. 137–154. Reproduced by permission of American Geophysical Union.]

between recording points and shotpoint intervals of 50–80 km. The shots were recorded in the offset range between 50–70 km and 220–320 km (Guterch, 1974, 1977).

The interpretation of the new LT-profiles included the data of the recently completed international profiles VII and VIII.

The densely spaced recordings allowed a reliable phase correlation of different waves, and resulted in a detailed picture of crustal structure in Poland (Fig. 7.2.7-03). The most important outcome of the interpretation was the detection of a 50 km deep Moho trough along the Tornquist-Teisseyre tectonic zone

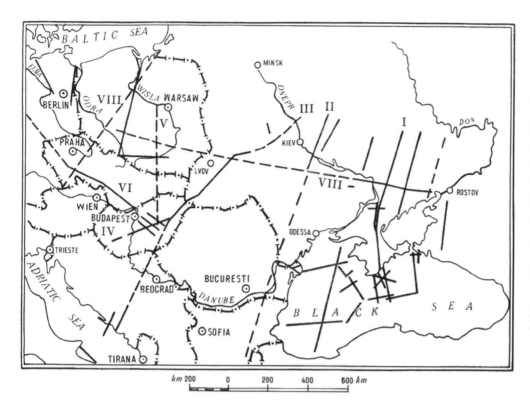

Figure 7.2.7-01. Location of deep seismic profiles and corresponding profiles in southeastern Europe (from Sollogub, 1969, fig. 1). Solid lines: completed in the 1960s, dashed lines: planned and partly completed in the 1970s. Remark by the authors: The NE-SW running profile through Poland and Czechoslovakia is Profile VII. It was erroneously named VIII. [*In* Hart, P.J., ed., The Earth's crust and upper mantle: American Geophysical Union, Geophysical Monograph 13, p. 189–195. Reproduced by permission of American Geophysical Union.]

Figure 7.2.7-02. Location of deep-seismic sounding profiles in Poland around the Tornquist-Teysseire Zone (from Guterch et al., 1991a, fig. 1). [Publs. Institute of Geophysics, Polish Academy of Science, A-19 (236), p. 41–61. Reproduced by permission of Instytut Geofizyki Akademii Nauk, Warsaw, Poland.]

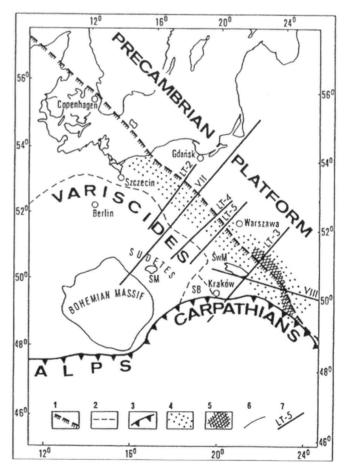

bordered by a 30-km-thick crust under the Hercynian terrain to the southwest and a 40-km-thick crust under the East European Platform to the northeast.

As already mentioned in Chapter 6, in the middle of the 1960s, a special investigation of the fore-Sudetic region had started, aiming primarily to study the sedimentary cover for location of possible oil reserves. The investigation extended over many years into the 1970s, when the interpretation of the complete network became possible (Toporkiewicz, 1986). On selected lines recordings were made up to 90 km distances, thus allowing to study the whole crust (Toporkiewicz, 1986), resulting in a network of deep seismic sounding profiles M1–M13 in Poland (Fig. 7.2.7-04) with a total length of 1500 km. This survey continued into the 1970s and its interpretation was not completed before 1978. The traveltime branches of the crustal phases could be traced up to 60–90 km distance ranges and the resulting models are shown in Figure 7.2.7-04 (Guterch et al., 1991a).

Finally, at the end of the decade, in 1979, the FENNOLORA project prompted the observation of two long-range profiles through eastern Europe (see Fig. 7.2.6-05); one of the lines ran in

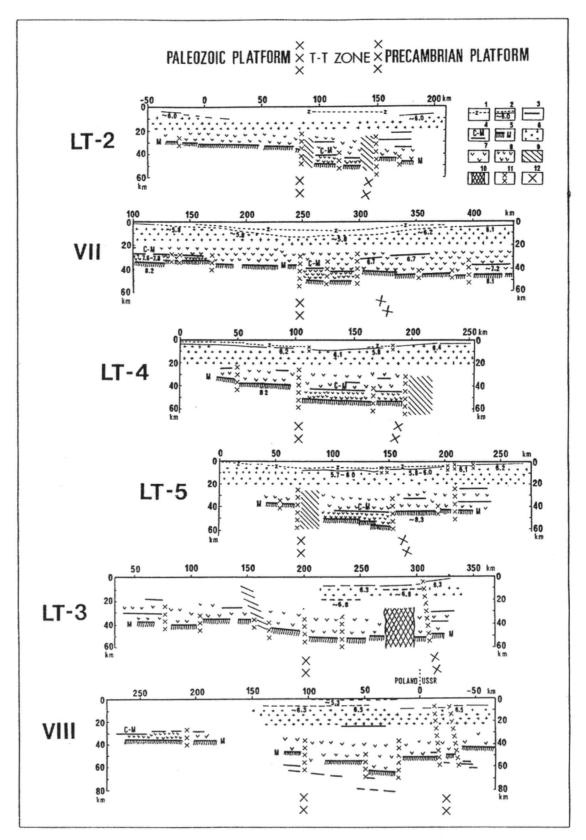

Figure 7.2.7-03. Crustal cross sections through Poland around the Tornquist-Teysseire Zone as interpreted in the 1970s 1Reproduced by permission of Instytut Geofizyki Akademii Nauk, Warsaw, Poland.]

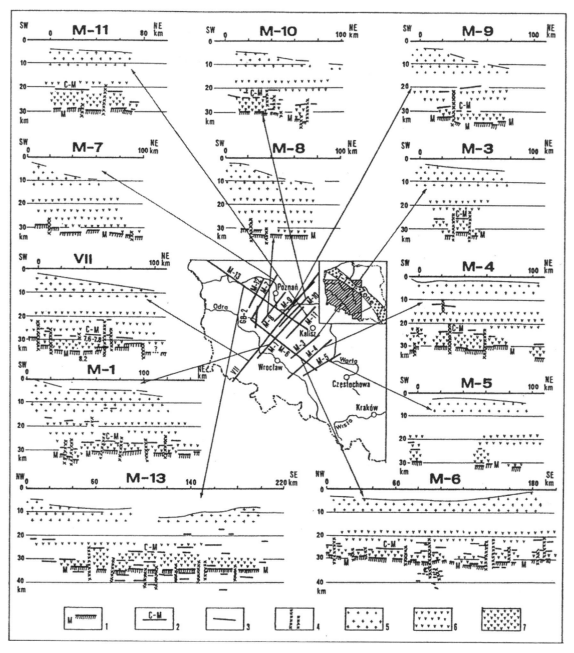

Figure 7.2.7-04. Location of seismic profiles and corresponding crustal cross sections in the fore-Sudetic region, southwestern Poland (from Guterch et al., 1991a, fig. 3). [Publs. Institute of Geophysics, Polish Academy of Science, A-19 (236), p. 41–61. Reproduced by permission of Instytut Geofizyki Akademii Nauk, Warsaw, Poland.]

a NW-SE direction through Poland into the Ukraine. Some of the FENNOLORA shots here were also successfully recorded, and for crustal control, additional shotpoints were organized.

In adjacent Czechoslovakia, after completion of the two International Profiles V and VI mentioned in Chapter 6 (for location see Figs. 6.3.1-01 and 7.2.7-01), the International Profile VII was completed in 1970 (Beránek et al., 1973). The previous profile VII was continued into southeastern Germany to the margin of the eastern Alps by the University of Munich, using the shots along profile VII as energy sources (Miller and Gebrande, 1976). Several seismic regional profiles were recorded between 1975 and 1980 investigating the deep structure of the Western Carpathians, and in 1979, shots of the international project FENNOLORA were also recorded on a N-S line through eastern Germany and through the westernmost Bohemian Massif of Czechoslovakia (Mayerova et al., 1994).

In Romania (Fig. 7.2.7-05), seismic investigations, along the 390-km-long International Profile XI, continued until 1974, and

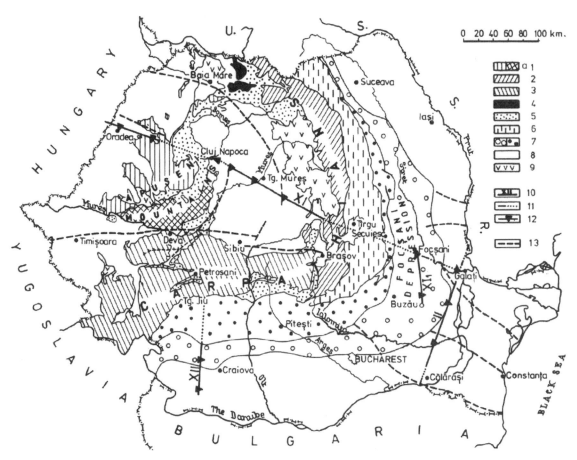

Figure 7.2.7-05. Seismic observations in Romania to the end of the 1970s (from Cornea et al., 1981, fig. 1). Legend: 1–9—geological units, 10—continuous seismic recording, 11—discontinuous seismic recording, 12—shotpoints, 13—tectonic fracture zones. [PAGEOPH, v. 119, p. 144–1156. Reprinted by permission from Birkhaeuser Verlag, Basel, Switzerland.]

were complemented by observations in the Focsani depression (profiles I_1 and II), as well as in the southern Carpathian foreland (line XII). Data examples and a detailed interpretation were published by Radulescu et al. (1976) and Cornea et al. (1981). The cross section along profile XI (Fig. 7.2.7-06) shows the crustal structure from the Pannonian basin in the west to the Moesian platform in the east.

Similar to the interpretation of the seismic lines through Poland by Guterch et al. (1991a), the Romanian crust was divided into crustal blocks, which were separated by deep-reaching fracture zones. The crust thickened to more than 40 km under the Eastern Carpathians and both the easternmost section of the Eastern Carpathians and the Focsani depression were underlain by a 20 km deep sedimentary trough.

Figure 7.2.7-06. Bouguer gravity and topography (top lines) and seismic crustal cross section through Romania along the international profile XI (from Cornea et al., 1981, fig. 9). 1—sediments, 2—granitic layer, 3—basaltic layer, 4—sediment-basement boundary, 5—Conrad discontinuity, 6—Moho, 7—Eastern Carpathian nappes, 8—volcanic rocks, 9—fracture zones, 10—crustal and subcrustal earthquake hypocenters. [PAGEOPH, v. 119, p. 144–1156. Reprinted by permission from Birkhaeuser Verlag, Basel, Switzerland.]

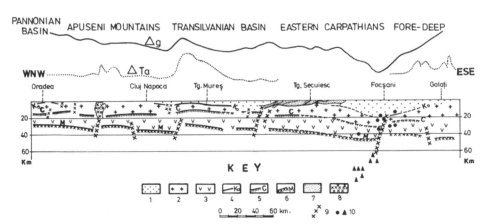

7.3. DEEP-SEISMIC SOUNDING IN EASTERN EUROPE AND ADJACENT ASIA (USSR)

In the USSR, deep seismic sounding operations were drastically reduced in the 1970s. It was a decade when new methods and interpretation programs were developed, and when the methods of field experimentation were also changed.

In the southwestern USSR, the network of international and national profiles, which had been initiated and pushed forward by V.B. Sollogub in the 1960s, was supplemented in the early 1970s by some new measurements, and individual interpretations were prepared (see, e.g., Grad and Tripolsky, 1995).

Since 1972, the Voronezh State University conducted seismic investigations across the Voronezh shield which represents a 300–500-km-wide horst of Archean and Proterozoic highly dislocated supracrustal and magmatic rocks of complex tectonic nature (Tarkov and Basula, 1983). Commercial quarry blasts were used for the investigation. The seismic signals were tape-recorded using 6-channel telemetric arrays of type Taiga with 150 m geophone spacing at an average station spacing of 5–10 km. Several profiles were recorded throughout the area, all recording wide-angle reflections from the Moho at distances between 90 and 190 km. The interpretation revealed an upper crust of 12–15 km in thickness, the middle crust consisted of one or two layers with velocities from 6.35 to 6.85 km/s and the lower crustal velocities ranged from 6.8 to 7.95 km/s within a depth range from ~32 km to 43–45 km (Moho depth).

The Novosibirsk group led by N.N. Puzyrev applied, for example, the new method of differential time fields to explore almost inaccessible areas of Siberia and the Baikal rift zone. Here the source and several arrays of receivers moved along a profile, keeping the distance between them optimal for waves reflected from basic crustal boundaries (for references, see Pavlenkova, 1996).

An overview of seismic observations made in Eastern Siberia around Lake Baikal in the late 1960s and early 1970s (Fig. 7.3-01) was compiled by Olsen (1983). The dashed lines in Figure 7.3-01

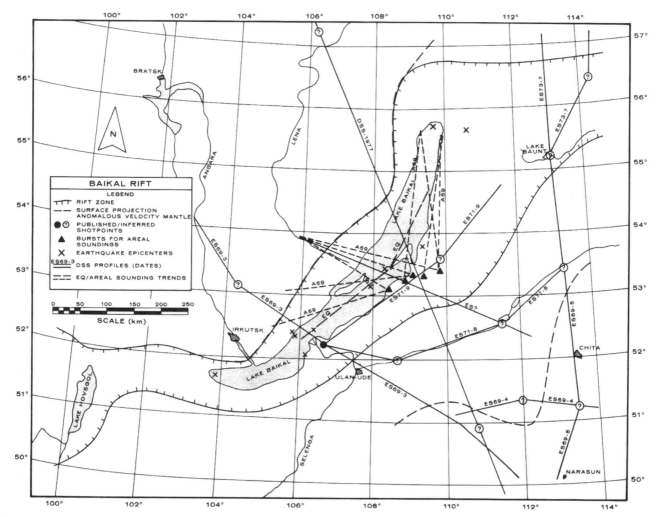

Figure 7.3-01. Map of the Lake Baikal area of eastern Siberia showing the approximate boundary of the Baikal rift zone, the surface projection of anomalous mantle material near Moho depths, and published seismic refraction lines employing both explosion and earthquake sources (from Olsen, 1983, fig. 4). [Tectonophysics, v. 94, p. 349–370. Copyright Elsevier.]

point from sources east of Lake Baikal to receivers which were distributed fan-like in order to study the three-dimensional structure beneath central and northern Lake Baikal. In this review by Olsen (1983), the original references can also be found.

Based on powerful explosions on land (nuclear devices, as became known after 1990), a number of profiles of 1000–2000 km in length were completed (Zverev and Kosminskaya, 1980; Vinnik and Ryaboi, 1981; Egorkin and Pavlenkova, 1981; see also Fuchs and Vinnik, 1982; Fuchs et al., 1987; Mechie et al., 1993), which were recorded by multichannel instruments with an average spacing of ~20 km.

A major part of these profiles was carried out throughout northwest Asia, across the West Siberian plate and the East Siberian platform as well as through central Turkmenia, central Kazakhstan, and other high-mountain areas of the USSR (Fig. 7.3-02). Both Kosminskaya and Pavlenkova (1979) and Zverev and Kosminskaya (1980) have published a representative collection of crustal and uppermost mantle cross sections for various

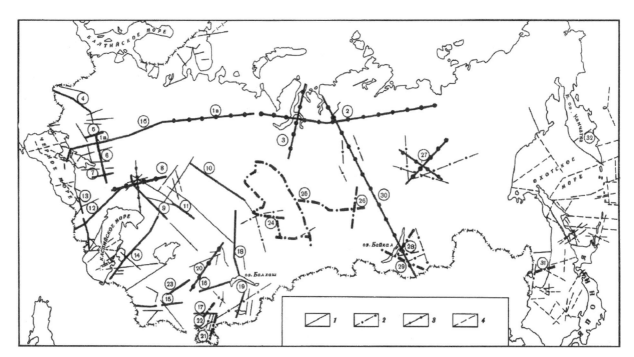

Figure 7.3-02. Location map of the main profiles recorded in the USSR between 1960 and 1978 (from Zverev and Kosminskaya, 1980, fig. 1). For references see Zverev and Kosminskaya (1980, table 1). [Publ. House NAUKA, Moscow, 180 p. Reproduced by permission of the Schmdt Institute of Physics of the Earth RAS.]

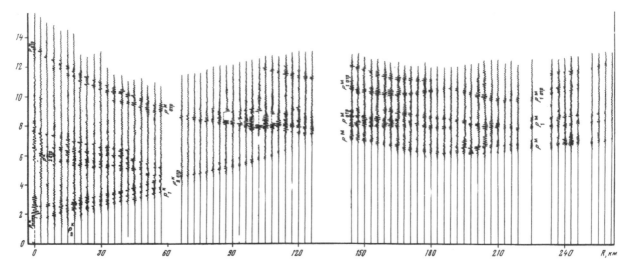

Figure 7.3-03. Data example of a profile recorded in 1970 through the Urals (from Zverev and Kosminskaya, 1980, fig. 50). [Publ. House NAUKA, Moscow, 180 p. Reproduced by permission of the Schmdt Institute of Physics of the Earth RAS.]

regions of the USSR. Data and a cross section of a profile recorded in the Urals in 1970 and published by Zverev and Kosminskaya (1980) are shown in Figures 7.3-03 and 7.3-04. In this context, a 700-km-long line across the Barents Sea also was recorded in 1976 (Davydova et al., 1985) to the northeast from Murmansk which consisted of a system of reversed and overlapping 200–400-km-long profiles.

The analysis of the long-range profiles not only revealed crustal structure information, but also showed pronounced differences in the structure of the lower lithosphere. Egorkin and Pavlenkova (1981) published a complete data set of a long-range profile across the East European platform with recordings up to 2800 km and discussed the inhomogeneous structure of the subcrustal lithosphere using the example of a profile through Kazakhstan (Fig. 7.3-05).

Examples of subcrustal velocity-depth functions for eastern Europe and northern Asia published until 1982 (Fig. 7.3-06) have been compiled, e.g., by Prodehl (1984; see also Fuchs et al., 1987).

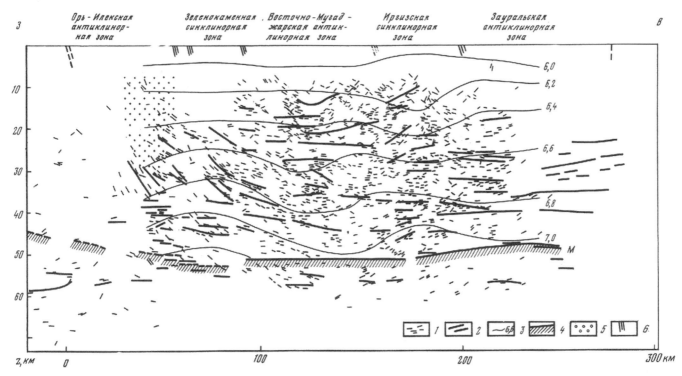

Figure 7.3-04. Crustal cross section along profile no 11 (Fig. 7.3-02) recorded in 1970 through the Urals (from Zverev and Kosminskaya, 1980, fig. 53). For references see Zverev and Kosminskaya (1980, Table 1). [Publ. House NAUKA, Moscow, 180 p. Reproduced by permission of the Schmdt Institute of Physics of the Earth RAS.]

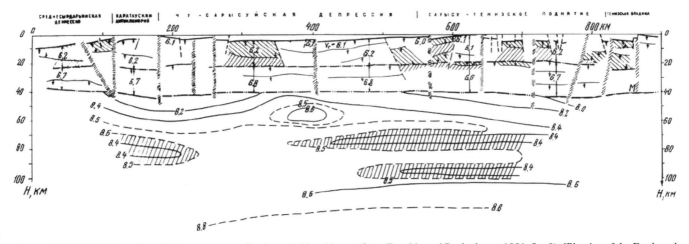

Figure 7.3-05. Seismic section of a long-range profile through Kazakhstan (from Egorkin and Pavlenkova, 1981, fig. 9). [Physics of the Earth and Planetary Interiors, v. 25, p. 12–26. Copyright Elsevier.]

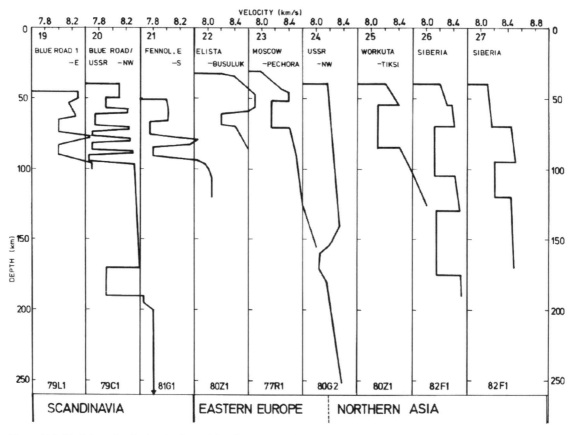

Figure 7.3-06. Selected velocity-depth models of upper-mantle structure under Scandinavia and the USSR (from Fuchs et al., 1987, fig. 2). [*In* Fuchs, K., and Froidevaux, C., eds., Composition, structure and dynamics of the lithosphere-asthenosphere system: American Geophysical Union Geodynamics Series 16, p. 137–154. Reproduced by permission of American Geophysical Union.]

7.4. CONTROLLED-SOURCE SEISMOLOGY IN NORTH AMERICA

7.4.1. Crustal Seismic-Refraction Work in the United States

Following the intensive seismic-refraction research activities in the 1960s in North America, the detection and development of plate tectonics turned much of the seismic and seismological activities at universities toward the oceans, and the interest in explosion seismology on continents in North America dropped. For example, the move of A.L. Hales from Dallas to Australia shifted the interest of the remaining scientists into other fields of geophysics. Furthermore, the major seismic activities of the U.S. Geological Survey, from 1966 onward, had moved into earthquake research with emphasis on the tectonically active west.

The interpretation of seismic data, which the U.S. Geological Survey had assembled throughout the United States in the 1960s, as described in Chapter 6 continued. Both for the Rocky Mountain and the Appalachian surveys, new

models were prepared (Prodehl and Pakiser, 1980; Prodehl et al., 1984), first versions of which had been compiled (Fig. 7.4.1-01) in a foregoing publication as a synthesis of crustal data (Prodehl, 1977).

Only a few centers remained active. Here, however, active seismology was pursued with much emphasis. For example, at the University of Wisconsin, e.g., R.P. Meyer and his group continued their activities on controlled-source seismology in the Andean regions of South America. In Cornell University at Ithaca, New York, J.A. Oliver, S. Kaufman, J.A. Brewer, L. Brown, and others started a continent-wide deep seismic-reflection program, an approach which was also pursued by M.J. Berry and co-workers at the Earth Physics Branch of the Department of Energy, Mines and Resources at Ottawa, Canada.

At the University of Utah at Salt Lake City, R.B. Smith with his co-workers, L.W. Braile, G.R. Keller, and others had a particular interest in the exploration of the crust and mantle in the tectonically active western United States with emphasis on the Basin and Range province, the Colorado Plateau, and the middle Rocky Mountains. A major source for the investigation of the crustal structure of the Wasatch Mountains and Wasatch Front

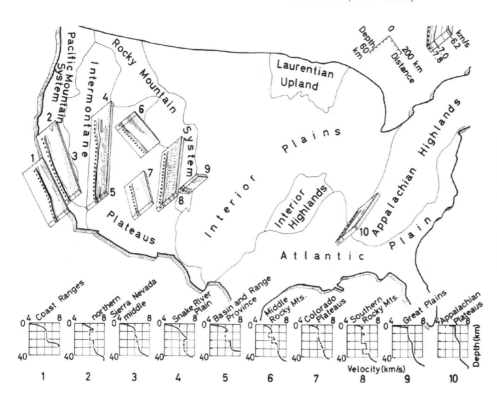

Figure 7.4.1-01. Compilation of selected crustal cross sections and representative velocity-depth functions for the U.S. Geological Survey seismic refraction data of the 1960s (from Prodehl, 1977, fig. 7). [*In* Heacock, J.G., ed., The Earth's crust: American Geophysical Union Geophysical Monograph 20, p. 349–369. Reproduced by permission of American Geophysical Union.]

was the regular quarry blasts in the Bingham copper mine (A in Fig. 7.4.1-02). In 1971 and 1972, several seismic-refraction profiles could be completed (lines A–C and A–B in Fig. 7.4.1-02; Braile et al., 1974; Keller et al., 1975; Appendix A7-5-1).

Finally, Yellowstone Park and the adjacent eastern Snake River Plain, which R.B. Smith claimed to be an active hot spot, became the focus of a major research project (e.g., Smith and Christiansen, 1980, Smith and Braile, 1994).

The seismic equipment available to universities in the late 1970s, throughout the United States, however, was not sufficient and not available in numbers great enough for large-scale experiments. Also, the powerful equipment of the U.S. Geological Survey of the 1960s was aging due to its extensive use.

So, in 1977, R.B. Smith asked his colleagues in Europe for help and in 1978, the first U.S.-European cooperation started with a joint venture of the University of Utah, Salt Lake City, and Purdue University in West Lafayette, Indiana, with the ETH Zurich, Switzerland, and the University of Karlsruhe, Germany.

The Europeans shipped their MARS-66 equipment to Salt Lake City to increase the number of portable recording stations and thus to enable effective seismic-refraction work in the Yellowstone–Snake River Plain area. Fifteen shots were recorded that provided coverage to distances of 300 km (Fig. 7.4.1-03).

The fieldwork of 1978 was followed by a second stage in 1980 (Braile et al., 1982; Smith et al., 1982; Sparlin et al., 1982).

Most of the data showed extremely clear wide-angle reflections, both in the eastern Snake River Plain (Fig. 7.4.1-04) and in the Yellowstone Park area (Fig. 7.4.1-05). More data and other details of the experiment are provided in Appendix A7-5-2. The

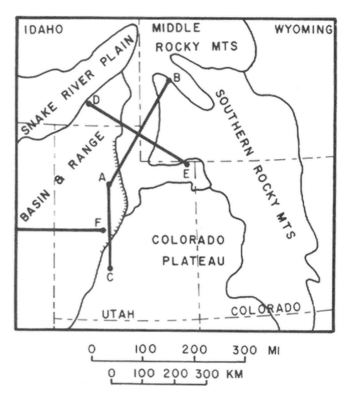

Figure 7.4.1-02. Location of seismic refraction profiles around the Wasatch Front, western United States (from Braile et al., 1974, fig. 1). A—Bingham Copper Mine, B—Fremont Lake, C—Piute Reservoir, D—American Falls Reservoir, E—Flaming Gorge Reservoir, F—Delta, Utah. [Journal of Geophysical Research, v. 79, p. 2669–2677. Reproduced by permission of American Geophysical Union.]

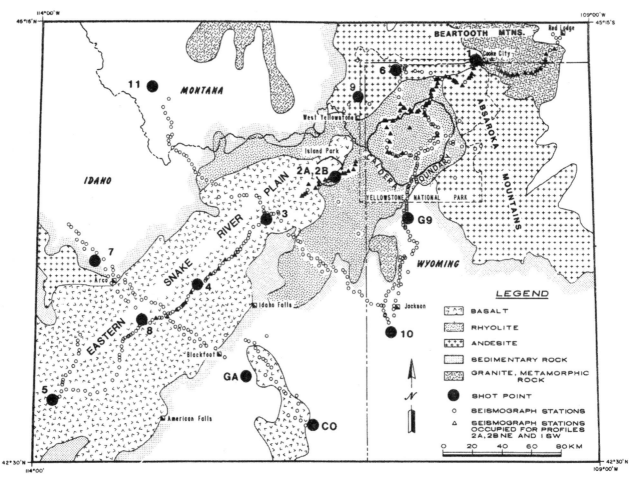

Figure 7.4.1-03. Location of the seismic refraction observations in the Yellowstone–Snake River Plain (from Smith et al., 1982, fig. 1). [Journal of Geophysical Research, v. 87, p. 2583–2596. Reproduced by permission of American Geophysical Union.]

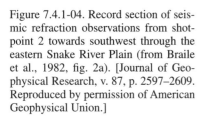

Figure 7.4.1-04. Record section of seismic refraction observations from shotpoint 2 towards southwest through the eastern Snake River Plain (from Braile et al., 1982, fig. 2a). [Journal of Geophysical Research, v. 87, p. 2597–2609. Reproduced by permission of American Geophysical Union.]

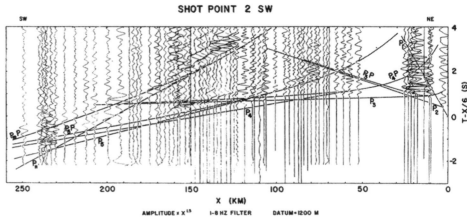

interpretation showed for the upper 10 km of the crust strong lateral inhomogeneities, probably reflecting the effects of a major lithospheric anomaly, which is evidenced by large volumes of volcanic rocks, and the systematic progression of the silicic volcanism along the eastern Snake River Plain to its present position beneath Yellowstone. The intermediate and lower crustal layers proved to be more homogeneous, and the total crustal thickness underneath the Island Park–Yellowstone Plateau–Beartooth Mountains was ~43 km, which increased slightly into the Snake River Plain to the southwest (Fig. 7.4.1-06).

Increased interest in the nature of the Rio Grande rift led to a first seismic-refraction study in 1976, organized and interpreted

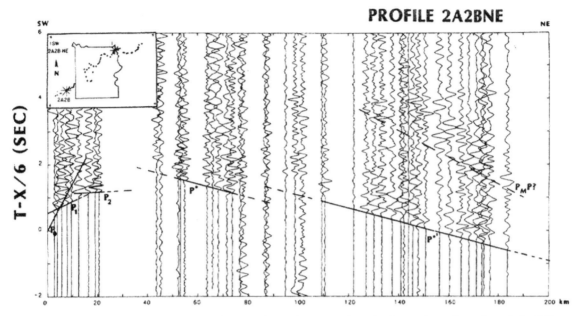

Figure 7.4.1-05. Record section of seismic refraction observations from shotpoint 2 towards northeast through the Yellowstone Park (from Smith et al., 1982, fig. 3a). [Journal of Geophysical Research, v. 87, p. 2583–2596. Reproduced by permission of American Geophysical Union.]

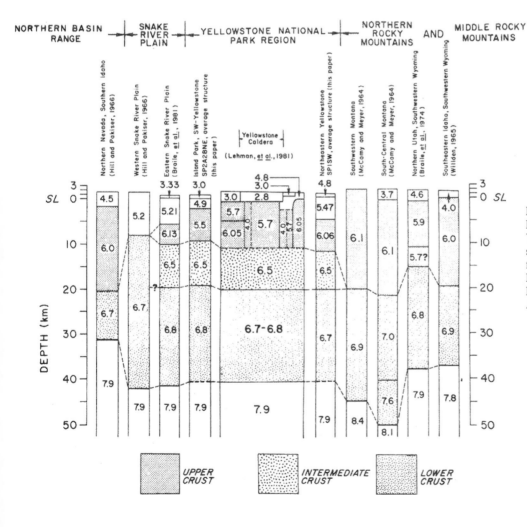

Figure 7.4.1-06. Generalized P wave crustal model for the Yellowstone–Snake River Plain survey (from Smith et al., 1982, fig. 8). [Journal of Geophysical Research, v. 87, p. 2583–2596. Reproduced by permission of American Geophysical Union.]

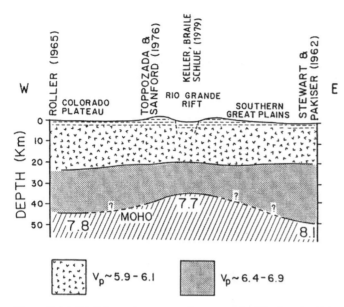

Figure 7.4.1-07. Schematic cross section across the Rio Grande rift in the Santa Fe–Albuquerque area (from Keller et al., 1979, fig. 6). [*In* Riecker, R.E., ed., Rio Grande rift; tectonics and magmatism: Washington, D.C., American Geophysical Union, p. 115–126. Reproduced by permission of American Geophysical Union.]

by K. Olsen of the Los Alamos National Laboratory, using the shotpoint DICE THROW north of Albuquerque, New Mexico, for a north-directed, single-ended profile along the eastern graben margin (Olsen et al., 1979, 1982; Olsen, 1983). Though only recorded with a large station separation of ~10 km, this profile (Appendix A7-5-3) provided information that the crust beneath the axis of the rift, with a thickness of 34 km, was ~10–15 km thinner than under the adjacent Great Plains and Colorado Plateau.

Also in the southern Rio Grande Rift, an unreversed seismic-refraction line was observed (Cook et al., 1979b; Appendix A7-5-3). It was recorded from mine blasts near Santa Rita, New Mexico, across the rift at ~32.9° latitude in a west-east direction and reached a maximum distance of 220 km.

From the data available, Keller et al. (1979) sketched an upward warping of the Moho underneath the Rio Grande rift (Fig. 7.4.1-07). In 1978, new data became available for the Basin and Range area in New Mexico and Arizona to the west of the Rio Grande rift (Gish et al., 1981; Sinno et al., 1981), which provided confirmation, e.g., of an earlier interpretation of crustal data around the Tonto Forest Seismological Observatory by Warren (1969), and showed that the crust thins rapidly from near 37 km along the Mogollon rim, the border of the Colorado Plateau toward the south, to less than 25 km south of Phoenix.

Involving all seismic-refraction and seismic-reflection data available by the early 1980s in the United States, including the data of the 1981 CARDEX project in the Rio Grande rift (Ankeny et al., 1986; Sinno et al., 1986; see Chapter 8), Prodehl and Lipman (1989) compiled a north-south section through the Rocky Mountains and Rio Grande rift (Fig. 7.4.1-08).

As mentioned above, scientists of the U.S. Geological Survey in the 1970s were mainly involved in earthquake studies with emphasis on active regions, California in particular. Over the years, the local seismic network along the San Andreas fault zone had become so dense that it became feasible to use these stations, in combination with suitable local earthquakes, as quasi-refraction lines (Fig. 7.4.1-09).

A successful approach in this direction was undertaken for the Coast Ranges south of San Francisco. Prodehl (1976) had compiled the recordings of a series of medium-size (magnitude 2.7–3.8) local earthquakes of 1975 as record sections. This collection was the base for a crustal study by Blümling and Prodehl (1983; Appendix A7-5-4) who used part of the existing seismic-refraction data (Fig. 7.4.1-10), which was recorded in 1967 (Stewart, 1968a) and which had been reinterpreted by Walter and Mooney (1982), and interpreted them together with some of the local earthquake data (Fig. 7.4.1-11).

A particular result of Blümling and Prodehl (1983) for both the refraction and the near-earthquake profiles was a velocity inversion in the lowermost crust between 21 and 26 km depth with velocities as low as 5.3. km/s overlying a 3–5-km-thick mantle transition zone with 7.6 km/s average velocity on top of the Moho at 29–30 km depth with 8.2 km/s.

Figure 7.4.1-08. Simplified cross section through Rocky Mountains and Rio Grande rift between 33° to 49°N latitude and 105° to 115°W longitude (from Prodehl and Lipman, 1989, fig. 23). [*In* Pakiser, L.C., and Mooney, W.D., eds., Geophysical framework of the continental United States: Geological Society of America Memoir 172, p. 249–284. Reproduced by permission of the Geological Society of America.]

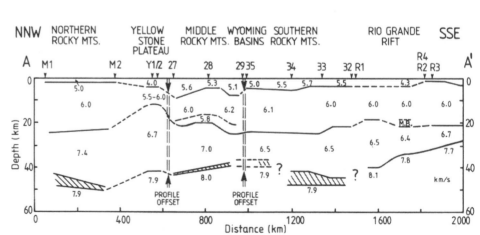

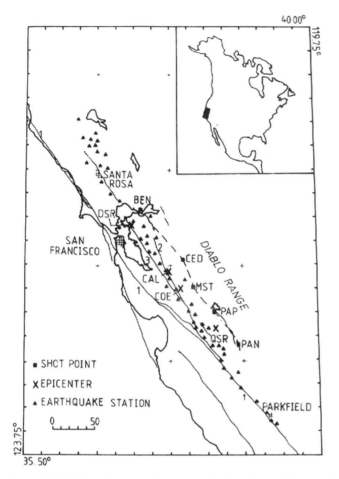

Figure 7.4.1-09. Location map of seismic refraction (dashed line) and near-earthquake (solid line) observations in the Diablo Range of central California (from Blümling and Prodehl, 1983, fig. 1). Squares: shotpoints of the 1967 refraction survey; triangles: permanent stations of the U.S. Geological Survey Central California Microearthquake Network. [Physics of the Earth and Planetary Interiors, v. 31, p. 313–326. Copyright Elsevier.]

Also north of the San Francisco Bay, a seismic-refraction survey was undertaken to study the Geysers–Clear Lake geothermal area in northern California (Warren, 1981).

Another project combining earthquake and active shot data dealt with the investigation of the crust in the western foothills of the Sierra Nevada (Spieth et al., 1981), where an earthquake near Oroville Dam, Oroville, California, prompted a very intense study. Record sections were compiled from both aftershock recordings up to 300 km and an active seismic-refraction experiment with distances up to 80 km (Appendix 7-5-4). The interpretation indicated a localized, shallow, high-velocity layer with 7 km/s at only 5 km depth, which had restricted extension and was underlain by normal 30-km-thick upper crust. The total crustal thickness was estimated at 39 km, based on a combined evaluation of Bouguer gravity and earthquake data recordings.

The surveillance of the Long Valley area, on the eastern side of the Sierra Nevada, employed seismic-refraction profiling in 1973. The goal was to define the upper 10 km of the crust in order to understand the nature and development of a resurgent caldron, and to investigate if the recorded wave forms might provide evidence for the presence of a geothermal reservoir within the caldera or a possible magma chamber at depth (Hill, 1976).

7.4.2. New Seismic-Recording Equipment Opens a New Stage of Crustal Research

A major step to reactivate active seismology was undertaken by J.H. Healy, whose approach led to major renewed research activity in controlled-source seismology in the 1980s, when W.D. Mooney became responsible for projects on crustal and upper-mantle studies. As the seismic equipment of the U.S. Geological Survey of the early 1960s was practically worn out, J.H. Healy started to develop a new seismic-refraction system, inspired by previous visits to Europe where he had watched the highly successful MARS-66 equipment.

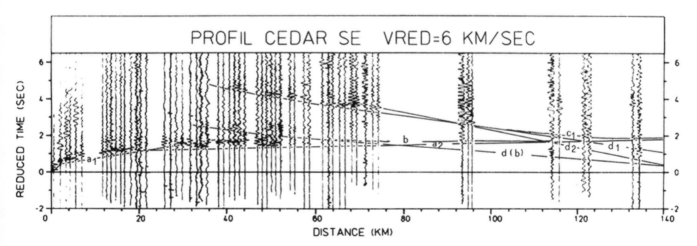

Figure 7.4.1-10. Example of seismic refraction observations in the Diablo Range of central California (from Blümling and Prodehl, 1983, fig. 5). [Physics of the Earth and Planetary Interiors, v. 31, p. 313–326. Copyright Elsevier.]

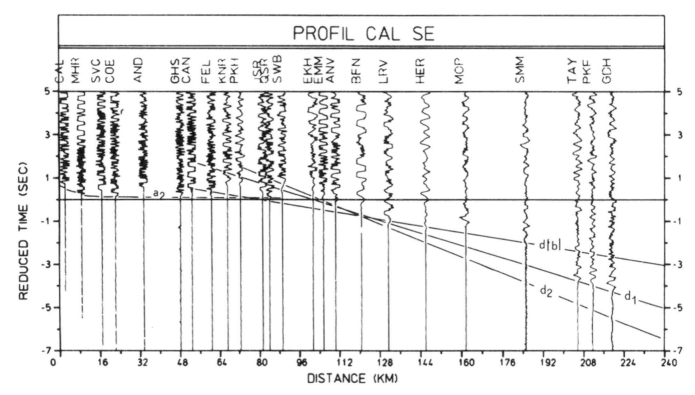

Figure 7.4.1-11. Examples of near-earthquake observations in the Diablo Range of central California (from Blümling and Prodehl, 1983, fig. 11). [Physics of the Earth and Planetary Interiors, v. 31, p. 313–326. Copyright Elsevier.]

This new system would fulfill several conditions which were not easily met. The instrument had to be small, with the ability to be built in large quantities, with at least 100 pieces. It would also have to be easily operated by untrained personnel, be able to record mostly automatically, i.e., to be switched on and off by a sophisticated clock of high accuracy, and be able to run unattended for several days.

In addition, the playback of the recorded data should be fast and easily gained for field control; i.e., within a few hours, the data of field tapes from 100 stations should be available on a field computer immediately after pick up, raw record sections be plotted on the spot, and the instruments would have to be ready for the next deployment within a very limited time period.

The rapid development of such instrumentation became possible by a grant and order from the Saudi Arabian government to perform a long-range seismic project in Saudi Arabia, which will be described in section 7.5, "The Afro-Arabian Rift System." By the beginning of 1978, 100 of these so-called cassette recorders were ready to be shipped to Saudi Arabia (Fig. 7.4.2-01). A detailed description of the recording system was prepared by Murphy (1988; Appendix A7-5-6).

Later in the same year, when R.B. Smith and colleagues accomplished their fieldwork in the Yellowstone–Snake River Plain area, the U.S. Geological Survey moved in and recorded an additional profile with these cassette recorders. It was this in-

strumentation which led to a renewed strong activity of the U.S. Geological Survey in the following decade.

Two more major projects using this new equipment followed at the end of the 1970s to investigate the crustal structure with much more detail than had been possible before. One project was the investigation of the northern Mississippi embayment (Hamilton, 1986; Mooney et al., 1983). In 1978 and 1979, thirteen deep seismic-reflection profiles with 3 s two-way traveltime had been recorded in the New Madrid, Missouri, seismic hazard zone. Subsequently, oil exploration in northern Arkansas provided 6400 km of reflection data, of which 350 km were acquired by the U.S. Geological Survey (Hamilton, 1986).

In 1979 and 1980, an extensive seismic-refraction program followed, providing velocity and structural information down to the crust-mantle boundary (Ginzburg et al., 1983; Mooney et al., 1983). The large number and the small size of the new cassette recorders now available and the instruments' refined clock system allowed a high station density as well as the layout of a series of profiles with a total of nine shotpoints across the embayment, within a reasonably short time frame (Fig. 7.4.2-02), and with a reasonably small number of people. Figure 7.4.2-03 shows the high data density and high data quality obtained by only one shot with the new equipment (Appendix A7-5-5). The important result of this study was the confirmation and delineation of a 7.3 km/s layer in the lower crust (see top section of

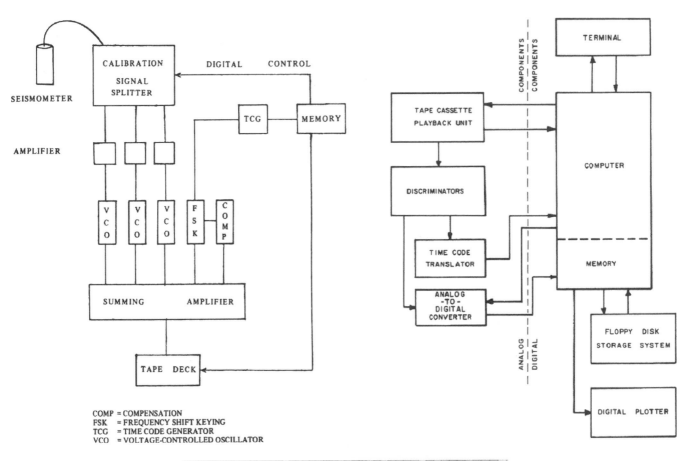

COMP = COMPENSATION
FSK = FREQUENCY SHIFT KEYING
TCG = TIME CODE GENERATOR
VCO = VOLTAGE-CONTROLLED OSCILLATOR

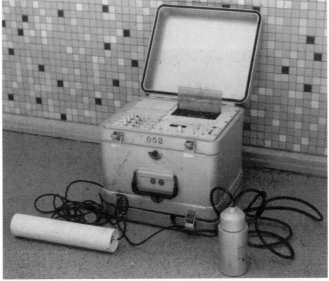

Figure 7.4.2-01. Cassette recorders developed by J.H. Healy and co-workers during 1976–1978. Schematic diagrams (from Healy et al., 1982, figs. 13 and 16). Left: Recording unit. Right: Seismic data processing system. [Final Project Report, Saudi Arabian Deputy Minister of Mineral Resources, Open-File Report, USGSOF-02-37, 429 p., including appendices; also U.S. Geological Survey, Open-File Report, 83-390.] Bottom: Cassette recorder with seismometer (courtesy of G. Fuis).

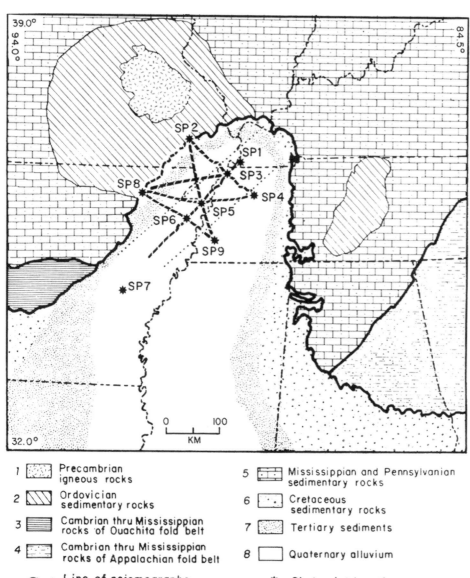

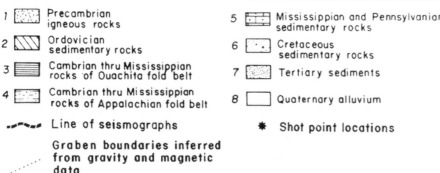

Figure 7.4.2-02. Location map of the seismic-refraction survey in the Mississippi embayment (from Mooney et al., 1983, fig. 1). [Tectonophysics, v. 94, p. 327–348. Copyright Elsevier.]

Fig. 7.4.2-05) which had already been suggested from previous studies, implying that the lower crust had been altered by injection of mantle material.

Another project aimed to make a detailed investigation of the Imperial Valley and its surroundings in southern California (Fuis et al., 1984; Kohler and Fuis, 1988; Appendix A7-5-6).

Forty shots at seven shotpoint sites, ranging in size from 400 to 900 kg, were detonated and recorded by 100 portable vertical seismic instruments, with a spacing of 0.5–1 km (Fig. 7.4.2-04). Usually in one night, three shots were recorded on a line. Data were primarily collected in January to March 1979, but, following the October 1979 Imperial Valley earthquake, additional data were collected (Fuis et al., 1984; Kohler and Fuis, 1988; Appendix A7-5-6).

Following the interpretation of the data in the Mississippi embayment in their publication, Mooney et al. (1983) added a discussion of young continental rift structures in central Europe, East Africa, and North America. All models show an anomalous transition zone above the Moho with velocities between 7 and 8 km/s. Their compilation of models, published by various authors, also contains velocity-depth cross sections through the Mississippi embayment (top section of Fig. 7.4.2-05) and through the Imperial Valley in the Salton Trough (bottom section of Fig. 7.4.2-05).

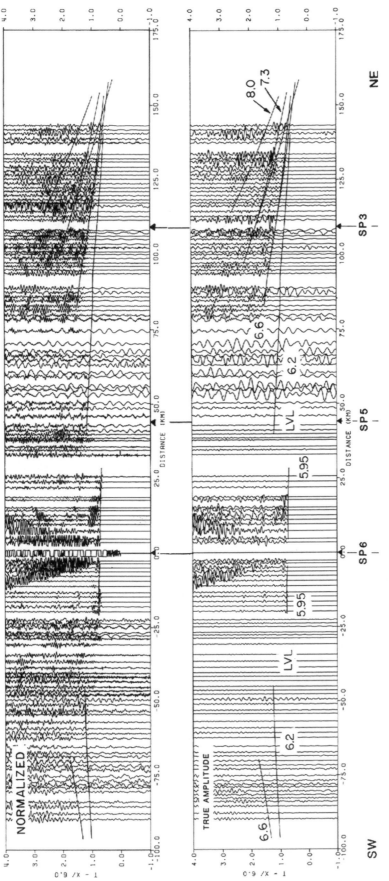

Figure 7.4.2-03. Data recorded from shotpoint 6 in SW and NE direction of the seismic-refraction survey in the Mississippi embayment (from Mooney et al., 1983, fig. 4). Top: record section plotted with normalized amplitudes. Bottom: with true amplitudes. [Tectonophysics, v. 94, p. 327–348. Copyright Elsevier.]

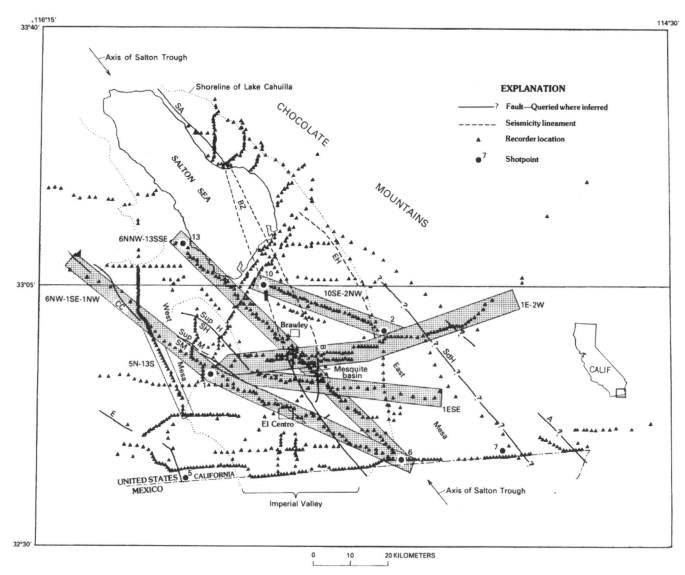

Figure 7.4.2-04. Location map of the seismic-refraction survey in the Imperial Valley, southern California (from Fuis et al., 1984, fig. 1). [Journal of Geophysical Research, v. 89, p. 1165–1189. Reproduced by permission of American Geophysical Union.]

7.4.3. The First Large-Scale Continental Reflection Profiling Project (COCORP and U.S. Geological Survey)

A completely different approach to study the Earth's crust in much more detail was started when at Cornell University (Ithaca, New York), a large-scale seismic-reflection program, the Consortium for Continental Reflection Profiling, or COCORP, was initiated. J.E. Oliver, who had already been one of the first promoters of plate tectonics in its early stages and was actually one of the first who discussed this theory in public, had thought about the complexity of the Earth's crust for a long time (Oliver, 1996). He realized how little science had hitherto contributed to the study of the continental crust in much more detail than large-scale seismic-refraction studies had been able to.

Pointing to successful deep seismic-reflection studies in the 1960s in Canada, the United States, Germany, the USSR, and Australia, he circulated the basic idea amongst the leading seismologists in the United States, which finally led to the establishment of COCORP. Together with S. Kaufman, whom he persuaded to leave the exploration industry, he developed the idea to study the Earth's crust by carrying out large-scale reflection seismology in applying commercially common methods of fieldwork and interpretation. In 1975, COCORP was created (Brewer and Oliver, 1980; Brown et al., 1986).

A commercial reflection company was hired, and the first fieldwork (Fig. 7.4.3-01) was started in 1975 in Hardeman County, Texas (Oliver et al., 1976). The Vibroseis system was used, because it offered high-quality data while maximizing

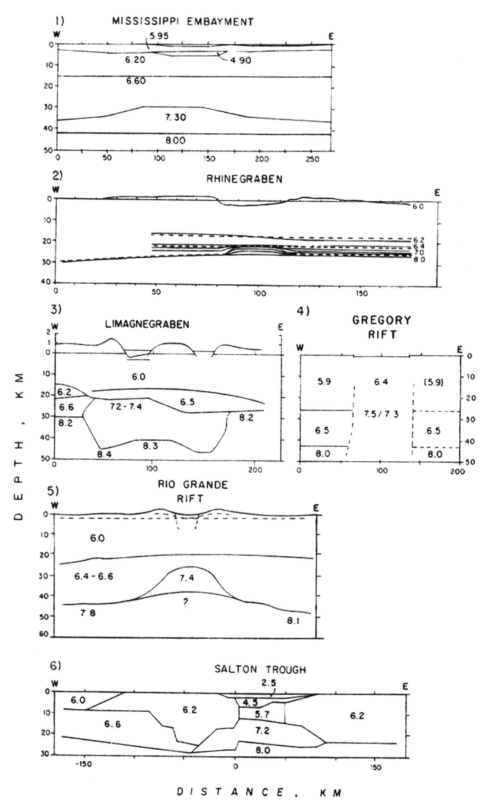

Figure 7.4.2-05. Continental rift structures (from Mooney et al., 1983, fig. 11). (1) Mississippi em-bayment (Mooney et al., 1983), (2) Rhinegraben, Germany (Edel et al., 1975), (3) Limagnegraben (Hirn and Perrier, 1974), (4) Gregory Rift, East Africa (Long et al., 1973), (5) Rio Grande Rift (Keller et al., 1979), (6) Imperial Valley (Fuis et al., 1984). [Tectonophysics, v. 94, p. 327–348. Copyright Elsevier.]

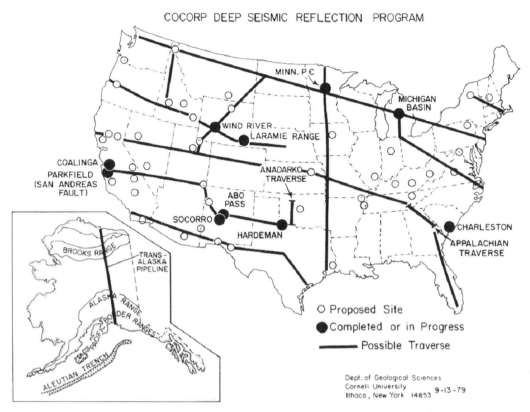

COCORP DEEP SEISMIC REFLECTION PROGRAM

Figure 7.4.3-01. Location map of COCORP deep seismic reflection studies accomplished until 1979 (from Brewer and Oliver, 1980, fig. 2). [Annual Reviews of Earth and Planetary Science, v. 8, p. 205–230. Copyright Annual Reviews.]

efficiency and minimizing environmental impact. COCORP used five vibrating trucks which transmitted signals of 8–40 Hz bandwidth into the ground. The returning echoes were collected by 6–10-km-long linear arrays of geophones connected to a 96-channel truck-mounted recording system. About 24 geophones per channel were used. The data were collected and computer processed to produce a 24-fold common-depth point stacked seismic section, representing a cross section through the Earth's crust in terms of traveltime of the seismic waves. These sections served to provide important preliminary geological interpretations, before depth conversion and migration were applied for detailed studies (Brewer and Oliver, 1980). Phinney and Roy-Chowdhury (1989) have discussed in much detail the basics of seismic-reflection techniques and the long procedure leading from the raw data to the seismic-reflection sections used for final geological interpretation. It will be described in some detail at the beginning of Chapter 8.3.

Soon, other surveys followed, using the same techniques as applied for the Hardeman county project (Fig. 7.4.3-01). The Rio Grande rift, New Mexico, was the site of the first full-scale investigation of a specific geological problem (Brown et al., 1979; Oliver and Kaufman, 1976). The objective was to study the structure of an active rift, including the geometry of faulting, the nature of the crust-mantle boundary, and the possible

existence of magma bodies at depth. Here, in the Socorro area of the Rio Grande rift, where a magma body had been inferred from local earthquake data, in 1975–1976 COCORP collected 155 km of 24-fold reflection data transverse and parallel to the rift. The most prominent coherent reflections were identified to correspond to a depth of ~20 km, corresponding to the top of the postulated magma body (Fig. 7.4.3-02).

In 1976, a 27-km-long seismic line was recorded across the San Andreas fault near Parkfield, California, where the San Andreas fault bifurcates to form a thin sliver. The line was short and crooked and the data quality was less good as elsewhere, but showed several interesting features indicating a zone of brittle fracturing limited to the upper 15 km of the crust, where generally the earthquakes occur, and a vertical blank zone of 4 km width beneath, possibly created by ductile flow (Long et al., 1978; Brewer and Oliver, 1980).

The seismic-reflection data, collected in 1977 in the Wind River Mountains and Laramide Range, Wyoming, on 158 km of seismic profiles, showed a pronounced set of dipping events originating at the surface at the position of the Wind River thrust and trending into the crust at a moderate angle (Brewer et al., 1980, 1982; Smithson et al., 1979), which could be traced to 24 km depth for certain and possibly further to 36 km depth.

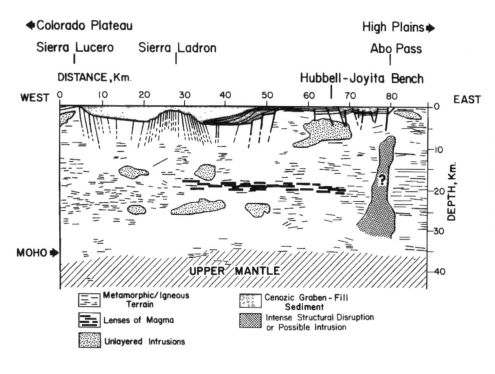

Figure 7.4.3-02. Schematic cross section of COCORP across the Rio Grande rift near Socorro, New Mexico (from Brown et al., 1980, fig. 14). [Journal of Geophysical Research, v. 85, p. 4773–4800. Reproduced by permission of American Geophysical Union.]

Other surveys were performed in the Midwest and eastern United States. In 1978, the Michigan Basin in Michigan was investigated (Jensen et al., 1979); reflection surveys in the Wichita uplift, Oklahoma (Brewer and Oliver, 1980) and in the Archean Province in Minnesota (Gibbs et al., 1984) followed in 1979. The seismic-reflection survey of 1980 in the Adirondack Mountains, New York, aimed to investigate an extensive dome of high-grade metamorphic rocks where seismic-reflection patterns against rocks once buried at mid-crustal depths but now exposed were to be calibrated (Brown et al., 1983a, 1986).

The southern Appalachian COCORP traverse of 1978 and 1979 was initiated to investigate the Brevard zone, an enigmatic fault zone in the Inner Piedmont. The interpretation of the seismic-reflection data implied a thin-skinned tectonics: that the crystalline southern Appalachians were allochthonous, vary in thickness between 6 and 15 km, and were thrust over a passive continental margin with little deformation of the underlying continental shelf and slope sediments (Cook et al., 1979a; Brewer and Oliver, 1980).

Besides intensifying its crustal seismic-refraction research by the new seismic cassette recorders, the U.S. Geological Survey also started a major seismic-reflection survey program which was to be continued in the 1980s. From 1973 to 1979, a major program of multichannel seismic exploration of the U.S. Atlantic continental shelf was carried out on the Long Island Platform (Phinney, 1986). It will be described in more detail in Chapter 7.8.3. In 1979, the first project on land was a survey conducted across the Grandfather Mountain, the line extending from Tennessee to North Carolina through the metamorphic sequences of the Appalachian orogen (Harris et al., 1981; for location, see line GMP in Fig. 8.5.3–24).

7.4.4. Canada

In 1970, geophysical studies were performed in the Baffin Bay, northern Canada (Keen et al., 1972).

In Canada, already in 1972, near-vertical reflection techniques were applied to record deep reflections along three profiles in the Canadian Cordillera (Mair and Lyons, 1976; Berry and Mair, 1977; Fig. 7.4.4-01). Up to six shotpoints were spaced ~1.5 km apart in each of three lakes, from where shots with charge sizes of 50 kg were recorded by closely spaced geophone arrays extended in a line along a road. The recording sites were spaced ~250 m apart, reaching maximum offsets of 20 km. This geometry allowed up to fivefold common reflection point (CDP) stacking. Only one relatively continuous event was recorded at ~11 s TWT, which, assuming an average crustal velocity of 6.3 km/s, was interpreted originating at a 34-km-deep reflector.

In 1973, a Vibroseis survey was added to obtain a roughly similar subsurface coverage. The experiment proved to be successful and reflections could be correlated throughout the crust and from the crust-mantle boundary.

In 1977, 1979, and 1981, experiments were conducted over the northern part of the Williston Basin as part of the COCRUST investigations (Fig. 7.4.4-02) which will be discussed in Chapter 8. The north-south and east-west inline profiles of 1977 and 1979 were supplemented in 1981 by three profiles arranged as a triangle which recorded all shots arranged at its edges. The data were interpreted several times (e.g., Hajnal et al., 1984; Kanasewich et al., 1987), indicating in particular a crustal thickening from 40 km in the Superior Province to 45–50 km under the Williston Basin.

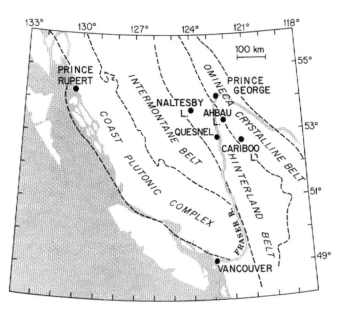

Figure 7.4.4-01. Location of Naltesby, Ahbau, and Cariboo Lakes in the Canadian Cordillera, shot sites for reflection profiles (from Berry and Mair, 1977, fig. 9) [American Geophysical Union, Geophysical Monograph 20, p. 319–348. Reproduced by permission of American Geophysical Union.]

7.5. THE AFRO-ARABIAN RIFT SYSTEM

7.5.1. The Northern End of the East-African Rift System

Following the joint effort of investigating the eastern Mediterranean area under Greece, a major expedition was planned and organized to investigate the northernmost part of the East African rift system in Ethiopia under the leadership of H. Berckhemer and J. Makris. The investigation aimed for the deep structure of the Afar region and part of the adjacent Ethiopian Highland by joint gravity and seismic-refraction measurements (Berckhemer et al., 1975).

In parallel, French scientists carried out a major crustal study of the French territory of Djibouti, adjacent to the Red Sea and Gulf of Aden (Ruegg, 1975). The investigations were embedded in an international research program on the Afar depression of Ethiopia within the Upper Mantle project mainly involving earth scientists from Britain, France, Germany, and Italy (Pilger and Rösler, 1975, 1976).

While quite conventional in principle, the actual field observations were rather unconventional due to the extreme and adverse environmental conditions. Five profiles, between 120 and 300 km long, were laid out in different parts of the Afar depression and one

Figure 7.4.4-02. Location of a triangular array of receivers used for the first COCRUST refraction/wide-angle reflection survey in the Williston Basin (from Green et al., 1986, fig. 4). Letters refer to geological details which are of no relevance here. [*In* Barazangi, M., and Brown, L., eds., Reflection seismology: a global perspective. [American Geophysical Union, Geodynamics Series, v. 13, p. 85–97. Reproduced by permission of American Geophysical Union.]

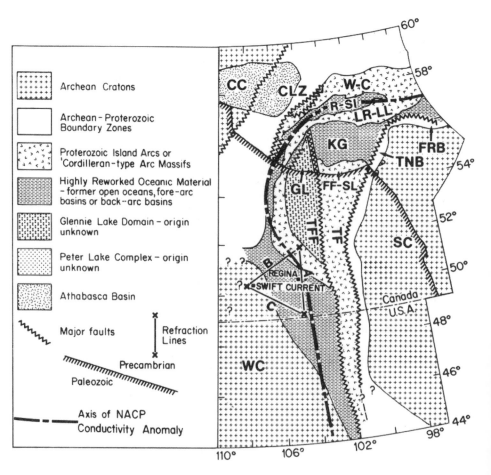

profile was arranged on the Ethiopian Plateau (Fig. 7.5.1-01). All profiles were chosen along roads or tracks that could be reached under favorable weather conditions. All data of the Afar 1972 expedition were reproduced in Appendix A2-1 (p. 89–90).

Seismic shots were generated in lakes or rivers preferably with charges between 300 and 1600 kg. A particular shooting technique to yield high seismic efficiency even from explosions in very shallow water was applied (Burkhardt and Vees, 1975). Also, by chance, a local earthquake occurred during the time of recording along profile V, which produced excellent S-wave arrivals at distances from 70 to 120 km.

The record section of profile I (Fig. 7.5.1-02, bottom) recorded on the Ethiopian Plateau was similar to data recorded on 40-km-thick continental crust elsewhere, while the first arrivals

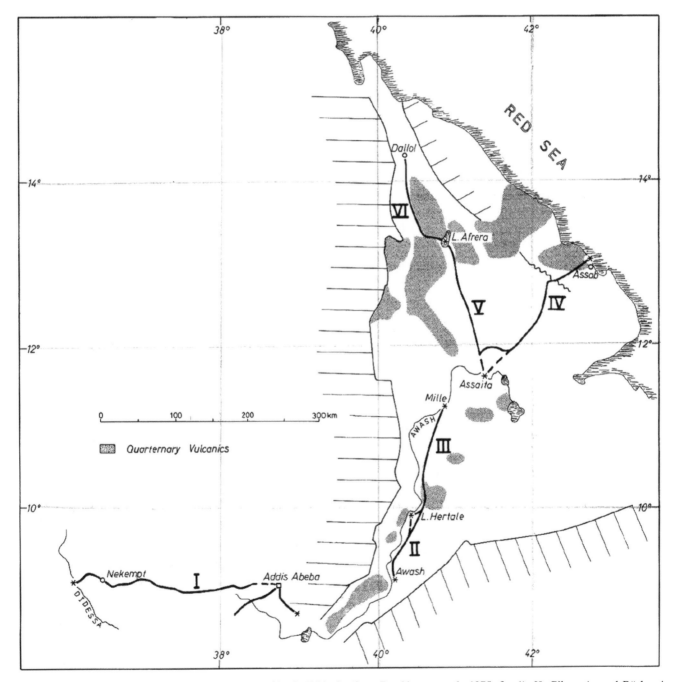

Figure 7.5.1-01. Location of deep seismic sounding profiles in Ethiopia (from Berckhemer et al., 1975, fig. 1). [*In* Pilger, A., and Rösler, A., eds., Afar Depression of Ethiopia: Stuttgart, Schweizerbart, p. 89–107. Reproduced by permission of Schweizerbart, Stuttgart, Germany (www.schweizertbart.de).]

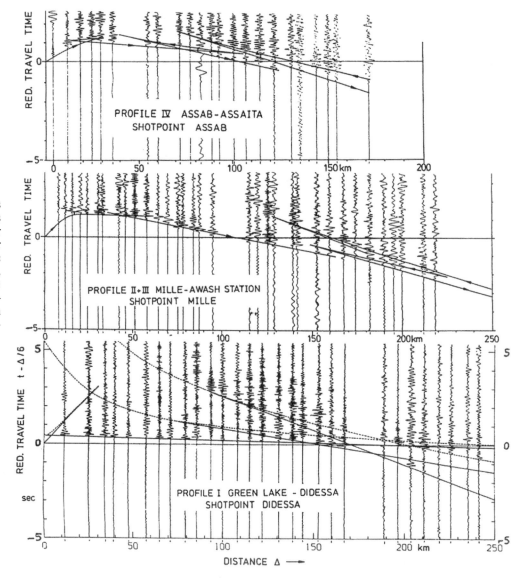

Figure 7.5.1-02. Record sections of deep seismic sounding profiles in Ethiopia (from Berckhemer et al., 1975, figs. 2b, 2a, 2e). For location, see Fig. 7.5.1-01. [*In* Pilger, A., and Rösler, A., eds., Afar Depression of Ethiopia: Stuttgart, Schweizerbart, p. 89–107. Reproduced by permission of Schweizerbart, Stuttgart, Germany (www.schweizertbart.de).]

of the other profiles (Fig. 7.5.1-02, top and center) reflected basic material with velocities of 6.6–6.8 km/s near the top of the crust.

The seismic investigation of Djibouti was carried out in 1971 by the Operation Grands Profils Sismiques de l'I.N.A.G. (Institut de Physique du Globe of Paris). Profiles were recorded along the coast of the Gulf of Tadjura making use of sea shots, while the inland profiles recorded sea shots, shots in shallow lakes, and borehole shots (Fig. 7.5.1-03).

The main result was that the bulk of the Afar crust, whose thickness changed from almost 26 km in the south (Fig. 7.5.1-02, center) to almost 16 km in the north (Fig. 7.5.1-02, top), appeared to have average P-wave velocities of 6.6–6.8 km/s and was underlain by a gradual transition into upper-mantle material with anomalously low velocities between 7.3 and 7.7 km/s under all of Afar (Fig. 7.5.1-04).

The authors saw evidence that this anomalous mantle was limited in depth to an extent of nearly 40 km (Berckhemer et al.,

1975). Under the adjacent Djibouti region to the east, Ruegg (1975) postulated an oceanic crust of thickness ~6–10 km, which was underlain by an anomalous upper mantle with P-wave velocities as low as 6.7–6.9 km/s increasing gradually to 7.4 km/s at a depth of 20 km.

7.5.2. First Crustal Studies of the Dead Sea Rift, around the Red Sea, and in Iran

In 1977, the first deep seismic sounding experiment was carried out in the Jordan–Dead Sea rift. Initiated by A. Ginzburg and promoted by geologists such as Z. Garfunkel, B. Freund, and others, the Geophysical Departments of the universities of Tel Aviv, Israel, as well as of Hamburg and Karlsruhe, Germany, carried out a seismic-refraction survey on the western side of the rift (Fig. 7.5.2-01) with side profiles toward the Mediterranean Sea (Ginzburg et al., 1979a, 1979b).

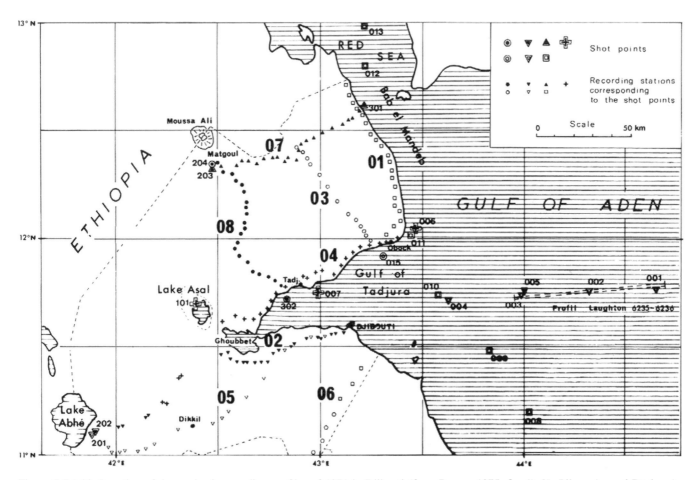

Figure 7.5.1-03. Location of deep seismic sounding profiles of 1971 in Djibouti (from Ruegg, 1975, fig. 1). [*In* Pilger, A., and Rösler, A., eds., Afar Depression of Ethiopia: Stuttgart, Schweizerbart, p. 120–134. Reproduced by permission of Schweizerbart, Stuttgart, Germany (www.schweizertbart.de).]

Shotpoints were in the Dead Sea, the Gulf of Aqaba, and the Mediterranean Sea (Fig. 7.5.2-01). Recording distances along most of the lines were up to 240 km, but along the rift a maximum distance of 400 km could be reached, due to the optimal damming of the shots in the Dead Sea (Fig. 7.5.2-03). The data (see Appendix A2-1, p. 82–83) suggested a transition zone between the lower crust and the upper mantle, which evidently was restricted to the rift area proper. The crust thickened from 20 km near the Mediterranean coast to ~30 km under the rift. The side profiles indicated a further crustal thickening toward 40 km under the Sinai Peninsula.

In 1978, a seismic-refraction line from Cyprus to Israel was added (Fig. 7.5.2-02). Thirty-three sea shots were recorded both by ocean-bottom seismometers along the line and by land stations in Cyprus and Israel (Makris et al., 1983a; Appendix A7-6-1).

Also in 1978, and in 1981, seismic profiles were recorded from the northernmost Red Sea into Egypt (Rihm et al., 1991) and a seismic-refraction profile was recorded from the Red Sea into northern Saudi Arabia (Makris et al., 1983b). Mechie and

Prodehl (1988) have compiled these observations together with other surveys accomplished in the early 1980s in Jordan and the Red Sea area and have published the corresponding crustal cross sections and references. Locations and a summary of results will be jointly discussed in Chapter 8, where data of the 1980s in this area is described.

Farther south, another long-range profile was realized in 1978. It traversed the Arabian Shield of Saudi Arabia over a distance of nearly 1000 km from the southern Red Sea to the neighborhood of Riad (Fig. 7.5.2-04). This survey was carried out by the U.S. Geological Survey, using for the first time the newly built mobile array of 100 cassette recorders mentioned above which had been designed by J.H. Healy and co-workers (Healy et al., 1982; Murphy, 1988; see above; Fig. 7.4.2-01; Appendix A7-5-6). In total, 100 portable seismographs were available; they were successively deployed in each of five 200-km recording spreads. A portable computer center, including field playback, digitizer, and plotting system, was moved along the profile to provide rapid feedback of data quality and content, and to allow for preliminary assessment of scientific results as the experiment

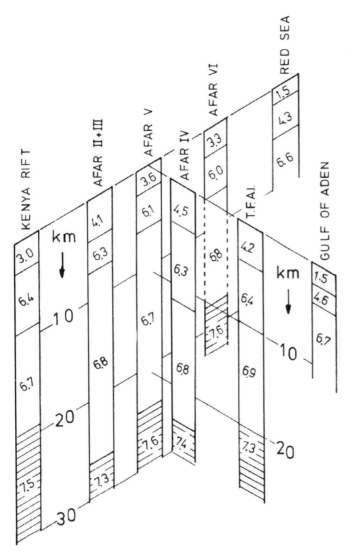

Figure 7.5.1-04. Simplified crustal models for Afar and adjacent rift zones (from Berckhemer et al., 1975, fig. 6). T.F.A.I. = French territory of Djibouti. [*In* Pilger, A., and Rösler, A., eds., Afar Depression of Ethiopia: Stuttgart, Schweizerbart, p. 89–107. Reproduced by permission of Schweizerbart, Stuttgart, Germany (www.schweizertbart.de).]

progressed (Gettings et al., 1986; Mooney et al., 1985; Healy et al., 1982; Appendix A7-6-3).

The data was of extremely good quality (Fig. 7.5.2-05; see Appendix A2-1, p. 86–88; Appendix A7-6-2 and A7-6-3), and was, in the course of a seismic workshop, interpreted by several other authors (e.g., Milkereit and Flueh, 1985; Prodehl, 1985; Mechie et al., 1986). This workshop will be discussed in some detail at the end of this chapter in the section "Advances of interpretation methodology." The overall structure (Fig. 7.5.2-06) shows a rather uniform 40-km-thick crust under the Arabian Shield, which thins more or less abruptly west of the escarpment under the coastal plain to less than 10 km in the Red Sea (Mooney et al., 1985; Gettings et al., 1986).

Also in 1978, a reconnaissance seismic-refraction survey was undertaken in southern Iran to study the deep structure of the Zagros Belt (Fig. 7.5.2-07). The project was a joint venture of the Geophysical Institutes of the Tehran University of Iran and the Free University of Berlin and the University Hamburg of Germany (Giese et al., 1984). Using 12 MARS66 seismic recording stations and a PCM equipment, commercial explosions at the mines Bafq and Sar-Chesmeh were used and recorded along two N-S directed profiles, the first one of 180 km length within the Lut block between the two mines, and the second one of 215 km length, running southward into the Zagros belt. The quality of the data was sufficient, but the density of the data obtained was relatively poor. Therefore only a preliminary interpretation with many open questions could be provided. The political circumstances prohibited any further research. Data and results, as published by Giese et al. (1984) are shown in Appendix A7-6-4.

7.6. DEEP SEISMIC SOUNDING STUDIES IN SOUTHEAST ASIA

7.6.1. India

Explosion seismology started in India at around 1972, when a three-year contract was signed by India and the USSR, and the first deep seismic sounding profiles were recorded. Field technology and interpretation was guided by Russian scientists, such as V.B. Sollogub, who was one of the initiators of the joint research program (e.g., Kaila et al., 1978, 1979, 1981a, 1981b; Mahadevan, 1994). A map of India showing all deep seismic lines shot between 1972 and 1992 is shown in the next Chapter 8 (Fig. 8.7.1-01).

The first 600-km-long profile crossed the Indian Shield in the southern part of India. It was carried out from 1972 to 1975. The profile started near Kavali, at the Bay of Bengal in the east, and ended near Udupi, at the Arabian Sea in the west, and resulted in a 40-km-thick crust (Kaila et al., 1979; Sarkar et al., 2001).

In the Deccan trap area of the Indian Shield south of Bombay, two 200-km-long profiles were recorded near the Koyna dam, and 60 km further north (Kaila et al., 1981b). A third N-S–trending profile was recorded in the Cambay basin, north of Bombay and east of the Gulf of Cambay, where 5-km-thick sediments overlie the Deccan traps (Kaila et al., 1981a). Later an E-W profile was added to the west of the Gulf of Cambay. Detailed crustal cross sections indicate deep faults and fracture zones throughout the crust down to a depth of ~50–60 km. Moho depths vary along these lines between 35 and 40 km, but the crust thins to 30 km and less toward coastal regions. The data were later reinterpreted by Rao and Tewari (2005).

Finally, an international Pamir-Himalaya project revealed, on average, a 60-km-thick crust in the Kashmir valley, but with strong local variations ranging from as low as 45 km to as deep as 76 km (Kaila et al., 1978, 1979).

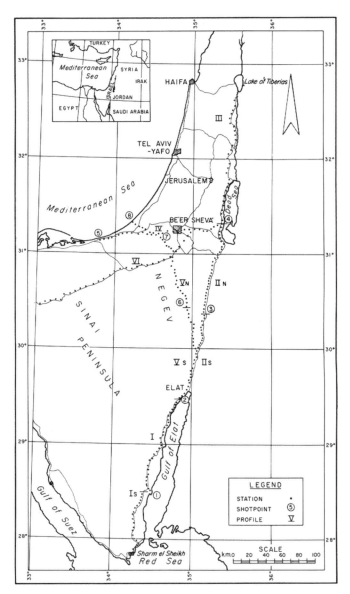

Figure 7.5.2-01. Location map of the seismic refraction survey of 1977 in Israel (from Ginzburg et al., 1979a, fig. 1). [Journal of Geophysical Research, v. 84, p. 1569–1582. Reproduced by permission of American Geophysical Union.]

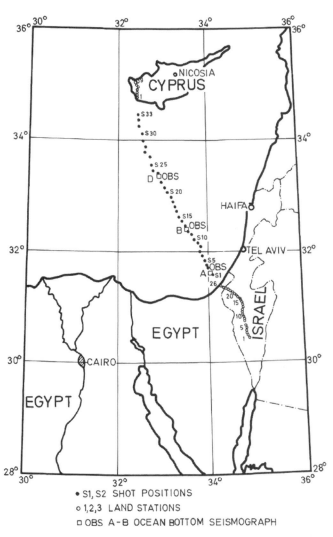

Figure 7.5.2-02. Location map of the seismic refraction survey of 1978 between Cyprus and Israel (from Makris et al., 1983a, fig. 1). [Geophysical Journal of the Royal Astronomical Society, v. 75, p. 575–591. Copyright John Wiley & Sons Ltd.]

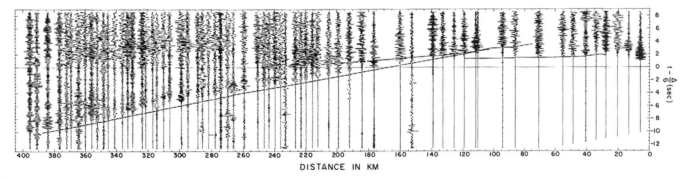

Figure 7.5.2-03. Record section, observed from the Dead Sea towards south to the southern end of the Sinai Peninsula (from Ginzburg et al., 1979a, fig. 4b). For location, see Fig. 7.5.2-01. [Journal of Geophysical Research, v. 84, p. 1569–1582. Reproduced by permission of American Geophysical Union.]

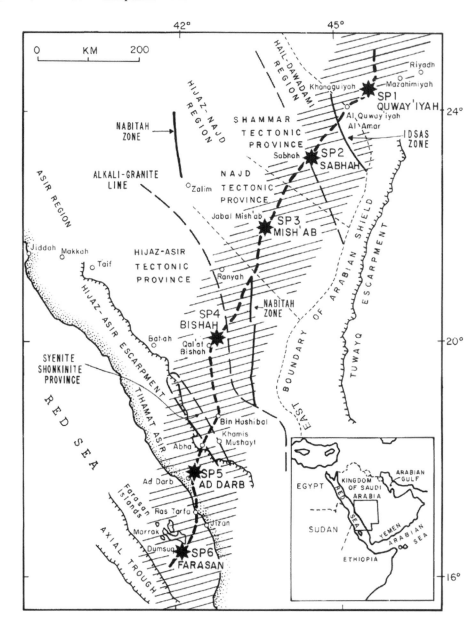

Figure 7.5.2-04. Location map of the long-range seismic refraction survey of 1978 through Saudi Arabia (from Mooney et al., 1985, fig. 1). [Tectonophysics, v. 111, p. 173–246. Copyright Elsevier.]

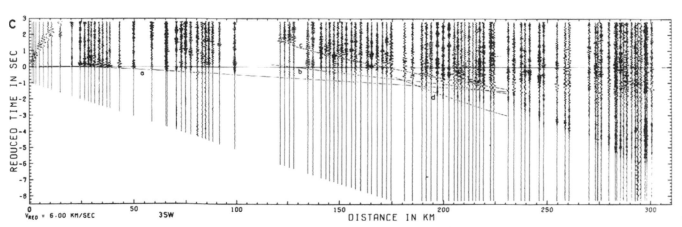

Figure 7.5.2-05. Record section of the Saudi Arabia 1978 experiment, observed across the Arabian Shield (from Mechie et al., 1986, fig. 3c). For location, see Fig. 7.5.2-04. [Tectonophysics, v. 131, p. 333–352. Copyright Elsevier.]

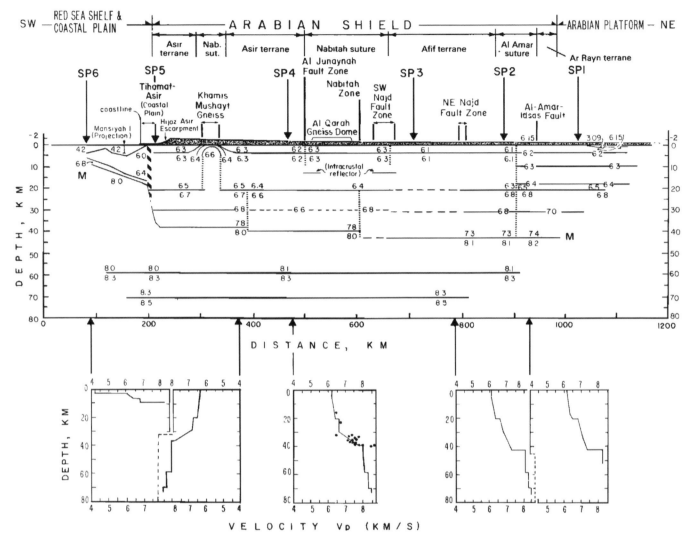

Figure 7.5.2-06. Crustal and upper mantle structure for southwestern Saudi Arabia under the Arabian Shield between the Red Sea and the Arabian platform. Velocity-depth functions for the major tectonic provinces are shown below (from Gettings et al., 1986, fig. 3). [Journal of Geophysical Research, v. 91, p. 6491–6512. Reproduced by permission of American Geophysical Union.]

7.6.2. China

In China, crustal studies were started moderately in the late 1970s and were connected with oil prospecting research in deep basins. Only one of these seismic surveys, carried out in the Qaidam basin in the 1970s, reached Moho level at 51 km depth (Teng et al., 1974; Teng, 1979). Explosion seismic studies of 1975–1977 in Tibet were continued in 1981 and 1982 (see Chapter 8.7.2) and the results, revealing a 75 km maximum thickness of the crust, were summarized by Teng (1987). In southeastern China in 1978, a 1000 ton explosion (Yongping explosion) ~500 km SW of Shanghai enabled the recording of a network of unreversed profiles which extended for 340–380 km toward the SE, NE, and SW and to a distance of 560 km toward the NW. A fence diagram shows a 3-layer crust of ~35 km thickness (United Observing Group of the Yongping Explosion, SSB, 1988; Yuan et al., 1986).

7.6.3. Japan

Japanese research activities in the 1970s were characterized by two directions. Big offshore shot experiments were carried out in 1974–1976 in the Pacific Ocean and the Sea of Japan, and onshore seismic-refraction surveys started in 1979 under the framework of the national project of earthquake prediction. One of these profiles, the Misima-Shimoda profile, was recorded across the Izu peninsula in central Japan (Asano et al., 1982; Yoshii et al., 1985). Another line was recorded on Shikoku Island (Ikami et al., 1982). The onshore activities were continued up to 2003, providing various crustal-scale heterogeneities existing within NE and SW Japan arcs (e.g., Asano et al., 1982; Yoshii, 1994).

The offshore research, carried out under the International Geodynamics Project, elucidated lateral variations in the upper-

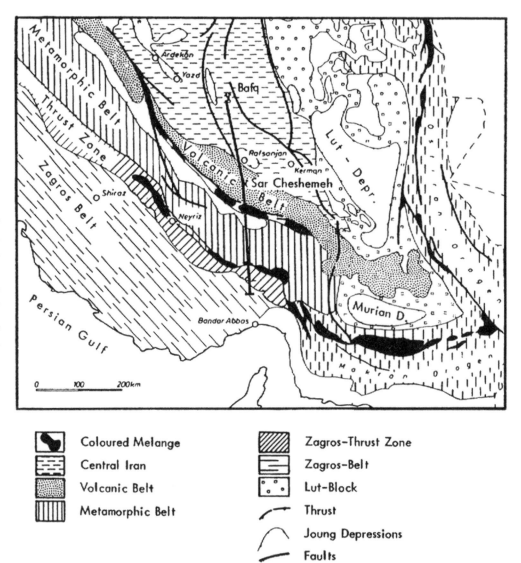

Figure 7.5.2-07. Tectonic map of Iran showing the position of the mines Bafq and Sar Cheshmeh and the layout of the seismic-refraction lines (from Giese et al., 1984, fig. 1). [Neues Jahrbuch für Geologie und Paläontologie Abhandlungen, v. 168, p. 230–243. Reproduced by permission of Schweizerbart, Stuttgart, Germany (www.schweizertbart.de).]

most mantle velocity from the trench area to the back arc basin (Research Group for Explosion Seismology, 1975; Okada et al., 1978, 1979; Yoshii et al., 1981). It is discussed in more detail in subchapter 7.8.1 (Pacific Ocean).

The seismic-refraction observations carried out in the 1970s in Japan, confirmed the basic results of the previous work summarized by the Research Group for Explosion Seismology (1973). These observations showed that the crust thickens from the south coast into the center of the island of Honshu, but that the total crustal thickness, as well as the thickness of the lower crust, proved to be substantially greater than in the previous models (Ikami, 1978; Kaneda et al., 1979).

From data obtained offshore, with ocean-bottom seismometers (Okada et al., 1979), and from previous results (Yoshii and Asano, 1972), e.g., compiled by Research Group for Explosion Seismology (1977), Asano et al. (1981) published a cross section from the Sea of Japan to the Pacific Ocean through northeast-

ern Honshu (Fig. 7.6.3-01). For northeastern Japan (Asano et al., 1979), the upper-mantle velocity is considerably lower (7.5 km/s) than in the other models (7.9 km/s).

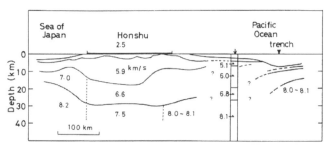

Figure 7.6.3-01. Crust and upper mantle structure from the Sea of Japan to the Pacific Ocean through northeastern Honshu (from Asano et al., 1981, fig. 9). Location of line is shown in Fig. 9.6.3-01, line A–A'. [Journal of the Physics of the Earth, v. 29, p. 267–281. Published by permission of the Seismological Society of Japan.]

7.7. CONTROLLED-SOURCE SEISMOLOGY IN THE SOUTHERN HEMISPHERE

7.7.1. Australia and New Zealand

The deep seismic-reflection experiments of the Bureau of Mineral Resources of the 1960s were continued in the 1970s. Extended records were successfully obtained at projects in the Galilee Basin, Queensland, in 1971 and in the Officer Basin, West Australia, and Murray Basin, New South Wales, in 1972 (Moss and Dooley, 1988), using large charges and in some cases an equipment which was specially modified for 24 s recording. Seismic-reflection investigations continued throughout the 1970s (e.g., Mathur, 1983, 1984). Where more comprehensive surveys were made, the events were consistent with deep reflections, though they lacked continuity in some cases. Finally, in 1976, the Bureau of Mineral Resources introduced digital recording and successfully carried out several deep crustal surveys between 1976 and 1978. Good quality reflection data up to 16 s were obtained at several sites in the southeastern Georgina, western Galilee and Bowen Basins (for locations, see Fig. 8.8.1-03) on short continuous profiles less than 15 km long (Dooley and Moss, 1988; Moss and Mathur, 1986; Finlayson, 2010; Appendix 2-2).

Long-range seismic probing of the crust and upper mantle in Australia, which was started in the 1960s, was continued. The ORD RIVER experiment with a series of shots in 1970 and 1971 (Denham et al., 1972) was already mentioned in Chapter 6.7.1.

In the following years, an increasing number of recording devices allowed the reduction of the spacing of stations considerably. It was particularly the group at the Bureau of Mineral Resources with D. Denham and D.M. Finlayson which took the lead in crustal and upper-mantle research. For investigations of the in-

land areas of Australia, the sources were mainly quarry blasts or drill-hole shots for mining exploration, but there was also marine work that was involved, e.g., around Tasmania and in the Papua New Guinea area and adjacent island chains.

With the move of A.L. Hales from Dallas, Texas, to the Australian National University at Canberra, active seismology research in Australia became a new component at the university. In particular, upper-mantle studies were the focus of this research. In subsequent years, K.J. Muirhead (Muirhead and Simpson, 1972) constructed long-recording devices (Figs. 7.7.1-01 and 7.7.1-02; Appendix A7-7-1; see also Finlayson 2010; Appendix A2-2) which were positioned on long-range profiles throughout the whole continent to record large explosions and the numerous earthquakes, which occurred frequently and with considerable magnitudes in the Southeast Asian island arcs.

Figure 7.7.1-01. Photograph of three-quarter watt seismic station, designed by Muirhead and Simpson (1972, fig. 3): complete field system with clock set-up box connected at right. [Bulletin of the Seismological Society of America, v. 62, p. 985–990. Reproduced by permission of the Seismological Society of America.]

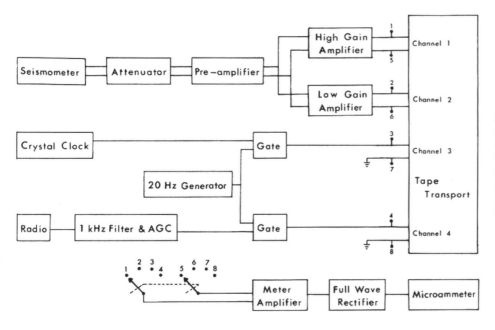

Figure 7.7.1-02. Three-quarter watt seismic station, designed by Muirhead and Simpson (1972, fig. 1): complete field system with clock set-up box connected at right. [Bulletin of the Seismological Society of America, v. 62, p. 985–990. Reproduced by permission of the Seismological Society of America.]

By these means, over the years, N-S–trending long-range lines were assembled (Fig. 7.7.1-03) which allowed scientists to derive refined models of the upper mantle down to several hundred kilometers in depth, showing, for example, discontinuities at 75 km and 200 km and below (e.g., Hales and Rynn, 1978; Hales et al., 1980; Leven, 1980). Data examples for the long-distance ranges of 500–2000 km and 1500–3100 km are shown in Figures 7.7.1-04 and 7.7.1-05 (see also Appendix A7-7-2).

A major survey designed to obtain upper-mantle arrivals was the Trans-Australia seismic survey (TASS) in 1972 was made possible by the Australian mining industry, which provided large explosions at Kunanalling, West Australia (80 tonnes), and Mount Fitton, South Australia (80 tonnes),

as well as in the Bass Strait where a 10 ton underwater shot was detonated at 250 m depth (Finlayson et al. (1974). Most of the data was recorded in the range from 200 to 1400 km. According to the position of the shotpoints, three main lines were established (Fig. 7.7.1-06): an east-west line through the Nullarbor Plain, a north-south line toward Alice Springs and a NW-SE line across the Snowy Mountains. For the interpretation (Finlayson et al., 1974; Muirhead et al., 1977; Appendix A7-7-2), the data of earlier explosions such as those from the 1956 Maralinga atomic bomb explosion and the 1970–1971 Ord River explosions were also included. In contrast to similar experiments in Europe, the existence of a low-velocity zone in the subcrustal lithosphere was not evident. Dooley (1979)

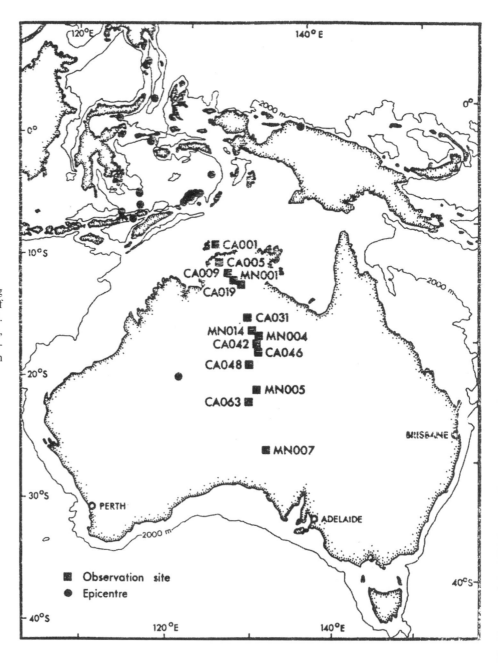

Figure 7.7.1-03. Location of recording stations in and of epicenters north of Australia (from Hales et al., 1980, fig. 1). [Research School of Earth Sciences, Australian National University Publication no. 1280. Published by permission of the Australian National University.]

compiled all available geophysical data along a geotraverse at 29°S (see Appendix A7-7-3); his crustal model from seismic data is shown in Figure 7.7.1-07.

A major effort was undertaken to study the crust under the Precambrian shield of western Australia with the Pilbara shield in the north and the Yilgarn shield in the south separated by the Capricorn orogenic belt. Two projects in 1977 and 1979 were carried out using quarry blasts of open-cut iron-ore mines (Drummond, 1979, 1981). A data example is shown in Figure 7.7.1-08 (for

location and more data, see Appendix A7-7-4). A fairly thin crust of 28 km thickness resulted for the northernmost Pilbara craton thickening gradually southwards to 33 km and more underneath the Capricorn orogen, and finally reached a crustal thickness of between 35 and 40 km under the Yilgarn craton in the south (Drummond, 1988; Dentith et al., 2000).

In 1973, a crustal study of the Bowen Basin in Queensland with its extensive coal measures was conducted (Collins, 1978) where a 35-km-thick crust, with surprisingly high velocities both

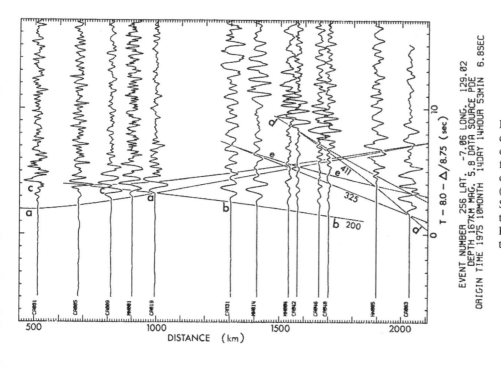

Figure 7.7.1-04. Record section of an event with focal depth of 167 km north of Australia (from Hales et al., 1980, fig. 14). The numbers denote the depth of the corresponding reflector in model 8 (see Appendix 7-7-2, fig. 9). [Research School of Earth Sciences, Australian National University Publication no. 1280. Published by permission of the Australian National University.]

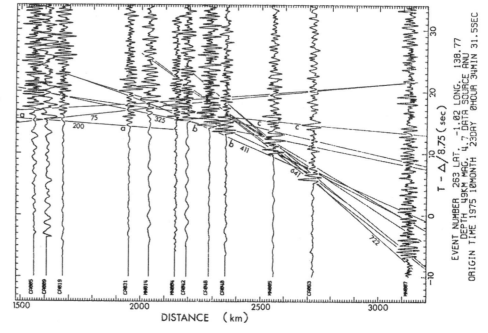

Figure 7.7.1-05. Record section of an event with focal depth of 49 km north of Australia (from Hales et al., 1980, fig. 20). The numbers denote the depth of the corresponding reflector in model 8 (see Appendix 7-7-2, fig. 9). [Research School of Earth Sciences, Australian National University Publication no. 1280. Published by permission of the Australian National University.]

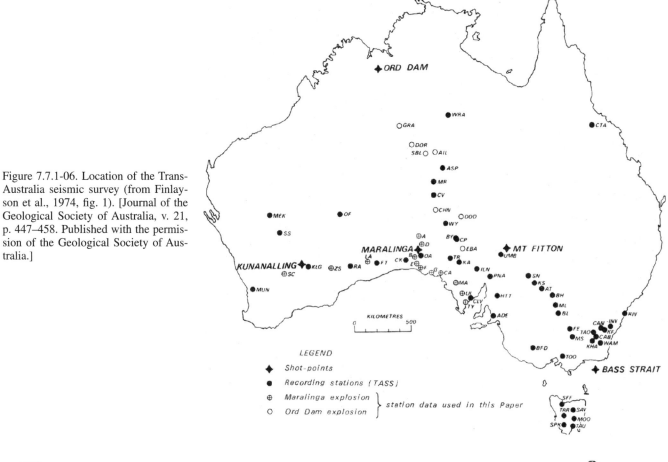

Figure 7.7.1-06. Location of the Trans-Australia seismic survey (from Finlayson et al., 1974, fig. 1). [Journal of the Geological Society of Australia, v. 21, p. 447–458. Published with the permission of the Geological Society of Australia.]

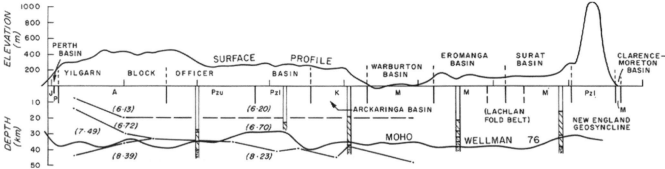

Figure 7.7.1-07. Crustal model along the Trans-Australia seismic survey (from Dooley, 1979, fig. 2c). [BMR Journal of Australian Geology and Geophysics, v. 4, p. 353–359. Copyright by Commonwealth of Australia (Geoscience Australia).]

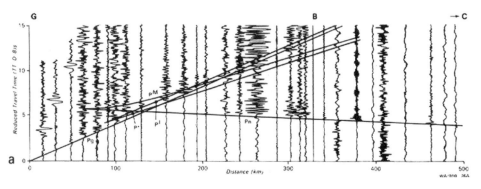

Figure 7.7.1-08. Record section of a profile through the Pilbara-Yilgarn cratons in northwestern Australia (from Drummond, 1979, fig. 3a). [BMR Journal of Australian Geology and Geophysics, v. 4, p. 171–180. Copyright by Commonwealth of Australia (Geoscience Australia).]

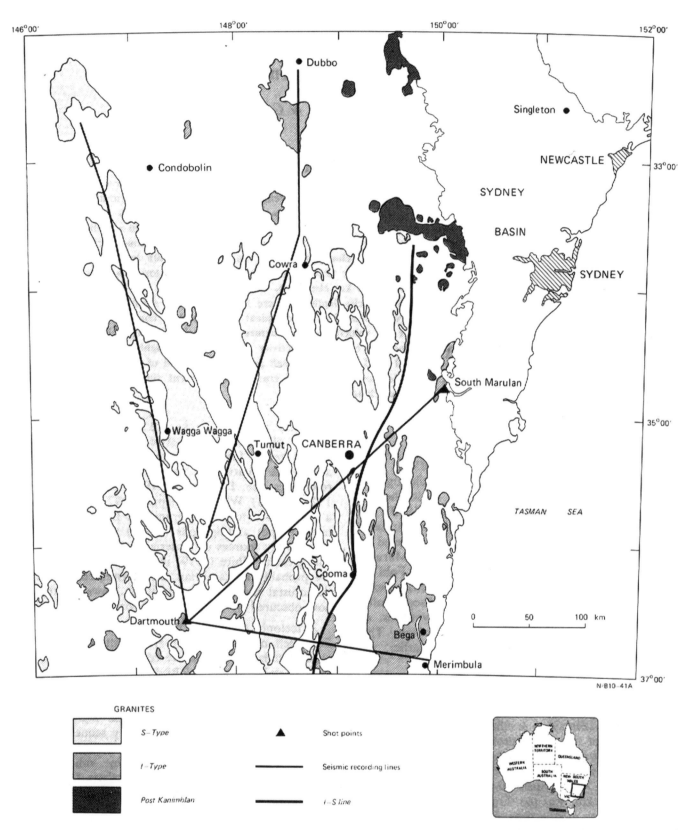

Figure 7.7.1-09. Location of seismic profiles through the Lachlan foldbelt in southeastern Australia (from Finlayson et al., 1979, fig. 1). [BMR Journal of Australian Geology and Geophysics, v. 4, p. 243–252. Copyright by Commonwealth of Australia (Geoscience Australia).]

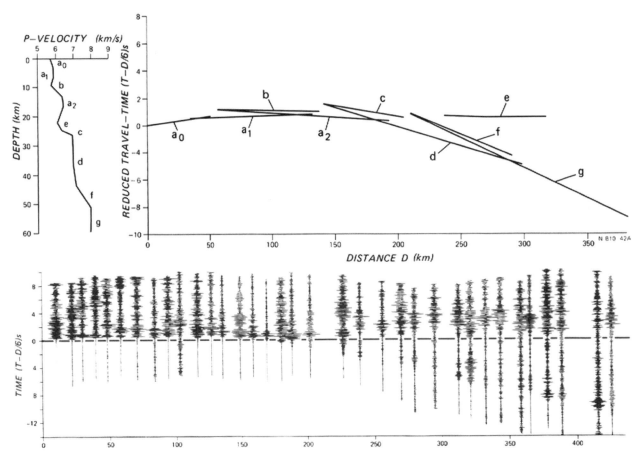

Figure 7.7.1-10. Top: Basic traveltime curve arrangement and model (from Finlayson et al., 1979, fig. 2). Bottom: corresponding data, here record section from Dartmouth towards NNW to Condobolin (from Finlayson et al., 1979, fig. 3a). [BMR Journal of Australian Geology and Geophysics, v. 4, p. 243–252. Copyright by Commonwealth of Australia (Geoscience Australia).]

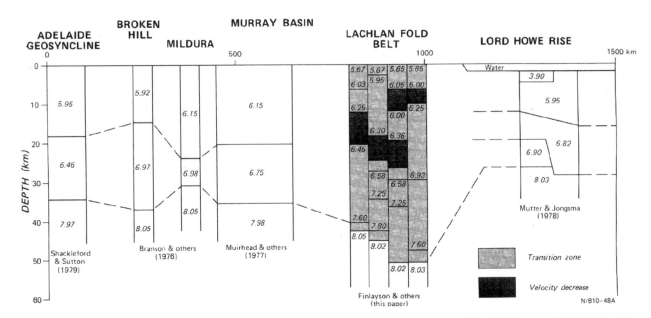

Figure 7.7.1-11. Crustal structure across southeastern Australia and its eastern continental margin (from Finlayson et al., 1979, fig. 8). [BMR Journal of Australian Geology and Geophysics, v. 4, p. 243–252. Copyright by Commonwealth of Australia (Geoscience Australia).]

in the upper and lower crust, was modeled (see Appendix A7-7-5 for data and model).

During the 1970s, a more detailed crustal research project was accomplished in New South Wales, southeastern Australia, in the Lachlan Fold Belt. Here, several lines, all based on quarry blast observations, were assembled between 1976 and 1978 (Finlayson et al., 1979; Fig. 7.7.1-09). In 1978, the survey was extended northward (Finlayson and McCracken, 1981). The interpretation of the data resulted in a thick continental crust averaging 40 km in thickness, but exceeding 50 km under the region of highest topography (Finlayson et al., 1979; Finlayson and McCracken, 1981; Appendix A2-1, p. 104–105, and Appendix A7-7-6). Also, the average crustal velocity was high and the interfaces were best modeled by thick transition zones (Fig. 7.7.1-10).

The cross section through southeastern Australia (Fig. 7.7.1-11) assembled by Finlayson et al. (1979) comprises earlier work in South Australia in the west (Branson et al., 1976; Muirhead et al., 1977; Shackleford and Sutton, 1979) and includes the Lord Howe Rise (Mutter and Jongsma, 1978) to the east of the southeastern Australian margin.

In 1979, a 500-km-long east-west line was recorded between the Mount Isa Mine, Queensland, using a 40 t rock breaking shot as the eastern source and the abandoned Skipper Mine near Tennant Creek, Northern Territory, shooting a 5 t charge at 40 m depth at the western end. In addition, on 9 August 1979, the stations along the profile recorded unexpectedly the New Britain earthquake (Finlayson, 1982).

In New Zealand in the 1970s, two offshore-onland seismic-refraction experiments were carried out. In 1974 and 1975, marine shots in fjords on the South Island were recorded by stations on land along several lines of up to 100 km length (for location, see Fig. 8.8.1-05). The objective was to verify an inferred shallow Moho under a high gravity anomaly (Davey and Broadbent, 1980; Priestley and Davey, 1983). The structure derived from the traveltime data was markedly different from the nearby stable areas of South Island. They gave no indication of a typical 6.0 km/s upper crustal layer, but substantiated the existence of high-velocity material typical of lower crust at shallow depths beneath western Fjordland and supported the idea that western Fjordlands represents a section of uplifted lower crust. The data did not succeed in defining the position of the Moho. At least there was no evidence that the Moho was shallower than 15 km.

The project WHUMP (Wellington–Hikurangi Upper Mantle Profile) of 1979, a SW-NE–directed crustal seismic-reflection–seismic-refraction experiment, was centered on Wellington (for location see Fig. 8.8.1-05). The project aimed to trace the extent of a strong reflector occurring at a depth of ~20 km beneath Wellington and to investigate if this reflector coincides with the interface between the lithosphere of the Indian plate and the underlying lithosphere of the Pacific plate, which is currently being subducted at the Hikurangi Trough in the east and dips under North Island, New Zealand, to the west (Davey and Smith, 1983). The measurements were made along a 50-km-long profile along the southwestern edge of the North Island between the Palliser Bay

in the southeast and the Cook Strait in the northwest recording shots from two shotpoints. A series of shots fired 500 m offshore at 10 m water depth on the seafloor near Wellington were recorded at 1 km intervals in northwest and southeast direction. The northwest sector of the profile was partly reversed by a series of shots at sea at the northwestern end of the line.

7.7.2. South Africa

In the University of Witwatersrand at Johannesburg, South Africa, R.W. Green had developed instrumentation which allowed him to carry out one-man seismic-refraction surveys (Green, 1973). He would install his recording devices (BPI) and let them run in a continuous mode for up to a week, and in between he would organize shotpoints, load them, and detonate the charges whenever ready. This worked very well for small-scale surveys.

At the same time, in southwest Africa, present-day Namibia, a special geological research project was carried out to study the Damara orogen with stratigraphic, tectonic, and petrographic

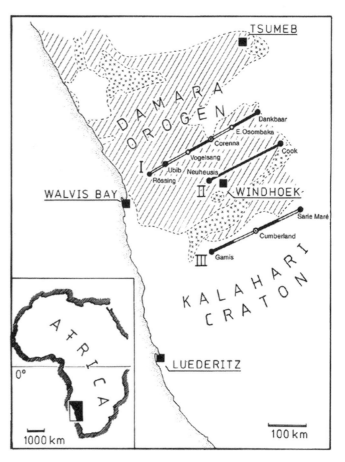

Figure 7.7.2-01. Simplified map of the Damara orogen in Namibia and location of the seismic-refraction lines (from Baier et al., 1983, fig. 1). [*In* Martin, H., and Eder, F.W., eds., Intracontinental Fold Belts: Berlin-Heidelberg, Springer, p. 885–900. Reproduced with kind permission of Springer Science+Business Media.]

investigations undertaken by the University of Goettingen, Germany, and funded by the German Research Society.

Within this research project, the crustal structure was also to be studied, for which purpose a cooperation of German geophysical institutes with the University of Witwatersrand at Johannesburg was initiated (Baier et al., 1983; Appendix A2-1,

p. 101–103, and Appendix A7-8). The deep seismic sounding research was carried out in 1975. It involved three profiles of 225–350 km length, with shotpoints at the ends, and with some in between (Fig. 7.7.2-01). All three of these profiles ran in a WSW-ENE direction. Two of them traversed the Damara orogen (Fig. 7.7.2-02A) and the third traversed the Kalahari craton in the

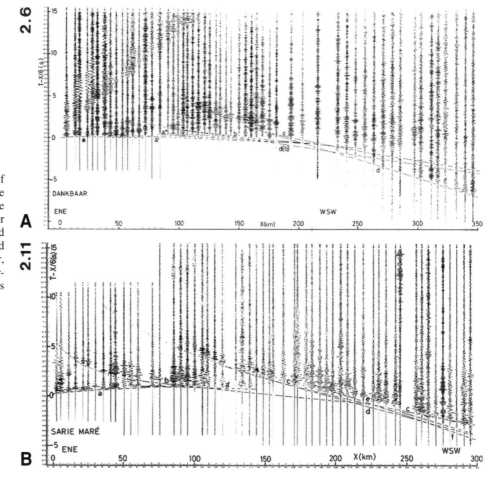

Figure 7.7.2-02. Record sections of the Namibia 1975 project. (A) Profile through the Damara orogen. (B) Profile through the Kalahari craton (from Baier et al., 1983, fig. 2). [*In* Martin, H., and Eder, F.W., eds., Intracontinental Fold Belts: Berlin-Heidelberg, Springer, p. 885–900. Reproduced with kind permission of Springer Science+Business Media.]

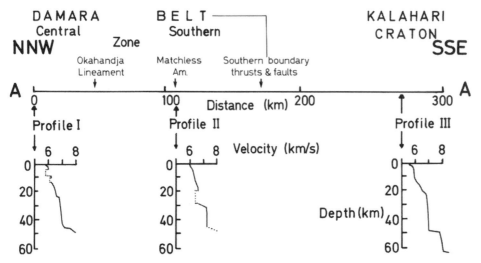

Figure 7.7.2-03. Velocity-depth sections through the Damara orogen and adjacent Kalahari craton, approximately 380 km NNE of the Atlantic coast (from Baier et al., 1983, fig. 4). [*In* Martin, H., and Eder, F.W., eds., Intracontinental Fold Belts: Berlin-Heidelberg, Springer, p. 885–900. Reproduced with kind permission of Springer Science+Business Media.]

south (Fig. 7.7.2-02B). With the exception of the westernmost shot of the northern profile I, which was a commercial mine blast, all shots were drill-hole shots arranged by teams of the participating institutions. Fifteen MARS-66 stations and 15 BPI stations were available for recording. The results were interpreted as a 47-km-thick crust, both underneath the Damara orogen and the adjacent Kalahari craton (Fig. 7.7.2-03).

7.7.3. South America

Crustal research in South America in the 1970s concentrated on the Andean region of Colombia and Ecuador, as well as the adjacent Pacific Ocean. In 1973, the Narino III project was launched between latitudes 2° and 4°N and was jointly organized by the University of Wisconsin under the direction of R.P. Meyer and the University of Texas at Dallas under the direction of A.L. Hales. Furthermore, the Carnegie Institution of Washington and the University of Kiel, Germany, joined in, arranging and recording quarry blast shots. The aim of the Narino III project was to investigate the crustal and upper-mantle structure in the area of

the subducting Pacific plate of oceanic character underneath the South American continent (Meyer et al., 1976). The project involved a series of sea shots along lines away from the Pacific coast and up to 400 km long, as well as parallel to the coast, which were recorded by stations on land. The additional quarry blasts were recorded along a north-south line along the eastern edge of the western Cordillera (Fig. 7.7.3-01). Some of the results are shown in Figures 7.7.3-02 and 7.7.3-03.

As part of the Narino project, the University of Kiel, Germany, also was involved. The crustal cross sections derived for the southern part of Colombia by Meissner et al. (1976b) revealed an average thickness of the crust of 20 km in the oceanic and western part of the line and of 30–40 km underneath the continental part.

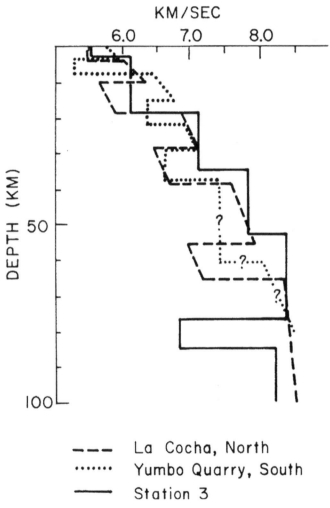

Figure 7.7.3-02. Velocity-depth functions for quarry-blast profiles running north-south along the eastern edge of the western Cordillera in comparison with Narino results (station 3) (from Meyer et al., 1976, fig. 24).). [*In* Sutton, G.H., Manghnani, M.H., Moberly, R., and McAffe, E.U., eds., The geophysics of the Pacific Ocean basin and its margin: American Geophysical Union Geophysical Monograph 19, p. 105–132. Reproduced by permission of American Geophysical Union.]

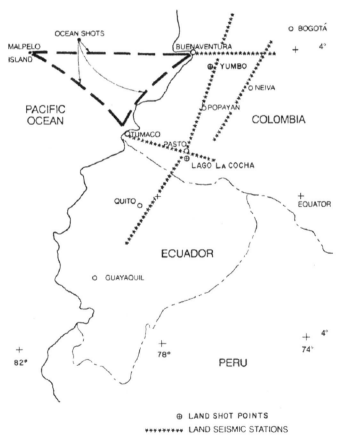

Figure 7.7.3-01. Sketch map of the Narino project in Colombia and Ecuador (from Meyer et al., 1976, fig. 1). [*In* Sutton, G.H., Manghnani, M.H., Moberly, R., and McAffe, E.U., eds., The geophysics of the Pacific Ocean basin and its margin: American Geophysical Union Geophysical Monograph 19, p. 105–132. Reproduced by permission of American Geophysical Union.]

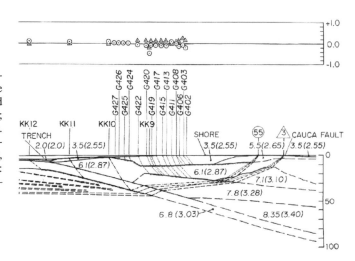

Figure 7.7.3-03. True-scale model derived from Narino data and gravity with shotpoints G402-G427 and KK12–KK9 (KK19–KK13 are farther west), ray paths and stations 55 and 3. Dotted and dash-dotted paths are critically refracted along the continental and oceanic mantle; dashed paths simulate diving waves (from Meyer et al., 1976, fig. 25). The numbers in the model are P-wave velocities; those in brackets densities. [*In* Sutton, G.H., Manghnani, M.H., Moberly, R., and McAffe, E.U., eds., The geophysics of the Pacific Ocean basin and its margin: American Geophysical Union Geophysical Monograph 19, p. 105–132. Reproduced by permission of American Geophysical Union.]

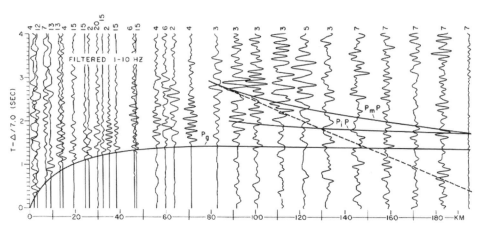

Figure 7.7.3-04. Record sections of the Colombia 1976 project. Yumbo-North profile (from Mooney et al., 1979, fig. 7a). [Bulletin of the Seismological Society of America, v. 69, p. 1745–1761. Reproduced by permission of the Seismological Society of America.]

In 1976, additional quarry blasts at one of the shotpoints of 1973 were arranged and recorded in the Western Cordillera of the Colombian Andes up to 200 km along strike as well as 60 km toward the west. In particular, the data of the quarry Yumbo, recorded to the north, were of very good quality (Fig. 7.7.3-04). A crustal thickness of 30 km resulted for this area (Mooney et al., 1979; Appendix A7-9-1).

In 1978, an experiment was carried out farther north, around latitude 5.5°N, with a series of shots at sea and a 350-km-long line crossing the Serra de Baudo near the coast and the Western and Central Cordilleras of northwestern Colombia (Flueh et al., 1981; Appendix A7-9-2). Again there was an international cooperation, involving U.S., German, Spanish, and Colombian scientists. Here, the Moho depth increases from 17 km near the coast to 30 km under the Serra de Baudo, and remains at this depth under the Western Cordillera and finally increases further to 45 km under the Central Cordillera.

7.7.4. Antarctica

A very detailed seismic survey by the Institute of Geophysics of the Polish Academy of Sciences was carried out in 1979 and 1980 in the southernmost end of the Pacific Ocean. The survey

was to explore the structure underneath West Antarctica, and it covered the area from 61° to 65°S and 56° to 66°W.

The seismic measurements reached from the southern Shetland Islands to the Antarctic Peninsula (Fig. 7.7.4-01) and included reflection seismic profiling along shelf geotraverses, shallow seismic-refraction soundings on shelves, deep seismic soundings of the whole crust, and magnetic profiling (Guterch et al., 1985).

The deep seismic sounding profiles reached a total length of 2000 km. The interpretation of their results was as follows: the crust is 32 km thick in the region of the South Shetland Islands, and thickens to 40–45 km in the region of the Antarctic Peninsula. A preliminary geodynamic model by Guterch et al. (1985) is shown in Figure 7.7.4-02.

7.8. OCEANIC DEEP STRUCTURE RESEARCH

7.8.1. Introduction

Until the mid-1960s, all offshore seismic exploration work used exclusively chemical explosions as energy sources. For deep crustal studies, this remained so in the 1970s. In particular for offshore-onshore projects, when shots at sea were both recorded

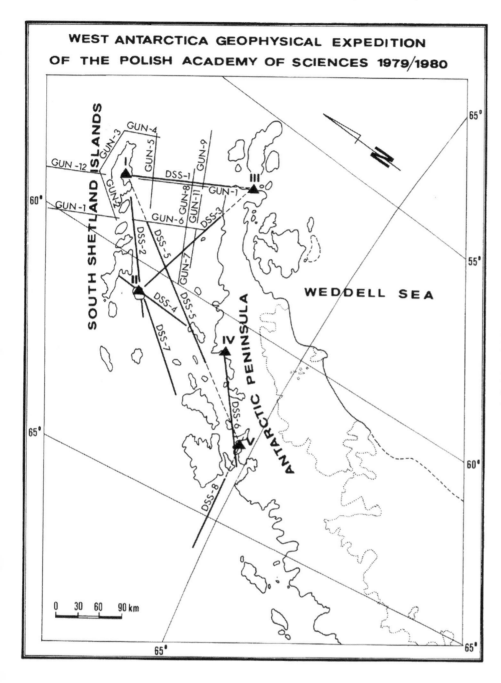

Figure 7.7.4-01. West Antarctica Polish expedition 1979–1980: Location of seismic profiles. DSS—deep seismic sounding profiles; GUN—reflection profiles; I, II, III, IV, V—position of seismic stations (from Guterch et al., 1985, fig. 1). [*In* Husebye, E.S., Johnson, G.L., and Kristoffersen, Y., eds., Geophysics of the polar regions: Tectonophysics, v. 114, p. 411–429. Copyright Elsevier.]

by recording devices on research vessels and on land, as well as for the 1000–2000-km-long land profiles in western Europe, small to large depth charges up to several tons of dynamite were detonated in shallow shelf waters wherever available. For the LISPB project of 1974, for example, Brian Jacob had carried out pre-studies to optimize the shots by splitting the single charges into smaller units and obtaining both a smaller quantity of explosives and a shallower optimum depth for underwater shots carrying sufficient energy over 1000 km distances (see subchapter 7.2.4). The development of airguns offered the chance to repeat shots much more rapidly than was possible with chemical explosions. Furthermore, the shallow depth of Moho under the deep

oceans did not require profile lengths of several 100 km as was needed for land profiles. So, from the mid-1970s onward, marine crustal data could very efficiently and at a fast speed be collected in large quantities and the oceanic structure could now be unraveled with a much higher precision. But in many deep-reaching scientific research projects explosives were still preferred as the most powerful energy source.

Dobrin (1976) provided background information on basic multichannel seismic theory and techniques of computer processing. Stoffa and Buhl (1979) discussed in detail the value of expanding spread profiling and constant offset profiling for deep crustal studies. They acquired expanding spread profile data

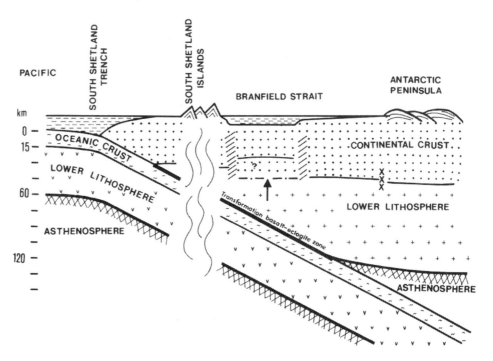

Figure 7.7.4-02. Preliminary geodynamic model of West Antarctica (from Guterch et al., 1985, fig. 12). [*In* Husebye, E.S., Johnson, G.L., and Kristoffersen, Y., eds., Geophysics of the polar regions: Tectonophysics, v. 114, p. 411–429. Copyright Elsevier.]

in the western Pacific, in the Caribbean, and in the Norwegian Sea and discussed in detail the accuracies and advantages of this specific techniques. They concluded that in the late 1970s, these experimental techniques were the only method available for a detailed continuous mapping of lateral crustal variations from non-reflecting horizons.

The 1970s were also the decade, when ocean-bottom seismometers (OBS) and ocean-bottom hydrophones were developed in various institutions and successfully tested in experiments around the world (Asada and Shimamura, 1976, 1979; Avedik et al., 1978; Carmichael et al., 1973; Rykunov and Sedov, 1967; Whitmarsh and Lilwall, 1983). Another example is "a new and inexpensive pop-up ocean-bottom hydrophone recorder" which had been developed in 1978 at the Bullard Laboratories at Cambridge, UK, for use in seismic-refraction experiments, in order to supplement or to replace the existing recording sonobuoys. The first "field" test at sea could be started in 1979 (Sinha et al., 1981a); in 1980 it was regarded to be ready.

These ocean-bottom instruments are self-contained recording units which are brought back to the surface on a line or after ballast weights are discarded by an acoustic command from the retrieving vessel (pop-up systems). They offered important advantages over free-floating sonobuoys. They could be placed at ranges beyond the maximum attainable with telemetering sonobuoys and could therefore be used for long offset seismic probing of several tens of kilometers below the seafloor. P-waves from shallow structural levels were recorded as first arrivals. The noise level was considerably lower, because the high-energy environment of the upper-water column was removed. The uncertainties in interpreting seismic traveltimes were considerably reduced, because the instruments were fixed on the ocean bottom. In an ocean-bottom seismograph both vertical and horizontal component seismometers and additional hydrophones could be incorporated.

From the numerous new investigations carried out in the 1970s, using airguns and expendable sonobuoys, a more detailed picture of the crust under the ocean basins could be obtained. Figure 7.8.1-01 presents a world map showing all locations of the Lamont-Doherty Laboratory where recordings with sonobuoys were made until the mid-1970s and which were used for a special investigation for the upper crustal structure (Houtz and Ewing, 1976). Following the summaries of Ewing and Houtz (1979), Jones (1999) and Minshull (2002), due to new and improved data, in many areas the seismic layer 2 had now to be split into two units, 2A and 2B, with average velocities of 3.5 km/s and 5.2 km/s. Between layers 2B and 3, refracted waves revealed a zone 2C with velocities of 6.0–6.2 km/s. In some regions, basement with velocities of 7.2–7.4 km/s was seen to occur just above Moho. This more refined crustal structure had the result, that Moho depth was ~1 km deeper than calculated from the earlier deep-sea refraction data. A correlation of these finer subdivisions of the oceanic basement with the P-wave velocity structure of ophiolites on land appeared satisfying.

The subdivision of layer 2 into layers 2A and 2B was originally based on slope-intersect interpretations of sonobuoy profiles, where layer 2A was recognized as a thin layer with low velocities at the top of the crust which did not generate refracted arrivals. The introduction of such a layer became necessary, because the layer-2 arrivals were not tangential to the top-basement reflections, while layer 2B was the deeper higher-velocity layer which generated the observed refracted arrivals (Houtz, 1976; Houtz and Ewing, 1976; Carlson, 1998; Minshull, 2002).

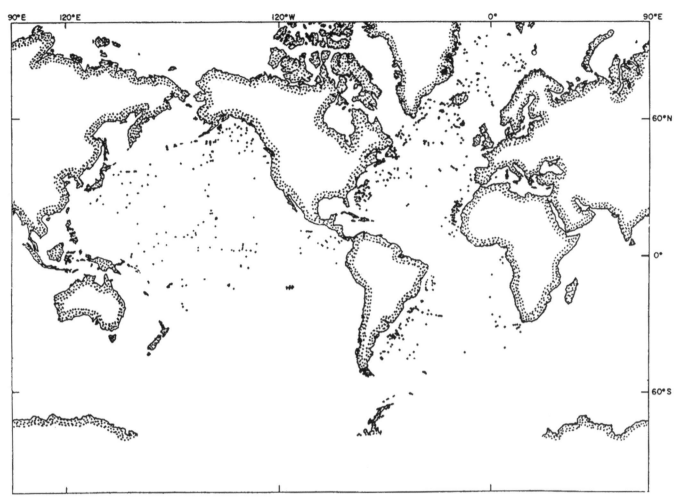

Figure 7.8.1-01. Location of Lamont-Doherty sonobuoy stations (from Houtz and Ewing, 1976, fig. 1). [Journal of Geophysical Research, v. 81, p. 2490–2498. Reproduced by permission of American Geophysical Union.]

Layer 2A has attracted geoscientists in connection with plate movements because of its supposed age-dependent properties. Therefore Houtz and Ewing (1976) and later in a subsequent study Houtz (1976) have analyzed all Lamont-Doherty sonobuoy data around the world as far as they were located on well-dated crust (Fig. 7.8.1-01). As a result, layer 2A was thicker on the Mid-Atlantic Ridge crest (1.5 km) than on the East-Pacific Ridge crest (0.7 km). In both oceans, layer 2A thinned gradually away from the ridge to ~100 m thickness, which was reached in the Atlantic at 60 m.y. old crust, in the Pacific at 30 m.y. old crust. This was attributed to the much faster spreading rate in the Pacific. Furthermore, regressions on refraction velocities in layer 2A as a function of age showed that its velocity increased from ~3.3 km/s at the ridge crests to that of layer 2B on crust ~40 m.y. old. In none of the deeper layers could a corresponding increase of velocity with age be detected.

The application of the reflectivity method to calculate synthetic seismograms for deep-ocean data proved to be most powerful to unravel details of the basement structure (Spudich and Orcutt, 1980). It was shown that the structure was better represented by velocity gradient zones than by discrete constant-velocity layers. Figure 7.8.1-02 shows a record section from observed data. Depending on the gradient, reflections were strong or weak. Figure 7.8.1-03 shows synthetic seismograms for an ocean-crust model with a velocity gradient above Moho, instead of Moho being a first-order discontinuity, which would not explain the observed data. The use of record sections and the reanalysis of the early data by modern synthetic seismogram analysis (White et al., 1992) showed that the slope-intercept time method systematically underestimated the thickness of the oceanic crustal layers, because secondary arrivals had not been recognized as reflections, but, if at all, had been interpreted as strong refracted arrivals. As Minshull (2002) points out, on modern record sections plotted from the early data, a strong layer-2 arrival, a weak layer-3 arrival, a strong Moho reflection, and weak mantle arrivals could be correlated and compared with synthetic seismograms.

There was also research work going on to determine if seismic anisotropy could be observed in the crust. In the 1960s,

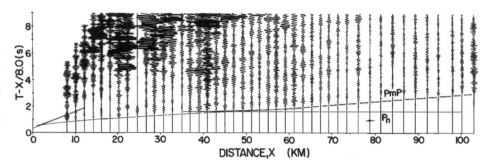

Figure 7.8.1-02. Observed data highlighting the P$_M$P phase, reflected from Moho, and the P$_n$ phase, refracted in the uppermost mantle layer (from Spudich and Orcutt, 1980, fig. 4). [Reviews of Geophysics and Space Physics, v. 18, p. 627–645. Reproduced by permission of American Geophysical Union.]

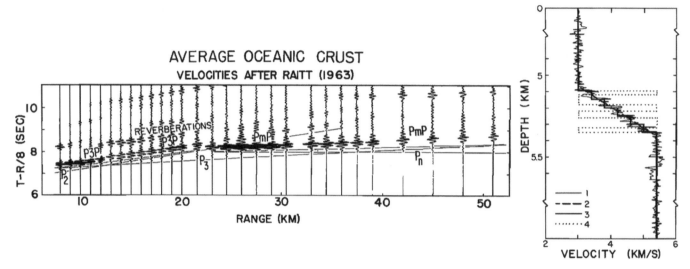

Figure 7.8.1-03. Left: Synthetic seismograms from the average oceanic crust model after Raitt (1963). P$_2$, P$_3$, P$_n$ = head waves in layers 2, 3, mantle (discontinuities at 5 km (bottom of water column), 7 km (layer 2–3 border) and 13 km (Moho) depth). P3P, PmP = reflections from tops of layer 3 and mantle (from Spudich and Orcutt, 1980, fig. 6). Right: Four possible models, highlighting a gradient or lamellar transition above Moho (from Spudich and Orcutt, 1980, figs. 6 and 8). [Reviews of Geophysics and Space Physics, v. 18, p. 627–645. Reproduced by permission of American Geophysical Union.]

special deep seismic sounding projects in the Pacific Ocean had investigated the question of anisotropy in the subcrustal lithosphere (Keen and Barrett, 1971; Morris et al., 1969; Raitt et al., 1969, 1971), but they concentrated on possible anisotropy in the uppermost mantle. In the 1970s, also on land, evidence of anisotropy was detected in the uppermost mantle (Bamford et al., 1979; Fuchs, 1975, 1977, 1983), but only in areas of extensional tectonics. Investigations of anisotropy in the crust on land, however, were contradictory and did not allow researchers to convincingly conclude that anisotropy was present in the crystalline crust. Bibee and Shor (1976) investigated a large number of standard marine-refraction studies to determine the general relationship of crustal and mantle velocity to spreading direction and age in the Pacific. They concluded that anisotropy in the crust is insignificant, as is variation of lower crustal velocity with age. Mantle velocities, however, exhibited a high correlation with both age and azimuth, indicating an increase of velocity with age and ~5% anisotropy with highest velocity in the direction perpendicular to the local magnetic anomalies. The situation seemed to be different to Stephen and co-workers for the oceanic layers (Stephen

et al., 1980; Stephen, 1981, 1985) and therefore in 1977, the specially designed Oblique Seismic Experiment was carried out in a borehole in the western Pacific 400 miles north of Puerto Rico. The experiment had been proposed to increase the usefulness of IPOD (International Phase of Oceanic Drilling) crustal holes as a means to investigate layer 2 of oceanic crust. However, the effect was quite small and it seems that for the usual deep seismic sounding experiments which were collected in our history, the possible neglecting of anisotropy in anyone of the crustal layers has not negatively influenced the interpretations of crustal structure and conclusions drawn from their results.

Christensen and co-workers studied rock samples from the Bay of Islands ophiolite complex of western Newfoundland (Salisbury and Christensen, 1978; Christensen and Salisbury, 1982) and from the Oman Mountains, where a continuous ophiolite succession is exposed (Christensen and Smewing, 1981). Based on their measurements of compressional and shear wave velocities to confining pressures of 6 kbar, they concluded that the seismic structure of the ophiolite sequences is remarkably similar to that of the oceanic crust and upper mantle. Kempner and Gettrust (1982a, 1982b) have calculated synthetic seismograms both for observed marine seismic data and for velocity profiles derived from laboratory measurements of rock samples from ophiolite sequences and proved that the composition of oceanic crust and upper mantle as derived from marine seismic data can directly be compared with exposed ophiolite sequences. A similar investigation was repeated in the 1980s by Collins et al. (1986).

Already in the 1960s and continuing in the 1970s, the main interest of oceanic crust and upper mantle research had shifted from the investigation of crustal and upper mantle structure under ocean basins to special structures with anomalous properties such as mid-ocean ridges, hotspots, ocean islands, and ocean-continent transition zones. Ewing and Meyer (1982) emphasized how the rapid evolvement of experimental and analytical techniques in marine seismology permitted a more flexible view of the variation of velocity with depth due to the employment of OBSs and due to the development of methods of analysis (e.g., Helmberger, 1968; Fuchs and Müller, 1971; Bessonova et al., 1974; Kennett, 1977). This greatly altered the conception of the structure of the oceanic crust and upper mantle and emphasized the need for detailed studies of specific areas and problems rather than well directed surveillance expeditions.

For example, in OBS refraction experiments in the 1970s on the northern East Pacific Rise, a fast-spreading ridge, a localized low-velocity zone was detected for the first time (Orcutt et al., 1976; Minshull, 2002) which was interpreted as resulting from the presence of a magma chamber containing partially molten rocks. Many seismic studies in the 1970s were concentrated on the Mid-Atlantic Ridge, a slow-spreading ridge. Early OBS refraction experiments on the Mid-Atlantic Ridge found neither a low-velocity zone corresponding to magma chamber nor a strong velocity contrast at Moho depths. Instead, the crust-mantle boundary was defined by a gradual increase in velocities (Fowler, 1976; Minshull, 2002). By the end of the 1960s, the traditional model of the hotspot Iceland consisted of four layers (Palmason, 1971): layer 0—unconsolidated material; layers 1 and 2—upper crust; layer 4—anomalous upper mantle with velocity of 7.4 km/s. At the end of the 1970s, a large-scale seismic experiment created a new model with a 15-km-thick two-layer crust and a mantle with gradual increase of velocity from 7.0 to 7.6 km/s at 50–60 km depth, with intermittent high-velocity sills near 30 km depth (Gebrande et al., 1980; see Fig. 7.8.4-08). Several aseismic ridges which are widespread in the oceans and oceanic plateaus representing vast outpourings of magma, but on a larger scale than aseismic ridges (Minshull, 2002) became the focus of seismic projects in the 1970s, such as the Iceland-Faeroe Ridge (Bott and Gunnarson, 1980), the Azores-Biscay Rise (Whitmarsh et al., 1982), or the Madagascar Ridge (Goslin et al., 1981; Sinha et al., 1981b) and the Agulhas Plateau off Cape Horn (Tucholke et al., 1981). Finally, continental shelves and adjacent oceanic areas were investigated in numerous projects along all continents.

An enormous amount of shallow seismic data was also collected in the 1970s in connection with the Deep Sea Drilling Project (DSDP). By 1968 the Joint Oceanographic Institutions for Deep Earth Sampling (JOIDES) had secured long-term funding from the National Science Foundation and leased a custom drillship from Global Marine Development Inc. named *GLOMAR Challenger*. Using powerful thrusters to stay in one spot even in heavy seas, the ship could lower a drill string through 7 km of water, then drill 1700 m and more into the sediment and rock of the seabed. For the next 15 years the *Challenger* crisscrossed the seas, making 96 separate voyages or "legs" and drilling at 624 sites on the seafloor. During that time, five other nations joined the program (Alden, 2010).

While the number of land projects of crustal and upper-mantle dimensions in a decade is limited due to the long precursory times of preparation, organization, and final fieldwork, marine projects, once ship time was available, could be achieved in much shorter time spans. Therefore, numerous oceanic research projects could be carried out in the 1970s and the following decades, many of them dealing exclusively with upper crustal studies. It would be beyond the scope of this history to mention them all. Therefore we have also not included the DSDP research work. Also, many projects were not published in easily accessible geophysical journals, but exclusively in special report volumes or journals specializing on general oceanic research topics other than geophysics and are therefore nearly impossible to find. We therefore restricted ourselves to a few and hopefully typical marine deep seismic projects in the following subchapters. Most projects we describe dealt with the crust and underlying uppermost mantle and were carried out in particular areas of tectonic unrest, such as along the mid-ocean ridges and along continental margins. Furthermore, similar to the continental efforts, some experiments aimed specifically to penetrate the subcrustal lithosphere by long-range seismic-refraction profiling. In the following description of active-source experiments carried out in the 1970s in the Pacific, Indian, and Atlantic Oceans, we will follow a more or less geographic order.

7.8.2. The Pacific Ocean

The knowledge on the crustal structure of the Pacific Ocean in the early 1970s was summarized, for example, by Woollard (1975), who had examined ~400 seismic-refraction measurements in the Pacific basin and presented them as block diagrams along selected lines (Fig. 7.8.2-01).

On a line through the northern Pacific from the Japan trench to the Gulf of Alaska (Fig. 7.8.2-02A), the crustal thickening beneath the Shatsky Rise was the dominant feature. On another line through the central Pacific from the Bismarck Archipelago to the Juan de Fuca Rise, the crust is 35–40 km thick underneath the Bismarck archipelago and adjacent Ontong-Java Rise, then thins to normal oceanic crustal thickness along the profile through the Gilbert islands, Darwin Rise, and Line Islands, then thickens substantially beneath the Hawaiian Islands, and thins again when traversing the Murray and Mendocino fracture

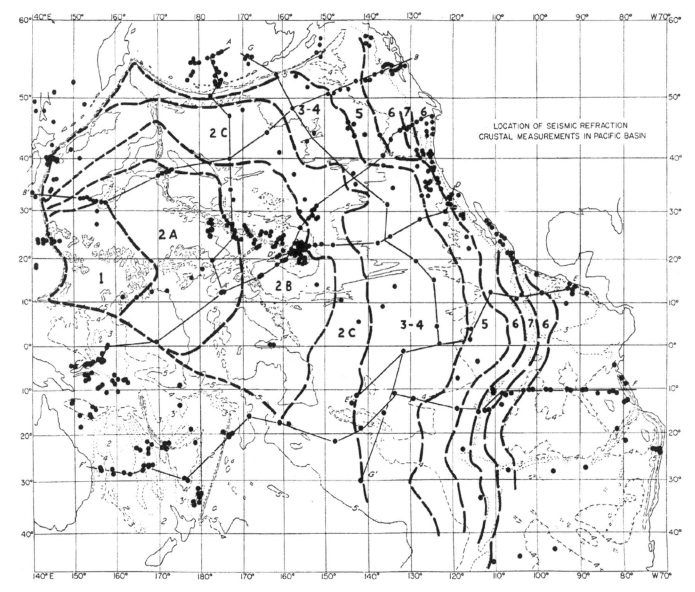

Figure 7.8.2-01. Location map of seismic refraction data measurements in the Pacific Ocean and location of profiles compiled by Woollard (from Woollard, 1975, fig. 2). [Reviews of Geophysics and Space Physics, v. 13, p. 87–137. Reproduced by permission of American Geophysical Union.]

zones as well as the Juan de Fuca Rise. A third line from the Queensland basin eastward (Fig. 7.8.2-02B) shows crustal thickening beneath the Lord Howe Rise (25 km), a slight thickening beneath the Tonga Trench and then a generally thin oceanic crust crossing the Tuamotu Islands and the East Pacific Rise until reaching the Peru-Chile trench.

Numerous investigations were performed by cruises of Soviet research vessels. In 1970 and 1971, the 49th cruise of the oceanographic research vessel *Vitiaz* covered a track length of 3500 miles of continuous reflection profiling using a 1-km-long multichannel streamer. The average penetration depth was 5–7 km; in some places up to 10 km. Deep seismic sounding observations were made along two polygons, which were located in the western part of the Pacific platform (151°W 29°S) and on the

Eauripik Upland (Micronesia) near drill site 62 of the JOIDES Deep Sea Drilling Project.

At the first location—in cooperation with the U.S. research vessel *Mahi* from Honolulu, Hawaii—velocity anisotropy at the top of the mantle was investigated. At the second site—in cooperation with the Japanese vessel *Hokuku Maru*—subcritical reflections from deep boundaries were observed with three OBSs.

In the late 1970s, a large-scale investigation of the Pacific Ocean was undertaken by the Shirshov Institute of Oceanology (Neprochnov, 1989). From 1977 to 1979, basins and rises of the central Pacific were the goal of several cruises of their research vessel. Many deep seismic sounding profiles were recorded using OBSs and big airguns. Figure 7.8.2-03 shows numbers and locations, and Figure 7.8.2-04 shows averaged crustal columns.

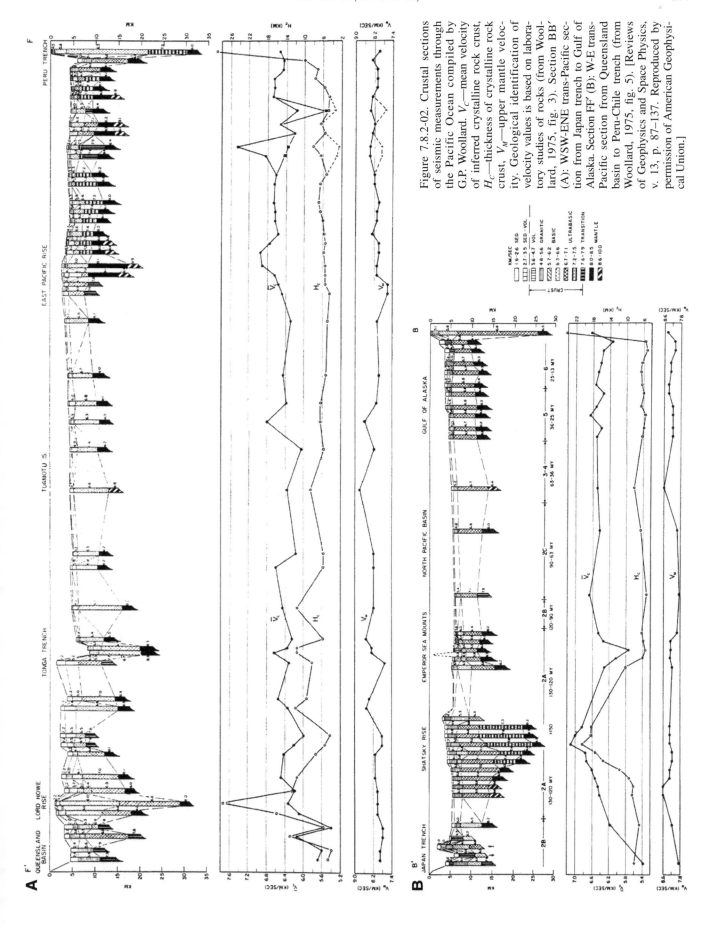

Figure 7.8.2-02. Crustal sections of seismic measurements through the Pacific Ocean compiled by G.P. Woollard. V_C—mean velocity of inferred crystalline rock crust, H_c—thickness of crystalline rock crust, V_M—upper mantle velocity. Geological identification of velocity values is based on laboratory studies of rocks (from Woollard, 1975, fig. 3). Section BB' (A): WSW-ENE trans-Pacific section from Japan trench to Gulf of Alaska. Section FF' (B): W-E trans-Pacific section from Queensland basin to Peru-Chile trench (from Woollard, 1975, fig. 5). [Reviews of Geophysics and Space Physics, v. 13, p. 87–137. Reproduced by permission of American Geophysical Union.]

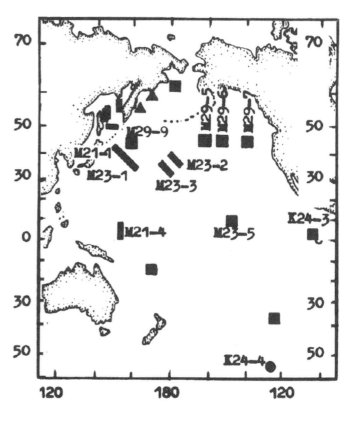

Figure 7.8.2-03. Location of seismic surveys during the various cruises of R/V *Kana Keoki* and R/V *Dmitry Mendeleev* from 1977 to 1979 (from Neprochnov, 1989, fig. 2). [*In* Mereu, R.F., Mueller, St., and Fountain, D.M., eds., Properties and processes of earth's lower crust: American Geophysical Union, Geophysical Monograph 51, p. 159–168. Reproduced by permission of American Geophysical Union.]

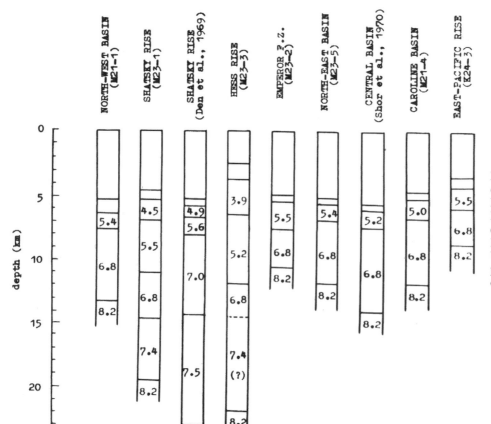

Figure 7.8.2-04. Averaged crustal columns for the Pacific region (from Neprochnov, 1989, fig. 3). [*In* Mereu, R.F., Mueller, St., and Fountain, D.M., eds., Properties and processes of earth's lower crust: American Geophysical Union, Geophysical Monograph 51, p. 159–168. Reproduced by permission of American Geophysical Union.]

In the following, we have selected a few projects to demonstrate the wealth of new data which became available with advanced technology gaining in particular from the development of OBSs and their gradual improvement both in quality and quantity. The majority of these marine experiments aimed for selected problems in tectonically anomalous areas.

In 1976, a marine geophysical survey was conducted in the region of the Explorer spreading center located west of Vancouver Island on Canada's west coast (Cheung and Clowes, 1981). It was an 80-km-long seismic-refraction line using an array of three OBSs and was located northwest of the northern tip of Explorer Ridge and parallel to Revere-Dellwood Fracture Zone in the northeast Pacific Ocean (Fig. 7.8.2-05). Eighty-one explosive charges, alternating large and small shots, were fired, resulting in one reversed profile and two split-spread profiles which overlapped the longer profiles. In addition, airgun shots were fired at

2 minute intervals to run 30-km profiles centered over each OBS in direction and across the explosion line, to add more detailed upper-crustal information. Figure 7.8.2-06 presents the record section obtained from the recordings of OBS1 (see also Appendix A7-10-1). Basically, a normal crustal thickness of 6.5 km resulted, but with an anomalously low P_n velocity of 7.4 km/s. A particular result was the absence of distinct structural discontinuities, even at the Moho.

During June 1974, a seismic-refraction and gravity survey of the East Pacific Rise was carried out near the Siqueiros Fracture Zone at 8°N. Three OBSs recorded shots along two N-S directed profiles west of (crossing the Fracture Zone at the 2.9–5 m.y. boundary) and along the crest of the East Pacific Rise (Orcutt et al., 1976). The charge sizes varied from 0.45 to 436 kg of TOVEX explosives. For the profile along the crest of the East Pacific Rise, the P-velocity increased quite

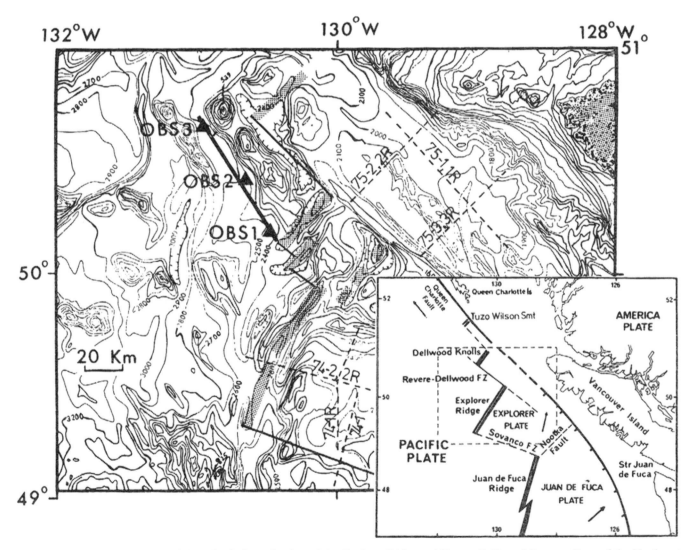

Figure 7.8.2-05. Location map of the seismic investigation of the Explorer Ridge and Revere-Dellwood Fracture Zone of the Northeast Pacific (from Cheung and Clowes, 1981, fig. 1). [Geophysical Journal of the Royal Astronomical Society, v. 65, p. 47–73. Copyright John Wiley & Sons Ltd.]

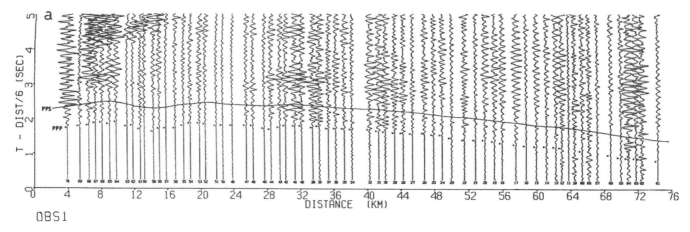

Figure 7.8.2-06. Record section of OBS1 of the Northeast Pacific experiment (from Cheung and Clowes, 1981, fig. 7). [Geophysical Journal of the Royal Astronomical Society, v. 65, p. 47–73. Copyright John Wiley & Sons Ltd.]

rapidly from 5 km/s at the seafloor to 6.7 km/s at the lid of a low-velocity zone. Mantle arrivals from below this zone indicated low P-velocities of 7.6 km/s. A second profile on 2.9 m.y. old crust showed a shallow region of strong velocity gradients which graded into velocities typical of the "oceanic" layer without any clear stratification and again a low mantle velocity of 7.6 km/s. The third profile along 5 m.y. old crust showed more distinct layering, but also velocity gradients within the layers. Here the velocities were more typical for Pacific refraction profiles and a mantle velocity of 8 km/s resulted. Orcutt et al. (1976) predicted that the reflector at 2 km depth beneath the crest of the East Pacific Rise might correspond to the top of a low-velocity magma chamber.

This was also postulated by Rosendahl et al. (1976), who from the same two-ship sonobuoy refraction study in this area concluded that this low-velocity zone is thickest under the summit of the rise, and then thins outward. Radical changes in the structure of the crust would occur within a few million years in this region, and stratification would become more pronounced with age.

In 1976, Lamont-Doherty Geological Observatory performed another investigation of the East Pacific Rise near the Siqueiros Fracture Zone near 9°N (Herron et al., 1978). They acquired ~900 km of digital 24-channel seismic-reflection data with large airguns on three ships crossings of the East Pacific Rise. The multichannel profiles were made aboard the vessel *Conrad* through the use of a 24-group hydrophone streamer, 2400 m long, and four airguns towed behind the ship. The airguns were fired every 50 m. The interpretation revealed three distinct crustal layers emerging as coherent events across the rise crest. The base of the third layer appeared at a depth of 2 km below the seafloor and was interpreted as possible top of a low-velocity magma chamber, as postulated by Orcutt et al. (1976). This layer seemed to thicken away from the rise crest toward the edge of the raised axial block that characterizes the East Pacific Rise according to Anderson and Noltimier (1973).

From a study of the northern Cocos plate, between the East Pacific Rise and the Middle America trench at 13°–15° N latitude, Lewis and Snydsman (1979) concluded that the crust thickens systematically with increasing age, caused by a gradual transformation of the top 2 km of the upper mantle into material having crustal velocities. The corresponding velocity-depth cross section showed a more or less constant crustal body of 5 km thickness slightly dipping toward the east underlain by a transformed mantle wedge changing gradually from 0 km thickness at 100 km distance from the rise axis, where Moho depth is 8 km, to 2 km thickness at ~600 km distance, resulting in a Moho depth of 10.5 km. The experiment was conducted in 1973 and 1974 on the Cocos Plate between the Orozco and Clipperton fracture zones and extended from the East Pacific Rise to the Middle America Trench (12.5°–15°N, 99°–104.5°W). Most of the data were obtained on five long-range telemetering buoys which were generally deployed as free-floating units ~2 km long. In addition, OBS were deployed twice, first near the trench and later near the rise axis.

In 1975, Orcutt and Dorman (1977) performed a long-range profile in the Pacific Ocean. The line was 600 km long (Fig. 7.8.2-07 and Appendix A7-10-1) and was laid out in the northeastern Pacific between the Clarion and Molokai fracture zones. In contrast to similar long-range observations in the western Pacific (see below), Orcutt and Dorman (1977) could not establish any velocities higher than 8.4 km/s down to 60 km depth, but they indicated the possible existence of a velocity inversion, with no less than 8.0 km/s, at around 50 km depth, similarly as had been detected by continental long-range investigations in western Europe (e.g., Hirn et al., 1973, 1975; see above).

In 1979, the Rivera Ocean Seismic Experiment (ROSE; Fig. 7.8.2-08) investigated the crust on the Orozco Fracture Zone of the East Pacific Rise at ~12° N (Ewing and Meyer, 1982; Lewis and Garmany, 1982). It was intended to become a combined sea and land seismic program to study a number of features of the structure and evolution of a mid-ocean ridge, a major oceanic

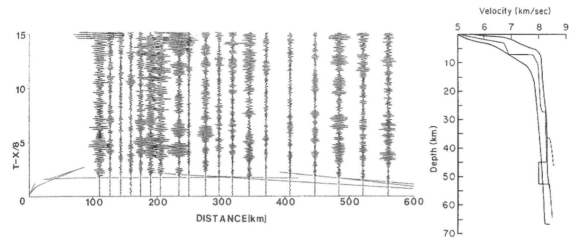

Figure 7.8.2-07. Left: Record section of a long-range seismic refraction profile in the Pacific Ocean (from Orcutt and Dorman, 1977, fig. 5). Right: Velocity-depth function derived from the long-range seismic observations on the Pacific Ocean (from Orcutt and Dorman, 1977, fig. 4). [Journal of Geophysics, v. 43, p. 257–263. Reproduced with kind permission of Springer Science+Business Media.]

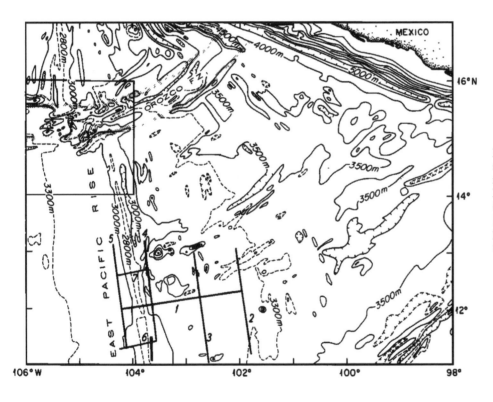

Figure 7.8.2-08. Location map of the seismic Rivera Ocean Seismic Experiment (ROSE) centered on the Orozco Fracture Zone of the East Pacific Rise (from Ewing and Meyer, 1982, fig. 1). [Journal of Geophysical Research, v. 87, p. 8345–8357. Reproduced by permission of American Geophysical Union.]

fracture zone, and the transition region between ocean and continent. Due to the lack of a permit to work within Mexico's 200-mile limit, the project had to be separated into three independent parts: an experiment to study the East Pacific Rise south of the Orozco Fracture Zone primarily using ocean bottom recording and explosive sources, a seismicity program at the Orozco, and a land-based program of recording natural events along the coastal region of Mexico. Due to the need to shift the marine part away from Mexican waters, only one strong event, the 7.6 Petatlan

earthquake of 14 March 1979, at 17.5°N, 101.5°W, and some of its fore- and aftershocks could be recorded jointly by both land and marine instruments.

As part of the project, a series of multichannel seismic-reflection profiles were run, and along these lines, military sonobuoys were routinely deployed and recorded on a digital seismic recording system along the reflection profiling. The seismic energy was produced by an array of airguns fired every 20 s (ca. 50 m). These typically provided seismogram data to a range of ~20 km from which crustal structure to a subseafloor depth of 2.5 km could be deduced. In addition to the sonobuoy data sets, three expanded spread profiles were collected. In these experiments involving three ships, which were alternatively shooting or receiving, explosive charges were fired on a 5 minute schedule (ca. 750 m). Data were recorded up to 50 km distance allowing researchers to deduce crustal structure to a depth of 5 km below the seafloor. The data were analyzed separately for details of the upper crust (Ewing and Purdy, 1982; Purdy, 1982) and for variations in crust and upper mantle structure (Gettrust et al., 1982). A detailed reinterpretation followed in 1988 (Vera and Mutter, 1988). In general, the overall structure of the crust consists of a rapid increase in velocity from ~2.5–6 km/s within the uppermost 1 km of crust (layer 2A), a zone of maximum 1 km thickness with little or no velocity gradient (layer 2B), a thin transition layer with a relatively rapid increase of velocity from ~6.2–7 km/s (layer 2C), and a thick nearly homogeneous basal layer with velocities around 7.2 km/s (layer 3). The reanalysis of these data (Vera and Mutter, 1988) allowed researchers to visualize aspects of structural variability of the lower oceanic crust which were not apparent in earlier interpretations (e.g., Gettrust et al., 1982).

Spudich and Orcutt (1980) presented a new picture of the oceanic crust based on progress in instrumentation and experimental design, as well as on advances in mathematical and computational seismology. In their new picture, layer 2 appeared as a region where velocity increases rapidly with depth. Layer 3 seemed more homogeneous in the vertical direction than layer 2, with gentle velocity gradients and occasional velocity inversions. Although it was observed at several sites, they doubted the widespread existence of a high-velocity basal crustal layer. The seismic picture of the crust appeared in good agreement with seismic velocities of rocks gained from direct drilling and laboratory velocity studies on both oceanic and ophiolitic samples.

An international conference, held in 1974 in Honolulu, Hawaii, to honor G.P. Woollard, led to the publication of a special volume on the geophysics of the Pacific Ocean basin and its margins (Sutton et al., 1976). The papers presented there dealt in particular with the continent-ocean transition which, both in the east under South America and in the west under Japan, appeared as the subduction of an oceanic plate underneath a continent.

In the South Pacific, a major project dealing with the subduction of the Pacific Ocean under the Colombian Andes of the South American continent was Narino III, an onshore-offshore experiment launched in 1973 between latitudes 2° and 4°N (Meyer et al., 1976). Another land-sea operation across the coast of Colombia followed in 1978 between the latitudes 5° and 6°N (Flueh et al., 1981). Details of both projects were discussed above in subchapter 7.7.3 on South America.

Hussong et al. (1976) dealt with the crustal structure of the Peru-Chile trench. Five standard two-ship reversed explosion lines defined the crustal velocity structure, and also served as end points for a 375-km-long refraction line comprised of 11 extended overlapping split profiles. Furthermore, a commercial reflection line was available, and twelve Airgun Sonobuoy Precision Echo Recorders (ASPER) supplemented the survey carried out from 1972 to 1973 (Fig. 7.8.1-09). Typical Moho depths near 10 km were obtained west of the trench, while under the continental shelf a crustal thickness of 30 km was found. The data were used by the authors to develop a complex model of the convergence of the Nazca plate with the South American continental margin (Fig. 7.8.1-10), with the Nazca plate appearing with higher velocity crust than typically found for other oceanic plates.

Two crustal investigations dealt with the Hawaiian archipelago, where the island of Hawaii with its active volcano Kilauea was investigated in detail. In 1976 and 1978, the U.S. Geological Survey established two seismic-refraction profiles on and off the Hawaii Islands with offshore shots of 100 km length each (Fig. 7.8.2-11). The first profile was directed southeastward normal to

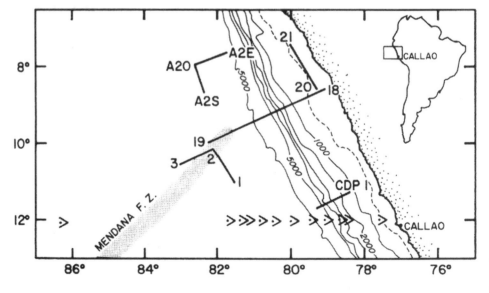

Figure 7.8.2-09. Location of seismic profiles around the Peru-Chile trench between 8° and 12°S latitude (from Hussong et al., 1976, fig. 1). The dot and arrow symbols at 12°S latitude assign the ASPER stations. [In Sutton, G.H., Manghnani, M.H., Moberly, R., and McAffe, E.U., eds., The geophysics of the Pacific Ocean basin and its margin: American Geophysical Union, Geophysical Monograph 19, p. 71–85. Reproduced by permission of American Geophysical Union.]

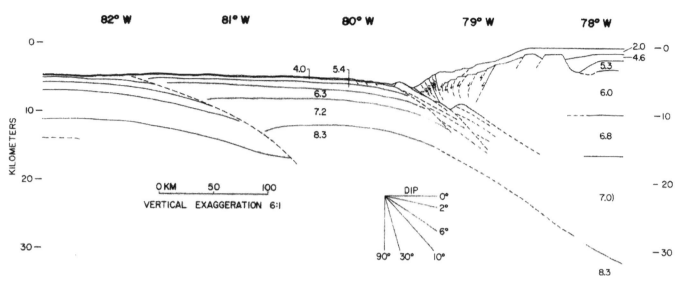

Figure 7.8.2-10. Composite crustal section across the Peru-Chile trench (from Hussong et al., 1976, fig. 12). [*In* Sutton, G.H., Manghnani, M.H., Moberly, R., and McAffe, E.U., eds., The geophysics of the Pacific Ocean basin and its margin: American Geophysical Union, Geophysical Monograph 19, p. 71–85. Reproduced by permission of American Geophysical Union.]

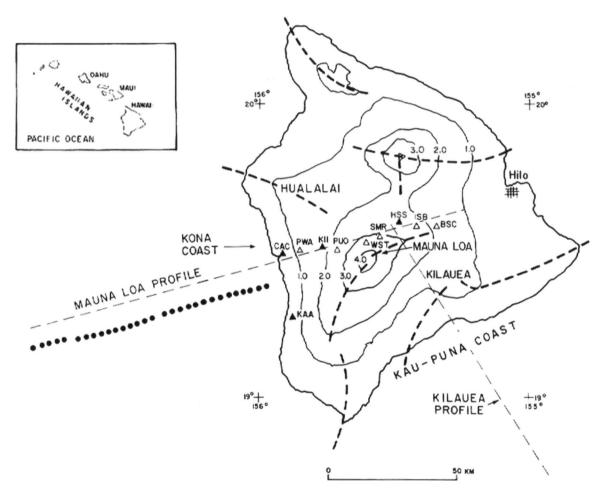

Figure 7.8.2-11. Location of the 1976 Kilauea and 1978 Mauna Loa profiles (from Zucca et al., 1982, fig. 1). Dots and triangles are shots at stations of the 1978 experiment. [Bulletin of the Seismological Society of America, v. 72, p. 1535–1550. Reproduced by permission of the Seismological Society of America.]

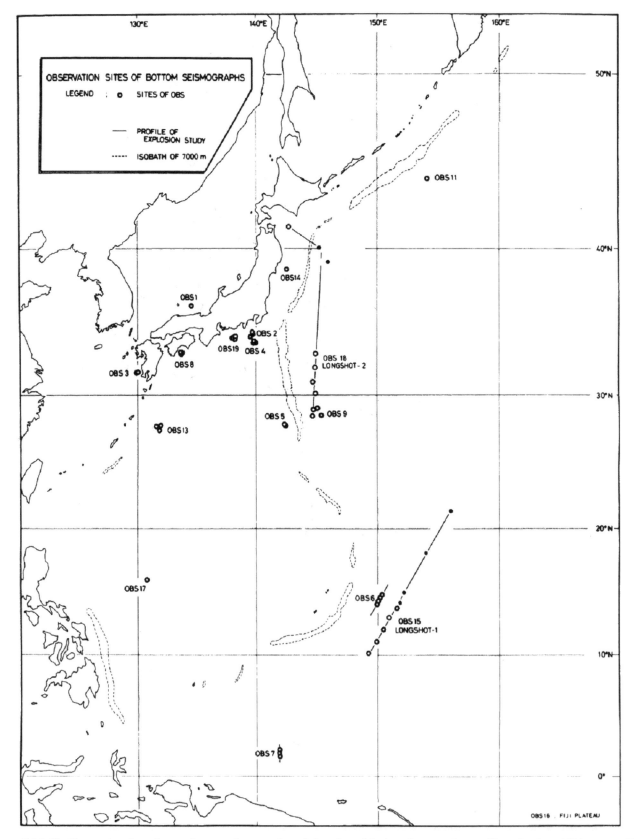

Figure 7.8.2-12. Observation sites of ocean-bottom seismographs and shot sites of the longshot experiments (from Asada and Shimamura, 1976, fig. 1). [*In* Sutton, G.H., Manghnani, M.H., Moberly, R., and McAffe, E.U., eds., The geophysics of the Pacific Ocean basin and its margin: American Geophysical Union, Geophysical Monograph 19, p. 135–154. Reproduced by permission of American Geophysical Union.]

the southeast coast across the submarine flank of Kilauea volcano (Zucca and Hill, 1980), with 43 explosive charges oriented along an offshore line roughly 100 km long and oriented normal to the southeast flank of Kilauea.(Kilauea profile in Fig. 7.8.2-11).

The second profile (Mauna Loa profile in Fig. 7.8.2-11) was directed toward WSW normal to the Kona coast and contained 30 explosive charges evenly distributed along the 100-km-long

offshore part (Zucca et al., 1982). Both profiles were recorded by the permanent seismic network, by portable 5-day recorders, and by OBSs (for details, see Appendix A7-10-2).

The objectives were to provide an improved resolution of the structure of the Mauna Loa and Kilauea volcanoes with emphasis on the nature of the transition from the "normal" oceanic crust to the massive volcanic pile. On both profiles a similar pattern

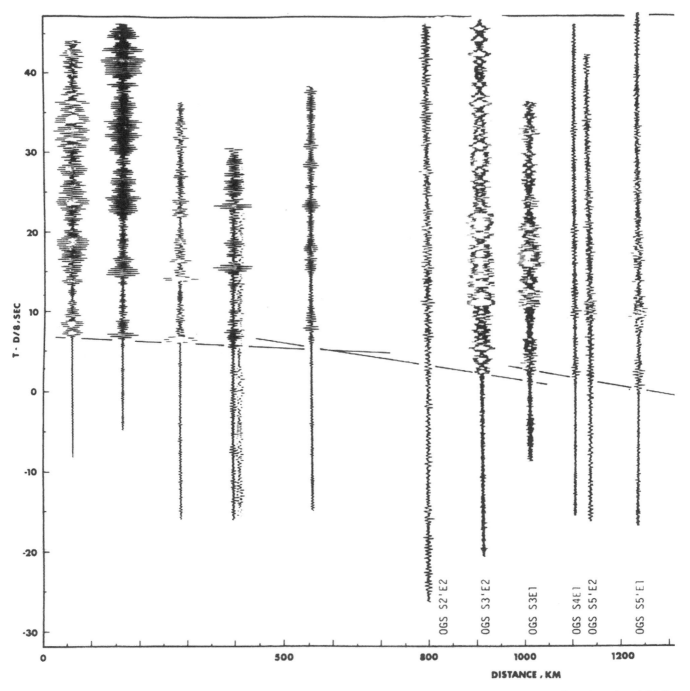

Figure 7.8.2-13. Record section of the longshot experiment (from Asada and Shimamura, 1976, fig. 9). [*In* Sutton, G.H., Manghnani, M.H., Moberly, R., and McAffe, E.U., eds., The geophysics of the Pacific Ocean basin and its margin: American Geophysical Union, Geophysical Monograph 19, p. 135–154. Reproduced by permission of American Geophysical Union.]

could be seen: The oceanic crust begins to thicken landwards at the break of the slope between the volcanic pile and the ancient sea floor. From here, the Moho dips gently toward the land. At the coast, the Moho is 14 km deep and it reaches a maximum depth of 17–18.5 km under the center of the volcanoes.

Japanese research activities in the 1970s involved large off-shore shot experiments in the western Pacific Ocean and in the Sea of Japan, carried out within the International Geodynamic Project. The aim was to elucidate lateral velocity variations in the upper mantle from the trench area to the back arc basin (Research Group for Explosion Seismology, 1975; Okada et al., 1978, 1979; Yoshii et al., 1981).

Two Longshot experiments were carried out in 1973 and in 1974 in the western Pacific basin. They aimed mainly for the structure of the subcrustal lithosphere and less for details of crustal structure (Asada and Shimamura, 1976, 1979).

For their seismic investigation of the structure of the western Pacific and its relation to Japan, Asada and Shimamura (1976, 1979) used large underwater explosions which were re-

corded by OBSs up to distances of 1300 km (Figs. 7.8.2-12 and 7.8.2-13). Since the ambient noise in deep oceanic basins turned out to be greater in the optimum frequency range (below 10Hz), 1.5-ton shots proved to be too small for obtaining clear signals, but 5- to 7-ton shots proved to be adequate for the Longshot experiments. The ocean-bottom seismographs, being used by the Japanese institutions since 1969, were designed to be self-contained in a small capsule and to withstand a depth of 9000 m.

Their preferred model 115 (Fig. 7.8.1-14) shows, underneath a 17-km-thick crust, a structure of the subcrustal lithosphere, similar to models obtained for long-range profiles in western Europe. From 1974 to 1980, further intensive long range experiments were undertaken in the northwestern Pacific, which detected large-scale lithospheric structure and anisotropy in the upper mantle (Asada and Shimamura, 1979; Asada et al., 1983; Shimamura et al., 1983).

Two other long-range explosion experiments using OBSs were carried out in the East Mariana Basin in 1973 and 1976 (Fig. 7.8.2-15). Twenty-five OBS stations were deployed along

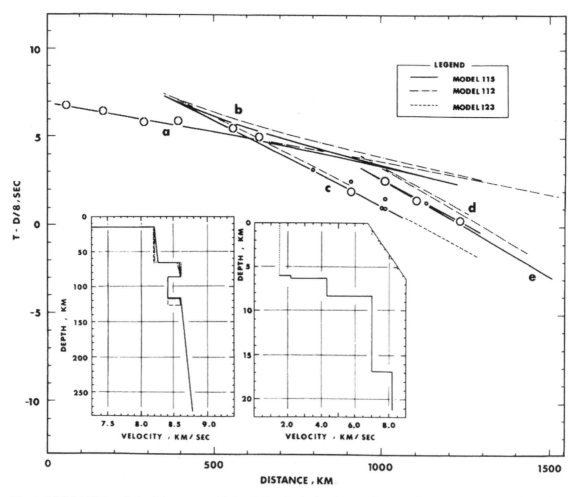

Figure 7.8.2-14. Reduced travel time curves of observed and calculated values for models shown below (from Asada and Shimamura, 1976, fig. 9). [*In* Sutton, G.H., Manghnani, M.H., Moberly, R., and McAffe, E.U., eds., The geophysics of the Pacific Ocean basin and its margin: American Geophysical Union, Geophysical Monograph 19, p. 135–154. Reproduced by permission of American Geophysical Union.]

an 800-km-long line and seven large (1.5–8.5 tons) shots and several tens of small shots were fired. The maximum recording range was 1900 km.

The interpretation of the data indicated significant differences in the structure of the lower lithosphere. The attenuation of seismic waves appeared to be larger than east of Honshu and the low-velocity layer below the Moho seemed more accentuated. Nagumo et al. (1981) confirmed the result that the sub-Moho velocity layer, with 8.4 km/s velocity, existed, but it appeared to be extremely thin. Alternating high-velocity and low-velocity layers down to 90 km depth were modeled. Possible models for the upper low-velocity zone and the corresponding data are shown in Figure 7.8.2-16 (see also Appendix A7-10-1).

In the southwestern Pacific in 1973, the seismic investigation of the major structures of the Papua Ultramafic Belt, which

had started in the 1960s, was continued. The ophiolite suite of rocks was widely considered to be a classic example of fragment of oceanic crust and upper mantle, obducted on to a continental mass during orogenesis in an area of continent/island arc convergence (e.g., Moores, 1973).

The intensive seismic investigation of the Papuan peninsula and its adjacent seas consisted of offshore-onshore operations in the Gulf of Papua and both the Coral Sea and Solomon Sea with recording distances up to 300 km and more (Drummond et al., 1979; Finlayson et al., 1976, 1977; Fig. 7.8.2-18; Appendix A7-10-3). Forty-two seismic stations were deployed in Papua New Guinea, and 111 shots were fired from the surrounding seas (Fig. 7.8.2-17). The shots had charges of 180 kg or 1000 kg and were detonated 100 m below the water surface, wherever water depth permitted. During the 1973 survey, a nuclear explosion on

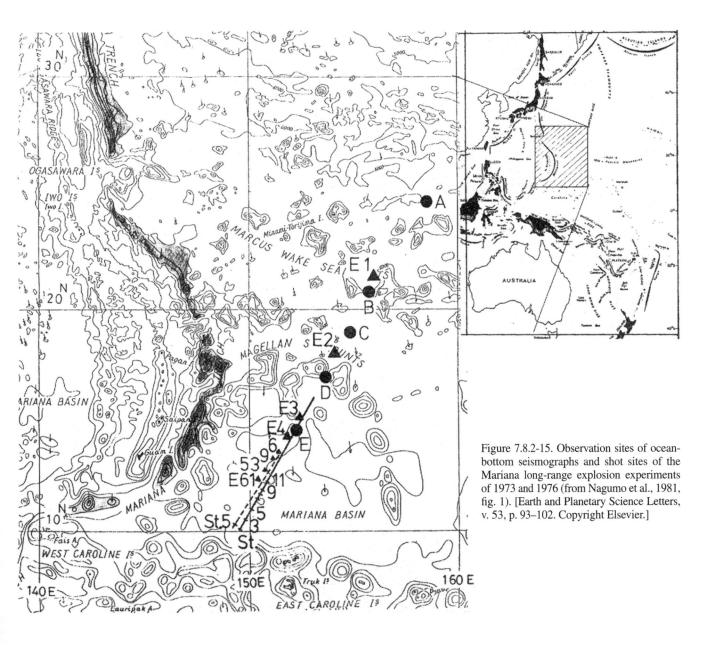

Figure 7.8.2-15. Observation sites of ocean-bottom seismographs and shot sites of the Mariana long-range explosion experiments of 1973 and 1976 (from Nagumo et al., 1981, fig. 1). [Earth and Planetary Science Letters, v. 53, p. 93–102. Copyright Elsevier.]

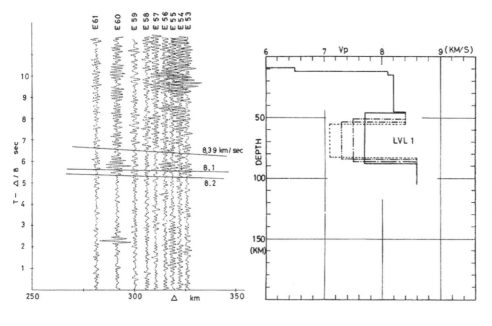

Figure 7.8.2-16. Part of record section (left) and models for upper low-velocity zone in the subcrustal lithosphere (right) of the Mariana long-range explosion experiments (from Nagumo et al, 1981, figs. 3 and 4a). [Earth and Planetary Science Letters, v. 53, p. 93–102. Copyright Elsevier.]

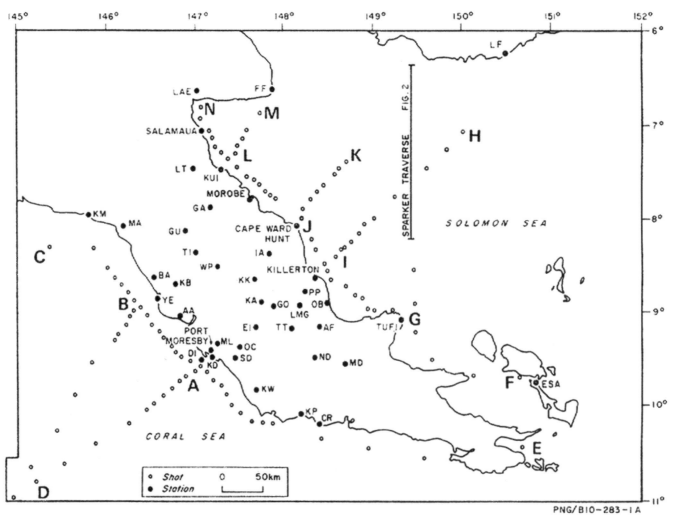

Figure 7.8.2-17. Location of shots (open circles) and recording sites (full circles) of the 1973 East Papua crustal survey (from Finlayson et al., 1976, fig. 3). [Geophysical Journal of the Royal Astronomical Society, v. 44, p. 45–60. Copyright John Wiley & Sons Ltd.]

Novaya Zemlya also was successfully recorded by 26 stations in east Papua (Finlayson, 1977).

The thickness of crustal material offshore ranges from 33 km in the western Solomon Sea (area L-M-N in Fig. 7.8.2-17) to 13 km in the central Solomon Sea (line H–I in Fig. 7.8.2-17). Along the southwest Papuan peninsula coast, a crustal thickness of 27–29 km was interpreted, which increased only slightly inland by 3–5 km until underthrusting takes place in the area of the Papuan Ultramafic Belt. The Moho appeared to shallow under the Moresby Trough of the Coral Sea (southwest of line A–B in Fig. 7.8.2-17) to 19 km and to deepen to 25 km under the Eastern Plateau in the center of the Coral Sea (around point D in Fig. 7.8.2-17). The results suggested a continental crust under the Papuan peninsula and the Eastern Plateau of the central Coral Sea, but an oceanic crust in between, underneath the Moresby trough (Fig. 7.8.2-19).

In 1978, a map was published showing the locations of marine crustal studies throughout the East and Southeast Asian Seas and the velocity-depth columns obtained. The map covered the area between latitudes 15°S and 45°N and longitudes 90°E and 150°E. The velocity columns contain velocity and thickness of both sedimentary layers and the layers of the crystalline crust and uppermost-mantle velocities as far as reached by the individual surveys. The map was compiled by Hayes et al. (1978).

Two major crustal surveys investigated the continental margins and adjacent seas of Australia. In 1976, the Scripps Institute of Oceanography, USA, conducted a seismic-reflection and wide-angle seismic survey in the Banda Sea north of Darwin using explosive seismic sources (Jacobson et al., 1979; Finlayson 2010; Appendix 2-2). The interpretation showed a crust of more than 30 km thickness north of Melville Island, decreasing slightly toward north to 30 km near the Timor Trough. The survey included observations across the Australian continental shelf between Darwin and Timor. In addition, Rynn and Reid (1983) recorded the shots fired in the Timor Sea along a 300-km-long offshore line on the continental shelf using OBSs. They determined a crustal thickness of 34 km. The offshore seismic shots were also recorded along a land profile between Darwin and Tennant Creek providing an upper-mantle velocity of 8.2 km/s and a crustal thickness of 45 km under the Pine Creek Geosyncline of northern Australia (Hales and Rynn, 1978; Hales et al., 1980).

The second marine survey of 1976 touched the margin of southern Australia, investigating the crustal structure within the Magnetic Quiet Zone of the Great Australian Bight. The

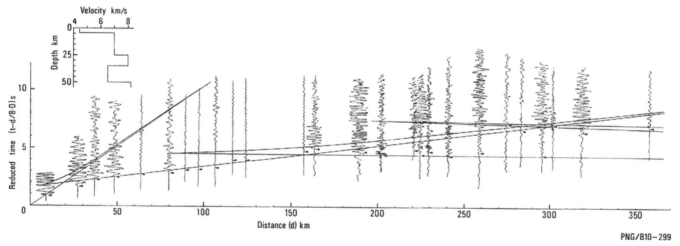

PNG/B10-299

Figure 7.8.2-18. Record section of shots along the northeastern coast of the Papuan peninsula recorded between stations LAE and TUFI during the East Papua crustal survey (from Finlayson et al., 1976, fig. 6). [Geophysical Journal of the Royal Astronomical Society, v. 44, p. 45–60. Copyright John Wiley & Sons Ltd.]

Figure 7.8.2-19. Seismic model of the Papuan ultramafic belt along line D-H of Figure 7.8.2-17 (from Finlayson et al., 1977, fig. 9a). The location of the line D-H of Fig. 7.8.2-17 was renamed by Finlayson et al. (1977) into A–B. [Physics of the Earth and Planetary Interiors, v. 14, p. 13–29. Copyright Elsevier.]

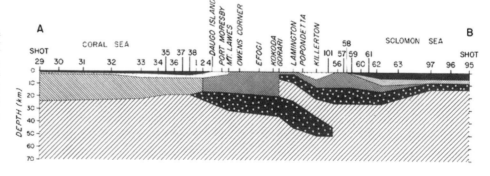

marine seismic work was carried out by the Lamont-Doherty research vessel RV *Vema* (Talwani et al., 1979; Finlayson 2010; Appendix 2-2). Using sonobuoys, both refraction and reflection data were recorded applying both air gun and explosive sources with charges up to 30 kg. The resulting crustal thickness was 12–13 km, but the velocities were typical of oceanic crust.

7.8.3. The Indian Ocean

From the available and known literature, it appears that in the 1970s, only a limited amount of research was carried out in the Indian Ocean. Most of this work concentrated on the southwestern part adjacent to Africa. Some work had been carried out by Hales and Nation (1973b) in 1968 beneath the Transkei–Natal plateau and the Mozambique plateau (see Chapter 6.8.3).

In the eastern Indian Ocean, Kieckhefer et al. (1980) did a detailed crustal structure survey in 1977 which extended from the Sunda trench axis, across Nias island ridge, to the forearc basin near the coast of Sumatra. Six refraction lines were shot parallel to the regional topography (Fig. 7.8.3-01), with one line in the

trench axis, two lines on the inner slope, and one line each on the Nias Ridge, the Nias Basin, and close to the coast of Sumatra. Both explosive charges and airguns provided energy recorded by arrays of sonobuoys.

A data example of long-range shots, recorded at station 1308 (Fig. 7.8.3-01) is shown in Figure 7.8.3-02. More data are shown in Appendix 7-10-4. The main feature of the interpretation of Kieckhefer et al. (1980) is a wedge of 4.7–4.9 km/s material seaward of the Nias Island ridge which underlies low-velocity sediments and thickens from 1.2 km beneath the trench axis to 13.5 km immediately seaward of the ridge (Fig. 7.8.3-03).

Tucholke et al. (1981) reported on new wide-angle reflection and refraction measurements in 1978 on the southern Agulhas Plateau 500 km off Cape Horn in the southwestern Indian Ocean, using an ocean bottom hydrophone, commercial long-range sonobuoys, and short-range military sonobuoys. Both explosives ranging from 2.5 to 100 kg and airgun shots were recorded up to distances of 72 km. Two reversed and six unreversed profiles were recorded over both smooth and irregular basement areas. Velocities of the deepest crustal layer

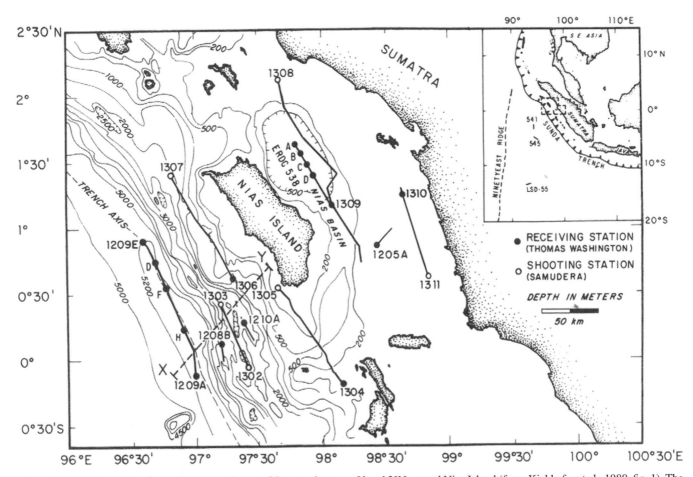

Figure 7.8.3-01. Location of seismic lines southwest of Sumatra between 0° and 2°N around Nias Island (from Kiekhefer et al., 1980, fig. 1). The contour interval of the bathymetric map is 500 nominal meters. [Journal of Geophysical Research, v. 85, p. 863–889. Reproduced by permission of American Geophysical Union.]

beneath the southern plateau ranged from 6.9 to 7.1 km/s and agreed with the basal crustal layer (6.6–7.0 km/s) of earlier models of the plateau and the adjacent Transkei Basin (e.g., Hales and Nation, 1973b). The thickness of this oceanic layer averaged 4 km or less in the Transkei Basin, but was at least 7–9 km thick under the Agulhas Plateau.

Goslin et al. (1981) investigated the anomalous crust of the Madagascar ridge south of Madagascar and the submarine Crozet plateau in 1978, located at ~45°S beyond the southwest Indian Ridge. Charges shot at intervals of 1–5 km and ranging from 25 to 400 kg were recorded on two seismic-refraction profiles of ~160 km length along the longitude of 45°E and north and south of latitude 32°S. On the Crozet Plateau, located to the south near 44°S, a 40-km-long refraction profile was recorded with shot intervals of 400 m. The deep structure of the Madagascar ridge resembled that of the submarine *Crozet Plareau*, corresponding to a strongly anomalous oceanic crust, Moho depths were 22–25 km under the northern and 14.5 km under the southern profile along the Madagascar ridge and 17.5 km under the Crozet Plateau; the P_n velocity was 8.0–8.1 km/s.

7.8.4. The Atlantic Ocean

As mentioned in the Introduction (section 7.8.1), seismic research in the 1970s concentrated mainly on particular structures, such as ridges, island arcs, continental margins, etc. This is evidently true also for the Atlantic Ocean.

Several surveys were carried out to study island chains off the western coast of Africa. The Cape Verde islands and emerged portions of a Mesozoic-Cenozoic volcanic accretion along fracture zones converging from the mid-Atlantic ridge toward Africa were both investigated by Dash et al. (1976).

The continental slope and the Cape Verde rise off Mauretania, an inactive margin with a sedimentary cover of 6 km and more, were the targets of a marine crustal survey by the University of Hamburg (Weigel and Wissmann, 1977).

The structure of the Canary Islands was investigated by a joint Spanish–Swiss research project (Banda et al., 1981a). Several *Meteor* (1964) expeditions—M33 in 1974 (Hinz et al., 1974), M39 in 1975 (see also Chapter 7.2.3; Makris et al., 1985), and M46 in 1977 (Sarnthein et al., 2008)—explored the continent-ocean transition west off Africa.

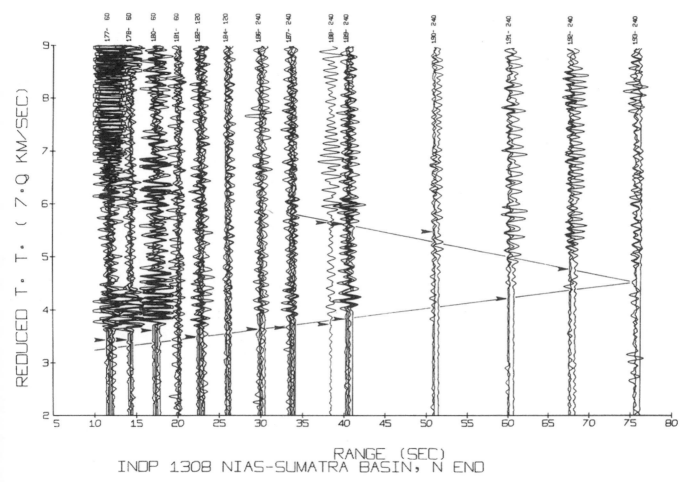

Figure 7.8.3-02. Record section for long-range shots (from Kiekhefer et al., 1980, fig. 4), recorded by array station 1308 (location shown in Figure 7.8.3-01). [Journal of Geophysical Research, v. 85, p. 863–889. Reproduced by permission of American Geophysical Union.]

The Mid-Atlantic Ridge was the goal of various seismic surveys, only a few of which are mentioned here. In 1977, a seismic-refraction experiment studied the crustal structure of the Kane Fracture Zone near 24°N, 44°W (Detrick and Purdy, 1980). Shooting lines and receivers formed a T configuration across the fracture zone, with two receivers located ~50 km apart in the fracture zone trough and the remaining eight receivers positioned 25–30 km apart on either side of the fracture zone (dashed lines in Fig. 8.9.4-01, showing both the 1977 and 1982 surveys; see Chapter 8.9.4). The data did not reveal any significant difference in crustal thickness or velocities on both sides

of the fracture zone and were in the range typically observed for oceanic crust. Anomalously thin crust of only 2–3 km thickness, however, was found beneath the Kane fracture zone trough proper, extending over a width of at least 10 km. Further investigations were added in 1982 (see Fig. 8.9.4-01), when three more lines were shot across the Kane fracture zone (Cormier et al., 1984; see Chapter 8.9.4).

In 1973 within the Franco-American Mid-Ocean Undersea Study (FAMOUS) during cruise 54 of RRS *Discovery*, four seismic-refraction profiles were recorded along, across, and parallel to the median valley at 37°N (Whitmarsh, 1975; Fig. 7.8.4-01;

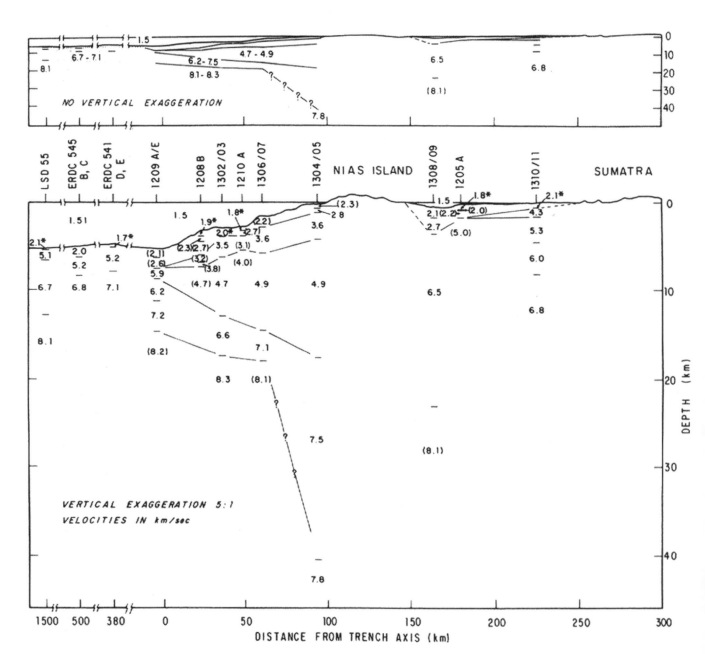

Figure 7.8.3-03. Summary cross section from the trench axis to Sumatra (from Kiekhefer et al., 1980, fig. 8), using depths of the midpoints of the reversed refraction profiles. [Journal of Geophysical Research, v. 85, p. 863–889. Reproduced by permission of American Geophysical Union.]

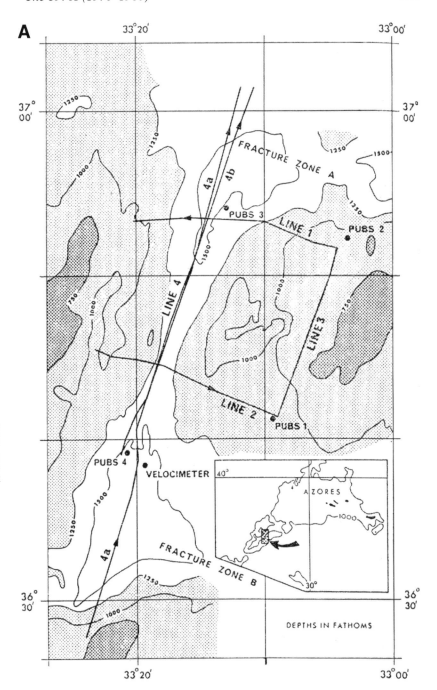

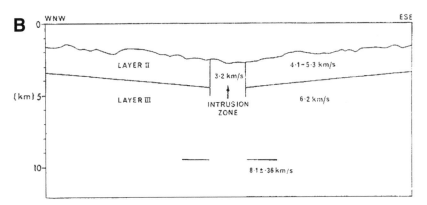

Figure 7.8.4-01. Location of seismic lines across the Median Valley of the Mid-Atlantic Ridge at 37°N (A) and model (B) along line 1 (from Whitmarsh, 1975, figs. 1 and 18). [Geophysical Journal of the Royal Astronomical Society, v. 42, p. 189–215. Copyright John Wiley & Sons Ltd.]

Appendix 7-10-5). Three of the lines had shots every 250 m. The shots were either explosive charges ranging from 2.5 to 100 kg, or were provided by a 1000 cubic inch airgun fired at 15 m depth every 2 minutes. As receivers, pop-up bottom seismic recorders were used, which recorded by means of a hydrophone and a tape recorder. Most arrivals were from a main refraction of apparent velocity of 5.4–6.3 km/s, only beyond a 35 km shotpoint distance were faster arrivals observed from an 8.1 km/s refractor. The main refractor corresponded in depth to approximately layer 3 of ocean basins, but its velocity was significantly less, possibly due to dip. A low-velocity layer with 3.2 km/s was found by lines 1 and 2 to underlie the whole width of the median valley floor. A problem remained as to how to reconcile the observation of

the 6.3 km/s refractor along line 4 with the presence of the low-velocity zone as indicated by lines 1 and 2 (Fig. 7.8.4-01). Details of this survey were also described by other authors (Poehls, 1974; Fowler, 1976).

From 40° to 43°N near the Azores, four lithospheric profiles were recorded in 1974 and 1975 to unravel the structure of the lithosphere near the Mid-Atlantic Ridge (Steinmetz et al., 1977; Fig. 7.8.4-02). Crustal structure was obtained from an east-west profile which crossed the ridge axis.

Although energy was propagated across the ridge axis within the crust, the axial region marked a clear barrier to waves propagating in the mantle. Energy was successfully recorded up to 450 km distance in the strike direction of the ridge on a profile

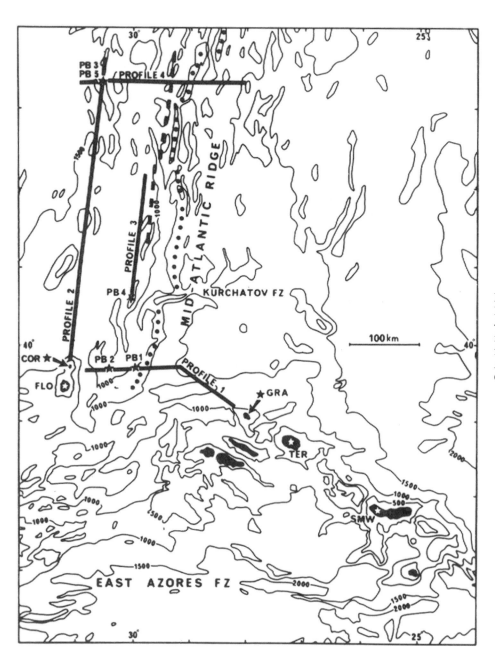

Figure 7.8.4-02. Location of seismic lines around the Mid-Atlantic Ridge at 40° to 43°N to the northwest of Azores islands (from Steinmetz et al., 1977, fig. 1). [Geophysical Journal of the Royal Astronomical Society, v. 50, p. 353–380. Copyright John Wiley & Sons Ltd.]

parallel to the axis along the 4 m.y. isochron. Here, below a layer with velocity 7.6 km/s, a low-velocity zone was found which was underlain by an 8.3 km/s refractor at 9 km depth below the sea bed (Fig. 7.8.4-03; see also Appendix 7-10-5). On another profile parallel to the rift axis, recorded along the 9 m.y. isochron, a velocity of 8.3 km/s was found at a much greater depth of 30 km.

Another survey investigated the crust along the Azores–Biscay rise (Whitmarsh et al., 1982). In 1978, a set of seismic-reflection and -refraction data was obtained in the vicinity of the rise. The seismic-refraction profiles were shot using pop-up bottom seismographs. Out to a range of 25 km, airgun firing provided shots every 2 minutes. At greater ranges, explosive charges of 11, 45, and 136 kg were used and recorded up to ~85 km distance. On all profiles, layer 2 with velocities of 5.7–6.0 and layer 3 with an average velocity of 6.7 km/s were underlain by an upper mantle with velocities of 7.8–8.2 km/s, but under one profile, a low-velocity zone on top of the mantle was found. Moho depth averaged around 13–14 km.

Fowler (1978) presented models for the crust beneath the Mid-Atlantic Ridge near 45°N. The above-mentioned 1973 seismic study of the FAMOUS area (37°N; Fowler, 1976) was continued in 1975 on RRS *Discovery* cruise 73 at 45°N. The purpose was to investigate the deep structure of the ridge crest. Seismographs were deployed west of the median valley, and shots were fired up to 120 km east of the valley. The experiment followed a refraction experiment shot by Keen and Tramontini (1970) in 1968 on the western flanks of the Mid-Atlantic Ridge at 45°N. They had obtained a "normal" oceanic crust yielding velocities of 4.6, 6.6, and 8.1 km/s for the first P-wave arrivals within ~30 km of the rift axis, but at 10 km from the western edge of the median valley, the first arrivals resulted in a 7.5 km/s velocity (see Chapter 6.8.2). During the experiment of 1975, shots of 50, 100, and 150 kg were shot along three lines and recorded by five OBSs deployed along and west of the median valley. The first line was

shot along the median valley and was 75 km long. A second line, also 75 km long, was recorded diagonally to the other two lines. For both lines, at the northern end of the median valley four internally recording sonobuoys were laid at 2 km intervals. The third line crossed the rift perpendicular to the rift axis; the majority of the shots were located east of the rift axis out to distances of near 120 km. On this line, an additional three sonobuoys were deployed at 2 km intervals at the eastern end of the line. The resulting traveltime data obtained from this experiment agreed well with those of Keen and Tramontini (1970), confirmed the presence of a normal oceanic upper-mantle velocity within a few million years of the ridge axis, and also seemed to confirm the presence of an anomalous upper mantle with lower velocities close to the ridge axis. The crustal structure appeared to be similar to the FAMOUS area at 37°N. At the ridge axis, an absorptive zone in the upper mantle was apparent; the depth below the seafloor to the top of this zone was determined to be ~6 km. Away from the rift axis, a positive velocity gradient of 0.04–0.05 km/(skm) was seen in the top 5–8 km of the upper mantle.

In the same area, at 45°N of the Mid-Atlantic Ridge, White and Whitmarsh (1984) reported on an investigation of seismic anisotropy due to cracks in the upper oceanic crust. The experiment was conducted in 1978 in an area centered ~28 km east of the median valley of the Mid-Atlantic Ridge at 45°25′N. The receivers were three 3-component ocean-bottom seismographs with hydrophones. The seismic source, two 16-l airguns, were towed along a shooting track, which gave an even azimuthal coverage at ranges of 4–10 km and 13–17 km, producing P-waves from layer 2 and from the lower crust, respectively. The result was a normal oceanic crustal velocity structure which could be explained by decreasing numbers of open cracks with increasing depth through the upper 2 km of the crust.

Seismic-refraction profiles across the continent-ocean transition, at the northern margin of the Bay of Biscay, identified an

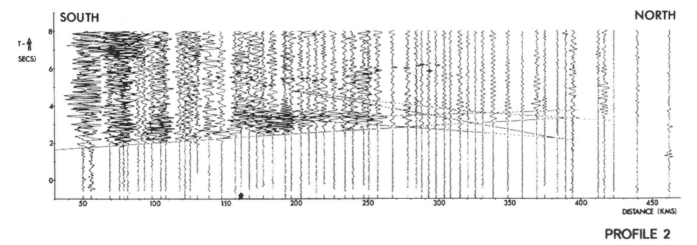

PROFILE 2

Figure 7.8.4-03. Record section of the north-south directed profile 2 west of the Mid-Atlantic Ridge at 40° to 43°N (from Steinmetz et al.,1977, fig. 5a). Data were observed on the island of Flores (FLO in Fig. 7.8.3-02). [Geophysical Journal of the Royal Astronomical Society, v. 50, p. 353–380. Copyright John Wiley & Sons Ltd.]

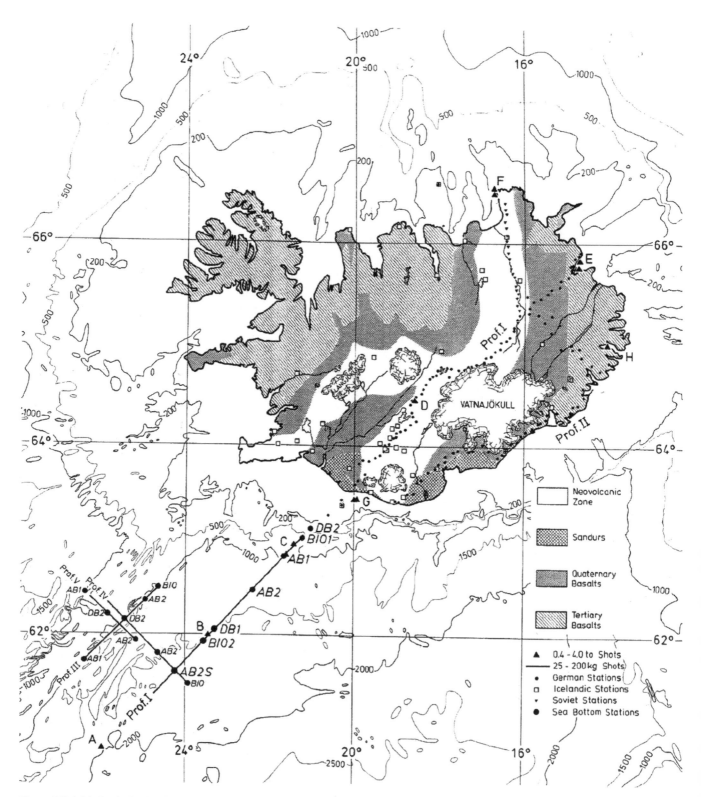

Figure 7.8.4-04. Geologic sketch of Iceland and bathymetric chart of the surrounding ocean showing shotpoints and recording sites of the RRISP project (from RRISP Working Group, 1980, fig. 1). [Journal of Geophysics, v. 47, p. 228–238. Reproduced with kind permission of Springer Science+Business Media.]

8-km-wide progressive thinning of the continental crust from 33 km thickness to ~5 km. In 1975 and 1976, several cruises were undertaken into the Bay of Biscay to shoot profiles, successfully using the newly constructed ocean-bottom seismographs under rough field conditions. More tests followed in 1976 and 1977 in the Norwegian Sea (Avedik et al., 1978).

The Hebridean Margin Seismic Project (Bott et al., 1979) targeted the crust west of northern Scotland and north of Ireland. During summer 1975, a line of large shots was fired across the continental margin between the Rockall Trough and the Hebridean shelf along latitude 58°N. Another line of large shots extended in an approximately N-S direction around longitude 8°W from the 58°N line southward to northern Ireland. The shots were recorded both across Scotland and in northwestern Ireland. The data indicated a 27-km-thick continental crust beneath the Hebridean shelf and a depth to Moho of 18 km beneath the Rockall Trough at 58°N.

Of particular interest to earth scientists in the 1970s was evidently the northernmost part of the Mid-Atlantic Ridge area around Iceland. Following Talwani's (Talwani et al., 1971) investigations of the Reykjanes Ridge between 60° and 63°N, several expeditions have dealt with the ridge between 59°N and Iceland.

In 1977, for example, between 60° and 63°N, three new along-strike profiles were recorded over crust $0.3–9 \cdot 10^6$ years old (Bunch, 1980). The profiles were 120 km long and were located at the southwestern limit of the section of the Reykjanes Ridge with no median valley where the ridge breaks up into sections whose axes lie obliquely to the overall spreading axis of the ridge. The detailed structures obtained indicated that with increasing age, the 4.6 km/s layer thins, and the crust with velocities 6.6–7.1 km/s thickens and increases its mean velocity. The velocity of the deepest layer seen in this experiment increased with age from 7.1 km/s at 0 m.y. to 8.2 km/s at 9 m.y.

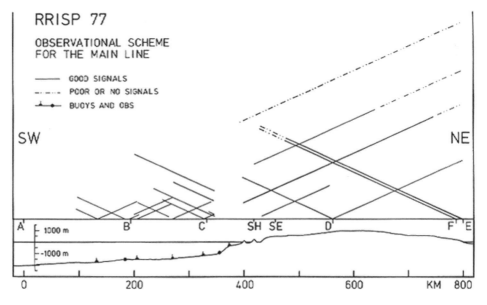

Figure 7.8.4-05. Layout of the RRISP project (from RRISP Working Group, 1980, fig. 3). [Journal of Geophysics, v. 47, p. 228–238. Reproduced with kind permission of Springer Science+Business Media.]

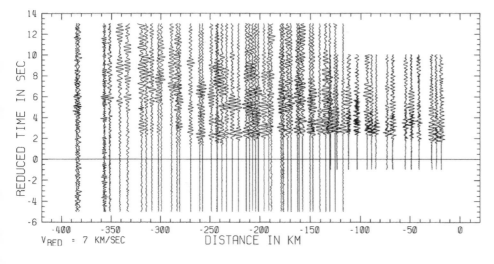

Figure 7.8.4-06. Record section along the land segment of the main profile, shots F1 to F4 (from RRISP Working Group, 1980, fig. 4). [Journal of Geophysics, v. 47, p. 228–238. Reproduced with kind permission of Springer Science+Business Media.]

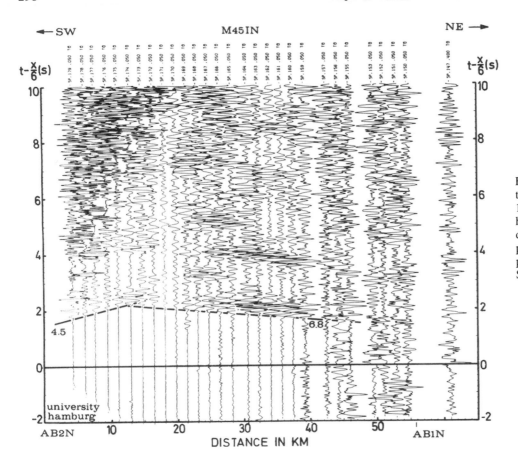

Figure 7.8.4-07. Marine record section (from RRISP Working Group, 1980, fig. 5), recorded by the ground hydrophone of buoy ABS2N (water depth 1389 m). [Journal of Geophysics, v. 47, p. 228–238. Reproduced with kind permission of Springer Science+Business Media.]

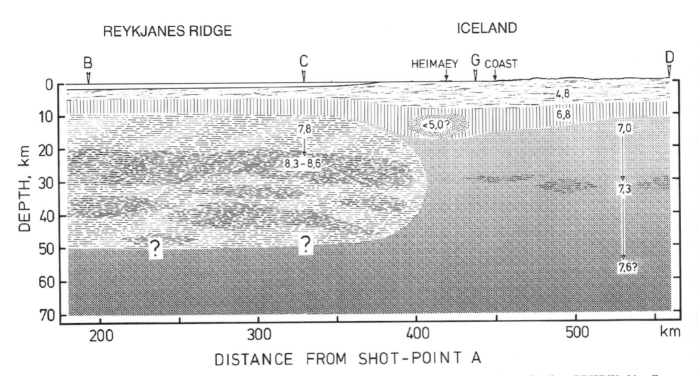

Figure 7.8.4-08. Generalized crustal and upper mantle cross section of the central part of the RRISP77 main profile (from RRISP Working Group, 1980, fig. 7). Letters indicate positions of large shots. [Journal of Geophysics, v. 47, p. 228–238. Reproduced with kind permission of Springer Science+Business Media.]

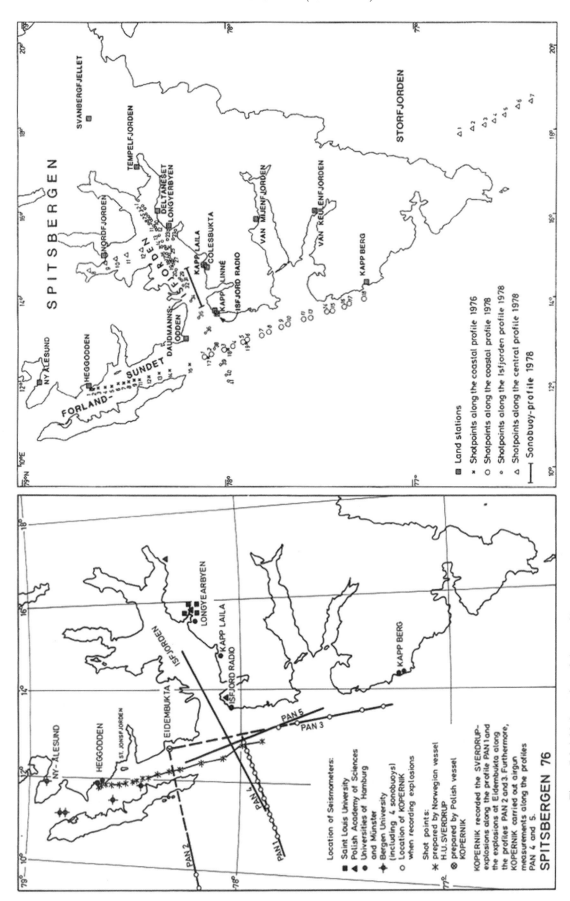

Figure 7.8.4-09. Location of shot profiles, sonobuoys and land stations seismic observations in and around Spitsbergen. Left: 1976 survey (from Guterch et al., 1978, fig.IV-2). Right: 1978 survey (from Sellevoll, 1982, fig. 4.1). [University of Bergen, Seismology Observatory, Bergen, 62 p. Published by permission of the author.]

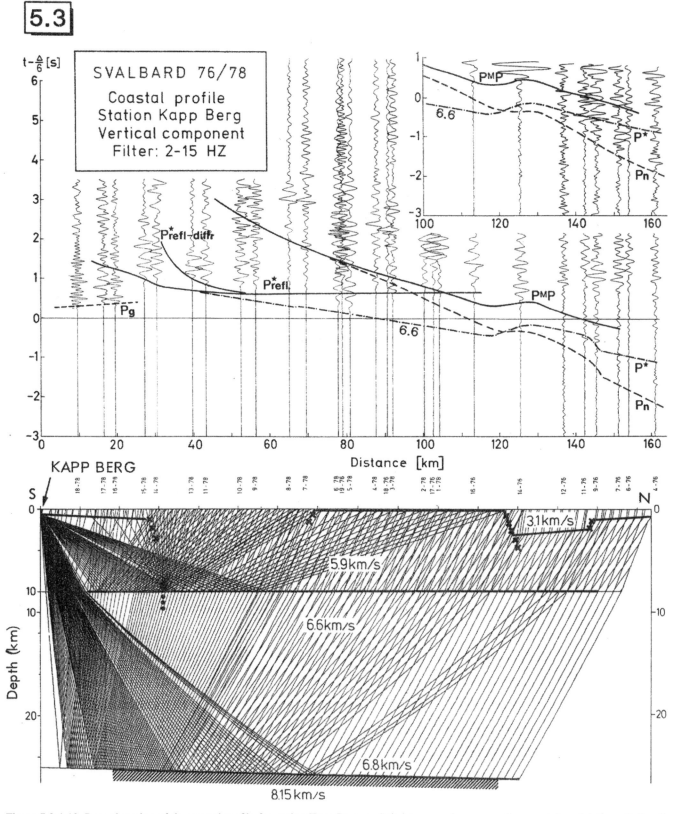

Figure 7.8.4-10. Record section of the coastal profile for station Kapp Berg on Spitsbergen and corresponding crustal model (from Sellevoll, 1982, fig. 5.3). [University of Bergen, Seismology Observatory, Bergen, 62 p. Published by permission of the author.]

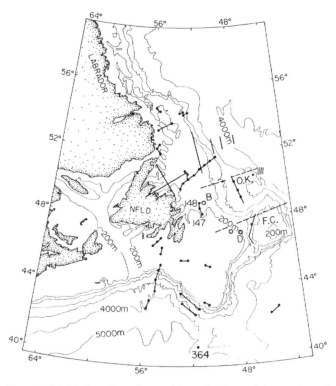

Figure 7.8.4-11. Location of crustal investigations of the continental margin off eastern Canada (from Keen and Barrett, 1981, fig. 1). Solid lines—earlier and the 1977–1978 seismic refraction lines; dashed lines—multichannel seismic reflection lines; open circles—deep exploratory wells; 111—DSDP drilling site; O.K.—Orphan Knoll; F.C.—Flemish Gap. [Geophysical Journal of the Royal Astronomical Society, v. 65, p. 443–465. Copyright John Wiley & Sons Ltd.]

The *Meteor* (1964) expedition M45 in 1977 aimed in particular to investigate the Reykjanes Ridge and to support the onshore-offshore Reykjanes Ridge Seismic Project (RRISP Working Group, 1980).

This was a major long-range seismic-refraction experiment which was realized in 1977 along an 800-km-long line along the southeastern flank of the Reykjanes Ridge, with the aim of resolving both crust and upper mantle to greater depths than previously possible, and of studying the transition from oceanic to the Icelandic structure (Fig. 7.8.4-04). Figure 7.8.4-05 presents the layout of the project. Shots on Iceland and at sea were recorded by up to 90 stations from Iceland, Germany, and the USSR, and by seven ocean-bottom seismographs from Germany and Canada (RRISP Working Group, 1980). In addition, detailed marine seismic investigations were carried out during cruise M45 of the German research vessel *Meteor* (1964). Data examples are shown in Figures 7.8.4-06 and 7.8.4-07 (see also Appendix A7-10-6).

Goldflam et al. (1980) reported normal, but thickened (8 km), oceanic crust for the flanks of the Reykjanes Ridge south of Iceland, thicker than elsewhere on mid-oceanic ridges. The unconsolidated sedimentary layer thickens when approaching the Iceland plateau, while the deeper crustal layers show an almost horizontal layering and no along-strike changes in crustal thickness were observed. Below the Moho, at 9 km depth below sea level, the upper-mantle velocity of 7.8 km/s was shown to increase gradually and reach 8.2 km/s at ~16 km depth and more.

Similar results were shown by Gebrande et al. (1980) and the RRISP Working Group (1980) for the Reykjanes Ridge flank, as well as for Iceland itself, which also sits on top of the active

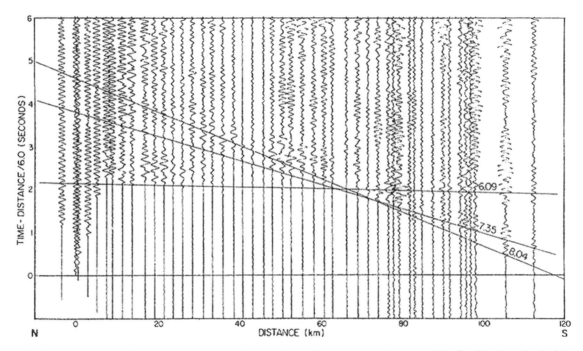

Figure 7.8.4-12. Record section of OBS 7-8-3, line 1, near Orphan Knoll (from Keen and Barrett, 1981, fig. 5b). [Geophysical Journal of the Royal Astronomical Society, v. 65, p. 443–465. Copyright John Wiley & Sons Ltd.]

spreading center of the Atlantic Ocean. Their structural model of Iceland beneath the neo-volcanic zone shows a generalized two-layered crust of variable (10–14 km) thickness underlain by an anomalous mantle with P-wave velocities of 7.0 km/s at the base of the crust increasing to 7.4 km/s at 30 km depth and 7.6 km/s at 50–60 km depth (Fig. 7.8.4-08). This anomalous mantle is evidently confined to Iceland, and a sharp transition resulted for the area of the shelf edge where normal oceanic lithosphere replaces the upward-doming asthenosphere (Gebrande et al., 1980).

In addition to the seismic-refraction measurements, reflection surveys were accomplished in the axial zone of southwestern Iceland, as well as in the flood basalts of northern Iceland (Zverev et al., 1980a, 1980b), to study the upper crust in detail. Reflectors with steep dip could be traced to depths of 15 km, while

for northern Iceland, a seismically homogeneous body without reflectors seems to be present at 10–15 km depth.

Bott and Gunnarson (1980) interpreted the data of a project on the Iceland-Faeroe Ridge, which connects the Iceland block on the active spreading ridge with the Faeroe block which may be of continental character. They interpreted the data to demonstrate a thick oceanic crust of 30–35 km under the ridge to be separated by a true continental margin from the Faeroe block, and underlain by continental crust beneath the lava cover (Bott et al., 1974, 1976). Also a *Meteor* (1964) expedition (cruise M48) in 1978 explored details along the Iceland-Faeroe Ridge (Sarnthein et al., 2008).

Spitsbergen (also named Svalbard) located close to the northern end of the Mid-Atlantic became the target of an inter-

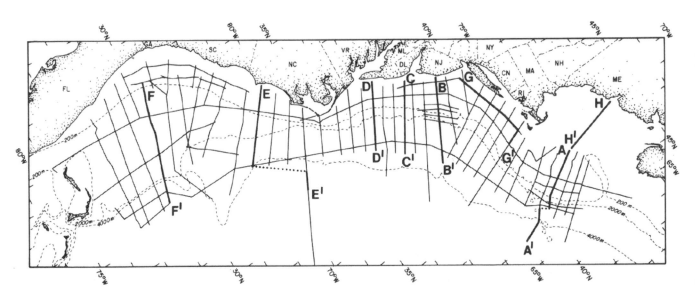

Figure 7.8.4-13. Location of multichannel seismic reflection profiles collected by the U.S. Geological Survey during the 1970s on the Atlantic continental margin (from Trehu et al., 1989, fig.7). [*In* Pakiser, L.C., and Mooney, W.D., eds., Geophysical framework of the continental United States: Geological Society of America Memoir 172, p. 349–382. Reproduced by permission of the Geological Society of America.]

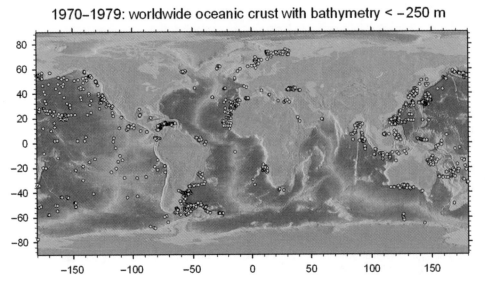

1970–1979: worldwide oceanic crust with bathymetry < −250 m

Figure 7.8.5-01. Seismic refraction measurements in the Atlantic, Pacific, and Indian Oceans performed between 1970 and 1979 (data points from papers published until 1979 in easily accessible journals and books).

national research project of institutions from Germany, Norway, Poland, and the United States under Polish leadership (Guterch and Sellevoll, 1981). Two expeditions were performed in 1976 and 1978. Shot profiles, sonobuoys, and stations on land provided a sophisticated system of observations along the western half of Spitsbergen and its adjacent oceanic margin (Fig. 7.8.4-09). The data were of good quality, but were only published in internal reports (Guterch et al., 1978; Sellevoll, 1982; Appendix 7-10-7). In 1976, the maximum observation distance along the N-S line was

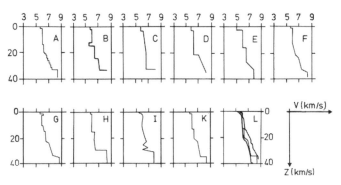

Figure 7.9-01. Comparison of interpretations of data set 2 by various authors (from Ansorge et al., 1982, fig. 24). [Journal of Geophysics, v. 51, p. 69–84. Reproduced with kind permission of Springer Science+Business Media.]

180 km, allowing to record P_MP and P_n arrivals from the Moho (Fig. 7.8.4-10), with a preliminary interpretation giving a crustal thickness of ~25 km. Further inland crustal depths up to 35 km resulted (Faleide et al., 1991; Sellevoll et al., 1991).

The sedimentary and crustal structure of the western margin of the Atlantic Ocean bordering the coast of North America was the target of detailed seismic-reflection campaigns.

The margin of eastern Canada (Fig. 7.8.4-11) had been intensively studied in the 1960s (Dainty et al., 1966). In the 1970s, multichannel seismic-reflection lines were added (Grant, 1975) and in 1977 and 1978, two deep crustal seismic-refraction measurements were carried out (Fig. 7.8.4-11; lines close to O.K. and F.C.), using OBSs (Keen and Barrett, 1981). The aim of this particular project was to determine the nature of the crust and the depth to Moho beneath Orphan Basin and Flemish Gap which region was characterized by the existence of continental fragments according to results of DSDP drilling at site 111 on Orphan Knoll (Ruffman and van Hinte, 1973) and other deep exploratory wells (Keen, 1979), for which locations are also shown in Figure 7.8.4-11. Data were recorded up to 120 km distance, (Fig. 7.8.4-12; Appendix A7-10-8) and the resulting crustal thickness ranged around 20 km both under the Flemish Gap and under the Orphan Basin (Keen and Barrett, 1981).

From 1973 to 1979, a major seismic-reflection program was carried out by the U.S. Geological Survey along the U.S. Atlantic

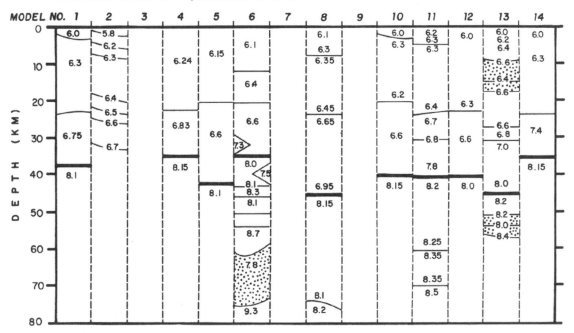

Figure 7.9-02. Comparison of interpretations of the same reversed data sets by various authors, here for shotpoints 3 and 4 (from Mooney and Prodehl, 1984, fig. 72). [Proceedings of the 1980 Workshop of the IASPEI on the seismic modeling of laterally varying structures: Contributions based on data from the 1978 Saudi Arabian refraction profile. U.S. Geological Survey Circular, 937, p. 140–153. .]

continental margin to evaluate the sedimentary basins of the shelf, slope, and rise for their hydrocarbon potential.

Within this program, a major portion was run on the Long Island Platform (Fig. 7.8.4-13, line LIP on Figure 8.5.3-24). The resultant CDP sections were processed to 12 seconds two-way traveltime and provided seismic sections through the cratonized crust of the southern New England Appalachians and its transition from continental thickness on the Long Island Platform to oceanic crust across the passive rifted margin (Hutchinson et al., 1986; Phinney, 1986; Phinney and Roy-Chowdhury, 1989).

Also in the 1970s, the surroundings of the site of the 1886 Charleston earthquake were investigated by the U.S. Geological Survey with regard to the likelihood of future earthquakes of similar size. Within this program, multichannel reflection work was started in 1979 on land and offshore in the Charleston area. The survey was continued in 1981 and was discussed in detail by Behrendt (1986), in conjunction with the 1978–1979 COCORP lines in Georgia and in the Charleston area (Cook et al., 1979a, 1981). A complete overview (Fig. 7.8.3-13) was published by Trehu et al. (1989b).

7.8.5. Summary of Seismic Observations in the Oceans in the 1970s

One of the highlights of the 1970s was the attempt to record super-long profiles not only on the continents, but also in the oceans, that penetrated well below Moho. In the Pacific Ocean, e.g., a 600-km-long line was laid out in the northeastern Pacific between the Clarion and Molokai fracture zones (Orcutt and Dorman, 1977). In the western Pacific basin, explosions were recorded by OBSs up to distances of 1300 km during two Longshot experiments in 1973 in the East Mariana basin (Asada and Shimamura, 1976). From 1974 to 1980, further intensive long-range experiments were undertaken in the northwestern Pacific, which detected large-scale lithospheric structure and anisotropy in the upper-mantle (Asada and Shimamura, 1979; Asada et al., 1983; Shimamura et al., 1983).

An excellent summary of marine crustal studies for the east and southeast Asian Seas was published as a map by Hayes et al. (1978) compiling all data available at that time between Japan in the northeast and Australia in the south, also covering part of the Indian Ocean west of Thailand and Indonesia.

Dealing also with the lower lithosphere, from 40° to 43°N a long-range profile was recorded near the Azores (Steinmetz et al., 1977). Another long-range seismic-refraction experiment in 1977 aimed to resolve both crust and upper mantle to greater depths than previously possible by establishing an 800-km-long line along the southeastern flank of the Reykjanes Ridge (RRISP Working Group, 1980).

An overview of seismic measurements in the oceans, compiled from the database assembled at the U.S. Geological Survey in Menlo Park, California, shows all surveys that were published until 1979 in generally accessible journals and books for the 1970s (Fig. 7.8.5-01).

The map shows, similar to the map for the 1960s, numerous observations in continental shelf and adjacent deep-ocean areas of North America and eastern Asia. There are only a few data points in the center of the Atlantic Ocean, and none in the Indian Ocean, but much research was conducted in the Pacific Ocean. A few points also touch the Antarctica. New are numerous proj-

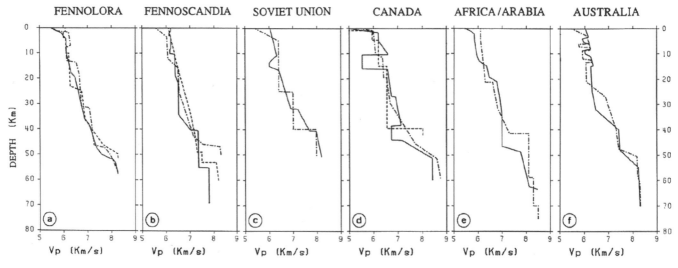

Figure 7.10-01. Representative velocity-depth functions of the crust of different shield areas (from Guggisberg et al., 1991, fig. 26). In corresponding order: solid line, dashed line, dashed-dotted line: (a) FENNOLORA 1979: E-south, C-north, G-south; (b) SVEKA 1981, block III (Luosto et al., 1984), BALTIC 82, block III (Luosto et al., 1985), Blue Road 1972, 5-west (Lund, 1979); (c) Russian platform (Pavlenkova, 1979), West Siberian Craton (Puzyrev and Krylov, 1971), (d) Grenville province, 2-west (Berry and Fuchs, 1973), Churchill province (Green et al., 1980), Superior province, 3-east (Berry and Fuchs, 1973); (e) Kalahari Craton, profile III (Baier et al., 1983), Arabian Shield, shotpoint 3 (Mooney et al., 1985); (f) North Australian Craton TCMI-3 west and TCMI-2 east (Finlayson, 1982). [Tectonophysics, v. 195, p. 105–137. Copyright Elsevier.]

ects which explored the continental shelves and adjacent ocean areas along eastern South America, northwestern Africa, the Norwegian coast and the Rockall trough, the Melanesian and Indonesian archipelagos, and southern Australia. The Mid-Atlantic Ridge in the northern Atlantic Ocean and the Hawaii Islands in the Pacific Ocean were of special interest.

7.9. ADVANCES IN INTERPRETATION METHODOLOGY

Throughout the 1970s, the interpretations of the many seismic-refraction experiments concerning the crust and upper mantle were mainly performed using traveltime routines for which many different computer programs had been developed over the years. In 1968, Helmberger (1968) had published the theory to calculate synthetic seismograms for layered transition zones. In addition, the reflectivity method published by Fuchs and Müller (1971) was applied all over the world. There is hardly an interpretation published during the 1970s in which the model has not been checked and verified by calculating synthetic seismograms and comparing them with the main observed phases. Braile and Smith (1975) published a whole series of synthetic record sections calculated for typical crustal models. Bessonova et al. (1974) had derived the tau method to invert seismic wide-angle reflection traveltimes.

However, a serious drawback of this methodology was that a horizontally homogeneous layered structure was required. Several, and not always successful, attempts were undertaken

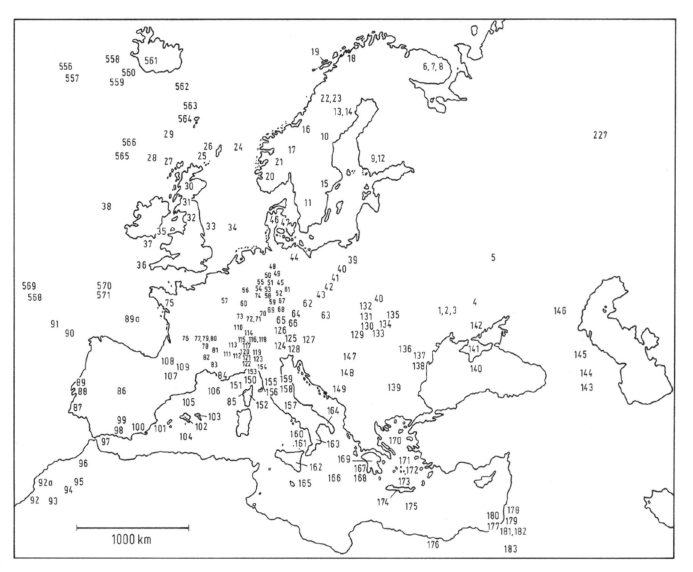

Figure 7.10-02. Map of Europe showing points where seismic data on crustal structure were available by 1980–1981 (from Prodehl, 1984, fig. 11). [*In* K.-H. Hellwege, editor in chief, Landolt Börnstein New Series: Numerical data and functional relationships in science and technology. Group V, Volume 2a: K. Fuchs and H. Soffel, eds., Physical properties of the interior of the earth, the moon and the planets: Berlin-Heidelberg, Springer, p. 97–206. Reproduced with kind permission of Springer Science+Business Media.]

to overcome this problem. For example, B. Kennett, amongst others, modified the reflectivity method to allow slight variations in structure on the shot side (Kennett, 1974, 1983).

The problem of laterally homogeneous structures was only solved when the first ray tracing approaches were published. In Munich, H. Gebrande was amongst the first who developed this approach which finally was successfully applied on Alpine data (Gebrande, 1976; Will, 1976). A breakthrough of this approach was made when Červený and co-authors published their book and later results on ray theory (Červený et al., 1977; Červený, 1979; Červený and Horn, 1980). A corresponding computer program for ray tracing had been prepared and made available for all interested users around the world, before it was finally published by Červený and Pšenčik (1984c). This procedure would become the major interpretation tool of the following decades. Červený's method and the computer routines developed on its basis also allowed researchers to compute synthetic seismograms. Other groups were also writing synthetic-seismogram programs on the

basis of ray theory which were widely applied, for example, the program by McMechan and Mooney (1980) in the United States or the program by Spence et al. (1984) in Canada and elsewhere.

In the USSR, when deep seismic sounding operations were drastically reduced in the 1970s, the method of interpretation of deep seismic data underwent a period of heated discussions (Pavlenkova, 1996). Up until the 1960s, the correlated later-arrival phases had been interpreted as head waves, and only gradually were the phases correlated and recognized to be reflected waves. In the 1970s, new methods and interpretation programs were developed, and the methods of field experimentation were also changed. Numerous papers on wave dynamics and the nature of the recorded waves had already appeared in the 1960s, and the discussion continued in the 1970s. I.P. Kosminskaya had published a book on seismic methods in 1968, which was translated into English by G.V. Keller in 1971 (Kosminskaya, 1971). In the 1970s, new facilities for processing wave fields were applied to reinterpret a large part of the old deep seismic sounding material.

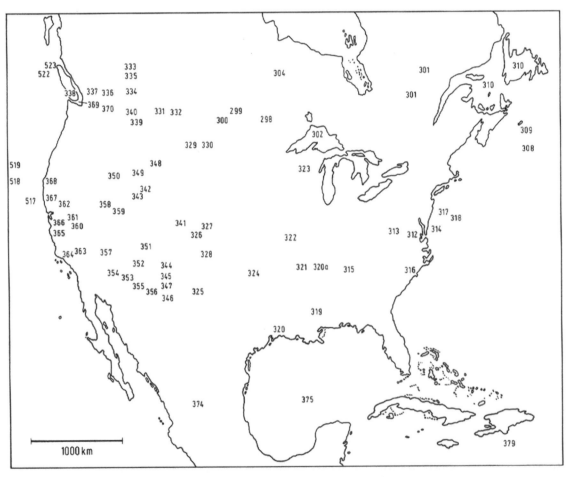

Figure 7.10-03. Map of the United States of America showing points where seismic data on crustal structure were available by 1980–1981 (from Prodehl, 1984, fig. 12). [*In* K.-H. Hellwege, editor in chief, Landolt Börnstein New Series: Numerical data and functional relationships in science and technology. Group V, Volume 2a: K. Fuchs and H. Soffel, eds., Physical properties of the interior of the earth, the moon and the planets: Berlin-Heidelberg, Springer, p. 97–206. Reproduced with kind permission of Springer Science+Business Media.]

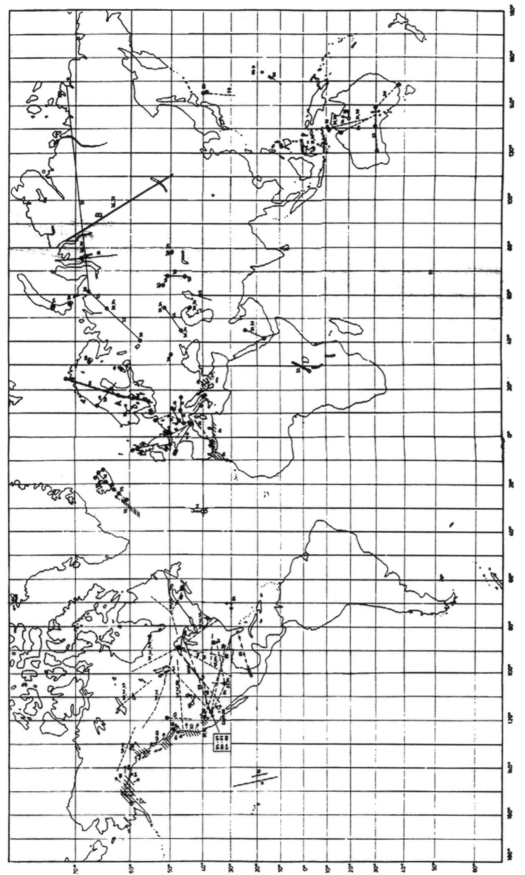

Figure 7.10-04. Enlarged section of the world map of Figure 2.7-01 showing long-range profiles until 1980 (from Fuchs et al., 1987, figure 1). [*In* Fuchs, K., and Froidevaux, C., eds., Composition, structure and dynamics of the lithosphere-asthenosphere system: American Geophysical Union Geodynamics Series 16, p. 137–154. Reproduced by permission of American Geophysical Union.]

In their book *Seismic Models of Basic Geostructures of the Territory of the U.S.S.R.*, Zverev and Kosminskaya (1980) presented a comprehensive set of reinterpreted—and representative—cross sections for various tectonic provinces of the USSR. As the scientists of the USSR at that time had very few opportunities to exchange ideas with their western colleagues, many approaches, algorithms, programs and methods of ray and wave field calculations in two- or three-dimensional media, as well as inversion methods for one- or two-dimensional media were developed in parallel and sometimes synchronously to those developed in the western world by, e.g., Alekseev, Bessonova, Matveeva, Michailenko, and others (for references, see Pavlenkova, 1996).

The problem of interpreting seismic-refraction data was internationally recognized. A number of workshops tried to bring together experimental and theoretical seismologists to discuss the various interpretation procedures and compare their results. Here, we mention only a few such workshops, where scientists of the whole world met to view the data and discuss the interpretation methods and their problems. So, in 1977, J.G. Heacock arranged a workshop on the Earth's crust at Vail, Colorado (Heacock, 1977). Czech scientists organized a series of workshops of differing character in 1977 and 1978 on problems of crustal studies and interpretation methods at the castle of Liblice near Prague. A year later, in 1978, I.P. Kosminskaya, N.I. Pavlenkova, and others arranged a workshop at Yalta on deep seismic sounding data and interpretation methods.

A special series of workshops starting in 1977 under the auspices of the IASPEI, specialized on problems of controlled-source seismology. These so-called CCSS workshops continue at non-regular intervals. For these workshops, data are distributed to the participants several months beforehand, and the scientists are requested to interpret the data with their presently used methodology. Thus, the various methods can be tested and verified during the discussions at these workshops.

The first CCSS workshop of this kind was organized in Karlsruhe, Germany, in 1977. Here, data from some unreversed seismic-refraction crustal profiles had been distributed to the participants, who were asked to bring their modeling results to the workshop (Fig. 7.9-01). One of the results was that the main differences between the models were not due to the interpretation method, but to phase correlations. The varying approaches and resulting models were finally published by Ansorge et al. (1982).

As data and modeling were tested using an one-dimensional approach only (not using any reversed shots), a second CCSS workshop followed three years later, organized by R.B. Smith at Park City, Utah, in August 1980. As a database for this workshop, the data of the 1978 refraction campaign of the U.S. Geological Survey in Saudi Arabia were distributed to the participants. This data allowed a two-dimensional approach to the interpretation. The project was already discussed in more detail above (see Figs. 7.5.2-04 to 7.5.2-06 and Appendix A7-6-3). Fourteen different models were presented at the workshop and were published in a U.S. Geological Survey Circular (Mooney and Prodehl, 1984). As an example, the interpretation between shotpoints 3 and 4 is shown (Fig. 7.9-02). Several of the models were also published independently in *Tectonophysics* (e.g., Mooney et al., 1985; Milkereit and Flueh, 1985; Prodehl, 1985).

7.10. THE STATE OF THE ART AT THE BEGINNING OF THE 1980s

For the first time, in the 1970s, seismic near-vertical incidence reflection experiments were organized in large scales covering distances of 100 km or more which brought new insight into details of crustal structure and composition of the Earth's crust down to Moho. Sedimentary basins on the continents and ocean basins at sea had already been studied in much detail, the former by petroleum seekers since the 1930s and the latter by marine scientists largely during the decades following World War II. For the study of sedimentary basins, the seismic-reflection profiling method had been highly developed over the years by the petroleum industry. Consequently, academic researchers took advantage of the expertise, techniques, and equipment of the industry to study the whole crystalline crust and obtain not only geophysical models with velocity, density, and attenuation, but also determine the extent and the configuration of reflecting horizons (Oliver, 1986).

So, in the 1970s, the first national large-scale seismic-reflection programs were initiated. COCORP in the United States (e.g., Oliver et al., 1976) was soon followed by the Canadian equivalent COCRUST (Mereu et al., 1989), a combination of large-scale crustal reflection and refraction studies. Similar large-scale seismic near-vertical incidence reflection profiles, accompanied by wide-angle reflection observations, were recorded in Germany (Bartelsen et al., 1982; Meissner et al., 1980).

Seismic-refraction work in the 1970s was for a large part devoted to research projects dealing with the subcrustal lithosphere, but also much emphasis was laid on crustal research in tectonically anomalous regions. For western Europe, Mostaanpour (1984) compiled the record sections of seismic-refraction profiles of the past 30 years, mapped the characteristic parameters, and plotted a series of contour maps covering mainly Germany, the Alps, France, and Italy, following a working scheme applied by Giese and Stein (1971) for western Germany.

A representative selection of lithospheric velocity-depth sections from data obtained during the 1970s was compiled by Guggisberg et al. (1991) when they were discussing the results of the 2000-km-long FENNOLORA profile through Scandinavia (Fig. 7.10-01).

Figure 7.10-05. Selected velocity-depth models of upper-mantle structure below Moho for Europe and Asia (from Prodehl, 1984, fig. 15). [*In* K.-H. Hellwege, editor in chief, Landolt Börnstein New Series: Numerical data and functional relationships in science and technology. Group V, Volume 2a: K. Fuchs and H. Soffel, eds., Physical properties of the interior of the earth, the moon and the planets: Berlin-Heidelberg, Springer, p. 97–206. Reproduced with kind permission of Springer Science+Business Media.]

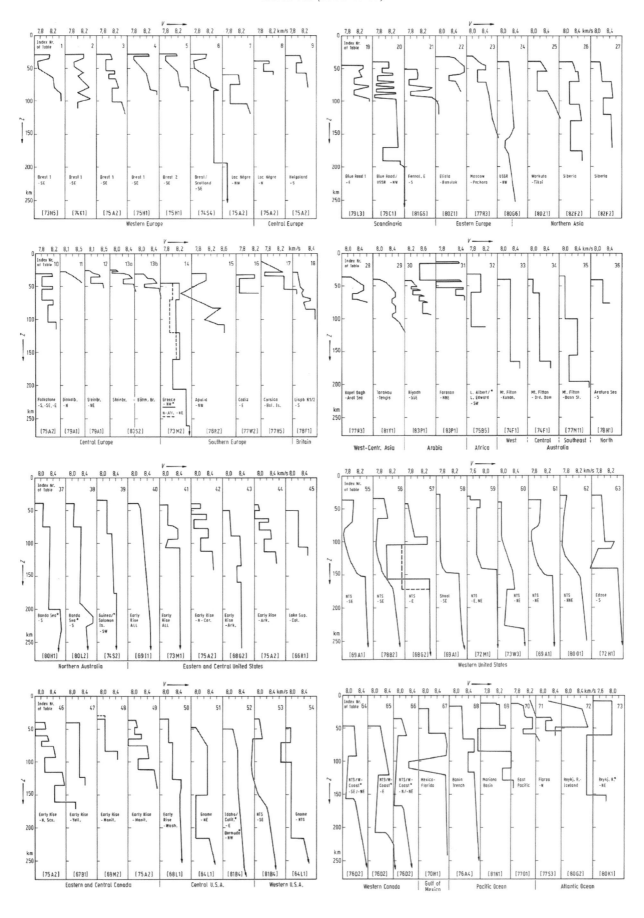

In 1981, a global crustal thickness map was published by Soller et al. (1981), based on the results of 273 publications.

An overview of crustal and upper-mantle studies by seismic-refraction experiments has been compiled by Prodehl (1984; see Appendix A7-1), essentially representing the state of the art by 1980–1981. The summary shows representative velocity-depth functions for the crust in tabular form and the corresponding locations on maps for Europe, North America, and the rest of the world. The map for the whole world (see Fig. 2.7-03 in Chapter 2) as well as the more detailed maps for Europe (Fig. 7.10-02) and the United States of North America (Fig. 7.10-03) show very well where data on crustal structure was available at the time. The second part of Prodehl's (1984) compilation also dealt with the subcrustal data (Fig. 7.10-04) and the models (Fig. 7.10-05), as they were published by 1980–1981.

The 1980s (1980–1990)

8.1. THE DECADE OF LARGE-SCALE REFLECTION SEISMOLOGY AND SEISMIC RAY TRACING

During the course of the 1960s, the technology of controlled-source seismology was very much improved so that in the beginning of the 1970s, this technique would not only obtain average properties of crustal structure, but could resolve major anomalous features. Furthermore, in the 1970s, improved shooting techniques allowed scientists to obtain resolvable energy up to distances of 1000 km or more. So, the fine structure of the lower lithosphere became one of the major goals of seismic-refraction projects including a rather good knowledge of details in the structure of the overlying crust. At the same time, the resolution was gradually increased by decreasing the interval distances between recording stations, and by increasing the number of shots along a profile. Also at this time, the interest in resolving the crustal structure in much more detail than hitherto led vertical-incident reflection seismic investigations into new dimensions. Large-scale reflection experiments such as COCORP in the United States, and special surveys in Europe which penetrated the whole crust were started and could be extended to continuous profiling up to 100 km profile length. In Europe, in most cases, these projects were supported by refraction piggy-back experiments to include the wide-angle reflection distance range.

The success of controlled-source seismology continued in the 1980s, but with increased interest as to the details of the crust. In Europe, a variety of seismic-refraction surveys was initiated in the frame of large-scale interdisciplinary projects, e.g., the establishment of the European Geotraverse from the North Pole to Tunisia, North Africa, and the increasing interest in deep continental drilling. Further stimulation was created by the establishment of large-scale seismic-reflection profiling leading to the establishment of national crustal-scale reflection programs such as BIRPS in Britain, DEKORP in Germany, or ECORS in France. In North America, the new seismic-refraction equipment of the U.S. Geological Survey led to considerably increased activities, parallel to the efforts of COCORP, and the Canadian geophysical institutions formulated subsequently the joint programs COCRUST and LITHOPROBE for a systematic crustal and upper-mantle investigation of the Canadian territory by a joint application of seismic-refraction and reflection methodology.

8.2. CONTROLLED-SOURCE SEISMOLOGY WORKSHOPS

The increasing wealth of explosion seismic data allowing the studying of the crust and the uppermost mantle in much detail, and in areas where strong anomalies were to be expected, led to an increasing need to improve the existing interpretation methods. As described in detail at the end of the foregoing Chapter 7, a first workshop of the CCSS (Commission on Controlled Source Seismology) working group had been held in 1977, at Karlsruhe, to discuss and compare current interpretation methods at the example of 1-D seismic-refraction data, which had been distributed beforehand to the participants of that workshop (Ansorge et al., 1982). The workshop had shown that the existing traveltime routines, and the application of the reflectivity method, led to very satisfying results (see Fig. 7.9-01), and the comparison of the different methods had proved that the interpretation results did not depend on the method used, but rather on the interpretation of seismic phases correlated in the raw data. Furthermore, most methods used had required horizontally homogeneous stratified media. However, it had become apparent that methods which allowed 2-D and 3-D approaches were needed. Already in the 1970s, the basic theory of the ray method in seismology had been published (Červený et al., 1977), and first computer routines to calculate traveltimes and ray paths for lateral inhomogeneous media (e.g., Červený and Horn, 1980; Gebrande, 1976; Will, 1976) as well as to compute ray theoretical seismograms (Červený, 1979; Červený and Pšenčik, 1984a; McMechan and Mooney, 1980; Spence et al., 1984) had been published. The following decade of the 1980s saw the systematic application of ray tracing methodology for 2-D interpretations of seismic-refraction surveys, which covered areas where the approach with 1-D modeling would have led to major errors.

As described in more detail at the end of Chapter 7, in 1980 another CCSS workshop, this time held at Park City, Utah, had the particular aim to study interpretation methods which could be applied to a long-range profile of several 100 km length with varying crustal structure and with a substantial number of shotpoints along the line. For data, the observations of a long-range profile through Saudi Arabia, recorded by the U.S. Geological Survey in 1978 (Gettings et al., 1986), could be used, which again were distributed to the participants beforehand and for which 17 separate interpretations (see Fig. 7.9-02) had been delivered (Mooney and Prodehl, 1984). It turned out that the individual interpretations showed only minor differences for the area of the Saudi Arabian shield and platform where lateral variations in structure were almost negligible. However, for the westernmost part of the line, the transition zone from the Arabian shield to the Red Sea, the individual models varied considerably.

As a consequence, and to ensure that differences in phase correlation could be ruled out, it was decided to hold another workshop, for which synthetic seismogram sections with clear secondary arrivals (Fig. 8.2-01), would be computed for a known

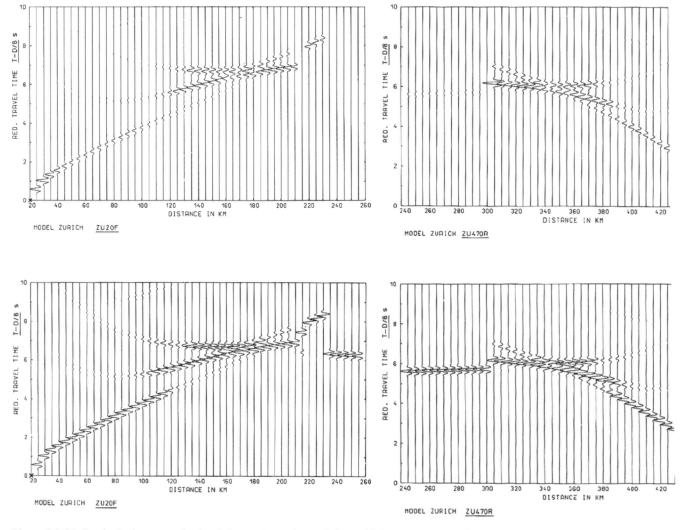

Figure 8.2-01. Synthetic data examples handed out prior to the workshop which were computed from the model shown in Figure 8.2-02 from the model sources at distances 20 km (left) and 470 km (right), normalized according to a distance scaling factor (top) and with equal maximum amplitude on all traces (bottom) (from Červený and Pšenčik, 1984a, Figs. 5 and 10). [*In* Finlayson, D.M., and Ansorge, J., 1984, Workshop proceedings: interpretation of seismic wave propagation in laterally heterogeneous structures: Bureau of Mineral Resources, Geology & Geophysics, report 258, Canberra, Australia, p. 3–14. Copyright by Commonwealth of Australia (Geoscience Australia).]

realistic structure and would be distributed to the participants for interpretation without knowing the original model (Fig. 8.2 02). The data preparation was taken care of by V. Červený and I. Pšenčik at Prague, Czechoslovakia, and the workshop was held in 1983, at Einsiedeln, Switzerland (Finlayson and Ansorge, 1984). Twenty-one interpretations of data set I—model Zurich—were presented at the workshop. In conclusion, the application of a variety of existing methods on these data led to a very satisfying result and ruled out that differing results were dependent on the particular method used, but were exclusively dependant on phase correlations.

Having ensured that the available methods would not lead to differing results, the emphasis was turned over to the joint interpretations of near-angle and wide-angle seismic-reflection and seismic-refraction data recorded along the same line. The problem of whether or not near-angle reflection data would see the same interfaces as wide-angle reflection and refraction data had already been approached at the CCSS workshop of 1983 at Einsiedeln, Switzerland, but would become the main focus of the next CCSS workshop held in Susono, Shizouka, Japan, in August 1985. In 1983, a data set III, consisting of seismic-reflection and refraction data obtained in 1978 in the area of the geothermal anomaly Urach, Germany (Bartelsen et al., 1982), and prepared by E.R. Flueh at Kiel, Germany, had been distributed and worked on by only three workshop attendees.

In 1985, the SJ-6 seismic-reflection/-refraction profile (Figs. 8.2-03 and 8.2-04), recorded through south central California perpendicular to and crossing the San Andreas Fault zone south of Parkfield, California, served as the only data set (Walter and Mooney, 1987; Appendix A8-3-3). One of the 11 separate

ZURICH DATA SET - 1

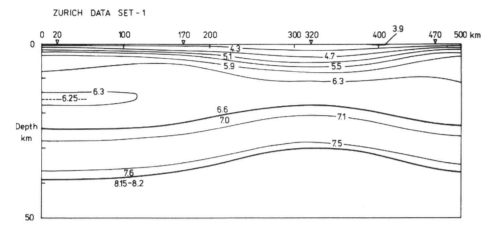

Figure 8.2-02. Model for which synthetic data (see examples in Fig. 8.2-01) handed out prior to the Workshop were computed (from Červený and Pšenčik, 1984b, fig. 1). [*In* Finlayson, D.M., and Ansorge, J., 1984, Workshop proceedings: interpretation of seismic wave propagation in laterally heterogeneous structures: Bureau of Mineral Resources, Geology & Geophysics, report 258, Canberra, Australia, p. 3–14. Copyright by Commonwealth of Australia (Geoscience Australia).]

models derived by workshop attendees and presented by Walter and Mooney (1987) is shown in Figure 8.2-05.

For the subsequent CCSS workshop, held in 1987 at Whistler near Vancouver, British Columbia, participants had been invited to work on one or more of the following topics: (1) a combined interpretation of multichannel seismic-reflection and seismic-refraction/wide-angle reflection data across Vancouver Island and the adjacent continental margin, collected by COCRUST and LITHOPROBE between 1981 and 1985, (2) enhanced processing of stacked data, and/or (3) processing of unstacked data from a land multichannel seismic-reflection profile collected on Vancouver Island. Data, results, and problems were discussed during the workshop (Green et al., 1990a, 1990b).

Data set (1) comprised the data of profile I, extending across Vancouver Island from near the volcanic arc to the deep ocean, and profile IV, shot along the length of Vancouver Island parallel to the continental margin, both recorded in 1980 (Fig. 8.2-06). Profile I comprised 32 land seismographs, deployed on a 160-km-long line on Vancouver Island, on islands in the Strait of Georgia, and on the British Columbia mainland, and it included four ocean-bottom seismometers (OBS) in the offshore region, of which three, shown in Figure 8.2-06, were of concern. Record sections from shotpoint J1 and from the shotpoints P8, P13, and P19, as well as the record sections recorded at the three OBS sites 1, 3, and 5 (Fig. 8.2-06), were distributed. In addition a reflection section of the marine profile 85-1, recorded in 1985, was handed out for the workshop.

8.3. CONTROLLED-SOURCE SEISMOLOGY IN EUROPE

The great success of explosion seismology in the 1970s was due to close cooperation of research institutions on an international scale. This cooperation allowed scientists to explore the entire lithosphere in great detail. This effort to cooperate was continued in the 1980s on an even broader base. Furthermore, details of crustal structure became the main focus of explorations. Three major research projects were formulated almost simultaneously, involving not only seismologists, but the whole range of earth sci-

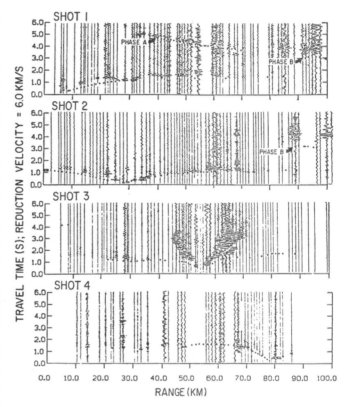

Figure 8.2-03. True-amplitude record sections of the of the original refraction data set handed out prior to the workshop (from Trehu and Wheeler, 1987, fig. 4a). [*In* Walter, A.W., and Mooney, W.D., 1987, Commission on Controlled Source Seismology (CCSS) proceedings of the 1985 workshop interpretation of seismic wave propagation in laterally heterogeneous terranes, Susono, Shizouka, Japan, 15–18 August 1985: Interpretations of the SJ-6 seismic reflection/refraction profile, south central California, USA: U.S. Geological Survey Open-File Report 87-73, p. 91–104.]

entists: (1) national deep-seismic-reflection programs in western and central Europe; (2) detailed seismic-reflection and seismic-refraction research within the frame of continental deep drilling programs successfully carried out in Germany by the German KTB project (Kontinentale Tief-Bohrung = continental deep drill-

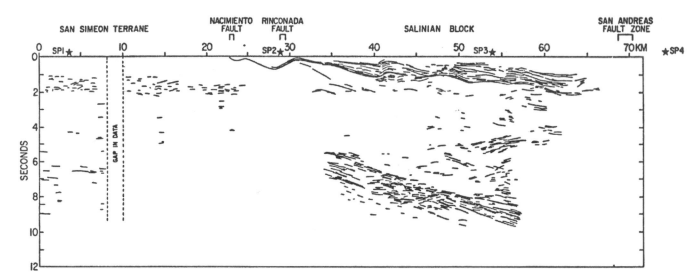

Figure 8.2-04. Line drawing of the original reflection data set handed out prior to the workshop (from Trehu and Wheeler, 1987, fig. 2). [*In* Walter, A.W., and Mooney, W.D., 1987, Commission on Controlled Source Seismology (CCSS) proceedings of the 1985 workshop interpretation of seismic wave propagation in laterally heterogeneous terranes, Susono, Shizouka, Japan, 15–18 August 1985: Interpretations of the SJ-6 seismic reflection/refraction profile, south central California, USA: U.S. Geological Survey Open-File Report 87-73, p. 91–104.]

Figure 8.2-05. Velocity model derived by 2-D ray tracing from the data set handed out prior to the workshop (from Trehu and Wheeler, 1987, fig. 4b). [*In* Walter, A.W., and Mooney, W.D., 1987, Commission on Controlled Source Seismology (CCSS) proceedings of the 1985 workshop interpretation of seismic wave propagation in laterally hetero-geneous terranes, Susono, Shizouka, Japan, 15–18 August 1985: Interpretations of the SJ-6 seismic reflection/refraction profile, south central California, USA: U.S. Geological Survey Open-File Report 87-73, p. 91–104.]

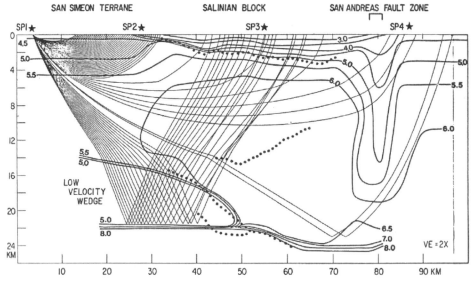

hole) as well as in Russia by the Kola deep drillhole on the Baltic Shield (which will be discussed further below); and (3) the European Geotraverse with major seismic-refraction surveys.

8.3.1. Large-Scale Seismic-Reflection Surveys in Western Europe

In contrast to most of the early near-vertical incidence seis-mic-reflection work in the 1950s and 1960s in Canada, Germany, the former USSR, and in Australia, the initiators of COCORP had hired a commercial reflection company for the fieldwork and also applied their processing and interpretation techniques (Oliver et al., 1976) on the new data involving the whole crust. The same

philosophy was also applied by the initiators of the national European large-scale seismic-reflection programs. Such national programs were BIRPS (British Institutions Reflection Profiling Syndicate) in Britain, ECORS (Etude de la Croûte Continentale et Océanique par Réflexion et Réfraction Sismique) in France, BELCORP (Belgian Continental Reflection Program) in Belgium, DEKORP (Deutsches Kontinentales Reflexionsseismisches Programm) in Germany, NFP (Swiss National-Fond Project) in Switzerland, CROP (Crosta Profonda) in Italy, and others.

Phinney and Roy-Chowdhury (1989) have described in much detail the basic seismic-reflection techniques and the long procedure leading from the raw data to the seismic-reflection sections used for final geological interpretation which we will

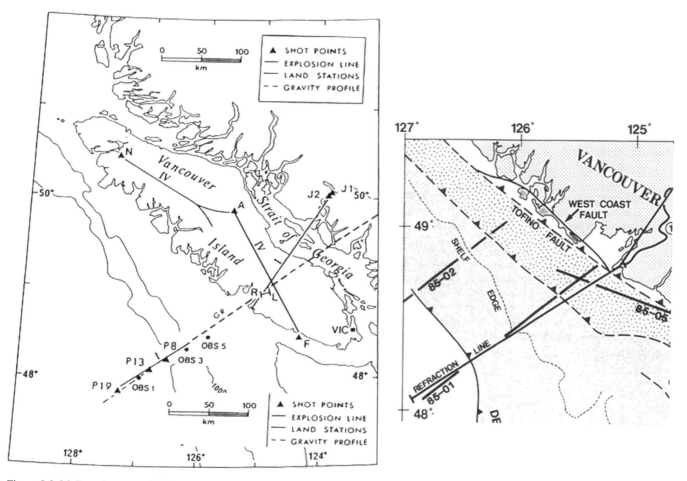

Figure 8.2-06. Location map of COCRUST and LITHOPROBE crustal study profiles across Vancouver Island (from Green et al., 1990b, figs. 4 and 1). Left: Map showing location of data handed out to workshop participants. Right: Location of reflection line 85-1 handed out to workshop participants. [Geological Survey Canada, paper 89-13: 3-25. Reproduced with the permission of Natural Resources Canada 2009, courtesy of the Geological Survey of Canada.]

briefly review. The conventional geometry for reflection profiling is an array of geophone groups, cabled together, to record seismic signals from a shotpoint or vibrator point within or slightly off-end the array. The seismic traces recorded from one such source point constitute a source gather. The arrivals in a source gather may include reflections, refractions, ground-roll, shear waves, air waves, converted waves, multiples and the like. The collection of all source gathers forms a 3-D data set (source, receiver group, time). The goal of seismic data processing is to convert this raw 3-D data manifold into a 2-D cross-sectional representation of subsurface structure. Phinney and Roy-Chowdhury (1989) have summarized the usual conventional seismic processing steps in table 2 of their publication. These steps are, in detail:

(1) involving production of shot gathers (demultiplex and Vibroseis correlation),

(2) initial clear-up (frequency filtering, resampling, trace editing, true amplitude recovery, datum statics, removal of shear and air waves and ground roll);

(3) preparation of common-depth point (CDP) gathers (rearrangement of traces, deconvolution, etc.);

(4) CDP processing (suppression of refraction arrivals, velocity analysis, residual statics, normal moveout, stack);

(5) post-stack processing (deconvolution, frequency filter);

(6) imaging the zero-offset wavefield (automatic gain control, coherency filter, convert digital data to graphic display); and

(7) imaging the subsurface (migration, inversion, convert subsurface image to graphic display).

Some of the steps described are mandatory, some optional. Of all steps described above, migration and stacking are the most important ones. Migration places the reflectors into their spatially correct locations. The final graphic display serves the same purpose as do record sections in seismic-refraction interpretation, where phases are to be correlated over larger distance ranges and identified as refractions or wide-angle reflections from discontinuities. The processed near-vertical incidence seismic-reflection sections show similarly reflecting horizons by correlating reflections over larger distance ranges and identifying them as geological features.

For the interpretation of seismic-reflection surveys it has become common use to present either the graphic display of a profile or corresponding line drawings as the final product and interpret the most evident features with depth indications in s TWT (two-way traveltime) and not, in many cases due to missing velocity information, to draw corresponding velocity-depth sections as is used in the interpretation of seismic-refraction surveys.

8.3.1.1. British Isles (BIRPS and Other Seismic-Reflection Surveys)

The continental region of the British Isles adjoins the Northeast Atlantic margin and includes the onshore regions of the British Isles as well as the stretched continental crust beneath the seas to the south and west of Ireland and northern Norway and beneath the North Sea.

Between 1980 and 1982, the Deep Geology Unit of IGS (Institute of Geological Sciences) continued their land-based seismic-reflection surveys which they had started in 1979 (Whittaker and Chadwick, 1983). In 1980, three surveys reached the base of the crust: the IGS 1980 Wiltshire and Dorset Survey, the IGS/NERC 1980 Hampshire Survey and the IGS/Leicester University 1980 Survey. For the interpretation of the Dynamite Survey 1981–1982 East Yorkshire and 1982 South Yorkshire data, the LISPB (Lithospheric Seismic Profile through Britain) 1974 data could be used to produce velocity profiles. Striking horizontal deep reflections around 11 s TWT were also seen on the Dynamite Survey 1982 Dorset. The preliminary interpretation of the geographically scattered data revealed a certain consistency of crustal structure for those areas with good quality dynamite lines. Beneath the sedimentary sections with flat-lying reflectors, the upper to mid-crustal zone was interpreted to be characterized by strongly tectonized rocks with brittle forms of thrusting and faulting. The lower crust appeared to support the concept of deformation by a dominantly ductile strain release mechanism. Discrete high-amplitude, subhorizontal events occurring between 8 and 11 s TWT were attributed to reflectors at or below the base of the crust (Whittaker and Chadwick, 1983).

A major impact on seismic studies in the UK came from the marine work undertaken by the group of scientists at Cambridge under the leadership of Drummond Matthews. Plans to investigate the Earth's crust in the British Isles at a large scale in much more detail than was possible by seismic-refraction work were born when Drummond Matthews visited COCORP to gain insight into the procedures of deep seismic reflection and how such a group functioned, and so to initiate deep seismic-reflection work.

The fact that the British Isles are surrounded by sea offered a relatively economical way to apply the reflection method to long distance ranges of several 100 km. BIRPS—the British Institutions Reflection Profiling Syndicate—was conceived in 1976 by a Royal Society working group and was first funded in 1981 by the British government through the Natural Environment Research Council to serve the interests of UK university groups and the British Geological Survey.

From its formation, BIRPS was pushed forward by the Cambridge group. Derek Blundell of London also became heavily involved. From its very beginning, it was supported by British-based oil companies. It was accepted that the objectives of reflection and refraction were different, the former providing near-vertical traveling ray-paths and highlighting structure, while the latter provided subhorizontal traveling ray-paths providing well-constrained velocities for the mid- and lower-crust and mantle, so highlighting composition and physical state.

The basics of BIRPS acquisition were summarized by Klemperer and Hobbs (1991) in their introduction to the BIRPS Atlas. The energy source used by BIRPS was based on airgun technology. For deep seismic profiling, the source had to produce in excess of 100 bar m in the bandwith from 5 to 80 Hz. It had to be rich in low frequencies, because frequencies below 20 Hz are best transmitted through the crystalline crust and get least attenuated. Typically, the source used by BIRPS consisted of 36 airguns which were distributed among six sub-arrays and were towed at ~7 m depth in an areal pattern of up to 175 m length and up to 100 m width behind the ship. The shot interval varied from 50 m for 15-s records to 100 m for 30-s records. The streamer was the same as used by seismic contractors for oil prospecting and was constantly exchanged whenever new technology became available. The streamer length was ~3000 m, consisted of 60 groups of hydrophones, and was towed at 15 m water depth. The data were recorded in digital format on magnetic tape. Each hydrophone group was recorded on one seismic data channel (Klemperer and Hobbs, 1991). The processing of the data followed the scheme described above by Phinney and Roy-Chowdhury (1989).

From early 1981 to midsummer 1984, a total of 3000 km of deep seismic-reflection profiles were acquired north, west, and south of Britain recording to 15 s TWT (Fig. 8.3.1-01), all being located within 500 km of the rifted margin of the continental shelf of NW Europe. Furthermore, Figure 8.3.1-01 shows 2600 km of seismic-reflection lines acquired in the North Sea by September 1984 and a few deep lines on land in Britain shot and observed by the British Geological Survey (Whittaker and Chadwick, 1983). Klemperer and Hobbs (1991) have compiled an atlas of 100 sections of the BIRPS surveys.

The first line, the Moine and Outer Isles Seismic Traverse (MOIST) was shot and processed in 1981. Figure 8.3.1-02 shows the MOIST traverse as a typical example of a BIRPS record. The reflections sketched have been interpreted as Mesozoic sedimentary basins and as subhorizontal layering in the lower crust extending from ~6 s to a relatively abrupt cut-off at ~10 s, the two-way reflection time to the Moho. Reflections dipping at ~30° from near-surface have been interpreted as low-angle normal faults, being reactivated Caledonian and Variscan thrust faults, merging with and being lost in the layering of the lower crust. Occasionally dipping events reach into the upper mantle as, e.g., the Flennan thrust in Figure 8.3.1-02. In order to see how far down the Flennan thrust penetrates, in 1984 the short line DRUM (Deep Reflections

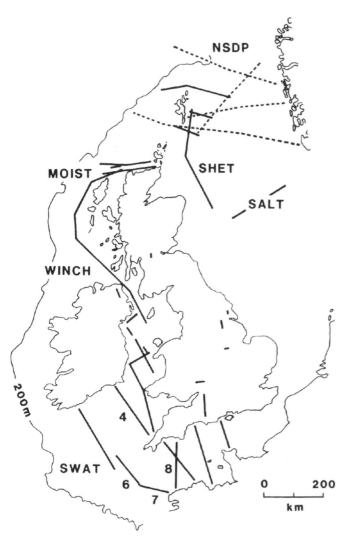

Figure 8.3.1-01. BIRPS deep seismic profiles completed prior to June 1984 over the continental shelf of NW Europe between Ireland, Britain, France, and Norway (from Matthews and Cheadle, 1986, fig. 1). [*In* Barazangi, M., and Brown, L., eds., Reflection seismology: a global perspective: American Geophysical Union, Geodynamics Series, v. 13, p. 5–19. Reproduced by permission of American Geophysical Union.]

from the Upper Mantle) was recorded to 30 s TWT parallel to the MOIST traverse. The Flennan thrust could be traced to 28 s as a nearly continuous bright reflector, corresponding to 80 km depth and 50 km below Moho (Blundell and Docherty, 1987).

The WINCH lines (Western Isles and North Channel traverse) followed in 1982 and confirmed the reality of the sub-Moho Flennan thrust. Deep seismic-reflection observations on the western side of Britain continued through the SWAT profiles (South-West Approaches Traverse), acquired jointly by BIRPS and the French group ECORS in late 1983 and processed in 1984. Thus, a more-or-less continuous set of deep seismic-reflection lines could be revealed from Scotland to France crossing the main faults of the Caledonian and Variscan orogenic belts, including the Variscan Front and several Permian and Mesozoic extensional basins. These were discussed and interpreted by Matthews and Cheadle (1986) who also refer to the corresponding original sources. Also jointly funded by BIRPS and ECORS, in 1985 a 600-km-long line WAM (Western Approaches Margin) was recorded off Cornwall in west-southwesterly direction to determine whether the layering of the lower crust could be related to extension or not (Peddy et al., 1989). The line started at the offset of the SWAT 6 line (Fig. 8.3.1-01) and ran out to the strongly stretched continental margin and further across the slope into the deep ocean up to magnetic anomaly 34. The Moho could be seen under the deep ocean, but could not be traced onto the continent (Blundell and Docherty, 1987).

In the North Sea, the line SHELL-UK82-101 was acquired in 1982 as a commercial profile by Shell UK showing the transition from the southern North Sea basin to the London-Brabant massif (Klemperer and Hurich, 1990). Another industrial line made available to BIRPS was the single deep profile SNST83-07 (Southern North Sea Tie 1983) recorded in 1983 in the southern North Sea, which provided a regional transect from East Anglia to Denmark across the southern part of the Central Graben and the Horn graben (Holliger and Klemperer, 1990). BIRPS reflection lines SALT (named for the Zechstein salt) were recorded in 1983, coinciding with the location of a wide-angle seismic experiment recorded in 1980 and 1981 across the Central Graben

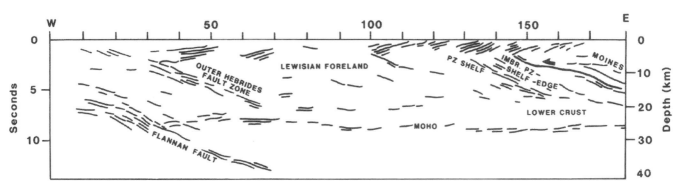

Figure 8.3.1-02. Line drawing of the BIRPS deep seismic line MOIST (from Matthews and Cheadle, 1986, fig. 3a). [*In* Barazangi, M., and Brown, L., eds., Reflection seismology: a global perspective: American Geophysical Union, Geodynamics Series, v. 13, p. 5–19. Reproduced by permission of American Geophysical Union.]

of the North Sea (Barton and Wood, 1984, Barton, 1986). Other BIRPS reflection surveys shot in 1984 in the North Sea, covering the northwest shelf of Europe, were SHET (Shetland survey), recorded in August 1984 along the shelf of northeastern Scotland (McGeary, 1987), and NSDP (North Sea Deep Profile), recorded in October 1984 between northern Scotland and southern Norway (Matthews and Cheadle, 1986).

Matthews and Cheadle (1986) argued that the BIRPS results in general support the idea of a ductile lower crust, showing numerous short reflections with a sharp downward cutoff at the Moho and that its flat-lying foliation is related to Mesozoic extension of the lower crust preceding the opening of the Atlantic. Dipping events reaching into the upper mantle are interpreted as imbricate faulting of a brittle layer in the uppermost mantle.

From 1985 to 1988, more surveys were carried out by BIRPS around the British Isles (Blundell and Docherty, 1987; Snyder and Hobbs, 1999). All surveys were recorded to 15 s TWT. The NEC (North East Coast line) project of 1985 was designed to confirm the north-dipping reflector seen on WINCH, which had been interpreted as the Iapetus suture (Freeman et al., 1988). NSDP was extended in 1985 by another 1600 km. GRID was a grid of lines shot in 1986 and 1987 north of Scotland around MOIST and DRUM to map the 3-D shape of the reflectors. The recording times varied between 16 and 48 s. SLAVE (Synthetic Large Aperture Velocity Experiment) in 1987 consisted of two ships towing airguns and 3.2 km digital streamers which steamed at constant offset to collect all the ray paths of a 16 km streamer. WIRELINES (West of Ireland Lines) in 1987 comprised 1265 km of reflection profiling crossing Caledonian structures to the northwest of Ireland and the Variscan front off the Shannon estuary. The survey extended a study of the deep crust to the Porcupine Bank, the westernmost part of Europe. Three short lines north of Ireland were designed to provide 3-D control on structures in the mantle and successfully imaged the northward-dipping structure of the Iapetus suture. MOBIL (Measurements Over Basins to Image Lithosphere) in 1987 was an extensive program in the southern North Sea. In addition, four onshore UK seismic recording stations extended the experimental network with wide-angle recordings. The three northern lines were designed to study the Iapetus suture in detail, the southern lines to investigate the gas basins. Finally WISPA (Weardale Integrated S- and P-wave Analysis) in 1988, a 20 km on-land acquisition program, was BIRPS' first onshore deep seismic-reflection experiment (Ward and Warner, 1991). Two crossing lines were shot across the Weardale granite, west of Durham, where in 1986 the British Geological Survey, using vibrators, had obtained good P-wave reflections down to the Moho. The BIRPS project was to provide corresponding S-waves applying a three-hole shooting technique and 3-component recordings. In 1990, the WESTLINE profile crossed the Rockall Trough west of Ireland (England and Hobbs, 1995, Snyder and Hobbs, 1999). It was a 450-km-long, normal-incidence, deep seismic-reflection profile which was shot perpendicular to the edges of the trough in order to image its conjugate margins.

8.3.1.2. ECORS (France)

The French national deep seismic-reflection program ECORS (Etude de la Croûte Continentale et Océanique par Réflexion et Réfraction Sismique) was designed to investigate the deep geology of the continental crust in France and to improve current seismic methodology both in fieldwork and in data processing (Bois et al., 1986, 1989). It aimed in particular to study deep basins which were of prime interest to the participating oil industry. The program started in 1983 and was jointly carried out by the Institut Français de Pétrole, Elf Aquitaine, and the Institut National d'Astronomie et de Géophysique (INAG) and assisted by the Centre National pour l'Exploitation des Océans for offshore operations. The first project in 1983 was cooperation with BIRPS in the Channel as described above (BIRPS and ECORS, 1986).

On land, more than 15 regional lines had been planned across France, of which fieldwork in northern France was completed by 1984 (double hatched line in Fig. 8.3.1-03). This first ECORS profile was 230 km long, and its goal was to study the Variscan belt below the Mesozoic and Tertiary cover of the Paris basin (Figs. 8.3.1-03 and 8.3.1-04). The originality of this first ECORS land profile was its combination of reflection, refraction and wide-angle reflection seismic surveying on the same con-

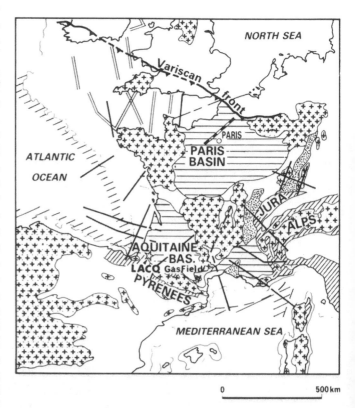

Figure 8.3.1-03. The ECORS project in France (from Bois et al., 1986, fig. 1a). [*In* Barazangi, M., and Brown, L., eds., Reflection seismology: a global perspective: American Geophysical Union, Geodynamics Series, v. 13, p. 21–29. Reproduced by permission of American Geophysical Union.]

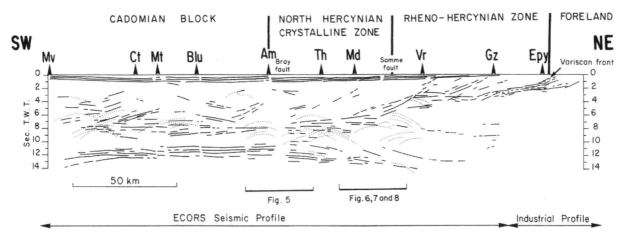

Figure 8.3.1-04. Line drawing of the deep seismic reflection profile in northern France (from Bois et al., 1986, fig. 4). The symbols on top mark borehole positions drilled down to the Paleozoic basement. [*In* Barazangi, M., and Brown, L., eds., Reflection seismology: a global perspective: American Geophysical Union, Geodynamics Series, v. 13, p. 21–29. Reproduced by permission of American Geophysical Union.]

tinuous profile, providing both resolution and velocity data. A 15-km-long geophone spread of 192 traces and 80 m separation recorded both sweeps from heavy vibrators placed every 6.6 m, six explosive shots at offsets of 0, 15, and 30 km on both sides of the spread and two more shots at offsets of 90 and 105 km to the north. Having thus recorded near-angle reflections, direct and reverse refraction data and direct wide-angle reflection data, the 15 km spread moved on and the procedure was repeated. The length of recording was 20 s. In addition, all waves from vibrators and explosive shots were also recorded by portable self-recording INAG stations which were distributed on both sides of the profile with offsets of 90–100 km. A line-drawing of the deep reflection profile in northern France is shown in Figure 8.3.1-04.

In 1984, a seismic-reflection survey was conducted in the Bay of Biscay along the Aquitaine shelf, with participation of FSH Hispanoil. The 300-km-long offshore seismic-reflection line ran in north-south direction (Marillier et al., 1988), and several perpendicularly oriented seismic-refraction lines were added. The long reflection line included a zero-offset and a 7.5 km constant offset vertical seismic-reflection profile and six expanding spread profiles with offsets larger than 100 km. As energy source an airgun array was used with very high energy in the 7–60 Hz frequency range. With a 3000 m streamer a 30-fold coverage was achieved. A second ship behind the first one also recorded the shots with a 2400 m streamer (Pinet et al., 1987).

After the completion of the detailed ECORS line through the Paris Basin, other lines followed in the east and south of France, partly again as international projects. Two profiles traversing the Central European Rift System in the Rhinegraben area will be described below in connection with the German deep seismic-reflection program.

The Bressegraben (Bergerat et al., 1990; Guellec et al., 1990; Truffert et al., 1990) was part of a long-range deep seismic-reflection line starting in the Massif Central and crossing the Alps toward Torino (Bayer et al., 1987), a project which was jointly

executed by ECORS and CROP (Crosta Profonda), a deep reflection program initiated by Italian institutions.

The Aquitaine Basin was studied in conjunction with a line through the Pyrenees (ECORS Pyrenees Team, 1988) and extended the above mentioned 1984 line in the Gulf of Biscay onto land.

Overviews of all ECORS activities in the 1980s were published by Bois et al. (1987) and Bois and ECORS Scientific Parties (1991).

8.3.1.3. CROP (Italy)

Crosta Profonda (CROP), the Italian deep seismic-reflection program, was planned between 1982 and 1984 and finally launched in 1986, supported by the Italian National Research Council (CNR) and the two leading companies in the energy sector, ENI-AGIP and ENEL (Bernabini and Manetti, 2003). The main goal of CROP was to investigate the crustal structure of Italy by near-vertical reflection seismic profiling, similar to the seismic-reflection programs in the United States and western European countries. From the very beginning, CROP had two components: onshore and offshore investigations (for location, see Fig. 9.2.3-05). The initial phase lasted from 1985 to 1988 and comprised two land projects in the Alps and one marine project in the western Mediterranean Sea, the second phase with the majority of acquired land and sea profiles was realized in the 1990s (see Chapter 9.2). The data of all acquired land and marine lines were assembled as sections in the CROP Atlas (Scrocca et al., 2003).

The initial phase was focused on the acquisition of crustal seismic profiles in the Alps, in cooperation with working groups from surrounding countries that were working on the same type of projects. The first task was, in cooperation with the French group ECORS (see section 8.3.1.2), to explore the Western Alps on a profile from Torino to Geneva. The fieldwork was carried out in 1986, when the 93-km-long Italian part of the 300-km-long line was recorded, crossing the 2999-m-high Col de la Galise (for location, see Figs. 8.3.1-03 and 9.2.3-05).

The second project was achieved from September to December 1988 in collaboration with the Swiss NFP 20 (see section 8.3.1.4), to explore the Central Alps on three lines in the area of Mount Generoso, Valle dello Spluga, and Val Brembana. In total, 102 km of seismic profiles were recorded.

The third project was the first marine project. In collaboration with ECORS, a seismic profile CROP-MARE I was recorded in the western Mediterranean from the Provence to NW Sardinia across the Sardinian-Balearic-Provencal Basin. It was part of the French-Italian ECORS-CROP project and consisted of a 570-km-long ECORS part and a 210-km-long Italian part, acquired in October 1988. It represents the Italian transect of a discontinuous line linking the continental inner shelf of the Gulf of Lyon and the northwestern Sardinian slope.

The data of all three projects were processed and interpreted by joint working groups and the results presented in crustal cross section along the complete lines. From the interpretation of the data of the ECORS-CROP traverse, two conclusions were drawn: (1) the existence of mantle fragments at different structural levels demonstrated that the mantle was also involved in the orogenesis, and (2) the predominant west-vergence of major crustal discontinuities confirmed that the European plate is being subducted to the southeast in the Western Alps (Bayer et al., 1987; Nicolas et al., 1990; Thouvenot et al., 1990; Roure et al., 1996; Bernabini et al., 2003). The data of the NFP20-CROP profiles through the Central Alps were interpreted jointly with the Swiss colleagues and partly combined with the interpretation of the seismic-refraction data of the European Geotraverse (e.g., Valasek et al., 1991; Ye, 1991; Ansorge et al., 1992). The Italian part contributed especially to the conclusion that the South Alpine upper crust appeared to be rootless and cut off along a basal thrust and could be linked with the south-verging Milano belt, while on the other hand the deeper Adria lithospheric elements wedged northward in the opposite direction into the remnants of the accretionary prism and into the European crust (Montrasio et al., 2003). The Mediterranean traverse indicated the complete absence of a spreading center or ridge in its oceanic sector. The structural pattern of the opposite margins appeared to be asymmetrical (de Voogd et al., 1991; Gorini et al., 1993; Fanucci and Morelli, 2003).

8.3.1.4. Switzerland

Deep seismic-reflection work in Switzerland started in 1982 when a total of 180 km Vibroseis lines were recorded in northern Switzerland to investigate the suitability of the crystalline basement as host rock for the storage of highly radioactive waste. Most of the lines were set up to resolve details in the uppermost 4 s; i.e., the upper crust, but part of the lines was selected for a reprocessing procedure to extend the correlated recording length to a maximum of 14 s (Finckh et al., 1984, 1986). Another, but only shallow detailed reflection survey dealt with the structure of the Swiss Rhône valley (Finckh and Frei, 1991).

The most important project was the project NFP 20 (Swiss National-Fond Project) with activities from 1986 to 1990. In connection with the European Geotraverse project, three deep seismic-reflection lines were established (Fig. 8.3.1-05). They crossed the Swiss Alps in a north-south direction (Heitzmann et al., 1991; Stäuble and Pfiffner, 1991; Valasek et al., 1991). What 20 years of seismic-refraction work in the Alps did not find, the Swiss reflection surveys in the 1980s showed clearly: the European subduction. The program also included the Swiss part of the European Geotraverse refraction program, described in more detail below. The combination of both reflection and refraction seismology contributed essentially to our knowledge of crustal structure and tectonics of the Alps.

Valasek et al. (1991) have outlined a crustal framework based on the initial results of the seismic-reflection profiles across the central Alps. It involves northward indentation of the Adriatic hinterland into the subducting European foreland. This occurs beneath the collapsed oceanic basins of the Penninic allochthon which is defined by highly reflective crystalline nappes. The present crustal thickness within the Alpine collision zone has doubled to ~60 km by both vertical and horizontal displacements along an inferred complex detachment system controlled by the Adriatic wedge. The south-plunging European foreland implies that portions of the thickened crust have been subducted into the upper mantle. Figure 8.3.1-06 shows NFP seismic-reflection data recorded in Switzerland in 1986 and 1988 (Valasek et al., 1991) along the European Geotraverse (EGT). The distances were measured from the seismic-refraction shotpoint E, also shown are locations of EGT shotpoints D and C (shotpoints CE, CD, and AC in Figure 8.3.4-02 in section 8.3.4.3 below).

8.3.1.4. DEKORP (Germany)

In Germany, deep seismic-reflection investigations had been pursued consistently since the end of the 1950s. It had started with long-time recording of oil prospecting reflection seismic investigations (Dohr, 1957, 1972), later specially planned seismic-reflection experiments were carried out in Southern Germany and the Rhenish Massif (Meissner, 1966; Dürbaum et al., 1971; Meissner et al., 1981), and in the mid-1970s a special working group "Deep Reflections" had been established. Following the example of COCORP, finally, in 1983 a ten-year program, DEKORP (Deutsches Kontinentales Reflexionsseismisches Programm = German continental reflection seismic program), was founded and funded by the German Ministry for Research and Technology.

Its basic scientific aim was to explore the deep geology of the German Variscides which had not been touched by commercial reflection seismic investigations (Dürbaum et al., 1997). Basically the Vibroseis technique was to be applied, and only occasionally some explosive sources should be added in particular for piggyback wide-angle investigations. For planning individual projects and their subsequent data interpretation, Regional Working Groups would be responsible, the final decision and application for funding would be made by the DEKORP steering

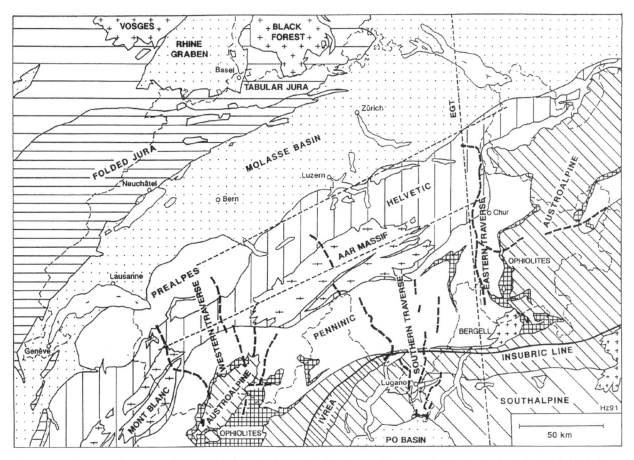

Figure 8.3.1-05. Location map of NFP seismic reflection and refraction profiles recorded in Switzerland in 1986–1990 (from Heitzmann et al., 1991, fig. 1). [*In* Meissner, R., Brown, L., Dürbaum, H.-J., Franke, W., Fuchs, K., and Seifert, F., eds., 1991, Continental lithosphere: deep seismic reflections: American Geophysical Union, Geodynamics Series, v. 22, p. 161–176. Reproduced by permission of American Geophysical Union.]

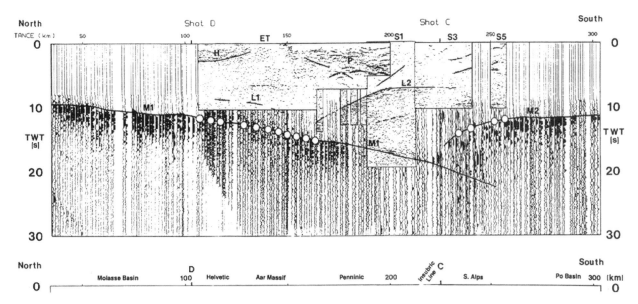

Figure 8.3.1-06. NFP seismic reflection data recorded in Switzerland in 1986 and 1988 (from Valasek et al., 1991, fig. 9) along the European Geotraverse (EGT). Distance measured from seismic-refraction shotpoint E, also shown are locations of EGT shotpoints D and C. [Geophysical Journal International, v. 105, p. 85–102. Copyright John Wiley & Sons Ltd.]

committee, the fieldwork would be submitted to a contractor, and the data would be processed in the DEKORP processing center at the University of Clausthal. Thus, in the following years from 1983 to 1993 a number of long-range seismic-reflection lines were realized (Fig. 8.3.1-07).

In total, seven individual projects were carried out until 1990 (Meissner et al., 1991a, 1991b). The first line, DEKORP-2 South (D2S in Fig. 8.3.1-07), was carried out in 1984. It crossed southern Germany between the rivers Main and Danube (DEKORP Research Group, 1985). In the upper crust down to 5–6 s TWT reflections were limited to occasionally dipping events, which were interpreted as thrusts, mostly marking boundaries between the Moldanubian, the Saxothuringian, and the Rhenohercynian. They transformed into flat-lying reflections when penetrating the lower crust. The lower crust was full of reflecting lamellae, most of them predominantly subhorizontal, which ended abruptly at the Moho. The distance range along which reflections could be correlated was generally less than 4 km (Meissner et al., 1987a).

Following DEKORP 2-S the seismic-reflection profiles K84 in the Black Forest (Lueschen et al., 1987, 1989) were recorded and funded by the KTB (continental deep drilling) project, which will be described in the next section. Also in the framework of KTB, in 1985 the DEKORP-4 line (D4) in northeastern Bavaria was recorded, together with some additional lines KTB 85/1 to KTB 85/6 (K85) around the proposed superdeep drillhole site in the Oberpfalz, and some 90 kg explosions for wide-angle seismics and expanding spread velocity recording (DEKORP Research Group, 1988). Finally, in 1989, a 3-D seismic-reflection survey was added (Dürbaum et al., 1992).

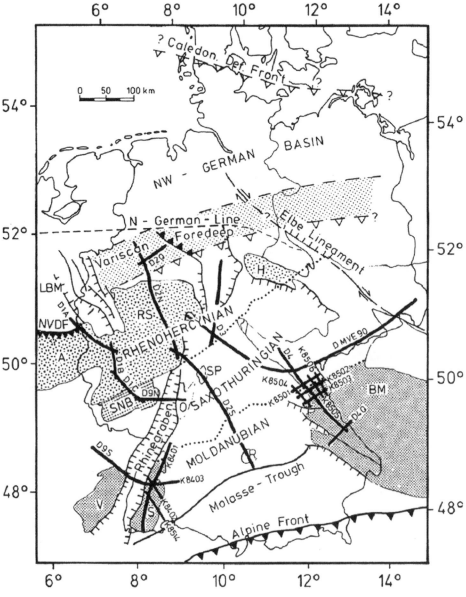

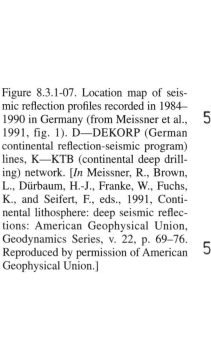

Figure 8.3.1-07. Location map of seismic reflection profiles recorded in 1984–1990 in Germany (from Meissner et al., 1991, fig. 1). D—DEKORP (German continental reflection-seismic program) lines, K—KTB (continental deep drilling) network. [*In* Meissner, R., Brown, L., Dürbaum, H.-J., Franke, W., Fuchs, K., and Seifert, F., eds., 1991, Continental lithosphere: deep seismic reflections: American Geophysical Union, Geodynamics Series, v. 22, p. 69–76. Reproduced by permission of American Geophysical Union.]

Closely connected with the Black Forest KTB profiles, the Rhinegraben and its flanks became the special target of two seismic-reflection lines. Both lines were cooperative projects with ECORS. DEKORP-9 South (D9S) traversed the southern Rhinegraben near Strasbourg where both Vibroseis technique and borehole shots were applied (Brun et al., 1991). DEKORP-9 North (D9N), starting in the Odenwald, traversed the Rhinegraben near Worms and led into the Saar-Nahe trench (Wenzel et al., 1991). Under the flanks of the Rhinegraben, the lower crust was characterized by strong reflectivity interpreted as a layered lower crust. The Moho was marked as a strong reflector at ~8 s below the eastern flank and the graben, but its depth increased gradually to the west to ~10 s TWT. East-dipping reflectors in the middle crust were attributed to Variscan features. In general, the Rhinegraben proved to be markedly asymmetric (Brun et al., 1991).

DEKORP-9 North (D9N) connected with DEKORP-1 (D1A–D1C) in the Saar-Nahe trench. DEKORP-1 crossed the Rhenish Massif west of the Rhine and the Aachen-Midi thrust zone and led as BELCORP line, organized by the Belgian Geological Survey, into southeastern Belgium (DEKORP Research Group, 1991). The aim of this 220-km-long line was to investigate the crustal structure of the western part of the Rhenish Massif. The results indicated the presence of NW-vergent tectonics of various styles which could often be traced into deeper parts of the crust. For the northern part Variscan compression playing a dominant role was inferred, while post-Variscan exten-

sion was seen to dominate in the Saar-Nahe trough. In the north, a deep extension of the Aachen thrust (Faille du Midi) was clearly observed. This prominent thrust with its characteristic ramp and flat structure could be followed over a distance range of 100 km down to 15 km depth. It contrasted sharply with the very complex deep fault system in the south, separating the post-orogenic Saar-Nahe trough from the Hunsrück Mountains (DEKORP Research Group, 1991).

East of the Rhine, the line DEKORP-2 North (D2N) was realized, traversing the Rhenish Massif and the Münsterland basin to the north (data example in Fig. 8.3.1-08), supplemented by DEKORP-2 NQ (D2Q), a perpendicularly oriented line (Franke et al., 1990). This line revealed an almost complete cross section through the Rhenohercynian zone. In the upper crust, many strong south-dipping reflections could be correlated with thrust faults known from surface geology. At depth, the thrusts flatten out in a relatively transparent zone between 3 and 5 s TWT. This zone could be correlated with high-conductivity layer detected in a magnetotelluric survey (Jödicke et al., 1983). The lower crust is relatively transparent in the north, but farther south, it is increasingly reflective, consisting of curvilinear thrust-related reflections which are cut by a conjugate set of weaker N- and S-dipping reflectors. The Moho was found to rise from north to south from ~11 to 8.5 s TWT (Franke et al., 1990).

Finally, in 1990, DEKORP-3 (D3) was realized, traversing the Hessen depression from Göttingen to the Main River in

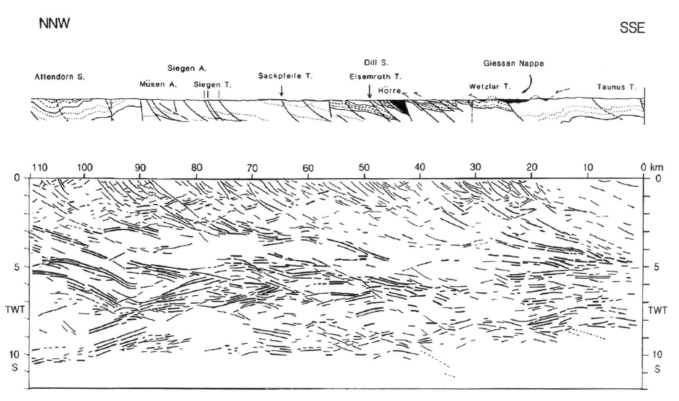

Figure 8.3.1-08. Line drawing of part of the deep seismic reflection profile DEKORP-2 North in western Germany (from Franke et al., 1990, fig. 6). [Geologische Rundschau, v. 79, p. 523–566. Reproduced with kind permission of Springer Science+Business Media.]

a north-south direction and then bending toward the east-northeast, to be continued as DEKORP-MVE-90 line D MVE90) into eastern Germany traversing the gneiss massif of Münchberg, the earthquake swarm area of the Vogtland and the Erzgebirge north of the Czech-German border (DEKORP Research Group, 1994). This line will be discussed in Chapter 9.2.2.

Based mainly on the dense deep-seismic network of DEKORP, Meissner and Sadowiak (1992) discussed the terrane concept as an important extension of plate tectonics and applied it successfully to the Variscides with their wide range of collisional belts. The DEKORP data reveal certain reflectivity patterns and enable to map old and new deep fault zones between the Variscan terranes which seem to consist of continental crust only, partly exclusively of rigid upper crust, and which are rooted in the ductile lower crust (Meissner and Sadowiak, 1992).

A systematic comparative study of the reflectivity of the continental crust was undertaken by Sadowiak et al. (1991), using all deep seismic-reflection data available in western, northern, and central Europe until 1988 (see Fig. 2.8-01). At almost all seismic-reflection profiles of BIRPS, ECORS, and DEKORP, as far as they were recorded over Caledonian and Variscan basement, the Moho is clearly identified as the lowest boundary of the laminated lower crust. They observed that the reflectivity changed dramatically from one seismic line to the next, and they classified the different seismic reflectivity patterns and established correlations between these seismic patterns and specific tectonic units. Lamellae and bands of reflections in the lower crust were widespread in post-orogenic extensional areas, "crocodiles" seemed to represent compressional zones that were occasionally accompanied by duplex structures. The "fishbone" pattern, consisting of many "crocodiles" was characteristic for the old London-Brabant Massif. The "ramp and flat" structure displayed a thin-skinned tectonics of the North Variscan Deformation Front over a length of 2000 km. Diffraction clusters in the lower crust accompanied by a dipping reflection in the upper crust was observed close to thick-skinned deformation fronts. Diffractions were also present in rift zones as in the North Sea. Some regions, where a decreasing reflectivity with depth, a concentration of reflections in the upper crust, and no Moho reflections were observed, were correlated to areas with Precambrian crust (Sadowiak et al., 1991).

A later review of the seismic-reflection profiles throughout central Europe was prepared by Meissner (2000) where he discusses the various reflectivity patterns and strongly varying characteristics with respect to the crustal evolution of central Europe by collision and collapse of continental terranes, considering the whole area of Western and Central Europe as a continental margin. In a similar way, England (2000) discusses the extensive grid of BIRPS offshore deep seismic-reflection profiles over the North Sea and the continental margins of Northwest Europe.

8.3.1.6. Czechoslovakia

Two deep seismic-reflection lines were observed in the 1980s in Czechoslovakia. One profile was 60 km long and was recorded through the Vienna and Pannonian basins in a NW-SE direction nearly parallel to the Austrian-Hungarian border, crossing the Little Carpathians north of Bratislava. The other profile ran through the Inner West Carpathians in a NNW-SSE direction from the Polish to the Hungarian border and consisted of two ~60-km-long parts. Cross lines were also recorded. All shots were dynamite charges. A 96-channel recording system was used with 50 m spacing on the western line and 80 m on the eastern line producing central offsets up to 2.4 km in the west and 3.8 km in the east. On both lines nominal 24-fold data resulted. Standard processing procedures, as described in the introduction above were applied.

The Czech profiles through the West Carpathians were the first systematic deep survey within the whole active Alpine-Himalayan mountain belt using near-incidence vertical seismic-reflection techniques (Tomek et al., 1987). On the western line, deep reflections up to 11.5 s were recorded underneath the Little Carpathians north of Bratislava, while at the southeastern end in the Danube Basin slightly westward dipping events at 10–11 s were observed. From the line through the Inner West Carpathians, only the northern portion was published by Tomek et al. (1987). The upper crust appeared highly reflective along the whole line. Underneath the northernmost 20 km section of the line also the lower crust between 7 and 12 s TWT was highly reflective expressed by a band of strong reflectors dipping southward. Farther south, the middle and lower crust were almost featureless; only at the southern end, under the center of the Carpathians, a short band of reflections appeared again from ~9–12 s TWT dipping in the reverse direction.

8.3.1.6. BABEL (Scandinavia)

Finally, in 1989 an international large-scale seismic-reflection experiment, BABEL (Baltic and Bothnian Echoes from the Lithosphere), was carried out as a joint venture by British, Danish, Finnish, German, and Swedish scientists from 12 research institutions in the Baltic Sea and the Gulf of Bothnia, parallel to the FENNOLORA (Fennoscandian Long Range Project) seismic-refraction survey of 1979 (BABEL Working Group, 1991, BABEL Working Group, 1993a, 1993b). It consisted of nine marine deep seismic-reflection profiles totaling 2268 km of high-quality data and formed a 1350 km traverse from the edge of the Archean nucleus to Phanerozoic western Europe. Lines 1–7 and line C (Fig. 8.3.1-09) followed more or less the Gulf of Bothnia, while lines B and A were planned parallel to the direction of the southeastern coasts of Sweden and Denmark. The BABEL lines were all located on the Precambrian Baltic Shield, with the exception of Line A which crossed the Tornquist Zone (TZ in Fig. 8.3.1-09) near Bornholm Island and continued into the North German Basin, underlain by Caledonian basement. In the same year (1989), the Institute of Oceanology of the Russian Academy of Sciences organized a 480-km-long profile, running NNE from the Bay of Gdansk ~50–100 km off the coast of the Baltic states and nearly parallel to the line B of the BABEL survey (Ostrovsky, 1993). The seismic signals of a powerful airgun (120 L) were recorded at five points by ocean-bottom seismometers equipped with hydrophones.

Figure 8.3.1-09. Location map of seismic reflection profiles of BABEL (from BABEL Working Group, 1991, fig. 1). Thick lines—reflection profiling; triangles—seismometer arrays or stations, stippled patterns—tectonic boundaries. A1—Aland islands, Bo—Bornholm island; Ö—Öland island; TZ—Tornquist Zone. [*In* Meissner, R., Brown, L., Dürbaum, H.-J., Franke, W., Fuchs, K., and Seifert, F., eds., 1991, Continental lithosphere: deep seismic reflections: American Geophysical Union, Geodynamics Series, v. 22, p. 77–86. Reproduced by permission of American Geophysical Union.]

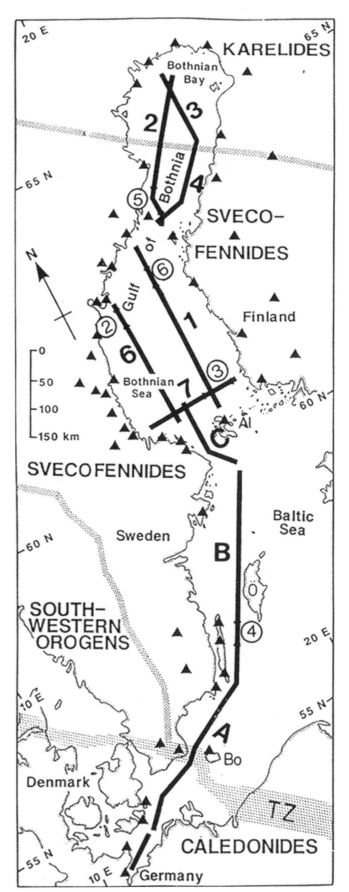

The BABEL survey not only acquired deep seismic-reflection data, but provided also coincident refraction/wide-angle reflection data to study the crustal structure of the Baltic Shield and its transition to the Phanerozoic crust of western Europe across the Tornquist Zone. For this purpose the 12-l airgun array used for this profiling was also recorded by 64 land stations to provide coincident refraction profiles, fan-spreads and 3-D seismic coverage of much of the Gulf of Bothnia and the Baltic Sea area adjacent to southern Sweden and Denmark.

Shot-to-receiver offsets ranged from 200 m to 350 km. In one case, even offsets of up to 700 km proved successful. The near offsets of 200–3200 m were used for traditional deep seismic profiling with specifications which the British BIRPS group had developed over the years. So, hydrophone groups in a 3-km-long streamer were towed behind a seismic vessel which produced a multiplicity of coverage up to 24-fold. The far offsets were obtained using the airgun array towed from the ship as the seismic source and recording on land at 64 points around the Gulf of Bothnia and the Baltic Sea. The geometry of the coastlines and islands allowed some land stations to be located on sites which were at each ends of offshore seismic lines and so provided a reversed refraction profile. In other cases, fan observations resulted. The success of the survey was based on the short shot-spacing of 62–75 m that made the resulting high-density profiles so effective. A great diversity of types of seismometers and geophone configurations was used on land, ranging from three-component seismometers with 1–2 Hz geophones to single-component outstations connected with a central recording site to linear arrays with 48 vertical seismometer groups at 50 m spacing recorded as traditional land array data for stacking.

In the Bothnian Bay (for locations, see Fig. 8.3.1-09), a south-dipping, non-reflective zone was seen to coincide with a conductive Archean-Proterozoic boundary onshore in Finland. Between the Bothnian Bay and the Bothnian Sea observed reflectivity geometries and velocity models reaching Moho depths suggested structures inherited from a 1.9 Ga subduction zone, the upper crust having anomalously low velocities. From wide-angle reflections, the metasedimentary crust of the Bothnian Bay seemed to be 10 km thicker than the adjacent Svecofennian subprovinces.

The line drawing of line A (Fig. 8.3.1-10) shows a variable character of deep reflections. South of the Tornquist Zone, a band of reflections is clearly visible in the lower crust, while the crust north of Tornquist Zone is dominated by diffractions. In

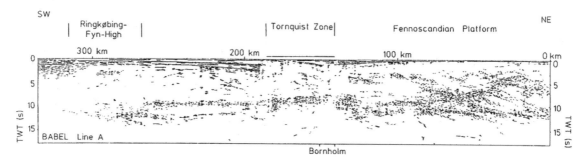

Figure 8.3.1-10. Line drawing of BABEL profile A (from BABEL Working Group, 1991, fig. 7). The line extends from the Caledonian German-Danish Basin in the SW across the Tornquist Zone (TZ) near Bornholm island into the Fennoscandian Platform in the NE. [*In* Meissner, R., Brown, L., Dürbaum, H.-J., Franke, W., Fuchs, K., and Seifert, F., eds., 1991, Continental lithosphere: deep seismic reflections: American Geophysical Union, Geodynamics Series, v. 22, p. 77–86. Reproduced by permission of American Geophysical Union.]

the shield parts of the profiles northeast of the Tornquist Zone, the Moho was found between 40 and 48 km, corresponding to 12–15 s TWT. From the wide-angle data, a three-layer crust was deduced with velocities of ~6.1–6.4, 6.6, and 6.9–7.2 km/s. South of the Tornquist Zone, a highly reflective lowermost crust between 8 and 10 s TWT corresponded to a depth range of 25–31 km and a high velocity gradient (6.7–7.1 km/s), but the wide-angle data did not show evidence for intracrustal discontinuities. The Ringkoebing-Fyn basement high could also be located in the data, and the crust underneath the North German lowlands was found underneath 10-km-thick post-Caledonian sediments, being 20 km thick with velocities between 6.0 and 6.9 km/s.

8.3.2. Seismic-Refraction Studies in the British Isles and Ireland

As mentioned above, when discussing the deep seismic-reflection surveys of BIRPS, the continental region of the British Isles includes the onshore regions of the British Isles as well as the stretched continental crust beneath the adjoining seas including the North Sea. Purely offshore marine surveys on the Northeast Atlantic margin to the south and west of Ireland, however, we will discuss in the section 8.9.4 (Atlantic Ocean).

In 1980 and 1981, a wide-angle offshore seismic line was shot across the Central Graben of the North Sea running from southern Scotland toward the southern tip of Norway. The location of its central part is identical with the location of the BIRPS line SALT (see Fig. 8.3.1-01). The object of the project was to test the lithospheric stretching model for the formation of sedimentary basins (MacKenzie, 1978). An array of twelve pull-up shallow-water seismometers was laid in the Central Graben and shots were fired in a line perpendicular to the axis of the basin (Barton and Wood, 1984; Barton, 1986). The data were analyzed using ray-tracing and amplitude modeling of the seismograms described in detail by Barton and Wood (1984). A model of the crust was developed with 2-D variations in both seismic velocity and structure. Seismic gradients were simulated by dividing the model into 1 km² elements, each of which contained a single

velocity value. The model, containing ~20,000 such elements, was well constrained by the numerous crossing ray paths which connected twelve shots and twelve seismometers. The velocity-depth-distance model was then converted to a normal incidence TWT profile of the Moho. Traveltimes were calculated by integrating the traveltime contributions from each of the constant velocity elements vertically from the surface to the Moho and back. The data were then compared with the two deep normal incidence BIRPS profiles SALT mentioned above in section 8.3.1.1 which were recorded along the seismic-refraction line, each being 70 km long and 15 s deep. The experiment proved that the time-converted Moho trace from the wide-angle model falls clearly at the base of the layering in the lower crust.

In the southwest of Scotland, the Western Isles Seismic Experiment of 1979 (see Chapter 7.2.4) was continued in 1981. Explosive and airgun shots were recorded along a second profile between Mull and Kintrye aiming to map the Lewisian basement from the foreland to the Caledonian mountain belt (Summers et al., 1982).

In 1984, using two 90-km-long seismic-refraction lines, MAVIS (the Midland Valley Investigation by Seismology) aimed to study the structure of the Carboniferous basin of West Lothian in the Midland valley of Scotland (Conway et al., 1987; Dentith and Hall 1990). The data revealed details of the Paleozoic trench sequence and the depth to basement, but did not penetrate into deeper levels of the crystalline crust.

In 1982, under the direction of Martin Bott, a British long-range seismic-refraction experiment of the University of Durham in 1982 aimed to study the crustal structure underneath the Caledonian suture zone separating Variscan from Caledonian basement in northern England with very closely spaced shots and stations (Bott et al., 1983, 1985). Their line, named CSSP (Caledonian Suture Seismic Project) started in the North Sea, crossed the north of England and ended with a line of shots terminating in the Irish Sea near the coast of Ireland (Fig. 8.3.2-01). It was a unique sea-to-land wide-angle explosion seismology project that produced data of unprecedented quantity and quality. The principal innovation was the use of closely spaced shots, mainly at sea, and closely spaced stations, mainly on land, yielding a data

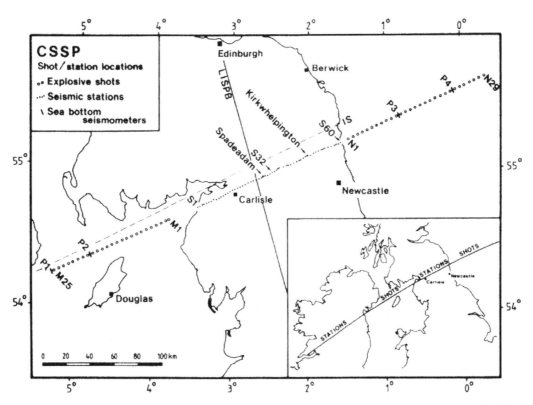

Figure 8.3.2-01. Location map of Caledonian Suture Seismic Project (CSSP) (from Bott et al., 1985, fig. 1), also showing the LISPB line of 1974 (see Chapter 7). [Nature, v. 314, no. 6013, p. 724–727. Reprinted by permission from Macmillan Publishers Ltd.]

set that permitted arrangement of the data in a variety of ways for phase correlation and modeling, such as in constant offset and common mid-point displays.

The results for this region differed radically from those of the earlier LISPB experiment (Bamford et al., 1978). The data allowed recognizing a well-defined pre-Caledonian metamorphic basement beneath most of the line at relatively shallow depth (Fig. 8.3.2-02). A well-defined mid-crustal refractor was also detected at ~16 km depth beneath the North Sea and northern England which was not recognized beneath the Irish Sea (Bott et al., 1985). Al-Kindi et al. (2003), concentrating on the lower crust beneath the Irish Sea, remodeled the data and identified a

Figure 8.3.2-02. Crustal structure along the Caledonian Suture Seismic Project (CSSP) line through Northern England (from Bott et al., 1985, fig. 4). *a*—shallow crustal structure, *b*—speculative cross section through northern England, perpendicularly crossing the CSSP line. Crustal structure under the Midland Valley is from LISPB (Bamford et al., 1978). [Nature, 314, no. 6013, p. 724–72. Reprinted by permission from Macmillan Publishers Ltd.]

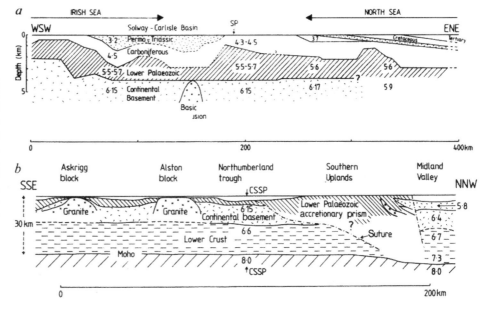

high-velocity crustal layer. They interpreted this layer as underplate resulting from a mantle thermal anomaly and causally related to Cenozoic uplift beneath this part of the British Isles.

The shots of the British CSSP project motivated Brian Jacob at the Dublin Institute for Advanced Studies (DIAS) to plan the first seismic crustal profile through Ireland. The CSSP line of shots in the Irish Sea was extended toward the southwest by the University of Durham to include five widely spaced shots with the outermost shot as close as possible to the Irish coast. This provided a starting point for the 250-km-long line of recording stations through Ireland along the Caledonian strike and roughly parallel to the postulated Caledonian suture zone (Appendix A2-1, p. 60–61). In total, there were 31 CSSP shotpoints in the Irish Sea. At the southwestern end of the line, the Irish group fired two additional shots in the Shannon river estuary and off the coast. These were small dispersed (50–100 kg) offshore shots (one at optimum depth) from which seismic energy was recorded at distances of almost 200 km. A quarry in the southwest of Ireland was also used as an additional shotpoint (Jacob et al., 1985). A total of 51 instruments from DIAS, Ireland, and Karlsruhe University, Germany, were deployed at 5 km spacing parallel to the postulated trace of the suture zone (Fig. 8.3.2-03).

The majority of the data is displayed in Appendix A2-1 (p. 60–61). The middle and lower part of Figure 8.3.2-04 demonstrate that the crustal structure is well constrained by the good coverage of shotpoints and recording stations.

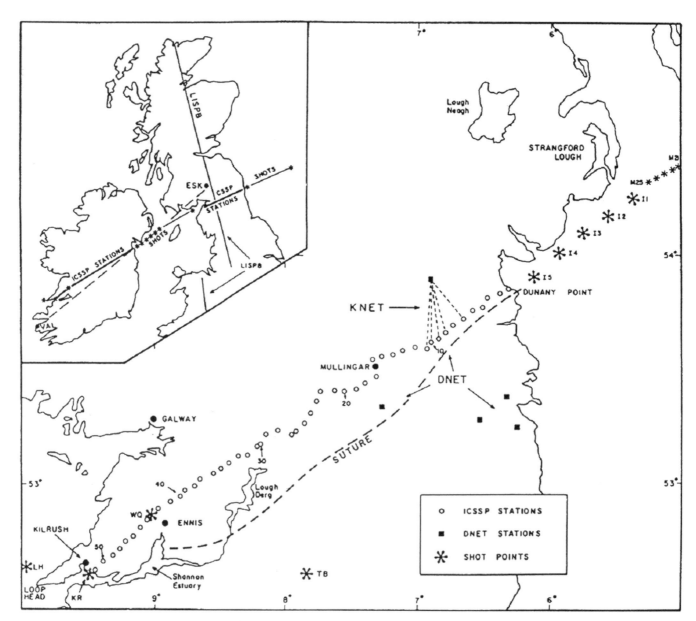

Figure 8.3.2-03. Location map of the Irish Caledonian Suture Seismic Project (ICSSP) project of 1982 (from Jacob et al., 1985, fig. 1). [Tectonophysics, v. 113, p. 75–103. Copyright Elsevier.]

This first seismic project in Ireland was soon to be followed by major activity to study the crust underneath Ireland and its surrounding seas. Figure 8.3.2-05 summarizes all seismic crustal observations in and around Ireland from the 1960s to the end of the 1990s.

In 1985 the Caledonian Onshore-Offshore Lithospheric Experiment (COOLE) was organized (Appendix A2-1, p. 62–63). It consisted of two separate seismic-refraction studies carried out onshore and offshore southern Ireland. The onshore part consisted of a 280-km-long N-S profile (COOLE'85-1 in Fig. 8.3.2-05) stretching from the southern coastline across the Irish Midlands up to Donegal Bay (Lowe and Jacob, 1989). A total of 6 shots was recorded by 59 seismic receivers deployed at ~3 km spacing. The profile was approximately perpendicular to the ICSSP profile of 1982 and the proposed surface trace of the suture zone. The main objectives of this project were to investigate the crustal structure across the suture zone and to provide seismic control for the very dense gravity data onshore Ireland.

The offshore study of the COOLE experiment concerned the investigation of the Celtic Sea to the southwest of Ireland. This will be discussed in more detail together with other offshore experiments in the subchapter 8.9.4, Atlantic Ocean.

In 1987, a 2-ton explosive source which was fired in the North Sea off Scotland, was recorded in southeastern Ireland on the so-called BB 1987 profile (BB'87 in Fig. 8.3.2-05) oriented in a SSW direction (Bean and Jacob, 1990; Jacob et al., 1991). Combined with the far distant shots of the CSSP project recorded along the ICSSP line in Ireland, a long-range line of ~600 km length resulted.

Thus together with the LISPB observations (Faber and Bamford, 1979), three long-range and intersecting refraction profiles resulted in Ireland and Britain which provided detailed information about the velocity-depth structure of the lower lithosphere and sampled the upper mantle along different azimuths. The observation that upper-mantle velocities were different for the CSSP and LISPB profiles, and that P-wave velocities of 8.6 km/s at 85 km depth were well in excess of the predicted isotropic velocities of 7.85 km/s (taking 80% olivine, 20% pyroxene and a calculated geotherm for the region), led Bean and Jacob (1990) to the conclusion that seismic anisotropy should be present in the lithospheric mantle.

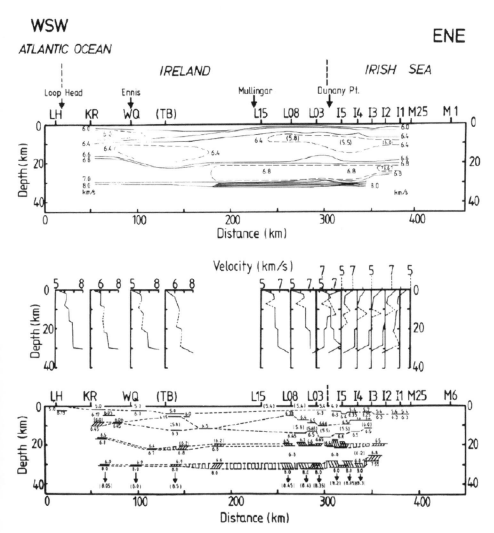

Figure 8.3.2-04. Crustal cross section resulting from the ICSSP project of 1982 (from Jacob et al., 1985, fig. 8). Upper part: Cross section with lines of equal velocity, middle part: 1-D velocity-depth functions, lower part: Position of zones where crustal reflected phases originate. [Tectonophysics, v. 113, p. 75–103. Copyright Elsevier.]

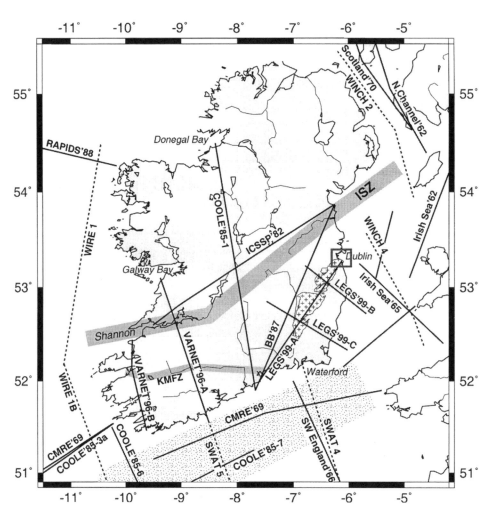

Figure 8.3.2-05. Seismic refraction and reflection profiles onshore/offshore Ireland recorded in the period of 1969–1999 (from Landes et al., 2005, fig. 1). Full lines: Irish Sea 1962, 1965 (Willmore, 1973), SW England 1966 (Willmore, 1973), CMRE 1969 (Bamford, 1971), ICSSP 1982 (Jacob et al., 1985), COOLE 1985 (Makris et al., 1988; Lowe and Jacob, 1989; O'Reilly et al., 1991), BB 87 RAPIDS 1988 (O'Reilly et al., 1995), VARNET-1996 Lines A and B (Landes et al., 2000; Masson et al., 1998), LEGS 1999 Lines A, B and C (Hodgson, 2001). Dashed lines: normal incidence reflection data by BIRPS: WIRE, WINCH, SWAT (Klemperer and Hobbs, 1991). Shaded: proposed locations of the Iapetus Suture Zone and the Killarney Mallow Fault Zone (ISZ, KMFZ). Dotted area offshore southern Ireland represents outlines of the North Celtic Sea Basin. Area with crosses in SE Ireland resembles surface outcrop of the Leinster Granite batholith (modified after Hodgson, 2001). [Terra Nova, v. 17, p. 111–120. Copyright Wiley-Blackwell.]

8.3.3. The German Deep Drillhole Project KTB

Following the successful completion of the super-deep drillhole KOLA SG3 which was located on the Baltic Shield of Russia and had penetrated 11 km of pre-Cambrian crust on the Kola peninsula (Kozlovsky, 1988), another super-deep drillhole project was planned. In 1981, the German geoscientific community started a similar project to unravel the petrology of the uppermost 10–15 km of the Hercynian crust in central Europe (Emmermann and Wohlenberg, 1989). A first funding for pre-investigations was provided in late 1981 and the active drilling and interpretation phase lasted from 1986 to 1996. At its first planning stage four locations were taken into consideration, but it was soon decided that the efforts of detailed seismic and other geophysical pre-investigations should be concentrated on two regions only: the central Black Forest around Haslach, and the Oberpfälzer Wald near Windischeschenbach, a small village close to the Czech border (Emmermann and Wohlenberg, 1989). In both areas, detailed seismic-refraction and seismic-reflection surveys were undertaken, partly within the framework of DEKORP, which will be discussed in more detail below.

8.3.3.1. Black Forest Location

For the Black Forest, a north-south line of 170 km in length was chosen following more or less the axis of the mountain range. Both a long-range near-incident seismic-reflection line (Lueschen et al., 1987, 1989) and a detailed seismic-refraction line (Gajewski and Prodehl, 1987; Gajewski, 1989) were planned and carried out (line S in Fig. 8.3.3-01). For logistical reasons, the refraction (full line with shotpoints S1 to S5) observations (Gajewski and Prodehl, 1987; Gajewski, 1989) were not completely coincident with the reflection line (double dashed lines), but workers made sure that the depth penetration of both surveys covered the same area. The favored proposed drill location was located near Haslach (about half-way between shotpoints S2 and S3 in Fig. 8.3.3-01) in the central part of the Black Forest within the central Black Forest gneiss complex. The reflection survey, organized by Ewald Lueschen, therefore included two side lines which crossed the main line in the area where the deep drillhole might be located, and where the W-E–directed side line connected with the Urach reflection survey of 1978 (Bartelsen et al., 1982) east of the refraction line Z. The goal of the second NW-SE–directed side line further south was

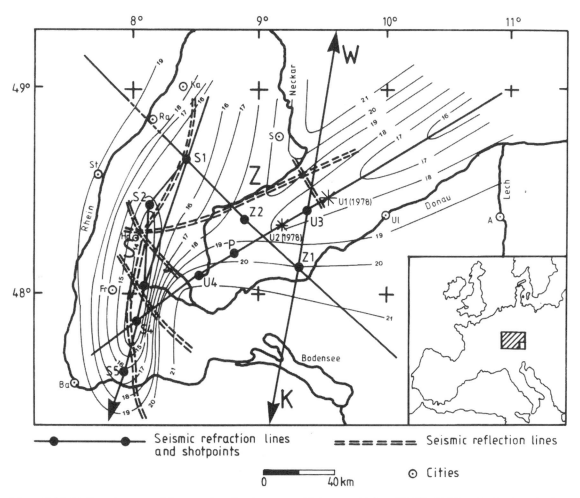

Figure 8.3.3-01. Location map of the seismic-refraction survey "Schwarzer Zollern-Wald" of 1984 and seismic-reflection surveys of 1978 (Urach) and 1984 (Black Forest), overlain on a depth contour map of the intracrustal boundary separating upper and lower crust (from Gajewski et al., 1987, fig. 11). [Tectonophysics, v. 142, p. 49–70. Copyright Elsevier.]

the investigation of the Badenweiler-Lenzkirch zone, bounding the central Black Forest gneiss complex in the south (Lueschen et al., 1987, 1989).

The refraction project, initiated and headed by Dirk Gajewski and Claus Prodehl, gained from additional projects. First, during several weekends in 1982, the German army organized a series of test explosions within the troop training area "Wildflecken" in the Rhön Mountains east of Frankfurt. Seventeen shots with charges of 450–1780 kg detonated in near-surface pits and shot times were arranged such that these explosions could be used to establish a series of fan profiles throughout southern Germany, the westernmost one being directed toward the Black Forest, the easternmost one toward the Bohemian Massif. (For location, see Fig. 8.3.4-08, no. 4; in the next section on the European Geotraverse, one of the lines is line W in Fig. 8.3.3-01, which is almost identical to the location of the European Geotraverse, recorded in 1986.) Forty recording stations were grouped such that the individual explosion were simultaneously recorded on all six lines at equivalent distances.

The largest recording distance on each line was 240 km, and the average station spacing ranged from 2.5 to 5 km (Zeis et al., 1990; data shown in Appendix A2-1, p. 27–28, and Appendix A8-1-1, Gajewski et al., 1989).

Second, the Black Forest refraction program could be jointly organized together with another research program of the University of Karlsruhe so that additional refraction lines were established crossing the Black Forest in a northeasterly (line U in Fig. 8.3.3-01) and southeasterly (line Z in Fig. 8.3.3-01) direction (Gajewski et al., 1987). One of the goals was to investigate in more detail (line Z) the Hohenzollerngraben, the site of an anomalous earthquake activity and where therefore another originally proposed drillhole site was located. Furthermore, line U along the Swabian Jura had already been the target of the Urach geothermal project in 1978 (Gajewski and Prodehl, 1985) as well as of the Rhinegraben seismic-refraction survey of 1972 (Edel et al., 1975), and it was the aim of this 1984 project to complete the already existing data along that profile. Finally, lines W and K were established, of which line W reversed one of the fan profiles from the troop training area "Wildflecken."

The seismic-refraction data of all profiles were of high quality and allowed workers to recognize many details. An example is shown in Figure 8.3.3-02, the complete data set is shown in Appendix A2-1 (p. 19–26) and Appendix A8-1-1 (Gajewski et al., 1988).

The combined interpretation of seismic-reflection and refraction data revealed a lot of details which each method alone could not have resolved (Fig. 8.3.3-03). The lower part of the upper crust proved to be a low-velocity channel and is almost without any reflectivity except for a bright spot in the Haslach area. Below 15 km depth a highly reflective lower crust was identified, interpreted by a strong and laterally consistent lamination with alternating high and low P-wave velocities differing by ~10% and a vertical layering on a scale of 100 m.

The seismic-refraction data also contained a wealth of shear-wave data so that for all lines cross sections could be established for the P-wave, S-wave and Poisson's ratio distribution. Surprisingly, the S-wave data neither supported the existence of a low-shear-wave velocity zone in the upper crust nor did they show the existence of any lamination for shear-waves in the lower crust (Fig. 8.3.3-04).

Consequently Holbrook et al. (1988) proposed a petrological model, based on laboratory-measured rock velocities, indicating a probably granitic composition for upper and middle crust

with a high quartz content in the middle crust (the P-wave low velocity zone) and a granulitic composition of the lower crust (Fig. 8.3.3-04 and Table 8.3.3-01). The subsequent calculation of synthetic seismograms showed a fair agreement between observed and synthetic data (Fig. 8.3.3-05). The horizontal-component synthetic seismograms predicted from the 1-D model in Figure 8.3.3-04 and Table 8.3.3-01 successfully predicted strong P_MP, S_MS, P_MP phases and P-wave reverberations from the lower crust (Pc) as well as weak S-wave lower crustal reverberations (Holbrook et al., 1988).

8.3.3.2. *Oberpfalz Location*

The second location for a super-deep drill-hole in the Oberpfälzer Wald in northeastern Bavaria was planned near the geological boundary between two structural units of the Variscan basement: the Moldanubian (MN in Fig. 8.3.3-06A) in the south and the Saxothuringian (ST in Fig. 8.3.3-06A) in the north which boundary was interpreted as a cryptic suture. This site was also investigated by detailed combined deep seismic-reflection and refraction studies.

First, the long-range seismic-reflection profile DEKORP-4 was arranged such that it passed the proposed drillhole site (Fig. 8.3.3-06, D4 in Fig. 8.3.1-07). Second, a number of 50–100-km-long crisscrossing seismic profiles were arranged like a net across the area (8501–8506 in Fig. 8.3.3-06, K85 in Fig. 8.3.1-07), organized by Helmut Gebrande. The whole reflection

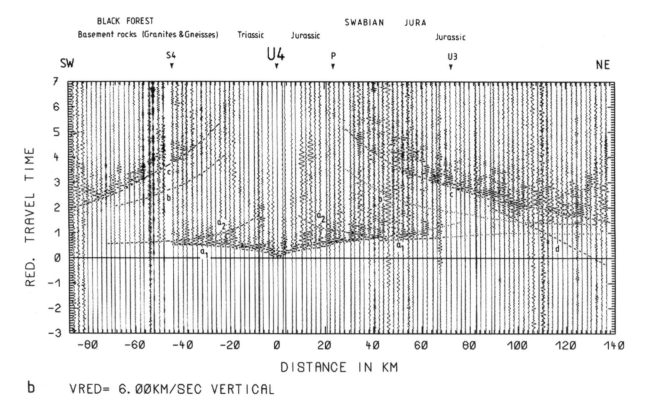

Figure 8.3.3-02. Data example of the seismic-refraction survey "Schwarzer Zollern-Wald" of 1984, record section of shot U4 on line U (from Gajewski et al., 1987, fig. 6b). [Tectonophysics, v. 142, p. 49–70. Copyright Elsevier.]

survey was carried out with the Vibroseis technique (Schmoll et al., 1989). Third, along the northwestern part of the DEKORP-4 line, 96 borehole shots with charges of 90 kg and 1 km spacing were fired and recorded threefold (Fig. 8.3.3-07): (1) for mapping wide-angle reflections from the upper crust by the contractor's 200-channel reflection spread at 42–58 km offset, (2) for mapping lower crustal and Moho wide-angle reflections by a mobile array of 24 MARS-66 stations at 60–90 km offset, and (3) for two expanding spread experiments with the midpoints in the Moldanubian and in the Saxothuringian zone, re-

spectively. Seismic-refraction data recorded between 1968 and 1983 added to the amount of available data (Gebrande et al., 1989, 1991). Fourth, a 3-D vertical incidence seismic-reflection survey was undertaken (Stiller, 1991).

Similarly, as in the Black Forest, the superimposition of the velocity-depth interpretation of the refraction data onto the seismic-reflection data resulted in a perfect agreement as shown by Gebrande et al. (1989, Fig. 8.3.3-08). In contrast to the Black Forest site, however, the upper crust in the area of the proposed drillsite appeared more reflective than the lower crust. The

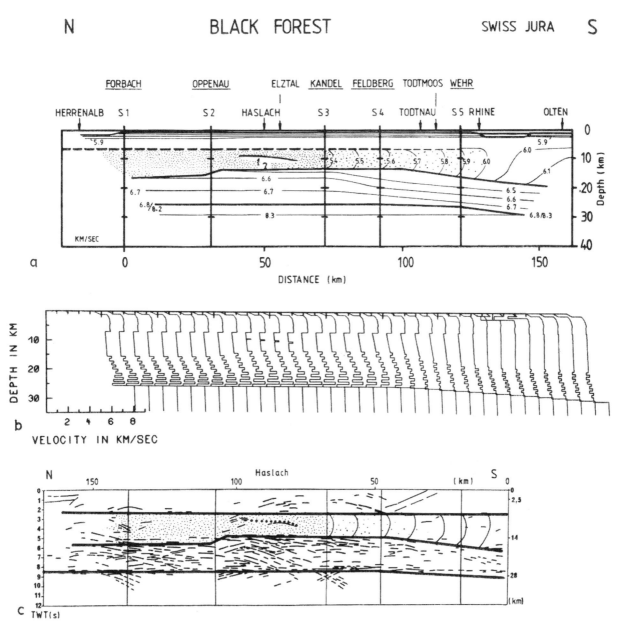

Figure 8.3.3-03. Crustal model of the Black Forest (from Gajewski and Prodehl, 1987, fig. 8). (a) Model with lines of equal velocity, derived from seismic-refraction data. (b) Velocity-depth sections along line S, derived from seismic-refraction data. (c) Line drawing of the seismic-reflection section along the Black Forest (simplified after Lueschen et al., 1987). Two-way traveltimes as calculated from model (a) are plotted as thick lines in (c). [Tectonophysics, v. 142, p. 27–48. Copyright Elsevier.]

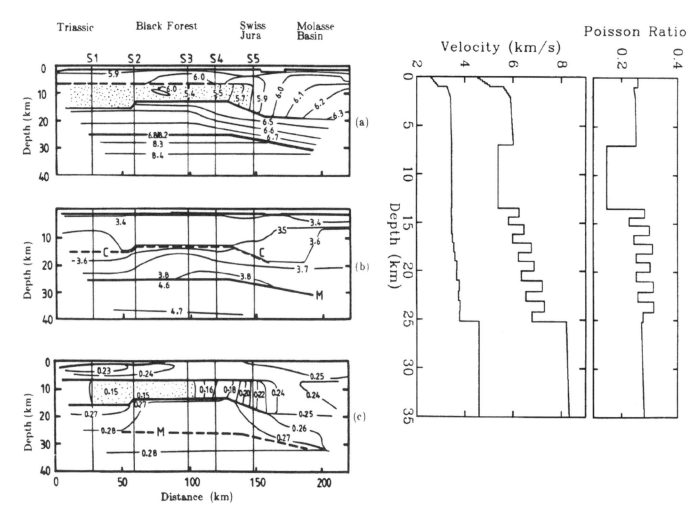

Figure 8.3.3-04. 3-component interpretation of the Black Forest data (line Z in Figure 8.3.3-01): Left: (a) P-wave velocity model, (b) S-wave velocity model, (c) Poisson's ratio model (from Holbrook et al., 1988, fig. 12). Right: 1-D depth functions of S-velocity (left), P-velocity (center), Poisson ration (right), which agree with the laboratory-measured rock velocities compiled in Table 8.3.3-01 (from Holbrook et al., 1988, fig. 16) and which were used to calculate synthetic seismograms. [Journal of Geophysical Research, v. 93, p. 12,081–12,106. Reproduced by permission of American Geophysical Union.]

TABLE 8.3.3-01. TWELVE-LAYER PETROLOGICAL MODEL OF THE BLACK FOREST LOWER CRUST
(FIG. 8.3.3-04) USED TO CALCULATE REFLECTIVITY SYNTHETIC SEISMOGRAMS (FIG. 8.3.3-05)

Layer	h (km)	v_p (km/s)	v_s (km/s)	σ	ρ (g/cm³)	Mineralogy
1	0.9	6.27	3.45	0.28	2.71	Granite (qtz-biot-fspar)
2	0.8	5.83	3.46	0.23	2.63	Granite (fspar-qtz)
3	1.0	6.47	3.46	0.30	2.73	Gabbro-anorthosite granulite
4	0.9	6.00	3.49	0.24	2.74	Diorite
5	1.0	6.78	3.55	0.31	2.78	Gabbro-anorthosite granulite
6	0.9	6.24	3.59	0.25	3.03	Gabbroic granulite (plg-cpx-biot)
7	1.0	6.89	3.67	0.30	3.08	Hornblende granulite (hbl-plg-cpx)
8	1.0	6.36	3.66	0.25	2.93	Quartz diorite
9	1.1	7.23	3.77	0.31	3.08	Garnet gabbroic pyriclasite(plg-px-gnt)
10	1.0	6.55	3.74	0.26	2.95	Al-rich gneiss (qtz-gnt-sil)
11	1.1	7.31	3.80	0.31	2.99	Gabbro (plg-epx-opx)
12	1.0	6.80	3.78	0.28	3.07	Gabbroic granulite (plg-cpx-opx)
Total	11.7	6.60	3.63	0.28	2.90	

Note: h—thickness of layer; v_p—P velocity; v_s—S velocity; σ—Poisson's ratio; ρ—density. For references, see Holbrook et al. (1988). [Journal of Geophysical Research, v. 93: 12,081–12,106. Reproduced by permission of American Geophysical Union.]

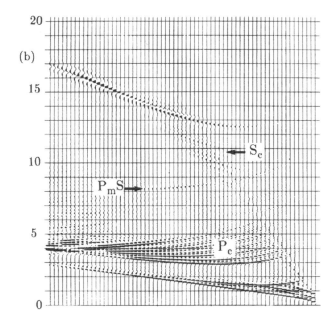

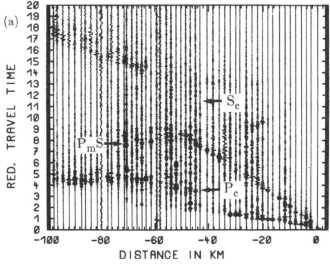

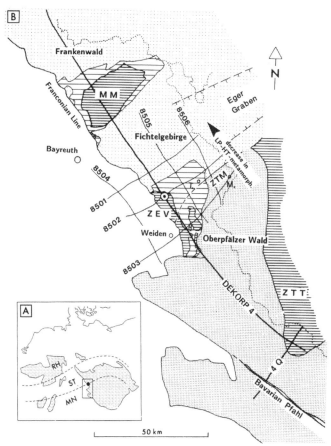

Figure 8.3.3-06. Geological sketch map of the western rim of the Bohemian Massif in NE Bavaria, Germany, showing the location of the DEKORP 4 and KTB (lines 8501 to 8506) seismic network (from Schmoll et al., 1989, fig. 1). Dot in (A) and circled dot in (B) is KTB (continental drillhole) site. [*In* Emmermann, R., and Wohlenberg, J., eds., 1989, The German continental deep drilling program (KTB)—site-selection studies in the Oberpfalz and Schwarzwald: Berlin-Heidelberg, Springer, p. 99–149. Reproduced with kind permission of Springer Science+Business Media.]

Figure 8.3.3-05. Synthetic and observed data (from Holbrook et al., 1988, fig. 17). Top: horizontal-component synthetic seismograms predicted from the 1-D model in Figure 8.3.3-04 and Table 8.3.3-01. Bottom: Transverse-component data from shotpoint S3 to south, reduction velocity = 7 km/s. [Journal of Geophysical Research, v. 93, p. 12,081–12,106. Reproduced by permission of American Geophysical Union.]

authors concluded that below a poorly defined upper crustal low velocity zone a marked velocity increase to 6.3–6.4 km/s was to be expected at ~7–8 km depth. The most spectacular feature, however, was a high-velocity zone detected at ~12–14 km.

At the end of 1985, a two-day meeting was held to review the data and results around the two drillhole sites in the Black Forest and in the Oberpfalz. Eighty-five projects had been supported until that time. The decision was finally made to choose the site in the Oberpfalz in NE Bavaria, and drilling started there on 18 September 1987.

8.3.4. The European Geotraverse

Following the tradition of European-wide international cooperation of geophysical research institutions, in 1980 the European Geotraverse was formulated. Based on the experience and excellent results of the FENNOLORA project of 1979 (Guggisberg et al., 1991; Guggisberg and Berthelsen, 1987), it was proposed in 1980 to elucidate the structure, composition, and dynamics of the continental lithosphere along a 4600-km-long and 100-km-wide corridor throughout Europe. The corridor would traverse all age provinces of Europe from Archean in Scandinavia through the Caledonian and Hercynian parts of central Europe to the Alpine orogenic belts of the Mediterranean area. This European Geotraverse would extend from the North Cape to Tunisia (Fig. 8.3.4-01). As a backbone of the European Geotraverse project, a comprehensive seismic-refraction investigation was planned

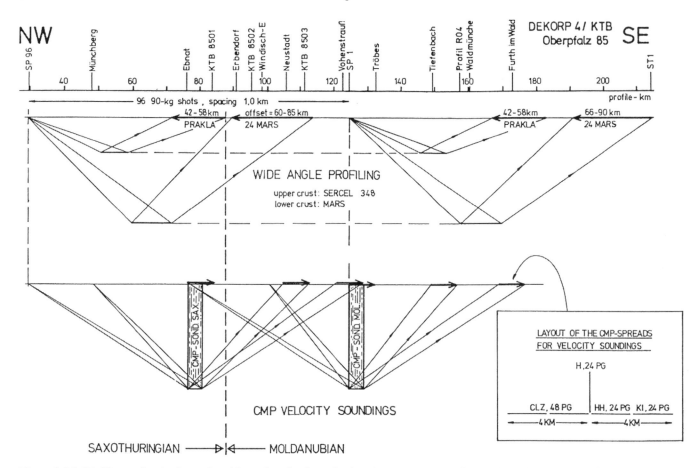

Figure 8.3.3-07. Observational scheme for wide-angle reflection seismics along DEKORP-4 (from Gebrande et al., 1989, fig. 4). [*In* Emmermann, R., and Wohlenberg, J., eds., 1989, The German continental deep drilling program (KTB)—site-selection studies in the Oberpfalz and Schwarzwald: Berlin-Heidelberg, Springer, p. 151–176. Reproduced with kind permission of Springer Science+Business Media.]

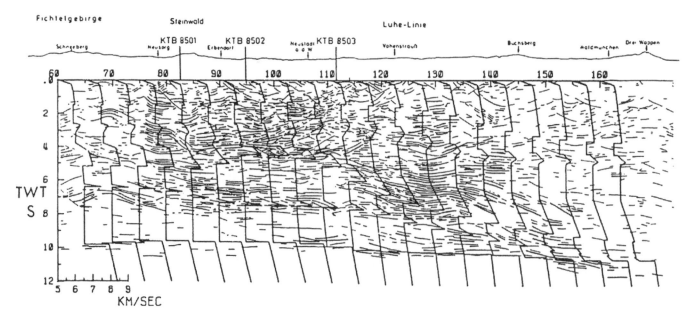

Figure 8.3.3-08. Line drawing of part of the DEKORP-4 near-vertical reflection data and superimposed velocity-depth functions, converted into two-way traveltimes (from Gebrande et al., 1989, fig. 13b). [*In* Emmermann, R., and Wohlenberg, J., eds., 1989, The German continental deep drilling program (KTB)—site-selection studies in the Oberpfalz and Schwarzwald: Berlin-Heidelberg, Springer, p. 151–176. Reproduced with kind permission of Springer Science+Business Media.]

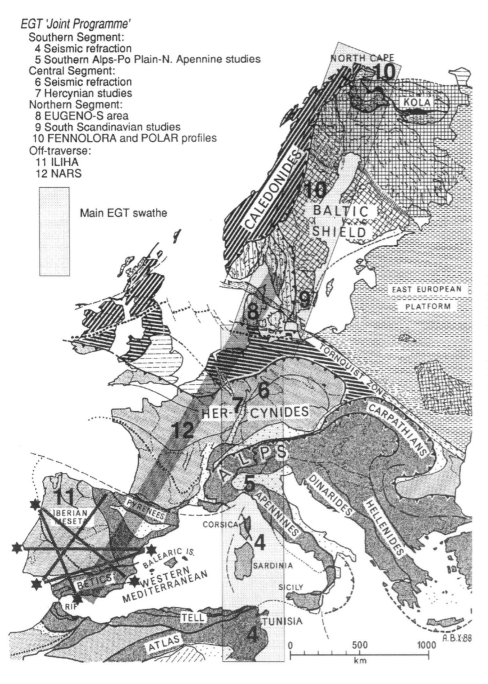

EGT 'Joint Programme'
  Southern Segment:
    4 Seismic refraction
    5 Southern Alps-Po Plain-N. Apennine studies
  Central Segment:
    6 Seismic refraction
    7 Hercynian studies
  Northern Segment:
    8 EUGENO-S area
    9 South Scandinavian studies
    10 FENNOLORA and POLAR profiles
  Off-traverse:
    11 ILIHA
    12 NARS

Main EGT swathe

Figure 8.3.4-01. Tectonic map of Europe outlining the joint programme of the European Geotraverse (EGT, from European Science Foundation, 1990, fig. 1). [European Geotraverse Project (EGT) 1983–1990, final report: European Science Foundation, Strasbourg, 67 p. Reproduced by kind permission of the European Science Foundation, Strasbourg, France.]

in order to study the detailed crustal and upper-mantle structure along the entire line.

The realization of the European Geotraverse was enabled by approval of and subsequent coordination by the European Science Foundation (ESF). A working group composed of scientists and administrators nominated by the participating organizations from 16 countries was set up.

In total, 13 major projects, and numerous geographically regional studies, were approved by ESF and, under the framework of the European Geotraverse, were recommended for funding to the appropriate national research foundations.

For logistical reasons, the Geotraverse was subdivided into three segments: The northern segment (nos. 8–10 in Fig. 8.3.4-01) comprised the already existing seismic-refraction line FENNOLORA of 1979 through Scandinavia, several projects in Finland (recorded in1981–1985; not shown in Fig. 8.3.4-02) and a transition zone between northern and central Europe in Denmark and southernmost Sweden (EUGENO-S of 1983). The central section of 1986 (nos. 6–7 in Fig. 8.3.4-01) would cover the Hercynian part of central Europe. For logistical reasons, the seismic-refraction line across central Europe also included the Alps and the Po Plain of northern Italy (no. 5 of Fig. 8.3.4-01). The

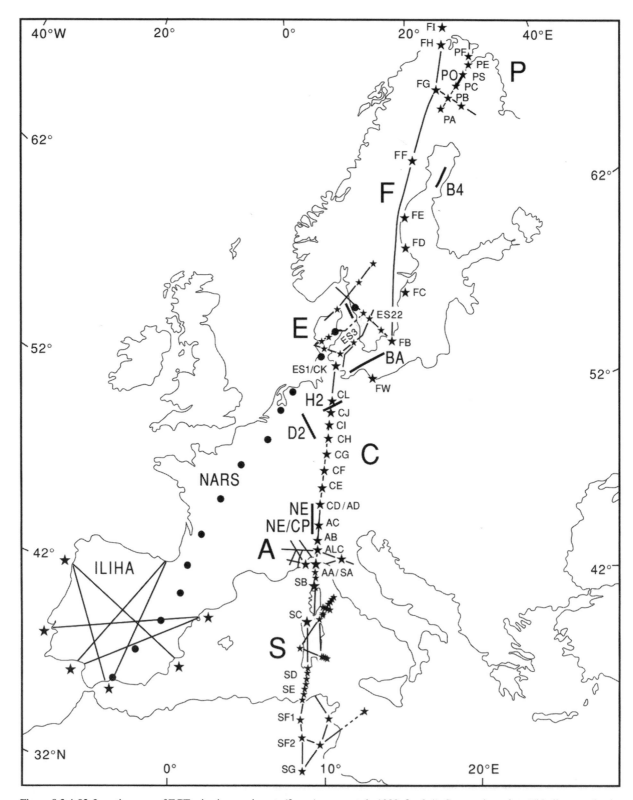

Figure 8.3.4-02. Location map of EGT seismic experiments (from Ansorge et al., 1992, fig. 3-1). Stars—shotpoints. Thin lines—seismic-refraction profiles: P—POLAR profile, F—FENNOLORA, E—EUGENO-S, C—EUGEMI, A—EGT-S86, S—EGT-South, ILIHA—Iberian Lithospheric Heterogeneity and Anisotropy project. Thick lines—deep seismic-reflection profiles: B4—BABEL line 4, BA—BABEL line BA, H2—North German basin, D2—DEKORP-2N, NE-NE/CP—NFP-20/CROP eastern and southern traverses. Filled circles—NARS (Network of Autonomously Recording Stations) array stations. [*In* Blundell, D., Freeman, R., and Mueller, S., eds., 1992, A continent revealed—the European Geotraverse: Cambridge University Press, p. 33–69. Copyright Cambridge University Press.]

southern segment (nos. 4–5 in Fig. 8.3.4-01) finally reached from the Alps to northern Africa, traversing northern Italy (1983), crossing the western Mediterranean Sea including the islands of Corsica and Sardinia (1983) and ending in northern Tunisia (1985). In addition, a large-scale anisotropic seismic-refraction experiment was planned on the Iberian peninsula (no. 11 in Fig. 8.3.4-01) and a wide-aperture seismological array NARS (no. 12 in Fig. 8.3.4-01) was to be installed to collect data on the upper mantle along the traverse Göteborg-Malaga (Nolet et al., 1986).

Being the backbone of the European Geotraverse project and other EGT-related seismic projects, for the EGT seismic-refraction experiments a comprehensive map (Fig. 8.3.4-02), showing all shotpoints, was compiled by Ansorge et al. (1992) in Chapter 3 of the major research publication *A Continent Revealed— The European Geotraverse* edited by Blundell et al. (1992). The joint program (European Science Foundation, 1990) also included experiments not specifically identified in Figure 8.3.4-01: simultaneous geomagnetic observations along the EGT (experi-

ment no. 1), mapping of the resistosphere and conductosphere along the EGT (experiment no. 2), mapping of the lithosphere-asthenosphere system along the EGT (experiment no. 3), and integrated geothermal studies along the EGT (experiment no. 13). In an interdisciplinary approach Blundell et al. (1992) reviewed all geological and geophysical data collected in the framework of the European Geotraverse.

In this review, a special section dealt with a compilation of all seismic surveys along and around the European Geotraverse (Ansorge et al., 1992) which was republished in a compressed scale (Fig. 8.3.4-03) by Mooney et al. (2002).

All projects, including the various deep seismic sounding projects, were carried out in the 1980s depending on the availability of financial resources. The results were discussed at seven workshops, concentrating on specific areas and published shortly afterwards in internal report volumes by the European Science Foundation for immediate use by the scientists involved: (1) the northern segment (Copenhagen, Denmark, 28–30 October 1983),

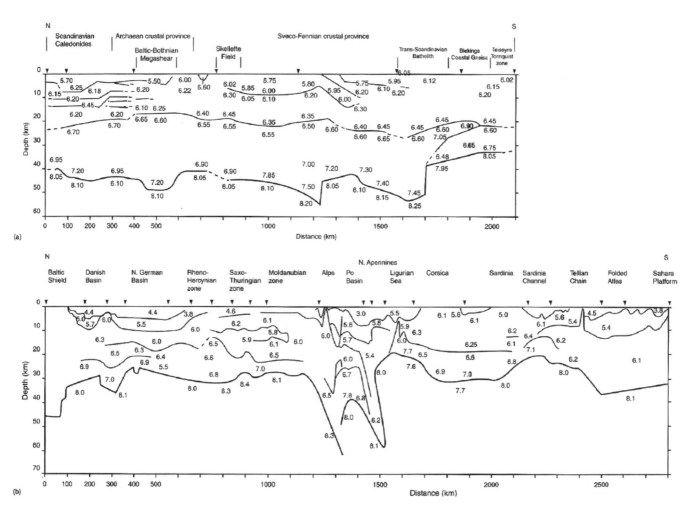

Figure 8.3.4-03. Crustal cross section through Europe along the EGT (from Mooney et al., 2002, fig. 9, after Ansorge et al., 1992, Figs. 3-4, 6, 9, 11, 12, 15, 16). Depth versus horizontal distance is 10:1. (a) Northern segment at approximately 20°E, (b) Central and Southern Segments at approximately 10°E. [*In* Lee, W., Kanamori, H., Jennings, P.C., and Kisslinger, C., eds., 2002, International Handbook of Earthquake and Engineering Seismology, Part A: Academic Press, Amsterdam, p. 887–910. Copyright Elsevier.]

(2) the southern segment (Venice, Italy, 7–9 February 1985), (3) the central segment (Bad Honnef, Germany, 14–16 April 1986), (4) the upper mantle (Utrecht, Netherlands, 11–12 March 1988), (5) the Iberian peninsula (Estoril, Portugal, 11–12 November 1988)), (6) data compilations and synoptic interpretation (Einsiedeln, Switzerland, 29 November to 5 December 1989), and (7) integrative studies (Rauischholzhausen, Germany, 26 March to 7 April 1990). In addition, other workshops for planning as well as for data interpretation of individual projects had been organized, for which only internal protocols were distributed.

All seismic-refraction data were collected in Open-File reports: the Northern Segment FENNOLORA by Stangl (1990; see also Appendix A2-1, p. 69–71; Appendix A7-4-2, p. A3–A58), its southern extension by Gregersen et al. (1987; see also Appendix A8-1-2), the Central Section by Aichroth et al. (1990; see also Appendix A2-1; p. 29–31, and Appendix A8-1-3) and Maistrello et al. (1991; see also Appendix A8-1-4), the Southern Segment by Egger (1990, 1992; see also Appendix A8-1-4) and Maistrello et al. (1990).

Most results were subsequently published in eight special issues of *Tectonophysics* entitled "The European Geotraverse": (1) 126, no. 1 (1986), (2) 128, nos. 3-4 (1986), (3) 142, no. 1 (1987), (4: Eugeno-S) 150, no. 3 (1988), (5: The Polar Profile) 162, nos. 1-2 (1989), (6) 176, nos. 1-2 (1990), (7) 195, nos. 2-4 (1991), (8) 207, nos. 1-2 (1992), and in a special issue on "Seismic studies of the Iberian peninsula": 221, no. 1 (1993). Finally, a comprehensive book was edited by Blundell et al. (1992) including an atlas of 13 maps and a database on CD-ROM, compiled by Freeman and Mueller (1992).

The individual seismic projects will be described here in geographical order, independent from the year of operation.

8.3.4.1. The Northern Segment

The already existing seismic lines FENNOLORA, extending from southeastern Sweden through northern Finland and northernmost Norway to the North Pole, and FINLAP, an unreversed lateral profile into Finland, both recorded in 1979, formed the core of the northern segment (Fig. 7.2.6-05 and no. 10 in Fig. 8.3.4-01). For details the reader is referred to section 7.2.6. Along the FENNOLORA line (Fig. 7.2.6-05 and F in Fig. 8.3.4-02 and *a* in Fig. 8.3.4-04) shots had been recorded up to 2000 km distance thus penetrating to a depth of 450 km (Guggisberg and Berthelsen, 1987; Guggisberg et al., 1991; Appendix A2-1, p. 69–71). Furthermore, in Finland and the adjacent USSR, a variety of additional seismic experiments had been carried out since the early 1980s. In their discussion of the POLAR profile (part of no. 10 in Fig. 8.3.4-01 and P in Fig. 8.3.4-02), Luosto et al. (1989) showed all modern deep seismic sounding profiles dealing with the lithospheric structure of the northern Baltic Shield (Fig. 8.3.4-04).

The first line in Finland was the 320-km-long SVEKA profile recorded in 1981 in central Finland. Forty-five shots with charges of 100–1000 kg were detonated in five lakes, ~80 km apart, and recorded by a continuously moving array of

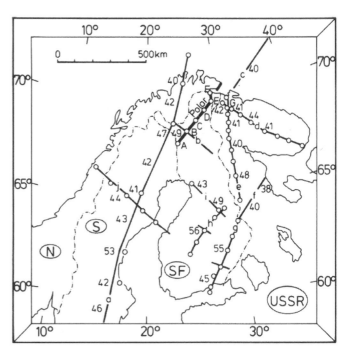

Figure 8.3.4-04. Modern deep seismic sounding profiles on the northern Baltic Shield (small letters *a* to *j*, from Luosto et al., 1989, fig. 2): *a*—FENNOLORA profile, *b*—FINLAP profile, *c*—Barents Sea profile, *d*—Nickel-Umbozero profile, *e*—Pechenga-Kostamuksha profile, *f*—Kem-Tulos profile, *g*—BALTIC profile, *h*—SVEKA profile, *I*—quarry blast line, *j*—BLUE ROAD profile. Average crustal thickness (in km) is indicated. A to G: large explosions along the POLAR profile. [Tectonophysics, v. 162, p. 51–85. Copyright Elsevier.]

19 recording units with 2 km station separation (Luosto et al., 1984; Luosto and Korhonen, 1986). In 1982 the 430-km-long BALTIC profile followed with 7 shotpoints in southeastern Finland (Luosto and Korhonen, 1986; Luosto et al., 1990). The seismic experiment along the BALTIC profile also included a very detailed seismic-reflection survey across the Granulite belt (Behrens et al., 1989).

The Russian part of the Baltic Shield was the target of two seismic lines recorded in 1981–1983: the Nickel-Umbozero profile recorded with 7 shotpoints through the center of the Kola Peninsula and the Pechenga-Kostamuksha profile recorded with 10 shotpoints in a north-south direction close to the Finnish border (Azbel et al., 1989). Davydova et al. (1985) described a seismic line recorded in 1976 in the Barents Sea, where shots of 135 kg were set off at 90 m depth and at 3–5 km intervals and recorded by OBSs on 19 sites with an approximate spacing of 50 km.

Finally, in 1985, the POLAR profile of 440 km length was recorded with 2 km station spacing to resolve the crustal structure of the northern Baltic Shield in northern Finland and northeastern Norway (Freeman et al., 1989; Luosto et al., 1989; Von Knorring and Lund, 1989; Walther and Flueh, 1993). Six shotpoints with large explosions of 200–1680 kg and three shotpoints with smaller explosions of 80 kg provided high-quality data. For the Finnish part, data were compiled in Appendix A2-1 (p. 72–81).

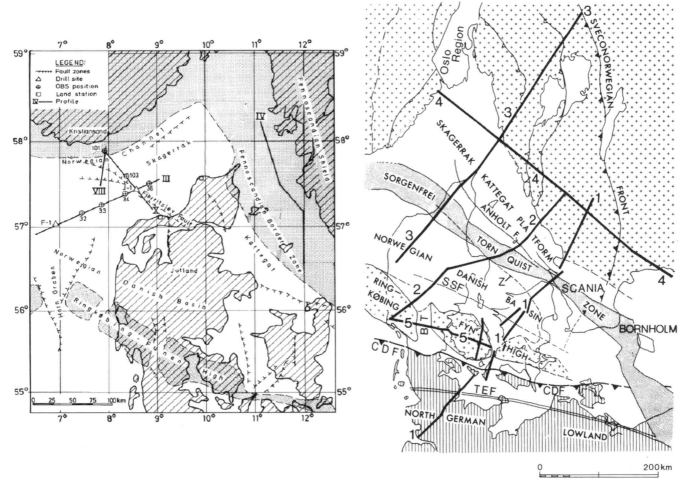

Figure 8.3.4-05. Left: *Meteor 66* cruise seismic reflection and refraction offshore lines of 1983 (from Behrens et al., 1986, fig. 1). [Tectono-physics, v. 128, p. 209–228.]. Right: EUGENO-S profiles of 1984 (from EUGENO-S Working Group, 1988, fig. 5). [Tectonophysics, v. 150, p. 253–348. Copyright Elsevier.]

In 1983, a multidisciplinary study of the Tornquist-Teisseyre Line, which connects to the FENNOLORA line of 1979, was conducted. The study of the Tornquist-Teisseyre Line, otherwise known as the contact zone between Precambrian and Hercynian Europe, was conducted using a network of seismic-refraction profiles (Fig. 8.3.4-05) from northern Germany through Denmark, and from Denmark to southern Sweden (EUGENO-S Working Group, 1988). The study also involved a marine survey with offshore and onshore observations (Behrens et al., 1986).

The offshore survey of the *Meteor 66* cruise in 1983 involved a streamer, OBSs, and seismic land stations which enabled both vertical-incidence and wide-angle reflection observations. Seismic energy was provided by an airgun array fired at two-minute intervals, thus providing for the vertical-incidence reflection observations one-fold coverage only, but a suitable energy source for crustal refraction measurements out to distances of more than 100 km. Thus, both the sedimentary horizons could be studied in some detail and a model for the whole crust could be obtained.

Figure 8.3.4-06 shows a data example of line IV where reflections from intracrustal boundaries and from the crust-mantle boundary near 40 km depth are clearly visible.

The EUGENO-S (European Geotraverse North–Southern extension) was achieved by a cooperative effort of scientists and funding from Denmark, Sweden, Norway, Finland, Germany, Switzerland, Great Britain, and Poland. The seismic project was carried out in summer 1984 and consisted of 5 intersecting lines covering southern Sweden and Denmark and adjacent areas with a total length of 2100 km. The spacing between shotpoints varied from 50 to 200 km, the station spacing on land being between 1.5 and 3.0 km. In total, 51 explosions were fired at 20 locations. Most of the shots were fired in boreholes, 5 shotpoints were in small lakes, and two shotpoints were in the Baltic Sea; charges ranged from 50 to 1200 kg. Most of the recording stations were MARS 66 stations (50 units) and Finnish SN-PCM-80 stations (4 units). Furthermore, 7 OBSs bridged the gap across the intervening marine part of the Kattegat.

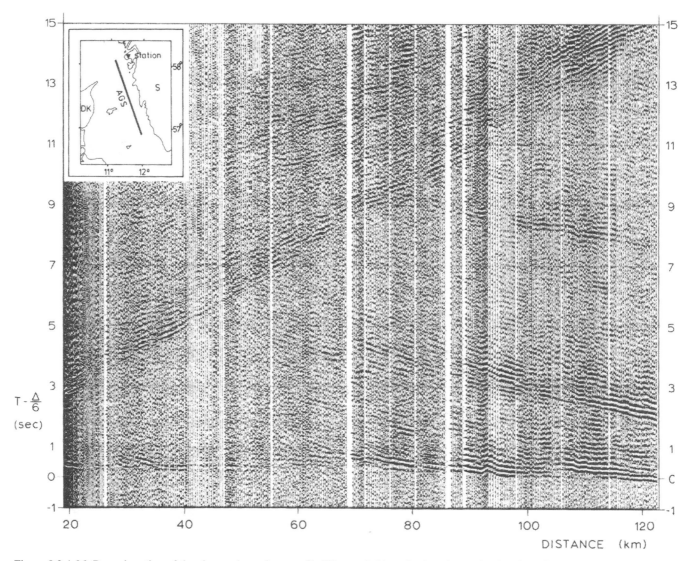

Figure 8.3.4-06. Record section of the air gun shots along profile IV recorded by a land station on hard rock onshore on the Tjörn peninsula, Sweden (from Behrens et al., 1986). [Tectonophysics, v. 128, p. 209–228. Copyright Elsevier.]

Along the offshore parts of the profiles, airgun shooting was also undertaken, which was not only recorded by the OBSs but also by land stations at up to 250 km in Sweden. The airgun array consisted of 4 airguns of 8 L each, fired simultaneously. Airgun shots were fired every 2 min corresponding to an average shot spacing of 300 m. Many of the data were published by EUGENO-S Working Group (1988) and are reproduced in Appendix A8-1-2.

Most of the interpretation was achieved during a 2-week interdisciplinary study center organized by the editorial team and supported by the European Science Foundation (Stege on Mon, Denmark, 4–21 November 1985); the results were subsequently published by EUGENO-S Working Group (1988). One of the principal results was the Moho map (Fig. 8.3.4-07), which was based on the EUGENO-S seismic results, but for which the gravity map was also used to infer the trends of the contours.

8.3.4.2. The Central Segment

Much of the area of the Central Segment of the European Geotraverse had already seen earlier seismic-refraction work in the 1970s (e.g., nos. 1–3 in Fig. 8.3.4-08, described in Chapter 7) and in the early 1980s. This data added substantially to the success of the EGT seismic survey of 1986. In 1982 and 1984, two major surveys in southern Germany had been performed in the framework of a special research project of the University of Karlsruhe: In 1982, a series of explosions on the troop training area "Wildflecken" in the Rhön Mountains in south-central Germany enabled scientists to simultaneously record a fan-like series of profiles throughout southern Germany (no. 4 in Fig. 8.3.4-08: profiles radiating from EGT-shotpoint H, Zeis et al., 1990). In 1984, a detailed crustal survey covered southwestern Germany ("Black Zollern-Forest," no. 5 in Fig. 8.3.4-08) in a triangle between Rhinegraben, the Swabian Jura and Lake Constance

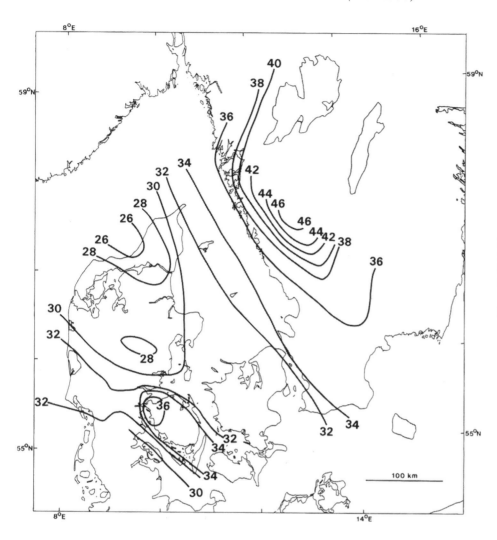

Figure 8.3.4-07. Depth of Moho (in km) in the transition zone between the Baltic Shield and Caledonian–Variscan Europe (from EUGENO-S Working Group, 1988, fig. 40). [Tectonophysics, v. 150, p. 253–348. Copyright Elsevier.]

(Gajewski et al., 1987). This survey had served to investigate the crustal structure underneath proposed sites for a deep drillhole in Germany and was already described in more detail in the previous section.

In North Germany, the hydrocarbon industry had continued to acquire an extensive seismic-reflection database in the North German Basin near the Elbe lineament between Hamburg to the south, Kiel to the north, the North Sea to the west and the Baltic Sea to the east (Dohr et al., 1989; fig. 2 of Yoon et al., 2008) The results of three more profiles, recorded in 1981 and 1984, were reprocessed (Yoon et al., 2008). Similar to the first line, recorded in 1974 (see Chapter 7.2.5), the data had been obtained with dynamite explosive sources recorded by 120 geophones with 20–100 m receiver spacings and 80–120 m shotpoint spacings down to 15 s TWT, resulting in mean common mid-point folds of ~20. Total profile lengths ranged from 30 to 100 km. The reprocessed sections showed an improved image quality at all time levels. Especially the reflections from the salt events and the Moho at 11.5–12 s TWT could be enhanced (Yoon et al., 2008).

The central segment of the European Geotraverse, organized by scientists from Germany (Beate Aichroth and Claus Prodehl), Switzerland (Jörg Ansorge), and Italy (Carlo Morelli, R. Cassinis, S. Scarascia, and others), was covered by a detailed seismic-refraction survey in 1986 (Fig. 8.3.4-08; EUGEMI Working Group, 1990; Aichroth et al., 1992; Prodehl and Aichroth, 1992; Appendix A2-1, p. 29–31, and Appendix A8-1-3; Aichroth et al., 1990).

This EGT project covered a distance of 1200 km and reached from its northern shotpoint in the Baltic Sea, off Northern Germany, across the whole of Germany (no. 6 in Fig. 8.3.4-01, C in Fig. 8.3.4-02), into Switzerland and northern Italy, thus covering the central Alps and the northern Po Plain (no. 5 in Fig. 8.3.4-01, A in Fig. 8.3.4-02). It was here that it overlapped with the northern end of the southern EGT segment. Its southernmost shotpoint was located in the Gulf of Genova and several fan profiles were arranged in the Apennines and the Western Alps to study lateral variations in crustal structure.

In total, 235 recording stations, most of them (~200) being MARS-66 equipment, were recording during three deployments. The stations recorded 15 shots in the southern part (Italy and

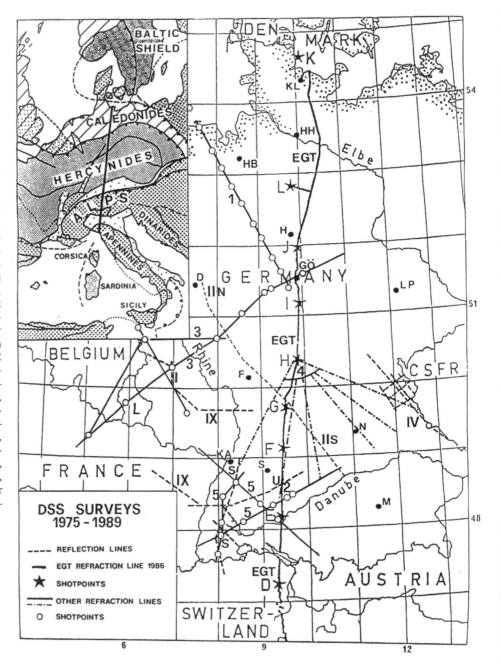

Figure 8.3.4-08. Location map of the 1986 seismic-refraction survey along the EGT Central Segment and other detailed seismic-refraction and -reflection lines in western Germany (from Prodehl and Aichroth, 1992, fig.1). EGT-86 shotpoints are denoted by stars and large capital letters D to K, observation sites by dots. Dashed lines represent seismic-reflection lines of DEKORP (I to IX) and other reflection surveys (U—Urach 1978, S—KTB Black Forest 1984). Single fat lines represent densely observed seismic-refraction surveys: 1—North Germany 1975/1976, 2—Urach 1978, 3—Rhenish Massif 1979, 4—"Wildflecken" 1982 (dash-dotted lines), 5—"Black Zollern-Forest." Cities (small capital letters): D—Dortmund, F—Frankfurt, G—Göttingen, H—Hanover, HB—Bremen, HH—Hamburg, KA—Karlsruhe, KL—Kiel, LP—Leipzig, M—München, N—Nürnberg, S—Stuttgart. The inset shows the tectonic sketch map of Europe (Berthelsen, 1983). [In H. Kern and Y. Gueguen, eds., Structure and composition of the lower continental crust: Terra Nova, v. 4, p. 14–24. Copyright Wiley-Blackwell.]

Switzerland), 10 shots in the central part (Hercynian part of Germany), and 5 shots in the northern part (North German lowlands). Most of the shots were drillhole shots; only at the southernmost and northernmost ends were underwater shots used. Participants came from Denmark, Germany, Finland, France, Ireland, Italy, Spain, Sweden, Switzerland, and the UK. A special seismic workshop from 27 February to 4 March 1989 at Karlsruhe, Germany, and an interdisciplinary EGT Study Center from 25 March to 5 April 1989 at Rauischholzhausen, Germany, served in particular to model and to interpret the data of the central segment of the EGT, leading to the seismic model and its petrological interpretation shown in Figure 8.3.4-09.

The data, obtained in 1986 and 1987 in the Alpine area around the European Geotraverse, were separately interpreted (Fig. 8.3.4-10). Ye (1991) interpreted the seismic-refraction data of the central segment from shotpoint F near the northern border of the Bavarian Molasse Basin to shotpoint A in the Ligurian Sea (data presentation in Appendix A8-1-4), in conjunction with a reinterpretation of other existing seismic-refraction profiles in the central Swiss Alps (Fig. 8.3.4-10, top), including a line along the northern border of the central Swiss Alps which was recorded from Jaun Pass to Saentis in 1987 (Maurer and Ansorge, 1992, also in Appendix A8-1-4). Valasek et al. (1991) discussed the model in conjunction with the deep seismic-reflection line

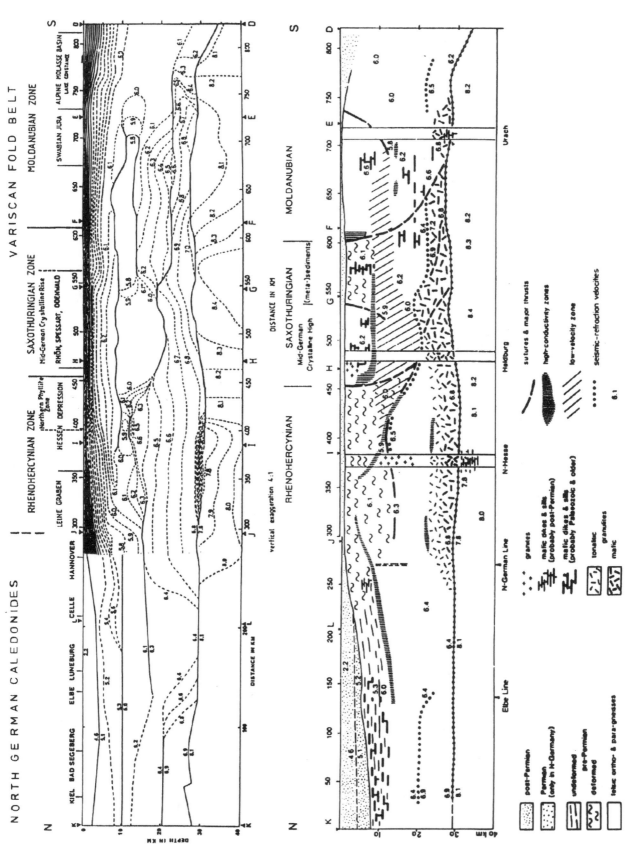

Figure 8.3.4-09. Crustal cross section of the EGT central segment from the Baltic Sea to Lake Constance (from Prodehl and Aichroth, 1992, fig. 3). Depth versus distance is exaggerated by 4:1. Top: Seismic velocity model. Bottom: Petrological interpretation. [*In* H. Kern and Y. Gueguen, eds., Structure and composition of the lower continental crust: Terra nova, v. 4, p. 14–24. Copyright Wiley-Blackwell.]

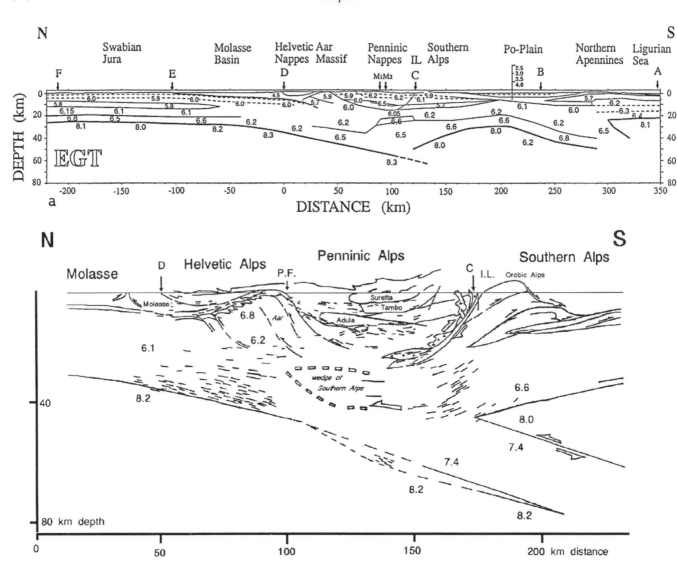

Figure 8.3.4-10. Top: Crustal cross section of the Alps along EGT from seismic-refraction data (from Ye, 1991, fig. 7-1a). Bottom: Geological-geophysical cross section of the Alps along EGT from EUGEMI, NFP 20 and other data (from European Science Foundation, 1990, fig. 8). [Ph.D. Thesis, Swiss Federal Institute of Technology Zürich (ETH), 114 p.] [European Geotraverse Project (EGT) 1983–1990, final report: European Science Foundation, Strasbourg, 67 p. Reproduced by kind permission of the European Science Foundation, Strasbourg, France.]

NFP 20 (Fig. 8.3.4-10, bottom), recorded in 1986 across the eastern part of the Swiss Alps which was already discussed in more detail in the previous section, "Large-Scale Seismic-Reflection Surveys in Western Europe" (for location, see Fig. 8.3.1-05).

As was discussed in subchapter 8.3.1.4, only by the combination of the seismic-refraction data with the much more detailed and with higher frequency recorded seismic-reflection data of the NFP 20 project was the detection of the European subduction shown in Figure 8.3.4-11 enabled.

In conjunction with the seismic exploration of the southern segment in 1983, which is discussed below, in northern Italy sea shots at shotpoint locations A1 to A4 shots not only provided energy for the main N-S line, but also, together with other borehole shots, served to record several perpendicularly oriented 30–80-km-long reversed and unreversed side lines (area A of Fig.

8.3.4-02). The goal of this special study was to obtain a detailed upper-crustal structure of the northern Apennines part of the EGT (Cassinis, 1986; Biella et al., 1987).

The complex crustal structure of the region between Alps and Ligurian Sea was discussed by a variety of authors. As an example, the interpretation of P. Giese and co-workers is shown in Figures 8.3.4-11 and 8.3.4-12, demonstrating the complex structure of the crust-mantle boundary in this area (Giese, 1985; Giese et al., 1982).

8.3.4.3. The Southern Segment and Other Investigations in Northwest Africa

The southern segment reached from northern Italy to northern Tunisia, crossing Corsica and Sardinia, and was organized by scientists from Italy (Trieste and Milano), Switzerland (ETH

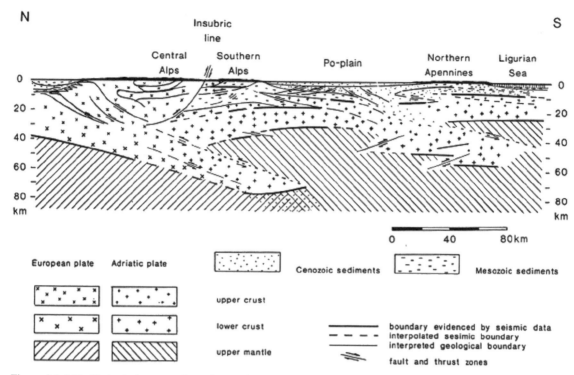

Figure 8.3.4-11. Tectonic interpretation of crustal structure along the EGT from the Alps to the Ligurian Sea, after P. Giese (from European Science Foundation, 1990, fig. 9). [European Geotraverse Project (EGT) 1983–1990, final report: European Science Foundation, Strasbourg, 67 p. Reproduced by kind permission of the European Science Foundation, Strasbourg, France.]

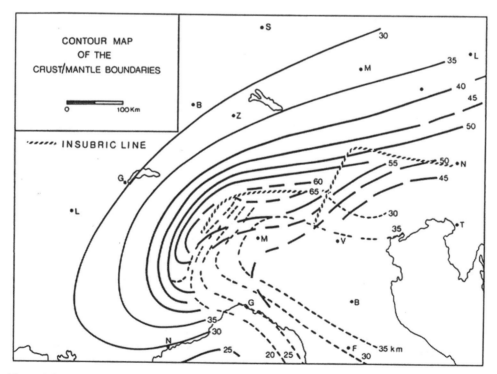

Figure 8.3.4-12. Moho surfaces in the region of Alps to Northern Apennines (from Giese, 1985, fig. 9). [*In* Galson, D.A., and Mueller, S., eds., 1985, Proceedings of the Second Workshop on the Geotraverse Project, the southern segment: European Science Foundation, Strasbourg, p. 143–153. Reproduced by kind permission of the European Science Foundation, Strasbourg, France.]

Zürich, and Germany (FU Berlin). It was investigated by two major seismic-refraction campaigns, performed in 1983 and 1985. The 1983 campaign covered northern Italy, Corsica, and Sardinia.

Following marine test profiles around the Corsica-Sardinia block in 1982, a marine survey, conducted in September 1983 in conjunction with the main experiment of 1983, covered the Ligurian Sea between Genova and Corsica (Ginzburg et al., 1986), using 45 small dynamite charges of 75–100 kg each and spaced 2–5 km apart. They were recorded by 10 OBSs, placed in the northern half of the line, and by land stations east of Genova and on northwestern Corsica, resulting in a number of partially overlapping reversed profiles.

The main profile of 1983 aimed to study the lithosphere under Corsica and Sardinia, using major depth charges of 1000–1125 kg, which were shot at four positions A to D (Fig. 8.3.4-13,

SA to SD in Fig. 8.3.4-02; Appendix A8-1-4). MARS-66 land recording stations were positioned on both islands (Egger et al., 1988; Egger, 1990, 1992; see Appendix A8-1-4). The interpretation (Fig. 8.3.4-14) included earlier crustal surveys of 1974 around Corsica, of 1979 in Sardinia, and of 1982, as well. Their positions are included in Figure 8.3.4-13.

The southernmost segment in Tunisia (Fig. 8.3.4-15) was a separate project, carried out in 1985. It consisted of marine surveys in the Sardinia Channel, between Sardinia and Tunisia as well as in the Pelagian Sea, off Tunisia toward northeast (Morelli and Nicolich, 1990), and of a land survey in Tunisia. The land survey in Tunisia used two offshore and five land shotpoints. Recording was achieved by 120 MARS-66 stations, arranged as a network of nine reversed profile segments (Research Group for Lithospheric Structure in Tunisia, 1992; Appendix A8-1-4), of

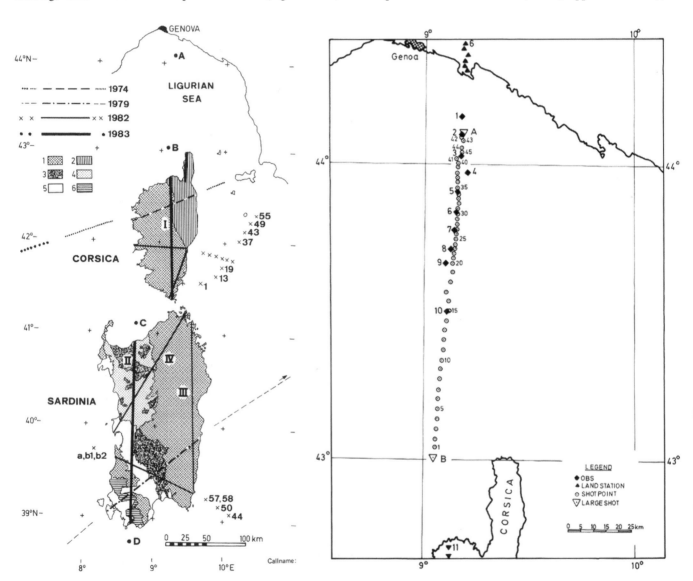

Figure 8.3.4-13. Left: Shotpoints and seismic profiles recorded on Corsica and Sardinia (from Egger et al., 1988, fig. 1). [Tectonophysics, v. 150, p. 363–389.]. Right: Location of ocean-bottom seismometers, land stations and shotpoints of the Ligurian Sea EGT segment (from Ginzburg et al., 1986, fig. 1). [Tectonophysics, v. 126, p. 85–97. Copyright Elsevier.]

which the main north-south line comprised 435 km on land, two sea shots at position SE and three land shots at positions SF1, SF2, and SG (Fig. 8.3.4-02).

Morelli and Nicolich (1990) have compiled a simplified cross section of the lithosphere from the Alps to Tunisia, summarizing the results from the individual sections along the southern segment (Fig. 8.3.4-16).

Also in North Africa, but not under the auspices of the European Geotraverse, a major seismic survey, carried out by the Free University of Berlin, Germany, was performed in 1983 and 1986 (Wigger et al., 1992; Appendix A8-1-5) in Morocco, Northwest Africa (Fig. 8.3.4-17, thick lines). The investigations added new information on the crust to the east of the crustal survey, which had been performed by the University of Hamburg (Fig. 8.3.4-17, thin lines) in 1975 (Makris et al., 1985).

The seismic investigations made use of commercial blasts of a phosphate mine near Oued Zem with charges between 4600 and 12600 kg and two specially arranged shot sites. In a lake near Zaida charges of 500 kg and in an abandoned shaft near Taklimt charges of 250 and 1000 kg were detonated. A trial experiment along the main line had been made already in 1983, using small charges at a then-active lead mine near Zaida (Wigger and Harder,

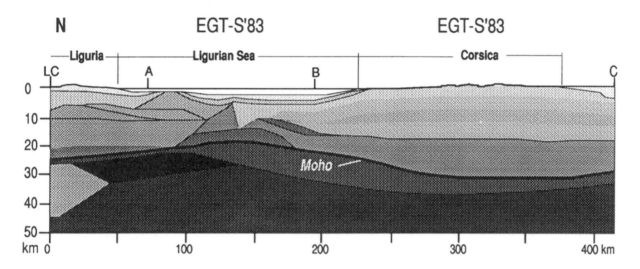

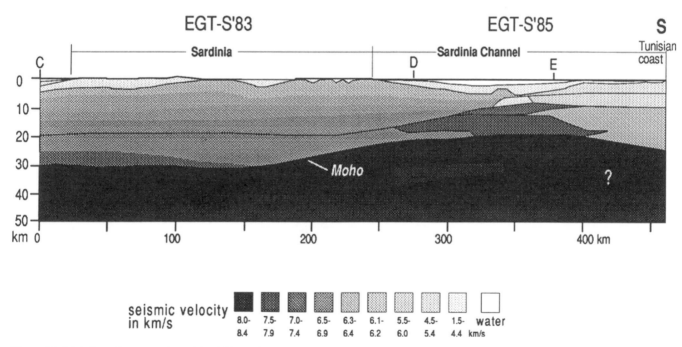

Figure 8.3.4-14. Crustal cross section of the EGT southern segment from the Ligurian Sea to Tunisian coast line (from European Science Foundation, 1990, fig. 10). Depth versus distance is exaggerated by 2:1. [European Science Foundation, Strasbourg, 67 p. Reproduced by kind permission of the European Science Foundation, Strasbourg, France.]

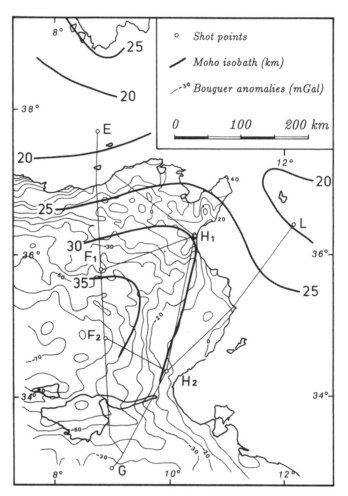

Figure 8.3.4-15. Location of shotpoints and profiles in and around Tunisia, and derived Moho depths overlain on a Bouguer gravity map (from Research Group for Lithospheric Structure in Tunisia, 1992, fig. 23). [Tectonophysics, v. 207, p. 245–267. Copyright Elsevier.]

1986). For recording, 30 stations were available, the station spacing ranged from 5 to 20 km. In total, three profiles were established, the longest line being a 350-km-long traverse between the Rif and the Anti Atlas, crossing the Middle and High Atlas mountains. The interpretation yielded several velocity inversions and gave an average Moho depth of 35 km, which under the High Atlas increased slightly to 40 km (Wigger et al., 1992).

8.3.4.4. Iberian Lithosphere Heterogeneity and Anisotropy Project

Finally, a project was designed to study possible anisotropy of the Hercynian lithosphere in Europe. Already in the 1970s, detailed seismic-refraction studies had explored the crust in southern and central Spain. In the early 1980s, the seismic exploration activity on land was focused in the Iberian Massif. The NW corner of Iberia was studied in a widespread survey, involving station spacing of 1–2.5 km (e.g., Cordoba et al., 1987).

As only the Iberian Peninsula had dimensions where large sea shots could be recorded up to 600–800 km distance on reversed long-range profiles on more or less homogeneous Hercynian crust, the Iberian Lithosphere Heterogeneity and Anisotropy (ILIHA) Project was designed as project no. 11 of the European Geotraverse. Following initial ideas formulated as early as 1984, during several special meetings in 1986 and 1987, a proposal to conduct a large-scale seismic anistropy experiment on the Iberian Peninsula was prepared and finally approved by the European Community, Directorate General XII, in 1988.

In 1989, the ILIHA deep seismic sounding experiment was conducted. Large sea shots of 500–1000 kg at positions B to F and X off the coasts of Portugal and Spain and land shots at positions G and P were recorded with 140 mobile recording stations along four reversed and two unreversed long-range profiles of 600–800 km length (Fig. 8.3.4-18) across the Iberian Peninsula (ILIHA DSS Group, 1993a, 1993b; Arlitt et al., 1993; Banda et al., 1993; Appendix A8-1-6). Unfortunately, the proposed shot at site A could not be realized.

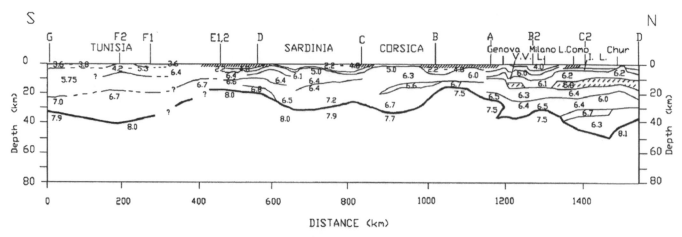

Figure 8.3.4-16. Crustal sketch along the EGT southern segment (from Morelli and Nicolich, 1990, fig. 4). [Tectonophysics, v. 176, p. 229–243. Copyright Elsevier.]

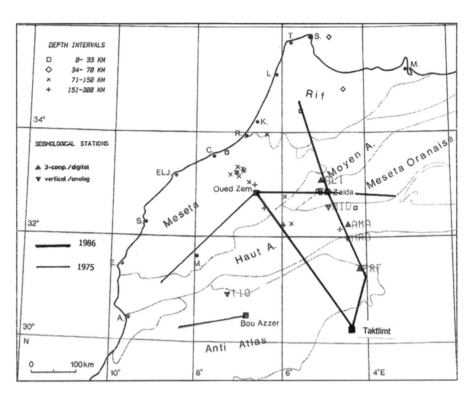

Figure 8.3.4-17. Location of seismic-refraction lines and seismological stations and epicenters in Morocco (base map with seismological information from Wigger and Harder, 1986, fig. 1). [Berliner Geowissenschaftliche Abhandlungen, v. A66, p. 273–288. Published by permission of Institut für Geologische Wissenschaften, Freie Universität Berlin.]

The ILIHA shots were also used for a special investigation of the Betic Cordillera (Banda et al., 1993; Appendix A8-1-6) where additional six 1500 kg borehole shots on land and three quarry blasts (4000 kg each) served as energy sources for two profiles, one across the Cordillera with 250 km length reaching the coast near Almeria, and one parallel to the Alboran Sea from Malaga to Alicante (profiles I and II in Fig. 8.3.4-18, right). The

shots were recorded by 90 analog seismic stations of type LOBS of the University of Hamburg.

The data of the successful seismic-refraction experiment were interpreted by several working groups. Data examples of the long-range profiles are shown in Figure 8.3.4-19. Besides the seismic-refraction observations, the ILIHA project also involved a broadband seismology investigation as well as other seismo-

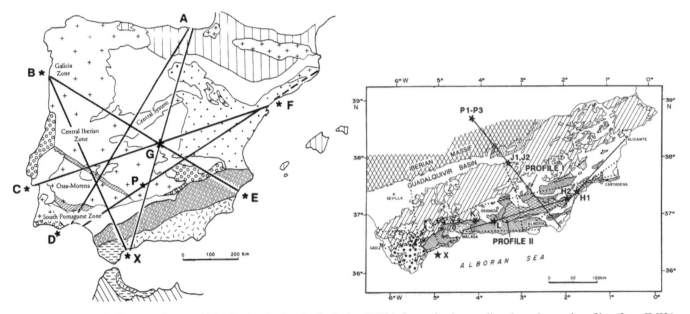

Figure 8.3.4-18. Left: Geotectonic map of the Iberian Peninsula displaying ILIHA deep seismic sounding shotpoints and profiles (from ILIHA DSS Group, 1993a, fig. 1). [Tectonophysics, v. 221, p. 35–51]. Right: Seismic survey of the Betic Cordillera, southern Spain (from Banda et al., 1993, fig. 1). For reference see position of shotpoint X. [Tectonophysics, v. 221, p. 35–51. Copyright Elsevier.]

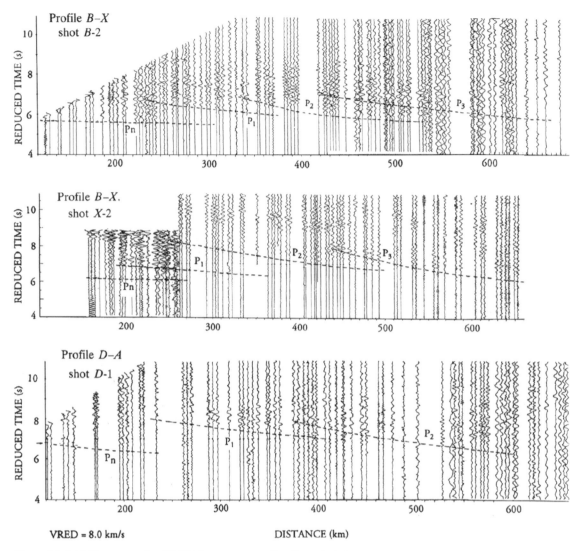

Figure 8.3.4-19. Data example of ILIHA long-range profiles (from ILIHA DSS Group, 1993a, Figs. 5, 6, and 8). [Tectonophysics, v. 221, p. 35–51. Copyright Elsevier.]

logical and geodynamical topics such as gravity, Rayleigh wave, focal mechanism, and delay time tomographic investigations which, together with 10 contributions on the deep-seismic sounding project, were jointly published in a special monograph edited by Mezcua and Carreno (1993).

The modeling of the mantle data allowed researchers to obtain fine structure of the lower lithosphere down to 90 km. The authors conclude that the seismic structure of the lithosphere is more uniform than expected, as significant lateral heterogeneities were not recognized within the lower lithosphere as would be expected if the frozen-in Hercynian structure was evident at these depths. However, there is an incompatibility between shear-wave velocities derived from Rayleigh and Love waves, suggesting anisotropy at depths greater than the 100 km penetrated by the ILIHA DSS experiment. Under the Betic Cordillera, a distinct crustal thickening to 35 km was modeled.

Following the overview of Díaz and Gallart (2009), one year before the ILIHA experiment, in 1988, the Valencia trough was explored within the VALSIS experiment. Up to 200 km of multichannel seismics were acquired including common depth point, common offset and expanding spread profiles (Pascal et al., 1992; Torné et al., 1992). The shots from some of these profiles were recorded onshore, providing the first onshore-offshore transects in Iberia (Gallart et al., 1990). The same area was explored further in 1989 with a wide-angle experiment using explosive sources, recorded by 110 land stations and 10 OBS (Danobeitia et al., 1992).

8.3.5. Deep-Seismic Sounding Projects in Eastern Europe

In the 1980s, only minor fieldwork in Poland was accomplished, but the interpretation of the whole set of existing profiles was pushed forward (Fig. 8.3.5-01).

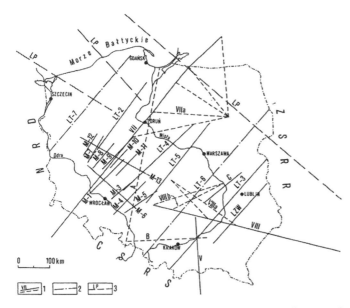

Figure 8.3.5-01. Location of seismic profiles and corresponding crustal cross sections in the fore-Sudetic region, southwestern Poland (from Guterch et al., 1986, fig. 1). [Publications of the Institute of Geophysics, Polish Academy of Sciences, A-17 (192), p. 3–83. Reproduced by permission of Instytut Geofizyki Akademii Nauk, Warsaw, Poland.]

A summary was published by Guterch et al. (1986). The location map shows the dense coverage of Poland with DSS lines, consisting of the first lines A, B, and C, accomplished in the early 1960s, the International Profiles V, VII, and VIII, completed by the early 1970s, and the sets of national profiles, the M-lines concentrating on the Sudetic foredeep, and the LT-lines, concentrating on the Tornquist-Teisseyre Zone in central Poland. In 1986, the LT-7 line in northwestern Poland could be added (Guterch et al., 1991b; Appendix A8-1-7), in cooperation with Finnish institutions. This line was both recorded as a seismic-refraction line, with 2 km station spacing and 5 shotpoints with charges between 100 and 900 kg, and as a steep-angle reflection line. This set of observations became the target of a special workshop, held in Warsaw in November 1993 (Guterch et al., 1994).

Also in the fore-Sudetic region, the first near-vertical reflection profiles were observed on profile A (Fig. 8.3.5-01, named GB2 in Fig. 7.2.7-04) in 1987, using 48-channel equipment and 24 10-Hz geophones per channel. Energy was produced by 15–30 kg borehole explosions and recorded at up to 18 s and 4 km maximum distance, and the set of M-profiles of 1966 was reinterpreted (Guterch et al., 1991a).

In Czechoslovakia a 150-km-long deep seismic-reflection transect was observed through the Slovakian West Carpathians. In 1983–1985, three lines were shot consecutively, using dynamite and recording with a 96-channel recording system. The deepest reflector was recorded continuously over 40 km distance from the Central (Apulian) West Carpathians to the Inner (Tethyan) West Carpathians, dipping toward SSE from 6 s TWT to ~9 s (Tomek et al., 1987; Tomek, 1993; for more details, see section 8.3.1.6).

8.4. DEEP SEISMIC SOUNDING PROJECTS IN EASTERN EUROPE AND ADJACENT ASIA (USSR)

The third period of Russian deep seismic sounding (DSS) investigations started at the end of the 1970s and continued through the 1980s. It was a time of stabilization of DSS methods, and an optimal system of observation and data interpretation was developed, including a program for a uniform study of the whole USSR (Benz et al., 1992; Pavlenkova, 1996). Considerable progress was made due to the efforts of the activities of the Special Regional Geophysical Expedition (SRGE; renamed GEON in the early 1990s) of the Ministry of Geology of the USSR (Egorkin and Chernyshov, 1983; Egorkin et al., 1987, 1991). The network of seismic profiles of the SRGE covered almost the whole territory of the USSR (Fig. 8.4-01).

This seismic research included three-component magnetic recordings of shots of varying sizes recorded by up to 300 stations on profiles with 2500–3000 km length (Fig. 8.4-02). Two types of energy sources were used. Chemical explosions with charges up to 5000 kg loaded in a series of boreholes 100–150 m apart allowed recording distances of 300–400 km. For distances up to 3000 km so-called "industrial" (nuclear) explosions were specially arranged for these investigations (later named PNE—peaceful nuclear explosions; dots in Fig. 8.4-02). Two to four of such shots, spaced 1000–1500 km apart, were arranged on several profiles (double lines in Fig. 8.4-01, lines in Fig. 8.4-02).

The long-range profiles allowed in particular the recording of reflections from the transition zone between upper and lower mantle. Most of its data were reprocessed and reinterpreted in the 1990s by several authors (e.g., Egorkin, 1999; Mechie et al., 1993; Morozova et al., 1999; Ryberg et al., 1998). The observation scheme and a data example from the "Quartz" profile are shown in Figures 8.4-03 and 8.4-04.

One of these projects, using large chemical explosions, aimed to investigate the crustal and uppermost mantle structure around the Mirnyi kimberlite field in Siberia (Suvorov et al., 2006). In 1981 and 1983, two profiles of nearly 400 km in length were recorded, perpendicular to each other (Fig. 8.4-05).

Profile 1, 370 km long, crossed the kimberlite pipe in a NNW-SSE direction. It was covered by 41 analogue Taiga seismographs, which recorded 21 explosions along the line including one offset shot. Profile 2, 340 km long, ran in a SW-NE direction and was located ~30 km to the south of the Mirnyi kimberlite field. The profile was covered by 45 seismographs which recorded 22 chemical explosions, including 4 off-end shots. Charges were between 1.5 and 6.0 tons of TNT, which were distributed in 100–200 kg individual charges in shallow water (less than 2 m deep). The resulting data are of high quality with high signal/noise ratio to the farthest offsets (Fig. 8.4-06; Appendix A8-1-8).

The interpretation of Suvorov et al. (2006) showed a normal cratonic crust of 45 km thickness with only slight Moho undulations (Fig. 8.4-07). The average velocity of the upper crust to 25 km depth was 6.3 km/s, but in an ~60-km-wide zone around the kimberlite pipe relatively small seismic velocities were

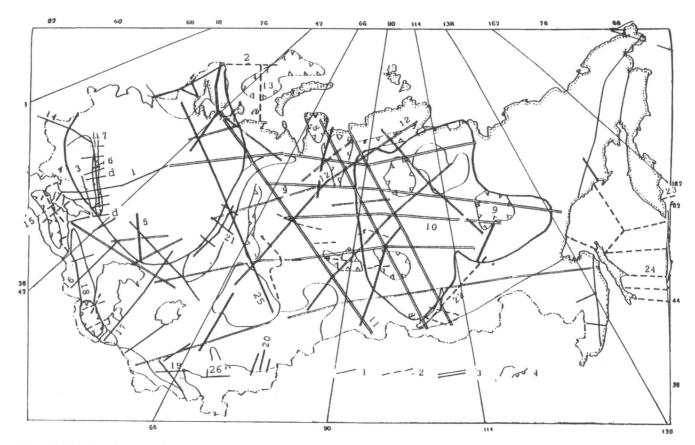

Figure 8.4-01. Location map of deep seismic sounding profiles in the USSR observed by SRGE (Special Regional Geophysical Expedition) in the 1970s and 1980s (from Pavlenkova, 1996, fig. 7). [Advances in Geophysics, v. 37, p. 1–133. Copyright Academic Press, Elsevier.]

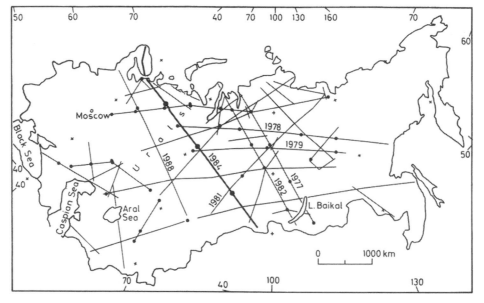

Figure 8.4-02. Location map of long-range deep seismic sounding profiles in the USSR using PNE (dots—peaceful nuclear explosions) as energy sources for distance ranges of several 1000 km (from Mechie et al., 1993, fig. 1). The observation scheme and data example shown in Figs. 8.4-03 and 8.4-04 is from the "Quartz" profile (thick line). [Physics of the Earth and Planetary Interiors, v. 79, p. 269–286. Copyright Elsevier.]

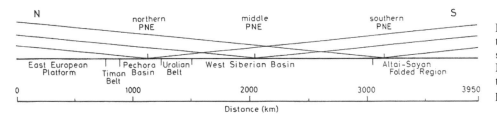

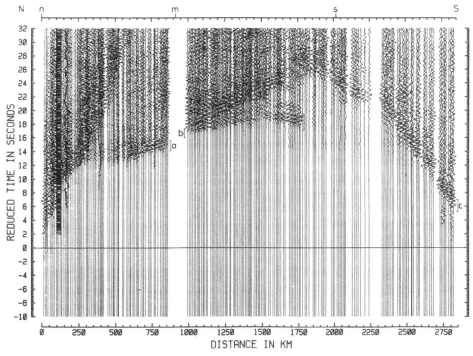

Figure 8.4-03. Observation scheme and tectonic units along the superlong deep seismic sounding profile "Quartz" (from Mechie et al., 1993, fig. 2). [Physics of the Earth and Planetary Interiors, v. 79, p. 269–286. Copyright Elsevier.]

Figure 8.4-04. Data example of the superlong deep seismic sounding profile "Quartz" in the USSR observed in 1984 (from Mechie et al., 1993, fig. 3b). [Physics of the Earth and Planetary Interiors, v. 79, p. 269–286. Copyright Elsevier.]

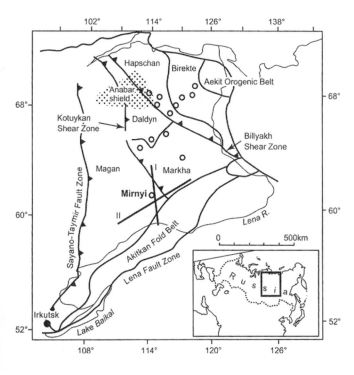

Figure 8.4-05. Location map of deep seismic sounding profiles (thick straight lines I and II) near the Mirnyi kimberlite field in the eastern Siberian craton (from Suvorov et al., 2006, fig. 1). Circles—kimberlite fields. [Tectonophysics, v. 420, p. 49–73. Copyright Elsevier.]

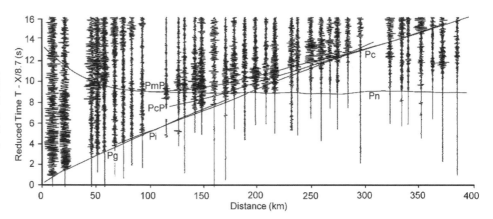

Figure 8.4-06. Data example showing record section of shotpoint 1 of DSS line I across the Mirnyi kimberlite field in the eastern Siberian craton (from Suvorov et al., 2006, fig. 9b). Reduction velocity is 8.7 km/s. [Tectonophysics, v. 420, p. 49–73. Copyright Elsevier.]

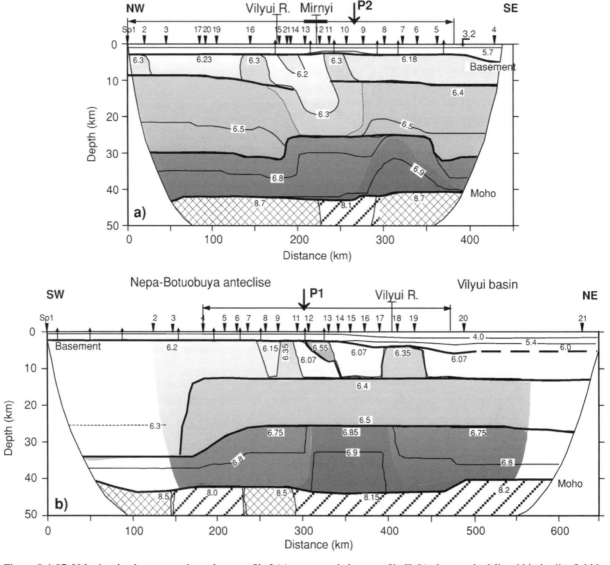

Figure 8.4-07. Velocity-depth cross sections along profile I (a) across and along profile II (b) close to the Mirnyi kimberlite field in the eastern Siberian craton (from Suvorov et al., 2006, fig. 7). [Tectonophysics, v. 420, p. 49–73. Copyright Elsevier.]

detected, surrounded by elevated average velocity of the upper crust (6.3 km/s). The lower crust, in contrast, had constant velocities of 6.8–6.9 km/s and appeared relatively unaffected by the presence of the kimberlite field. Furthermore, along both profiles extremely large sub-Moho mantle P-velocities of greater than 8.5–8.7 km/s were detected, except for a 70-km-wide zone with a "normal" P_n velocity of 8.1 km/s below the kimberlite field on profile 1 and below an ~100-km-wide zone at the southwestern end of profile 2, which again was close to the kimberlite field. The difference in the velocities in the two profile directions indicated anisotropy, but the effect of unusual rock compositions, e.g., from a high concentration of garnet, could not be excluded by the authors.

Near-vertical reflection studies were also performed in the USSR in the 1980s, but at a much smaller scale than in the United States and in western Europe. On the Ukrainian Shield and in the Urals, records were obtained up to 6 s TWT, allowing the study of the upper 15 km of the crust in some detail. In Kazakhstan, deep seismic-reflection records up to 12 s were recorded, and in Belorussia CDP profiles were obtained (Pavlenkova, 1996).

8.5. CONTROLLED-SOURCE SEISMOLOGY IN NORTH AMERICA

8.5.1. Instrumentation

The seismic experiments of the 1980s in North America differed substantially from seismic experiments in earlier decades. In reflection seismic surveys, long lines that were several 100 km in length were being realized. Furthermore, starting in 1978, seismic-refraction projects employing a rapidly growing number of instruments came into use. The requirements in the new development of portable seismic recording instruments had been: the instrument had to be small and available in large quantities (at least 100 pieces); it should be easily be operated by untrained personnel; it should record mostly automatically (i.e., to be switched on and off by a sophisticated clock of high accuracy); it should run unattended for several days; and the playback of the recorded data should be fast and easily gained for field control.

By the beginning of 1978, the U.S. Geological Survey, under the initiative of J.H. Healy, had developed new recording equipment (see Fig. 7.4.2-01), the so-called Seismic Cassette Recorder (SCR), and had built 100 individual units. The instrument was a single-component device consisting of a Mark Products L-4A 2-Hz vertical-component geophone, a set of three parallel amplifier boards with adjustable gain settings, a temperature-compensated oscillator (TCXO) providing the time standard for each unit, a voltage-controlled oscillator (VCO), and a cassette recorder. The three parallel amplifier boards with overlapping dynamic gain ranges allowed a variable total dynamic range. These three data channels and the time code signal were frequency modulated and all four frequencies plus a tape-speed compensation carrier frequency were summed and recorded on cassette tape. During the digitizing process, the cassette tapes were played

back and the signals demultiplexed and demodulated (Murphy, 1988; Appendix A7-5-6).

Over the following 10 years, other sets of new instrumentation were being developed and successively introduced into the fieldwork of the North American research groups. In Canada, a digital instrument, the Portable Refraction Seismograph (PRS1), was being developed by the Geological Survey of Canada and afterwards built by EDA Instruments Ltd. By the end of the 1980s, the instrument was ready for use (Asudeh et al., 1992). This unit had been built in large quantities so that by 1989, ~200 individual units were available for seismic fieldwork.

During the second half of the 1980s, digital equipment had also been developed by U.S. companies and was ready for use by the scientific community at the end of the decade, replacing the analogue systems. One of the first experiments where digital equipment came into use was the 1990 TACT survey in Alaska. The first digital U.S. equipment for scientific use was a donation of 200 units of Seismic Group Recorders (SGR) by the oil industry to Stanford University. Another development in the late 1980s ready for use by 1990 was the RefTek 72A-02 instrument which had been developed by RefTec Ltd. and which was being bought by PASSCAL (Program for Array Seismic Studies of the Continental Lithosphere), funded by the U.S. National Science Foundation for the use by North American university groups. For the fieldwork in Alaska in 1990, where all four types of instruments described above were in the field, 35 RefTek's were available (see section 8.5.4; Fuis et al., 1997). Both the SGRs and the RefTeks will be described in more detail in Chapter 9.

8.5.2. Canada

Over ~10 years starting in the late 1970s, a series of long-range seismic-refraction and wide-angle reflection experiments was conducted by a consortium of Canadian university and government crustal seismologists (COCRUST). Figure 8.5.2-01 shows the location of seismic profiles and the main tectonic features of interest (Mereu et al., 1989). In each experiment, between 8 and 23 shots were fired and recorded up to distances between 150 and 400 km, mainly on vertical instruments. In total, 59 inline record sections were obtained. Mereu et al. (1989) describe all COCRUST experiments in some detail, show a series of record sections as data examples (Appendix A8-3-1), and discuss the crustal complexities arising from these data and the nature of the P_MP phase reflected from the Moho.

The first project was carried out in 1977, 1979, and 1981 across southern Saskatchewan and Manitoba and was conducted over the northern part of the Williston Basin and a transition zone between the Churchill Geological Province and the older Superior Province. The north-south and east-west inline profiles of 1977 and 1979 were supplemented in 1981 by three profiles arranged as a triangle which recorded all shots arranged at its edges. The data were interpreted several times (e.g., Hajnal et al., 1984; Kanasewich et al., 1987), indicating in particular a crustal

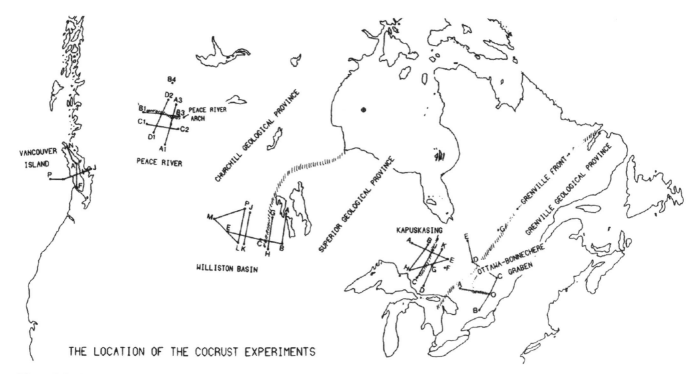

THE LOCATION OF THE COCRUST EXPERIMENTS

Figure 8.5.2-01. Location of COCRUST seismic refraction surveys in Canada and main tectonic features of interest (from Mereu et al., 1989, fig. 1). [*In* Mereu, R.F., Mueller, St., and Fountain, D.M., eds., Properties and processes of earth's lower crust: American Geophysical Union, Geophysical Monograph 51, p. 103–119. Reproduced by permission of American Geophysical Union.]

thickening from 40 km in the Superior Province to 45–50 km under the Williston Basin.

The second COCRUST experiment was the Vancouver Island project in 1980. These data were later (in 1984) complemented with a series of onshore and offshore near-vertical deep reflection lines, the first project of LITHOPROBE, which will be discussed below (Fig. 8.5.2-02). The main results showed that the continental crust overlying a subducting plate has a very complex crustal structure (e.g., Ellis et al., 1983; Green et al., 1986; Clowes et al., 1987a). Part of the data was used as basic data for a CCSS Workshop, which was discussed in the first section of this chapter (Green et al., 1990a).

The Ottawa-Bonnechere Graben-Grenville Front experiment of 1982 aimed to study the rather complex region of the Canadian Shield in eastern Ontario and western Quebec. The seismic lines were oriented parallel and perpendicular to the strike of the Ottawa-Bonnechere Graben and were interpreted by Mereu et al. (1986). The crust in the region of the graben appeared to be extremely heterogeneous, and a significant thickening of the crust along the

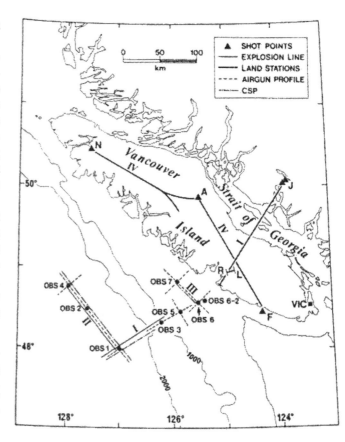

Figure 8.5.2-02. Location map of the COCRUST and LITHOPROBE surveys around Vancouver Island (from Green et al., 1990b, fig. 4). OBS—ocean bottom seismometer locations. [Geological Survey Canada, 89–13. 3–25. Reproduced with the permission of Natural Resources Canada 2009, courtesy of the Geological Survey of Canada.]

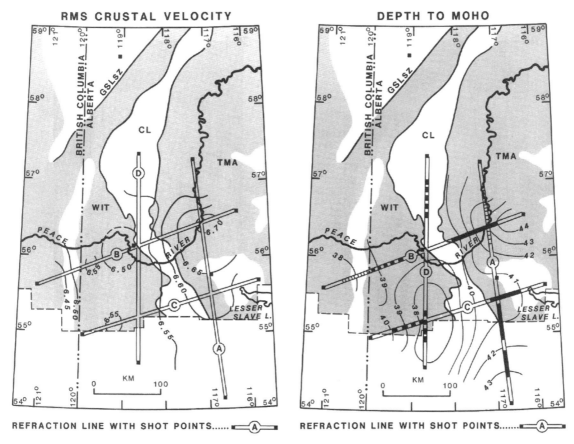

Figure 8.5.2-03. Location of the Peace River Arch experiment (from Stephenson et al., 1989, Figs. 4 and 5) superimposed on major aeromagnetic domains (shaded areas = positive residual magnetic anomalies). Left: Mean crustal velocities. Right: Moho depth contours. [Bulletin of Canadian Petroleum Geology, v. 37, p. 224–235. Permission granted by Bulletin of Canadian Petroleum Geology.]

Grenville front indicated that the front is a deep seated tectonic feature. These data across the Abitibi-Grenville region which later became one of the LITHOPROBE transects (AG in Fig. 8.5.2-05) and the data of the 1988 Ontario–New York refraction experiment (GRAP-88) were discussed and reinterpreted by Mereu (2000a) and White et al. (2000) in the context with new seismic-reflection and refraction data of the 1992 LITHOPROBE experiment (see Fig. 9.4.1-06 and corresponding text in Chapter 9.4.1).

The 1984 Kapuskasing experiment was carried out in northern Ontario over a gravity high, a zone of high-grade metamorphic rocks within the Superior Geological Province, and consisted of five seismic-refraction lines (Fig. 8.5.2-01) sampling the crust both parallel and perpendicular to the structure (Mereu et al, 1989). The investigations continued as the LITHOPROBE Kapuskasing Structural Zone Transect, and in 1987–1988, some 350 km of reflection data were recorded over the uplift (Percival et al., 1991; Leclair et al., 1994). The crust along the axis of this structure is extremely complex with a crust-mantle transition zone occurring from 38 to 48 km depth (Percival et al., 1991; Percival, 1994).

Finally, in 1985, the Peace River arch experiment was carried out in central Alberta consisting of four seismic lines of 290–330 km length, shot parallel and perpendicular to the arch structure (Fig. 8.5.2-03) and interpreted by Stephenson et al. (1989). All profiles were reversed, two of them had 3 km station spacing and an additional mid-point shot, the other two had 5 km station spacing and no mid-point shot. The most significant feature of the data was a very clear P_MP phase reflected from a very sharp Moho at 38–45 km depth. The Moho depth contour map also indicates the character of the observed Moho reflections. Solid black bars indicate a sharp Moho; cross-hatched bars indicate a more diffuse crust-mantle transition.

In 1984, the Geological Survey of Canada established a permanent base camp on the northeastern part of the Canadian Arctic continental shelf on a floating ice island that had broken off an ice shelf on Ellesmere Island in 1983 (Forsyth et al., 1990). In 1985 and 1986, air-supported seismic-refraction surveys were carried out on the adjacent ice pack, and in 1985, 1986, and 1988, the ice island was used as a platform for a floating seismic-reflection array. The refraction lines were up to 40 km long, and shots of 136–653 kg charges were detonated at depths of 100 m below the sea ice. Location of shotpoints and profiles and a velocity model fence diagram are shown in Figure 8.5.2-04.

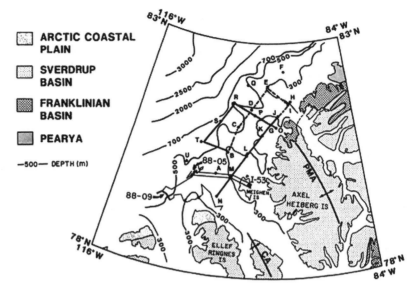

Figure 8.5.2-04. Ice island refraction work on the northeastern part of the Canadian Arctic continental shelf (from Forsyth et al., 1990, fig. 2). Top: location map (Letters: shotpoint locations), Bottom: crustal structure fence diagram. [*In* Pinet, B., and Bois, C., eds., The potential of deep seismic profiling for hydrocarbon exploration: Editions Technip, Paris, p. 225–236. Copyright Editions Technip, Paris.]

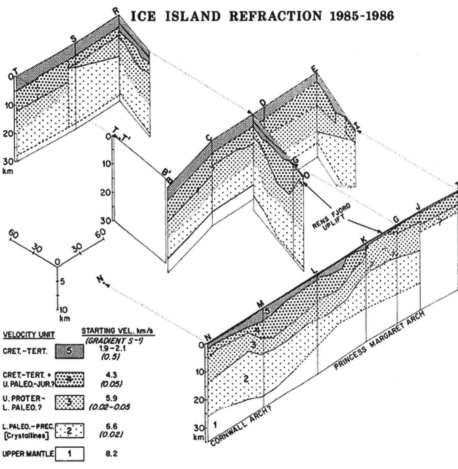

On the basis of the successful COCRUST operations and their results, in 1981, first discussions were started to involve the whole earth science spectrum. Soon thereafter, a LITHO-PROBE (probing the Earth's lithosphere) steering committee was founded, involving representatives from universities, government agencies and petroleum and mining industries. LITHOPROBE was designed as a coordinated yet highly decentralized research program. Its scientific and operational components were built around a series of transects or study areas (Fig. 8.5.2-05).

Each transect was to represent globally geotectonic processes and addressed the problems of continental evolution by undertaking geological, geochemical, and geophysical surveys and experiments in a variety of tectonic settings representing a wide range of geological time periods. The program was to be spearheaded

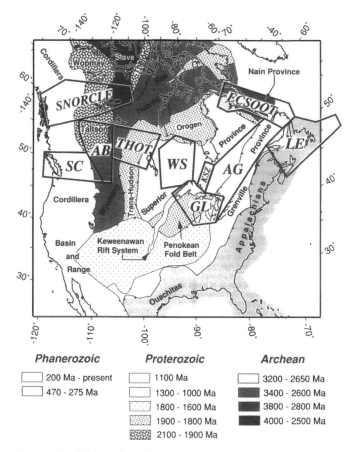

Figure 8.5.2-05. Location of LITHOPROBE transects (study areas) on a simplified tectonic map (from Clowes, 1997, fig. 1.1-1). SC—Southern Cordillera, AB—Alberta Basement, SNORCLE—Slave Northern Cordillera Lithospheric Evolution, THOT—Trans-Hudson Orogen Transect, WS—Western Superior, KSZ—Kapuskasing Structural Zone, GL—Great Lakes International Multidisciplinary Program on Crustal Evolution (GLIMPCE), AG—Abitibi-Grenville, LE—LITHO-PROBE East, ECSOOT—Eastern Canadian Shield Onshore-Offshore Transect. [LITHOPROBE Phase V proposal—evolution of a continent revealed. LITHOPROBE Secretariat, University of British Columbia, Vancouver, B.C., 292 p. Reproduced by permission of R. Clowes.]

by the seismic-reflection method, as reflection images provide the most directly applicable information at depth. For this reason, the LITHOPROBE Seismic Processing Facility was established consisting of a central site at the University of Calgary and an associated national network of seismic research nodes at several universities and at the Geological Survey of Canada. In particular, it provided a central facility for storage and rapid access of LITHOPROBE and related digital seismic-reflection and refraction as well as magnetotelluric data (Clowes, 1997; Clowes et al., 1984).

The first project was funded in 1984–1985 for a one-year Phase I preliminary program with primary scientific activity on Vancouver Island including a detailed seismic-refraction and reflection survey (Clowes et al., 1986, 1987a, 1987b), part of the Southern Cordillera Transect, which was continued in Phase II from 1987 to 1990, involving multichannel seismic-reflection acquisition (Kanasewich et al., 1994; O'Leary et al., 1993; Zelt et al., 1992, 1993). In 1985, five marine multichannel seismic-reflection profiles (lines 85-1 and 85-2 are shown on Fig. 8.2-06 in Chapter 8.2) totaling 520 km were recorded across the western Canada convergent margin where the Juan de Fuca plate is subducting beneath North America (Clowes et al., 1987b). The primary objectives were the definition of the offshore accretionary structures and clarification of the convergent interaction between the two plates. The seismic source was a 50 airgun array tuned to concentrate energy in the band below 50 Hz. Signals were recorded by a 3000 m, 120 channel streamer. The top of the subducted crust was clearly resolved and correlated with the lowestmost reflector seen on Vancouver Island on the 1980 COCRUST data described above (Clowes et al., 1987a).

Part of Phase II was SCoRE (Fig. 8.5.2-06), the Southern Cordillera Seismic Refraction Experiment, which was carried out in 1989 and 1990 (Zelt et al., 1992, 1993; Clowes et al., 1995; Spence and McLean, 1998) and which provided seven 350-km-long profiles over the southern Canadian Cordillera. SCoRE 89 included the first six lines and was conducted primarily across the Intermontane and Coast belts, including Vancouver Island, while during SCoRE 90, three lines were recorded from the Intermontane belt to the Foreland belt east of the Cordillera and one line was recorded along strike in the Coast belt. Kanasewich et al. (1994) discussed in detail line 8, which followed the strike direction of the Omineca Belt of the Canadian Cordillera. The corresponding crustal cross section is depicted in Figure 8.5.2-06.

In total, SCoRE comprised more than 50 individual shot gathers (Clowes et al., 1995). Interpreting the longest profile, the 450-km-long SW-NE–directed profile 2, which traversed the southern Cordillera from Vancouver Island through the Coast belt into the Intermontane belt, Spence and McLean (1998) concluded that throughout the crust, seismic velocities were in general high beneath the Insular belt, low beneath the coast and western Intermontane belts, and intermediate beneath the eastern Intermontane belt. The Moho depth increases only slightly along the line from near 33 km in the SW to ~38 km in the NE.

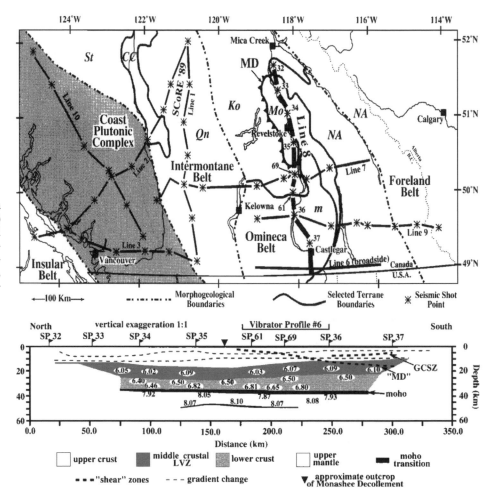

Figure 8.5.2-06. Top: Southern Cordillera Refraction Experiment (SCORE 89–90) shot and receiver lines). Bottom: Crustal cross section of Line 8 (from Kanasewich et al., 1994, figs. 1 and 7b). [Journal of Geophysical Research, v. 99, p. 2653–2670. Reproduced by permission of American Geophysical Union.]

Prior to SCoRE, in 1985 LITHOPROBE had recorded nearly 270 km of crustal seismic-reflection data across the eastern part of the southern Canadian Cordillera (Cook et al., 1987). Reflections near 12 s TWT were interpreted to possibly originate at the crust-mantle boundary which was estimated to lie at ~35 km depth.

In 1984–1985, a marine seismic-reflection survey, to become part of the LITHOPROBE East Transect (LE in Fig. 8.5.2-05), was recorded off the northeast coast of Newfoundland. Further seismic-reflection data were added in 1986, and a seismic-refraction survey was performed in 1988 and interpreted at a later stage together with data of the 1990s (Marillier et al., 1994; see also Fig. 9.4.1-04), which will be discussed in Chapter 9.4.1. Another seismic-reflection survey collected 600 km of data in 1989 in two main corridors, both directed in a NW-SE direction in the southwestern half of Newfoundland, which data were reprocessed and reinterpreted later by Van der Velden et al. (2004).

A fourth transect of LITHOPROBE, where work was started in the 1980s was the Great Lakes zone (GL). Here (Fig. 8.5.2-07) in 1986 the GLIMPCE experiment (Great Lakes International Multidisciplinary Program on Crustal Evolution) was undertaken in a cooperative effort of the Geological Survey of Canada and the U.S. Geological Survey (USGS). The experiment was a combined on-ship seismic-reflection and onshore seismic-refraction experiment to determine the structure of the crust under the Great Lakes, with the main tectonic targets being the Midcontinent Rift System, the Grenville Front, the Penokean and Huronian Fold Belts, and the Michipicoten Greenstone Belt (Green et al., 1989; Behrendt et al., 1989, 1990; Epili and Mereu, 1989; Hall and Quinian, 1994; White et al., 2000).

In total, ~1350 km of multichannel seismic-reflection data and an equivalent amount of seismic-refraction data have been collected across the North American Great Lakes. Across the Midcontinent System in Lake Superior and Lake Michigan alone, 700 km of 120-channel deep seismic-reflection data were collected. Refraction observations were made at the ends of most lines (stars in Fig. 8.5.2-07), and five OBSs recorded wide-angle information at equally spaced intervals along line A. The Moho reflections indicated varying crustal thickness from 37 to 55 km depth along the rift and volcanic and interbedded sediments of the rift to depths as great as 32 km, suggesting the greatest thickness of intracratonic rift deposits ever found on earth.

Taken together, by the end of LITHOPROBE, a series of ten transects, planned to provide a 3-D structure for each of the

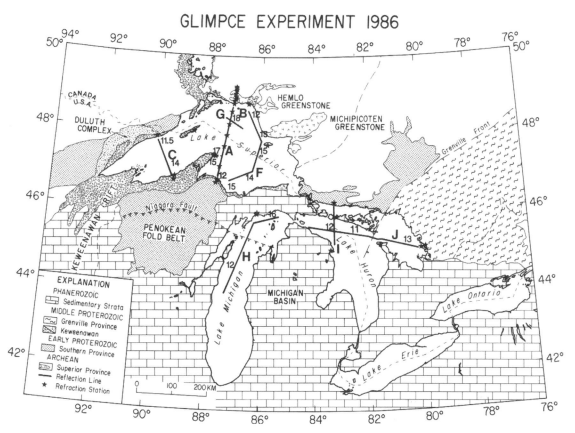

Figure 8.5.2-07. Location of the 1986 GLIMPCE experiment (from Behrendt et al., 1990, fig. 1b). [Tectonophysics, v. 173, p. 595–615. Copyright Elsevier.]

ten key areas (Fig. 8.5.2-05), should provide a nearly continuous coverage of the northern part of the North American continent. Results available up to late 1990 were summarized by Clowes et al. (1992) and Clowes (1993).

8.5.3. Continental United States

8.5.3.1. Seismic-Reflection Surveys

Following the successful start of COCORP in the second half of the 1970s, COCORP surveys continued in the 1980s. Brown et al. (1986) summarized the efforts and published a map showing the location of COCORP surveys completed until 1985 (Fig. 8.5.3-01). Since the beginning of the 1980s, a large amount of new data had been assembled which can best be recognized when comparing Figure 7.4.3-01 and Figure 8.5.3-01. Some of these operations involved some very long lines, such as the traverses through the northwestern Cordillera, through the northern Basin and Range province and through the Appalachians.

In the west, the new data comprised deep seismic-reflection surveys in the northwestern Cordillera from Washington to Idaho (Potter et al., 1986, 1987) along 48.5° latitude; across and east of the Cascades in Oregon; from the Sierra Nevada in northern California to the northwestern Colorado Plateau in western Utah along a 1000-km-long 40°N seismic-reflection transect re-

corded (see also Fig. 8.5.3-02) in 1982–1984 with its main emphasis on the Basin and Range province (Allmendinger et al., 1987, Klemperer et al., 1986); and in the Mojave desert and in Death Valley in southern California (de Voogd et al., 1988). Also in 1986, COCORP recorded a seismic-reflection line along the PACE lines through Arizona (Hauser et al., 1987a) and in 1989, a COCORP survey in Montana was added (Latham et al., 1988).

The COCORP seismic-reflection traverses of the U.S. Cordillera at 48.5°N and 40°N latitude revealed both fundamental similarities and differences in reflection patterns. On both traverses, autochthonous crust beneath thin-skinned thrust belts of the eastern part of the Cordillera is unreflective, immediately to the west the Cordilleran interior was very reflective above a flat prominent reflection Moho. The prominent reflection Moho was confined to areas which have undergone large Cenozoic post-thrusting extension. The main differences between the two traverses were seen in the reflection patterns of the middle and lower crust in the Cordilleran interior (Potter et al., 1987).

The COCORP 48.5° transect had a total length of 296 km and was acquired in 1984. It formed the initial part of a planned Cordilleran transect at 48–49° latitude, and it consisted of six reflection lines (Washington lines 1–5 and Idaho line 1). The data were collected using Vibroseis technique to produce a correlated record length of 16 s TWT corresponding to a maximum depth

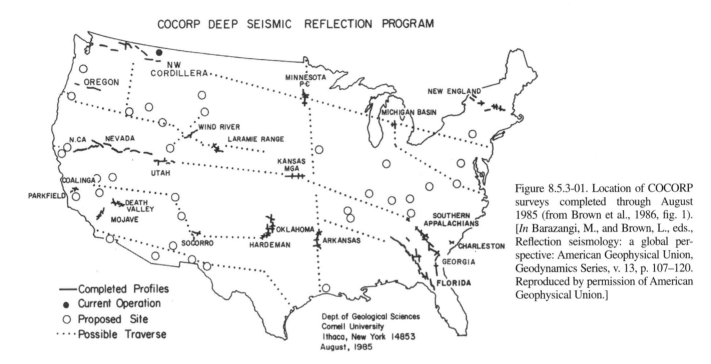

Figure 8.5.3-01. Location of COCORP surveys completed through August 1985 (from Brown et al., 1986, fig. 1). [*In* Barazangi, M., and Brown, L., eds., Reflection seismology: a global perspective: American Geophysical Union, Geodynamics Series, v. 13, p. 107–120. Reproduced by permission of American Geophysical Union.]

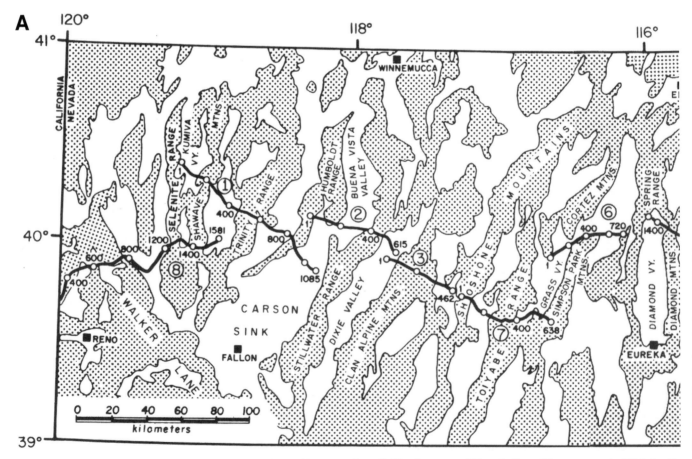

Figure 8.5.3-02 (*on this and following page*). (A) COCORP 40° transect from California to central Nevada (from Klemperer et al., 1986, fig. 1). [Geological Society of America Bulletin, v. 97, p. 603–618. Reproduced by permission of the Geological Society of America.]

of 50 km. The Moho was evident by strong horizontal reflections at 10.5–11.0 s TWT and was recognized as a flat surface underlying a complexly deformed crust, suggesting that it developed in Cenozoic times during widespread magmatism and extension possibly related to the generation of the Columbia River basalts (Potter et al., 1986).

The COCORP 40° transect (Fig. 8.5.3-02) followed the King Survey of the 40° parallel of 1867–1872 which had provided the first systematic reconnaissance traverse of the Cordillera of western North America. It was chosen because a large amount of subsequent research had shown that this transect contained one of the most complete records of Cordilleran evolution in the United States (Allmendinger et al., 1987). It was thus chosen as the site of the first deep seismic-reflection profile across the orogen where it crossed tectonic features ranging in age from Proterozoic to Recent and provided an acoustic cross-section of a complex orogen affected by erosion, compression, magmatism, and terrane accretion.

Allmendinger et al. (1987), for example, described the key features of the 40°N transect as follows The part of the transect traversing the Basin and Range province showed asymmetric seismic fabrics with west-dipping reflections in the east and mainly subhorizontal reflections in the west. The Moho was

clearly determined by pronounced reflections at 30 km depth, also at 34 km locally. No clear sub-Moho reflections were seen. In contrast, under the Sierra Nevada and under the Colorado Plateau, complex-dipping reflections and diffractions occurred locally as deep as 48 km depth. The features seen in the eastern part of the transect, which is underlain by Precambrian crystalline basement, were interpreted to be related to the entire geological history of the orogen.

In the western part, in contrast, most reflectors were interpreted to be not older than Mesozoic, indicating a strong Cenozoic overprint which is characterized by asymmetric half-grabens, low-angle normal faults, and a pervasive subhorizontal system of reflections in the lower crust.

The Death Valley COCORP seismic sections also showed a series of subhorizontal events that characterized the lower crust and abruptly ended at 11 s TWT. Between 10 and 11 s, a reflecting horizon was correlated and interpreted as Moho at ~30 km depth. Beneath the reflective lower crust, the upper mantle appeared to be seismically transparent (de Voogd et al., 1988).

The COCORP Arizona transect of 1986 crossed the transition zone between the Basin and Range province and the Colorado Plateau. The Moho reflections under the Basin and Range province contrasted with the non-reflective Moho under the

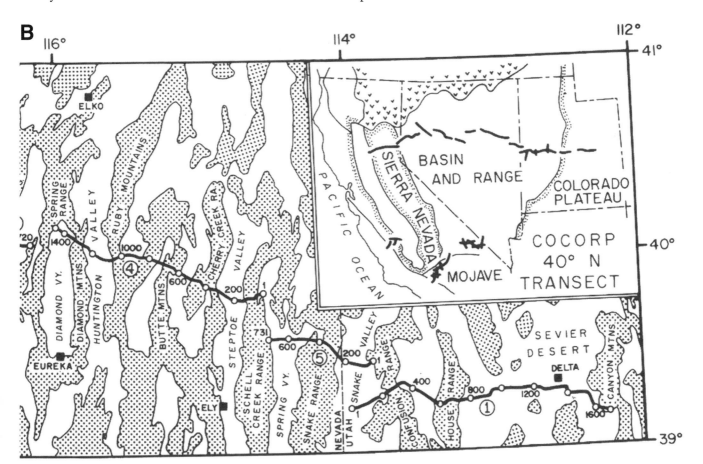

Figure 8.5.3-02 (*continued*). (B) COCORP 40° transect from central Nevada to Utah. The inset shows also the location of the COCORP surveys in southern California: Mojave Desert, Death Valley and San Andreas fault (Parkfield).

Colorado Plateau. This observation was interpreted that the development of reflectors at the Moho was clearly influenced by extension and associated igneous processes (Hauser et al., 1987a).

The seismic-reflection profiles recorded by COCORP in 1987 in the Montana plains between the Rocky Mountains and the Williston basin imaged the crystalline continental basement of the Archean Wyoming cratonic province. The crust appeared to be reflective throughout its entire thickness. West of the Williston basin, the Moho was not marked by the presence of any distinct reflections, but was defined to be located at the base of the reflective crust. The lowermost crust underneath the Williston basin, however, was characterized by a prominent laterally extensive zone of relatively high-amplitude reflections (Latham et al., 1988).

Besides COCORP, the USGS also performed a major study of the Earth's crust by seismic-reflection profiling (for location, see Fig. 8.5.3-24). Already since the early 1970s, a series of multichannel seismic-reflection projects had been supported and organized by the USGS (Hamilton, 1986). They included ~115,000 km of marine multichannel seismic-reflection data in a continuing program to study the geological framework of the U.S. continental margins of the Pacific Ocean off the west coast and Alaska as well as of the Atlantic Ocean off the east coast.

In 1982, the USGS received funds for a new program specifically designed to study the deep structure of the continent primarily by the use of seismic-reflection profiling (Hamilton, 1986). The first project was a study in central California from the Coast Ranges to the Sierra Nevada foothills. In 1984, work began in Alaska to carry out a crustal transect along the pipeline route. In 1985, studies were undertaken along a transect in southern California and southwest Arizona.

In the Midwest (for location, see Fig. 8.5.3-24), COCORP recorded seismic-reflection transects in 1981 both in Kansas and in Arkansas.

The COCORP survey in northeastern Kansas was part of a major east-west traverse across the mid-continent geophysical anomaly (Brown et al., 1983b). The seismic-reflection sections revealed clear indications of complex structure in the mid-to-lower crust which contrasts with the simplicity of the overlying sedimentary cover revealing numerous dipping and arcuate reflections and diffractions in the deeper crust. The basement above shows significantly fewer reflections, which was interpreted as being granitic terrane. The expected Moho at ~36 km depth was less characterized by specific reflections but more by an apparent decrease in the density and number of reflections (Brown et al., 1983b).

The Arkansas transect crossed the Ouachita Mountains in western Arkansas. The data indicated that Carboniferous foreland basin deposits within the Arkoma basin in the north thickened dramatically toward the south reaching more than 12 km thickness. Beneath the Benton uplift in the center of the line a broad antiform could be defined cresting at ~7 km depth. Beneath the southern Ouachitas, south-dipping stratified events were observed to depths in excess of 14 km. At the southern end

of the line beneath the northern coastal plain/southern Ouachitas, a prominent gently northward-dipping reflection occurred at ~22 km depth. The major structural features observed on this line were tentatively interpreted as having formed during a "simple" arc/continent collision, in which the subduction zone dipped south (Nelson et al., 1982).

In the eastern United States (for location, see Fig. 8.5.3-24), COCORP recorded the Adirondack–New England transect in 1980–1981, consisting of six lines and traversing the Grenville craton, the Adirondacks, to the west and the New England Appalachians to the east. Across the southern Adirondacks, a striking band of high reflectivity between 6 and 8 s TWT was seen which was interpreted to originate at 18–26 km depth. The Green Mountains to the east were identified as an imbricated thrust slice obducted above the lower crustal penetrating ramp (Brown et al., 1983a). Other intrabasement features were major east-west–dipping reflections, interpreted as possible faults, arched reflections that were interpreted as evidence for folds, a well-defined featureless zone, and possible scattered Moho reflections at times of at least 11 s which, according to Brown et al. (1983a), corresponds to Katz's (1955) estimated Moho depth.

In addition to the southern Appalachian traverse of 1978–1979, in 1983–1984 and 1984–1985, COCORP recorded a second line through the Appalachians and southeastern coastal plains from Georgia to Florida (Nelson et al., 1985). Reflections associated with the Appalachian detachment were the most prominent features of the northwestern portions of both the eastern and western Appalachian COCORP traverses. On both traverses these reflections comprised a prominent, ~0.5 s thick, gently southeast-dipping horizon in the upper crust, which could be traced from the level of detachment within the Valley and Ridge province southeastward for a considerable distance beneath the crystalline interior of the orogen. Beneath the coastal plains of western Georgia, a broad crustal-penetrating zone of dipping reflections was found, probably marking the Alleghanian suture in the southeastern United States. The COCORP data further southeast elucidated the internal structure of the Mesozoic Georgia basin within the Alleghanian orogen showing several large half-grabens separated by intervening highs. Beneath the well-developed Appalachian detachment reflections the Grenville basement exhibited few intracrustal reflections and no obvious reflections from Moho, but under the Georgia and Florida lines, numerous lower crustal reflections and locally prominent reflections at Moho level were evident (Nelson et al., 1987).

In 1983, the USGS shifted part of its seismic-reflection program to the eastern United States, starting with a project in Maine across the northern Appalachian Mountains and continuing in 1984 (QMT in Fig. 8.5.3-24). This project was closely connected with wide-angle profiling and will be described in more detail in subchapter 8.5.3.4. In 1984, a seismic-reflection survey in eastern Pennsylvania was also undertaken to study the seismic-reflection geometry of the Newark basin margin (Ratcliffe et al., 1986).

Another major goal of the USGS program was to study earthquake hazard zones where bedrock is covered by a thousand or more meters of poorly consolidated sediments (Hamilton, 1986), such as the New Madrid, Missouri, zone in the Northern Mississippi Embayment, which was accomplished in the late 1970s (see Chapter 7) and the Charleston, South Carolina, zone in the Atlantic Coastal Plain (Gohn, 1983; Behrendt, 1986), an investigation of which started in 1979 with seismic-reflection lines on land up to 3 s TWT and offshore with lines up to 12 s. The project was continued in 1981 with profiles of 1350 km total length through South Carolina and Georgia recorded to 6–8 s TWT (Hamilton, 1986). The USGS Interstate 64 seismic-reflection line (JRC in Fig. 8.5.3-24) was another relatively long line conducted in 1980 and 1981 that extended from north-central Virginia to near the Virginia coast (Lampshire et al., 1994; Pratt et al., 1987, 1988). The stacked section showed a highly reflective upper crust and also a sequence of reflections at ~9–12 s, indicative of lower crustal layering ~5–10 km in thickness, the base of which coincided with the Moho at 55 km depth under the Valley and Ridge province interpreted from earlier seismic-refraction work (James et al., 1968; Pratt et al., 1988).

As part of the ADCOH (Appalachian Ultradeep Core Hole, line ADC in Fig. 8.5.3-24) project site investigations of 1985, a seismic-reflection Vibroseis survey was carried out along regional lines in North and South Carolina and Georgia that extended for ~180 km across part of the southern Appalachians and, due to the experimental design and careful processing, collected seismic-reflection data of extremely high quality (Coruh et al.,1987; Costain et al., 1989a, 1989b; Hatcher et al., 1987; Hubbard et al., 1991). The ADCOH data were originally recorded with 14–56 Hz bandwith and 8 s length, but an extended Vibroseis correlation was used to produce 17 s data length. The upper crust to 9 km depth could be clearly imaged; below 8 s reflections were weak, but observable to as late as 13 s. However, these Moho reflections were generally short segments (Coruh et al., 1987). During the project, wide-angle and expanding spread data were also recorded (Hatcher, 2010, personal commun.).

Complete overviews of all deep seismic-refraction and reflection programs completed until late 1987 can be found in a summary volume *Geophysical Framework of the Continental United States*, edited by Pakiser and Mooney (1989). In their specific review of seismic-reflection programs in the eastern and central United States, Phinney and Roy-Chowdhury (1989) compiled a table containing the responsible organization, year, and specific references for all programs, which can be found at the end of subchapter 8.5.3.4 (see Fig. 8.5.3-24). Smithson and Johnson (1989) published a similar approach for the western United States.

8.5.3.2. Seismic-Refraction Projects in the Pacific States

The new equipment—100 portable Seismic Cassette Recorders (SCR)—which the USGS had acquired in the late 1970s (Healy et al., 1982; Murphy, 1988; Appendix A7-5-6) enabled a new phase of activity in the 1980s. Following the first projects in the late 1970s (Saudi Arabia, Mississippi embayment, Imperial Valley; for references see Chapter 7), under the leadership

of Walter Mooney and Gary Fuis, detailed crustal investigations started in 1981 throughout the United States. Some of the surveys aimed for upper-crustal studies only as, e.g., a detailed investigation of the Long Valley caldera in 1982 and 1983 (Meador and Hill, 1983; Meador et al., 1985; Appendix A8-3-2), and will not be discussed here in more detail, but most projects involved recording distances of 100 km and beyond and allowed study of the whole crust and uppermost mantle.

The beginning of the 1980s saw major activity in California. The western part of the Mojave Desert between the Garlock and San Andreas faults was the goal of a 200-km-long north-south profile with additional lines in 1980 and 1981 involving 12 shotpoints (Harris et al., 1988; Appendix A8-3-3).

From 1980 to 1982, projects followed in the Santa Cruz Mountains and in the Livermore area in the Diablo Range southeast of San Francisco (Meltzer et al., 1987; Williams et al., 1999; Appendix A8-3-3) and in the Great Valley of California (Holbrook and Mooney, 1987; Murphy, 1990; Appendix A8-3-3). The Great Valley project involved six profiles; the three major ones had three shotpoints each, were 180 km long, and were directed parallel to the tectonic strike direction. Furthermore, in 1982, a long line perpendicular to the tectonic strike direction was recorded with 7 shotpoints from Morro Bay at the Pacific coast into the foothills of the Sierra Nevada (Murphy and Walter, 1984; Appendix A8-3-3) across the San Andreas fault zone in the Parkfield-Cholame area and across the Great Valley (line SJ-6 in Figs. 8.5.3-03 and 8.5.3-04).

In 1983, an eighth shotpoint was added and part of the profile reoccupied with a much closer station spacing. The data of this project (see Fig. 8.2-03) together with seismic-reflection data collected by Western Geophysical in 1981 and purchased by the USGS in 1983 (see Fig. 8.2-04) served as basic material for the CCSS Workshop held in Susono, Shizouka, Japan in August 1985 (Walter and Mooney, 1987; Appendix A8-3-3) as was mentioned in the first section of this chapter. In 1986, a northwest-southeast–directed line of 86 km length, centered at San Luis Obispo, was recorded with 120 stations perpendicular to the line SJ-6 with a central shotpoint at the crossing point and two other in-line shots northwest and southeast. The survey was complemented by several fan shots located on land to the northeast and offshore in Morro Bay and farther south. The marine shots were airgun shots (Sharpless and Walter, 1988; Appendix A8-3-3).

Trehu (1991) investigated large-aperture data that had been collected in 1986 by seismometers deployed offshore (OBS) and recording airgun shots spaced ~150 m along a deep crustal seismic profile that was shot off Morro Bay toward southwest across the central California margin. This so-called Pacific Gas and Electric EDGE experiment (Howie et al., 1993) involved two parallel 190-km-long NE-SW onshore-offshore wide-angle reflection/refraction lines (PG&E lines 1 and 3) and three MCS (multichannel seismic) marine reflection lines. Two MCS lines were identical with the wide-angle lines; the third was a tie line. The above-mentioned PG&E-3 transect extended from the San Andreas fault (70 km inland) to

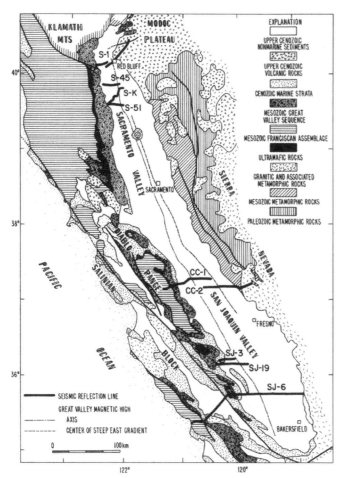

Figure 8.5.3-03. Regional geological map of California showing location of seismic reflection lines (from Zoback and Wentworth, 1986, fig. 1). [*In* Barazangi, M., and Brown, L., eds., Reflection seismology: a global perspective: American Geophysical Union, Geodynamics Series, v. 13, p. 183–196. Reproduced by permission of American Geophysical Union.]

the coast at Pismo Beach and then across the entire continental shelf and slope 120 km offshore.

The occurrence of a M 6.7 earthquake on 2 May 1983, near Coalinga, California, triggered a further detailed seismic investigation of crustal structure (Figs. 8.5.3-03 and 8.5.3-04) across the Mesozoic ocean-continent boundary beneath the California Coast Ranges and Great Valley which involved both seismic-reflection and refraction investigations (Colburn and Walter, 1984; Macgregor-Scott and Walter, 1985, 1988; Walter, 1990; Wentworth et al., 1983, 1984, 1987; Appendix A8-3-3). Some reflection lines were purchased and partly recorrelated from 6 to 12 s (lines SJ). The east-west–directed COCORP line C-1 (Fig. 8.5.3-04) was collected 12 km south of Coalinga.

Farther north, a 15-s reflection line (CC-1 and CC-2 in Fig. 8.5.3-03) was collected under contract. This 170-km-long line extends westward from the Sierran foothills across the northern San Joaquin Valley and into the Diablo Range of the central California

Coast Ranges. The interpretation of the seismic-reflection data was coordinated with two seismic-refraction lines (Colburn and Walter, 1984; Walter, 1990) shot after the Coalinga earthquake and crossing each other close to the epicenter area. The east-west line was 83 km long (shotpoints 9–12 in Fig. 8.5.3-04); the northwest-southeast line was 102 km (shotpoints 13–17 in Fig. 8.5.3-04). As the seismic-refraction lines were installed immediately after the 2 May 1983 $M_L = 6.7$ earthquake, the seismic-refraction data were further complemented by aftershock recordings along both lines (Macgregor-Scott and Walter, 1985, 1988).

For the Coalinga east-west profile, only the upper crust was modeled (Fig. 8.5.3-05). The interpretative west-east structural cross section of Wentworth and Zoback (1990) at the bottom of Figure 8.5.3-05 runs parallel to the west-east seismic-refraction line (top of Fig. 8.5.3-05) interpreted by Walter (1990) and follows approximately the reflection line SJ-19. For comparison, a projection of the seismicity superimposed into the west-east line is shown which concentrated on both the Cretaceous Great Valley sequence (Kg) and the Jurassic and Cretaceous Franciscan assemblage (KJf) overlying the basement (B). The model of the NW-SE profile of Macgregor-Scott and Walter (1988) covers the whole crust and shows underneath the 6.6-km/s layer a 6-km-thick lower crust with velocities from 7.1 to 7.4 km/s and the Moho at 27 km depth with a low uppermost mantle velocity of 7.95 km/s. Similar results have been published for the Great Valley profiles (Holbrook and Mooney, 1987) and by most of the CCSS Workshop participants for the Morro Bay–Sierra Nevada line (Walter and Mooney, 1987) mentioned above.

In northeastern California, the USGS conducted seismic-refraction experiments (Fig. 8.5.3-06) in order to characterize the crustal structure of and boundaries between the Klamath Mountains, Cascade Range, Modoc Plateau, and Basin and Range geological provinces (Zucca et al., 1986). The experiments took place in 1979, 1981, 1982, and 1985 (Kohler et al., 1987; Appendix A8-3-3; Zucca et al., 1986; Fuis et al., 1987). A total of 55 shots was detonated at 29 shotpoints; the number of stations varied between 60 and 120, with intervals varying between 0.5 and 1.75 km. The surveys comprised north-south lines both in the Klamath Mountains and the Modoc Plateau; northwest-southeast lines centered on Mount Shasta and Medicine Lake volcanoes, and a west-east line from the Klamath Mountains to the Basin and Range province linking the other profiles. All lines involved a 1 km station spacing and at least three shotpoints over 80 km distance. Along the 300-km-long west-east line (see also Fig. 8.5.3-08, line 9) six shotpoints were placed. One of the northwest-southeast lines across Mount Shasta and Lassen Peak was also 300 km long, had eight shotpoints, and reached to the northern end of the Sierra Nevada. The Medicine Lake Volcano was of particular interest with specially designed recording schemes.

The crustal cross section in Figure 8.5.3-07 by Fuis et al. (1987; Mooney and Weaver, 1989) shows that the Klamath Mountains are underlain by a stack of oceanic layers, while the Modoc Plateau is inferred to be underlain by crystalline igneous and metamorphic rocks beneath volcanic and sedimentary

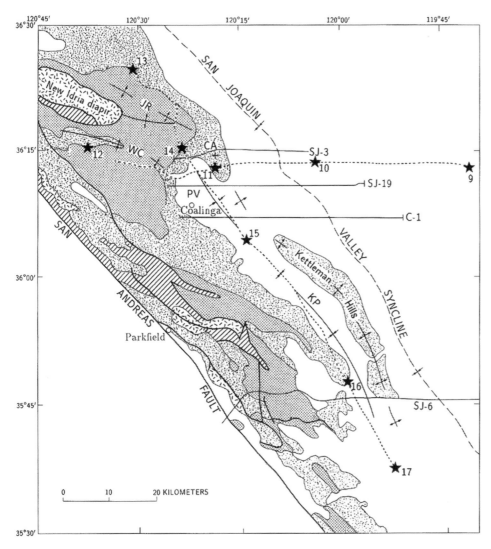

Figure 8.5.3-04. Location of seismic profiles around Coalinga, central California, extending from the southern Diablo Range into San Joaquin Valley (from Walter, 1990, fig. 3.1). +—Main shock (M_L=6.7). Stars—shotpoints, full lines: seismic profiles recorded before 2 May 1983. Dotted lines: seismic profiles after 2 May 1983. SJ—industrial seismic reflection lines; C-1—COCORP line. [U.S. Geological Survey Professional Paper 1487, p. 23–39. Copyright U.S. Geological Survey.]

rocks. The Cascade Range, in between, is a complex suture region currently intruded by magmas. The base of the model in Figure 8.5.3-07 at 14 km consists of a 7.0-km/s layer that extends to unknown depths. The total crustal thickness under the southern Cascades has not reliably been determined; it is estimated to be around 38-40 km. The crustal structure beneath the Modoc Plateau, however, is well known from the seismic-refraction profiles; the Moho lies in ~38 km depth.

Several seismic-refraction investigations were also performed by the USGS farther north in Oregon and Washington (Fig. 8.5.3-08). The Oregon Cascades were the target of a 275-km-long seismic-refraction profile (line 7 in Fig. 8.5.3-08) with three shotpoints and 100 recording stations along the north-south axis of the Cascades from Mount Hood to Crater Lake. The profile was located 20–30 km west of the crest of the High Cascades and nearly coincided with the contact between the High and the Western Cascades. The interpretation of Leaver et al. (1984) includes the observations of three regional earthquakes. Two of the events occurred to the north in southern Washington (Elk Lakes

event and Goat Rocks event), and the third event (Stephen's Pass earthquake) occurred in the south in northern California. They had been recorded by local networks of the USGS and the University of Washington and record sections had been prepared. The interpretation of the data resulted in a total crustal thickness of ~45 km underneath the Oregon Cascades, subdivided into a 10-km-thick upper crustal layer and underlain by an intermediate crust with 6.5–6.6 km/s from 10 to 30 km depth and a lower crust with a mean velocity of 7.1 km/s (Leaver et al., 1984).

The investigation of the Oregon Cascades was continued in 1983 by a 180-km-long west-east profile across east-central Oregon (line 8 in Fig. 8.5.3-08) which reached from the High Cascades 30 km west of Newberry crater (triangle on line 8) to the eastern High Lava Plains (Catchings and Mooney, 1988b; Cotton and Catchings, 1989; Appendix A8-3-4). The target was the arc to backarc transition zone between the volcanic Cascade Mountains and the actively extending Basin and Range province. Two-hundred forty recording sites recorded shots from six sites, of which five shotpoints were located in the west around Newberry

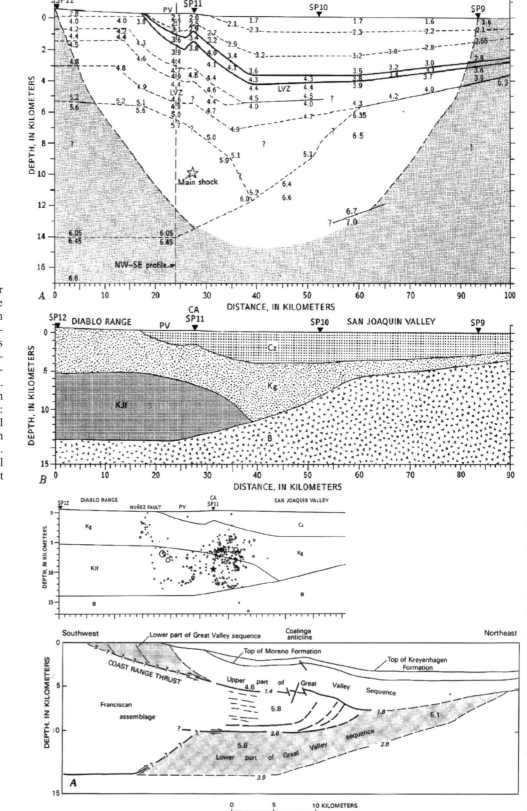

Figure 8.5.3-05. Top and upper center: Upper crustal model of the Coalinga east-west profile (from Walter, 1990, Figs. 3.3). Cz—Cenozoic strata, Kg—Cretaceous Great Valley sequence, KJf—Jurassic and Cretaceous Franciscan assemblage, B—basement. Lower center: Seismicity (from Walter, 1990, fig. 3.9). Bottom: Interpretative west-east structural cross section (from Wentworth and Zoback, 1990, fig. 4.6). [U.S. Geological Survey Professional Paper 1487, p. 23–39. Copyright U.S. Geological Survey.]

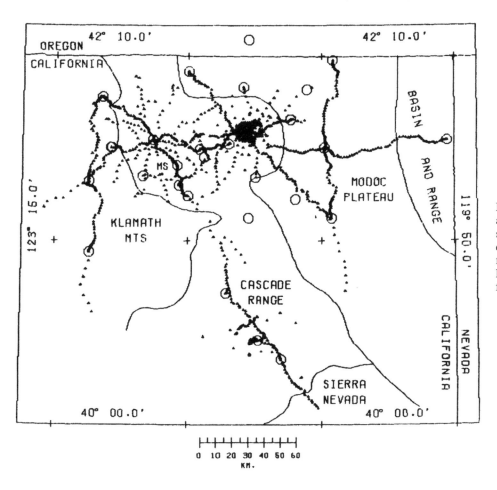

Figure 8.5.3-06. Location of shotpoints and recording sites of the 1979–1985 seismic-refraction surveys in northeastern California (from Kohler et al., 1987, fig. 2). [U.S. Geological Survey Open-File Report 87-625, Menlo Park, California, 99 p.]

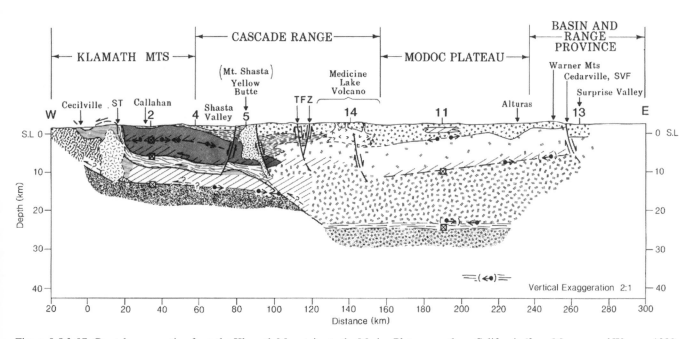

Figure 8.5.3-07. Crustal cross section from the Klamath Mountains to the Modoc Plateau, northern California (from Mooney and Weaver, 1989, fig. 12). [*In* Pakiser, L.C., and Mooney, W.D., eds., Geophysical framework of the continental United States: Geological Society of America Memoir 172, p. 129–161. Reproduced by permission of the Geological Society of America.]

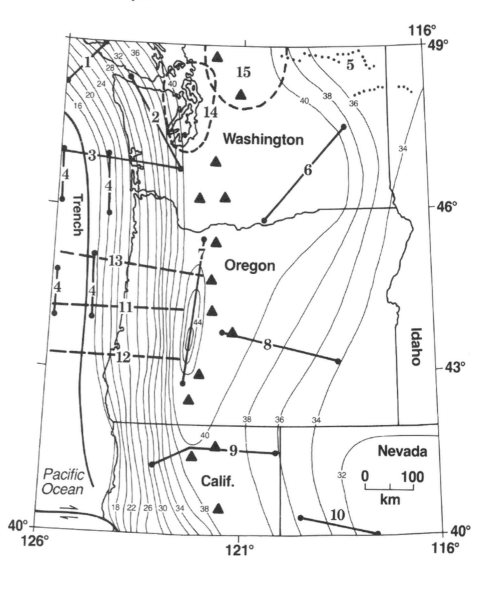

Figure 8.5.3-08. Location map of seismic-refraction (solid lines), seismic-reflection (dotted lines), and other geophysical (dashed lines and circles) surveys in the northwestern United States (from Mooney and Weaver, 1989, fig. 10). 1—Vancouver Island, Canada (Clowes et al., 1986); 2, 3—Washington continental margin (Taber and Lewis, 1986); 4—offshore seismic continental margin studies (Shor et al, 1968); 5—COCORP reflection survey across northwestern Cordillera (Potter et al., 1986); 6—Columbia Plateau (Catchings and Mooney, 1988a); 7—Oregon Cascades (Leaver et al., 1984); 8—Newberry volcano, east-central Oregon (Catchings and Mooney, 1988b); 9—northern California (Zucca et al., 1986); 10—northwestern Basin and Range province (PASSCAL Working Group, 1988; Holbrook, 1990; Catchings and Mooney, 1991); 11–15—other geophysical surveys. Triangles—volcanoes; thin lines—contours of crustal thickness. [*In* Pakiser, L.C., and Mooney, W.D., eds., Geophysical framework of the continental United States: Geological Society of America Memoir 172, p. 129–161. Reproduced by permission of the Geological Society of America.]

volcano and spaced ~15 km apart and one shotpoint was located at the eastern end of the line.

To image any extant small silicic magma chambers, the investigation of the upper crust beneath Newberry volcano was continued in 1984 by a dense 13-km-diameter 120-element array of vertical seismographs with average site spacing of 1.07 km and 16 shotpoints. For the interpretation a high-resolution active-source seismic-tomography method was applied (Achauer et al., 1988; Dawson and Stauber, 1986; Appendix A8-3-4). The upper and middle crust (Fig. 8.5.3-09) is similar to that of the Cascade Range but the Moho at the bottom of the lower crust of 7.4 km/s is only 35 km deep (Catchings and Mooney, 1988b).

Mooney and Weaver (1989) compiled a summary of upper crustal structure underneath the most spectacular volcanoes in the Cascade Range (Fig. 8.5.3-09). Seismic velocities of ~6 km/s are found at shallow depths beneath the stratovolcanoes (A, B, C, D, G in Fig. 8.5.3-09), but are not attained until a depth of 8 km for Newberry volcano (E) and 5 km for Medi-

cine Lake Caldera (F) leading to the conclusion that the crustal structure underneath the stratovolcanoes does not differ significantly from that of the overall crustal structure along the axis of the High Cascades.

In 1989, an offshore-onshore seismic experiment was initiated with the aim to investigate the subduction zone underneath the Cascade Mountains in central Oregon (Trehu and Nakamura, 1993; Brocher et al., 1993a) which was followed by the 1991 Pacific Northwest Experiment (Trehu et al., 1993). Both projects are jointly discussed in Chapter 9.4.2.2.

In conjunction with the search of a nuclear waste facility planned in the northwestern United States, in 1984, a 260 km long-range seismic-refraction/wide-angle reflection survey was conducted across the central Columbia Plateau in northeasterly direction from northern Oregon into central Washington (line 6 in Fig. 8.5.3-08) to evaluate its crustal and upper mantle structure and the nature of the Columbia River basalt group (Cotton and Catchings, 1988; Appendix A8-3-4; Catchings and Mooney,

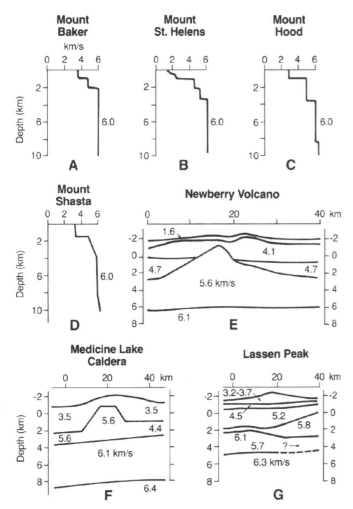

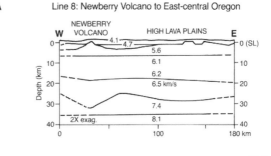

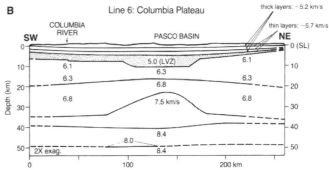

Figure 8.5.3-10. Crustal structure of east-central Oregon and the Columbia Plateau (from Mooney and Weaver, 1989, fig. 14), corresponding to lines 8 and 6 respectively in Figure 8.5.3-08. [*In* Pakiser, L.C., and Mooney, W.D., eds., Geophysical framework of the continental United States: Geological Society of America Memoir 172, p. 129–161. Reproduced by permission of the Geological Society of America.]

Figure 8.5.3-09. Upper crustal velocity structure of the stratovolcanoes and off-axis basaltic centers Newberry and Medicine Lake volcanoes (from Mooney and Weaver, 1989, fig. 13). [*In* Pakiser, L.C., and Mooney, W.D., eds., Geophysical framework of the continental United States: Geological Society of America Memoir 172, p. 129–161. Reproduced by permission of the Geological Society of America.]

1988a). The survey was centered on the Hanford site in the Pasco Basin of south-central Washington, a proposed high-level nuclear waste repository and the site of existing nuclear facilities. Four shotpoints were approximately located at profile-kms 60, 115, 160, and 200, and shots of 900–2200 kg sizes were detonated in 50 m deep drillholes. The profile contained 240 recording sites spaced 900 m apart between the shotpoints and 1300 m outside. Though vertical-incidence seismic-reflection profiling has not been successful due to the 3–6-km-thick basalt cover of the area, the wide-angle seismic operation succeeded in obtaining clear seismic signals from the crust and upper mantle beneath the Columbia Plateau.

An interpretation by Catchings and Mooney (1988a) is shown in Figure 8.5.3-10. The crust is 40 km thick, in the center of the line, the lower crust consists of two layers with 6.8 and 7.5 km/s. Beneath the Pasco Basin, the lower basal layer with

7.5 km/s thickens and shows geometry similar to continental rifts. From the seismic-reflection data across the Okanagon Highlands (line 5 in Fig. 8.5.3-08; see also Fig. 8.5.3-01) a highly reflective, complexly deformed crust and a relatively flat Moho at 33–35 km depth have been identified (Potter et al., 1986).

8.5.3.3. Seismic-Refraction Projects in Basin and Range and Rocky Mountains

In 1986, a multi-method seismic experiment was conducted in Nevada (Fig. 8.5.3-11) in the northwestern Basin and Range province (Catchings and PASSCAL Working Group, 1988; Holbrook, 1990; Catchings and Mooney, 1991; Whitman and Catchings, 1987; Appendix A8-3-5) under the auspices of the Program for Array Seismic Studies of the Continental Lithosphere (PASSCAL). The aim was to collect and integrate new wide-angle and near-vertical incidence explosion seismic data to gain a detailed picture of the structure and velocity distribution.

The 280-km-long "west-east" line had 7 shotpoints with large shots, 45 km apart, and four smaller shots in the center, and the 200-km-long "north-south" profile had 5 shotpoints, 50 km apart, which shots were all recorded simultaneously on both lines. In total, 28 shots were detonated in boreholes with charges ranging from 230 to 2700 kg which were recorded by three deployments of 120 one-component and 40 three-component stations. Receiver spacing was 900 m for the center 100 km of the west-east line and 1400–1500 m elsewhere. Furthermore, three

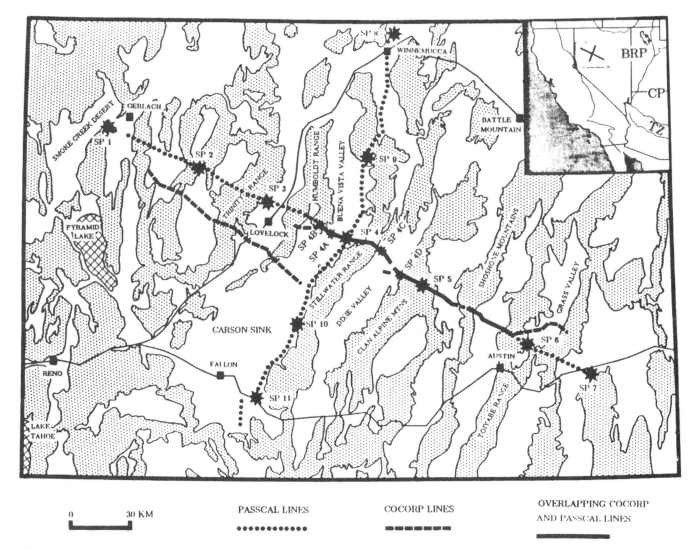

Figure 8.5.3-11. Location of the 1986 PASSCAL Basin and Range lithospheric seismic experiment. Stars—shotpoints (from Holbrook, 1990, fig. 1). Also shown is the COCORP investigation of 1982–1984 (see Fig. 8.5.3-01, western half). [Journal of Geophysical Research, v. 95, p. 21,843–21,869. Reproduced by permission of American Geophysical Union.]

48-channel seismic-reflection recording systems were involved which also were deployed three times.

The seismic-refraction experiment was placed as close as possible at the same position as the western half of the COCORP experiment in 1982–1984 (COCORP 40° transect in Fig. 8.5.3-02, western half). Due to the propagation path of waves to be recorded at wide-angle distance ranges, the seismic-refraction line had to be as straight as possible and therefore the northwestern half and the southeasternmost end deviate from the COCORP reflection lines. The experiment included the objective to compare the entire wavefield recorded coincidently by the "refraction method" with 2-Hz geophones at 1 km spacing and the "reflection method" with 10-Hz geophones and 100 m spacing.

For comparison the main results of the seismic-reflection COCORP 40° transect and the PASSCAL seismic-refraction proj-

ect are shown (Figs. 8.5.3-12 to 8.5.3-14) and briefly discussed here. In general, the Moho is 30 km thick (Fig. 8.5.3-12). This is also evident in the reflection data where the Moho is identified as the bottom of the highly reflective lower crust (Fig. 8.5.3-13). Middle and lower crust with 6.6 and 7.4 km/s (Fig. 8.5.3-12) more or less agree with the highly reflective lower crust of Figures 8.5.3-13 and 8.5.3-14.

The new data confirm the results of the VELA UNIFORM project in the 1960s, but present a much higher degree of accuracy and details. For example, the interpretation of Catchings and Mooney (1991) contains a lower crustal layer (shaded in Fig. 8.5.3-12) which in an earlier interpretation was identified as crust-mantle transition zone (Prodehl, 1979; see Fig. 6.5.1-09).

More local investigations in the Basin and Range Province of Nevada aimed for detailed crustal studies around Yucca Mountain near Beatty, Nevada. Here the USGS conducted seismic

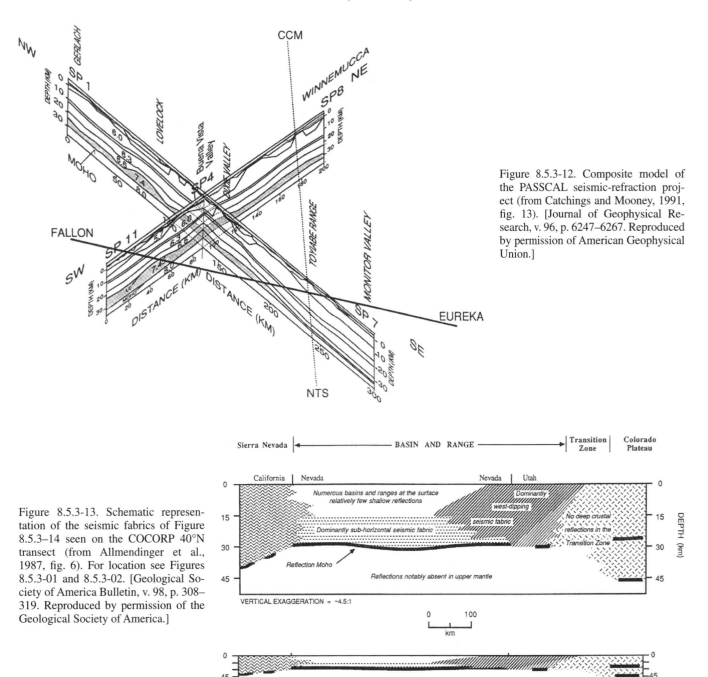

Figure 8.5.3-12. Composite model of the PASSCAL seismic-refraction project (from Catchings and Mooney, 1991, fig. 13). [Journal of Geophysical Research, v. 96, p. 6247–6267. Reproduced by permission of American Geophysical Union.]

Figure 8.5.3-13. Schematic representation of the seismic fabrics of Figure 8.5.3–14 seen on the COCORP 40°N transect (from Allmendinger et al., 1987, fig. 6). For location see Figures 8.5.3-01 and 8.5.3-02. [Geological Society of America Bulletin, v. 98, p. 308–319. Reproduced by permission of the Geological Society of America.]

studies to aid an investigation of the regional crustal structure at a possible nuclear waste repository site near Yucca Mountain. Initial seismic-refraction studies had already been carried out in 1980 and 1981 (Hoffman and Mooney, 1983), recording three nuclear events at two regional and one N-S profile up to 110 km offsets. Furthermore, an E-W profile had been established in 1982, recording up to 65 km offset from a shotpoint SE of Beatty, Nevada, across Yucca Mountain to the Nevada Test Site area. The data showed prominent reflections from a mid-crustal boundary at 15 km depth, identified other mid-crustal boundaries at 24

and 30 km depth, and revealed a total crustal thickness of 35 km (Hoffman and Mooney, 1983).

In 1988, a test experiment was carried out south of Yucca Mountain, to see if the seismic-reflection method would be feasible for a larger seismic-reflection crustal study in this area (Brocher et al., 1990, 1993b). In this test, both large Vibroseis arrays (more than 72,200 kg) and large chemical explosive sources (charges ranging from 23 to 182 kg) were recorded by a large number (480) of closely spaced (25 m) receiver channels. The recording line was 27 km long and a 60-fold coverage was

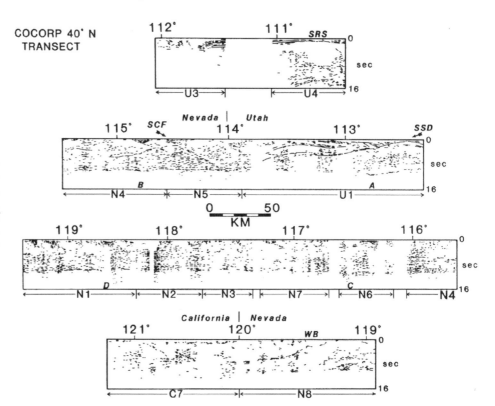

Figure 8.5.3-14. Simplified line drawings of the entire COCORP 40°N transect (from Allmendinger et al., 1987, fig. 5). For location see Figures 8.5.3-01 and 8.5.3-02. [Geological Society of America Bulletin, v. 98, p. 308–319. Reproduced by permission of the Geological Society of America.]

acquired using a 12-km-long symmetric split spread (Brocher et al., 1990). It turned out that the explosive sources produced a higher-quality image of the lower crust (5–10 s). The base of the crust at 35 km depth was estimated from the absence of high-amplitude reflected arrivals below 9–10 s confirming the result obtained already by the 1980–1982 seismic-refraction study described above.

The Long Valley Mono Craters volcanic complex at the western margin of the Basin and Range province in eastern California had already been the target of a major seismic project in the 1970s (Hill, 1976). New seismic-refraction data were obtained in the summers of 1982 and 1983 as part of a study to evaluate the region as a possible site for deep scientific drilling in a thermal regime. One hundred stations in 1982 and 120 in 1983 recorded 1000 kg charges fired in 50–60 m deep boreholes. The profiles sampled the upper 7–10 km of the crust within the western half of the Long Valley caldera, the Mono craters ring fracture system and the "normal" crust north and northeast of Long Valley. At a depth of 2 km beneath the topographic surface, the P-wave velocity is uniformly 5.6 km/s beneath which the velocity increases with a gradient of ~0.1 s^{-1}. Clear secondary arrivals in the western part of the caldera supported evidence from 1972 seismic-refraction profiles for a reflecting boundary at a depth of 7–8 km beneath the west margin of the resurgent dome which may represent the top of a magma chamber (Hill et al., 1985).

A special goal was to study the evolutionary history of metamorphic core complexes of the western Cordillera (Fig. 8.5.3-15). In particular, information on the thickness of the crust, the geom-

etry of the Moho, and the composition of the middle and lower crust that underlie metamorphic core complexes (McCarthy et al., 1991; Wilson et al., 1991) was needed.

For this reason the USGS chose to undertake another multimethod seismic experiment: PACE (Pacific to Arizona Crustal Experiment), a major seismic-refraction/wide-angle reflection study of the core complex belt along the Colorado River in southeastern California and western Arizona. The study of 1985 (profiles 2 and 3) and 1987 (profile 1) was conducted as part of PACE (Fig. 8.5.3-15) and was designed to complement vertical incidence seismic-reflection profiles collected by COCORP (Consortium for Continental Reflection Profiling) and by CALCRUST (California Consortium for Crustal Studies).

The project involved three profiles across the Chemehuevi (A, profile 3), Whipple (B, profile 2), and Buckskin-Rawhide (C, profile 1) mountains metamorphic core complexes. Lines 2 and 3, recorded in 1985 (Wilson and Fuis, 1987; Appendix A8-3-5), were each 130 km long and a total of 30 shots from 20 shotpoints was recorded. Including offset shots, the maximum offsets along profiles 2 and 3 was extended up to 170 km. Profile 1, recorded in 1987 (Larkin et al., 1988; Appendix A8-3-5) with 27 shots from 19 shotpoints, was 180 km long, with maximum offsets extended to 210 km. During both experiments, higher-resolution reflection arrays were deployed within the center of the lines (Henyey et al., 1987).

Crustal thickness is near 30 km and fairly uniform throughout the region. The prominent feature in the model of Figure 8.5.3-16 is a thick midcrustal layer centered beneath or just

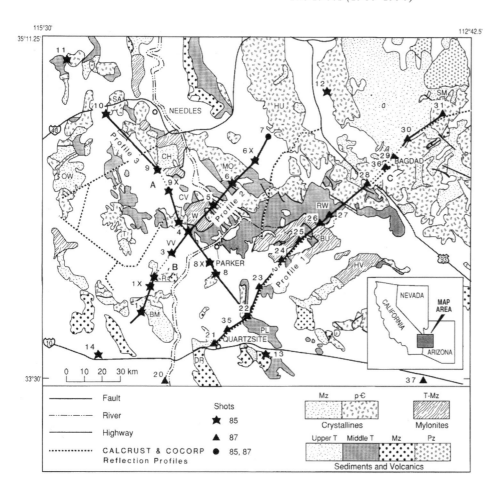

Figure 8.5.3-15. PACE seismic-refraction survey of 1985 and 1987 in southeastern California and western Arizona (from McCarthy et al., 1991, fig. 2). BM—Big Maria Mountains; BU—Buckskin Mountains; CH—Chemehuevi Mountains; DR—Dome Rock Mountains; HU—Hualapai Mountains; HV—Harcuvar Mountains; MO—Mohavi Mountains; OW—Old Woman Mountains; PL—Plomosa Mountains; R—Riverside Mountains; RW—Rawhide Mountains; SA—Sacramento Mountains; T—Turtle Mountains; VV—Vidal Valley; W—Whipple Mountains. [Journal of Geophysical Research, v. 96, p. 12,259–12,291. Reproduced by permission of American Geophysical Union.]

west of the metamorphic core complex (McCarthy et al., 1991; Wilson et al., 1991).

The arch-like shape of its upper boundary was regarded to possibly correlate with the large amount of uplift noted at the surface. For its interpretation the authors offered various explanations, from a magmatically thickened crust composed of diorite or sill-like interfering gabbro and silicic country rock to low-velocity felsic material laterally flowing in from less extended regions and thereby inflating the middle crust (McCarthy et al., 1991).

In 1989, the Pacific to Arizona Crustal Experiment (PACE) was conducted across the northeastern Transition zone and the southwestern margin of the Colorado Plateau. Two profiles were acquired (Fig. 8.5.3-17, McCarthy et al., 1994; Appendix A8-3-5). The PACE Colorado Plateau profile P1 extended 150 km from the northeastern end of the 1987 PACE study (profile 1 in Fig. 8.5.3-15) across Chino Valley to the western edge of the Navajo Indian Reservation near Cameron, Arizona, and had 24 shots with an average spacing of 10 km. The Grand Canyon profile P2 (Wolf and Cipar, 1993) was located entirely within the Colorado Plateau. It was also 150 km long and had inline 10 shots, one of them offset resulting in a maximum recording distance of 225 km. The location map (Fig. 8.5.3-17) also shows the position of earlier seismic-refraction lines obtained in the 1960s (for details and references, see Chapter 6).

The data of profile 1 were inverted together with the 1987 data. Crustal thickness increases from 30 km in the southwest to almost 40 km where profile 1 enters the Colorado Plateau, but decreases to 35 km at the northeastern end of the line (Parsons et al., 1996). Coincident with the thickening of the crust is the thickening of the lower crustal layer (6.5–6.6 km/s). For the Grand Canyon profile a 35-km-thick crust was obtained near Meteor Crater at the southeastern end, but when reaching the Grand Canyon, the crust thickens to almost 50 km (Wolf and Cipar, 19923; Parsons et al., 1996).

In 1986 COCORP (Hauser et al., 1987a; Hauser and Lundy, 1989) recorded 550 km of vertical incidence seismic-reflection data along a transect in Arizona, coincident with the PACE profiles and in areas to the east. The COCORP vertical incidence seismic-reflection data had obtained a gradual increase in reflectivity between 16 and 20 km depth. Moho reflections were generally between 9 and 10 s TWT in the Basin and Range province, corresponding to 27–30 km depth, and were dipping gently eastward beneath the central transition zone, but also showed distinct offsets and reached up to 12 s corresponding to ~36 km depth (Hauser et al., 1987a; Hauser and Lundy, 1989).

Further east, in the area of the Southern Rocky Mountains, two new experiments in 1981 comprised the CARDEX (Caldera

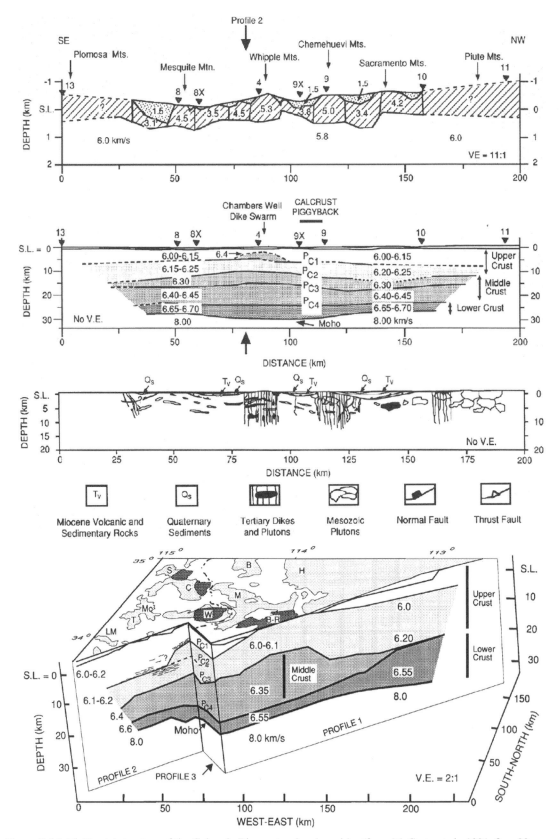

Figure 8.5.3-16. Crustal structure of the Colorado River extensional corridor (from McCarty et al., 1991, figs. 23 and 26, and Wilson et al., 1991, fig. 6). Abbreviations, see Figure 8.5.3-15. [Journal of Geophysical Research, v. 96, p. 12,259–12,291. Reproduced by permission of American Geophysical Union.]

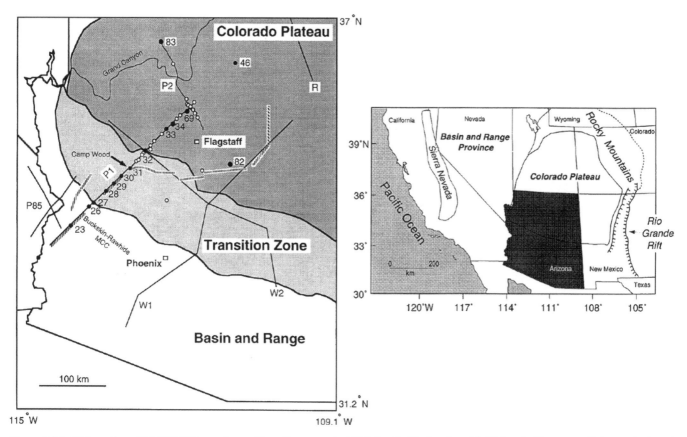

Figure 8.5.3-17. Location of PACE 1989 experiment (from Parson et al., 1996, fig. 1). Circles along P1 (Colorado Plateau profile) and P2 (Grand Canyon profile) are shotpoints. Older lines: R—Hanksville-Chinle profile (Roller, 1965); W1—Gila Bend–Surprise profile (Warren, 1969); W2—Blue Mountain Bylass profile (Warren, 1969). [Journal of Geophysical Research, v. 101, p. 11,173–11,194. Reproduced by permission of American Geophysical Union.]

and Rift Deep Seismic Experiment) project and added to the knowledge of the Rio Grande rift.

In 1981, the Valles caldera within the Jemez Mountain volcanic field at the western flank of the Rio Grande rift in north central New Mexico (Fig. 8.5.3-18; see also Fig. 9.4.2.24 for location of the caldera) was investigated by 72 portable recording units and 10 permanent telemetered stations of the Los Alamos Seismic Network providing radial profile and fan-beam coverage by recording six shots with maximum offsets of 100 km (Olsen et al., 1986). By applying a simultaneous inversion of seismic-refraction and earthquake data, the project elucidated upper crustal structure beneath the caldera providing evidence of a low-velocity body presumed to correlate with the so-called Bandelier magma chamber (Ankeny et al., 1986; see Appendix A8-3-6).

The second experiment targeted the southern Rio Grande rift (Sinno et al., 1986; Appendix A8-3-6), when three interlocking regional profiles were recorded between 1980 and 1983: a 200-km-long reversed SE-NW cross line across the rift, an unreversed 210-km-long axial N-S line, and an unreversed NW-SE line, named Pipeline Road Profile, which was only recorded between offsets of 80 and 195 km (Fig. 8.5.3-19; for data and models see Appendix A8-3-6). The first and second profiles re-

corded preannounced shots on White Sands Missile Range, while for the second and third profiles quarry blasts from the Tyrone Mine west of the Rio Grande rift were used. The average station spacing on all lines was 4 km.

The reversed SW-NE reversed cross line showed no significant dip of the 32 km deep Moho, while along the axial line a significant dip of Moho indicated a crustal thinning by 4–6 km to the south with respect to the northern rift and thus confirmed a similar upward warping of Moho (see Fig. 7.4.1-07), as sketched earlier by Keller et al. (1979) and as shown by Prodehl and Lipman (1989) in their generalizing NNW-SSE–trending crustal cross section of the Rocky Mountains and adjacent Rio Grande Rift (see Fig. 7.4.1-08).

8.5.3.4. Seismic Projects in the Eastern United States

In 1983, a joint consortium of geoscientists from Canada and the United States arranged a nearly continuous 1000-km-long seismic-reflection survey traversing the northern Appalachians from Quebec to Maine along a line extending from near the edge of the Precambrian Shield in Quebec to the eastern limit of the continental margin bordering the ocean basin (Fig. 8.5.3-20; QMT in Fig. 8.5.3-24).

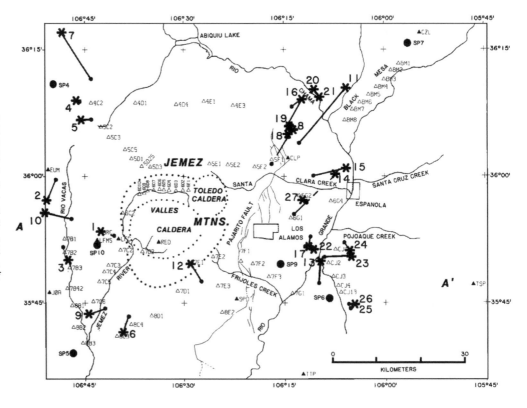

Figure 8.5.3-18. Location of the 1981 CARDEX (Caldera and Rift Deep seismic Experiment) refraction shotpoints (large solid dots), seismograph stations (triangles) and earthquakes (dots and asterisks = initial and relocations) in the Valles caldera within the Jemez Mountain volcanic field (from Ankeny et al., 1986, fig. 1). [Journal of Geophysical Research, v. 91, p. 6188–6198. Reproduced by permission of American Geophysical Union.]

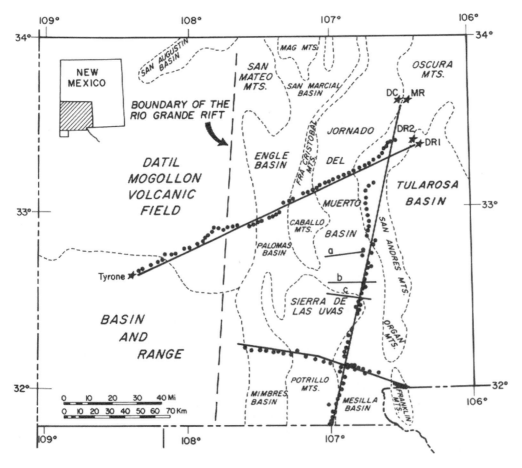

Figure 8.5.3-19. Location of the seismic lines recorded from 1980 to 1983 in the southern Rio Grande Rift in southeastern New Mexico (from Sinno et al., 1986, fig. 2). [Journal of Geophysical Research, v. 91, p. 6143–6156. Reproduced by permission of American Geophysical Union.]

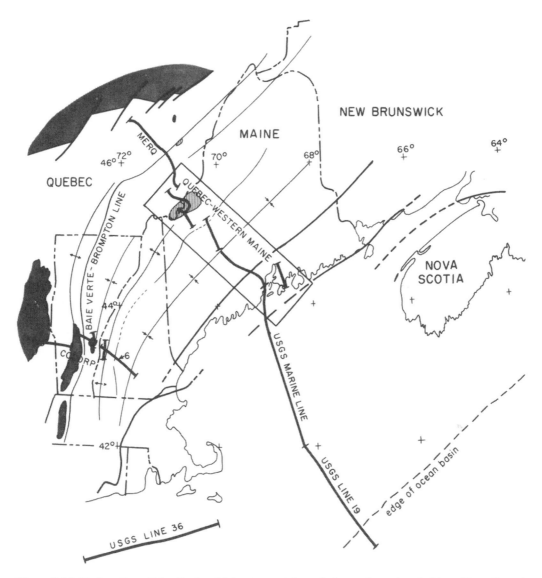

Figure 8.5.3-20. Location of the Quebec-Maine transect through the northern Appalachians (from Stewart et al., 1986, Fig 1). The map also shows a COCORP line through New England (Brown et al., 1986) and one of line 36 of the Long Island platform marine survey obtained in the 1970s (Hutchinson et al., 1986). [*In* Barazangi, M., and Brown, L., eds., Reflection seismology: the continental crust: American Geophysical Union, Geodynamics Series, v. 14, p. 189–199. Reproduced by permission of American Geophysical Union.]

The Canadian section in Quebec was completed in 1983 collecting 219 km of multichannel reflection data for 15 s TWT using Vibroseis sources. Variable upsweeps from 7 to 45 Hz with a 12 km spread, 30 m group intervals, and 90 m vibration points were used (Stewart et al., 1986). It was originally obtained for oil and gas assessment and was recorded to 4–6 s TWT (Green et al., 1986; Stewart et al., 1986). The continuation of this survey into the United States through Maine to the sea coast followed in 1983 and 1984 (Hamilton, 1986) and included a detailed seismic-refraction investigation (Spencer et al., 1987, 1989). The reflection line was extended by a marine profile across the Gulf of Maine to the continental margin (GMM in Fig. 8.5.3-24). Along the marine line in 1984 a 6000 in³ tuned airgun array was fired

every 50 km and recorded by 3 km streamer with 120 channels spaced 25 m apart with a record length of 15 s with a 0.004 s sampling interval, resulting in 30-fold data (Hutchinson et al., 1987).

The seismic-refraction survey was carried out in 1984 and involved eight profile lines, designed to complement the seismic-reflection survey. A 300-km-long traverse consisting of three collinear profiles crossed the Appalachians between Quebec and the Atlantic coast. Three shorter crossing profiles examined individual Appalachian terranes and provided 3-D control for the primary traverse (no. 1 in Fig. 8.5.3-21). A final 145-km-long profile examined the Avalon terrane of southeastern Maine (no. 2 in Fig. 8.5.3-21 and MRP in Fig. 8.5.3-24). (Murphy and Luetgert, 1986, 1987; Appendix A8-3-7; Luetgert

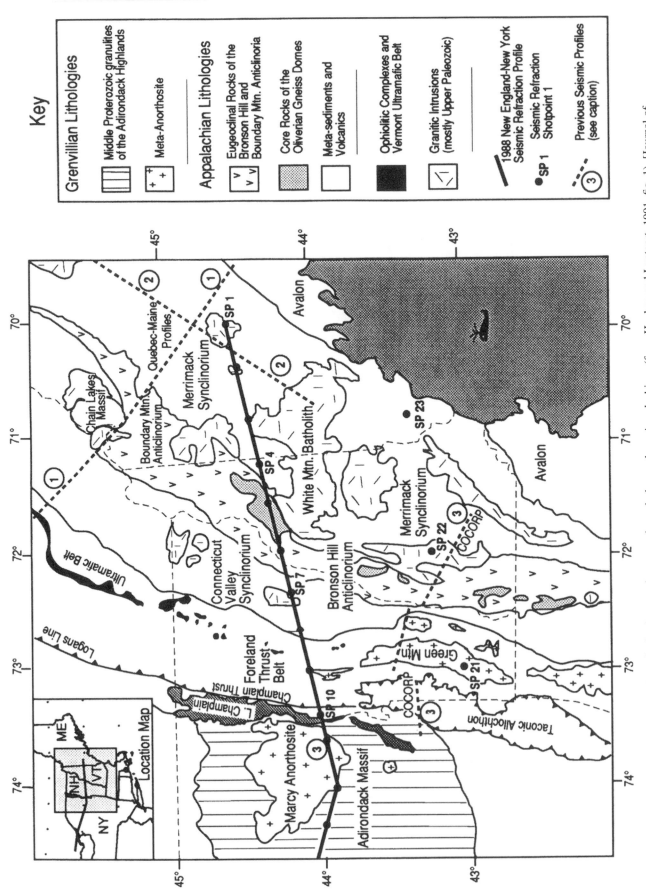

Figure 8.5.3-21. Location of seismic crustal surveys through the northern Appalachians (from Hughes and Luetgert, 1991, fig. 1). [Journal of Geophysical Research, v. 96, p. 16,471–16,494. Reproduced by permission of American Geophysical Union.]

et al., 1987; Spencer et al., 1989; Hughes et al., 1994). A linear array of 120 portable cassette recorders with 2 Hz geophones were deployed at a nominal spacing of 800 m along each profile. A total of 48 shots with an average shot spacing of 30 km was detonated. In addition to the profile data, several shots were fired broadside to the recording arrays and provided detailed wide-angle data. Also shown in the location maps are the New England COCORP lines (Brown et al., 1986; see Fig. 8.5.3-01; no. 3 in Fig. 8.5.3-21; NEC and ADK in Fig. 8.5.3-24) and one of the lines of a marine survey of the USGS investigating the Long Island Platform (LIP in Fig. 8.5.3-24) recorded in 1979 (Hutchinson et al., 1986; Fig. 7.4.3-01).

In 1988, a seismic-refraction/wide-angle reflection experiment crossed the northern Appalachians and the Adirondack Massif in the eastern United States and ended in the Grenville province of the North American craton in Canada (thick line in Fig. 8.5.3-21), leading from New England through upstate New York into southern Ontario (Luetgert et al., 1990; Appendix A8-3-7; Hughes and Luetgert, 1991; Hughes et al., 1994). The main line saw 3 deployments of 300 recording stations and 35 shots at 22 shotpoints, two of them off line.

Furthermore in New England near the Vermont–New Hampshire border a 150-km-long southeastward directed side line and a special high-density wide-angle reflection deployment with 100-m station spacing were added.

The geological interpretation (Fig. 8.5.3-22) of Hughes and Luetgert (1991) shows a midcrustal ramp separating the high-velocity Grenvillian crustal block from the accreted lower-velocity subhorizontally layered Appalachian crust. The lower crust is shown as a continuous layer, but may be divided into discrete Grenville and Appalachian units.

Hughes and Luetgert (1991) compiled a comparison of the main results of the experiments carried out in this region in the 1980s (Fig. 8.5.3-23) which all cross the Appalachian-Grenville boundary. For the Maine-Quebec profile (Fig. 8.5.3-23A), Spencer et al. (1989) have interpreted a "décollement" surface which separates the allochthonous upper crustal units of the Appalachians from the autochthonous Grenvillian crust which underlies much of the western Appalachians. This "décollement" is very similar to the ramp structure seen by Hughes and Luetgert (1991) in the New England–New York seismic experiment (Fig. 8.5.3-23B). For the COCORP New England seismic-reflection line Brown et al. (1983a) have imaged a "thin-skin" detachment near 5 km depth which extends in the form of a steep "steplike" thrust imbricated structure to a depth of 30 km (Fig. 8.5.3-23C).

The New England–New York experiment had not only also produced good P-arrivals, but also excellent S-data from which V_p/V_s ratios Musacchio et al. (1997) inferred the composition of the crust in the Grenville and Appalachian provinces.

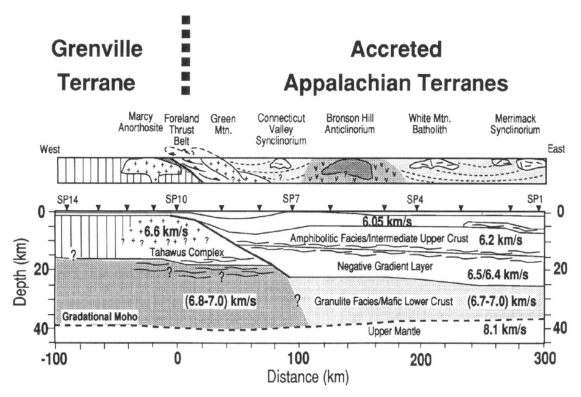

Figure 8.5.3-22. Crustal structure of the 1988 seismic-refraction survey through the northern Appalachians (from Hughes and Luetgert, 1991, fig. 13). [Journal of Geophysical Research, v. 96, p. 16,471–16,494. Reproduced by permission of American Geophysical Union.]

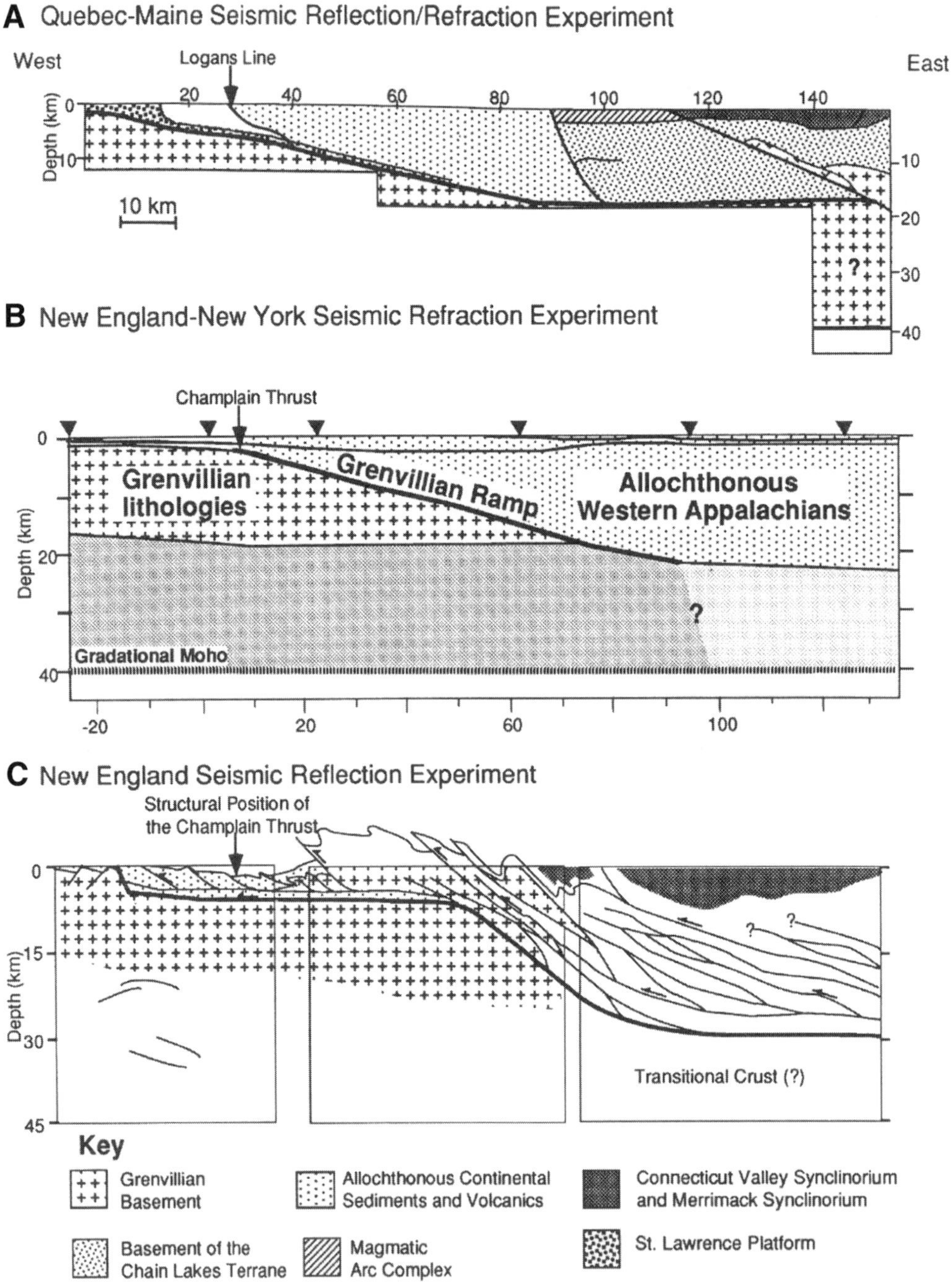

A Quebec-Maine Seismic Reflection/Refraction Experiment

B New England-New York Seismic Refraction Experiment

C New England Seismic Reflection Experiment

Key

Grenvillian Basement	Allochthonous Continental Sediments and Volcanics
Basement of the Chain Lakes Terrane	Magmatic Arc Complex
	Connecticut Valley Synclinorium and Merrimack Synclinorium
	St. Lawrence Platform

Figure 8.5.3-23. Comparison of the seismic experiments across the Appalachian-Grenville terrane boundary in New England (from Hughes and Luetgert, 1991, fig. 14). [Journal of Geophysical Research, v. 96, p. 16,471–16,494. Reproduced by permission of American Geophysical Union.]

For an overview of all deep seismic-refraction and reflection programs completed until late 1987, the reader is referred to the summary volume *Geophysical Framework of the Continental United States*, edited by Pakiser and Mooney (1989) in which, for the eastern and central United States, Phinney and Roy-Chowdhury (1989) compiled a map (Fig. 8.5.3-24) and a table containing the responsible organization, year, and specific references for all programs, with the exception of the 1988 seismic-refraction project across the northern Appalachians and the Adirondack Massif (Hughes and Luetgert, 1991). For the Appalachians and the adjacent Grenville Province in Canada, Taylor (1989) compiled all available data into a geophysical framework. Smithson and Johnson (1989) published a similar approach for the western United States.

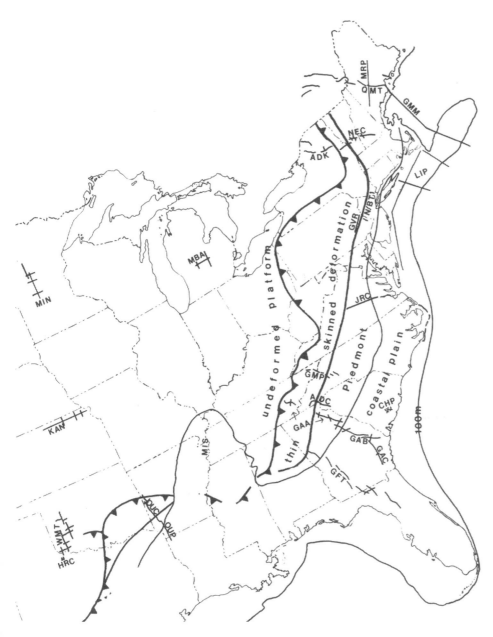

Figure 8.5.3-24. Seismic reflection lines in the eastern United States (from Phinney and Roy-Chowdhury (1989, fig. 1). QMT—Quebec-Maine reflection (USGS 1984–1985), MRP—Maine refraction (USGS 1985), GMM—Gulf of Maine marine reflection (USGS 1985), ADK—Adirondack + NEC-New Hampshire (COCORP 1980–1981), LIP—Long Island Platform marine (USGS 1979), NBT—Newark Basin rift Tectonics (USGS 1984), GVR—Great Valley wide angle (Princeton 1982–1984), JRC—James River Corridor, Virginia (USGS 1980–1981), GMP—Grandfather Mountain, Tennessee–North Carolina (USGS 1979), GAA + GAB + GAC—Southern Appalachians reflection (COCORP, 1978 and 1979, and USGS 1981), CHP—Charleston (COCORP 1978, and USGS 1981), GFT—Georgia-Florida Transect (COCORP 1983–1985), ADC—ADCOH site study (crystalline overthrust belt, 1985), MIS—Mississippi embayment (USGS 1978–1979), OUC + OUP—Ouachita transect (COCORP 1981 reflection and PASSCAL 1986 imaging experiment), HRC—Hardeman Basin (COCORP 1975), WMT—Wichita Mountains (COCORP 1979–1980), KAN—Kansas (COCORP 1981), MIN—Minnesota (COCORP 1979), MBA—Michigan (COCORP 1978). [*In* Pakiser, L.C., and Mooney, W.D., eds., Geophysical framework of the continental United States: Geological Society of America Memoir 172, p. 613–653. Reproduced by permission of the Geological Society of America.]

8.5.4. Alaska

In 1984, a major seismic crustal study was launched in Alaska: the Trans-Alaska Crustal Transect project (TACT; Fig. 8.5.4-01). Seismic-refraction work started in 1984 along a 320-km-long transect line in southern Alaska (Fig. 8.5.4-02), crossing the Prince William, Chugach, Peninsular, and Wrangelia terranes (Daley et al., 1985; Flueh et al., 1989; Fuis et al., 1991; Meador et al., 1986; Appendix A8-3-8).

About 500 km of seismic-refraction data were collected. The shots were recorded by four deployments, each ~140 km long and consisting of 120 1-component recording instruments equipped with 2-Hz seismometers, spaced by 1 km. Two north-south segments ran along the Richardson Highway from Valdez in the south to the Denali fault in the north with 13 shotpoints, on average 25 km apart. Four shotpoints were used for both deployments. The two west-east deployments ran roughly parallel to the major geologic structures; the northern one started northeast of Glenallen, extended ~160 km westward along Glenallen Highway and was provided with 5 shotpoints, two of them offset to the east-northeast. The southern line extended 136 km along the trend of the Chugach Mountains from west of Valdez to Lake Bremner in the east and had also 5 shotpoints, one of them offset to the east. Charges ranged from 500 kg to 2730 kg.

In 1985 the seismic-refraction survey in south-central Alaska was continued (Wilson et al., 1987; Appendix A8-3-8). A total of 27 shots with charges ranging in size from 245 to 1850 kg was recorded. The northern west-east line was supple-

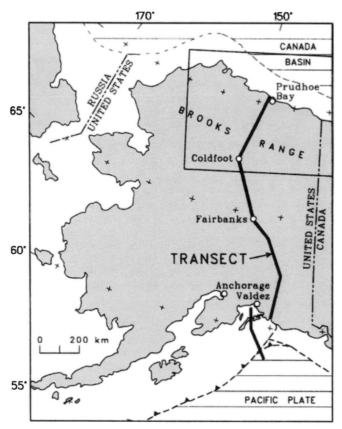

Figure 8.5.4-01. Map of Alaska (from Plafker and Mooney, 1997, fig. 1) showing the location of the Trans-Alaska Crustal Transect project (TACT). [Journal of Geophysical Research, v. 102, p. 20,639–20,643. Reproduced by permission of American Geophysical Union.]

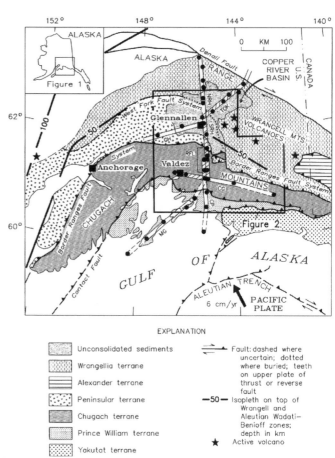

EXPLANATION

Unconsolidated sediments
Wrangellia terrane
Alexander terrane
Peninsular terrane
Chugach terrane
Prince William terrane
Yakutat terrane

Fault: dashed where uncertain; dotted where buried; teeth on upper plate of thrust or reverse fault
—50— Isopleth on top of Wrangell and Aleutian Wadati-Benioff zones; depth in km
★ Active volcano

Figure 8.5.4-02. Map of south-central Alaska (from Fuis et al., 1991, fig. 1) showing terranes and TACT seismic-refraction lines with shotpoints (double full lines and dots). Dashed lines indicate directions to offset shots. [Journal of Geophysical Research, v. 96, p. 4187–4227. Reproduced by permission of American Geophysical Union.]

mented for near-surface control west of Glenallen and extended by the so-called Tok Cutoff line to the east-northeast by 170 km with 7 shotpoints east of Glenallen and two offset shots 50 and 100 km off to the west. Furthermore, the 130-km-long southwestern line (Fig. 8.5.4-02), starting at Cordova south of Valdez, was observed with 4 underwater shotpoints along the line and two 50-km offset shots at both ends sampling the deep structure of the Prince William terrane. Finally the Cordova Peak deployment was located in the Chugach Mountains east of Valdez, partly overlapping with the 1984 north-south line and extending it further south. Here 7 shotpoints were used, two of them offset to the north and to the south.

In 1988, the seismic survey of the TACT was extended into the Prince William Sound and the northern Gulf of Alaska (Brocher and Moses, 1990; Appendix A8-3-8) by a marine multichannel seismic-reflection and wide-angle reflection/refraction survey (Fig. 8.5.4-03). Six lines were laid out resulting in a total of 1100 km marine multichannel data acquired by over 65 wide-angle seismic profiles, having either in-line or fan geometry. Airgun signals, provided by a large airgun array (7770 in³) and generated at 50 m interval, were recorded by a temporary array of 18 seismic recorders along the reflection lines as wide-

angle reflections and refractions. Two marine stations designed to reverse the ray coverage were located offshore and used GEOS instruments to record expendable sonobuoys from a nearby ship dedicated to that purpose. Sixteen land stations using three-component five-day recorders (Criley and Eaton, 1978) were deployed in the vicinity of the Prince William Sound and on some islands and continuously recorded the airgun signals.

A third phase of the TACT land-seismic investigation followed in 1987 with three 130-km-long seismic-refraction/wide-angle reflection profiles and a 17-km-long reflection profile (Beaudoin et al., 1989; Goldman et al., 1992; both Appendix A8-3-8), extending the investigation to the north of the Denali fault (Fig. 8.5.4-04).

The southernmost refraction profile, the Alaska Range deployment, followed the Richardson Highway between Paxson and Delta Junction and was centered on the Denali fault, thus overlapping with the 1984 survey. The central profile, the Fairbanks South deployment, started near Delta Junction and extended to Fairbanks, following the Trans-Alaska Oil Pipeline.

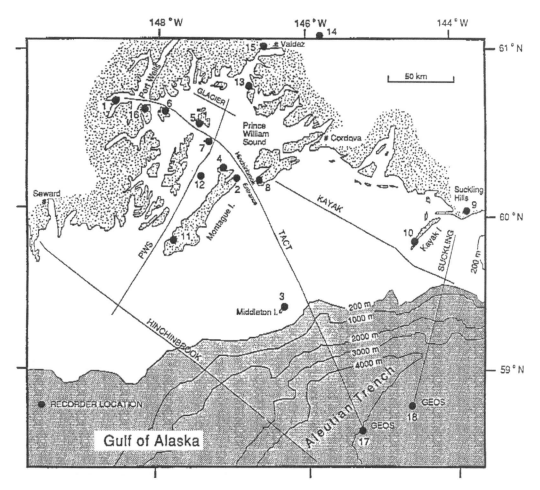

Figure 8.5.4-03. Map of Prince William Sound and the northern Gulf of Alaska showing the locations of the TACT marine reflection lines and temporary seismic recorders (from Brocher and Moses, 1990, fig. 1). [U.S. Geological Survey Open-File Report 90-663, Menlo Park, California, 40 p.]

The northernmost profile, the Fairbanks North deployment, followed the Elliot and Dalton highways and the pipeline to the Yukon River. Each deployment had 140 recording instruments with 1 km spacing. The 17-km Olnes reflection profile had 60 m spacing of the 140 recorders and was coincident with the southern end of the Fairbanks North profile. A total of 48 shots with charges from 227 to 2725 kg was recorded.

The last experiment along the TACT was conducted in 1990 across the Brooks Range, Arctic Alaska (Fig. 8.5.4-05), in a collaboration of the U.S. Geological Survey, Rice University, the Geological Survey of Canada, and IRIS/PASSCAL (Fuis et al., 1997; Levander et al., 1994; Murphy et al., 1993; Appendix A8-3-8). It was the first experiment in which refraction and reflection were simultaneously merged in the TACT investigations. A 700-channel seismograph system recorded 63 shots from 44 shotpoint locations. The instruments were deployed five times (see Fig. 8.5.4-05) in abutting and overlapping arrays, producing a 315-km-long profile with nominally 100 m instrument spacing. Both small (50–300 kg) and large (750–2000 kg) charges were fired to produce a vertical incidence to

wide-angle refraction-reflection data set with continuous offset coverage from 0 to 200 km. The project was the first USGS experiment which involved first generations of digital equipment. The instruments used included 35 PASSCAL RefTek 72A-02s, each recording 6 channels, 190 Stanford University Seismic Group Recorders (SGRs), 120 USGS Seismic Cassette Recorders (SCRs) and 180 Geological Survey of Canada Portable Refraction Seismographs (PRS1s).

In the vertical-incidence data of the marine TACT line (Fig. 8.5.4-06) of Brocher et al. (1991a), a prominent mid- to lower-crustal reflector was seen between 6 and 8 s TWT and a landward 9° dip of Moho was indicated in the refraction data. Discontinuous weak reflections might define the top of the subducting oceanic plate to as deep as 10 s TWT. However, the depth to Moho at the northern end of the TACT line was neither constrained in the reflection nor in the refraction data (Brocher et al., 1991a).

Several models have been proposed by Fuis et al. (1991). The preferred model in Figure 8.5.4-07 shows a horizontal velocity discontinuity at 9 km depth underneath the northern Chugach Mountains and southern Copper River basin, overlain by a low-

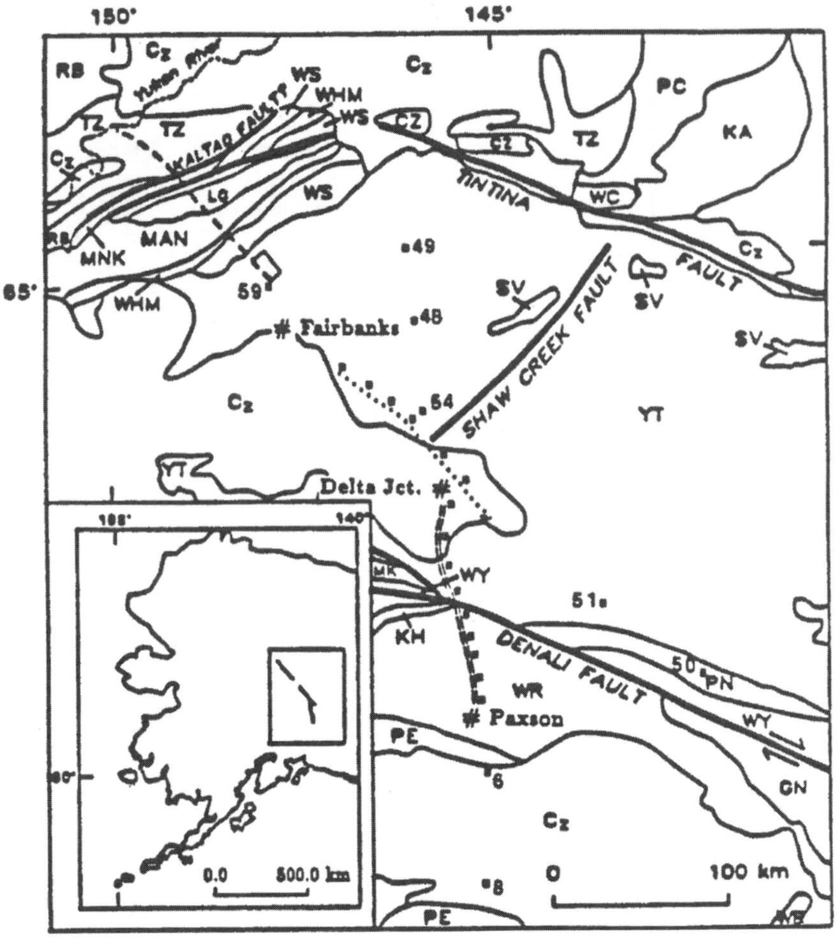

388   *Chapter 8*

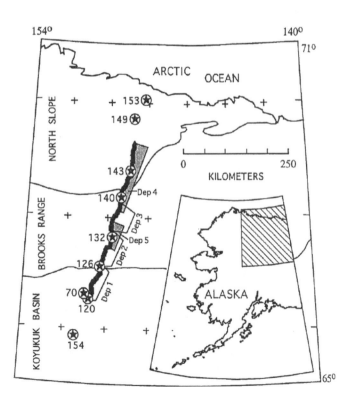

Figure 8.5.4-04. Map of central Alaska (from
Beaudoin et al., 1989, fig. 1) showing the location
of the 1987 TACT seismic survey. [U.S. Geological Survey Open-File Report 89-321, Menlo Park,
California, 114 p.]

Figure 8.5.4-05. Map of Arctic Alaska (from
Murphy et al., 1993, fig. 1) showing the loca-
tion of the 1990 TACT seismic survey. [U.S.
Geological Survey Open-File Report 93-265,
Menlo Park, California, 128 p.]

velocity zone, both of which extend across the deep projection of the Border Ranges fault system. To the south the model shows gently north-dipping layers of alternating high and low velocities that seem to be truncated in the south by the contact fault.

At depth, a subducting plate north-dipping from 20 to 25 km to at least 30–35 km is inferred under the Chugach ter-

rane consisting of a 3–6 km lower crust over a clearly defined Moho. Beneath the northern Copper River basin subhorizontal low and high velocity layers in the middle crust appear to extend across the West Fork fault system and a deep Moho is seen rising from more than 55 km northwards to at least 45 km depth. Below 20 km depth, there is a blank region caused by uncertainty

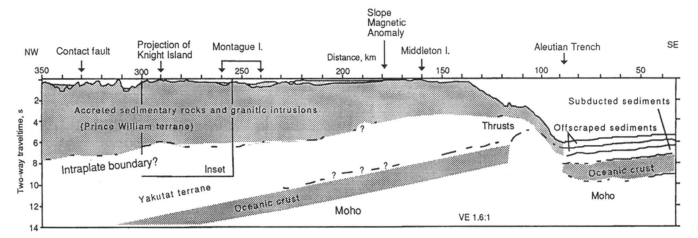

Figure 8.5.4-06. Interpreted line drawing of the unmigrated TACT line (see Fig. 8.5.4-03) showing prominent mid-crustal reflections and reflections from Moho (from Brocher et al., 1991, fig. 2). [*In* Meissner, R., Brown, L., Dürbaum, H.-J., Franke, W., Fuchs, K., and Seifert, F., eds., Continental lithosphere: deep seismic reflections: American Geophysical Union, Geodynamics Series, v. 22, p. 241–246. Reproduced by permission of American Geophysical Union.]

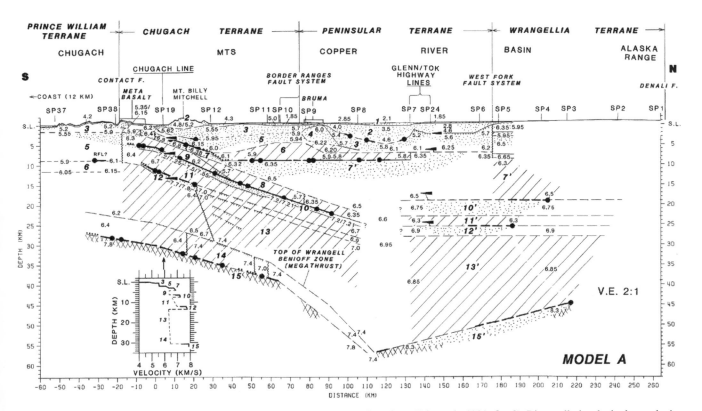

Figure 8.5.4-07. Preferred velocity-depth model for the north-south transect line (from Fuis et al., 1991, fig. 3). Diagonally hatched—low-velocity regions; stippled—regions controlled by bottoming refractions; large dots—subsurface points of critical reflections; cross-hatched—mantle. [Journal of Geophysical Research, v. 96, p. 4187–4227. Reproduced by permission of American Geophysical Union.]

in correlation of features between the south and north halves of the model. A fence diagram shown in Figure 8.5.4-08 shows the inferred structure of the crust for all three lines recorded in south-central Alaska.

The continuation of the seismic observations into central Alaska across the Denali fault toward Fairbanks enabled Brocher et al. (2004a) to study in particular the Denali fault zone in some detail. As a result, the fault zone proper might extend to 30 km depth and, including its nearby associated fault strands, appeared to cause a 5-km-deep low-velocity region, extending up to 40 km width in a north-south direction. From the wide-angle reflection the crustal thickness underneath the Denali fault zone appeared as a transition between the 60-km-thick crust under the Alaska Range to the south and the extended 30-km-thick crust under the Yukon-Tanana terrane to the north. Figure 8.5.4-09 is a compilation of compressional wave velocity models across the Denali fault (Fuis et al., 1991; Beaudoin et al., 1992; Brocher et al., 2004a).

The data of the TACT section recorded northwest of Fairbanks (see Fig. 8.5.4-04) were interpreted by Beaudoin et al.

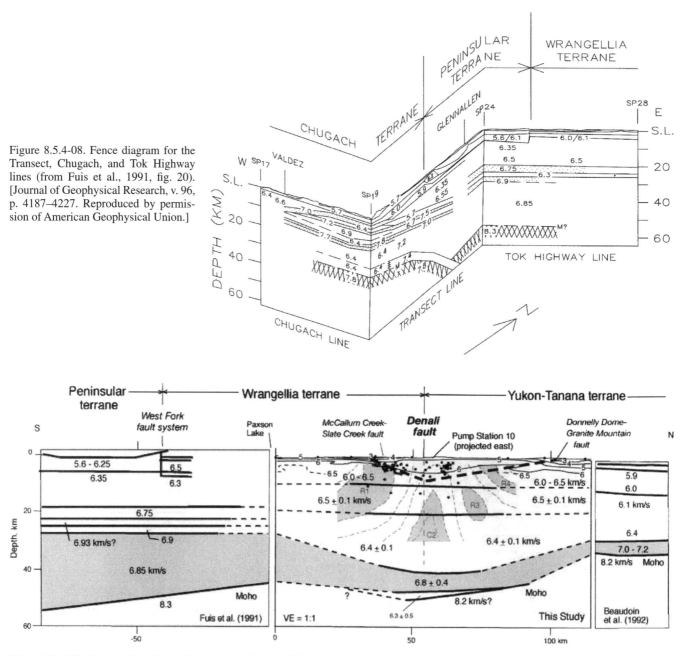

Figure 8.5.4-08. Fence diagram for the Transect, Chugach, and Tok Highway lines (from Fuis et al., 1991, fig. 20). [Journal of Geophysical Research, v. 96, p. 4187–4227. Reproduced by permission of American Geophysical Union.]

Figure 8.5.4-09. Compilation of crustal models near the Denali fault, central Alaska (from Brocher et al., 2004a, fig. 12). [Bulletin of the Seismological Society of America, v. 94, p. S85-S106. Reproduced by permission of the Seismological Society of America.]

(1994). Their crustal model underneath the Yukon-Tanana upland shows the Moho at ~30–35 km depth (Fig. 8.5.4-10). For the lower crust (layers 2–4) the authors have proposed several tectonic interpretations, one of which is reproduced here.

The seismic-refraction and -reflection data along the transect through the Brooks Range in northern Alaska have been interpreted by Fuis et al. (1997) and Wissinger et al. (1997). Figure 8.5.4-11 shows a velocity-depth interpretation by Wissinger et al. (1997).

According to a tectonic model by Fuis et al. (1997), tectonic wedges are characteristic for the crust under northern Alaska, mostly north-vergent. There is high reflectivity throughout the crust above a basal décollement which deepens southward from 10 km to ~30 km under beneath the Brooks Range. Below this décollement there is low reflectivity which Fuis et al. (1997) interpreted as a southward-verging tectonic wedge (Fig. 8.5.4-12).

8.6. THE AFRO-ARABIAN RIFT SYSTEM

8.6.1. Crustal Studies of the Dead and Red Sea Areas

Following the first deep seismic sounding experiment carried out in 1977 on the west side of the Jordan–Dead Sea rift (see Chapter 7.5.2), a second experiment followed in 1984, this time on the eastern flank of the rift in Jordan, initiated by Zuhair El-Isa of the University of Amman and promoted by scientists of Hamburg (Jannis Makris) and Karlsruhe (Jim Mechie and Claus Prodehl), Germany (El-Isa et al., 1987a, 1987b). The location map of Figure 8.6.1-01 shows both surveys.

Parallel to the rift profile of 1977, a long-range profile was organized in the 1984 experiment in a north-south direction along the eastern flank of the rift passing the cities of Amman and Maán (lines I, III, and IV in Fig. 8.6.1-01) with land shotpoints (1–4)

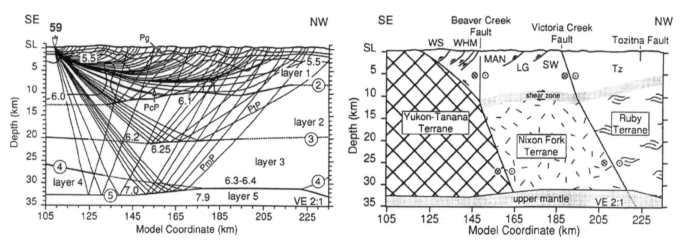

Figure 8.5.4-10. Crustal model of the northern Yukon-Tanana upland, central Alaska: Left: Ray diagram for shotpoint 59. Right: A possible tectonic interpretation (from Beaudoin et al., 1994, figs. 4 and 11). [Geological Society of America Bulletin, v. v. 106, p. 981–1001. Reproduced by permission of the Geological Society of America.]

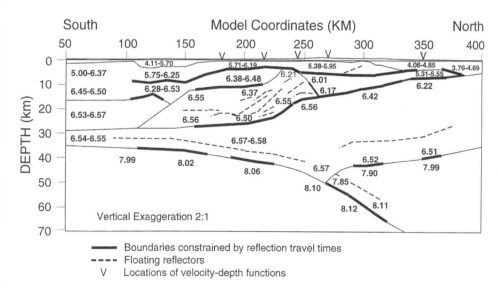

Figure 8.5.4-11. Velocity-depth model for the Brooks Range, Arctic Alaska (from Wissinger et al., 1997, fig. 4a). [Journal of Geophysical Research, v. 102, p. 20,847–20,871. Reproduced by permission of American Geophysical Union.]

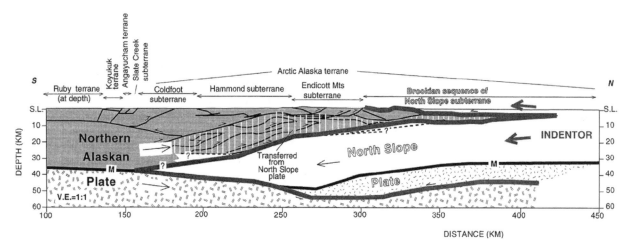

Figure 8.5.4-12. Inferred model for the Brooks Range, Arctic Alaska (from Fuis et al., 1997, fig. 6). Crust is gray (indented plate) or blank (indentor plate); mantle has double-tick pattern (indented plate) or single-tick pattern (indentor plate). Heavy gray line separates indented northern Alaskan plate from North Slope plate. Rock that has been clearly transferred from indentor plate to indented plate is indicated with vertical white lines. [Journal of Geophysical Research, v. 102, p. 20,873–20,896. Reproduced by permission of American Geophysical Union.]

and an underwater shotpoint in the Gulf of Aqaba (no. 5 in Fig. 8.6.1-01) providing energy up to 200 km offsets. A third profile II of 170 km length was recorded to the east of the main line. Here, to obtain information on the near-surface structure, in addition to large shots at shotpoint 3, several small shots (6–10 in Fig. 8.6.1-01) were detonated in shallow drillholes providing energy up to 30 km offset. In total, 28 shots, of which 3 were underwater shots, were recorded by 20 mobile seismic stations of type MARS-66 (for instrument details, see Chapter 6.2.1), recording three components of ground motion and an external time signal of radio Moscow, all frequency modulated, on single-track analogue magnetic tape. Station spacing averaged 5 km along the main lines I, III, and IV and 1–2 km along line II.

All three components were interpreted (Fig. 8.6.1-02) and subsequently published (El-Isa et al., 1987a, 1987b). The upper crust including sediments has a Poisson's ratio of 0.25 except in northwestern Jordan where it gets as high as 0.32. The lower crust has a higher Poisson's ratio of 0.29–0.32 for which the authors have offered two possibilities, either the lower crust possesses high feldspar and low quartz content or fluid phases exist in the form of penny-shaped inclusions. El-Isa et al. (1987a) have also combined their results with those of Ginzburg et al. (1979b) and have compiled a west-east crustal cross section through the Jordan–Dead Sea rift and its flanks (Fig. 8.6.1-03). Under the rift proper crustal thickness is ~30 km. Different from the western flank, the elevated crust under the rift proper extents eastward for at least 40 km away from the rift underneath the eastern flank and only then gradually starts to thicken toward the thick Arabian Shield crust.

In 1984, deep seismic-reflection studies were started in Israel with a SE-NW–directed, 90-km-long line from near the southern end of the Dead Sea to the Mediterranean coast (Rotstein et al., 1987) for a deep crustal study. Two other lines in the Mediterranean coastal area were derived by recorrelation of oil explora-

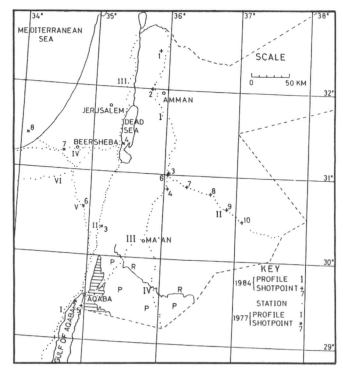

Figure 8.6.1-01. Location map of seismic-refraction lines in the Jordan–Dead Sea rift and on its flanks (from El-Isa et al., 1987b, fig. 1). [Geophysical Journal of the Royal Astronomical Society, v. 90, p. 265–281. Copyright John Wiley & Sons Ltd.]

tion lines. In continental inner Israel a reflective lower crust is apparent between a transparent upper crust and an almost transparent upper mantle, in agreement with the seismic-refraction results. Near the coast, strong reflections from 11 to 12 s TWT characterize the upper mantle, which cannot be associated with the crust-mantle boundary, because the conversion of the Moho

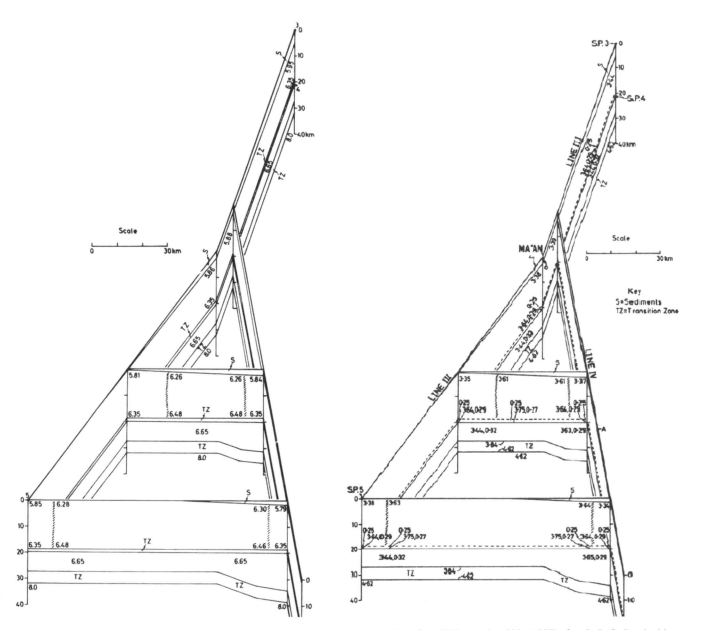

Figure 8.6.1-02. Three-dimensional model of crustal structure beneath southern Jordan (from El-Isa et al., 1987a, 1987b, fig. 6). Left: P-velocities, right: S-velocities and Poisson's ratio. [Tectonophysics, v. 138, p. 235–253; Geophysical Journal of the Royal Astronomical Society, v. 90, p. 265–281. Copyright Elsevier.]

depth derived from seismic-refraction data into TWT results in only approximately 10 s.

In 1986, a French research group investigated the Gulf of Suez and the Egyptian part of the northern Red Sea (Fig. 8.6.1-04) by detailed marine seismic profiling called the "Minos" cruise (Gaulier et al., 1988). Fifteen expanding spread profiles (ESP; see Fig. 8.9.1-01 below) and a seismic wide-angle reflection/refraction line were observed by a two-ship operation using four 16.4 airguns as sound source on one vessel and a 96-channel 2.4-km-long streamer towed by the second vessel.

Three unreversed seismic wide-angle reflection/refraction lines (Avedik et al., 1988) were obtained with a self-contained

seismic receiving and recording system, an ocean-bottom vertical seismic array (OBVSA in Fig. 8.6.1-04). In addition, three seismic profiles were recorded on land while the ship shot parallel to the shore.

Most profiles recorded good crustal reflection and refraction arrivals, and Moho arrivals were obtained at offsets of 80–100 km. Crustal thickness decreases from 30 km underneath the shore of Egypt and Sinai to near 10 km in the center of the northern Red Sea (Fig. 8.6.1-05).

Parallel to the French group, the University of Hamburg under the leadership of Jannis Makris was extremely active in the Red Sea area. The location map (Fig. 8.6.1-06) shows all seismic-refraction observations where the University of Hamburg was partner or leader (Rihm et al., 1991). The seismic surveys of the 1970s in Afar and Israel and those of 1978 and 1981 in Egypt, the northern Red Sea, and Saudi Arabia (1979 Egypt and 1981 "Stefan E" in Fig. 8.6.1-06) were already mentioned in Chapter 7. Also shown in Figure 8.6.1-06 is the location of the Jordan 1984 experiment described above in some detail. The 1984 *Conrad* cruise experiment partly overlaps with the French survey of Gaulier et al. (1988). The results for the northern Red Sea and

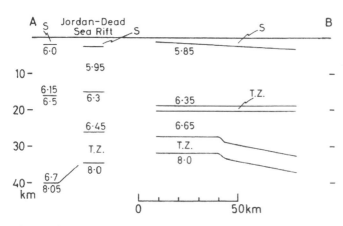

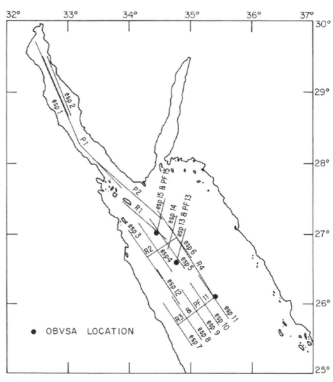

Figure 8.6.1-03. Crustal cross section across the southern Jordan–Dead Sea rift (from El-Isa et al., 1987a, fig. 13). [Tectonophysics, v. 138, p. 235–253. Copyright Elsevier.]

Figure 8.6.1-04. Location map of seismic lines in the Gulf of Suez and northern Red Sea (from Avedik et al., 1988, fig. 1). OBVSA—positions of ocean-bottom seismometer array. [Tectonophysics, v. 153, p. 89–101. Copyright Elsevier.]

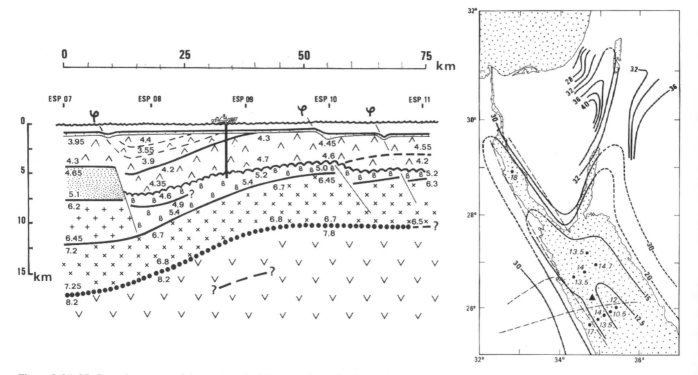

Figure 8.6.1-05. Crustal structure of the northern Red Sea area (from Gaulier et al., 1988, figs. 13 and 17). Left: Crustal cross section through the southern five expanding spread profiles (Figure 8.6.1-04). Right: Moho map. [Tectonophysics, v. 153, p. 55–88. Copyright Elsevier.]

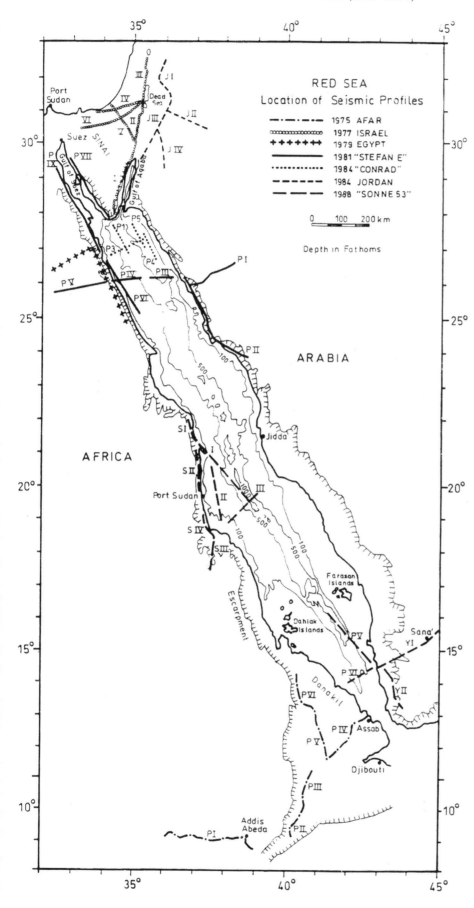

Figure 8.6.1-06. Location map of seismic-refraction lines in the Red Sea in the 1970s and 1980s by or under participation of the University of Hamburg, Germany (from Rihm et al., 1991, fig. 1). [Tectonophysics, v. 198, p. 279–295. Copyright Elsevier.]

the transition toward Egypt are similar to those of the French research group and are incorporated into the Moho map of Gaulier et al. (1988; Fig. 8.6.1-05). Under Egypt, the crustal thickness decreases slightly from 30 km near the coast to 28 km further west.

In 1988 the activities moved into the central and southern Red Sea (Egloff et al., 1991). During the cruise 53 of *Sonne* (Fig. 8.6.1-06) seismic lines were shot in the central Red Sea off the coast of Sudan (lines I to III) and into Sudan (lines SI to SIV) as well as in the southernmost Red Sea (lines PV and PVI) and into Yemen (lines YI and YII). Egloff et al. (1991) have compiled several crustal cross sections from which an example for the Red Sea off Sudan and another one for the line PVI-Y1 from the southern Red Sea into Yemen are reproduced here (Fig. 8.6.1-07). Underneath the axis of the central Red Sea around latitude 19.5°N Moho is found at 5 km depth, while underneath the axis of the southernmost Red Sea near latitude 14°N the Moho depth has increased to 13 km depth. When approaching the coastal waters, the Moho depth increases drastically, but evidently in several steps. At the central Red Sea off Sudan the Moho is at almost 15 km depth, while at the coast off Yemen it is as deep as 20 km, but depth increases further inland to 30 km under the escarpment, and decreases again further to the northeast.

8.6.2. The First Project of Kenya Rift International Seismic Project

In 1985, American, British, German and Kenyan scientists started the first phase of a seismic investigation of the East African rift in Kenya: the Kenya Rift International Seismic Project (KRISP). In Britain the long research tradition in Africa of the University of Leicester with its scientists Don Griffiths, Roy King, Aftab Khan, and Peter Maguire was resumed. In Germany, the University of Karlsruhe with its scientists Karl Fuchs and Claus Prodehl was interested in the Afro-Arabian rift system within the scope of its interdisciplinary collaborative research center "Stress and Stress Release in the Lithosphere" which had already supported the investigations in Jordan. In the United States, it was Randy Keller of the University of Texas at El Paso and Larry Braile of Purdue University at West Lafayette, Indiana, who became the leading U.S. scientists in KRISP because of their major interest in rifts around the world (see, e.g., Keller et al., 1991).

The KRISP project of 1985 was intended as a test phase to explore the propagation of seismic waves within and outside the East African rift in Kenya as a basis for a future major investigation. The project led to the observation of two perpendicularly oriented seismic-refraction profiles (KRISP Working Group, 1987). The north-south line was 300-km-long, extended along the axis of the rift from south of Lake Naivasha to Lake Baringo and was served by five shotpoints, three of them underwater shots in lakes. The second line traversed the rift in east-west direction and had only borehole shots (Fig. 8.6.2-01).

Only half of the data was recorded digitally on U.S. equipment, the rest on frequency-modulated analogue systems (European MARS and UK Geostores). It turned out that, due to the volcanic deposits covering the central section of the rift, only the lake shots were efficient and therefore the east-west line was not successful. However, the north-south profile was a full success.

A data example from shots in Lake Baringo and a preliminary interpretation is shown in Figure 8.6.2-02.

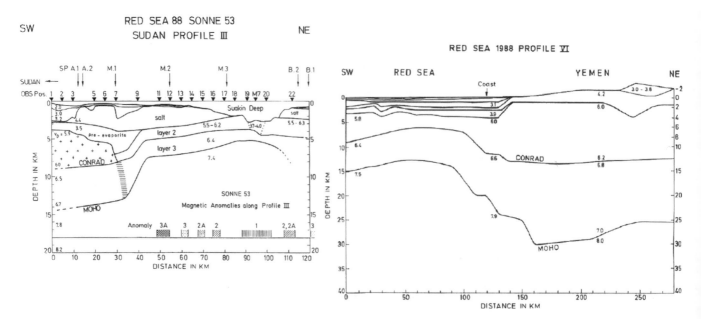

Figure 8.6.1-07. Crustal cross sections through the Red Sea area (from Egloff et al., 1991, figs. 6 and 18). Left: Sudan profile III in the central Red Sea between 19° and 20°N. Right: Profile VI: southern Red Sea into Yemen, passing Sana, between latitudes 14° and 16°N. Note the different scales and different vertical exaggerations! [Tectonophysics, v. 198, p. 329–353. Copyright Elsevier.]

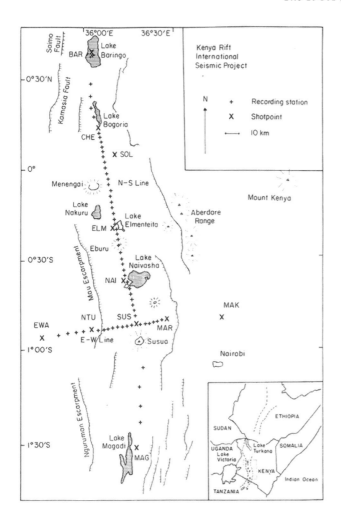

Figure 8.6.2-01. Location of the KRISP85 seismic experiment (from KRISP Working Group, 1987, fig. 1). [Nature, v. 325, no. 6101, p. 239–242. Reprinted by permission from Macmillan Publishers Ltd.]

Surprisingly, Moho depth resulted as deep as 35 km (Henry et al., 1990; Khan et al., 1989). The outcome of this test phase was extremely valuable for the design of the follow-up experiments of KRISP in 1990 and 1994, which is discussed in detail in Chapter 9.5.1. In a review of geophysical experiments and models of the Kenya rift until 1989, the KRISP85 results were discussed in the context with earlier data (Swain et al., 1994).

8.7. DEEP SEISMIC SOUNDING STUDIES IN SOUTHEAST ASIA

8.7.1. India

Since the initiation of the deep seismic sounding technique by an Indo-Soviet collaboration agreement in 1972, as described in Chapter 7.6.1, a great number of seismic-refraction profiles were covered during the 1970s, but extensive deep seismic sounding studies also continued in the 1980s (Mahadevan, 1994), with emphasis placed on basin structures (e.g., Kaila et al., 1987a, 1987b, 1990a, 1990b).

By the end of the 1980s, a total of 6000 km long-range refraction and wide-angle reflection observations had been obtained by deep seismic sounding surveys throughout India along 20 profiles (Fig. 8.7.1-01).

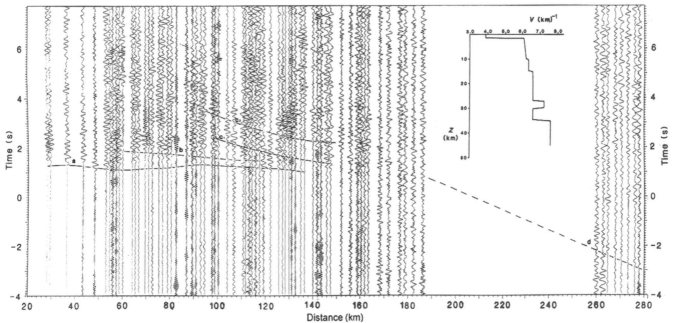

Figure 8.6.2-02. Record section from the Lake Baringo shots with phase correlation and velocity-depth function (from KRISP Working Group, 1987, fig. 2). [Nature, v. 325, no. 6101, p. 239–242. Reprinted by permission from Macmillan Publishers Ltd.]

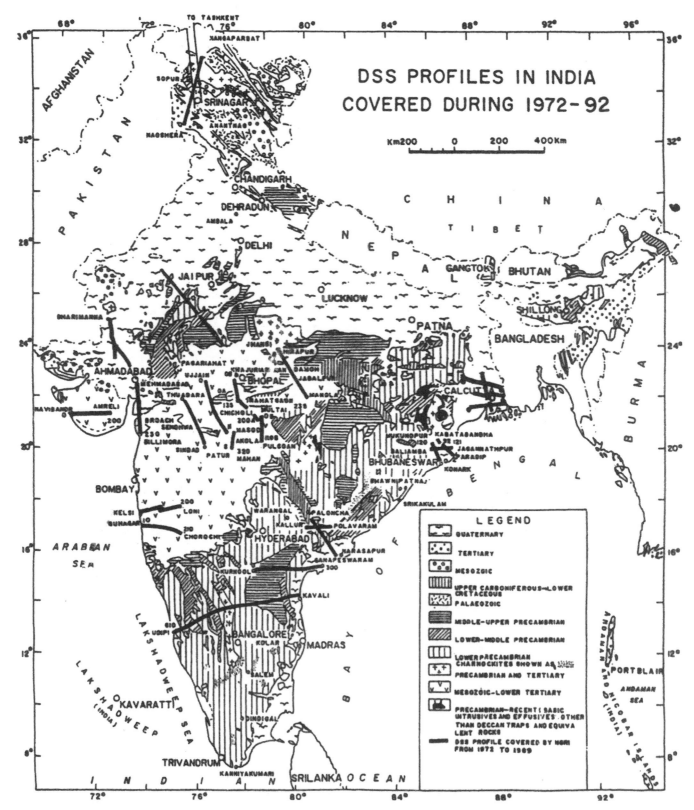

Figure 8.7.1-01. Location map of deep seismic sounding profiles in India observed by NGRI (National Institute of Geophysical Research) between 1972 and 1992 (from Mahadevan, 1994, fig. 2.4). [Geological Society of India, Memoir 28, Bangalore, 569 p. Published by permission of the Geological Society of India.]

For example, to investigate the deep structure of Gondwana basins, three seismic profiles were recorded across the Mahanadi delta 200 km SSW of Calcutta (Kaila et al., 1987b) and another two deep seismic sounding profiles were recorded across the Godavari basin further southwest east of Hyderabad (Kaila et al., 1990a). Another example is the crustal structure study of the Narmada-Son lineament, a continental rift system in the western part of central India.

Here a continuous profiling system was adopted over major portions of four north-south–oriented deep seismic sounding profiles across this lineament (Kaila, 1986; Kaila et al., 1985). The shotpoints were 20–40 km apart and were recorded to maximum distances of 200 km. For this purpose, a multichannel cable of 11.6 km length with geophone groups every 200 m was laid out and connected to two 48 channel magnetic tape systems. The 11.6 km spread was shifted along the line and shots were repeated along the profile until 200 km distance for any shotpoint was reached, shot sizes ranging from 50 to 1000 kg depending on the corresponding recording distance. The interpretation included the data obtained in the Cambay basin to the

west described already in Chapter 7. The data allowed a detailed interpretation of the shallow crust, but also revealed a fairly detailed picture of the whole crust. A Moho contour map shows variations of the crustal thickness between 34 and 42 km (Kaila, 1986; Kaila et al., 1985, 1990b).

An example of a crustal cross section along the Paloncha-Narsapur DSS profile running from SE to NW through the Godavari basin and graben in southeastern India is presented (Fig. 8.7.1-02). The data were recorded in the 1980s and their interpretation published by Kaila et al. (1990a).

In the west Bengal basin bordering Bangladesh, in 1987–1988 two wide-angle seismic-refraction lines were recorded in an east-west direction, north (180 km long) and south (140 km long) of Calcutta (Kaila et al., 1992; Fig. 8.7.1-01). Along the two profiles, 1953 line-km of traveltime data were observed from 22 shotpoints at a geophone spacing of 200 m. Moho depths of 32–34 km resulted for the central and western parts of the lines, while to the east and southeast considerable crustal thinning was revealed. In 1988–1990 another east-west line, 120 km long, was added in between, and a 227-km-long north-

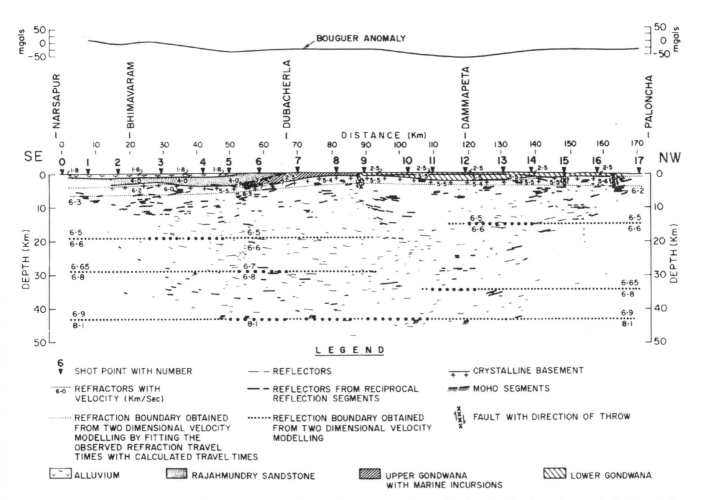

Figure 8.7.1-02. Crustal cross section of a deep seismic sounding profile through the Godavari basin and graben in southeastern India (from Kaila et al., 1990a, fig. 5). [Tectonophysics, v. 173, p. 307–317. Copyright Elsevier.]

south line crossed all the three east-west lines. These new lines, also shown in Figure 8.7.1-01 were recorded digitally with 80 m sensor spacing (Kaila et al., 1996), but only shallow structures were revealed in much detail.

Another line, shot in 1983–1984 between Hirapur and Mandla in central India (Kaila et al., 1987c; Fig. 8.7.1-01; ~500 km to the SSE of Delhi) dealt mainly with the shallow velocity structure and was later interpreted in much detail by Sain et al. (2000).

In 1986–1987, three seismic lines covering 350 km were recorded in the northern Cambay and Sanchor sedimentary basins in western India, north of Mehmadabad (Kaila et al., 1990b; Fig. 8.7.1-01). This project continued the seismic work carried out in the 1970s in the southern Cambay basin (Kaila et al., 1981a).

8.7.2. China

Explosion seismology research in China was considerably increased in the 1980s. The main tectonic provinces are shown in Figure 8.7.2-01.

In 1981 and 1982, a joint Sino-French explosion seismology study was carried out in Tibet investigating the Himalayan border and the adjacent Lhasa block to the north (Hirn and Sapin, 1984; Hirn et al., 1984a, 1984b, 1984c; Teng et al., 1983a; Yuan et al., 1986) which on the French side was organized by Alfred Hirn, University of Paris, France.

In 1981, a 500-km-long east-west line was recorded between the High Himalayas and the Yarlung Zangbo Jiang as part of a French-Chinese geoscientific exploration of southern Tibet (solid line in Fig. 8.7.2-02). A system of reversed and overlapping profiles and some auxiliary profiles were established using 40 portable tape recording seismic stations of type MARS-66 and recorded seven charges ranging from 2000 to 10,000 kg detonated in two lakes and at one drillhole. Spacing of seismographs was 10 km. Twelve additional stations stationed in Nepal tested the energy propagation through the High Himalayas.

Crustal reflections at critical wide-angle distance from the Moho were exceptionally clear and the crust-mantle boundary

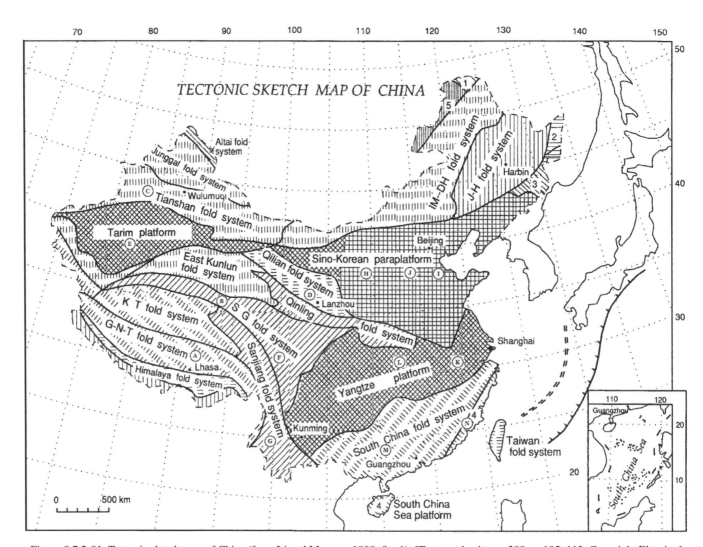

Figure 8.7.2-01. Tectonic sketch map of China (from Li and Mooney, 1998, fig. 1). [Tectonophysics, v. 288, p. 105–113. Copyright Elsevier.]

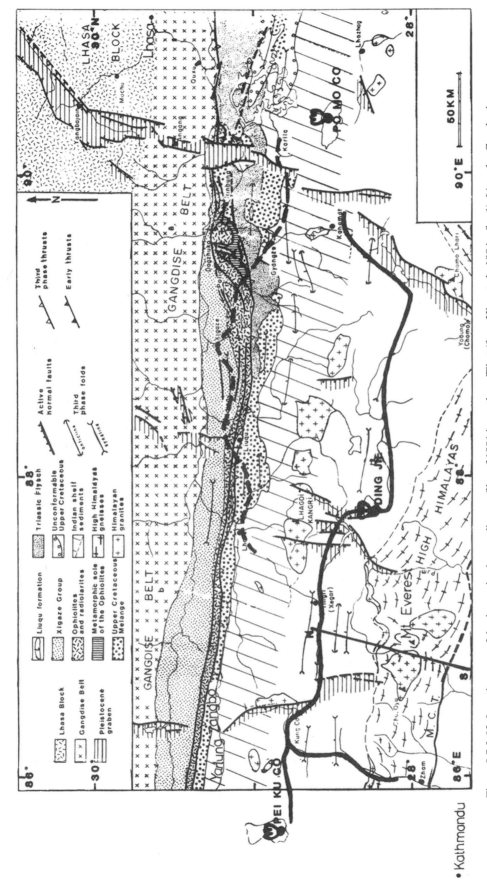

Figure 8.7.2-02. Location map of the deep seismic sounding profiles of 1981 in southern Tibet (from Hirn et al., 1984a, fig. 1). [Annales Geophysicae, v, 2, p. 113–118. Reproduced by permission of the author.]

was modeled as a 12-km-thick transition zone at 75 km depth beneath the present-day surface at 4500 m mean elevation above sea level (see top record section of Fig. 8.7.2-03). The second record section (Fig. 8.7.2-03, lower left) shows the recordings from the center shotpoint Ding Jie along the fan profile running south from the western shotpoint Pei Ku Ko.

The crustal cross section (Fig. 8.7.2-03, lower right) is the result of the fan profile interpretation plotted half way between recording line and shotpoint along the straight line N-S in Figure 8.7.2-02. It shows an intracrustal boundary at 40 km depth and the Moho at 55 km, offset to greater depth up to 70 km and more north of the High Himalayas.

Following this study, in 1982, the fan line was continuously extended toward the north, crossing several fold systems and the eastern part of the Tarim basin and ending in the Junggar basin as can be viewed on the tectonic map (Fig. 8.7.2-01). In this second project in 1982 (Hirn et al., 1984b), a 500-km-long Moho traverse (solid line in Fig. 8.7.2-04) was laid out.

It crossed the Yarlung Zangbo Jiang suture zone, which separates the Lhasa block from the Tibetan part of the Indian continental plate in southernmost Tibet, then traversed the Lhasa block and continued further through the Bangong-Nujiang suture, an earlier suture which limits the Lhasa block 300 km to the north. Most seismograms were obtained at constant distance, 200–250 km from shotpoints; in some parts, however, the seismograms had to be corrected for larger offset distances (Fig. 8.7.2-05). Both sutures appeared as vertical 20-km steps in the Moho and were interpreted by Hirn et al. (1984b) as possible loci for eastward strike-slip motion of the Tibetan lithosphere. Also in the early 1980s, an E-W profile was shot in the Tibetan plateau in the northern part of the Lhasa block (Sapin et al., 1985), south of the Banggong-Nujiang Suture (for location, see Fig. 9.6.2-05A).

Following the French-Chinese investigations, many other seismic-refraction surveys followed. The location of the seismic-refraction/wide-angle reflection profiles in China from 1958 to 1990 is summarized in Figure 8.7.2-06. Some 250 standardized instruments for deep seismic sounding had been distributed within China among various research groups belonging to the State Seismological Bureau. The instruments recorded on two-channel magnetic tape cassettes, one for the seismometer signals, and the other for timing control. The seismic channel was frequency modulation multiplexed at two or three levels to obtain a wide dynamic range. The sensors were

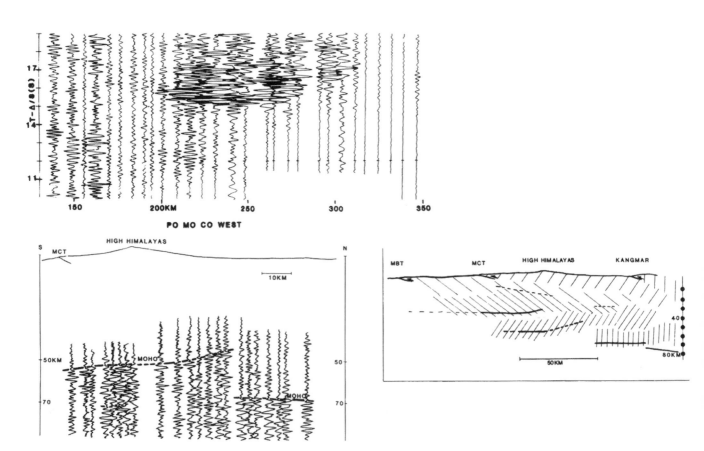

Figure 8.7.2-03. Top: Shots Po Mo Co recorded on the east-west line (from Hirn et al., 1984c, fig. 2). Bottom: Fan profile to the west from shotpoint Ding Jie (left) and seismic structure across the High Himalayas along N-S line of Figure 8.7.2-02 (from Hirn et al., 1984c, figs. 3 and 4). MBT—Main Boundary Thrust, MCT—Main Central Thrust. [Nature, v. 307, no. 5946, p. 25–27. Reprinted by permission from Macmillan Publishers Ltd.]

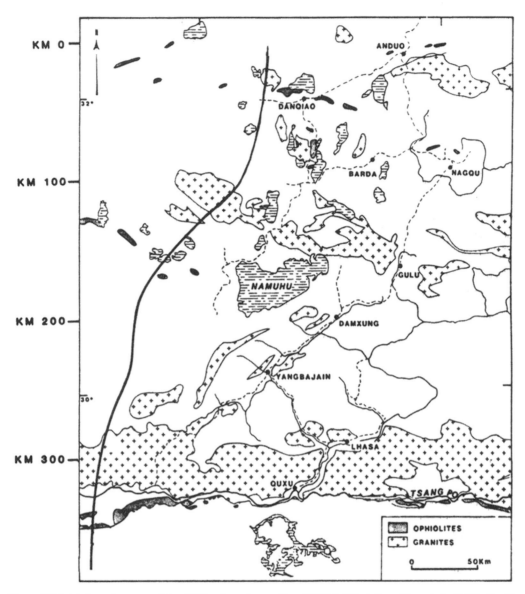

Figure 8.7.2-04. Location map of the 500 km long fan profile of 1982 in Tibet (from Hirn et al., 1984b, fig. 1). [Nature, v. 307, no. 5946, p. 23–25. Reprinted by permission from Macmillan Publishers Ltd.]

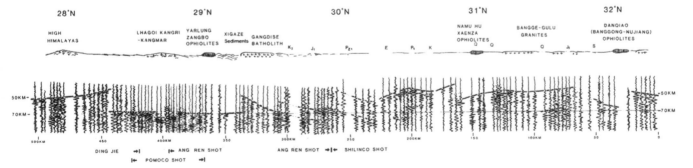

Figure 8.7.2-05. Time section of Moho reflections at constant offset of 250 km (critical to wide-angle range) along the north-south fan profile of 1982 through Tibet (from Hirn et al., 1984b, fig. 2). [Nature, v. 307, no. 5946, p. 23–25. Reprinted by permission from Macmillan Publishers Ltd.]

2-Hz vertical seismometers. Li and Mooney (1998) have published a brief overview of the seismic crustal studies which may have started in 1958, but only in the 1970s was the first modern seismic survey performed (see Chapter 7.6.2). The main activity of seismic research, however, started in the 1980s and continued since then.

An overview of seismic crustal studies in Tibet was published by Teng (1987). He reviewed the explosion seismic studies carried out on the Qinghai-Xizang (Tibet) plateau from 1975 to 1977 and 1981–1982. From shots fired in five lakes and three borehole sites, five seismic profiles had been completed. Moho depths reached 75 km under the Tibetan plateau.

Many seismic-refraction lines were established between longitudes 98° and 106°E from the Qinling fold system in central China (Chen et al., 1988; Zhang et al., 1988) to the Sanjiang fold system in south-central China (Kan et al., 1986, 1988) and the transition into the Yangtse platform where the Moho increases rapidly from 30 to 40 km under the Yangtse platform in the east to

more than 70 km depth under the various fold systems in the west (for references see Table 1 of Li and Mooney, 1998, and Table 1 of Zhang et al., 2005).

A dense network of seismic profiles was recorded in the Bejing area in the North China plain and adjacent regions (Sun et al., 1988). To the south a few seismic lines were observed in the eastern part of the Qinling fold belt and adjacent areas of the North China plain and the Yangtse quasi-platform (Hu et al., 1988; Ding et al., 1988; Chen and Gao, 1988). Finally, adding to the observations from the Yongping explosion in the 1970s (United Observing Group of the Yongping Explosion, SSB,, 1988) the South China foldbelt was covered by three more lines (Lhiuzou Explosion Research Group, 1988; Liao et al., 1988). The training of Chinese scientists and their interpretation of seismic data was strongly supported by the USGS under the initiative of Walter Mooney.

One of these projects, which was carried out in cooperation with the USGS, is discussed here in more detail. To define the crustal structure in an area of active tectonics on the southern

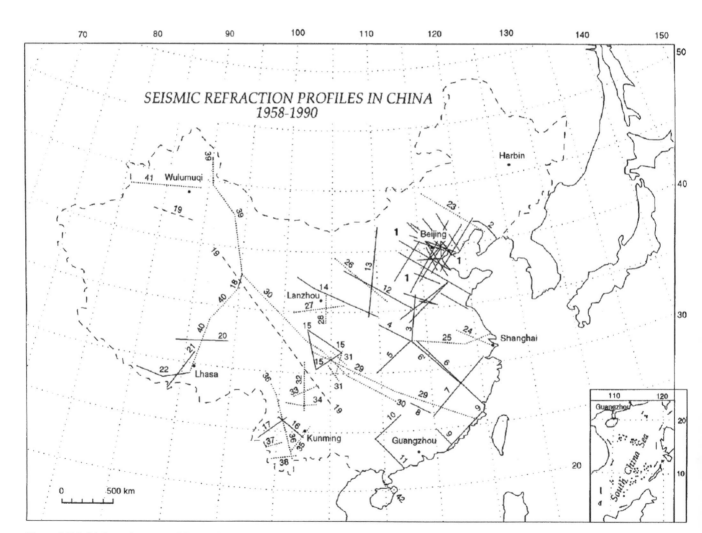

Figure 8.7.2-06. Location map of deep seismic sounding profiles in China observed between 1958 and 1990 (from Li and Mooney, 1998, fig. 2). Solid lines—completed until 1986 and published; dotted lines—completed from 1987 to 1990; dashed line—profile with sparse observational points. [Tectonophysics, v. 288, p. 105–113. Copyright Elsevier.]

end of the Himalaya-Burma arc, in 1982 a seismic project was carried out in the Yunnan province in southwestern China to the southwest of the city of Kunming (Kan et al., 1986, 1988).

It consisted of three 300–400-km-long seismic-refraction profiles with a total length of 1070 km (nos. 16, 17, and 35 in Fig. 8.7.2-06). Four shotpoints were organized along each line and 196 instruments enabled an average spacing of 2 km. The resulting crustal thickness varies from 38 to 46 km. For lines 16 and 17 a fence diagram is displayed in Figure 8.7.2-07 showing some details of the interpreted crustal structure.

Another 1100-km-long seismic-refraction survey was conducted in 1988 across northwest China (Wang et al., 2003). Seismic energy was provided by 12 chemical explosive shots fired in boreholes (Fig. 8.7.2-08). The charges ranged in size from 1500 to 4000 kg provided sufficient energy for first arrivals clearly seen to a maximum distance of 300 km. The spacing between the three-component recording stations was 2–4 km; the spacing between shotpoints was 63–125 km. Both P- and S-wave phases could be well correlated to determine a detailed crustal model (Fig. 8.7.2-09). The data are shown in Appendix A9-4-2, together with other data of western China recorded in the 1990s. The three-layer crust is more than 40 km thick, which was interpreted as being due to subsequent Indo-Asian collision. According to the authors the variation of the average velocity along the line is due to the non-uniform composition of the accreted blocks which comprise continental and oceanic terranes, accretionary prismas, and magmatic arcs.

8.7.3. Japan

Since the early days of seismology, Japanese seismologists continuously exchanged their scientific results with their colleagues worldwide and were actively involved in the foundation of IASPEI (International Association of Seismology and Physics of the Earth's Interior) which provides a platform for communication by its biannual meetings. This concerned also the special field of explosion seismology. For example, as mentioned above, in 1985 they hosted one of the CCSS (Commission on Controlled Source Seismology) workshops where interpretation methods were being discussed and compared.

After intensive investigations of crustal structure underneath the Japanese islands in the 1960s and 1970s, active explosion seismology research in Japan in the 1980s concentrated mainly on offshore research of the adjacent Pacific Ocean. A corresponding offshore investigation by Suyehiro and Nishizawa (1994) will be described below in subchapter 8.9.1. On land most explosion seismic studies were carried out under the national Earthquake Prediction Program and focused on the upper crustal structure, searching particularly for major fault zones and tectonic lines. In effect, every year a seismic-refraction survey was recorded (Ikami et al., 1986; Iwasaki et al., 1998; Matsu'ura et al., 1991; Moriya et al., 1998; Research Group for Explosion Seismology, 1992a, 1992b; Sasatani et al., 1990; Yoshii et al., 1985).

Only two research projects, carried out in 1989 and 1990 within the scope of the Japanese Earthquake Prediction Program,

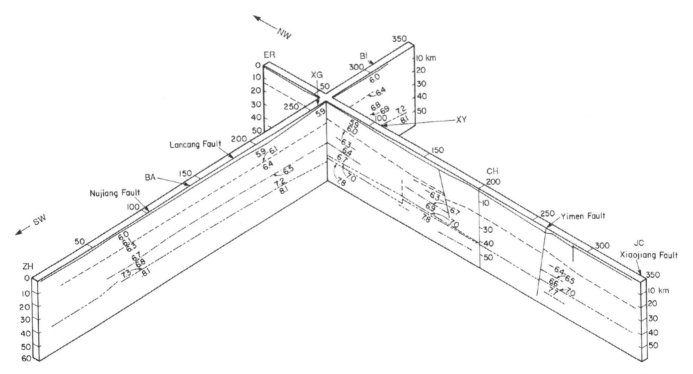

Figure 8.7.2-07. Fence diagram showing crustal structure in the Yunnan province, southwestern China (from Kan et al., 1986, fig. 3). [Science, v. 234, p. 433–437. Reprinted with permission from AAAS.]

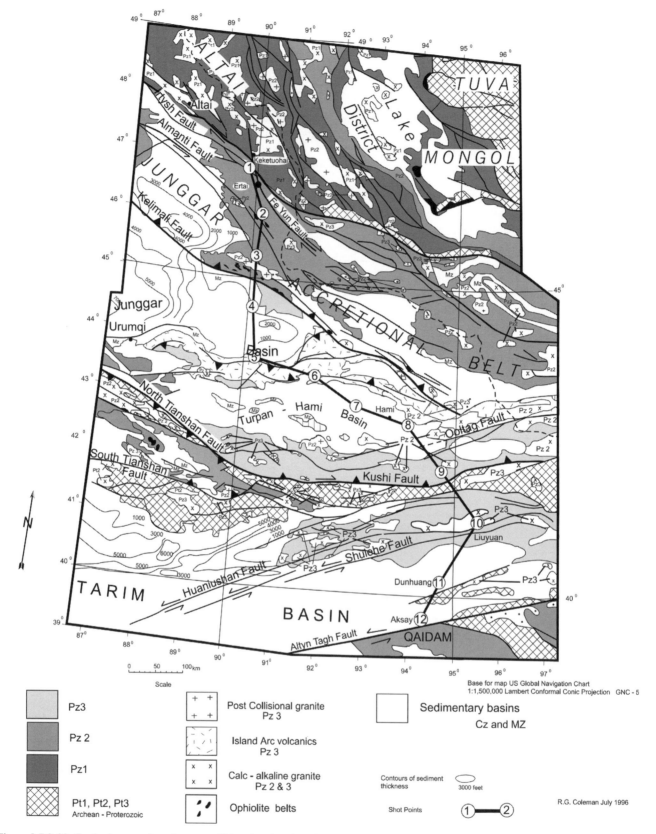

Figure 8.7.2-08. Geologic map of northwestern China showing the seismic refraction survey of 1988 between the Altai Mountains and the Altyn Tagh fault (from Wang et al., 2003, fig. 2). [Journal of Geophysical Research, v. 108, B6, 2322, doi: 10.1029/2001JB000552, ESE 7-1–7-15. Reproduced by permission of American Geophysical Union.]

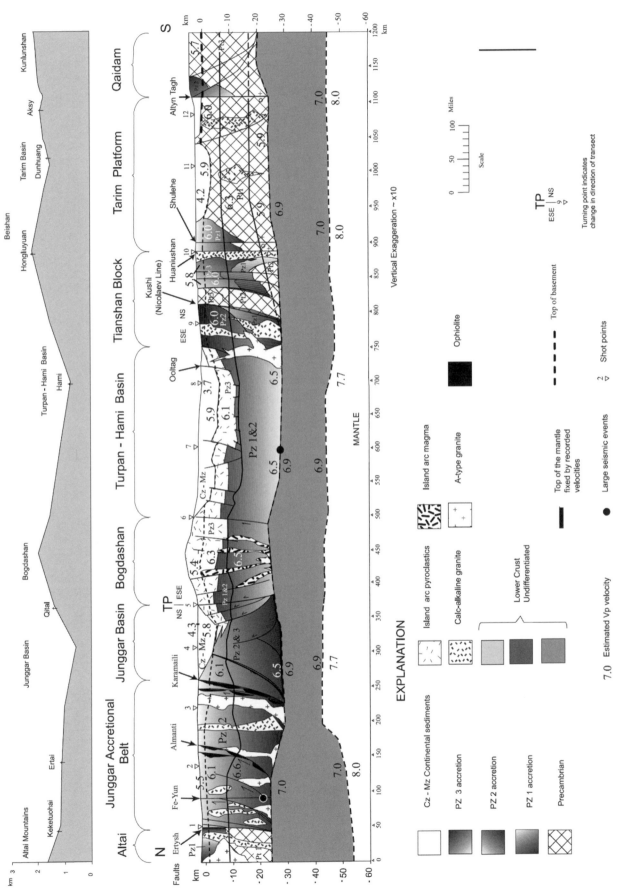

Figure 8.7.2-09. Geologic section through northwestern China showing the crustal structure from the Altai Mountains to the Altyn Tagh fault (from Wang et al., 2003, fig. 10). [Journal of Geophysical Research, v. 108, B6, 2322, doi: 10.1029/2001JB000552, ESE 7-1–7-15. Reproduced by permission of American Geophysical Union.]

both on the island of Honshu, may be mentioned here in more detail. The project of 1989 was a 220-km-long profile between Fujihashi and Kamigori in the western part of Honshu. It involved four shotpoints with charges of 500–800 kg and 137 recording sites (Research Group for Explosion Seismology, 1995).

The second seismic-refraction experiment, again with four explosive sources, was 194 km long and was conducted in 1990 in the Kitakami Massif (Research Group for Explosion Seismology, 1992b). The target area was located in the eastern part of northern Honshu, one of the typical island arcs forming the western rim of the Pacific Ocean, located above the westward subduction zone of the Pacific plate starting beneath the Japan Trench (Fig. 8.7.3-01). The interpretation by Iwasaki et al. (1994) revealed a detailed crustal model with a 32–35-km-thick crust and

its P and S velocity structure (Fig. 8.7.3-02) with an upwarping of Moho beneath the Hayachine tectonic belt. The data also allowed studying the corresponding seismic attenuation (Q-structure). Atypically the P/S ratio resulted to be fairly constant throughout the whole crust.

8.8. CONTROLLED-SOURCE SEISMOLOGY IN THE SOUTHERN HEMISPHERE

8.8.1. Australia and New Zealand

In Australia, seismic-reflection recording using digital equipment started in 1976. The positive outcome of experiments of the late 1970s encouraged Australian scientists such

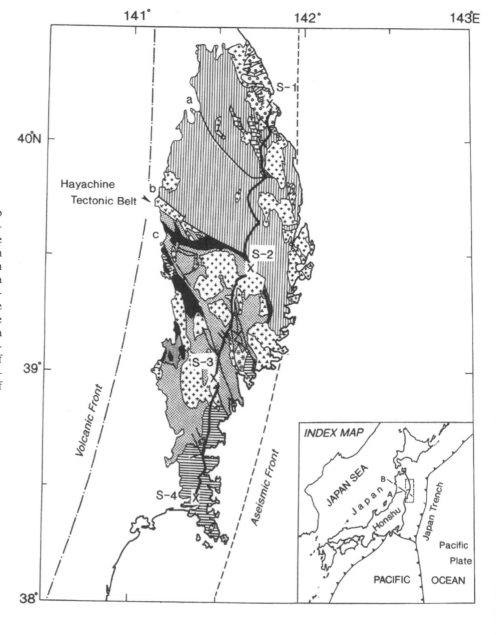

Figure 8.7.3-01. Geological sketch map of the Kitakami region, northern Honshu, Japan, showing the location of the seismic-refraction experiment (from Iwasaki et al., 1994, fig. 1). Refraction line and shotpoints are indicated by a heavy solid line and crosses, respectively. Dashed lines in the index map are earlier seismic refraction lines. A—the Isibuchi-Kamaishi profile (Matsuzawa et al., 1959), B—Oga-Kesennuma profile (Hashizume et al., 1968). [Journal of Geophysical Research, v. 99, p. 22187–22204. Reproduced by permission of American Geophysical Union.]

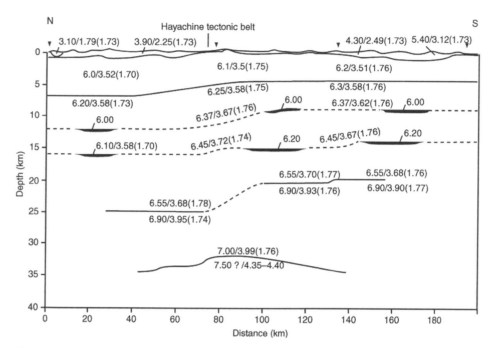

Figure 8.7.3-02. Cross section through the Kitakami region, northern Honshu, Japan (from Mooney et al., 2002, fig. 14, modified from Iwasaki et al., 1994). Indicated are P-velocity/S-velocity, and their ratio. [*In* Lee, W., Kanamori, H., Jennings, P.C., and Kisslinger, C., eds., International Handbook of Earthquake and Engineering Seismology, Part A. Academic Press, Amsterdam, p. 887–910. Copyright Elsevier.]

as Doug Finlayson, Clive Collins, John Moss, Jim Leven, and others to continue with 20 second record length from 1980 onward. Thus, during 1980–1982 good deep seismic-reflection data were obtained in the central Eromanga basin in southwestern Queensland, resulting in 1400 km of traverse in a regional grid on continuous profiles up to 270 km long (Moss and Mathur, 1986; Finlayson et al., 1989). In 1984, a long west-east traverse was added extending from the central Eromanga basin eastward across several basins to near the coast (Fig. 8.8.1-01).

The quality of reflections from the deep crust was generally very good, but the signals from explosive sources had a generally better penetration than the Vibroseis signals. Wide-angle seismic profiling was also conducted along numerous lines coincident with the near-vertical incidence profiles in the Central Eromanga Basin, the Surat Basin, and Clarence-Moreton Basin (e.g., Finlayson and Mathur, 1984; Finlayson 2010; Appendix 2-2).

Overall in the Eromanga basin a highly reflective lower crust between 8 and 12 s TWT was separated from reflective basin sequences in the uppermost crust (1–2 s) by an upper crust with little to no reflectivity. The deep crustal features correlated across a network of traverses at 20–40 km depth were discussed in detail by Finlayson et al. (1989). Figure 8.8.1-02 shows how these lower crustal reflections may be bundled and explained as lower crustal lenticles.

Following these successful studies, a cooperative seismic program with federal and state government surveys and academic institutions was initiated (Fig. 8.8.1-03) in which government and academic institutions were engaged together with industry

with the aim to record deep reflection profiling across many parts of Australia (Moss and Mathur, 1986).

Within this program, in 1983 the Australian Geological Survey Organization (AGSO) investigated the southwestern Yilgarn Craton by recording wide-angle seismic profiles along two transects at right angles to each other, both of ~400 km in length. Energy sources were special seismic shots with charges of 800–2800 kg and coal mine shots at the Collie open pit mine. The original two-layer crust of Drummond (1988) was later reinterpreted and a high-velocity layer (7.25–8.10 km/s) in the lower crust added. The resulting Moho depth increases from 35 km in the east to 38 km in the west (Dentith et al., 2000; Finlayson 2010; Appendix 2-2).

In 1984, a wide-angle seismic survey was undertaken across the New England Batholith in eastern Australia. Moho depths were 34–35 km, under the northern part of the batholith a sill-like feature of high velocity was found at 21–24 km depth (Finlayson and Collins, 1993; Finlayson 2010; Appendix 2-2).

In 1985, two major north-south–oriented deep seismic-reflection surveys were conducted in central Australia (Fig. 8.8.1-03) by the Australian Geological Survey Organization (AGSO). Four hundred eighty-six line kilometers were acquired across the Arunta block and the Amadeus basin (Wright et al., 1990; Korsch et al., 1998). The seismic sections were recorded up to 20 s TWT, equivalent to a depth to 60 km assuming an average crustal velocity of 6 km/s. Underneath the Arunta block, the middle to lower crust appeared moderately to strongly reflective, and a series of planar reflections dipping gently to the north were

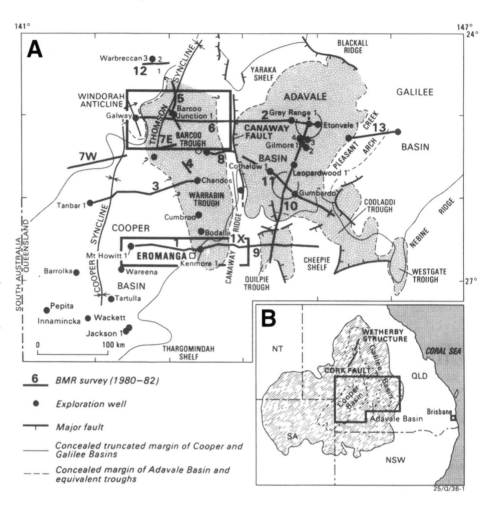

Figure 8.8.1-01. Central Eromanga basin study area, eastern Australia (from Finlayson et al., 1989, fig. 1). (A) Location of Bureau of Mineral Resources (BMR) deep seismic reflection traverses. (B) Outline of the Jurassic-Cretaceous Eromanga basin. [*In* Mereu, R.F., Mueller, St., and Fountain, D.M., eds., Properties and processes of earth's lower crust: American Geophysical Union, Geophysical Monograph 51, p. 3–16. Reproduced by permission of American Geophysical Union.]

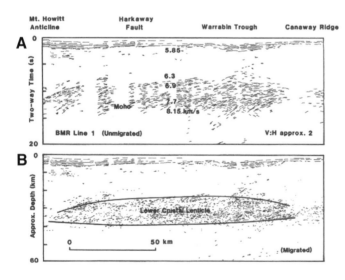

Figure 8.8.1-02. Digitized line diagram of reflectors along BMR line 1 (from Finlayson et al., 1989, fig. 10; for location see Fig. 8.8.1-01). (A) Unmigrated data with velocities derived from coincident seismic refraction data. (B) Migrated data converted to depth using an average crustal velocity of 6 km/s. The migrated data emphasize the reflections in a lower crustal lenticle. [*In* Mereu, R.F., Mueller, St., and Fountain, D.M., eds., Properties and processes of earth's lower crust: American Geophysical Union, Geophysical Monograph 51, p. 3–16. Reproduced by permission of American Geophysical Union.]

interpreted as a series of parallel, planar, north-dipping, thick-skinned thrust faults of which most could be traced at the surface.

In the Amadeus basin beneath the strongly reflective sub-horizontal reflections of the basin sediments, the crust showed low reflectivity in its upper part, but again moderate to strong reflectivity beneath. Only one thrust fault has been recognized in the basin at the southern end of the seismic section (Fig. 8.8.1-04). The crustal thickness across central Australia was interpreted to be between 45 and 50 km and the Mesoproterozoic crustal architecture dominated by planar crustal scale structures were interpreted as ancient thrust faults that had been reactivated by later orogenies.

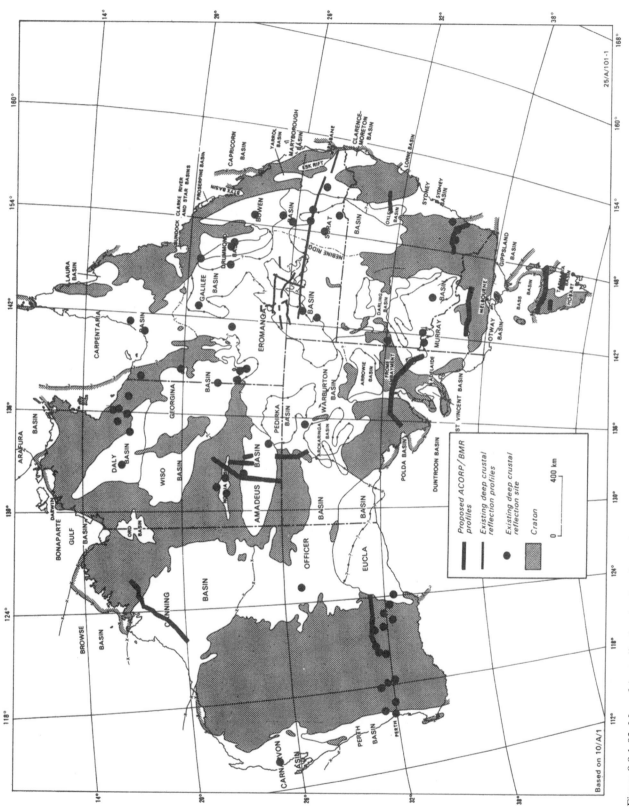

Figure 8.8.1-03. Map of Australia showing locations where deep seismic reflection surveys have been recorded and the planning stage of the Australian Continental Reflection Profiling Program (ACORP) (from Moss and Mathur, 1986, fig. 1). [*In* Barazangi, M., and Brown, L., eds., Reflection seismology: a global perspective: American Geophysical Union, Geodynamics Series, v. 13, p. 67–76. Reproduced by permission of American Geophysical Union.]

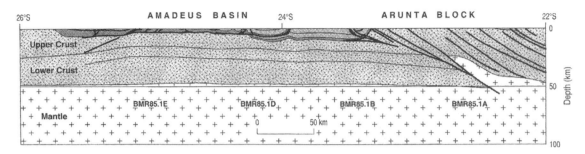

Figure 8.8.1-04. Transect through central Australia showing the crustal architecture based on seismic-reflection, seismic-refraction, and teleseismic investigations (from Korsch et al., 1998, fig. 7b). [Tectonophysics, v. 288, p. 57–69. Copyright Elsevier.]

Finally, from 1987 to 1989, the Australian Geological Survey Organization (AGSO) undertook a number of small deep seismic sounding projects across the Lachlan Orogen in cooperation with various state geology agencies and universities (Finlayson, 2010). In 1989 a seismic traverse involving 294 km of seismic-reflection data was recorded across the northern Bowen basin and the western margin of the New England orogen (e.g., Korsch et al., 1992; Finlayson 2010; Appendix 2-2).

The achievements of seismic and other related studies on the Australian lithosphere were discussed at an interdisciplinary symposium on the Australian lithosphere (Drummond, 1991).

New Zealand is located near the Indian-Pacific plate boundary with a relative motion of oblique convergence. Therefore a wide variety of crustal and upper mantle structures is observed. Beneath the North Island, oceanic lithosphere of the Pacific plate is being subducted, resulting in a configuration of accretionary prism, forearc basin, volcanic arc and backarc spreading basin. In the central South Island there is continental lithosphere on both sides of the plate boundary and mountain building and shear are the mechanisms of deformation (Stern et al., 1986).

The Northland seismic survey of 1982–1983 consisted of marine explosives shots into a 600-km-long line of land stations along the Auckland peninsula. The Central Volcanic Region seismic-refraction profile was recorded through the center of North Island (Fig. 8.8.1-05) with shots in Lake Taupo and in the vicinity of White Island (Stern, 1985). The interpretation of the data along this 200-km-long line resulted in crustal thickness of 14.9 km and an upper-mantle velocity of 7.4 km/s underneath the Central Volcanic Region (Stern, 1985; Stern and Davey, 1987; Stern et al., 1986, 1987).

In 1984 and 1985, a detailed seismic-reflection profile (dashed lines in Fig. 8.8.1-05) was recorded from the Pacific Ocean across Hikurangi Trough up to the east coast of North Island and continued from the west coast of southern North Island across the South Wanganui Basin (Davey et al., 1986; Davey, 1987; Davey and Stern, 1990). The eastern profile of 1984 across the Hikurangi Trough was 250 km long and was recorded up to 20 s TWT. It consisted of a deep-water section of ~100 km length and a slope-shelf segment of ~150 km length.

Its objective was to obtain a crustal cross section across the fore-arc region of the Hikurangi subduction zone. It delineated thrust wedges and back-tilted basins of the accretionary wedge which overlies a detachment zone marking the top of the subducted Pacific plate. No Moho reflections could be identified (Davey et al., 1986). The western profile recorded in 1985 across the South Wanganui Basin defined a broad crustal downwarp in the backarc region of the Pacific-Indian plate boundary overlying the 20–50 km deep subduction zone. The goal was to investigate the region behind the Hikurangi margin, which changes from an extensional backarc basin under the central North Island to a postulated crustal downwarp under the southern North Island. The profile was run across this postulated downwarp.

The data showed discontinuous coherent reflectors dipping westward at the east end of the profile, interpreted as marking the top of the subducting Pacific plate, with overlying east dipping reflectors, which were interpreted as the base of the Australian plate crust, from depths of 9–15 s. Farther to the west a generally transparent middle crust and a reflective lower crust were evident in the reflection section. In 1989 a parallel 110-km-long profile, 150 km north of the western part of the previous profile, was acquired with recording times up to 15 s TWT. The goal was to obtain a crustal cross section across the backarc platform of the Hikurangi subduction zone (Davey, 1987; Davey and Stern, 1990).

The South Island was investigated in 1983. Large marine explosive shots were detonated off each coast and in Lake Tekapo and recorded on 19 seismographs along a 150-km-long profile

Figure 8.8.1-05. Overview of crustal seismic lines recorded between 1952 and 1985 and some results in New Zealand (from Stern et al., 1986, fig. 2). Wellington profile of 1952 is at southern end of North Island; the Fjordland survey of 1974–1975 is at the SW end of South Island; WHUMP of 1979 is a short line near Wellington across the southern end of North Island; Northland line of 1982–1983 is a 600 km line along Auckland peninsula (northern extension of North Island); Central Volcanic Region profile of 1982–1983 is at center of North Island (Lake Taupo in center and White Island off the north coast); solid line through South Island–Central South Island line of 1983; dashed lines around southern end of North Island mark location of multichannel seismic reflection lines of 1984–1985. [*In* Barazangi, M., and Brown, L. eds.: Reflection seismology: a global perspective: American Geophysical Union, Geodynamics Series, v. 13, p. 121–132. Reproduced by permission of American Geophysical Union.]

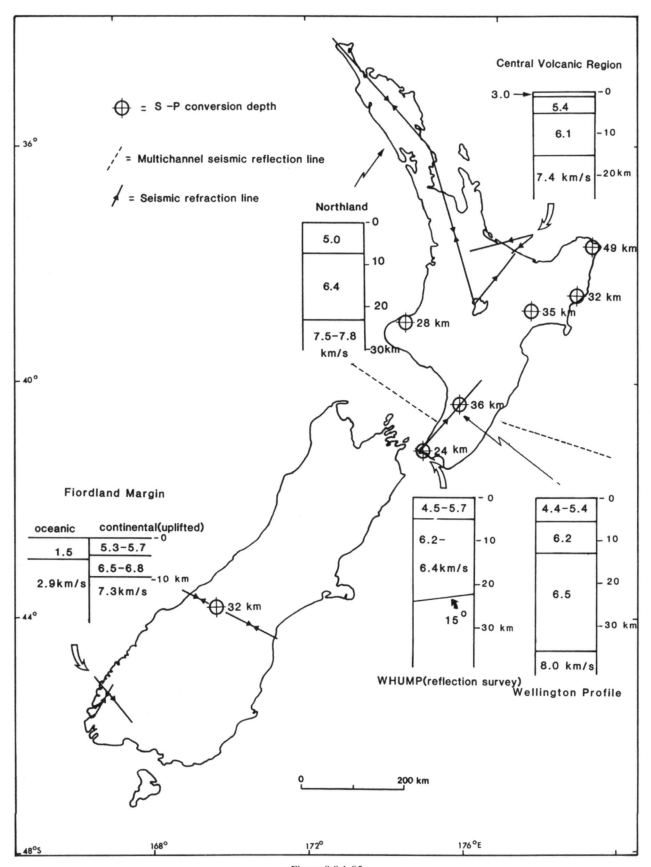

Figure 8.8.1-05.

(Fig. 8.8.1-05), extending across the central South Island from Timaru in the SW to the Pukaku Seismograph Network in the NE (Smith et al., 1995). Strong secondary arrivals were seen from both offshore shots and interpreted by a layer at 24 km depth with a true velocity of 7.2 km/s at the east coast dipping inland at 3.4°. Arrivals from a deeper layer with 8.3 km/s were only evident on the West Coast shot records. Assuming a true velocity of 8.0 km/s as determined from earthquake P_n data, the 7.2 km/s layer would be 11 km thick and its base dip with 2.9°. The extrapolated crustal thickness under the east coast would be 35 km.

8.8.2. South Africa

South Africa has seen some limited controlled-source seismology research in the 1980s. Durrheim (1986) reports on two shallow test seismic-reflection surveys up to 4 s TWT and a short 36-km-long refraction profile in 1982 in the Witwatersrand basin of South Africa that demonstrated successfully that the structure of supracrustal rocks, Proterozoic metamorphosed sediments and lavas, could be mapped and that the boundary between upper and middle crust could also seismically be recognized, being a relatively sharp transition zone (Durrheim,

1986). In 1988, a 112-km-long seismic-reflection profile was recorded across the Witwatersrand Basin that reached from the Ventersdorp dome at the northeastern edge across the Potchefstroom syncline to the center of the Vredefort dome in the center of the basin (Durrheim et al., 1991). The recording time was 16 s TWT. On the basis of contrasting seismic fabrics the crystalline basement could be subdivided into three domains, and it was suggested that a systematic mapping of such basement domains by reflection profiling might provide insights regarding processes responsible for localizing stratified basins. The northwestern portion of the profile showed a change in reflective character at ~12 s TWT, which was interpreted to mark the crust-mantle transition. The absence of a distinct "reflection Moho" suggested a smooth transition from crust to mantle of a depth range of a few kilometers. A seismic-reflection line recorded in 1985 off the Indian Ocean coast (Durrheim, 1987) is described in more detail in Chapter 8.9.3.

In 1987 and 1988, a reconnaissance seismic survey in Botswana (Fig. 8.8.2-01) was carried out by Petro Canada International Corporation on behalf of the government of Botswana (Wright and Hall, 1990). Shots were dynamite charges of 2–4 kg in each hole drilled 6 m deep in a six-hole pattern. Shotpoints

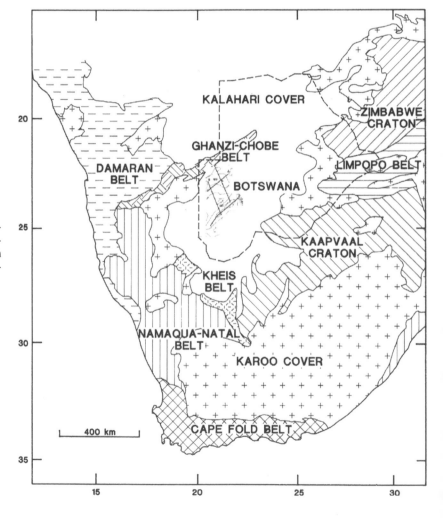

Figure 8.8.2-01. Map of South Africa showing location of deep seismic-reflection survey in the Kalahari desert in western Botswana (compiled from Wright and Hall, 1990, figs. 1 and 2). [Tectonophysics, v. 173, p. 333–343. Copyright Elsevier.]

were spaced such that a 15-fold common mid-point coverage could be obtained. The processed seismic sections covered a time range up to 15 s. In total 600 km of 12 fold data were obtained which showed good-quality reflections between 5 and 15 s TWT. An example is shown in Figure 8.8.2-02. The Moho ("base of crust" in Fig. 8.8.2-02) was tentatively traced piecewise on all lines at ~14 s TWT at the base of a reflective lower crust.

8.8.3. South America

In 1982, crustal research of the Andes in South America received a new impetus. Under the leadership of Peter Giese at the Free University of Berlin, Germany, the geoscientific interdisci-

plinary research group "Mobility of Active Continental Margins" was established, funded by the German Research Society from 1982 to 1989. This project centered on the Andes of Chile and enabled a variety of onshore and offshore seismic experiments.

The map in Figure 8.8.3-01 shows the location of seismic-refraction lines established in northern Chile and adjacent Bolivia and Argentina in the 1980s. In the first two expeditions in 1982 and 1984, the large explosions in the copper mine Chuquicamata were used as the energy source (Wigger, 1986) and three profiles with recording distances of 240–260 km to SW, S, and SE were successfully recorded in northern Chile. On a fourth profile to the east, reaching into southern Bolivia, seismic energy disappeared beyond 100 km distance.

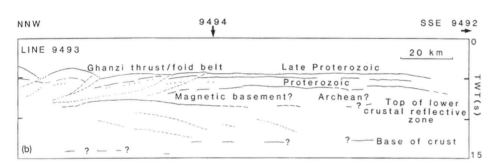

Figure 8.8.2-02. Schematic crustal cross section through the Kalahari Desert in western Botswana (from Wright and Hall, 1990, fig. 6) along line 9493, the eastern NNW-SSE deep seismic-reflection line. Line 9494 is the northern SW-NE line (see Fig. 8.8.2-01). [Tectonophysics, v. 173, p. 333–343. Copyright Elsevier.]

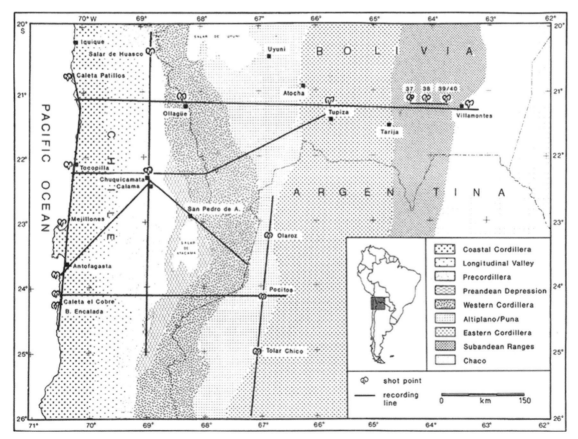

Figure 8.8.3-01. Map of northern Chile and Bolivia showing location of deep seismic refraction surveys from 1982 to 1994 between latitudes 20–25°S and 71–68°W (from Patzwahl, 1998, fig. 2.5). [Berliner Geowissenschaftliche Abhandlungen, Reihe B30: 150 p. Published by permission of Institut für Geologische Wissenschaften, Freie Universität Berlin.]

In 1987 and 1989, additional shotpoints were arranged, and additional profiles were recorded (Wigger et al., 1994). The aim was to obtain as many reversed observations as possible. Energy was primarily obtained by using large quarry blasts from various copper mines, but also some self-organized borehole shots were added. Such land shots, with charges between 500 and 2000 kg, were arranged in Chile as well as in Argentina and were detonated in salt lakes in patterns of shallow boreholes (10–30 m). In Bolivia, borehole shots in the eastern Cordillera and further east were fired.

Furthermore, underwater shots in the Pacific Ocean were arranged, fired by the Chilean navy close to the Chilean coast. The charges, ranging from 50 to 250 kg, were placed on the sea bottom at water depths between 57 and 80 m. The resulting pattern of profiles is shown in Figure 8.8.3-01. In total, 35 recording instruments were available; the average station spacing was between 5 and 10 km.

The data of the various projects carried out from 1982 to 1989 were collectively interpreted (Schmitz, 1993; Wigger et al., 1994). An E-W cross section for the Chuquicamata W-profile (Fig. 8.8.3-02) shows a complicated crustal structure for the forearc region, characterized by a sequence of high- and low-velocity regions, underlain by a dipping Moho from 40 km depth near the coast to more than 70 km depth under the Western Cordillera.

The Central Andean Transect (Wigger et al., 1991, 1994) at latitude 22°10′S displays the crustal and uppermost mantle structure from the Nazca Plate in the southwestern Pacific to the Chaco Plains in northern Argentina (Fig. 8.8.3-03).

8.8.4. Antarctica

With improved technology, extremely hostile areas such as Antarctica also came into reach of extensive seismic research projects. The seismic surveys of the Institute of Geophysics of the Polish Academy of Sciences, which had started in 1979 and 1980 to explore the structure underneath West Antarctica, con-

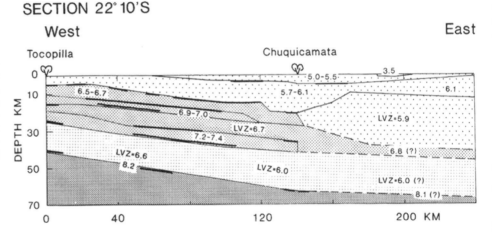

Figure 8.8.3-02. 2-D west-east crustal cross section at latitude 22°10′S of the forearc region in northern Chile (from Wigger et al, 1994, fig. 6). [*In* Reuter, K.-J., Scheuber, E., Wigger, P. eds.: Tectonics of the Southern Central Andes: New York, Springer, p. 23–48. Reproduced with kind permission of Springer Science+Business Media.]

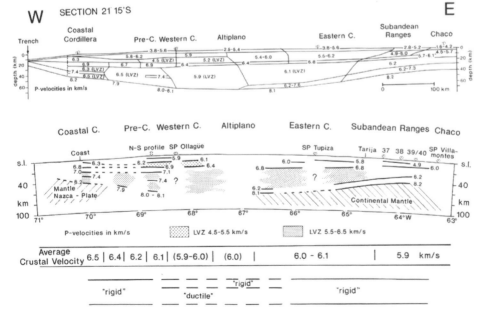

Figure 8.8.3-03. 2-D west-east crustal and upper mantle cross section at latitude 21°15′S from the Pacific Ocean through the Andes to the Chaco in Argentina, showing seismic discontinuities and P-wave velocities (from Wigger et al, 1994, fig. 18). [*In* Reuter, K.-J., Scheuber, E., Wigger, P. eds., Tectonics of the Southern Central Andes: New York, Springer, p. 23–48. Reproduced with kind permission of Springer Science+Business Media.]

tinued in 1984–1985 and in 1987–1988 (Fig. 8.8.4-01; Janik, 1997). The area of the Bransfield Strait between the South Shetland Islands and the Trinity Peninsula was of particular interest, but a series of profiles was also recorded farther south along the western margin of the Antarctic Peninsula. Land stations on the islands and on the peninsula in the Bransfield Strait recorded shots along various profiles.

The shots, ranging in size from 16 to 144 kg, were detonated at 60 m water depth and were usually 5–7 km apart. Recording stations on land had 3–6 seismometers, spaced by 100–200 m. The resulting modeled velocities were high throughout the crust (Fig. 8.8.4-02). The so-called upper crust yielded already velocities up to 6.8 km/s. A particular result of the interpretation of the data was a high-velocity lower crustal layer with velocities above 7.4 km/s beneath the central subbasin of the Bransfield Strait, which could be separated from the remaining pre-rift mafic lower crust with velocities from 6.8 to 7.3 km/s. Total crustal thickness varied from 36 to 42 km in the coastal area and decreased to ~25–28 km toward the Pacific Ocean (Sroda et al., 1997).

The goal of the U.S. Louisiana State University since the early 1970s was to investigate the east-west boundary of McMurdo Sound at the southern end of the Ross Sea (McGinnis et al., 1985). Following a series of shallow seismic-refraction, magnetic and gravity studies, in 1980 an E-W reversed seismic-

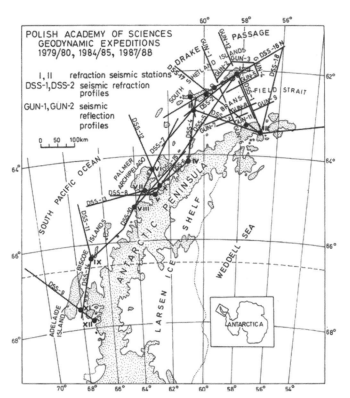

Figure 8.8.4-01. Location of deep seismic sounding lines (DSS), observed in the area of the Antarctic Peninsula from 1979 to 1988 (from Birkenmajer et al., 1990, fig. 2). [Polish Polar Research, v. 11, p. 241–258. Reproduced with permission of Polish Academy of Sciences, Warsaw, Poland.]

refraction profile was recorded between the Antarctica mainland and Ross Island, approximately from 164°30′ to 166°30′E and along latitude 77°45′S parallel to the ice shelf. The observations with maximum shot-to-receiver distances of 40 km provided the velocity distribution in the upper crust. Along this line, a common-depth point reflection profile was also obtained. Layered reflectors in the upper crust dipped and thickened to the east, away from the coast, down to a depth of ~7 s TWT, corresponding to 14–16 km depth. In 1981, a 200-km-long reversed seismic-refraction N-S–directed profile parallel to the coast followed. The interpretation resulted in a 5.0 km/s basement and a 6.5 km/s intracrustal unit, underlain by the mantle at ~21 km depth with a velocity of 8.2 km/s.

8.9. DEEP SEISMIC SOUNDING STUDIES IN THE OCEANS

8.9.1. Introduction

From the early 1980s onward investigation of the crustal structure focused on large dynamic ranges and dense spatial sampling. For this reason, experiments were designed for obtaining large offsets and large-aperture seismic-refraction/wide-angle reflection data as well as near-vertical incidence reflections. Possible configurations of sources and receivers to collect offshore seismic data are shown schematically in a sketch (Fig. 8.9.1-01) by Trehu et al. (1989b).

In the 1980s the development of non-explosive sources had become effective enough to be successfully recorded over long ranges of several 100 km at sea and also, in onshore-offshore experiments, on land. In particular, reflection experiments dealing with the sedimentary structure at convergent margins were important for the interpretation concerning sediment subduction, subduction erosion, and the growth of the continental crust (von Huene and Scholl, 1991).

The technique was described in much detail by Jones (1999) in a textbook which corresponds to the technical status available by the end of the 1990s. However, airguns of great power were in use since the early 1980s, in the earlier years possibly with fewer technical refinements than in the 1990s, depending on the actual state of technical development.

Airguns produce a wide range of pulse shapes and source spectra. A most important feature of any energy source is the repeatability of its signature over many firings which turned out to be a major advantage with ongoing progress in data processing. The outputs can be monitored with a deep-towed hydrophone or, more conveniently, with near-field hydrophones, ~1 m from the gun, whose outputs allow the calculation of the far-field waveform. To provide broadband output signals every 10–15 s, airguns of various sizes were grouped in arrays. Typically, two source arrays are towed at depths of 3–10 m and are alternately fired. The gun positions are monitored by hydrophones which receive pulses from a high-frequency transducer (50–100 kHz) close to the ship and/or, later in the 1990s, by a GPS receiver in a head buoy. Several tens of airguns with a total volume of 100 l or more were used for deep seismic profiling, for example, by BIRPS, ECORS, BABEL and CROP (Avedik et al., 1993; BABEL Working Group, 1991; Klemperer and Hobbs, 1991).

Nevertheless, for many marine projects, the use of explosives was not yet out of range. Some projects used both energy sources either to ensure general success or to use the explosive shots to provide enough energy for the larger distance ranges. Also, tests have been made to apply a special technique in which explosives are used with a much less damaging effect. Orcutt (1987) has described new experiments where, e.g., ocean bottom seismometers and seismic sources were placed near the seafloor

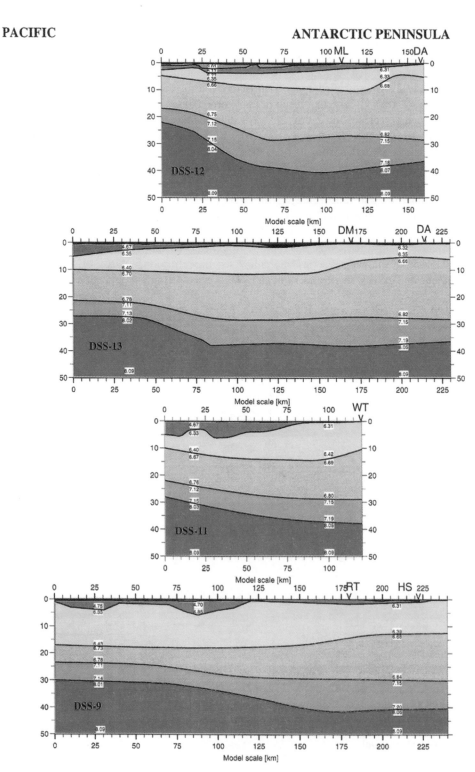

Figure 8.8.4-02. Crustal structure of profiles DSS-12, -13, -11, and -9 crossing the northwestern margin of the Antarctic Peninsula (from Guterch et al., 1998, fig. 6). [Polish Polar Research, v. 19, p. 113–123. Reproduced with permission of Polish Academy of Sciences, Warsaw, Poland.]

(e.g., Purdy, 1986) or multichannel streamers and low frequency sources towed near the seafloor (e.g., Fagot, 1986).

Kirk et al. (1991) have also devised a technique for firing arrays of bottom shots on the ocean bed in depths of 4000 m and more. Ten-kg charges were dropped from the surface while the shooting ship was navigated acoustically. The charges were fired at preset times by an electric timer which initiates an electrical detonator, detonating cord and cast TNT explosives. The ranges to OBS could be calculated from the arrival-time difference of the direct and surface-reflected water waves. Experiments were carried out in 1989 at six sites in the Norwegian Sea at depths of 1500 m and 3900 m. It was possible to obtain regularly spaced traces 200–300 m apart along bottom profiles up to 4 km in length. It was possible to determine the shot-receiver ranges with an accuracy of better than 22 m and shot instants to within 14 msec.

Of greatest importance for the advance of marine deep seismic sounding was the fact that gradually over the years more and more ocean bottom seismographs came into use, giving much better signal-to-noise ratios than was possible with strings of hydrophones which suffered also from the noise produced by the towing ship. While in the 1970s, the use of ocean-bottom seismometers very often was part of the title of a subsequent publication, in the 1980s, ocean-bottom seismographs became practically a standard of deep seismic sounding projects at sea (e.g., Kirk et al., 1982; Whitmarsh and Lilwall, 1983; Duschenes et al., 1985; Nakamura et al., 1987; Moeller and Makris, 1989; Jones, 1999).

Nevertheless, for the exploration of the sedimentary part of the oceanic crust, reflection measurements by a string of towed hydrophones has become the standard practice of any commercial or scientific marine experiment. The hydrophones are packed in streamers or cables and form a hydrophone array. Most towed arrays are multichannel receivers. They are divided into sections which are recorded separately. With developing techniques, the number of channels in a streamer increased, and streamer or cable lengths up to several kilometer lengths could be used to be towed by a ship (Jones, 1999). Techniques were also developed to place multichannel hydrophone arrays on the sea bottom (e.g., Powell et al., 1986).

The Ocean Drilling Project (ODP) continued into the 1980s. The *Challenger* was retired in 1983, and for the succeeding Ocean Drilling Program JOIDES, again with NSF funding, a more advanced drillship was converted from industrial use and rechristened to *JOIDES Resolution*. From 1985 to 2003 this vessel travelled almost nonstop around the world drilling 650 more holes on 110 more legs. By ODP's end the program included more than 20 nations (Alden, 2010).

The theory for the calculation of synthetic seismograms was also developed further. A short overview was published by Chapman and Orcutt (1985). The use of synthetic seismograms in the interpretation of refraction data, also for marine seismic surveys, had been introduced in the 1970s and was well established in the 1980s. Following the pioneering papers of Helmberger (1968) and Fuchs and Müller (1971), the Cagnard-de Hoop-Pekeris and reflectivity methods were widely used to compute synthetic seismograms. Seismic models were iteratively refined to obtain a reasonable fit between the synthetics and the observed data. The procedure was also extremely successful for marine seismic data (Spudich and Orcutt, 1980), but was usually tedious and expensive. Subsequently, attempts were made to automate the inversion process (Chapman and Orcutt, 1980, 1985; Mellman, 1980; Given and Helmberger, 1983; Shaw and Orcutt, 1985). At the same time, the basic theory of the ray method in seismology had been formulated (Červený et al., 1977), and first computer routines to calculate traveltimes and ray paths for lateral inhomogeneous media (e.g., Červený and Horn, 1980; Gebrande, 1976; Will, 1976) as well as to compute ray theoretical seismograms (Červený, 1979; Červený and Pšenčik, 1984a; McMechan and

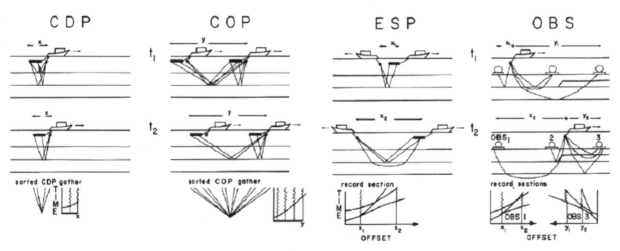

Figure 8.9.1-01. Source-receiver configurations to collect offshore seismic data (from Trehu et al., 1989b, fig. 4). CDP—common depth point, COP—constant offset profile, ESP—expanding spread profile, OBS—ocean bottom seismometer. [*In* Pakiser, L.C., and Mooney, W.D., eds., Geophysical framework of the continental United States: Geological Society of America Memoir 172, p. 349–382. Reproduced by permission of the Geological Society of America.]

Mooney, 1980; Spence et al., 1984) had been published. The following decade of the 1980s saw the systematic application of ray
tracing methodology for 2-D interpretations of seismic-refraction
surveys used for marine seismic data as well.

A similar approach, as was attempted by Kempner and
Gettrust (1982a, 1982b) was published by Collins et al. (1986).
They calculated 2-D synthetic seismogram profiles of the inferred fossil oceanic crust-mantle transition observed in the Bay
of Islands ophiolite complex of western Newfoundland (Salisbury and Christensen, 1978; Christensen and Salisbury, 1982).
To simulate a seismic-reflection experiment, they calculated
near-vertical incidence seismograms at a horizontal spacing of
500 m for three separate sections of the ophiolite totaling 64 km
in length. Multichannel seismic data from both the western Pacific and the western North Atlantic showed Moho traveltimes
similar to those observed in the synthetic profiles, suggesting that
the structures observed in the inferred fossil crust-mantle transition of the ophiolite were characteristic of oceanic lithosphere.

With advanced techniques, the number of marine seismic
experiments carried out in the past three decades since the beginning of the 1980s literally exploded. As we have pointed
out in Chapter 1, it is far beyond our ability to aim for a similar completeness of marine seismic projects as we may have
achieved for land-based deep seismic crustal projects. Rather,
we will restrict ourselves to a selected number of marine seismic projects and will describe them in some detail in the following subchapters.

Furthermore, many marine seismic projects were already
described in previous sections of this chapter. This concerns in
particular the large national marine seismic-reflection campaigns
of BIRPS, ECORS, BABEL, and CROP in subchapter 8.3 which
have exclusively investigated continental crust covered by shallow waters. For other marine seismic projects which were carried
out in the Mediterranean and in the Red Sea as well as in the
North and Baltic Seas, the reader is referred to the corresponding
sections. Furthermore, some of the projects which investigated
continental margin areas around Africa, Canada, Japan, New
Zealand, Australia, South America, or Antarctica, were already
described above in the corresponding sections of subchapters 8.3
to 8.8, if they were connected with a strong land component. In
the following description of active-source experiments carried
out in the 1980s in the Pacific, Indian, and Atlantic Oceans, we
will follow again a more or less geographic order.

8.9.2. Pacific Ocean

Besides local investigations in specially defined areas, which
will be described further below, Neprochnov (1989) has reported
on a bulk of seismic data which were assembled at a worldwide
study of the lower crust and upper mantle using OBSs and big
airguns, performed by the Shirshov Institute of Oceanology in
the period from 1977 to 1984 (Fig. 8.9.2-01). While the first
cruises from 1977 to 1979 had concentrated on the central Pacific
Ocean (K24, M21 to M23 in Fig. 8.9.2-01), the cruise 29 in 1982

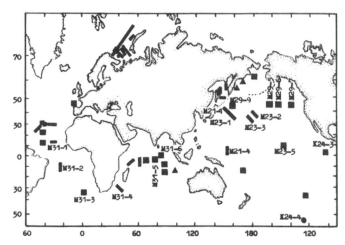

Figure 8.9.2-01. World map showing locations of deep seismic sounding experiments made in expeditions of the Shirshov Institute of Oceanology using OBS and big airguns (from Neprochnov, 1989, fig. 2).
Triangles—single OBS, rectangles—several OBS along one profile,
square—system of profiles in different directions, long solid lines—
geotraverse with a number of OBS. [*In* Mereu, R.F., Mueller, St., and
Fountain, D.M., eds., Properties and processes of earth's lower crust:
American Geophysical Union, Geophysical Monograph 51, p. 159–
168. Reproduced by permission of American Geophysical Union.]

emphasized research in the North Pacific Ocean (M29 in Figs.
8.9.2-01 and 8.9.2-02).

The Juan de Fuca Ridge, located off western North America
between latitudes 44° and 52°N and separating the large, mod-
estly sedimented Pacific plate from the small, broken up Juan de
Fuca plate, was the subject of several projects in the 1980s.

In 1981, multichannel seismic-reflection profiles were recorded at the southern Juan de Fuca Ridge (Morton et al., 1987).
Two profiles perpendicular to the axis crossed the rift at latitudes
44°40′N and 45°05′N. A third line ran along the ridge axis from
latitude 45°20′N to south of the intersection with the Blanco
Fracture Zone. A weak reflection centered beneath the axis was
found at a depth similar to seismically detected magma chambers
on the East Pacific Rise (e.g., Detrick et al., 1987).

Later in the 1980s a multichannel seismic-reflection survey
(Rohr et al., 1988) collected data by means of a 3000 m long,
120 channel streamer and a 50-l airgun array. As supposed, the
deep crustal structure proved to be highly asymmetric. Moho
was found to exist at ~6.6 km depth below the sea floor within
4 km of the rift valley on the Pacific plate, in contrast to the Juan
de Fuca plate, which showed a reflector at intermediate depths
dipping away from the rise to reach Moho depth 12 km from
the rift valley.

In 1990, the Juan de Fuca Ridge was investigated by a seismic-refraction experiment conducted with airguns and OBSs at
45°N at the northern Cleft segment and the overlapping rift zone
between the Cleft and Vance segments (McDonald et al., 1994).
Four OBSs, deployed twice, recorded airgun shots along 700 km
of track lines, divided up into three ridge-parallel, one diagonal,
and three ridge-perpendicular lines. Seismic velocity anisotropy

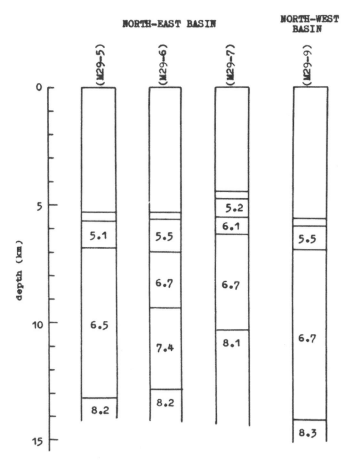

Figure 8.9.2-02. Averaged crustal columns for seismic surveys during the 29th cruise of R/V *Dmitry Mendeleev* (from Neprochnov, 1989, fig. 4). [*In* Mereu, R.F., Mueller, St., and Fountain, D.M., eds., Properties and processes of earth's lower crust: American Geophysical Union, Geophysical Monograph 51, p. 159–168. Reproduced by permission of American Geophysical Union.]

was observed, being confined to the upper 500 m or less of the oceanic crust, the ridge-parallel velocity being ~3.35 km/s and the ridge-perpendicular velocity ~2.25 km/s.

An overview on shallow seismic-reflection surveys performed along the continental margins of California and Baja California as well as in the Gulf of California until the end of the 1980s is contained in the Memoir of Dauphin and Simoneit (1991).

One of the deep sea boreholes, drilled by the Deep Sea Drilling Project (DSDP) and Ocean Drilling Program (ODP), the Hole 504B on the southern flank of the Costa Rica rift in the eastern equatorial Pacific, provided an opportunity to directly establish the geological nature of the boundary between seismic layers 2 and 3 in the oceanic crust. Since 1979, drilling on seven different DSDP and ODP legs enabled the progressive deepening of hole 504B, to 2111 m below the sea floor and more than 1800 m into the igneous crust, placing it near the expected depth of the layer 2/3 boundary. Seismic experiments were carried out in and around the hole over 15 years since drilling had started. Sonic velocity logs resulted in P-wave velocities from 4 to 6 km/s in the upper 600 m and an increase to more than 6 km/s near 1000 m depth. By 1.3–1.4 km depth velocities reached 6.5 km/s. Based on these observations the top of the seismic layer 3 was placed just above the base of the hole at 1.8 km depth. The available seismic-refraction data from around this hole were reexamined by Detrick et al. (1994). The authors concluded that the seismic layer 2/3 boundary lies within a sheeted-dyke complex and is associated with a gradual downhole change in crustal porosity and alteration and is not a lithological transition from sheeted dykes into gabbro.

Various experiments carried out in the 1970s on the Pacific Rise crest had provided the first geophysical data which could be explained in terms of a crustal magma chamber associated with the genesis of oceanic crust (e.g., Orcutt et al., 1975). Therefore, in order to unambiguously constrain the depth and lateral extent of the crustal magma chamber, in 1982 an extensive seismic-refraction experiment (MAGMA) was conducted on the East Pacific Rise at 12°50′N (Orcutt et al., 1984). The MAGMA experiment utilized OBSs at three sites to record the near-surface explosive sources. The sites were located on the ridge (age = 0), 6 km to the east (age = 0.11 Ma), and 16 km east (age = 0.30 Ma). In a first phase, a grid of short, 15–17-km-long lines were shot using small charges (0.5–1.5 kg) and a shot spacing of ~230 m, to provide a dense areal coverage of the upper crust at the ridge axis and adjacent areas. In a second phase, larger shots (8–150 kg) were placed at 900 m spacing covering an expanded 2-D grid. Investigating the refracted arrivals parallel to the rise crest, Orcutt et al. (1984) attributed the attenuation of seismic arrivals beyond a range of 10–12 km to a shadow zone created by the interaction of seismic waves with a low-velocity zone and interpreted it as evidence for a crustal magma chamber at most 1.5 km below the seafloor.

Detrick et al. (1987) reported on a new experiment in 1985 across the East Pacific Rise near 9°N, based on multichannel seismic imaging of a crustal magma chamber along the East Pacific Rise by Herron et al. (1978), because the shape, longevity, and along-strike variability of ridge-crest magma chambers had remained the subject of considerable controversy. This experiment was a much more extensive two-ship multichannel seismic survey of the East Pacific Rise between 8°50′N and 13°30′N and was designed to provide new information on the shape and dimensions of the axial magma chamber and to investigate how the magma chamber varies along the rise axis over distances of tens to hundreds of kilometers. More than 3500 km of conventional, 48-channel common-depth point (CDP) seismic-reflection data were obtained including 30 new profiles across and a series of CDP reflection lines along the crest of the East Pacific Rise. The seismic velocity structure of the crust and upper mantle in the axial region was obtained using the two-ship expanding spread profiling (ESP) techniques. The ESPs were oriented parallel to the rise and were shot using airguns as well as explosives. In the 9°N area, ESPs were shot on the rise crest as well as 5 km away on either side of the crest. In the 13°N area, ESPs were located within 3.6 km of the rise axis. As a result, a reflection observed on the multichannel profiles along and across the East Pacific Rise was interpreted as to arise from the top of a crustal magma chamber located 1.2–2.4 km below the seafloor. The magma chamber appeared to be quite narrow (less than 4–6 km wide), but could be traced as a nearly continuous feature for tens of kilometers along the rise axis (Detrick et al., 1987). The data were further interpreted in more detail in subsequent papers (Collier and Singh, 1997, 1998; Kent et al., 1993; Vera et al., 1990; Vera and Diebold, 1994). Orcutt (1987) has published a review on experiments in the Pacific Ocean dealing with various aspects of marine crustal research.

Toomey et al. (1990) described a marine seismic survey investigating the 3-D seismic velocity structure of the East Pacific Rise near latitude 9°30′N, encompassing a 12-km-long linear rise segment trending north-south. The project involved 15 ocean-bottom instruments which recorded energy from 480 explosive shots and from several airgun profiles, yielding more than 12,000 seismic records. They found pronounced heterogeneity over distances of a few kilometers. A linear, 1–2-km-wide high-velocity anomaly, centered on the rise axis, was seen to be restricted to the uppermost 1 km of the crust. Consistent with a zone of higher crustal temperatures an axial low velocity zone was found at 1–3 km depth which they interpreted as injections of mantle derived melt.

A two-ship seismic experiment near the Hawaiian islands Oahu and Molokai investigated the lithospheric flexure across the Hawaiian-Emperor seamount chain in 1982 (Watts et al., 1985). The single-ship multichannel reflection profiling techniques was used to determine the configuration of individual crustal layers and the Moho; the two-ship techniques served to obtain the detailed velocity structure of the crust and upper mantle. Both large airgun and explosive shots were fired into the streamers of both ships. The seismic experiment involved normal incidence CDP data, constant offset profiles (COPs), and eleven ESPs obtained across a 600-km-long, 100-km-wide transect of the Hawaiian ridge centered near Oahu. While shooting the COP, the two vessels maintained a constant distance apart of 3.6 km and both ships alternately fired their airguns every minute. During the ESP, the two vessels separated from other, one ship shooting the airgun, the other receiving up to the end of the 65–90-km-long lines and then reversed shooting and receiving when returning along the same profile. During the ESP shooting, at the same time a CDP profile was obtained by one of the ships which was equipped with a 3.6-km-long seismic engineering multichannel streamer with 48 active channels. In a following second part of the experiment a similar sequence of COP/ESPs was completed between the Molokai channel and the arch to the north. The principal result was a total crustal thickening from 6.5 to 6.8 km beneath the Hawaiian arch to 18–19 km beneath the ridge, and that the flexed oceanic crust underneath the Hawaiian Emperor seamount chain is underlain by a 4-km-thick deep crustal body which was interpreted as a deep crustal sill complex associated with the tholeiitic stage of volcano building along the chain (Watts et al., 1985). Some of the ESP record sections were interpreted by Lindwall (1988) in detail to obtain 2-D models of the entire crust of the volcanoes and the underlying oceanic crust near Oahu, of the southeastern submarine flank of Kauai, and of the ridge axis between Oahu and Molokai.

Other experiments dealt with the verification of thin oceanic crust under fracture zones (e.g., Cormier et al., 1984; see subchapter 8.9.4).

Data of the 1983 Ngendei experiment (Orcutt et al., 1983) in the South Pacific (23.82°S, 165.53°W) helped Shearer and Orcutt (1985) investigate the problem of anisotropy in the oceanic crust. The site was ~1000 km east of the Tonga trench and 1500 km WSW of Tahiti. The oceanic lithosphere here is estimated to be ~140 m.y. old, one of the oldest areas in the Pacific. The Scripps Institution of Oceanography shot four split refraction profiles with azimuth spacing of 45° and a circular line of 10 km radius. All of the lines were recorded by at least two ocean-bottom seismometers, and two of the lines were recorded by a borehole seismometer, emplaced 124 m in DSDP Hole 595B by the *Glomar Challenger* on DSDP Leg 91. The analysis of the data indicated anisotropy at two levels in the oceanic lithosphere. In the upper mantle, P-wave velocities varied between 8.0 and 8.5 km/s, with the fast direction at N30°E. Crustal anisotropy was seen within layer 2 with velocity differences of 0.2–04 km/s, with the fast direction at N120°E, orthogonal to the mantle anisotropy.

In 1982 and 1984, a bright midcrustal reflector was seismically imaged beneath an active backarc spreading center, the Valu Fa Ridge, in the Lau Basin of the SW Pacific (Morton and Sleep, 1985), between the Fiji Islands and the Tonga Trench. The data consisted of a grid of seismic normal incidence and wide-angle reflection profiles located at 22°10′–22°30′S, 176°35′–176°50′W. The bright reflector was coincident with a velocity inversion at a depth of 3.2 km below the seafloor and was observed on every one of the 40 across-axis profiles. It was interpreted as the top of a crustal magma chamber. The widest magma chamber reflector occurred beneath the overlapping spreading center, where it extended up to 4 km and was imaged beneath both ridges and the overlap basin (Collier and Sinha, 1992).

A project dealing with the investigation of the Society Islands hotspot in 1989 (Grevemeyer et al., 2001b) is described in the context of follow-up projects in Chapter 9.8.2.

Various projects have dealt with the ocean-continent transition at continental margins. In particular, around Japan various marine projects were carried out. Seismic profiling crossing the trench axis was started in 1983 at the southernmost part of the Kuril trench off Hokkaido, Japan, from which a detailed subduction structure of the Pacific plate was delineated for the first time (Nishizawa and Suyehiro, 1986; Iwasaki et al., 1989). In 1984 and 1988, a German-Japanese cooperative project with refraction profiling was undertaken in the northern part of the Ryukyu trench area. The development of the pop-up type ocean bottom seismographs led to dense receiver spacing (10–20 km) along 190- and 295-km profile lines. This enabled delineation of subduction and accretion structure near the trench axis and also proved that crustal thinning was associated with the backarc spreading of the Okinawa trough (Iwasaki et al., 1990; Kodaira et al., 1996). These results provided the breakthrough for imaging a subducted plate boundary beneath the NE and SW Japan arcs.

In the vicinity of the northern Japan Trench, seismic reflection, two-ship refraction, and OBS refraction data, mostly by airgun shooting, were collected in particular in the 1980s (Fig. 8.9.2-03) and a summary of their interpretation of the seismic velocity structure beneath the forearc of the Tohoku subduction zone was published by Suyehiro and Nishizawa (1994). The area of study extended from the trench axis to the Tohoku coast, ~200 km from the trench. The interpretation revealed that the Pacific crust gradually increases its dip to ~7°, 110 km landward of the trench axis. The basic results are shown in Figure 8.9.2-04. Low-velocity material with 3 km/s is likely to the present-day

accretionary prism consisting of unconsolidated sediments. A corresponding onshore investigation by Iwasaki et al. (1994) is described above in the subchapter 8.7.3 (Japan).

The continental margins around Australia became a special target of marine seismic profiling from 1985 to 1989 by the research ship *Rig Seismic*. The vessel had been leased in 1984 by the Australian government to conduct seismic profiling and other geological research of the structure and architecture of the offshore sedimentary basins. At that time *Rig Seismic* was one of a few vessels worldwide which was capable of undertaking a full range of geoscientific data collection including multichannel seismic data (Finlayson 2010; Appendix 2-2).

The first cruise of *Rig Seismic* in 1985 targeted the Lord Howe Rise in the central Tasman Sea, and in cooperation with

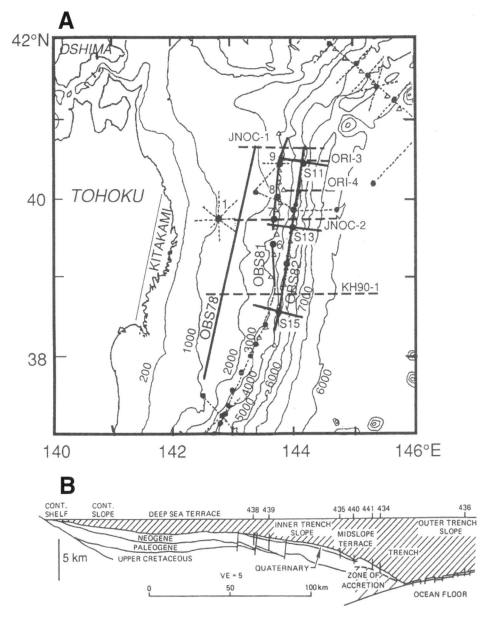

Figure 8.9.2-03. Japan Trench area (from Suyehiro and Nishizawa, 1994 Figure 2). (A) Map showing seismic lines observed from 1978 to 1990. Thick dashed lines—multichannel seismic recording lines, thick solid lines—OBS lines, discussed in detail by the authors, thin dashed lines—other OBS data, solid circles—OBS locations, open triangles—dynamite shotpoints. (B) Schematic explanatory section across northern Japan and adjacent trench. Numbers 436 to 441—DSPD drilling sites. [Journal of Geophysical Research, v. 99, p. 22,331–22,347. Reproduced by permission of American Geophysical Union.]

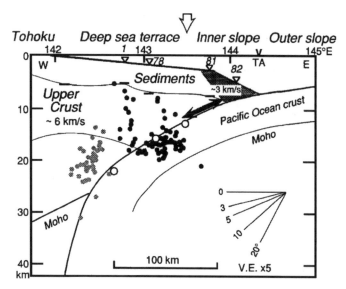

Figure 8.9.2-04. Cross section of the northern Japan Trench area with important structural boundaries and seismicity (from Suyehiro and Nishizawa, 1994, fig. 15). [Journal of Geophysical Research, v. 99, p. 22,331–22,347. Reproduced by permission of American Geophysical Union.]

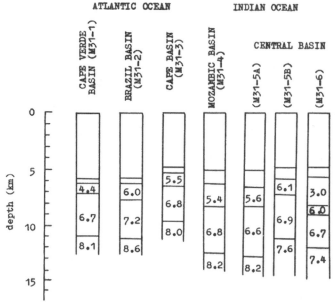

Figure 8.9.3-01. Averaged crustal columns for seismic surveys in the Indian and Atlantic Oceans during the 31st cruise of R/V *Dmitry Mendeleev* (from Neprochnov, 1989, fig. 5). [*In* Mereu, R.F., Mueller, St., and Fountain, D.M., eds., Properties and processes of earth's lower crust: American Geophysical Union, Geophysical Monograph 51, p. 159–168. Reproduced by permission of American Geophysical Union.]

the German R/V *Sonne*, 1250 km of seismic-reflection data were acquired ~200 km southeast of Lord Howe island. In 1985 and 1987, the cooperative two-ship research of *Rig Seismic* and *Sonne* continued in the offshore part of the Otway Basin off southern Victoria, Australia, collecting ~3500 km of multi-channel seismic data (e.g., Williamson et al., 1989). In 1986, the Great Australian Bight off Victoria and South Australia was investigated by two cruises, collecting 3500 km of seismic profiling data. Several cruises of *Rig Seismic* targeted the continental margins of northeast Queensland in 1985, 1987, and 1989, and from 1987 to 1989 the Lord Howe Rise was again visited, collecting seismic data over the Gippsland Basin, Lord Howe Rise, and Tasman Basin. In 1988, 1750 km of multichannel seismic data were also collected along a survey targeting the western margin of Tasmania and the deep abyssal plain of the Southern Ocean (Finlayson, 2010; Appendix 2-2). Other cruises of the AGSO vessel *Rig Seismic* investigating the western continental margins of Australia are being discussed in the following sub-chapter 8.9.3.

8.9.3. Indian Ocean

Rounding South Africa, cruise 31 of R/V *Dmitry Mendeleev* of the Shirshov Institute of Oceanology from 1983 to 1984 covered both the South Atlantic Ocean and the Indian Ocean. In 1984, it had reached the Indian Ocean (Neprochnov (1989). Here, one profile with three OBS was shot in the Mozambique basin and five profiles with 12 OBS deployments in the central basin south of India (see Fig. 8.9.2-01). The results are shown as crustal columns in Figure 8.9.3-01.

East of the Cape on the Agulhas Bank off the Indian Ocean coast of South Africa, in 1985, a 12 s TWT seismic-reflection profile of 46 km length was surveyed (Durrheim, 1987). In contrast to a typical BIRPS profile around the British Isles, the crust underneath the Agulhas Bank off South Africa is characterized by strong upper crustal reflectors ascribed to Proterozoic sediments, while in the lower crust, only a few reflectors were visible. The Moho proper was evident by a zone of strongly reflecting elements at 9.5 s TWT.

Off the coast of Oman, a coincident normal incidence and wide-angle seismic experiment was carried out in 1986 during cruise 18/86 of RRS *Charles Darwin* across the east Oman continental margin (Fig. 8.9.3-02) north of the Masirah Island ophiolite (Barton et al., 1990). The normal incidence reflection profile served to define the upper crustal structure, while the wide-angle reflection profile was obtained with ten digital OBSs recording 110 explosive shots. A steep landward-dipping reflector underneath the continental slope was interpreted by Barton et al. (1990) as Moho (Fig. 8.9.3-03). Beyond the slope, under the deep offshore Masirah basin, the crustal thickness resulted to be as small as 5 km. Depth to Moho increases further out to 14 km under the Masirah ridge, but thins to ~6 km under the oceanic crust beginning 40 km west of Owen Ridge (Fig. 8.9.3-02).

In 1985, the Australian vessel *Rig Seismic* conducted a continental margins cruise to the Kerguelen Plateau in the central Indian Ocean around Heard Island. The cruise included deep seismic profiling to determine the nature and extent of sedimen-

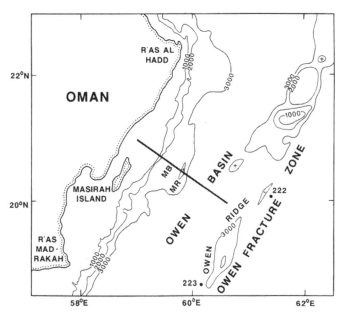

Figure 8.9.3-02. Bathymetric map of the east Oman offshore area with position of wide-angle seismic line (from Barton et al., 1990, fig. 3). MB—Masirah basin, MR—Masirah ridge, 222, 223—DSDP (Deep Sea Drilling Project) holes. [Tectonophysics, v. 173, p. 319–331. Copyright Elsevier.]

tary basins across the region and the underlying crustal architecture. About 5600 km of 48 channel (12-fold) seismic-reflection data were collected (Ramsay et al., 1986; Finlayson 2010; Appendix 2-2).

In another major project, the *Rig Seismic*, in cooperation with the Lamont vessel *Conrad*, investigated in 1986 the margin of offshore northwestern Australia (Mutter et al., 1989; Lorenzo et al., 1991; Finlayson 2010; Appendix 2-2). Here the Exmouth Plateau forms an unusually broad region of continental crust which was deformed during a Jurassic rifting period that preceded Early Cretaceous sea-floor spreading in the adjacent Indian Ocean (e.g., Veevers and Cotterill, 1978). The data included two-ship reflection and refraction measurements. Several ESPs were acquired along each reflection line. Together with velocities obtained by the analysis of CDP gathers, velocity-depth structures were derived and ages assigned to the major tectonic-seismic units using drilling results from near-by ODP Site 763. For some of the lines the whole crust was sampled, for others

the upper 5 km of sedimentary section down to an inferred extensional detachment surface was modeled. The refraction data resolved the presence of a 5–10-km-thick 7.3 km/s high-velocity layer in the lower crust beneath a strike-slip deformation zone. The total crustal thickness increased from ~8 km underneath the Cuvier Basin in the southwest, which is overlain by 5 km of water, reached up to 25 km under the transitional strike-slip deformation zone and decreased to ~20 km under the Exmouth Plateau. The crust under the Exmouth Plateau was modeled to be subdivided into a 10-km-thick crystalline crust and 8–10-km-thick detachment complexes. From this result it was inferred that the evolution of the transform was attended by intense magmatism (Lorenzo et al., 1991).

In 1986 and 1988, the North and South Perth Basins were investigated by cruises of the Australian vessel *Rig Seismic*. In the North Perth Basin, 2370 km of 48-channel seismic-reflection profiling data together with sonobuoy data were collected from four traverses in 1986. In the South Perth Basin, a total of 4000 km of multichannel seismic data were acquired in 1988, mostly on the continental shelf but with several lines extending offshore to the deep abyssal plain of the Indian Ocean (Finlayson 2010; Appendix 2-2).

8.9.4. Atlantic Ocean

While the investigations of the Earth's crust in the Pacific and Indian Ocean were carried out exclusively by institutions specializing in marine research work, in the Atlantic Ocean, groups whose expertise had been gained by research work on land were also active.

In the South Atlantic Ocean, the above mentioned cruise 31 of R/V *Dmitry Mendeleev* of the Shirshov Institute of Oceanology had mainly served the purpose of investigating oceanic basins worldwide (Neprochnov (1989). The expedition started in 1983 in the Cape Verde basin with one profile and five OBSs, continued southward to the Brazil basin where again one profile with three OBS was shot, and reached the Cape basin west of southernmost Africa in early 1984 where two profiles with five OBSs were recorded (see Figs. 8.9.2-01 and 8.9.3-01).

The North Atlantic Transect (NAT Study Group, 1985; McCarthy et al., 1988; Mithal and Mutter, 1989) had provided a major improvement on the knowledge of the structure of the oceanic crust and its variability on a large regional scale. In an effort to define the seismic properties of the oceanic crust in detail

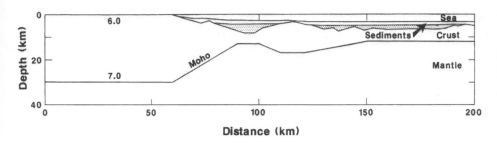

Figure 8.9.3-03. Crustal model of Oman continental margin (from Barton et al., 1990, fig. 11). [Tectonophysics, v. 173, p. 319–331. Copyright Elsevier.]

and on a regional scale, two-ship multichannel seismic data acquisition techniques as described by Stoffa and Buhl (1979) were employed along a transect across the western North Atlantic from the North American continental margin to the Mid-Atlantic Ridge near 23°N. The experiment crossed oceanic crust spanning almost 200 m.y. in age. The two-ship acquisition phase resulted in 3880 km of wide-aperture, CDP data and eleven ESPs at selected intervals (NAT Study Group, 1985).

Other spectacular images of the internal structure of the oceanic crust, showing widespread occurrence of intracrustal reflectivity in the western Central Atlantic Ocean had been published by White et al. (1990). Seismic-reflection studies of the Blake Fracture Zone had provided a complex detailed structure, and crustal thinning underlain by an anomalous upper mantle across this fracture zone had been interpreted by Mutter et al. (1985) which was confirmed later by new seismic data and velocity control from ESPs (see Fig. 8.9.1-01; White et al., 1990).

In 1982, the Kane fracture zone, offsetting the Mid-Atlantic Ridge rift valley near 24°N ~160 km in a left-lateral sense, was studied again by three seismic-refraction experiments (Cormier et al., 1984). They were positioned (Fig. 8.9.4-01) so as to extend and complement the 1977 experiment (Detrick and Purdy, 1980) which had been conducted at ~44°W (dashed line in Fig. 8.9.4-01). The westernmost line was 150 km long and was shot down the median valley of the Mid-Atlantic Ridge south of the Kane fracture zone. The two easternmost lines were shot to investigate whether the anomalously thin crust of only 2–3 km, found in the 1977 experiment along 44°W (Detrick and Purdy, 1980), was present only adjacent to the Kane fracture zone ridge or if it extended also into the older parts of the fracture zone. Shots of 14.5 kg and sometimes 112.6 kg were fired along the lines with a spacing of 1.5 and 4.5 km, respectively. Some of the lines were reshot with an airgun, using a shot spacing of 100 m to improve the resolution of the shallow crust.

The data analysis showed strong lateral variations in crustal thickness and velocities. Along most of the eastern fracture zone trough, upper-mantle velocities were found at depths of only 2–3 km below the seafloor, less than half the typical depth to Moho

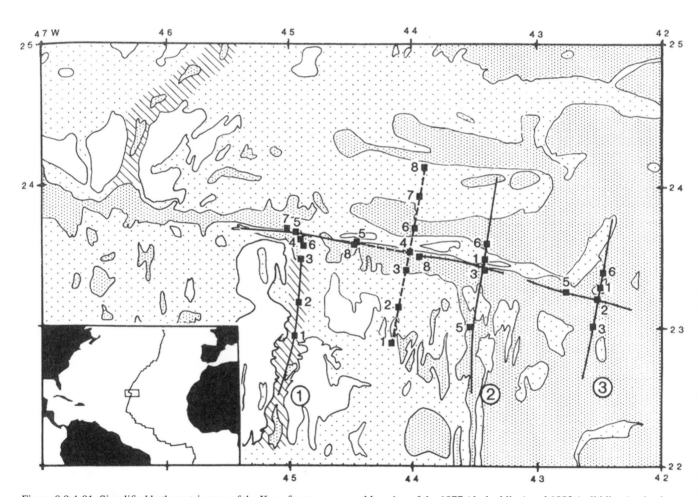

Figure 8.9.4-01. Simplified bathymetric map of the Kane fracture zone and location of the 1977 (dashed line) and 1982 (solid lines) seismic refraction experiments. Black squares—receiver positions; diagonal lining—Mid Atlantic Ridge rift valley; densely stippled areas—depths greater than 4000 m; lightly stippled areas—depths from 3000–4000 m (from Cormier et al., 1984, fig. 1. [Journal of Geophysical Research, v. 89, p. 10,249–10,266. Reproduced by permission of American Geophysical Union.]

in the ocean basins. This anomalous fracture zone crust was generally characterized by low P-wave velocities, relatively high velocity gradients and a distinct crust-mantle boundary. The very thin crust appeared to be confined to the deepest parts of the Kane fracture zone (less than 10 km wide), although a more gradual crustal thinning seemed to extend up to several tens of kilometers from the fracture zone.

Inactive sections of the Tydeman fracture zone at latitude 36°N with approximate crustal ages of 59 Ma and 71 Ma were the target of two separate experiments in 1982 (Calvert and Potts, 1985). The first experiment consisted of two two-ship ESPs located at ~23°30'W. In the second experiment two OBS were laid ~50 km apart in the fracture valley and RRS *Shackleton* fired 24 explosive charges varying in size from 12.5 to 250 kg. The shooting track was centered on 36°N, 26°W, running from west to east along the strike of the fracture zone and passing over the receivers. The resulting crustal velocities were low compared with "normal" oceanic crust; there was a marked absence of an identifiable layer 3, the depth to Moho was shallow and anomalously low velocities of 7.2–7.5 km/s were observed. Using these results and those from other surveys, Calvert and Potts concluded that the seismic velocity in the upper mantle beneath fracture zones systematically decreases with increasing age.

A review of seismic studies on crustal structure of North Atlantic Fracture Zones conducted in the 1970s and 1980s was prepared and published by Detrick et al. (1993b).

A detailed and extensive survey along the Atlantic continental margin of Africa was undertaken by project PROBE, an academia-industry project aimed at understanding the evolution from continental rift to passive rift margin, and from passive continental margin to oceanic crust (Rosendahl et al., 1991). Near the end of the 1980s a grid of multifold seismic-reflection profiles was acquired in the Atlantic Ocean between the Cameroon volcanic line and the coasts of Cameroon, Equatorial Guinea, and Gabon (Fig. 8.9.4-02).

The PROBE lines crossed several major sedimentary basins which all formed as continental rift basins during the early Cretaceous opening of the southern and equatorial Atlantic Ocean. For the data acquisition a 6000 m long cable with a 100–1000 m offset and a tuned airgun array were used. The six-source subarrays had a width of 75 m and a length of up to 100 m. The nominal fold coverage was 60, and common-depth spacing was 25 m. The record lengths were 20 s. Over most of the survey area the reflection Moho was clearly imaged at depths ranging from 7 to 10 s TWT over most of the survey area east of the Cameroon volcanic line. The Moho reflection usually consisted of a 2–4 Hz event and was in many areas the highest amplitude event. On many profiles, the Moho could be continuously traced from oceanic crust to the shoreward ends of the lines, well beyond the assumed transition from oceanic to continental crust. In many cases, the two-way traveltimes remained fairly constant from oceanic to continental crust. The spatial variation in thickness of the igneous component of the oceanic crust was extremely large; even doubling of the crust occurred

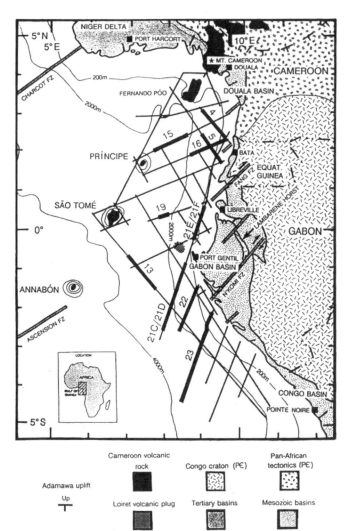

Figure 8.9.4-02. Location of the PROBE survey grid along the African coast between 4°N and 5°S (from Rosendahl et al., 1991, fig. 1). [Geology, v. 19, p. 291–295. Reproduced by permission of the Geological Society of America.]

over distances of a few kilometers. Oceanic basement and the depth to the Moho was greater south of latitude 1°S than north of this latitude. In some profiles, but not on all, the igneous crust appeared to thicken at the transition from continental to oceanic crust (Rosendahl et al., 1991).

The data were later reprocessed and a tectonic model of the continental margin between Cameroon and southern Gabon was developed (Rosendahl and Groschel-Becker, 2000). Under the North Douala Basin, north of the Kribi Fracture Zone, a 75-km-wide transform fault trending NE-SW and intersecting with the coastline at 2–3°N, oceanic crust extended essentially to the coastline. The reflection Moho was relatively weak but continuous here and the oceanic crustal thickness under the North Douala Basin appeared rather uniform, averaging ~1.75 s in TWT. South of the Kribi Fracture Zone, under the Gabon Basin, oceanic crust appeared to be offset ~350 km to the southwest, resulting in a broad rift margin off Gabon. Here, a single continuous event representing the reflection Moho could not be delineated.

Also the *Meteor* (1964) cruise M67 in 1984 under the direction of W. Weigel (University of Hamburg) and K. Hinz (BGR Hannover) provided an extended crustal seismic survey off Northwest Africa (Sarnthein et al., 2008). Banda et al. (1992) had carried out a geophysical study of the oceanic crust in the eastern Central Atlantic Ocean, in the Canary basin west of the Canarian Archipelago (Fig. 8.9.4-03) which is shown here as an example. The crust is ~9–10 km thick, and the multichannel reflection data showed a similar complex internal structure as reported by White et al. (1990) for the western Atlantic Ocean (Fig. 8.9.4-04).

In 1988, an extensive geophysical data set was collected over the Josephine Seamount, located at the NE end of the Madeira-Tore Rise in the eastern North Atlantic. A 275-km-long seismic-refraction line, running in a NW-SE direction between 38°11′N, 14°53′W and 35°57′N, 13°17′W, crossed the Josephine Seamount and the northwestern Horseshoe Abyssal Plain (Peirce and Barton, 1991). Six DOBS (digital ocean-bottom seismometers) were deployed with both 3-component geophone packages and hydrophones and recorded 110 explosive charges (25 and 100 kg) at ~1.8 km intervals. In addition an airgun array fired at 60 s intervals provided a shot spacing of ~150 m. The crust at either side of the seamount was typically oceanic in character. Beneath the seamount, however, a region of anomalously high velocity and crustal thickening to a depth of 17–18 km was found.

Various investigations of fracture zones offsetting slow-spreading ridges, particularly in the North Atlantic Ocean (e.g., Cormier et al., 1984; Fowler, 1976, 1978; Mutter et al., 1985; White and Whitmarsh, 1984) prompted Whitmarsh and Calvert (1986) to conduct in 1982 a seismic experiment in the Charlie-Gibbs Fracture Zone at 52°–53°N which consists of two narrow, deep E-W transform valleys which together offset the median valleys of the Reykjanes and Mid-Atlantic Ridge by 350 km. The objectives of the experiment were to determine the crustal structure within one of the transform valleys and to measure the

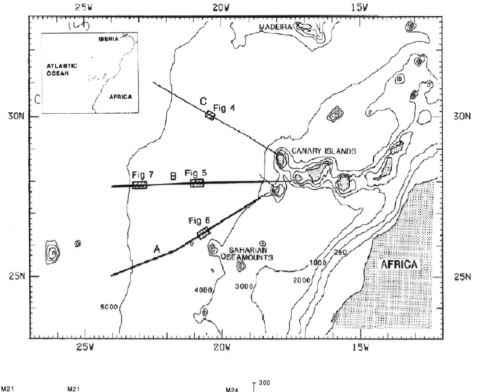

Figure 8.9.4-03. Location of seismic profiles in the Canary basin (from Banda et al., 1992, fig. 1). [Geological Society of America Bulletin, v. 104, p. 1340–1349. Reproduced by permission of the Geological Society of America.]

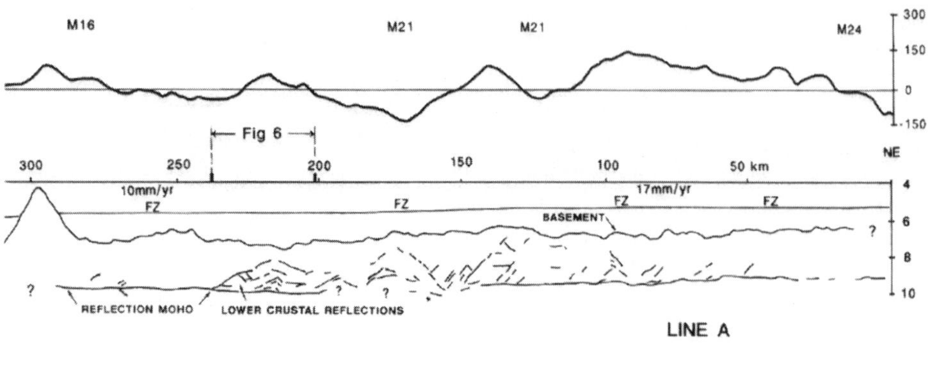

Figure 8.9.4-04. Crustal cross section of the Canary basin along line A, northeastern part (from Banda et al., 1992, fig. 3). [Geological Society of America Bulletin, v. 104, p. 1340–1349. Reproduced by permission of the Geological Society of America.]

rate of crustal thinning normal to the valley axis. Ocean-bottom seismometers recorded airgun shots along two perpendicular profiles, an 84-km-long refraction line along the valley axis and a 94-km-long line N-S profile through the center of the E-W line. A third E-W line, 90 km long, was recorded south of the fracture zone over normal oceanic crust. The crust in the fracture zone appeared to be anomalously thin (4 km); it thinned along the N-S line over a distance of 40 km from 8 to 5 km toward the fracture zone where the belt of anomalous crust was 12 km wide.

Three other groups may be mentioned in particular. In the UK, university research centers and the British Geological Survey connected land and sea crustal research work out of which the BIRPS program developed. In Germany, the University of Hamburg under Jannis Makris had been active both on land and at sea since the early 1970s. In Ireland, it was the Dublin Institute for Advanced Studies at Dublin, Ireland, under Brian Jacob which did major crustal research in continental areas as well as in the surrounding seas.

The 200-km-long Goban Spur continental margin southwest of Britain was investigated in 1985 by the Western Approaches Margin (WAM) deep seismic-reflection profile, recorded jointly by BIRPS and ECORS (Peddy et al., 1989; Pinet et al., 1991). The profile was shot perpendicular to the margin (for location, see Fig. 8.3.1-01) and was aimed to image the structure of the thinned continental crust. Beyond the shelf edge, it was only possible to image the layered lower crust clearly above the first seabed multiple. In 1987, a seismic-refraction experiment was added on the Goban Spur margin to define the crustal structure and the extent of igneous rocks in order to complement the information provided by the deep seismic-reflection data (Horsefield et al., 1994). The project consisted of three seismic-refraction lines, one along the WAM line and two perpendicular to the WAM line, parallel to the margin. Several DOBS were deployed on each line which recorded airgun shots, spaced ~300 m. The project also included 500 km of four-channel normal incidence seismic-reflection profiles to provide control of the sediment structure and depth to the acoustic basement. From traveltime and amplitude modeling of the wide-angle OBS profiles, Horsefield et al. (1994) concluded that the continental crust thins in a seaward direction to ~7 km before it breaks to allow a new oceanic spreading center to develop. They also stated that the thinning was accompanied by only limited igneous intrusion and extrusion of the upper crust and that the adjacent oceanic crust is abnormally thin (5.4 km).

In 1985 a two-ship Synthetic Aperture Profile (SAP-5) was shot from the Rockall Bank into the center of the Rockall Trough (Joppen and White, 1990). The 1985 seismic-reflection operation included a single ship reflection profile N-11 and two expanding spread profiles (ESP 11 and 12) which centered on the intersection of the lines SAP-5 and N-11. In 1990, BIRPS recorded the WEST-LINE profile. This line ran parallel to the SAP-5 line and also crossed the Rockall Trough (England and Hobbs, 1995; Snyder and Hobbs, 1999). It was a 450-km-long, normal-incidence, deep seismic-reflection profile which was shot perpendicular to the edges of the trough in order to image its conjugate margins.

The offshore study of the COOLE experiment (for locations, see Fig. 8.3.2-05), a joint project of the Dublin Institute for Advanced Studies and the University of Hamburg, comprised three seismic-refraction profiles offshore southern Ireland (Makris et al., 1988; O'Reilly et al., 1991). Profile COOLE'85-7, along the axis of the North Celtic Sea Basin, comprised ten OBS stations, which were deployed at 15–20 km intervals over a distance of 160 km ~40 km south of the position of the CMRE 1969 profile (CMRE'69 in Fig. 8.3.2-05; lines A in Fig. 6.2.3-01; Bamford, 1971). Sea shots were fired every 2.5 km. The transverse profile COOLE'85-6 comprised eleven OBS stations deployed at 14–16 km intervals over a distance of 140 km, with airguns used as a seismic source. The main objectives were to investigate lithospheric structures related to rifting/stretching during the development of the North Celtic Sea and Porcupine basins. In addition, profiles COOLE'85-3a and -3b were laid out in a NE-SW direction (Makris et al., 1988). Profile COOLE'85-3a (for location, see Fig. 8.3.2-05) served to determine the structure of the transition region between the Porcupine Seabight and SW Ireland. Profile COOLE'85-3b was recorded in the prolongation of COOLE'85-3a farther southwest between longitudes 13°W and 15°W and approximately at latitude 50°N; i.e., it reached from the Porcupine basin (for location, see Fig. 8.9.4-05) into the Porcupine abyssal plain to the west.

Another enterprise of the Dublin Institute for Advanced Studies and the University of Hamburg was the RAPIDS (Rockall and Porcupine Irish Deep Seismic) project, shot in 1988 and 1990 (Figs. 8.3.2-05 and 8.9.4-05). It consisted of two orthogonal wide-angle seismic profiles totaling 1600 km.

Individual lines were typically 200–250 km long. Ocean-bottom seismometers were deployed every 7–10 km and explosive charges of 25 kg were fired at optimum depth of 65 m over intervals of ~1 km. Energy propagated usually farther than 150 km. Data examples are shown in Appendix A8-5-1.

The model along the 1000-km-long east-west profile from Ireland to the Iceland Basin shows the complex structure when crossing the various troughs and basins and intervening banks (Fig. 8.9.4-06). The low-velocity mantle with velocities of 7.5–7.8 km/s underneath the Rockall Trough was interpreted as serpentinized mantle, the still lower velocity of 7.2 km/s west of the Hatton basin as underplated body (Fig. 8.9.4-06). The Moho appeared as a transition zone and is shallow under the basins (with 12–15 km depth, Figs. 8.9.4-06 and 8.9.4-07), but reaches continental dimensions under the Rockall bank (Shannon et al., 1994, 1999; Hauser et al., 1995; O'Reilly et al., 1995, 1996; Vogt et al., 1998).

Prior to the RAPIDS project, in 1985 Cambridge, Durham and Birmingham universities had undertaken a detailed geophysical study of the Hatton bank volcanic margin over a SE-NW–trending line of ~150 km length between the Rockall plateau and the Iceland basin, centered around 59°N and 18.5°W (Morgan and Barton, 1990). The experiment included expanding spread profiles, parallel and synthetic aperture profiles, a wide-angle ocean-bottom seismometer and a multichannel refraction line

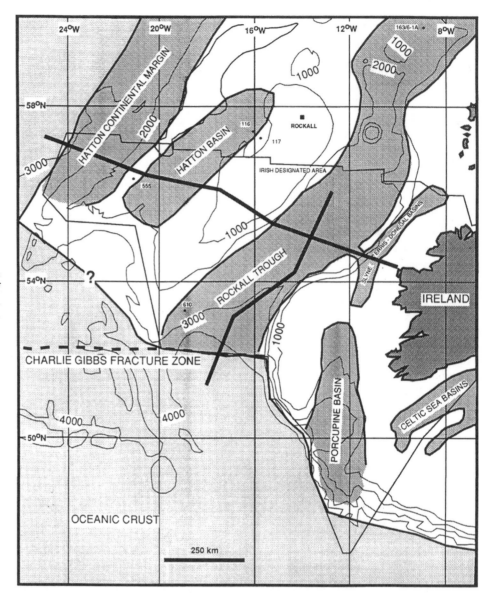

Figure 8.9.4-05. Location of seismic profiles in the Atlantic Ocean west of Ireland (from Shannon et al., 1994, fig. 1). [First Break, v. 12, p. 515–522. Copyright EAGE.]

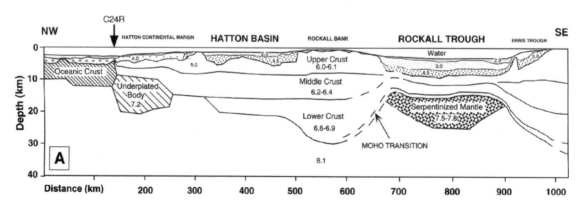

Figure 8.9.4-06. WNW-ESE–directed crustal cross section from Ireland across Rockall Trough, Hatton Basin to the Iceland Basin, Atlantic Ocean (from Shannon et al., 1999, fig. 2A). For location see Fig. 8.9.4-05. [*In* Fleet, A.J., and Boldy, S.A.R., eds., Petroleum Geology of Northwest Europe: Proceedings, 5th Conference, p. 421–431, Petroleum Geology, v. 86, Geological Society of London. Reproduced by permission of Geological Society Publishing House, London, U.K.]

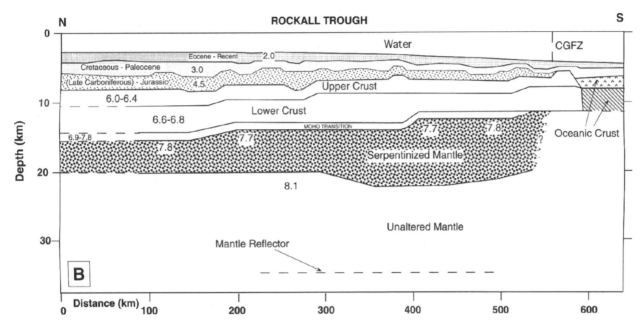

Figure 8.9.4-07. NNE-SSW directed crustal cross section through the Rockall Trough, Atlantic Ocean, west of Ireland (from Shannon et al., 1999, fig. 2A). For location see Fig. 8.9.4-05. [*In* Fleet, A.J., and Boldy, S.A.R., eds., Petroleum Geology of Northwest Europe. Proceed. 5th Conf.: p. 421–431. Petroleum Geology 86, Geological Society of London. Reproduced by permission of Geological Society Publishing House, London, U.K.]

perpendicular to the margin, and a multichannel refraction line through the midpoints of the expanding spread profiles and 1400 km of normal incidence seismic observations. The resulting velocity model showed a gradual depth decrease of Moho from ~27 km under the Rockall plateau to 15 km toward the Iceland basin.

Farther north, off the coast of Norway on the Voering plateau, north of the Jan Mayen fracture zone, a wide-aperture CDP line was recorded across the midpoints of a series of ESP profiles (Fig. 8.9.4-08; see Fig. 8.9.1-01 for definitions of CDP and ESP) using the two-ship multichannel seismic techniques. The crustal structure was determined from a series of expanding spread profiles with midpoints along a wide-aperture CDP transect which also intersected two deep-drilling sites (DSDP) and was plotted by Zehnder et al. (1990) as a velocity contour diagram (Fig. 8.9.4-09). The data were interpreted to imply the presence of a 15–20-km-thick igneous crust that was emplaced along the margin during the initiation of seafloor spreading. Seaward, the igneous crust diminishes in thickness.

At the same time, in 1988, farther northeast a major seismic-reflection/-refraction experiment including long-offset refraction profiles along the Lofoten passive margin was started. It was a cooperative project among Norwegian, German, and Japanese institutes to investigate the detailed structure of the continental breakup. In particular, the OBS method was applied (Mjelde et al., 1992; Goldschmidt-Rokita et al., 1994; Kodaira et al., 1995). The program was continued in the 1990s, also covering the Voering marginal high and amongst other lines also re-shooting the line of Zehnder et al. (1990). In Chapter 9.8.4,

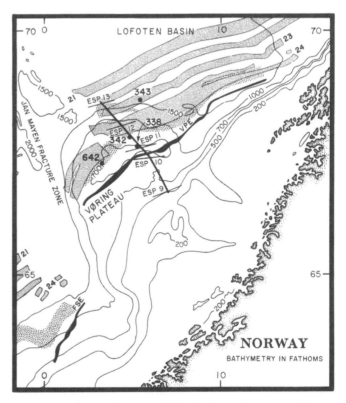

Figure 8.9.4-08. Location map of seismic observations on the Voering plateau off Norway (from Zehnder et al., 1990, fig. 2). VPE is Voering plateau escarpment; dots are deep sea drilling sites; dense dot pattern identifies seafloor magnetic lineations. [Tectonophysics, v. 173, p. 545–565. Copyright Elsevier.]

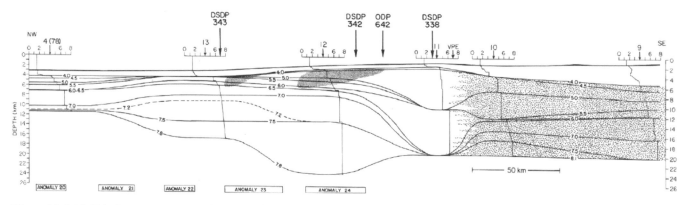

Figure 8.9.4-09. Velocity contours across the Norwegian continental margin (from Zehnder et al., 1990, fig. 6). VPE—Voering plateau escarpment. [Tectonophysics, v. 173, p. 545–565. Copyright Elsevier.]

measurements and results of this survey are summarized (see, e.g., Figs. 9.8.4-18 and 9.8.4-22).

In 1987, the Aegir rift was investigated by two crossing profiles (Grevemeyer et al., 1997). It is located half-way between Norway and Iceland and to the north of the Faeroe Islands at 3°–5°W, 65°–67°N and trends in a SW-NE direction (for location, see Fig. 9.8.4-18). The Aegir rift is a graben structure and centered between the Jan Mayen Ridge and the Faeroe-Shetland Escarpment southwest of Norway and is regarded as an extinct spreading center. Ocean-bottom hydrophones and seismographs were deployed along two reversed lines. One profile of 90 km length with five OBSs was shot along the rift's median valley; the other one was 200 km long and oriented perpendicular to the strike and was occupied with ten OBSs. The shooting interval of the airgun array was 2 min; spacing between the shots ~330 m. In addition, explosives were detonated at 2 km intervals. The crust within the extinct spreading center was found at 10 km depth to be thinner and of lower velocity compared to the crust sampled off-axis at 11 km depth below sea level, water depth being ~3 km. Upper mantle velocities were 8 km/s.

In 1985, deep seismic sounding measurements were performed in the transition zones west and north of Spitsbergen (also named Svalbard) by a Polish expedition (Czuba et al., 1999). The recordings were made by stationary land stations (stars in Fig. 8.9.4-10, upper left), which recorded all 105 underwater shots detonated along five profiles. Two shot profiles were recorded to the west toward the Knipovich Ridge (K1 and K2 in Fig. 8.9.4-10) and two profiles to the north toward the Yermak Plateau (C1 and C2 in Fig. 8.9.4-10). Furthermore a shot profile C1 was recorded along the west coast of Spitsbergen, where in 1976 and 1978 the first experiments had been carried out. For comparison the map shows Moho depths by large numbers in quadrangles as obtained by Czuba et al. (1999) and Moho depths by small numbers in small circles and triangles as had been obtained by the earlier expeditions (Faleide et al., 1991; Sellevoll et al., 1991).

Eastern Greenland and its margin on the western side of the northernmost Atlantic Ocean became the goal of detailed seismic mapping since the end of the 1980s which continued in the 1990s

(for locations, see Fig. 9.8.4-30). In 1988 and 1989 deep seismic-reflection data were acquired on Jameson Land in connection with oil exploration activities in eastern Greenland providing the first seismic depth information on the deeper crustal structure (Mandler and Jokat, 1998).

During 1988 the first combined marine-land refraction experiment was carried out in East Greenland (Weigel et al., 1995). The primary goal was to map the continent-ocean transition east of Scoresby Sund, hereby deploying six recording stations on the southern coast of the Sund. The results indicated strong variations in crustal thickness from at least 30 km in the western part of the Scoresby Sund to less than 20 km at its eastern termination (Weigel et al., 1995).

The western side of Greenland also became the goal of a seismic investigation in 1989 (Clement et al., 1994; Gohl et al., 1991; Gohl and Smithson, 1993). The survey was conducted by the University of Wyoming and consisted of a marine airgun array recorded by land-based receivers (Fig. 8.9.4-11). Fifteen PASSCAL RefTek 2-Hz three-component seismometers were placed on 35 locations along the coast, fjords, and farther inland. An array of five airguns (6000 in³) was fired with 100–150 m shot spacing along a north-south–striking offshore line, covering offsets of up to 350 km, and along the Godthaab and Ameralik fjords to obtain data farther inland (Gohl et al., 1991). An excellent quality of first and secondary-arrival P-phases was obtained, and strong S-phases were recorded. For the P-phases Moho depths between 30 and 40 km resulted (Gohl et al., 1991; Gohl and Smithson, 1993). From the S-wave analysis (Clement et al., 1994), the continental crust appeared to be clearly seismically anisotropic (Fig. 8.9.4-12).

Several projects dealt with the Arctic Ocean. The Canadian expedition CESAR (Canadian Expedition to Study the Alpha Ridge) explored the Alpha Ridge complex in the Amerasia basin in 1983 with a seismic-refraction survey (Asudeh et al., 1988; Forsyth et al., 1986; Jackson et al., 1986; Weber, 1986, 1990). The Amerasia basin is dominated in the north by the Alpha-Mendeleev Ridge complex which is the most prominent and least understood of the Arctic submarine ridges. The ice station

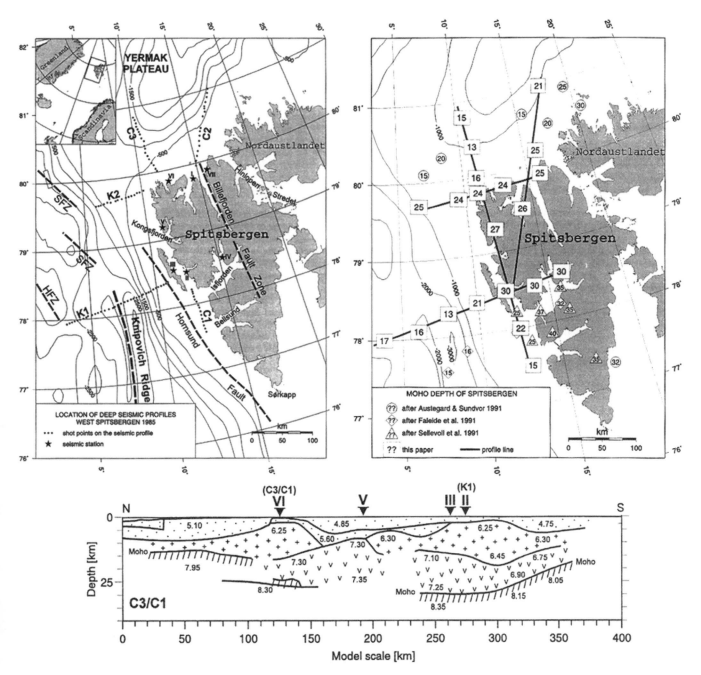

Figure 8.9.4-10. Top: Locations and Moho depths of the 1985 Spitsbergen survey (from Czuba et al., 1999, figs. 1 and 11). Bottom: Crustal cross section along the profiles C1/C3 (from Czuba et al., 1999, fig. 9). [Polish Polar Research, v. 20, p. 131–148. Reproduced with permission of Polish Academy of Sciences, Warsaw, Poland.]

CESAR was set up on the drifting Arctic sea ice in the vicinity of the Alpha Ridge and experiments were conducted from March to May 1983. A refraction line, 210 km long, was shot along the crest of the Ridge. The resulting crustal model showed a nearly 40-km-thick crust. A velocity of 6.45 km/s with a steep gradient was measured below layer-2 at the ends of the line and in the central portion an additional velocity of 6.8 km/s was obtained. Beneath this, a high-velocity layer of 7.3 km/s, 10 and 16 km

thick, occurred along the entire line. The layer-2 and -3 velocities appeared typical of oceanic crust but several times thicker than in ocean basins. The model for the Ridge suggested many similarities with the region of Iceland.

In 1989 a seismic-refraction experiment was carried out in the Makarov Basin of the Arctic Ocean (Sorokin et al., 1999). Ten 6-channel analog recorders with array lengths of 200–500 m were placed on drifting ice by helicopter with intervals of 10–

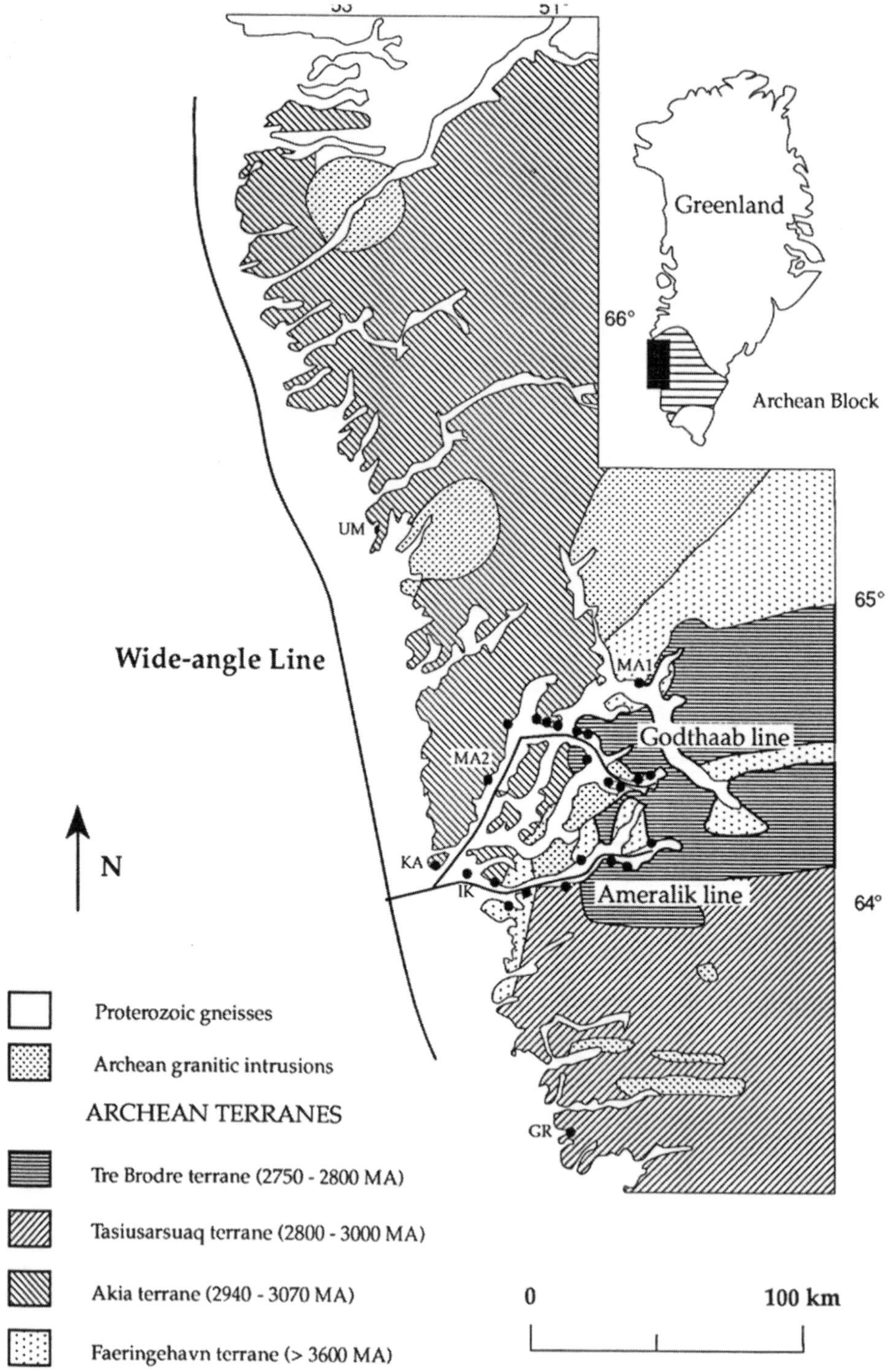

Figure 8.9.4-11. Location of the 1989 survey along the southwest coast of Greenland (from Clement et al., 1994, fig. 1). Solid lines—positions of source array for wide-angle recordings; dots—land-based receiver stations. GR—Graedefjord; IK—Ikaarisat; KA—Kangeq; MA—Manistog; UM—Uman. [Tectonophysics, v. 232, p. 195–210. Copyright Elsevier.]

15 km. Five underwater shots were fired at a depth of ~50 m. Charge sizes varied from 100 to 500 kg. A split spread of 220 km was obtained with a total of 45 seismic traces with a high signal to noise ratio. The sedimentary and upper crustal arrivals were not constrained due to the paucity of shots near the origin; however, the lower crust and M discontinuity were resolved. Velocities of 6.7 km/s were determined for the lower crust and 8.0 km/s

for the upper mantle The 6.7 km/s velocity layer was interpreted as oceanic crust layer 3. The well-resolved thickness of this layer of 15 km was substantially greater than the average thickness of 4 km for layer 3.

In 1987–1989 the Soviet ice-station North Pole-28 (NP-28) drifted across the Arctic Ocean (for location, see Fig. 9.8.4-39). It started its crossing above the Podvodnikov Basin at ~81°N.

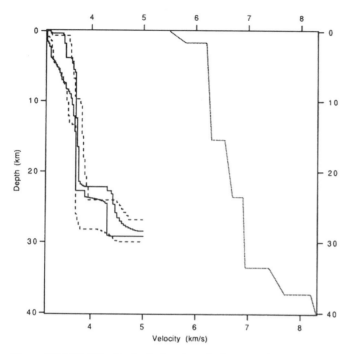

Figure 8.9.4-12. Velocity-depth functions for P-wave (dotted line) and S-wave (radial—solid line; transverse—dashed line) modeling of the 1989 survey along the southwest coast of Greenland (from Clement et al., 1994, fig. 4). [Tectonophysics, v. 232, p. 195–210. Copyright Elsevier.]

It passed near the North Pole and was finally abandoned above the Yermak Plateau (~81°N) north of Spitsbergen. NP-28 crossed the Lomonossov Ridge three times; on the second and third times it also traversed across the Marvin Spur (Lebedeva-Ivanova et al., 2006). The total length of the profile was ~4000 km. The reflection experiment was performed on drifting ice, with an ice-drift of 5–10 km/day, its direction being variable and unpredictable. Seismic receiver arrays were set up along two lines, perpendicular to each other in order to attenuate out-of-plane waves. The arms of the array were 545 m long. Receivers were placed at 50 m interval at each of the two arms. Shots were provided by 3–5 detonators in the vertex of the array at a water depth of 8 m and fired every 2–4 h, resulting in a shot rate of ~0.5 km intervals. Data were recorded for 12 s on magnetic analog tape. For the interpretation, reflection lines of the Alfred-Wegener Institute (Bremerhaven, Germany) were also used, which were recorded in 1990 and 1991 across the Makarov Basin and around site of the ACEX (Arctic Coring Expedition) drillholes, drilled in 2004, coring the first 430 m section on the Lomonossov Ridge near the North Pole. The data provided a basis for correlating a highly reflective sedimentary package and the basement along the Lomonossov Ridge.

In the western North Atlantic Ocean, most deep seismic research projects concentrated on the complex structure of the continent-ocean transition of North America. Large dynamic range and dense spatial sampling were aimed for.

The continental margin of Canada around Newfoundland was investigated from 1983 to 1987. In 1984 and 1986, Hall

et al. (1990) investigated dipping shear zones and the base of the marine part of the Appalachian crust west and north of Newfoundland as part of the LITHOPROBE East transect (see Chapter 8.5.2).

In 1985 and 1987 a series of deep seismic-reflection profiles was recorded on the continental shelf to the east of Newfoundland (de Voogd et al., 1990; Keen et al., 1990). The goal was the crustal structure underneath a major transform margin southwest of the Grand Banks, off Newfoundland (Fig. 8.9.4-13). A standard method of data acquisition was applied, but the 127-l airgun source was larger than normally used and the record length was 20 s. Near the edge of the shelf the Moho was well defined near 13 s TWT (de Voogd et al., 1990), converted to ~28 km depth (Keen et al., 1990). A line drawing of the data recorded along line 87-5, i.e., perpendicular to the transform trend, was discussed by Keen et al. (1990) in some detail (Fig. 8.9.4-14). It shows a reflective lower crust near the edge of the shelf (D) overlying the well defined Moho (C). Below a region where a band of reflections (F, G) could be correlated the Moho deepens by 7 km (E), but toward the landward northeastern end of the line it shallows again to 30 km. Under the continental slope to the southwest (I in Fig. 8.9.4-14) the Moho shallows rapidly within the continental shear zone (J). Seaward of the Fogo seamount the top of the oceanic basement is relatively flat (L), and ~6-7 km below the oceanic Moho (M) is found. K in Figure 8.9.4-14 may represent the top of one of the volcanic Fogo seamounts.

The marine database was later enlarged by additional profiles, one of which (line 7) coincided with LITHOPROBE line 85-2 and was discussed by Reid (1994) in some detail. After 2000, the project SCREECH (e.g., Funck et al., 2003; Lau et al., 2006a) added more data to the Flemish Cap margin (see Chapter 9).

A major seismic survey was undertaken across the southern edge of the Grand Banks off eastern Canada, south of Newfoundland (Reid, 1988; Todd et al., 1988). In 1983–1984, a SSW-NNE–directed refraction line was shot as a split spread to a closely spaced array of OBSs along the base of the continental slope at the southwest Newfoundland transform margin, a 900-km-long segment of the ocean-continent boundary south of the Grand Banks between 42 and 44°N, 51°–54°W. A second combined reflection and refraction line was added at right angles, together with some reflection profiles across the margin (Todd et al., 1988).

In 1984, this work was extended to a systematic refraction experiment across the southwest Newfoundland transform margin. Seven OBS were deployed and a pattern of split-spread refraction lines was shot subparallel to structure in order to obtain a detailed coverage of the ocean-continent transition. The seismic source consisted of two 32-l airguns which were fired every 60 s, giving a shot spacing of ~200 m. For the reflection profiles, for which data were acquired by a single-channel streamer, the shot rate was increased and the source reduced. The continental crust with an average P-velocity of 6.2–6.5 km/s with 30 km thickness beneath the southern Grand Banks thinned oceanward to a

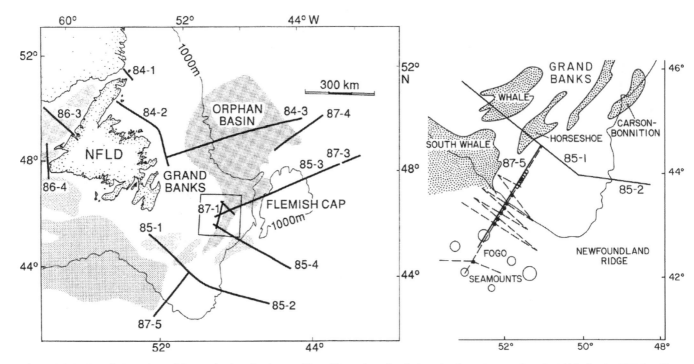

Figure 8.9.4-13. Left: Location of deep seismic reflection profiles off Newfoundland, Canada (from de Voogd et al., 1990, fig. 1). NFLD—Newfoundland. Right: Detailed geography for line 87-5 (from Keen et al., 1990, fig. 1). [Left: Tectonophysics, v. 173, p. 527–544. Copyright Elsevier. Right: Geological Society of America Bulletin, v. 100, p. 1550–1567. Reproduced by permission of the Geological Society of America.]

Figure 8.9.4-14. Line drawing of migrated seismic reflection data recorded along line 87-5 (from Keen et al., 1990, fig. 3) between the Grand Banks and the Fogo seamounts (see Fig. 8.9.4-13 for location). [Tectonophysics, v. 173, p. 527–544. Copyright Elsevier.]

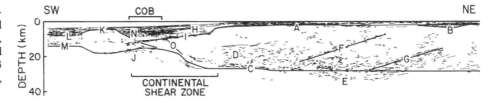

25-km-wide transition zone, where the Paleozoic basement of the Grand Banks (5.5–5.7 km/s) was interpreted to be replaced by a basement of oceanic volcanics and synrift sediments (4.5–5.5 km/s). Seaward of the transition zone the crust appeared oceanic in character, with a velocity gradient from 4.7 to 6.5 km/s and a thickness of 7–8 km. Oceanic layer 3 was not found, and no significant intermediate velocity material (larger than 7 km/s) was seen at the ocean-continent transition which was interpreted to mean that no underplating of continental crust has taken place. The continent-ocean transition across the transform margin appeared much narrower than across rifted margins.

Along the Atlantic continental margin of the United States one of the first of these experiments was the 1981 Large Aperture Seismic Experiment (LASE) during which ESPs and COPs were shot across the Baltimore Canyon (Fig. 8.9.4-15, top section) following approximately line B–B' in Figure 7.8.4-13 (LASE Study Group, 1986; Diebold et al., 1988; Trehu et al., 1989b).

Another large-aperture experiment in the Gulf of Maine in 1984–1985 (Hutchinson et al., 1987; line GMM on Fig. 8.5.3-24)

followed line H–H' (Fig. 7.8.4-13) and the large-aperture experiment of 1985 followed approximately line E–E' in Figure 7.8.4-13 in the Carolina Trough (Fig. 8.9.4-15, lower section). Line G–G' in Figure 7.8.4-13 traversed the Long Island Platform (Trehu et al., 1989a, 1989b). OBS profiles were also laid out, e.g., in 1985, both in the Gulf of Maine and above the Carolina Trough where a series of lines was shot parallel to the structure which complemented the CDP lines (Trehu et al., 1989b). Examples of interpretive cross sections are shown for the Baltimore Canyon Trough and the Carolina Trough (Fig. 8.9.4-15).

In 1988 a deep-penetrating multichannel seismic-reflection (MCS) survey was performed in the Southeast Georgia Embayment on the U.S. Atlantic continental margin consisting of six profiles totaling 1200 km in length (Fig. 8.9.4-16), using a 6-km-long, 240-channnel streamer (Oh et al., 1991). Ocean-bottom seismometers were deployed along the 250-km-long lines BA-3 and BA-6, of which line BA-3 was located entirely on the shelf (Lizarralde et al., 1994), while profile BA-6 aligned across the Carolina trough (Holbrook et al., 1994c). Over 5000

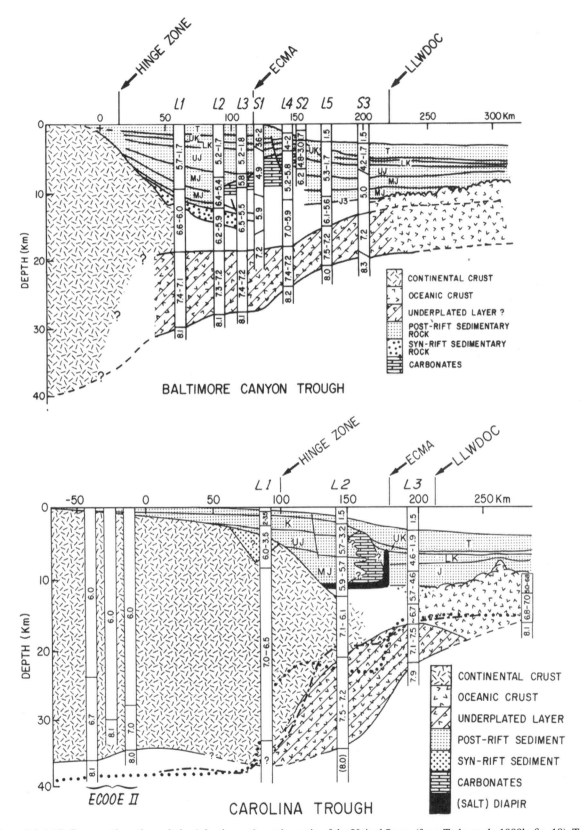

Figure 8.9.4-15. Cross sections through the Atlantic continental margin of the United States (from Trehu et al., 1989b, fig. 18). Top: Baltimore Canyon Trough obtained by LASE Study Group (1986) following line B–B' in Figure 7.8.4-13. Bottom: Carolina Trough based on data of the 1985 U.S. Geological Survey large-aperture experiment (Trehu et al., 1989a) following line E–E' in Figure 7.8.4-13. [Journal of Geophysical Research, v. 94, p. 10,585–10,600. Reproduced by permission of American Geophysical Union.]

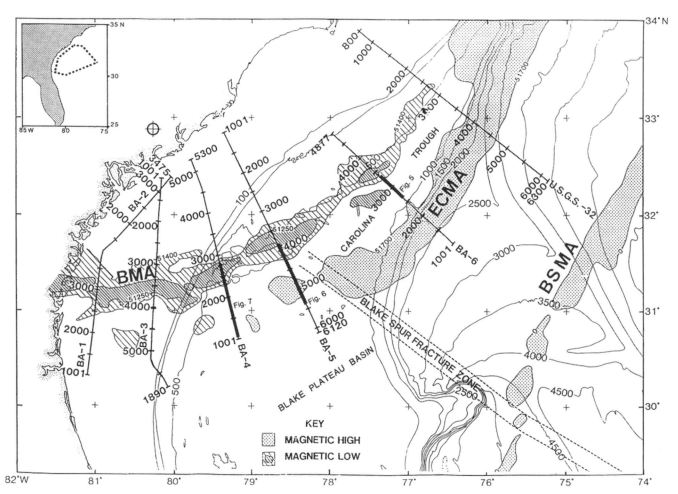

Figure 8.9.4-16. Location of deep-penetrating MCS reflection lines from the Southeast Georgia Embayment (from Oh et al., 1991, fig. 1). Prominent magnetic anomaly trends: BMA—Brunswick; BSMA—Blake Spur; ECMA—East Coast. The USGS line 32 was recorded in the 1970s (see Fig. 7.8.4-13). [*In* Meissner, R., Brown, L., Dürbaum, H.-J., Franke, W., Fuchs, K., and Seifert, F., eds., Continental lithosphere: deep seismic reflections: American Geophysical Union, Geodynamics Series, v. 22, p. 225–240. Reproduced by permission of American Geophysical Union.]

airgun shots were recorded, fired by an airgun array at 50 m intervals. The interpretation of line BA-6 (Fig. 8.9.4-17) shows the structural change from a rifted 34-km-thick continental crust through a 70–80-km-wide transitional zone, where 12 km of postrift sedimentary rocks overlie a 10–24-km-thick subsedimentary crust, to oceanic crust, which comprises 8 km of sedimentary rocks overlying an 8-km-thick crystalline crust (Holbrook et al., 1994c).

The EDGE Mid-Atlantic multichannel seismic experiment was launched to provide more and better images of the deep velocity structure of rifted continental crust on the Atlantic continental shelf of North America (Holbrook et al., 1992b, 1992c). In 1990, data were collected off northern Virginia, United States (Fig. 8.9.4-18). Three marine profiles were recorded, two across the margin (801 and 803) and one parallel to the coast on the continental shelf (802). Shots from a 36-element airgun array were fired at constant distance intervals of 50 m and recorded by a 240-channel, 6-km-long, hydrophone streamer and ten OBS. In

addition, shots were recorded by an onshore array of ten RefTeks that extended the aperture of the experiment by ~250 km. Average spacing on line 801 was 13 km on the continental shelf and upper slope and 28 km on the rise.

The crustal cross section of Figure 8.9.4-18 shows the line drawing of the multichannel data along line MA802 (Sheridan et al., 1991) with corresponding crustal velocities from wide-angle data (Holbrook et al., 1992b). The crust appeared to be highly reflective and the Moho was observed between 11 and 12 s TWT corresponding to a depth of ~30 km (Fig. 8.9.4-18).

A preliminary interpretation of line 802 revealed a 34-km-thick crust with four layers beneath the sediments which detailed velocity structure that appeared indistinguishable from the velocity structure found in the 1980s beneath several Appalachian terranes in the northern Appalachians of New England (Holbrook et al., 1992b). Across the margin strong lateral changes in deep-crustal structure were observed. Lower-crustal velocities varied from 6.8 km/s in the rifted continental crust,

Distance (km)

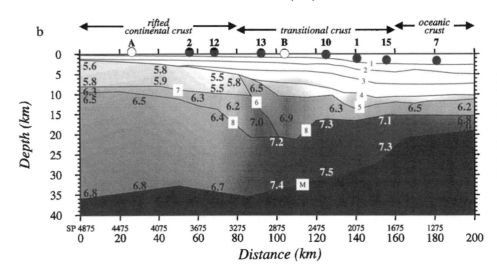

Figure 8.9.4-17. Crustal cross sections through the Carolina Trough of the Atlantic continental margin of the United States (from Holbrook et al., 1994c, fig. 4). [Journal of Geophysical Research, v. 99, p. 9155–9178. Reproduced by permission of American Geophysical Union.]

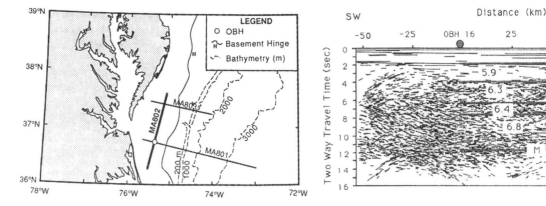

Figure 8.9.4-18. Left: Location of EDGE mid-Atlantic seismic experiment off the coast of northern Virginia (from Holbrook et al., 1992a, fig. 1). Right: Line drawing of multichannel seismic data along line MA802 with corresponding velocities from coincident wide-angle data (from Holbrook et al., 1992a, fig. 5). [Geophysical Research Letters, v. 19, p. 1699–1702. Reproduced by permission of American Geophysical Union.]

to 7.5 km/s beneath the outer continental shelf, to 7.0 km/s in the oceanic crust.

High-velocity lower crust and the seaward-dipping reflections comprised a 100-km-wide, 25-km-thick ocean-continent transition zone. The boundary between rifted continental crust and the seaward thick igneous crust was abrupt, occupying only 20 km of the margin, and the Appalachian intracrustal reflectivity largely disappeared across this boundary as the velocity increased from 5.9 to greater than 7.0 km/s, implying that the reflectivity is disrupted by massive intrusion and that very little continental crust persists seaward of the reflective crust (Holbrook et al., 1994c). The inclusion of the onshore data showed a gradual thickening of the crust from the continental margin westward across the coastal plain, where a uniform thickness of 35 km was observed, to ~45 km under the Blue Ridge section of the southern Appalachians (Lizarralde and Holbrook, 1997).

The Gulf of Mexico also saw increased activities. As an example, the results for a series of experiments carried out after 1983 along the Gulf of Mexico using large airgun sources and dense shot spacing is shown in Figure 8.9.4-19 (Ebeniro et al., 1988; Trehu et al., 1989b).

8.9.5. Summary of Seismic Observations in the Oceans in the 1980s

In the oceans a worldwide study of the lower crust and upper mantle using OBS and big airguns was performed by the Shirshov Institute of Oceanology (Neprochnov, 1989) in the period from 1977 to 1984 (see Fig. 8.9.2-01). The North Atlantic Transect (NAT Study Group, 1985) provided a major improvement on the knowledge of the structure of the oceanic crust and its variability on a large regional scale. Other spectacular images of the internal structure of the oceanic crust were published by White

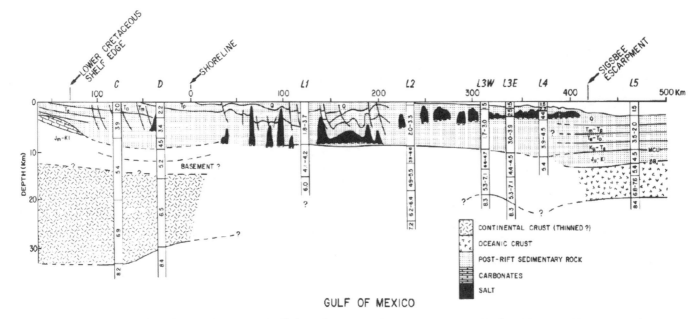

Figure 8.9.4-19. Cross sections through the northern Gulf of Mexico margin (from Trehu et al., 1989b, fig. 18C). [*In* Pakiser, L.C., and Mooney, W.D., eds., Geophysical framework of the continental United States: Geological Society of America Memoir 172, p. 349–382. Reproduced by permission of the Geological Society of America.]

et al. (1990). They showed widespread occurrence of intracrustal reflectivity in the western Central Atlantic Ocean. The research project RAPIDS (Rockall and Porcupine Irish Deep Seismic) of the Dublin Institute for Advanced Studies and partners consisted of two orthogonal wide-angle seismic profiles totaling 1600 km. The individual lines were typically 200–250 km long and produced a 1000-km-long east-west seismic profile from Ireland to the Iceland Basin crossing various troughs, basins and intervening banks (e.g., Hauser et al., 1995).

Various projects have dealt with the ocean-continent transition at continental margins. For example, seismic surveys were carried out in the vicinity of the northern Japan Trench (Suyehiro and Nishizawa, 1994) or across the east Oman continental margin north of the Masirah Island ophiolite (Barton et al., 1990). An extended offshore marine survey targeted the crustal structure off Norway on the Voering plateau (Zehnder et al., 1990) and along the Lofoten margin (Mjelde et al., 1992). Detailed seismic investigations aimed for the deep structure in the transition zones west and north of Spitsbergen (Czuba et al., 1999) as well as eastern Greenland and its margin on the western side of the northernmost Atlantic Ocean (e.g., Mandler and Jokat, 1998). The 1981 Large Aperture Seismic Experiment (LASE) was one of the first experiments along the Atlantic continental margin of North America (e.g., Trehu et al., 1989b).

Increased research activity is shown for the Polar Regions, both north and south. The summary map for the 1980s in Figure 8.9.5-01, which lists only data points published until 1989, indicates that apparently less activity happened in the 1980s. This is not quite true. Many projects of the 1980s which were described in Chapter 8.9 do not show up on the map produced

from data in the USGS database. None of the marine components of the European large-scale seismic-reflection surveys like BIRPS, ECORS, CROP, or BABEL appear on the map. This is due to the fact that in the database all these surveys were classified as land projects similar to what we have done in this volume. In other cases, as is probably true, for example, for the projects northwest of Ireland on the Rockall Trough, the corresponding publications may have appeared only in the following decade. Finally, few data points do not always mean little activity. Concentration of projects on a specific area may have led to an accumulation of data points and caused them to appear as one data point only. In spite of these discrepancies, which are evident especially for the 1980s, we have included this map in our summary, as it is consistent with the summary maps for the other decades.

8.10. ADVANCES OF INSTRUMENTAL AND INTERPRETATION METHODOLOGY AND THE STATE OF ART AT THE END OF THE 1980s

The interpretation of the fast-growing amount of data in crustal and upper mantle research work was carried out more and more by applying ray theory (Červený et al., 1977). On that basis computer programs for ray tracing and calculating ray-theoretical synthetic seismograms had been developed in the 1970s and early 1980s which allowed interpreting complicated structures in two or three dimensions (e.g., Červený, 1979; Červený and Horn, 1980; Červený and Pšenčik, 1984a, 1984b; Gajewski and Pšenčik, 1987, 1988; McMechan and Mooney, 1980; Sandmeier and Wenzel, 1986; Spence et al., 1984). With

1980-1989: worldwide oceanic crust with bathymetry < −250 m

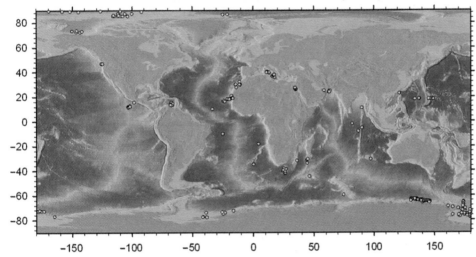

Figure 8.9.5-01. Seismic refraction measurements in the Atlantic, Pacific and Indian Oceans performed between 1980 and 1989 (data points from papers published until 1989 in easily accessible journals and books).

this technique, the requirement to obtain a denser data coverage, in particular in tectonically complicated regions, grew rapidly.

Consequently, the 1980s saw a transition in the development of new instrumentation. It had become clear that the station spacing of 5 km or more for detailed crustal structure studies, which had been regarded as sufficient up to the 1970s, did not provide the resolution required to answer many scientific questions. Rather, smaller shotpoint and station spacing was required. By the end of the 1970s, the development of the cassette recorders at the USGS (Murphy, 1988; for details see Appendix A7-5-6) was the first attempt to fill this gap. A large number of easily maintained instruments with fast play-back facilities became available, which led to an increasing number of fine-tuned experiments in North America. The increasing demand for more recording equipment sped up the developments of a new generation of recording units, and by the end of the 1980s, both in Canada and in the United States not only a large number of instruments became available, but also the age of digital recorders had started, pushed forward in particular in the United States by the foundation of IRIS in 1984 with its sub-organization PASS-CAL (Program for Array Seismic Studies of the Continental Lithosphere).

In Europe, the MARS-66 analogue system experienced a late peak, and by the early 1980s, ~200 units were available in continental Europe. Station spacing also decreased to 2 km in most seismic-refraction experiments in Europe. The search for suitable super-deep drill sites and the idea of the European Geotraverse spanning the whole of Europe added to the activities in crustal research. The development of a new digital generation of seismic instruments, however, was slower than in North America, partly caused by lack of a Europe-wide organization to enable special funding. Only in the first half of the 1990s would the time be ripe for the test and subsequent purchase of newly developed digital equipment. An exception was the University of Hamburg where Jannis Makris had collected sufficient fund-

ing to push for an internal development which finally enabled the building of a large quantity of recording instruments. The development of this analogue device was such that the individual units could either be built into ocean-bottom devices (OBS) or be used on land (LOBS) accompanied by a computerized fast play-back facility. In Britain, the Cambridge university group had undertaken another approach. Here a new digital recording system had been developed, named DOBS, especially designed for controlled-source work, both to be used at sea or on land allowing up to six hours total recording duration (Owen and Barton, 1990). The system was successfully tested in a marine survey off Oman.

On the other hand, crustal research in Europe received a new impulse from reflection seismology by the rapid formation of national groups, such as BIRPS, ECORS, DEKORP, and other national groups, favoring vertical-incidence reflection work in a big style and following the ideas developed by COCORP in North America in the late 1970s. In a review paper, Mooney and Meissner (1992) describe the development and the main achievements of reflection seismology on the research of continental lower crust and Moho.

In the beginning of COCORP, wide-angle piggyback experiments were rare. The different techniques and different frequency ranges of the seismic signals of near-incidence reflection research work and of wide-angle reflection profiling led to quite different presentations of crustal structure, and it took a while before the different philosophies were jointly discussed. On the other hand, in central Europe, close cooperation between the "reflection" and "refraction" groups started from the very beginning. ECORS profiling was always accompanied by simultaneous wide-angle operations (Bois et al., 1986) as were the first long reflection profiles accompanying the search for deep drill sites in southern Germany (e.g., Wenzel et al., 1987). From 1985 onward, the CCSS workshops aimed particularly at a joint interpretation of seismic-reflection and refraction data (Walter and Mooney, 1987; Green et al., 1990a). Mooney and Brocher (1987) have compiled a global review on coincident seismic-reflection/-refraction studies of the continental lithosphere.

Comprehensive reviews of the seismic velocity structure of the deep continental lower crust that represent the state of art by the end of the 1980s were published (e.g., Mooney, 1987). For example, Holbrook et al. (1992a) included a table of lower-crust velocities and corresponding references typical for different tectonic environments (see also Christensen and Mooney, 1995). Mooney and Meissner (1992) discussed the multi-generic origin of crustal reflectivity of the lower crust in various environments.

Other compilations concern summaries of crustal and upper mantle structure based on controlled-source seismology observations for different continents or large parts of those continents, e.g., Meissner et al. (1987b) for Europe, Pakiser and Mooney (1989) for North America, Pavlenkova (1996) for the territories of the former Soviet Union, and Mechie and Prodehl (1988) and Prodehl and Mechie (1991) for the Afro-Arabian rift, while in Olsen (1995), e.g., reviews for continental rifts around the world were compiled by different authors based on a series of discussion meetings of experts.

⌒ CHAPTER 9 ⌒

The 1990s (1990–2000)

9.1. THE FIRST DECADE OF DIGITAL RECORDING

With advanced technology, controlled-source seismology had already reached a stage in the 1980s where its models were no longer schematic diagrams, but approached realistic cross sections and sometimes even 3-D models, which attracted geoscientists from other fields and led them to accept seismic models as a real picture of geologic-tectonic features. In particular, the recently obtained reflection seismograms of large dimensions covering the whole Earth's crust along hundreds of kilometers down to the Moho were simulating the complicated realistic features of the real Earth. At the same time, however, seismic projects had become more and more expensive, using multiple energy sources and a large number of sophisticated recording devices, with the change from analogue to digital recording at the turn of the decade causing a major breakthrough toward modern recording and interpretation techniques.

The combination of these factors meant that most projects were no longer those of individual researchers or individual research institutions, but became mostly embedded in large-scale research programs which involved a multitude of cooperating institutions and an interdisciplinary cooperation among scientists of various geoscientific fields. Starting in the 1980s and continuing in the 1990s, this was reflected, e.g., in the formulation of large reflection programs on national scales such as COCORP (Phinney and Roy-Chowdhury, 1989; Smithson and Johnson, 1989), BIRPS (Klemperer and Hobbs, 1991), ECORS (Bois and ECORS Scientific Parties, 1991), DEKORP (Meissner and Bortfeld, 1990), etc., which was followed by the foundation of IRIS/PASSCAL (Incorporated Research Institutions for Seismology/Program for Array Seismic Studies of the Continental Lithosphere) in the United States and LITHOPROBE in Canada, and by international and interdisciplinary geoscientific priority programs in Europe, dealing with large-scale tectonic topics.

Here, in 1992, supported by the European Science Foundation, EUROPROBE was founded, a Lithosphere Dynamics program, concerned with the origin and evolution of the continents (Gee and Stephenson, 2006). EUROPROBE's focus was mainly, but not restricted, on Europe and was particularly driven to emphasize and encourage East-Central-West European collaboration. It was dedicated to enable the realization of major projects, which aimed to investigate the whole lithosphere and which required close multinational cooperation of geologists, geophysicists and geochemists. Particular EUROPROBE projects, which involved major controlled-source seismic experiments,

were the "TESZ" project, investigating the Trans-European fault zone Tornquist-Teysseire Zone (TTZ) in Scandinavia and Poland, the "Uralides" with a major seismic campaign in the Urals, "Georift," emphasizing the Dniepr-Donets basin, "Eurobridge," establishing a seismic crustal traverse from Lithuania to the Ukraine, and "Pancardi," an interdisciplinary geoscientific frame for large-scale investigations in the Carpathian area, in particular focusing on the deep-earthquake region of Vrancea in the Romanian Carpathians. Other large-scale interdisciplinary projects in Europe were the finalization of the deep-drilling project in Germany, a seismic-reflection traverse through the Eastern Alps, a German priority program focusing on the tectonics of the Central European Variscides, and a major research program studying the Variscan front in Ireland, to name only a few examples.

An important role for successful worldwide cooperation in seismic projects was the development of new digital equipment which had started by the end of the 1980s and continued in the 1990s both in North America and in Europe. In particular, the North American equipment not only stimulated and enabled new large-scale seismic-refraction -reflection experiments in Canada and in the United States, but would also enable many large-scale projects in Europe. As will be described in detail in the individual subchapters of this chapter, e.g., the IRIS/PASSCAL instrument pool, starting with the U.S. Geological Survey–Stanford instrumentation Seismic Group Recorders and continuing with the powerful RefTek generation, was fundamental for successful joint European-U.S. projects in Kenya, France, and Spain in the early 1990s. The Canadian equipment PRS-1 would be the fundamental equipment for seismic-reflection work in Hungary in 1992. Finally, at the end of the 1990s and in the beginning of the 2000s, large-scale projects in Eastern Europe and in Ethiopia would not have been possible without the IRIS/PASSCAL instrument pool. Vice versa, European participation supported several seismic projects in North America.

The continuing rapid progress of data collection and interpretation required the continuation of continuous scientific exchange which was only partly met by the annual meetings of the various national and international geoscientific organizations, such as, e.g., the American Geophysical Union or the European Geophysical Union. Other special meetings of earth scientists were also required.

Since the 1960s, a special subcommission of the European Seismological Commission, which meets every two years, dealt in particular with exchange of knowledge concerning fieldwork, interpretation methods, and results of explosion seismic data. In the early 1980s, the series "International Symposia on Deep

443

Seismic Reflection Profiling of the Continental Lithosphere" was started, which continued into the 1990s and beyond, with meetings at Banff, Canada, in 1992 (Clowes and Green, 1994), Budapest, Hungary, in 1994 (White et al., 1996), Asilomar, California, in 1996 (Klemperer and Mooney, 1998a, 1998b), Platja d'Aro, Spain, in 1998 (Carbonell et al., 2000a), Ulvik, Norway, in 2000 (Thybo, 2002), and in Taupo, New Zealand in 2003 (Davey and Jones, 2004). Also the irregularly scheduled CCSS (Commission on Controlled-Source Seismology) workshops brought together scientists to discuss and practice interpretation methods on pre-distributed data sets. They were organized under the umbrella of IASPEI (International Association of Seismology and Physics of the Earth's Interior) since the late 1970s and continued in the 1990s, e.g., in 1993 in Moscow, Russia (Pavlenkova et al., 1993), in 1996 in Cambridge, UK (Snyder et al., 1997), and in Dublin, Ireland, in 1997 (Jacob et al., 1997) and 1999 (Jacob et al., 2000).

Figure 9.1.1-01. Seismic Group Recorders being prepared at a warehouse in Nairobi for the KRISP90 experiment (Prodehl et al., 1994b) in Kenya. (Photograph by C. Prodehl.)

9.1.1. The North American Controlled-Source Seismology Community

The development of digital instrumentation for the controlled-source seismology community in North America had started in the mid-1980s. It had been initiated by increased crustal and upper mantle research activity of the U.S. Geological Survey (USGS) and the Geological Survey of Canada, and, involving all North American university groups, by the initiation of the LITHOPROBE project in Canada, and by the founding of IRIS in the United States in 1984 with its sub-organization PASSCAL, which was funded by the U.S. National Science Foundation. So, at the beginning of the 1990s a large set of digital instrumentation was available.

The Portable Refraction Seismograph (PRS-1), which, as part of LITHOPROBE, was developed and acquired by the Geological Survey of Canada and afterwards built by EDA Instruments Ltd., was the first digital recording device, which came into use in a large quantity in the late 1980s. Its involvement in large cooperative U.S.–Canadian projects was already described in Chapter 8. The PRS-1 was a single-channel instrument that used a Mark Products L-4A 2-Hz vertical-component geophone. Automatic gain-ranging from 1 to 1024 in binary steps allowed a total dynamic range of 132 dB. Seismic data were sampled at 120 samples per second by a 12-bit A/D board and stored in memory until the data were uploaded to a computer (Murphy et al., 1993; see also Appendix A8-3-8, OFR 93-265). This type of instrument was later complemented by a three-component version, the PRS-4 unit.

Toward the end of the 1980s, the oil industry (AMOCO) had donated 200 units of Seismic Group Recorders (SGR) to Stanford University (Figs. 9.1.1-01 and 9.1.1-02). This SGR III was a single channel, digital seismic recorder with a dynamic range of 156 dB. Data were sampled at 500 samples per second by a 12-bit A/D board with gain ranging from 0 to 90 dB in six steps. Originally these units, manufactured by Globe Universal

Figure 9.1.1-02. Seismic Group Recorders at a field test during the KRISP90 experiment (Prodehl et al., 1994b) in Kenya. (Photograph by C. Prodehl.)

Sciences, Inc. (a division of Grant-Norpac, Inc.), were designed to record seismic data on cartridge tape under the control of a radio signal.

In partnership with the USGS and IRIS/PASSCAL, the Stanford SGRs were modified by adding a timer board that could trigger the recording of data in place of the radio control signal. Each timer board had a precision crystal clock and was loaded with a schedule of trigger times from a PC or laptop to allow programmable turn-on as well as radio-triggering. The clock, a temperature-compensated internal oscillator (TXCO), was synchronized to a master clock prior to deployment, so that the units could individually be installed in the field. The fixed sample rate of 2 ms, maximum record length of 99 s, and maximum recording capacity per cartridge of ~20 min was ideal for then-conventional refraction recording, but greatly limited applicability for roll-along reflection experiments, or recording of offshore airgun sources.

Each SGR III unit was connected to a single string of 6 modified Mark Products L-10B vertical-component geophones (8 Hz) connected in series. Later in the 1990s, 2-Hz Mark Product geophones were also used (Murphy et al., 1993; for more details, see Appendix A8-3-8, OFR 93-265).

In the 1990s, until 1998, the SGR became a workhorse instrument for the community thanks to cooperation between Stanford University, the USGS, and PASSCAL. Figures 9.1.1-01 and 9.1.1-02 show the SGRs at a survey in Kenya (Prodehl et al., 1994b).

The latest development in the late 1980s ready for use by 1990 was the RefTek 72A-02 instrument which was developed by RefTec Inc. and was being bought by PASSCAL. These instruments were six-channel digital seismic data acquisition systems with a 200 Mbyte data disk and an OMEGA clock. The instruments had a variable-gain preamplifier which could be set between a gain of 1 and a gain of 8192. During the 1990 Alaska campaign, for example, the gain was set to 256. Cables were available to be connected to six sets of Mark Products L-10B 8-Hz vertical-component geophone strings or to L-4 2-Hz vertical- or three-component geophones (Murphy et al., 1993; see also Appendix A8-3-8, OFR 93-265). For the fieldwork in Alaska in 1990, where, besides the analogue cassette recorders of the USGS, all three digital types of instruments described above were in the field, 35 RefTeks were available (see Chapter 8, section 8.5.4; Fuis et al., 1997).

Recognizing the controlled-source community's need for a simple instrument for their experiments, PASSCAL worked with RefTek to design a three-component version of the main RefTek instrument (Fig. 9.1.1-03). However, this instrument required heavy batteries, external GPS clocks for many applications, and was far from simple. To complement the PASSCAL passive-seismology instrument center at Lamont, an active-seismology PASSCAL center was set up at Stanford University, from 1991 (50 RefTeks, 1 staff member) to 1998 (320 RefTeks, 3 60-channel Geometrics, 4 staff members plus interns). Though not a simple instrument, the RefTeks were reliable, and data recovery

Figure 9.1.1-03. RefTek 3-component equipment in the field during the CD-ROM experiment (Rumpel et al., 2005; Snelson et al., 2005) in the Rocky Mountains. (Photograph by C. Prodehl.)

was routinely >90%, in cases >98%, even on international, multi-deployment experiments. More than 20 RefTek experiments a year were supported from this center. In 1998, PASSCAL relocated both instrument centers to New Mexico Tech in a new purpose built facility that has grown to manage the Earthscope U.S. Array instrumentation (Keller et al., 2005b).

By pooling various combinations of the Canadian instruments, the RefTeks, the Seismic Group Recorders (SGR), and the Seismic Cassette Recorders (SCR), a number of large experiments were carried out. In the early 1990s, the typical large experiment was 200 SGR, 200 PRS-1, and 200 RefTek (e.g., Mendocino 1993). The SSCD (Southern Sierra Continental Dynamics), DELTA FORCE and DEEP PROBE experiments were the last to use the SGRs (Keller et al., 2005b).

A new generation of recorders designed primarily for use in controlled-source experiments became available in 1999. The RefTek 125 (Texan) instruments became available via a combination of grants to the University of Texas at El Paso. Though only a one-component instrument, due to its small size and weight it has become a very powerful tool for fast-moving large-scale experiments (Fig. 9.1.1-04). Its weight is less than 1.4 kg, the power consumption is small (150 mW in recording and 25 mW in sleeping mode), allowing to be operated by 2 D-cell batteries for 5–8 days, it has an adequate memory (the initial standard was 32 Mbyte) to record continuously for at least 32 h at a 10 ms sampling interval, and it has a high-precision clock (0.1 ppm accuracy temperature-compensated crystal oscillator) for accurate timing of reflection-seismic data. Experience showed that the clock drift rate was 5–7 ms per day.

The initial grant was from the Texas Higher Education coordinating board and funded collaboration between the University of Texas at El Paso and RefTek to design this instrument. Broad input from the international community and PASSCAL was sought during this process. Thanks to an MRI grant from NSF, 440 instruments were available by mid-1999, and during this year, 5 large experiments (3 with 400 instruments) were undertaken (Keller et al., 2005b).

Figure 9.1.1-04. RefTek 1-component equipment ("Texan") during the VRANCEA-2001 experiment (Hauser et al., 2002) in Romania. (Photograph by C. Prodehl.) The transport boxes (right) contain 15 recorders.

9.1.2. The European Controlled-Source Seismology Community

The European Geotraverse was the last large international experiment in Europe that had used MARS-66 analogue equipment on a large scale. However, also in Europe, controlled-source seismic projects as well as a gradually increasing number of tele-seismic projects for studies of crust and upper-mantle structure called for new instrumentation which would also allow to record over a great length of time. However, in contrast to North America, and in spite of the previous success of the European-wide distributed MARS66 system, a new European initiative did not get established. Thus, individual national initiatives were started, but for many years with very limited success. For example, since the early 1980s, in Germany a working group with members from several German universities had started to design the needs for a modern recording device as a replacement for the worn-out analogue MARS66 system, and a first digital system was being developed, but never came to completion. Various companies offered the so-called PCM recording devices which had a great storage capacity and could be left in the field for several days and record earthquake activity. However, these instruments were complicated to use and expensive, so that only very limited numbers of instruments could be purchased, mainly for seismology purposes and teleseismic tomography projects. It was finally the North American initiative, which brought about various products becoming available for the European community by the beginning of the 1990s. In 1991, for example, hardware and software tests of different equipment, offered by the companies Lennartz (Germany) and EDA (Canada) as well as by the U.S. companies Geotech-Teledyne, Kinemetrics, RefTec Inc., and Sprengnether, were arranged at Munich, and a recommendation was made as to which instrument might best serve the German research community. However, it took another couple of years before a major number of new digital instruments became available.

After the unification of Germany, the former East German "Central Institute of Earth Science" at Potsdam had been overtaken by the Federal Republic of Germany and reorganized into the "GeoForschungsZentrum Potsdam" (GFZ, GeoScience-Center Potsdam). It was this new research facility, which finally obtained sufficient funding to establish a new German digital equipment pool which, similar to the PASSCAL pool, would become available for all research institutions in Germany. In 1993, GFZ had purchased a larger number of PDAS equipment from Teledyne and was going to purchase a larger set of RefTek equipment, so that the first research projects with new GFZ digital recording equipment could go ahead from 1993 onward.

In 1993, a first cooperative project INSTRUCT of GFZ, DEKORP, and the University of Kiel was undertaken around the deep continental drilling site KTB in Germany with 14 PDAS. However, the first major projects with the new pool were the PISCO experiment in South America (60 PDAS; see section 9.7.3 below), the FRAC project of the University of Bochum with 25 PDAS, both in 1994 and, in 1995, the GRANU-project with 126 stations (61 PDAS and 65 RefTeks; Fig. 9.1.2-01 and 9.1.2-02; see section 9.2.1 below).

In 1995, a working group was established consisting of GFZ and university delegates who would meet semiannually to deal with the use of the equipment on application by German and other researchers. At that time, 59 PDAS-100 (Fig. 9.1.2-01; Table 9.1.2-01), 66 RefTek-72A/07 (Table 9.1.2-01), and 5 MARS-88 digital recording devices had become available, and 107 short-period three-component 1Hz MARK-L4 seismometers and 21 broadband sensors had been purchased.

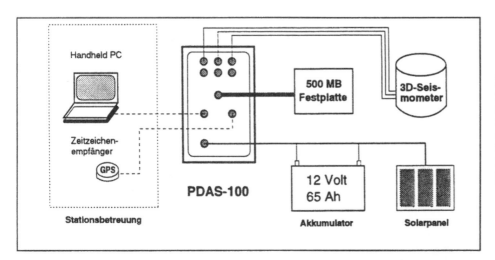

Figure 9.1.2-01. PDAS-100 digital recording device of GeoForschungsZentrum (GFZ) Potsdam (from Lessel, 1998, fig. 4.4), used both for controlled-source and seismology experiments. [Berliner Geowissenschaftliche Abhandlungen, Reihe B, Band 31, 185 p. Published by permission of Institut für Geologische Wissenschaften, Freie Universität Berlin.]

Figure 9.1.2-02. PDAS-100 digital recording device of GeoForschungs-Zentrum (GFZ) Potsdam, being prepared for fieldwork at the GRANU-95 experiment (Enderle et al., 1998) in eastern Germany. (Photograph by C. Prodehl.)

TABLE 9.1.2-01. TECHNICAL SPECIFICATIONS OF THE SEISMIC-REFRACTION EQUIPMENT FROM THE GEOPHYSICAL INSTRUMENT POOL OF THE GEOFORSCHUNGSZENTRUM (GFZ) POTSDAM

Model:	RefTek 72A-07	PDAS-100
Description	Digital recording systems for standalone use, time window and trigger mode, final storage on hard disk	
Manufacturer	Refraction Technology, Dallas, Texas	Teledyne Brown Engineering, Dallas, Texas
Availability	Geophysical instrument pool of the GFZ Potsdam, available for academic use	
Sample rate	20, 25, 40, 50, 100, 125, 200, 250, 500 or 1000 samples per second, adjustable	0.1, 0.2, 0.5, 1, 2, 5, 10, 20, 50, 100, 200, 500, 1000 samples per second, adjustable
Max. recording	Variable length	
Anti-aliasing	6pole Butterworth, 200 Hz cut-off	
Digital filter	Linear phase finite impulse response; pass band up to 82% Nyquist, −130 dB at Nyquist	Linear phase finite impulse response, pass band up to 80% Nyquist, −120 dB at Nyquist
Pre-amp	Unity and 32, adjustable settings	Unity, 10 and 100, adjustable
Programmability	99 time windows of different length	10 time windows of different length, expandable to 9999 repetitions each
Clock drift	Max. 5 ms due to sync. by GPS each hour, $10e-6$ or about, 100 ms per day if GPS fails	$7*10e-8$ (DCXO type), $2*10e-7$ (TCXO type), <10 ms due to sync. by GPS every 4 hours
Power consumption	2.7 to 3 Watts (3 channels, 100 sps each)	3 to 3.6 Watts (3 channels, 100 sps each)
Disk capacity	500 MB and 1 GB	540 MB
Seismometers	Mark L-4C-3D, 1Hz, three components	
Operating temperature	−20 to +60 °C	−20 to +65 °C

Note: From Geophysics Journal International, v. 133, p. 245–259. Copyright John Wiley & Sons Ltd.

Concerning the new Texan instruments, of which the PASSCAL pool and the University of Texas at El Paso in the United States had a total of 840 by 2000, the University of Copenhagen, Denmark, purchased 100, the Earth Science Research Institute TUBITAK-MAM in Gebze, Turkey, acquired 60, and other groups in Central Europe obtained another quantity of ~40 units (Guterch et al., 2003a).

Also in Britain, PDAS and RefTek instrumentation was acquired by the NERC Geophysical Equipment Pool and by various universities, e.g., Leicester University, University of East Anglia, and Cardiff University. Some years later, Leeds University purchased the Orion system. In 2000, finally, the British seismic community (Leicester, Cambridge, Leeds, and Royal Holloway) was awarded a grant to acquire large numbers of Guralp 6TDs and 40Ts, together with a number of the Guralp 3Ts, bringing the British scientists to the forefront of observational broadband seismology.

9.2. CONTROLLED-SOURCE SEISMOLOGY IN EUROPE

9.2.1 The British Isles and Ireland

The fruitful cooperation of the Dublin Institute for Advanced Sciences with German partners from Hamburg and Karlsruhe, which had led to major seismic surveys in Ireland and the adjacent seas continued in 1995 with a 3-year-program named VARNET (Variscan Front Network) The project was funded by the European Community and aimed to study the deep structure of the Variscan Front in southern Ireland by seismic, seismological and magnetotelluric studies. For this purpose, in 1996, a seismic-refraction project was designed in southwestern Ireland, which crossed the Variscan-Caledonian boundary, the supposed Variscan Front (Fig. 9.2.1-01).

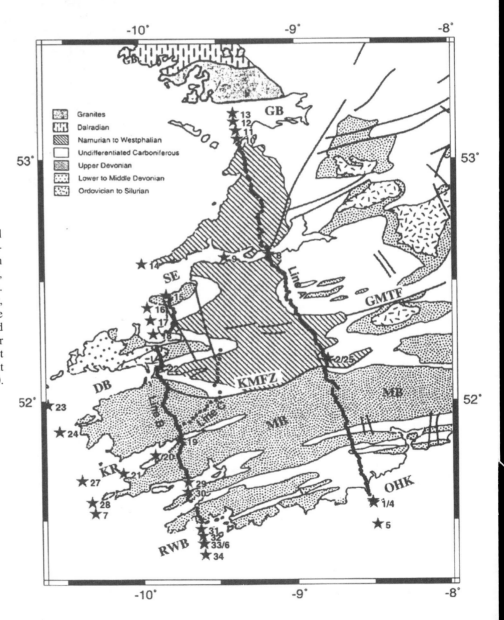

Figure 9.2.1-01. Simplified geological map of southwestern Ireland and location of the VARNET seismic-refraction profiles (from Landes et al., 1990, fig. 1). Stars—shotpoint locations. DB—Dingle Bay, GB—Galway Bay, SE—Shannon Estuary, KR—Kenmare River, MB—Munster Basin, OHK—Old Head of Kinsale, RWB—Roaringwater Bay, GMTF—Galtee Mountains Thrust Fault, KMFZ—Killarney-Mallow Fault Zone. [Terra Nova, v. 17, p. 11–120. Copyright Wiley-Blackwell.]

During this experiment, 34 shots were recorded by 170 stations deployed mainly along two approximately N–S–directed profiles of 140 and 200 km in length, extending from the southern coastline toward the Shannon Estuary and Galway Bay, respectively. The shotpoint geometry was designed to allow for both inline and offline fan shot recordings on the profiles to extend the 2-D interpretation of the inline data to three dimensions (Landes et al., 2000, 2003, 2005; Masson et al., 1998).

A series of shots at small intervals were fired at the northern and southern ends of lines A and B as well as on the western extension of line C. Along the western line B, a series of sea inlets made it possible to fire quite small shots (25 or 50 kg) in water in between the various land segments, so that interleaving of shots and stations provided a multicoverage of observations. An additional "land" shotpoint was added in the center of line A, using a lake in a drowned quarry. The P-wave data of the in-line shots recorded on lines A and B are shown in Appendix A2-1 (p. 64–68) and Appendix A9-1-1.

The large number of shots at the ends of line A allowed a well constrained model for this line (Landes et al., 2000), where particularly in the area of the Shannon estuary at the proposed site of the Iapetus Suture Zone (ISZ) a pronounced thickening of the middle crust resulted. For line B, a rather detailed picture of the upper and middle crust was established by Masson et al. (1998), while the data for the lower crust and upper mantle were rather scarce, due to the insufficient length of the profile. Therefore the Moho was not modeled for line B in that first interpretation. The models shown in Figures 9.2.1-02 and 9.2.1-04 are from a more recent compilation by Landes et al. (2005). The overall crustal thickness reached from 28 km near the coast in

the south and west to 33 km in the center of southwestern Ireland (Landes et al., 2003).

In 1999, the latest onshore seismic-refraction experiment, LEGS (Leinster Granite Seismics), was carried out in the southeast of Ireland (Hodgson, 2001; Hodgson et al., 2000; for location, see Fig. 8.3.2-05 and inset of Fig. 9.2.1-04). Its main aim was the determination of the dimensions of a major granite batholith, outcropping as the Leinster Granite in this area, which according to gravity data should have a much larger subsurface extension toward the southwest. A total of 19 shots was recorded simultaneously along three profiles and recorded by 300 TEXAN stations of PASSCAL and the University of Texas at El Paso. The axial profile, which extended in the NE-SW direction, was, including its off-end shots, 230 km long. Two perpendicular profiles were each 110–140 km long. Station spacing was 1.0–1.5 km.

The resulting depth to Moho showed a decrease from 32 km to 28 km underneath the center of the main granite body, which itself appeared to be between 2 and 5 km thick (Figs. 9.2.1-03 and 9.2.1-04).

Landes et al. (2003, 2005) compiled all crustal investigations in Ireland and the surrounding seas into 3-D block diagrams (Fig. 9.2.1-04). The review of Landes et al. (2005) also includes seismic-tomography studies, the first of which was an outcome of the Varnet-96 program, as mentioned above, when some earthquakes and a nuclear event had also been recorded during the recording windows set for the controlled-source seismic experiment. These data could be used to model the probable trace of the Iapetus suture in the upper mantle just south of the Shannon river estuary between 30 and 110 km depth (Masson et al., 1999).

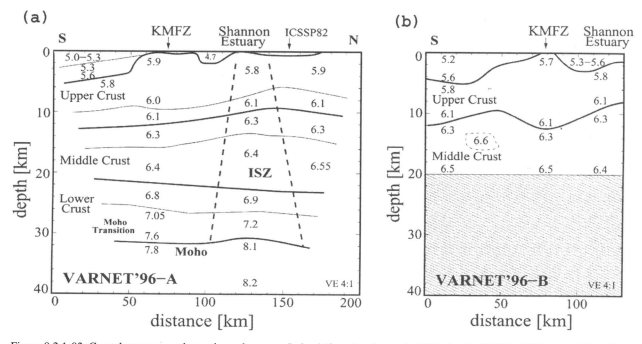

Figure 9.2.1-02. Crustal structure underneath southwestern Ireland (from Landes et al., 2005, fig. 6). KMFZ—Killarney-Mallow Fault Zone; ISZ—Iapetus Suture Zone. [Terra Nova, v. 17, p. 111–120. Copyright Wiley-Blackwell.]

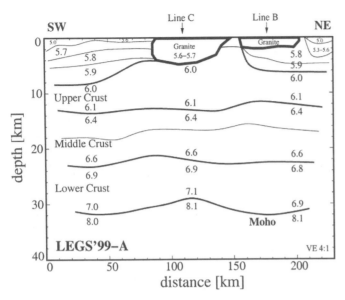

Figure 9.2.1-03. Crustal structure underneath the Leinster Granite region (from Landes et al., 2005, fig. 9): LEGS'99 profile A, modified after Hodgson (2001). [Terra Nova, v. 17, p. 111–120. Copyright Wiley-Blackwell.]

9.2.2. Central Europe

In Germany, research interest in the first half of the 1990s concentrated on two areas: the Varisticum in Saxony, eastern Germany, and the southern end of the Rhinegraben, in southwestern Germany. In the second half of the 1990s, major seismic research projects were undertaken in northeastern Germany and the adjacent Baltic Sea (DEKORP-BASIN) as well as in the Eastern Alps (TRANSALP).

In 1990, immediately following the approach of West and East Germany toward unification, geophysicists agreed to shoot a modern seismic-reflection line across the former border, exploring the Saxothuringian area of the central-European Hercynian mountain system from the Hessen depression in the west to the Erzgebirge in the east. The survey was named DEKORP 3/MVE-90 and covered a total of 600 km deep reflection profiling (Fig. 9.2.2-01). DEKORP 3B had already been recorded at an earlier stage of DEKORP (see Chapter 8.3.1.5), as one of three nearly parallel seismic profiles crossing the Rhenohercynian and the Saxothuringian Zone of the western central part of the Federal Republic of Germany nearly perpendicularly to the Variscan strike. DEKORP 3-A was planned to connect 3-B to oil industry

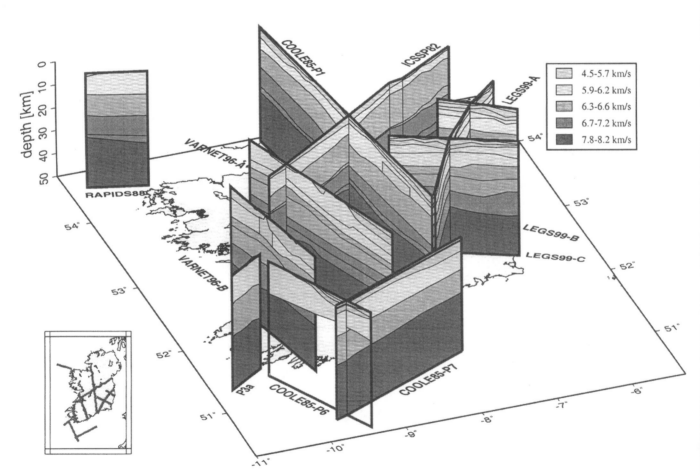

Figure 9.2.1-04. Crustal structure underneath Ireland and adjacent seas (from Landes et al., 2005, fig. 8). [Terra Nova, v. 17, p. 111–120. Copyright Wiley-Blackwell.]

seismic profiles which had been recorded in the Leinegraben area (DEKORP Research Group, 1994a).

DEKORP 3-A and MVE-90 were both carried out in late 1990. The same field parameters and processing procedures were applied as had been used for DEKORP 3-A and 3-B. For logistic reasons, the MVE part was divided into a western and eastern part and the results were published as two separate papers in the same volume (DEKORP Research Group, 1994a, 1994b).

In addition, wide-angle observations accompanied the reflection survey: expanding-spread surveys and a seismic-refraction line were performed simultaneously, while the recording equipment, the 16-km-long, 320-channel receiver spread of the contractor, moved along the line. To obtain reliable and detailed velocity information down to the Moho, two large-scale expanding-spread experiments with shot-to-receiver distances up to 110 km were incorporated into the program. They were centered at two locations with particular gravity anomalies. While the reflection unit moved gradually during the course of the standard Vibroseis reflection survey from SW to NE, two series of properly timed borehole shots were placed symmetrically NE of the sounding points. Charges of 90 kg provided satisfactory signal/noise ratios up to 110 km offset.

The twofold reversed and partly overlapping seismic-refraction line was observed along the 200-km-long eastern portion of MVE-90 (Fig. 9.2.2-01) and was designed in particular to provide wide-angle reflections. The experiment was performed by recording repeated and well-timed explosions at three shotpoints using the 16-km-long receiver spread of the standard reflection survey. The three shotpoints were equally spaced with mutual offsets of ~100 km. Shot charges varied with increasing distance to the receiver spread between 30 and 150 kg. The maximum observation distances were 100 km from the central shotpoint in both directions, and 120 km from the other two shotpoints.

Data and interpretation were published in a special edition of the *Zeitschrift für Geologische Wissenschaften* (*Journal for Geological Sciences*) including an appendix in which the final stacks of the deep reflection seismic data were reproduced (DEKORP Research Group, 1994). The interpretation of the seismic data also included existing gravity data and resulted in a 4-layer crustal model.

The NE-SW–striking section of the MVE-90 profile of 385.3 km length revealed different features in comparison with other, mainly NW-SE–trending DEKORP profiles. While in the NW-SE–striking seismic DEKORP lines, south- and north-dipping reflections appeared widespread, reflective elements of that scale were absent on the MVE-90 profile. This general pattern showed an upper crust down to 8 km depth, characterized by relatively poor reflectivity. In the middle crust, to a depth of

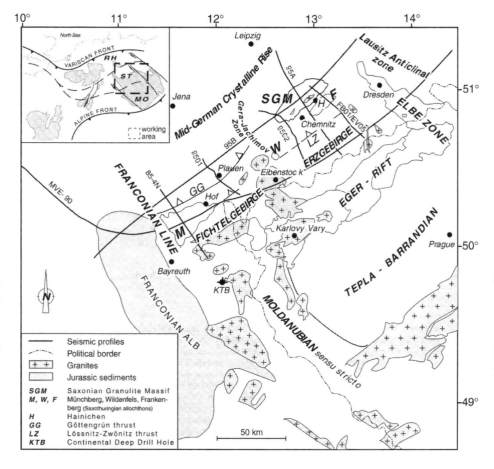

Figure 9.2.2-01. Simplified geological map of southeastern Germany, showing seismic profiles of 1990–1995, investigating the Saxuthuringian part of the Variscides in Saxony, Germany (from Krawczyk et al., 2000, fig. 1). Inset shows the major divisions of the German Variscides into Rhenohercynian, RH, Saxuthuringian, ST, and Modanubian, MO. MVE-90, 85-4N, 9501, 9502 are DEKORP seismic-reflection lines, 95A and 95B are seismic-refraction lines (project GRANU-95), FB01/EV05 are older seismic-reflection lines, reprocessed in the nineties. [*In* Franke, W., Haak, V., Oncken, O., and Tanner, D., eds., Orogenic processes: quantification and modelling in the Variscan belt: Geological Society of London Special Publication 179, p. 303–322, 2000. Reproduced by permission of Geological Society Publishing House, London, U.K.]

18 km, flat and lense-like structures became evident by higher reflectivity. Underneath, again, a poorly reflective crust was seen with sloping short reflective elements dipping to the SW. This general pattern along the MVE-90 line deviated substantially, however, where the line crossed the Münchberg Gneiss Complex, the Vogtland and the Elbe Zone. The lower crust between 24 km depth and the Moho emerged by high reflectivity and horizontal to scarcely dipping structures. The geometry of the Moho showed, different than seen on other DEKORP profiles, an uneven topography varying between 28 and 30 km depth. Under the eastern Erzgebirge an updoming of the Moho was interpreted from both the recording of near-vertical and wide-angle reflections (DEKORP Research Group (B), 1994b).

In 1992, the German Research Society established a new priority program, "Orogenic processes: their quantification and simulation at the example of the Variscides." One of the highlights was the investigation of the so-called Saxonian Granulite Mountains (SGM in Fig. 9.2.2-01), a metamorphic core complex within the Saxuthuringian section of the Variscides (see inset of Fig. 9.2.2-01), located 50–100 km north of the seismic-reflection line MVE-90 which had targeted the Erzgebirge.

In 1995, a major seismic survey targeted this area with both a seismic-refraction survey, GRANU-95 (Enderle et al., 1998a, 1998b) with two lines crossing each other at the center of the Saxonian Granulite Mountain metamorphic core complex, and two seismic-reflection lines, one, 9502, crossing the same structure, the other, 9501, outside to the southwest. Furthermore, the data of an older seismic-reflection survey, FB01/EV05, located to the northeast, and the data of DEKORP 4 (see Chapter 8.3.1.5) and the above described MVE-90 line (lines 85-04 and MVE-90 in Fig. 9.2.2-01), were incorporated in the final interpretation (Krawczyk et al., 2000).

The seismic-refraction project GRANU-95 (Fig. 9.2.2-02) involved 126 recording stations and two deployments. In the first deployment, stations were installed along the 90-km-long line 95-A, running from Leipzig to the Czech border, and along the perpendicular line 95-B around the crossing point D, and 4 shots, A1 to A4, along line 95-A, were recorded. In the second deployment each second station of line 95-A was removed and shifted to line 95-B, which extended over a length of 260 km from north of Dresden, Saxony, to Bamberg, Bavaria, and 8 shots, B to I, more or less equally spaced along line 95-B, were recorded.

Depending on the distance range to be covered, shot sizes varied from 100 kg to 900 kg. The P-wave data of the two seismic-refraction profiles are shown in Appendix A2-1 (p. 32–35) and Appendix A9-1-2 (Enderle et al., 1998b). Furthermore, seismic-reflection equipment with 96 channels, allowing geophone spacing of 80 m, was installed at shotpoint F and recorded all shots.

Southwest of a region with exposed granulites, a highly reflective zone at 1.5 s TWT (two-way traveltime) beneath the reflection lines 9501 and 9502 corresponded well with the top of a

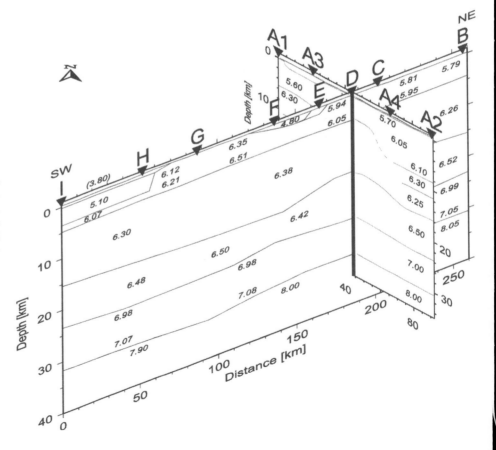

Figure 9.2.2-02. 3-D model of the seismic-refraction project GRANU-95 (from Enderle, 1998, fig. 2.28). [Ph.D. thesis, University of Karlsruhe. Published by permission of Geophysical Institute, University of Karlsruhe, Germany.]

high-velocity zone under line 95-A at ~4 km depth. In this region, high velocities of 6.5–6.6 km/s were also seen at 15–17 km depth (Enderle et al., 1998a; DEKORP and OROGENIC PROCESSES Working Groups, 1999). A lower crust with 7 km/s between 24 and 30 km depth could be well established. Figure 9.2.2-03 gives a composite picture of seismic-reflection boundaries, magneto-telluric anomalies, and Bouguer gravity.

In 1996, DEKORP continued its activities in northeast Germany. The DEKORP-BASIN Group, centered at GFZ Potsdam, undertook a major seismic-reflection survey in the southeastern Baltic Sea and in the lowlands of East Germany. In the Baltic Sea, a grid of marine seismic-reflection profiles was recorded (Fig. 9.2.2-04), covering the area between southern Sweden, Denmark, the island of Bornholm, and the region of northeastern Germany between the Bay of Kiel and the island of Ruegen (DEKORP/BASIN Research Group, 1998, 1999; Krawczyk et al., 1999, 2002; Bleibinhaus et al., 1999).

The marine survey was extended on land into northeastern Germany (Fig. 9.2.2-04) by a detailed seismic-reflection profile BASIN 9601 along a 330-km-long, SSW-directed line from the island of Ruegen to the Harz Mountains (Bayer et al., 1999) and a short cross line BASIN 9602. The land survey consisted of two identical reflection seismic profiles: First, a Vibroseis line was recorded for imaging the upper crust, aiming for depth coverage from the surface to 40 km depth, but concentrating on details of the sedimentary cover and the upper crystalline crust.

Second, an explosive seismic survey was added to ensure imaging of the lower crust and the Moho, aiming for the depth interval between 15 and 40 km (Fig. 9.2.2-05). The observations were complemented by wide-angle recordings for mapping velocities in the deeper crust and at Moho level.

In general, the reflection Moho in the Baltic Sea was found at 28–35 km depth. The data also showed that north of the island of Rügen, the ambiguous picture was characterized by a sequence of northward-dipping reflectors extending from the Moho into the upper mantle. On land, the Moho appeared as a broad band throughout the basin. The continuation of the Moho from its northern boundary into the Baltic Sea, however, was less obvious, and the authors discussed two possible solutions: either the Moho might rise to ~28 km or deepen to ~38 km. Below the basin the Moho appeared flat at a constant depth of 30–32 km depth, but somewhat uplifted south of the southern basin margin to 28 km depth and finally dipping to ~37 km below the Harz Mountains.

In summary, the crystalline crust was found to thin from 32 km at the basin margins to ~22 km below the basin center (derived by subtracting the sediments) and a high-velocity lower crust was detected below the basin. The thickness of the lower crust approached 18 km below the basin center, and thus the middle crust was thinned to only a few kilometers. Nevertheless, the Moho appeared extremely flat and continuous and was well defined as a first-order velocity contrast.

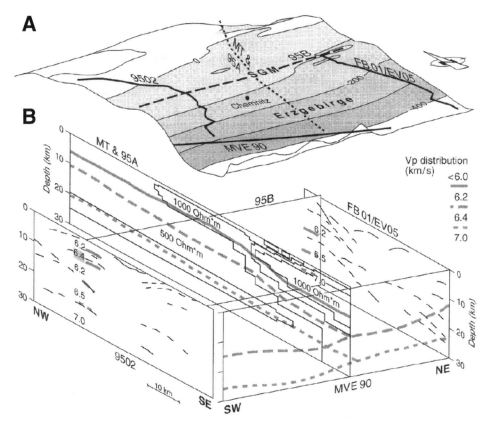

Figure 9.2.2-03. (A) Bouguer gravity contours. (B) 3-D model of the Saxonian Granulite Mountain area comprising seismic, magnetotelluric and gravity results (from Krawczyk et al., 2000, fig. 6). [*In* Franke, W., Haak, V., Oncken, O., and Tanner, D., eds., 2000, Orogenic processes: quantification and modelling in the Variscan belt: Geological Society of London Special Publication 179, p. 303–322. Reproduced by permission of Geological Society Publishing House, London, U.K.]

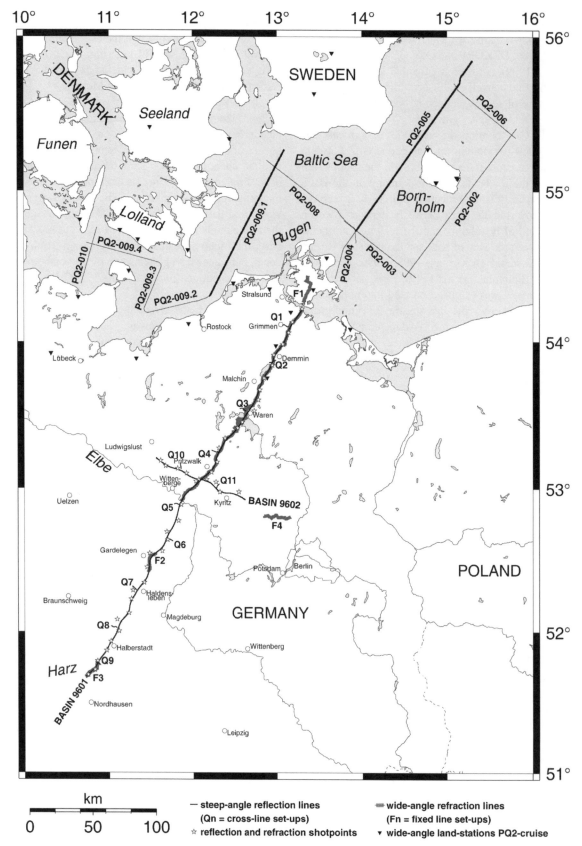

Figure 9.2.2-04. Marine (PQ2-lines) and land (BASIN 9601) seismic reflection lines of the DEKORP-BASIN campaign in 1996 (from Krawczyk et al., 1999, fig.1). [Tectonophysics, v. 314, p. 241–253. Copyright Elsevier.]

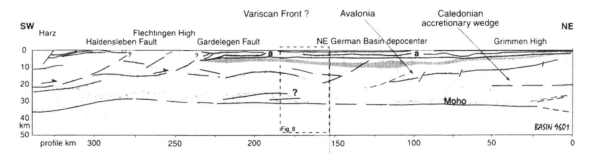

Figure 9.2.2-05. Depth-converted seismic reflection line BASIN 9601 (from Krawczyk et al., 1999, fig. 3). [Tectonophysics, v. 314, p. 241–253. Copyright Elsevier.]

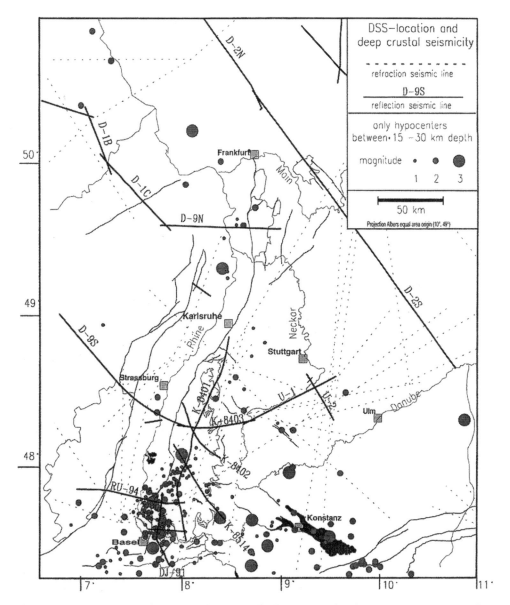

Figure 9.2.2-06. Location of seismic-reflection (solid lines) and refraction (thin dotted lines) profiles in the southern Rhinegraben (from Mayer et al., 1997, fig. 2). Also shown are deep crustal events with hypocenters between 15 and 30 km depth (after Bonjer, 1997). [Tectonophysics, v. 275, p. 15–40. Copyright Elsevier.]

Almost simultaneously with the DEKORP activity in eastern Germany in 1990, the University of Karlsruhe undertook a new initiative to explore the southernmost end of the Rhinegraben (Mayer et al., 1997). From seismology observations, it was long known that this part of the Rhinegraben had an unusual asymmetric distribution of local earthquakes which only on the eastern flank north of Basel occurred at depths of 15–30 km (Bonjer, 1997; Figs. 9.2.2-06 and 9.2.2-07).

Therefore, in 1990 and 1991, two seismic-reflection lines, D-90 and DJ-91, were recorded across the region of deep-earthquake activity, across the Dinkelberg fault block, which is located at the southeasternmost edge of the Rhinegraben (Fig. 9.2.2-07).

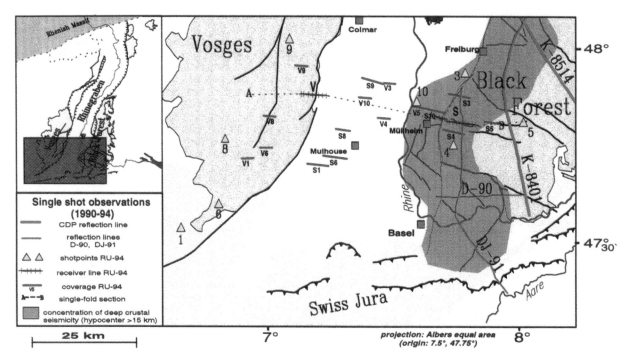

Figure 9.2.2-07. Location of reflection observations in the southern Rhinegraben (from Mayer et al., 1997, fig. 9). RU-94—reflection experiment of 1994 with single-shots (yellow triangles) and recordings (short violet lines) ranging from near-vertical to near-critical reflection distances. D—reflection lines of 1990 and 1991 across the Dinkelberg area. K (thick red lines)—multifold KTB reflection lines of 1984–85. Green area—region of deep earthquake locations. [Tectonophysics, v. 275, p. 15–40. Copyright Elsevier.]

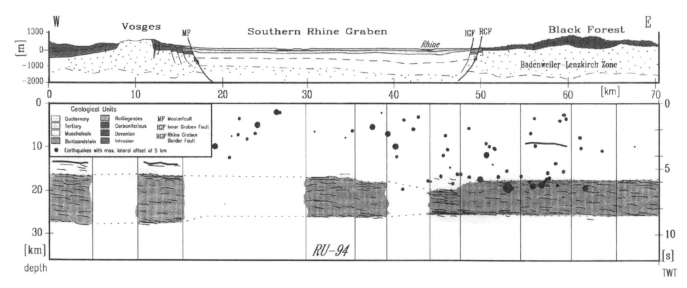

Figure 9.2.2-08. Crustal cross section through the southern end of the Rhinegraben, extending from the Vosges Mountains in the west to the Black Forest in the east, showing local seismic events and a reflective lower crust (from Mayer et al., 1997, fig. 12). [Tectonophysics, v. 275, p. 15–40. Copyright Elsevier.]

Furthermore, a near-vertical to wide-angle survey, partly under-shooting the Rhinegraben and traversing it at its southern end, was added in 1994 (RU-94). It had source positions on both rift shoulders and receiver positions both on the rift shoulders and in the graben proper (Fig. 9.2.2-07).

The resulting cross section RU-94 (Fig. 9.2.2-08) was compiled from single-shot observations and shows the top of the lower crust at an almost constant depth of 5.7–6.0 s TWT, corresponding to 17 km depth. Only beneath the inner graben fault a local increase in depth to 19 km was observed.

The Moho was located at the lower end of the reflective lower crust, at 9 s TWT, which is ~27 km depth and, beneath the Vosges, dips slightly toward the west. The maximum depth of hypocenters of local events, projected into the cross section, was found to increase from west to east, reaching the deepest position close to the Rhinegraben border fault and seemed to coincide with the upper boundary of a thinned lower crust.

Also in 1991 and 1992, a second initiative to obtain more details about the crustal and upper-mantle structure of the Central European Rift System was undertaken in central France by a joint French-German experiment in the Limagne graben and surrounding Massif Central. The project consisted of two parts (Fig. 9.2.2-09). First, a teleseismic tomography study involved a six-month period in 1991 and 1992, during which 79 mobile short-period stations and 14 permanent stations recorded teleseismic events (Granet et al., 1995).

The second part was a special seismic-refraction survey in the central part of the Central Massif, observed in late 1992 (Zeyen et al., 1997). The seismic-refraction survey was performed with 150 mobile stations, which recorded seven borehole shots with charges of 400 and 800 kg along two ESE-WNW–trending traverses. As the new equipment pool of GFZ Potsdam had not yet materialized, the recording equipment was borrowed from PASSCAL: 150 SGR stations including a field computer were shipped from the United States to France, and personnel from PASSCAL and the University of Texas, El Paso, helped to run the equipment in the field and produce a complete data tape immediately at the end of the project.

One hundred and five stations were deployed on the 230-km-long southern line crossing two of the neogene volcanic fields and the southern end of the Limagne graben. The other 45 stations were placed on a parallel profile, 70 km farther north, traversing the Limagne graben at its central part, with a gap in the recording sites in the noisy graben proper. Station spacing was generally between 2 and 2.5 km. Due to the simultaneous recording on both profiles and the 70 km distance between the two lines (the expected critical distance for P_MP reflections), reversed fan-type data were obtained, which aimed in particular for a 3-D coverage of the Moho. The P-wave data of the two seismic-refraction profiles are shown in Appendix A2-1 (p. 49–53).

Figure 9.2.2-10 gives an idea which phases were recorded at which locations, assuming horizontal reflectors, and Figure 9.2.2-11 shows the resulting cross section for the southern line.

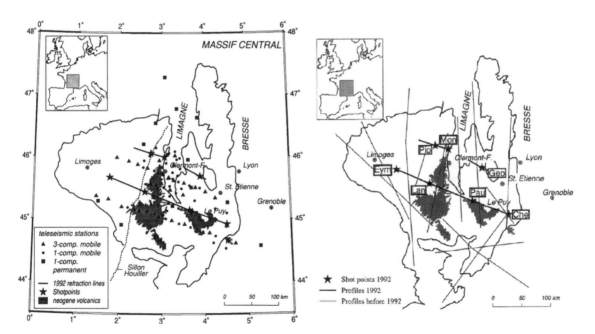

Figure 9.2.2-09. Left: Map of the Central Massif of France, showing the location of teleseismic stations (points and triangles) and the seismic-refraction lines (stars—shotpoints). Also shown is the location of the Limagnegraben and the distribution of neogene volcanics (from Prodehl et al., 1995, fig. 4-4).). Right: Map of Central Massif of France, showing the location of all seismic-refraction lines since 1969. Thick lines design project of 1992 (from Zeyen et al., 1995, fig. 2). [Left: Olsen, K.H., ed., Continental rifts: Evolution, structure, tectonics: Amsterdam, Elsevier, p. 133–212. Right: Tectonophysics, v. 275, p. 99–118. Copyright Elsevier.]

The position where the southern end of the Limagne graben is located clearly showed a drastically decreased velocity to 7.5–7.7 km/s below the general Moho depth range of 28 km.

The resulting crustal thickness varied between 25 km depth under the Limagne graben and 30 km depth under most of the Massif Central. The contemporary teleseismic tomography study resulted in a 3-D velocity structure model of the upper mantle

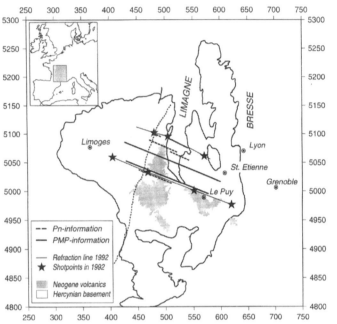

Figure 9.2.2-10. Map of the Central Massif of France, showing the location of the two refraction lines and the resulting positions where P_MP and P_n information is expected assuming horizontal reflectors (from Zeyen et al., 1997, fig. 8). [Tectonophysics, v. 275, p. 99–118. Copyright Elsevier.]

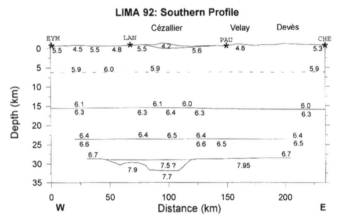

Figure 9.2.2-11. Crustal cross section through the neogene volcanic fields of Cantal (south of Clermont Ferrand) and le Deves (around Le Puy) and the southern end of the Limagne graben in between. (from Zeyen et al., 1997, fig. 7). [Tectonophysics, v. 275, p. 99–118. Copyright Elsevier.]

down to 180 km depth, whereby the upper 60 km of the lithosphere displayed strong lateral heterogeneities and showed a remarkable correlation between the volcanic provinces and the observed negative velocity perturbations (Granet et al., 1995).

The deep seismic sounding (DSS) observations of the early 1990s had gathered a great wealth of new high-quality and dense data sets. A critical review of the then-utilized interpretation procedures led Karl Fuchs and co-workers to a new view of the dynamic properties of seismic wave propagation through the crust-mantle transition. In various papers (e.g., Enderle et al., 1997; Tittgemeyer et al., 1996), the wave propagation was tested by modeling the scattering in the upper mantle and by comparing the resulting synthetic seismograms with observed record sections.

The abrupt termination of near-vertical reflections at the Moho and the coincident presence of a strong supercritical P_MP reflection coda in record sections of continental seismic wide-angle refraction experiments led to the vision of a hitherto unrec-

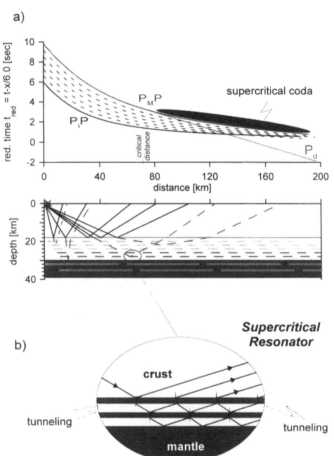

Figure 9.2.2-12. Scattering of energy between P_IP and P_MP phases, caused by internal structure of the lower crust from 18 to 30 km model-depth, is followed by supercritical reverberations following P_MP (supercritical coda), caused by internal structure of the topmost mantle below 30 km model-depth, which has a different scale of structural dimensions (from Enderle et al., 1997, fig. 4a). [Tectonophysics, v. 275, p. 165–198. Copyright Elsevier.]

ognized property of the crust-mantle boundary and a new picture of the lithosphere. It was proposed (Enderle et al., 1997) that the Moho formed a sandwiched mix of crust-mantle material between the lower crust and uppermost mantle in combination with a stepwise increase in mean velocity, and that at Moho level a significant change occurred in the scale of the structural dimensions and of velocity variance (Fig. 9.2.2-12). A similar approach was later used by Mereu (2000a, 2000b) to interpret LITHOPROBE data obtained in the Abiti-Grenville province of Canada (see subchapter 9.4.1.3).

Another advance in interpreting the large amount of recent seismic data was developed at the ETH Zurich. Here, F. Waldhauser undertook a new approach to determine the 3-D topography and lateral continuity of seismic interfaces using 2-D–derived controlled-source seismic reflector data (Waldhauser et al., 1999). He applied his method on more than 250 controlled-source seismic-reflection and -refraction profiles in the greater Alpine region (Fig. 9.2.2-13) to obtain structural information of the crust-mantle boundary (Moho).

The resulting 3-D model of the Alpine crust-mantle boundary showed two offsets that divided the interface into a European, an Adriatic, and a Ligurian Moho, with the European Moho subducting below the Adriatic Moho, and with the Adriatic Moho underthrusting the Ligurian Moho (Fig. 9.2.2-14). Each subinterface depicted the smoothest possible (i.e., simplest) surface, fitting the reflector data within their assigned errors. The results were consistent with previous studies for those regions with dense and reliable controlled-source seismic data and confirmed

earlier models proposed, e.g., by P. Giese and co-workers (see Chapter 8.3.4). The newly derived Alpine Moho interface, however, surpassed earlier studies by its lateral extent over an area of ~600 km by 600 km, by quantifying reliability estimates along the interface, and by obeying the principle of being consistently as simple as possible.

The great success of the Swiss NFP 20 (Swiss National-Fond Project) deep seismic-reflection program at the end of the 1980s, when a seismic-reflection survey crossed the entire Swiss Alps in the north-south direction (Fig. 8.3.1-05), had led to the foundation of an international Working Group aiming to establish a second north-south geotraverse through the Eastern Alps. By the end of the 1990s, from 1998 to 2001, a detailed seismic-reflection program, TRANSALP, targeted the Eastern Alps along a north-south–directed geotraverse between Munich, Germany,

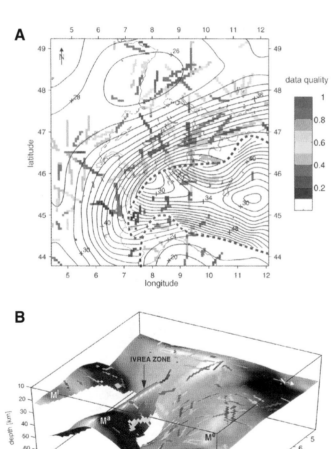

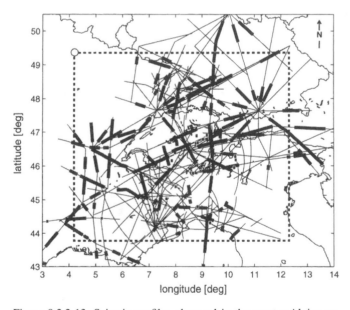

Figure 9.2.2-13. Seismic profiles observed in the greater Alpine region in the past decades (from Waldhauser et al., 1999, fig. 1). Superimposed as thick lines are locations of 2-D–migrated Moho elements from published interpretations of controlled-source seismic data. The dashed box shows the extent of Waldhauser's model in Figure 9.2.2-16. [Geophysical Journal International, v. 135, p. 264–278. Copyright John Wiley & Sons Ltd.]

Figure 9.2.2-14. 3-D model of the Alpine Moho (from Waldhauser et al., 1999, fig. 13). (A) Alpine Moho interface contoured at 2-km intervals derived by the smoothest interpolation of the 3-D–migrated controlled-source data. Data quality of the 3-D–migrated database, i.e. reliability of the original models, is indicated by colors. (B) Perspective view toward the SW on the Alpine Moho. [Geophysical Journal International, v. 135, p. 264–278. Copyright John Wiley & Sons Ltd.]

and Venice, Italy (TRANSALP Working Group, 2001, 2002; Lueschen et al., 2004, 2006; Gebrande et al., 2006). It was a cooperative program of Germany, Austria, and Italy (Fig. 9.2.2-15).

The main data acquisition was divided into four different phases. The first phase of 1998 accomplished 120 km and reached from the northern foreland near Freising, Bavaria, across the Northern Alps to the Inn valley in Austria. The second phase, during the winter of 1998–1999, covered 50 km in northern Italy, between Belluno and Treviso, and was devoted to the southernmost Alps and the adjacent Po plain. In 1999, the main phase was realized, covering 170 km of the Central Alps between the Inn valley, Austria, and Belluno, Italy. Finally, in 2001, additional explosive measurements were undertaken in the northernmost Italian section because of noise problems and some bad shots in this particular area during the 1999 campaign.

For the first time, a continuous, 340-km-long seismic-reflection section could be achieved that included the complete orogen at its broadest width, as well as the two adjacent peripheral foreland basins (Fig. 9.2.2-15). The design of the experiment was very complex. Vibroseis near-vertical seismic profiling formed the core of the field-data acquisition. It was complemented by explosive near-vertical profiling and cross-line recording for 3-D control. Furthermore, wide-angle recording of Vibroseis and explosive sources were undertaken by a stationary array for velocity control and active-source tomography, and finally, for another 9–11 months, another stationary array recorded local and teleseismic events to enable passive tomography studies (Lueschen et al., 2004, 2006; Millahn et al., 2006).

The Vibroseis survey was designed to achieve high resolution and penetrate the upper and middle crust. The explosive seismic survey was to provide low-fold, but high-energy signals from the deeper parts of the crust. Shots of 90 kg charges, with 5 km spacing, were fired in 30-m-deep boreholes, when the Vibroseis rolling spread arrived at both the north and south off-

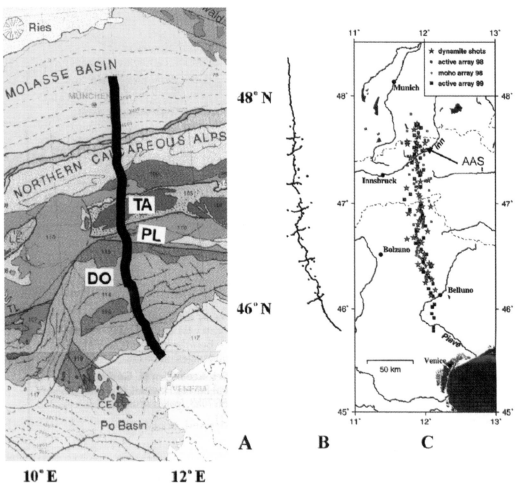

Figure 9.2.2-15. Location map of TRANSALP (from TRANSALP Working Group, 2001, figs. 1 and 3). (A) Tectonic map. (B) Location of explosive sources (charges of 90 kg in three 30-m-deep shot holes) recorded off-end with up to 30-km-long spreads. (C) Stationary, continuously running network of up to 50 three-component units recording all Vibroseis and explosive sources. [Eos (Transactions, American Geophysical Union), v. 82, p. 453, 460–461. Reproduced by permission of American Geophysical Union.]

end configuration. By this procedure, Vibroseis and explosive sources were recorded by the same recording unit and produced data simultaneously.

The resulting data were of excellent quality, with the different types of data complementing each other in depth penetration and resolution characteristics. The northernmost and southernmost parts of the 300-km-long Vibroseis section displayed the stratified Molasse basins, the Tertiary base in the Bavarian Molasse being the most prominent reflection. The section also showed the transition from unfolded Molasse sediments at maximum 6 km depth to the folded Molasse sediments and Flysch zone, characterized by a chaotic signature, and into the Northern Calcareous Alps at maximum 11 km depth. Several onlap structures beneath the Alpine front and a sudden displacement of the Tertiary base by 4–5 km were also visible. At ~9–10 km depth, the Northern Calcareous Alps and their substrata are bounded by a basal, almost horizontal reflection pattern above a relatively transparent upper crystalline crust.

Farther to the south, a prominent south-dipping reflection pattern, outcropping at the Inn valley, was interpreted as a thrust fault system along which the Northern Calcareous Alps were overthrusted by their former basement, the Greywacke Zone. Beneath the axis of the Central Alps the profile shows a bivergent asymmetric structure of the crust. Here, the crust reaches a maximum thickness of 55 km, and two 80–100-km-long transcrustal ramps could be traced in the data: north of the axis a southward dipping "Sub-Tauern-Ramp" and south of the crest a northward-dipping "Sub-Dolomites-Ramp" (Fig. 9.2.2-16, Lueschen et al., 2004). The two models, shown in the center and bottom of Figure 9.2.2-16, illustrate clearly the main results of the TRANSALP seismic survey, but express different viewpoints of the tectonic interpretation. For comparison, the predicted model of TRANSALP Working Group (2001) is shown in Figure 9.2.2-16 (top cross section). A more refined interpretation of the wide-angle data was prepared by Bleibinhaus and Gebrande (2006).

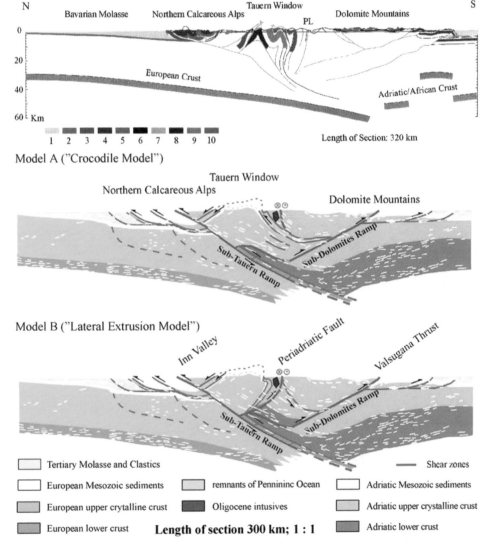

Figure 9.2.2-16. Top: Pre-experiment crustal cross-section along the TRANSALP traverse. The section is mainly based on geological mapping, deep seismic refraction velocity models of the 1970s, and inspired from "indenter tectonics" applied for the Western Alps (from TRANSALP Working Group, 2001, fig. 2). Center and Bottom: Cartoon of two alternative structure and evolutionary models, based on the data of TRANSALP (from Lueschen et al., 2004, fig. 10). [Tectonophysics, v. 388, p. 85–102. Copyright Elsevier.]

9.2.3. Southern Europe

Following the last stage of the European Geotraverse, the ILIHA project on the Iberian Peninsula in 1989, in the early 1990s the focus shifted to multichannel seismic-reflection profiles. In 1990 the Spanish Research and Development Plan initiated a deep vertical seismic-reflection program named ESCI (Estructura Sismica de la Corteza Iberica). Three key areas were selected: the northwest Hercynian Peninsula, the northeast Spain-Valencia trough, and the Betics. Subsequently 450 km of terrestrial and 1325 km of offshore profiles in northern, eastern, and southern Iberia were recorded (Díaz and Gallart, 2009).

Crustal research activities started in 1991 in northern Spain. In 1991 the project ESCIN (Estudio Sismico de la Corteza Iberica Norte) was launched. It consisted of several land and marine seismic-reflection profiles (dotted lines ESCIN-1 to -4 and IAM-12 in Fig. 9.2.3-01). Two of the ESCIN lines (1 in 1991, 2 in 1993) were shot and observed on land (Pulgar et al., 1996). The other two lines ESCIN-3 and -4 were marine profiles investigating the continental margin and heading into the abyssal plain of the Bay of Biscay (Alvarez-Marron et al., 1996, 1997; Ayarza et al., 1998).

In 1992, the scientists of Spain made use of the 150 PASSCAL recording units and its computing facility that had been shipped to France for the seismic investigation of the Central Massif before (see above) for a detailed land survey of the Cantabrian Mountains and the adjacent Duero Basin. Five seismic lines were recorded in the Cantabrian Mountains (bold lines in Fig. 9.2.3-01), one in the east-west direction along the strike, and four in the north-south direction into the Duero Basin to the south (Perez-Estaun et al., 1994; Fernandez-Viejo et al., 2000).

In 1993, the Iberian Atlantic Margins European project (IAM), a marine seismic project, focused on the Atlantic margin of Iberia (for location, see Fig. 9.8.4-15). Up to 3500 km of multichannel profiles were acquired in the North Iberian and Atlantic margins, the Gorringe bank and the Gulf of Cadiz (Banda et al., 1995). The shots were also recorded by ocean-bottom seismometers (OBSs) and by a variety of land stations (Fig. 9.2.3-01).

One profile, IAM12, was oriented in N-S direction north of La Coruna. Three profiles, IAM 3, 9, and 11, were continued on land by short in-line profiles providing images of the continent-ocean transition. IAM 11 ran in an east-west direction from the Galician coast across the Galicia margin, IAM 9 also ran east-west farther south in front of the central Portuguese coast, and IAM 3 ran NE-SW in front of Canbo San Vicente. In the follow-

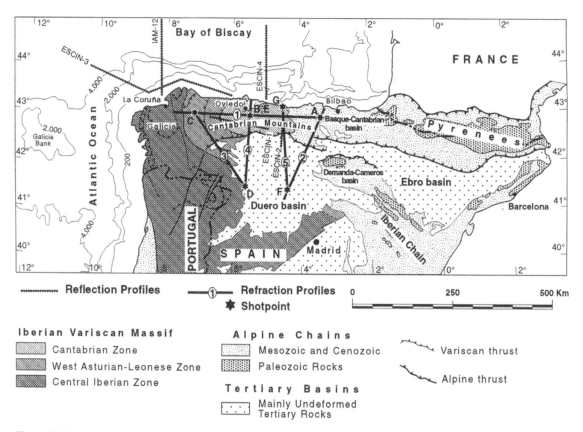

Figure 9.2.3-01. Simplified tectonic map of the northern Iberian Peninsula and adjacent North Iberian margin showing seismic profiles for crustal studies (from Fernandez-Viejo et al., 2000, fig. 1). Bold lines—crustal survey of 1992, dotted lines: projects ESCI and IAM, also recorded in 1992. [Journal of Geophysical Research, v. 105, p. 3001–3018. Reproduced by permission of American Geophysical Union.]

ing years, additional marine experiments were implemented in Atlantic Margins. More details will be discussed in subchapter 9.8.4.3 (North Atlantic).

Combining the data of ESCIN and one of the new lines of 1992, Pulgar et al. (1996) compiled a 350-km-long transect across the Cantabrian Mountains, which consisted of an onshore part in the south and an offshore part in the north (Fig. 9.2.3-02A). The onshore section showed gradual thickening of the crust from 30 km underneath the Duero Basin in the south to ~35 km at the southern edge of the Cantabrian Mountains. Underneath the mountains, the crust thickens drastically northward to almost 60 km at the shoreline. The northern marine cross section, viewed from north to south, started with 16 km crustal thickness under

the abyssal plain of the Bay of Biscay. It showed a smooth gradual thickening across the northern Iberian continental margin to near 30 km at the shoreline, overlapping the root zone, seen from the south (Fig. 9.2.3-02A). Fernandez-Viejo et al. (2000) interpreted the whole set of 5 profiles recorded in 1992 and compiled a block diagram (Fig. 9.2.3-02B) showing the thickening of the crust from the Duero Basin into a pronounced crustal root zone underneath the Cantabrian Mountains.

For the second key area, northeastern Spain and the adjacent Valencia trough, a 50-km-long seismic-reflection profile was set up on land between the southern end of a 250-km-long north-south–striking ECORS profile, recorded in the late 1980s, and the Mediterranean coast, and continued toward the south-

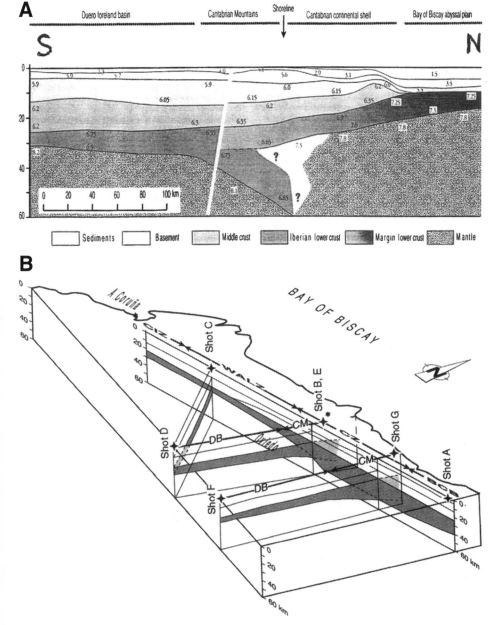

Figure 9.2.3-02. (A) Schematic cross section and (B) fence diagram illustrating the Alpine crustal thickening along the Cantabrian Mountains obtained from the seismic refraction data in the northern Iberian Peninsula (from Pulgar et al., 1996, fig. 11 [Tectonophysics, v. 264, p. 1–19. Copyright Elsevier] and Fernandez-Viejo et al., 2000, fig. 13). CM—Cantabrian Mountains; OB—Duero basin. [Journal of Geophysical Research, v. 105, p. 3001–3018. Reproduced by permission of American Geophysical Union.]

east with a 400-km-long marine profile extending in NW-SE direction across the Valencia trough and continuing south of Mallorca (Gallart et al., 1994, 1995a). The marine airgun shots were also recorded at far offsets by six stations on the Iberian mainland and by two stations on Mallorca (Fig. 9.2.3-03). Also shown on the map are two earlier marine seismic-reflection profiles (VALSIS) which were recorded parallel and ~30 km off the coast line. The ESCI deep seismic-reflection profile showed a highly reflective lower crust of ~15 km thickness overlying the Moho at 29 km depth. The far offset recordings of the marine shots indicated that a major crustal thinning into the Valencia trough of as much as 10 km over a distance of 60 km starts close to the shoreline.

In 1995, the French project LISA (Ligurian Sardinian margins project) sampled different areas of the western Mediterranean from seismic multichannel profiling (Nercessian et al., 2001). Five marine profiles were observed near the eastern end of the Pyrenees. The airgun shots were also recorded by 5 por-

table land stations (Gallart et al., 2001). This addition aimed to investigate the transition between the Pyrenees affected by the Alpine compression and the Mediterranean, an area of Neogene extension.

In southern Spain, two seismic-reflection profiles ESCI-B1 and ESCI-B2 (Fig. 9.2.3-04) were recorded in the course of the ESCI (Estudio Sismico de la Corteza Iberica) project, traversing the Betic Cordillera, with ESCI-B1 in NW-SE direction, sampling mostly the external Betics, and ESCI-B2 in NE-SW direction, through the inner Betics (Garcia-Duenas et al., 1994; Carbonell et al., 1998a). A third marine seismic-reflection profile, ESCI-A1, extended the second line ESCI-B2 into the Alboran Sea (Comas et al., 1995). The land and marine surveys were connected by a piggy-back wide-angle recording of the airgun shots at sea by portable land stations (Gallart et al., 1995b). Another offshore line, approximately W-W directed, connected the Alboran Sea and the South Balearic Basin (Booth-Rea et al., 2007).

On average the Iberian crust appeared to be seismically featureless. On ESCI-B1 a high-amplitude reflection sequence was seen between 20 and 30 km, partly down to 40 km, and the Moho was imaged as a 3–5-km-thick double band dipping southwards. ESCI-B2, providing a transect through the Sierra Nevada core, showed a highly reflective deep crust, overlying a subhorizontal Moho, but a fairly transparent upper crust and upper mantle (Martínez-Martínez et al., 1997). The highly reflective structure, identified as the crust-mantle boundary at 28–30 km depth differed from tomographic modeling, based on the 21 seismic stations of the Spanish National Seismic Network, also shown in Figure 9.2.3-04, which suggested the existence of a 34–36 km crustal root (Carbonell et al., 1998a).

In Italy, the deep seismic-reflection program CROP (Crosta Profonda) continued with its second phase from 1989 to 1997 (Bernabini and Manetti, 2003). Four seismic-reflection surveys were accomplished on land (red lines in Fig. 9.2.3-05).

CROP 04 covered the southern Apennines with an acquisition length of 154 km. Fieldwork lasted from December 1989 to April 1990. It was recorded using explosives or Vibroseis as energy sources. The acquisition parameters and processing sequence was the same as had been applied in the Alps. However, the results were less favorable, probably due to the different geological subsurface conditions. The first seismic sections showed almost no reflections. Therefore, the line was completely reprocessed in the time interval 0–10 s TWT. The reprocessed section is now comparable with high-quality commercial lines in the region (Scandone et al., 2003). But while commercial lines normally do not exceed 5–6 s TWT suitable for geologic interpretation, the reprocessed CROP 04 stacks show continuous and well-structured events up to 8–9 s TWT. Reflectors evident at 6–8 s TWT have been interpreted as important constrain for the sole thrust of the Apenninic tectonic wedge that in the region of the Alborni Mountains reaches a depth of 20–25 km. This depth coincides with the depth of the so-called "Tyrrhenian Moho" revealed by seismic-refraction experiments in the 1970s (e.g., Colombi et al., 1973; Giese et al., 1973).

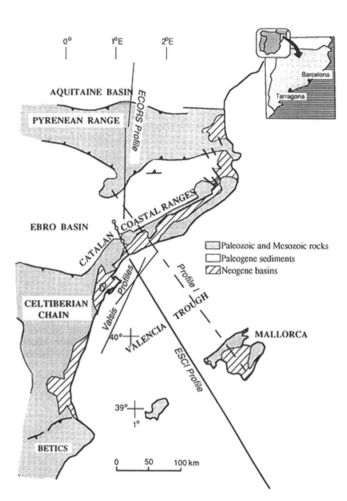

Figure 9.2.3-03. Location map of the ESCI profile extending from the northeastern Iberian Peninsula into Valencia trough and beyond (from Gallart et al., 1994, fig. 1). Open circles mark the land stations recording the marine shots. Profile 1: a seismic-refraction line not yet performed by 1994. [Tectonophysics, v. 232, p. 59–75. Copyright Elsevier.]

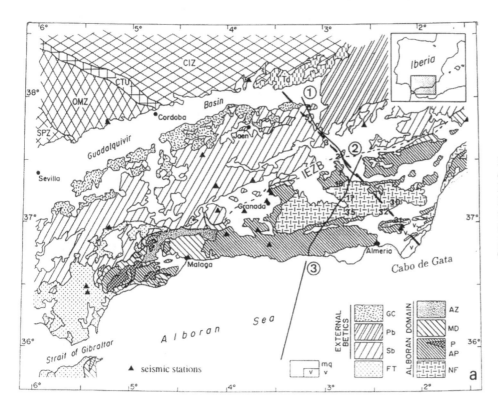

Figure 9.2.3-04. Geologic map of southern Spain, showing the location of the ESCI-profiles: 1—ESCI-B1, 2—ESCI-B2, 3—ESCI-A1, triangles—stations of the Spanish National Seismic Network (from Carbonell et al., 1998, fig. 1a). [Tectonophysics, v. 288, p. 137–152. Copyright Elsevier.]

In 1992–1993, the 220-km-long line CROP 03 was recorded across the northern Apennines, oriented in WSW-ENE direction between Punta Ala near Grosseto on the Tyrrhenian side, opposite of the island Elba, and Gabicce near Pesaro on the Adriatic side, and crossing the crest of the Apennines between the cities Siena-Arezzo to the north and Perugia to the south. The acquisition parameters were adjusted, and so events up to 15 s TWT are visible in the sections. Wide-angle reflections were also recorded along the line with large shots. The crustal structure could be well resolved, both for the extensional Tyrrhenian domain, where the Moho is supposed to be very young and is located at a depth of 22–25 km, and of the compressional Adriatic domain, where the compression is still active and where the well-stratified crust reaches a depth of ~35 km (Barchi et al., 2003).

CROP 18 was shot in 1995 and was recorded in a NW-SE direction in southern Tuscany between the geothermal areas of Ladarello and Mount Amiato which is an area with exceptionally high heat flow of 120 mW/m² and local peaks of up to 1000 mW/m². Line 18 crosses line 03 near its western end (Fig. 9.2.3-05). The area was affected first by the convergence and subsequent collision of the European and African margins in Late Cretaceous and early Miocene and second by the extensional tectonics in the inner part of the Apennines since the Miocene with the emplacement of mixed crustal and mantle magmas. The reflection line was acquired primarily for geothermal purposes. A regional mid-crustal reflector, referred to as K-horizon, and some bright spots are the dominant features in the reflection section. The crust-mantle boundary was located at 22–25 km depth (Batini et al., 2003).

CROP 11 crosses central Italy between Marina di Tarquinia at the Tyrrhenian side near the city of Civitavecchia and Vasto at the Adriatic coast. The line was arranged oblique, but as perpendicular as possible to the strike of the Apennines. It was acquired in two stages, the first 109 km in 1996, and the second 156 km in 1999. The deep crustal structures are well imaged in correspondence to the Apulian foreland. At around 7 s TWT, a reflector could be recognized, dipping gently toward the west. It could be easily followed to the Bomba ridge where complex structures and velocity changes near the surface and at depth were encountered. Other deeper and subhorizontal reflective sections were images at 10 s and at 12 s TWT. They were assigned to the 2.5 s thick lower crust of ~9 km thickness. The Moho was identified with 12.5 s reflection time at 32 km depth. The corresponding reflections deepened westward reaching 13–14 s TWT beneath Mount Maiella.

The final land project of CROP was the participation at the international TRANSALP project (see subchapter 9.2.2; TRANSALP Working Group, 2001, 2002; Lueschen et al., 2004, 2006; Gebrande et al., 2006), a cooperative program of Germany, Austria, and Italy (Fig. 9.2.2-15) carried out from 1998 to 2001. The Italian seismic-reflection part was accomplished in 1998 and 1999 and covered a distance range of 327 km. In addition to the main profile, six short profiles, orthogonal to the main line, were recorded. Both Vibroseis and explosive sources were used.

At the end of CROP Phase II, plans were under way to continue the seismic-reflection program in the 2000s with profiles through Calabria and Sicily.

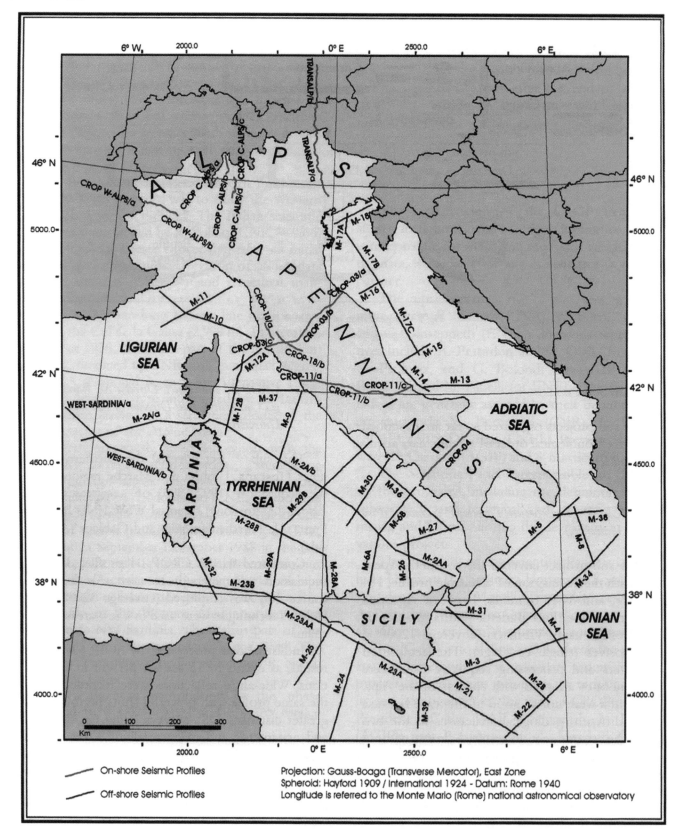

Figure 9.2.3-05. Location map of CROP seismic reflection lines on land and at sea (from Bernabini and Manetti, 2003, fig. 2) [*In* Scrocca et al., eds., 2003, CROP Atlas: seismic reflection profiles of the Italian crust: Memorie Descrittive della Carta Geologica d'Italia, Roma, v. 62, 194 p. Copyright CROP.]

Besides the onshore program, a large marine program was realized in two phases. During CROP MARE 1, in 1991 ~3400 km of seismic profiles were recorded in the Ligurian, Tyrrhenian, and Ionian Sea (Fig. 9.2.3-05). In the second phase, CROP MARE 2 in 1993–1994, over 5000 km marine seismic profiles were acquired in the southern Tyrrhenian Sea, in the Sardinia Channel, and in the Ionian and Adriatic Seas (Fig. 9.2.3-06). The profiles of both

phases were acquired by the oceanographic vessel *Explora*, using airgun sources (Bernabini and Manetti, 2003; Finetti, 2003). The data quality was fair to good in foreland regions, such as the Adriatic and Pelagian Seas, and in parts of the deep basins, but in the areas with a more complex tectonic setting, such as the Tyrrhenian Sea margins, the Corso-Sardinian block, and the Calabrian arc, the data quality was strongly variable (Finetti, 2003)

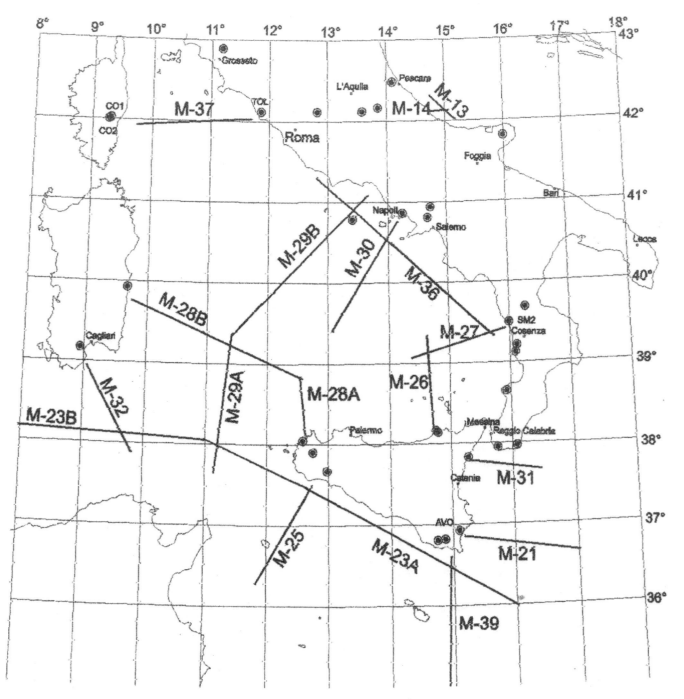

Figure 9.2.3-06 Location map of wide-angle seismic profiles recorded during the CROP MARE 2 phase (from Caiello et al., 2003, fig. 2) [*In* Scrocca et al., eds., 2003, CROP Atlas: seismic reflection profiles of the Italian crust: Memorie Descrittive della Carta Geologica d'Italia, Roma, v. 62, 194 p. Copyright CROP.]

During the acquisition of the marine data, low-frequency seismographs recorded the offshore shots on land (Fig. 9.2.3-06). These data provided additional high-density wide-angle reflection/refraction profiles in some offshore-onshore configurations along the Italian coast (Caielli et al., 2003). Depending on the local conditions, a shooting interval of 20 s provided at least 300 and up to 3000 traces of wide-angle data with trace spacing of 50 m, covering maximum recording distances between 60 and 200 km.

While the originally planned CROP lines in southern Italy could not be realized in the CROP Phases 1 and 2, in the early 1990s, southern Italy became the target of some other seismic projects, most of them combined onshore-offshore experiments, targeting in particular the Ionian Sea between southern Italy and Greece.

The first project was a small-scale land-seismic experiment. In southern Calabria, southern Italy, crystalline basement is exposed, which is interpreted as lower-crust material and therefore had been defined as a key area for an investigation of the structure, composition, and evolution of the Hercynian lower crust in Europe (Schenk, 1990).

Therefore, in early 1990, a particular experiment was carried out by an Italian-German collaboration (Lueschen et al., 1992). A small-scale seismic-reflection experiment aimed to compare in situ seismic properties with petrological data from field mapping and from laboratory measurements of seismic properties in the same rock units (Fig. 9.2.3-07, top). In this respect, the experiment was going to complement super-deep continental drilling programs, which aim for upper and mid-crustal levels. The seismic profile was located in the center of the Serre Mountains and crossed the main lithological units, following a plateau in a N-S direction. The basic principle used in the field layout was to deploy a fixed, densely spaced receiver spread with all available recording channels and fire all shotpoints along the complete line, then move the spread to the next of a total of 3 sectors and repeat all 9 shots (Fig. 9.2.3-07, bottom).

Additionally, four short E-W transverse profiles ~1 km long (with off-end shots of small 5 kg charges) were recorded within the main lithological units in order to detect azimuth-dependent seismic velocities. Recording was done with three-component geophones at 80 m spacing. The experiment gave reasonable results (Fig. 9.2.3-07, center), however, the reflectivity of the outcropping lower crustal units was lower than theoretically predicted, and the analysis of refracted-wave velocities revealed values systematically lower by up to 30% than laboratory data on rock samples or calculated data from modal analysis. The authors attributed the main discrepancy to large-scale alteration of the rocks due to Apennine tectogenetic events (Lueschen et al., 1992).

In 1992, the STREAMERS project was launched, comprising more than 500 km of seismic-reflection profiles throughout the Ionian Sea between southern Italy and Greece (Hirn et al., 1996).

A series of six offshore lines, using a 180-channel 4.5-km-long streamer and airgun sources was recorded to the south and east of Calabria into the western Ionian Sea (Cernobori et al., 1996). On land recording stations were positioned at the site TS-1 and along a profile named Serre (Fig. 9.2.3-08). The map

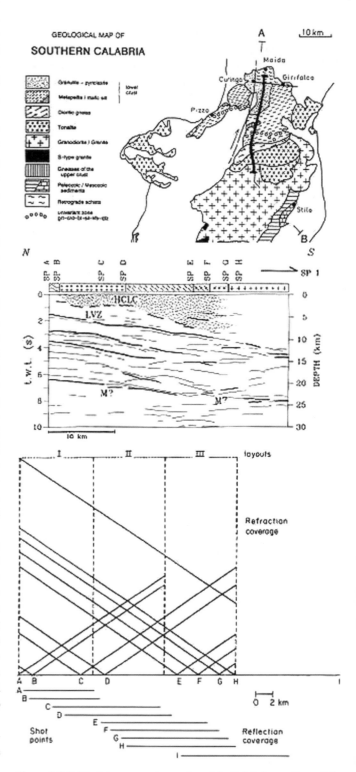

Figure 9.2.3-07. A lower-crust experiment in Calabria, southern Italy (from Lueschen et al., 1992, figs. 1, 4, and 6). Top: Geologic background. Center: Line drawing. Bottom: Layout. [Terra Nova, v. 4, p. 77–86. Copyright Wiley-Blackwell.]

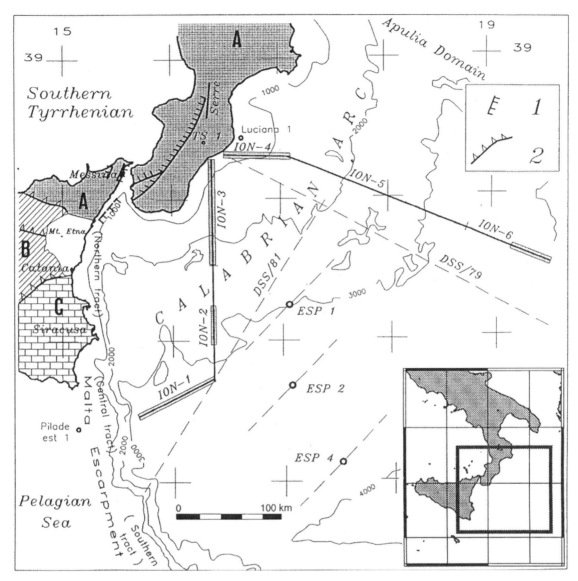

Figure 9.2.3-08. Location map of the STREAMERS profiles (ION-1 to -6) in the western Ionian Sea (from Cernoberi et al., 1996, fig. 1). [Tectonophysics, v. 264, p. 175–189. Copyright Elsevier.]

also shows the sites of the earlier seismic investigations of the area (Ferrucci et al., 1991; Makris et al., 1986; de Voogd et al., 1992), carried out in 1979 and 1981 (DSS79, DSS81, and ESP [expanding spread profile] locations). One of the principal features of the multichannel reflection data beneath the Ionian basin was a band of "layered" high-amplitude reflections near the base of the crust, whereby the traveltimes increased gradually from the basin toward the southern and eastern margins of Calabria.

A more detailed investigation with additional shorter lines (Fig. 9.2.3-09) west and north of the long offshore lines ION-1, -2, and -3 of the 1992 survey followed in 1993, using industrial-grade reflection profiling with improved marine sources, and aiming to investigate the offshore region east of Sicily near Mount Etna (Nicolich et al., 2000). The inferred Moho depth

(Fig. 9.2.3-09) decreases from near 25 km at the shore lines to 17–18 km in the Ionian basin.

In the eastern Ionian Sea, a 180-km-long seismic-reflection profile, augmented by a few wide-angle seismometer stations on land, has been recorded to the west of the Peleponnesus, Greece (Fig. 9.2.3-10), using a powerful airgun source (Hirn et al., 1996), in order to investigate the crust in the area where the Ionian Sea plate is being subducted beneath the western Hellenides. This profile, ION-7, extended from the deep Ionian basin into the Gulf of Patras. East of the Ionian Islands, the reflectivity pattern suggested a rather thin continental-type crust of ~23 km thickness.

Under the western slope of the islands, a major normal-incidence reflector dipping eastward was found. In particular, a bright reflector at 13 km depth was detected, which had

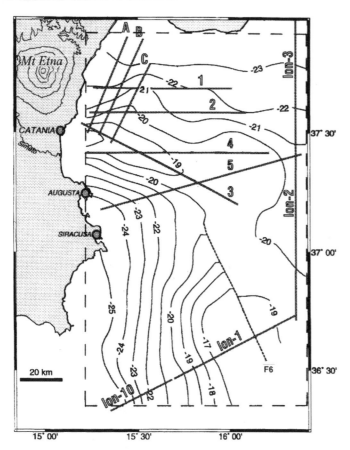

Figure 9.2.3-09. Location of seismic lines and Moho contours in the western Ionian Sea east of Sicily (from Nicolich et al., 2000, fig. 7). [Tectonophysics, v. 329, p. 121–139. Copyright Elsevier.]

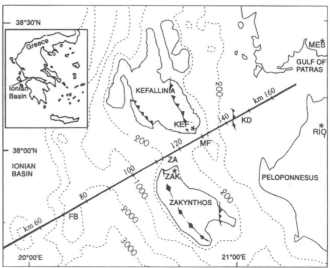

Figure 9.2.3-10. Location map of the STREAMERS profiles ION-7 in the eastern Ionian Sea west of the Peloponnesus, Greece (from Hirn et al., 1996, fig. 1). [Tectonophysics, v. 264, p. 35–49. Copyright Elsevier.]

been suggested as being the interface along which the western Hellenides override the African plate.

To investigate this structure in more detail, in 1997 the French-Greek project SEISGRECE was performed around the Ionian islands west of the Peloponnesus, consisting of a number of marine multichannel reflection profiles and OBS and land-based stations to record refraction and wide-angle reflection data (Clément et al., 2000).

From three of the multichannel seismic profiles of the projects STREAMERS and SEISGRECE, Laigle et al. (2000) determined an active fault which they could image to 10 km depth.

During the SEISGRECE project of 1997, an airgun profile was also recorded with shots every 50 m in the part of the Gulf of Corinth to the east of Aigion (Sachpazi et al., 2003; Clément et al., 2004). Along the N 120° E axis of the Gulf of Corinth, a multichannel reflection seismic profile was shot in the 0.9 km deep marine basin and OBS refraction lines were recorded that penetrated the basement. Furthermore, multichannel seismic reflection profiles, which had been shot earlier with a Maxipulse explosive source at 25-m intervals into a 96-channel, 2.4-km-long streamer, were reprocessed for a joint interpretation. The latter two profiles were NNW-SSE transects which crossed

almost orthogonally the deep marine basin of the Gulf (Sachpazi et al., 2003). Using an array of 14 airguns, a source with maximum efficiency became available, because a signal was provided that is peaked in the rather low-frequency 12–20 Hz but has a duration short enough for acceptable resolution for seismic normal-incidence reflections. Ocean bottom seismometers and land stations, offset at either end of the shot line along the axis of the Gulf of Corinth recorded clear seismic waves out to the maximum recording distance of 105 km. The Moho was found at 40 km depth under the western Gulf north of Aigion and 32 km under its north coast, north of Corinth (Clément et al., 2004).

Another shallow-marine seismic survey was performed in the eastern Mediterranean between south of Cyprus and the Syrian coast revealing a thick sedimentary sequence in the Levantine Basin with the base of the Mesozoic sedimentary unit at 7.8–8.0 s TWT, being interpreted as top of the crystalline basement (Vidal et al., 2000).

9.2.4. Eastern Europe

When the EUROPROBE program started in 1993, the investigation of the Trans-European Suture Zone, separating the mobile western Hercynian Europe with its predominantly 30-km-thick crust from the East European Platform and adjacent Baltic Shield with 40–50-km-thick crust, became one of its earliest international collaborative research projects (Thybo et al., 1999). While the transition from the Baltic Shield to Hercynian Europe had been one of the objectives of the European Geotraverse in the 1980s, the deep structure of the border zone between Hercynian Europe and the East European platform in Poland, which had always been a target of Polish DSS research since the late 1960s, was now, in the second half of the 1990s, addressed with renewed power.

In 1997, the Polish Academy of Sciences and the University of Warsaw, under the leadership of A. Guterch and M. Grad, organized a large seismic-refraction program, POLONAISE (Polish Lithosphere Onsets—An International Seismic Experiment), to investigate the deep structure of the lithosphere under the Trans-European Suture Zone in east-central Europe and the southwestern portion of the Precambrian East European craton in the area of northwestern Poland (Fig. 9.2.4-01).

The seismic-refraction wide-angle reflection experiment was conducted in May 1997 and included contributions from the geophysical and geological communities in Poland, Denmark, United States, Germany, Lithuania, Finland, Sweden, and Canada (Guterch et al., 1999). The project involved two deployments of 613 refraction instruments, the majority of which came from the University of Texas and the IRIS/PASSCAL pool, and five multichannel (90 or 120 channels) seismic-reflection stations to record shots along five interlocking profiles with a total length of ~2000 km, covering 800 sites, which, on average, were spaced 1.5 km on profile 4 and 3 km on the other lines. By this simultaneous recording of shots on several profiles, a good 3-D coverage was ensured. Sixty-four explosions were fired in 40 m

deep boreholes, each loaded with 50 kg of TNT. The total charges varied from 100 to 1000 kg. Distances between shotpoints were 15–35 km. The shot efficiency was excellent. In general, clear first arrivals and later phases were obtained to the farthest offsets of ~600 km from the shotpoints. As the two sample record section demonstrate, the data showed striking differences between the timing and appearance of seismic phases, depending on whether the shots were located southwest (Fig. 9.2.4-02) or northeast (Fig. 9.2.4-03) of the Trans-European Suture Zone (Guterch et al., 1999; see also Appendix A9-1-3).

An example of the modeling procedure is shown in Figure 9.2.4-04 for profile 4. This line started in Germany, ran through the whole of Poland, and extended through most of Lithuania, covering a distance of ~900 km. The model was obtained by forward ray-tracing using the SEIS83 package of Červený and Pšenčik (1984c). It clearly shows the stepwise change of crustal thickness, from near 30 km underneath the Paleozoic platform of Hercynian Europe, from where the Moho increases drastically to almost 45 km under the Trans-European Suture Zone and farther to near 50 km under the Precambrian East European Platform. Figure 9.2.4-04 also demonstrates the enormous thickness

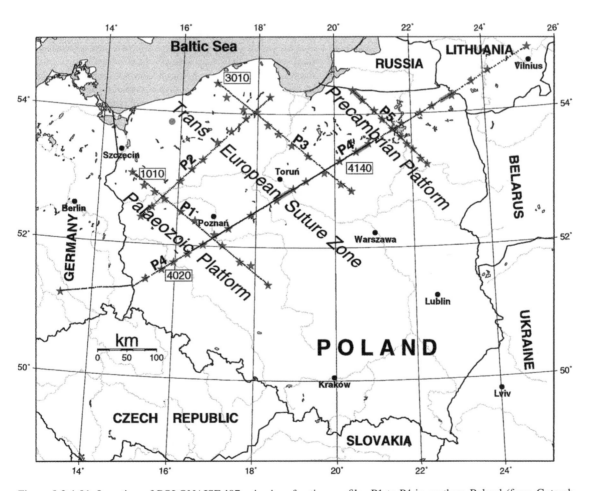

Figure 9.2.4-01. Location of POLONAISE '97 seismic refraction profiles P1 to P4 in northern Poland (from Guterch et al., 1999, fig. 2). [Tectonophysics, v. 314, p. 101–121. Copyright Elsevier.]

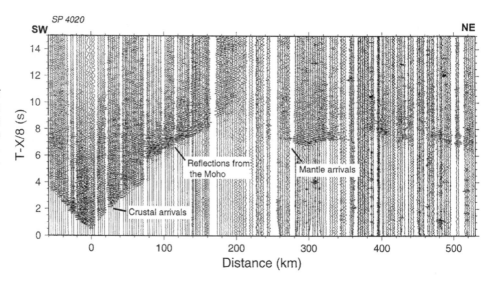

Figure 9.2.4-02. Data example of POLONAISE '97 seismic refraction profile P4, with shotpoint (SP 4010) location SW of Trans-European Suture Zone (from Guterch et al., 1999, fig. 11). [Tectonophysics, v. 314, p. 101–121. Copyright Elsevier.]

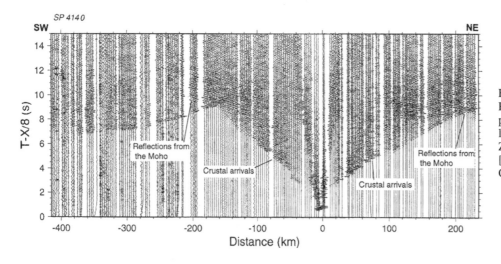

Figure 9.2.4-03. Data example of POLONAISE '97 seismic refraction profile P4, with shotpoint (SP 4140) location NE of Trans-European Suture Zone (from Guterch et al., 1999, fig. 12). [Tectonophysics, v. 314, p. 101–121. Copyright Elsevier.]

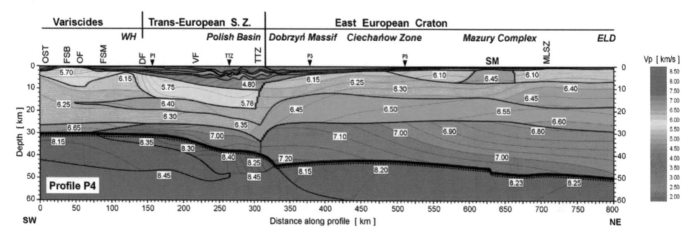

Figure 9.2.4-04. Two-dimensional P-wave crustal model along the POLONAISE profile P4 (from Dadlez et al., 2005, fig. 2). [Tectonophysics, v. 411, p. 111–128. Copyright Elsevier.]

of sediments, filling the Polish Basin above the Trans-European Suture Zone, with velocities of less than 6 km/s. The basin has maximum depths of 16–20 km.

In only another three years, the organizers of POLONAISE undertook a project of even larger size, the project CELEBRATION 2000 (Central European Lithospheric Experiment Based on Refraction). Scientific organizations in Poland, Hungary, the Czech Republic, Slovakia, Austria, Russia, Belarus, and Germany provided the sources in their countries. This project involved the participation of 28 institutions from Europe and North America and targeted the Trans-European Suture Zone region, as well as the southwestern portion of the East European craton (southern Baltica), the Carpathian Mountains, the Pannonian basin, and the Bohemian massif (Fig. 9.2.4-05). The profiles covered southeastern Poland, the Czech Republic, Slovakia, and Hungary, and reached into Austria in the southwest, into Germany in the west, and into Belarus and Russia in the northeast (Guterch et al., 2001).

The layout of the CELEBRATION 2000 experiment was organized similarly to POLONAISE 1997. The project comprised a network of interlocking seismic-refraction profiles (Fig.

9.2.4-05) with a total length of ~8900 km. Station spacing along the profiles was either 2.8 or 5.6 km. One hundred forty-seven shots, ranging in size from 90 to 15,000 kg, were fired along most of the profiles, so that ~5400 km of traditional profile data were obtained. In addition, a large array of 3-D coverage resulted. To accomplish this network of profiles, three deployments were required, which were completed within a period of one month (June 2000). In total, ~100,000 usable seismograms were obtained. About 840 of the 1230 instruments were RefTek

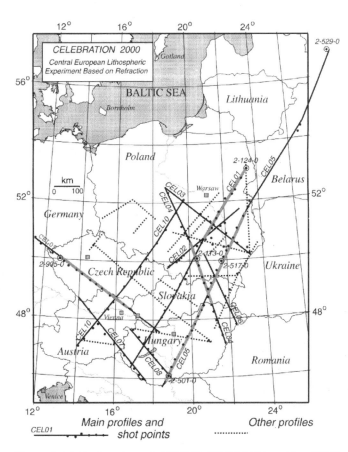

Figure 9.2.4-05. Location of CELEBRATION 2000 profiles (CEL) in central-eastern Europe (from Guterch et al., 2003a, fig. 2). [Studia Geophysica et Geodaetica, Academy of Science, Czech Republic, Prague, v. 47, p. 659–669. Permission granted by Studia Geophysica et Geodaetica, Prague, Czech Republic.]

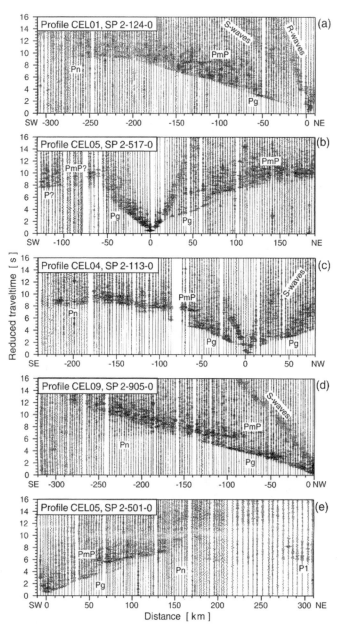

Figure 9.2.4-06. Record sections from different tectonic regions traversed by CELEBRATION 2000 profiles (CEL) in central-eastern Europe (from Guterch et al., 2001, fig. 3). [Eos (Transactions, American Geophysical Union), v. 82, p. 529, 534–535. Reproduced by permission of American Geophysical Union.]

125 ("Texan") recorders, 640 of them from the joint pool of
IRIS/PASSCAL and the University of Texas at El Paso, and
200 from European partners: Austria—15, Denmark—100, Finland—10, Poland—15, and Turkey—60 (Guterch et al. (2003b).
In addition, 50 three-component RefTek 72 instruments of the
IRIS/PASSCAL pool were deployed in the Czech Republic, and
another 20 three-component stations in Poland (Polish MK-4P
units) for one month to record shots and earthquakes. Figure
9.2.4-06 shows record sections recorded in different tectonic
regimes: (a) East European craton, (b) Trans-European Suture
Zone region, (c) Carpathian Mountains, (d) Bohemian Massif,
and (e) Pannonian Basin (see also Appendix A9-1-4).

The longest line, CEL05, was 1400 km long, traversed
the East European craton, the Trans-European Suture Zone,
the Carpathians, and the Pannonian basin, and provided the
key information on the lithosphere (Guterch et al., 2003b). For
the first modeling, a tomographic approach after Hole (1992),
using first arrivals only, was applied (Guterch et al., 2003b).
Later forward ray-tracing modeling (Fig. 9.2.4-07) used the full
P-wave field (Grad et al., 2006).

The resulting crustal thickness along this line varied from
43 to 50 km beneath the East European craton to more than
50 km under the southernmost unit of the Trans-European Su-

ture Zone (crossing point of lines CEL03 and CEL05; see Fig.
9.2.4-08), and decreased to 30 km under the Carpathians and less
than 23 km under the Pannonian Basin. While under the East
European craton a three-layer crust prevailed, the crust under the
Carpathians and the Pannonian basin appeared to consist of only
two distinct layers with relatively low velocities (Fig. 9.2.4-07).
The uppermost mantle velocity also varied, from 8.25 km/s and
more under the craton to 8.0 km/s and less under the Carpathian-
Pannonian area.

Another profile, CEL03, 700 km long, targeted central Poland
along the Trans-European Suture Zone, a key target of the Polish
crustal structure investigations already in earlier decades, also
named TTZ after its discoverers Teisseyre and Tornquist in the
nineteenth century (Teysseire and Teysseire, 2003). The detailed
interpretation of profile CEL03 revealed that the Trans-European
Suture Zone is not uniform, but has to be subdivided into several
units (Fig. 9.2.4-08) with crustal thickness increases from near
30 km near the Baltic Sea in the northwest to 50 km and more
near the Ukrainian border in the southeast (Janik et al., 2005).

The 450-km-long line CEL09 (Fig. 9.2.4-05) served to investigate the deep structure of the Bohemian Massif (Hrubcova
et al., 2005). Its northwestern end in eastern Germany crosses
the German seismic lines GRANU'95 and MVE'90 and runs

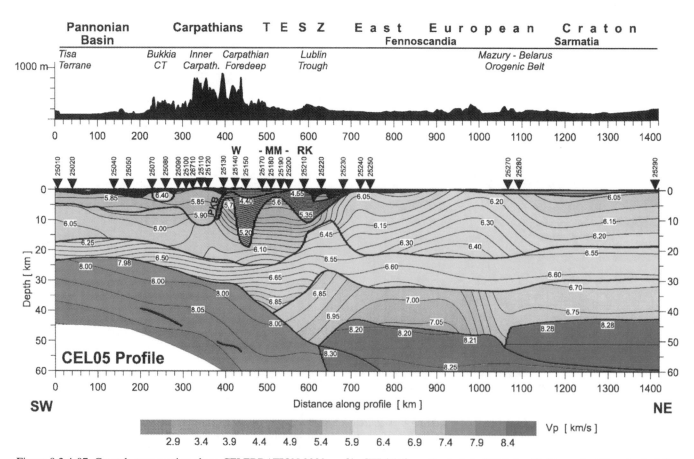

Figure 9.2.4-07. Crustal cross section along CELEBRATION 2000 profile CEL05 (from Grad et al., 2006, fig. 7). [Journal of Geophysical Research, v. 111: B03301, doi:10.1029/2005JB003647. Reproduced by permission of American Geophysical Union.]

almost parallel to the reflection line 9HR (Tomek et al., 1997) within the Czech Republic. The highly reflective lower crust under the Saxothuringian zone in the northwest terminates at the Moho at 26–35 km depth, while the central Moldanubian crust reaches up to 39 km in thickness and is underlain by a pronounced Moho with the strongest velocity contrast. To the southeast, a thick crust-mantle transition zone ranges from 23 to 40 km depth (Fig. 9.2.4-09).

One of the particular tectonic features in central-eastern Europe is the Pannonian basin, with an extremely thin crust of 20–25 km which is surrounded by orogenic regions of much thicker crust. Its thick Neogene cover has been the goal of an ongoing program for deep exploration. The PANCARDI project within the EUROPROBE program in the 1990s became a driving force for a detailed review of existing and new reflection seismic profiles within the Pannonian basin (Horvath et al., 2006).

Here, in 1992, with the support of Canadian scientists, the Pannonian Geotraverse (Fig. 9.2.4-10) was launched as an extension of previous investigations (Hajnal et al., 1996). Using a special observation scheme for recording units, normally used in large-scale refraction profiling, special seismic-reflection measurements were arranged. Expecting narrow-band, low-frequency signals, a 100-km-long seismic-reflection profile was

designed, using 200 Canadian PRS units and 12 Swiss recorders. A mobile reflection spread comprised 195 of the PRS units, while the remaining stations were permanently placed at the ends of the line. Fifty-kg explosive sources, placed at 65 m depth, generated excellent record quality, in which the Moho reflections were generally visible already in the field monitors. The results confirmed the existence of a thin crust with 25–27 km thickness, with the crust being decoupled by well-defined detachment faults and the Moho dipping gently eastwards.

A more extended but less detailed coverage of the Pannonian basin followed in 2000 with CELEBRATION 2000, which has been described above, and in 2002 and 2003, during the long-range refraction profiling programs ALP 2002 and SUDETES 2003, which will be described in Chapter 10.

While CELEBRATION 2000, ALP 2002, and SUDETES 2003 targeted also the western Carpathians in Poland, Slovakia, and Hungary, the Carpathians of Romania became the goal of a joint German-Romanian research program, aiming to study the seismically high-risk Vrancea area at the center of the southeastern Carpathians and the causes of the strong and deep-seated (around 100 km depth) intermediate-upper-mantle earthquakes.

Seismic-refraction studies and a large-scale tomography study of southeastern Romania were part of an interdisciplinary

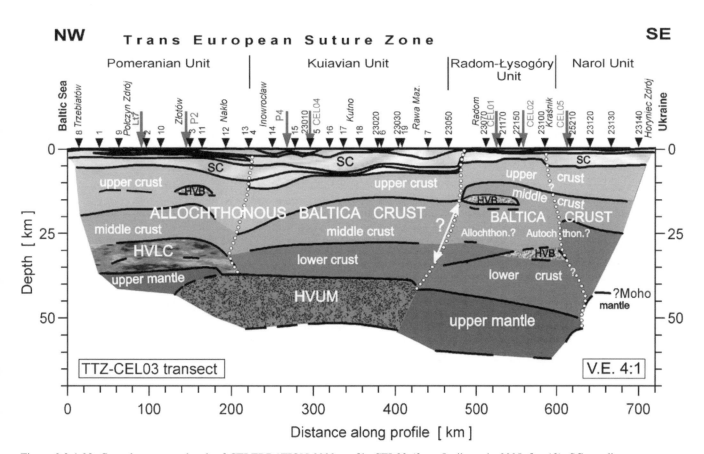

Figure 9.2.4-08. Crustal structure sketch of CELEBRATION 2000 profile CEL03 (from Janik et al., 2005, fig. 12). SC—sedimentary cover, HVB—high-velocity body, HVLC—high-velocity lower crust, HVUM—high-velocity upper mantle. [Tectonophysics, v. 411, p. 129–156. Copyright Elsevier.]

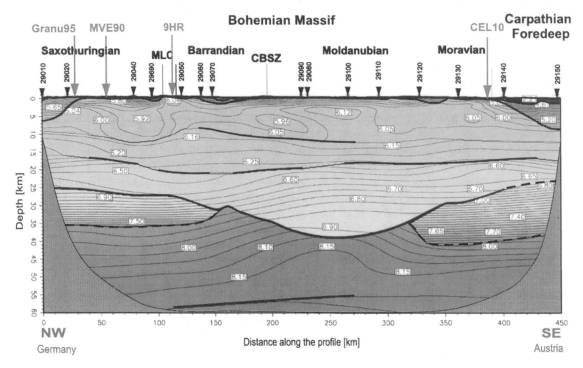

Figure 9.2.4-09. Crustal structure along CELEBRATION 2000 profile CEL09 traversing the Bohemian Massif (from Hrubcova et al., 2005, fig. 6) [Journal of Geophysical Research, v. 110: B11305, doi:10.1029/2004JB003080. Reproduced by permission of American Geophysical Union.]

research program, carried out by the Collaborative Research Center 461 (CRC 461) "Strong Earthquakes—a Challenge for Geosciences and Civil Engineering" at the University of Karlsruhe (Germany) and the Romanian Group for Vrancea Strong Earthquakes (RGVE) at the Romanian Academy in Bucharest (Wenzel, 1997; Wenzel et al., 1998a, 1998b; Martin et al., 2005, 2006).

Two major active-source seismic-refraction experiments were carried out in the eastern Carpathians of Romania in 1999 and 2001 (Fig. 9.2.4-11) as a contribution to this research program (Hauser et al., 2001, 2002, 2007a). They were designed to study the crustal and uppermost mantle structure to a depth of ~70 km underneath the Vrancea epicentral region and were jointly performed by the Geophysical and Geological Institutes of the University of Karlsruhe (Germany), the National Institute for Earth Physics in Bucharest (Romania) and the University of Bucharest (Romania).

The VRANCEA'99 seismic-refraction experiment (Hauser et al., 2001) was a 300-km-long refraction profile and was

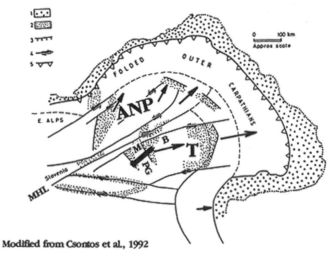

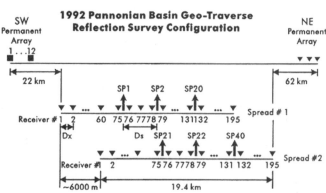

Figure 9.2.4-10. Top: Tectonic position of the Pannonian Basin, PG—location of the 1992 Pannonian Geotraverse (from Hajnal et al., 1996, fig. 3). ANP—Alpine North Pannonian unit, T—Tisza unit, MHL—Mid Hungarian lineament, M—Mako trough, B—Bekes basin. Bottom: Survey configurations of the 1992 Pannonian Geotraverse data acquisition program (from Hajnal et al., 1996, fig. 4). Dx—100 m, Ds—300 m. Total number of shots: 289, record length: 60 s, total coverage 95 km, nominal fold: 24. [Tectonophysics, v. 264, p. 191–204. Copyright Elsevier.]

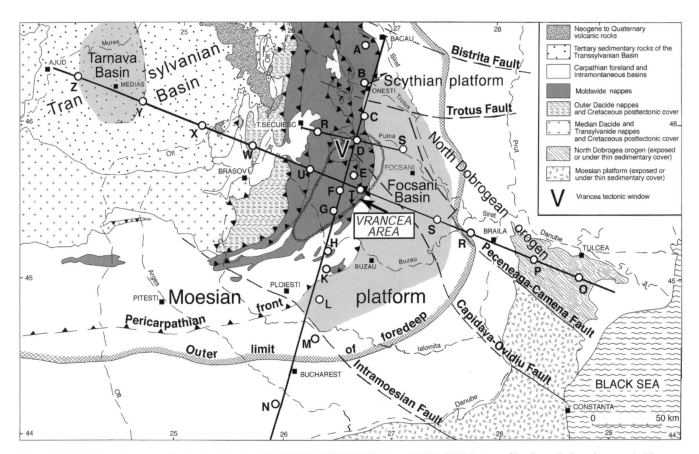

Figure 9.2.4-11. Location map of seismic-refraction VRANCEA 1999 (N-S line) and 2001 (W-E line) profiles through the seismogenic Vrancea zone in the eastern Carpathians of Romania (from Hauser et al., 2002). V—Vrancea seismogenic zone with strong intermediate-depth earthquakes between 70 and 160 km depth. [Eos (Transactions, American Geophysical Union), v. 83, p. 457, 462–463. Reproduced by permission of American Geophysical Union.]

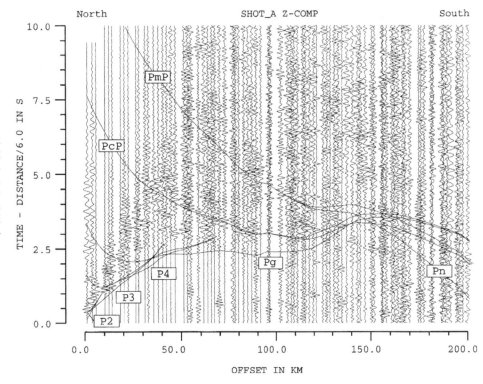

Figure 9.2.4-12. Record section from the northernmost shotpoint A along the north-south profile VRANCEA 1999 (from Hauser et al., 2001, fig. 5a). P1–P4—first arrivals from sedimentary cover; Pg, PcP, PmP—reflections from top basement, top lower crust and Moho; [Tectonophysics, v. 340, p. 233–256. Copyright Elsevier.]

recorded between the cities of Bacau and Bucharest with 12 shot-points A to N, traversing the Vrancea epicentral region in a NNE-SSW direction (Fig. 9.2.4-11). It was complemented by a short perpendicular line, which ran transverse to the geological structures along the Putna Valley, north of the Vrancea seismogenic zone (V in Fig. 9.2.4-11), intersected the main refraction line at shotpoint D, and had two additional shotpoints R and S (north of V in Fig. 9.2.4-11).

Along these two segments, a total of 140 recording sites were occupied with an average station spacing of 2 km. Seismic recording equipment for the experiment was provided by the GeoForschungsZentrum Potsdam (Germany), by Leicester University (UK), and by the NERC geophysical equipment pool (UK). The spacing of shotpoints ranged from 12 to 30 km, with an average of 22 km. The charge sizes varied from 300 kg to 900 kg with the larger shots at the end points of the main line (Fig. 9.2.4-11). All shots were recorded simultaneously along the main line and the transverse line. Hauser et al. (2000; Appendix A9-1-5) and Landes et al. (2004b) compiled data and other technical details in data reports. A data example is shown in Figure 9.2.4-12 (for more data, see Appendix A9-1-5; Hauser et al., 2000, 2001).

The P-wave interpretation (Fig. 9.2.4-13) showed up to 10-km-thick sedimentary sequences. Both the sedimentary and crystalline upper-crustal layers appeared to thicken underneath the seismogenic Vrancea zone and the adjacent southern margin of the Carpathians. As a result, the intra-crustal discontinuity varies strongly in depth, between 18 km at both ends of the line and 31 km at its center. Strong wide-angle P_MP reflections indicated the existence of a first-order Moho, the depth of which also varies from 30 km near the southern end of the line to 41 km near the center and to 39 km near the northern end. Within the uppermost mantle, a low-velocity zone was interpreted from a reflection, named P_LP, which defined the bottom of this low-velocity layer at a depth of 55 km (Hauser et al., 2001). The data also showed reasonable appearance of S-waves, which were incorporated into the P-wave model (corresponding velocities are shown in brackets in Fig. 9.2.4-13) and Poisson's ratios were calculated for the individual crustal layers (Raileanu et al., 2005).

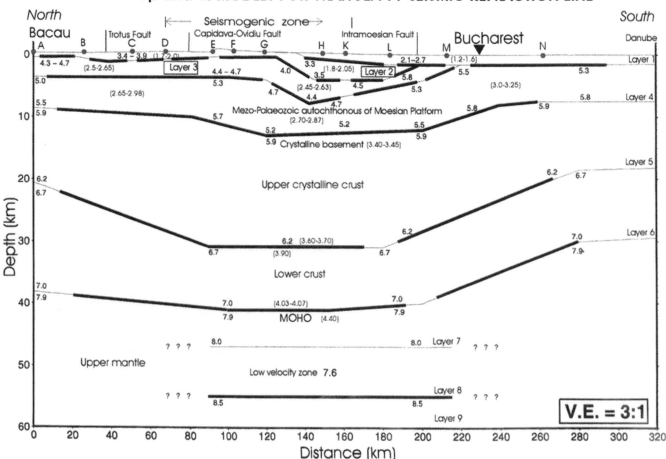

Figure 9.2.4-13. Crustal P-wave cross section along the north-south profile VRANCEA 1999 (from Hauser et al., 2001, fig. 9). Velocities in brackets are corresponding S-velocities. [Tectonophysics, v. 340, p. 233–256. Copyright Elsevier.]

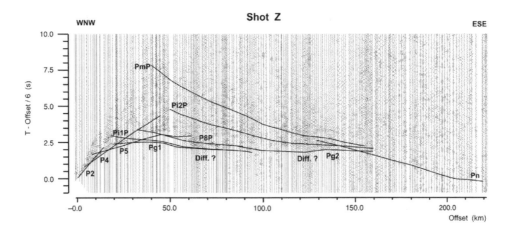

Figure 9.2.4-14. Record section from the westernmost shotpoint Z in Romania along the west-east profile VRANCEA 2001 (from Hauser et al., 2007a, fig. 9). P1–P5—first arrivals from sedimentary cover; Pgx, PxP—refracted/reflected phases from crustal discontinuities; PmP, Pn—reflected/refracted phases from Moho. [Tectonophysics, v. 430, p. 1–25. Copyright Elsevier.]

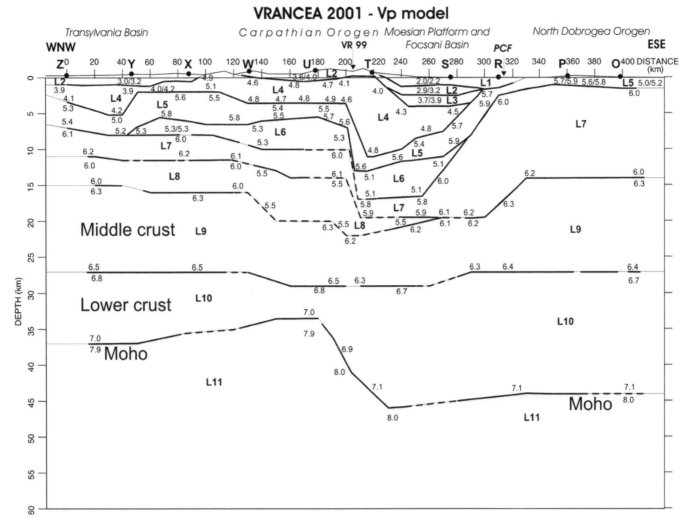

Figure 9.2.4-15. Crustal P-wave cross section along the west-east profile VRANCEA 2001 (from Hauser et al., 2007a, fig. 11). [Tectonophysics, v. 430, p. 1–25. Copyright Elsevier.]

The second experiment across the Vrancea zone, VRANCEA 2001, was recorded in August/September 2001. Because of its close relation to the VRANCEA'99 project, its details will be described here in Chapter 9. VRANCEA'2001 was a 700-km-long WNW-ESE–trending seismic-refraction line in Romania with a short extension into Hungary (Hauser et al., 2002, 2007a). The main part (Fig. 9.2.4-11) ran from the Transylvanian Basin across the East Carpathian Orogen and the Vrancea seismic region to the foreland areas with the very deep Neogene Focsani Basin and the North Dobrogea Orogen near the Black Sea. Ten large drillhole shots with 300–1500 kg charges (O to Z in Fig. 9.2.4-11) were fired in Romania between the town of Aiud at the western margin of the Transylvanian Basin and the Black Sea, which resulted in an average shotpoint spacing of 40 km. To the west an additional shot (500 kg) was fired in Hungary. These shots generated 11 seismogram sections each with almost 800 traces (data examples in Fig. 9.2.4-14 and Appendix A9-1-5; Hauser et al., 2007a). The spacing of the geophones was variable. It was around 1 km from the eastern end to Aiud (~450 km length), 6 km from Aiud to Oradea (Romanian-Hungarian border), and ~2 km on the Hungarian territory. Between shotpoints T and U, the geophones were deployed at a spacing of 100 m (Panea et al., 2005). In total, 790 recording instruments were available. The small 640 one-component geophones (Texan type) were mostly deployed in the open field outside of localities, while, for safety reasons, the 150 three-components geophones (RefTek and PDAS type) were deployed in guarded properties within towns and villages.

The experiment was jointly performed by the research institutes and universities of Germany, Romania, the Netherlands, and the United States. They included the University of Karlsruhe and The GeoForschungsZentrum Potsdam, Germany, the Free University of Amsterdam, Netherlands, the National Institute for Earth Physics and the University of Bucharest, Romania, and the Universities of El Paso and South Carolina, United States. Field recording instruments were provided by the University of El Paso and PASSCAL, United States (Texan one-component stations), and the GeoForschungsZentrum, Germany (three-component RefTek and PDAS stations).

The data interpretation for the eastern 450-km-long segment (Hauser et al., 2007a) indicated a multi-layered structure with variable thicknesses and velocities. The sedimentary stack comprised up to 6 layers (L1-L6 in Fig. 9.2.4-15) with seismic velocities of 2.0–5.9 km/s and reached a maximum thickness of more than 15 km within the Focsani Basin area. The underlying crystalline crust (L7–L10 in Fig. 9.2.4-15) showed considerable thickness variations in total as well as in its individual subdivisions. The model (Fig. 9.2.4-15) shows the lateral velocity structure of these blocks along the seismic line which remained almost constant with ~6.0 km/s along the basement top and with 7.0 km/s above the Moho. Under the Transylvanian basin, the crust appeared to be 34 km thick with low-velocity zones in its uppermost 15 km. Under the Carpathians, the Moho deepens to beyond 41 km, reaching 46 km under the Focsani basin. However, the crystalline crust does not exceed 25 km in thickness and is covered by up to 15 km of sedimentary rocks. The North

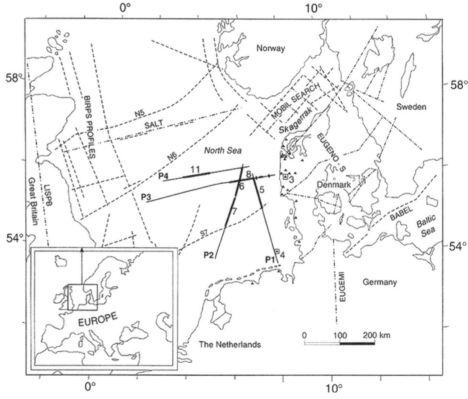

Figure 9.2.5-01. Location map of MONA LISA seismic normal incidence profiles and land recording sites (solid lines P1 to P4 and triangles) and former seismic investigations (dashed lines) in the North Sea area (from MONA LISA Working Group, 1997, fig. 1). [Tectonophysics, v. 269, p. 1–19. Copyright Elsevier.]

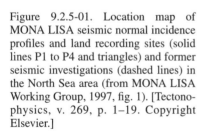

Dobrogea crust reaches a thickness of ~43 km and was interpreted as thick eastern European crust overthrusted by a thin 1–2-km-thick wedge of the North Dobrogea Orogen (Hauser et al., 2007a).

The two seismic-refraction campaigns and the teleseismic tomography survey of 1999 (Wenzel et al. 1998b; Martin et al., 2005, 2006) produced a wealth of data, concentrating on the Vrancea zone proper, and allowed for a 3-D tomographic interpretation of the crustal data, which were presented in contour maps at various depth levels (Landes et al., 2004a).

9.2.5. North Sea and Northern Europe

After many successful marine seismic-reflection and -refraction investigations of the North Sea and the adjacent Skagerrak in the 1980s (see Fig. 9.2.5-01), as, e.g., BIRPS, SALT, MOBIL SEARCH, and others (e.g., Klemperer and Hobbs, 1991; Barton and Wood, 1984; Husebye et al., 1988), two major seismic projects were added in the 1990s: in 1992, a BIRPS experiment and in

1993, the MONA LISA project (Marine and Onshore North Sea Acquisition for Lithospheric Seismic Analysis).

The BIRPS (British Institutions Reflection Profiling Syndicate) project of 1992 was a two-ship coincident near-vertical and wide-angle experiment to determine the continental structure of the central North Sea (Neves and Singh, 1996; Singh et al., 1998a). Its location was within the "BIRPS profiles" (Fig. 9.2.5-01) parallel to and halfway between the SALT line and the next WSW-ENE–oriented line to the south shown in Figure 9.2.5-01. It was a 106-km-long expanding spread profile. It was obtained by the steaming of two vessels at a constant speed toward a central midpoint, with one of them shooting every 75 m with simultaneous recording on a 114-channel 2.7-km dual streamer, and the other one receiving with a 240-channel 6-km streamer. In addition, seven ocean-bottom hydrophones (OBHs) were deployed at 20-km intervals along the profile. In a second stage, the two vessels steamed in the same direction at constant offsets corresponding to critical distances for the top of the lower crust and the crust-mantle transition. The principal results along the

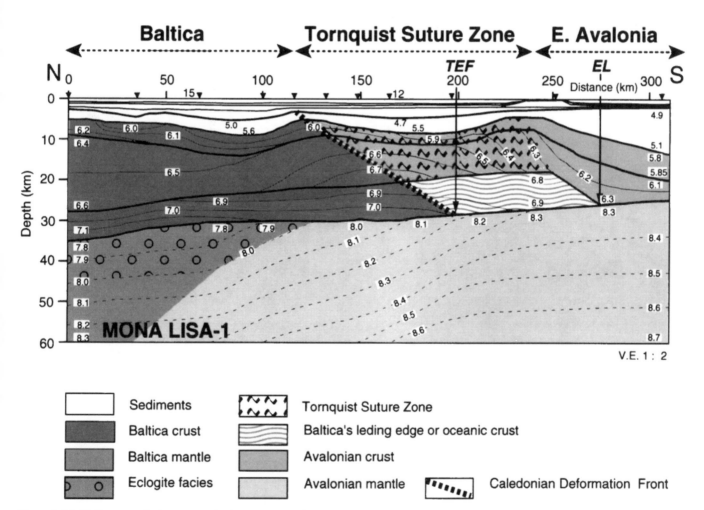

Figure 9.2.5-02. P-wave velocity model of MONA LISA profile 1 (from Abramovitz et al., 1999, fig. 2). Small triangles indicate the position of OBHs. TEF—Trans-European Fault, EL—Elbe Lineament. [Tectonophysics, v. 314, p. 69–82. Copyright Elsevier.]

135-km-long profile were an upper crustal thickness of 19 km and a Moho depth of 32–34 km (Singh et al., 1998a).

The main aim of the MONA LISA project (Fig. 9.2.5-01) was to image plate collision structures and reflectivity characteristics of the crust and uppermost mantle in order to map the Caledonian Deformation Front and possible remains of the

Tornquist and Iapetus lithospheric plates (MONA LISA Working Group, 1997). Another aim was to identify rifting processes by profiles crossing the Central Graben and the Horn Graben. Here, attempts were made to avoid anticipated problems due to strong attenuation of seismic signals in the graben sediments by locating the seismic lines primarily across structural highs

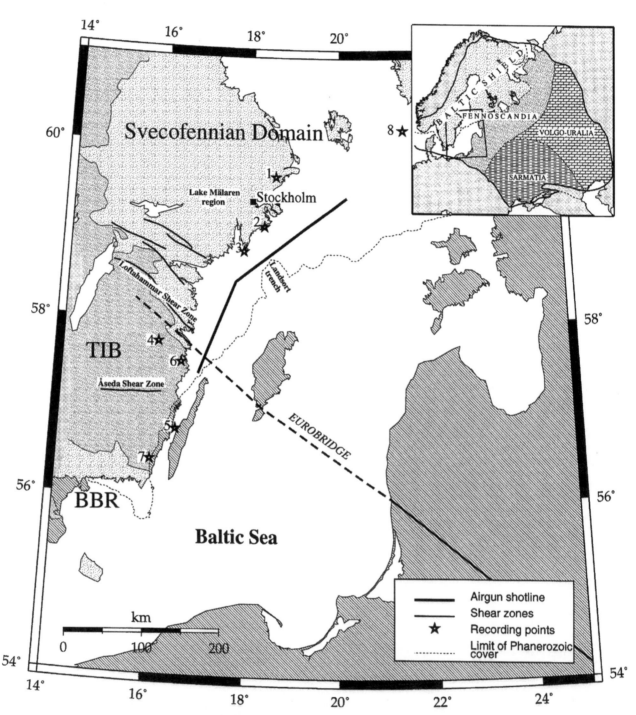

Figure 9.2.5-03. Location map of the Coast Profile transect along the southeastern coast of Sweden (from Lund et al., 2001, fig. 1). Dashed line—EUROBRIDGE line. TIB—Transscandinavian Igneous Belt; BBR—Blekinge-Bornholm region. [Tectonophysics, v. 339, p. 93–111. Copyright Elsevier.]

and by the use of a very powerful airgun array tuned to low-frequencies.

The fieldwork was achieved in 1993 by the cooperation of scientific and commercial institutions of Denmark, Germany, Great Britain, and Sweden. The four profiles comprised a total length of 1112 km of normal incidence seismic data and the recording of the airgun signals at 26 onshore and 2 offshore locations, allowing for successful recording distances of up to 400 km. Shots at 75 m intervals were recorded by a 180-channel streamer of 4500 m length, towed at 15 m depth which provided 30 fold data with 26 s record length and 12.5 m CMP spacing. In 1995, additional wide-angle profiling with OBH stations was carried out.

The normal-incidence reflection profiles showed an unusually low intracrustal reflectivity, in particular a poorly reflective lower crust. The Moho could be identified on all four profiles, but also, for the first time, dipping and subhorizontal reflections from the mantle up to 24 s TWT could be correlated in the normal-incidence reflection sections. For the Central Graben a reduction in crustal thickness by ~5 km was observed (MONA LISA Working Group, 1997). The following interpretation (Abramovitz et al., 1998) revealed a three-layered Baltica crust in the north thinning from 34–35 km to 29–30 km at the Tornquist Suture Zone and changing into a 24–25-km-thick two-layered East Avalonia crust (see, e.g., Fig. 9.2.5-02) in the south.

In Scandinavia, after the thorough investigation of the Scandinavian crust by a large series of seismic-refraction campaigns in the 1980s, which concentrated particularly on the crust underneath the Baltic Shield in Sweden, Finland, and under the Baltic Sea, only one crustal survey was added in the 1990s. This was a marine study in 1995 along the southeastern coast of Sweden named Coast Profile (Lund et al., 2001). The shooting vessel, employed from the Polar Geosurvey Expedition in St. Petersburg, Russia, fired airgun shots every two minutes for 38 h along the shooting line, resulting in distances of 270 m between consecutive shotpoints.

Eight recording stations had been deployed on land along the Swedish coast, of which six provided high-quality data ranging from 25 km to 327 km airgun-receiver offsets (Fig. 9.2.5-03). The data were interpreted together with data of the FENNOLORA project of 1979 resulting in a crustal model along the southeastern Swedish coast where Moho depths varied from 36 to 52 km (Fig. 9.2.5-04). The results were later discussed in the context of EURO-BRIDGE which will be discussed in subchapter 9.3, "Deep Seismic Sounding in Eastern Europe and Adjacent Asia (Former USSR)."

Otherwise, seismic projects in Scandinavia in the 1990s were devoted primarily to upper mantle studies. Already in the 1980s, some of the long-range seismic-refraction profiles had enabled the detection of details of sub-Moho upper-mantle structure (see, e.g., Guggisberg and Berthelsen, 1987).

In Russia, the reinterpretation of super-long profiles of several 1000 km length, based on the use of nuclear explosions in the late 1970s, had revealed details of the upper-mantle structure to several 100 km depth. In Scandinavia, such strong energy sources were not available, therefore in the 1990s, several major projects were organized, using the methodology of teleseismic tomography. According to the scope of this publication, however, these projects will not be discussed in much detail.

Examples are the Teleseismic Tomography Tornquist (TOR) project, investigating the deep lithospheric structure of the Trans-European Suture Zone between Phanerozoic and Proterozoic Europe by a teleseismic array of 120 stations, recording for half a year in 1996 and 1997 and simulating a 100-km-wide and 900-km-long profile from northern Germany through Denmark to southern Sweden and reaching a depth range of 300 km (Gregersen et al., 2002). Some details of crustal S-wave structure were obtained by applying the time-domain inversion method to the receiver functions to compute S-wave velocities in the crust and uppermost mantle beneath each station down to 60 km depth (Wilde-Piorko et al., 2002).

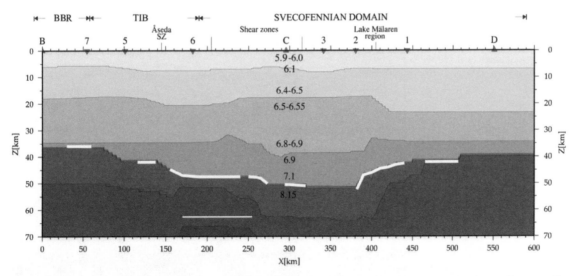

Figure 9.2.5-04. P-wave velocity model of the Coast profile 1 (from Lund et al., 2001, fig. 7). Numbers—Coast profile recording sites; letters—FENNOLORA shotpoints. [Tectonophysics, v. 339, p. 93–111. Copyright Elsevier.]

The second large project was the Svecofennian-Karelian-Lapland-Kola Transect (SVEKALAPKO) project, which included near-vertical and wide-angle seismic-reflection and refraction work, large-scale magnetotelluric studies, geothermal surveys, and an intensive study of teleseismic data with a network of 144 temporary and permanent seismic stations (Bock et al., 2001). The temporary seismic station array was operated between August 1998 and May 1999 covering the Proterozoic and Archean crust in Finland and Russian Karelia.

9.3. DEEP SEISMIC SOUNDING IN EASTERN EUROPE AND ADJACENT ASIA (FORMER USSR)

Also located on the Baltic Shield, in 1992, a seismic-reflection line was recorded across the Kola Superdeep Borehole (Fig. 9.3-01), which had been cored to a depth of 12.25 km and had attracted worldwide attention, because, due its large depth, hypotheses on the nature of crustal reflectivity could be tested against in situ geological control (Kozlovsky, 1988).

Single-fold reflection shooting and DSS had been conducted in the SG-3 region before, but it had lacked deep-crustal, common-depth-point (CDP) seismic profiling (Ganchin et al., 1998). Near-vertical reflection profiling, based on the "common midpoint" seismic method (CMP), began in the Kola Peninsula in the beginning of the 1990s. The Russian state geophysical company EGGI recorded two profiles in the northwestern part with a record length of 15s (Kostyuchenko et al., 2006).

In 1992, an international group acquired a 25s TWT line across the Kola super-deep borehole (Ganchin et al., 1998). The experiment of 1992 comprised the 38-km-long CDP profile across, and two deep (6 km) and four shallow (0.5 km) VSP (vertical seismic profiles) within the borehole. The CDP profiling was arranged along a SW-NE–directed line, starting ~38 km south of the borehole and ending just north of the borehole. It provided a 20-fold coverage and was recorded to 15 s TWT (Smythe et al., 1994; Ganchin et al., 1998). It resulted in an image of a highly reflective section: Dipping reflections could be correlated with lithological contrasts within the supracrustal series and at shear

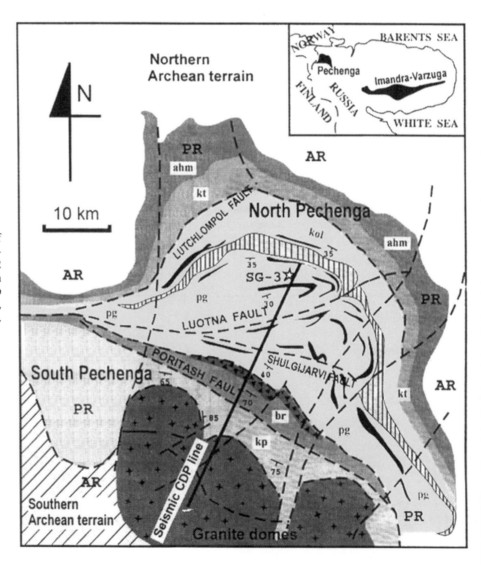

Figure 9.3-01. Geological sketch map of the Pechenga region on the Kola Peninsula, showing the location of the Kola-92 seismic-reflection line (seismic common depth point line) and the Kola Superdeep Borehole SG-3 (from Ganchin et al., 1998, fig. 1). [Tectonophysics, v. 288, p. 1–16. Copyright Elsevier.]

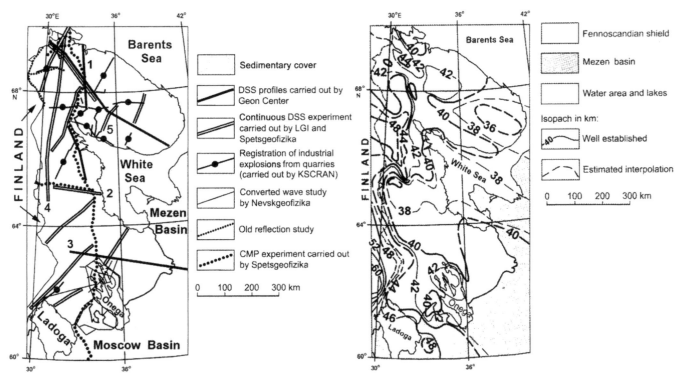

Figure 9.3-02. Left: Location of regional deep seismic sounding (DSS) profiles across the eastern part of the Fennoscandian Shield (from Kostyuchenko et al., 2006, fig. 4). Numbered single lines—DSS profiles performed by GEON; numbered double lines—continuous DSS profiles performed by Spetsgeofizika. Right: Depth to Moho under the northeastern part of the Fennoscandian Shield (from Kostyuchenko et al., 2006, fig. 10). [Gee, D.G., and Stephenson, R.A., eds., European lithosphere dynamics: Geological Society of London Memoir 32, p. 521–539. Reproduced by permission of Geological Society Publishing House, London, U.K.]

zones. The results suggested the presence of fluids down to a depth of at least 12 km in the upper crust.

From 1997 to 2000, Spetsgeofisika completed two profiles (dotted lines in Fig. 9.3-02), including a long CMP profile from the Kola borehole in the north to Petrozavodsk in the south and a second profile in W-E direction from the Finnish border to the White Sea (Berzin et al., 2002).

From 2000 to 2002, Spetsgeofisika conducted a Vibroseis CMP experiment across the western and central Mezen Basin and along a line across the Timan Range (dotted lines 3 and 4 in Fig. 9.3-03). The total length of available CMP profiles (both state and industry) amounted to ~2000 km. Figures 9.3-02 and 9.3-03 show only the state-financed CMP lines. The maximum source-receiver distance was 10 km. The energy was provided by 4 or 5 10-t vibrators spread over a base of 50 m. Groups of 12

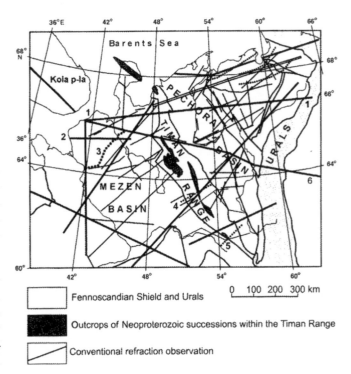

Figure 9.3-03. Location of deep seismic sounding (DSS) profiles across the Mezen Basin, Timan Range and Pechora Basin (from Kostyuchenko et al., 2006, fig. 5). Thick lines—CMP profiles; thin lines—older conventional DSS lines. [Gee, D.G., and Stephenson, R.A., eds., European lithosphere dynamics: Geological Society of London Memoir 32, p. 521–539. Reproduced by permission of Geological Society Publishing House, London, U.K.]

geophones were spaced every 44 m. The record length was 25 s TWT and a 50-fold stack was obtained. Figure 9.3-03 also shows older conventional DSS lines which had been recorded by the former Ministry of Geology.

During these studies, 48–60 channel recording instruments had been used, usually spaced every 100–200 m. Seismograms were recorded on paper by the wiggle-trace method. Shotpoint spacing was from 10–20 km to 50–70 km and shooting was direct and reversed. Refraction observations were shot with maximum offset up to 100–150 km (Kostyuchenko et al., 1999, 2006). The interpretation of these data resulted in several cross sections (Kostyuchenko et al., 2006, figs. 7–9) and two Moho maps (Figs. 9.3-02 and 9.3-04), showing on average an overall 38–42-km-thick crust. Only underneath the Mezen and Pechora basins, crustal thinning to less than 36 km was observed. The Moho contour maps also show a sudden crustal thickening along the Finnish border as well as to the east of the Pechora basin (Figs. 9.3-02 and 9.3-04).

A link between the super-deep SG-3 well on the Kola peninsula to the base Hayes-1 well on Franz-Joseph land was established in 1995, when geophysical investigations in the southern part of the AP-1 geotraverse across the Barents Sea (Fig. 9.3-05) were completed (Sakoulina et al., 2000). The integrated seismic experiment included offshore and onshore wide-angle reflection/refraction observations along a 700-km-long profile (Fig. 9.3-05). The offshore recording devices comprised 41 three-component OBSs with 5–20 km spacing and a multi-channel reflection study at different offset ranges. Onshore, eight seismometers had been installed. The seismic energy was provided by powerful 80-l array airgun shots with a 250 m interval. They produced clear arrivals to 150–180 km distances, and in some cases even up to 280 km. Intensive reflections from the Moho at 35–40 km depth were almost continuously recorded (Fig. 9.3-06). The seismic investigations continued into 2001 (Roslov et al., 2009). The AP-2 profile with a total length of 935 km crossed the center of the Barents Sea, Novaya Zemlya Island and the Kara Sea up to the Yamai Peninsula of western Siberia.

To investigate the deep lithospheric structure of the East European Craton between the exposed Proterozoic and Archean complexes of the Baltic and Ukrainian Shields, in 1994 the project EUROBRIDGE was established within the frame of EURO-PROBE and aimed for an integrated geoscience research of a ~2000-km-long swath of the lithosphere, extending from Scandinavia to the Black Sea (Bogdanova et al., 2001, 2006). The main part of the EUROBRIDGE project was a major seismic-refraction and wide-angle reflection experiment. It was conducted in several legs from 1994 to 1997, commencing in the north and terminating in the south, in the Ukraine (Fig. 9.3-07).

The first leg was a marine project across the Baltic Sea (EUROBRIDGE'95 Seismic Working Group, 2001) from Västervik (Sweden) to Shventoji (Lithuania), recorded in 1994. For its northwestern end at the Swedish coast, the Coast Profile (Lund et al., 2001), discussed further above, added important information (Fig. 9.3-07; see also Figs. 9.2.5-03 and 9.2.5-04).

It was followed by three seismic-refraction surveys on land, carried out in 1995, 1996, and 1997, establishing a

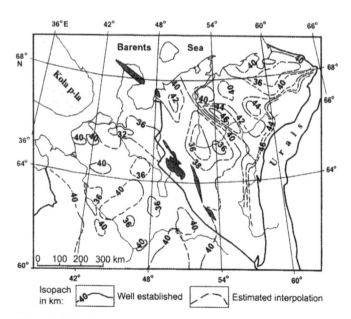

Figure 9.3-04. Depth to Moho under the Mezen Basin, Timan Range and Pechora Basin (from Kostyuchenko et al., 2006, fig. 10). [Gee, D.G., and Stephenson, R.A., eds., European lithosphere dynamics: Geological Society of London Memoir 32, p. 521–539. Reproduced by permission of Geological Society Publishing House, London, U.K.]

Figure 9.3-05. Tectonic sketch map of the Barents Sea showing the location of the AP-1-95 geotraverse (from Sakoulina et al., 2000, fig. 1). [Tectonophysics, v. 329, p. 319–331. Copyright Elsevier.]

DISTANCE (km)

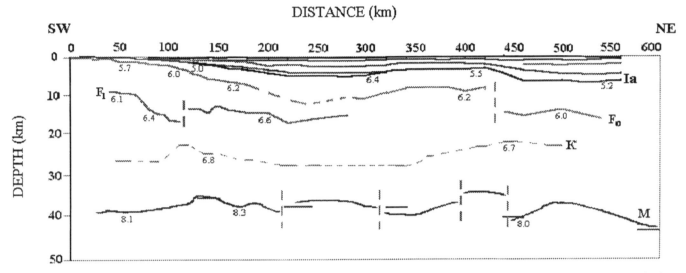

Figure 9.3-06. Crustal cross section through the Barents Sea along the AP-1-95 geotraverse (from Sakoulina et al., 2000, fig. 6). [Tectonophysics, v. 329, p. 319–331. Copyright Elsevier.]

line running from NW to SE, which extended from Sweden across the Baltic Sea and farther on land from the coast of Lithuania through Belarus into the Ukraine (Figs. 9.3-07 and 9.3-08). The 1995–1996 line crossed the northeastern end of POLONAISE profile P4 and ran parallel to P5. Interpretations of profiles P4 and P5 are included in the summary of Bogdanova et al. (2006, Fig. 8).

The first onshore stage, EUROBRIDGE'95, was 280 km long and was recorded on the Lithuanian part of the East European Platform, traversing the Proterozoic West Lithuanian Granulite Domain and East Lithuanian Belt terranes (Fig. 9.3-08). Explosive shots of up to 1000 kg TNT were detonated at 10 shotpoints (SP01-SP10) at intervals of ~30 km, and an offshore shot (SP00) was fired in the Baltic Sea close to Gotland. 76 three-component stations allowed an average station spacing of 3.5 km (EUROBRIDGE'95 Seismic Working Group, 2001).

The second onshore stage, EUROBRIDGE'96, was 544 km long and covered the area (Fig. 9.3-08) from the East Lithuanian Belt to the Ukrainian Shield. The southernmost part was recorded in the Pripyat Trough (see Fig. 9.3-10). During EUROBRIDGE'96, 14 shots were fired at 30 km intervals from new shotpoints 11–24, and two shots from shotpoints of 1995, nos. 08 and 10. The shots were recorded in two deployments by 114 three-component stations, allowing a station spacing between 3 and 4 km (EUROBRIDGE Seismic Working Group, 1999). A record section for shotpoint 24 is shown in Figure 9.3-09, demonstrating the energy distribution for crustal phases up to 300 km recording distance (more data are shown in Appendix A9-1-6).

The third onshore stage, EUROBRIDGE'97 (Fig. 9.3-10) covered Ukrainian territory and was recorded in 1997 (Thybo et al., 2003). The line was 530 km long and crossed the EUROBRIDGE'96 profile in southern Belarus. Data acquisition was obtained in two deployments with overlapping coverage in the center of the line. Seismic energy came from 18 borehole shots

with charges ranging from 250 to 1000 kg, which were recorded by 120 mobile three-component seismographs with a typical station spacing of 3–4 km (for data, see Appendix A9-1-6).

The interpretation of EUROBRIDGE'95 and '96 (Fig. 9.3-11) revealed that the crust consists generally of three layers, but the middle crustal layer is much thinner in the northwest. The Moho in the northwest of Lithuania (West Lithuanian Domain in Fig. 9.3-11) and the adjacent Baltic Sea is ~44 km deep, but its depth increases toward southeast under the East Lithuanian Belt to 50 km and the crust remains 50 km thick below southern Lithuania and Belarus, with Moho elevations of a few kilometers in between. Uppermost-mantle velocities immediately beneath the Moho are generally 8.2–8.4 km/s. A lower lithosphere reflector was found at 65–70 km depth. High lower-crustal velocities and a crustal thickness of ~50 km were observed throughout the EUROBRIDGE' 96 profile. The boundary between the East Lithuanian Belt and West Lithuanian Granulite Domain was associated with pronounced crustal velocity changes, and a thinning of the crust toward the northwest (EUROBRIDGE Seismic Working Group, 1999; EUROBRIDGE'95 Seismic Working Group, 2001).

The interpretation of the EUROBRIDGE'97 data (Fig. 9.3-12) revealed for most of the line (Thybo et al., 2003) a similar velocity-structure model as shown above for the southernmost part of EUROBRIDGE'96, with the exception that the upper part of the lower crust had velocities less than 7.0 km/s and was separated from the lowermost crust by a distinct velocity jump near 35–40 km depth. In the model of Thybo et al. (2003) the lowermost crust, defined by velocities greater than 7 km/s, gradually dissolves toward the south underneath the Ukrainian Shield.

While EUROBRIDGE covered the westernmost part of the Dniepr-Donets rift system, DOBRE (Donbas Refraction and Reflection), a multinational study group, concentrated on the Donbas Foldbelt of the Ukraine (Fig. 9.3-13), the uplifted and

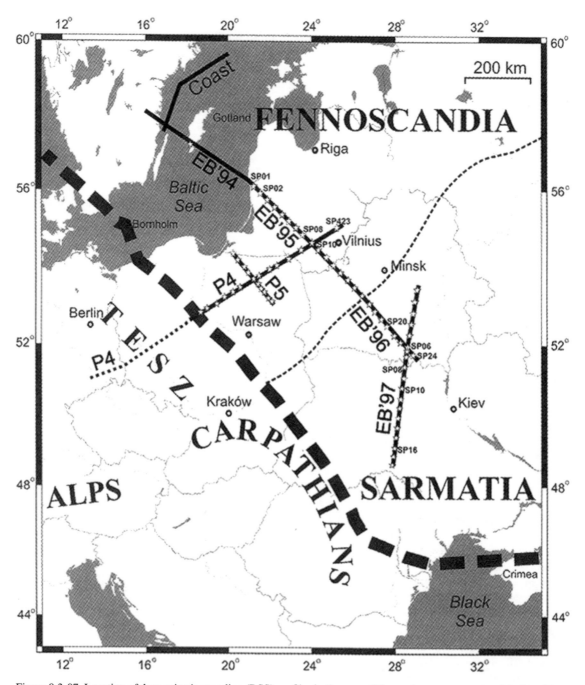

Figure 9.3-07. Location of deep seismic sounding (DSS) profiles in the area of the southwestern margin of the East European Craton (from Bogdanova et al., 2006, fig. 5). EB'94 to EB'97—EUROBRIDGE 1994–1997; Coast—Coast Profile 1995; P4, P5—Polonaise 1997; TESZ—Trans-European Suture Zone. [Gee, D.G., and Stephenson, R.A., eds., European lithosphere dynamics: Geological Society of London Memoir 32, p. 599–625. Reproduced by permission of Geological Society Publishing House, London, U.K.]

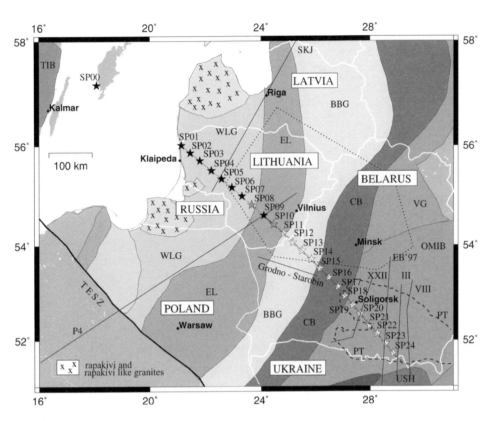

Figure 9.3-08. Location map of EURO-BRIDGE '95 and '96 seismic-refraction campaigns (from EUROBRIDGE Working Group, 1999, fig. 2), showing major regional tectonic units. Black and grey stars—shotpoints of 1995; white and grey stars—shotpoints of 1996; thin black lines—other seismic refraction lines; Roman numbers—former lines of 1960s and 1970s; EB'97—EUROBRIDGE 1997; P4—Polonaise line 4; SKJ—seismic line Sovietsk-Kohtla-Jarve; TESZ—Trans-European Suture Zone; dashed area, PT—Pripyat Trough; USH—Ukrainian Shield; other signatures—regional tectonic units. [Tectonophysics, v. 314, p. 193–217. Copyright Elsevier.]

deformed part of the more than 20-km-thick Dniepr-Donets Basin that formed due to Late Devonian rifting of the East European Craton in the eastern Ukraine and in southern Russia (DOBREflection-2000 and DOBREfraction '99 Working Groups, 2002; DOBREfraction '99 Working Group, 2003; Stephenson et al., 2006).

In 1999 a seismic-refraction/wide-angle reflection survey was performed across the Donbas Foldbelt. A total of 261 one-component stations was supplied by the University of Texas at El Paso and the University of Copenhagen, and 20 three-component stations by the Polish Academy of Sciences. The project consisted of a main line with 245 recording stations and a 190-km-long subsidiary line with 36 stations (line DOBRE'99 in Fig. 9.3-13), and 11 shots were placed in line every 25–30 km along the main line. In 2000, a seismic-reflection survey, using Vibroseis and explosive sources, along a 133-km-long part of the 1999 refraction line was added between shotpoints 3 and 7, reaching from the Azov Massif of the Ukrainian Shield northward across the axial part of the Donbas Foldbelt (DOBREflection-2000 and DOBREfraction '99 Working Groups, 2002; DOBREfraction '99 Working Group, 2003; Appendix A9-1-7; Stephenson et al., 2006).

The resulting DOBREfraction '99 P-wave model (Fig. 9.3-14) showed a remarkable basin underneath the 110-km-wide Donbas Foldbelt with up to 20-km-thick sediments. The Moho showed some topography, but was found on average at ~40-km depth. Along the reflection profile this generally corresponded to a broad reflective band which could be traced along the whole reflection profile at ~14 s below the Ukrainian Shield and at 12 s

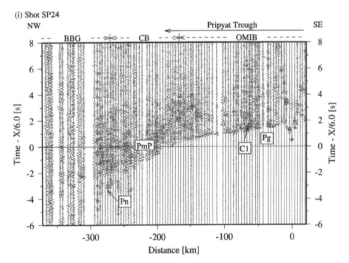

Figure 9.3-09. Record section of EUROBRIDGE '96 shotpoint 24 (from EUROBRIDGE Seismic Working Group, 1999, fig. 4i), showing Pg and Pn (refracted first arrivals from crystalline basement and Moho) and reflections: C1 from a mid-crust layer, PmP from Moho. For location, see Fig. 9.3.08). [Tectonophysics, v. 314, p. 193–217. Copyright Elsevier.]

below the Donbas Foldbelt (DOBREflection-2000 and DOBREfraction '99 Working Groups, 2002; Stephenson et al., 2006).

For the Peri-Caspian basin a N-S oriented crustal cross section along longitude 50.7°E (Figs. 9.3-15 and 9.3-16) was compiled from unpublished data (Brunet et al., 1999; Stephenson

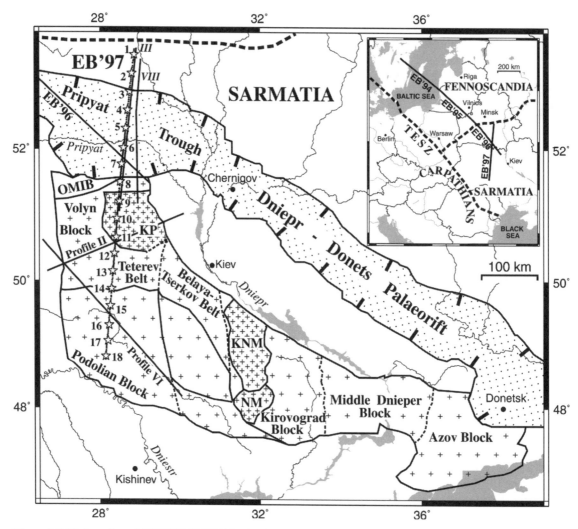

Figure 9.3-10. Location of EUROBRIDGE '97 seismic-refraction line (from Thybo et al., 2003, fig. 1). Nos. 1 to 18: shotpoint locations. Profiles III, VI, VIII: earlier deep seismic sounding profiles, but numbered differently by Sollogub et al. (1969, see Fig. 6.3.1-01 and 1972, see Fig. 6.4-04). [Tectonophysics, v. 371, p. 41–79. Copyright Elsevier.]

et al., 2006), showing a 35–40-km-thick crust, underlain at its center, where the sediments reach a thickness of 15–20 km, by a high-velocity layer of 8.1 km/s which was discussed with some controversy as to whether it is part of the crust or not.

When EUROPROBE started in 1990, a close cooperation between Russian and German scientists was immediately agreed upon. As a result, when in 1991 an ~3000-km-long DSS profile, named GRANIT (Fig. 9.3-17), was recorded by Russian scientists from eastern Ukraine to the West Siberian basin (GRANIT transect, 1992; Juhlin et al., 1996), a sub-experiment, aiming for a detailed anisotropy investigation was arranged. It was one of several sub-experiments of the project GRANIT and was a large-scale seismic anisotropy experiment, carried out in 1991 across the Voronesh Massif, named ASTRA (Fig. 9.3-17, left). It consisted of 280-km-long seismic-refraction profiles, which crossed each other in a central point. The ASTRA experiment was characterized by observations with dense multi-azimuthal

more-component recordings in the near-vertical incidence to wide-angle distance range. This recording scheme (Fig. 9.3-17, right) allowed for investigation of a 5 million km² large block for signs of rock anisotropy, including shear-wave birefringence and azimuthal dependency (Lueschen, 1992).

The size of the borehole shots ranged between 200 and 300 kg. For recording the following devices were available: five 48-channel "Progress" systems and 10 three-component analog stations plus two digital PDAS from GFZ Potsdam. The operation shown in Fig. 9.3-17 was planned to be repeated for each line. However, for logistic reasons, only half of the experiment could finally be realized. Unfortunately, results of this project were never published in the international literature. The northeastern part of the profile GRANIT is shown in Fig. 9.3-18.

Being part of the international program EUROPROBE, and coincident with the 1991 GRANIT line across the Middle Urals,

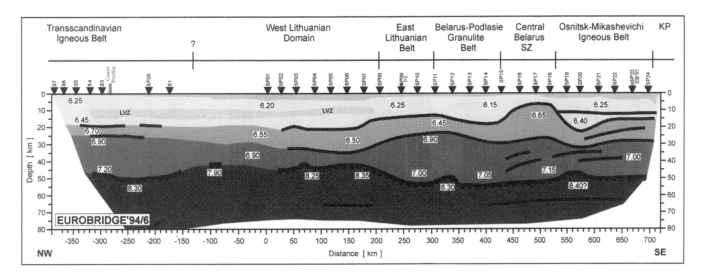

Figure 9.3-11. P-wave crustal model from Lithuania to Belarus from EUROBRIDGE '94 to '96 seismic-refraction campaigns (from Bogdanova et al., 2006, fig. 8). Triangles—shotpoint locations; KP—Korosten Pluton. [Gee, D.G., and Stephenson, R.A., eds., European lithosphere dynamics: Geological Society of London Memoir 32, p. 599–625. Reproduced by permission of Geological Society Publishing House, London, U.K.]

Figure 9.3-12. P-wave crustal model from southern Belarus to northwestern Ukraine from the EUROBRIDGE '97 seismic-refraction campaign (from Bogdanova et al., 2006, fig. 8). Triangles—shotpoint locations; dipping black lines—elements of seismic boundaries obtained from reflected and refracted waves; body with thin dashed lines—high-velocity zone; zones with white bars—zones of high reflectivity. [Gee, D.G., and Stephenson, R.A., eds., European lithosphere dynamics: Geological Society of London Memoir 32, p. 599–625. Reproduced by permission of Geological Society Publishing House, London, U.K.]

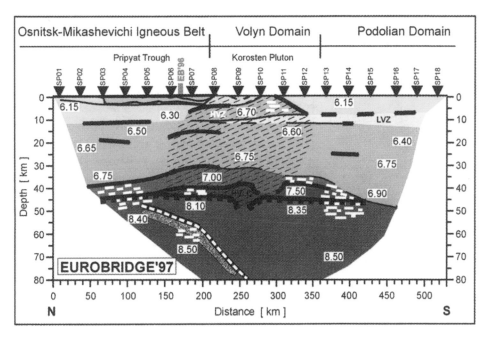

in 1993 the ESRU (EUROPROBE Seismic Reflection profiling in the Urals) project started in a joint effort of Russian, Swedish, U.S., and German scientists. Until 1999, it was conducted in five seismic-reflection experiments in the Middle Urals (Figs. 9.3-18 and 9.3-19), approximately along the latitude 58°N, and was in total ~335 km long (Juhlin et al., 1995, 1996; Kashubin et al., 2006). The ESRU line crossed the orogen close to the Urals Deep Borehole SG4.

ESRU93 was oriented in a SW-NE direction and was 60 km long. ESRU95 consisted of two profiles, running in W-E and in N-S directions, which crossed each other at the deep borehole SG-4. ESRU96 was an extension of the ESRU95-WE line.

ESRU98 overlapped the eastern end of ESRU96 before running toward the NE over Mesozoic strata of the West Siberian Basin. A 2 km gap remained between ESRU98 and ESRU 99, which continued the whole ESRU line to the east. The line was extended to the west in 2001–2003 reaching a total length of 440 km (see Chapter 10.2.3).

In addition to the ESRU-SB and Michailovsky profiles, data of the R17 profile (Juhlin et al., 1996) in the northern part of the area and the Alapaevsk and Chernoistoichinsk profiles (Steer et al., 1995), just south of the ESRU profile, were relevant for the large-scale structural interpretation of the Middle Urals. The 70-km-long seismic-reflection profile R-17 in the northern Middle

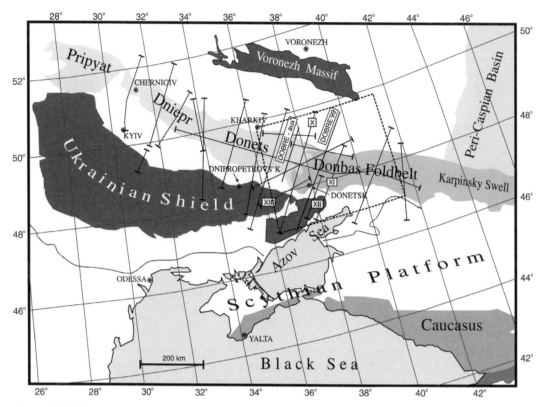

Figure 9.3-13. Regional tectonic setting of the Dniepr-Donets Basin and its contiguous inverted segment Donbass Foldbelt, showing location of DOBRE '99 (from DOBREfraction '99 Working Group 2003, fig. 1) and earlier seismic lines (see also Figure 6.4-04, corresponding profiles X, XI, XII re-numbered 10, 11, 12 by Sollogub et al., 1972). [Tectonophysics, v. 371, p. 81–110. Copyright Elsevier.]

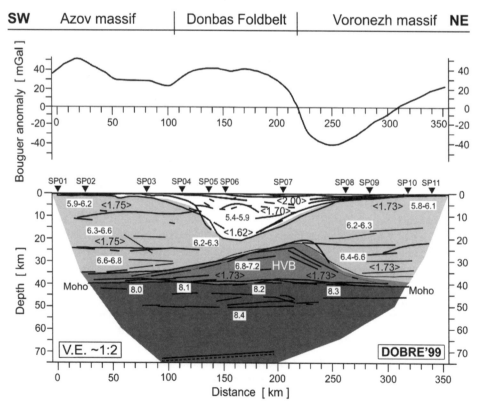

Figure 9.3-14. Top: Bouguer gravity anomaly along DOBREfraction '99 line. Bottom: DOBREfraction '99 P-wave model (from DOBREfraction '99 Working Group, 2003, fig. 11). Also shown are V_p/V_s ratios (in brackets). HVB—high velocity body. [Tectonophysics, v. 371, p. 81–110. Copyright Elsevier.]

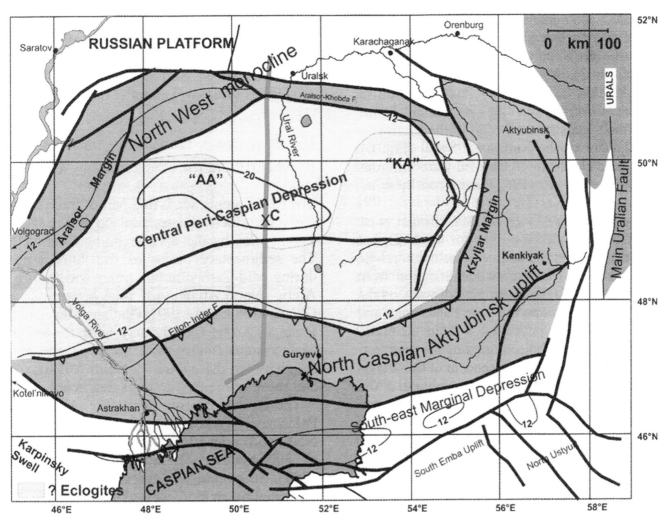

Figure 9.3-15. Crystalline basement contour map of the Pre-Caspian Basin showing the location of the cross section in Figure 9.3-16 (from Stephenson et al., 2006, fig. 6, *after* Brunet et al., 1999, fig. 2). [Gee, D.G., and Stephenson, R.A., eds., European lithosphere dynamics: Geological Society of London Memoir 32, p. 599–625. Reproduced by permission of Geological Society Publishing House, London, U.K.]

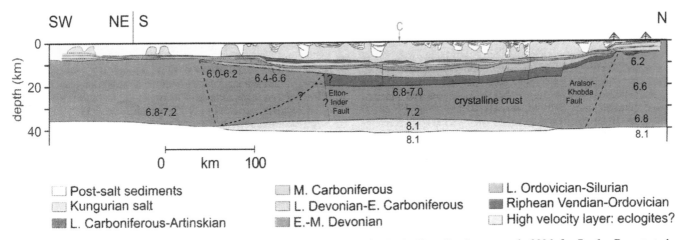

Figure 9.3-16. Simplified sketch along a N-S transect through the Peri-Caspian Basin (from Stephenson et al., 2006, fig. 7, *after* Brunet et al., 1999, fig. 5). [Gee, D.G., and Stephenson, R.A., eds., European lithosphere dynamics: Geological Society of London Memoir 32, p. 599–625. Reproduced by permission of Geological Society Publishing House, London, U.K.]

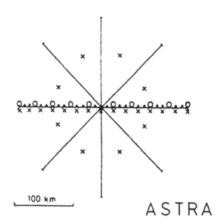

ASTRA

Figure 9.3-17. Left: Location of the ASTRA sub experiment (from Lueschen, 1992). Right: Observation scheme. Crosses are borehole shots with 200–300 kg charges, 20 km apart. Circles are 48-channel digital recording instruments of type <Progress>, which were spaced 30 km apart. Dots are 3-component analogue recording devices, which were 10 km apart. [DGG-Mitteilungen (News of the German Geophysical Society), v. 3, p. 24–39. Reprinted with permission of German Geophysical Society (DGG).]

Urals (Fig. 9.3-18) was recorded also in 1993 in conjunction with a standard oil prospecting survey in the West Siberian basin.

Every eighth shot (~400 m) along the profile was recorded up to 20 s TWT, which was also available for the reinterpretation of Juhlin et al. (1996), who interpreted the Moho to be at the base of a zone of continuous reflections in the lower crust between 11 and 14 s TWT. Finally, the data of the Michailovsky profile, recorded in the southern Middle Urals, were available for the interpretation (line M in Fig. 9.3-19).

Juhlin et al. (1995, 1996) have described and interpreted the first phase of ESRU93. The reinterpretation of the corresponding wide-angle part of the GRANIT line together with the ESRU data by Juhlin et al. (1996) showed comparatively high uppermost-mantle velocities of 8.4–8.5 km/s and a lower-crustal root with its deepest point along the ESRU profile and directly below the surface expression of the Main Uralian fault (Fig. 9.3-20).

Similar Moho depths were reported for the recent wide-angle UWARS experiment, which line is located 100 km farther south (Fig. 9.3-18), but the thickest crust there was interpreted to be farther to the east (Thouvenot, et al., 1994). Together with older DSS and peaceful nuclear explosions data, Juhlin et al. (1996) saw convincing evidence for the presence of an anomalously thick root under the Urals from ~45–55 km depth with low reflectivity below the Urals.

A presentation of the complete ESRU data set and its interpretation (Fig. 9.3-21) is described by Kashubin et al. (2006). A reflective zone of ~12–15 km thickness, deepening to the east, had been interpreted already by Juhlin et al. (1995) as East European Craton crystalline lower crust. This eastward thickening was also recognized farther south on the Michailovsky profile (line M in Fig. 9.3-19). Below the strong mid-crustal reflectivity, west-dipping reflections were seen, beginning at 60 km depth (km –30) and extending to ~25 km depth (km +100), indicating the deepest part of the Urals crustal root under the Kvarkush Anticline. Toward the east (east of km +50), underneath the hinterland east of the Main Uralian Thrust Zone in the Middle Urals, which according to Kashubin et al. (2006) appears to be dominated by Paleozoic island-arc terranes, the reflectivity of the lower crust is variable, but the reflective

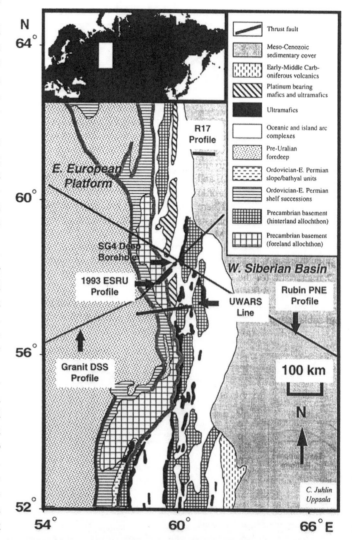

Figure 9.3-18. Seismic profiles in the Middle Urals (from Juhlin et al., 1996, fig.1). The RUBIN PNE profile was recorded in the 1970s. [Tectonophysics, v. 264, p. 21–34. Copyright Elsevier.]

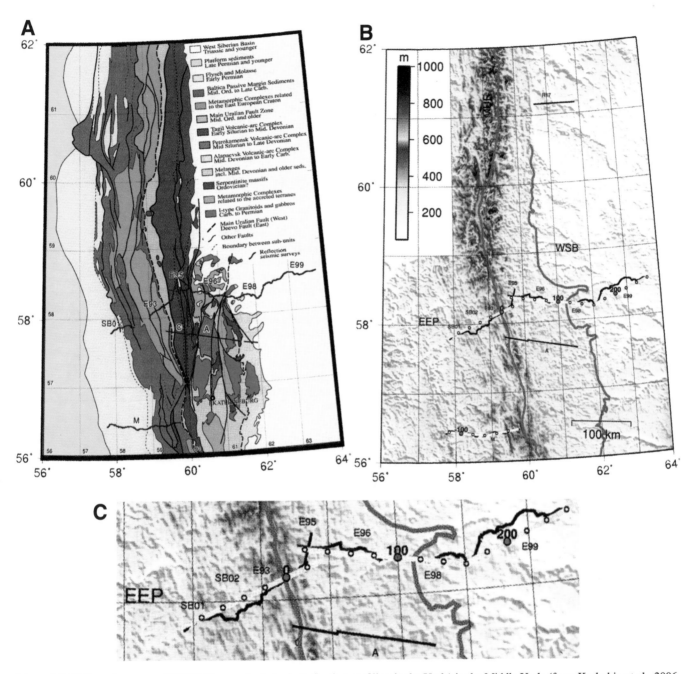

Figure 9.3-19. Location maps of ESRU (Europrobe Seismic Reflection profiling in the Urals) in the Middle Urals (from Kashubin et al., 2006, fig.1). (A) Geological map. (B) Physiographic map of the Middle Urals (EEP—East European Platform; WSB—West Siberian Basin). E93 to E99—ESRU; SB—Serebrianka-Beriozovka profile; M—Michailovsky profile. (C) Enlarged section of (B) showing location of ESRU. [Gee, D.G., and Stephenson, R.A., eds., European lithosphere dynamics: Geological Society of London Memoir 32, p. 427–442. Reproduced by permission of Geological Society Publishing House, London, U.K.]

Moho is still reasonably well defined, is generally flat and continuous at 40–45 km depth (Kashubin et al., 2006, 2009).

The largest international seismic project, carried out in Russia after 1990, was the URSEIS'95 project (Berzin et al., 1996, Carbonell et al., 1996, Knapp et al., 1998), which crossed the Urals at ~53°N, ~400 km farther south than the lines discussed above. URSEIS'95 was an integrated seismic experiment de-

signed by scientists from Russia, Germany, the United States, and Spain to reveal the detailed crustal and upper-mantle structure of the Urals. The transect, which was chosen for this purpose, extended from the East European platform in the Uralian foreland to the West Siberian basin (Fig. 9.3-22). The seismic survey comprised three parts: (*i*) a 465-km-long near-vertical incidence, Vibroseis-source reflection survey, (*ii*) a coincident

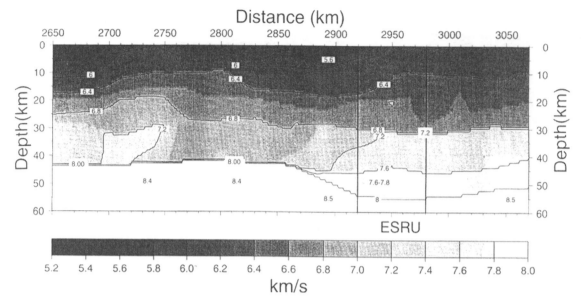

Figure 9.3-20. Velocity model based on a 400 km segment of the GRANIT deep seismic sounding profile (from Juhlin et al., 1996, fig.4). [Tectonophysics, v. 264, p. 21–34. Copyright Elsevier.]

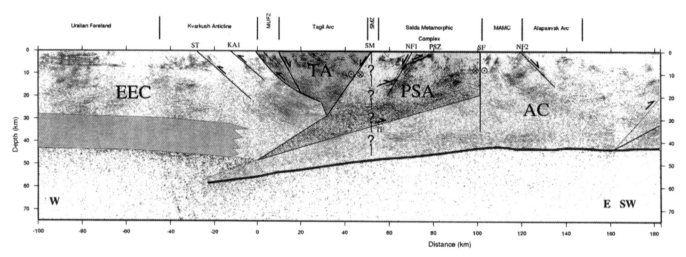

Figure 9.3-21. Merged migrated line drawing with interpretation along the entire ESRU profile from km −100 to km +200 (from Kashubin et al., 2006, fig.5). MTFZ—Main Uralian Thrust Fault (km 0); SMZ—Serov-Mauk Zone; MAMC—Murzinka-Adui Metamorphic Zone; EEC—East European Craton; TA—Tagil Arc; PSA—Petrokamensk-Salda crust; AC—Alapaevsk Crust. [Gee, D.G., and Stephenson, R.A., eds., European lithosphere dynamics: Geological Society of London Memoir 32, p. 427–442. Reproduced by permission of Geological Society Publishing House, London, U.K.]

near-vertical incidence, explosive-source survey, and (iii) a 340-km-long explosive-source, wide-angle reflection and refraction survey, including a cross line and two off-line fan-recording shots (Fig. 9.3-22).

From six shotpoints (Fig. 9.3-22), a total of 33,000 kg of explosives distributed among 15 shots, with charge sizes that ranged between 1500 and 3000 kg was used for the seismic-refraction/wide-angle reflection survey. The seismic energy was recorded by 50 three-component digital recording instruments (RefTek and Lennartz) in successive deployments. The difficult logistics forced station spacings that ranged from 1 to 2.5 km (Carbonell et al., 1996, 1998b, 2000b; Stadtlander et al., 1999).

The operation required six deployments. During the first deployment, a N-S line along the 60°E meridian was completed. A shot was fired at shotpoint S6 and recorded southwards from the cross-point with the E-W main line for ~120 km. With the second and third deployments, fan profiles were recorded from shotpoints S5 and S6, located 60 km north of the main line, and recorded along an E-W line, arranged 60 km south of the main line (Fig. 9.3-22).

The fans were designed so that the reflection points would lie below the E-W main line. The major effort of the wide-angle experiment involved a 335-km-long E-W main profile (Fig. 9.3-22). This profile was coincident with the eastern 335 km of

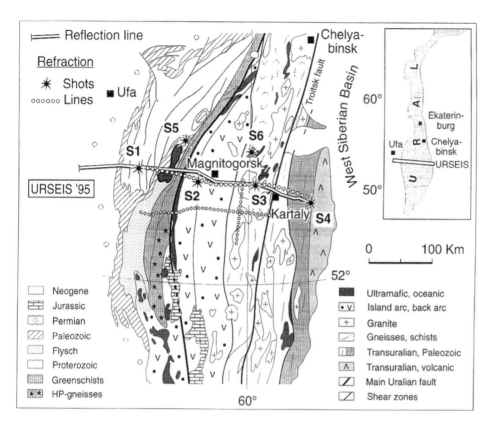

Figure 9.3-22. Generalized tectonic map of the southern Urals, showing the location of the seismic-reflection and refraction surveys (from Berzin et al., 1996, fig. 1). [Science, v. 274, p. 220–221. Reprinted with permission from AAAS.]

the 465-km-long URSEIS'95 near-vertical incidence reflection profile. The profile was completed in deployments four to six, and an average instrument spacing of 2.3 km was realized along the whole line for shotpoints S2, S3, and S4. For shotpoint S1 an average spacing of 2.3 km was achieved between shotpoints S1 and S2, while between shotpoints S2 and S4 the average spacing was 4.6 km. Carbonell et al. (2000b) show both reflection and refraction data sets displayed as CMP stacks (Figs. 9.3-23B and 9.3-23C). The P-wave data of shotpoints 1–4 are shown in detail in Appendix A9-1-8.

The major aim of the E-W main line was to delineate the Moho (crust-mantle boundary) structure beneath the line and thus identify the presence of a crustal root beneath this part of the mountain belt and, if present, to determine the depth of such a root. The recording with three-component instruments allowed to determine both P- and S-wave velocities and thus to place constraints on the rock types beneath the orogen and in particular in the crustal root, if present. At the same time the main profile served to provide velocity control for the near-vertical incidence reflection profile. The main purpose of the N-S cross-line was to determine crustal velocities and structure parallel to the trend of the mountain belt. The fan profiles were planned to detect major E-W variations in the main structural interfaces and in the Moho.

Carbonell et al. (2000b) and Stadtlander et al. (1999) have both interpreted the P- and S-wave data of the seismic-refraction/ wide-angle reflection experiment of the URSEIS'95 seismic project (see, e.g., Fig. 9.3-23A). Both author groups found the presence of a crustal root at least 10 km thick beneath the central part of the orogen, where the crustal thickness increased from 43 to 45 km at the margins of the transect up to 53–56 km beneath the central part of the profile. However, the center of this crustal root appeared to be displaced by 50–80 km to the east of the present-day maximum topography. Also, an upper crustal body from 10 to 30 km depth on average, with a high P-wave velocity of 6.3 km/s, was found and interpreted as consisting of mafic and/or ultramafic rocks. Another major feature of the seismic models showed the presence of high P- and S-wave velocities (7.5 and 4.2 km/s, respectively) at the base of the crustal root. P-wave velocities of 8.0–8.2 km/s and S-wave velocities of 4.5–4.7 km/s were found as characteristic features of the upper mantle.

The Vibroseis reflection profiling component of URSEIS'95 provided a high-resolution crustal-scale image of the unextended southern Uralide orogen. The section showed a marked lateral and vertical variation of reflectivity throughout the entire crust which was interpreted to differentiate the former margin of the East European craton to the west from the accreted terranes to the east. Continuous reflections and reflective domains correlated with major tectonic features interpreted from surface geology and defined a bivergent orogen in which the Paleozoic collision structure was largely preserved. At depth a less reflective zone between the East European platform and the accreted terrane regions coincided with the location of the Magnitogorsk volcanic arc and corresponded to the crustal root zone visible in the refraction data.

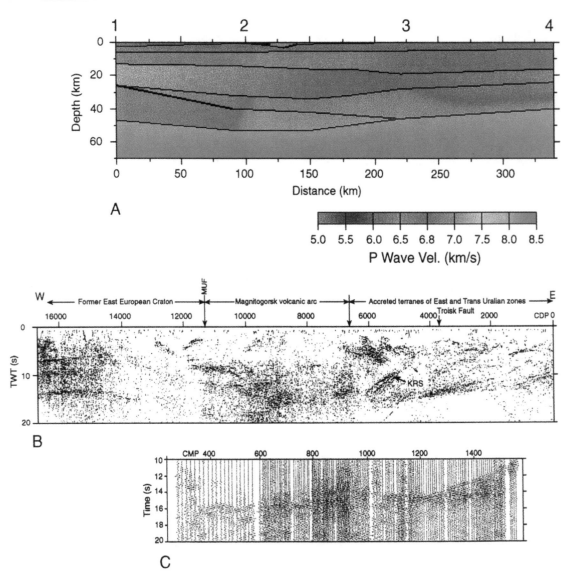

Figure 9.3-23. URSEIS '95 (from Carbonell et al., 2000, plates 1 and 2). (A) P-wave crustal cross section from seismic-refraction data. (B) Normal incidence stack of the explosion seismic-reflection common depth point data set. (C) Stack of the seismic wide-angle reflection data. [Journal of Geophysical Research, v. 105, p. 13,755–13,777. Reproduced by permission of the American Geophysical Union.]

9.4. CONTROLLED-SOURCE SEISMOLOGY IN NORTH AMERICA

9.4.1 Canada

First established in 1984, LITHOPROBE, Canada's national collaborative Earth science research program, was established to develop a comprehensive understanding of the evolution of the North American continent. It was active in the following 20 years with various phases, and aimed to present 3-D structure and the past geotectonic processes which were responsible for the present structure. For this purpose a series of ten study areas (named transects) was defined, each of which presented a representative

geological key target. In total, the transect provided a nearly continuous coverage across the northern part of the North American continent, including an active margin on the west coast, the Cordillera, platform and shield regions, the Appalachians, and the passive margin on the east coast (Clowes et al., 1996, 1999). Phase I started in 1984, Phase II followed from 1987 to 1990. A summary of results to late 1990 was provided by Clowes et al. (1992) and Clowes (1993; see corresponding part of Chapter 8).

In the 1990s, from 1990 to 1998, Phases III and IV followed, and finally, LITHOPROBE's Phase V ran from 1998–1999 to 2002–2003, and, prolonged into 2004–2005, presented a concluding phase of an extensive, Canada-wide project. In Phase IV, four transects were completed (Fig. 9.4.1-01): Abitibi-Grenville

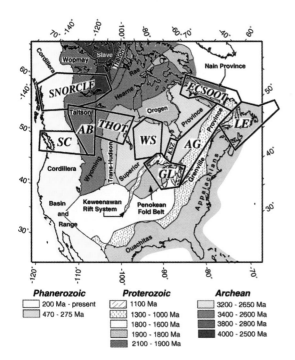

Figure 9.4.1-01. Location of LITHOPROBE transects (study areas) on a simplified tectonic map (from Clowes, 1999, fig. 1). SC—Southern Cordillera; AB—Alberta Basement; SNORCLE—Slave Northern Cordillera Lithospheric Evolution; THOT—Trans-Hudson Orogen Transect; WS—Western Superior; KSZ—Kapuskasing Structural Zone; GL—Great Lakes International Multidisciplinary Program on Crustal Evolution (GLIMPCE); AG—Abitibi-Grenville; LE—LITHOPROBE East; ECSOOT—Eastern Canadian Shield Onshore-Offshore Transect. [Episodes, v. 22, p. 2–20. Reproduced by permission of IUGS Publications Committee, Calgary, Alberta, Canada.]

(AG; Hynes and Ludden, 2000), Trans-Hudson Orogen (THOT; Hajnal et al., 2005a), Alberta Basement (AB; Ross, 1999, 2002), and Eastern Canadian Shield Onshore-Offshore (ESCOOT; Wardle and Hall, 2002a, 2002b), while the transects WS (Western Superior) and SNORCLE (Slave Northern Cordillera Lithospheric Evolution) (Cook and Erdmer, 2005) were started in Phase IV, but completed in Phase V. For each transect (Fig. 9.4.1-01), data analyses and interpretations, preparation of syntheses and compilation of a digital database and archive were completed. The main purpose of Phase V, however, was to develop a pan-LITHOPROBE synthesis from all ten LITHOPROBE transects and to complete the LITHOPROBE data archives. In all transects, the primary geophysical data acquired by LITHOPROBE comprised seismic-reflection, seismic-refraction and magnetotelluric data.

Clowes et al. (1996, 1999) presented summaries for various LITHOPROBE transects, as far as available at the time being. Clowes et al. (1996) discussed in particular seismic-reflection lines located in Precambrian regions. Almost all the data was acquired in the early 1990s: a marine example of ECSOOT, recorded in 1992 along the Labrador coast, new reflection data along two corridors in the east-central Grenville province (A-G), recorded in 1993, and data of the western Grenville province (GLIMPCE), recorded in the 1980s (Fig. 9.4.1-02). Furthermore, they discussed seismic-reflection lines located in central to western Canada: profiles recorded in 1991 across the Trans-Hudson orogen (THOT) and profiles recorded in 1992 across AB—central Alberta (Fig. 9.4.1-03). On all data reflectivity persisted throughout the crust, and the Precambrian Moho could be well defined.

In another summary paper, Mandler and Clowes (1998) discussed a number of multichannel seismic-reflection surveys, which were carried out in western and central Canada (Fig.

9.4.1-03) and which showed bright extensive reflectors between 2 and 3 s TWT. Such data were recorded in a number of different locations, as, e.g., in the surveys CAT92, PRAISE94, THOT94, and SALT95, whereby the Winagami reflection sequence was the most extensive one, extending over an estimated area of 120,000 km^2 in the Paleoproterozoic basement of Alberta, and the Wollaston Lake Reflector stretched continuously over 160 km in the western hinterland of the Paleoproterozoic Trans-Hudson Orogen.

In the following, starting at the east coast of North America, only the major seismic-refraction and reflection studies in Canada will be discussed, as they were performed within the specific transect regions following the definition of LITHOPROBE's division (Fig. 9.4.1-01).

9.4.1.1. LITHOPROBE East

Already in the 1980s several seismic-reflection and refraction surveys had been performed in the area of the LITHOPROBE East transect (Fig. 9.4.1-04) to investigate the structural architecture of the Northern Appalachians.

In 1991 another large survey, the LITHOPROBE East 91 (LE91) onshore-offshore seismic-refraction/wide-angle reflection experiment added a new wealth of data. The observations comprised closely spaced receivers on land and closely spaced shots offshore. The principal elements of the LE91 experiment were: (1) a marine component north, east, and south of Newfoundland consisting of five profiles using OBSs and an airgun source; (2) a land component within central Newfoundland consisting of three profiles using explosive shots and land seismographs; and (3) a series of combined offshore-onshore recordings (Marillier et al., 1994). More than 400,000 seismic traces were generated from ~800 land sites and 36 OBS stations (Fig. 9.4.1-04).

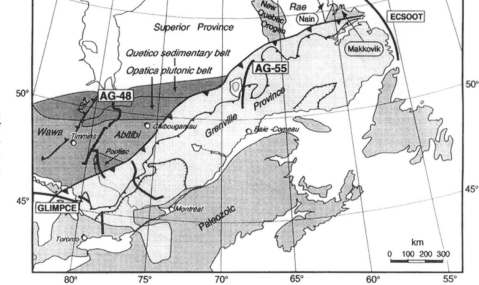

Figure 9.4.1-02. Simplified tectonic map of eastern Canada with location of LITHOPROBE seismic-reflection profiles (from Clowes, et al. 1996, fig. 2). [Tectonophysics, v. 264, p. 65–88. Copyright Elsevier.]

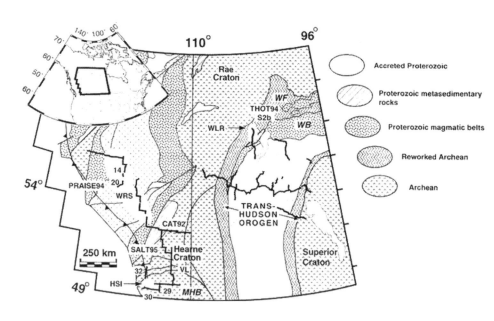

Figure 9.4.1-03. LITHOPROBE seismic reflection lines in western Canada (from Mandler and Clowes, 1998, fig. 1). THOT94 were recorded in the Trans-Hudson Orogen (see also Figure 9.4.1-08). CAT92, PRAISE94 and SALT95 were recorded in the Alberta Basement Transect region (AB; see also Figure 9.4.1-09). [Tectonophysics, v. 288, p. 71–81. Copyright Elsevier.]

One of the results is the thinning of a crustal unit in the west from the Laurentian plate margin toward the east from 44 to 40 km. In central Newfoundland, underneath the central mobile belt, the crust thins to a maximum of 35 km with average continental crustal velocities. Farther east, the crust thickens again to 40 km (Marillier et al., 1994).

9.4.1.2. LITHOPROBE Eastern Canadian Shield Onshore-Offshore Transect

In 1992, 1250 km of normal incidence seismic profiles and, in 1996, seven wide-angle seismic profiles were recorded (Fig. 9.4.1-05) across Archean and Proterozoic rocks of Labrador, northern Quebec, and the surrounding marine areas (Wardle

and Hall, 2002a, 2002b; Hall et al., 2002) to study the crustal and upper mantle structure within the Eastern Canadian Shield Onshore-Offshore Transect (ECSOOT). The seismic-refraction data were collected by variable combinations of marine shooting and recording and land recording.

The interpretation revealed a 33–44-km-thick Archean crust with P-wave velocities increasing downwards from 6.0 to 6.9 km/s. A moderate crustal reflectivity was observed, but the reflection Moho remained unclear (Hall et al., 2002). The Proterozoic crust appeared similar, partly with greater thickness and with variable lower crustal velocities, but the crustal reflectivity was strong and the Moho could be clearly defined.

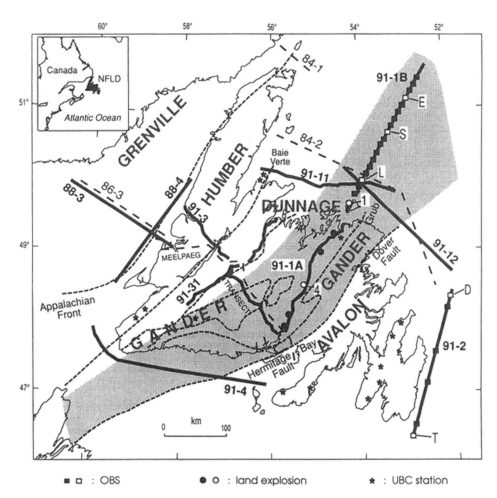

Figure 9.4.1-04. Location map for the Newfoundland Appalachians, showing LITHOPROBE East 1991 and 1988 wide-angle/refraction profiles (from Marillier et al., 1994, fig. 1). Continuous black lines: seismic refraction/wide-angle reflection profiles. Dashed lines: near-vertical incidence reflection profiles. [Tectonophysics, v. 232, p. 43–58. Copyright Elsevier.]

■ □ : OBS   ● ○ : land explosion   ★ : UBC station

Strong crustal variations appeared in particular along line 5: While under the Ungara Bay the data revealed a 38–40-km-thick crust, under the Torngat Orogen (for location, see Fig. 9.4.1-05) a crustal root down to 55 km depth was found. Farther to the east, under line 5E, the Moho depth decreased in two steps; first to a plateau of 40 km depth under the onshore part of line 5E and then, offshore, almost discontinuously to ~20 km depth (Hall et al., 2002).

9.4.1.3. LITHOPROBE Abitibi-Grenville Transect

The early 1990s saw major seismic activities in the area of the LITHOPROBE Abitibi-Grenville Transect (AG). Both in 1991 and 1993, high-resolution near-vertical incidence seismic-reflection lines were recorded (Clowes et al., 1992, 1996) and in 1992 a major seismic-refraction/wide-angle reflection survey took place (Fig. 9.4.1-06), the results of which were discussed together with previous observations carried out in the area in the 1980s (Winardhi and Mereu, 1997; Mereu, 2000a; White et al., 2000).

The seismic-reflection surveys of the LITHOPROBE studies in the Neoarchean southeastern Superior Province and the Meso-proterozoic Grenville Province in the southeastern Precambrian Canadian Shield included more than 1800 km of near-vertical incidence reflection data.

The 1992 LITHOPROBE Abitibi-Grenville seismic-refraction experiment comprised four profiles across the Grenville and Superior provinces of the southeastern Canadian Shield (Fig. 9.4.1-06). Several hundred kilometers coincided with or were parallel to the 1991 LITHOPROBE reflection lines (Clowes et al., 1992).

Three deployments of 417 instruments, 47 of those three-component seismographs with an average spacing of 1–1.5 km, provided more than 17000 seismograms produced by 44 shot-points with drillhole shots of 400–800 kg charges. The shot spacing averaged 30 km. The longest line (G-M in Fig. 9.4.1-06) extended in N-S direction over 620 km across the Grenville Front, supported by a 210-km-long cross line (E-F) in strike direction of the Abitibi Greenstone Belt. Its southern end intersected one of the 1986 GLIMPCE experiment lines and the 1988 GRAP line (named "1988 Ontario–New York refraction experiment" in Fig. 9.4.1-06; Hughes and Luetgert, 1992). The two shorter lines to the west (A-B and X-Y) covered 280 km in the N-S direction and 170 km in the W-E direction and were dedicated to study in particular the Sudbury Impact Structure.

The detailed controlled-source seismic surveys were complemented by a seismic tomography study based on data from a teleseismic experiment which comprised a N-S array of 28

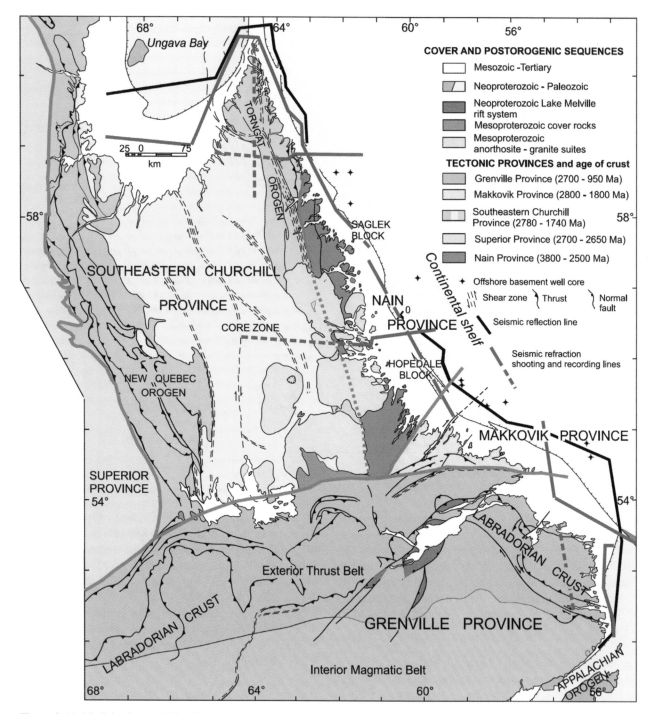

Figure 9.4.1-05. Seismic acquisition lines in the ECSOOT Transect area (from Wardle and Hall, 2002, fig. 2). Straight black lines: marine seismic lines. Straight dashed lines: onshore refraction recording lines. [Canadian Journal of Earth Science, v. 39, p. 563–567. Reproduced by permission of NCR Research Press/National Research Council, Canada.]

broadband seismographs deployed for half a year in 1996 along a 600-km-long line along the Quebec-Ontario border across the Grenville and Superior Provinces of the Canadian Shield (Rondenay et al., 2000).

Complete overviews of the geoscientific investigations including the seismic data and their interpretations were published in special review papers and special volumes (Ludden, 1995; Ludden and Hynes, 2000a, 2000b). In the summary volume on the Abitibi-Grenville transect, Ludden and Hynes (2000a, 2000b) have compiled two composite cross sections based on more than 1800 km of seismic-reflection data, complemented by a seismic-refraction experiment to constrain the velocity structure of the

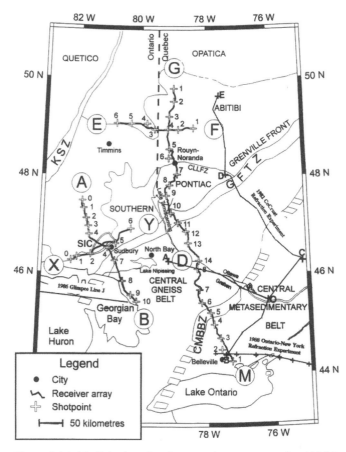

Figure 9.4.1-06. Seismic refraction experiments across the Abitibi-Grenville region (LITHOPROBE AG, from Mereu, 2000a, fig. 1). The location map includes observation lines recorded from 1982 to 1992. GFTZ—Grenville Front Tectonic Zone. [Canadian Journal of Earth Science, v. 37, p. 439–458. Reproduced by permission of NCR Research Press/National Research Council, Canada.]

crust and upper mantle. They have discussed how the different styles of crust formation and tectonic accretion in the Neoarchean and Mesoproterozoic have been defined by these data.

In their interpretation of both the reflection and refraction data, Winardhi and Mereu (1997) have noted the Grenville Front Tectonic Zone (GFTZ in Fig. 9.4.1-06) as a southeast-dipping region of anomalous velocity gradients extending to the Moho. From the wide-angle P_MP reflections, they inferred that the Moho is 4–5 km deeper south of the Grenville Front and varies from a sharp discontinuity to a rather diffuse and flat boundary under the Abitibi-Greenstone Belt north of the Front.

This was essentially confirmed by White et al. (2000) who have seen additional structural boundary zones in the same data. Mereu (2000a, 2000b) has tried to reconcile the differences between seismic models which usually occur when interpreting either reflection or refraction data. He analyzed crustal coda waves that follow the first arrivals and explained the coda by a class of complex heterogeneous models in which sets of small-scale, high-contrast sloping seismic reflectors are "embedded" in

a uniform seismic velocity gradient field. A similar approach had first been attempted by Sandmeier and Wenzel (1986) and later on by Enderle et al. (1997) and Tittgemeyer et al. (1996) when reinterpreting seismic-refraction data in Central Europe (see Fig. 9.2.2-12, subchapter 9.2.2).

9.4.1.4. Western Superior Transect

In 1996, LITHOPROBE acquired two ~600-km-long seismic-refraction/wide-angle reflection profiles oriented parallel (line 1, 600 km, running in an E-W direction north of latitude 50°N) and orthogonal (line 2, 500 km, running in a N-S direction west of longitude 90°W) to the regional geologic strike of the Archean western Superior Province of the Canadian Shield (Fig. 9.4.1-07).

Five hundred fifteen seismographs were deployed at intervals of 1–1.5 km. Shot sizes ranged from 1000 to 3000 kg, with shots being at 30–50 km intervals along each line. The 515 recording devices deployed consisted of: 170 PRS-1 seismographs with 2Hz vertical-component seismometers and 35 PRS-4 with three-component 4Hz seismometers from the Geological Survey of Canada; 170 SGR seismographs with 2Hz vertical-component seismometers or 8Hz vertical-component geophone strings from the USGS; and 140 RefTek seismographs with three-component geophones from IRIS/PASSCAL. The interpretation of the data by Musacchio et al. (2004) showed an intermediate middle crust with 6.6–6.7 km/s velocity, extending to 25–30 km depth and its top varying in depth from 12 to 20 km. Large variations were determined for the lower crust (6.7–7.5 km/s) and upper-mantle (8.0–8.8 km/s) velocities as well as for the Moho depths (32–45 km). From wide-angle reflections a deep boundary was modeled at 110–120 km depth along both profiles. For parts of the lower crust the authors saw a V_P anisotropy of 8%, with the fast propagation normal to the strike, as well as in a 15–20-km-thick layer in the upper mantle (>6%, fast propagation parallel to strike), which dips northwards from 50 to 75 km depth at ~ 10°.

9.4.1.5. Trans-Hudson Orogen Transect

The investigation of the deep structure underneath the Trans-Hudson Orogen Transect (THOT) of Manitoba and Saskatchewan started in 1991 (Lucas et al., 1994; Hajnal et al., 2005b), when LITHOPROBE acquired ~1100 km of near-vertical incidence Vibroseis seismic-reflection data (gray lines in Fig. 9.4.1-08). The E-W profiles showed strong, E-dipping reflections extending throughout the crust, while the N-S profiles outlined antiformal and synformal structures (Lucas et al., 1994). A second phase was implemented in 1994 with acquisition of 590 km of regional seismic high-resolution reflection profiling with 32 km in the western part plus 370 km in the eastern part of the orogen. This Phase II was the first LITHOPROBE survey to use a 480-channel recording system (Nemeth and Hajnal, 1998; Hajnal et al., 2005b).

The Trans-Hudson Orogen seismic-refraction experiment (THORE) followed in 1993. It consisted of three long-range pro-

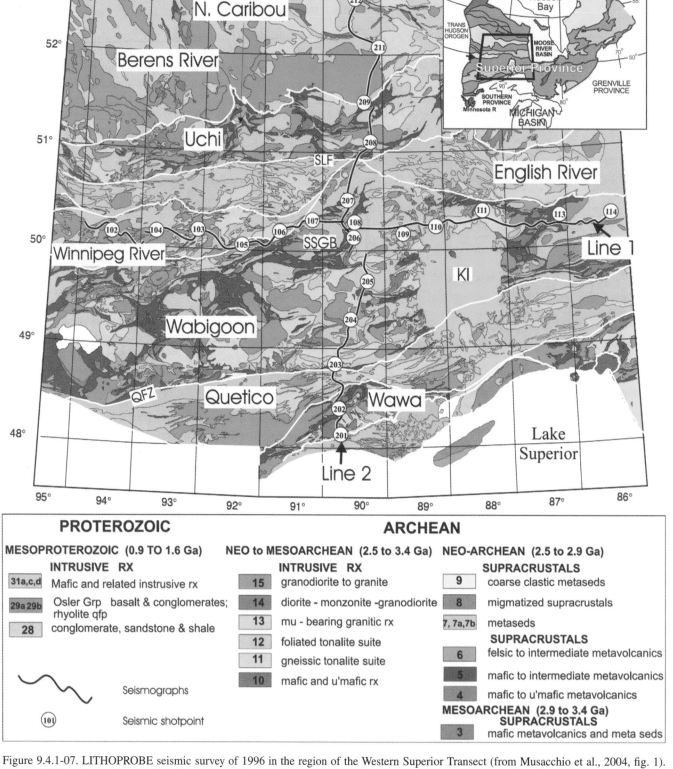

Figure 9.4.1-07. LITHOPROBE seismic survey of 1996 in the region of the Western Superior Transect (from Musacchio et al., 2004, fig. 1). [Journal of Geophysical Research, v. 109, B3, B03304, 10.1029/2003JB002427. Reproduced by permission of American Geophysical Union.]

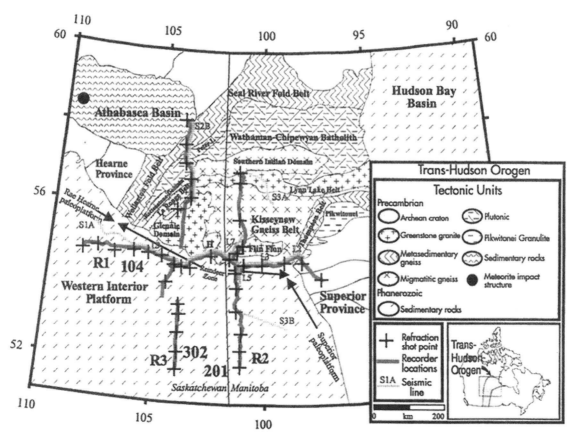

Figure 9.4.1-08. Seismic reflection and refraction lines and shotpoints in the Trans-Hudson Orogen Transect region (LITHOPROBE THOT, from Nemeth and Hajnal, 1998, fig. 1). [Tectonophysics, v. 288, p. 93–104. Copyright Elsevier.]

files, two of them 750 km long and one 500 km long (black lines in Fig. 9.4.1-08). A total of 525 seismographs was involved: 185 PRS-1 (vertical-component Portable Refraction Seismographs with 2Hz seismometers) and 35 three-component PRS-4 recorders with 4Hz seismometers, both from the Geological Survey of Canada; 185 USGS instruments (135 vertical-component Seismic Group Recorders [SGR] with 2Hz seismometers and 50 SGRs with strings of 8Hz geophones); and 120 three-component RefTek recorders with 2Hz seismometers from IRIS/PASSCAL. A maximum of 505 recorders with a spacing of 1.0–1.5 km was deployed along each line. The additional (three-component) systems were placed on a selected broadside profile. Thirty-nine borehole shots (crosses in Fig. 9.4.1-08), with charge sizes from 800 to 3000 kg and an average shot spacing of 50 km, were detonated during acquisition (Nemeth et al., 1996, 2005; Nemeth and Hajnal, 1998). All THOT seismic-reflection lines are also shown in Figure 9.4.1-03.

Generally, crustal velocity structure did not correlate with the location and extent of major geological domains. Crustal thickness variations were substantial. The generally thick crust, 40–45 km, thickened at some locations up to 55 km, and a small root down to ~50 km depth was observed below the Sask craton. However, the crustal thickness variations seemed not to be correlated with major crustal features or boundaries, such as suture zones or paleosubduction zones, suggesting post-collisional variations (Nemeth et al., 2005; White et al., 2005).

Bezdan and Hajnal (1998) discussed, in particular, the results and advantages of four expanding spread profiles that had been recorded in various tectonic domains along line R3, which had been selected on the basis of high crustal reflectivity identified during the regional survey and which provided more detailed images of the crust below 6 s TWT.

9.4.1.6. Alberta Basement Transect and "Deep Probe"

The Alberta Basement region was of particular relevance for the oil and gas exploration industry and many seismic-reflection surveys were performed which partly were also relevant for the crust underlying the sediments. Two special issues of the *Canadian Journal of Earth Sciences*, edited by G.M. Ross, provided a synthesis and new results for the sub-sedimentary region of the transect (Ross, 1999, 2000, 2002).

In 1992, an innovative 3-D seismic-reflection experiment (CAT92) was undertaken by LITHOPROBE (for location, see Fig. 9.4.1-03) as a feasibility study to study the influence of the basement structure beneath the Western Canada Sedimentary Basin on the sedimentary section above (Kanasewich et al.,

1995) and to study the crustal structure underneath the Alberta Basement Transect. Two-dimensional reflection data sets were recorded on N-S and E-W profiles and totalled 513 km. A 3-D recording was conducted in the region of the intersection of lines 2 and 3 (Kanasewich et al., 1995, fig. 1). The Moho was found at the base of a generally strong reflective crust at 12–14 s TWT, corresponding to ~35–40 km depth, but its sharpness varied.

In 1994 and 1995, LITHOPROBE added further reflection lines to the CAT92 survey (for location, see Fig. 9.4.1-03), including PRAISE94 toward the NW and SALT95 (Southern Alberta Lithospheric Transect) toward the S and SE (Mandler and Clowes, 1998). Furthermore, in 1995, the project VAULT (Vibroseis Augmented Listen Time) was conducted (Ross et al., 1997), another seismic-reflection experiment across the northern margin of the Medicine Hat Block, a region with anomalous sedimentation in southern Alberta and northern Montana, ~250 km south of CAT92 (no. 32 in Fig. 9.4.1-03).

Again, exceptional reflectivity in the crust was recorded, but also good reflectivity within the uppermost mantle down to ~20 s TWT (Ross et al., 1997). The SALT95 data in southern Alberta, ~100 km south of VAULT, were later re-investigated by Welford and Clowes (2004) in conjunction with an extended exploration survey to longer traveltimes, performed in 1999 by a partnership group of Canadian oil industry partners, with respect to an approach to investigate upper-middle crustal structure to depths of ~20 km using industry-style 3-D data.

All seismic-reflection lines, conducted between 1991 and 1995 in the region of the Alberta Basement Transect, are shown as dashed, thin full, or thick full lines in Figure 9.4.1-09 (Clowes et al., 2002) and will be partly included in the following discussion of the South Alberta Seismic Refraction Experiment.

SAREX, LITHOPROBE's Southern Alberta Refraction Experiment, was the corresponding seismic-refraction experiment to the seismic-reflection surveys discussed above. It was conducted in 1995 in connection with the long-range upper-mantle project "Deep Probe" and aimed to provide crustal velocity structure in southern Alberta as part of the Alberta Basement Transect (Clowes et al., 2002; Gorman et al., 2002). The line extended over 800 km from east-central Alberta to central Montana (Fig. 9.4.1-09). Charges, ranging from 800 to 1300 kg, were detonated at ten shotpoints in Canada at a spacing of 50 km and were recorded by 521 seismographs deployed at 1 km intervals. An additional 140 seismographs recorded these shots along the extension of the line into Montana with 1.25–2.50 km intervals. Data from the Deep Probe shotpoint 49 (large star in Fig. 9.4.1-09), a 5000 kg shot co-located within 300 m of shotpoint 1, supplemented the SAREX data for the region south of the Canadian-U.S. border.

The interpretation (Fig. 9.4.1-10) by Clowes et al. (2002) showed a strong relief of the Moho, with its depth within the Canadian territory gradually increasing from ~35 to 45 km near the Canadian-U.S. border and then suddenly increasing to almost 60 km under Montana. Also, the upper-mantle velocity increased

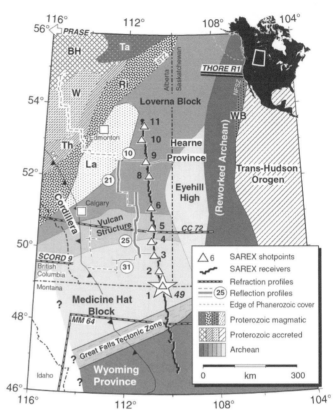

Figure 9.4.1-09. LITHOPROBE's Southern Alberta Refraction Experiment (SAREX) and location of seismic-reflection lines in the Alberta Basement Transect region (from Clowes et al., 2002, fig. 1). Also shown for comparison are the locations of the western end of the THOT-R1 line and SCORD9 (Southern Cordillera Refraction Experiment (Zelt and White, 1995). [Canadian Journal of Earth Science, v. 39, p. 351–373. Reproduced by permission of NCR Research Press/National Research Council, Canada.]

in the same direction from 8.05 to 8.2 km/s. The most prominent feature in the crust was a thick (10–25 km) lower-crustal layer (LCL in Fig. 9.4.1-10) with high velocities ranging from 7.5 to 7.9 km/s, located under the Medicine Hat block of southern Alberta and the Wyoming block of Montana.

The "Deep Probe" experiment (Fig. 9.4.1-11) followed immediately after the SAREX crustal investigation. It targeted the velocity structure from the base of the crust to depths as great as the mantle transition zone near 400 km, i.e., to expose the roots of western North America in terms of both lithospheric structure and constructional history (Deep Probe Working Group, 1998; Gorman et al., 2002; Snelson et al., 1998). Originally, it was planned only for the region of Alberta, but could be substantially enlarged by cooperation with the United States institutions and scientists, enlarging the number of seismographs and in particular providing a number of large shots throughout the United States. As a result, the project spanned a distance range of 3000 km from the southern Northwest Territories to southern New Mexico. The study corridor followed approximately the 110th meridian. Ten shots, which were up to 10 times larger than

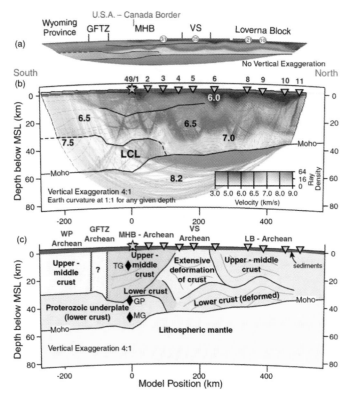

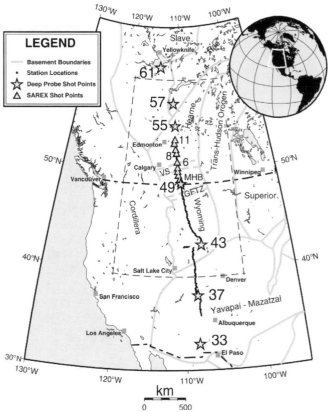

Figure 9.4.1-10. Crustal cross section along LITHOPROBE's SAREX line (from Clowes et al., 2002, fig. 8). LB—Loverna Block; GFTZ—Great Falls Tectonic Zone; MHB—Medicine Hat Block; WP—Wyoming Province; VS—Vulcan Structure. Numbers in circles: crossing of seismic reflection profiles. Diamonds identify locations of xenolith samples. [Canadian Journal of Earth Science, v. 39, p. 351–373. Reproduced by permission of NCR Research Press/National Research Council, Canada.]

Figure 9.4.1-11. Location map showing the Deep Probe corridor in western North America (from Gorman et al., 2002, fig. 1). Stars: Deep Probe shotpoints. Triangles: SAREX shotpoints. Small circles (overlapping to a line): seismograph locations. [Canadian Journal of Earth Science, v. 39, p. 375–398. Reproduced by permission of NCR Research Press/National Research Council, Canada.]

those used for conventional crustal studies were recorded every 1–2 km to maximum source-receiver offsets of ~2500 km (Gorman et al., 2002). Three of these shotpoints had already served as shotpoints for the SAREX line.

A first lithospheric model displayed a colored mantle cross section reaching from Tyrone, southern New Mexico, to shotpoint 55 in the Alberta plains north of Edmonton (Deep Probe Working Group, 1998, fig. 3), while the analysis of Gorman et al. (2002) focused on the region between Deep Probe shots 43 and 55. Both interpretations resulted in a continental-scale model of the velocity structure of the lithosphere of platformal western Laurentia to a depth of 150 km.

A particular detail of the cross section of the Deep Probe Working Group (1998) is the indication of a low-velocity zone within the uppermost mantle between 50 and 80 km depth. At the top of this zone, the velocity decreased from 7.9–8.0 km/s to 7.6–7.7 km/s. This low-velocity zone gradually thins out to the north, disappearing in Wyoming north of the Cheyenne Belt, and is absent under Canada. A further interpretation (Snelson et al., 1998) concerns mainly the territory of the United States and will be discussed in section 9.4.2.5 (see Fig. 9.4.2-25).

9.4.1.7. Slave Northern Cordillera Lithospheric Evolution

The Slave geological province is a relatively small craton in NW Canada and includes the five oldest rocks in the world. Its southern portion was already addressed in the 1980s during Phases I and II. Phase I in 1984 and 1985 dealt mainly with an investigation of Vancouver Island, and Phase II in 1987–1990 investigated the Southern Cordillera Transect. In particular, in 1989 and 1990, major seismic-refraction work had been carried out by SCoRE (Southern Cordillera Seismic Refraction Experiment) which was described in Chapter 8.5.2 (Clowes et al., 1995).

The ACCRETE seismic experiment in 1994 (thin lines south of line 22 in Fig. 9.4.1-12), in fact a U.S.-based project (Hammer et al., 2000; Morozov et al., 1998, 2001), was the first project executed in the area of the Northern Cordillera Transect. Using coastal fjords for access along Portland Inlet and adjoining waterways, 1700 km marine multichannel reflection data were acquired. The tuned airgun array also served as the seismic source for a seismic-refraction/wide-angle reflection survey, a land-based seismograph array deployed inline with Portland Inlet from north of Prince Rupert to the north around Stewart (see Fig. 9.4.1-12). Thirty-seven instruments were positioned along the

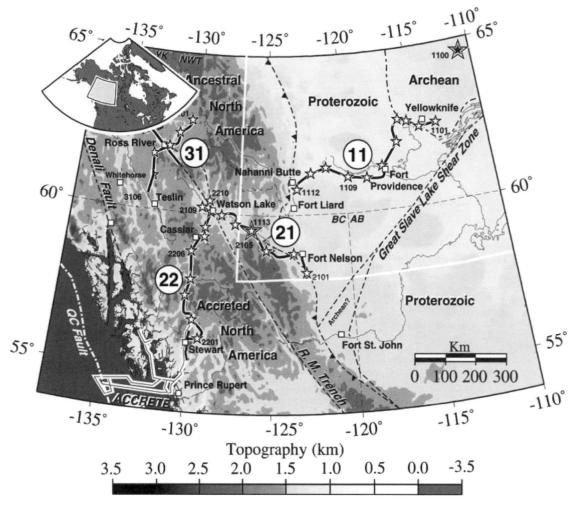

Figure 9.4.1-12. Location map of the SNoRE seismic refraction project within the SNORCLE transect (from Fernandez-Viejo and Clowes, 2003, fig. 1). [Geophysical Journal International, v. 153, p. 1–19. Copyright John Wiley & Sons Ltd.]

eastern shore of the 200-km-long fjord, and other instruments were deployed along a 105-km-long profile extending NE through the Coast Mountains.

The next project was a multichannel seismic-reflection survey, which was carried out in 1996 along line 11 (Fig. 9.4.1-12; Cook et al., 1998, 1999), following a sequence of roads from east of Yellowknife to west of Nahanni Butte in the southwestern part of the Northwest Territories.

In 1997, the SNoRE97 seismic-refraction/wide-angle reflection survey was initiated to investigate the lithosphere of the entire Slave Northern Cordillera Lithospheric Evolution (SNORCLE) Transect (Clowes et al., 2005; Fernández-Viejo and Clowes, 2003). It included four individual lines along three transect corridors (Fig. 9.4.1-12). A total of 37 shots, with a nominal spacing of 60 km and with charges between 800 and 3000 kg, was recorded by 594 recording stations, 226 of them being three-component seismographs. The resulting station spacing was 0.8–1.2 km for vertical and 1.6–2.4 km for three-component recordings.

The easternmost line 11 was 720 km long and ran along the same roads as the 1996 reflection survey. Only along line 11 were all available seismographs exclusively deployed in-line. In addition, two shots 200 km off either end of line 11, with charges of 5000 and 10,000 kg, were recorded along the line (stars in Fig. 9.4.1-12) to determine structure to the base of the lithosphere. Lines 21 and 22 were observed simultaneously, and during the recording of line 31, a number of stations were installed up to 100 km off-line along a fault running perpendicular to the main line.

Together with earlier studies, a 2000-km-long lithospheric velocity model resulted that extended from the Archean Slave craton to the present Pacific basin (Clowes et al., 2005). In Figure 9.4.1-13, we show the 800-km-long part of lines ACCRETE and SNORE 22. On a large scale, the seismic data identified a variety of orogenic styles. The Moho remained remarkably flat and shallow (33–36 km) across the majority of the transect, despite the variety of ages, orogenic styles, and tectonomagmatic deformations, which were crossed by the seismic corridors. Laterally variable crustal velocities were consistently slower

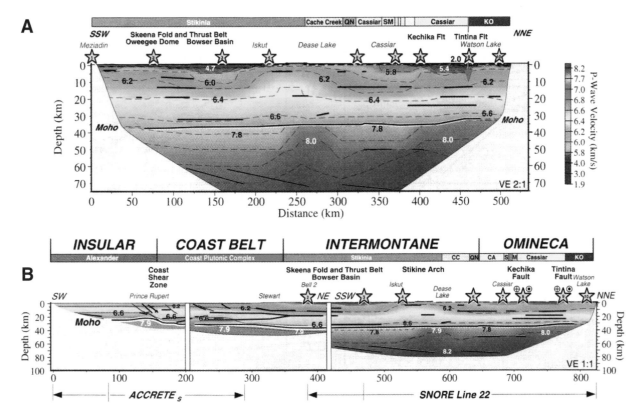

Figure 9.4.1-13. Crustal cross section (A) along SNoRE line 22 vertical exaggeration 2:1), (B) along lines AACRETE and SNoRE 22 (no vertical exaggeration). Stars—shotpoints (from Hammer and Clowes, 2004, fig. 5). [Journal of Geophysical Research, v. 109, B06305, doi:1029/2003JB002749, 19p.] Reproduced by permission of American Geophysical Union.]

beneath the Cordillera than beneath the cratonic crust. Beneath much of the Cordillera slow upper mantle velocities (7.8–7.9 km/s) were observed.

More details of the various lines were published separately. Fernández-Viejo and Clowes (2003) concentrated on line 11, including the reflection data. Lines 21 and 22 were interpreted by Welford et al. (2001) and Hammer and Clowes (2004), respectively. The latter tied their interpretation of the data of line 22, generated by explosive shots, to data of the earlier overlapping project ACCRETE of 1994 (Fig. 9.4.1-13). Creaser and Spence (2005) finally dealt with line 31. An alternative interpretation of part of the seismic-reflection data was prepared by Evenchick et al. (2005).

9.4.2. United States

The successful start of IRIS in the United States in the 1980s, with its sub-organization PASSCAL had led to a continuous funding by the U.S. National Science Foundation in the 1990s and the following decade until today. Today, PASSCAL is one of two major instrumentation programs of IRIS, with the second one being the Global Seismic Network. PASSCAL instruments and support are available to the academic research community according to the rules and policies set by the IRIS Executive Committee. It nowadays operates a pool of over 1000 portable seismic instruments to record active source reflection data, active source refraction data or natural sources, e.g., earthquakes. The instrumentation is supported by an instrument center which at present is based at the New Mexico Institute of Mining and Technology at Socorro, New Mexico.

With its large set of digital instrumentation, for which small research teams with limited funding can apply, it enables a large number of North American university large-scale research projects and also many major projects of the USGS. On request, IRIS/PASSCAL also provides technicians for individual fieldwork and other technical support, and it takes care that all instruments are properly maintained and replaced, when needed. A special study group also ensures that the equipment is properly modernized, whenever advanced technology provides new generations of equipment. The rules of IRIS/PASSCAL also request from their customers that copies of the data that are obtained by their equipment are provided to be stored at a joint data center.

9.4.2.1. Alaska

The experiments for deep seismic exploration in mainland Alaska had reached a preliminarily final stage with the northernmost segment of TACT (Trans-Alaska Crustal Transect), namely the seismic exploration of the Brooks Range in 1990 (Fuis et al.,

1997) which was already described in Chapter 8.5.4. A synthesis of all data obtained in the 1980s and 1990s along the 1350 long TACT corridor was finally compiled by Fuis et al. (2008). The most distinctive crustal structures and the deepest Moho along the transect were found near the Pacific and Arctic margins. Near the Pacific margin, a stack of tectonically underplated oceanic layers was inferred which were interpreted as remnants of the extinct Kula (or Resurrection) plate. Continental Moho just north of this underplated stack was found more than 55 km deep. Near the Arctic margin, the Brooks Range is underlain by large-scale duplex structures that overlie a tectonic wedge of North Slope crust and mantle. There, the Moho has been depressed to nearly 50 km depth. In contrast, the Moho of central Alaska is on average 32 km deep (Fuis et al., 2008).

Approximately parallel to the TACT survey, in 1989 the EDGE Alaska transect had been acquired over the shelf off Alaska using a 4-km streamer and airgun sources (Moore et al., 1991). Along this transect, the top of the subducting oceanic

plate, representing the North America–Pacific plate boundary, had been traced to more than 30 km depth beneath the inner shelf 250 km landward of the trench. In 1994, a deep-crustal seismic-refraction experiment was carried out along the same transect using an airgun source with a high shot density and closely spaced OBHs. This experiment aimed to verify the previously observed deep crustal structure, to increase the depth range, and to obtain velocity information for an improved image reconstruction (Ye et al., 1997). In addition, across EDGE 302 several perpendicular profiles were shot. Eleven OBHs were used to record the shots, which had an average spacing of 90 m, along six single-line segments, with two each along EDGE 301 and 302 and two perpendicular crossing EDGE 302. The new experiment improved in particular the data along EDGE 302 (Fig. 9.4.2-01).

In the 1990s, experimental work continued in the marine areas around mainland Alaska. A large-scale active-source seismic experiment (Fig. 9.4.2-02) in the region of the Aleutian vol-

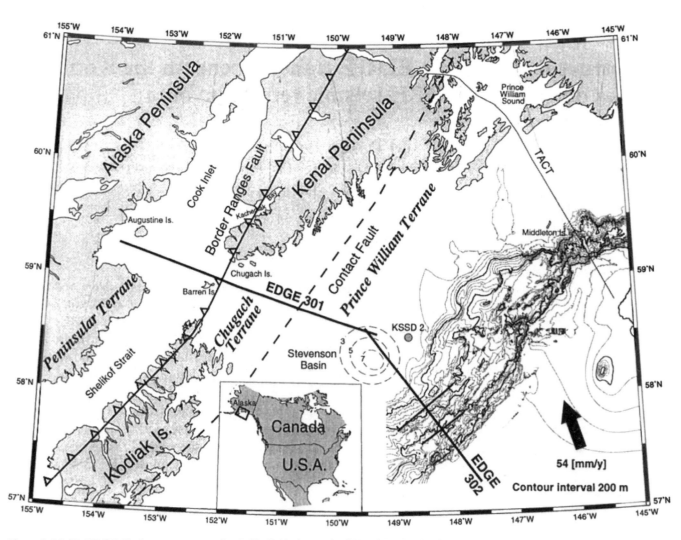

Figure 9.4.2-01. EDGE Alaska transect over the shelf off Alaska south of Kenai Peninsula (from Ye et al., 1997, fig. 1). [Geophysical Journal International, v. 130, p. 283–302. Copyright John Wiley & Sons Ltd.]

canic arc and the back-arc Bering shelf was carried out in July 1994 (Fliedner and Klemperer, 1999). The first component of the experiment was a multichannel reflection acquisition along the ship tracks shown in Figure 9.4.2-02, of which tracks A1, A2, and A3 were used by Fliedner and Klemperer (1999) for their interpretation.

The second component was a wide-angle reflection survey, for which OBS/OBH recorders and 27 portable three-component stations on islands and on the Alaskan peninsula were deployed (Fig. 9.4.2-02). These seismometers recorded the airgun shots of the reflection survey up to distances of 450 km (300 km in the island are section). Stations 1–15 recorded all shots along ship

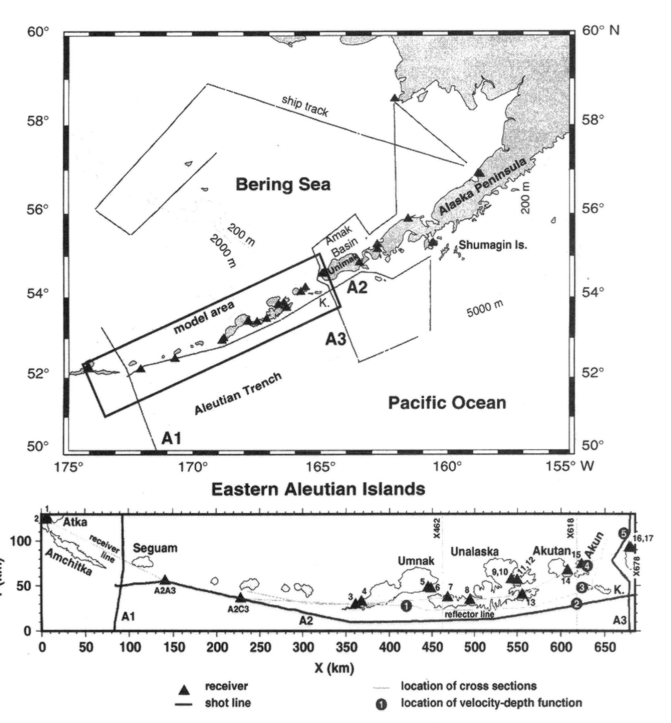

Figure 9.4.2-02. Seismic experiment in southwestern Alaska (from Fliedner and Klemperer, 1999, fig. 1) showing location of the stations and ocean-bottom seismometers (triangles). [Journal of Geophysical Research, v. 104, p. 10,667–10,694. Reproduced by permission of American Geophysical Union.]

lines A2 and A3, while usable data from shots of line A1 were only recorded by OBS/OBH stations and two land stations 1 and 2 on the island of Atka (Holbrook et al., 1994a). Stations to the east of track line A3 recorded only shots from this line but not from line A2.

From their interpretation, Fliedner and Klemperer (1999) concluded that the Aleutian crust, with the Moho at ~30 km depth, is almost continental in thickness and consists of ~20% preexisting oceanic crust (Fig. 9.4.2-03). The average crustal velocity of 6.7 km/s was substantially higher than usually found for continental crust, and the uppermost-mantle velocity beneath the Moho was generally 7.8 km/s, increasing gradually to 8 km/s near 50 km depth.

One month later, in August 1994, two long deep-crustal seismic-reflection profiles were acquired along the continental shelf with airgun shots at intervals of 50 m or 75 m (thick lines in Fig. 9.4.2-04). The profiles had a total of 3754 km and on each line ~40,000 shots were fired. The marine reflection data were obtained by a 4.2 km, 160-channel digital streamer, with record lengths up to 16–23 seconds. From these marine airgun sources in the Bering and Chukchi Seas, 14 deep-crustal wide-angle seismic-reflection and -refraction profiles were recorded (Brocher et al., 1995a; Appendix 9-2-1). This was achieved by an onland array of twelve three-component stations (quadrangles in Fig. 9.4.2-04) which continuously recorded the airgun shots on line EW94-10, where the ship steamed northward.

Seven of these stations also recorded the airgun shots fired during the southward cruise along the USGS/Stanford line. The map also shows the ship tracks and stations of the July 1994 survey (thin dashed lines and quadrangles in Fig. 9.4.2-04), part of which was interpreted by Fliedner and Klemperer (1999).

Airgun sources were recorded as far as 400 km, which enabled to image reflectors throughout the crust, to determine crustal and upper mantle refractor velocities, and to provide con-straints on the geometry of the Moho. The recorded crustal and upper mantle refractions indicated substantial variations in the crustal thickness along the transect (Brocher et al., 1995a).

9.4.2.2. Pacific Northwest

As a consequence of its position along the Cascadia sub-duction zone, where the Juan de Fuca plate of the Pacific Ocean is being subducted underneath the North American continent, western Washington and adjacent British Columbia are regarded as having a potential for seismic hazards, as was recognized by earthquakes, which have been observed within the subducting Juan de Fuca plate since 1940, and by crustal faults within the Puget Lowlands capable of large earthquakes (Fisher et al., 1999; Parsons et al., 1998).

On the Canadian side, the deep crustal structure had been investigated intensely by LITHOPROBE already in the 1980s, the Southern Cordillera Transect being the first target of LITHO-PROBE. Its Phase I had started in 1984 with primary scientific activity on Vancouver Island including a detailed seismic-refrac-tion and -reflection survey (Clowes et al., 1986, 1987a) and was continued in Phase II from 1987 to 1990, involving SCoRE, the Southern Cordillera Seismic Refraction Experiment, which was carried out in 1989 and 1990 (Clowes et al., 1995) and which provided long profiles over the southern Canadian Cordillera, including lines across the Intermontane and Coast belts and Van-couver Island (see Chapter 8.5.2).

In the adjacent United States, the Pacific Northwest and the Cascadia margin were investigated in great detail in the 1990s to better understand crustal structure, tectonics, and earthquake and volcanic hazards. Five major projects were carried out. (1) The first project was conducted in 1991 and concentrated on the area south and east of Puget Sound. (2) Close to this survey in 1995 another offshore-onshore experiment investigated again the seismic properties of the subducting Juan de Fuca plate. (3) The

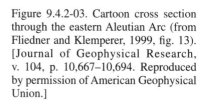

Figure 9.4.2-03. Cartoon cross section through the eastern Aleutian Arc (from Fliedner and Klemperer, 1999, fig. 13). [Journal of Geophysical Research, v. 104, p. 10,667–10,694. Reproduced by permission of American Geophysical Union.]

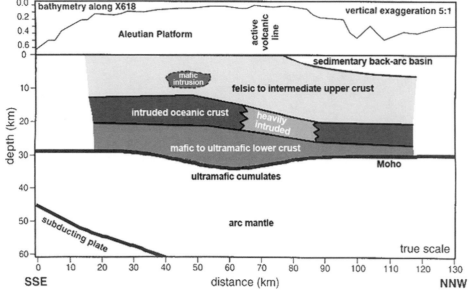

marine waterway system of Puget Sound and its outlets into the Pacific Ocean around Vancouver Island, the Strait of Juan de Fuca in the south, and the Strait of Georgia in the north, were the target of the first part of SHIPS 98 (Seismic Hazards Investigation of Puget Sound, also called "Wet SHIPS") in 1998 (Brocher et al., 1999). (4) SHIPS 99 (also called "Dry SHIPS") was an E-W–directed onshore-offshore survey across the Puget Sound targeting the Seattle Basin (Brocher et al., 2000). (5) The Cascadia margin of southern Oregon was the goal of an onshore-offshore seismic experiment in 1994.

The 1991 Pacific Northwest experiment investigated the Cascade Range and the adjacent Puget Lowland of Washington and was a major seismic-refraction/wide-angle reflection survey. It was preceded by an offshore-onshore seismic experiment in 1989 (Fig. 9.4.2-05) which collected both seismic-reflection and -refraction data along an east-west line in central, coastal Oregon (Brocher et al., 1993a, Trehu and Nakamura, 1993; Appendix 9-2-2). An airgun array served as a seismic source for 8 OBSs positioned on the continental shelf, slope, and adjacent abyssal plain. The airgun shots were also successfully recorded by 10 land stations, consisting of two one-component stations and 8 three-component five-day recorders which were deployed in coastal central Oregon and recorded continuously the airgun signals up to recording distances of 180 km.

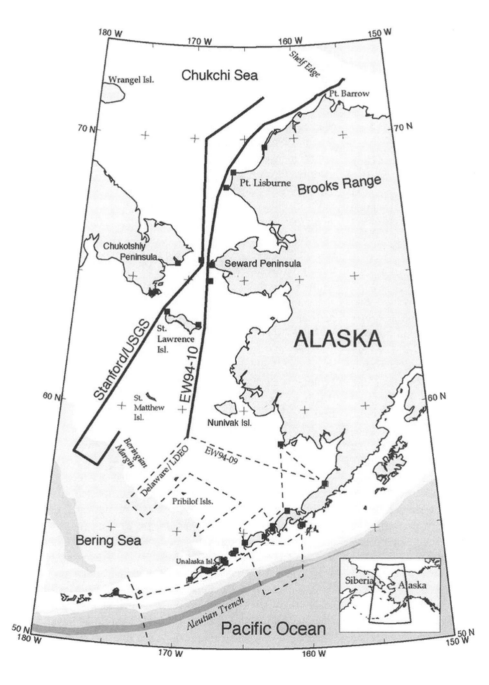

Figure 9.4.2-04. Seismic reflection and refraction survey in the Bering and Chukchi Seas and recording sites in Alaska and Siberia (from Brocher et al., 1995a, fig. 1). [U.S. Geological Survey Open-File Report 95-650, 57 p.]

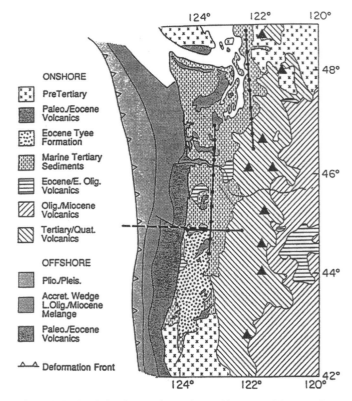

ONSHORE

PreTertiary	
Paleo./Eocene Volcanics	
Eocene Tyee Formation	
Marine Tertiary Sediments	
Eocene/E. Olig. Volcanics	
Olig./Miocene Volcanics	
Tertiary/Quat. Volcanics	

OFFSHORE

Plio./Pleis.	
Accret. Wedge L.Olig./Miocene Melange	
Paleo./Eocene Volcanics	

△—△ Deformation Front

Figure 9.4.2-05. Seismic experiment in Washington and Oregon (from Luetgert et al., 1993, fig. 1) showing the location of the 1991 seismic refraction lines (dots—shotpoints) and the 1989 marine seismic reflection profile (dashed line). Triangles—Cascade volcanoes. [U.S. Geological Survey Open-File Report 93-347, 71 p.]

The 1991 Pacific Northwest experiment comprised two N-S directed profiles, 270 km and 330 km long, and a shorter E-W line, which were recorded in the Puget Basin and the Willamette Valley of western Washington (Fig. 9.4.2-05). Along each of the N-S lines 10 shots of 900–1800 kg charge size at intervals of ~30 km were recorded by ~500 seismic recorders, spaced 600–700 m (Luetgert et al., 1993). The E-W line, located at 44.8–44.9°N, crossed the western 330 km long N-S profile, here 7 shots of 1000 kg charge size were recorded by 420 stations at 350 m spacing (Trehu et al., 1993; Appendix 9-2-2). This line followed and complemented a marine seismic-reflection profile and complementary onshore-offshore recordings (dashed line in Fig. 9.4.2-05), collected in 1989, to the east to the Cascade foothills (Trehu et al., 1990; Trehu et al., 1993; Appendix 9-2-2). Prior to this study crustal studies had only been performed in the Columbia Plateau and the central Cascade Range of Oregon (Catchings and Mooney, 1988a, 1988b; Leaver et al., 1984), and Schultz and Crosson (1996) had investigated the seismic velocity structure across the central Washington Cascade Range from refraction interpretation with earthquake sources.

Interpretations of these data were published separately for the individual lines. Miller et al. (1997) investigated the eastern line which reached farther north and concentrated on the eastern part

of the Puget Sound Basin and adjacent Cascade Range. Trehu et al. (1994) investigated the western N-S line in the Willamette Lowland and the W-E cross line across the Cascade Range (Fig. 9.4.2-05). Along the western N-S profile, a 45–50-km-thick crust was obtained, whereby the lower 5–8 km were interpreted as oceanic crust of the subducted Juan de Fuca plate (Trehu et al., 1994). For the east-west profile, the authors suggested a dip of the Moho from 22 km beneath the outer shelf to 30 km beneath the coast (Trehu et al., 1994). The resulting crustal thickness along the eastern line (Miller et al., 1997) ranged from 42 km in the north to 47 km in the south, where the lower-crustal layer appeared as a 2–8 km transitional layer with velocities of 7.3–7.4 km/s above a low-velocity upper mantle (7.6–7.8 km/s).

In 1995, wide-angle seismic data were collected along a 325-km-long E-W profile near 46.5°N (Fig. 9.4.2-06), which crossed southern Washington from Willapa Bay to the Columbia plateau (Parsons et al., 1998, 1999). About 1500 stations spaced at 200 m intervals recorded 17 large explosions. The wide-angle reflections of these explosion data showed images of the subducting slab and provided continuous first arrivals to 230 km offsets (Parsons et al., 1999).

In 1996, the German research vessel *Sonne* conducted an extensive investigation of the offshore Oregon and Washington margins (Flueh et al., 1997; Parsons et al., 1999). Offshore Washington (Fig. 9.4.2-06), a total of more than 14,500 airgun sources (50–150 m spacing) was fired, with just over 6000 detonated along lines instrumented with a cumulative total of 53 OBSs on the seafloor. The balance of the airgun sources was fired for marine multichannel profiles. On land, 44 stations distributed along three profiles (triangles in Fig. 9.4.2-06) continuously recorded all the airgun sources. The OBSs and on-land profiles provided continuous phase coverage across the margin and enabled a comparison of the structure from north to south.

The 3-D crustal and upper mantle model (Parsons et al., 1998; Fig. 9.4.2-07) was the basis for an extended 3-D modeling of coastal Washington (Parsons et al., 1999), in which surface geology, well data and seismic data were combined to resolve details of the individual crustal layers/terranes at the transition from ocean to continent.

The experiment offshore central Oregon (Fig. 9.4.2-08, Flueh et al., 1997; Gerdom et al., 2000) comprised two N-S directed offshore strike lines over the continental shelf, on each of which six OBHs and one OBS were deployed and recorded the airgun shots of both lines. On the E-W dip line, located at ~44.7°N, 14 OBHs and six OBSs were deployed. Additionally, the airgun shots were recorded by a linear array of 29 seismometers located across central western Oregon, located near 44.7°N, slightly south of the 1991 Pacific Northwest E-W line described by Trehu et al. (1993, 1994). Along the E-W line, the subducting oceanic crust of the Juan de Fuca plate could be traced from the trench to some 10 km landward of the coast with an increasing dip angle changing gradually eastward from 1.5° west of the trench to 16° under the coast. The Moho at the coast line was found at 30 km depth (Gerdom et al., 2000).

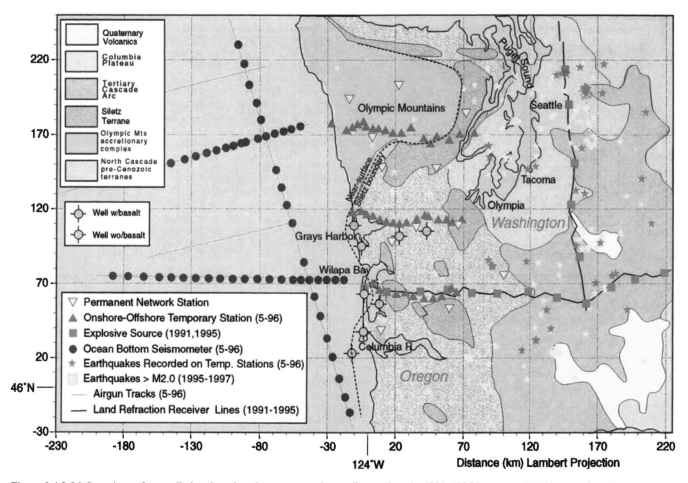

Figure 9.4.2-06. Locations of controlled and earthquake sources and recording stations in 1991–1996 in western Washington (from Parsons et al., 1999, plate 3). The southern E-W and the eastern N-S lines include the locations of the 1991 explosive sources of Figure 9.4.2-05. [Journal of Geophysical Research, v. 104, p. 18,015–18,093. Reproduced by permission of American Geophysical Union.]

The 1998 SHIPS seismic project (Fig. 9.4.2-09) used airgun shots and acquired 1000 km of deep-crustal multichannel reflection profiles and 1300 km of wide-angle lines in northwestern Washington and southwestern British Columbia (Brocher et al., 1999; Appendix 9-2-2). In total, nearly 33,300 airgun shots were released. The experiment was carried out in several stages. First, nearly 13,000 airgun shots were fired by one of the ships exclusively for wide-angle recording, while the ship, without towing a streamer, sailed from Lake Washington near Seattle up and down through the Puget Sound and the narrow Hood Canal and then continued into the Strait of Juan de Fuca, returned and went on through the Strait of Georgia to Texada Island, Canada.

For the airgun shot profiles 257 land stations, 95 of them equipped with three-component seismometers, and 15 OBSs were deployed which allowed recordings to far offsets. Most of the land stations were deployed along the shores of the waterways to provide quasi–2-D lines. Three-dimensional coverage with land stations was achieved east of Puget Sound in the Seattle area and in the southeastern edge of Vancouver Island. Linear arrays were deployed SSE-ward from Tacoma, Washington, and

along the Fraser River north of Vancouver, British Columbia. In addition, the seismic stations of the permanent earthquake monitoring arrays in Washington and British Columbia recorded all the shots. During the deployment, 30 local earthquakes were also recorded.

In the second part of the experiment (Fig. 9.4.2-09), the ship towed a 2.4-km-long, 96-channel digital seismic streamer which reduced the size of the airgun array. With this equipment, reflection lines were acquired along the three main waterways. The near-trace of the seismic streamer was 200 m and the far-trace 2575 m behind the ship, resulting in a group interval of 25 m. Using a second ship, towing two short streamers of 300 m length, it was furthermore possible to record constant offset profiles (COPs), thus prolonging the recording aperture of the first ship. Also, 12 expanding spread profiles were acquired as the two ships sailed apart from each other in opposite directions at the same speed, maintaining a common midpoint.

Only a limited amount of wide-angle data could be collected. Most stations provided usable data to offsets of 40–50 km, although airgun shots can be recorded to much greater ranges if

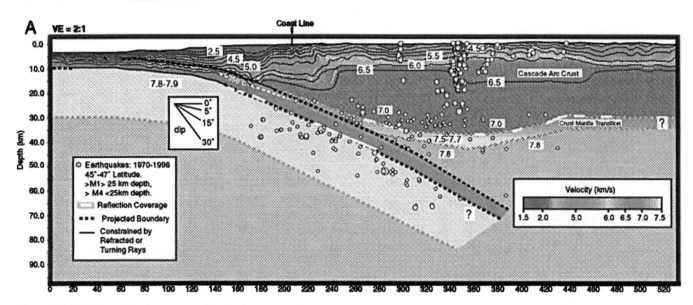

Figure 9.4.2-07. Crustal and upper mantle velocity cross section across southern Washington (from Parsons et al., 1999, plate 2A). [Journal of Geophysical Research, v. 104, p. 18,015–18,093. Reproduced by permission of American Geophysical Union.]

Figure 9.4.2-08. Location of seismic refraction lines of 1996 across the Oregon continental margin (from Gerdom et al., 2000, fig. 2). [Tectonophysics, v. 329, p. 79–97. Copyright Elsevier.]

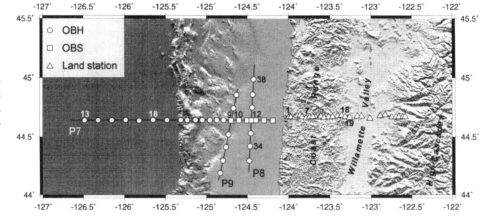

recorded on remote bedrock sites. One of the results from this experiment was a tomographic model which imaged the upper crust of the Puget Lowland to a depth of ~11 km (Brocher et al., 2001).

As a sequel to the 1998 airgun experiment "Wet SHIPS," the 1999 SHIPS experiment, nicknamed "Dry SHIPS," acquired a 112-km-long E-W–trending multichannel seismic-reflection and -refraction line in the Seattle Basin (Fig. 9.4.2-10). A total of 1008 seismographs were deployed at an average spacing of 100 m, and 29 shotpoints were organized at ~4 km intervals along the seismic line. To bridge the gap in line 2 caused by the Puget Sound, eight of the seismographs were OBSs, deployed in the Sound at water depths ranging from 50 to 250 m. In addition to the active source data, 35 continuously recording three-component stations, deployed at 4 km intervals, and the OBSs sampled data from 26 regional earthquakes and blasts as well as 17 teleseismic events (Brocher et al., 2000; Appendix 9-2-2). The receiver geometry provided both a low-fold reflection cover-

age and a high-fold refraction coverage through the center of the basin. Due to the geometry of the local waterways and the availability of public lands, the line had to be "broken" into lines 1 (39 km long) and 2 (88 km long) which overlapped for 12 km on the Kitsup Peninsula (Fig. 9.4.2-10). Lines 3–6 (22–25 km long) are N-S–trending fan lines to provide 3-D control in the eastern part of the Seattle Basin. A total of 38 shots was detonated at the 29 shotpoints, all in drill holes, with charges ranging from 11 to 1136 kg. At nine shotpoints, shots were repeated to allow stacking of the data.

A subsequent experiment, the "Kingdome SHIPS" phase, followed in early 2000 with 206 stations deployed with a nominal spacing of 1 km in a hexagonal grid in the city of Seattle to look at the site response within the upper 2 km of the Seattle Basin. They recorded the implosion (100,000 kg delayed) of the Kingdome sports arena and four additional 68-kg shots at the corners of the grid, recording usable energy up to 10 km (Snelson, 2001).

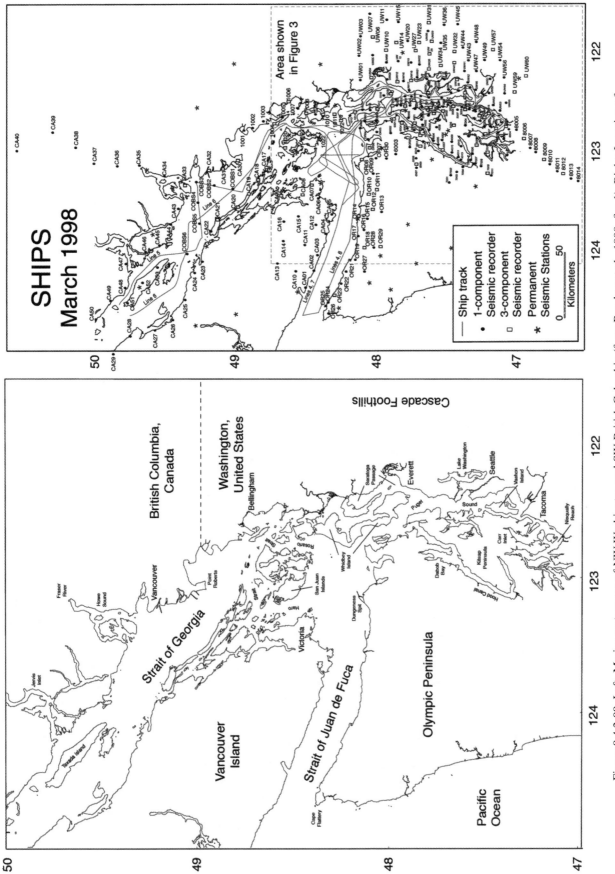

Figure 9.4.2-09. Left: Marine waterways of NW Washington and SW British Columbia (from Brocher et al., 1999, fig. 1). Right: Location of ship tracks (seismic lines) and recorders of SHIPS 98 seismic project (from Brocher et al., 1999, fig. 2). [U.S. Geological Survey Open-File Report 99-314, 123 p.]

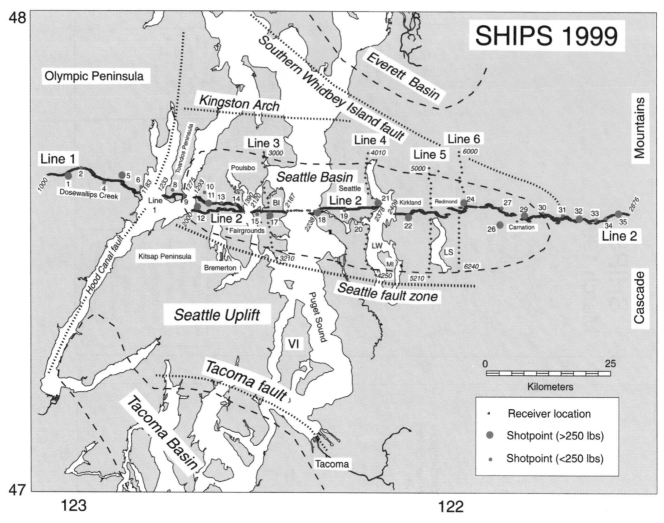

Figure 9.4.2-10. Location of Dry SHIPS 99 seismic shots and recorders in the Puget Lowland (from Brocher et al., 2000, fig. 1). [U.S. Geological Survey Open-File Report 00-318, 81 p.]

As a result of a 3-D P-wave tomography study of the data of the 1999 SHIPS experiment, the basin is ~70 km wide and contains sediments with velocities from 1.8–4.5 km/s. The basement rocks underneath the Seattle basin, characterized by a rapid increase in velocity from 4.5 to 5.0 km/s, are found at 6–7 km depth at the center of the seismic line, consistent with the 1998 N-S reflection line recorded along the Puget Sound (Snelson, 2001; Snelson et al., 2007).

The southernmost seismic project in the coastal area of Washington and Oregon was carried out in late 1994 in southern Oregon in the vicinity of Cape Blanco between 42°N and 43.5°N. It was a deep-crustal wide-angle seismic-reflection and -refraction experiment to study the lateral variations in crustal structure along the Cascadia margin (Brocher et al., 1995b; Appendix A9-2-2). The aim of the project was to image the lower crustal structure, observe significant crustal differences across a major E-W–trending shear zone near Cape Blanco, and to image the subducting Gorda and Juan de Fuca plates.

Two linear arrays of land stations were deployed along E-W transects across the Oregon Coast Range, each array having 10 three-component recording stations and stretching landward to 80 km from the coast (Fig. 9.4.2-11). Nearly 12,300 airgun shots were recorded by both arrays shot in-line along lines 5 (86 km long) and 6 (81 km long), while the shots of the other offshore lines were recorded onshore in a fan geometry. Airgun signals were recorded from 8 km to 160 km offsets on both deployments. The resulting data quality was extremely satisfying. Coherent arrivals could be traced as far as 170 km, P_n arrivals, e.g., were visible over intervals of 30 km (Brocher et al., 1995b).

The marine data collection comprised 760 km of multi-channel seismic-reflection profiles, collected on the continental margin of southern Oregon by recording airgun shots on a 4.2-km-long, 160-channel digital streamer. Seven profiles were obtained along strike-lines parallel and dip-lines perpendicular to the Oregon coast (Fig. 9.4.2-11).

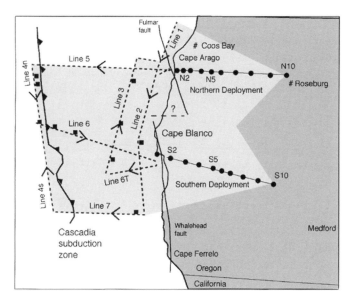

Figure 9.4.2-11. Location of onshore-offshore seismic wide-angle reflection lines near Cape Blanco, southern Oregon (from Brocher et al., 1995b, fig. 1). [U.S. Geological Survey Open-File Report 95-819, 69 p.]

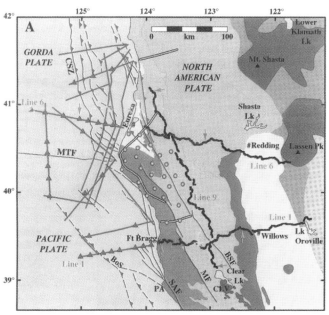

Figure 9.4.2-12. 1994 offshore multichannel seismic profiles (straight blue lines); 1994 OBS/OBH positions (triangles); and 1993 onshore profiles (crooked black and dark blue (reoccupied in 1994) lines) of the Mendocino Triple Junction experiment in northern California (from Trehu and Mendocino Working Group, 1995, fig. 1). Red lines: faults. [Eos (Transactions, American Geophysical Union), v. 76, no. 38, p. 369, 380–381. Reproduced by permission of American Geophysical Union.]

9.4.2.3. San Andreas Fault Region

9.4.2.3.1 Mendocino Triple Junction in northern California. The Mendocino Triple Junction in northern California has raised the particular interest of geoscientists, because here the active tectonic regime of northwestern California changes abruptly from transform motion along the San Andreas fault zone to subduction north of the Mendocino fault zone. Since Atwater (1970) proved the influence of plate tectonics on the geological history of western North America, it has become evident that three plates meet at the Mendocino Triple Junction (Fig. 9.4.2-12).

The North American and Pacific plates slide against each other onshore and offshore along the complex San Andreas fault system, while the Pacific and Gorda plates slide against each other offshore along the Mendocino transform fault. North of the Triple Junction, the Gorda plate is being subducted obliquely beneath the North American continent along the megathrust of the Cascade subduction zone (Trehu and Mendocino Working Group, 1995).

To study the crust and upper-mantle structure underneath this complex region of northern California and the adjacent continental margin, in 1993 and 1994 a network of large-aperture seismic profiles was arranged (Fig. 9.4.2-12). Approximately 650 km of onshore seismic-refraction and reflection data and 2000 km of offshore multichannel seismic (MCS) reflection data were collected during this survey. Furthermore, simultaneous onshore and offshore recordings of the MCS airgun shots were made (Godfrey et al., 1995; Appendix 9-2-3; Godfrey et al., 1998; Trehu and Mendocino Working Group, 1995).

The onshore survey of 1993 comprised two seismic lines of 200 km in length, where along each line 570 stations, spaced on average at 350 m, recorded 8 shots, spaced 25 km apart, with

charges of 400–2000 kg (lines 1 and 6). The third profile (line 9) was 250 km long. Here 11 shotpoints were realized and the station separation was near 400 m. The offshore MCS data of 1994 were obtained by airgun shots, fired at 50 m intervals and recorded on a 4-km streamer up to 14 s TWT. The survey was complemented by OBS/OBH deployments at 34 seafloor sites and by the reoccupation of 477 onshore sites. In addition, a 1993 shotpoint near the intersection of lines 1 and 9 was revived (Godfrey et al., 1995, 1998; Trehu and Mendocino Working Group, 1995).

Interpretations of the individual data sets were published (e.g., Godfrey et al., 1998; Leitner et al., 1998; Henstock et al., 1997; Henstock and Levander, 2000). A cartoon of Leitner et al. (1998), interpreting MCS data in combination with gravity data, indicated a crustal thinning from 20 km near the coast to less than 11 km at 100 km distance from the mainland. Interpreting in particular the data of line 1 (Fig. 9.4.2-12), the southernmost offshore-onshore line, Henstock and Levander (2000) found that the top of the lower crust, a 5–10-km-thick high-velocity layer (6.4–7.2 km/s), deepened from 7 km beneath the Pacific Ocean basin at the west end to 23 km at the east end of line 1, being offset at the San Andreas and the Maacama faults (SAF and MF in Fig. 9.4.2-12) by up to 4 km, down to the east. The Moho was found to be similarly deformed, although by only 2 km.

9.4.2.3.2. San Francisco Bay Area. While the San Andreas fault zone throughout most of northern California, with a few exceptions, runs offshore close to the coast, it enters the mainland in the area of the city of San Francisco. The San Francisco Bay Area is one of the most seismically active metropolitan areas in the United States. Here, the fault zone is spread over a broad, 100-km-wide zone, encompassing several major active faults, such as the San Gregorio, San Andreas, Hayward and other faults. Southwards, in Central California, the San Andreas fault juxtaposes the Cretaceous granitic Salinian terrane to the west and the Late Mesozoic/Early Tertiary Franciscan Complex to the east side (Fig. 9.4.2-13). On the San Francisco Peninsula,

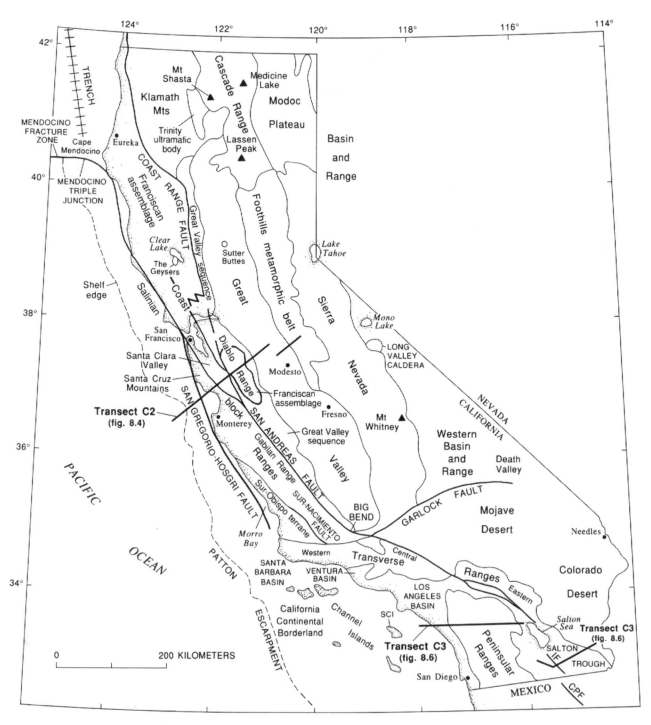

Figure 9.4.2-13. Map of California, showing place names, geologic provinces, selected geologic units, and locations of the crustal transects summarizing the deep structure of the crust known until 1990 (from Fuis and Mooney, 1990, fig. 8.2). [U.S. Geological Survey Professional Paper 1515, p. 207–236.]

however, the present-day San Andreas fault is completely within a Franciscan terrane, while the Salinian-Franciscan boundary is located southwest, marked by the subparallel Pilarcitos fault (Parsons and Zoback, 1997).

Due to the major earthquake hazard, the area contains one of the most densely occupied seismic networks in the world. However, until the early 1990s, little was known about the details of crustal structure which is important in understanding the tectonic processes that produce the seismicity and in accurately locating regional earthquakes. Sparsely sampled 1967 refraction lines south of the San Francisco Bay area (Walter and Mooney, 1982) and 1976 lines north of the Bay (Warren, 1981) provided first-order information on either side of the fault. Several tomography studies made use of the dense local network of the permanent Northern California Seismic Network (NCSN), which was started in 1966 and has since then been successively enlarged to its present-day status. However, none of the models focused on the entire Bay Area region. Fuis and Mooney (1990) have published two transects summarizing the deep structure of the crust along the San Andreas fault zone based on the knowledge available until 1989.

It was only in 1990, following the Loma Prieta magnitude 7.1 earthquake of 17 October 1989, that detailed crustal studies started. The first project was a marine seismic-reflection investigation of the central California margin and its prolongation on land. The marine seismic-reflection investigation of the central California margin was conducted along two 60-km-long parallel profiles running offshore from the San Francisco peninsula toward the SW. The airgun array of the offshore reflection lines was used as a sound source for recording with three-component seismic five-day recorders deployed onshore along two profiles, continuing the offshore lines on land (Brocher et al., 1992; Appendix A9-2-4).

The main recorder array consisted of 13 stations and stretched from the coast across the epicentral region of the Loma Prieta earthquake over a distance of 92 km (profile line identical with line FG in Fig. 9.4.2-14). The second array of six stations was located 25 km to the northeast over a distance of 70 km (Brocher et al., 1992, 2004a; Page and Brocher, 1993). On the basis of this data set, Brocher et al. (2004a) concluded that the hypocenter of the Loma Prieta earthquake was located at the base of a wedge of Salinian rocks beneath the continental margin, but above oceanic crust. The Loma Prieta line was reversed in 1993 (Brocher, 1994; Appendix A9-2-4) when 33 stations (line FG in Fig. 9.4.2-14) recorded some of the chemical explosions set off during the 1993 experiment which is described further below.

To obtain fundamental information on the crustal structure and fault geometries that underlie the Bay Area, in the fall of 1991 a major seismic-reflection investigation was conducted utilizing the marine waterway system that dissects the area (Figs. 9.4.2-15 and 9.4.2-16). The project BASIX (Bay Area Seismic Imaging Experiment) comprised three complementary seismic methods: high-resolution reflection profiling, wide-angle reflection/refraction profiling, and multichannel seismic profiling (McCarthy and

Hart, 1993; Brocher et al., 1991b, 1994; Holbrook et al., 1996; Hole et al., 2000). The high-resolution shallow-penetration data were acquired to constrain near-surface faulting, while the wide-angle reflection/refraction data were essential to explore the deep-crustal velocity structure of the crust down to the Moho (McCarthy and Hart, 1993; Appendix A9-2-4).

The experiment involved an airgun array towed at an average depth of 7–8 m. For the multichannel seismic-reflection profiling the seismic energy was recorded by hydrophones installed at the edge of the dredged shipping channel and radioed back to a 120-channel recording system on the ship. Each day the hydrophones were shifted to new positions along the profiles, and each night the ship steamed back and forth along the hydrophone array while firing the airguns. Thus, the nominal shot interval was 50 m on the inland and Golden Gate profiles and 75 m on the offshore profile. The extensive marine waterway system that dissects the San Francisco Bay Area allowed recording of a total of 140 km of multichannel reflection data (track lines in Fig. 9.4.2-15), which was achieved by acquiring the data in 13 separate, overlapping segments, each ~25 km long. More than 15,000 separate airgun shots were recorded during BASIX.

Furthermore, wide-angle reflection profiles were recorded (Brocher and Moses, 1993; Appendix A9-2-4) from the airgun shots by land stations, recording the airgun shots of the 13 reflection lines (Fig. 9.4.2-16). The majority of these stations were deployed along the E-W–trending line from the Sacramento River to the Golden Gate, resulting in offsets from 1 km to 160 km. The NW-SE line consisted of a mixture of wide-angle in-line and fan recording, spanning a length of 49 km. In addition to the 13 reflection lines, the airguns were shot along two other lines solely for wide-angle recording: along the NW-SE–trending line along the San Francisco Bay with a shot interval of 50 m and along the E-W–trending line on the continental margin with a shot interval of 100 m. Here, five OBSs were successfully deployed for each line separately (Fig. 9.4.2-16). The source-receiver offsets along the offshore line ranged from 11 to 259 km. As a result, the base of the upper crust with seismic velocities from 5.5 to 6.0 km/s was marked by high-amplitude reflections at a TWT of 6 s, corresponding to a depth of 15 km. The upper crust showed no internal discontinuities and was interpreted as the Franciscan assemblage, which is exposed throughout the Bay Area. For the lower crust high seismic velocities between 6.5 and 7.2 km/s and a thickness between 8 and 11 km east of the San Andreas fault resulted (Brocher et al., 1994).

A subsequent interpretation (Holbrook et al., 1996) confirmed the thickness of the upper crust to be 15–18 km, but divided the lower crust into two layers with 6.4–6.6 and 6.9–7.3 km/s velocities (Fig. 9.4.2-17). The authors saw a thickening of the crust from 12 km at the southwestern end of line OBS-2-25 km under the Bay Area. They also indicated a similar offset of the top and the bottom of the lower crust at the San Gregorio and San Andreas faults to the NE (SGF and SAF in Fig. 9.4.2-17, bottom section), as was noted farther north near the Mendocino Triple Junction by Henstock and Levander (2000).

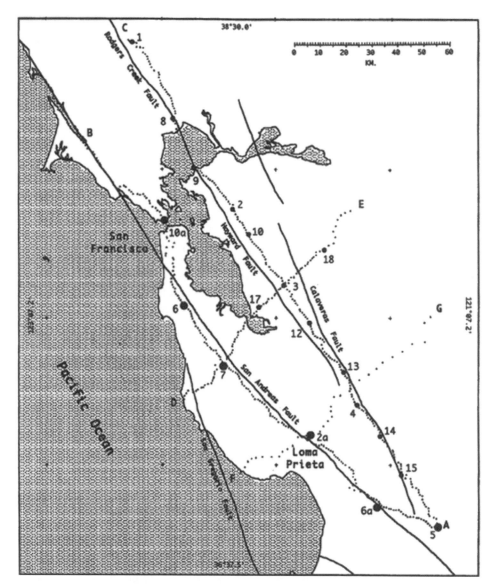

Figure 9.4.2-14. Map of the San Francisco Bay Area showing onshore seismic-refraction lines (from Kohler and Catchings, 1994, fig. 2). Line AB recorded along the San Andreas fault in 1991 (shotpoints 2a, 6a, 10a) and reshot in 1993; line AC recorded along the Hayward fault; line DE cross line; line FG Loma Prieta line reversing the 1991-airgun line-38 onshore-extension. [U.S. Geological Survey Open-File Report 94-241, 71 p.]

A tomography approach providing a 3-D model of the seismic velocity structure of the San Francisco Bay Area was applied by Hole et al. (2000), showing lateral variations of the velocity contrast from upper to lower crust which were correlated with surface faults, indicating changes in the depth of the mafic lower crustal layer. A comprehensive review was prepared by Parsons (2002) for seismic data obtained between 1991 and 1997, with the exception of the seismic-refraction survey of 1993, described further below.

In May 1991, a 180-km-long seismic-refraction line was recorded with 1 km spacing from five drillhole shotpoints (line AB in Fig. 9.4.2-14). The line started north of 38°N near Point Reyes, paralleled the San Andreas fault, passed through the San Francisco area (Murphy et al., 1992; Appendix A9-2-4) and ended south of 37°N ~40 km southeast of San Jose (Murphy et al., 1992). The line was extended by 20 km toward the south

and re-shot in 1993 with the addition of a sixth shotpoint, when refraction profiling was carried out both along the Hayward and the San Andreas faults (Kohler and Catchings, 1994). The Moho was found at 30–32 km depth under the southeastern end of both lines and decreased toward the NW to 22 km under the city of San Francisco (Catchings and Kohler, 1996).

During this project, two more lines were recorded: the 220-km-long Hayward fault line (line AC in Fig. 9.4.2-14) and a 100-km-long cross line (line DE in Fig. 9.4.2-14) from the Pacific coast into the Diablo Range east of the Calaveras fault (Kohler and Catchings, 1994). The instrument spacing was 1.5 km. Along line AC, four OBSs provided recordings from the bottom of the San Pablo Bay. The 1993 experiment comprised a total of 17 shots. In the first deployment, lines AB and DE were occupied and five shots fired at shotpoints 5, 6, 7, 17, and 18. In the second deployment, lines AC and CD were occupied and 12

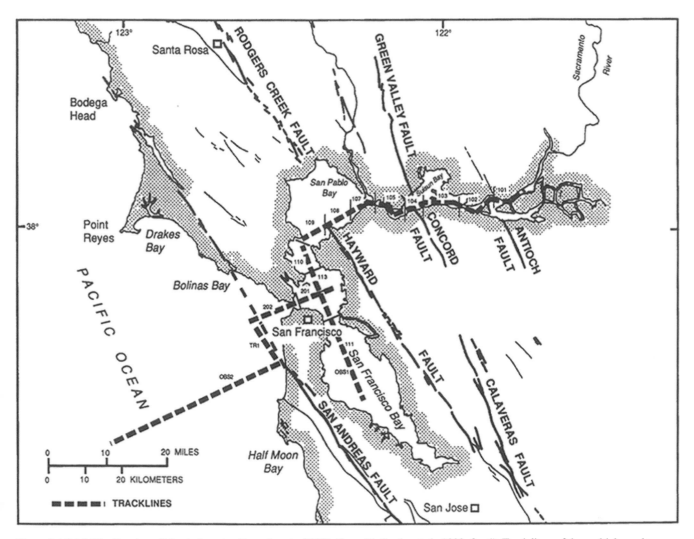

Figure 9.4.2-15. The Bay Area Seismic Imaging Experiment of 1991 (from McCarthy et al., 1993, fig. 4). Track lines of the multichannel survey. [U.S. Geological Survey Open-File Report 93-301, 26 p.]

shots at all remaining shotpoints were fired. A comprehensive interpretation of these data has not been published, but Boatwright et al. (2004) have collected the first-arrival traveltimes and geometry data and calculated tomographic velocity models for each line.

In 1995, 183 stations on the San Francisco peninsula recorded seven in-line and four fan shots along a SW-NE–trending linear array of 9 km length with an instrument spacing of 50 m. Reflections up to 14 s TWT could be correlated. Here the Moho was inferred at the lower end of middle and lower crustal reflectivity at 9s TWT, which was converted into 27–28 km depth (Parsons, 1998).

9.4.2.3.3 Central California. The crustal investigation of the surroundings of the San Andreas fault was shifted southwards in 1995, when an array of 48 seismic instruments, which had been installed already by the end of 1994 to record local earthquakes for half a year, was used for an active-source experiment (Fig. 9.4.2-18). Thirteen shots of 200–300 kg charge size were fired

into the array and additional stations, forming three profiles that paralleled and crossed the San Andreas fault (Thurber et al., 1996). An important result of this project was the confirmation that there was a sharp lateral velocity contrast at the surface trace of the San Andreas fault and that the San Andreas fault bounds the southwestern edge of a 6–7-km-wide trough of low P-wave velocity that extends to 3–4 km depth.

The most recent study of the San Andreas fault zone finally targeted the Parkfield area, halfway between the densely populated areas of San Francisco and Los Angeles. A series of magnitude-6 events with very similar characteristics (Brown et al., 1967) had led to a detailed geophysical investigation of this area (Bakun and Lindt, 1985) which should finally lead to the deep-drilling project SAFOD (San-Andreas-Fault-Observatory-at-Depth; Hickman et al., 1994). The main activities, the realization of a deep drillhole and a major seismic-refraction/-reflection project, however, were only started after 2000 and will therefore be described in Chapter 10.

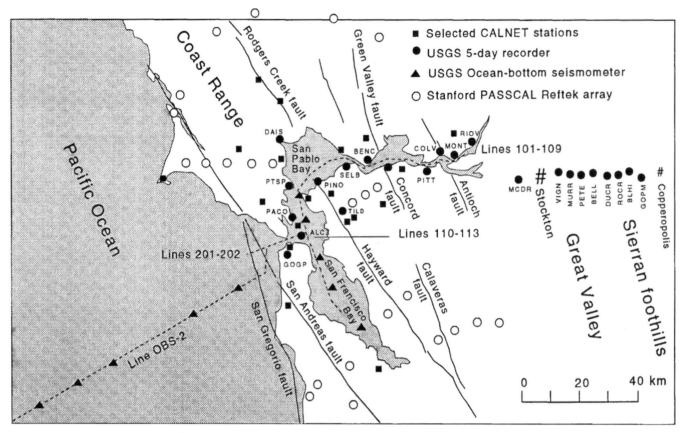

Figure 9.4.2-16. The Bay Area Seismic Imaging Experiment of 1991 (from Brocher and Moses, 1993, fig. 1): Basic seismic reflection lines (dashed lines) and location of wide-angle recorders. [U.S. Geological Survey Open-File Report 93-276, 89 p.]

9.4.2.3.4. Southern California. Similar to the San Francisco Bay area in central California, also in the densely populated area of Los Angeles–San Diego in southern California, the strong earthquake activity in the immediate neighborhood of the active San Andreas fault system causes a severe threat. A dense network of permanent seismic stations is used to analyze the statistics of the earthquake activity, and major events are being systematically studied (e.g., Ellsworth, 1990). The lithospheric cross section through southern California by Fuis and Mooney (1990) indicates, however, how little was known about the crust and upper mantle up to 1990. The only detailed crustal seismic investigation had been carried out in the Imperial Valley in 1979 south of the Salton Sea (Fuis et al., 1984), described in Chapter 7.4.2.

Following several major earthquakes in the Los Angeles area which were associated with buried faults, the need for detailed knowledge of subsurface structures became obvious and the project LARSE (Los Angeles Region Seismic Experiment) was born with the aim to obtain a very detailed structural crustal model for the Greater Los Angeles Area along selected traverses reaching from the coastal regime across its basins and mountain ranges into the Mojave desert. The planning of LARSE-I concentrated mainly on line 1 (Fig. 9.4.2-19), but later on, follow-

ing the Northridge event of 1991, in 1999 a detailed land survey (LARSE-II) was added on line 2 (Fig. 9.4.2-19).

LARSE-I started in 1993 with a passive study along line 1, with the goal to collect seismic waveform data from local and teleseismic earthquakes to refine 3-D images of the lower crust and upper mantle in southern California. In October 1994 an active seismic source experiment was conducted. It consisted of an onshore part along line 1 and an onshore-offshore survey. The onshore part, which followed the marine airgun shooting survey, concentrated on line 1 which started at Seal Beach, crossed the Los Angeles Basin and the San Gabriel Mountains and ended in the Mojave Desert (Fuis et al., 1996). Sixty explosions, with charges ranging from 5 kg to 3000 kg, were recorded by a linear array of 640 seismographs, with the majority of drillhole shots being set off in the San Gabriel Mountains (Fig. 9.4.2-19).

During LARSE-I, in October 1994, also deep-crustal multichannel seismic-reflection data were recorded offshore (Brocher et al., 1995c; ten Brink et al., 1996; Appendix A9-2-5). Three onshore-offshore wide-angle reflection lines, all centered on the Los Angeles basin, were recorded, each 200–250 km long in total, with the offshore portions being between 90 and 150 km long (Fig. 9.4.2-20). Using an airgun array as the seismic energy source, nearly 24,000 airgun shots were done. Wide-angle data

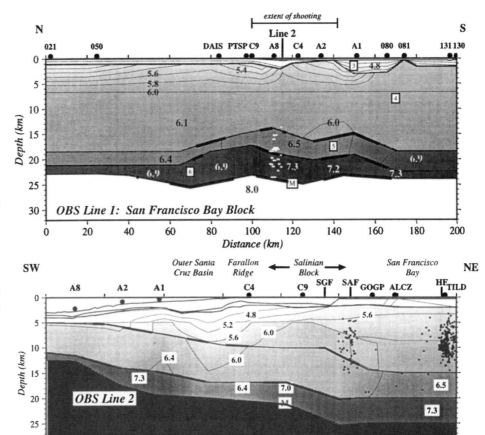

Figure 9.4.2-17. Crustal and upper mantle velocity cross sections across the San Francisco Bay Area (from Holbrook et al., 1996, figs. 2 and 16). Top: N-S section through the San Francisco Bay area. Bottom: SW-NE section across the continental margin to the Hayward fault (HF) east of the San Francisco Bay. [Journal of Geophysical Research, v. 101, p. 22,311–22,334. Reproduced by permission of American Geophysical Union.]

were recorded by 170 land stations deployed at 2 km intervals along all three profiles, and at sea by two sonobuoys and by 10 OBSs, deployed along lines 1 and 2 (details in ten Brink et al., 1996; Appendix A9-2-5). Three airgun lines TR1 to TR3 connected the main lines 1, 2, and 3 and were recorded by the on-shore stations in an oblique fan geometry. Three more lines, 4–6, were solely multichannel seismic-reflection offshore lines. All multichannel reflection data were recorded by a 160-channel, 4.2 km-long digital streamer (Brocher et al., 1995c; Appendix A9-2-5). Airgun signals were visible at OBSs as far as 70 km, and in some cases also up to 140 km. At onshore sites the airgun energy carried over distances of 200 km and was even visible in the Mojave Desert.

In October 1999, LARSE-II was conducted along line 2 (Fuis et al., 2001a, 2001b; Appendix A9-2-5, Murphy et al., 2002). As on line 1, prior to the active-source experiment local and distant earthquakes were recorded along line 2 for a period of 6.5 months. The active-source survey complemented a 1994 onshore-offshore (airgun) survey along the same corridor. Chief imaging targets included the Santa Monica, San Gabriel, and San Andreas faults, blind thrust faults (including the Northridge fault), and the depths and shapes of the sedimentary basins in the San Fernando Valley and Santa Monica areas.

Ninety-three borehole explosions were detonated along the main line 2, which was 150 km long and extended from Santa Monica Bay northward into the western Mojave Desert, crossing the Santa Monica Mountains, the San Fernando Valley (through the Northridge epicentral area), the Santa Susana Mountains, and the western Transverse Ranges (Fig. 9.4.2-19), and along 5 auxiliary lines in the San Fernando Valley and Santa Monica areas, ranging in length from 11 to 22 km. The explosions were recorded by ~1400 seismographs, simultaneously on all lines. The goal of 1 km shotpoint spacing and 100 m geophone spacing for the main line was nearly achieved for the first 79 km from the coast to the southern Mojave Desert. Farther north, shotpoint spacing averaged 2.75 km from 79 to 101 km, and the remaining shotpoint was located at 136 km, while station spacing increased stepwise from 90 km to the end of the line at 150 km. The shot size, as far as open space was available, was planned to be a mix from small (near 100 kg) to large (~450 kg), in order to investigate reflective features with differing frequency return. However, in the densely populated areas the charge sizes depended on the location of the shotpoints and were determined from ground-shaking data obtained during LARSE-I and from calibration shots (Fuis et al., 2001a, 2001b).

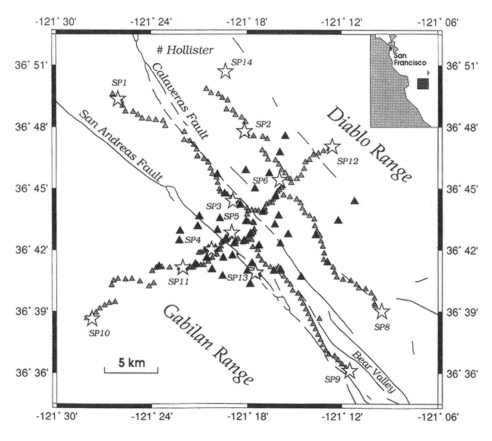

Figure 9.4.2-18. Location of the 1995 seismic experiment (from Thurber et al., 1996, fig. 2). Passive array (large triangles) and in-line active-source recording stations (small triangles) and shotpoints (stars). [Eos (Transactions, American Geophysical Union), v. 77, p. 45, 57–58. Reproduced by permission of American Geophysical Union.]

LARSE-I imaged the bottom of the Los Angeles basin for the first time (Fuis et al., 1996, 2001b). A velocity model (Lutter et al., 1999) combined with the line drawing of CMP (crustal mid-point) data by Ryberg and Fuis (1998) and a subsequently developed schematic block diagram of Fuis et al. (2001b) illustrate the tectonics which could be interpreted from the seismic results of the LARSE-I project (Fig. 9.4.2-21). One important first result was the discovery that areas of high seismicity correlated well with areas of increased velocities (Fuis et al. (1996). Beneath the San Gabriel Mountains, bright reflective zones (A and B in Fig. 9.4.2-21) were discovered along line 1 in the middle crust at ~20 km depth that appeared to connect with the San Andreas fault in the north and with compressional faults in the Los Angeles basin to the south (Ryberg and Fuis, 1998).

Furthermore, the explosion seismic data showed a dip of the Moho from 30 km under the Mojave Desert to a maximum of 38 km under the San Gabriel Mountains (bottom of the reflective lower crust in Fig. 9.4.2-21). Toward the south the data could not be continuously correlated.

From teleseismic observations a root zone thickened by 15 km under the San Gabriel Mountains was deduced (Kohler and Davis, 1997). Hauksson and Haase (1997) have obtained velocity-depth cross sections to 20 km depth by a 3-D inversion using all available explosion seismic and earthquake data. This model, however, was detailed enough only along line 1 where from the explosion seismic data the traveltimes between sources and receivers were

sufficiently accurately known. Here an intrusion of basement rocks into the sediments of the basin was interpreted as a likely origin of the Whittier thrust fault (also seen in Fig. 9.4.2-21).

Tomographic modeling of the first arrival data of line 2 (Lutter et al., 2004) produced a similar model for line 2 as for line 1 (Lutter et al., 1999). Along line 2 prominent zones of dipping reflectors, which were superimposed onto the tomographic model, could be correlated. The most prominent feature is a north-dipping reflective zone which extends downward from near the hypocenter of the 1971 San Fernando earthquake at ~13 km depth (brown circle in Fig. 9.4.2-22) to ~30 km depth in the vicinity of the San Andreas fault in the central Transverse Ranges. A fainter south-dipping reflective zone extends from the 1994 Northridge hypocenter at ~16 km depth (blue circle in Fig. 9.4.2-22) to 20 km depth beneath the Santa Monica Mountains. These reflective bands could be correlated with aftershock series of the 1971 San Fernando and the 1994 Northridge earthquakes (brown and blue dots, respectively, in Fig. 9.4.2-22), which were projected from a zone 10 km east of line 2 (Fuis et al., 2001b, 2003).

9.4.2.4. Sierra Nevada and Basin and Range Province

In summer 1993, the Southern Sierra Nevada Continental Dynamics (SSCD) project recorded two seismic-refraction lines in central California, which addressed the Sierra Nevada root question (Wernicke et al., 1996; Ruppert et al., 1998; Fliedner et al., 2000). The first line was 400 km long and ran in a W-E

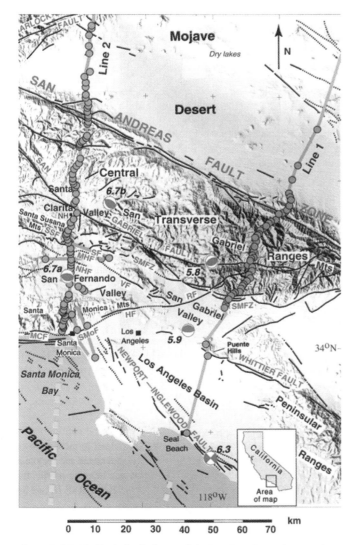

Figure 9.4.2-19. Shaded relief map of the Los Angeles region, southern California, showing location of the LARSE active source seismic profiles (from Fuis et al., 2003, fig. 1). Small circles—shotpoints; colored lines—seismic recording stations; yellow dashed lines—airgun pulses; large red circles—epicenters of large earthquakes (with focal mechanisms and magnitudes). [Geology, v. 31, p. 171–17. Reproduced by permission of the Geological Society of America.]

at Owens Lake. Near Death Valley, some additional 680-kg shots were added for increased resolution of upper-crustal features (Wernicke et al., 1996).

A small crustal root (40–42 km thick) centered ~80 km west of the Sierran topographic crest, low average P_n velocities of 7.6–7.9 km/s under a laminated transitional Moho under the Sierra Nevada batholith, and slightly higher P_n velocities of 7.9–8.0 km/s under a sharp first-order Moho under the Basin and Range province were the main results of a 3-D interpretation (Ruppert et al., 1998). Including earlier refraction and earthquake data, a subsequent interpretation revealed a 3-D velocity model of the crust (Fliedner et al., 2000). Crustal thickness increased from less than 25 km at the coast to 35–42 km in the western Sierra Nevada and decreased to 30–35 km in the Basin and Range province. The authors also saw a systematic thinning of the crust in the Basin and Range province from north to south in concordance with a decrease in elevation, and both the Basin and Range province and the Sierra Nevada batholith showed average crustal velocities of 6.0–6.2 km/s well below the worldwide continental average of 6.45 km/s.

In the Basin and Range Province of Nevada the USGS had first conducted seismic-refraction studies in 1980–1982 (Hoffman and Mooney, 1983) to aid an investigation of the regional crustal structure at a possible nuclear waste repository site near Yucca Mountain. In 1988, another test experiment had been carried out south of Yucca Mountain (Brocher et al., 1990, 1993b).

On the basis of the previous studies, in 1994 a detailed seismic-reflection survey along two lines of 11 and 26 km length followed, using hybrid seismic sources including vibrators, surface-mounted explosive charges, and conventional shallow minihole and deep shot holes for data collection. Besides a detailed image of the upper crust, the interpretation revealed a reflective lower crust between 15 km depth and the Moho between 27 and 30 km depth along the longer profile and excluded the existence of a lower-crustal magma chamber (Brocher et al., 1996; Appendix A9-2-6).

Another seismic-reflection survey, combining normal-incidence and wide-angle reflection data, was arranged by a joint project of the Universities of Arizona, Wyoming, and Georgia, and concentrated on the Ruby Mountains metamorphic core complex southeast of Elko, Nevada (Satarugsa and Johnson, 1998). The focus was to prove or disprove the statement that the Ruby Mountains had a flat Moho as had been postulated by the COCORP 40° transect (Allmendinger et al., 1987; Hauser et al., 1987b) which crossed the Ruby Mountains at its southern end, where topography is relatively low. The fieldwork was carried out in 1992 and 1993. The 1992 survey consisted of a 90-km-long N-S–trending line along the eastern flank of the mountains and three 3–9-km-long cross lines. The southern end of the long line coincided approximately with location 1000 of the COCORP 40° transect (see Fig. 8.5.3-02, eastern half, profile 4 at position 1000). Twelve shots had charges from 250 to 1080 kg and an average shot spacing of 20 km, and the in-line receiver spacing varied between 70 and 100 m with one-,

direction from the Coast Ranges across the Sierra Nevada into Death Valley, in the western Basin and Range province. The second line was 325 km long and ran in a N-S direction along Owens Valley east of the Sierra Nevada to the south across the Garlock fault (Fig. 9.4.2-23). A third line, unreversed, spanned 450 km from Beatty, Nevada, across the Sierra Nevada into the Coast Ranges and ran along the first line, recording an explosion of 2 million pounds of chemicals, the Department of Energy's Nuclear Non-Proliferation Experiment (NPE). Up to 675 seismic recorders had been installed for each seismic profile, with a nominal spacing of 500 m along the first two lines and 1500 m along the third line. Shot sizes ranged from 3600 kg for the end shots of the first two lines to 1360 kg at the profiles' intersection

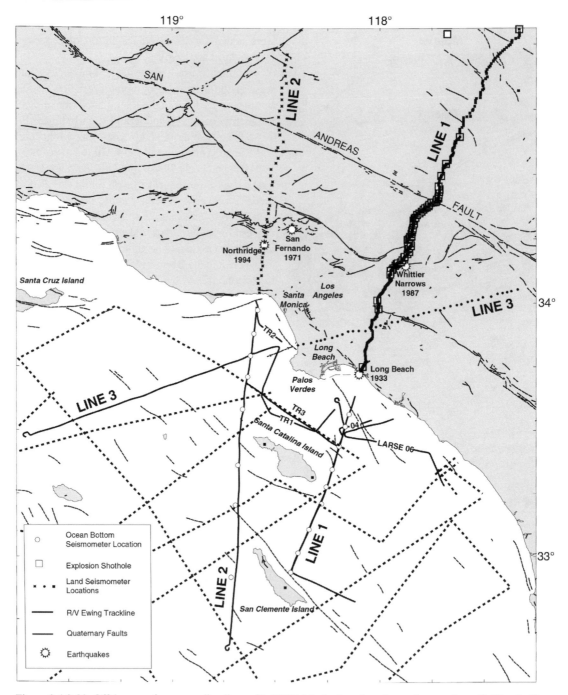

Figure 9.4.2-20. Offshore-onshore recording lines of LARSE-I in the Los Angeles region, southern California (from Brocher et al., 1995c, fig. 1), showing locations of the LARSE airgun lines and locations of recording stations (OBS and land stations). Also shotpoints of the LARSE-I land phase along line 1 are shown. [U.S. Geological Survey Open-File Report 94-241, 71 p.]

two-, and three-component geophones totaling 1622 recording channels. The line was divided into two 45-km-long segments which recorded the shots along the corresponding segment. The 1993 survey consisted of a 60-km-long in-line profile and two 8–9-km-long cross profiles which recorded 5 shots and provided infill subsurface coverage.

The data revealed that the depth to Moho along the eastern edge of the Ruby Mountains varied irregularly between 30.5 and 33.5 km and thus did not reflect local relief, concluding that isostatic balance is achieved by modifications of the crust during extension by intracrustal processes but not by a local crustal root (Satarugsa and Johnson, 1998, 2000).

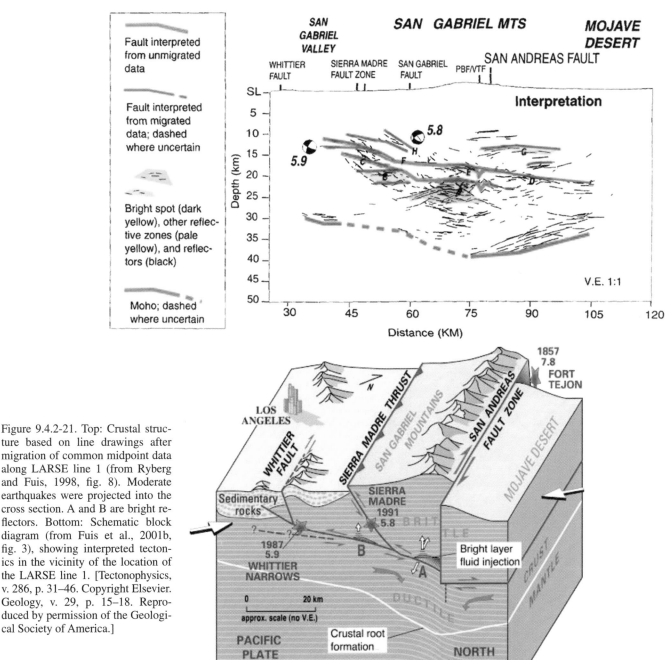

Figure 9.4.2-21. Top: Crustal structure based on line drawings after migration of common midpoint data along LARSE line 1 (from Ryberg and Fuis, 1998, fig. 8). Moderate earthquakes were projected into the cross section. A and B are bright reflectors. Bottom: Schematic block diagram (from Fuis et al., 2001b, fig. 3), showing interpreted tectonics in the vicinity of the location of the LARSE line 1. [Tectonophysics, v. 286, p. 31–46. Copyright Elsevier. Geology, v. 29, p. 15–18. Reproduced by permission of the Geological Society of America.]

The 1993 Delta Force seismic experiment aimed to develop a regional lithospheric structure model of southern Nevada, southwestern California, and western Arizona portions of the Basin and Range province and its transition into the Colorado Plateau. The project was designed as a cost effective, regional seismic experiment by taking advantage of the previous results from higher resolution seismic experiments such as the SSCD (Southern Sierra Continental Dynamics) and PACE (Pacific to Arizona Crustal Experiment) carried out in 1985, 1987, 1989,

and 1992 (Hicks, 2001). The Delta Force experiment employed 4 seismic sources and 474 digital recording instruments laid out in a series of three intersecting recording lines, providing in-line and fan-type coverage of the study area.

9.4.2.5. The Rocky Mountain Region

In spite of great geologic interest and abundant natural resources, relatively little was known about the crust and upper mantle structure of the Rocky Mountains until in the 1990s, when several major seismic experiments were started. Summaries of early surveys, all prior to 1980, were published by Keller et al. (1998), Prodehl and Lipman (1989), Smithson and Johnson (1989), and Prodehl et al. (2005). It was only in the 1990s, that three major seismic projects were undertaken: the Jemez Tomography Experiment dealing with the crustal structure of the Jemez Mountains in New Mexico (Baldridge et al., 1997); the Deep Probe experiment, with a major crustal component in Canada as described above, and dealing exclusively with the upper-mantle structure of the Rocky Mountain area in the United States (Deep Probe Working Group, 1998; Snelson et al., 1998; Henstock et al., 1998); and the CD-ROM experiment in 1999, a 1000-km-long lithospheric transect around 106°W, covering the Southern

Rocky Mountains from Wyoming to New Mexico (Keller et al., 1999; Karlstrom and Keller, 2005).

The Jemez Tomography Experiment targeted the crustal structure beneath a large continental silicic magmatic system in the Jemez Mountains of New Mexico (Baldridge et al., 1997). The project centered upon the Valles Caldera, which at the same time was being geothermally explored, and followed an experiment conducted in 1981 (Olsen et al., 1986; Ankeny et al., 1986) which had elucidated upper crustal structure beneath the caldera providing evidence of a low-velocity body presumed to correlate with the so-called Bandelier magma chamber. The seismic component involved both active and passive sources and aimed to constrain the structure and composition of the lower crust and

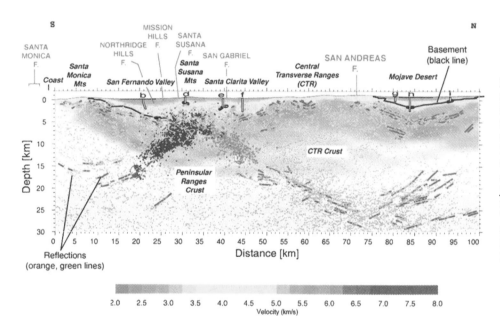

Figure 9.4.2-22. Tomographic velocity model and superimposed line drawings of CMP data along LARSE line 2 (from Ryberg and Fuis, 2004, written commun.). The 1971 San Fernando and 1994 Northridge main shocks (big circles) and relocated aftershocks (brown and blue dots, respectively) were projected into the cross section. [internal report, compiled from Fuis et al., 2001b, Tectonophysics, v. 286, p. 31–46. Copyright Elsevier. Geology, v. 29, p. 15–18. Reproduced by permission of the Geological Society of America.]

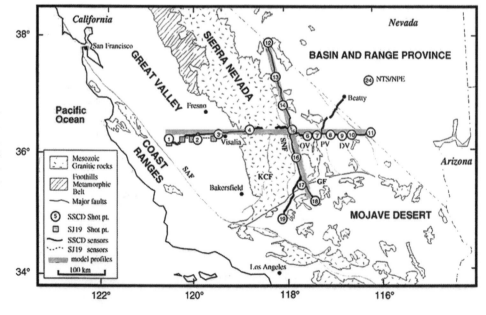

Figure 9.4.2-23. Location map of seismic profiles through the Sierra Nevada, recorded by SSCD (from Ruppert et al., 1998, fig. 1). NTS/NPE—Department of Energy's Nuclear Non-Proliferation experiment. [Tectonophysics, v. 286, p. 237–252. Copyright Elsevier.]

upper mantle, to delineate the geometry and physical state of the magma chamber, and to investigate the geometry of the caldera structure and its fill.

Though a comparatively modest experiment, the active component had to deal with unusual public oppositions and social issues in permitting which were described in much detail by Baldridge et al. (1997), before the three lines which crossed each other in the Valles caldera could be realized (Fig. 9.4.2-24).

The first line could be recorded in 1993, with the compromise that shotpoints in the caldera were dropped and replaced by Vibroseis trucks, while the other two lines, because of the severe permitting problems, had to be delayed until 1995. The three profiles finally realized were three nearly in-line arrays, 150–180 km long and incorporated 300 seismometers. The lines extended from the Colorado Plateau across the Rio Grande rift and Southern Rocky Mountains into the Great Plains.

The first line in 1993 was 180 km long and comprised 150 three-component and 135 one-component recorders with a nominal spacing of 300 m in the middle of the line and 1600 m near the ends. The simulation of the inner shotpoints was realized by three 20,000 kg truck-mounted vibrators with sweeps ranging from 5 to 30 Hz over 30-s intervals and repeated every two minutes, so that up to 150 sweeps could be stacked later on. The second phase in 1995 added the other two lines with similar instrument spacing, but with the addition of 23 seismographs placed in a 3-D array within the caldera and left in place for both lines. Again vibration points were added in the caldera to compensate for the loss of the originally planned shots.

The experiment confirmed the existence of a strong 12-km-thick and 10-km-wide mid-crustal low-velocity zone, interpreted as a magma chamber coincident with the above mentioned Bandelier magma chamber. The data also evidenced a deeper low-velocity zone extending from the lower crust into the uppermost mantle, which was broadest and strongest near the assumed Moho depth of 35 km.

The passive component extended over two years. It started also in 1993 with 22 instruments, placed along two profiles with a 3 km spacing, and recording for 3 months. A second phase in 1994 recorded for 2 months with a larger array of 49 stations deployed in and adjacent to the caldera (Lutter et al., 1995).

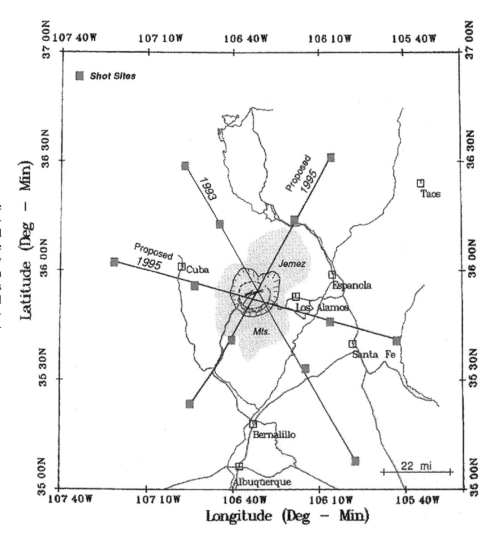

Figure 9.4.2-24. Layout of the active component of JTEX (Jemez Tomography Experiment). Seismic lines extend from the Colorado Plateau across the Rio Grande rift and Southern Rocky Mountains into the Great Plains (from Baldridge et al., 1997, fig.2). [Eos (Transactions, American Geophysical Union), v. 78, p. 417, 422–423. Reproduced by permission of American Geophysical Union.]

Some information on deep crustal structure farther north in the Southern Rocky Mountain region in southern Colorado was obtained by reprocessing data of a seismic-reflection survey, completed by Chevron USA Inc. in 1984 near latitude 38°N in the San Luis Valley, the northernmost part of the Rio Grande rift (Tandon et al., 1999). As a result, the trace of the main bounding fault of the basin against the Sangre de Cristo Range could be correlated to a wide zone of dipping reflections which soled out at lower crustal level at 26–28 km depth. However, unequivocal Moho reflections below the San Luis basin could not be identified, possibly due to a gradational nature of the Moho.

The "Deep Probe" experiment of 1995 followed approximately the 110th meridian (for location, see Figs. 9.4.1-11 and 9.4.2-26) and spanned a distance range of 3000 km from the southern Northwest Territories to southern New Mexico. It targeted the velocity structure from the base of the crust to depths as great as the mantle transition zone near 400 km depth. Details of the project were described in section 9.4.1.6, in particular those dealing with the Canadian side. On the U.S. side, only a few large shots were organized so that only the gross structure of the crust could be revealed.

The interpretation of the data resulted in a 2400-km-long lithospheric cross section (Fig. 9.4.2-25) reaching from the Colorado Plateau in southern New Mexico, United States, into the Alberta plains in Alberta, Canada (Snelson et al., 1998; Gorman et al., 2002). Two outstanding features were a high-velocity zone in the lower crust under Wyoming and Montana and indications

for a deep-reaching suture zone connected with the Uinta uplift in Colorado. Under the Colorado Plateau, the crustal thickness increases gradually from ~35 km in southern New Mexico to ~45 km near the suture zone, where the crust thickens to 50 km.

Farther north beyond the Wind River uplift and basin with only 40 km thickness, the model shows a sudden increase to more than 50 km and then, underneath the Archean Wyoming Province, a northward gradual decrease to ~35 km under the Alberta Plains. Upper-mantle velocities are on average 7.9 km/s under the Proterozoic crust of the Colorado Plateau and 8.1 km/s under the Archean crust of the Wyoming Province. A low-velocity zone with 7.6–7.7 km/s between ~50 and 80 km depth in the uppermost mantle underneath a high-velocity lid of 7.9–8.0 km/s at Moho level was shown in the colored mantle cross section of Deep Probe Working Group (1998), but was not included in the more detailed interpretation of Snelson et al. (1998, Fig. 9.4.2-25).

In the same year (1995), when the Deep Probe experiment was conducted, 14 U.S. institutions combined their efforts and created the interdisciplinary geoscience project CD-ROM (Continental Dynamics–Rocky Mountains) with the aim of realizing a detailed crustal and upper mantle interdisciplinary investigation of the Southern Rocky Mountains from central Wyoming to central New Mexico (Fig. 9.4.2-26), involving tectonics, structure geology, regional geophysics, geochemistry, geochronology, xenoltih studies, and seismic studies (Keller et al., 1999, 2005a; Karlstrom and Keller, 2005).

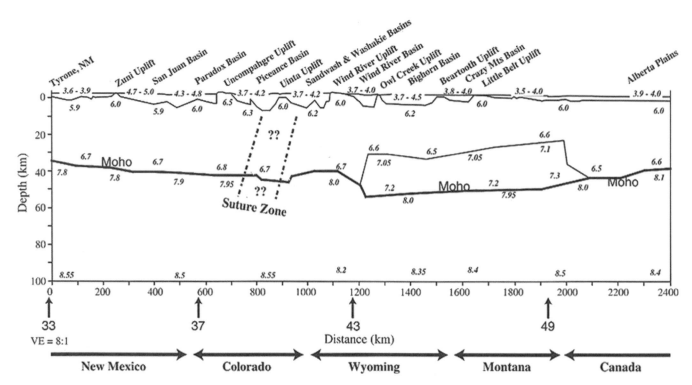

Figure 9.4.2-25. Velocity model for the Deep Probe transect (from Snelson et al., 1997, fig.2). For location, see Fig. 9.4.1-11. [Rocky Mountain Geology, v. 33, p. 181–198. Copyright Rocky Mountain Geology.]

Its main aim was to study the Proterozoic assembly of southwestern North America which has been overprinted by Phanerozoic orogenic processes such as the Ancestral Rocky Mountain orogeny (Kluth and Coney, 1981), the Laramide orogeny (e.g., Erslev, 2005), and the formation of the Rio Grande rift (Keller and Baldridge, 1999). After an initial workshop in 1995, CD-ROM was funded from 1997 to 2002 by the Continental Dynamics Program of the National Science Foundation, with supplemental funding from the German Research Society to the University of Karlsruhe.

The seismic investigations in particular comprised passive-source seismic studies (e.g., Sheehan et al., 2005, Burek and Dueker, 2005) and two active-source experiments (Figs. 9.4.2-26 and 9.4.2-27), consisting of two seismic-reflection profiles ~200 km long (Morozova et al., 2005, Magnani et al., 2005), and a 950-km-long seismic-refraction profile (Rumpel et al., 2005, Snelson et al., 2005, Levander et al., 2005).

The 950-km-long seismic-refraction/wide-angle reflection line was arranged perpendicular to the strike of the Proterozoic assembly of southwestern North America and therefore followed a NNW-SSE direction. It extended from central Wyoming into central New Mexico. Originally it had been planned to establish a seismic-reflection survey along the entire transect following ex-

actly the seismic-refraction line (dashed line in Fig. 9.4.2-26), but for financial reasons only 400 km of seismic-reflection recordings could be realized, with the lines deviating partly from the refraction line for logistic reasons.

The northern reflection section crossed the Cheyenne belt (Morozova et al., 2005), an assumed Archean-Proterozoic suture zone, and the southern reflection section investigated the Yavapai-Mazatzal transition zone and the Jemez lineament in northern New Mexico (Magnani et al., 2005). To compensate at least in part for the central segment of the traverse in Colorado not being covered by reflection profiling, more shotpoints and reduced

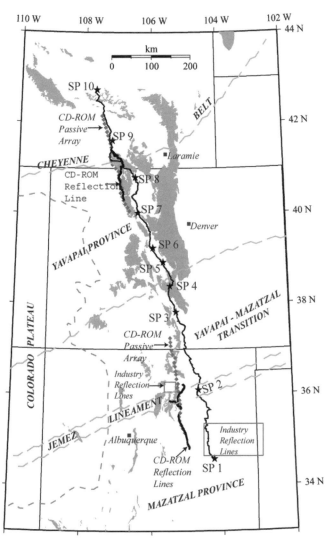

Figure 9.4.2-27. Location map of the CD-ROM seismic lines (from Rumpel et al., 2005, fig.1). Thick black lines—seismic-reflection profiles; thin black line—seismic-refraction profile; stars—CD-ROM refraction shotpoints; boxes—areas where industrial seismic reflection profiles were recorded; grey areas—exposed Precambrian basement. [The Rocky Mountain Region: An evolving lithosphere—tectonics, geochemistry, and geophysics: American Geophysical Union Monograph 154, Washington, D.C., p. 257–269. Reproduced by permission of American Geophysical Union.]

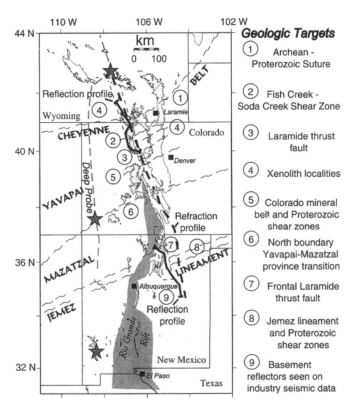

Figure 9.4.2-26. Targets of CD-ROM and preliminary location map of the CD-ROM transect (from Keller et al., 1999, fig.1). Thick full lines—seismic-reflection profiles; thick dashed line—seismic-refraction transect; thin dashed line—Deep Probe profile; stars—Deep Probe shotpoints 43, 37, and 33 (from N to S). [Rocky Mountain Geology, v. 33, p. 217–228. Copyright Rocky Mountain Geology.]

station spacing was established which became possible by the additional funding and participation from Germany.

The line started at the Deep Probe shotpoint 43 in the Gas Hills region of central Wyoming (shotpoint 10 in Fig. 9.4.2-27; see also Figs. 9.4.1-11 and 9.4.2-26), where a charge of 4530 kg could be detonated without involving ripple firing. From the next shotpoint (no. 9), coinciding with the northern end of the northern seismic-reflection profile, the seismic line continued through the Sierra Madre in Wyoming, crossed the North and South Parks, located to the west of the Front Range, and the Wet Mountains in SSE direction and left the Rocky Mountains south of the Wet Mountains near shotpoint 3 and terminated 350 km farther south in the Great Plains of central New Mexico. With the exception of shotpoint 10, the charge sizes ranged between 170 and 2720 kg. Shot spacing varied from 200 km to 41 km, with an overall average spacing of 106 km, and was smallest in the central section between shotpoints 3 and 8, where an average shot spacing of 60 km resulted. Approximately 600 seismographs were deployed twice along the entire profile, resulting in an average receiver spacing of 800 m for all shots. Three-component stations occupied the central section between shotpoints 3 and 8, and one-component stations were used for the northern and southern sections. Energy was well recorded up to 350 km offsets, but in some cases greater distances were reached.

The data preparation (Appendix A9-2-7) and first modeling was achieved simultaneously at El Paso, Texas, and Karlsruhe, Germany. A data example is shown in Figure 9.4.2-28, presenting the record section from the southernmost shotpoint near Gardner, with correlations by Snelson et al. (2005). The data were described and interpreted in detail by Rumpel (2003) and Snelson (2001), and the results were presented in three publica-

tions. Rumpel et al. (2005) and Snelson et al. (2005) primarily used a ray-tracing methodology and divided their attention in particular on details of the upper and middle crust (Rumpel et al., 2005) and on the lower crust and uppermost mantle (Snelson et al., 2005), while Levander et al. (2005) used an inversion and tomographic approach.

Crustal thickness varies substantially along the entire line (Figs. 9.4.2-29 and 9.4.2-30). Moho depth is near 50 km at the northern end, increases gradually toward the south, and reaches a maximum depth of ~57 km underneath the North and South Parks west of the Front Range in central Colorado, where the Colorado Mineral belt traverses the Southern Rocky Mountains. A local Moho high of ~50 km was interpreted under the Wet Mountains, followed by a crustal thickness increase of ~5 km around shotpoint 3 where the line enters the Great Plains. Southward from here Moho depth decreases gradually to near 40 km depth at the southern end of the line.

As mentioned above, the seismic-reflection investigations concentrated on the two major suture zones of the Proterozoic assembly of southwestern North America, the Cheyenne belt in southern Wyoming/northern Colorado and the Jemez lineament in north-central New Mexico, while the central Proterozoic boundary, coinciding with the present-day Colorado Mineral belt, had been covered by the more densely recorded central section of the seismic-refraction profile (Figs. 9.4.2-25 and 9.4.2-26).

The northern seismic-reflection line was recorded in 1999 under the leadership of scientists from the Universities of Wyoming and Arizona. It followed the axis of the Sierra Madre in northern Colorado and southern Wyoming in an overall N-S orientation, parallel to the strike of the Laramide uplifts, and was the first reflection survey recorded on land within the United

Figure 9.4.2-28. Record section of the CD-ROM seismic-refraction profile from SP1 Gardner with travel time correlations (from Snelson et al., 2005, fig.2). [The Rocky Mountain Region: An evolving lithosphere—tectonics, geochemistry, and geophysics: American Geophysical Union Monograph 154, Washington, D.C., p. 257–269. Reproduced by permission of American Geophysical Union.]

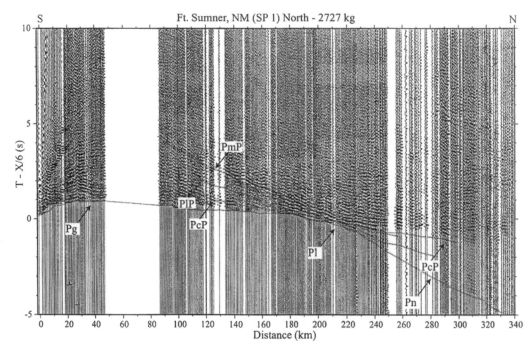

States since the last COCORP survey 8 years previously. It was a Vibroseis survey with four 25,000 kg-force vibrators, 25 m station and 100 m source intervals and a 4 ms sample interval into 1000 channels resulting in 25-km-long recording spreads, giving a nominal fold of 125. Record lengths were 25–30 s. The new reflection data, which were modeled together with the northern part of the refraction data, offered a dramatic new view of the geometry and tectonic history of the Cheyenne belt suture which was imaged as a complex series of wedges resulting in an interleaving of the Archean crust in the north and the Proterozoic crust in the south (Morozova et al., 2005).

The southern seismic-reflection line, organized by scientists from Rice University at Houston, Texas, and the University of Texas at El Paso, was also recorded in 1999. It extended for a total of 170 km north and south of the town of Las Vegas, New Mexico, which is located at the southern edge of the Jemez lineament. It was again a Vibroseis survey, and the data were acquired with 1001 active recording channels along a 25-km-long seismograph array, with 25 m group intervals and a 100 m source interval. The resulting record length was 25 s.

The data, crossing the Cenozoic Jemez lineament in northern New Mexico, showed impressive reflectivity throughout the crust. The middle-lower crust, from 10 to 45 km depth, evi-

denced oppositely dipping reflections that converged roughly at the center of the profile at 35–37 km depth. They were interpreted as Proterozoic-age crustal structures forming a doubly vergent suture zone that has survived subsequent tectonism (Magnani et al., 2005).

Keller et al. (2005a) compiled the results of pre-CD-ROM and CD-ROM results (Fig. 9.4.2-29) and summarized them in a Moho contour map (Fig. 9.4.2-30) which also included results obtained from receiver functions (Sheehan et al., 2005; Burek and Dueker, 2005). They also illustrated the geologic evolution of the crust-mantle boundary in the Southern Rocky Mountain region in time from the early Proterozoic until today and its connection in time with the evolution of surface elevation and petrologic crustal modifications (Keller et al., 2005a; Rumpfhuber and Keller, 2009; Rumpfhuber et al., 2009).

9.4.2.6. The Southern Appalachian–Atlantic Coast Region

In a comparison of the tectonic history of the Appalachian-Ouachita orogen and the Trans-European suture region, Keller and Hatcher (1999) have examined all large-scale geophysical studies conducted in the U.S. portion of the Appalachians. As since the late 1980s, onshore geophysical studies in the southern Appalachians have been modest in scope, they had to rely

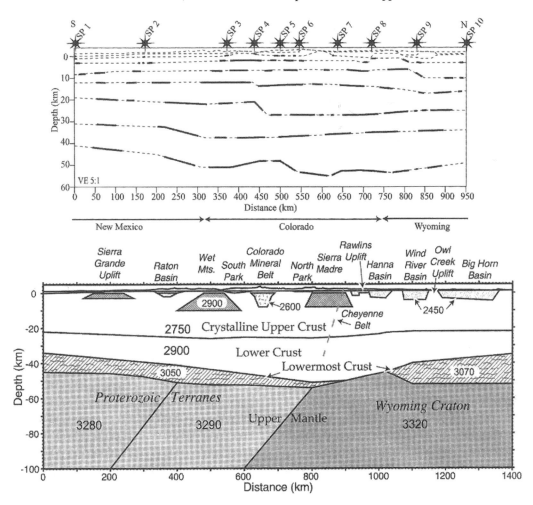

Figure 9.4.2-29. Interpretation of the CD-ROM seismic-refraction data (from Snelson et al., 2005, figs. 11c and 12d). Top: Model based on first arrivals and wide-angle reflections (from Snelson et al., 2005, fig. 11c). Bottom: 2.5-D density model along the CD-ROM transect. The distance scale corresponds to that of the seismic model (from Snelson et al., 2005, fig.12d). [The Rocky Mountain Region: An evolving lithosphere—tectonics, geochemistry, and geophysics: American Geophysical Union Monograph 154, Washington, D.C., p. 271–291. Reproduced by permission of American Geophysical Union.]

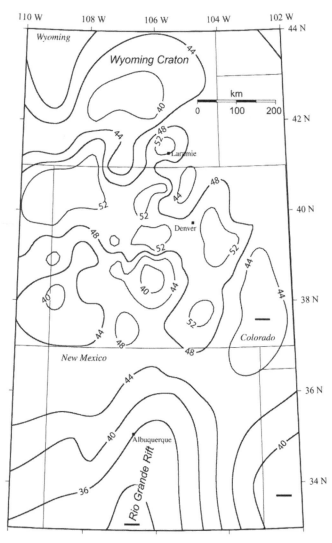

Figure 9.4.2-30. Moho contour map of the Southern Rocky Mountains derived by gridding and contouring the results of previous studies of seismic-refraction data and receiver functions in the area with the results of the CD-ROM seismic experiment (from Keller et al., 2005). [The Rocky Mountain Region: An evolving lithosphere—tectonics, geochemistry, and geophysics: American Geophysical Union Monograph 154, Washington, D.C., p. 271–291. Reproduced by permission of American Geophysical Union.]

on summaries prior to the late 1980s, provided, e.g., by Taylor (1989) and Phinney and Roy-Chowdhury (1989). Besides a local seismic-refraction survey in the New Madrid seismic zone, investigating the shallow upper crustal structure to 12 km depth (Hildenbrand et al., 1995), only one major active-source onland crustal investigation was carried out in the eastern United States in the 1990s in South Carolina (Luetgert et al., 1992, 1994). It was a seismic-refraction/wide angle reflection profile in the Atlantic coastal plain of South Carolina (Luetgert et al., 1992; Appendix 9-2-8).

The recording line of 1991 in South Carolina crossed the coastal plain from southeast to northwest over a length of 120 km.

Portable vertical seismographs of the type SGR III were installed every 1 km, in addition horizontal seismographs were placed at 2 km spacing. Five shotpoints with 1000 kg charges were located at intervals of 30 km (Fig. 9.4.2-31).

The model of Luetgert et al. (1994) showed that the Coastal Plain sediments thicken from a few tens of meters to more than 1 km. Large lateral velocity variations were found for the upper 6–7 km of the crystalline crust. Around shotpoint 3 the velocity in the upper part of the basement (1–7 km depth) increased almost abruptly from 5.7 to 5.8 km/s in the east to 6.1–6.3 km/s in the west. Below 7 km depth no lateral velocity changes were evident. The mean crustal velocity increases only slightly from 6.4 to 6.5 km/s in the middle crust to 6.6–6.7 km/s in the lower crust. The Moho dips from ~32 km depth in the east to 37 km in the west (Luetgert et al., 1994).

In 1996 and 1997, Hawman et al. (2001) recorded quarry blasts in the area of the Cumberland Plateau Seismic Observatory, where the USGS had observed two 300-km-long crossing profiles in 1965 (Prodehl et al., 1984). The study was undertaken to evaluate the feasibility of using relatively small, delay-fired quarry blasts as energy sources for wide-angle reflection profiling of the middle and lower crust beneath the southern half of the eastern Tennessee seismic zone. Timed blasts at four crushed-stone quarries were recorded by a portable array of 19 stations with three-component geophones at roughly 200 m interval. Source-receiver offsets ranged from 16 to 67 km.

The blasts ranged from 75 to 542 ms duration and from 259 to 942 lbs explosives per delay. The first tests of the complicated interpretation procedure due to the extended source signatures produced by the ripple firing showed that usable signal levels at frequencies up to 40 Hz at distances to at least 65 km could be obtained and that also nowadays, since quarry blasts use exclusively ripple firing, useful results for deep crustal studies can be expected by a systematic observation of commercial quarry blasts.

The EDGE Mid-Atlantic seismic experiment along the east coast, which was performed in 1990 and collected multichannel and wide-angle seismic data offshore and onshore of Virginia (Holbrook et al., 1992b, 1994b, Lizarralde and Holbrook, 1997), was discussed in Chapter 8.9.4 in connection with marine seismic investigations in the Atlantic Ocean. The inclusion of the onshore data (Fig. 9.4.2-32) showed a gradual thickening of the crust from the continental margin westwards across the coastal plain, where a uniform thickness of 35 km was observed, to ~45 km under the Blue Ridge section of the southern Appalachians (Lizarralde and Holbrook, 1997).

9.5. THE AFRO-ARABIAN RIFT SYSTEM

9.5.1. The East African Rift System in Kenya

The Afro-Arabian rift system has attracted geoscientists since the early 1900s, but hostile environmental and problematic political conditions have set severe restrictions until today. Modest

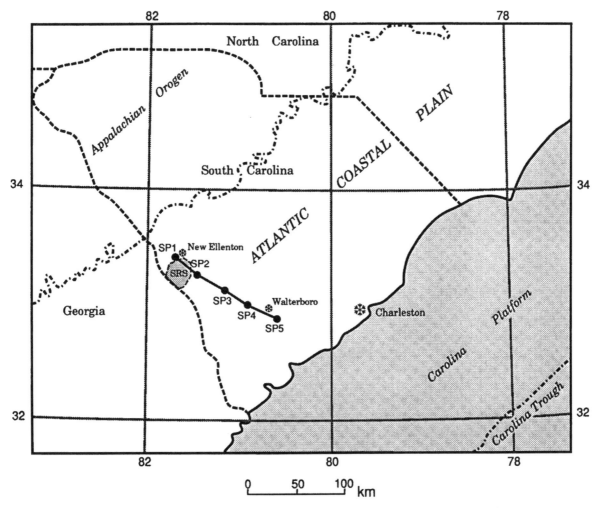

Figure 9.4.2-31. Location map of the 1991 South Carolina seismic experiment (from Luetgert et al., 1992, fig.1). SRS—Savanna River Site. [U.S. Geological Survey Open-File Report 92-723, 35 p.]

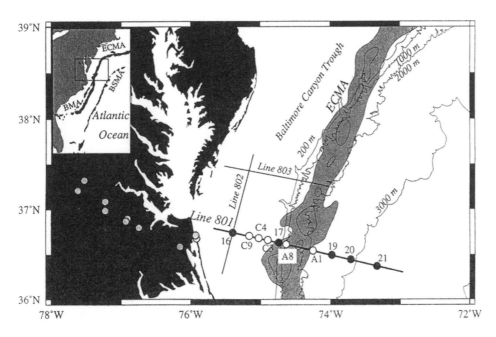

Figure 9.4.2-32. Location map of the EDGE Mid-Atlantic seismic experiment (from Holbrook et al., 1994b, fig.1). Black and white circles—locations of OBS; shaded circles—onshore seismometers. [Journal of Geophysical Research, v. 99, p. 17,871–17,891. Reproduced by permission of American Geophysical Union.]

attempts to study the crustal structure by controlled-source seismology had started in the late 1960s in the East African rift system of Kenya and continued in the 1970s, until a first major project was undertaken in 1985. The Ethiopian part of the East African rift system saw a major experiment in 1972, and the Dead Sea rift was studied in much detail at the end of the 1970s and in the mid-1980s (see corresponding sections in Chapters 6 to 8).

Based on the experiences collected by the earlier expeditions, major efforts were undertaken in the 1990s and in the first years of the twenty-first century to unravel the details of crustal structure of the East African rift system. In 1990 and 1994, two major international campaigns explored the crust and upper mantle under the East African rift of Kenya (KRISP Working Party, 1991; Keller et al., 1992; Prodehl et al., 1994a; KRISP Working Group, 1995a; Fuchs et al., 1997). In 2001, the northern end of the East African Rift system was the target of a major seismic experiment (Maguire et al., 2003; see Chapter 10), and the Dead Sea transform saw the beginning of a major international seismic campaign in 2000, covering both sides of the rift in Jordan and Israel (DESERT Team, 2000; DESERT Group, 2004), followed by a second cross-rift seismic profile in 2004 (ten Brink et al., 2006; see Chapter 10) and a third one in 2006 (Weber et al., 2007; see Chapter 10).

The seismic investigation of the East African Rift in Kenya of 1990 by KRISP (Kenya Rift International Seismic Project) contained, similar to the preceding project of 1985, two components (Prodehl et al., 1994b). An extended seismic-refraction/wide-angle reflection survey and a teleseismic tomography experiment (Fig. 9.5.1-01) were jointly undertaken to study the lithospheric structure of the Kenya rift down to depths of greater than 200 km.

During the teleseismic survey of 1989–1990, an array of 65 seismographs was deployed (circles in Fig. 9.5.1-01) to record teleseismic, regional and local events for a period of 7 months. The elliptical array spanned the central portion of the rift, with Nakuru at its center, and covered an area of ~300 × 200 km, with an average station spacing of 10–30 km (Green et al., 1991; Achauer et al., 1992, 1994; Slack et al., 1994).

The controlled-source seismic part consisted of three profiles. All P-wave data, except for line A, are presented as record sections in Appendix A2-1 (p. 91–96) and Appendix A9-3 (KRISP Working Group 1991). The axial profile ran in N-S direction from Lake Turkana in the north to Lake Magadi in the south (Fig. 9.5.1-01). A total of 206 mobile vertical-component seismographs were deployed three times and, with the exception of the first deployment, recorded the energy of underwater and borehole explosions, with an average spacing of 2 km, to distances of ~550 km.

The first deployment served as a test for instrumentation and shooting techniques in a hostile environment as well as to resolve fine structure of the crust for the part of the rift where a larger number of shots could most economically be arranged. It covered 140 km along the western shore of Lake Turkana. Average station spacing was 0.7 km and small underwater shots of 100–400 kg

were recorded from five positions in Lake Turkana (LT1 to LT5). To decrease the recording interval to 350 m, at three positions two shots, separated by 350 m, were fired (Gajewski et al., 1994).

For the main part of the rift, based on the experience of the 1985 experiment, the majority of shots were also underwater shots, which were fired twice, once for each deployment, in Lake Turkana near its northern end (LTN, Fig. 9.5.1-02) and in the center (LTC), in Lake Baringo (BAR), in Lake Bogoria (BOG), and in Lake Naivasha (NAI) with charges ranging from 400 to 1200 kg. Only one drillhole shot was detonated near Lokoro (LKO), halfway between Lake Turkana and Lake Baringo (Mechie et al., 1994; Keller et al., 1994a).

The cross profile was 450 km long and traversed the rift in W-E direction from Lake Victoria in the west to Chanler's Falls on the Ewaso Ngiro River in the east, 50 km east of Archers Post. The line crossed the rift at Lake Baringo and was recorded with one deployment. Here only two underwater shot positions were available in Lake Victoria (VIC, Fig. 9.5.1-03) and Lake Baringo (BAR), where charges of ~1000 kg were set off. For security reasons, the easternmost shotpoint CHF (Chanler's Falls), planned as an underwater shot in the river, could not be realized for this line. Instead, several borehole shots were arranged which were located at positions with shallow groundwater table near the towns of Barsalingo (BAS), Tangulbei (TAN) and Kaptagat (KAP). Charges of 1000–2200 kg were set off here. Station spacing was 1 km within the rift proper, 2 km on both flanks to ~120 km east and 170 km west of Lake Baringo, and 5 km to both ends of the line (Maguire et al., 1994; Braile et al., 1994).

A third flank profile explored the eastern flank of the rift and ran in NW-SE direction for a distance of 300 km from Lake Turkana to Chanler's Falls. The purpose of this line was to obtain a model for an area which was not influenced by the processes shaping the Kenya rift. Except for the two shots in Lake Turkana (LTC and LTS), where underwater shots of 350 kg were set off, borehole shots with charges between 400 and 2000 kg had been organized along the remainder of the line near the villages of Illaut (ILA) and Laisamis (LAI), and at Chanler's Falls (CHF) on the Ewaso Ngiro River, where the only shot possible in this area was used for this flank line toward the northwest. Station spacing was 1.5 km with two major gaps at both ends (Prodehl et al., 1994c).

The KRISP90 field experiment required a careful organization and permissions, which lasted for several years and was finally accomplished by a close cooperation with Kenyan scientists who provided substantial support in all circumstances. Prodehl et al. (1994b) have summarized all technical details from the first planning to the final realization of the fieldwork. Jacob et al. (1994) had carefully investigated how to optimize wide-angle seismic signal-to-noise ratios and P-wave transmission. In total, 34 shots were fired from 19 shotpoints with charges ranging from 100 kg to 2000 kg. All lake shots had produced excellent recordings, limited in the rift to 400 km offsets and on the flanks to 450 km. Also the borehole shots, except LKO, provided suitable energy to at least 250 km distance. In total, 1063 recording

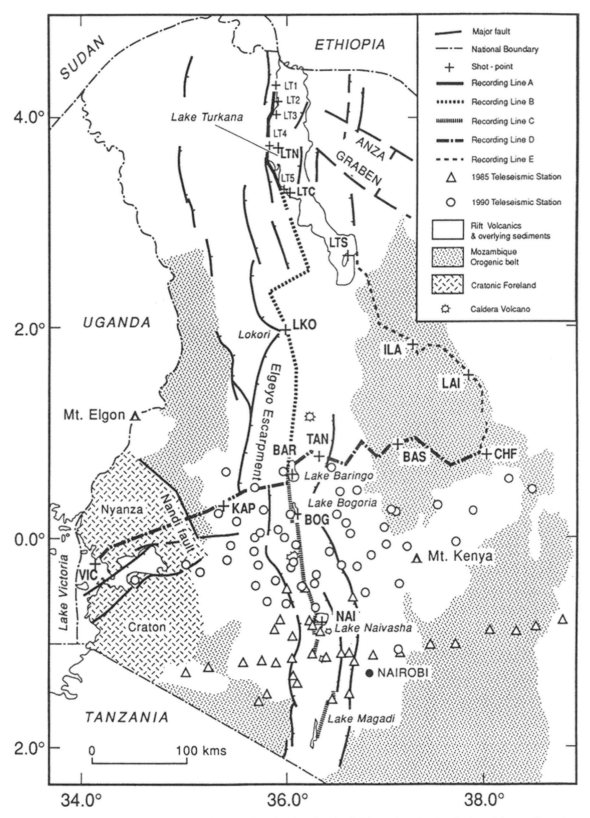

Figure 9.5.1-01. KRISP90 location map showing the seismic refraction/wide-angle reflection lines and the configuration of the teleseismic networks in 1985 and 1989–1990 (from Keller et al., 1994b, fig. 1). The 1985 refraction lines extended from Lake Baringo to Lake Magadi and across the rift valley south of Lake Naivasha, parallel to the 1985 teleseismic cross line (white triangles). [Tectonophysics, v. 236, p. 465–483. Copyright Elsevier.]

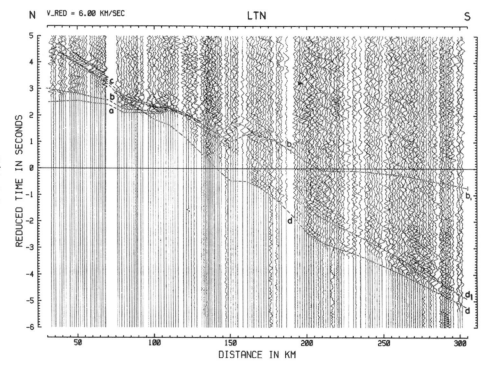

Figure 9.5.1-02. KRISP90 data example for the axial line. Record section of shotpoint LTN (from Mechie et al., 1994, fig. 3). [Tectonophysics, v. 236, p. 179–200. Copyright Elsevier.]

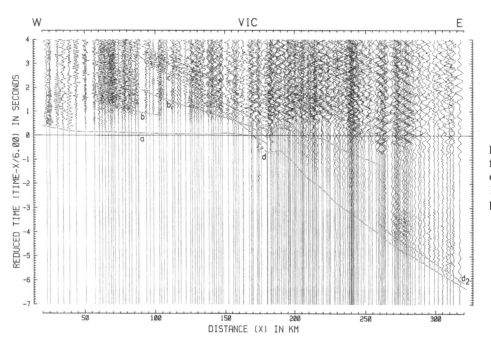

Figure 9.5.1-03. KRISP90 data example for the cross profile. Record section of shotpoint VIC (from Mechie et al., 1997, fig. 4). [Tectonophysics, v. 278, p. 95–119. Copyright Elsevier.]

sites were occupied. All sites had been located beforehand in pre-site surveys with many positions being determined by GPS measurements. Seventy-two scientists divided their tasks in self-sustaining recording and shooting parties and technical services at a mobile headquarters where batteries were recharged, instruments repaired and tapes collected to be transported to a temporary processing center at Egerton University near Nakuru.

The data were jointly processed, and their interpretation was coordinated by several workshops, leading to a special publication volume (Prodehl et al., 1994a). Concerning the controlled-source data, several contributions were prepared to deal with individual aspects and lines. The main results (Figs. 9.5.1-04 and 9.5.1-05) were summarized by Keller et al. (1994b) and Mechie et al. (1997) including a cross section through the southern part

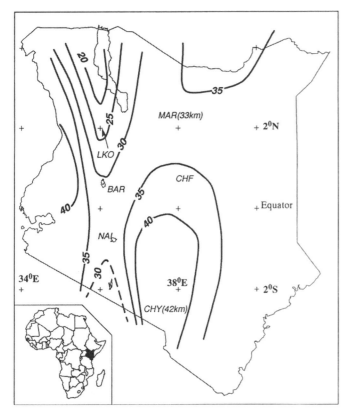

Figure 9.5.1-04. Crustal thickness map for the Kenya rift region (from Keller et al., 1994b, fig. 3). MAR and CHY indicate estimates of crustal thickness from xenoliths in the Marsabit and Chyulu Hills areas, respectively. [Tectonophysics, v. 236, p. 465–483. Copyright Elsevier.]

of the Kenya rift provided by teleseismic delay time studies (Achauer et al., 1992, 1994).

The most significant discovery from KRISP90 was that the crustal thickness along the rift varies from as little as 20 km beneath Lake Turkana to ~35 km under the culmination of the Kenya dome near Lake Naivasha. The transition between Lokori and Lake Baringo occurs over a horizontal distance of 150 km and resolves any discrepancies of earlier models. The limited data south of Lake Naivasha suggested a decrease in crustal thickness southwards toward Lake Magadi, where in 1990 no shotpoint could be realized. This was later accomplished in the KRISP94 experiment described further below. While under the rift uppermost mantle velocities near the Moho ranged from 7.5 to 7.7 km/s, under the flanks normal continental uppermost mantle velocities of at least 8.1 km/s were found. Here also the thickest crust of 40 km was found, but restricted to the surroundings of the southern portion of the rift (Keller et al., 1994a, Mechie et al., 1997).

Last but not least, during the active phase of KRISP90, the teleseismic array also recorded all shots. The data were used to interpret the P-wave first arrivals in order to determine lateral crustal inhomogeneities with a 3-D approach. Several high-velocity bodies found within the crust were attributed to possible mafic cumulates or magma chambers, which evolved during the

major magmatic episodes and which are typical indicators for an active rift (Ritter and Achauer, 1994).

The KRISP89–90 project also unraveled details of the upper mantle structure beneath the Kenya rift and its flanks. The resolution of the controlled-source seismic data was restricted concerning the upper-mantle structure below the Moho. Only for the axial and the cross profiles some structure in the uppermost mantle could be revealed (Keller et al., 1994a; Maguire et al., 1994), indicating a low-velocity zone within the uppermost mantle below 55 km depth under the northern rift, and an overall velocity increase to 8.3–8.4 km/s under the northern rift and under both flanks at 60–65 km depth (Fig. 9.5.1-05).

In agreement with the model for the southern portion of the Kenya rift derived from the active-source seismic data of the axial profile, the teleseismic results (Achauer et al., 1994; Slack et al., 1994) indicated that below the southern part of the rift, a low velocity body exists which extends from the Moho to at least 165 km depth. While it was shown to be confined to the width of the rift proper in its upper portion, it widened at depths greater than 125 km. In a subsequent summary on the deep structure of the East African rift system, the teleseismic velocity perturbations in the upper mantle from 40 to 160 km depth, which had been published by Slack et al. (1994) as relative velocity variations in % (see Fig. 9.5.1-05), had been combined with the seismic-refraction cross section of Braile et al. (1994) and converted into absolute velocities (Braile et al., 1995; see unpublished report [IRIS, 2000] and figure by Braile [1997] in Appendix A9-3).

The continuation of the deep structure of the Kenya rift southward had remained an open question from KRISP90. Therefore, beginning in 1993 and continuing into early 1995, the KRISP Working Group began a new series of multidisciplinary field investigations in the rift in southern Kenya (KRISP Working Group, 1995a; Prodehl et al., 1997b). The research program involved six experimental efforts: (1) a teleseismic survey of the area of the Chyulu Hills located on the eastern flank of the rift, ~100 km to the southeast of Lake Magadi (Fig. 9.5.1-06; Ritter et al., 1995; Ritter and Kaspar, 1997); (2) a seismic-refraction/wide-angle reflection survey across southern Kenya extending from the Indian Ocean near Mombasa to Lake Victoria (Fig. 9.5.1-06); (3) the installation of a temporary seismological network in southern Kenya; (4) a local earthquake recording survey around Lake Magadi (Fig. 9.5.1-06) which was continued in 1997 and 1998 (Ibs-von Seht et al., 2001); (5) a detailed gravity survey along the active-source seismic lines (Birt et al., 1997); and (6) a magnetotelluric study at selected sites along the same seismic lines (Simpson et al., 1997).

The seismic-refraction/wide-angle reflection survey across southern Kenya consisted of two profiles which were in fact two deployments. Line F, the eastern line, extended from the eastern flank of the rift near Magadi to the Indian Ocean near Mombasa (Fig. 9.5.1-06). Line G, the western line, extended from the northwestern end of the Chyulu Hills across the southern rift and the game park Masai Mara to Lake Victoria near the border of Kenya and Tanzania. In total, 10 borehole shots were fired with

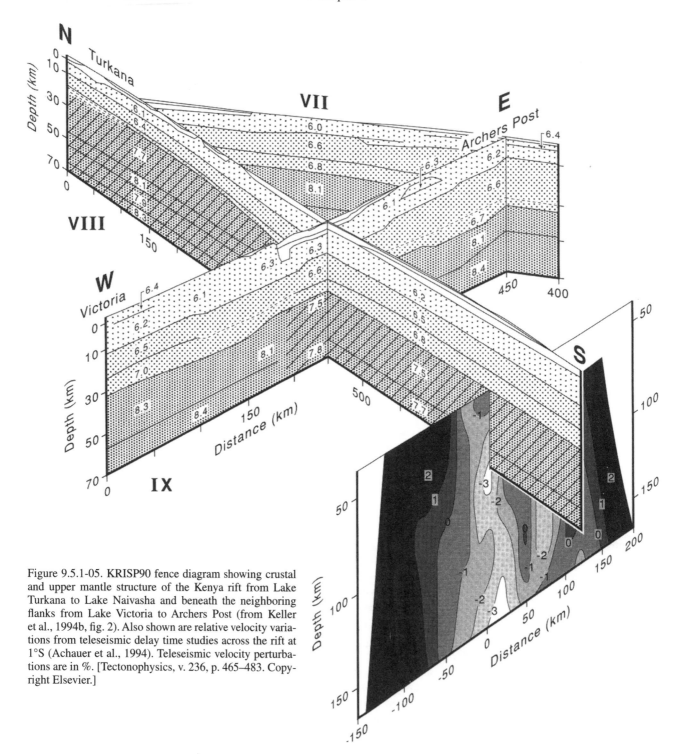

Figure 9.5.1-05. KRISP90 fence diagram showing crustal
and upper mantle structure of the Kenya rift from Lake
Turkana to Lake Naivasha and beneath the neighboring
flanks from Lake Victoria to Archers Post (from Keller
et al., 1994b, fig. 2). Also shown are relative velocity variations from teleseismic delay time studies across the rift at
1°S (Achauer et al., 1994). Teleseismic velocity perturbations are in %. [Tectonophysics, v. 236, p. 465–483. Copyright Elsevier.]

charges from 450 to 2080 kg, loaded in one or two boreholes. At
the ends, underwater shots, where the charge was subdivided into
smaller, separated units, could be arranged. Two shots in Lake
Victoria had total charges of 900 kg each (VIS, Fig. 9.5.1-07).
In the Indian Ocean, using the support by a Kenyan Navy vessel,
only one small charge of 300 kg could be fired (IND). Furthermore, two quarry blasts (KIB) with 1000 kg charges near Kibini

could be timed in accordance with the layout of the two lines and
fired without delays between detonators.

To ensure an overlapping energy coverage along both lines,
shots at CHN, MOR, and in Lake Victoria (VIS) were executed
twice and recorded during both deployments. The energy distribution was almost perfect for all shots. In particular, the shots
along the western part MOR and VIS) gave good arrivals to 450

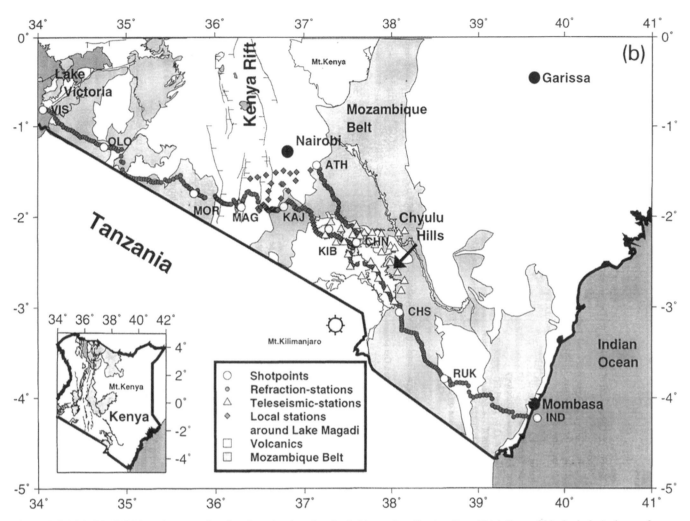

Figure 9.5.1-06. KRISP94 location map showing the seismic-refraction/wide-angle reflection lines (thick lines of black circles), the configuration of the Chyulu Hills teleseismic network of 1993–1994 (white triangles), and the local earthquake recording network (black quadrangles) around Lake Magadi (from Prodehl et al., 1997b, fig. 1b). [Tectonophysics, v. 278, p. 121–147. Copyright Elsevier.]

and 600 km distances, respectively (Fig. 9.5.1-07). All P-wave data, except for line A, are presented as record sections in Appendix A2-1 (p. 97–100) and Appendix A9-3 (KRISP Working Group, 1995b).

The details of crustal structures were interpreted separately for each line, the western line by Birt et al. (1997) and the eastern line by Novak et al. (1997a). Furthermore, Byrne et al. (1997) concentrated on the interpretation of mantle phases along both lines. A summary model for both lines F and G is included in a fence diagram covering the seismic-refraction/wide-angle reflection interpretations for both KRISP90 and KRISP94 (Khan et al., 1999; Fig. 9.5.1-08). Beneath both lines the crystalline crust could be divided into three principal layers with major boundaries at 10–15 km and 20–30 km depth. In the upper crust the P-wave velocities were found to gradually increase from 6.0 to 6.2 km/s under the western flank to 6.3–6.4 km/s underneath the area of the rift. The middle crust with velocities of around 6.5 km/s appeared thickest under the

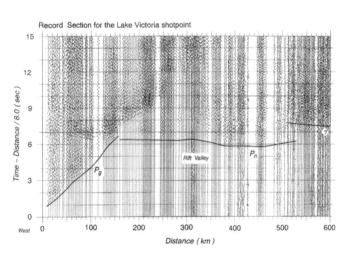

Figure 9.5.1-07. KRISP94 data example. Record section of shotpoint VIS (from Prodehl et al., 1997b, fig.5). [Tectonophysics, v. 278, p. 121–147.] Copyright Elsevier.

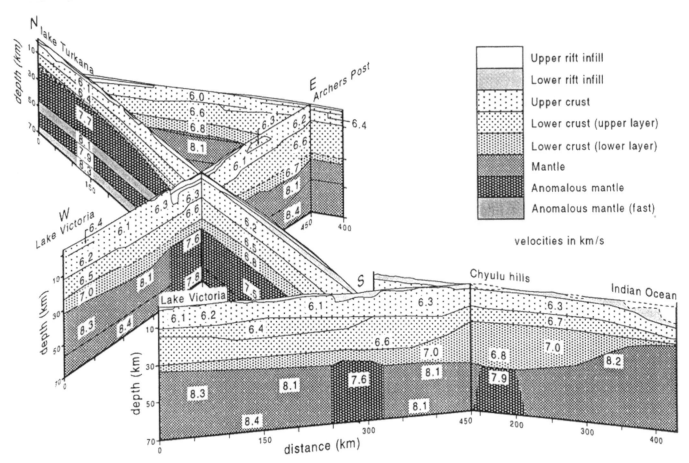

Figure 9.5.1-08. KRISP90/94 fence diagram showing crustal and upper mantle structure of the Kenya rift from Lake Turkana to Lake Magadi and beneath the neighboring flanks from Lake Victoria to the Indian Ocean (from Khan et al., 1999, fig. 4). [*In* Mac Niocaill, C., and Ryan, P.D., eds., Continental tectonics: Geological Society of London Special Publication 164, p. 257–269. Reproduced by permission of Geological Society Publishing House, London, U.K.]

western flank, where it might be subdivided into two layers. The lower mafic crust with velocities of around 7 km/s was thin near Lake Victoria, while under Lake Magadi the seismic velocity structure appeared similar to that found under the northern portion of the Kenya dome.

The dramatic thickening of the lower-crustal layer to more than 20 km eastward underneath the Chyulu Hills was unexpected, where it constitutes about half of the whole crust which here thickened to 44 km. However, the Moho here appeared as a diffuse crust-mantle transition zone, the lower limit of which could hardly be defined. Therefore, in agreement with the teleseismic modeling, a low-velocity zone was introduced into the lowermost section of the lower crust (Figs. 9.5.1-09 and 9.5.1-10). Furthermore, Novak et al. (1997b) interpreted the S-wave data and calculated V_p/V_s ratios for all layers. Finally, Novak et al. (1997b) developed an integrated model for the deep structure of the Chyulu Hills volcanic field in southeastern Kenya, based on the seismic-refraction P- and S-data and on the tomographic modeling of the teleseismic data, as well as on available petrological data (Fig. 9.5.1-09).

The state of the art of crust and upper mantle research for the Dead Sea transform, Red Sea area, and the East African rift system was summarized in a compilation of all crustal cross sections published until 1994 (Prodehl et al., 1997a) from all controlled-source seismic experiments since the late 1960s touching the Afro-Arabian rift system. A collection of representative crustal columns indicated crustal thickness variations viewed as a function of rift development (Fig. 9.5.1-11).

9.5.2. The East African Rift System in Tanzania

Following the successful completion of crustal and upper mantle research in the eastern branch of the East African rift system in Kenya, the international Working Group discussed various possibilities how to continue their research work into Tanzania and Malawi.

This was partly stimulated by successful research work of the shallow subsurface structures in East African lakes by Duke University under the leadership of Bruce Rosendahl and co-workers in the 1980s (see, e.g., Rosendahl et al., 1989), but

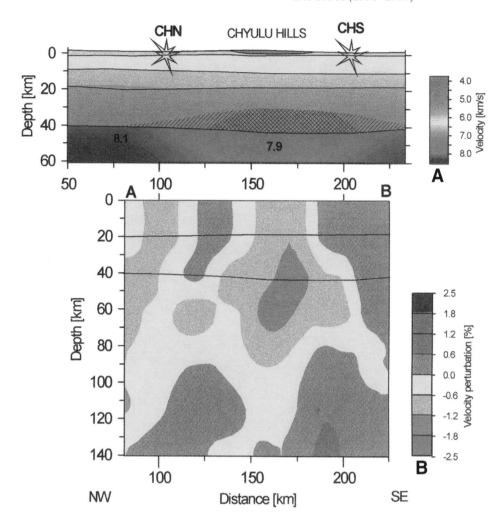

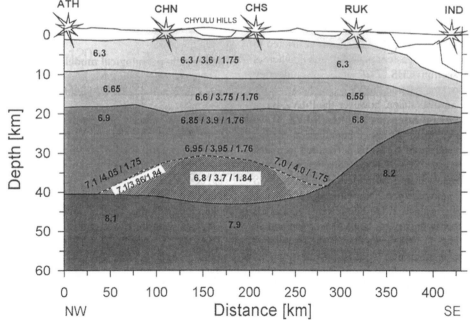

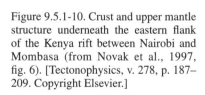

Figure 9.5.1-09. Crust and upper mantle structure underneath the Chyulu Hills volcanic field (from Novak et al., 1997, fig. 7). (A) Part of the crustal model along line F (see Fig. 9.5.1-06). (B) Cross section through the teleseismic tomography model of Ritter and Kaspar (1997). [Tectonophysics, v. 278, p. 187–209. Copyright Elsevier.]

Figure 9.5.1-10. Crust and upper mantle structure underneath the eastern flank of the Kenya rift between Nairobi and Mombasa (from Novak et al., 1997, fig. 6). [Tectonophysics, v. 278, p. 187–209. Copyright Elsevier.]

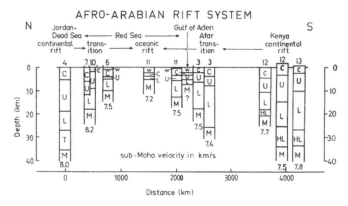

Figure 9.5.1-11. Evolutionary sequence illustrating the variation in crustal thickness under the Afro-Arabian rift system from the Dead Sea transform system through the Red Sea, Gulf of Aden and Afar triangle to the southern end of the Kenya rift (from Prodehl et al., 1997a, fig. 4). Depths refer to sea level. W—water; C—cover rocks; U—upper crust; L—lower crust; HL—high-velocity lower crust; T—crust-mantle transition rocks; M—Moho. [Tectonophysics, v. 278, p. 1–13. Copyright Elsevier.]

finally plans were dropped due to the lack of likely sources for funding and the nonexistence of local scientists or institutions as in Kenya. Nevertheless, experimental work in East Africa continued with large-scale teleseismic tomography surveys. Using the IRIS/PASSCAL broadband seismic array, in 1994 and 1995, teleseismic experiments were undertaken in Tanzania, in 2001–2002 in southern Kenya, and in 2000–2002 in Ethiopia (Fig. 9.5.2-01). These surveys, however, did not resolve any details of crustal and uppermost mantle structure but provided a telescoping view of the East African rift within this suspected plume province (Nyblade and Langston, 2002). In Tanzania, broadband stations were spread along two lines throughout the country (Nyblade et al., 1996). The array in southern Kenya followed approximately the KRISP94 crustal refraction survey (Nyblade and Langston, 2002; Fig. 9.5.2-01).

A small-scale experiment should also be mentioned, when from 1992 to 1994 a seismic network of five digital stations operated along the Rukwa trough in southwestern Tanzania near the border to Malawi, recording deep crustal earthquakes (Camelbeeck and Iranga, 1996). One hundred ninety-nine events were successfully recorded, and record sections were constructed for parts of the data. A crustal thickness of 42 km was determined from the arrivals identified as P_MP reflections.

9.5.3. The Jordan–Dead Sea Transform

Crustal structure studies in the Jordan–Dead Sea transform, which had first been undertaken independently on the western side in Israel in 1978 (Ginzburg et al., 1979a) and on the eastern side in Jordan in 1984 (El-Isa et al., 1987a) and for which results for both sides had been summarized by Mechie and Prodehl (1988), were revived in 2000. For the first time a seismic profile could be established which covered both sides of the transform (Fig. 9.5.3-01).

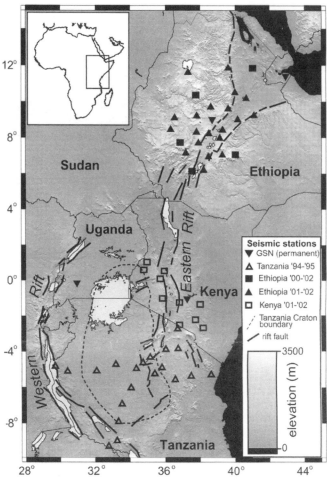

Figure 9.5.2-01. Topographic map of East Africa, showing the major rift faults of the East African rift system and the locations of broadband seismic stations (from Nyblade and Langston, 2002, fig. 1). [Eos (Transactions, American Geophysical Union), v. 83, no. 37, p. 405, 408–409. Reproduced by permission of American Geophysical Union.]

The seismic line started at the Mediterranean Sea south of Gaza, crossed the Jordan–Dead Sea transform in the Arava Valley south of the Dead Sea, and terminated in southern Jordan southeast of Ma'an (DESERT Team, 2000; DESERT Group, 2004; Mechie et al., 2005).

The project consisted of two parts: a seismic-refraction/wide-angle reflection line of 260 km length and a near-vertical incident seismic-reflection line of 100 km length. The project started with the refraction line. The 260-km-long NW-SE–trending line passed through Palestine, Israel, and Jordan and crossed the Arava fault ~70 km south of the Dead Sea (Figs. 9.5.3-01 and 9.5.3-02).

Thirteen shots, including two quarry blasts (nos. 10 and 31 in Fig. 9.5.3-01) were executed. The quarry blasts had charges of 8500 kg and 12000 kg. Five of the 11 borehole shots ranged from 720 to 950 kg, and the other six shots were smaller with charges of 30–80 kg. All shots were recorded by 99 three-component instruments, spaced at 1 km in the Arava Valley, at

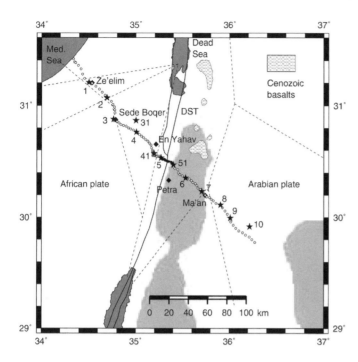

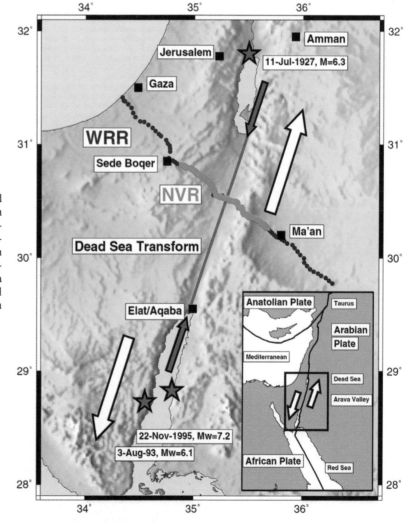

Figure 9.5.3-01. Location map of DESERT 2000 (from Mechie et al., 2005, fig. 1). Stars—shotpoints; circles—seismic recording sites. Dashed lines are earlier seismic-refraction lines (see Mechie and Prodehl, 1988). [Geophysical Journal International, v. 160, p. 910–924. Copyright John Wiley & Sons Ltd.]

Figure 9.5.3-02. Topographic map of the Jordan–Dead Sea rift showing the location of DESERT 2000 (from DESERT Group, 2004, fig. 1a). Light blue dots—near-vertical seismic reflection line; dark blue dots—extension of reflection line for wide-angle reflection/refraction observations. White arrows indicate the relative movement of plates; red line and arrows indicate the Dead Sea Transform between Dead and Red Seas. [Geophysical Journal International, v. 156, p. 655–681. Copyright John Wiley & Sons Ltd.]

2.5 km on the adjacent flanks between Sede Boqer and Ma'an, and 4–4.5 km apart along the remaining parts of the line. All shots were recorded by 125 vertical-component geophone groups with 100 m spacing along a 12.5-km-long section of the profile on the Jordanian side of the Arava Valley between shots 5 and 51. A data example is shown in Figure 9.5.3-03 (for more data, see Appendix A9-3).

The study provided the first whole-crustal image across the Dead Sea transform (Fig. 9.5.3-04B). While for the seismic basement an offset of several kilometers was interpreted underneath the Arava fault, the main fault of the Dead Sea transform, the Moho depth increased steadily from ~26 km at the Mediterranean Sea to ~39 km under the Jordan highlands, except for a small asymmetry under the Arava valley (DESERT Group,

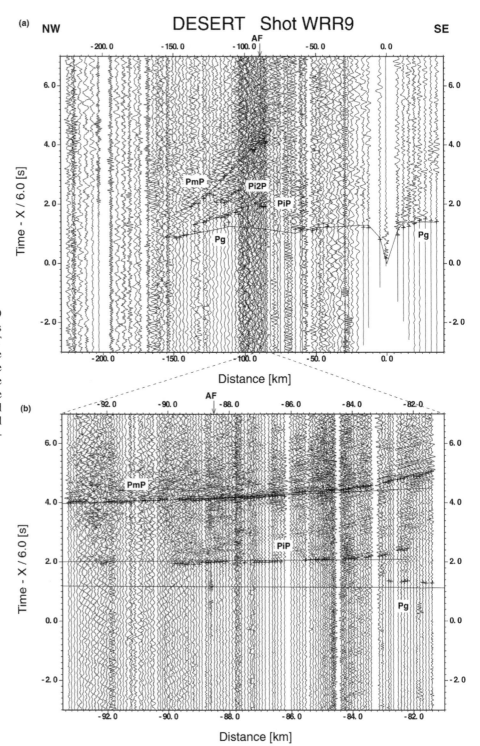

Figure 9.5.3-03. Record section of shot 9 along the DESERT WRR profile across the Jordan–Dead Sea rift (from DESERT Group, 2004, fig. 8). For location, see Fig. 9.5.3-01. Top: P-wave data along the whole length of the line. Bottom: P-wave data along the 12.5-km-long segment in the Arava valley with densely spaced vertical geophone groups. [Geophysical Journal International, v. 156, p. 655–681. Copyright John Wiley & Sons Ltd.]

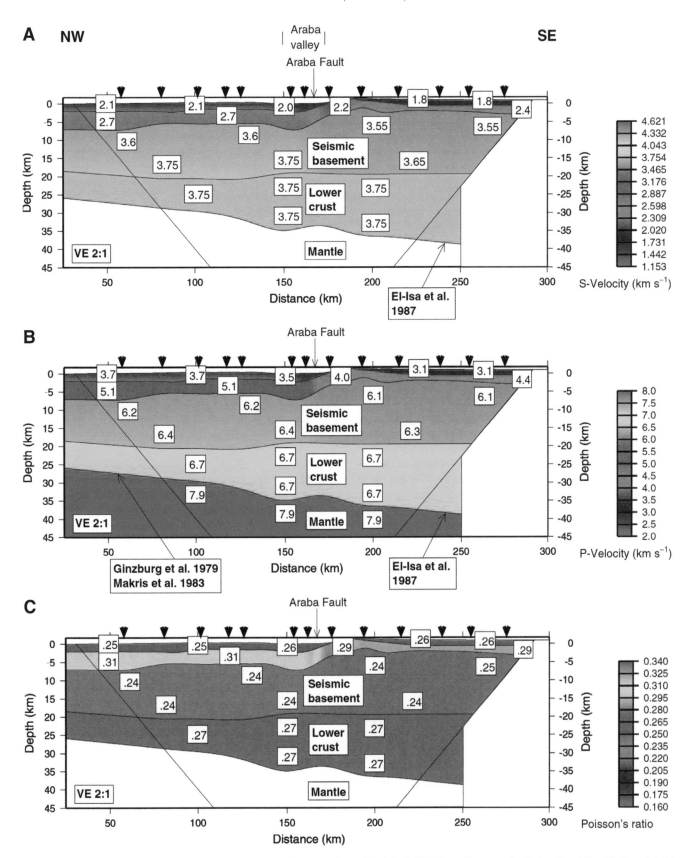

Figure 9.5.3-04. (A) S-wave model. (B) P- wave model. (C) Poisson's ratio model of the DESERT profile across the Jordan–Dead Sea rift (from Mechie et al., 2005, fig. 10). For location, see Fig. 9.5.3-02. [Geophysical Journal International, v. 160, p. 910–924. Copyright John Wiley & Sons Ltd.]

2004). Under the Arava Valley, no significant uplift of the Moho was noted. From the near-vertical data it could be inferred that the Arava fault cuts through the crust, develops into a broad zone in the lower crust and reaches down to Moho level.

Based on the excellent quality of the observed P- and S-wave data, a more detailed interpretation (Mechie et al., 2005) specialized on the S-wave data of the seismic-refraction/wide-angle profile (Fig. 9.5.3-04A), resulting in Poisson's ratios of 0.24 for most of the upper crust and 0.27 for the lower crust (Fig. 9.5.3-04C).

9.6. DEEP SEISMIC SOUNDING STUDIES IN SOUTHEAST ASIA

9.6.1. India and Indonesia

Crustal and upper mantle research in India has been almost exclusively performed by the National Geophysical Research Institute at Hyderabad. Scientists from this institution have visited all leading research institutions around the world and over a few decades a very powerful research team has evolved.

Following the example of COCORP in the United States and equivalent organizations in many West European countries in the 1980s promoting deep crustal reflection work in large dimensions, seismic-reflection investigations on a large scale were performed in India (Rajendra Prasat et al., 1998; Reddy et al., 1995; Tewari et al., 1995, 1997), aiming to unravel the structure and tectonics of the Aravalli-Delhi fold belt in northwestern India. For example, a 200-km-long deep seismic-reflection profile across the northwestern Indian shield is described in more detail (Fig. 9.6.1-01).

The seismic-reflection profile ran from the Marwar basin across the Proterozoic Delhi Fold Belt and gneissic complexes into the Vindhyan basin. Field parameters included a 20 s TWT record with 60-fold theoretical CDP coverage using 120-channel data at a 4 ms sampling rate and geophone interval/ shot interval of 100 m. The actual fold coverage, however, varied between 40 and 45. The energy source was 30–35 kg of explosives loaded at 15–20 m depth. The Delhi Fold Belt appeared to be a zone of thick (45–50 km) crust that might have been generated due to the doubling of the lower crust during compressional Proterozoic tectonics. This resulted in

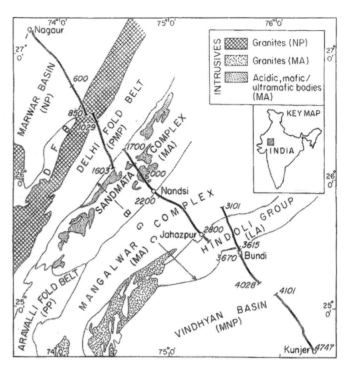

Figure 9.6.1-01. Location map of the Nagaur-Kunjer deep reflection profile, traversing the northwestern Indian shield (from Rajendra Prasat et al., 1998, fig. 1). [Tectonophysics, v. 288, p. 31–41. Copyright Elsevier.]

three Moho reflection bands, two of which dip SE from 12.5 to 15.0 s TWT and from 14.5 to 16.0 s TWT. Another band of subhorizontal Moho reflections at ~12.5 s TWT was attributed to the post-Delhi tectonic orogeny. The signatures of the Proterozoic collision have been preserved in the form of strong SE-dipping reflections in the lower crust and Moho. The lower crust appeared fairly reflective in the Delhi Fold Belt, but poorly reflective under the gneissic complexes, where the base of the crust was estimated to be at 50 km, based primarily on gravity and not on reflection data (Tewari et al., 1997).

The deep-crustal reflection data recorded along the traverse were projected onto a cartoon (Fig. 9.6.1-02) imaging the two sets of oppositely dipping reflection bands from upper to lower

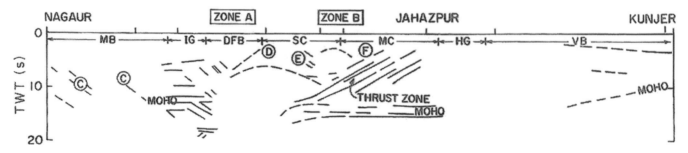

Figure 9.6.1-02. Cartoon showing the prominent reflections along the Nagaur-Kunjer deep reflection profile, traversing the northwestern Indian shield (from Rajendra Prasat et al., 1998, fig. 3a). [Tectonophysics, v. 288, p. 31–41. Copyright Elsevier.]

crustal levels correlated with the collision leading to the development of the Aravalli-Delhi fold belt in the Proterozoic (Rajendra Prasat et al., 1998).

The Indian crustal structure and Moho configuration beneath the Himalaya was investigated with teleseismic recording experiments (Mitra et al., 2005; Rai et al., 2006) the locations of which are shown in the next subchapter (Fig. 9.6.2-02). Mitra et al. (2005) reported on the project NE INDIA in northeastern India, while the NW Himalaya was studied by the project HIMPROBE extending from India into Tibet (Rai et al., 2006). A teleseismic receiver function analysis of seismograms recorded on a 700-km-long profile of 17 broadband seismographs traversing the NW Himalaya showed a progressive northward deepening of the Indian Moho from 40 km beneath Delhi south of the Himalayan foredeep to 75 km beneath Taksha at the Karakoram fault (Rai et al., 2006). Similar studies by Wittlinger et al. (2004) to the north of the Karakoram fault indicated that the Moho may continue to deepen to as much as 90 km depth beneath western Tibet before shallowing substantially to 50–60 km at the Altyn Tagh fault, which, however, is not shown in the Moho contour map of China (see Fig. 10.5.2-01), compiled by Li et al. (2006).

In southern India, new deep seismic-reflection and refraction data did not become available. However, following the devastating 1993 Latur earthquake in southern India (~18°N, 76°30′E), a mobile network of 45 seismic stations was operated for aftershock activity in cooperation with the GeoForschungsZentrum Potsdam, Germany. The data could be used to plot record sections of well-located aftershocks out to 80 km offsets and deduce details of crustal structure. The result was a sequence of alternating high- and low-velocity layers for P- and S-waves in the upper crust at depths of 6.5–9 km and 12.3–14.5 km with ~7% velocity reduction and a low-velocity layer at 24–26 km depth (Krishna et al., 1999).

In 1998, a small-scale seismic project was undertaken on Java, Indonesia, to investigate the structure of an active volcano. The German institution GFZ Potsdam studied the seismic structure of the stratovolcano Merapi (Java, Indonesia) by means of a controlled-source seismology experiment, using 90 three-component stations and a specially devised airgun source. The project concentrated on the heterogeneous structure of the volcano within a radius of 5 km from the active dome, where the sources of most of the natural volcanic seismic events are located, because the seismic properties of the propagation media in an active volcano is important to understand the natural seismic signals used for eruption forecasting. Three 3-km-long seismic profiles, each consisting of up to 30 three-component seismometers with a station interval of 100 m, were deployed in an altitude range between 1000 m and 2000 m above sea level. The distance from the highest stations to the active summit region at an altitude of almost 3000 m was ~5 km. As energy sources airgun sources were used at the lower end of each of the three profiles. The sources consisted of water basins equipped with a 2.5-l mudgun. Airgun shots were released every 2 min, so that at each seismometer the repeated shots were recorded between 20

and 120 times. The modeling of the data was described in detail by Wegler and Lühr (2001). The observed seismograms showed a spinlep-like amplitude increase following the direct P-phase. This shape of the envelope could be explained by the diffusion model. According to this model there were so many strong inhomogeneities that the direct wave could be neglected and all energy was regarded as being concentrated in multiple scattered waves. The natural seismic signals at Merapi volcano showed similar characteristics to the artificial shots. The first onsets had only small amplitudes and the energy maximum arrived delayed compared to the direct wave. Therefore it can be concluded that these signals are also strongly affected by multiple scattering.

9.6.2. China

The large-scale seismic investigations of the crust and upper mantle of China with explosion seismology of the 1980s were continued even more intensively in the 1990s. Summary papers representing the current state of the art were published from time to time (e.g., Li and Mooney, 1998; Li et al., 2006) presenting DSS profile locations in mainland China (Fig. 9.6.2-01), many of which were completed by the Research Center of Exploration Geophysics (RCEG) and China Earthquake Administration (CEA, previously known as the China Seismological Bureau or CSB).

Besides the location map, a list of all profiles, shown as red lines in Figure 9.6.2-01, contains details of the RCEG/CEA lines (Li et al., 2006), such as location names, length in km, azimuth, number of shots, and year. From this list it can be noted that the RCEG/CEA lines 1–25 were completed between 1980 and 1989, except for nos. 9–10 and 12–14 which were already observed from 1977 to 1979, while lines 26–46 were shot in the 1990s including the year 2000, and lines 47–50 in the early 2000s. The list also contains information on the seismic profiles completed by other institutions, such as location names and corresponding references for each profile A1 to A41 shown in Fig. 9.6.2-01 as green lines. The corresponding tectonic sketch map of the main geological provinces was also shown (see, Figs. 8.7.2-01 and 10.5.2.04). Furthermore, Li and Mooney (1998) and Li et al. (2006) have summarized the main results in crustal columns with average crustal structure for the main tectonic areas as well as in contour maps, showing crustal thickness (Moho depths) and uppermost-mantle velocities (P_n-velocities).

Figure 9.6.2-01 does not show, however, lines which had been organized by international collaborations in Tibet such as the Sino-French lines recorded in the 1980s or all the INDEPTH lines observed in the 1990s (Figs. 9.6.2-02 and 9.6.2-03).

It was Tibet in particular that attracted scientists from all over the world. In the 1980s, major collaborations of Chinese and French scientists had stimulated major seismic-refraction investigations (e.g., Hirn et al., 1984c; Sapin et al., 1985), but with modern technology, even major seismic-reflection work became possible in the 1990s.

For the area of Tibet and surroundings, Klemperer (2006) has investigated all major seismic and magnetotelluric experiments

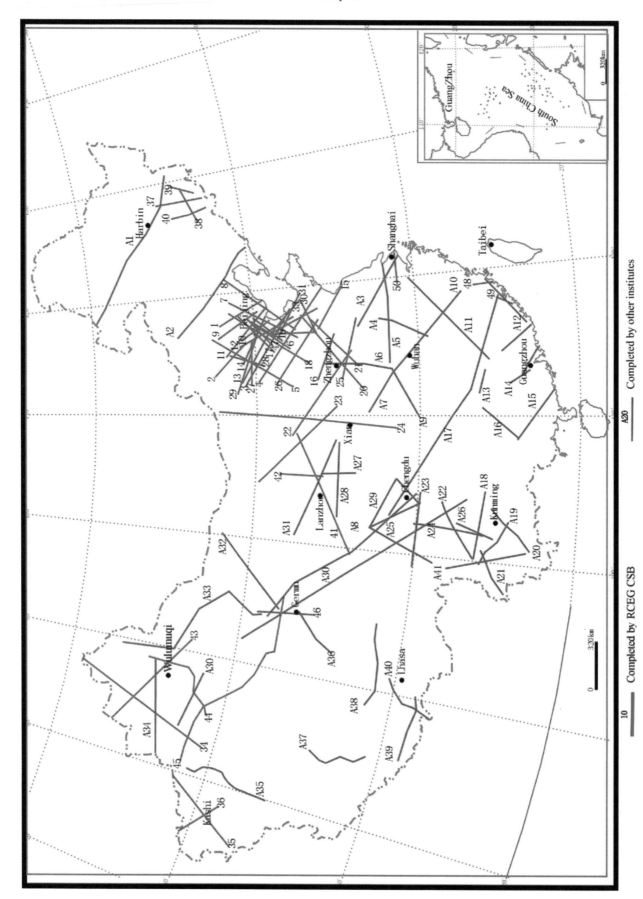

Figure 9.6.2-01. Location map of deep seismic sounding profiles in mainland China (from Li et al., 2006, fig. 1). Profiles 1–45 were completed by RCEG and CEA, profiles A1–A41 were completed by other institutions. [Tectonophysics, v. 420, p. 239–252. Copyright Elsevier.]

Figure 9.6.2-02. Location map of main geophysical experiments in and around Tibet (from Klemperer, 2006, fig. 1). Heavy black lines and upper-case names are generalized locations of seismic experiments. Heavy dotted lines are magnetotelluric profiles. Thin lines are major faults and sutures. Thin dotted line is 3000 m elevation contour. INDEPTH: IN-1 (Zhao et al., 1993; Makovsky et al., 1996), IN-2 (Nelson et al., 1996; Makovsky and Klemperer, 1999), IN-3 (Zhao et al., 2001; Ross et al., 2004). Sino-French: SF-1 (Hirn et al., 1984c), SF-2 (Zhang and Klemperer, 2005), SF-3 (Hirn et al., 1995; Galvé et al., 2002), SF-4 (Herquel et al., 1995; Vergne et al., 2003), SF-5 (Wittlinger et al., 1998), SF-6 (Galvé et al., 2002; Vergne et al., 2003), SF-7 (Wittlinger et al., 2004). PASSCAL (e.g., McNamara et al., 1996): WT—West Tibet (Kong et al., 1996), TB—Tarim Basin (Gao et al., 2000; Kao et al., 2001), NB—Nanga Parbat (Meltzer et al., 2001). HIMPROBE (Rai et al., 2006): TQ—Tarim-Qaidam (Zhao et al.,

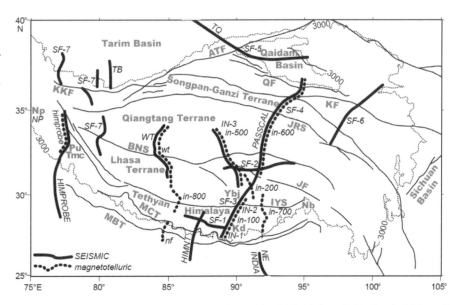

2006), HIMNT—Himalaya-Nepal-Tibet experiment (Schulte-Pelkum et al., 2005), NE INDIA (Mitra et al., 2005). [*In* Law, R.D., Searle, M.P., and Godin, L., eds., Channel flow, ductile extrusion and exhumation in continental collision zones: Geological Society of London Special Publication 268, p. 39–70. Reproduced by permission of Geological Society Publishing House, London, U.K.]

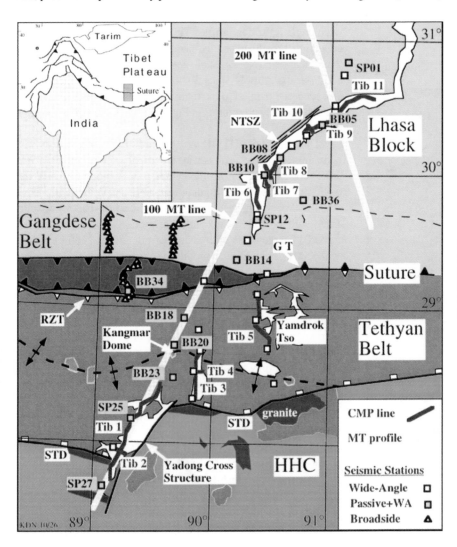

Figure 9.6.2-03. INDEPTH I-II location map (from Nelson et al., 1996, fig. 1). SP—short period stations, BB—broadband stations for both active-source wide-angle and passive-event recording. [Science, v. 274, p. 1684–1688. Reprinted with permission from AAAS.]

with respect to crustal flow in Tibet and has published a location map of all experiments in and around Tibet including the corresponding references (Fig. 9.6.2-02). From the experiments cited, IN-1, IN-2, and locally along IN-3 were near-vertical reflection profiles, while controlled-source wide-angle profiles were SF-1, SF-2, SF-3, SF-6, IN-1, IN-2, IN-3, WT, and TQ. Teleseismic recording experiments were PASSCAL, IN-2, IN-3, SF-3, SF-4, SF-5, SF-6, SF-7, TB, NP, HIMPROBE, HIMNT, and NE INDIA.

One of the largest cooperative seismic projects in Tibet was INDEPTH (Figs. 9.6.2-03 and 9.6.2-05), involving both near-vertical incident and wide-angle seismic-reflection methodology. This project was accomplished with major U.S. and German participation in several phases, starting in 1992 (e.g., Brown et al., 1996; Makovsky et al., 1996; Nelson et al., 1996). Its southern part (IN-1, IN-2) is shown by line A40 of Figure 9.6.2-01. The cooperation with French institutions was also continued in northeastern Tibet (e.g., Galvé et al., 2002). Major cooperative work of Chinese institutions was accomplished and results published together with scientists of the USGS, e.g., in the Dabie Shan orogenic belt in eastern China (Wang et al., 2000; A33 in Fig. 9.6.2-01), or at the northern and northeastern margin of the Tibetan plateau (Zhao et al., 2006; Liu et al., 2006).

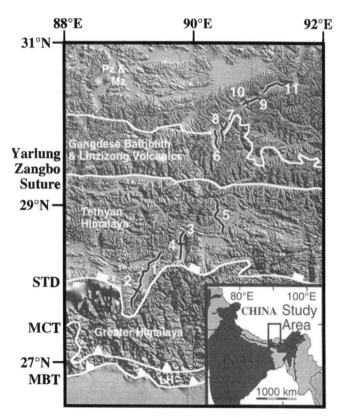

Figure 9.6.2-04. Topographic map showing the location of INDEPTH I and II seismic lines and major structural boundaries of the India-Tibet collision system (from Alsdorf et al., 1998, fig. 1). [Journal of Geophysical Research, v. 103, p. 26,993–26,999. Reproduced by permission of American Geophysical Union.]

The project INDEPTH (International Deep Profiling of Tibet and the Himalaya) was a geoscientific project to understand the crustal and upper mantle structure beneath the Himalaya and the Tibetan Plateau. It was carried out under the direction of a collaborative geoscience team involving scientists from the Chinese Academy of Geological Sciences and several North American and European institutions. The project acquired multichannel seismic-reflection (common midpoint or CMP data), wide-angle reflection/refraction, broadband earthquake, magnetotelluric, and surface geological data. In the first two phases, data were collected along a 400-km-long corridor, extending from the crest of the Himalaya to the center of the Lhasa block, and most of the data were collected within the Yadong-Gulu rift in southern Tibet (Figs. 9.6.2-03 and 9.6.2-04).

INDEPTH consisted of four phases. INDEPTH I was conducted in 1992 when pilot near-vertical incidence CMP reflection profiles of 100 km length in southern Tibet (Tib-1 and Tib-2) were recorded, with 50-m receiver-group spacing and 50-s listening time (Zhao et al., 1993). Along these profiles, companion wide-angle data were also collected.

INDEPTH II was carried out in 1994 and 1995 and involved a larger program with the acquisition of CMP reflection and wide-angle reflection profiles Tib-3 to Tib-11 in 1994 as well as collecting broadband earthquake, magnetotelluric, and surface geological data (Nelson et al., 1996; Alsdorf et al., 1998). As an addition, the densely spaced shots of the reflection lines were also recorded parallel to the main lines at offsets between 70 and 130 km (broadband stations in Fig. 9.6.2-03; GEDEPTH location in Fig. 9.6.2-05B).

During INDEPTH III, in 1998 a 400-km-long SSE-NNW wide-angle seismic profile was recorded in central Tibet (Fig. 9.6.2-05), extending from the central Lhasa block in NNW direction to the central Qiangtang terrane (Zhao et al., 2001; Ross et al., 2004). It consisted of a closely spaced (5–10 km) seismic array with 60 broadband and short-period stations which recorded 11 large shots with charges from 180 to 1160 kg (Zhao et al., 2001). In addition, a mobile 60-channel array also recorded at near-vertical angles of incidence four in-line groups and one cross-line group of 160 small shots with charges of 6–50 kg at variable offsets up to 16 km, two of them sampling the Lhasa block south of the Banggong-Nujiang suture, and two of them sampling the Qiangtang terrane farther north (Ross et al., 2004). Following the active-source experiment, 57 three-component stations, including eight offline stations, were left until 1999 to record local, regional, and far-distant events. This phase also included a magnetotelluric experiment.

INDEPTH IV was designed to study the northeastern boundary of the Tibetan Plateau as represented by the Kunlun Mountains and the Qaidam Basin. It started in spring 2007 and was planned to continue into 2009 (Brown, 2006).

The most prominent laterally extensive feature of the INDEPTH I seismic-reflection section was a "mid-crustal" reflection dipping gently to the north from 9 s to 13 s TWT. Using a range of velocities of 6.0–6.4 km/s as was provided by the

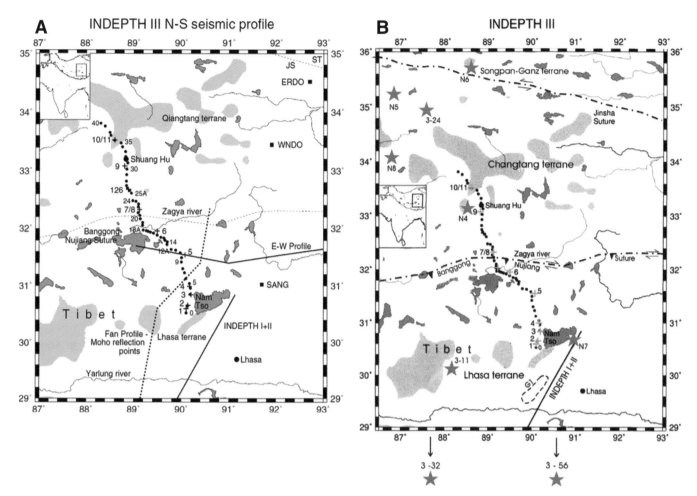

Figure 9.6.2-05. (A) INDEPTH III location map (from Zhao et al., 2001, fig. 1). Also shown are the locations of INDEPTH I and II and earlier profiles of Sino-French campaigns in the eighties: an E-W profile (Sapin et al., 1985) and Moho reflection points of a fan profile (dotted line, Hirn et al., 1984c; Fig. 9.6.2-04). [Geophysical Journal International, v. 145, p. 486–498. Copyright John Wiley & Sons Ltd.] (B) INDEPTH III and earthquake location map (from Meissner et al., 2004, fig. 1). G1—location of the wide-angle GEDEPTH recording area during INDEPTH I and II. [Tectonophysics, v. 380, p. 1–25. Copyright Elsevier.]

French-Sino refraction experiment in the Himalaya (Hirn et al., 1984b), for the southern end of the line a depth of 28 km and for the north a depth of 40 km was calculated. Wide-angle reflections recorded by RefTek stations deployed farther north, allowed to extend this horizon in the subsurface by at least 30 km beyond the north end of the reflection line. Comparison with existing earthquake and geological data let assume that this reflection could be interpreted as marking the thrust fault—termed the Main Himalayan Thrust—along which India is presently underthrusting Tibet. Another notable feature of this first INDEPTH project was the occurrence of coherent reflections at ~23 and 24 s TWT. These were interpreted as originating from the Moho at the base of the Indian continental crust. Using an average velocity of 6.3 km/s, a Moho depth of ~75 km results.

A particular emphasis of all INDEPTH projects was its near-vertical incidence reflection component. So far the most significant results of this component were summarized by Ross et al. (2004):

(1) The reflection Moho was found at 23 s TWT at the southern end of the transect (Zhao et al., 1993). (2) A prominent reflection marks the Main Himalayan Thrust, the active décollement along which the Indian plate is underthrusting Asia. (3) A series of extremely high-amplitude reflections or "bright spots" was detected at 15 km depth beneath the northern portion of the INDEPTH II profile (Brown et al., 1996; Alsdorf et al., 1998; Makovsky and Klemperer, 1999). (4) The short seismic-reflection sections from the four locations of INDEPTH III, spanning the Jurassic Banggong-Nujiang Suture, showed thin (<2 km) sedimentary cover sequences, an unreflective upper crust down to ~25 km depth, a strongly reflective lower crust down to 65 km depth (22 s TWT), and a distinct change in seismic character from layered reflectivity to an unreflective domain, that Ross et al. (2004) interpreted as marking the reflection Moho.

In order to collect wide-angle seismic reflections during INDEPTH I and II, the densely spaced shots of the reflection line were recorded at offsets up to 300 km along and parallel to the main line and thus provided an image of the crust straddling the Indus-Yarling suture. The major features were prominent reflections at ~20 km depth beneath and extending to ~20–30 km north and south of the surface expression of the suture, and north-dipping reflections north of the suture (Makovsky et al., 1999; Zhao et al., 1997).

The excellent data along the wide-angle seismic profile of INDEPTH III in central Tibet (Fig. 9.6.2-06; Appendix A9-4-1) revealed a crustal thickness of 65 km beneath the line (Zhao et al., 2001), in agreement with 70 km depth found by Sapin et al. (1985) just south of the Banggong-Nujiang suture along the E-W line shown in Figure 9.6.2-05A. However, the 20 km-step in the Moho postulated by Hirn et al. (1984c) from the fan-recordings, could not be confirmed by the new data.

Another major result of the INDEPTH III line was the evidence for a 25–30-km-thick lower crust with high velocities ranging from 6.5 to 6.6 km/s at its top to 7.2–7.3 km/s at its base and a velocity contrast of 0.7–0.8 km/s across the Moho (Fig. 9.6.2-07).

Zhao et al. (2001) have summarized all data available at that time in a sketch showing how Moho depths in Tibet decrease from south to north between 28°N and 36°N (Fig. 9.6.2-08). Further details of structural and tectonic features were detected by Meissner et al. (2004), who reinterpreted the INDEPTH III wide-angle data in conjunction with the observations of eight earthquakes (data examples in Appendix 9-4-1), recorded during the same time.

The Sino-French cooperation of the 1980s was continued at the end of the 1990s with a network of seismic-refraction profiles in northeastern Tibet on the Tibetan Plateau (Galvé et al., 2002; Appendix 9-4-2).

Six borehole shots with charges of 3–5 tons were recorded in three deployments of 250 seismometers with 5–10 km spacing on profiles reaching offsets of more than 250 km. The system of reversed and overlapping profiles extended from the North Kunlun crustal block to the Qang Tang terrane of central Tibet (Fig. 9.6.2-09).

The interpretation (Fig. 9.6.2-10) showed a drastic change of crustal structure when crossing the Kun Lun fault. In the south, a Moho-like high-velocity lid was obtained at 35 km depth, overlying a 37-km-thick lower crust with an average velocity of 6.5 km/s and a Moho at 72 km depth. Under the North Kunlun block the "35-km" reflection was absent. Here, a very low average velocity resulted for the more or less featureless crust and Moho depth was found at 64 km (Galvé et al., 2002). The data were further interpreted a few years later (Jiang et al., 2006; Galvé et al., 2006).

The northern margin of the Tibetan Plateau was investigated by an active-source seismic-refraction profile across the Altyn Tagh Range (Zhao et al., 2006). The line crossed the Qaidam Basin in E-W direction and continued in northwesterly direction into the Tarim Basin (Fig. 9.6.2-11). Ten shotpoints separated by 60–320 km provided the seismic energy by 2-ton-charges which were recorded by 240 three-component seismographs spaced at intervals between 1.5 and 3.0 km (Fig. 9.6.2-12; Appendix A9-4-2). The southeasternmost shotpoint (no. 10 in Fig. 9.6.2-11) was located ~200 km south of the southernmost shotpoint of the seismic-refraction survey of 1988 (Wang et al., 2003; see Chapter 8.7.2 and Appendix A9-4-2) which had been

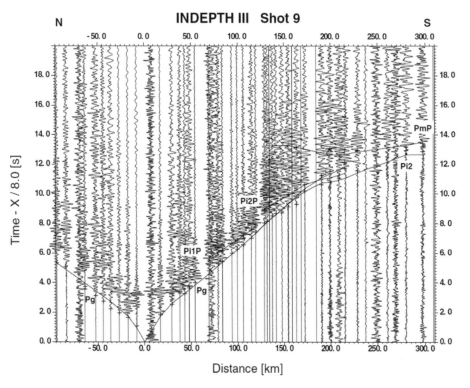

Figure 9.6.2-06. Data example of the INDEPTH-III profile across central Tibet, shot 9 (from Zhao et al., 2001, fig. 8). For location, see Fig. 9.6.2-05. [Geophysical Journal International, v. 145, p. 486–498. Copyright John Wiley & Sons Ltd.]

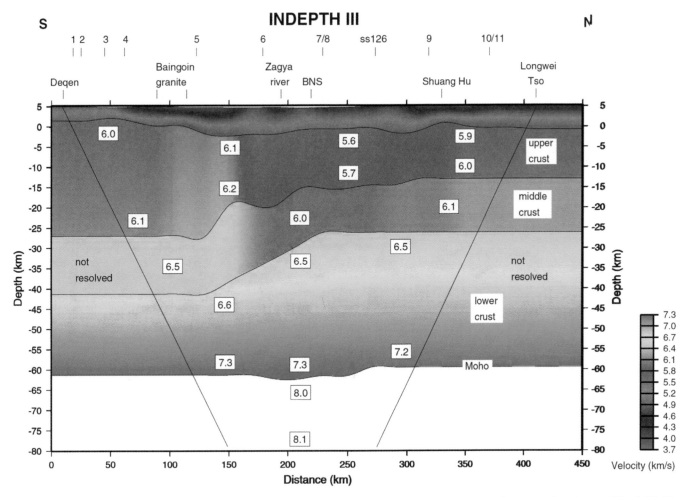

INDEPTH III

Figure 9.6.2-07. P-wave model of the INDEPTH-III profile across central Tibet (from Zhao et al., 2001, fig. 7). For location, see Fig. 9.6.2-05. [Geophysical Journal International, v. 145, p. 486–498. Copyright John Wiley & Sons Ltd.]

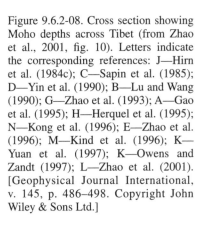

Figure 9.6.2-08. Cross section showing Moho depths across Tibet (from Zhao et al., 2001, fig. 10). Letters indicate the corresponding references: J—Hirn et al. (1984c); C—Sapin et al. (1985); D—Yin et al. (1990); B—Lu and Wang (1990); G—Zhao et al. (1993); A—Gao et al. (1995); H—Herquel et al. (1995); N—Kong et al. (1996); E—Zhao et al. (1996); M—Kind et al. (1996); K—Yuan et al. (1997); K—Owens and Zandt (1997); L—Zhao et al. (2001). [Geophysical Journal International, v. 145, p. 486–498. Copyright John Wiley & Sons Ltd.]

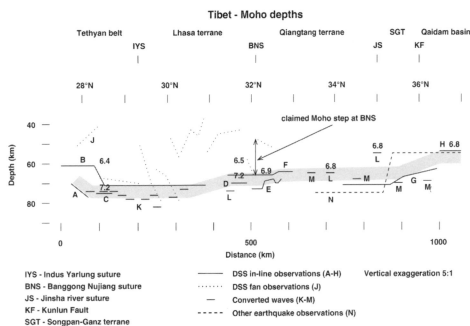

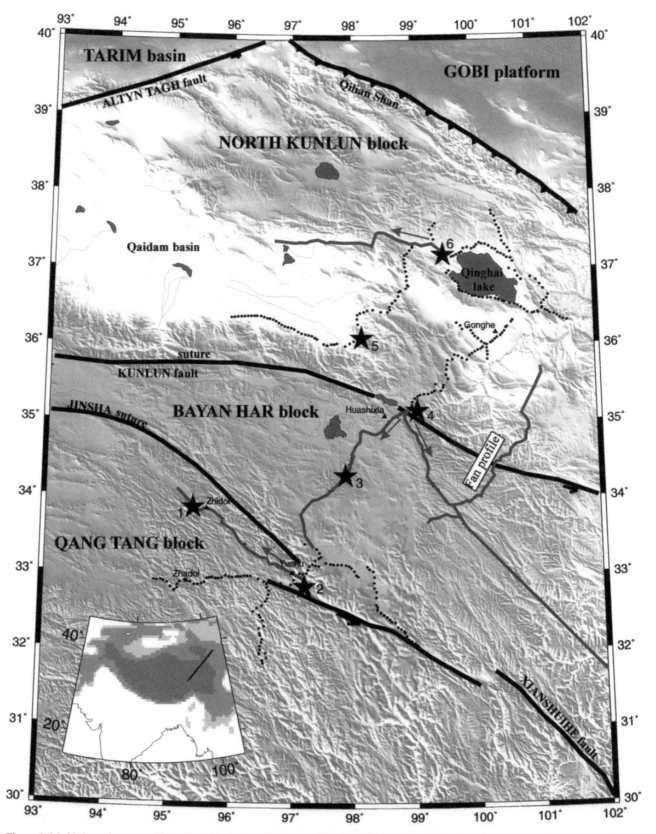

Figure 9.6.2-09. Location map of Sino-French seismic refraction profiles (from Galve et al., 2002, fig. 1). Stars—shotpoints; full and dotted lines—discussed and not discussed, respectively, in detail by Galvé et al. (2002). [Earth and Planetary Science Letters, v. 203, p. 35–43. Copyright Elsevier. .]

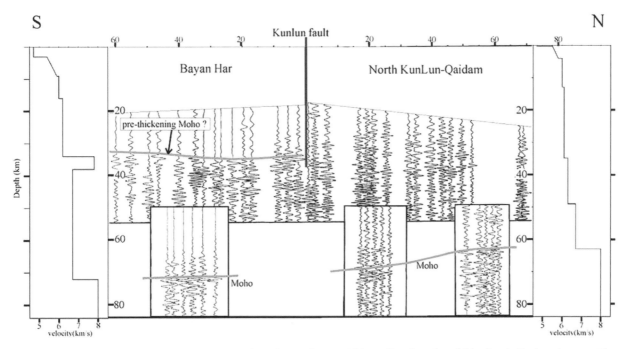

Figure 9.6.2-10. Depth-converted record section of the fan profile at 155 km offset from broadside shot 4 (for location, see Fig. 9.6.2-09) and depth section spanning 140 km across the Kun Lun fault (from Galvé et al., 2002, fig. 4). The insets show depth-converted seismograms from offsets of 190–240 km. [Earth and Planetary Science Letters, v. 203, p. 35–43. Copyright Elsevier.]

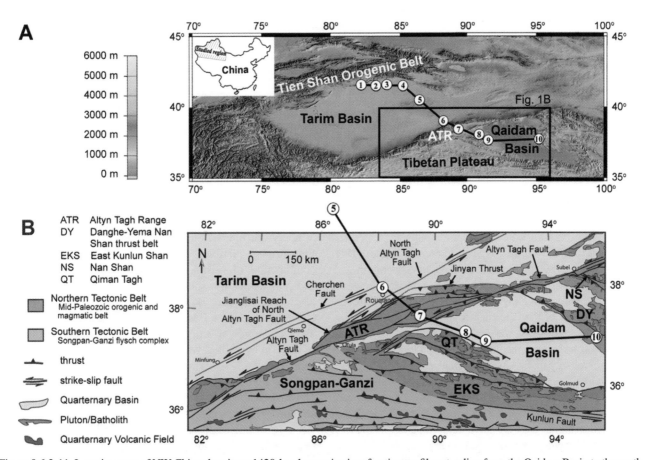

Figure 9.6.2-11. Location map of NW China showing a 1420-km-long seismic refraction profile extending from the Qaidam Basin to the northern margin of the Tarim Basin (from Zhao et al., 2006, fig. 1). [Earth and Planetary Science Letters, v. 241, p. 804–814. Copyright Elsevier.]

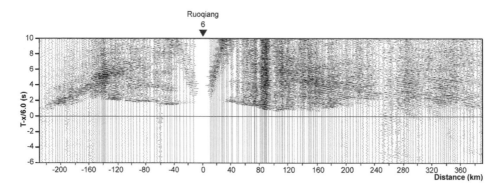

Figure 9.6.2-12. P-wave record section of shot 6 of the 1420-km-long seismic refraction profile across the Altyn Tagh Range, NW China (from Zhao et al., 2006, fig. 2A). For location, see Fig. 9.6.2-11. [Earth and Planetary Science Letters, v. 241, p. 804–814. Copyright Elsevier.]

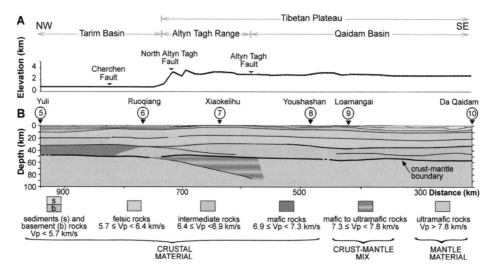

Figure 9.6.2-13. P-wave model across the Altyn Tagh Range, NW China (from Zhao et al., 2006, fig. 7). For location, see Fig. 9.6.2-11. [Earth and Planetary Science Letters, v. 241, p. 804–814. Copyright Elsevier.]

recorded in NNW direction from the Altyn Tagh fault to the Altai Mountains (see Fig. 8.7.2-08).

Similar to the Kun Lun fault farther south, the Cherchen fault seemed to mark a drastic change in crustal structure (Fig. 9.6.2-13). The interpretation was carried out in cooperation with the USGS (Zhao et al., 2006). The resulting model showed a platform-type crust with a high-velocity lower crust beneath the Tarim Basin north of the Cherchen fault (for location, see Fig. 9.6.2-09). This lower crust was not found south of the fault. In contrast, the crust beneath the northern Tibetan Plateau evidently lacks a high-velocity mafic lower crustal layer.

The velocity-depth structure was found to be similar under the Altyn Tagh Range and the Qaidam Basin to the southeast. However, the seismic model of Zhao et al. (2006) also indicated that the high topography of the Altyn Tagh Range of ~3000 m is supported by a wedge-shaped region underlying the crust with seismic velocities of 7.6–7.8 km/s that extends to 90 km depth and which the authors defined as crust-mantle mix.

Another 1000-km-long seismic profile (Liu et al., 2006) investigated the northeastern margin of the Tibetan Plateau to study the interaction between the Tibetan Plateau and the Sino-Korean platform. The joint China-U.S. project was a combination of seismic refraction, teleseismic observations and magnetotelluric soundings. The line started to the southeast of the Qaidam Basin in the Songpan-Ganzi-terrane and ended in the Ordos Basin, part of the Sino-Korean platform. Twelve charges of 2 tons were fired at nine shotpoints, eight of them in boreholes, while shotpoint no. 1 was an underwater shot (Fig. 9.6.2-14). The seismic energy was recorded by 200 portable digital seismographs, DAS-1, developed by the Geophysical Exploration Center of the China Earthquake Administration. The spacing of the three-component receivers was 1.5–2 km, enabling the interpretation of both P- and S-waves (Fig. 9.6.2-15; Appendix A9-4-2).

The interpretation revealed significant variations of crustal velocities along the profile and a decrease of average crustal thick-

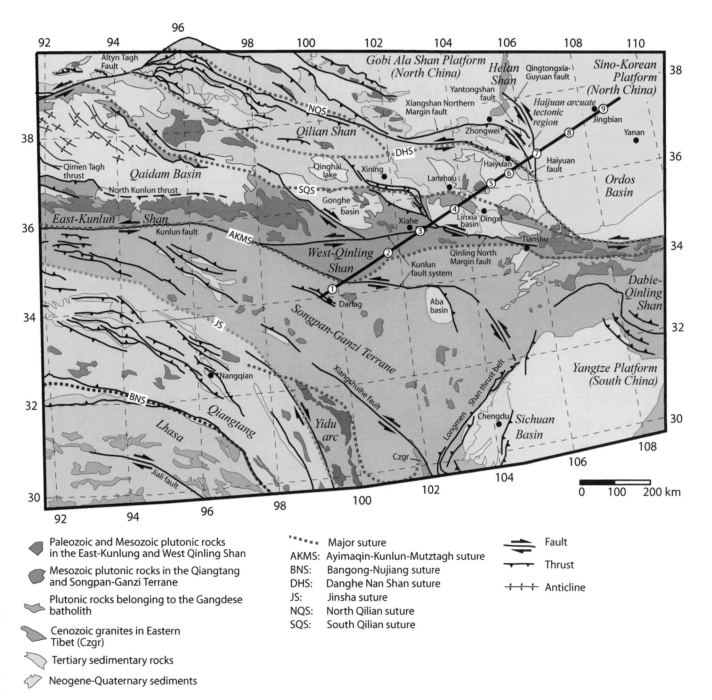

Figure 9.6.2-14. Location map of north central China showing a 1000-km-long seismic refraction profile extending from the Songpan-Ganzi terrane into the Sino-Korean platform (from Liu et al., 2006, fig. 1). [Tectonophysics, v. 420, p. 253–266. Copyright Elsevier.]

ness from around 60 km in the Songpan-Ganzi terrane at the northeastern margin of the Tibetan Plateau to 40–45 km under the Sino-Korean platform (Fig. 9.6.2-16). In particular, the lower crust varies in thickness from 38 km in the southwest to 22 km in the northeast, but in general, relatively low velocities are predominant in the lower crust (Liu et al., 2006).

Another example of a cooperative project between Chinese and U.S. scientists was a seismic-refraction project through the

Dabie Shan in eastern central China (Wang et al., 2000), shown as project A33 on Fig. 9.6.2-01. The project was carried out in 1994 and consisted of a 400-km-long, N-S–directed line through the E-W–trending ultra-high pressure Dabie Shan orogenic belt in east central China (Fig. 9.6.2-17).

Eight shots (seven borehole shots; one underwater shot) of 1000–2000 kg, spaced ~50 km apart, were recorded by a total of 150 analog recorders (Fig. 9.6.2-18; Appendix A9-4-2).

Chapter 9

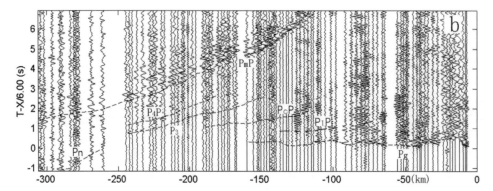

Figure 9.6.2-15. P-wave record section of shotpoint 4 of the seismic refraction profile extending from the Songpan-Ganzi terrane into the Sino-Korean platform (from Liu et al., 2006, fig. 3B). [Tectonophysics, v. 420, p. 253–266. Copyright Elsevier.]

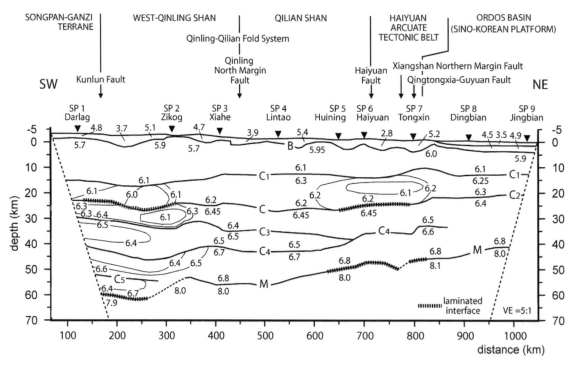

Figure 9.6.2-16. P-wave model between the Tibetan Plateau and the Sino-Korean platform, north central China (from Liu et al., 2000, fig. 6a). For location, see Fig. 9.6.2-14. [Tectonophysics, v. 420, p. 253–266. Copyright Elsevier.]

The recorders were spaced 1 km apart inside the orogen and 3–4 km outside.

The interpretation (Fig. 9.6.2-19) revealed that the cratonal blocks north and south of the orogen were composed of a 35-km-thick, three-layered crust with average seismic velocities of 6.0, 6.5, and 6.8 km/s. The crust thickened underneath the Dabie Shan orogenic belt and reached a maximum thickness of 41.5 km beneath the northern margin (right half of Fig. 9.6.2-19) of the orogen. Here, also a Moho offset was observed, caused possibly by an active strike-slip fault (Wang et al., 2000).

The recording with three-component seismographs allowed an interpretation of both P- and S-waves and the determination

of Poisson's ratio resulting in values of 0.23–0.25 for the upper crust, 0.24–0.27 for the middle crust, and 0.27–0.29 for the lower crust. In the upper crust, the lowest Poisson's ratios were observed underneath the orogen, whereas in the middle and lower crust, Poisson's ratios were highest in the south and decreased northward (Wang et al., 2000).

Li and Mooney (1998) and Li et al. (2006) have summarized the main results of crustal studies in China in crustal columns with average crustal structure for the main tectonic areas (see Fig. 10.5.2-03) as well as in contour maps, showing crustal thickness (Moho depths, see Fig. 10.5.2-01) and uppermost-mantle velocities (P_n-velocities).

Explanation

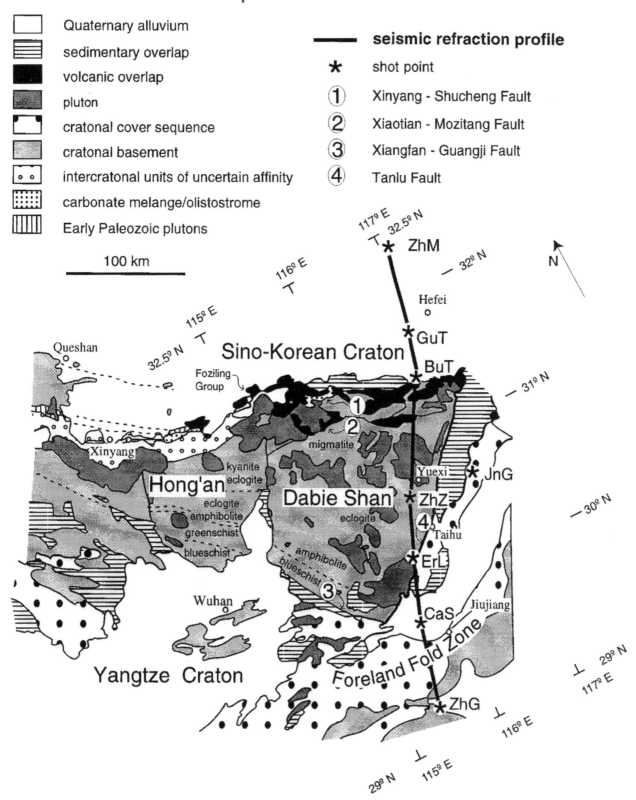

Figure 9.6.2-17. Location map of the Dabie Shan orogen, central eastern China, showing the location of the 400-km-long N-S–directed seismic refraction profile (from Wang et al., 2000, fig. 2). [Journal of Geophysical Research, v. 105, p. 10,857–10,869. Reproduced by permission of American Geophysical Union.]

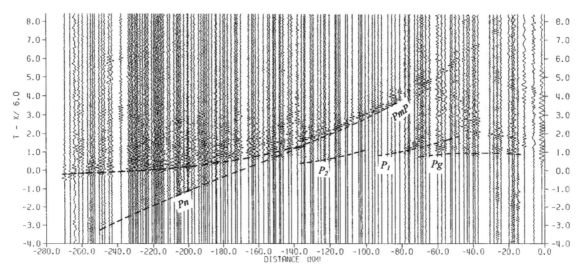

Figure 9.6.2-18. P-wave record section of shotpoint ZhM of the seismic refraction line across the Dabie Shan orogen, central eastern China (from Wang et al., 2000, fig. 3a). [Journal of Geophysical Research, v. 105, p. 10,857–10,869. Reproduced by permission of American Geophysical Union.]

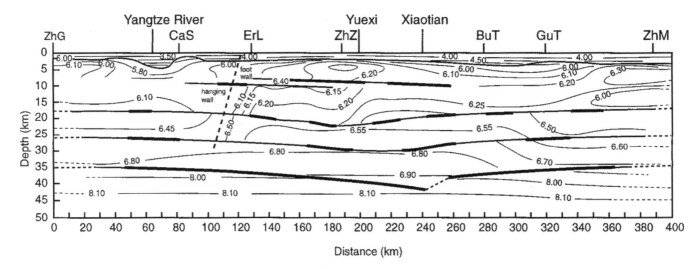

Figure 9.6.2-19. P-wave model across the Dabie Shan orogen, central eastern China (from Wang et al., 2000, fig. 9a). For location, see Fig. 9.6.2-17. [Journal of Geophysical Research, v. 105, p. 10,857–10,869. Reproduced by permission of American Geophysical Union.]

9.6.3. Japan

A summary of crustal studies in Japan performed in the 1990s also contained a review of results of the pre-1990s studies (Iwasaki et al., 2002). The tectonic map shows all seismic profiles performed in and around Japan until the end of the 1990s (Fig. 9.6 3-01). Lines A–A′, B–B′, and C–C′ are profiles which had already been obtained in the 1960s (see Chapter 6).

Controlled-source seismology in Japan was activated in the 1990s as part of the Japanese Earthquake Prediction Program. The first projects took place on the island of Honshu in 1989 and 1990. A 200-km-long profile was shot in 1989 with four shotpoints between Fujihashi and Kamigori in central-western Honshu (named

"SW Japan" in Fig. 9.6 3-01), and a project in the Kitakami region was accomplished in 1990 (line *a* in Fig. 9.6 3-01). These projects were described in Chapter 8.7.3 (Iwasaki et al., 1993, 1994; Research Group for Explosion Seismology, 1992b, 1995).

Other seismic-refraction studies followed. Some of these projects were described in Japanese with only some technical details and local maps provided in the English abstracts. The general scheme of most projects, described in the following, involved four shotpoints and 170–190 recording sites along profiles which usually were ~180 km long.

In 1991 a 180-km-long, E-W–directed, line with 169 recording sites was observed in central Japan (*1* in Fig. 9.6.3-01) between Agatsuma and Kanazawa (Research Group for Explosion

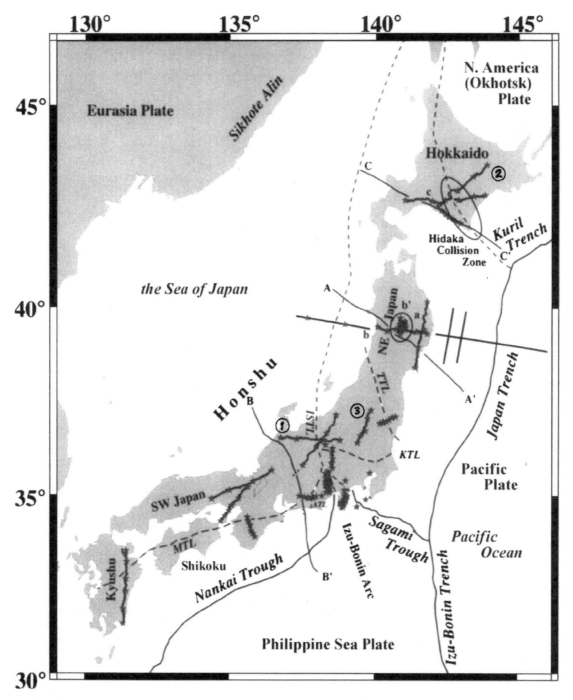

Figure 9.6.3-01. Seismic-refraction profiles in and around Japan (modified from Iwasaki et al., 2002, fig. 1 [Tectonophysics, v. 355, p. 53–66.]). A–A′: line of cross section by Asano et al. (1981) shown in Figure 7.6.3-01. B–B′: Atsumi-Noto profile, central Japan (Aoki et al., 1972). C-C′: southern Hokkaido (Okada et al., 1973).

Seismology, 1994). In 1992, the central part of Hokkaido was the target of a 178-km-long profile (*2* in Fig. 9.6.3-01) with 184 sites between Tsubetsu and Hikada Monbetsu (Research Group for Explosion Seismology, 1993). In 1993 a 105-km-long profile (*3* in Fig. 9.6.3-01) with five shotpoints and 195 recording sites was recorded in central Japan in the northern part of the Kanto district between Shimogo and Kiryu (Research Group for Explosion Seismology, 1996; Ozel et al., 1999).

Following the destructive Kobe earthquake of 17 January 1995, seismic-refraction (Research Group on Underground Structure in the Kobe-Hanshin Area, 1997; Research Group for Explosion Seismology, 1997; Ohmura et al., 2001) and -reflection (Sato et al., 1998) measurements were carried out in the surroundings of Kobe. Four refraction lines covered the Kobe area (Fig. 9.6.3-02), and more than 150 recorders were deployed. Six major shots (S1 to S6, some outside the map area in Fig. 9.6.3-02), with charges between 350 and 700 kg, provided information on the whole crust. In addition, smaller shots were fired to study small-scale crustal heterogeneities in the area of Awaji Island (Demachi et al., 2004), while shot U1 served in particular to study the basement structure under the city of Kobe. To reveal the structure of the seismogenic and other associated active faults, a seismic-reflection line (Sato et al., 1998) was shot across the northern part of Awaji Island and the neighboring offshore region. On land, four Vibroseis trucks were in operation, while the offshore operation used airguns with a shot spacing of 50 m. The total length of the line was 42 km.

In the 1990s, the Nankai Trough to the south of Shikoku Island and Kii peninsula on Honshu Island as well as the Japan Trench east of Honshu (for geographic locations of the plate structure around Japan, see Fig. 9.8.2-13) were investigated by many marine deep seismic wide-angle reflection surveys, some of which also had a land component. These projects will be discussed in subchapter 9.8.2.

In 1997 a new interdisciplinary project was started to elucidate deformation processes of island arc crust in NE Japan, starting with an onshore-offshore wide-angle seismic experiment

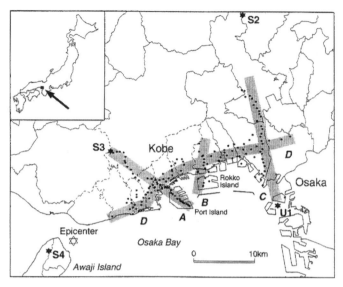

Figure 9.6.3-02. Seismic-refraction profiles in the Kobe-Hanshin area (from Research Group on Underground Structure in the Kobe-Hanshin Area, 1997, fig. 1). [Bulletin of the Earthquake Research Institute, University of Tokyo, v. 72, p. 1–18 (in Japanese). Published by permission of Earthquake Research Institute, University of Tokyo.]

across the northern part of Honshu Island. 293 stations were placed along a 150-km-long E-W line (line *b* in Fig. 9.6.3-01), perpendicular to and crossing the 1990 N-S line through the Kitakami Mountains (line *a* in Fig. 9.6.3-01), and recorded 10 onshore (100-kg and 500-kg shots, data example in Fig. 9.6.3-03) and two offshore shots to the west in the Sea of Japan (Research Group for Explosion Seismology, 1997; Iwasaki et al., 2001a, 2002). The crustal cross section of the onshore profile showed a thickening of the crust from 27 km under its western end to 32 km under the Kitakami Lowland and a thinning again farther east to near 28 km under the Kitakami Mountains (Fig. 9.6.3-04).

From the offshore data, a further crustal thinning toward the east could be proved underneath the continental margin, until 80 km off the coast the continental upper-mantle wedge

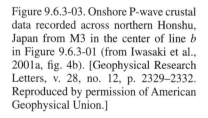

Figure 9.6.3-03. Onshore P-wave crustal data recorded across northern Honshu, Japan from M3 in the center of line *b* in Figure 9.6.3-01 (from Iwasaki et al., 2001a, fig. 4b). [Geophysical Research Letters, v. 28, no. 12, p. 2329–2332. Reproduced by permission of American Geophysical Union.]

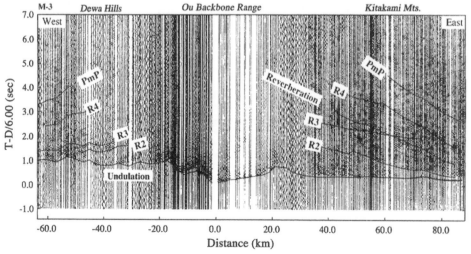

1997 Kamaishi-Iwaki Profile

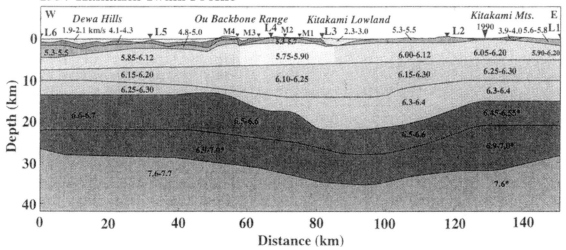

Figure 9.6.3-04. Onshore P-wave model across northern Honshu, Japan (from Iwasaki et al., 2001a, fig. 3; line *b* in Figure 9.6.3-01). [Geophysical Research Letters, v. 28, no. 12, p. 2329–2332. Reproduced by permission of American Geophysical Union.]

converges into the oceanic crust and mantle (Fig. 9.6.3-05). The data also allowed mapping the subduction of the oceanic crust and mantle underneath the Japan island arc, with the subduction angle changing rapidly in E-W direction from 2 to 4° near the trench axis to almost 20° under Japan.

As already mentioned, also Hokkaido Island, northern Japan, was seismically investigated. Following the first seismic-

refraction experiment along the southern shore in 1984, in 1992 a 178-km-long line was recorded across the center of the island (Fig. 9.6.3-06), in order to investigate the crustal-scale structural variation and deformation associated with the accretion and collision tectonics in central Hokkaido (Iwasaki et al., 1998).

Here, from 1994 to 1997, the first deep seismic-reflection surveys in Japan were carried out in the southern part of the Hikada

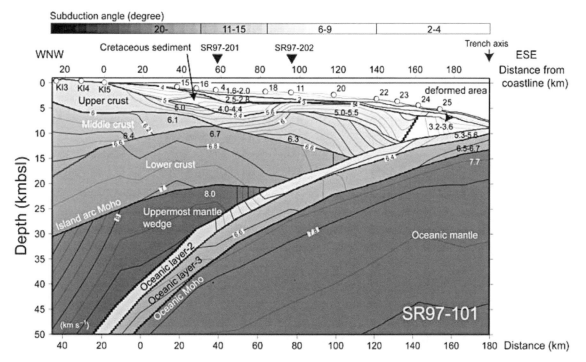

Figure 9.6.3-05. Continuation of the onshore P-wave model across northern Honshu, Japan, into the Pacific Ocean (from Takahashi et al., 2004, fig. 7a; line *b* in Figure 9.6.3-01). [Geophysical Journal International, v. 159, no. 1, p. 129–145. Copyright John Wiley & Sons Ltd.]

---

568 Chapter 9

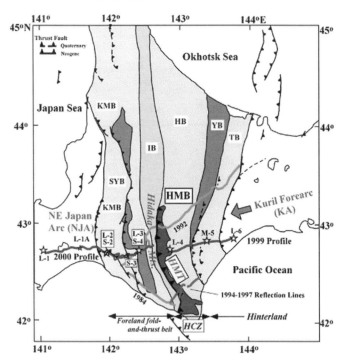

Figure 9.6.3-06. Tectonic map of central Hokkaido with all seismic refraction/reflection profiles and shotpoints (from Iwasaki et al., 2004, fig. 2). [Tectonophysics, v. 388, p. 59–73. Copyright Elsevier.]

Three successive experiments followed from 1998 to 2000. In 1998 (Iwasaki et al., 2001b), two short profiles of 10–20 km length were shot, using Vibroseis and dynamite sources. Additional offline recorders mapped the deeper parts of the crust. In 1999, another reflection line was observed, which, together with the 1998 observations, resulted in 83 km total length of reflection observations, where 1567 receivers with spacings of 50–150 m recorded 74 Vibroseis and 30 explosive shots with charge sizes ranging from 40 to 500 kg (Iwasaki et al., 2002). These reflection experiments of 1998–1999 aimed to investigate the deeper structure of the northern part of the collision zone.

In 1999 (Research Group for Explosion Seismology, 2002a), a refraction experiment followed along a 227-km-long east-west–running profile with 297 stations and 6 shots with charge sizes of 100–700 kg. In 2000, the eastern 114 km of the 1999 profile were again occupied by 327 stations and additional 4 shots with charges of 100–300 kg to constrain the crustal structure which had not been resolved by the 1999 project (Research Group for Explosion Seismology, 2002b; Iwasaki et al., 2001b, 2002, 2004). The aim of the wide-angle seismic experiment was to obtain an entire crustal model from the hinterland to the forearc fold-and-thrust belt crossing the Hikada collision zone. Figure 9.6.3-01 and, more detailed, Figure 9.6.3-06 show the location of these experiments. A structural model of upper and middle crust across southern Hokkaido (Fig. 9.6.3-07), obtained from the refraction/wide-angle reflection data was compiled by Iwasaki et al. (2004). In this section, the most important results of the reflection studies were integrated, namely two remarkable reflections with an eastward dip at depths of 15–22 km and 17–26 km. However, the down-

collision zone (Arita et al., 1998, Ito, 2000, 2002; Tsumura et al., 1999), applying Vibroseis techniques. The crustal section based on the 1994–1997 Vibroseis reflection surveys imaged direct evidence for the delamination of the Kuril forearc, whereby the wedge of the NE Japan arc was seen as intruding eastward into the delaminated Kuril forearc (Iwasaki et al., 2002).

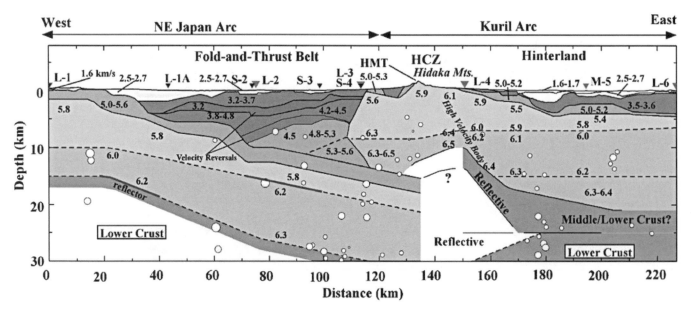

Figure 9.6.3-07. Upper and middle crustal structure model across southern Hokkaido (from Iwasaki et al., 2004, fig. 4). Open circles—hypocenters relocated within 10 km from the seismic profile. [Tectonophysics, v. 388, p. 59–73. Copyright Elsevier.]

going lower crust could not be imaged by the seismic data (Iwasaki et al., 2002, 2004).

In 2000, the Japan Marine Science and Technology Center (JAMSTEC), in cooperation with the University of Tokyo, performed an onshore-offshore experiment along a 492 km profile from the Sea of Okhotsk in the NNW to the Pacific Ocean in the SSE, crossing the northeastern coastal part of Hokkaido and the southwestern Kuril Arc-Trench system (Kurashimo et al., 2007; Nakanishi et al., 2009). In the Okhotsk Sea to the NNW of Hokkaido, 10 OBSs and in the Pacific to the SSE, 30 OBSs were deployed. Energy was supplied by a large airgun array and fired every 150 m. For the onshore part four explosive charges with 300 kg charge size (J1 to J4 in Fig. 9.6.3-08) were detonated and 74 seismic stations were deployed with 1.5 km spacing along the 96-km-long onshore line. The three types of observation—shot gathers airgun shots–OBS, airgun shots–land stations, and land shots–OBS—provided detailed data which also allowed revealing the structure around the onshore-offshore boundary. The interpretation revealed that the crust in the Kuril back-arc region has a thick accretionary complex with velocities of 5.8–6.2 km/s and was considered to continue into the Cretaceous accretionary belt observed in the Hikada Belt based on onshore studies (Iwasaki et al., 1998). The lower crust could not be clarified in this study, but it was concluded that an extremely thick middle/lower crust should occupy ~70% of the entire crust (Nakanishi et al., 2009).

In a piggyback experiment, the land shots of this experiment were also recorded along a 162-km-long line along the northeastern coast of Hokkaido (Taira et al., 2002), using 75 recorders (Fig. 9.6.3-08). On this line shots were recorded in one direction only. The profile was located above the Tokoro belt and aimed to investigate the deep structure of the back arc of the Kuril Arc. The high-quality data allowed estimating the depth to the top of the lower crust and of the Moho (Taira et al., 2002).

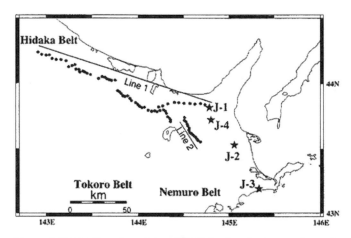

Figure 9.6.3-08. Location of the 2000 Okhotsk seismic experiment in northeastern Hokkaido showing recording sites and shotpoints (from Taira et al., 2002, fig. 1). [Bulletin of the Earthquake Research Institute, University of Tokyo, v. 77, p. 225–230 (in Japanese). Published by permission of Earthquake Research Institute, University of Tokyo.]

9.7. CONTROLLED-SOURCE SEISMOLOGY IN THE SOUTHERN HEMISPHERE

9.7.1. Australia and New Zealand

9.7.1.1. Australia

As in the decades before, in the 1990s and 2000s the main activities concerning deep crustal research in Australia with seismic-reflection and -refraction methods remained with the Bureau of Mineral Resources, Geology and Geophysics at Canberra, renamed in the 1990s to the Australian Geological Survey Organisation (AGSO). The Research School of Earth Sciences of the Australian National University at Canberra (ANU), in contrast, concentrated its deep earth research on the upper mantle and below, applying mainly passive seismological methods using earthquake sources and broadband recording systems. The 1992 SKIPPY project—a broadband recording program across the Australian continent—was initiated by ANU and was the first of many field deployments following between 1993 and 1996 (Van der Hilst et al., 1994, 1998; Zeilhuis and van der Hilst, 1996; Clitheroe et al., 2000). The main features of the recording systems used in Australia were summarized by Finlayson (2010; Appendix 2-2).

Since 1985, AGSO has run a large program of deep seismic-reflection surveys offshore and onshore Australia (Fig. 9.7.1-01), which until 1992 were summarized in a table by Goleby et al. (1994).

The onshore group targeted basins, fold belts or Archean basement with petroleum, coal or mineral deposits and operated a 120-channel telemetric acquisition system, using buried explosive charges as energy sources. The data were typically collected using a 40–60 m station interval and appropriate shot intervals to give an 8- to 18-fold data redundancy. The offsets were up to 4 km and record lengths were usually 20 s. Between 1989 and 1992, the yearly data collection amounted to 450–500 km (Goleby et al., 1994). Throughout the 1990s, onshore DSS investigations were focused on the major mineral provinces of Broken Hill, the eastern Lachland Orogen, the goldfields of western Victoria, the Yilgarn Craton, Tasmania, the Hamersley Basin, and the New England Orogen (Finlayson 2010; Appendix 2-2). Surveys which were connected with seismic-refraction/wide-angle reflection investigations will be discussed in some detail.

Together with the deep marine seismic-reflection profiling, wide-angle seismic data were obtained from OBSs along the Vulcan transect in the Timor Sea and the Petrel transect in the Bonaparte basin off northern Australia (Petkovic et al., 2000) in 1996 and 1998. Additionally, a few land stations recorded the airgun shots (Fig. 9.7.1-02). The Vulcan line, e.g., was 336 km long and shot spacing was 100 m.

From the data of the Vulcan transect, the crustal and upper mantle architecture between the Precambrian Australian craton and the Timor Trough was determined. Linking the data with earlier deep seismic marine profiling, the crustal architecture across the major boundary between the Australian and SE Asian

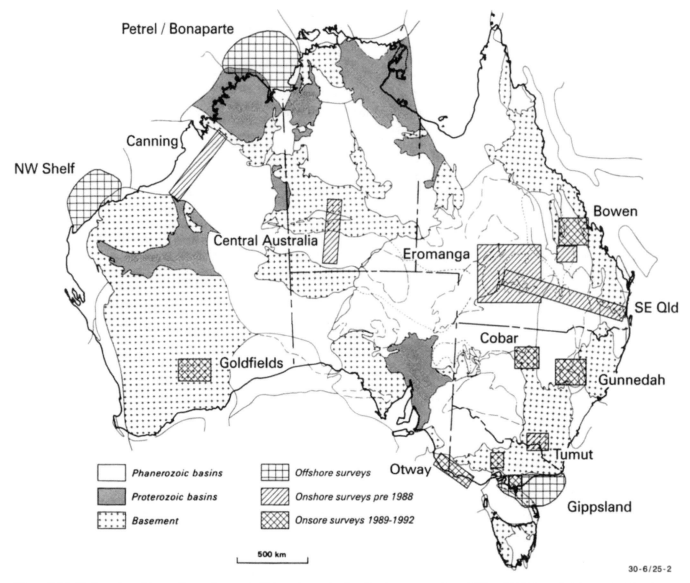

Figure 9.7.1-01. Location of Australian onshore and offshore deep seismic reflection surveys from 1979 to 1992 (from Goleby et al., 1994, fig. 1). [Tectonophysics, v. 232, p. 1–12. Copyright Elsevier.]

plates could be established (Petkovic et al., 2000). Near the Australian coast, a crustal thickness of 35 km resulted. The crust then thinned northwestward to 26 km under the outer shelf of the Timor Trough. While the Paleozoic/Mesozoic basin sequences thickened from almost zero to more than 10 km, the Precambrian basement rocks suffered attenuation from 35 km to 13–14 km across the margin.

Following the 1989 seismic-reflection traverse across the northern Bowen basin, discussed in Chapter 8.8.1, in 1991 a 253-km-long seismic-reflection profile was recorded along a transect on the western margins of the New England batholith from the Gunnedah basin (for location, see Fig. 9.7.1-01) across the Tamworth belt to the Bundarra pluton west of Armidale, Queensland (Korsch et al., 1997; Finlayson 2010; Appendix 2-2).

Another seismic transect was recorded farther north through Queensland, crossing the Mount Isa inlier in North Australia (Drummond et al., 1998). The seismic survey was commissioned in 1994 within a multidisciplinary study of the inlier and its mineral deposits. The reflection line was 250 km long and was overlapped and extended at each end by a refraction survey, 500 km long, to create sufficient offsets for energy penetrations into lower crust and Moho depth ranges (Fig. 9.7.1-03).

Underlain by a thin low-velocity zone, the interpretation (Fig. 9.7.1-04) showed a westward dipping high-velocity body with velocities of 6.9–7.3 km/s in the middle of the crust underneath the Eastern Fold Belt, which appeared to be interrupted toward the west but to continue west of the Mount Isa fault, and finally to merge at ~40 km depth, with the crust-mantle bound-

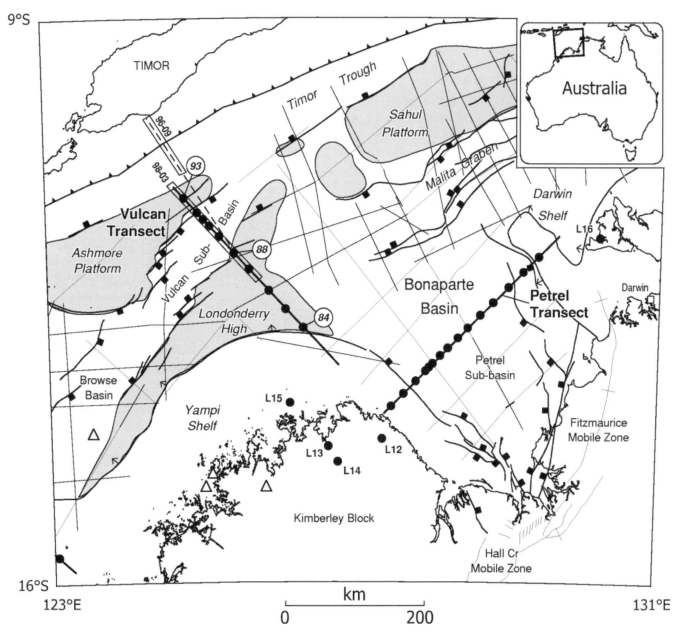

Figure 9.7.1-02. Location of deep seismic reflection surveys in the Timor Sea region (from Petkovic et al., 2000, fig. 1). Thick lines with dots—Vulcan and Petrel transects; black dots—OBS and land recording sites; thin lines—AGSO deep seismic profiling lines. [Tectonophysics, v. 329, p. 23–38. Copyright Elsevier.]

ary, which itself was modeled as a 15-km-thick transition zone between 40 and 55 km depth (Goncharov et al., 1998; Drummond et al., 1998).

In central Australia in 1985, 485 line kilometers of deep seismic-reflection profiling had already been acquired across the Arunta Block and Amadeus Basin, which were followed in 1993 by another 550 line kilometers across the southern Musgrave Block and adjacent Officer Basin (all located within the box labeled "Central Australia" in Fig. 9.7.1-01). All seismic sections were recorded to 20 s TWT. The resulting 900-km-long,

N-S–directed, crustal cross section displayed a crustal thickness of ~45 km underneath the Officer Basin in the south and ~50 km underneath the Amadeus Basin farther north and a disruption of the Moho underneath the Musgrave Block in between (Leven and Lindsay, 1995; Korsch et al., 1998).

The Goldfields Province of Western Australia (Fig. 9.7.1-01) was investigated several times. In 1991, the Kalgoorlie traverse EGF01 (Fig. 9.7.1-05) was shot (Goleby et al., 1994, Drummond et al., 2000a). In 1997, several high-resolution seismic-reflection lines were recorded north and south of Kalgoorlie, and in 1999, a

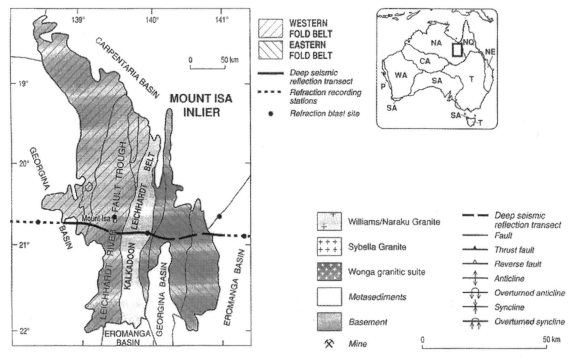

Figure 9.7.1-03. Location of the deep seismic reflection transect through the Mount Isa Inlier, North Australia (from Drummond et al., 1998, fig. 1 top). [Tectonophysics, v. 288, p. 43–56. Copyright Elsevier.]

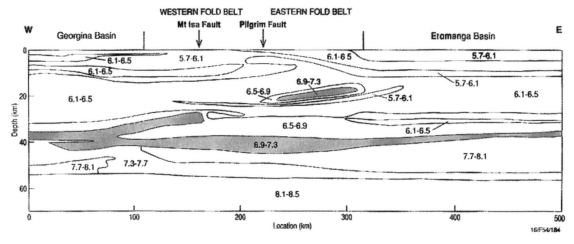

Figure 9.7.1-04. Crustal cross section through the Mount Isa Inlier, North Australia (from Drummond et al., 1998, fig. 4). [Tectonophysics, v. 288, p. 43–56. Copyright Elsevier.]

grid of deep seismic-reflection traverses was acquired to examine the 3-D geometry of the region. It was tied to the 1991 and 1997 profiles (Goleby et al., 2002; Finlayson 2010; Appendix 2-2). In 2001, another deep seismic-reflection survey 01AGSNY was recorded (Goleby et al., 2004), located ~200 km farther to the north-northeast (Fig. 9.7.1-06) using vibrator seismic sources. It will also be discussed briefly in Chapter 10.6.1.

The seismic images of the crust below the greenstone supra-crustal rocks of the Eastern Goldfields Province of the Archean Yilgarn Craton showed structures that Drummond et al. (2000a)

interpreted in terms of tectonic events. While the upper crust appeared to be largely unreflective, the geometry of the reflective middle crust implied thickening by west-directed thrust stacking, and the lower crust showed a fabric indicative of ductile deformation. Total crustal thickness was determined to be 32–34 km in the west and ~36–38 km in the east, deepening slightly east of the Ida fault (Goleby et al., 1994; Drummond et al. (2000a). The new data of the 2001 traverse confirmed the subdivision of the crust into three layers. A review of all investigations across the Yilgarn Craton considering wide-angle and reflection profiling as well as

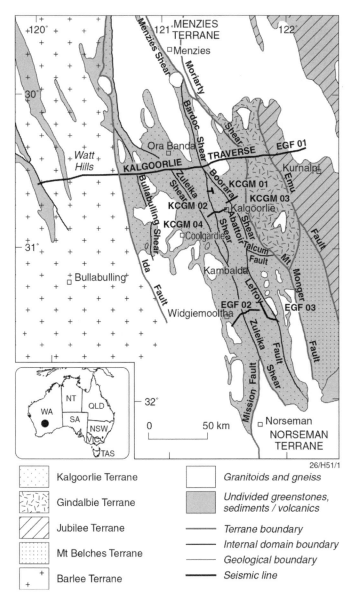

Figure 9.7.1-05. Location of the 1991 Kaloorlie Traverse EGF01 through the Eastern Goldfields Province, Western Australia (from Drummond et al., 2000a, fig. 1). [Tectonophysics, v. 329, p. 193–221. Copyright Elsevier.]

et al. (1996) to be 31 km (10.3–10.5 s TWT) near the northern basin margin, thinning to ~25 km (9 s TWT) seaward of the Tartwaup fault (Fig. 9.7.1-08; TWF in Fig. 9.7.1-07).

In late 1994 and early 1995, marine seismic profiling was added to explore the offshore region of the Otway Basin (Fig. 9.7.1-09) both by near-vertical and wide-angle incidence across the Otway Basin continental margin (Finlayson et al. (1998). For this survey, seismic stations were also established onshore to record the airgun signals to distances as far as 200 km. The right panel of Figure 9.7.1-09 shows the locations of lines and stations which were used for wide-angle recording. The composite result of the 1992 and 1994–1995 surveys as interpreted by Finlayson et al. (1998) shows that the Moho shallows from a depth of nearly 30 km onshore to 15 km depth at 120 km from the nearest shore, and farther to 12 km in the deep ocean at the end of the profile (Fig. 9.7.1-10).

The 1990s saw several other long-range seismic-reflection programs on continental Australia described in detail by Finlayson (2010; Appendix 2-2) which are only briefly mentioned here. In 1996 and 1997, a 295-km-long seismic-reflection line was recorded across the Broken Hill block in the west of New South Wales. In 1999, a deep seismic-reflection profile northeast of Broken Hill followed. In 1997, the Grampian Mountains and western Lachlan Orogen in western Victoria were studied by deep seismic-reflection lines (Korsch et al., 2002). In 1997, the Molong volcanic belt of the eastern Lachlan Orogen in New South Wales became the location of 105 km of deep seismic-reflection profiling (3 lines) and of a wide-angle seismic survey using explosive seismic sources aimed at determining possible crustal compositional variations across the region (Finlayson et al., 2002). In 1999, reflection profiling was performed farther west in the Junee-Naromine volcanic belt (Glen et al., 2002).

Finally, a seismic-reflection survey of AGSO in 1995 in and around Tasmania (Fig. 9.7.1-11), part of which was interpreted by Drummond et al. (2000b), should be mentioned. The program of deep seismic-reflection profiling was combined with a seismic tomography survey.

The offshore program covered the surrounding continental shelf areas of the entire island of Tasmania and was designed to map the large-scale structures of Tasmania at depth. The airgun seismic energy was also recorded by a number of onshore stations (dots in Fig. 9.7.1-11) to study structures which could not be imaged by the offshore profiling. A preliminary interpretation (Drummond et al., 2000b) revealed a crustal thickness of 33–35 km for data of the offshore line 1 along the eastern coast of Tasmania, which is about the same as derived from line 15 along the southern coastline (Fig. 9.7.1-12).

Several reviews have summarized the results of crust and upper-mantle studies for Australia (Collins, 1991; Collins et al., 2003; Clitheroe et al., 2000). Clitheroe et al. (2000) have based their crustal thickness map mainly on applying the receiver function techniques on events beyond 30° with sufficiently energetic P-wave arrivals recorded by an array of permanent and temporary

receiver function work included the compilation of a 3-D seismic model of crustal architecture (Goleby et al., 2006).

The seismic survey program of AGSO in 1992 in southeastern Australia targeted the Otway Basin along Australia's southern continental margin (Fig. 9.7.1-01). It consisted of seven regional deep seismic profiles which were shot across the onshore part of the basin (Fig. 9.7.1-07) using boreholes with 10-kg dynamite charges and recorded by 120-channel split spreads with group intervals of 50 m and 20-s record lengths (Finlayson et al., 1996). The longest line (AGSO line 5) was 101 km long and started in Paleozoic basement of the Delamerian Lachlan Orogen (Fig. 9.7.1-07). The crustal thickness was interpreted by Finlayson

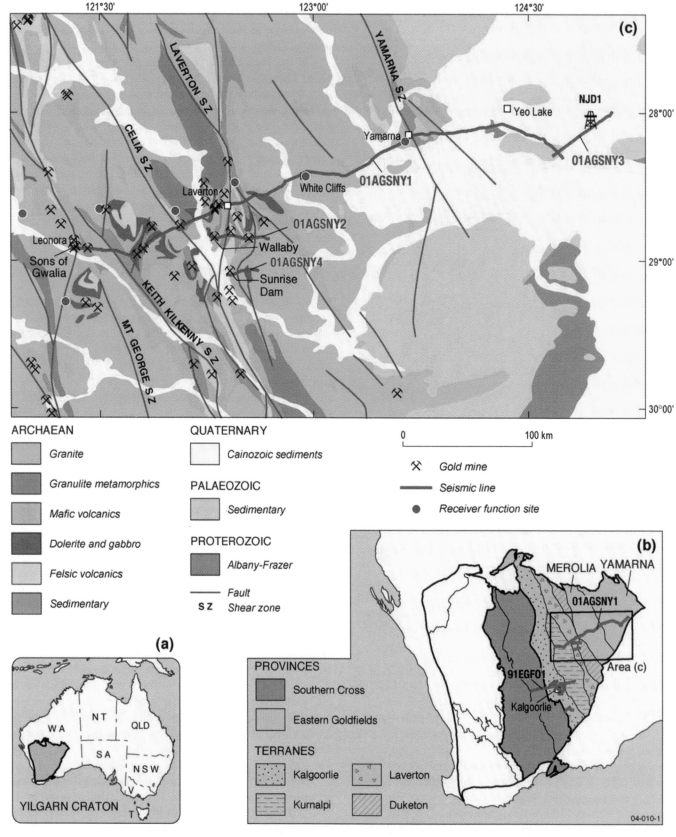

Figure 9.7.1-06. Location of the deep seismic reflection profiles through the Eastern Goldfields Province, Western Australia (from Goleby et al., 2004, fig. 1). (a) Index map. (b) Location of the 1991 and 2001 seismic surveys. (c) Detailed map of the 2001 reflection profiles. [Tectonophysics, v. 388, p. 119–133. Copyright Elsevier.]

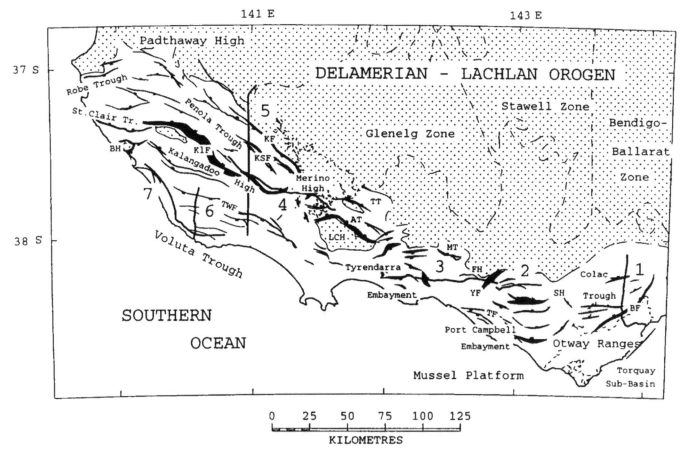

Figure 9.7.1-07. Location of the 1992 onshore seismic reflection profiles in SE Australia (from Finlayson et al., 1996, fig. 3). [Tectonophysics, v. 264, p. 137–152. Copyright Elsevier.]

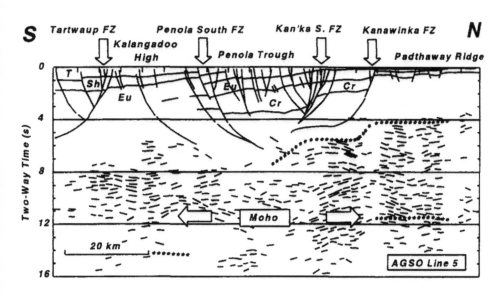

Figure 9.7.1-08. Line drawing and simplified basement features along 1992 AGSO line 5 in SE Australia (from Finlayson et al., 1996, fig. 8). [Tectonophysics, v. 264, p. 137–152. Copyright Elsevier.]

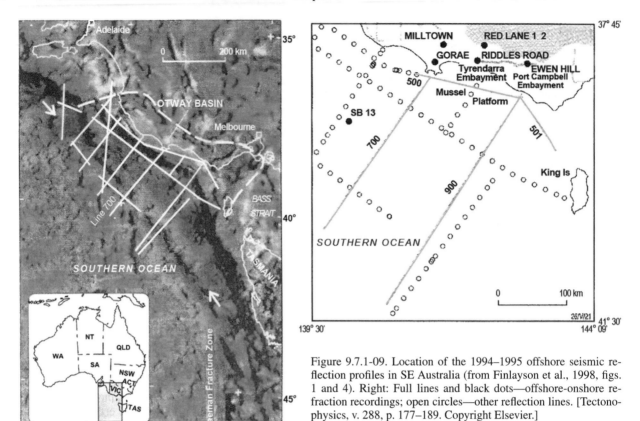

Figure 9.7.1-09. Location of the 1994–1995 offshore seismic reflection profiles in SE Australia (from Finlayson et al., 1998, figs. 1 and 4). Right: Full lines and black dots—offshore-onshore refraction recordings; open circles—other reflection lines. [Tectonophysics, v. 288, p. 177–189. Copyright Elsevier.]

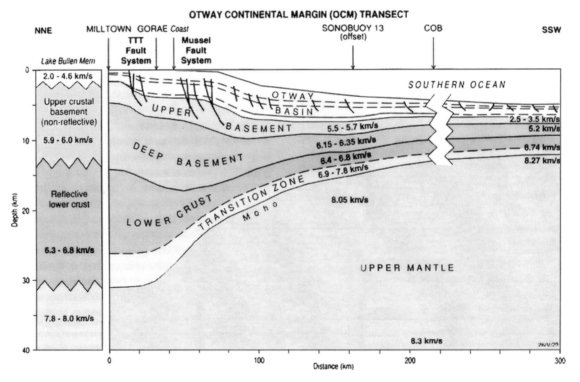

Figure 9.7.1-10. Cross section from the Otway Basin, SE Australia, across the continental margin based on the 1992 and 1994–1995 onshore and offshore seismic reflection profiles (from Finlayson et al., 1998, fig. 9). [Tectonophysics, v. 288, p. 177–189. Copyright Elsevier.]

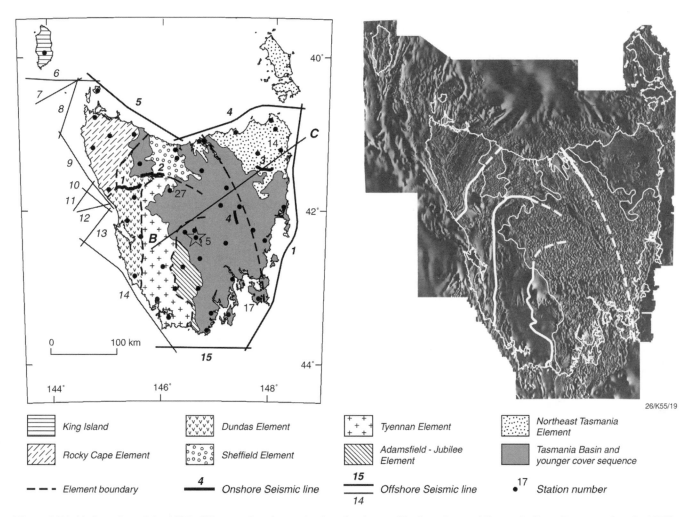

26/K55/19

	King Island		Dundas Element		Tyennan Element		Northeast Tasmania Element
	Rocky Cape Element		Sheffield Element		Adamsfield - Jubilee Element		Tasmania Basin and younger cover sequence
– – –	Element boundary	**4**	Onshore Seismic line	15 / 14	Offshore Seismic line	•17	Station number

Figure 9.7.1-11. Location of the 1995 offshore and onshore seismic reflection profiles in and around Tasmania (from Drummond et al., 2000b, fig. 1a). Stars—locations of onshore blasts; thick short lines 1–4—onshore profiles; thin lines 1–15—offshore seismic lines; dots—stations of a seismic tomography survey. [Tectonophysics, v. 329, p. 1–21. Copyright Elsevier.]

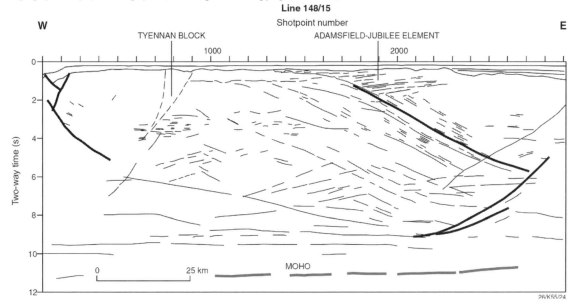

Figure 9.7.1-12. Line drawing of the 1995 offshore seismic reflection profile south of Tasmania (from Drummond et al., 2000b, fig. 8). [Tectonophysics, v. 329, p. 1–21. Copyright Elsevier.]

broadband stations. The latter ones were installed during two projects, where, from 1993 to 1996 during the SKIPPY project, a single array of 12 stations was moved across the continent in six deployments, and in the KIMBA project, an array of 10 stations was deployed in the Kimberley Craton in the northwest of Australia. Collins (1991) based his crustal thickness map primarily on the results of controlled-source seismology investigations, but the most recent, revised map (Collins et al., 2003) also included the results of Clitheroe et al. (2000). The depth to Moho under Australia shows significant variations (Fig. 9.7.1-13).

In general, the authors stated that within Archean regions of Western Australia the Moho appears to be relatively shallow with a large velocity contrast, while the Moho is significantly deeper under the Proterozoic North Australian platform, under central Australia and under Phanerozoic southeastern Australia. They also stated that there is a broad transition zone from crustal to mantle velocities where the Moho is deep (Collins et al.,

2003). While usually the Moho is mapped as a continuous sub-horizontal boundary, based on averaging wide-angle data, from seismic-reflection profiling, due to the smaller wavelengths, offsets were observed where major crustal faulting reached down to the Moho or intersected the crust-mantle boundary, as was shown, e.g., for seismic surveys in central Australia or in Tasmania. In conclusion, the crustal thickness under Australia varied considerably, ranging from ~24 km along the coastlines and continental shelves to 56 km under central Australia and underneath the Lachlan Foldbelt in southeastern Australia.

9.7.1.2. New Zealand

In contrast to the Australian continent, New Zealand lies on the Australian/Pacific plate boundary with Pacific oceanic crust being subducted under eastern North Island, a transform boundary which transects the South Island, and oceanic crust of the Australian plate being subducted under southwestern South

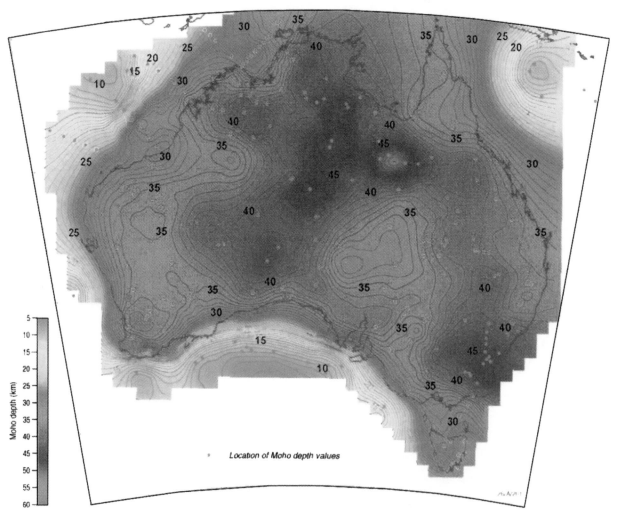

Figure 9.7.1-13. Crustal thickness map of Australia (from Collins et al., 2003, fig. 1). Contour interval—1 km. Red dots—location of depth values used to generate the contours. [Geological Society of Australia Special Paper 273, p. 121–128. Published with the permission of the Geological Society of Australia.]

Island (Sutherland, 1999). South Island is thus being deformed by the oblique collision of the oceanic Pacific plate, the eastern South Island and Campbell plateau to the southeast, with the continental Australian plate, the northwestern South Island and North Island and the Challenger Plateau to the northwest (Fig. 9.7.1-14). This deformation of South Island is marked by the rise of the Southern Alps. About 450 km of dextral strike-slip motion, 80 km of convergence and 20 km of uplift have occurred across this plate boundary since the Oligocene (e.g., Allis, 1986; Davey et al., 1998; Sutherland, 1999).

One of the main features of the North Island is the Central Volcanic Region which has been a zone of back-arc extension and arc volcanism associated with the subduction of the Pacific plate under the Australian plate along the Hikurangi margin. Its present activity is believed to be concentrated on the Taupo Volcanic Zone along its eastern margin (Davey and Lodolo, 1995).

As described in the previous chapters, detailed seismic surveys started in the 1970s and continued through the 1980s. In 1990 a crustal marine seismic-reflection profile of 360 km length was recorded along the north coast of North Island which crossed the Central Volcanic Region (Davey and Lodolo, 1995; Davey et al., 1995). The record length was 16 s TWT. The objective was to obtain a crustal section across the Hikurangi subduction zone and back-arc in the north of North Island. One major goal was

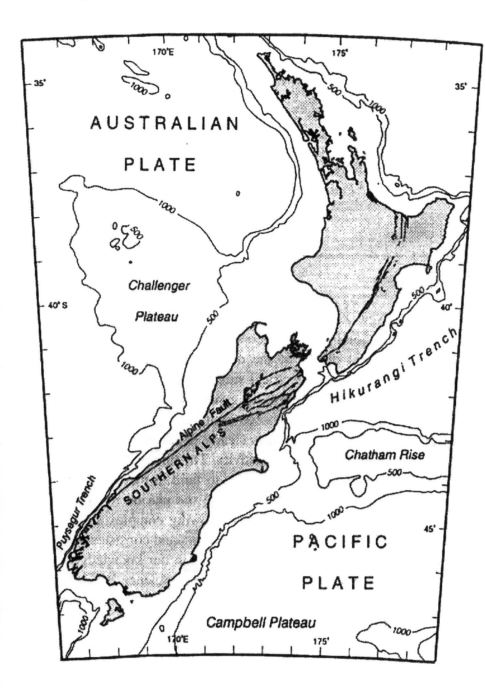

Figure 9.7.1-14. Location map of the New Zealand region (from Davey et al., 1998, fig. 1). [Tectonophysics, v. 288, p. 221–235. Copyright Elsevier.]

to delineate the near-surface structure and image the extensional tectonics inferred to form the Central Volcanic Region. A second goal was to image the lower crust–upper mantle structure, in particular to define the base of the crust which had been determined at 15 km depth by seismic-refraction work in the 1980s (Stern and Davey, 1987). The data have revealed mainly the complex tectonics of the near-surface structures. Under the presently active Taupo Volcanic Zone strong mid and lower crustal reflectors were found at depths of 4.5–5.5 s TWT under the volcanic ridge and western frontal graben which were associated with magma or volcanic sills. The deepest reflector was close to the refraction Moho of Stern and Davey (1987).

In 1994, another project, MOOSE (Marlborough Onshore Offshore Seismic Experiment), investigated the southern Hikurangi subduction zone at the northeastern part of South Island. It was a 300-km-long marine crustal reflection profile with 16 s record length, where normal-incidence and also sparse off- onshore wide-angle seismic data were collected (Henrys et al., 1995). A band of reflectors at 5–8 s TWT could be correlated with the base of a 12–15-km-thick upper crustal layer of velocity 5.0–5.5 km/s underlying part of South Island.

Also in 1994, a 27-km-long crustal seismic-reflection profile was recorded in the central South Island using explosive charges. Its aim was to obtain a crustal section over the eastern part of the Pacific-Australian plate boundary in central South Island (Kleffman et al., 1998).

A third project in 1994 was a marine seismic-reflection survey of 80 km in length in the area of Stewart Island, south of South Island. The recording times were 16 s TWT (Davey, 2005). These data were interpreted together in the context with the marine seismic-reflection profile along the eastern coast of South Island, which was part of SIGHT (see below) and a short third profile along the south coast of Stewart Island recorded in 1996. These seismic lines which partly overlapped provided a crustal seismic image from the southern end of South Island to south of Stewart Island (Davey, 2005). The data provided constraints on the evolution of the Gondwana margin in this region, because lower crustal and upper mantle seismic reflectivity off southeastern Stewart Island were interpreted in terms of a northeast-dipping paleosubduction zone.

The investigations in the area of Stewart Island were continued in 1996 (Melhuish et al., 1999). A marine seismic-reflection line of 610 km in length was recorded from the southern tip of Stewart Island toward WNW through the Solander Basin and across the Puysegur Bank and Trench south of South Island. In the Solander Basin, the seismic-reflection data showed well developed reflections to depths of 30–35 km (12–13 s TWT). On the basis of lower crustal reflectivity, beneath the adjacent Stewart Island shelf the base of the crust was determined at ~30 km depth (9 s TWT), rising westwards to ~20 km depth (8 s TWT) under the Solander basin. Prominent dipping reflections were interpreted as an active fault which could be traced to 30 km depth (~12 s TWT), where they merged into a zone of strong reflectivity in the upper mantle (Melhuish et al., 1999).

To study the characteristics of the Southern Alps orogen of New Zealand and to understand the collisional processes involved in continental collision, in 1995 and 1996, a joint U.S.–New Zealand geophysical project was undertaken which involved active source and passive seismology, magnetotelluric and geoelectric studies, petrophysics, geological mapping and gravity measurements (Okaya et al., 2007). The principal field activity during the southern summer was an integrated onshore and onshore-offshore seismic-refraction and wide-angle reflection experiment, carried out across the South Island and extending ~200 km offshore (Fig. 9.7.1-15) and named SIGHT (South Island Geophysical Transect). The advantage of the offshore-onshore geometry of SIGHT was the illumination of the lower crust and mantle by dense shots offshore into land stations from both sides of the orogen allowing high resolution of the crust and upper mantle structure (Davey et al., 1998; Godfrey et al., 2001; Okaya et al., 2002).

The experiment had two main components. The first was a wide-angle reflection-refraction experiment along the two land transects (profiles 1 and 2 in Fig. 9.7.1-15) across central South Island. Twenty-three chemical explosives with charges of 350–1200 kg were equally spaced, with 16 shots on the northern transect and 7 shots on the southern transect. Four hundred twenty recording instruments were deployed along each transect with a nominal spacing of 400 m. Data examples of line 2 are shown in Figure 9.7.1-16.

The second component consisted of three offshore-onshore transects, two along the profiles 1 and 2 and a third along southeastern South Island. This third transect supplied tie lines across the eastern parts of the two main transects. During this phase, data were recorded by 225 land stations, spaced at ~1.5 km intervals along the two profiles 1 and 2 and at ~10 km intervals along the tie lines of the third transect. Twenty ocean-bottom seismograph/hydrophone instruments (OBS/H) were deployed offshore, first on the west side of South Island, and airgun shots were fired at 50 m intervals along eight western profiles in total, each ~200 km long. The OBS/Hs were then redeployed on the east side of South Island when shooting the profiles on this side of the island. Energy from the onshore shots was clearly recorded to maximum distances of 160 km, while airgun shots reached to recording distances of 300 km. The vessel firing the shots also recorded near-vertical incidence multichannel seismic data to 16 s TWT along all eight profiles (Fig. 9.7.1-15).

To record both active (transect) and passive source data, SAPSE (Southern Alps Passive Seismic Experiment), in addition to 17 permanent New Zealand stations, broadly distributed 26 broadband and 15 short-period recording stations across the South Island, but centrally weighted toward the central Alpine fault and transect region (black and white circles in Fig. 9.7.1-15).

The preliminary modeling (Fig. 9.7.1-17) of both explosion and onshore-offshore data indicated a crust of fairly constant P-wave velocity overlying a lower crustal layer with seismic velocity of ~7 km/s and 5–10 km thick (Davey et al., 1998; Stern and McBride, 1998). The total crustal thickness varied from ~30 km

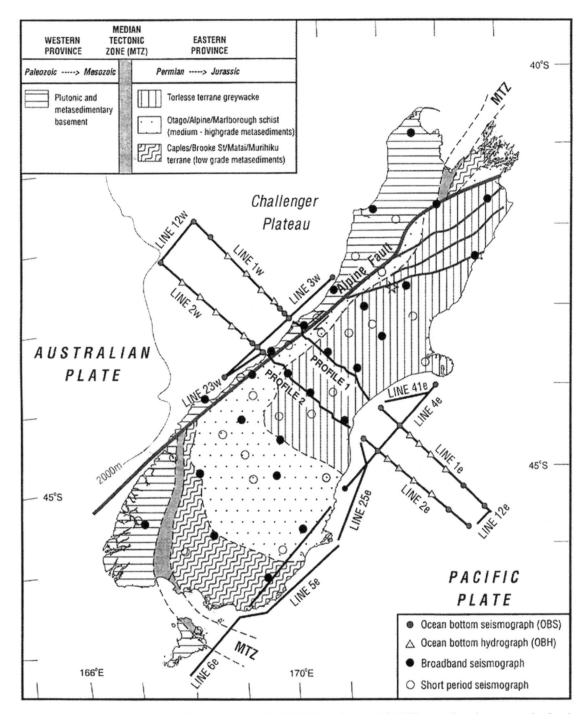

Figure 9.7.1-15. Location of the SIGHT transects, OBS/OBH locations and SAPSE recording sites across the South Island of New Zealand (from Davey et al., 1998, fig. 2). [Tectonophysics, v. 288, p. 221–235. Copyright Elsevier.]

at the east coast to ~42 km under the Southern Alps, whereby the deeper crustal layer was interpreted as possibly corresponding to old oceanic crust. The data also provided evidence of deep reflections from within the dipping Alpine fault zone and delayed lower crustal phases for west coast seismographs from shots east of the Southern Alps, interpreted as due to a low-velocity zone associated with the Alpine fault zone (Davey et al., 1998; Stern and McBride, 1998). In the marine data clear lower crustal-Moho reflections were imaged along most of the tracks, indicating that Moho lies at ~10 s TWT off the west coast and 8 s off the east coast.

The double-sided onshore-offshore seismic imaging of the South Island of New Zealand by "super-gathers" across a plate boundary was discussed in more detail by Okaya et al. (2002) with respect to its application in other areas around the world

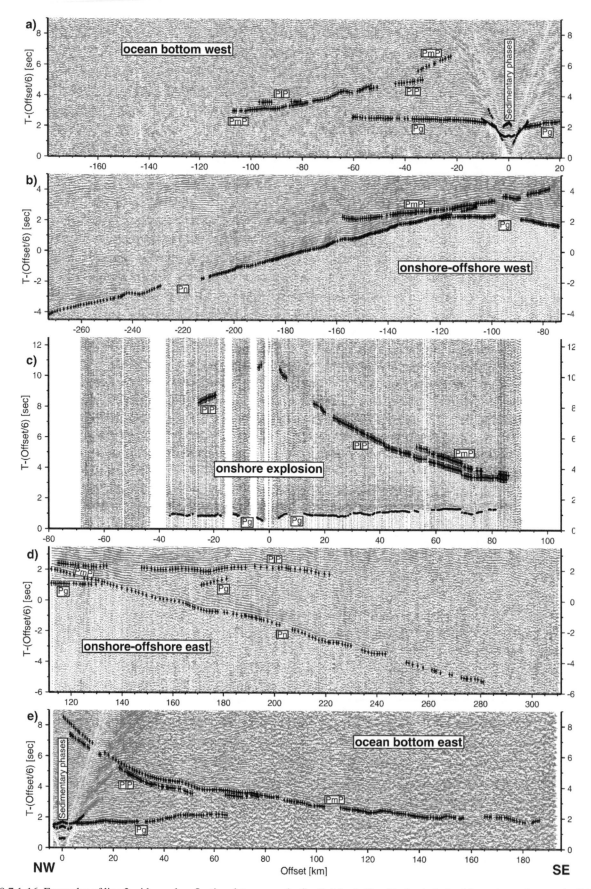

Figure 9.7.1-16. Examples of line 2 wide-angle reflection data across the South Island, New Zealand, derived from the onshore and offshore data (from Scherwath et al., fig. 2). [Journal of Geophysical Research, v. 108 (B12), 2566, doi: 10.1029/2002JB002286. Reproduced by permission of American Geophysical Union.]

where a transition between continental and oceanic environments coincides with the existence of islands or narrow landmasses allowing for a double-sided experiment.

A full interpretation of the two main transects (Scherwath et al., 2003; Van Avendonk et al., 2004) showed up to 17 km of crustal thickening to form a 44-km-deep crustal root that was offset 10–20 km to the southeast of the highest mountains, with a velocity reversal below the midcrust in the eastern part of the

crustal root and a strong thickening of the 7.0 km/s lower crust within the root (Fig. 9.7.1-18).

In 1998 a 65 km seismic-reflection transect (SIGHT98) was shot onshore on South Island across the MacKenzie Basin to Mount Cook Village to provide a detailed crustal image in the central Southern Alps. Fifty kg shots at 1 km intervals were recorded by an 800 channel array, providing a CDP survey with vertical seismic profiling in central South Island (Long et al.,

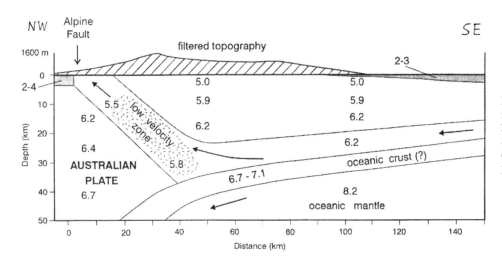

Figure 9.7.1-17. Preliminary crustal model across the South Island, New Zealand, derived from the onshore wide-angle data of profile 2 (from Davey et al., fig. 6). [Tectonophysics, v. 288, p. 221–235. Copyright Elsevier.]

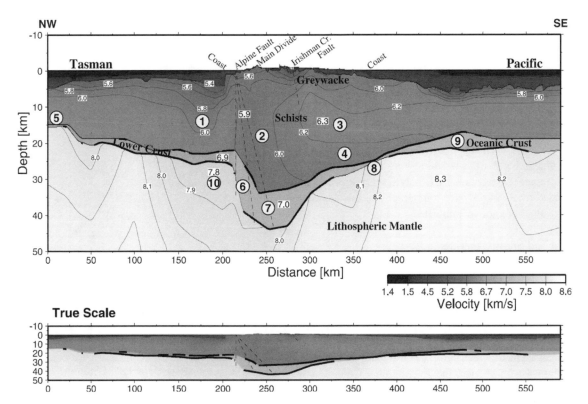

Figure 9.7.1-18. Final crustal model across the South Island, New Zealand, derived from the onshore and offshore data (from Scherwath et al., fig. 4). [Journal of Geophysical Research, v. 108 (B12), 2566, doi: 10.1029/2002JB002286. Reproduced by permission of American Geophysical Union.]

2003; Okaya et al., 2007). The maximum array aperture was nearly 35 km and the maximum shot-receiver offset ~30 km. The raw seismic data were recorded up to 30 s time and processed in two parts. For the first study, only the first 5 s TWT were processed with a maximum offset of 14 km and imaged the upper 12–15 km of the crust (Long et al., 2003). These data did not image any major regional-scale reflections or discontinuities in the uppermost 5 s. Instead, more subtle, smaller-scale features were present, interpreted as to mark faults, bedding, or lithological contacts. The deeper data imaged a band of reflectivity inferred to correspond to a lower crustal décollement that turned upwards at its western end to align with the Alpine fault at the surface (Stern et al., 2007).

9.7.2. South Africa

Besides the large-scale projects in the East African rift system (see subchapter 9.5.1), active-source seismic projects were also carried out in the 1990s in Namibia. In 1995, the project MAMBA (Geophysical Measurements Across the Continental Margin of Namibia experiment), a combined onshore-offshore multi-channel and wide-angle seismic survey, included a major land component along two transects (for location, see Fig. 9.8.4-02) traversing deeply eroded basement rocks of the late Precambrian Damara Orogen onshore and extending offshore out across the ocean-continent transition. Airgun shots were recorded simultaneously on land and at sea. For the interpretation receiver gathers were compiled for each station (Bauer et al., 2000). More details are described in subchapter 9.8.4.

A second project, ORYX of the German GeoForschungs-Zentrum Potsdam, was carried out in 1999. The project aimed to investigate details of the crustal structure crossing the Waterberg fault zone south of the Erongo Mountains in central Namibia. The special aim was a combined near-incidence and wide-angle seismic-reflection survey using 35 mobile six-component stations of GFZ Potsdam, each supplied with 6 vertical geophones (180 seismic channels) along two 18-km-long lines across the Waterberg fault zone. Conventional CDP-style seismic processing was extended by P- and S-wave tomography of the shallow subsurface. Results were not yet published.

9.7.3. Central and South America

Since the first beginnings of crustal structure studies in the central plateau of Mexico in the 1950s (Meyer et al., 1958, 1961a) and in the Andean region of Peru in the 1950s (Tatel and Tuve, 1958) and 1960s (Ocola and Meyer, 1972), Central and South America have been the target of continued international research projects, some of which continue until today.

In the southeast of Mexico, the International Continental Deep Drilling Program (ICDP) initiated deep crustal studies of the Chicxulub meteorite impact crater. The continental margin of Central America became the target of a German project, with an onshore-offshore project at the coast of Costa Rica. In the

north and northeast of South America, crustal structure studies were undertaken in Venezuela and Brazil. The majority of crustal investigations, however, concentrated on the region of the Andes, where since the 1960s the Free University of Berlin, Germany, had initiated detailed geological research work which in the 1980s and 1990s was finally complemented by intensive onshore and offshore geophysical crustal and upper-mantle studies.

9.7.3.1. Mexico and Costa Rica

The Chicxulub impact crater, located on the present coast of the Yucatan peninsula and expressed onshore as a semicircular ring of sinkholes with a diameter of ~160 km (Pope et al., 1996) was recognized as such in 1980 (Alvarez et al., 1980). Widely distributed drill holes sampled cores of the post-impact Tertiary sediments and showed that the sediments thicken to more than a kilometer in the interior of the basin (Ward et al., 1995).

In 1996, BIRPS (British Institutions Reflection Profiling Syndicate) acquired ~650 km of deep seismic-reflection profiles off the northern coast of the Yucatan peninsula in Mexico (Snyder et al., 1999; Christeson et al., 2001; Morgan et al., 2002).

The data were recorded to 18 s TWT on a 240-channel, 6-km-long streamer, using a 50 m shot spacing. The 13,000 airgun shots were also recorded at wide-angle on 34 OBSs, provided by the University of Texas at Austin, and by 91 land recorders supplied by PASSCAL (Fig. 9.7.3-01).

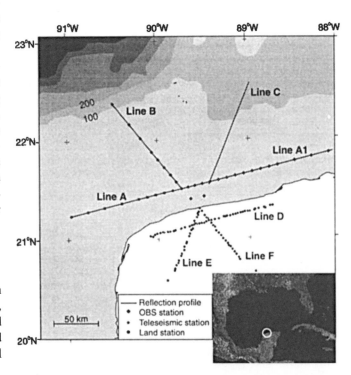

Figure 9.7.3-01. Location of the 1996 Chicxulub seismic experiment, showing BIRPS seismic reflection survey lines, OBS, and land stations (from Snyder et al., 1999, fig. 1). Inset—Gulf of Mexico, showing location of the Chicxulub impact crater on the Yucatan peninsula. [Journal of Geophysical Research, v. 104, p. 10,743–10,755. Reproduced by permission of American Geophysical Union.]

The shallow water limited the acquisition of marine seismic data to distances of not closer than 25 km from the crater center which is located close to the coastline.

The seismic-reflection data of the Chicxulub deep seismic-reflection profile A/A1 were converted to depth using the velocity information from wide-angle OBS data (Fig. 9.7.3-02), imaging the crater fill and impact lithologies and structures to the base of the crust at ~35 km depth (Christeson et al., 2001). Using in particular the higher-resolution data covering the uppermost 8 km (4-s record length), the new seismic offshore data revealed a ring around the peak and three annular zones where the style of deformation changed at discrete radial distances of 55–65 km and 85–98 km (Snyder et al.,1999).

Morgan et al. (2002) inverted the whole seismic-refraction data set in 3-D using first-arrival travel-time tomography and found a high-velocity body within the central crater region that most likely represents lower-crustal rocks which were stratigraphically uplifted during the formation of this complex crater. A suite of checkerboard tests confirmed that the observed velocity anomalies were real, in spite of the irregular experimental geometry of the refraction data.

A second seismic experiment took place in the beginning of 2005, organized by a team of scientists from Mexico, the United States, and the UK (Morgan et al., 2005). At this time, 1500 km of reflection profiles were acquired and 36,560 airgun shots were recorded on 28 ocean-bottom seismometers and 87 land seismometers (Fig. 9.7.3-03).

A subset of the seismic data constituted a site survey for two proposed Integrated Ocean Drilling Project (IODP) drill holes (Chicx-01A and Chicx-02A in Fig. 9.7.3-04).

The seismic data of the 1996 experiment had allowed the conclusion that the crater is 180–200 km in diameter, and in 2002 Chicxulub was drilled onshore at Yaxcopil-1 (Yax-1 in Fig. 9.7.3-04) under the auspices of the ICDP. Besides the proposals for the two offshore drill holes Chix-01A and Chix-02A, another onshore drill hole (ICDP-hole 2 in Fig. 9.7.3-04) has been proposed close to the crater center. Initial analyses of the new seismic data indicated that important information could be obtained of how large impact craters are formed. They also promise important constraints on changes in lithology across the crater (Morgan et al., 2005).

In Costa Rica, the convergent margin off the Nicoya Peninsula had been the site of single- and multichannel seismic-reflection surveys (von Huene and Flueh, 1995). Within the project TICOSECT (Trans Isthmus Costa Rica Scientific Exploration of a Crustal Transect) project in 1995 a major seismic-refraction survey was conducted offshore Costa Rica near the Nicoya

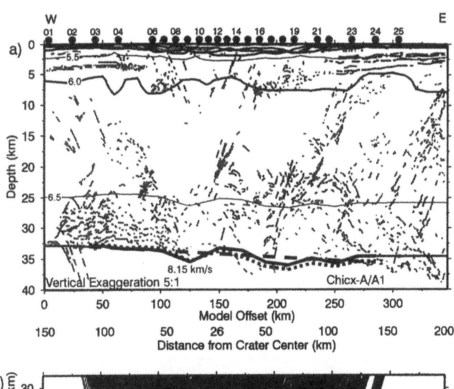

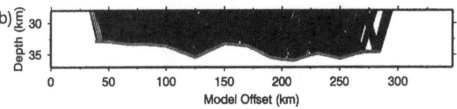

Figure 9.7.3-02. Line drawing for Chicxulub deep seismic reflection profile A/A1, converted to depth using the velocity information from wide-angle OBS data (from Christeson et al., 2001, fig. 9). [Journal of Geophysical Research, v. 106, p. 21,751–21,769. Reproduced by permission of American Geophysical Union.]

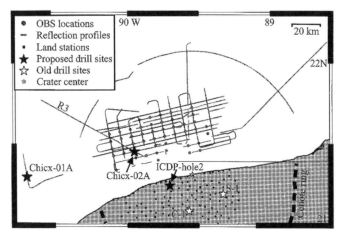

Figure 9.7.3-03. Location of the 2005 Chicxulub seismic experiment (from Morgan et al., 2005, fig. 1). [Eos (Transactions, American Geophysical Union), v. 86, no. 36, p. 325, 328. Reproduced by permission of American Geophysical Union.]

Peninsula. Using 22 ocean-bottom hydrophones and seismographs and a number of land stations as well, a dip profile and three strike profiles were recorded (Christeson et al., 1999). The land recorders extended the dip line for ~50–60 km onto the Nicoya Peninsula. The airgun profiles were carried out by R/V *Maurice Ewing* with a 20-gun, 138-L array and was recorded on a 100-m two-channel streamer. The dip line was shot twice, at both 50 m and 125 m shot spacing. The strike line closest to the shore, running approximately along the 1500 m bathymetric depth contour, was shot with 50 m shot spacing; the other two strike lines, running approximately above the 3000 m and 3500 m water depth contour lines, were shot with 125 m spacing.

The joint onshore-offshore data of the 1995 survey were used to construct a crustal structure model of the convergent margin from 20 km seaward of the Middle American Trench onto the Nicoya Peninsula (Christeson et al., 1999). At 60–80 km off the coastline the Moho was found at 11 km depth below sea level and then dipped gradually to ~25 km depth at ~10 km inland from the coastline. The top of the subducted slab was well resolved, it deepens from 5 km depth at the trench to 15–16 km depth at the Nicoya Peninsula coastline, the dip angle of the subducting plate increasing from 6° to 13° at a distance of ~30 km from the trench.

The project was continued in 1996 as part of the COTCOR (Comparative Transactions in Costa Rica) international collaboration project of GEOMAR (Kiel, Germany), ICE (San José, Costa Rica) and ICTJA-CSIC (Barcelona, Spain). Borehole shots with charges from 550 to 1230 kg TNT were fired twice at five locations and recorded by 60 portable stations, occupying alternatively one half of the 170-km-long profile, resulting in 120 recordings. The 1996 profile was located coincident with the 1995 onshore profile (Sallares et al., 1999). The resulting crustal model was combined with the 1995 results to a 250-km-long crustal section, from the outer rise to the back-arc basins. The oceanic Moho reaches 10 km depth west of the Middle American Trench, deepening progressively toward the isthmus to reach an angle of seduction of 35° down to 40 km depth. The obtained onshore model is characterized by significant changes in the surface velocity (2.0–5.0 km/s), an upper crust of 4 km with velocities of 5.3–5.7 km/s, a mid-crust with a mean thickness of 13 km and velocity variations of 6.2–6.5 km/s and an up to 20-km-thick lower crust with velocities between 6.9 and 7.3 km/s. The Moho is reached at 38–40 km depth.

9.7.3.2. Northeastern South America

In Venezuela, two deep wide-angle seismic projects were undertaken to investigate the crust and upper-mantle structure of the Guayana Shield and the Oriental Basin to the north.

The field campaign ECOGUAY (Estudios de la Estructura Cortical del Escudo de Guayana) in 1998 targeting the Guayana Shield comprised nine seismic-refraction lines of up to 320 km distance (Fig. 9.7.3-05), using blasts of iron mines, gold mines, and a quarry with ripple firing as energy sources. The profiles were partly reversed and were obtained by choice scrolling the available 13 seismic stations in up to four deployments for each shotpoint. The spacing of recording points varied between 5 and 10 km (Schmitz et al., 2002).

In 2001, a second experiment followed, this time targeting the Oriental Basin north of the Guayana Shield. The study area of ECCO (Estudio Cortical de la Cuenca Oriental) comprised a region from the Coastal Cordillera and Interior Ranges in the north, crossing the entire Oriental Basin (eastern Venezuela basin) up to the Guayana Shield in the south (Fig. 9.7.3-06). Also this project used mine blasts and borehole blasts as energy sources.

The north-south line (squares in Fig. 9.7.3-06) was 300 km long and ran from the Caribbean Sea across the Oriental Basin

Figure 9.7.3-04. Cartoon of the Chicxulub crater based on marine seismic reflection, offshore-onshore refraction, and drill hole data (from Morgan et al., 2005, fig. 3). [Eos (Transactions, American Geophysical Union), v. 86, no. 36, p. 325, 328. Reproduced by permission of American Geophysical Union.]

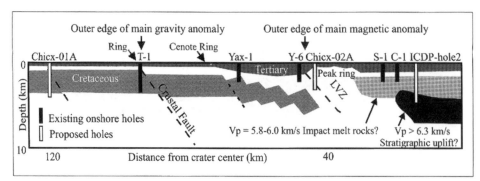

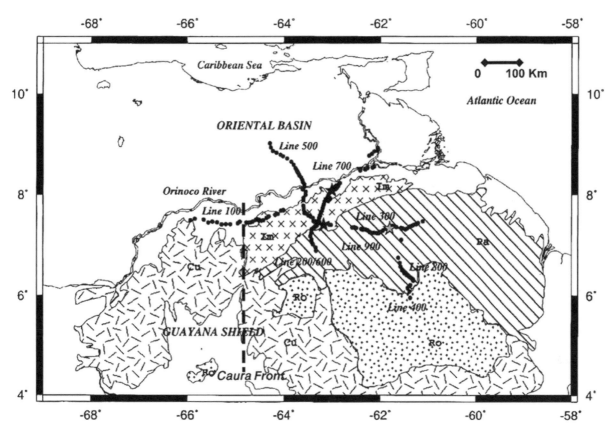

Figure 9.7.3-05. Location of the 1998 seismic experiment investigating the Guayana Shield in Venezuela (from Schmitz et al., 2002, fig. 1). [Tectonophysics, v. 345, p. 103–118. Copyright Elsevier.]

into the Guayana Shield. It comprised five shotpoints, spaced 50–100 km apart, with charges of 150–500 kg, which were recorded by 190 one-component stations (Texans) with a spacing between 1.5 and 2 km (Schmitz et al., 2005). Two other profiles of this project are shown by triangles and circles in Figure 9.7.3-06.

The interpretation of the 1998 data revealed a subdivision of the crust into a 20 km upper crust with velocities of 6.0–6.3 km/s, and a lower crust with velocities of 6.5–7.2 km/s below 20–25 km depth. The total crustal thickness of the Guayana Shield varied from west to east and ranged from 46 km in the west (Archean segment) to 43 km in the east (Proterozoic segment). The average crustal velocity was ~6.5 km/s (Schmitz et al., 2002).

Toward the north, underneath the Oriental Basin, a sedimentary thickness of up to 13 km was derived, while the total crustal thickness decreased from 45 km beneath the Guayana Shield to 39 km at the Orinoco River, and 36 km close to El Tigre in the center of the Oriental Basin (Fig. 9.7.3-07). The average crustal velocity decreased in the same direction from 6.5 to 5.95 km/s.

Starting in 1997, combined geological and geophysical studies aimed to investigate the deep structure of central Brazil. For this reason a seismic-refraction survey was planned. The project started with a trial experiment in late 1997 (Berrocal et al., 2004). It consisted of 20 digital recorders, deployed along a 260-km-long, N-S–oriented profile with two seismic sources with 800 kg

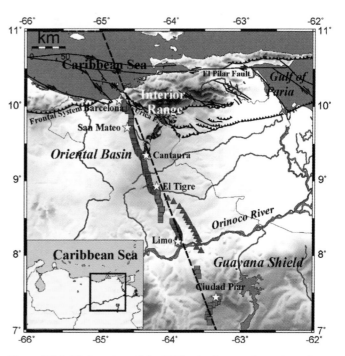

Figure 9.7.3-06. Location of the 2001 seismic experiment investigating the Oriental Basin in Venezuela (from Schmitz et al., 2005, fig. 1). [Tectonophysics, v. 399, p. 109–124. Copyright Elsevier.]

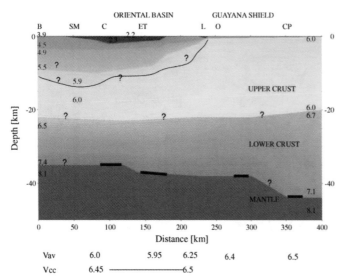

Figure 9.7.3-07. Crustal structure along a N-S section from the Oriental Basin to the Guayana Shield in Venezuela (from Schmitz et al., 2005, fig. 10). [Tectonophysics, v. 399, p. 109–124. Copyright Elsevier.]

charges at each end of the profile. Some of the stations were permanent seismographic stations of the Serra de Mesa network and Brasilia array. This experimental line served mainly to define the shot sizes for the main seismic-refraction lines. Following the successful test experiment, three main seismic profiles were recorded in central Brazil (L1, L2, L3 in Fig. 9.7.3-08).

The profiles L1 and L2, located in the central section of Tocantins Province, were each ~300 km long and were aligned in WNW-ESE direction. On each profile, 120 stations were deployed at 2.5 km spacing, and explosive sources were detonated in boreholes, spaced 50 km apart on each line. The charges used varied from 1000 kg at the end of the lines to 500 kg for the intermediate shots. The two lines overlapped each other by 50 km. Line L3 was located in the southeastern sector of Tocantins Province and ran in NE direction. It started in the SW in the Parana basin, crossed the Brasilia fold belt and ended on the Sao Francisco craton. On this line, only shots from the ends of the line and some middle shots were successful.

The interpreted model for lines L1 and L2 (Fig. 9.7.3-09) showed a crustal thickening from west to east, where the depth to the Moho varied from 32 to 43 km. The mean crustal velocity was around 6.3 km/s. For line L3 on the southeastern Tocantins Province a thick two-layer crust of 41–42 km was determined both under the Parana basin and the Sao Francisco craton, but a thinner crust of ~38 km was found under the Brasilia fold belt (Berrocal et al., 2004).

9.7.3.3. Southern Andean Region of South America

The successful research work of the geoscientific interdisciplinary research group "Mobility of active continental margins" of the Free University of Berlin, Germany, under the leadership of Peter Giese from 1982 to 1989 was only the beginning of a large interdisciplinary research of the Andes and the adjacent

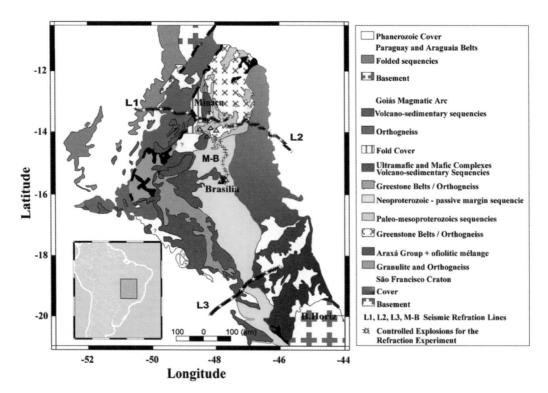

Figure 9.7.3-08. Geological map of Tocantins Province, central Brazil, with location of seismic-refraction lines (from Berrocal et al., 2004, fig. 1). [Tectonophysics, v. 388, p. 187–199. Copyright Elsevier.].

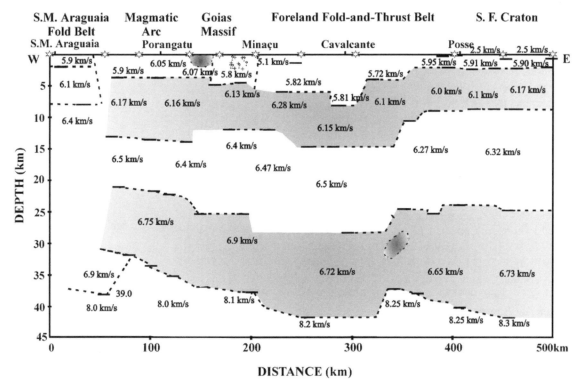

Figure 9.7.3-09. Seismic model of the central sector of Tocantins Province, central Brazil (from Berrocal et al., 2004, fig. 6). [Tectonophysics, v. 388, p. 187–199. Copyright Elsevier.].

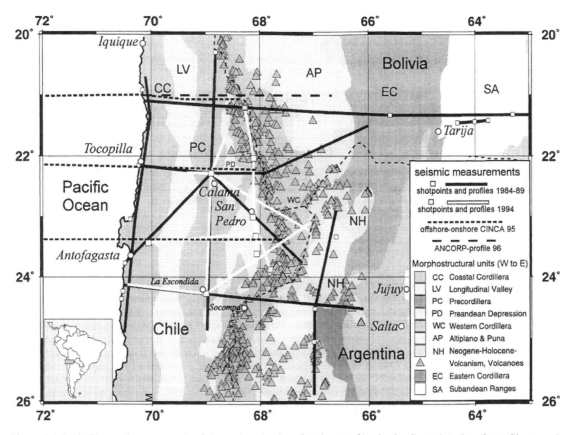

Figure 9.7.3-10. The onshore network of the main seismic refraction profiles in the Central Andes (from Giese et al., 1999, fig. 1). [Journal of South American Earth Science, 12, p. 201–220. Copyright Elsevier.]

Pacific Ocean in the 1990s which went on into the first years of the twenty-first century. The core of this research was formed by the foundation of the Collaborative Research Center 267 (CRC 267) "Deformation Processes in the Andes" at the Free University of Berlin and the University of Potsdam (Germany), which was funded by the German Research Society for 15 years. Soon after its foundation, the GeoScience Center at Potsdam, Germany, became a partner in this project and as such increased its research power. The new research center not only enabled extended seismic research, but involved all geoscientific branches available at the three research institutions and included furthermore a strong support from various South American research facilities.

In the years 1994–1996, three major seismic investigations were performed in northern Chile and adjacent parts of Bolivia and Argentina (Fig. 9.7.3-10): the project PISCO 94 (Proyecto de Investigaciones Sismologicas de la Cordillera Occidental 1994), the project CINCA 95 (Crustal Investigations off- and onshore Nazca plate/Central Andes 1995), and the project ANCORP 96 (Andean Continental Research Project 1996) after a successful test project PRECORP in 1995.

The project PISCO 94 (white lines in Fig. 9.7.3-10) was the first seismic campaign in the 1990s (Schmitz et al., 1999; Lessel 1998). The main profile of 320 km length extended in a N-S direction from shotpoints OLL to VAR (Fig. 9.7.3-11), approximately along the longitude 68°W, along the western border of the recent magmatic arc, the western Cordillera. In its southern part it crossed the Pre-Andean depression (Salar de Atacama). The station elevations ranged from 2300 m to 5000 m above NN; their spacing was on average 2.5 km, accomplished by 4 deployments recording shots from 5 shotpoints along the line. In the southern half, along the Salar de Atacama, with a sedimentary sequence of up to 8 km thickness, an additional N-S line with an additional shotpoint at its southern end was arranged where a station spacing of 1 km was achieved. In addition, the seismic profiles of the 1980s radiating from the mine Chucicamata (CHU in Fig. 9.7.3-11) were partly reoccupied, additional shotpoints were arranged, and a seismological network consisting of 31 stations distributed in an area of 200 × 200 km was in operation for 100 days recording more than 5300 local seismic events and also all the shots (Graeber and Asch, 1999). In total, 33 shots were recorded with charges ranging from 50 kg to 1300 kg, fired in patterns of up to 20 holes with depths from 10 to 50 m. The shotpoints were either located in salt lakes or in sedimentary basins. Data examples as published by Schmitz et al. (1999) are shown in Figure 9.7.3-12 and Appendix A9-5-1.

In 1995, a second project, CINCA 95, followed. It was a combined land-sea experiment (dashed lines in Fig. 9.7.3-10 and white lines in Fig. 9.7.3-14) and was jointly organized by the German Geological Survey (BGR Hannover), the German research institution GEOMAR (Kiel) and the CRC 267, the joint interdisciplinary research group of the Free University at Berlin and the GeoScience Center at Potsdam (Patzwahl, 1998; Patzwahl et al., 1999; Appendix A9-5-1). It comprised three west-east running lines crossing the Peru-Chile Trench offshore and

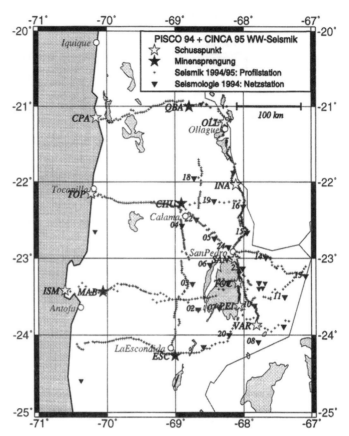

Figure 9.7.3-11. Location of the seismic refraction profiles of the projects PISCO'94 and CINCA'95 (stars—shotpoints, dots—recording sites) in the Central Andes onshore and offshore and the location of the seismological network PISCO'94 (triangles) in the Central Andes (from Lessel, 1998, fig. 4.3). [Berliner Geowissenschaftliche Abhandlungen, Reihe B, Band 31: 185 p. Published by permission of Institut für Geologische Wissenschaften, Freie Universität Berlin.]

reached to the magmatic arc of the western Cordillera in northern Chile. The BGR offshore reflection lines are shown as white lines in Figure 9.7.3-14.

Furthermore, in 1995, the PRECORP seismic-reflection experiment (Fig. 9.7.3-13) was carried out as a feasibility study for a longer deep seismic-reflection survey across the Central Andes. The 50-km-long profile was located at 22.5°S south of Calama, with shot and receiver locations extending in an E-W direction. Thirty-eight shot gathers were recorded using a maximum of 144 receivers with a spacing of 100 m. The maximum offset was 14.3 km. The shotpoint interval of the split-spread recording geometry was 2.4 km, providing twofold coverage (Yoon et al., 2003).

Its successful operation allowed a full-scale long seismic-reflection experiment of ~400 km length, the project ANCORP 96 (ANCORP Working Group, 1999, 2003). Its realization was enabled by a successful cooperation between the CRC 267 of Berlin-Potsdam, Germany and university and governmental research institutions of Chile and Bolivia. As shown in Figure 9.7.3-14, it continued the northern E-W offshore line of CINCA 95 at 21°S and extended across the Coastal Cordillera, the Longi-

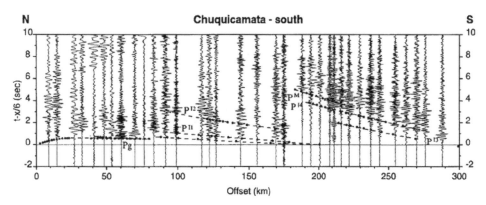

Figure 9.7.3-12. Record section of the profile Chuquicamata-S in the Precordillera (from Schmitz et al., 1999, fig. 6a). P_g, P^{I1}, P^{I2} are phases from the upper and middle crust, P^M branch is interpreted as reflection from the recent geophysical Moho. [Journal of South American Earth Science, 12, p. 237–260. Copyright Elsevier.]

tudinal Valley, the Precordillera, the western Cordillera, and the Altiplano (Figs. 9.7.3-13 and 9.7.3-14).

Recording was achieved by 56 independently operating 3- and 6-channel PDAS stations with strings of vertical geophones of 4.5 Hz natural frequency, resulting in 252 recording channels. With a spacing of 100 m a spread length of 25 km was achieved. During the roll-along operation, the back 6.25 km of the spread was removed daily and added to the front end.

The aim of the project was to provide an image of the deep lithosphere. Near-vertical profiling was performed with a relatively wide shot spacing of 6 km. Each shot was fired twice off-end into the 252 recording channels at 100 m spacing, resulting in a 50-km-long split spread when the spread had completely moved. The resulting subsurface coverage was fourfold. Shots of 90 kg were fired in 20 m deep boreholes, except for 5 shots in Bolivia that had a charge size of 300 kg. Figure 9.7.3-15 shows the automatic line drawing of the migrated ANCORP 96 reflection line data (see also Appendix A9-5-1).

From 10 locations along the ANCORP transect, repeated shots were fired into the 25-km-long geophone line to achieve seismic observations at large offsets. These shots for wide-angle recording, which were detonated each time, after the recording spread had obtained a full new position, were repeated up to 9 times with increasing charges of up to 450 kg and served in particular for velocity control. One of these wide-angle shotpoints was 6 km offshore, operated by the Chilean Navy, and one was at a mine using regular quarry blasts. The resulting 70 shot sections of 25 km length each reached maximum observation distances between 120 and 230 km.

The combined interpretation of the processed near-vertical incidence reflection data, the wide-angle refraction data, and the data of a subsequent seismological monitoring of local and teleseismic earthquakes by a widespread 30-station array for 3 months continuously allowed workers to derive a rather detailed cross section (Fig. 9.7.3-16a) and its geodynamic interpretation (Fig. 9.7.3-16b).

In an overview, Giese et al. (1999) summarized the seismic, seismological and other geophysical investigations and compiled a representative crustal and uppermost-mantle model for the Andean region of Chile and adjacent areas at latitude 21–22°S

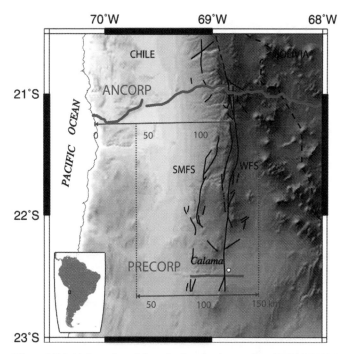

Figure 9.7.3-13. Location of the seismic reflection profiles PRECORP'95 and ANCORP'96 (from Yoon et al., 2003, fig. 1). [Geophysical Research Letters, v. 30: no. 4, 1160, doi: 10.1029/2002GL015848. Reproduced by permission of American Geophysical Union.]

(Fig. 9.7.3-17). In total, the cross section covered an area from the Pacific Ocean to the Brazilian craton, a distance range of more than 800 km. In the forearc and in the eastern backarc, the geophysical Moho was well defined by seismic-refraction data. Beneath the western Cordillera and the western Altiplano, the depth to the Moho transition zone could be derived from P to S converted waves (Giese et al., 1999; Scheuber and Giese, 1999).

Besides northern Chile, central Chile became the target of an expedition of the German *Sonne*. Also in 1995, within the multidisciplinary CONDOR (Chilean Offshore Natural Disaster and Ocean Environmental Research) project, an amphibic operation (Fig. 9.7.3-18) investigated the Valparaiso Basin offshore Valparaiso, central Chile, along two marine-seismic reflection and

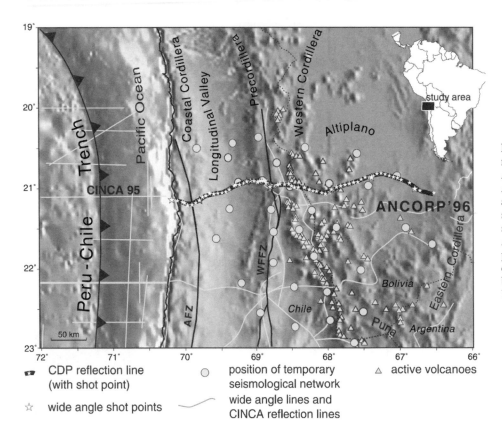

Figure 9.7.3-14. Location of the seismic reflection line ANCORP'96 (from ANCORP Working Group, 2003, fig. 1). Also shown are the CINCA'95 wide-angle and reflection lines onshore and offshore (white lines) and the ANCORP'96 seismological network in the Central Andes (white circles). [Journal of Geophysical Research, v. 108, B7, 2328, doi: 10.1029/2002JB001771. Reproduced by permission of American Geophysical Union.]

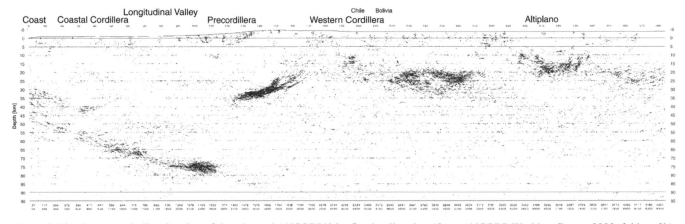

Figure 9.7.3-15. Automatic line drawing of the migrated ANCORP'96 reflection line data (from ANCORP Working Group, 2003, foldout 2b). Only laterally coherent phases were enhanced. [Journal of Geophysical Research, v. 108, B7, 2328, doi: 10.1029/2002JB001771. Reproduced by permission of American Geophysical Union.]

refraction profiles north and south of the latitude 33°S (Flueh et al., 1998a). The project was operated by the BGR (German Geological Survey, Hannover) and GEOMAR (Kiel).

The profiles crossed the continental margin and the Chile Trench, where thick trench sediments had been observed. Profile 1 crossed the rupture area of the 1985 Chile earthquake. Multichannel seismic-reflection data were obtained by a 3-km-long streamer recording airgun shots. Wide-angle measurements were performed by OBH recording devices (digital, high data ca-

pacity ocean-bottom recorders) and by land stations prolonging the 60–100-km-long marine profiles up to 200 km on land (data examples in Fig. 9.7.3-19 and Appendix A9-5-1). The crustal velocity models (Fig. 9.7.3-20) showed that the continental upper-plate crust extended to the middle-lower slope boundary and that the crustal structure of the oceanic plate appeared rather similar on both profiles (Flueh et al., 1998a).

In 2000, the Collaborative Research Center SFB 267 at Berlin and Potsdam, Germany, started a detailed research project in

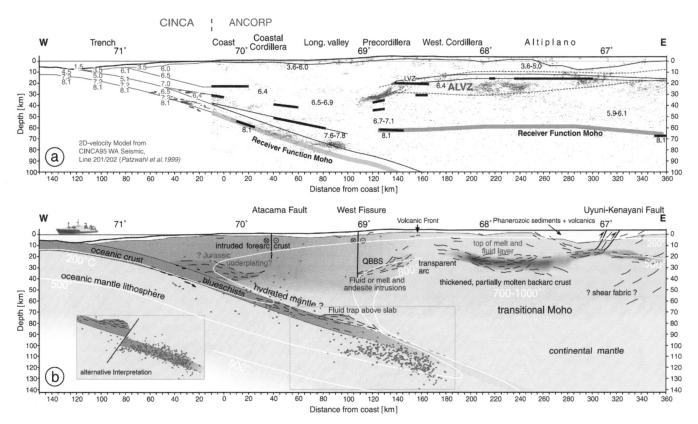

Figure 9.7.3-16. Seismic results and interpretation of CINCA'95 and ANCORP'96 along a cross section at 21°S (from ANCORP Working Group, 2003, fig. 7). (a) Automatic line drawing of depth-migrated ANCORP reflection data, including onshore wide-angle and receiver function results and merged with results from the offshore CINCA experiment and its onshore recordings. (b) Suggested geodynamic interpretation of the seismic results. [Journal of Geophysical Research, v. 108, B7, 2328, doi: 10.1029/2002JB001771. Reproduced by permission of American Geophysical Union.]

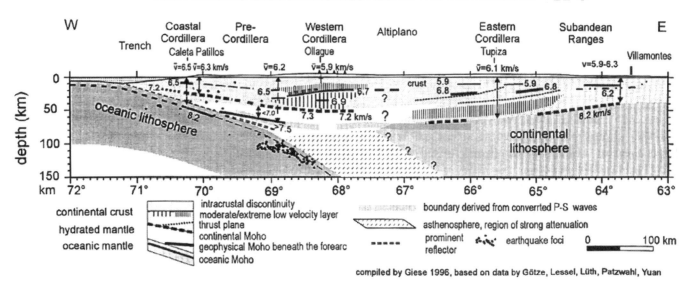

Figure 9.7.3-17. West-east cross section along 21°S through the Andean lithosphere (from Giese et al., 1999, fig. 2). The crustal structure is based on active and passive seismic measurements. [Journal of South American Earth Science, 12, p. 201–220. Copyright Elsevier.]

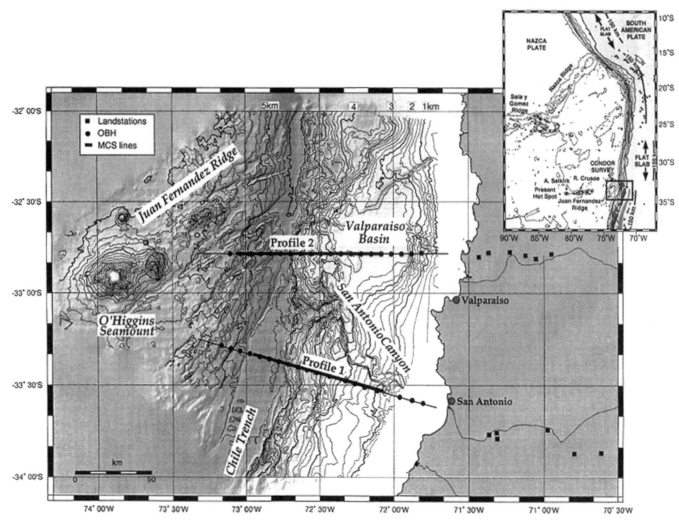

Figure 9.7.3-18. West-east offshore-onshore profiles near 33°S in the CONDOR study area in central Chile (from Flueh et al., 1998a, fig. 1). [Tectonophysics, v. 288, p. 251–263. Copyright Elsevier.]

southern Chile. The first component was the project ISSA 2000 (Integrated Seismological experiment in the Southern Andes) which consisted of a temporary seismological network and a seismic-refraction profile (Fig. 9.7.3-21). Before and after the installation of this network across the Southern Andes around 38°S, which recorded 333 local events over a period of 3 months, all 62 land stations were deployed along a profile at 39°S and five chemical explosions were recorded, one from a small lake in the Chilean Main Cordillera and four in the Pacific Ocean (Bohm et al., 2002; Lueth et al., 2003). Data examples are shown in Figure 9.7.3-22 and Appendix A9-5-1.

A typical continental Moho was not clearly observed in this region, crustal thickness was first inferred from Bouguer gravity to be ~40 km (Fig. 9.7.3-23). The oceanic Moho was observed down to ~45 km depth under the continental fore-arc.

This project was followed by the project SPOC (Subduction Processes off Chile) in 2001 (Krawczyk and the SPOC

Team, 2003, 2006). It consisted of three components. The first one was a shipborne geophysical experiment with the R/V *Sonne* in fall and winter 2001/2002 and was operated by the BGR (German Geological Survey, Hannover), GEOMAR (Kiel) and the SFB 267 (Berlin and Potsdam). The second and third components were predominantly land-based onshore-offshore experiments.

The second component was a 54-km-long near-vertical incidence reflection line (NVR) which was recorded in three spreads and recorded 14 shots fired at 10 locations. Furthermore, the airgun shots along the offshore profile SO161-38/42 (Fig. 9.7.3-24) were recorded along the first 18 km. The third component was a wide-angle seismic experiment where simultaneously the airgun pulses from R/V *Sonne* were recorded by 30 three-component stations deployed in an array, and 52 stations along three consecutive W-E profiles and 22 OBH/OBS stations along the offshore part of the 38°15′S line. Besides the airgun shots, 8 chemical

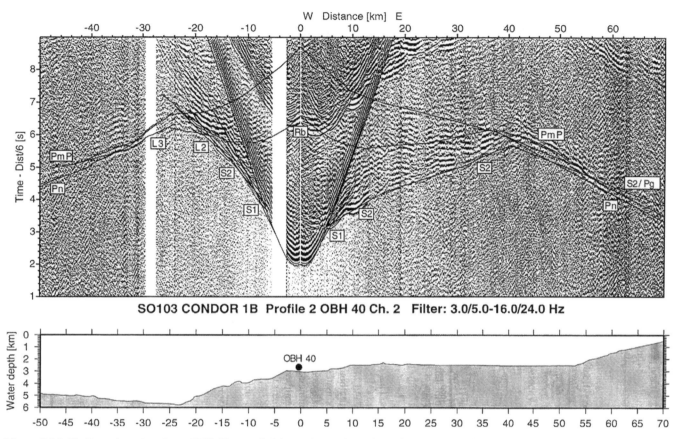

Figure 9.7.3-19. Record section from OBH 40 recorded from airgun shots along the west-east offshore-onshore profile 2 near 33°S in the CONDOR study area in central Chile (from Flueh et al., 1998a, fig. 5). [Tectonophysics, v. 288, p. 251–263. Copyright Elsevier.]

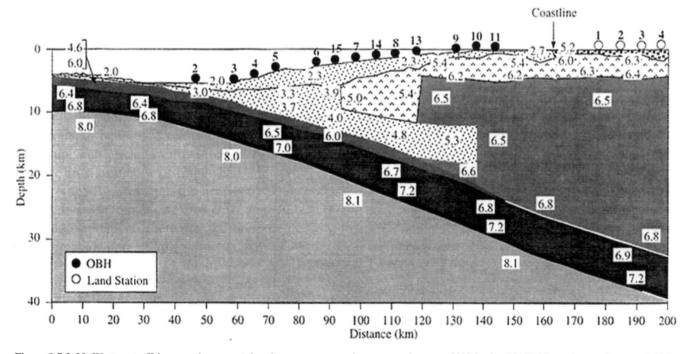

Figure 9.7.3-20. West-east offshore-onshore crustal and uppermost-mantle cross section near 33°S in the CONDOR study area in central Chile (from Flueh et al., 1998a, fig. 2, lower part). [Tectonophysics, v. 288, p. 251–263. Copyright Elsevier.]

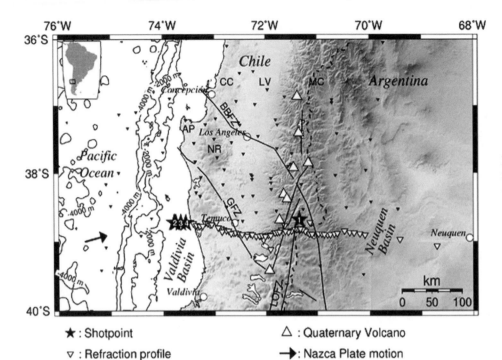

Figure 9.7.3-21. ISSA 2000 seismic measurements in southern Chile and Argentina (from Lueth et al., 2003, fig. 1). [Revista Geologica de Chile, v. 30, no. 1, p. 83–101. Reproduced with permission of Servicio Nacional de Geologia y Mineria, Chile, Andean Geology.]

★ : Shotpoint

▽ : Refraction profile

• : Seismological network

△ : Quaternary Volcano

➡ : Nazca Plate motion

— : Fault Zone

shots were fired, mainly at the ends of the three profiles (Fig. 9.7.3-24 and Fig. 9.7.3-25).

The whole array also recorded teleseismic, regional and local events. Furthermore, broadband stations were deployed along the ISSA 2000 line. The most comprehensive data set of the SPOC onshore experiment came from the southern profile along 38.25°S (Sick et al., 2006). Here, a wide-angle OBS/OBH profile was acquired offshore. Simultaneously, the airgun pulses were recorded by onshore receivers.

The near-vertical reflection seismic land experiment consisted of a 54-km-long receiver set, with three 18-km-long deployments of 180 geophone groups at 100 m spacing. In total, a series of 2 offshore and 12 onshore chemical shots, fired at 10 different locations, was recorded. Thus, a 72-km-long common-depth point line resulted, with twofold coverage in the innermost 45 km and singlefold coverage at the margins, covering the offshore-onshore transition along the same transect (Krawczyk et al., 2006).

The images of the near-vertical profile provided a complex reflection pattern with several distinct reflectors, but did not reveal structures below 35 km depth. The whole data set covered the crustal and upper-mantle structure from ~200 km offshore to ~240 km onshore (Fig. 9.7.3-26). On the SPOC data, the oceanic Moho was observed down to ~40 km depth under the Coastal Cordillera, and the continental crust from the trench to the Main Cordillera. The continental crust was characterized by gradually increasing P-wave velocities from the trench to the Main Cordillera, whereby the upper crust was ~10–15 km thick and the lower crust ~25 km thick (Krawczyk et al., 2006; Sick et al.,

2006). The continental Moho was imaged only along the northern SPOC profile. The combined results of ISSA 2000 and SPOC 2001 and teleseismic observations resulted in a Moho depth of 40 km beneath the Main Cordillera shallowing toward the east to ~35 km depth beneath the Neuquen basin.

The latest experiment in southern Chile was the TIPTEQ (from The Incoming Plate to Mega-Thrust EarthQuake Processes) array project of 2004 and 2005, which is described in Chapter 10.6.3.

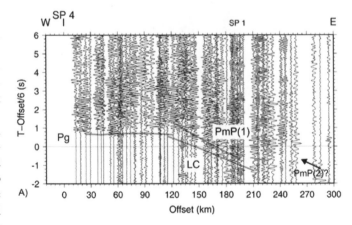

Figure 9.7.3-22. Record section of the ISSA 2000 seismic measurements in southern Chile and Argentina (from Lueth et al., 2003, appendix 1, fig. 4). [Revista Geologica de Chile, v. 30, no. 1, p. 83–101. Reproduced with permission of Servicio Nacional de Geologia y Mineria, Chile, Andean Geology.]

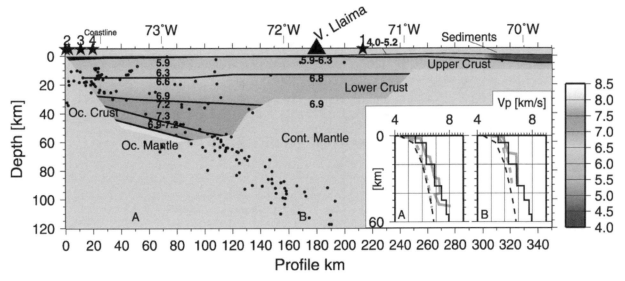

Figure 9.7.3-23. West-east onshore crustal and uppermost-mantle cross section near 39°S of the seismic-refraction line of the project ISSA 2000 in southern Chile (from Bohm et al., 2002, fig. 7). The thin black lines of the inset shows the detailed velocity-depth structure at the positions A and B of the cross section. Black dots indicate hypocenter locations of earthquakes observed by the ISSA seismological network. [Tectonophysics, v. 356, p. 275–289. Copyright Elsevier.].

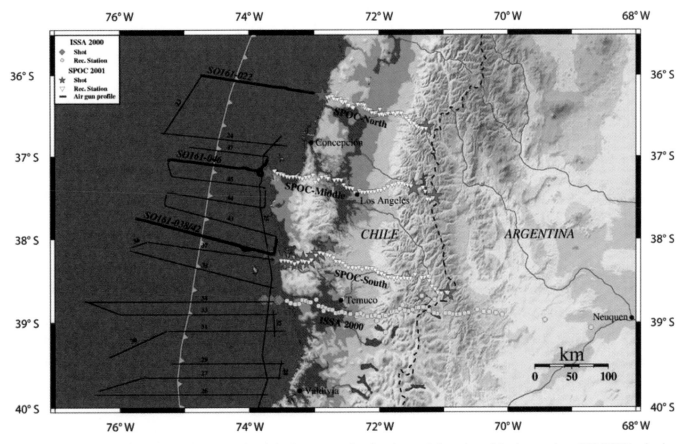

Figure 9.7.3-24. Map of SPOC 2001. White triangles and red stars—receiver locations and shotpoints of the three onshore SPOC 2001 seismic-refraction lines at latitudes 36°, 37°, and 38°S. The near-vertical reflection experiment covered the offshore-onshore transition at 38.25°S. Black lines—offshore airgun profiles. Grey circles and gray diamonds—receiver and shot locations of ISSA 2000 at 39°S. (from Sick et al., 2006, fig. 7–13). [Oncken et al., 2006, The Andes: Berlin-Heidelberg-New York, Springer, p. 147–169. Reproduced with kind permission of Springer Science+Business Media.].

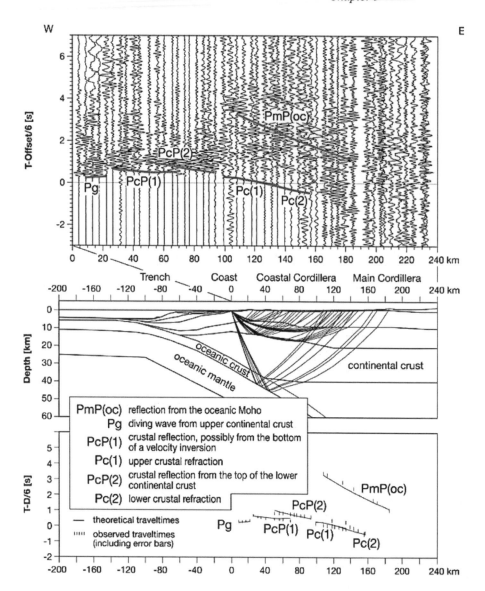

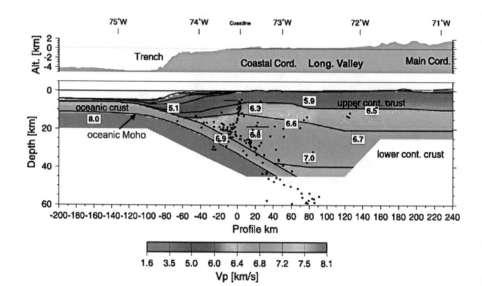

Figure 9.7.3-25. Seismic data, ray paths and traveltime curves for the western shotpoint of the south profile of the SPOC onshore survey (from Krawczyk et al., 2006, fig. 8-5). [Oncken et al., 2006, The Andes: Berlin-Heidelberg-New York, Springer, p. 171–192. Reproduced with kind permission of Springer Science+Business Media.]

Figure 9.7.3-26. P wave model along the SPOC 2001 south profile (from Sick et al., 2006, fig. 7-15). [Oncken et al., 2006, The Andes: Berlin-Heidelberg-New York, Springer, p. 147–169. Reproduced with kind permission of Springer Science+Business Media.].

9.7.4. Antarctica

In 1990 and 1991, Antarctica again saw a Polish research investigation, in cooperation with Japanese scientists (Grad et al., 1997; Janik, 1997). While the previous surveys were primarily based on shot profiles and isolated land stations and had covered the whole northwestern margin of the Antarctic Peninsula, this new survey added a new profile along the axis of the Bransfield Strait (DSS-20) between the northernmost tip of the Antarctic Peninsula and the South Shetland Islands and involved OBSs (Fig. 9.7.4-01). The OBSs were supplied by Hokkaido University. They were equipped with three components and their operation depth was up to 6000 m.

In total, four expeditions were organized (Guterch et al., 1998; Janik, 1997). Some of them covered in particular the area of the Bransfield Strait between the South Shetland Islands and the Trinity Peninsula. Some of the lines were observed twice. The measurements of 1979 and 1980 were repeated in 1990 and 1991, e.g., along profiles DSS-1, -3 and -4, and other profiles were added in 1990 and 1991, when the OBSs became available. A data example from profile DSS-6 is shown in Figure 9.7.4-02 (for more data, see Appendix A9-5-2).

While the crust along the margin of the Antarctic Peninsula possesses a normal continental crustal thickness, the internal structure is quite complicated (Fig. 9.7.4-03). Only beyond the South Shetland Islands toward the northwest, the crust gradually changes into oceanic crust (Fig. 9.7.4-04). The interpretation of the whole Polish West Antarctica DSS data, obtained between 1979 and 1991, was summarized in Figure 9.7.4-04, illustrating the Moho depths along the various profiles. A recent reinterpretation was published by Janik et al. (2006; Appendix A9-5-2).

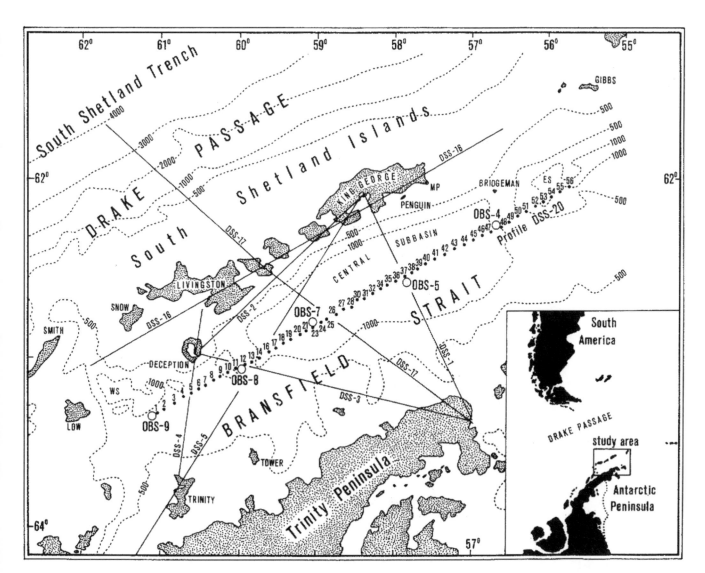

Figure 9.7.4-01. Location of deep seismic sounding lines (DSS) in the area of the Bransfield Strait, Antarctica (from Janik, 1997, fig. 11). [Polish Polar Research, v. 18, p. 171–225. Reproduced with permission of Polish Academy of Sciences, Warsaw, Poland.]

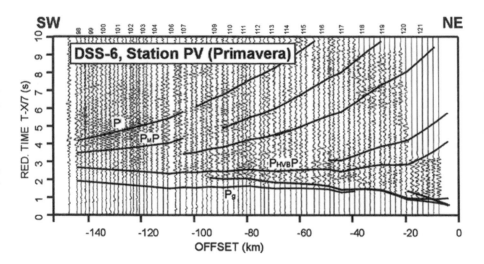

Figure 9.7.4-02. Record section of DSS-6 profile, Station PV, recorded along the Bransfield Strait, Antarctica (from Janik et al., 2006, fig. 5.2-2). [*In* Fütterer, D.K., Damaschke, D., Kleinschmidt, G., Miller, H., and Tessensohn, F., eds., Antarctica: contributions to global earth sciences: Berlin-Heidelberg-New York, Springer, p. 229–236. Reproduced with kind permission of Springer Science+Business Media.]

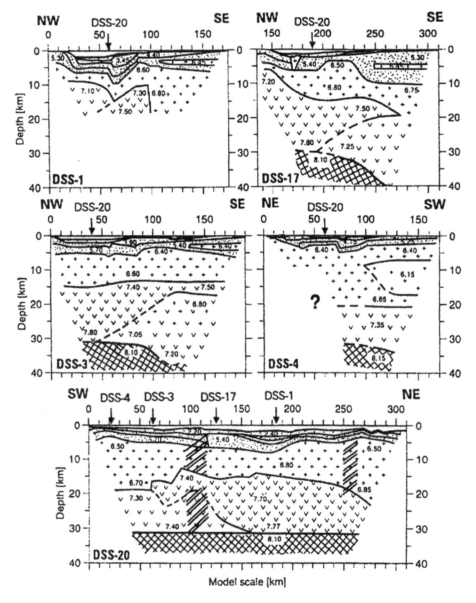

Figure 9.7.4-03. Crustal cross sections of DSS-1, -3, -4, -17 and -20 profiles through the Bransfield Strait, Antarctica (from Janik, 1997, fig. 31). [Polish Polar Research, v. 18, p. 171–225. Reproduced with permission of Polish Academy of Sciences, Warsaw, Poland.]

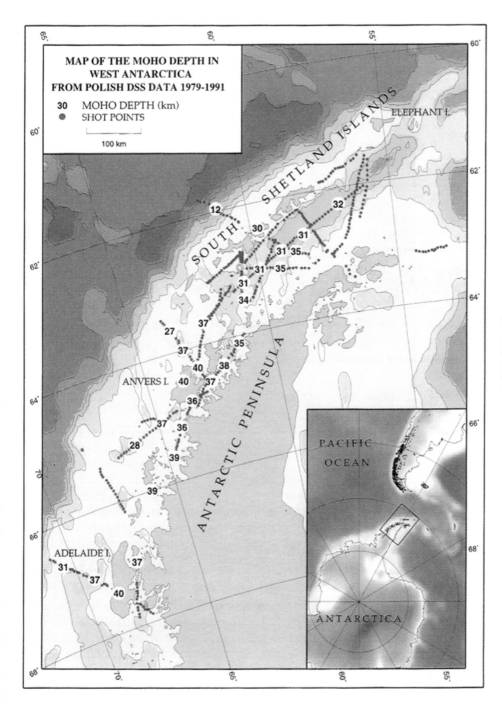

Figure 9.7.4-04. Moho depths in West-Antarctica (from Guterch et al., 1998, fig. 7). [Polish Polar Research, v. 19, p. 113–123. Reproduced with permission of Polish Academy of Sciences, Warsaw, Poland.]

SERIS (Seismic Experiment Ross Ice Shelf) was a large-scale modern multichannel seismic-reflection and -refraction experiment in the south summer of 1990–1991 which was carried out over the snow in Antarctica and aimed to examine the geometry of the boundary of the West Antarctica rift system, which includes the Ross Embayment and the East Antarctica shield as well as the causes for the uplift of the 3500-km-long and 4500-m-high Transantarctic Mountains (ten Brink et al., 1993). As access of East Antarctica from the Ross Ice Shelf, the Robb Glacier (about halfway between the McMurdo station and the South Pole) was

chosen and a 141-km-long line was laid out across the Ross Ice Shelf over the Robb Glacier into the Transantarctic Mountains, from ~171°E, 82°S to 166°E, 83°S. Two types of seismic data were collected along the traverse. Multichannel seismic-reflection data were recorded with a 1.5-km-long streamer, which was a towed seismic cable to which 60 groups of gimbaled geophones were connected at 25 m intervals. Wide-angle seismic-reflection and -refraction data were recorded up to 90 km maximum distance. The energy was provided by dynamite charges, shot in 17 m deep holes. In this way, images of the subsurface layers and

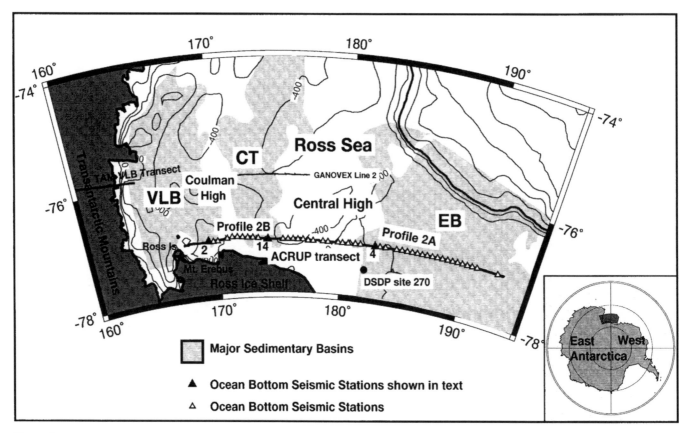

Figure 9.7.4-05. Location of the ACRUP transect in the Ross Sea, West Antarctica (from Trey et al., 1999, fig. 1). [Tectonophysics, v. 301, p. 61–74. Copyright Elsevier.].

images of the deep crust and velocities were provided. Because of logistical constraints, the wide-angle seismic part of the profile was split into a partially reversed 90-km-long profile within the Ross Embayment and a fully reversed 51-km-long part within the Transantarctic Mountains.

Underneath the ice shelf, an ~1-km-thick sedimentary layer was seen to overlie a low-grade basement or high-velocity sedimentary layer. A strong mid-crustal reflector was seen at 15 km depth. The crustal structure under the mountains was found to differ from that under the Ross Ice Shelf. Under the ice, no sedimentary layer was noticed and the upper crust appeared with a high velocity gradient. The midcrustal reflection could hardly be seen, despite recording good-quality Moho arrivals. Lower crustal velocities appeared to be typical for continental crust and lower than under the Ross Ice Shelf. The seismically determined Moho was found at similar depth under the Ross Ice Shelf as under the Ross Sea, dipping smoothly from the Ross Embayment toward the Transantarctic Mountains. Entering the mountains, where also the base of the crust dips from 15° to 20°, the crust gradually thickens from 22 to 30–35 km.

During the Australian summer of 1993–1994, the University of Hamburg, Germany, the USGS, and the Osservatorio Geofisico Sperimentale of Trieste, Italy, undertook the ACRUP (Antarctica Crustal Profile) project, during which a 670-km-

long transect was recorded across the southern Ross Sea (Fig. 9.7.4-05) to study the velocity and density structure of the crust and uppermost mantle of the West Antarctic rift system. Forty-two OBSs, consisting of 34 analog three-component systems and 8 digital one-component systems (single vertical seismometer plus hydrophone), recorded strong airgun signals to distances of 120 km, which were generated by the Italian research vessel *Explora* at ~250 m intervals along the entire profile. The shots were also recorded by a 1200-m-long, 48-channel streamer to provide near-vertical incidence reflection data for mapping sedimentary sequences.

The interpretation of the data (Trey et al., 1999) revealed that the three major sedimentary basins (early-rift grabens), the Victoria Land Basin, the Central Trough and Eastern Basin, are underlain by highly extended crust and shallow mantle (minimum depth of ~16 km). Beneath the adjacent basement highs, the Coulman High and Central High, the Moho was found to deepen and to be at a depth of 21 and 24 km, respectively. For the crustal layers velocities ranging from 5.8 to 7.0 km/s were found. A distinct reflection was observed on numerous OBSs from an intracrustal boundary between the upper and lower crust at 10–12 km depth. In summary, the Moho depths and crustal thicknesses were found to vary significantly across the West Antarctic rift system (Fig. 9.7.4-06).

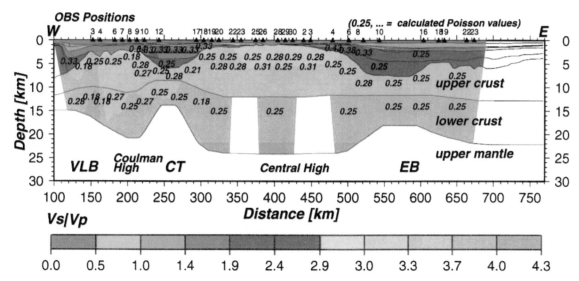

Figure 9.7.4-06. Crustal structure along the ACRUP transect in the Ross Sea, West Antarctica, showing Poisson's ratios (from Trey et al., 1999, fig. 6). [Tectonophysics, v. 301, p. 61–74. Copyright Elsevier.].

9.8. OCEANIC DEEP STRUCTURE RESEARCH

9.8.1. Introduction

Oceanic research in the 1990s and 2000s is characterized by two facts. In many projects, powerful airgun arrays replaced the conventional shooting technique with explosives and the analogue recording techniques was gradually replaced by digital equipment. The length of streamers, involving up to 300 channels, has reached several kilometers and for deep crustal research the use of large numbers of ocean-bottom instruments equipped with seismometers (OBSs) and/or hydrophones (OBHs) has become the rule. Since the introduction of digital equipment, storage of data is no longer a problem.

In particular, to concentrate the efforts of mid-oceanic ridge research worldwide, "InterRidge" was founded in 1993 to promote interdisciplinary, international studies of oceanic spreading centers through scientific exchange and the sharing of new technologies and facilities among international partners. Its members are research institutions dealing with marine research around the world. InterRidge is dedicated to sharing knowledge amongst the public, scientists, and governments and to provide a unified voice for ocean ridge researchers worldwide (www.interridge.org).

Many seaborne projects were conducted along the continental margins and were combined with land-based recording devices. Many of these projects were therefore included, as we have pointed out in Chapter 1, in the previous subchapters on onshore seismic projects, in particular those exploring the North Sea and Mediterranean Sea (e.g., many Spanish projects, all CROP surveys and surveys around Greece) and the continental margins of the British Isles, the Americas, Japan, Australia, New Zealand, and Antarctica.

The growing number of seaborne seismic projects which study crust and upper-mantle structure in oceanic areas makes it impossible, to include them all in the framework of this volume. As for the previous decades, only a limited number of examples can be presented with some detail.

9.8.2. Pacific Ocean

A purely marine wide-angle seismic study by three German institutions (GEOMAR, Kiel, GFZ Potsdam, and University of Hamburg) was carried out during the GEOPECO cruise of R/V *Sonne* in 2000 off the coast of Peru (Hampel et al., 2004; Krabbenhoeft et al., 2004). The aim of the survey was to study the area where the oceanic Nazca plate which originates at the East Pacific Rise is obliquely subducting underneath the continental South America plate (Figs. 9.8.2-01 and 9.8.2-02) along the Peru-Chile trench with water depths of more than 6000 m. The Nazca plate is segmented by fracture zones, e.g., the Viru Fracture Zone or the Mendana Fracture Zone (VFR and MFR in Fig. 9.8.2-01), leading to different crustal ages along the plate boundary.

The survey comprised six seismic wide-angle profiles (Figs. 9.8.2-01 and 9.8.2-02) with shots fired by an airgun array with intervals of 60 seconds and a ship speed of 4 knots, resulting in an average shot spacing of 120 m. The seismic-reflection and -refraction data were recorded by a total of 97 stations using the OBSs and OBHs of GEOMAR, Kiel, Germany (Krabbenhoeft et al., 2004). Simultaneously, multichannel seismic data were recorded with 150 m shotpoint distances by a 24-channel streamer, towed 50 m behind the research vessel.

As a result, the oceanic crust of the Nazca plate, divided into an upper and a lower crustal layer, was determined to be 6.4 km thick on average, with a pelagic sediment cover ranging from 0 to 200 m. In the area of the Trujillo Trough (TT in Fig. 9.8.2-01),

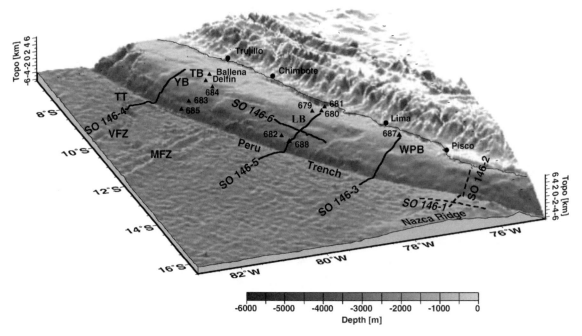

Figure 9.8.2-01. Location of the GEOPECO seismic survey off Peru, illuminated from the northeast (from Krabbenhoeft et al., 2004, fig. 2). [Geophysical Journal International, v. 159, p. 749–764. Copyright John Wiley & Sons Ltd.]

at 8°S, the crust thinned to 4 km. The underlying mantle had an average velocity of 7.9 km/s. Across the margin, the plate boundary, where the South American plate overrides the Nazca plate, could be traced down to 25 km depth (Fig. 9.8.2-03). Adjacent to the trench axis, a frontal prism, associated with the steep lower slope, was inferred from the velocity models (Krabbenhoeft et al., 2004). The data farther south, of profiles 146-1 and 146-2, were interpreted by Hampel et al. (2004).

Another project dealt with the seismic structure of the Galapagos volcanic province, located between 78°W and 95°W longitude and 3°S and 9°N latitude (Fig. 9.8.2-04). Three seismic experiments were carried out in this area, PAGANINI in 1999, G-PRIME in 2000, and SALIERI in 2001. Other seismic projects in this area had investigated the Cocos-Nazca Spreading Center (CNCS in Fig. 9.8.2-04) and the Galapagos Hot Spot (GHS in Fig. 9.8.2-04), described, e.g., by Canales et al. (2002) and Toomey et al. (2001), respectively.

The PAGANINI (Panama Basin and Galapagos Plume–New Investigations of Intraplate Magmatism) seismic project dealt in particular with detailed investigations of the Cocos and Malpelo volcanic ridges. The thickened oceanic crust found here was thought to be the result of an interaction between the Galapagos hot spot and the Cocos-Nazca Spreading Center during the last 20 m.y. (Sallares et al., 2003). To obtain accurate 2-D velocity models, three wide-angle seismic profiles were acquired across the two ridges (nos. 1–3 in Fig. 9.8.2-04). The airgun shots along the three lines had a shot spacing of ~140 m and were recorded by OBS and OBH over a length of 275 km, 260 km, and 245 km, respectively, with a receiver spacing rang-

ing from 1 to 5 km. In total, 21 German OBSs and OBHs and 13 French OBSs were available. Clear arrivals could be recorded on all profiles up to 150 km offsets, but P_n was only identified on profile 3. The main result (Sallares et al., 2003) was that the maximum thickness of the crust ranged between 16.5 km (southern Cocos) and 18.5–19.0 km (northern Cocos and Malpelo). While the upper layer of the crust (layer 2) was characterized by high-velocity gradients, the lower layer appeared in having nearly uniform velocities (layer 3).

The SALIERI (South American Lithosphere transects across volcanic Ridges) seismic experiment followed in 2001 (Fig. 9.8.2-04). The seismic data set obtained and used by Sallares et al. (2005) included 52 seismic sections recorded along two wide-angle profiles acquired by the German research vessel *Sonne*. Airgun shots were spaced 120 m apart, and 24 GEOMAR OBSs and OBHs and 13 French OBSs were available. The first profile 1 included 30 OBSs and OBHs with a receiver spacing between 7 and 15 km, which were deployed along a 350 km N-S–trending transect, which crossed the Carnegie Ridge at 85°W. Profile 2 comprised 22 instruments deployed along a 230 km N-S transect with a similar receiver spacing covering the northern flank of the ridge near the subduction zone at 82°W. From the interpretation (Sallares et al., 2005), a maximum crustal thickness of 19 km was obtained for the central Carnegie-1 profile, located over a 19–20-m.y.-old oceanic crust at about 85°W, while for the eastern Carnegie-2 profile, located over a 11–12-m.y.-old oceanic crust at about 82°W, a crustal thickness of only 13 km resulted. In comparison with the profiles of the PAGANINI-1999 project, the thickness of the oceanic layer 2 proved to be quite uniform along

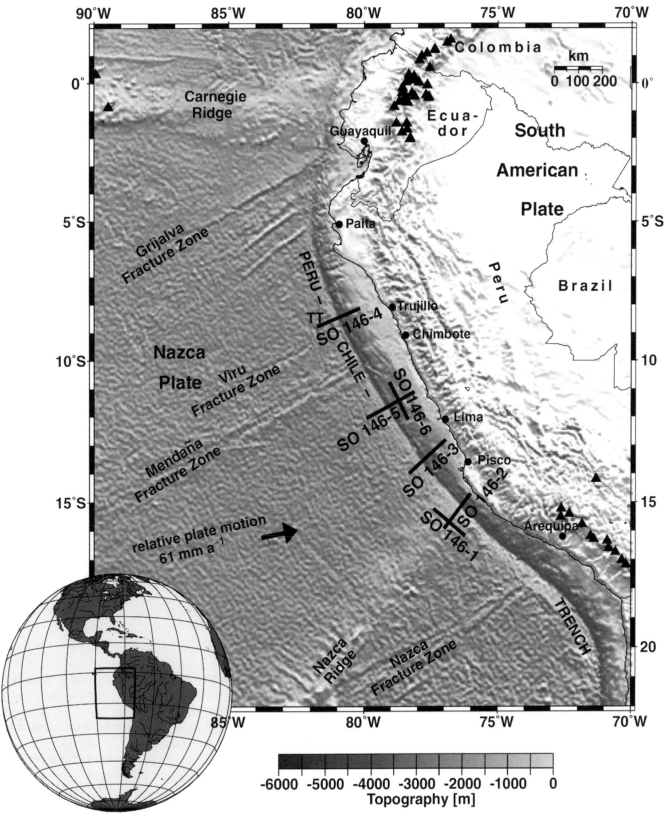

Figure 9.8.2-02. Location of the GEOPECO study area off Peru (from Krabbenhoeft et al., 2004, fig. 1). [Geophysical Journal International, v. 159, p. 749–764. Copyright John Wiley & Sons Ltd.]

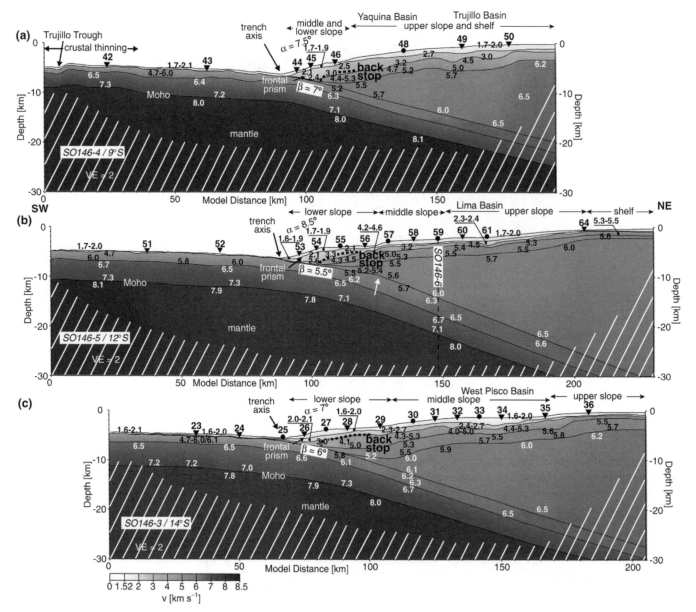

Figure 9.8.2-03. Models from N to S, obtained from the three seismic wide-angle profiles SO 160-4, -5, and -3, shown by thick black lines in Figure 9.8.2-01 (from Krabbenhoeft et al., 2004, fig. 10). [Geophysical Journal International, v. 159, p. 749–764. Copyright John Wiley & Sons Ltd.]

the five profiles, while crustal thickening was mainly accommodated by layer 3, the velocities of which appeared systematically lower where the crust thickened.

Farther west in the equatorial Pacific, at the western end of the Cocos-Nazca axial ridge and east of the Galapagos microplate (Fig. 9.8.2-05), the Hess Deep rift valley structure around 101.5°W, 2.1°N was investigated by a seismic-refraction experiment conducted in 1993 by Scripps Institution of Oceanography at San Diego, California (Wiggins et al., 1996). This experiment did not use energy sources near the sea surface, but seafloor shots of 30 kg charge size whose energy was coupled to the seafloor in a relatively small area. The shots were recorded by a linear array

of digital OBSs which contained three-component seismometers and a hydrophone (Fig. 9.8.2-06). The resulting velocity model differed from typical young fast-spreading East Pacific Rise crust by ~1 km in depth with slow velocities beneath the valley of the Hess Deep and a fast region forming the intrarift ridge.

In 1997, a 3-D MCS reflection and tomography experiment was carried out over the 9°03′N overlapping spreading center of the East Pacific Rise (Singh et al., 2006a). It has an offset of 8 km and the two limbs overlap for ~27 km. The data were acquired over a 20 km by 20 km area of seafloor covering both limbs of the overlapping spreading center. Over a distance of less than 7 km along the ridge crest, a rapid increase in two-way traveltime of

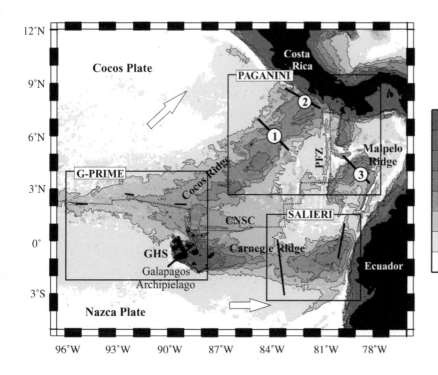

Figure 9.8.2-04. Location of the PAGANINI-1999, G-PRIME-2000, and SALIERI-2001 seismic projects in the Galapagos volcanic area (from Sallares et al., 2003, fig. 1). [Journal of Geophysical Research, v. 108, B12, 2564, doi: 10.1029/2003JB002431. Reproduced by permission of American Geophysical Union.]

seismic waves between the magma chamber and Moho reflections was observed

The Garrett Fracture Zone, located along a flow line on the eastern flank of the East Pacific Rise at 14°S, was the goal of the wide-angle reflection/refraction survey EXCO (Exchange between Crust and Ocean) of the Universities of Hamburg and Bremen, Germany, at the end of 1995 (Fig. 9.8.2-07). Six profiles were shot on 0.5–8.3-m.y.-old seafloor, using 17 OBHs, digital recording systems of GEOMAR, Kiel, and a single airgun source, towed at 10 m depth behind the ship. Shot spacing was 180 m and 250 m on the 8.3 my-line, and the maximum recording distances were 70 km (Grevemeyer et al., 1998). The result was a classical model of oceanic crust of ~6 km thickness. It comprised a 1.4–1.8-km-thick layer 2, with a low-velocity (2.9–4.6 km/s) surficial layer 2A at its top and bounded at its base by a sharp velocity increase to 5.5 km/s at the top of layer 2B, and a lower crustal layer 3 with velocities ranging from 6.8 to 7.2 km/s with a moderate velocity gradient.

The nearby East Pacific Rise, south of the Garrett Fracture Zone, had been investigated earlier, in 1991, by a U.S. team from Woods Hole Oceanographic Institution in Massachusetts, Scripps Institution of Oceanography at San Diego, California, and Lamont-Doherty Geological Observatory at Palisades, New York (Detrick et al., 1993a; Hussenoeder et al., 1996; Hooft et al., 1997). The seismic survey TERA was concentrated in three areas centered at 14°15′S, 15°55′S and 17°20′S (Fig. 9.8.2-08).

Three types of seismic data were obtained: common midpoint (CMP) reflection data, two-ship expanding spread profiles and ocean-bottom seismic refraction data. In each of the three areas, six OBSs were deployed. The reflection data were acquired by a 20-gun airgun array and a 4-km-long, 160-channel

digital streamer. In the 14°15′S and 17°20′S areas, 23 expanding profiles, oriented parallel to the rise axis, were shot. A series of reconnaissance reflection profiles was also obtained along and across the rise axis over a distance of more than 800 km from ~20.7°S to the Garrett transform (Fig. 9.8.2-08). One of the main results of the interpretation of the seismic data from the ultrafast-spreading southern part of the East Pacific Rise was that the rise was underlain by a thin extrusive volcanic layer (seismic layer 2A of less than 200 m thickness) that thickened rapidly off the ridge axis. A relatively continuous reflection at 5.5–6.0 s TWT and at 1.8 s below the sea floor, which could be traced within 2–3 km off the rise axis on some profiles, was interpreted as Moho. A more or less continuous reflection from the axial magma chamber (Fig. 9.8.2-09) could be traced along the entire reconnaissance survey, resulting in an average depth of 2.6 km below the seafloor along the rise axis between 14°S and 18°S, but dipping farther south to nearly 3 km depth at the end of the survey line (Detrick et al., 1993a). Parts of the data were later further interpreted in more detail (Hussenoeder et al., 1996; Hooft et al., 1997; Singh et al., 1998b).

A more detailed description of the experiment in the 14°15′S area and its interpretation was published by Kent et al. (1994). They stated that, where the Moho could be imaged, it lies at a nearly constant traveltime below the base of layer 2A which suggested that layers 2B and 3 were, to first order, of uniform thickness. Aside from the Moho reflections at an approximate traveltime of 1.4 s below the base of layer 2A, the lower crust appeared acoustically transparent indicative of a seismically homogeneous layer 3.

The East Pacific Rise at 9°–10°N which had been the focus of an extended multi-channel experiment in 1985 (Detrick et al.,

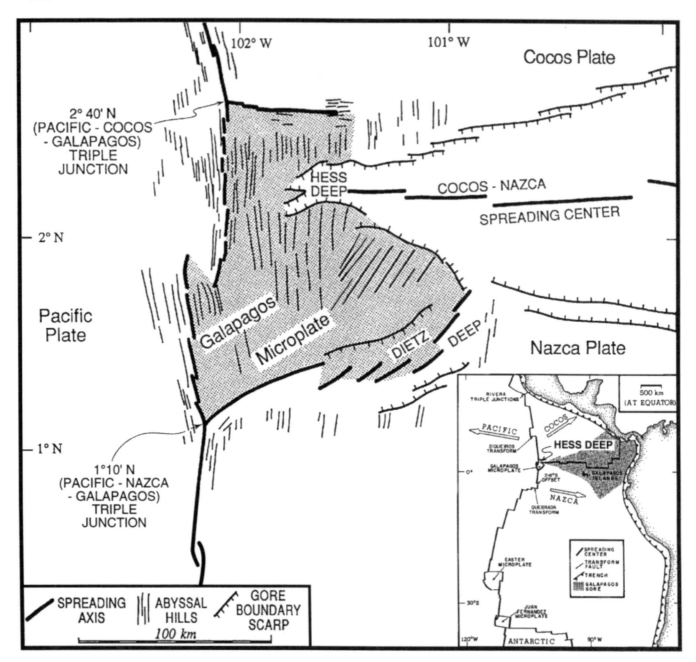

Figure 9.8.2-05. Location of the Hess Deep rift valley and its relation to the East Pacific Rise (from Wiggins et al., 1996, fig. 1). [Journal of Geophysical Research, v. 101, p. 22,335–22,353. Reproduced by permission of American Geophysical Union.]

1987), was again the focus of a large-scale experiment in 1993 (Christeson et al., 1997), this time using 59 OBSs deployments along six lines shot with a large airgun shot spacing of 250 m. The large number of OBSs allowed the study of regional compressional (P) and shear (S) velocity structure of seismic layer 2B and layer 3.

The MELT (Mantle Electromagnetic and Tomography) experiment in 1996 targeted the East Pacific Rise between 15° and 18°S and was designed to distinguish between competing models of magma generation beneath mid-ocean ridges (MELT Seismic

Team, 1998). The project comprised seismic and electromagnetic observations. Besides the 800-km-long two seismic-passive linear arrays of 51 OBSs, an active-source component was also involved to study the crustal structure in detail (Canales et al., 1998). Airgun shots, fired at a 100-s repetition rate, were recorded by 15 OBSs deployed along three seismic-refraction lines. The northern line along the second MELT array was 360 km long and crossed the East Pacific Rise a few kilometers north of the 15°55'S overlapping spreading center and included five OBSs spaced 40–90 km apart. The remaining ten OBSs were deployed

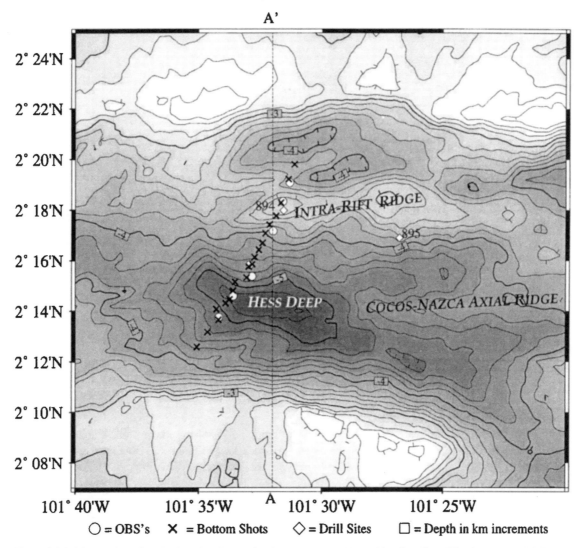

Figure 9.8.2-06. Location of a seismic-refraction profile through the Hess Deep rift valley (from Wiggins et al., 1996, fig. 2). [Journal of Geophysical Research, v. 101, p. 22,335–22,353. Reproduced by permission of American Geophysical Union.]

along the primary MELT array which intersected the rise axis near 17°15′S where a shallow axis was present. Four of these instruments, spaced 15–50 km apart, were located on the Nazca Plate to the east along a 160-km-long line. The other six instruments, spaced 15–80 km apart, were employed on the Pacific Plate along a 230-km-long line. The records showed clear crustal refractions (P_g) and Moho reflections (P_MP), but on only some of the receivers were upper-mantle refractions (P_n) observed. Below the top layer with velocities of 2.5–4.0 km/s, crustal velocities increased from 4.0 to 7.0 km/s at a sub-seafloor depth of 3.0–3.5 km. The lower crust was modeled by a 2.5-km-thick layer with P-velocities of 7.0–7.1 km/s on the Nazca Plate and up to 7.2 km/s on the Pacific Plate. Crustal thickness was 5.1–5.7 km at 17°15′S on both sides of the rift axis and was not resolvably different, but a small along-axis difference was found for similarly aged crust on the Nazca Plate, crustal thickness increasing slightly to 5.8–6.3 km at 15°55′S (Canales et al., 1998).

As example of a seismic project in the south central Pacific around 150°W, 17°S the investigation of the Society Island hotspot chain is described here (Fig. 9.8.2-10). The oceanic crust under the Society Islands volcanic chain, which is ~400 km long, ranges from 65 Ma in the south to 80 Ma in the north. The island chain is composed of five atolls and nine islands, one of them being Tahiti. The progressive decrease in age of volcanics eastwards implies that the chain was created by the passage of the Pacific plate over a hotspot source (Grevemeyer et al., 2001b).

The experimental part of the project had already been carried out in 1989. One profile approached Tahiti and covered the abyssal hill terrain to the east of Tahiti, whereas the other profiles covered oceanic crust north of the hotspot region. Twenty-six OBHs were deployed along four profiles. However, only 16 instruments provided seismic recordings useful for geophysical data analysis (Grevemeyer et al., 2001b). The instruments were

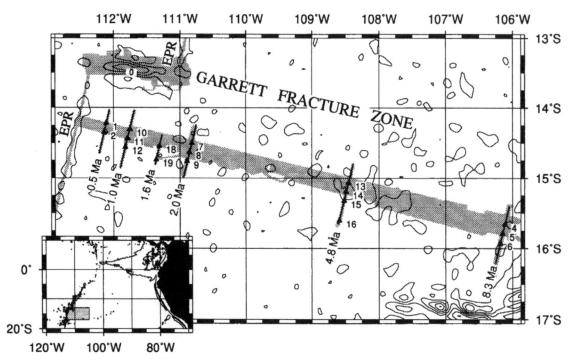

Figure 9.8.2-07. Location of the EXCO-1995 seismic project on the southern East Pacific Rise at 14°S. (from Greve-meyer et al., 1998, fig. 1). [Geophysical Journal International, v. 135, p. 573–584. Copyright John Wiley & Sons Ltd.]

analog pop-up systems from the University of Hamburg and digi-tal systems from IFREMER (Institut Francais de Recherche et d'Exploitation de la Mer) in Brest. An airgun array towed at 10 m depth provided the energy source with a shot spacing of 250 m. No onshore shots had been planned, but in addition to the OBS data, the airgun shots were also recorded at the seismic stations of the Polynesian Seismic Network onshore Tahiti and Moorea, from which picks of the onset times of the first breaks were pro-vided which could be used for the interpretation of profiles 2 and 3. Profile 2, for example, located offshore Tahiti, was ~280 km in length and airgun shots were recorded in-line by 9 OBHs and one land station.

The traveltime modeling provided pronounced variations in crustal structure along the hotspot chain (Grevemeyer et al., 2001b). After removal of the contribution of the uppermost low-velocity layer representing the volcanic edifices and debris, it was suggested that the islands and atolls sit on normal oceanic crust. In general, off the islands a nearly flat crust-mantle boundary at around 14 km depth below sea level was obtained, while under the islands, in response to loading, the crust has bent down and thickened. From amplitude modeling, Grevemeyer et al. (2001b) concluded that the crust on the hotspot trail had been modified with respect to the pre-hotspot crust and that the lower crust had been intruded by sills during the ongoing late-stage volcanic activity and that some melts were trapped in the crust-mantle boundary region.

The next project described in this subchapter was conducted off Taiwan on the Ryukyu margin, forming the boundary of the

Philippine Sea plate and the Chinese continental margin at 22°N between 122°30′E and 123°30′E, where a north-south–trending oceanic ridge, the Gagua ridge, is entering the subduction zone (Fig. 9.8.2-11). Here the TAICRUST project was initiated as a seismic imaging program, which utilized deep seismic-reflection profiles and wide-angle reflection and refraction data recorded by both ocean-bottom seismometers and onshore seismic stations. The aim of the project was to investigate the deep structure and geodynamic processes of the south Ryukyu margin and the Tai-wan arc-continent collision zone (Liu et al., 1995).

During this project, in 1995 three multichannel deep seismic-reflection profiles, located to the east, west, and north of the Gagua ridge (Fig. 9.8.2-11) were recorded by the R/V *Maurice Ewing* (Schnuerle et al., 1998). A 20-airgun array was fired at 20-s intervals and recorded up to 16 s by a 160-channel digital streamer of 4 km length. A sub-basement reflection was recorded on two of the profiles (EW9509-1 and EW9509-18) at ~1.5 s below the top of the oceanic basement (see, e.g., Fig. 9.8.2-12) as a high-amplitude long-wavelength reflector which could have been the Moho (Schnuerle et al., 1998).

However, if it were the Moho, the igneous oceanic crust between the basement and the above-mentioned sub-basement reflector would be extremely thin, i.e., less than 5 km, and even thinner toward the trench. Previously conducted seismic-refrac-tion surveys offshore south Ryukyu showed a strong velocity contrast between layer 2 and layer 3 (5.6/7.0 km/s) and an ab-normally thin layer 3. Therefore, Schnuerle et al. (1998) tended toward the interpretation that their continuous sub-basement re-

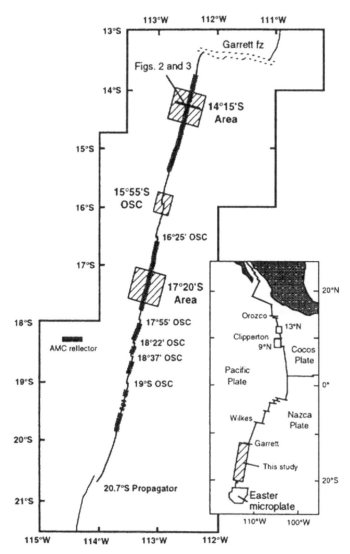

Figure 9.8.2-08. Location of the TERA-1991 seismic projects on the southern Pacific Rise axis south of the Cocos-Nazca-Pacific triple junction (from Detrick et al., 1993, fig. 1). Hatched boxes show areas of extensive reflection/refraction/tomographic surveys. Also shown are reconnaissance profiles and portions of the rise axis associated with an axial magma chamber (AMC) reflector. [Science, v. 259, p. 499–503. Reprinted with permission from AAAS.]

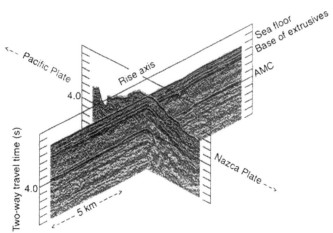

Figure 9.8.2-09. Two intersecting multichannel reflection profiles across and along the crest of the East Pacific Rise (from Detrick et al., 1993, fig. 2). AMC—reflection from top of axial magma chamber, Moho reflection at 5.5–6.0 s not shown. [Science, v. 259, p. 499–503. Reprinted with permission from AAAS.]

flector was rather the top of oceanic layer 3, whereas the Moho, which should be less than 1 s deeper, was not well imaged.

In continuation of the Ryuku Trench toward the northeast follow the Nankei Trough south of Shikoku Island, Japan, the Japan Trench east of Honshu, Japan, and the NE-trending Kuril Trench, east of Hokkaido, Japan (Fig. 9.8.2-13). Several projects were conducted in Japanese waters around Japan.

The instrumentation used in the 1990s and following 2000s was the same for most of the Japanese marine expeditions. The OBSs, designed originally by Shinohara et al. (1993) and converted into a digital system, had gimbal-mounted 4.5-Hz geophones and a hydrophone. Data were recorded by a digital

recorder in each OBS. The digital recorder had a 16-bit A/D converter and was able to record data for ~17 days on 4 channels with 100 Hz sampling. Experiments and the airgun array for large distances had a volume of up to 207 l. The design of the onland recording systems in onshore-offshore experiments was similar to the OBS system, with 4.5-Hz geophones and a digital recorder (Kodaira et al., 2002).

In 1994, 1995, and 1997, wide-angle OBS studies investigated the crustal structure of the adjacent Nankai Trough seismogenic zone (Kodaira et al., 2000). The Nankai Trough (Fig. 9.8.2-13; see also Fig. 9.6.3-01), southwestern Japan, is known as a vigorous seismogenic zone with well-studied historic earthquakes. Several OBS profiles were recorded from the shelf areas south of Shikoku Island and Kii peninsula toward the SSW across the Nankai trough into the Philippine Sea (Fig. 9.8.2-14). The 250-km-long line MO104 crossed the presumed seismic slip zone of the 1946 Nankaido earthquake (M_S = 8.2) and was recorded in conjunction with a network of multichannel reflection profiles. The resulting crustal model showed a gentle sloping subducting oceanic crust and a thick overlying sedimentary wedge. The normal oceanic crust was interpreted to consist of two oceanic layers 2 and 3 with velocities of 5.0–5.6 km/s and 6.6–6.8 km/s. The maximum of the sedimentary wedge was 9 km at 70 km to the north from the trench axis with velocities of 3.4–4.6 km/s in its deepest parts. The base of the ~5-km-thick subducting oceanic crust, seen at 10 km depth below sea level under the Philippine Sea, could be traced landward to a depth of 25 km. Under Shikoku Island it is estimated to be at 40 km depth.

In 1999, the Japan Marine Science and Technology Center (JAMSTEC), in cooperation with several universities, undertook another extensive experiment in the western Nankai Trough seismogenic zone and acquired onshore-offshore wide-angle as well as offshore multichannel seismic data (Kodaira et al., 2002).

612

Chapter 9

Along the 185-km-long offshore part of the wide-angle seis-mic profile, 98 OBSs were deployed into which an airgun array (207 l) fired shots every 200 m. The onland part contained 93 land stations which were deployed at 1–2 km spacing and extended across the eastern side of Shikoku Island and farther to the north onto Honshu Island. Two explosive 500-kg charges were fired at both ends of Shikoku Island and recorded both on- and offshore. Also the airgun shots were successfully recorded onland by the stations on Shikoku Island. The multichannel seismic-reflection data which covered the southernmost 135-km-long section of the profile provided a shallow image of the offshore profile, but also saw clear reflection events at 11–12 km depth which were inter-preted as the Moho of the subducting slab. From the wide-angle seismic data, a subducting seamount was found at the center of the proposed rupture zone (see Fig. 9.8.2-14) with dimensions of 13 km thick by 50 km width at 10 km depth and was inter-preted as now colliding with the Japanese island arc crust. Be-neath the subducted seamount a low P-wave velocity of 7.5 km/s was found (Kodaira et al., 2002).

Park et al. (2002) have interpreted additional multichannel seismic-reflection lines farther east across the Nankai Trough south and southeast of Kii Peninsula which revealed steeply land-ward-dipping splay faults in the rupture area of the 8.1 Tonankai earthquake of 1944. They were found to branch upward from the plate boundary interface, the subduction zone, at a depth of 10 km ~50–55 km landward of the trench axis, breaking through the upper crustal plate.

Already in 1992, an extensive marine seismic-reflection and -refraction survey was conducted at 32°15′N between 138° and 143° across the northern Izu-Ogasawara island arc system, which extends for more than 1000 km off the southern coast of Japan to the south, between the Izu-Ogasawara Trench and the Shikoku Basin south of Honshu Island (Suyehiro et al., 1996). Controlled sources and natural earthquakes were used to model the 3-D crustal and uppermost mantle structure and subduction slab geometry. The crust was thickest with ~20 km beneath the presently active rift zone. The middle crust with ~6 km/s was confined beneath the arc and comprised ~25% of the crustal vol-ume, the lower crust with 7.1–7.3 km/s velocities comprised an-other 30% of the crust and the Moho became obscure east of a north-south–trending zone of volcanoes.

In 2000, the Tsushima Basin, located in the southwestern Japan Sea north of southern Honshu, was targeted by a seismic velocity structure survey using OBSs and airguns (Sato et al.,

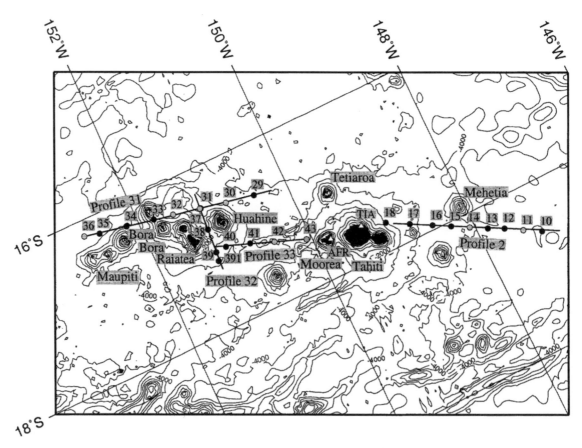

Figure 9.8.2-10. Location of the seismic investigation of the Society Islands hotspot chain in the south central Pacific (from Grevemeyer et al., 2001b, fig. 1). Solid lines are profile segments covered by airgun shots, black dots are OBH positions, and white triangles are land stations. [Geophysical Journal International, v. 147, p. 123–140. Copyright John Wiley & Sons Ltd.]

2006). The survey was an 89-km-long line extending from the southeastern Tsushima Basin toward the southwestern Japan Island Arc. Six OBSs were deployed with 15–20 km spacing, the shot interval was 80–100 s, corresponding to a spacing of ~200–270 m. The resulting crustal thickness was 17 km under the southeastern Tsushima Basin including a 5-km-thick sedi-

mentary layer, and 20 km including 1.5-km-thick sediments under its margin.

Off the coast of northern Honshu, from 36° (latitude of Tokyo) to 41°N and at 142° to 143°E, a series of MCS reflection surveys was performed in the years 1996–2001, all lines crossing the Japan Trench. In total, 3120 km MCS lines were obtained.

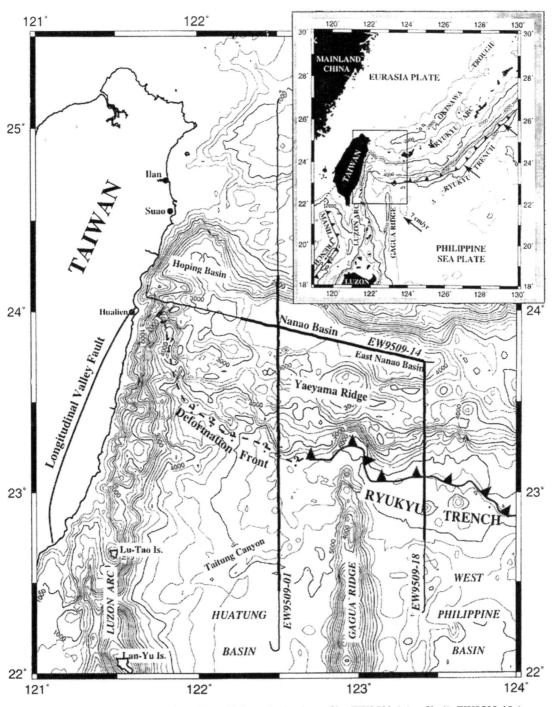

Figure 9.8.2-11. Location of the TAICRUST multichannel seismic profiles EW9509-1 (profile 1), EW9509-18 (profile 2) and EW9509-14 (profile 3) in the southern Ryukyu arc-trench system in the vicinity of the Gagua ridge (from Schnuerle et al., 1998, fig. 1). [Tectonophysics, v. 288, p. 237–250. Copyright Elsevier.]

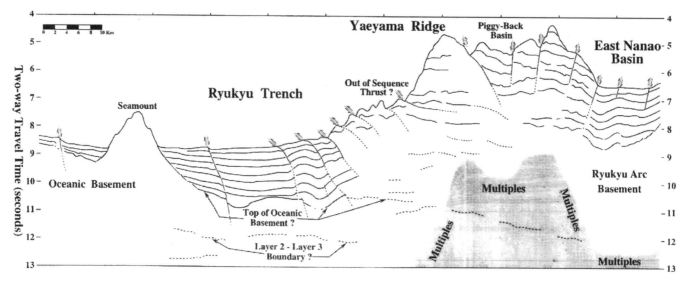

Figure 9.8.2-12. Seismic image along the TAICRUST multichannel seismic profile 2 (EW9509-18) showing at depth the top-basement and the sub-basement reflectors (from Schnuerle et al., 1998, fig. 4). [Tectonophysics, v. 288, p. 237–250. Copyright Elsevier.]

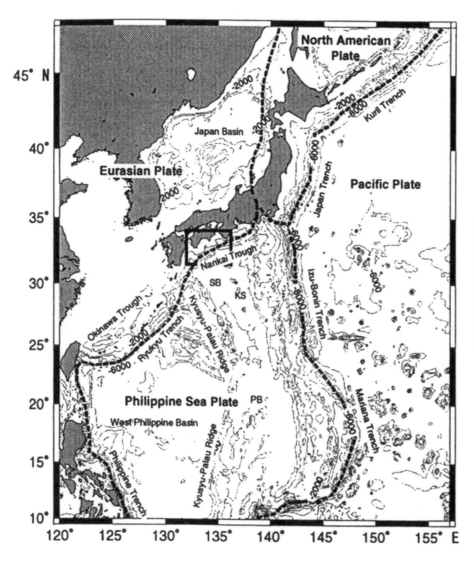

Figure 9.8.2-13. Map of the western Pacific and Philippine Sea showing the geographic location of oceanic features east of the Philippines and China and around Japan (from Kodeira et al., 2000, fig. 1). SB—Shikoku Basin, KS—Kinan Sea Mounts, PB—Parece Vera Basin. [Journal of Geophysical Research, v. 105, p. 5887–5905. Reproduced by permission of American Geophysical Union.]

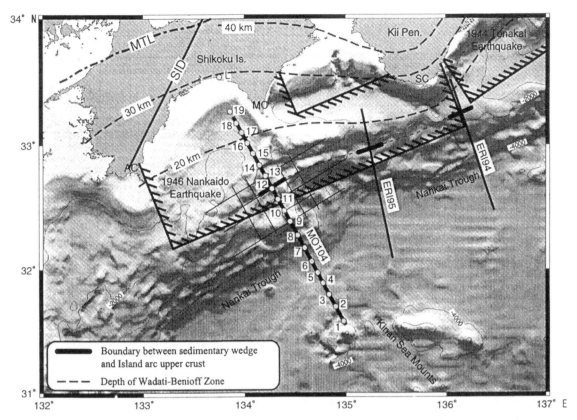

Figure 9.8.2-14. OBS seismic-refraction profiles MO108, ERI95 and ERI94 across the western Nankai Trough seismogenic zone south of Shikoku Island and Kii peninsula (from Kodeira et al., 2000, fig. 2). Also shown are the locations a network of multichannel seismic reflection lines (thin lines) and of the onshore wide-angle seismic profile on Shikoku Island. Thick dashed lines—isodepth contours of the Wadati-Benioff zone (after Hyndman et al., 1995). Short thick marks on the OBS profiles—boundary between sedimentary wedge and island arc upper crust. Rectangular areas—coseismic slip zones of the 1944 and 1946 8.2-earthquakes (after Hyndman et al., 1995). [Journal of Geophysical Research, v. 105, p. 5887–5905. Reproduced by permission of American Geophysical Union.]

For recording, a 3200–4100 m streamer cable and an airgun array of 66–197 l were used (Tsuru et al., 2002). The surveys revealed the tectonic pattern of the sedimentary layers and the top of the igneous crust, but did not penetrate into lower crustal levels.

In 1997, an offshore experiment was carried out in the Pacific Ocean to the east of northeastern Honshu, consisting of two parallel N-S–directed lines of 150–160 km length parallel to the Japan Trench axis farther east and furnished with seven OBSs each (Takahashi et al., 2000) and an E-W–directed line of 180 km length between the coast and the trench axis, along which another eight OBSs and three land stations were deployed (Takahashi et al., 2004). The shot interval of the airgun array along all lines was ~50 m. The OBSs were equipped with a hydrophone sensor and a three-component geophone. The location of these lines is shown in Figure 9.6.3-01 (north of line A–A′). Resulting from the interpretation of these data, the subduction angle increases from east to west, from 3° to 8° to 11°, with the top of the plate located at depths of 11 km at 40 km, 21 km at 120 km, and at 28 km at 140 km distance west from the trench. The Moho beneath the fore-arc was found at ~20 km depth, the velocity of the island arc

upper mantle is ~8 km/s. While the velocities in the Cretaceous sedimentary layer and in the upper crust showed large variations in E-W direction, the velocity in the lower crust with 6.7–7.0 km/s was relatively stable. A middle crustal layer, which was found between upper and lower crust on land in some locations, extended 40 km seaward beyond the coastline (Takahashi et al., 2004).

In 1998, a wide-angle survey was recorded in the southern Japan Trench fore-arc region, consisting of two cross lines, one line being 290 km long and running perpendicular to the trench axis (from ~37°N, 141°E to 36°N, 144°E), the other one being 195 km long and running parallel to the trench. OBSs were deployed at 20 km intervals along the trench-parallel profile, on the perpendicular line nine OBSs were deployed at 10 km intervals symmetrically to the crossing point. The airgun shooting interval was 50 m for the 120-channel streamer and the land stations prolonging the marine line into Honshu. The interpretation resulted in a 4-layer crust consisting of Tertiary and Cretaceous sedimentary rocks and the island arc upper and lower crust, underlain by the island-arc Moho at 18–20 km depth. The uppermost mantle of the island arc (mantle wedge) extends to 100 km landward of the trench axis, its velocity increasing from 7.4 km/s at the ocean-side corner to 7.9 km/s below the coastline (Miura et al., 2003).

In 1999, a seismic experiment was conducted farther north in the Miyagi fore-arc region. 36 OBSs were deployed at 3–5 km intervals along a 270-km-long east-west profile at ~38°N which reached from the coast across the Japan Trench to the Pacific plate. During the shooting with a 50 m shot interval a 12-channel streamer for multichannel seismic surveying was towed to determine the shallow structure (Miura et al., 2005). The depth to the Moho was determined to be 21 km at 115 km distance from the trench axis which increased progressively landwards. The P-wave velocity of the mantle wedge was 7.9–8.1 km/s, which the authors regarded as typical for uppermost mantle without large serpentinization. The dip angle increased from 5 to 6° near the trench axis to 23° 150 km landward from the trench axis. The P-wave of the oceanic mantle was as low as 7.7 km/s.

Finally, farther to the north, the Kuril Island Arc and Trench were investigated in 2000 (Kurashimo et al., 2007, Nakanishi et al., 2009) by an onshore-offshore experiment. Because of its large onshore component, it is described in some detail in subchapter 9.6.3.

As noted, since 1985, the Australian Geological Survey Organisation (AGSO) ran a large program of deep seismic-reflection surveys offshore and onshore Australia (Fig. 9.7.1-01), which until 1992 were summarized in a table by Goleby et al. (1994). The offshore program targeted, in particular, Mesozoic basins with hydrocarbon potential. It had started in 1985 with a seismic research vessel operating a 4800 m, 240 channel streamer, which was towed at depths of 10–12 m and a 3000 cubic inch, tuned, sleeve airgun array towed at 9 m depth. The data which were usually sixfold were recorded up to 16 s. The continental margin surveys off Australia continued in the 1990s (Goleby et al., 1994; Finlayson, 2010). A total of 64 research cruises were conducted in the period 1985–1998. A table showing all cruises of the AGSO vessel *Rig Seismic* was also published by Finlayson (2010; Appendix 2-2).

The projects concerning the margins of NW and S Australia will be described in the next subchapter 9.8.3 (Indian Ocean). On the Pacific side of Australia, an investigation of the Macquarie Ridge and the Australian/Pacific plate boundary, southwest of New Zealand, took place in 1994. In cooperation with French and U.S. agencies, the AGSO vessel *Rig Seismic* acquired ~600 km of seismic-reflection data across the ridge (Finlayson 2010; Appendix 2-2).

9.8.3. Indian Ocean

In this subchapter, a few projects are described as examples for seismic investigations of the Indian Ocean. Detailed investigations covered the margins of Australia and the Banda Sea east of Timor. Other projects dealt with the center and the western parts of the Indian Ocean.

The offshore program of AGSO in the 1990s mainly covered areas off NW and S Australia. From 1990 to 1993 along the continental margin of northwestern and northern Australia ~35,000 km of mostly deep—commonly 15 s record lengths—seismic-reflection data were acquired. Maps and other details were presented by Finlayson (2010; Appendix 2-2). These surveys included 1894 km of seismic profiling recorded in 1990 in and across the Vulcan graben, two sets of data recorded in 1990 and 1991 in the Arafura Sea, several surveys on the Exmouth Plateau in 1990–1992; another 2100 km of seismic-reflection data were recorded in 1991 and 1993 in the Bonaparte basin–Timor Sea area, ~3460 km of data across the Browse basin were acquired in 1993, and 4085 km of data across the offshore Canning basin were also recorded in 1993.

The deep marine seismic-reflection surveys in 1996 and 1998 along the Vulcan transect in the Timor Sea and the Petrel transect in the Bonaparte basin off northern Australia have already been discussed in subchapter 9.7.1.1 (for location, see Figs. 9.7.1-01 and 9.7.1-02; Petkovic et al., 2000). These surveys had also supplied wide-angle seismic data from OBSs and from a few land stations recording the airgun shots.

The marine seismic profiling of the offshore region of the Otway Basin in late 1994 and early 1995 (for location, see Fig. 9.7.1-09) was already discussed in subchapter 9.7.1 in connection with the near-vertical and wide-angle incidence observations on land across the Otway Basin continental margin (Finlayson et al. (1998).

During 1992, the British Institutions Reflection Profiling Syndicate (BIRPS) and the Indonesian Marine Geological Institute (MGI) jointly investigated the crustal architecture of the Banda Arc. The project involved two long seismic-reflection profiles east of Timor and were 350 km long (Finlayson 2010; Appendix 2-2).

In 1998, a large ocean-bottom seismometer and hydrophone experiment was conducted near 17°S near ODP site 757 on Ninetyeast Ridge, a major Indian Ocean hotspot track which was created in Late Cretaceous and Early Tertiary times by the Kerguelen hotspot. The ridge can be traced for ~5000 km from 30°S latitude, where it joins the Broken Ridge, northward to the Bay of Bengal, where it is buried beneath the Bengal fan (Fig. 9.8.3-01). The aim of the project was to study the regional structure of the hotspot trail and the small-scale structure of the volcanic edifice.

In total, ~2500 km of seismic-refraction/wide-angle and near-vertical incidence data were acquired by R/V *Sonne* in a joint experiment of three German institutions—University of Bremen, GEOMAR at Kiel, and BGR at Hannover (Flueh et al., 1998b, 1999; Grevemeyer et al., 2001a). More than 100 OBSs and OBHs had been deployed and successfully recovered. In addition, a three-channel ministreamer was used. The kernel was a network of profiles in an area of ~55 × 110 km around ODP site 757 (17°S, 88°E) and a roughly 550-km-long east-west transect at 17°S through its center (Fig. 9.8.3-02).

Along the 550-km transect, 60 ocean-bottom receivers provided data used for traveltime inversion and modeling. The transect extended from the Indian Ocean basin across the Ninetyeast Ridge into the Wharton basin. It was composed of three profiles,

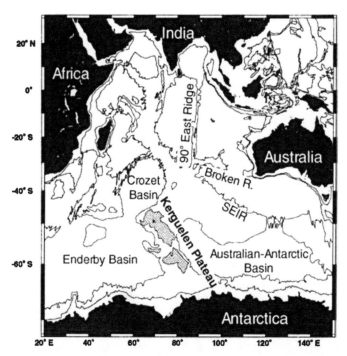

Figure 9.8.3-01. Map of the Indian Ocean, showing the bathymetric contours 2000 m and 4000 m (from Operto and Charvis, 1996, fig. 1). SEIR—South East Indian Ridge. [Journal of Geophysical Research, v. 101, B11, p. 25,077–25,103. Reproduced by permission of American Geophysical Union.]

suggest that the crust under the ridge was bent downward, and hotspot volcanism has underplated the preexisting crust, reaching a total crustal thickness of ~24 km (Fig. 9.8.3-03). This underplating also occurred to the east of the ridge under the Wharton basin (Grevemeyer et al., 2001a).

The Southwest Indian Ridge saw two seismic projects in the 1990s during the RRS *Discovery* cruise 208. A wide-angle seismic experiment investigated the Atlantis II Fracture Zone and was interpreted together with geochemical analyses of dredged basalt glass samples from a site conjugate to Ocean Drilling Program hole 735B and thus allowed the determination of the thickness and the most likely lithological composition of the crust beneath hole 735B (Muller et al., 1997). The wide-angle seismic data were recorded by OBH deployed in a grid pattern and with an instrument spacing of 20 km. Shot spacing from an airgun source was ~100 m. The largest offsets from which seismic arrivals from crust and upper mantle were recorded were 25–35 km. The Moho depth was well constrained by P_MP reflections along most of the N-S line. A seismic crustal thickness of 4 km was determined to the north and south of the Atlantis Platform on which hole 735B was located, while beneath the hole the Moho was found at 5 km depth beneath the seafloor. The base of the magmatic crust was estimated to be 2 km beneath the seafloor, underlain by a 2–3-km-thick layer of serpentinized mantle peridotite with a P-wave velocity of 6.9 km/s.

The E-striking Southwest Indian Ridge between 65°20′ and 66°30′E at latitude 27°40′S was the goal of an E-W–directed seismic profile recorded during the RRS *Discovery* cruise 208 (Muller et al., 1999). The line was directed approximately perpendicular to the spreading direction. By large changes in the crustal and seismic layer 3 thicknesses, three segments were defined, anomalously thin crust of 2–3 km being observed beneath the nontransform segment boundaries and 5–6 km beneath the segment midpoints. It was inferred that melt generation is focused on segment midpoints and varies between segments.

Another project investigated the Kerguelen Plateau to the south of the Ninetyeast Ridge (Fig. 9.8.3-01). Here, during

each being ~300 km long. To simulate a single refraction line, the profiles were shot with an overlap of 100–150 km; 25, 25, and 15 ocean-bottom stations were deployed on profiles 05, 06, and 31, respectively (Fig. 9.8.3-02). The seismic source was a tuned array of 20 airguns providing a total volume of 51.2 l. Shot interval was 60 seconds, resulting in an average shot spacing of 110 m.

The resulting models indicated normal oceanic crust to the west and east of the edifice. Crustal thickness was on average 6.5–7 km. Based on wide-angle reflections from both the pre-hotspot and the post-hotspot crust–mantle boundary, the authors

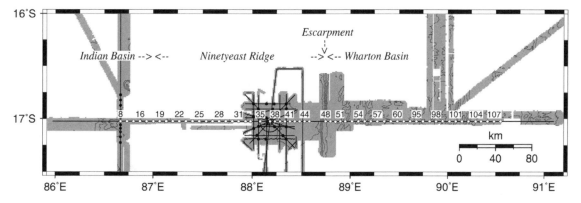

Figure 9.8.3-02. Location of the seismic survey lines along a 550-km-long transect across Ninetyeast Ridge (from Grevemeyer et al., 2001a, fig. 2). [Geophysical Journal International, v. 144, p. 414–431. Copyright John Wiley & Sons Ltd.]

the Kerguelen OBS experiment in early 1991, seven wide-angle seismic lines were shot at sea during the cruise of the M/V *Marion Dufresne* on the Kerguelen Plateau and in adjacent oceanic basins. Of those, two lines, lines 4 and 5, were shot in the Raggatt basin, along NNW-SSE and E-W directions, respectively (Fig. 9.8.3-04).

Operto and Charvis (1996) concentrated their discussion on lines 4 and 5 (lower part of Fig. 9.8.3-04). More than 960 shots were recorded along each seismic line with a maximum source-receiver offset of 165 km and an average spacing of shots of ~185 m. Five OBSs were evenly deployed along each line. The central OBS (OBS 3) was located at the cross point of lines 4

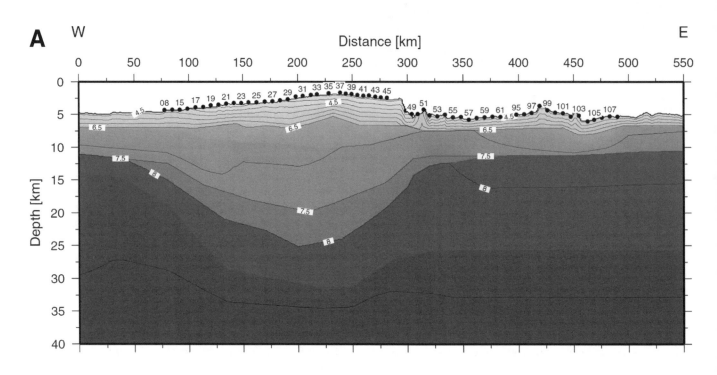

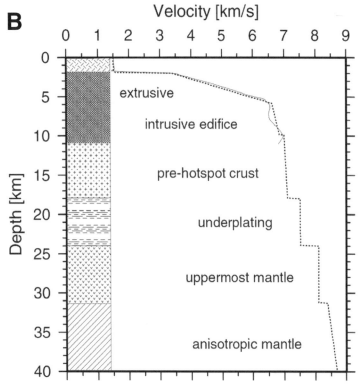

Figure 9.8.3-03. Crustal cross section (A) and velocity-depth model (B) along the 550-km transect from the Indian Ocean basin across Ninetyeast Ridge into the Wharton basin (from Grevemeyer et al., 2001a, fig. 11). [Geophysical Journal International, v. 144, p. 414–431. Copyright John Wiley & Sons Ltd.]

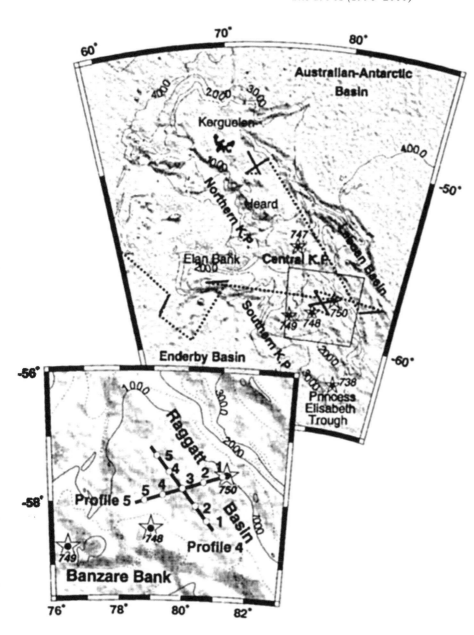

Figure 9.8.3-04. Location of the Kerguelen ocean bottom seismometer experiment (from Operto and Charvis, 1996, fig. 2). [Journal of Geophysical Research, v. 101, B11, p. 25,077–25,103. Reproduced by permission of American Geophysical Union.]

Figure 9.8.3-05. Crustal cross section along profile 4 of the Kerguelen ocean bottom seismometer experiment (from Operto and Charvis, 1996, fig. 4). [Journal of Geophysical Research, v. 101, B11, p. 25,077–25,103. Reproduced by permission of American Geophysical Union.]

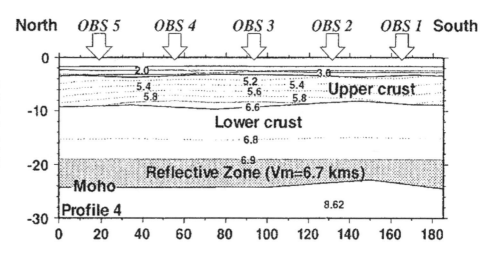

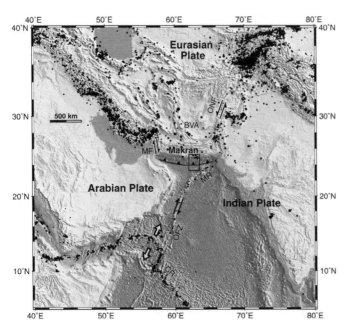

Figure 9.8.3-06. Location of the Makran accretionary wedge off Pakistan (from Kopp et al., 2000, fig. 1). MR—Murray Ridge; rectangle—research area. [Tectonophysics, v. 329, p. 171–191. Copyright Elsevier.]

and 5. The seismic source was an untuned array of eight airguns, 16 l, each fired simultaneously every 100 seconds. Receivers were three-component digital OBSs.

The result of the data interpretation of profile 4 was a 22-km-thick crust, subdivided into a 1.6-km-thick sedimentary cover, an upper crust of 5.3 km, a lower crust of 11 km thickness and a 4–6-km-thick reflective transition zone immediately above the Moho (Fig. 9.8.3-05). The structure along profile 5 was almost identical down to 19 km depth, but with some more relief. However, the transition zone appeared to be seismically transparent (Operto and Charvis, 1996).

The data of lines 1, 2 (the northern two cross lines at 50–51°S, 72–73°E), and 7 (the southwestern line at 57–58°S, 62–63°E) (Fig. 9.8.3-04) were interpreted by Charvis and Operto (1999). The crustal structure under the northern Kerguelen Plateau is almost identical to that shown in Figure 9.8.3-05 for profile 4. In the Enderby basin to the west, however, on line 7, the Moho was found at a shallower depth, dipping from ~15.5 km depth in the SW to 18 km in the NE, while the lower crust with substantially higher velocities of 6.8–7.2 km/s with an undulating upper boundary at 8.5 km depth in the SW and 6 km in the NE thickens in the same direction.

In the northwestern Indian Ocean, the detailed crustal structure of the Makran subduction zone off Pakistan (Fig. 9.8.3-06) was the goal of a cruise of the German R/V *Sonne* in 1997 in a cooperation of GEOMAR, Kiel, and BGR, Hannover in Germany, Cambridge University in Great Britain and the National Institute of Oceanography of Pakistan. The seismic experiment, performed in 1997, was part of a greater project MAMUT

(Makran-Murray Traverse) with the objective to study the evolution of recent processes at the Murray Ridge and the Makran accretionary wedge (Kopp et al., 2000). Five wide-angle seismic lines were shot across and along the Makran accretionary wedge (Fig. 9.8.3-07). The 160-km-long, N-S–trending, dip line 8 was chosen to be coincident with an MCS line collected by Cambridge University in 1986. Four E-W–trending strike-lines, 100–125 km long, were shot parallel to the deformation front. Between seven and nine OBHs were deployed along each profile, located between and parallel to adjacent ridges (Kopp et al., 2000).

The interpretation (Fig. 9.8.3-08) revealed that the Moho dipped gently from ~14 km depth to ~22 km depth below sea level along the dip line. A bright reflector was interpreted as a décollement within the turbiditic sedimentary sequence, where a sequence of sediments more than 3 km thick sediments have bypassed the first accretionary ridges and have been transported to greater depth. In the strike lines, low-velocity layers were seen landward of the deformation front. The oceanic crust was identified by a strong velocity contrast between layers 2 and 3 (Kopp et al., 2000).

9.8.4. Atlantic Ocean

9.8.4.1. South Atlantic Ocean

The seismic surveys we have collected for the South Atlantic deal mainly with surveys near the continental margins. Continental flood basalts on the conjugate margins of southern Africa and South America and the existence of paired hotspot tracks on the South Atlantic seafloor had been recognized as evidence that the South Atlantic is flanked by plume-related volcanic margins (Fig. 9.8.4-01). Shallow marine seismic data supported this interpretation with the recognition of seaward dipping reflector sequences on both sides of the South Atlantic (Light et al., 1993; Hinz et al., 1999). We first discuss projects exploring margin areas of Africa (Bauer et al., 2000; Edwards et al., 1997), and then projects investigating the offshore region of South America (Franke et al., 2006; Mohriak et al., 1998).

To obtain the hitherto missing deep-crustal seismic data, in 1995 the project MAMBA (Geophysical Measurements across the Continental Margin of Namibia experiment), a combined onshore-offshore multichannel and wide-angle seismic survey, was performed jointly by three German institutions, the BGR (Bundesanstalt fuer Geowissenschaften und Rohstoffe, Hannover), AWI (Alfred-Wegener-Institut, Bremerhaven) and GFZ (GeoForschungsZentrum, Potsdam). The seismic measurements were obtained along two transects (Fig. 9.8.4-02) offshore and onshore of Namibia, traversing deeply eroded basement rocks of the late Precambrian Damara Orogen onshore and extending offshore out across the ocean-continent transition. The experiment consisted of a two-ship marine survey using the Russian vessel R/V *Akademik Nemcinov* and the Norwegian vessel R/V *Polar Queen*. Airgun shots were fired at 60 s intervals, corresponding to a 140 m source spacing, and were recorded simultaneously by OBHs of AWI and by onshore seismometers of GFZ. Distances

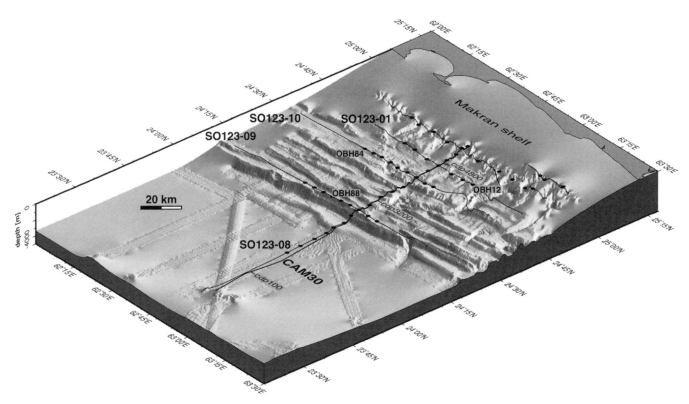

Figure 9.8.3-07. Bathymetric map of the Makran accretionary wedge with locations of seismic profiles and OBH locations (from Kopp et al., 2000, fig. 2). [Tectonophysics, v. 329, p. 171–191. Copyright Elsevier.]

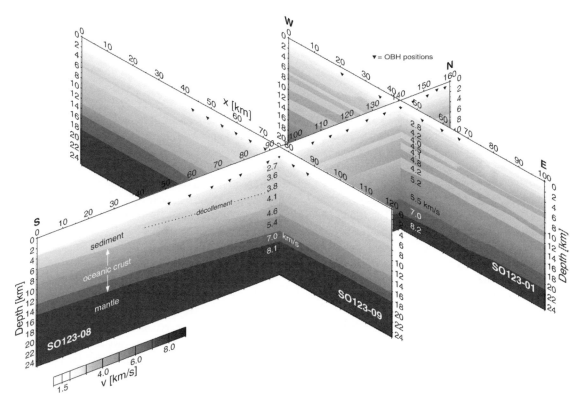

Figure 9.8.3-08. Fence diagram of the wide-angle seismic velocity models displaying shallow subduction beneath the Makran accretionary wedge (from Kopp et al., 2000, fig. 7). [Tectonophysics, v. 329, p. 171–191. Copyright Elsevier.]

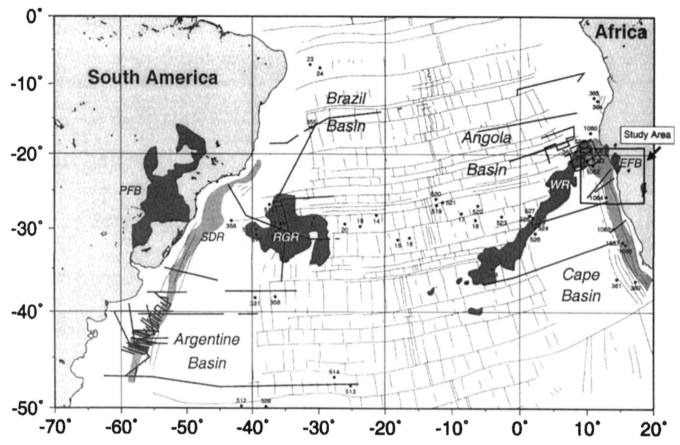

Figure 9.8.4-01. Reflection seismic lines of BGR in the South Atlantic Ocean (from Bauer et al., 2000, fig. 1) including transects 1 and 2 of Figure 9.8.4-02 (box "Study Area"). Grey areas—appearance of seaward dipping reflectors (SDR); black areas—magmatic features (PFP Parana flood basalts; RGR—Rio Grande Rise; WR—Walvis Ridge; EFB—Etendeka flood basalts) related to the Mesozoic opening of the South Atlantic. Numbers indicate DSDP/ODP drill sites. [Journal of Geophysical Research, v. 105, p. 25,829–25,853. Reproduced by permission of American Geophysical Union.]

between the OBHs varied between 40 and 70 km. On land, 25 recorders were spaced from 2.5 km (coastal region) to 15 km (NE end of the lines) apart. The refraction/wide-angle reflection data were compiled as receiver gathers for each OBH and land station (Bauer et al., 2000).

The interpretation (Fig. 9.8.4-03) shows a crustal thickness of ~10 km at the SW end of both lines. The Moho depths indicated a typically oceanic crust at the SW end of both lines to ~320 km distance from the shoreline. It was followed by a 150–200-km-wide zone of igneous crust of up to 25 km thickness off Namibia. The continental crust along the transects reflected differences in the regional geology. The uniform velocity structure of the southern transect showed similar values as derived by the 1975 Damara project (see Fig. 7.7.2-03) for the southern central zone of the Damara Belt (Baier et al., 1983).

In 1993, Edwards et al. (1997) explored the transform margin off Ghana. Six lines, between 80 and 120 km long, were shot parallel to the margin in SW-NE direction with ~10 km spacing, from the continental shelf down to the Guinea basin. A seventh line was normal to and bisected these lines. Airgun shots with 170 m

and 300 m spacing as well as explosive charges with 1.0–1.5 km spacing were recorded by digital OBS. The modeling of this suite of wide-angle seismic lines across the continental margin at the eastern end of the Romanche fracture zone off Ghana showed that the transition from continental to oceanic crust is evidently confined to a narrow, 6–11-km-wide zone, located at the foot of the steep continental slope. The crust of the ocean-continent transition proved to be characterized by high velocities (5.8–7.3 km/s). The continental crust appeared unaffected by the proximity of the adjacent oceanic crust, and there was no evidence for underplating of continental crust adjacent to the transition zone.

On the South American side of the South Atlantic Ocean, one project dealt with the broad shelf off Argentina, where in 1998, a 665-km-long onshore-offshore refraction profile was recorded along latitude 40°S across the Colorado Basin, a sedimentary basin separated from the deep-sea basin by a basement high paralleling the continental margin (Franke et al., 2006). Along the 457-km-long marine wide-angle seismic profile and accompanying MCS lines, nine OBHs systems were deployed at an average spacing of 40 km. In addition, five three-component

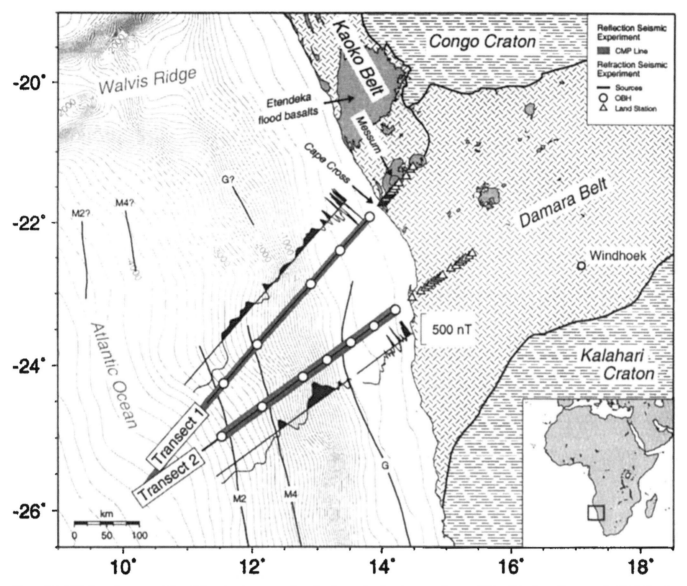

Figure 9.8.4-02. Location map of the offshore/onshore seismic reflection and refraction experiment off Namibia (from Bauer et al., 2000, fig. 2). The wiggle traces beside the seismic lines are simultaneously measured magnetic data. [Journal of Geophysical Research, v. 105, p. 25,829–25,853. Reproduced by permission of American Geophysical Union.]

seismometers were operated on land along the western prolongation of the refraction line BGR98-01 (Fig. 9.8.4-04).

The land stations were six-channel digital recorders (Teledyne PDAS 100), each with a three-component 1-Hz geophone and a six-4.5-Hz-geophones string sensor (SM-6). The seismic source was a tuned set of four linear subarrays with 32 airguns, fired at equidistant shot intervals of 175 m. The resulting velocity-depth cross section (Fig. 9.8.4-05) showed a relatively homogeneous three-layered crystalline crust with Moho depths around 30 km along the whole line.

Farther north, another seismic project concentrated on divergent margin basins located in the South Atlantic Ocean at the northeastern Brazilian margin aiming for petroleum explora-

tion. It consisted of the acquisition of regional seismic profiles in ultradeep waters, close to the boundary between continental and oceanic crust. Here, during an extensive program of deep reflection profiling, high-quality 18 s (two-way traveltime) seismic-reflection profiles were acquired in the Sergipe-Alagoas and the Jacuipe basins off NE Brazil in the period between 1989 and 1991 by Petrobras (Petroleo Brasileiro S.A.; Rio de Janeiro, Brazil). Three lines, 80–110 km long, extended from the coastline toward the boundary with the oceanic crust (profiles A to C in Fig. 9.8.4-06). Along these lines special seismic acquisition techniques were applied. The cable lengths varied between 4 and 6 km, serving 160–240 channels. Airgun shots were spaced at 50 m, and the sampling rate of the digital recording was 2 ms.

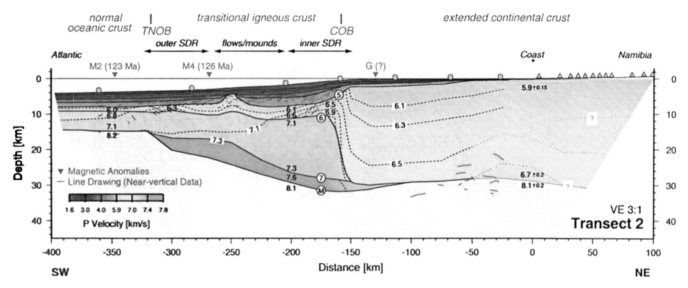

Figure 9.8.4-03. Crustal cross section along the southern Transect 2 of the MAMBA experiment (from Bauer et al., 2000, plate 1). [Journal of Geophysical Research, v. 105, p. 25,829–25,853. Reproduced by permission of American Geophysical Union.]

Other marine data with record lengths of 12 s were acquired in 1992 by the Brazilian Navy in cooperation with PETROBAS at a geophysical survey along the continental margin off northeastern Brazil (Gomes et al., 2000) to map the location of the ocean-continent transition. Known as EPLAC project, over 45,000 line-kilometers of multichannel seismic, gravity, magnetic and bathymetric data were collected along the Brazilian Continental Margin. Around the transition, ~50–100 km off the coast line, the Moho depth was found at a rather uniform depth of 15 km below sea level, the crystalline crust being ~8 km thick, overlain by 2–3-km-thick sediments.

Combined with potential field data and calibrated with the results from several exploratory boreholes, the seismic data provided evidence that the deep seismic reflectors in the deep-water region were related to crustal architecture rather than to sedimentary features. The record section of profile C is shown as an example (Fig. 9.8.4-07), showing the rapid decrease of crustal thickness from near 10 s to ~5 s over a distance range of 80 km. The section also shows a seaward dipping reflector wedge, which Bauer et al. (2000) also found on the Namibian side. Mohriak et al. (1998) pointed out that the deep-water province is characterized by wedges of reflectors that dip seaward marking the transition to a pure oceanic crust. A schematic model (Fig. 9.8.4-08) shows the gradational thinning of the crust (including sediments) from nearly 27 km thickness at the coast to ~10 km thickness at 100 km seaward distance and 4 km water depth, corresponding to a thinning of the crystalline crust from 18 to 9 km thickness.

A review of the major continental margin surveys performed in the 1980s and 1990s along the western borders of the Atlantic Ocean, both for South and North America, was prepared by Talwani and Abreu (2000) discussing the marine seismic-refraction and -reflection data in the context with other geophysical obser-

vations with the aim to compare and contrast the U.S. coast volcanic continental margin with the volcanic continental margins of the South Atlantic.

9.8.4.2. Mid-Atlantic Ridge

The Mid-Atlantic Ridge was repeatedly studied since the 1960s. In each following decade, new experiments followed with modernized instrumentation. A particular goal was to map the seismic properties of the slow spreading Mid-Atlantic Ridge in contrast to the fast-spreading East Pacific Rise.

In 2000, a grid of four wide-angle seismic profiles were shot during the second leg of the *Meteor* (1986) cruise M47, using airguns and OBS or OBH across interesting structures of a segment boundary at 5°S on the Mid-Atlantic Ridge. This region had been selected because satellite gravity data showed that this region was relatively undisturbed: The 5°S Fracture Zone separates relatively wide spreading segments and is itself characterized by a single structure with an offset of about 50 km. Two profiles crossed the median valley obliquely, the other two crossed the 5° transform fault and fracture zone.

In total, 77 successful deployments of the GEOMAR ocean-bottom hydrophones and ocean-bottom seismometers were carried out and provided a solid database to derive a detailed picture of the seismic velocity field. Shooting along the profiles was made using up to three 32-l airguns. Velocity models showed a strong velocity gradient in the first 2–3 km below the seafloor, underlain by a lower velocity gradient in the lower crust/upper mantle. In the Median Valley, the velocities were significantly reduced. The crust mantle boundary was not well defined in all regions, but crustal thickness varied between 3 and 6 km. In general, morphological highs were marked by elevated velocities below the seafloor, coincident with results from dredging that recovered gabbros (Borus et al., 2001).

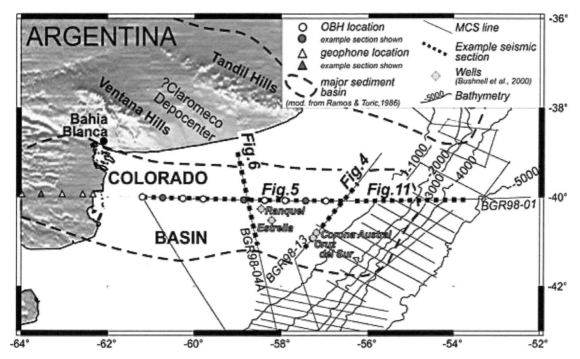

Figure 9.8.4-04. Location map of the offshore/onshore seismic reflection and refraction experiment off Argentina across the Colorado Basin (from Franke et al., 2006, fig. 2b). [Geophysical Journal International, v. 165, p. 850–864. Copyright John Wiley & Sons Ltd.]

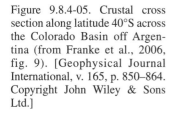

Figure 9.8.4-05. Crustal cross section along latitude 40°S across the Colorado Basin off Argentina (from Franke et al., 2006, fig. 9). [Geophysical Journal International, v. 165, p. 850–864. Copyright John Wiley & Sons Ltd.]

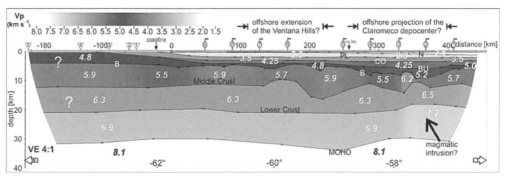

A group of projects (MARBE, Mid-Atlantic Ridge Bull's Eye) covered the Mid-Atlantic Ridge south of the triple junction west of the Azores. Another area of interest was a segment of the Reykjanes Ridge, investigated under the umbrella of BRIDGE (British Mid-Ocean Ridge Initiative). In 1996, the MARBE Experiment covered three spreading segments (Fig. 9.8.4-09) from ~33°30′ N to 35°30′ N longitude at around 37° W latitude (Canales et al., 2000; Hooft et al., 2000; Hosford et al., 2001). The 1996 cruise of R/V *Maurice Ewing* was the first major controlled-source seismic experiment on the northern Mid Atlantic Ridge south of the Reykjanes Ridge in over 15 years and was located ~450 km southwest of the Azores triple junction (Fig. 9.8.4-10A). Three seismic-refraction profiles followed the rift valleys (OH-1, OH-2 and OH-3 in Fig. 9.8.4-09), which are separated from each other by non-transform offsets (NTO-1 to NTO-3 in Fig. 9.8.4-09).

The shot interval was 90 seconds, resulting in an average shot spacing of ~200 m. The 90-km-long refraction profile along the northern segment, OH-1, comprised ten receivers, spaced 7–15 km apart, and 485 shots. Along OH-2 five stations, spaced 12–15 km apart, recorded 301 shots, and six stations along OH-3 also recorded 301 shots (Hooft et al., 2000). Furthermore, two off-axis refraction profiles were shot in segment OH-1. The western line was located in the western rift mountains (Canales et al., 2000) and crossed the non-transform offset NT01, following the eastern rift mountains of segment OH-2 (numbered line in Fig. 9.8.4-10). The line was 155 km long and comprised ten OBHs which recorded 615 airgun shots, fired at an interval of 120 seconds, providing a seismic trace spacing of ~250 m. The eastern line was 70 km long and was located in the eastern rift mountains of segment OH-1, 20 km away from the rift valley (Hosford et al., 2001). Here, eight OBHs were deployed at an

Figure 9.8.4-06. Location of three regional deep seismic profiles A, B, and C on the northeastern Brazilian margin on a regional map with tectonic elements along the South Atlantic Ocean (from Mohriak et al., 1998, fig. 1). [Tectonophysics, v. 288, p. 199–220. Copyright Elsevier.].

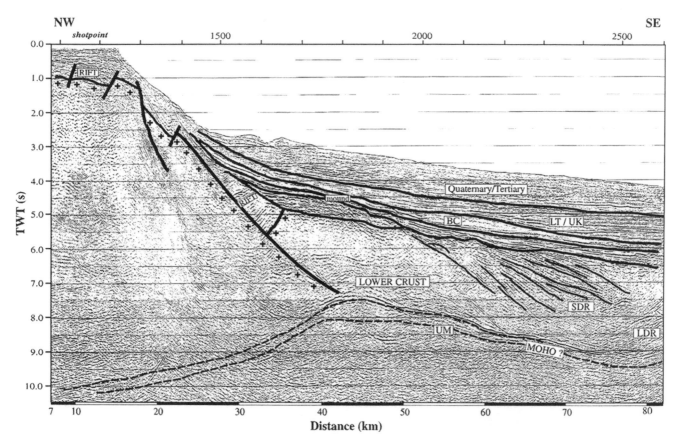

Figure 9.8.4-07. Deep seismic profile C in the Jacuipe basin on the northeastern Brazilian margin in the South Atlantic Ocean with interpretation and tectonic elements (from Mohriak et al., 1998, fig. 8). UM—upper mantle–lower crust transition; SDR—seaward dipping reflector wedge. [Tectonophysics, v. 288, p. 199–220. Copyright Elsevier.].

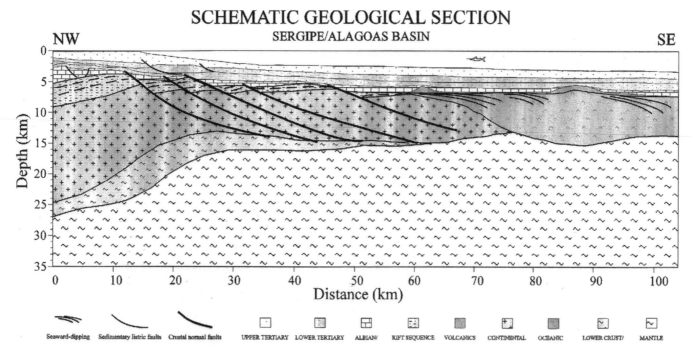

Figure 9.8.4-08. Schematic geological model of the Sergipe sub-basin on the Brazilian margin in the South Atlantic Ocean (from Mohriak et al., 1998, fig. 10). [Tectonophysics, v. 288, p. 199–220. Copyright Elsevier.].

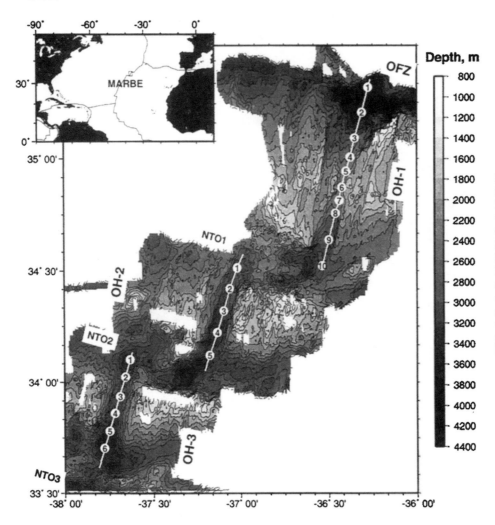

Figure 9.8.4-09. Bathymetric map along the Mid-Atlantic Ridge from 33°30′ N to 35°30′ N with the location of the three rift valley deep seismic profiles of the project MARBE (from Hooft et al., fig. 1). Lines—location of airgun shots; numbered circles—locations of ocean-bottom hydrophones; OH-1 to OH-3—spreading segments; NTO1 to NTO3—nontransform offsets; OFZ—oceanographer fracture zone. [Journal of Geophysical Research, v. 105, p. 8205–8226. Reproduced by permission of American Geophysical Union.]

average spacing of 9 km. They recorded 265 airgun shots with a 120 s firing rate, yielding an average shot spacing of 280 m (line C in Fig. 9.8.4-10).

Hosford et al. (2001) compiled P-wave velocity models for segment OH-1, both for the two rift mountain lines, imaging 2-m.y.-old crust, and the axial rift valley line. The seismic Moho along the lines was defined by the P_MP reflection points (white stars in Fig. 9.8.4-11). The main results were summarized by Hosford et al. (2001) as follows: the mean upper crustal velocity (layer 2) increases by 16% from 4.5 km/s at the axis to 5.2 km/s at the 2-m.y.-old crust on each flank. The maximum crustal thickness is 8–9 km beneath the centers of all three lines and the crust thins by up to 5 km toward the ends of the segment OH-1. A mid-crustal high-velocity body was spatially associated with clusters of large seamounts in the western and eastern rift mountains (Fig. 9.8.4-11).

Rather little was evidently hitherto known about the internal velocity structure of ocean islands and of seamounts which have not reached the sea surface. Good constraints were obtained (Minshull, 2002) for only a few oceanic intraplate volcanoes, such as the Jasper Seamount in the northeast Pacific (Hammer

et al., 1994) and the Great Meteor Seamount in the central Atlantic (Weigel and Grevemeyer, 1999).

Here the exploration of the Great Meteor shall be discussed in some detail. West from the triple junction of the Mid-Atlantic Ridge and the Azores-Gibraltar zone the Atlantis-Meteor seamount complex is located at 28°–30°W, 30°–34°N, a group of submarine volcanoes and guyots, ~700 km to the south of the Azores. As a continuation of the "Atlantische Kuppelfahrten" in the 1960s (Closs et al., 1968), the University of Hamburg, Germany, carried out a detailed geophysical survey in 1990, sampling the crustal structure of the four seamounts Meteor, Cruiser, Plato, and Atlantis. Weigel and Grevemeyer (1999) interpreted the data of the largest seamount, the Great Meteor. It had been discovered in 1938 and is a large-size, flat-topped guyot, rising from a depth of more than 4700 m to a crestal depth of 275 m. The top is covered by 200–900 m thick sediments which superimpose basement rocks with P-velocities of 5.9–6.0 km/s. Five OBHs were deployed along a 200 km line across the seamount. An airgun array of 76 l capacity was towed along the line and shot every 2 min. Two of the OBHs were positioned at the western and eastern flanks at 2930 m and 2700 m depth, respectively, two on the top

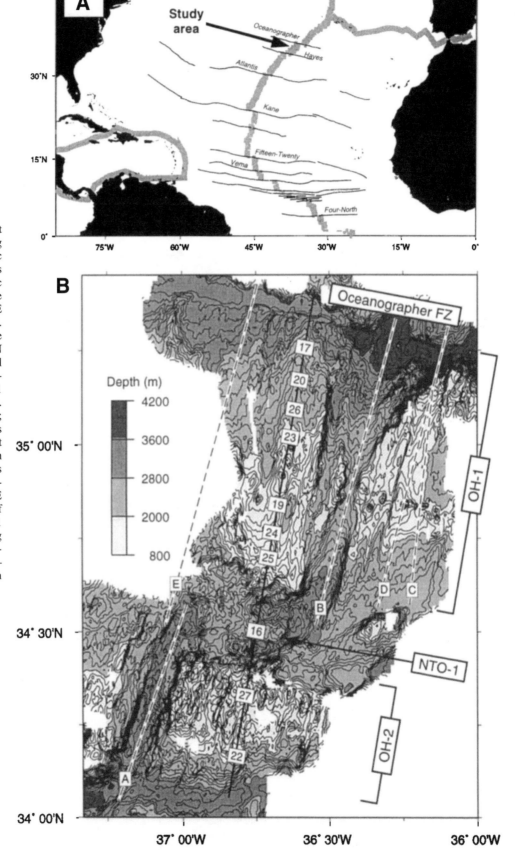

Figure 9.8.4-10. (A) Central part of the Atlantic Ocean showing the location of the Mid-Atlantic Ridge and other plate boundaries (shaded bands), the main Atlantic fracture zones (solid lines) and the location of the project MARBE (from Canales et al., 2000, fig. 2a). (B) Bathymetric map along the Mid-Atlantic Ridge from 34°00′ N to 35°30′ N with the location of all deep seismic profiles of the project MARBE along segment OH-1 (from Canales et al., 2000, fig. 2). C—eastern rift mountains line; B—rift valley line (corresponds to the seismic line along segment OH-1 in Figure 9.8.4-09); line with numbers—western rift mountains line (Canales et al., 2000; number denote locations of OBHs). E and D—earlier refraction lines of Sinha and Louden (1983, lines C1 and B); A—rift valley line along segment OH-2. [*Journal of Geophysical Research*, v. 105, p. 2699–2719. Reproduced by permission of American Geophysical Union.]

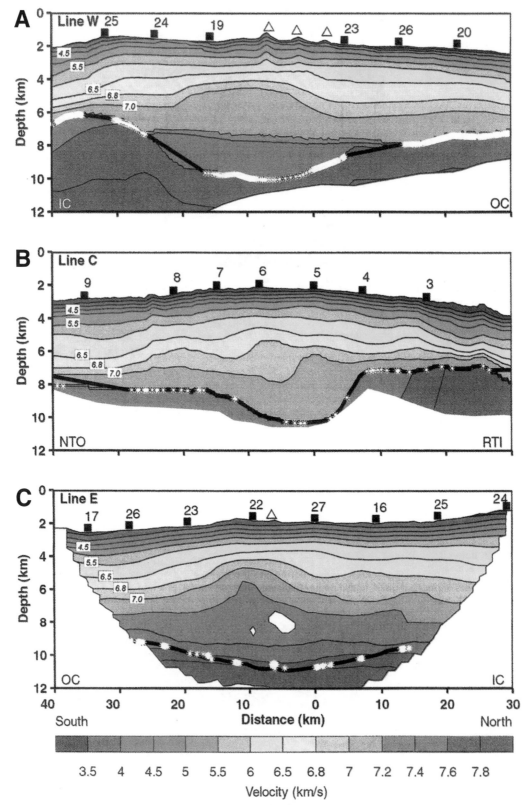

Figure 9.8.4-11. Final P-wave velocity models for segment OH-1 along the Mid-Atlantic Ridge from 34°15′ N to 35°30′ N (from Hosford et al., 2001, Plate 1). (A) Western rift mountains line W (= numbered line in Fig. 9.8.4-10). (B) Rift valley line C (= line B in Fig. 9.8.4-10). (C) Eastern rift mountains line E (= line C in Fig. 9.8.4-10). White stars—P_MP reflection points; full squares—OBH locations; open triangles—seamount clusters. [Journal of Geophysical Research, v. 106, p. 13,269–13,285. Reproduced by permission of American Geophysical Union.]

of the guyot at 300 and 440 m depth and the fifth OBH located 55 km off to the SE at 4700 m depth.

The interpretation was performed with a 2-D ray-tracing procedure. Starting from the 7-km-thick oceanic crust to the NW, with layer 2 at 5.0–8.5 km depth (velocities increasing gradually from top to bottom from 4.5 to 5.5 km/s) and layer 3 at 8.5–12.0 km depth (velocity of 6.5 km/s) crustal thickness increases underneath the edifice to a total thickness of ~17 km, with the top of the lower crustal layer upwarping to 2 km depth and the bottom downwarping to 17 km with a gradual velocity increase from 6.0 at the top to 7.5 km/s at the bottom. The upper-mantle velocity in the model was kept within 8.0 km/s. The upper crustal layer remains more or less constant in thickness, but its velocity is heavily decreased being upwarped along the flank of the sea mountain over the uprising and thickening lower crustal body (Weigel and Grevemeyer, 1999).

During the same expedition in April 1990 of the German vessel *Meteor* (1986), the Atlantis and Plato seamounts as well as the Great Meteor bank (27°–32°W, 28°–35°N) were targeted with 28 reflection seismic measurements and 40,000 shot positions deploying eight OBSs in two arrays of four along each line and using an airgun array (eight airguns deployed as two simul-

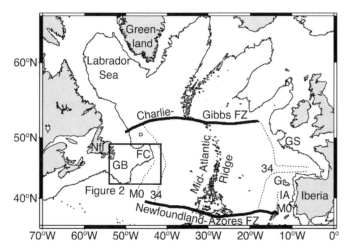

Figure 9.8.4-12. Location map of the mid–North Atlantic showing the Newfoundland, Iberia, and Irish margins (from Funck et al., 2003, fig. 1). Thick solid lines—fractures zones (FZ); thin solid line—2000-m bathymetric contour; dashed lines—M0 and 34 magnetic anomalies; FC—Flemish Cap; GB—Grand Banks; Nfl—Newfoundland; G—Galicia Bank; IA—Iberia Abyssal Plain; GS—Goban Spur. [Journal of Geophysical Research, v. 108 (B11), 2531, doi: 10.1029/2003JB002434. Reproduced by permission of American Geophysical Union.]

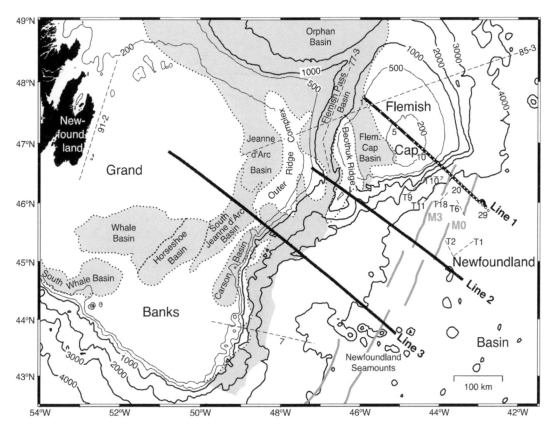

Figure 9.8.4-13. Bathymetric map of the Grand Banks and Newfoundland Basin east of Newfoundland, showing the location of the SCREECH experiment (from Funck et al., 2003, fig. 2). Thick solid lines—the three main lines; thin solid line—bathymetric contour lines; thin dashed lines—other seismic lines (e.g., for location of line 91-2 see also Figure 9.4.1-04); grey lines—M0 and M3 magnetic anomalies. [Journal of Geophysical Research, v. 108 (B11), 2531, doi: 10.1029/2003JB002434. Reproduced by permission of American Geophysical Union.]

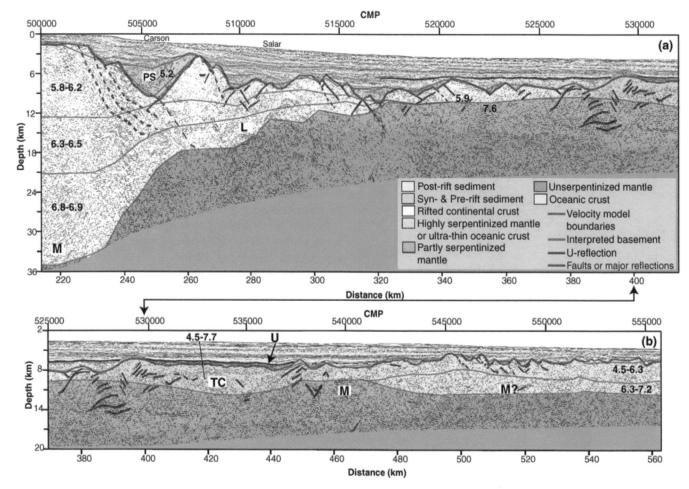

Figure 9.8.4-14. Velocity-depth model of SCREECH line 3: Comparison between reflectivity and the coincident velocity model derived from wide-angle data (from Lau et al., 2006a, fig. 8). L—landward-dipping reflections; TC—transitional crust; M—Moho reflections that coincide with the velocity model Moho; U—U-reflection; PS—package of pre-rift/syn-rift sediment within Carson basin. [Geophysical Journal International, v. 167, p. 157–170. Copyright John Wiley & Sons Ltd.]

taneously shot independent arrays of four airguns each with an airgun interval of 1 m at 5 m water depth) as energy source. Furthermore, along five refraction profiles 4000 shots were recorded from an airgun array of four 19-l-airguns. In total, 18 sites were occupied by five analogue OBSs, while for 11 sites, four digital OBSs were available (Wefer et al., 1991).

The Canary Islands, located at 14°–18°W, 28°–29°N to the west of the North African coast are an ocean island complex with anomalous structure of the oceanic lithosphere. Here a deep seismic experiment was carried out around the island Gran Canaria in 1993 during a cruise of the German research vessel *Meteor* (1986) (Schmincke and Rihm, 1994; Ye et al., 1999). Previous studies had been carried out on these islands in the 1960s and 1970s (e.g., Bosshard and MacFarlane, 1970; Banda et al., 1981a; see Chapter 7.8.4). The field work comprised three profiles radiating from Gran Canaria toward N, NNE, and NE. In addition, several transverse profiles and short profiles around the island were surveyed. Airguns were fired on each line at 1 or 2 min intervals and were recorded by eight onshore land sta-

tions and eight OBHs covering the N and NE sector off Gran Canaria up to a distance of 60 km from the coast. The interpretation showed that under the ocean basin adjacent to the volcanic edifice a 4-km-thick sedimentary sequence (average velocity 3.4 km/s) overlies the 7-km-thick igneous crust (velocity of 4.5 km/s). Beneath the island a pronounced lateral velocity variation takes place. Within the central volcanic edifice a 4–5-km-thick low-velocity zone was found, at roughly 4–12 km depth south of the island center. The massive volcanic island flank thins rapidly away from the island with velocities decreasing from 5.0 km/s near the coast to 3.5 km/s in the outermost part of the profiles, 50–60 km off the coast. A clear doming of lower crust material (>6.6 km/s) to 8–10 km depth was attributed to young mafic plutonic rocks. The Moho depth varies only slightly from 14 km north and northeast of Gran Canaria to 16 km depth under the center of the island (Ye et al., 1999).

In 1993, the RAMESSES study (Reykjanes Axial Melt Experiment: Structural Synthesis from Electromagnetics and Seismics) targeted an apparently magmatically active axial volcanic

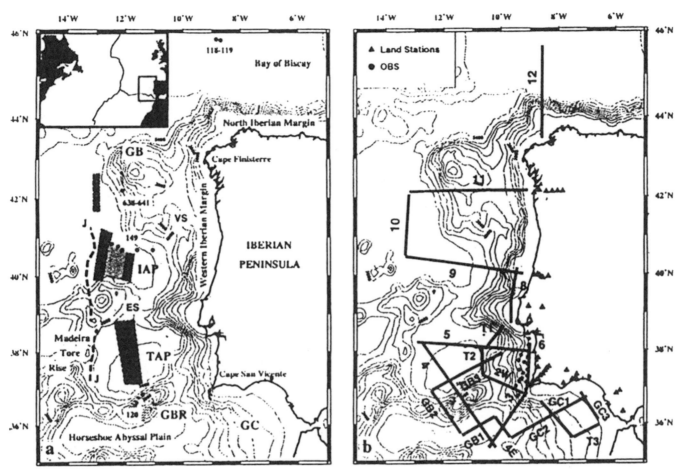

Figure 9.8.4-15. Bathymetry (500 m intervals) map of Iberian Atlantic margin (from: Banda et al., 1995, fig. 1). Left: Map of the IAM (Iberian Atlantic Margins European project) study area. Numbered solid circles—DSDP and ODP drill sites; ES—Extremadura Spur; GB—Galicia Bank; GBR—Gorringe Bank Region; GC—Gulf of Cadiz; IAP—Iberian Abyssal Plain; TAP—Tagus Abyssal Plain; VS—Vigo Seamount. Right: Location map of the IAM seismic reflection profiles. Triangles—land stations; circles—ocean-bottom seismometers. [Eos (Transactions, American Geophysical Union), v. 76, no. 3, p. 25, 28–29. Reproduced by permission of American Geophysical Union.]

ridge, centered on 57°45′N at the Reykjanes Ridge, with the aim of investigating the processes of crustal accretion at a slow-spreading mid-ocean ridge. As part of the project, airgun and explosive wide-angle seismic data were recorded by ten digital OBSs along profiles oriented both across- and along-axis. Co-incident normal-incidence and other geophysical data were also collected (Navin et al., 1998). At 57°45′N, the Reykjanes Ridge appeared to have a crustal thickness of ~7.5 km on-axis. Slightly off-axis, the crust was modeled to decrease in thickness, imply-ing off-axis extension. Modeling also indicated that the axial vol-canic ridge was underlain by a thin (~100 m), narrow (~4 km) melt lens some 2.5 km beneath the seafloor, which overlay a broader zone of partial melt ~8 km in width.

Under the auspices of the British Mid-Ocean Ridge Initia-tive (BRIDGE) a multicomponent geophysical experiment was conducted on a magmatically active, axial volcanic ridge seg-ment of the Reykjanes Ridge, centered on 57°43′N (Sinha et al., 1999). The site of the experiment was approximately located 840 km south of Reykjanes peninsula and over 1000 km off the

center of the Iceland plume. At the site a clear median valley is present which extends for more than 100 km northward toward Iceland. The project involved wide-angle seismic profiling with OBSs, seismic-reflection profiling, controlled-source electro-magnetic sounding and magnetotelluric sounding. The seismic part consisted of two lines where an array of eleven OBSs was deployed. The WNW-ESE line was 100 km long and line 2, 35 km long, ran perpendicular to line 1 along the axial volcanic ridge. From the seismic data it could be established that at dis-tances of 10 km away from the axis, a normal Atlantic oceanic crust of 6–8 km thickness existed. There the sedimentary cover was thin (up to 200 m) and discontinuous and the seismic veloc-ity at the top of the crystalline basement varied from 2.8–3.5 km/s. Following a steep velocity gradient at 2.0–2.5 km below the seafloor, a velocity of 6.5 km/s was reached, which depth was regarded as boundary between oceanic layers 2 and 3. With increasing depth the P-velocity increases more slowly, and just above the crust-mantle boundary at 6–8 km below the seabed a velocity of 7 km/s was seen. The uppermost mantle velocity was 7.9 km/s. The axial region had a quite different structure. The axial volcanic ridge lies within the shallow 30-km-wide and 800 m deep rift valley. Here, the P-velocities were between 1.0 and 1.5 km/s lower than the off-axis velocities at equivalent depths throughout most of layers 2 and 3. In agreement with the other data sets, the result was explained by the existence of an anomalous body, which was characterized by anomalously low P-velocity and low resistivity and was associated with a seismic reflector. This body was subsequently interpreted as melt body,

i.e., a shallow crustal magma chamber could be identified beneath a slow spreading ridge segment.

In 1995, a 310-km-long seismic-refraction profile, ICE-MELT, was recorded on land through central Iceland (Darbyshire et al., 1998). It traversed Iceland from the Skagi Peninsula on the north coast to the southeast coast, crossing central Iceland over the glacier Vatnajökull, below which the locus of the Iceland mantle plume is supposed to currently center. Up to 60 land-based instruments recorded each of six explosive shots, ranging

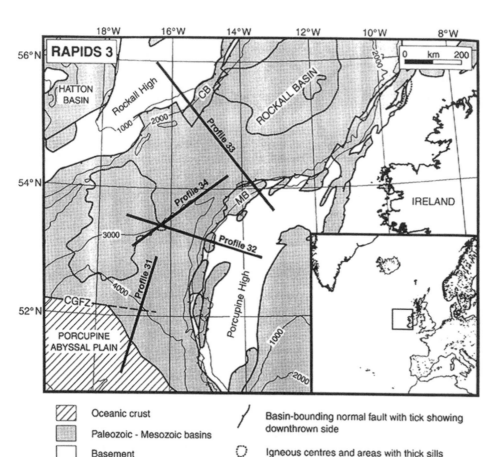

Figure 9.8.4-16. Location map of the RAPIDS 3 seismic profiles in the Hatton-Rockall area of the Northeast Atlantic margin (from Morewood et al., 2003, fig. 1). CGFZ—Charlie-Gibbs Fracture Zone; CB—Conall basin; MB—Macdara basin. [Eos (Transactions, American Geophysical Union), v. 84, no. 24, p. 225, 228. Reproduced by permission of American Geophysical Union.]

Figure 9.8.4-17. Preliminary P-wave velocity-depth model of RAPIDS 3 profile 33 across the Rockall basin (from Morewood et al., 2003, fig. 3a). [Eos (Transactions, American Geophysical Union), v. 84, no. 24, p. 225, 228. Reproduced by permission of American Geophysical Union.]

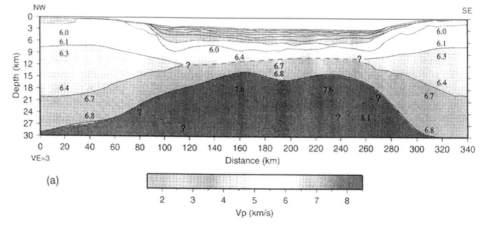

in size from 25 to 400 kg. Three shots were fired offshore (one in the north, two in the south) and three shots were in lakes in the central highlands of Iceland. Velocities in the upper crust ranged from 3.2 to 6.4 km/s. At both ends of the seismic line, the upper crust could be subdivided into two layers with a total thickness of 5–6 km. For the central highlands, only a single unit of upper crust was modeled, with seismic velocity increasing continuously with depth to almost 10 km below the surface. Below the central volcanoes of northern Vatnajökull, the upper crust was only 3 km thick. Below the base of the upper crust, velocities of

6.6–6.9 km/s were found and apparently increased with depth to 7.2 km/s. The seismic Moho was determined from the two large offshore shots at the profile ends. The resulting crustal thickness was 25 km at the north end and increased to 38–40 km beneath southern central Iceland. Mantle velocities could not be constrained (Darbyshire et al., 1998).

A second project (named B96), ~100 km farther east, was carried out in 1996 in northern Iceland (Menke et al., 1998). Along a north-south–striking array near the northern coast of 80 km length, three chemical underwater explosions were recorded, one

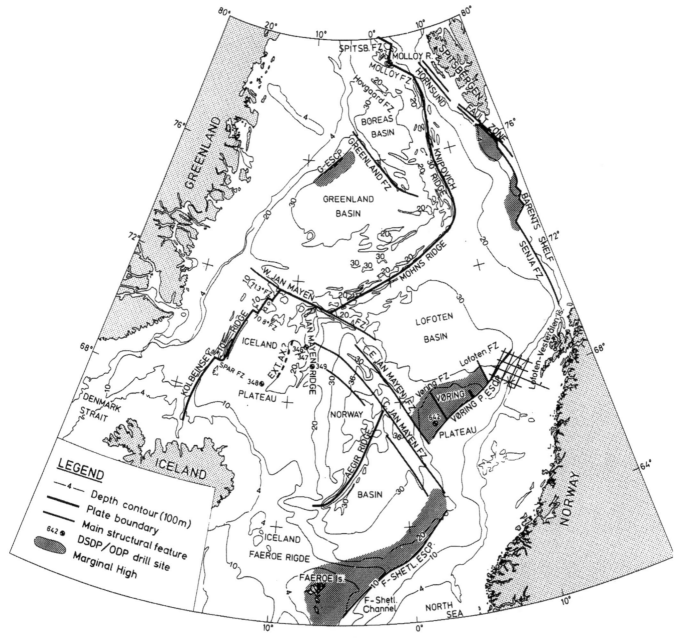

Figure 9.8.4-18. Regional physiography and structural basins and continental margins in the Norwegian-Greenland Sea. Also indicated are ODP drill sites and the 1988 survey off the Lofoten Islands, Norway (from Mjelde et al., 1992, fig. 2). [Tectonophysics, v. 212, p. 269–288. Copyright Elsevier.].

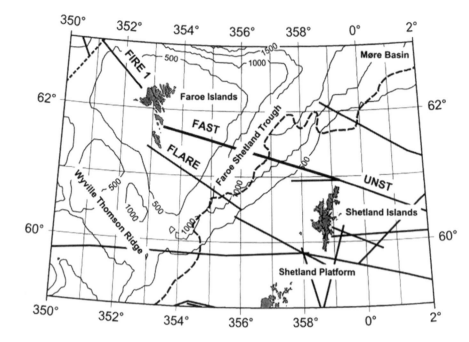

Figure 9.8.4-19. Location of FIRE 1, FAST-UNST and FLARE across the Faeroe-Shetland Trough (from England et al., 2005, fig. 1). [Journal of the Geological Society of London, v. 162, p. 661–673. Reproduced by permission of Geological Society Publishing House, London, U.K.].

Figure 9.8.4-20. Location of FIRE, the Faeroe-Iceland Ridge experiment (from McBride et al., 2004, fig. 1). EVZ, WVZ, NVZ—eastern, western, northern volcanic zones of Iceland. [Tectonophysics, v. 388, p. 271–297. Copyright Elsevier.].

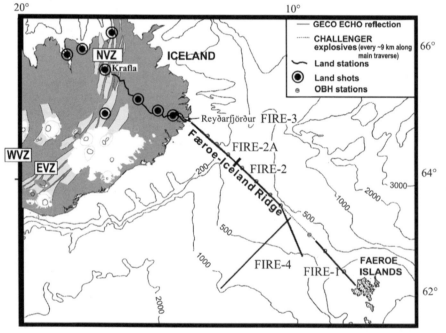

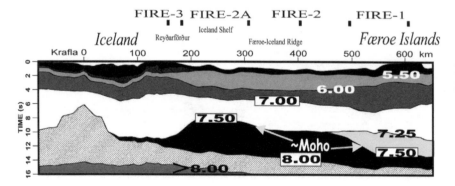

Figure 9.8.4-21. Velocity-depth model of the project FIRE between the Faeroe Islands and north-central Iceland (from McBride et al., 2004, fig. 2). [Tectonophysics, v. 388, p. 271–297. Copyright Elsevier.]

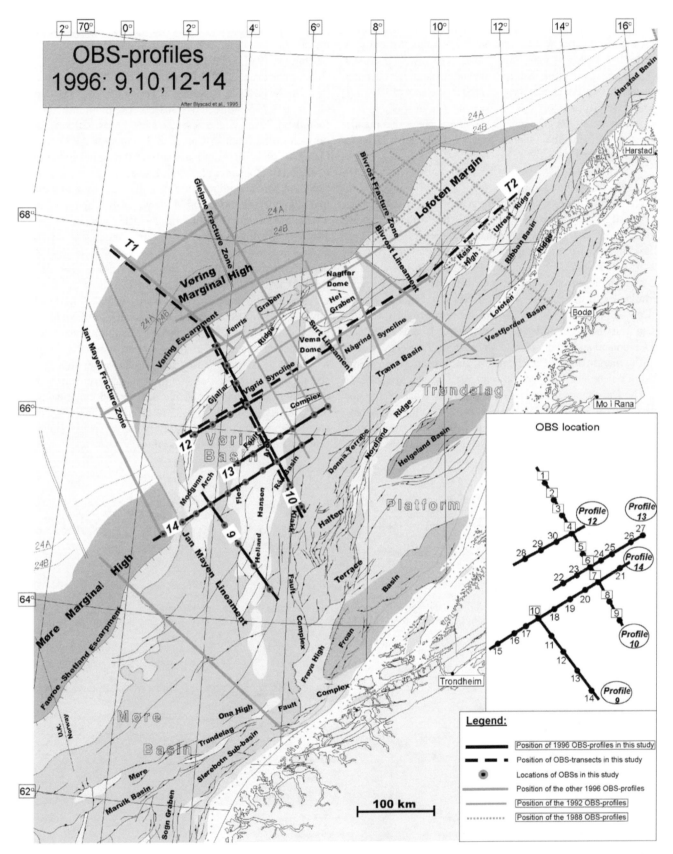

Figure 9.8.4-22. Location map of regional profiles at mid-Norwegian continental margin (from Raum et al., 2002, fig. 1). [Tectonophysics, v. 355, p. 99–126. Copyright Elsevier.].

shot at each end, and the third fan shot more than 100 km offline to the east. The array was operational for two months and also recorded a number of microearthquakes. The resulting crustal thickness was 25–31 km, with a southward thickening occurring in an abrupt step. An apparent rather high P_n velocity of 8.0 km/s was observed from recordings of an earthquake in southern Iceland, which the authors explained to be consistent with a mantle lid, a layer of subsolidus mantle separating the Moho from a deeper partial melt zone.

9.8.4.3. North Atlantic Ocean

The North Atlantic Ocean was the goal of many crustal structure investigations in the 1990s. We will start with the experiment SCREECH dealing with the margin areas off Canada (Fig. 9.8.4-12). Then we will jump to the other side of the Atlantic Ocean and continue with the project IAM (Iberian Atlantic Margin) and continue from there northwards to the microcontinents off Ireland and the margin areas off Norway. Finally, several projects will be discussed dealing with the northernmost Atlantic Ocean from the Shetland Island via the Faeroe Islands to Iceland and Greenland and finally close with an Arctic Transect.

The margin areas of the North Atlantic Ocean were investigated in much detail between 40° and 50°N latitude (Fig. 9.8.4-12). The Canadian margin in the west was the goal of a major experiment (SCREECH, Study of Continental Rifting and Extension on the Eastern Canadian Shelf) in 2000 across the Flemish Cap and into the deep basin east of Newfoundland (Funck et al.,

2003; Hopper et al., 2006; Lau et al., 2006a, 2006b). In the east, major experiments (RAPIDS, Rockall and Porcupine Irish Deep Seismic) in 1988, 1990, and 1999 covered the margin off Ireland (Mackenzie et al., 2002; Morewood et al., 2003).

The SCREECH experiment, performed in 2000, consisted of three NW-SE–directed transects reaching into the Newfoundland Basin (Fig. 9.8.4-13). Line 1 was 320 km long and ran across the Flemish Cap, line 2 started at the Outer Ridge Complex, and line 3 covered parts of the Grand Banks (Funck et al., 2003; Hopper et al., 2006; Lau et al., 2006a, 2006b). The refraction/wide-angle reflection seismic data were obtained in a two-ship operation, one ship deploying OBSs and the other one towing the airgun array. In total, 29 OBSs were deployed, with 14 of them being equipped with three-component 4.5 Hz geophones and one hydrophone, and the other 15 having a hydrophone component only. On line 1 the station spacing varied from 21 km in the shallow water of Flemish Cap to 8 km in the Newfoundland Basin. Shot spacing was 200 m. For line 3, only 24 OBSs were available, of which 11 were equipped with three-component 4.5 Hz geophones. Along this line, the instrument spacing ranged from 10 to 40 km. While the OBSs were recovered by the first vessel, the second vessel shot the line a second time with a denser shot interval, mostly 50 m, to collect coincident multichannel seismic data by a 6-km streamer with 480 channels, resulting in 12.5-m group spacing. Furthermore, supplementary multichannel data were collected along additional parallel and perpendicular lines.

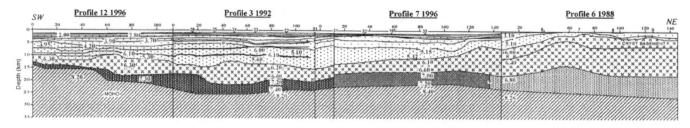

Figure 9.8.4-23. Crustal transect of the Voring margin off mid-Norway along the strike direction (from Raum et al., 2002, fig. 16). Explanation of symbols in Figure 9.8.4–24. [Tectonophysics, v. 355, p. 99–126. Copyright Elsevier.].

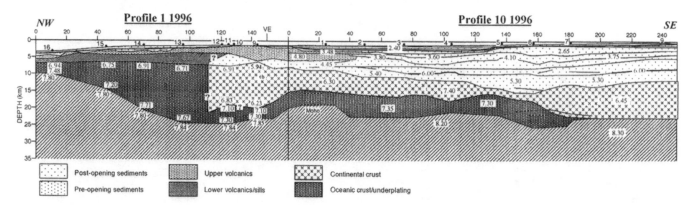

Figure 9.8.4-24. Crustal transect of the Voring margin off mid-Norway along the dip direction (from Raum et al., 2002, fig. 17). [Tectonophysics, v. 355, p. 99–126. Copyright Elsevier.].

The model for line 3 (Lau et al., 2006b) shows both the velocity-depth model, derived from the wide-angle data, and the coincident reflectivity pattern (Lau et al., 2006a, 2006b). The result (Fig. 9.8.4-14) was a 160–170-km-wide zone of continental rifted crust extending 120 km into the Newfoundland basin with abrupt thinning beneath the Carson basin from ~12–4 s TWT. It is separated from the oceanic crust in the east and southeast by a 70-km-wide transitional zone with a flat and unreflective basement in its central part (Lau et al., 2006b).

On the eastern side of the North Atlantic Ocean, many marine seismic projects were performed. They involved continental margin investigations off the west coast of Iberia, detailed crustal studies of the stretched continental crust west of Ireland, marine studies off the coast of Norway, and investigations of the northernmost areas around Spitsbergen, the Faeroes, Iceland and Greenland.

In 1993, the Iberian Atlantic Margins European project (IAM), a marine seismic project, focused on the Atlantic margin of Iberia (Fig. 9.8.4-15). Up to 3500 km of multichannel profiles were acquired in the North Iberian and Atlantic margins, the Gorringe bank and the Gulf of Cadiz (Banda et al., 1995). The shots were also recorded by OBS and by a variety of land stations. Three

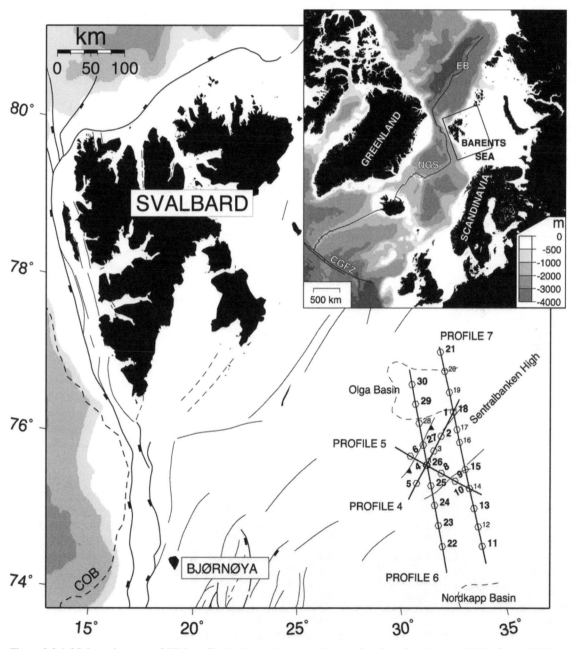

Figure 9.8.4-25. Location map of OBS profiles in the northwestern Barents Sea (from Breivik et al., 2002, fig. 1). COB—continent-ocean boundary west of Spitsbergen. [Tectonophysics, v. 355, p. 67–97. Copyright Elsevier.].

profiles, IAM 3, 9, and 11, were continued on land by short in-line profiles providing images of the continent-ocean transition.

The northernmost profile, IAM 12, was oriented in a N-S direction north of La Coruna (see also Fig. 9.2.3-01). IAM 11 ran in an E-W direction along latitude 42°N from the Galician coast across the Galicia margin, the E-W line IAM 9 ran approximately along latitude 41°N, and the N-S line IAM 10 surveyed the Iberian Abyssal Plain approximately along longitude 13.5°W. Lines IAM 8 and IAM 6 ran N-S parallel to the central and southern Portuguese coast and investigated the continental margin. The Tagus Abyssal Plain and the southwestern Iberian margin were surveyed by lines T1, 5, 2W, T2, and GB3. IAM 4 and some other lines investigated the Gorringe Bank region. IAM 3 ran NE-SW in front of Canbo San Vicente into the Horseshoe Abyssal Plain. Finally, four lines were recorded off southwestern Spain in the Gulf of Cadiz (GC1 to GC3 and T3).

The Gorringe Bank between the Tagus Abyssal Plain and the Gulf of Cadiz was of interest because of features resulting from the collision of the African and Eurasian plates including large-amplitude gravity anomalies. Line IAM 4 showed clear structural differences. To the north, a 1.5–2.0 s TWT thick sedimentary sequence overlay the oceanic crust of the Tagus Abyssal Plain; to the south at the Horseshoe Abyssal Plain the sediments are deformed by faulting and folding. Furthermore, a multicycle high-amplitude low-frequency reflector was located at ~8–9 s TWT. Along the IAM 9 seismic-reflection line in the southern Iberia Abyssal Plain between 40°N and 41°N, a wide transition zone was found, while the region of extended continental crust was relatively narrow. The transition zone along the IAM-9 profile was found to be a peridotite ridge, characterized by a thin upper layer, less than 3 km thick, and with velocities between 4.0 and 6.5 km/s with a high-velocity gradient. At depth followed a high-

velocity layer with ~7.6 km/s with a low velocity gradient. The Moho was absent or very weak.

In the following years, additional marine experiments were implemented in Atlantic margins. In 1995, a wide-angle profile was shot along IAM-9 which included the deployment of up to 16 OBSs (Chian et al., 1999; Dean et al., 2000) with the aim to investigate the transition zone in more detail. The resulting velocity structure allowed the distinction of four zones. From west to east followed each other a 6.5–7-km-thick oceanic crust, an overlap of two peridotite ridges, transition zone, and thinned continental crust. The peridotite ridges, which cover a distance of ~50 km width, differ from the rest of the transition zone by a reduced top basement velocity and an elevated middle crust velocity. The transition zone is 170 km wide and has a velocity structure which is neither oceanic nor continental. The continental crust with velocities ranging from 5.5 to 6.8 km/s thickens from 7 km in the west across the lower continental slope over a distance of 80 km to 28 km in the east.

The stretched continental crust west of Ireland was the goal of continuing research. Following the RAPIDS project, shot in 1988 and 1990 (see, e.g., Fig. 8.9.4-05), in 1999 RAPIDS 3 followed (Fig. 9.8.4-16), involving the acquisition of four wide-angle seismic-reflection/-refraction lines within the Irish sector of the Rockall Basin area of the northeast Atlantic margin, offshore of the Republic of Ireland. It involved 1038 km of seismic profiling (Mackenzie et al., 2002; Morewood et al., 2003).

A total of 197 ocean-bottom seismometers were deployed and successfully recovered by the vessel RN *Akademik Boris Petrov*, operated by GeoPro GmbH of Hamburg. A 50-l airgun array was used as a seismic source. Firing every minute this gave a shot spacing of 120 m. A second vessel, the RN *Celtic Voyager*, operated by the Irish Marine Institute, was employed

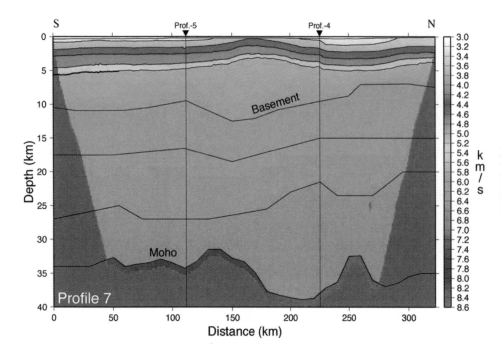

Figure 9.8.4-26. Model of OBS profile 7 in the northwestern Barents Sea (from Breivik et al., 2002, fig. 9b). [Tectonophysics, v. 355, p. 67–97. Copyright Elsevier.].

to fire 25-kg explosive shots as a secondary source. These were used as a backup to the airgun source, to ensure deep penetration of seismic energy into the crust and upper mantle. The quality of the airgun data was extremely good, with energy visible up to 140 km offset. Signal-to-noise ratios were high, and clear crustal and mantle phase arrivals were observed on the seismic sections.

Two profiles (32 and 33 in Fig. 9.8.4-16) were shot transverse to the trend of the Rockall Basin, thereby allowing the nature of the crustal thinning to be examined. Another profile (31) ran from the Rockall Basin and across the extension of the Charlie-Gibbs Fracture Zone. This profile was designed to allow examination of the crustal transition as the highly thinned continental crust of the Rockall Basin passes into the oceanic crust of the Porcupine Abyssal Plain. A fourth profile (34) extended approximately along the axis of the Rockall Basin.

The modeling resulted in a three-layer crust on the structural highs which flank the Rockall Basin. The crust appeared to be similar to that documented beneath onshore Ireland and the UK. A dramatic thinning of the crust was observed beneath the Rockall Basin. On profile 33 (Fig. 9.8.4-17), the crust thins from over 25 km beneath the Porcupine High to ~7 km beneath the Macdara Basin over a distance of ~60 km.

A similar thinning gradient was observed on profile 32. Crustal thinning at the northwestern end of profile 33 is more gradual, decreasing from over 25 km to ~6 km over a distance

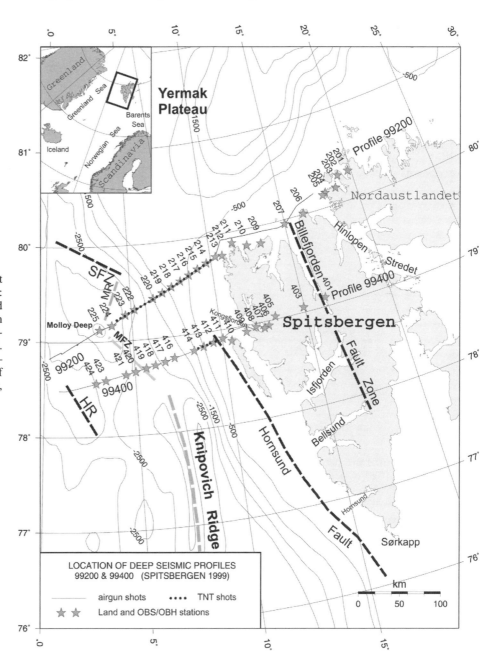

Figure 9.8.4-27. Location map of shot profiles and recording stations (stars: ocean-bottom seismometers and land stations) in the transition zone between the North Atlantic Ocean and Spitsbergen (from Czuba et al., 2004, fig. 1). [Polish Polar Research, v. 25, p. 205–221. Reproduced with permission of Polish Academy of Sciences, Warsaw, Poland.]

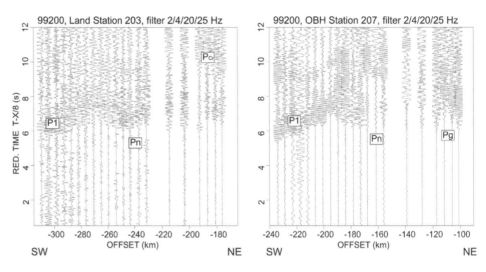

Figure 9.8.4-28. Examples of amplitude-normalized TNT seismic record sections (from Czuba et al., 2004, fig. 6) along line 99200, for stations RefTek 203 (left) and OBH 207 (right). Pg—first arrivals of crustal P-waves; Pcr—later crustal refracted P-waves; Pn—refracted P-waves beneath the Moho; P1—lower lithosphere reflections. [Polish Polar Research, v. 25, p. 205–221. Reproduced with permission of Polish Academy of Sciences, Warsaw, Poland]

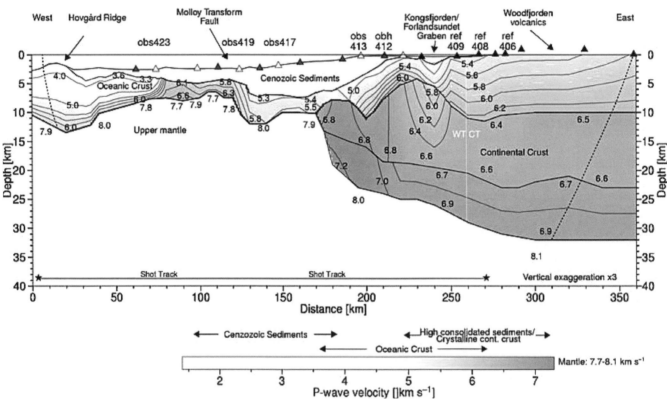

Figure 9.8.4-29. Crustal structure of the transition zone between the North Atlantic Ocean and Spitsbergen along line 99200 (from Ritzmann et al., 2004, fig. 7). [Geophysical Journal International, v. 157, p. 683–702. Copyright John Wiley & Sons Ltd.]

of ~120 km. Beneath the center of the Rockall Basin, a 6–7-km-thick sub-crustal layer was modeled, which thins toward the basin margins, with P-wave velocities lower than expected for normal upper mantle (Morewood et al., 2003).

South of the RAPIDS project, in 1997, data were obtained in the southern Porcupine Basin between 51° and 52°N. A 6-km-long, 240 channel digital streamer and a tuned sleeve-gun array were used with record length up to 9 s TWT (Reston et al., 2004). Five lines traversed the Porcupine Basin in an E-W direction; one line was shot along-strike in a N-S direction. Most prominent in the data was a bright reflection that was seen to cut down to the west from the base of the sedimentary section and in part to follow the top of a partially serpentinized mantle.

Various projects dealt with detailed seismic investigations of the northernmost part of the Atlantic Ocean, reaching from the Faeroe-Iceland area and the offshore area of Norway in the south to the Barents Sea, Spitsbergen and the offshore areas of Greenland in the north (Fig. 9.8.4-18).

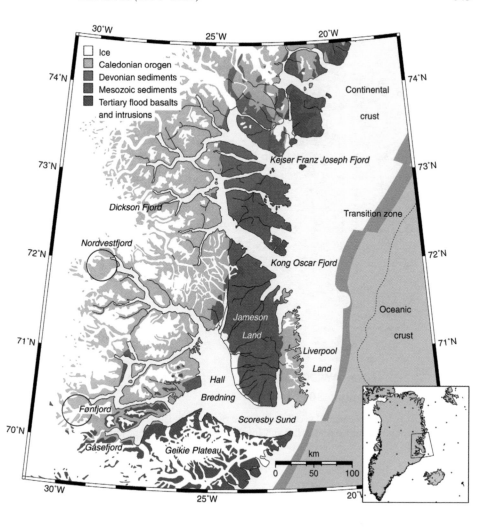

Figure 9.8.4-30. Simplified geological map of the East Greenland fjord region (from Schmidt-Aursch and Jokat, 2005, fig. 1). [Geophysical Journal International, v. 160, p. 736–752. Copyright John Wiley & Sons Ltd.]

Between the Faeroe Islands and the Shetland Islands, Geo-Prakla shot several seismic-reflection lines FLARE and FAST-UNST across the Faeroe-Shetland Trough (Fig. 9.8.4-19) using the same acquisition configuration which had previously been used by BIRPS (England et al., 2005). The aim was to study the crust beneath the Faeroe basalts. Bright reflections were seen at 7–9 s beneath the basalts, dipping westward in the opposite direction than the dip of the basalts and reflections within the basalts. The sub-basalt reflections were interpreted to originate from near the top of the basement. The Moho was not imaged beneath the basalts. The authors explained this as being due to missing impedance contrast at the base of the crust. The main result was a thinning of the basement to ~10 km thickness beneath the center of the trough, which was explained as being caused by more than one extension period.

The Faeroe–Iceland Ridge to the northwest of the Faeroe Islands was investigated in 1994 by the project FIRE (Faeroe–Iceland Ridge Experiment), a combined onshore-offshore seismic-reflection and wide-angle acquisition program (Fig. 9.8.4-20) which extended from the Faeroe Islands continental block to the mid-Atlantic spreading center of Iceland (Smallwood et al., 1999; McBride et al., 2004).

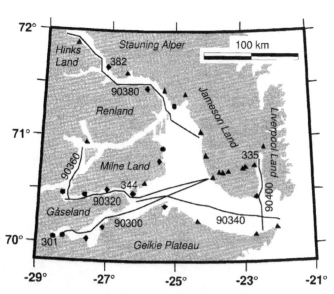

Figure 9.8.4-31. Location map of the 1990 seismic lines in the narrow fjords west of Scoresby Sund, to the east and south of Jameson Land, eastern Greenland (from Mandler and Jokat, 1998, fig. 2). [Geophysical Journal International, v. 135, p. 63–76. Copyright Wiley & Sons Ltd.]

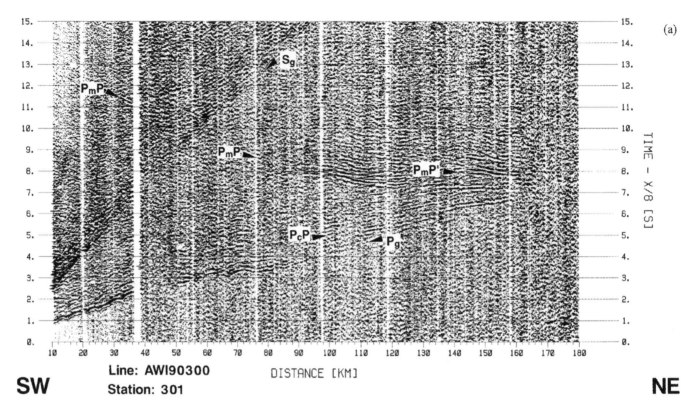

Figure 9.8.4-32. Seismic record section of station 301, line 90300, of the 1990 seismic survey in the Gaose fjord west of Scoresby Sund, eastern Greenland (from from Mandler and Jokat, 1998, fig. 3a). [Geophysical Journal International, v. 135, p. 63–76. Copyright John Wiley & Sons Ltd.]

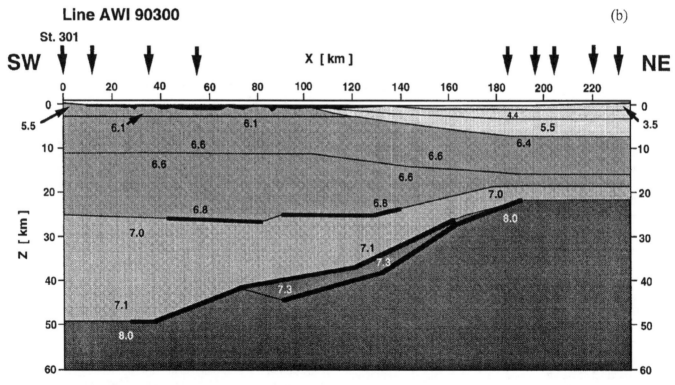

Figure 9.8.4-33. Crustal structure along line 90300 of the 1990 seismic survey in the Gaose fjord west of Scoresby Sund, eastern Greenland (from Mandler and Jokat, 1998, fig. 3b). [Geophysical Journal International, v. 135, p. 63–76. Copyright John Wiley & Sons Ltd.]

The project included a 440-km-long, conventional near-normal incidence multichannel seismic profile. The airgun shots were also recorded by land seismometers on both sides and by OBH. Over 6000 shots were fired at 75 m intervals. In addition, 55 explosive charges of 50 or 200 kg were detonated and recorded on the onshore seismometers, on the OBH and on a 6-km-long digital multichannel streamer with 240 receiver groups of hydrophones.

The land recording array on Iceland comprised 104 seismometers between the eastern coast and the axis of the present-day spreading center. The onshore array on Iceland also recorded explosive shots in lakes and fjords in Iceland. On the Faroe Islands seismometers were deployed at six sites.

The model presented in Figure 9.8.4-21 is based on interpretations of Richardson et al. (1998) for the Faeroes platform, Staples et al. (1997) for Iceland and the simultaneous modeling of the recordings from the FIRE shots along the arrays on Iceland and the Faeroe Islands and from the offshore OBH (Smallwood et al., 1999). The crust beneath the Faeroe Islands was estimated by McBride et al. (2004) to be ~46 km thick, and the Moho in Figure 9.8.4-21 was interpreted as the depth where velocity reached 7.5 km/s in the middle of a transition from 7.3 to 7.7 km/s.

From 1988 to 1996, a large number of seismic-refraction/wide-angle reflection profiles was recorded in the offshore area of central Norway off the Lofoten Islands (Figs. 9.8.4-18 and 9.8.4-22) by a consortium of Norwegian institutions, supported by Japanese scientists.

By introducing the OBS method, long-offset refraction studies were enabled, mapping the whole crust from upper sedimentary layers to Moho, because in small-offset reflection seismic surveys extrusive/intrusive rocks in the sedimentary layers caused seismic-imaging problems (Mjelde et al., 1992).

Following a first survey in 1988 (Mjelde et al., 1992), in 1992, an extensive OBS survey was performed in the central and northern part of the Voering Margin (Mjelde et al., 1997, fig. 1), comprising several experiments from regional to local-scaled acquisition. The success of the 1992 survey suggested a similar regional OBS survey to map larger areas of the margin.

In 1996, 15 profiles were acquired investigating the northern and southern part of the Voering Basin, seaward of the escarpment in the central part of the Voering Margin, as well as across the Jan Mayen Fracture Zone into the More Margin (Mjelde et al., 1998, 2001). Three profiles in the northern part of the Voering Basin were interpreted by Mjelde et al. (1998), while Raum

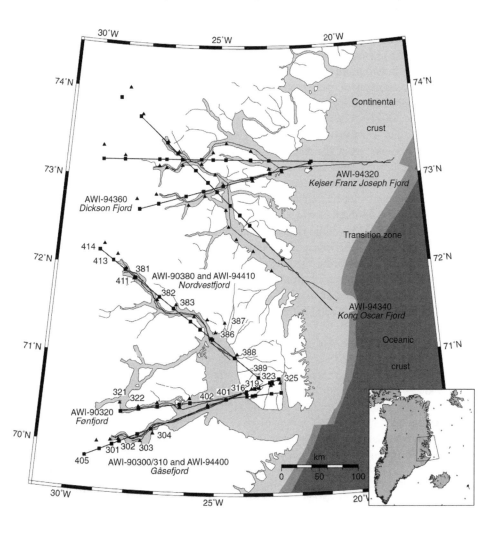

Figure 9.8.4-34. Location map of the 1990 and 1994 deep seismic lines in East Greenland (from Schmidt-Aursch and Jokat, 2005, fig. 2). [Geophysical Journal International, v. 160, p. 736–752. Copyright John Wiley & Sons Ltd.]

et al. (2002) modeled five profiles with a total length of 892 km
in the southern part, which comprised the data of 30 recovered
OBSs. The profiles were oriented in both strike and dip directions
(lines 9, 10, 12–14 in Fig. 9.8.4-22).

The crystalline basement was modeled as a 6+ km/s refractor. Along one of the strike profiles, the transition from continental to oceanic crust was modeled where P-velocity increased from
6.2 to 7.3 km/s. Finally, Raum et al. (2002) compiled summary
models of the 1988, 1992 and 1996 surveys along two transects,
in both the strike and dip directions (Figs. 9.8.4-23 and 9.8.4-24).

The areas around Spitsbergen in the northernmost Atlantic
Ocean had already seen several Polish-Norwegian activities in the
past decades. The research was continued in the 1990s. In 1998,
the northwestern Barents Sea southeast of Spitsbergen was the target of a wide-angle survey with OBSs (Breivik et al., 2002). Four
lines were shot as a network (Fig. 9.8.4-25), with airgun shots
every 200 m along each OBS profile. Two 160-km-long profiles
extended in a SW-NE direction, and two 300-km-long profiles
were recorded in N-S direction. The experiment was carried out
by the vessel R/V *Hakon Mosby* of the University of Bergen.

The observed data enabled the identification and mapping
of the depths to the crystalline basement and to the Moho by ray
tracing and inversion. Refractions from the top of the basement
together with reflections from the Moho constrained the basement P-velocity to increase from 6.3 to 6.6 km/s from the top to
the bottom of the crust for which an average thickness of 35 km
was obtained. On two profiles, the Moho deepened locally from
near 33 to ~38 km depth in root structures (Fig. 9.8.4-26), which
could be associated with high top-mantle velocities of 8.5 km/s
(Breivik et al., 2002).

In 1999, the continent-ocean transition of northwestern
Spitsbergen became again the target of a Polish expedition,
ARKTIS VV/2 (Czuba et al., 2004). Seismic energy from airgun
and TNT shots was recorded by onshore land stations, ocean-
bottom seismometers and hydrophone systems on two profiles
(Fig. 9.8.4-27). The northern profile 99200 was 430 km long and
ran from the Molley Deep in the northern Atlantic to Nordaust-
landet in northeastern Spitsbergen. The southern line 99400 was
360 km long and ran from the Hovgard Ridge to Billefjorden,
Spitsbergen. Good quality records from the airgun shots were observed up to 200 km distance at land stations and 50 km at OBSs,
while TNT shots reached to 300 km distance (Fig. 9.8.4-28 and
Appendix A9-6-1).

The crustal structure of the transition zone proved to be
very complex (Fig. 9.8.4-29), with Moho depths in the Atlantic
Ocean as shallow as 6 km and Moho depths around 30 km
under the center and the northern coast of Spitsbergen, thus
confirming the result of the 1985 expedition (Czuba et al.,
1999; Ritzmann et al., 2004).

Eastern Greenland and its margin on the western side of the
northernmost Atlantic Ocean had been the goal of deep seismic
investigations since the end of the 1980s (see Chapter 8; Mandler
and Jokat, 1998). The research in the fjords of the Scoresby Sund
(Fig. 9.8.4-30) was continued in 1990 by an expedition of the

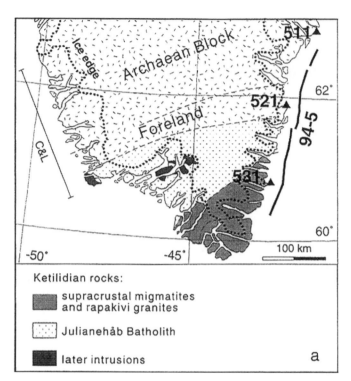

Figure 9.8.4-35. Location map of the offshore line 94-5 and three
land-based recording stations (triangles) along the southeastern coast
of Greenland (from Dahl-Jensen et al., 1998, fig. 1a). [Tectonophysics,
v. 288, p. 191–198. Copyright Elsevier.].

Alfred-Wegener-Institute for Polar and Marine Research (AWI,
Bremerhaven, Germany), offering the unique opportunity of applying marine seismic methods for the investigation of Greenland's continental crust up to 300 km west of the continent-ocean
boundary. In total, more than 2300 km of seismic-refraction
profile-km and 4000 km of normal-incidence reflection data were
collected during the cruise. Mandler and Jokat (1998) discussed
in particular data and results of profiles, obtained along the narrow western fjords crossing the southern part of East Greenland's
Caledonian province (Fig. 9.8.4-31).

Here, the lines totalled ~1000 km and had maximum shot-
receiver offsets of 230 km. The land-based recording stations
were deployed by helicopter with between 5 and 10 stations per
line, giving an average spacing of 25 km. The seismic source
was provided by a single powerful airgun, shooting at an average
spacing of 75 m. The record section shown in Figure 9.8.4-32
(see also Appendix A9-6-1) demonstrates the energy distribution
and possible phase correlations.

In the western area, which is part of the Caledonian mountains of East Greenland, the seismic velocities of crystalline rocks
increased continuously with depth from ~5.5 km/s at the surface to 6.6 km/s at 12 km. In the southwestern part of the region
(28°W), the seismic measurements revealed high values of up to
48 km for the total thickness of the crust. Toward the east, the
crustal thickness decreased rapidly, reaching a minimum of 22 km
under the late Paleozoic–Mesozoic sedimentary basin of Jameson

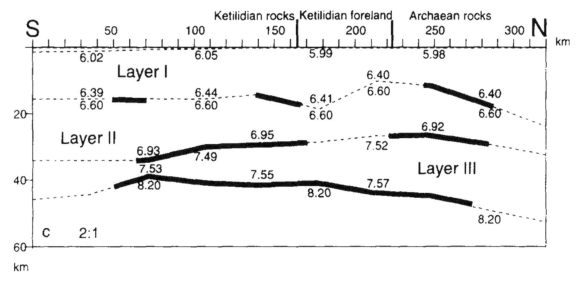

Figure 9.8.4-36. Crustal structure along the southeastern coast of Greenland (from Dahl-Jensen et al., 1998, fig. 3c). [Tectonophysics, v. 288, p. 191–198. Copyright Elsevier]

Land (Fig. 9.8.3-33). In the area of transition from thick to thin crust, the seismic data indicated a layered structure at the Moho (Mandler and Jokat, 1998).

The experiment of 1990 was continued in 1994 (Schmidt-Aursch and Jokat, 2005), re-shooting and adding to the East Greenland survey of 1990, and adding lines farther north around 73°N, some of which were recorded up to maximum distances of 325–375 km (Fig. 9.8.3-34).

In 1994, the Danish Lithosphere Center collected 1094 km of deep multichannel reflection seismic data along the SE Greenland coast south of 63°30′N, ~20 km offshore (Dahl-Jensen et al., 1998). The data were recorded on a 368-channel 4600-m-long streamer with a shotpoint interval of 75 m and 27 s recording time. Three land-based recording stations were set up along the coast and recorded wide-angle data during the shooting of line 94-5 (Fig. 9.8.4-35) and a margin-crossing line.

The data of the land station recordings were modeled by a three-layer crystalline crust, with layer III having high velocities of 7.4–7.5 km/s. The depth to Moho varies from 39 km in the south to 49 km in the north (Fig. 9.8.4-36).

In 1996, the R/V *Maurice Ewing* was used to conduct the SIGMA (Seismic Investigation of the Greenland Margin) experiment, a combined MCS reflection and wide-angle onshore/offshore seismic experiment across the southeast Greenland continental margin, organized jointly by the Woods Hole Oceanographic Institution (WHOI) and the Danish Lithosphere Center (Korenaga et al., 2000). Four transects were located at ~68°N, 66°N, 63°N and 59°N to systematically sample the structure of the margin with increasing distance from the Greenland-Iceland Ridge, the presumed Iceland hotspot track (Fig. 9.8.4-37).

Korenaga et al. (2000) focused their description on line 2, whose data comprised ~300 km of multichannel seismic-reflection data and coincident wide-angle seismic-reflection and

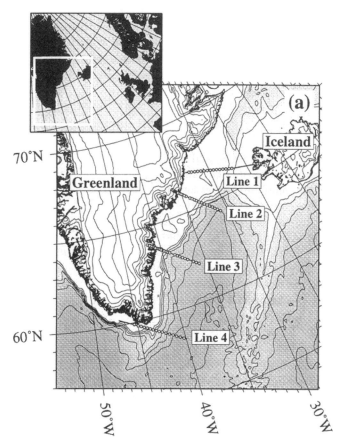

Figure 9.8.4-37. Location map of the 1996 SIGMA seismic experiment along the southeastern Greenland coast (from Korenaga et al., 2000, fig. 1a). Circles denote the locations of the onshore and offshore seismic instruments. Bathymetry is shown with 500-m contour interval. [Journal of Geophysical Research, v. 105, p. 21591–21,614. Reproduced by permission of American Geophysical Union.]

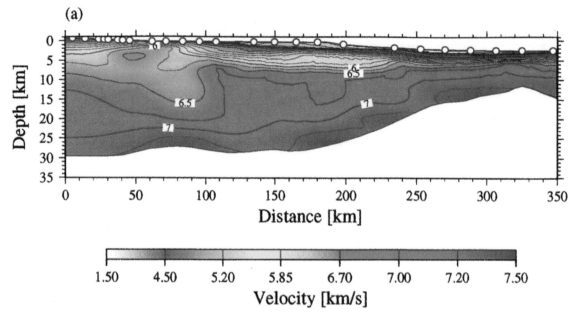

Figure 9.8.4-38. Model of transect 2 of the 1996 SIGMA seismic experiment along the southeastern Greenland coast (from Korenaga et al., 2000, plate 3a). [Journal of Geophysical Research, v. 105, p. 21591–21,614.] Reproduced by permission of American Geophysical Union.]

-refraction data recorded on 17 ocean-bottom and eight onshore seismometers. The multi-channel profiling was conducted using a 20-element, 8460 in³ airgun array fired every 50 m (20 s) and a 4.0-km hydrophone streamer with 160 channels. Wide-angle seismic data were recorded on 10 WHOI ocean-bottom hydrophones, eight USGS ocean-bottom seismometers, and eight on-land seismic recorders from PASSCAL, deployed along the transect. The total model distance of transect 2 was 350 km.

Most of the wide-angle data on transect 2 were of high quality. Airgun firing at a 20-s interval at 4.7 knots yielded densely spaced seismic traces at 50 m. The sampling interval was 10 ms, and a 5–14 Hz bandpass filter was applied to the data followed by predictive deconvolution and coherency weighting.

For the interpretation, a new seismic tomographic method was developed to jointly invert refraction and reflection travel-times for a 3-D velocity structure. The resulting P-wave velocity-depth model for the 350-km-long transect (Fig. 9.8.4-38) showed strong lateral heterogeneity and included (1) a 30-km-thick, undeformed continental crust with a velocity of 6.0–7.0 km/s near the landward end, (2) a 30–15-km-thick igneous crust within a 150-km-wide continent-ocean transition zone, and (3) a 15–9-km-thick oceanic crust toward the seaward end. The thickness of the igneous upper crust varied from 6 km within the transition zone to 3 km seaward and was characterized by a high-velocity gradient. The bottom half of the lower crust generally had a velocity higher than 7.0 km/s, reaching a maximum of 7.2–7.5 km/s at the Moho.

A very sophisticated project covered the Mendeleev Ridge at 82°N along the Arctic-2000 transect in the Arctic Ocean between 164°W and 165°E (Lebedeva-Ivanova et al., 2006), a distance

of 485 km (Fig. 9.8.4-39). It was the first ship-based Arctic expedition performing seismic investigations in the high Arctic Basin and was carried out by the ice-class vessel *Academian Fedorov*, accompanied through the pack-ice by the nuclear ice-breaker *Russia*. The seismic acquisition included a comprehensive DSS survey along most of the line and a simple reflection profiling along the entire line. The experiment was carried out in three long segments, each 125 km long, and a fourth shorter segment in the west, 32 km long. The digital three-component seismometers were deployed by two helicopters along each segment and 7–8 shots per segment with charges ranging from 100 to 1000 kg were detonated in water at water depths of 70–100 m.

The shot spacing was 40 km. For each segment, four shots were within the deployment range, while the other shotpoints were offset by up to 75 km on each side. Data, as published by Lebedeva-Ivanova et al. (2006; Appendix 1-4), are shown in Appendix A9-6-1.

Simultaneously, a reflection survey was performed, using the same equipment and the same deployment, by detonating 0.4 kg shots at each recording site, i.e., at 5 km spacing. The measurements were single-point recordings of up to 15 s TWT. They were made while picking up the instruments and were designed to acquire data on the depth to the seafloor and the structure and thickness of the sedimentary cover in the uppermost crust at each recording site.

The velocity-depth structural cross section (Fig. 9.8.4-40) revealed seven layers. The uppermost three layers represent 3.5-km-thick sediments. The character of layer IV was debated as to whether it was to be interpreted as carbonate sediments or crystalline basement rocks. Layers V and VI constituted the

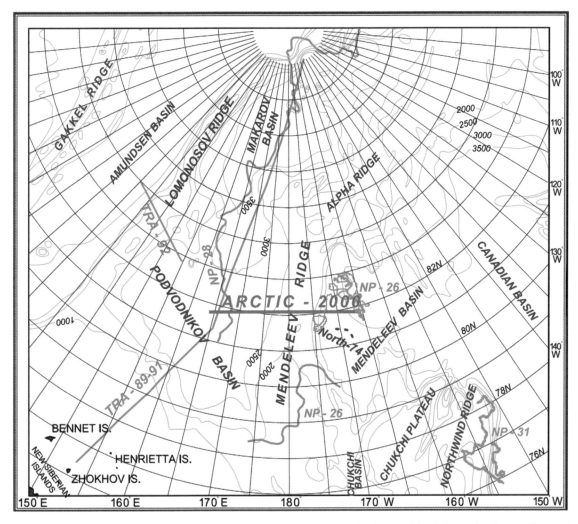

Figure 9.8.4-39. Location of the geotraverse "Arctic Transect-2000" and other combined deep seismic sounding and reflection geotraverses (TRA-92, TRA89-91) and reflection profiles (NP-26, NP-28, NP-31 and North-74) in the Arctic Sea (from Lebedeva-Ivanova et al., 2006, fig. 1). [Geophysical Journal International, v. 165, p. 527–544. Copyright John Wiley & Sons Ltd.]

crystalline crust, with the lower crust (layer VI) being 19–20 km thick. Layer VII with velocities of 7.4–7.8 km/s was interpreted as a crust-mantle mix, possibly resulting from underplating. The depth to the Moho along the Arctic Transect varied considerably, from 20 km under the Mendeleev Basin in the east and 13 km under the Podvodnikov Basin in the west to a maximum depth of 32 km in the center of the Arctic Transect. In conclusion, the authors favored the interpretation that the Mendeleev Ridge is composed of an attenuated underplated continental crust.

9.8.5. Summary of Seismic Observations in the Oceans in the 1990s

Most of the experiments we have described in more detail for the 1990s were largely devoted to the investigation of details of the mid-ocean rises, such as the East Pacific Rise.

Here, for example, projects around the Galapagos hot spot and the Cocos-Nazca Spreading Center (e.g., (Sallares et al., 2003) and the Garrett Fracture Zone (Grevemeyer et al., 1998) were undertaken. A large number of Japanese marine surveys, some of which also had land components, targeted the trench system in the Philippine Sea and northwestern Pacific Ocean around Japan. In particular, the Nankei Trough and the Japan Trench south and east of Honshu was intensively investigated (e.g., Kodaira et al., 2000, Miura et al., 2003, 2005, Takahashi et al., 2004, Tsuru et al., 2002).

In the Indian Ocean, the Ninetyeast Ridge was investigated (Grevemeyer et al., 2001a). In the Atlantic, the Mid-Atlantic Ridge (e.g., Hooft et al., 2000) but also the continent-ocean transitions were major targets. In particular, in the northern Atlantic Ocean, projects targeted the microcontinents and intervening basins off Ireland (e.g., Mackenzie et al., 2002), the Norwegian

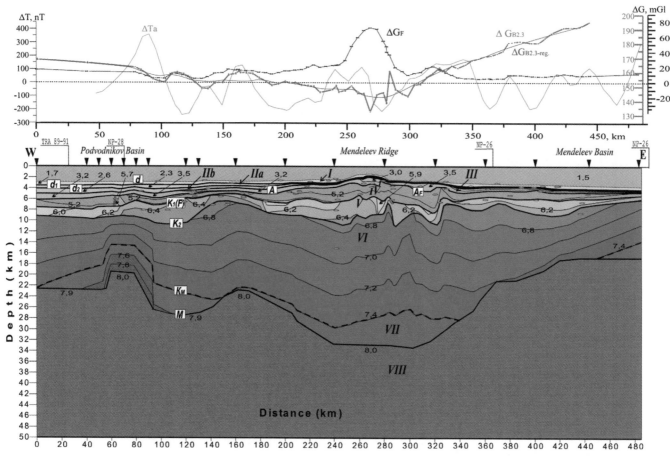

Figure 9.8.4-40. Crustal structure model along the Arctic Transect 2000 across the Mendeleev Ridge in the Arctic Ocean at 82°N (from Lebedeva-Ivanova et al., 2006, fig. 10). [Geophysical Journal International, v. 165, p. 527–544. Copyright John Wiley & Sons Ltd.]

margin (e.g., Raum et al., 2002), the Faeroe–Iceland Ridge (e.g., Smallwood et al., 1999), the northwestern Barents Sea southeast of Spitsbergen (Breivik et al., 2002), and eastern Greenland (e.g., Korenaga et al., 2000; Schmidt-Aursch and Jokat, 2005). The Arctic-2000 transect in the Arctic Ocean between 164°W and 165°E (Lebedeva-Ivanova et al., 2006) and the investigation of the North American margin off Canada (e.g., Funck et al., 2003) were other large marine projects.

The map of locations of experiments in the 1990s is based on data points of the USGS database which were published until 1999 in easily accessible journals and books (Fig. 9.8.5-01). It shows a concentration of projects on the Atlantic Ocean and the Polar Regions. The continental margins and adjacent deep-ocean parts of South America, South Africa, southern Australia, and New Zealand were covered by many new experiments. The map also indicates the activities along the East Pacific Rise and the Galapagos hot spot and special features in the southern Indian Ocean. The numerous marine projects of CROP MARE in the Mediterranean Sea, however, did not find their way into the database yet, as they were only published in connection with the CROP Atlas, which is little known.

9.9. ADVANCES IN INTERPRETATION METHODOLOGY

As mentioned at the beginning of this chapter, the early 1990s saw the gradual replacement of analogue recorders by digital recording instrumentation and the establishment of instrument pools, which allowed organizations to buy and to service large quantities of individual recording sets, and from which individual groups of scientists and/or institutions could loan any number. At a later stage, these instrument pools also started to request copies of the recordings and to archive large amounts of data for future use.

At sea, powerful airgun systems, replacing in many projects the destructive underwater explosions, as well as reliable ocean-bottom seismometer/hydrophone equipment and large-scale digital streamers of several kilometers length were developed and widely used in the following decade, supplying a vast amount of data which allows studying details of the oceanic crust and upper mantle structure.

Since large-scale international projects often needed more equipment than one instrument pool could provide, the equip-

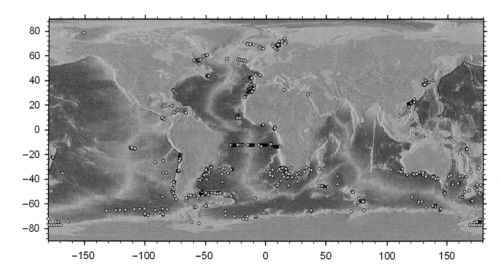

Figure 9.8.5-01. Seismic refraction measurements in the Atlantic, Pacific and Indian Oceans performed between 1990 and 2000 (data points from papers published until 1999 in easily accessible journals and books).

ment from various sources had to be handled. Here the problem of differing data recording formats soon became a time-consuming problem when trying to compile the various sub-data sets into a uniform format. To discuss this problem and seek a satisfactory solution, the International Lithosphere Program (ILP) and the International Association for Seismology and Physics of the Earth (IASPEI) formed the Mega Earth Mobile Seismic Array Consortium (MEMSAC), combining members of government agencies, universities and industry, which discussed the relevant problems at two meetings held in Germany and in California (MEMSAC Working Group, 1993).

In connection with large-scale seismic-refraction programs (so-called active studies, i.e., recording controlled sources, such as quarry blasts, borehole and underwater explosions, vibrators, or airguns), large-scale teleseismic tomography projects (so-called passive studies, i.e., long-term recording of earthquakes) were carried out since the early 1990s extending the crustal and uppermost mantle research to greater depth ranges. Examples, to name a few, are the KRISP and EAGLE investigations of the East African Rift System in Kenya (Prodehl et al., 1994a; Fuchs et al., 1997) and Ethiopia (Maguire et al., 2003), the onshore-offshore investigations of the Andean region in South America (e.g., ANCORP Working Group, 1999, 2003; Flueh et al., 1998a; Giese et al., 1999; Krawczyk and the SPOC Team, 2003, 2006; Rietbrock et al., 2005), the INDEPTH expeditions to Tibet (e.g., Brown et al., 1996; Nelson et al., 1996; Zhao et al., 1993, 1997, 2001), or the CD-ROM project in the southern Rocky Mountains (Karlstrom and Keller, 2005). For the interpretation of these data special inversion methods, used originally in passive experiments, were applied which in the course of time proved to be practical to apply to controlled-source seismology data also. In particular, the method for studies of crust and upper mantle with P-receiver functions, which was being developed since the early 1980s, and later S-receiver function studies in the late 1990s proved to be a very efficient and relatively cheap interpretation tool of the large-scale teleseismic data.

An overview on interpretation methods used in the 1990s to interpret in particular combined active and passive studies, was given at one of the latest CCSS (Commission on Controlled Source Seismology) workshop meetings, held in Dublin in 1999 (Jacob et al., 2000). Here, for example, Mechie (2000) referred to popular tools commonly used for interpreting seismic-refraction/wide-angle reflection data. Introduced in the 1970s, the ray-tracing method has remained an almost universal method for data interpretation (Červený and Pšenčík, 1984c), and the most commonly used programs include ray-theoretical and Gaussian-beam seismograms (Červený, 1985). Based on the theory of the finite-differences method for calculating the full wavefield in laterally inhomogeneous media, which was formulated already in the 1970s (e.g., Kelly et al., 1976; Reynolds, 1978), in the 1990s it became possible in practice to make such calculations for refraction/wide-angle reflection data on a crustal/lithosphere scale over several hundreds of kilometers and at realistic frequencies (e.g., Mechie et al., 1994). Since Vidale (1988) introduced ray-tracing using a finite-difference approximation of the eikonal equation, this method was improved in the 1990s (e.g., Podvin and Lecomte, 1991; Schneider et al., 1992) and also extended in a way that reflected arrivals and second arrival refractions from prograde traveltime branches could be calculated (Hole and Zelt, 1995). Also, commonly used methods in the processing of near-vertical incidence seismic-reflection data, such as normal moveout correction and migration, were applied to refraction/wide-angle reflection data (e.g., Lafond and Levander, 1995; Patzwahl, 1998).

Since the early 1990s, with the introduction of partial derivatives for the traveltimes with respect to the velocity field and layer boundaries, inversion of traveltime data in laterally heterogeneous media became possible (Lutter et al., 1990; Zelt and Smith, 1992; Hole, 1992). By the end of the 1990s, various tomographic methods existed (Hole, 1992; Zelt and Barton, 1998). Traveltime tomography became popular and evolved to a method widely applied as a first approach to model large amounts of data as well as a last check of the validity of a model (Zelt, 1998, 1999; Zelt and Barton, 1998; Zelt and White, 1995; Zelt et al., 1996, 1999, 2003; Hole et al., 2000, 2001). Nowadays there is hardly a publication on crustal and upper mantle interpretation which does not first apply a tomographic approach to the seismic data before a refined ray-tracing modeling is applied (for example, interpretations of the Polonaise and CELEBRATION 2000 data; see, e.g., Guterch et al., 2003a, 2003b; Grad et al., 2006; Janik et al., 2005). The irregularly scheduled CCSS workshops mentioned in the beginning of this chapter, which brought together scientists to discuss and practice interpretation methods on pre-distributed data sets, were continued in the 2000s (Hole et al., 2005).

9.10. THE STATE OF THE ART AROUND 2000

The large number of recording devices available by the year 2000 as well as the ability to record continuously over long time periods opened a new dimension. Projects could now be planned which extended seismic surveys into three dimensions. The availability of recording over long time periods has opened the possibility of avoiding organizing man-made sources such as quarry blasts, borehole or underwater explosions, but, at least for projects on land, to plan for seismic tomography surveys with teleseismic and/or local events as energy sources. As has been proved in particular by the examples of the Kenya Rift in East Africa and the CD-ROM project in North America, the combination of controlled-source seismology and teleseismic tomography has enlarged the areal coverage as well as the depth range to be explored.

Based on the incredible wealth of data gathered during the past 50 years on seismic crustal structure studies, Walter Mooney started to build up a database, which will be described in more detail in Chapter 10. Already its preliminary versions with data collected to the mid- and end-1990s enabled construction of a worldwide syntheses of crustal parameters (e.g., Christensen and Mooney, 1995; Mooney et al., 1998; Mooney, 2002) or, e.g., a synthesis of the seismic structure of the crust and uppermost mantle of North America and adjacent oceanic basins (Chulick and Mooney, 2002).

The Twenty-first Century (2000–2005)

10.1. THE SECOND DECADE OF DIGITAL RECORDING

The new decade brought continuing advances in recording techniques. Digital technology, which had taken over as the recording device of controlled-source seismology in the 1990s, was further improved. After many successful deployments, an updated design (RefTek-125A) for the Texan instrument was finalized in 2004. PASSCAL and the EarthScope program began to purchase these models immediately, and the UTEP (University of Texas at El Paso, USA) group focused on an upgrade path for the existing instruments. Via grants from several sources, UTEP obtained the funds needed to upgrade almost 400 of its units, and upgrades were under way by 2005 (Keller et al., 2005b).

The PASSCAL pool and UTEP in the United States had a total of 840 of the new Texan instruments by 2000. The University of Copenhagen, Denmark, purchased 100 units, the Earth Science Research Institute TUBITAK-MAM in Gebze, Turkey, acquired 60 units, and other groups in Central Europe obtained ~40 units (Guterch et al., 2003a).

In the UK, Leeds University had purchased the Orion system, and in 2000, the universities of Leicester, Cambridge, Leeds, and London Royal Holloway were awarded a grant to acquire large numbers of Guralp 6TDs and 40Ts, together with a number of Guralp 3Ts, bringing the British scientists to the forefront of observational broadband seismology. Following the end of the grant, the equipment was taken over by the NERC Geophysical Equipment Pool, who kept the name SEIS-UK for the seismological component of its structure. A total of 182 Guralp instruments were acquired at the start of SEIS-UK (10x3TD, 20x40TD and 152x6TD). The 2010 inventory included 28x3TD, 15x3T, 20x40TD, 35x3ESPDC (60sec) and 112x6TD.

In Germany in 2000, the GeoScience Center Potsdam started to gradually renew its instrument pool for short-period and broadband seismology and has now replaced the worn-out RefTek and PDAS data loggers with a new system, the Earth Data PR6-24. Up to mid-2007, almost 240 of these new data loggers were acquired and made available for the geophysical community, with an additional 10 broadband GURALP units for special purposes. A major cooperative effort between land and sea investigations promises a new instrumental pool, named DEPAS (Deutscher Pool für Aktive Seismologie). In early 2006, it consisted of 30 OBS and 65 GURALP seismometers plus Earth Data Loggers, which had been enabled by a joint venture of the German institutions GFZ (GeoForschungsZentrum Potsdam)

and AWI (Alfred-Wegener-Institut Bremerhaven) for amphibious seismology (Schulze and Weber, 2006; Schmidt-Aursch et al., 2006). When completed, 100 Earth Data Loggers PR6-24 and 100 GURALP CMG-3ESP compact units will be available for the land part of DEPAS. For the offshore part, 68 standard and 12 deep-sea OBS (Guralp CMG-40T seismometers plus hydrophones) and 5 OBH (hydrophones only) will be available.

Large-scale research programs that involved a multitude of cooperating institutions and interdisciplinary cooperation of various geoscientific fields continued to dominate the scene in the early 2000s. With the increasing number of recording devices, the possibility of covering large areas two-dimensionally enabled more and more the application of tomographic methodologies such as teleseismic tomography. Many earth scientists started to prefer long-term deployments using natural events as energy sources instead of short-term projects using expensive controlled sources.

Since the early 1990s, the organization of large controlled-source seismology projects has faced growing difficulties, which is another reason for the decline in controlled-source work. At sea, the use of explosives as a seismic source is now banned due to the impact on marine life from such effects as vibration and chemical contamination. Nowadays, modern marine projects have to rely mostly on powerful airgun sources. For land-based projects, more and more demanding environmental restrictions involving major environmental impact assessments have very nearly led in the recent past to the cancellation of experiments, in spite of the fact that in general, the real impact is pretty small. This and other organizational difficulties dealing with officials or with the problem of needing a lot of people for a short time, for example, are good reasons that scientists plan more and more for passive rather than active seismic-source experiments.

Nevertheless, in many cases, a combination of active- and passive-source experiments is being aimed for. IRIS/PASSCAL in the United States and LITHOPROBE in Canada continued to support large seismic (active) and seismological (passive) projects, occasionally combined with international and interdisciplinary geoscientific priority programs in Europe, dealing with large-scale tectonic topics. EUROPROBE, supported by the European Science Foundation, continued to support programs concerned with the origin and evolution of the continents (Gee and Stephenson, 2006) emphasizing East-Central-West European collaboration and close multinational cooperation of geologists, geophysicists, and geochemists.

Special sessions on large national and international seismic programs became important components of the annual meetings of the various national and international geoscientific organizations, such as, e.g., the American Geophysical Union or the European Geophysical Union, but special meetings of earth scientists at regular intervals also continued such as the biannual meetings of the special subcommission of the ESC's (European Seismological Commission) "Structure of the Earth's Interior," the CCSS (Commission on Controlled Source Seismology) workshops on interpretation methods (e.g., Hole et al., 2005), and the series of "International Symposia on Deep Seismic Reflection Profiling of the Continental Lithosphere" which continued into the twenty-first century with meetings at Ulvik, Norway, in 2000 (Thybo, 2002), in Taupo, New Zealand, in 2003 (Davey and Jones, 2004), in Mont-Tremblant, Quebec, Canada, in 2004 (Snyder et al., 2006), in Hayama, Japan, in 2006 (Ito et al., 2009a), in Saariselkä, Finland, in 2008, and in Cairns, Queensland, Australia, in 2010.

In particular, the Working Group on Seismic Imaging of the Lithosphere played and plays an important role for the development and improvement of seismic interpretation methods. Since 1968 the IASPEI CCSS had sponsored a series of "Deep Seismic Methods" workshops (see, e.g., Chapter 7.9—Ansorge et al., 1982; Mooney and Prodehl, 1984; Chapter 8.2—Finlayson and Ansorge, 1984; Walter and Mooney, 1987; Green et al., 1990a; and Chapter 9.1—Pavlenkova et al., 1993; Snyder et al., 1997; Jacob et al., 1997, 2000). The workshops focused on the methodological aspects of reflection and refraction seismology as applied to the crust and lithosphere, and the goal was to develop and assess improved acquisition, processing, modeling, inversion, and integration procedures. Prior to the workshops, data had been distributed to the participants with the request to interpret them with the available methodologies and discuss methods and results at the subsequent workshops. The CCSS Working Group was recently reorganized as the Working Group on Seismic Imaging of the Lithosphere, continuing the series of international CCSS workshops on interpretation methods (Hole et al., 2005). Just as in the workshop held at Einsiedeln, Switzerland, in 1983 (Finlayson and Ansorge, 1984), before the 2005 workshop, synthetic seismograms were computed from a given model, this time using a 2-D viscoelastic finite difference algorithm (Robertsson et al., 1994). Fifty sources were fired into the model and recorded by 2800 receivers to create a crustal-scale refraction survey geometry. For the interpretation without knowing the model, the current standard methodology was applied. The interpretation of the synthetic data with a first-arrival traveltime tomography gave a high resolution for the upper crust and a correct depth and long-wavelength structure for the Moho, but the resolution deteriorated in the lower crust. The model was subsequently improved by using refraction and reflection traveltimes from the Moho to invert for a discrete Moho and velocity structure below 15 km (Hole et al., 2005; Zelt et al., 2003).

The following description of major seismic projects undertaken between 2001 and the beginning of 2006 cannot be complete and will be rather limited. Due to the large number of available instruments and the resulting huge amount of data, the completion of large seismic projects, with major publications on the re-

sults available in print, usually takes at least 5 years. For many projects undertaken since 2001, only abstracts or short outlines have become available up to the end of 2005. In the following, we have concentrated in particular on those projects for which reasonable information has already been published. Nevertheless, we will also mention some other major projects which were undertaken in the first five years after 2000, for which either abstracts or internal reports at scientific meetings have become available to us.

10.2. CONTROLLED-SOURCE SEISMOLOGY IN EUROPE

10.2.1. Western and Southern Europe

The fruitful cooperation of the Dublin Institute for Advanced Sciences with German partners continued in 2002. During the recording windows set for the controlled-source seismic experiment of VARNET96 (see Chapter 9), some earthquakes and a nuclear event had been recorded. These data had been used to model the probable trace of the Iapetus suture in the upper mantle just south of the Shannon river estuary at between 30 and 110 km depth (Masson et al., 1999). In 2002 and 2003, a specially designed Irish Seismological Lithospheric Experiment (ISLE) was undertaken to investigate the nature and tectonic history of the deep lithospheric and asthenospheric structure of the Iapetus Suture Zone (Landes et al., 2004c, 2005).

On the Iberian peninsula in 2001, the IBERSEIS deep seismic-reflection transect across the SW Iberian Massif (Carbonell et al., 2004; Flecha et al., 2009; Palomeras et al., 2009) crossed three major geological zones. The Vibroseis data set provided a crustal image that revealed the geometry of the structures and average physical properties of the shallow crust, but not for deep crust and upper mantle. One of the most relevant results of the Vibroseis study was a high-amplitude, strongly reflective structure named the IBERSEIS Reflective Body, which is ~140 km long and located at mid-crustal level at ~12–14 km depth. However, due to the fact that the Variscan belt of SW Iberia is a transpressive orogen in which large strike-slip movements have been suggested, the two-dimensional view provided by the deep normal incidence transect IBERSEIS might not be sufficient, and some kind of 3-D control appeared to be needed for a reasonable structural interpretation. Therefore, also in 2001, two wide-angle seismic transects were acquired (Palomeras et al., 2009). The project aimed to provide velocity constraints on the lithosphere and to complement the previously acquired normal incidence seismic profile IBERSEIS (Carbonell et al., 2004). The wide-angle reflection lines were ~300 km long. They had up to 6 shotpoints with an approximate interval of 60 km, with charge sizes between 500 and 1000 kg. A total of 690 digital seismic recording instruments (650 1-component Texans and 40 RefTek 3-component units) from the IRIS/PASSCAL were used. With 400 m spacing along one line and 150 m on the other one, the acquisition parameters provided closely spaced seismic images of the lithosphere. The combined data sets indicated rocks of mafic composition in the middle crust and revealed new aspects related to the lithospheric evolution of this transpressive orogen (Palomeras et al., 2009).

The IBERSEIS line was extended by another 300 km to the northeast by the ALCUDIA project in 2007, providing a complete transect from the Gulf of Cadiz to the Hercynian domain of the Spanish Meseta (Díaz and Gallart, 2009).

A detailed review on crustal structure of the Iberian Peninsula was published by Díaz and Gallart (2009). Their location map shows all seismic profiles observed since the 1970s to present, both on land and in the surrounding waters. Several cross sections reaching from coast to coast, and a Moho depth contour map, accompanied by a topographic map showing the individual data points, demonstrate the gradual crustal thickening from near 10 km under the Atlantic margin areas to 30 km under the Variscan Iberian Massif. Maximum crustal thicknesses are reached with 50 km under the Pyrenees and with 40 km under both the Iberian Chain in eastern Spain and the Betic Chain in Southern Spain. Toward the east and south, the crust thins rapidly to 15 km under the Valencia Trough and the Alboran Sea. Under the Balearic Islands, Moho depth increases locally to ~25 km.

10.2.2. Northern Europe

In Denmark in 2004 and 2005, the project ESTRID (Explosion Seismic Transects around a Rift in Denmark) was launched to investigate primarily the lower crust and Moho around the Silkeborg gravity high in central Denmark underneath the Mesozoic sedimentary Danish Basin (Nielsen et al., 2006; Thybo et al., 2006; Sandrin and Thybo, 2008). This is located close to the margin of the Precambrian Baltic Shield within the former Baltica plate. Two seismic profiles in central Jutland were recorded. The first one in 2004 was a refraction/wide-angle survey in an E-W direction across the peninsula and of 160 km length, the other one was a 185-km-long combined reflection/refraction survey oriented in a N-S direction and performed in 2005 (Sandrin and Thybo, 2008). The reflection part of the 2005 line had 4×760 recording sites over 110 km profile length and used 94 shots of 15–25 kg. The data confirmed the presence of a high-velocity body with velocities greater than 6.5 km/s and a N-S width of 30–40 km extending from 10 to 12 km depth into the lower crust underneath Silkeborg. Along the E-W profile, the Moho showed a substantial relief, with its depth varying between 27 and 34 km (Thybo et al., 2006).

On the Baltic Shield, new seismic-reflection studies were added to the existing seismic observations. In Finland from 2001 to 2003, a large deep seismic-reflection survey consisting of four sections, FIRE 1–4, was performed throughout Finland to study the structural architecture and evolution of the Fennoscandian Shield. The profiles had a total length of 2165 km and transected all major tectonic units and boundaries of the Fennoscandian Shield in Finland. The data and an interpretation have not yet been published (FIRE consortium, 2006).

On the Kola Peninsula in 2000–2002, Spetsgeofisika conducted a Vibroseis common mid-point experiment across the western and central Mezen Basin and along a line across the Timan Range which was already mentioned in Chapter 9 and discussed in context with the data obtained from 1998 to 2000 (Kostyuchenko et al., 2006, see dotted lines 3 and 4 in Fig. 9.3-03).

10.2.3. Central and Eastern Europe

After the successful completion of TRANSALP from 1998 to 2001 (e.g., Lueschen et al., 2006; Gebrande et al., 2006), in 2002 an additional large-scale seismic-refraction experiment, ALP 2002, was carried out in the Eastern Alps (ALP 2002 Working Group, 2003; Brueckl et al., 2003, 2010, Fig. 10.2.3-01; Sumanovac et al., 2009). Its realization developed out of a large international cooperation, headed and initiated by the Polish scientists M. Grad and A. Guterch, who, in 1997 and 2000, had successfully completed the two large-scale seismic-refraction campaigns POLONAISE'97 and CELEBRATION 2000, which were described in more detail in Chapter 9.2.4. On the basis of these projects, they had initiated the third international campaign, ALP 2002, this time centered on the Carpathians and adjacent Alps (Fig. 10.2.3-01).

ALP 2002 was designed to meet the following criteria: the whole area of investigation should be covered by a 3D geometry of profiles, the CELEBRATION 2000 technique would be applied, and the outline of the survey would connect with the

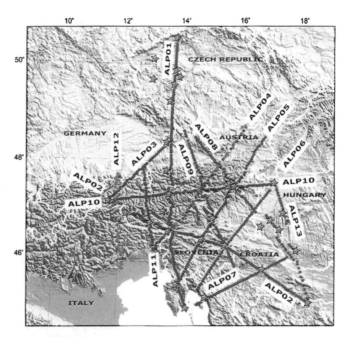

Figure 10.2.3-01. Location map of ALP 2002. Red stars—shotpoints; blue lines—seismic refraction profiles (from Brueckl et al., 2003, fig. 1). [Studia Geophysica et Geodaetica, Academy of Science, Czech Republic, Prague, v. 47, p. 671–679. Permission granted by Studia Geophysica et Geodaetica, Prague, Czech Republic.]

CELEBRATION profiles as well as with the TRANSALP profile. In total 40 shots were set off, 32 of them "strong" shots with 300 kg charges, distributed in 5–6 boreholes of 40–50 m depth. The remaining 8 shots had less than 100 kg and were shot in Hungary. Thirteen seismic-refraction lines were set up, using a total number of 926 Texan instruments (RefTek-1-component stations with 4.5 Hz geophones). Of these lines, line ALP04 overlapped with and extended the CELEBRATION 2000 line CEL10, and ALP12 followed the TRANSALP profile. In total, 4300 profile-km were covered. In addition a reflection recording spread was deployed in Austria and a dense local 3D-deployment was accomplished in Hungary. Furthermore, 70 stations were deployed along lines ALP04 and ALP12 for passive seismic monitoring. The ALP 2002 Working Group comprised scientists from Austria, Canada, Croatia, Czech Republic, Denmark, Finland, Germany, Hungary, Poland, Slovenia, and the United States (ALP 2002 Working Group, 2003; Brueckl et al., 2003). Recent interpretations were published in 2006, 2007, and 2010 (Bleibinhaus et al., 2006; Brueckl et al., 2007, 2010).

The large-scale seismic-refraction experiment ALP 2002 (ALP 2002 Working Group, 2003; Brueckl et al., 2003) also covered major parts of Eastern Europe outside the Alps. Some of the profiles, ALP01, ALP04, and ALP05 traversed the Bohemian

Massif in the Czech Republic, while profiles ALP13, ALP02, and ALP07 were centered in the Pannonian basin of Hungary, also touching the Dinarides of Slovenia and Croatia (Sumanovac et al., 2009; Figs. 10.2.3-01 and 10.2.3-02).

In 2003, the project SUDETES 2003, a 3-D seismic-reflection experiment with the aim to contribute to the investigation of deep crustal structure and geodynamics of the northern part of the Bohemian Massif, the largest outcropping part of the Late Paleozoic Variscan orogen in Central Europe, was launched. An overview of all long-range controlled-source seismic experiments in central and Eastern Europe from 1997 to 2003 was published by Guterch et al. (2007).

The layout of the field part of the project consisted of a network of profiles (Fig. 10.2.3-02), where besides in-line observations also off-line recordings were planned to obtain dense ray coverage for 3-D modeling (Guterch et al., 2003a; Majdanski et al., 2006).

Fifty-three shots with charges ranging from 50 to 1000 kg were fired into a network of six profiles, S01 to S06, which, with an average station spacing of 3–4 km, covered a total of 3450 km. The longest line, S04, ran from Germany in a NW-SE direction through the Czech Republic and Slovakia into the Pannonian Basin in Hungary, and was located ~100 km further north of and

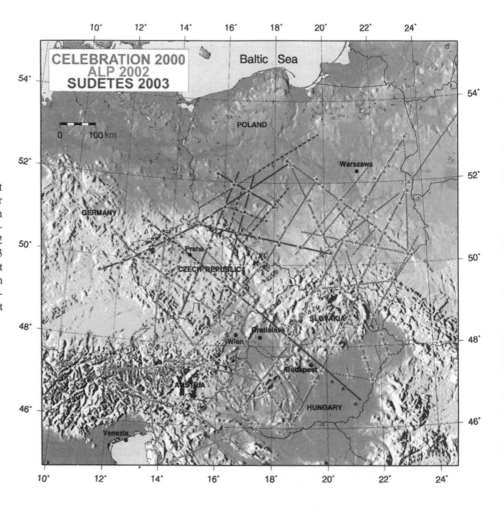

Figure 10.2.3-02. Map of Central-East Europe, showing location of major seismic refraction experiments (from Guterch et al., 2003a, fig. 1): CELEBRATION 2000 (red lines), ALP 2002 (light blue lines), and SUDETES 2003 (dark blue lines). [Studia Geophysica et Geodaetica, Academy of Science, Czech Republic, Prague, v. 47, p. 659–669. Permission granted by Studia Geophysica et Geodaetica, Prague, Czech Republic.]

parallel to line CEL09, described above. The second-longest line, S01, ran in a SW-NE direction through Czech territory parallel to the German MVE 90 line, south of the Erzgebirge, into Poland and ended on the East European craton.

Lines S02, S03, and S06 were jointly interpreted by Majdansky e al. (2006). The lines S02 and S03 ran almost parallel in an approximately N-S direction from Poland into the Czech Republic, and line S06 was a short, east-west–oriented cross line near latitude 52°N. As a result of the interpretation of the lines S02, S03, and S06 (Fig. 10.2.3-03; for more data examples see Appendix A10-1-1), the crust of the Bohemian Massif with a thickness of 33–35 km appeared slightly thicker than that of the Polish

fore-Sudetic area further north (Majdanski et al., 2006; Fig. 10.2.3-04). The lithospheric structure of the Bohemian Massif and adjacent Variscan belt in central Europe based on Profile S01 was interpreted by Grad et al. (2008).

Within the framework of the SUDETES 2003 project, the GRUNDY 2003 was performed with the aim of investigating the extent of the Variscan deformation front, which is one of the key problems of the regional geology of the Central European Permian Basin system, particularly in its Polish part (Malinowski et al., 2007), because conventional reflection seismics usually fail to produce a satisfactory image of the pre-Permian strata due to the shielding effect of Zechstein (Upper Permian) evaporites.

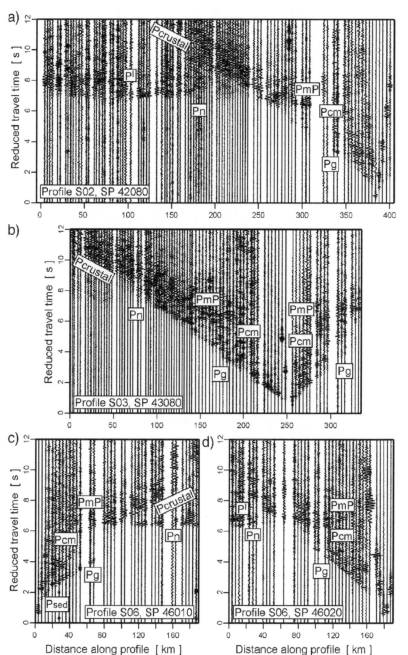

Figure 10.2.3-03. Examples of record sections for profiles S02, S03, and S06 of the SUDETES 2003 seismic project (from Majdanski et al., 2006, fig. 3). [Tectonophysics, v. 413, p. 249–269. Copyright Elsevier.]

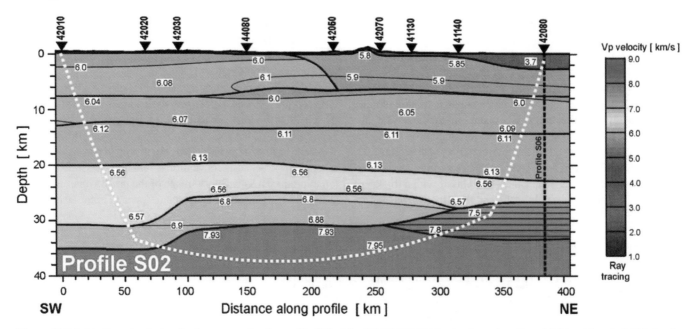

Figure 10.2.3-04. Crustal velocity-depth cross section for profile S02 of the SUDETES 2003 seismic project (from Majdanski et al., 2006, fig. 5). [Tectonophysics, v. 413, p. 249–269. Copyright Elsevier.]

Thus, Malinowski et al. (2007) used a novel seismic acquisition technique to study the base of the Permian complex and its Variscan basement. In the GRUNDY 2003 experiment, wide-angle reflection-refraction measurements were combined with the near-vertical reflection seismics by the use of the constant geophone array with dense (100 m) receiver spacing occupying a 50-km-long profile. The 3-D design of the experiment, covering a 50 × 10 km area, helped in eliminating the effect of out-of-plane propagations and local inhomogeneities. In the 50 × 10 km rectangular area, a total number of 786 single channel RefTek 125 "Texan" seismic stations equipped with 4.5 Hz geophones were deployed, forming a high-density central line (receiver spacing 100 m, 50 km long, referred to as G01 line) and additional 4 parallel profiles (G12–G15) with similar length but with mean receiver spacing of 600 m. Thirty shots were fired along the G01 profile, and 7 shotpoints were put in the side profiles. The mean charge of the shots was 50 kg of TNT explosives located in 2 drill holes 30–40 m deep. The data were recorded both inline and crossline, which allowed for 3D tomographic modeling of the whole target area and 2D reflection processing along G01 line. An effective integration of traveltime tomography, common-depth-point reflection processing, and prestack depth migration of wide-angle reflections applied to the data, allowed for a model in which the contact zone of the Variscan overthrust structure (Variscan front), with its molasse-filled foredeep, was deduced.

From 2001 to 2003, a passive teleseismic tomography project, BOHEMA, was added to the crustal and uppermost mantle investigations of the Bohemian Massif by controlled-source seismic experiments with the aim to develop a 3-D geodynamic model of the whole lithosphere-asthenosphere system (Babuska et al., 2003).

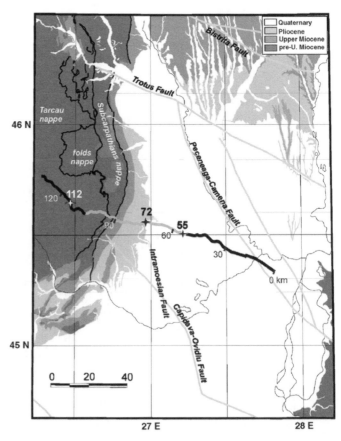

Figure 10.2.3-05. Map of southeastern Romania, showing the location of the DACIA PLAN seismic survey with the separate recorder deployments 1–3 (black-grey-black) indicated (from Panea et al., 2005, fig. 3). [Tectonophysics, v. 410, p. 293–310. Copyright Elsevier.]

In Romania in 2001, the second experiment across the Vrancea zone, VRANCEA 2001, was recorded. It was a 700-km-long WNW-ESE–trending seismic-refraction line and was carried out in August and September 2001 in Romania with a short extension into Hungary (Hauser et al., 2002, 2007a). The main part (for location see Fig. 9.2.4-11) ran from the Transylvanian Basin to the Black Sea coast, across the East Carpathian Orogen, the Vrancea seismic region, the foreland areas with the very deep Neogene Focsani Basin and the North Dobrogea Orogen. Landes et al. (2004b) have compiled data and other technical details in a data report. Due to its close relation to the VRANCEA'99 experiment, this project and its main results (see Fig. 9.2.4-15) have already been described in more detail in Chapter 9.

The VRANCEA 2001 project (Hauser et al., 2002, 2007a) also included a near-vertical incidence seismic-reflection part.

Between shotpoints T and U the geophones had been deployed at a spacing of 100 m. Immediately following the seismic-refraction work of VRANCEA 2001, two more deployments, using all available 640 one-component geophones (Texan type), were again laid out at 100 m spacing along the VRANCEA 2001 refraction line to the east of shotpoint T, traversing most of the Focsani basin, and shots were fired every 1 km (Fig. 10.2.3-05). This additional seismic-reflection program was enabled by the project DACIA PLAN (Danube and Carpathian Integrated Action on Process in the Lithosphere and Neotectonics). Panea et al. (2005) processed the data and published a first interpretation (Fig. 10.2.3-06).

In order to study in more detail a possible connection between crustal structure and the intermediate-depth seismicity of the Vrancea Seismogenic Zone (V in Fig. 9.2.4-11) with epi-

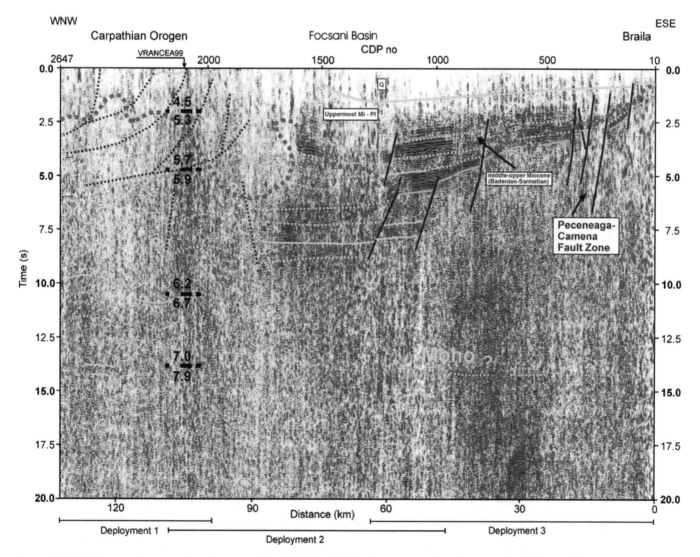

Figure 10.2.3-06. Interpreted time seismic section of the DACIA PLAN seismic survey (from Panea et al., 2005, fig. 10b). Dark blue velocity interfaces are from the VRANCEA99 model of Hauser et al. (2001), red dashed interface represents the 5.0 km/s velocity contour of the tomographic model of Bocin et al. (2005), blue and yellow lines—geological interpretation of the seismic-reflection data. [Tectonophysics, v. 410, p. 293–310. Copyright Elsevier.]

centers as deep as 160–200 km (Martin et al., 2005, 2006), the seismic-reflection project DACIA PLAN of 2001 was continued in 2004 with the project DRACULA (Deep Reflection Acquisition Constraining Unusual Lithospheric Activity). It consisted of three parts. DRACULA I was collected across the Transylvanian Basin and the SE Carpathian bend zone. DRACULA II was a 35-km-long E-W–directed seismic-reflection line and was located east of Onesti in the northern part of the Carpathian foreland. DRACULA III was also a 35-km-long seismic-reflection line which was recorded in the southern foreland south of Buzau, halfway between the Intramoesian and the Capadiva-Ovidiu Faults (Enciu et al., 2009).

In the SEISMARMARA French-Turkish program (Carton et al., 2007; Laigle et al., 2007), the N/O *Le Nadir* acquired multichannel seismic-reflection profiles in the North Marmara Trough in August 2001. This acquisition was restricted to the deep part of the Sea of Marmara, in water depths greater than 100 m. The profiles had an unprecedented depth-penetration due to the 4.5 km length of a 360-channel streamer and the strength of the low-frequency airgun arrays. Efficient penetration was obtained with the original "single-bubble" method (Avedik et al., 1996), which resulted from the modification of conventional airgun array shooting. The data allowed imaging the deep structure

of the marine North Marmara Trough on the strike-slip North Anatolian Fault west of the destructive Izmit 1999 earthquake. A reflective lower crust and the Moho boundary were detected. The OBS array data allowed especially constraining velocities for the deeper structural units, which ranged between 5.7 and 6.3 km/s for the upper crust and on the order of 6.7 km/s for the lower crust. Crustal thinning was evidenced by the seismic imaging of the reflective lower crust and the Moho boundary on an E-W profile with a shallowing of the lower crust from the south of the Central Basin toward the east into more internal parts of the deformed region.

In the Ukraine, after the successful investigations of the Donbas Foldbelt by both refraction and reflection profiling in 1999, in 2005 and 2006 the DOBRE-2 project extended the DOBRE-99 line (see Fig. 9.3-13) to the south across the Skythian Platform (Fig. 10.2.3-07). In 2005 common-depth-point profiling on land was completed with 100 km near the Azov Massif and 47 km through Crimea, while for 2006 the survey through the Azov Sea (160 km) and a short section (43 km) into the Black Sea had been planned using sea-bottom seismic stations (Starostenko et al., 2006).

A detailed investigation of the eastern Black Sea basin was performed in 2005 by the British universities at Southampton

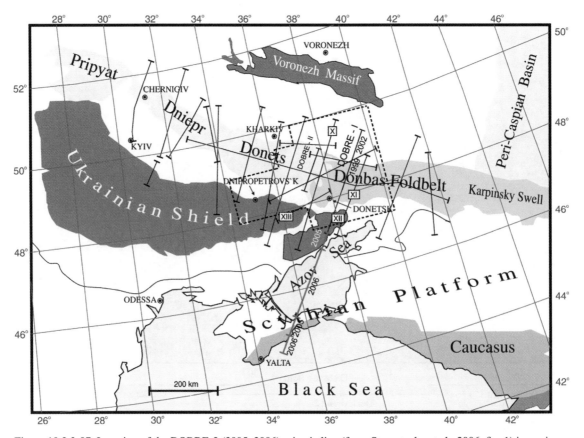

Figure 10.2.3-07. Location of the DOBRE-2 (2005–2006) seismic line (from Starostenko et al., 2006, fig. 1) in conjunction with DOBRE-1 (=DOBRE'99 in Figure 9.3-12). [*In* Grad, M., Booth, D., and Tiira, T., eds., European Seismological Commission (ESC), Subcommission D—Crust and Upper Mantle Structure, Activity Report 2004–2006.]

and Cambridge. A 500-km-long ESE-WNW–trending profile was crossed by three shorter perpendicular lines of up to 90 km length (Scott et al., 2006). A preliminary interpretation revealed an 8–10-km-thick sedimentary package on top of the crystalline crust which thins slowly from ~28 km at the eastern margin to 7–8 km in the center of the basin. One of the profiles crossed the Archangelsk Ridge where a maximum crustal thickness of 32 km was detected followed by a rapid decrease to 8 km in the center of the basin over a distance of only 30 km.

Knapp et al. (2004) discussed the reprocessing and reinterpretation of the two perpendicular ABSHERON 20 s deep seismic-reflection profiles in the South Caspian region, which were recorded with 70 km length in the deep water (200–700 m) of the South Caspian Sea, offshore Azerbaijan, at the Absheron Ridge with industry-like marine acquisition parameters. According to Knapp et al. (2004) these data provided evidence that the South Caspian basin may contain the thickest accumulation of sediments (26–28 km) in the world. The data also provided seismic reflections at all crustal levels, down to ~16–17 s and identified the Moho at 38–40 km depth, resulting in a 10-km-thick crystalline crust.

The ESRU (Europrobe Seismic Reflection profiling in the Urals) project, a EUROPROBE project which had started in 1993 (see Chapter 9, Fig. 9.3-19), was continued in 2001–2003, when the line was extended to the west in three campaigns along the Serebrianka-Beriozovka profile (SB), resulting in a total length of ESRU-SB of 440 km (Kashubin et al., 2006). In 2001, ~50 km were acquired west of ESRU93 in the Uralian foreland (SB01). In 2002, an additional 42 km of data were recorded on profile 2 (SB02) between the western end of ESRU93 and SB01. Finally, 73 km of profile SB03 were acquired in 2003 as a western extension of SB01.

10.3. NORTH AMERICA

10.3.1 The Pacific States of the United States

In 2002, the Georgia Strait Geohazards Initiative (SHIPS 2002) added new wide-angle recordings to the 1998 SHIPS experiment data (Brocher et al., 2003; Pratt et al., 2003). Eight hundred km of wide-angle recordings were made together with shallow penetration multichannel seismic-reflection profiles acquired by the Canadian Coast Guard Ship *Tully*, using a 1967-l airgun. Onland, 48 3-component RefTek stations recorded the airgun signals at wide offsets reaching ranges as far as 120 km (Fig. 10.3.1-01).

The active seismic experiment was preceded by a passive experiment Seattle SHIPS 2002, when 87 three-component stations were deployed for a period of three months, recording local and teleseismic events (Pratt et al., 2003). One set of 26 seismographs was deployed along an east-west line across the center of the basin and on bedrock at both ends, approximately coincident with the 1999 SHIPS experiment (see Fig. 9.4.2–10), while another set of 24 seismographs was deployed along a north-south line

through the basin center and the remaining 36 were distributed along two additional shorter north-south lines (Fig. 10.3.1-02). An example of the data quality is shown in Fig. 10.3.1-03, while more details of the projects and examples of data can be viewed in Appendix A10-2.

Starting in the 1990s, plans were developed to investigate the San Andreas fault zone by launching a deep-drilling project in central California named SAFOD (San Andreas Fault Observatory at Depth). Though only little populated, the Parkfield area, halfway between the densely populated areas of San Francisco and Los Angeles had attracted particular public interest. In 1966, a moderate earthquake had occurred with many aftershocks (Brown et al., 1967). At intervals of ~20 years, events of very similar characteristics (all with a magnitude of ~6) had repeatedly occurred in this area (1881, 1901, 1922, and 1934). This discovery led in the mid-1980s to the prediction of a similar event to occur in the near future, i.e., between 1985 and 1993. Therefore, in 1985, the U.S. Geological Survey initiated a large field experiment which should investigate the details of such an event (Bakun and Lindt, 1985). Even if the predicted event did not happen, the scientific interest in the Parkfield area has nevertheless remained extremely high. Thus, Parkfield became one of the target areas proposed as the location for a deep drillhole which should penetrate the San Andreas fault zone at various depth ranges by applying a particular drilling scheme proposed by Hickman et al. (1994), and which was finally named SAFOD. An initial shallow seismic experiment PSINE (Parkfield Seismic Imaging–Ninety Eight) was launched in 1998, comprising a 5-km-long high-resolution seismic-reflection/-refraction profile across the proposed drilling site (Catchings et al., 2002, 2003).

SAFOD is one of three major components of EarthScope, a National Science Foundation–funded initiative in collaboration with the U.S. Geological Survey, which is designed to investigate the powerful geological forces that shape the North American continent. The completion of SAFOD was foreseen for 2007. One of the prime goals of SAFOD is to be an earthquake observatory with instruments installed directly within the active fault where earthquakes occur. In July 2002, the first phase of this drilling project was realized when a test hole was drilled into the seismically active section of the San Andreas fault to a depth of 3 km.

The crustal structure in the surroundings of the proposed drilling site was only roughly known from large-scale investigations of the U.S. Geological Survey (for references see corresponding sections of Chapter 6 and Chapter 7 and Fuis and Mooney, 1990). The closest investigations had been the seismic-refraction survey in the surroundings of the Coalinga event of 1983 (for references see corresponding sections of Chapter 8) and the seismic-reflection and refraction line SJ-6, the data of which had been interpreted by a larger number of scientists within the frame of a CCSS (Commission on Controlled Source Seismology) Workshop (Walter and Mooney, 1987; for references, see corresponding sections of Chapter 8).

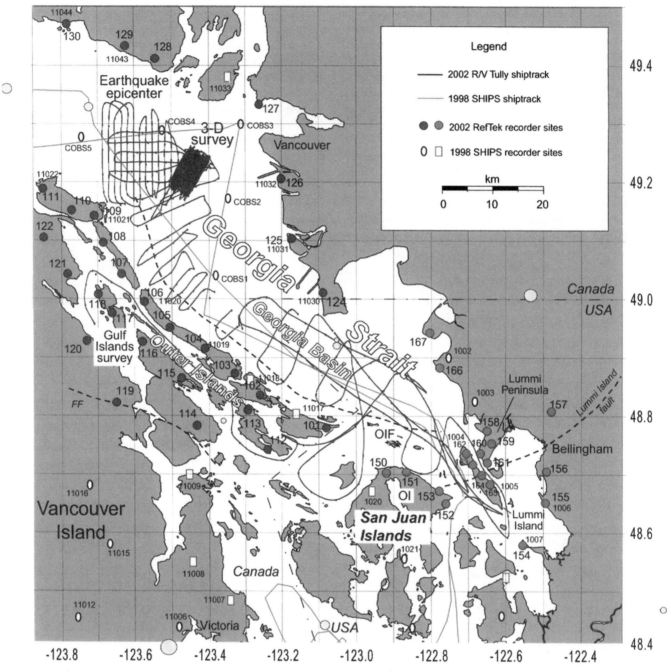

Figure 10.3.1-01. Locations of RefTek receivers (blue and red dots) and Tully tracklines (blue lines) of the active SHIPS 2002 project (from Brocher et al., 2003, fig. 1). Also shown are the 1998 SHIPS (see also Figure 9.4.2-09) recorder sites (open squares) and tracklines (red lines). Dashed lines—fault lines. [U.S. Geological Survey Open-File Report 03-160, Menlo Park, California, 34 p.]

Due to limited resolution of the seismic-refraction data, a detailed seismic-reflection line which crossed the SAFOD site had been planned (Hole et al., 2001) and was finally realized in 2004. The data were recorded along a 46 km profile centered on the SAFOD drill site and perpendicular to the San Andreas fault, with 62 shots (spacing 0.5–1.0 km) and 912 recorders (receiver spacing 25–50 m). The maximum depth penetration was 20 km. The interpretation of the data showed gently east- and west-dipping reflective bands within or below the granitic Salinian terrane at depths of 6–14 km, beginning at the San Andreas fault. Within the Franciscan terrane to the east of the fault, diffuse gently west-dipping reflectivity at 4–10 km depth was found (Ryberg et al., 2005; Hole et al., 2006; Bleibinhaus et al., 2007).

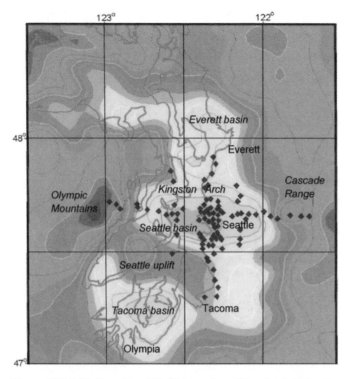

Figure 10.3.1-02. Location of RefTek receivers (blue dots) during the passive Seattle SHIPS 2002 project (from Pratt et al., 2003, fig. 1), superimposed on a tomography map showing isolines of speed at 2.5 km depth (VanWagoner et al., 2002). [Journal of Geophysical Research, v. 107, no. 12, p. 22-1–22-23. Reproduced by permission of American Geophysical Union.]

10.3.2. Basin and Range Province and Rio Grande Rift

In 2002, in the Walker Lane experiment, a new 450-km-long crustal refraction profile was collected between Battle Mountain, Nevada, and Auburn, California, traversing western Nevada, the Reno area, Lake Tahoe, and the northern Sierra Nevada (Fig. 10.3.2-01).

One hundred ninety-nine one-component instruments recorded mine blasts and earthquakes (Louie et al., 2004). Several ripple-fired mine blasts at the eastern end of the line, which were fired daily, produced arrivals visible up to distances of 300 km. From quarry blasts at the western end, energy was seen up to 600 km distance. A local earthquake, with an epicenter ~100 km SSE of Lake Tahoe, produced fan-shot data across the array. An unexpectedly deep crustal root of 50 km under the northern Sierra Nevada, centered west of the crest, and an anomalously thin crust with Moho depths of 19–23 km over a limited region near Battle Mountain in the Basin and Range province are the main results of this survey (Louie et al., 2004).

The northwestern margin of the Basin and Range province near the northern border of eastern California and western Nevada was the goal of an active-source seismic-refraction survey in 2004. It was part of the EarthScope Program investigating an area which is characterized by a transition from low-magnitude extension in northwestern Nevada to relatively unextended volcanic plateaus in northeastern California (Lerch et al., 2007). The experiment involved a 300-km-long east-west line with 1100 vertical seismometers which recorded five in-line shots and one off-line fan shot. Furthermore, ripple-fired mine blasts produced

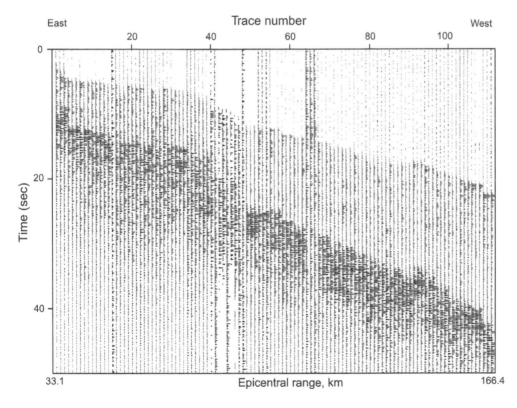

Figure 10.3.1-03. Three component recordings of a local M-2.7 earthquake with epicenter near Mount Vernon, recorded during the passive Seattle SHIPS 2002 project at distance ranges from 33 to 166 km (from Brocher et al., 2003, fig. 9). All three components of each station are plotted side by side. [U.S. Geological Survey Open-File Report 03-160, Menlo Park, California, 34 p.]

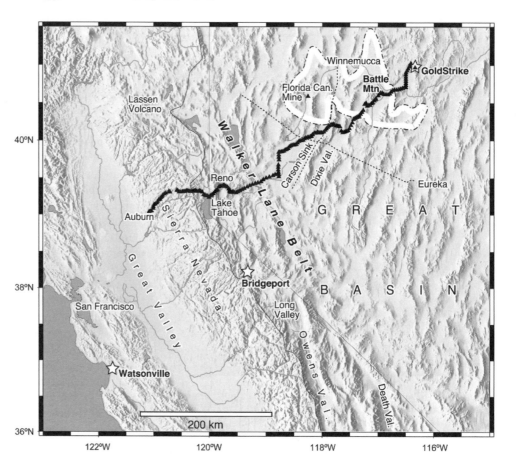

Figure 10.3.2-01. Location of receivers (triangles) and sources (stars) of the Walker Lane experiment 2002 (from Louie et al., 2004, fig. 1). Dotted lines: 1986 PASSCAL arrays (see Fig. 8.5.3-11). [Tectonophysics, v. 388, p. 253–269. Copyright Elsevier.]

useful first-arrival data. Altogether, average 300 m station spacing resulted. The interpretation of the data was able to document a 20% crustal thinning associated with Basin and Range extension from ~37 km crustal thickness under northeastern California to ~31 km under northwestern Nevada.

In 2003, the Potrillo Volcanic Field experiment was carried out in southern New Mexico and far west Texas. Its goal was to address the questions of how far to the south the Rio Grande rift extends and how far is it distinguishable from the Basin and Range province (Averill, 2007). The experiment consisted of a 205-km-long east-west–oriented line at ~32°N latitude with eight borehole shots with charges of 500–1000 kg. Almost 800 instruments (Texans) were deployed at variable spacing of 100 m, 200 m, and 600 m between Hachita, New Mexico, and Fort Bliss on the east side of El Paso, Texas. The western part of the line was located in the Basin and Range province. In the center of the line, an elevated platform of extensive volcanism occurs, making up the Potrillo Volcanic Field, while the eastern third of the line traversed the Rio Grande rift region. The most distinctive feature of the interpreted crustal model was an apparent thickening of the middle crust underlying the Potrillo Volcanic Field. In addition, an overall thinning of the crust from 35 km in the west to ~30 km in the east and a decrease in upper mantle velocities from 7.9 to 7.75 km/s provided evidence that the southern Rio Grande rift differs from the adjacent Basin and Range province.

10.4. THE AFRO-ARABIAN RIFT SYSTEM

10.4.1. The Jordan–Dead Sea Transform

In February 2002, the *Meteor* (1986) cruise M52 followed marine research projects in the Mediterranean, the Black Sea and the Red Sea. One of the projects was a seismic investigation offshore Israel. Twenty OBS/H were deployed along a 150-km-long profile, which was an extension of the DESERT 2000 line (Dead Sea Rift Transect). Signals of 4 vibrators ashore eastward of the Gaza Strip were recorded on land and by the OBH/S, followed by shooting with three airguns with a total volume of 100 l. The airguns were triggered every 60 s giving a shot distance of 120 m (Hübscher et al., 2003).

Four years after DESERT 2000, another seismic experiment was conducted across the Jordan–Dead Sea transform (ten Brink et al., 2006). The experiment consisted of two wide-angle seismic-reflection/-refraction profiles. The first line was a 250-km-long profile from the Gaza strip through the Dead Sea basin into eastern Jordan (Figs. 10.4.1-01A and 10.4.1-01B). The second line, 280 km long, ran along the international border between Jordan, Israel, and the West Bank at the center of the Dead Sea transform.

Only preliminary results of the first line were published hitherto (Figs. 10.4.1-01A and 10.4.1-01B). Along this profile,

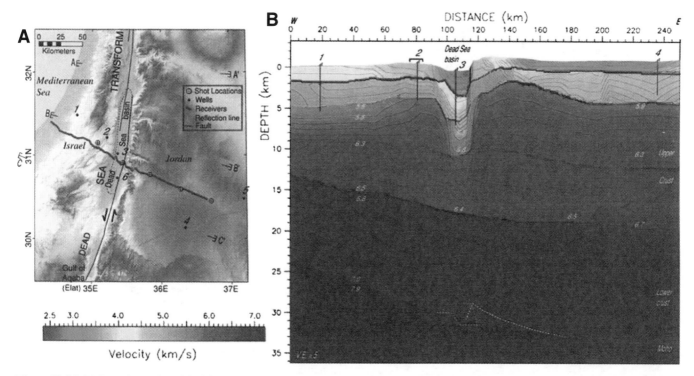

Figure 10.4.1-01. Location and model of the 2004 Dead Sea basin seismic refraction profile (from ten Brink et al., 2006, fig. 1). (A) Shaded relief map of the Jordan–Dead Sea rift showing the location of the Dead Sea basin profile of 2004. (B) P-wave velocity model along the seismic profile. [Geophysical Research Letters, v. 33, L24314. Reproduced by permission of American Geophysical Union.]

334 one-component stations (RefTek 125, Texans) were deployed at intervals of 0.65–0.75 km and recorded five one-ton explosions. Different from the DESERT 2000 line, crossing the Dead Sea transform ~70 km further south, was the observation of a narrow low-velocity zone within the basement underneath the Dead Sea basin proper, extending to 18 km depth (Figs. 10.4.1-01A and 10.4.1-01B). Moho depths varied from 35 km at the eastern end of the profile to 31 km under the rift valley and 24 km at the western end of the line.

Though only carried out after 2005, we want to mention a 235-km-long seismic wide-angle reflection/refraction profile across the Dead Sea Transform in the region of the southern Dead Sea basin (Mechie et al., 2009) which was completed in spring 2006 as part of the Dead Sea Integrated Research project (DESIRE). The DESIRE seismic profile crossed the Dead Sea Transform ~100 km north of where the DESERT 2000 seismic profile crossed the Dead Sea Transform. The wide-angle seismic-refraction measurements comprised 11 shots recorded by 200 three-component and 400 one-component instruments spaced 300 m to 1.2 km apart along the whole length of the E–W–trending profile. The depth to the seismic basement beneath the southern Dead Sea Basin was modeled at ~11 km below sea level beneath the profile, the interfaces below ~20 km depth, including the top of the lower crust and the Moho, appeared to show less than 3 km variation in depth beneath the profile as it crosses the basin.

10.4.2. The East African Rift System in Ethiopia

The investigation of the crust of the East African Rift System was revived in the beginning of the twenty-first century with a large international campaign in Ethiopia. With the funding of a new generation of mobile recording stations by NERC (National Environmental Research Council) to SEIS-UK (Maguire and SEIS-UK, 2002), funding for major crustal seismic research work in Ethiopia also became available.

Additional U.S. funding by the NSF (National Science Foundation Continental Dynamics Program) and other sources enabled to establish the Ethiopia Afar Geoscientific Lithospheric Experiment (EAGLE) under the heading of the Universities of Leicester, UK, and Texas at El Paso, United States, aiming to probe the crust and upper mantle structure between the continental main Ethiopian rift and Afar, an oceanic spreading rift. The region provided an ideal laboratory to examine the process of breakup as it is occurring (Maguire et al., 2003; Keranen et al., 2004).

EAGLE was a three-phase seismic experiment (Fig. 10.4.2-01), which started in October 2001 and was completed in February 2003. It was followed in March 2003 by a magnetotelluric project (Phase 4) and a variety of supporting efforts that collectively provided a diversity of data for integrated analysis.

The initial tomographic inversions of relative teleseismic residuals of Phase 1 showed low velocities underlying the

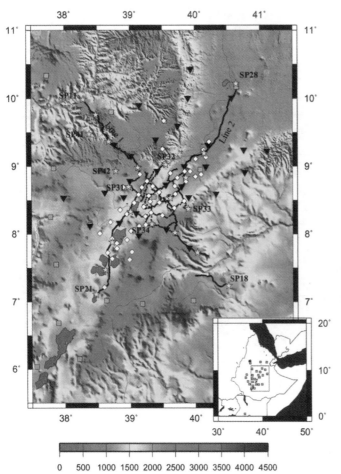

Figure 10.4.2-01. Topographic map of the central Ethiopian rift showing the locations of EAGLE (from Maguire et al., 2003, fig. 1): Phase 1 (inverted black triangles), Phase 2 (white diamonds), Phase 3 instruments (black dotted lines) and shots (yellow stars). Grey squares: locations of broadband seismic stations of Nyblade and Langston (2002, fig. 9.5.2-01). [Eos (Transactions, American Geophysical Union), v. 84, no. 35, p. 337, 342–343. Reproduced by permission of American Geophysical Union.]

Ethiopian rift down to depths of at least 200 km. The velocity anomaly appeared tabular in shape beneath the continental part of the rift in the southwestern region of the EAGLE-1 deployment, but was more diffuse and triangular in shape toward Afar in the northeast. An analysis of local events recorded during Phases 1 and 2 suggested a concentration of seismicity on the western margin of Afar.

The interpretation of the controlled-source seismic data (Fig. 10.4.2-02) involved forward and tomographic modeling techniques to obtain crustal and upper mantle models for the various lines (Maguire et al., 2006). The interpretation of the cross-rift profile of Phase 3 (line 1), e.g., resulted in a 2–5-km-thick layer of sedimentary and volcanic sequences across the entire region, underlain by a 40–45-km-thick crust under the central Ethiopian rift (Fig. 10.4.2-03) with a 15-km-thick high-velocity lower-

most crustal layer beneath the western plateau (Mackenzie et al., 2005). This lowermost crustal layer was not found under the eastern side, where the crust was modeled to be 35 km thick under the sediments. Beneath the rift a slight crustal thinning was observed, where P_n velocities indicated the presence of hot mantle rocks containing partial melt.

10.5. CONTROLLED-SOURCE SEISMOLOGY IN SOUTHEAST ASIA

10.5.1. India

The first five years did not see any major seismic investigation of the entire crust in India, as far as can be inferred from the accessible international literature. Rather, the research effort concentrated on the support of the hydrocarbon exploration geophysics. One example will be briefly mentioned here, in spite of our general aim to concentrate on projects dealing with basement features.

The project dealt with a detailed seismic investigation of the shallow structure of the Naga thrust and fold belt, located in the Assam province in northeastern India (Jaiswal et al., 2008). The deepening of the Naga thrust fault in the northeast caused particular challenges when drilling for hydrocarbons, traditionally guided by surface manifestations of the thrust fault. Therefore multichannel 2-D seismic data were collected along a line perpendicular to the trend of the thrust belt. The recorded line was 21 km long with 50 m shot and receiver spacing, involving 313 small borehole shots with 2.5–5.0 kg charges and maximum offsets of ~6 km. The result of the data interpretation was a detailed picture of the basement surface, where the Naga thrust belt could be traced at depth.

10.5.2. China

Two of the projects, the 320-km-long line through the Altyn Tagh Range (Zhao et al., 2006) and the 1000-km-long profile investigating the northeastern margin of the Tibetan Plateau (Liu et al., 2006) from the Qaidam Basin to the Ordos Basin, were described in the previous Chapter 9.6.2, because the authors did not indicate any date. However, the experiments may actually have been performed after the year 2000.

To summarize the multitude of new crustal and upper mantle structure results, obtained since the late 1970s until recently, Li et al. (2006) have created new contour maps for crustal thickness (Fig. 10.5.2-01) and P_n velocities (Fig. 10.5.2-02) as well as representative crustal thickness columns for the various tectonic regions of China (Fig. 10.5.2-03) which have been reproduced here.

In general, the crust thins from west to east. Beneath the Tibetan Plateau, it reaches a thickness of at least 74 km, which is the maximum value measured in the world from DSS data (Li et al., 2006). Along the east coast of China, in contrast, the crust is only 28–30 km thick. A strong lateral gradient in crustal thickness extends in the north-south direction throughout central China, approximately along 103–105°E (Fig. 10.5.2-01).

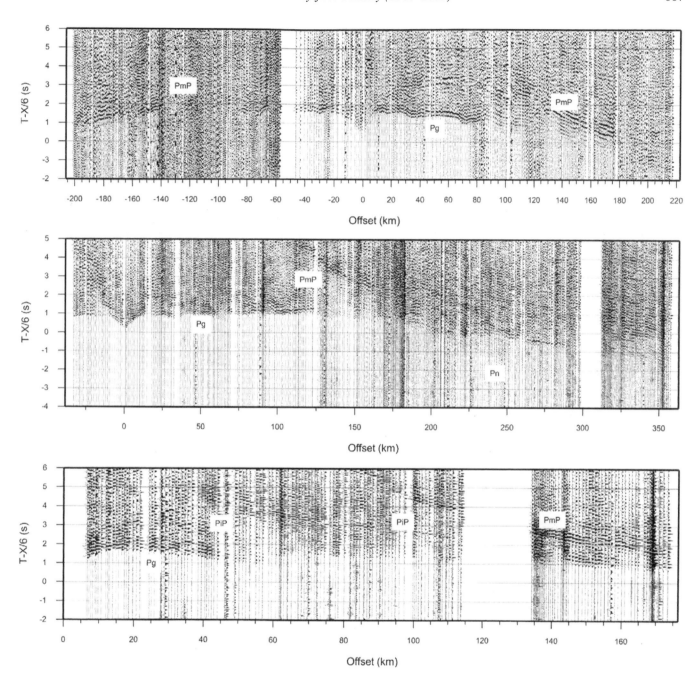

Figure 10.4.2-02. Data examples from the EAGLE 2002 project (from Maguire et al., 2003, fig. 3): Top: Line 1, central shotpoint 25. Center: Line 2, shotpoint 12 at NW flank of the rift. Bottom: Line 1 eastern half, central shotpoint 25. [Eos (Transactions, American Geophysical Union), v. 84, no. 35, p. 337, 342–343. Reproduced by permission of American Geophysical Union.]

The P_n velocity beneath China is relatively uniform, however, and was found to be between 7.9 and 8.0 km/s for the majority of observations, though the precision of P_n velocity determination is limited. The differences in crustal thickness are also evident in the crustal structure columns, shown in Figure 10.5.2-03 for selected tectonic regimes, shown in Figure 10.5.2-04, and which have been divided into a western China section with crustal thicknesses of 46 km and more and an eastern China section with crustal thicknesses usually less than 36 km.

10.5.3. Japan and South Korea

In Japan, a major crustal research project was launched in 2001 to complement a research project of 1985 (Iidaka et al., 2004).

A 260-km-long profile was recorded in southern Japan across south-central Honshu in N-S direction from coast to coast (Fig. 10.5.3-01). Three hundred ninety-one seismic stations with an average spacing of 670 m recorded five explosive sources with 500-kg charges and one 100-kg shot. sixty-three

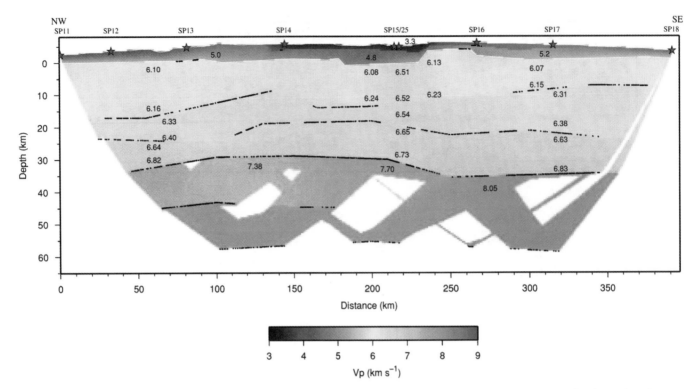

Figure 10.4.2-03. P-wave model across the central Ethiopian rift along EAGLE line 1 (from Mackenzie et al., 2005, fig. 6). For location, see Fig. 10.4.2-01. [Geophysical Journal International, v. 162, no. 3, p. 994–1006. Copyright John Wiley & Sons Ltd.]

of these stations were equipped with three-component geo-phones. The profile ran along line A–A′ in Figure 10.5.3-01 (identical with line B–B′ of Fig. 9.6.3-01) and had the same location as the Atsumi-Noto profile of Aoki et al. (1972). It crossed line B–B′ in Figure 10.5.3-01 which had been recorded in 1985 (Matsu'ura et al., 1991).

Figure 10.5.3-02 shows a record section of the 2001 survey (see also Appendix A10-3-1). The resulting crustal structure (Fig. 10.5.3-03) is similar to that in Figure 9.6.3-04, but shows less deepening of the Moho underneath the center of the line (Iidaka et al., 2004).

The most important experiments in 2001–2005 were intensive seismic-reflection/-refraction projects extending from the Philippine Sea across SW Japan to the Sea of Japan, which continued the investigations conducted during the 1990s (Kodaira et al., 2002; Ito et al., 2009b). Under the framework of disaster mitigation research program (the Special Project of Earthquake Disaster Mitigation in Urban Areas), a series of seismic investigations in and around Kanto plain (Sato et al., 2005), central Japan, and Kii peninsula and Kinki district, SW Japan (Ito et al., 2006; Sato et al., 2009) were performed.

In the Kanto region, four seismic profiles were recorded from 2002 to 2003 across the greater Tokyo region to image the Philippine Sea Plate which subducts northwestward underneath the Tokyo metropolitan region in central Japan (Sato et al., 2005, 2006). Seismic reflections from the upper surface of the Philippine Sea Plate were observed on all seismic sections and reached

from 6 to 26 km depths. They were nearly concordant with the slab geometry estimated from seismicity.

In 2004, a deep seismic-reflection profile was recorded in the Kinki metropolitan area between Osaka and Suzuka, using Vibroseis and dynamite shots (Ito et al., 2006). Wide-angle recording using widely spaced shots of 300 kg charge size and some sets of 100 stationary Vibroseis sweeps was also used. The main targets were to unravel the deep geometry of an active fault and the velocity structure. Beneath the mountain range separating the two basins with the cities Osaka in the west and Suzuka in the east, horizontal, coherent mid-crustal reflectors at 16 km depth were identified with the base of the seismogenic zone. Also in the seismogenic zone dipping reflectors were detected which were interpreted as possibly deeper extensions of active faults. Another set of reflectors was seen at 26 km depth.

An active-source onshore-offshore 485-km-long seismic profile extended from the Nankai Trough to the northern shore of central Honshu into the Sea of Japan (Kodaira et al., 2004). The wide-angle seismic data from onland explosions (maximum 500 kg) were recorded by 70 ocean-bottom seismometers (OBS) and 328 land-based stations. Furthermore, signals from a 197 l airgun array were recorded by all OBSs and for ~100 km onshore by 63 land stations. On the offshore part, multichannel seismic-reflection data were also recorded. The seismic velocity and reflectivity images showed several regions of crustal thickening down to a depth of 45 km under the offshore part which was interpreted as subducted ridges. The subducted crust beneath

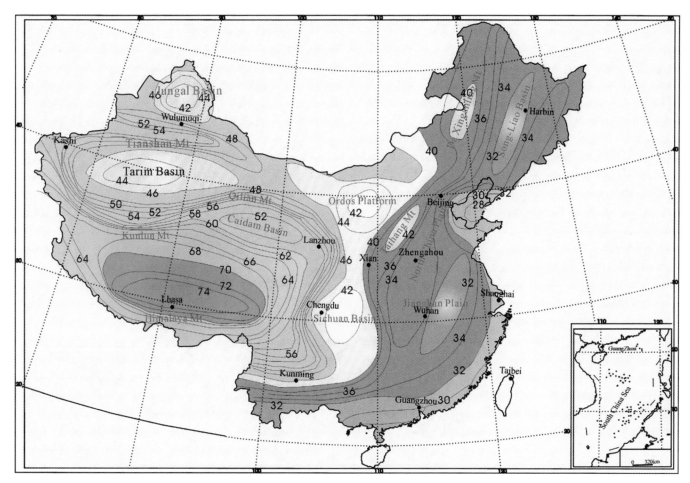

Figure 10.5.2-01. Contour map of crustal thickness in China from DSS data (from Li et al., 2006, fig. 4). For location of profiles, see Fig. 9.6.2-01 and Figure 10.5.2-02. [Tectonophysics, v. 420, p. 239–252. Copyright Elsevier.]

the onshore part appeared as a highly reflective interface from a depth of 25–45 km. Coincident in location with the highly reflective subducted crust was a landward-dipping zone recognized with a high Poisson's ratio of 0.34.

The most important results from these experiments were the detection of very strong reflections from the Philippine sea plate, which almost correspond to the aseismic part of the plate boundary. The seismogenic zone (asperity) along the plate boundary, on the other hand, appeared less reflective. Such structural difference was interpreted as probably being controlled by water dehydrated from the subducting oceanic lithosphere.

Furthermore, seismic-refraction/-reflection surveys were also conducted in inland active fault zones (e.g., the Itoigawa-Shizuoka tectonic line and Atototsugawa fault zone, central Honshu) (Iidaka et al., 2009; Ikeda et al., 2004, 2009; Sato et al., 2004).

In 2002, the Itoshizu seismic survey was recorded in the southwest of Honshu. It was a 68-km-long seismic line and was arranged perpendicular to the trend of major faults and folds across the Itoigawa-Shizuoka Tectonic Line (ISTL). For the western part of the line, a high-resolution image of the geometry

of the active fault system was acquired by a standard common mid-point reflection experiment using Vibroseis trucks and a digital telemetry cable system. For the eastern half, 600 channels of offline recorders were deployed. To obtain a deeper image by refraction/wide-angle reflection, four explosive sources and six Vibroseis locations were recorded. The interpretation suggested that the Miocene basin was formed by an east-dipping normal fault with a shallow flat segment to 6 km depth and a deeper ramp penetrating to 15 km depth.

The first seismic investigation of the Earth's crust in South Korea was undertaken in 2002 and 2004 (Kim et al., 2007). In 2002, a 294-km-long profile was recorded in the southern part of the Korean peninsula, installed in a WNW-ESE direction with two shotpoints, one at the WNW end and one in the center of the line, reaching from the Yellow Sea in the west to the East Sea in the east. A second line, 335 km long, followed in 2004 in a NNW-SSE direction with four shotpoints, one at each end and two in the center, and ran from near the border to North Korea more or less parallel to the coasts through the center of the peninsula to the southern coast bordering the East China Sea. Seismic waves were generated by 500–1000 kg borehole shots and recorded

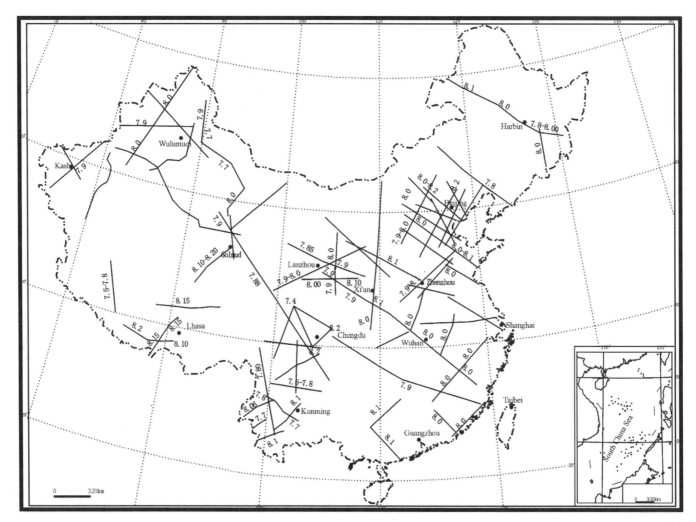

Figure 10.5.2-02. Seismic velocities of the uppermost mantle from P$_n$-waves (from Li et al., 2006, fig. 6). [Tectonophysics, v. 420, p. 239–252. Copyright Elsevier.]

along the lines by 170 PRS-1 portable seismographs at nominal intervals of 1.5–1.7 km. The interpretation indicated several crustal layers at depths of 2–3 km, 15–17 km, and 22 km. Moho was found at 37–39 km maximum depths with uppermost mantle velocities of 7.8–8.4 km/s. The Moho became shallower when approaching the Yellow Sea and the East Sea on the west and east coasts of the peninsula.

10.6. CONTROLLED-SOURCE SEISMOLOGY IN THE SOUTHERN HEMISPHERE

10.6.1. Australia and New Zealand

In western Australia in 2001, the deep seismic-reflection survey 01AGSNY was recorded (Goleby et al., 2004) as a continuation and supplement of the 1991 investigations of the Goldfields Province of Western Australia (Fig. 9.7.1-06). It comprised 430 km of deep seismic-reflection data across the northeastern

Eastern Goldfields Province portion of the Yilgarn Craton and was already discussed in Chapter 9.7.1, together with the 1991 survey. Including a number of additional high resolution deep seismic-reflection data obtained during 2002 and 2004 (Finlayson, 2010; Appendix 2-2), Goleby et al. (2006) compiled a 3-D seismic model of crustal architecture. The review included wide-angle and reflection profiling as well as receiver function work.

In 2003, a 193-km-long, N-S–directed, deep seismic-reflection traverse was acquired across the Archean-Proterozoic Gawler Craton in South Australia, centered on the Olympic Dam Mine. The survey was carried out using Vibroseis and collected data up to 18 s TWT. The interpretation revealed a Paleoproterozoic succession reaching to a depth of 14 km which was underlain by a 5-km-thick mid-crustal reflective layer. The Moho was seen at ~13 s TWT (Drummond et al., 2006).

With the same techniques, another 197-km-long, E-W–directed, seismic-reflection traverse was recorded in 2003–2004 from the Broken Hill block in New South Wales across the Curna-

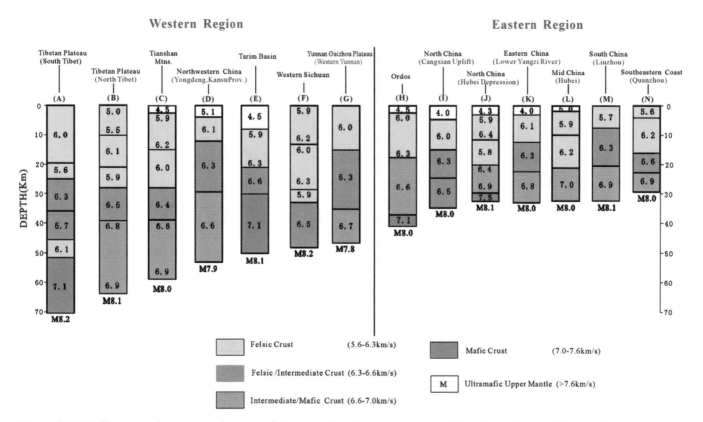

Figure 10.5.2-03. Representative seismic velocity-depth functions for 14 tectonic regimes of China (from Li et al., 2006, fig. 5). For location of the 14 regimes, see Fig. 10.5.2-04. [Tectonophysics, v. 420, p. 239–252. Copyright Elsevier.]

mona region of South Australia to the central Flinders Range to study the Meso- to Paleoproterozoic basement architecture. The line continued a line of a 1996–1997 survey across the Broken Hill block (Finlayson, 2010; Appendix 2-2).

An onshore-offshore operation in 2005 using the vessel *Pacific Sword* investigated the southwesternmost corner of Australia. The *Pacific Sword* was equipped with a 4900 cubic inch airgun array and a 6–8 km digital streamer. Marine seismic profiling off the southeastern and southwestern margins of the Yilgarn Craton intended to reveal the crustal architecture of the offshore frontier basins, acquiring a network of 2700 km of seismic data. During the marine work, seismic recorders were deployed onshore at 25–50 km intervals which recorded the airgun shots up to 400 km distance. A key feature of the modeling of the wide-angle data was a deep crustal root located ~100 km inland from the coast (Finlayson, 2010; Appendix 2-2).

In 2005, a 290-km-long north-south traverse was recorded across the boundary of the Thomson Orogen in western Queensland–northern New South Wales and the Lachlan Orogen in central and southern New South Wales. A 48-km-deep Moho underneath the Thomson Orogen and a 33-km-deep Moho under the Lachlan Orogen, separated by a major northward dipping fault that cuts the entire crust, was one of the major results. Also, the lower crust was more reflective under the Lachlan Orogen (Finlayson, 2010; Appendix 2-2).

A second project in 2005 collected 720 km of deep seismic-reflection data in a grid of four deep seismic traverses in the Tanami region of the Northern Territory. Associated with the reflection profiling was a large-scale passive listening refraction survey along a NW-SE transect to record wide-angle seismic data from the vibrator source and thus to provide the crustal velocity structure for the region and its relationship to lithology (Finlayson, 2010; Appendix 2-2). Moho was defined as the base of the reflective crust, dipping slightly from ~35 km depth in the NW to more than 50 km in the SE.

On the North Island of New Zealand, in 2001, a major seismic project was carried out. The Central Volcanic Region or Taupo Volcanic Zone occupying the northern half of the North Island of New Zealand is a region where onshore backarc extension within continental lithosphere resulting from the westward subduction of the Pacific oceanic plate could be studied. Therefore, in 2001 the NIGHT (North Island Geophysical Transect) project was initiated (Henrys et al., 2003a, 2003b; Harrison and White, 2006; Stratford and Stern, 2006). It involved both an active-source and a passive-source component and had also an offshore component (Fig. 10.6.1-01).

The active source part had two main deployments. First, in January-February 2001, 88 3-component stations (RefTeks), 200 1-component stations (Texans), and 15 OBSs were deployed with an average spacing of 1 km along a NW-SE profile and recorded

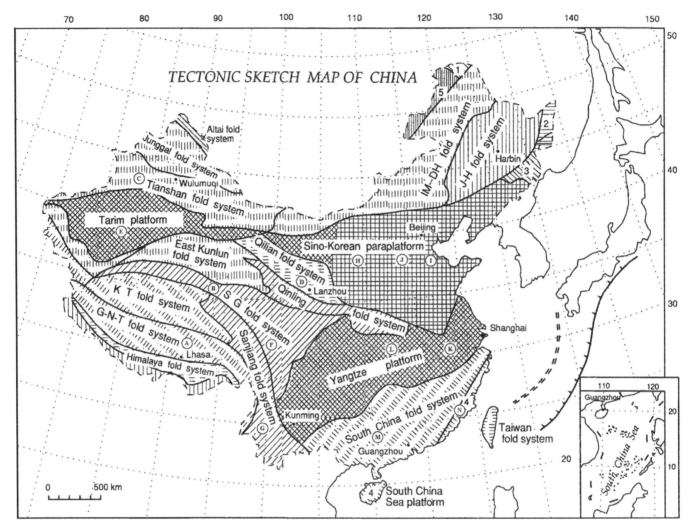

Figure 10.5.2-04. Tectonic map of China (from Li et al., 2006, fig. 2). The circled letters indicate the regimes for which representative crustal structure columns are shown in Fig. 10.5.2-03. [Tectonophysics, v. 420, p. 239–252. Copyright Elsevier.]

shots from M/V *Geco Resolution* (8200 cu in airgun array), that also recorded vertical incidence crustal multichannel seismic data (6000 m 480 channel array). The 288 land recorders were redeployed along the NW-SE profile and a N-S line through the back-arc Taupo Volcanic Zone, and nine 500 kg shots recorded (7 on the NW-SE profile and 2 offset onto the N-S line). The second main deployment, in December 2001, fired 3 shots into 200 1-component stations located on the N-S line and 100 1-component stations on the western half of the NW-SE line. Ninety-five stations were redeployed onto the eastern part of the NW-SE line, and two additional 500 kg shots were fired (Fig. 10.6.1-01). The NW-SE line was 250 km long and ran from coast to coast through the axis of the Taupo Volcanic Zone, while the N-S line was 120 km long and was located entirely within the Taupo Volcanic Zone.

To complement the active-source data, passive source (earthquake) data were recorded during additional deployments (Fig. 10.6.1-02). First, 63 short-period 3-component stations were de-

ployed at 2-km spacing along the NW-SE line for two months following the active source phase 1, and second, following active-source phase 2, a broadband array of 20 stations (Guralp) was deployed along the same line with 5-km spacing (Harrison and White, 2006). The location and timing of the local shallow (<10 km depth) and deep (>40 km depth) earthquakes recorded by these linear passive arrays was constrained by the New Zealand National Seismograph Network as well as by the CNIPSE array (Henrys et al., 2003a). Data examples of the active-source experiment and the subsequent ray trace modeling are shown in Figure 10.6.1-03.

The long-range seismic-refraction data showed velocities of 6 km/s and less within the top 15 km of the crust of the Taupo Volcanic Zone. At 15 km depth, the P-velocity increased to 6.8 km/s and then to 7.4 km/s at ~20 km depth. Rocks between 15 and 20 km depth were interpreted as new crust formed by underplating. The strongest reflection, interpreted as the reflection Moho, was observed from the top of the proposed underplated

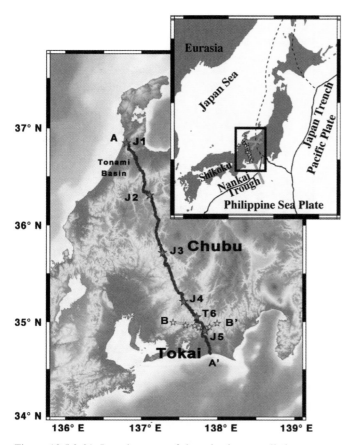

Figure 10.5.3-01. Location map of the seismic controlled-source experiment (line A–A') of 2001 across south-central Honshu, Japan (from Iidaka et al., 2004, fig. 1). Line B–B' was recorded in 1985 (Matsu'ura et al., 1991). Shotpoints are shown by stars. The orange lines in the inset (for location, see also Fig. 9.6.3-01) are profile lines analyzed by Takeda (1997). [Tectonophysics, v. 388, p. 7–20. Copyright Elsevier.]

layer at 15 km depth. No such distinct reflection was observed at 20 km depth. Rather, wide-angle reflection data showed a continuum of low-level reflectivity between 15 and at least 35 km depth. Thus, Stratford and Stern (2006) concluded that the transition from lower crust to upper mantle should be broad.

Harrison and White (2006) discussed two possible interpretations of the geophysical results. *Either* the crust beneath Taupo Volcanic Zone is 15 km thick with a huge layer of melt at the top of the mantle (Fig. 10.6.1-04), *or* the crust is 30 km thick with the lower half comprising extremely heavily intruded crust (Fig. 10.6.1-05). Harrison and White (2006) favored the latter interpretation.

The interpretation of the marine data which extended the line of observations toward southeast beyond the Hikurangi subduction zone showed the subducting crust as a shallow dipping (3°) strongly reflecting interface at 6 s TWT (two-way traveltime) which became less pronounced where the plate interface steepens landward at a depth of ~12 km, 120 km from the trench axis. From the velocity model, based on the wide-angle onshore-offshore refraction data and the inversion of arrival times from earthquakes recorded by the CNISPE array, the fore-arc had been seen as a region with low P-velocity (<5.5 km/s), high V_P/V_S (>1.85), high Poisson's ratio (>0.29) overlying subducting oceanic crust ($V_P = 7.0$ km/s) (Reyners et al., 2006). After a pre-stack depth migration a dominant reflector was observed which coincided with the top of the subducting plate and an increase of dip could be imaged as a pronounced step of the reflector by 5 km (Henrys et al., 2006).

Three more seismic-reflection surveys may be mentioned briefly. In 2001, 500 km of crustal reflection data were gathered across eastern Chatham Rise, recording a source of 8204 cu in along a 6000 m streamer up to 16 s TWT (Davy et al., 2008).

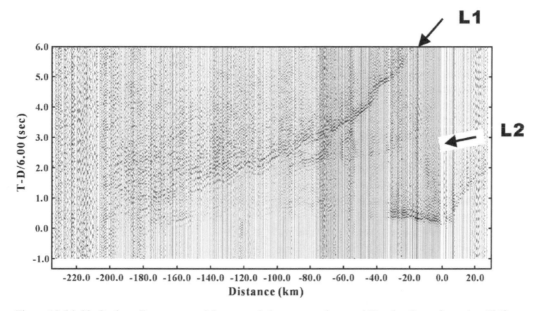

Figure 10.5.3-02. Onshore P-wave crustal data recorded across south-central Honshu, Japan from shot J5 (from Iidaka et al., 2004, fig. 2c). For location, see Fig. 10.5.3-01. [Tectonophysics, v. 388, p. 7–20. Copyright Elsevier.]

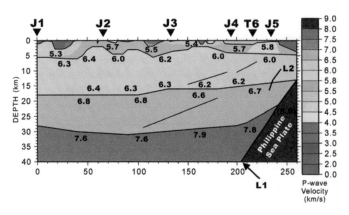

Figure 10.5.3-03. P-wave crustal model across south-central Honshu, Japan (from Iidaka et al., 2004, fig. 4). For location, see Fig. 10.5.3-01. [Tectonophysics, v. 388, p. 7–20. Copyright Elsevier.]

In 2003, two crustal transects were obtained east of South Island, recording a source of 3000 cu in by 25 OBSs and a 2250 m streamer. The first transect was observed across Bounty Trough and was 410 km long, the second transect extended for 500 km across Great South Basin and Campbell Plateau (Grobys et al., 2007, 2009). Finally, in 2005, 2950 km of marine reflection data were recorded off northeastern North Island to investigate the geometry of the Hikurangi subduction thrust and upper plate. Records of 12 s TWT were recorded by a 6–12-km-long streamer, using a source of 4140 cu in (Barker et al., 2009).

10.6.2. South Africa

In 2003, seismic investigations in Namibia supported an onshore-offshore experiment on the west coast of South Africa in order to reveal the structure of the oceanic-continent transition zone. The land part was jointly organized by the Council of Geoscience (RAS) and the GFZ Potsdam (Germany). The onshore part consisted of two profiles running almost perpendicular to the coastline with length of 100 km and three shots (100 kg, 50 kg, 100 kg) were fired on each line. The land stations also recorded airgun shots, fired along the offshore lines. One of the lines, the 500-km-long Springbok profile, was interpreted by Hirsch et al. (2009) and is described in more detail in subchapter 10.7.4.

A second seismic project of the GFZ Potsdam (Germany) was the 800-km-long Agulhas-Karroo transect, carried out in 2005 within the project INKABA ya Africa (Lindeque et al., 2007; Stankiewicz et al., 2007, 2008). It was a north-south offshore-onshore transect which ran over a distance of 400 km from the offshore Agulhas Plateau across the Agulhas Fracture zone and the continental margin onto to South African coast, and for other 600 km onland across the Cape Fold Belt, the Karoo Basin and into the Kapvaal Craton. The project also included the recording of 600 km of magnetotelluric data. For the offshore part the reader is referred to subchapter 10.7.4.

Within this frame, two wide-angle onshore seismic lines were collected. The lines ran parallel to each other, ~200 km

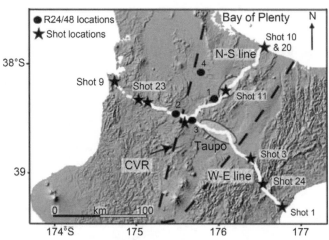

Figure 10.6.1-01. Location of the NIGHT active-source transects across the Taupo Volcanic Zone or Central Volcanic Region (CVR) on North Island of New Zealand (from Stratford and Stern, 2006, fig. 3). Black dots are the deployment locations of the R24/48-channel seismograph. [Geophysical Journal International, v. 166, p. 469–484. Copyright John Wiley & Sons Ltd.]

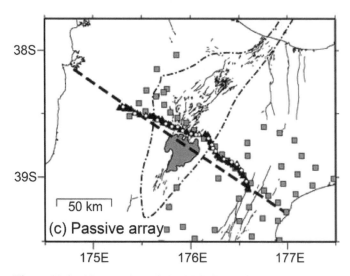

Figure 10.6.1-02. Location of the NIGHT passive-source transects across the Taupo Volcanic Zone on North Island of New Zealand (from Harrison and White, 2006, fig. 2c). Black triangles—linear short-period array of early 2001; white circles—linear broadband array of 2001–2002; grey squares—CNIPSE array. [Geophysical Journal International, v. 167, p. 968–990. Copyright John Wiley & Sons Ltd.]

apart, and were ~240 km long. At each line, 48 receivers recorded data from 13 shots fired in boreholes with charges between 75 and 125 kg. The average shot spacing was 20 km; that of the receivers was 5 km. The record sections from the land stations showed clear P_MP arrivals, both from the land shots and from part of the airgun shots, as well as a clear mid-crust reflection. In the joint model from airgun and land shots the Moho discontinuity was clearly visible as a high-velocity contrast. It was found at 40 km depth underneath the Karoo basin and slightly deeper

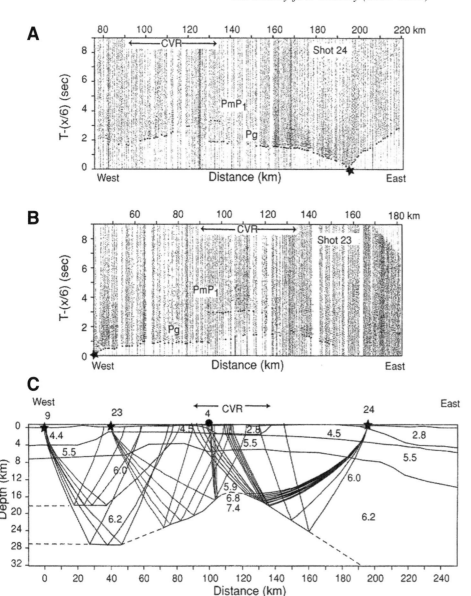

Figure 10.6.1-03. NIGHT active-source experiment across the Taupo Volcanic Zone on North Island of New Zealand (from Stratford and Stern, 2006, fig. 4). Data examples (A) shot 24, (B) shot 23, and (C) ray trace model. [Geophysical Journal International, v. 166, p. 469–484. Copyright John Wiley & Sons Ltd.]

(42 km) under the Cape Fold Belt. South of the Fold Belt, the Moho shallows abruptly to 30 km at the present coast.

Furthermore, a 100-km-long near-vertical seismic-reflection experiment yielded a high-quality seismic image of the crust and Moho across the southern Karoo basin in South Africa (Lindeque et al., 2007). The highly reflective crust comprised upper, middle, and lower layers. A well-defined middle crustal layer of 20 km thickness occurred below a seismically imaged unconformity. The underlying lower crustal layer is wedge shaped and 24 km thick in the north and decreases to 12 km thickness beneath the Cape Fold Belt. A clearly imaged undulating Moho occurs at a depth of ~43 km in the north with a nick point at 42–35 km depth and then deepens to 45 km in the south beneath the tectonic front of the Cape Fold Belt. A possible 1–2 km lowermost crustal layer of high seismic reflectivity overlies the Moho and was interpreted as underplated mafic material.

10.6.3. South America

In Venezuela, the ECCO (Estudio Cortical de la Cuenca Oriental) study of 2001 (Schmitz et al., 2005) targeted the Oriental Basin north of the Guayana Shield (see Fig. 9.7.3-06). It was a quasi-continuation of the 1998 ECOGUAY (Estudios de la Estructura Cortical del Escudo de Guayana) study targeting the Guayana Shield (Schmitz et al., 2002) and was therefore described already in Chapter 9.7.3.2.

The coastal area of Venezuela, representing the Caribbean–South America plate boundary, became the target of a large active-source land-sea seismic experiment in 2004 (Levander et al., 2006; Clark et al., 2008; Schmitz et al., 2008; Magnani et al., 2009). About 6000 km of marine multi-channel seismic-reflection data were collected offshore Venezuela. In addition five wide-angle seismic profiles were recorded both onshore and

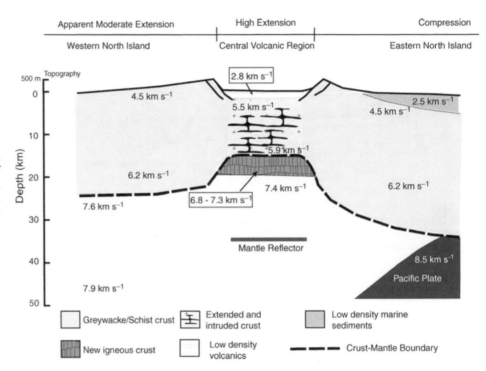

Figure 10.6.1-04. Model 1 across the Taupo Volcanic Zone on North Island of New Zealand (from Stratford and Stern (2006, fig. 10). [Geophysical Journal International, v. 166, p. 469–484. Copyright John Wiley & Sons Ltd.]

offshore Venezuela and the Antilles arc region (Fig. 10.6.3-01), spanning an area of almost 12 degrees of longitude and 5 degrees of latitude as part of the project BOLIVAR (Broadband Ocean Land Investigation of Venezuela and the Antilles Arc Region).

The 550-km-long N–S profile crossed the structures involved in the active 55-m.y.-long continent-arc oblique collision between the Caribbean and the South American plate. From the north to the south, these structures include the accretionary prism, the extinct volcanic arc (Leeward Antilles arc), the Tertiary Bonaire basin, the continental-size dextral strike-slip fault system (San Sebastián–El Pilar fault), the allochthonous exhumed terranes, and the autochthonous fold and thrust belt (Caribbean Mountain system) and foreland basin. The wide-angle data showed that these elements are characterized by different velocity structures and that they are separated by sharp lateral velocity variations. The Leeward Antilles island arc, being 27 km thick with a high velocity (7 km/s) crustal layer, exhibited a velocity structure similar to that of the Lesser Antilles active volcanic arc indicating that the extinct arc has not been modified by the collision with the South American plate.

The data showed an ~20 km change in crustal thickness across the San Sebastián fault, separating the 25-km-thick Bonaire crustal block with average crustal seismic velocities ranging between 5.5 and 6.7 km/s from the 45 km South American crustal block and suggesting that the dextral strike-slip fault is a crustal feature that likely continues in the mantle as a primary strand of the plate boundary between the South American and the Caribbean plates. South of the strike-slip fault and beneath the exhumed eclogitic terranes, the data imaged a north-dipping, high-velocity (>6.5 km/s) anomaly in the upper crust (3–11 km),

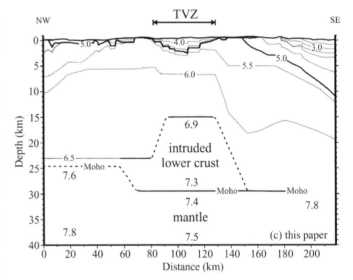

Figure 10.6.1-05. Model 2 across the Taupo Volcanic Zone (TVZ) on North Island of New Zealand (from Harrison and White, 2006, fig. 17). [Geophysical Journal International, v. 167, p. 968–990. Copyright John Wiley & Sons Ltd.]

indicating that high-pressure/low-temperature rocks are the likely lithologies responsible for the high seismic velocities and suggesting that exhumation of these assemblages is enabled by the strike-slip fault (Magnani et al., 2009).

Chile had become the target of a series of major research projects in the 1990s, such as PISCO94, CINCA95, CONDOR in 1995, ANCORP96, and ISSA2000. The research work con-

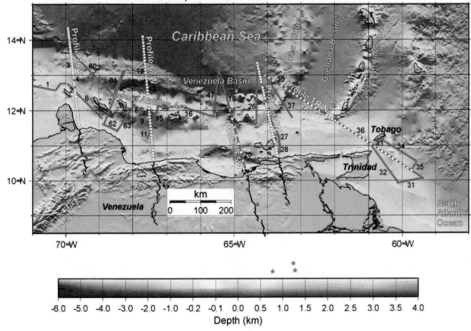

SE Caribbean Continental Dynamics Project:
Marine Reflection, OBS Profiles and Land Profiles

Figure 10.6.3-01. Location of the BOLIVAR controlled-source seismic project 2004 (from Magnani et al., 2009, fig. 1b). Red lines—seismic reflection lines; white circles—ocean-bottom seismometers; black lines—land-based seismometers; red stars—shotpoints. [Journal of Geophysical Research, v. 114, B02312, doi:10.1029/2008JB005817. Reproduced by permission of American Geophysical Union.]

tinued into the 2000s, of which the project SPOC (Subduction Processes off Chile) in 2001 (Krawczyk et al., 2003, 2006), due to its close relation to the foregoing project ISSA2000 (Bohm et al., 2002; Lueth et al., 2003), was already described in some detail in Chapter 9.7.3.3.

The latest experiment in southern Chile was the TIPTEQ (from The Incoming Plate to Mega-Thrust Earthquake Processes) array project of 2004 and 2005 (Fig. 10.6.3-02). The large-scale onshore-offshore array included 140 stations onshore with a dense station spacing (less than 7 km in the center) and 30 OBSs offshore. An active source, onshore seismic experiment was carried out in January 2005. The profile trended E-W at 38.2°S and ran from the Pacific coast for ~100 km inland, crossing the Coastal Cordillera and the western portion of the Longitudinal Valley. The profile thus crossed the area of the hypocenter of the great 1960 Chile earthquake (Mw = 9.5). The experiment was cored by a near-vertical incidence reflection survey with shots every 1.5 km and 3-component geophones every 100 m. With an active spread length of 18 km and a daily roll-along of 4.5 km, this resulted in an 8 fold common-depth-point coverage.

The near-vertical incidence reflection survey was supplemented by an expanding spread survey consisting of 15 extra shots with offsets up to almost 100 km. The expanding spread survey provided tenfold coverage of the middle portion of the profile. The project aimed to study the fine-scale structure around the 1960 Valdivia, Chile, earthquake, which was the largest earthquake ever recorded instrumentally, and to determine the key controlling parameters causing subduction zone earth-

quakes at convergent plate boundaries (Rietbrock et al., 2005; Groß et al., 2008).

Furthermore, a 90-day marine campaign, from the end of 2004 to the beginning of 2005, acquired a broad variety of geological and geophysical data offshore Chile between 35° and 48°S (Fig. 10.6.3-03) which included active and passive seismics. The experiment was set up to obtain data along corridors between fracture zones separating relatively small areas of distinct ages (Scherwath et al., 2006).

For the seismic wide-angle data acquisition, some 200 deployments of OBS and OBH were carried out along the TIPTEQ transects, each line with 30–50 stations at a nominal spacing of 5.5 km. A short seismic streamer towed by the research vessel complemented the data set. A first preliminary velocity model was published by Scherwath et al. (2006) for TIPTEQ corridor 3 (Fig. 10.6.3-04).

10.7. OCEANIC DEEP STRUCTURE RESEARCH

10.7.1. Introduction

The seismic structure of the oceanic crust and passive margins was reviewed by Minshull (2002) and published in Part A of the *International Handbook of Earthquake and Engineering Seismology*, edited by Lee et al. (2002). The structure of the oceanic crust, as known by 2000, is shown in chapters on P- and S-wave velocity structure, anisotropy and attenuation, and variations of crustal structure with spreading rate and with age. Other chapters deal with the seismic structure of the Moho and the uppermost

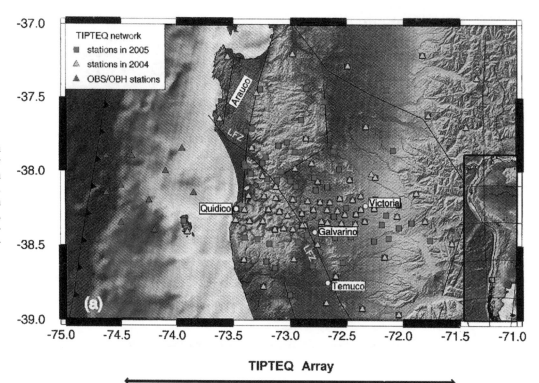

Figure 10.6.3-02. Location of the TIPTEQ seismic array project 2004–2005 (from Rietbrock et al., 2005, fig. 1). [Eos (Transactions, American Geophysical Union), v. 86, no. 32, p. 293, 298. Reproduced by permission of American Geophysical Union.]

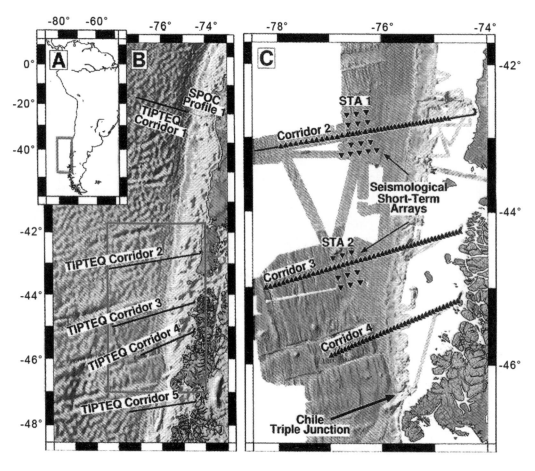

Figure 10.6.3-03. Location of the TIPTEQ project (from Scherwath et al., 2006, fig. 1). (A and B) Overview of all marine seismic corridors, 2004–2005, (C) Central transects with locations of ocean-bottom seismometers (triangles) along refraction profiles and the outer rise seismological networks STA1 and STA2. [Eos (Transactions, American Geophysical Union), v. 87, no. 27, p. 265, 269. Reproduced by permission of American Geophysical Union.]

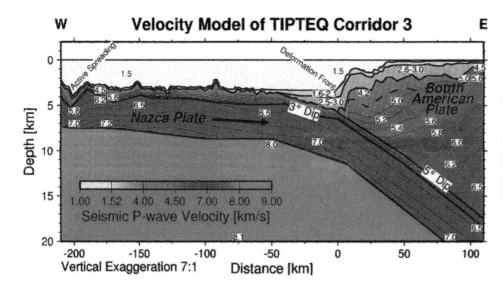

Figure 10.6.3-04. Preliminary structural velocity model of the TIPTEQ marine seismic corridor 3 (from Scherwath et al., 2006, fig. 3). Plate age is 6.5 Ma at the trench. [Eos (Transactions, American Geophysical Union), v. 87, no. 27, p. 265, 269. Reproduced by permission of American Geophysical Union.]

oceanic mantle, of mid-ocean ridges, of oceanic fracture zone and segment boundaries, and hotspots, ocean islands, aseismic ridges and oceanic plateaus. In the textbook *Marine Geophysics*, Jones (1999) has described in much detail the instrumentation and methodologies, as used in marine seismic exploration until the end of the 1990s.

Another summary on our knowledge on the structure of the lithosphere under the oceans, as obtained until 2000, was published in the *Encyclopedia of Sciences* (Steele et al., 2001), where individual papers inform on seismic structure (Harding, 2001), on mid-ocean ridges (Carbotte, 2001) and seismology sensors (Dorman, 2001) and other geophysical topics as, e.g., origin of the oceans (Turekian, 2001), geophysical heat flow (Stein and von Herzen, 2001), or magnetics (Vine, 2001).

It is interesting to compare the generalizing results of average oceanic crustal structure of Raitt (1963) and Ewing (1969) with those published by Harding (2001), who distinguishes between "traditional" and "modern." The "modern" summary of average oceanic crustal structure of Harding (2001) is based on models which were the result of synthetic seismogram analysis, where a range of typical velocities and velocity gradients characterizes layers 2 and 3 (Table 10.7.1-01).

For many projects carried out since 2000, full interpretations are still under way. Therefore, as for the continents, also for numer-

ous oceanic research projects in many cases only abstracts of presentations presented at international meetings or internal reports have become available and only for a limited number full interpretations have been published. We will present a few examples.

10.7.2. Pacific Ocean

The continental margin off the coast of Nicaragua and Costa Rica was the target of a marine seismic survey carried out in 2005 by the research vessel *Meteor* (1986) cruise 66-4a. Two wide-angle seismic profiles were conducted seaward of the trench in the outer rise area offshore of central Nicaragua. Instruments were deployed at ~5 km intervals (Ivandic et al., 2010). An airgun array fired shots every 60 s, corresponding to a shot interval of ~130 m. Both profiles were 120 km long. The first one ran along the trench axis and intersected the trench at 11°N, the second one had the same orientation, but was located ~60 km seaward of the trench axis. Seismic structure was characterized by low velocities both in the crust and upper mantle. The crustal thickness on both lines was ~5.6–5.8 km. Velocities in the uppermost mantle were found in the range of 7.3–7.5 km/s, interpreted as caused by serpentinization.

The SALIERI (South American Lithosphere Transects across Volcanic Ridges) seismic experiment in 2001 followed the

TABLE 10.7.1-01. COMPARISON OF AVERAGE VELOCITIES IN OCEANIC CRUST AS OBSERVED IN THE 1960S AND 2000S (WITHOUT ERROR INDICATIONS)

Year	Layer 2, igneous crust		Layer 3, igneous crust		Layer 4, upper mantle	
	Velocity (km/s)	Thickness (km)	Velocity (km/s)	Thickness (km)	Velocity (km/s)	Thickness (km)
1963	5.15	1.21	6.82	4.57	8.15	
1969	4.5–5.5	1.5	6.5–7.1	5	7.7–8.3	
2001 ("Traditional")	5.07	1.71	6.69	4.86	8.13	6.57
2001 ("Modern")	2.5–6.6	2.11	6.6–7.6	4.97	>7.6	7.08

PAGANINI and the G-PRIME experiments in 1999 and 2000 to investigate the seismic structure of the Galapagos volcanic province between 78°W and 95°W longitude and 3°S and 9°N latitude (see Fig. 9.8.2-04). It was already described in Chapter 9.8.2 (Sallares et al., 2003, 2005).

An experiment in the south-central Pacific, conducted in 2004, explored the region southwest of New Caledonia. It is centered on 164°E, 24°S and is already well documented. It was a deep seismic survey conducted within the western part of New Caledonia's Exclusive Economic Zone (EEZ) and revealed for the first time the thinned continental and oceanic natures of the crust beneath the eastern Tasman Sea. In a collaboration of French institutions with New Caledonia scientists, 2500 km of deep seismic-reflection and 60 km of wide-angle seismic data were collected within the western part of New Caledonia's EEZ (Fig. 10.7.2-01). Airgun shots were recorded by a 4.5 km, 360-channel digital streamer and 15 OBSs.

Since Cretaceous times, the southwest Pacific region has been dominated by an extensional episode which dismembered Gondwanaland and has led to a fragmentation of continental crust to form subparallel marginal basins such as the Tasman Sea basin and the New Caledonia basin. Today, e.g., the Tasman Sea basin is floored by oceanic crust. This and other basins isolated microcontinental fragments such as the Lord Howe Rise and the New Caledonia–Norfolk Ridge. The study area, located in the southwest Pacific east of both Australia and the oceanic Tasman Sea basin, comprises such continental fragments.

Onboard interpretation of seismic-reflection data and preliminary modeling of wide-angle data confirmed the continental nature of the Lord Howe Rise and the Norfolk Ridge and also revealed the continental-type seismic velocities and crustal thicknesses of the Fairway Ridge and basin systems (Fig. 10.7.2-02, top + bottom). To the east, wide-angle and near-vertical incidence reflection data showed the oceanic nature of the N-S central segment of the New Caledonia basin (Fig. 10.7.2-02, center + bottom), characterized by a shallow, 16 km deep Moho and velocities which are typical for 8-km-thick oceanic crust (Lafoy et al., 2005).

Twenty degrees farther east, around 178°–170°W, 18°–19°S, Crawford et al. (2003) have modeled the crustal structure across the Tonga–Lau arc–backarc system from the Lau-Ridge to the Pacific Plate, using data from an 840-km-long airgun

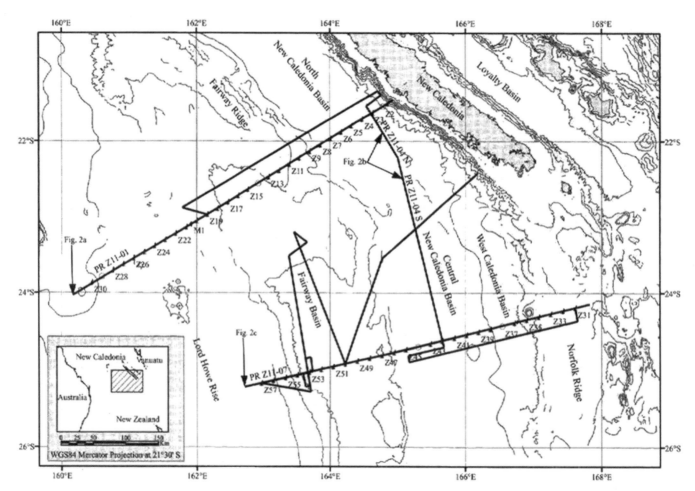

Figure 10.7.2-01. Ship tracks of the ZoNeCo11 deep seismic survey in the Pacific Ocean SW of New Caledonia (from Lafoy et al., 2005, fig. 1). [Eos (Transactions, American Geophysical Union), v. 86, p. 101, 104–105. Reproduced by permission of American Geophysical Union.]

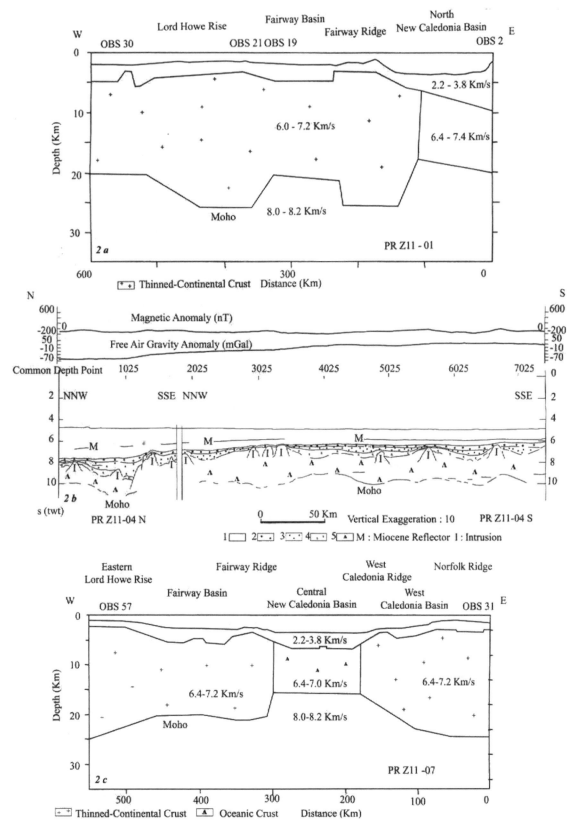

Figure 10.7.2-02. Preliminary models of the ZoNeCo11 deep seismic survey (from Lafoy et al., 2005, fig. 2). Top: preliminary velocity-depth model of the seismic refraction line Z11 01. Center: multichannel seismic reflection tie line Z11-04. Bottom: preliminary velocity-depth model of the seismic refraction line Z11 07. [Eos (Transactions, American Geophysical Union), v. 86, p. 101, 104–105. Reproduced by permission of American Geophysical Union.]

seismic-refraction line over 19 OBSs and one land station (Fig. 10.7.2-03). The airgun line consisted of 5148 shots at 90s interval (160 m) and the average spacing of the seismometers was 42 km.

The overall crustal model (Fig. 10.7.2-04) showed a 5.5-km-thick Pacific crust in the east, thickening to 9 km beneath the flank of the Capricorn seamount at 172°W and showing an up to 7.5-km-thick intermediate-velocity layer (6–7 km/s) beneath the Tonga Ridge. Further west, the crust is exceptionally thin at the boundary between the Lau Basin and Tonga Ridge, before thickening again to 6 km in the east part and 9 km in the west part of the Lau Basin.

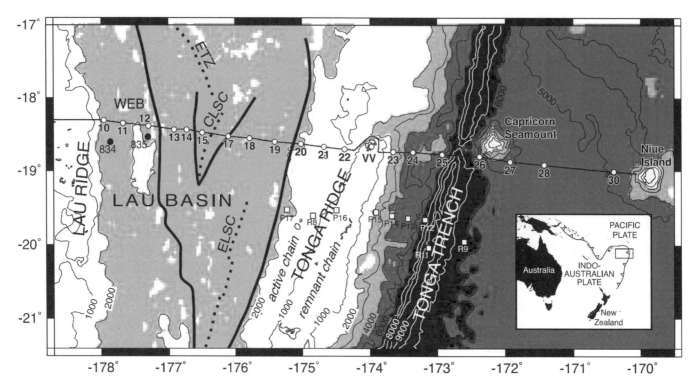

Figure 10.7.2-03. Location of an 840-km-long deep seismic airgun refraction line across the Tonga Ridge from the Pacific Ocean into the Lau Basin (from Crawford et al., 2003, fig. 1). WEB—Western Extensional Basin; CLSC—Central Lau Spreading Center; ELSC—Eastern Lau Spreading Center; ETZ—Extensional Transform Zone; 834 and 835 indicate ODP drill holes. Open squares: centerpoints of earlier 1-D surveys (R—sites from Raitt et al., 1955; P—sites from Pontoise and Latham, 1982). [Journal of Geophysical Research, v. 108, no. 4, doi:10.1029/2001JB001435. Reproduced by permission of American Geophysical Union.]

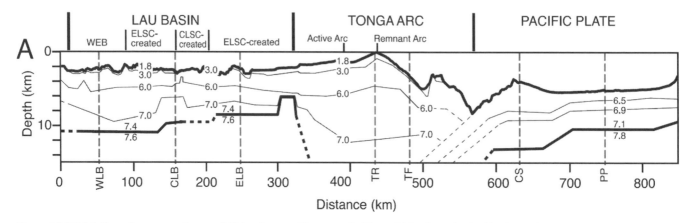

Figure 10.7.2-04. Crustal structure along an 840-km-long profile across the Tonga Ridge from the Pacific Ocean into the Lau Basin (from Crawford et al., 2003, fig. 8A). WEB—Western Extensional Basin; CLSC—Central Lau Spreading Center; ELSC—Eastern Lau Spreading Center; WLB, CLB, ELB—West, Central, East Lau Basin; TR, TF—Tonga Ridge, Forearc; CS—Capricorn Seamount; PP—Pacific Plate. [Journal of Geophysical Research, v. 108, no. 4, doi:10.1029/2001JB001435. Reproduced by permission of American Geophysical Union.]

Major research activities were performed in the sea areas around Japan in the late 1990s and early 2000s. To investigate megathrust earthquake and tsunami generation in a subduction seismogenic zone, controlled-source seismology experiments continued to be conducted in the Nankai Trough and the Japan Trench, where the recurrence interval between the great earthquakes and coseismic slip distribution are well studied (e.g., Baba and Cummins, 2005; Yamanaka and Kikuchi, 2004). Recent availability of a large number of OBSs, a large airgun array, and a long streamer cable for academics provided the background to obtain data which revealed several new results of lithospheric-scale structures in subduction zones. Some of the most striking findings in the Nankai trough were several scales of subducted seamounts/ridges (Kodaira et al., 2002, 2005) and splay fault branching from the subducting plate boundary (Park et al., 2002). These structures strongly controlled the rupture propagations of the 1944 Tonankai and the 1946 Nankai megathrust earthquakes. Recently (after 2005) a high velocity and density domed body was detected in the overriding plate above the deeper part of the segmentation boundary, and strongly coupled patches at the segmentation boundary (Kodaira et al., 2006). Structural variations such as different velocities of the mantle wedge and different degrees of development (thickness) of a low-velocity layer at the plate boundary were detected along the Japan Trench (Tsuru et al., 2002; Miura et al., 2003, 2005; Takahashi et al., 2000, 2004). These variations correspond to the difference in the rate of large earthquake (M > 7) occurrence along the Japan Trench.

Crossing the rupture segmentation boundary between the 1944 8.1-Tonankai and the 1946 8.4-Nankai earthquakes, three wide-angle seismic-reflection profiles were recorded running parallel to the Nankai trough axis off the coast of Kii peninsula. Ocean-bottom seismographs were deployed at 5 and 7 km intervals and a large airgun array (197 l) fired every 200 m (Kodaira et al., 2006). Along the 145-km-long line farthest away from the coast also multichannel reflection seismic data were recorded. The other two lines were 175 km long. The wide-angle seismic data revealed a fractured oceanic crust with a strike-slip fault system in the segmentation boundary and a shallow high-velocity body near Cape Shionomisaki of Kii peninsula forming a strongly coupled patch at the segmentation boundary. The thickness of the subducted oceanic crust at the far-off profile varied from 5–7 km in the southwest to 7–10 km in the northeast. The oceanic Moho depth below sea level increases only slightly toward the coast from ~15 to 18 km.

The Izu-Ogasawara (Bonin)-Mariana (IBM) arc is one of the typical intra-oceanic arc that has been geologically and petrologically studied from long ago to investigate formation of the continental crust. It is necessary to obtain the seismic velocity structure to verify several petrological models explaining the crustal growth of the intra-oceanic arc. Suyehiro et al. (1996) have for the first time revealed the existence of the andesitic rock like as continental crust and high-velocity lower crust (>7 km/s) from the whole crustal structure across the arc (see project description in Chapter 9.8.2).

Two active-source experiments were conducted to image the structure immediately beneath the fronts of the Izu arc in 2004 (Kodaira et al., 2007a, 2007b) and the Bonin arc (Kodaira et al., 2007b). In both experiments, a linear array of densely deployed OBSs with 5 km spacing and a large airgun array were used. The seismic data revealed marked structural differences between the Izu arc in the north (at 35°–30°N and 139°–140°E) and the Bonin arc to the south (at 30°–25°N and 140°–141°E). The thickest crust (32 km) was found under the Izu arc and the thinnest crust (10 km) under the Bonin arc. The 1000-km-long profile along the volcanic front in the Izu-Bonin arc provided new seismological constraints on growth of continental crust in an intra-oceanic arc, e.g., crustal growth corresponding to basaltic arc volcanoes, structural variation of arc crusts and rifted margins.

In 2003, an active-source seismic experiment was conducted farther south around 16.5°–17.5°N latitude along a 700-km-long line across the middle Mariana arc region in an approximately WNW-ESE direction. The line started in the Parece Vela basin in the west, crossed the West Mariana ridge, the Mariana trough, and the Mariana arc, and ended in the Mariana Trench (Takahashi et al., 2008). The seismic line was almost perpendicular to the strike of the Mariana Arc and the West Mariana Ridge. One hundred and six OBSs at intervals of 5.4 km or 10 km recorded shots from a large airgun array (197 l). Reflection records were also obtained using a towed 12-channel hydrophone streamer during the airgun shooting. The characteristic data showed two clear refractions from ocean layers 2 and 3, clear Moho reflections, and small-amplitude upper-mantle refractions. The resulting model of the Mariana arc-backarc system showed a 4–6-km-thick oceanic crust (sediments and oceanic layers 2 and 3) under the Parece Vela basin and the Mariana trough, corresponding to a Moho depth of ~10 km below sea level. Under the West Mariana ridge and under the Mariana arc the crust thickens abruptly to near 16 km total crustal thickness, corresponding to maximum Moho depths of 18–19 km below sea level. Toward the east under the Mariana Trench the crust appears to remain relatively thick. While oceanic layer 2 remains fairly constant in thickness and follows more or less the topography of the sea bottom, the thickening of the crust occurs in layer 3 which was subdivided into two layers with velocity gradients from 6.0 to 6.5 km/s and 6.7 to 7.3 km/s. The upper-mantle velocity varies from 7.6–7.7 km/s under the West Mariana ridge and Mariana arc to 7.9–8.0 km/s under the Parece Vela basin and the Mariana trough (Takahashi et al., 2008).

The efforts to investigate the nature and architecture of the continental margins around Australia continued in 2000–2004. In the program to map the limits of the continental shelves for the Australian submission to the corresponding United Nations Commission, Geoscience Australia (the former Australian Geological Survey Organization) prepared various seismic transects. On the Pacific side major surveys targeted the Three Kings Ridge east to the South Fiji basin and the Lord Howe Rise, both located east of Australia and north of New Zealand, as well as the South Tasman Rise to the south of Tasmania (Finlayson, 2010; Appendix 2-2).

10.7.3. Indian Ocean

The seismic investigations of Geoscience Australia in 2000–2004 on the Australian margins of the Indian Ocean side targeted the Great Australian Bight in the south and the Naturaliste Plateau as well as the Wallaby and Exmouth Plateaus to the west. In the northwest of Australia, the Argo Abyssal Plain was another goal (Finlayson, 2010; Appendix 2-2).

In the Indian Ocean, the Seychelles-Laxmi Ridge, recognized as a microcontinent, was the target of a major seismic experiment in 2003 (Fig. 10.7.3-01), carried out by scientists of the Universities of Southampton and Leeds, UK, and GFZ Potsdam, Germany. The project involved both a controlled-source and a passive seismic element (Collier et al., 2004).

The controlled-source part comprised a deep seismic-reflection/-refraction transect across the Seychelles/Laxmi Ridge

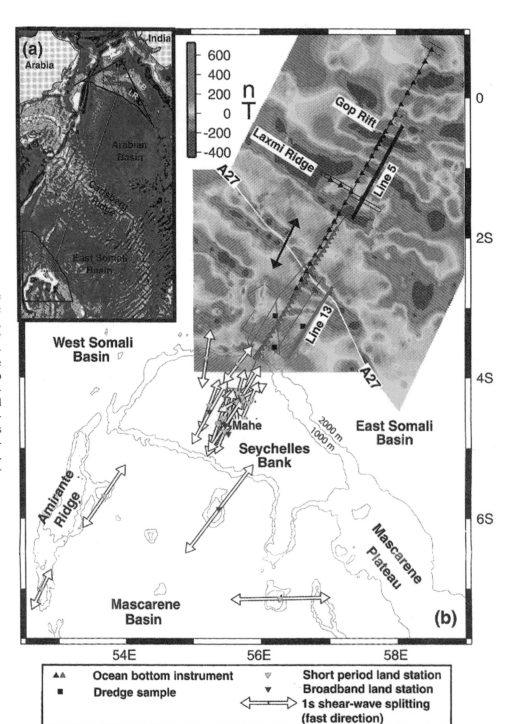

Figure 10.7.3-01. Location of the Seychelles-Laxmi Ridge seismic observations (from Collier et al., 2004, fig. 1). (a) Satellite gravity map of the NW Indian Ocean outlining the Seychelles and Laxmi Ridge margins. (b) Magnetic anomaly map with location of multichannel seismic reflection profiles 5 and 13 and locations of ocean-bottom instruments (black and red triangles). [Eos (Transactions, American Geophysical Union), v. 85, no. 6, p. 481, 487. Reproduced by permission of American Geophysical Union.]

margins (reconstructed to A27 in Fig. 10.7.3-01) using an airgun array, ocean-bottom seismographs, and land stations installed on the central Seychelles islands. At some of the land stations, P_n arrivals could be recorded out to shot-receiver offsets of 380 km.

After the controlled-source seismic survey was completed, the land seismometers were redeployed widely across the Seychelles islands and recorded earthquakes for another 10 months.

Preliminary P-wave models (Fig. 10.7.3-02) showed evidence that the rifted margins are highly asymmetric. The Laxmi Ridge was seen as wide and complex, whereas the Seychelles side appeared narrow and simple. At both, the Seychelles margin and the India/Pakistan margin, the crust thins drastically from ~26 km to 10–12 km depth below sea level (Collier et al., 2004).

10.7.4. Atlantic Ocean

For the South Atlantic, we have selected a few recent projects investigating the continental margin areas of South Africa and South America, while for the North Atlantic, we will deal with investigations west of Ireland and east of Greenland.

In 2003, the project MAMBA (Geophysical Measurements across the Continental Margin of Namibia Experiment) of 1995 (Bauer et al., 2000) was continued with a similar onshore/offshore experiment on the west coast of South Africa in order to reveal the structure of the oceanic-continent transition zone. The involved institutions were the Council of Geoscience (RAS), the BGR Hannover and the GFZ Potsdam (both Germany). The onshore/offshore part consisted of two profiles running almost perpendicular to the coastline with length of 100 km on land and ~300 km on sea. Each minute a 52 l airgun produced seismic signals, while on land three shots (100 kg, 50 kg, 100 kg) were fired on each line. Furthermore, two coast-parallel lines on land recorded the seismic signals from two shotlines in the sea which ran also parallel to the coast. The aim of this investigation was to obtain a P_n tomographic image. One of the lines, the 500-km-long Springbok profile, was described in detail by (Hirsch et al., 2009). It crossed the western margin of the Republic of South

Africa at 31°S. In the offshore part, the profile started at 3600 m water depth and ran through the central part of the Orange basin in the shelf region. The 100-km-long land part crossed the coast-paralleled Gariep belt, where the topography reaches maximum elevations of 1000 m. The offshore part was a combined transect of reflection and refraction seismic data. The velocity modeling revealed a segmentation of the margin into three distinct parts of continental, transitional and oceanic crust. The oceanic crust beneath the outer 50 km is subdivided into layers 2A and 2B, beneath of which the velocity increased to 7 km/s, interpreted as gabbroic layer 3. Along the following 200 km of the line transitional crust was identified where the middle and lower crust were characterized by high P-velocities of 6.9–7.4 km/s, as seen on many rifted volcanic margins. The remaining 250 km including the 100 km land section showed continental crust with 5.9–6.1 km/s near the surface. Here, the crust was characterized by a continuous velocity increase, but no significant velocity contrasts indicated a subdivision into an upper and a lower continental crust (Hirsch et al., 2009).

A second seismic project of the GFZ Potsdam (Germany) was the 800-km-long Agulhas-Karroo transect, carried out in 2005 within the project INKABA ya Africa (Parsiegla et al., 2007; Stankiewicz et al., 2007, 2008). It was a north-south offshore-onshore transect which ran over a distance of 400 km from the offshore Agulhas Plateau across the Agulhas Fracture zone and the continental margin onto to South African coast, and for other 600 km onland across the Cape Fold Belt, the Karoo Basin and into the Kapvaal Craton. The project also included the recording of 600 km of magnetotelluric data. The onshore part was already described in subchapter 10.6.2.

The offshore part (Parsiegla et al., 2007) consisted of 20 four-component (three-component seismometer and a hydrophone component) OBSs deployed over 400 km length. Eight G-airguns and one Bolt-airgun (volume of 96 l) were fired every 60 seconds during cruise SO-182 of the R/V *Sonne*, resulting in a shot spacing of about 150 m. As the onshore wide-angle and offshore parts were carried out simultaneously, all airgun

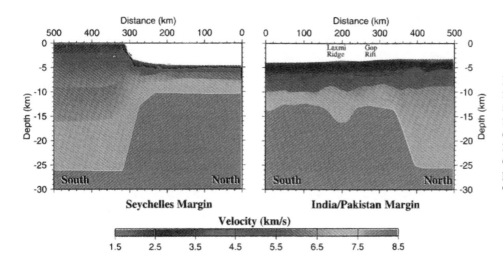

Figure 10.7.3-02. Preliminary P-wave models of the two halves of the transect indicated in Fig. 10.7.3-01 derived from wide-angle data by forward modeling and inversion (from Collier et al., 2004, fig. 3). [Eos (Transactions, American Geophysical Union), v. 85, no. 46, p. 481, 487. Reproduced by permission of American Geophysical Union.]

shots were also recorded by the land receivers. The Moho, found at 30 km depth at the present coast, starts to shallow gradually seaward for a distance of 250 km across the continental margin to 20 km depth at the Agulhas-Falkland Fracture Zone which marks the transition zone between continental and oceanic crust. Here the Moho depth of 20 km is reduced to 12 km over a horizontal distance of 50 km and remains at 11–12 km depth, i.e., 6–7 km crust under 5 km of ocean water (Parsiegla et al., 2007; Stankiewicz et al., 2008).

A detailed investigation with numerous reflection and refraction profiles of the margin off Uruguay was carried out in 2004 (Temmler et al., 2006) by a joint venture of the BGR (Bundesanstalt für Geowissenschaften und Rohstoffe, Hannover, Germany) and SOHMA (Servicio de Oceanografie, Hidrografia y Meteorologia da la Armada, Uruguay) with 70 OBSs along profile lengths exceeding 250 km. The interpretation of Temmler et al. (2006) resulted in a gradual decrease of Moho depth from 27 to 20 km from 50 to 140 km distance from the coast and a further decrease to 15 km, and finally a crustal thickness of 10 km beyond 170 km.

Another marine survey along the Atlantic side of South America was performed offshore French Guiana in 2003 (Greenroyd et al., 2006). The survey collected coincident multichannel seismic reflection and wide-angle seismic-refraction data along two transects across the margin. The wide-angle refraction data were derived from 39 OBS and nine land-based instruments and could be modeled along a 535 km profile across the Demerara Plateau and along a 427 km profile across the margin, 175 km to the south. Crustal velocities were generally lower than 7 km/s and both profiles showed abnormally thin oceanic crust of 3.5–5 km thickness. Furthermore two contrasting margin structures, separated by 175 km distance, were derived from the data: along the northern profile the continental crust thinned gradually over 300 km distance, while the southern profile showed a more rapid thinning over only 60 km distance (Greenroyd et al., 2006).

Also the coastal area of Venezuela, representing the Caribbean–South America plate boundary, became the target of a large active-source seismic experiment in 2004 (e.g., Levander et al., 2006; Schmitz et al., 2008). About 6000 km of marine multichannel seismic-reflection data were collected offshore Venezuela. In addition, five wide-angle seismic profiles were recorded both onshore and offshore Venezuela and the Antilles arc region, spanning an area of almost 12 degrees of longitude and 5 degrees of latitude as part of the project BOLIVAR (Broadband Ocean Land Investigation of Venezuela and the Antilles Arc Region; for more details, see subchapter 10.6.3).

The Mid-Atlantic Ridge area was visited by the *Meteor* (1986) cruise M62-4 in October 2004. The target of a seismic project was the region of the Ascension Transform system at ~7°S and the spreading segment immediately south of this transform. The Ascension Transform is actually a "double transform fault" consisting of two parallel transform fault/fracture zone systems sandwiching a very short segment. OBS/H were deployed along a WSW-ENE–trending profile (~120 km long between 11.8°W and 13.5°W) within the Ascension Fracture zone system and along several shorter profiles and clusters to the south of it at ~13.5°W (Reston et al., 2009).

Farther north, the Mid-Atlantic Ridge was studied in 2005 at the Lucky Strike Volcano at ~32°16′W, 37°17′N with a network of 39 cross lines of 19 km length and one along-strike profile (Singh et al., 2006b). The seismic-reflection data were acquired using a 4.5-km-long digital streamer with 12.5 m receiver intervals towed at 15 m depth. The cross lines were 100 m apart; the airgun shot interval was 37.5 m. The along-axis line was shot at an interval of 75 m. With a width of 6 km and a length of 15 km, the Lucky Strike Volcano is one of the largest volcanoes at the Mid-Atlantic Ridge. The existence of an axial magma chamber was inferred, the reflection from the top of the crustal magma chamber being ~3 km beneath the seafloor and 3–4 km wide and extending 7 km along-axis.

At the eastern side of the southern Atlantic, in the second half of 2004, the *Meteor* (1986) cruise M62-3 investigated the causes for a prominent bathymetric swell around the Islands of Cape Verde, a feature caused by hotspot volcanism. To study the mechanical properties of the lithosphere over the hotspot, a roughly 500-km-long, N-S–directed, deep seismic line was shot across the swell between the islands of Fogo and Santiago to image tomographically the crustal structure of the swell. Forty OBS/H stations were deployed at 1 km intervals to sample the seismic wave field from a 64-l array of airguns fired at 90 s intervals. Detailed seismic investigations in the area of the volcanically active islands of Fogo and Brava were aimed to reveal magma reservoirs and the magma plumbing system of the islands. Three profiles were obtained, one running from the island of Santiago toward Fogo, around the island and continuing roughly 100 km to the west, providing a 170-km-long profile across Fogo. Two shorter lines—between 60 and 80 km long—provided additional deep seismic data from the Fogo and Brava area. The recorded data were of high quality. On line 2, seismic energy was recorded to offsets of 120 km; shots from profile 3 were recorded on the other side of the island of Fogo. A land station on Fogo recorded many shots, and P_g, P_MP, and P_n phases could be identified in the data. While the ocean bottom instruments were on the seafloor, some stations did record signals associated with volcanic activity, including earthquakes and volcanic tremors (Grevemeyer et al., 2009).

Seismic projects of the twenty-first century in the waters surrounding the Iberian Peninsula and corresponding references were compiled by Díaz and Gallart (2009). Most of them, however, deal with sedimentary and uppermost crust only. The TASYO project aimed to investigate the tectonic structure of the Gulf of Cadiz and comprised two E-W–oriented multichannel profiles. In 2002, the VOLTAIRE project followed in the same area with more than 1000 km multichannel seismic profiles. In 2004, the TECALB project sampled the Eastern Alboran Sea. In 2006, the West-Med experiment acquired wide-angle data in the Alboran Sea and the transition to the South Balearic Basin, using a network of OBS and land stations, and the MARSIBAL cruise of 2006 added multichannel seismic data.

In 2002, RAPIDS (Rockall and Porcupine Irish Deeps Seismic wide-angle seismic experiment) was extended by a new profile RAPIDS-4 (O'Reilly et al., 2006), following the RAPIDS 1 and 2 projects of 1988 and 1990 (see Fig. 8.9.4-05; Shannon et al., 1994, 1999; Hauser et al., 1995; O'Reilly et al., 1995, 1996; Vogt et al., 1998), and RAPIDS 3 of 1999 (see Fig. 9.8.4-16; Mackenzie et al., 2002; Morewood et al., 2003). The new profile RAPIDS-4 (Fig. 10.7.4-01) was a 230 km east-west–oriented profile across the Porcupine Arch, a deep structural feature in the Porcupine Basin. Sixty-five OBSs with 4 components (one hydrophone and 3 orthogonally directed geophones) were deployed at 3–4 km interval and seismic sources were fired every 120 m across the deployment line. The experiment gave the first well-resolved results of P-wave velocity variations in the deeper part of the Porcupine Basin (Fig. 10.7.4-02). The crystalline crust thins rapidly from almost 30 km under the Porcupine High in the west and Celtic Platform in the east to less than 2 km under the Porcupine Basin, overlain by a sedimentary basin of up to 10 km thickness and underlain by an upwarping mantle. Also the uppermost mantle velocity decreases from normal (8 km/s) to anomalously low values as low as 7.2 km/s.

Farther northwest of Ireland in the North Atlantic, in 2002 the Hatton Deep Seismic (HADES) project was designed to investigate the crust of the Hatton Basin and the Hatton continental

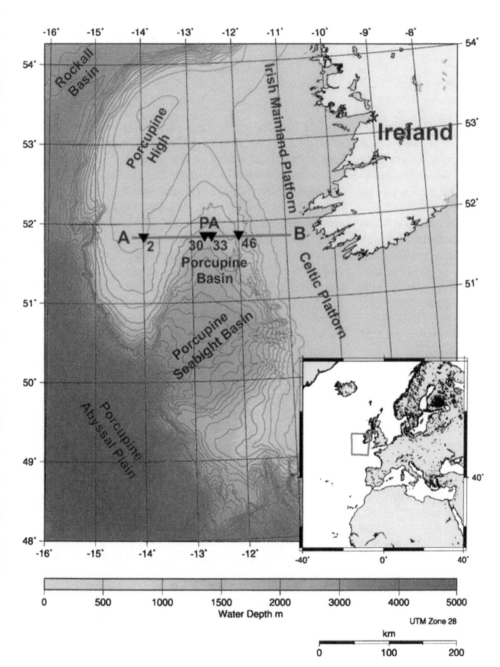

Figure 10.7.4-01. Location of the RAPIDS-4 profile in the northern part of the Porcupine Basin southwest of Ireland (from O'Reilly et al., 2006, fig. 1). [Journal of the Geological Society, v. 163, p. 775–787. Reproduced by permission of Geological Society Publishing House, London, U.K.]

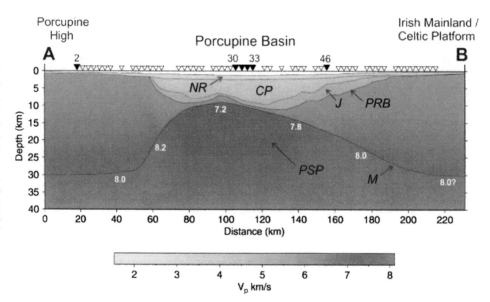

Figure 10.7.4-02. Crustal structure along the RAPIDS-4 profile in the northern part of the Porcupine Basin southwest of Ireland (from O'Reilly et al., 2006, fig. 10). NR, CP—Neogene to Recent, Cretaceous to Paleogene post-rift sequences; J—predominantly Jurassic syn-rift sediment sequence; PRB—pre-rift basement; M—Moho; PSP—partially serpentinized mantle peridotite. [Journal of the Geological Society, v. 163, p. 775–787. Reproduced by permission of Geological Society Publishing House, London, U.K.]

margin (Chabert et al., 2006, Smith et al., 2005). This is located to the northwest of the Rockall Bank at ~16–20°W, 56–58°N in the North Atlantic. The project provided an exceptionally large amount of wide-angle seismic data along the axis of the basin. Along a 363-km-long NE-SW profile, 100 OBS were deployed and airgun shots were fired at 130 m intervals.

In 2004, a joint project of GEOMAR (Kiel, Germany) and DIAS (Dublin, Ireland) re-investigated the southern Porcupine Basin to the southwest of Ireland along marine profiles along which in 1997 data had been obtained using a 6-km-long, 240 channel digital streamer and a tuned sleeve-gun array with record length up to 9 s TWT (Reston et al., 2004). This time, during the *Meteor* (1986) cruise M61-2 in 2004, 25 OBS/H stations were deployed along each line and additionally seven land stations recorded all shots (Fig. 10.7.4-03). The energy was generated by two to three 32-l airguns which fired over 2000 shots per profile. The water depth ranged from 200 m near the coast of Ireland to 1600 m in the center of the basin. It shallowed to ~400 m under the Porcupine Ridge. The seven land stations recorded all shots during the month-long duration of the marine experiment. On the better-quality record sections from the onshore stations clear primary and secondary phases could be seen out to 180 km offsets. The data quality apparently depended primarily on the number of airguns used for shooting (Hauser et al., 2007b; Reston et al., 2006).

Two major goals were set. First, both sides of a rift basin occurring in close proximity to each other could be studied here, allowing questions about the symmetry of extension to be addressed by several east-west profiles parallel to the direction of extension. Second, the amount of extension increases from north to south, so a series of east-west cross sections on different latitudes provided information on crustal structure during variable extension. The spatial changes between these sections also represent the temporal development of the rift through continued extension. The data quality was excellent in general; clear arrivals were apparent to offsets of over 70 km on most instruments, giving a complete coverage of the crust beneath the basin. In places, clear refraction-reflection-refraction triplications were observed for several interfaces (Reston et al., 2006).

Within the framework of the EUROMARGINS project, the investigations in the North Atlantic around Greenland of 1990 and 1994 (Schmidt-Aursch and Jokat, 2005; Fig. 9.8.4-30) were extended in 2003 east of Greenland toward the Greenland Basin (Voss et al., 2006), exploring the continent-ocean transition between the Jan Mayen and Greenland Senja fracture zones off East Greenland (Fig. 10.7.4-04). The investigations were carried out by the Alfred Wegener Institute for Polar and Marine Research during an expedition with the research vessel R/V *Polarstern*. The main aim of the project was to investigate the crustal architecture and the evolution of the conjugate volcanic margins off mid-Norway and East Greenland in the context of rifting.

Deep seismic-refraction data were acquired on four 300–450-km-long profiles (Fig. 10.7.4-04). On each profile, a total of 25 OBSs were deployed. On the northernmost profile, a typical oceanic crust, 6 km thick, was found further offshore. The other three profiles were extended onshore by six land stations each. The present interpretation revealed a division along these profiles into three parts: continental crust with a thickness of 29–32 km was found at the western ends of the profiles, followed by a diffuse continent-ocean transition zone, the width of which increased from 70 km in the north to 190 km in the south, and an oceanic crust whose thickness decreased from 16 km near the continent-ocean transition zone to 5–6 km at the eastern ends.

The complete data set provided new insights into the formation of the entire segment of the continental margin between the Jan Mayen and the Greenland fracture zones. The deeper structure of the East Greenland margin was of special interest, because it is conjugate to the Vøring Plateau off Norway (Voss et al., 2006).

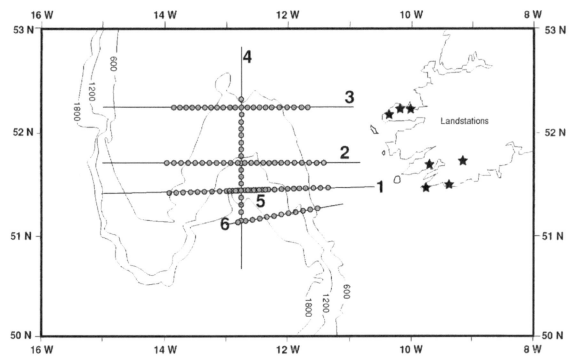

Figure 10.7.4-03. Map of a seismic survey in 2004 in the Porcupine Basin of the Atlantic Ocean southwest of Ireland. (from Hauser et al., 2007b). Straight lines and circles—shot profiles and ocean bottom seismometer/hydrophone stations of GEOMAR; stars—land stations operated by DIAS; thin lines—bathymetric contours (600 m interval). [50th Annual Irish Geological Research Meeting, School of Environmental Sciences, University of Ulster, Coleraine, Northern Ireland, 23–25 February (abstract). Reproduced by kind permission of the authors.]

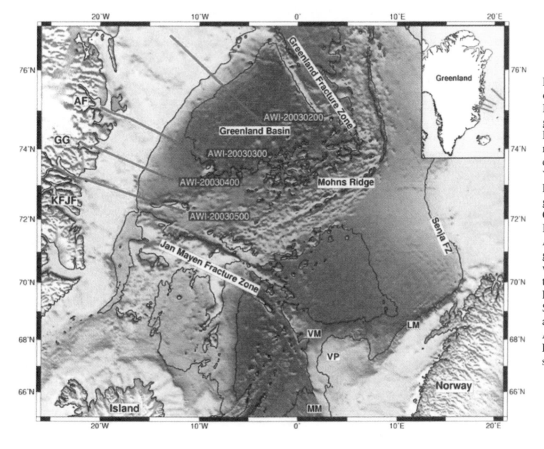

Figure 10.7.4-04. Overview of the East Greenland and Norwegian conjugate margins. Red lines represent the locations of the AWI seismic refraction profiles (from Voss et al., 2007, fig. 1). VM—Vøring margin; VP—Vøring Plateau; MM—Møre margin; LM—Lofoten Margin; GG—Godthåb Gulf; KFJF—Kejser Franz Joseph Fjord; AF—Ardencaple Fjord. Background is bathymetry grid with 1500 m and 3000 m contour lines. [European Seismological Commission (ESC), Subcommission D—Crust and Upper Mantle Structure, Activity Report 2004–2006. Reproduced by kind permission of the authors.]

Geophysical surveys at the Norwegian Vøring and Lofoten margins (Mjelde et al., 1992, 1998, 2001; Raum et al., 2002) were described earlier in Chapter 9.8.4, revealing important vertical and lateral variations in crustal structure and composition.

The results of seismic experiments on velocities and associated crustal thickness along the entire East Greenland margin were compiled by Voss et al. (2009) in two new wide-angle seismic transects. In addition, maps were compiled for the East Greenland margin covering a distance range of ~500 km distance from the southeastern and central coast of East Greenland reaching as far as Iceland. The maps show the depth to the crystalline basement, the depth to Moho, the thickness of the crystalline crust and the thickness of a high-velocity layer with velocities above 7 km/s.

10.7.5. Summary of Seismic Observations in the Oceans from 1950 to 2005

In Figure 10.7.5-01, the approximate locations of marine experiments from 2000 until 2005 or even more recently were plotted as far as results were published until 2010. For a number of projects, however, publications may not yet be available. Other projects may have been published in conjunction with land seismic projects and do not appear as marine projects in the database. As was mentioned for previous decades, another reason for the fact that some of the projects which were discussed in the individual "Oceans" subchapters are not seen on this map may also be due to the database containing only projects which were published in easily accessible journals or books.

Evident on the map for the 2000s is a concentration of projects on the line Norway–Faeroe Islands–Iceland. Deep-ocean trenches and adjacent continental margins (Indonesia and South America) have also attracted increased activities.

With the large number of recording devices available, projects can now be planned on land and at sea which extend seismic surveys into three dimensions. Large-scale research programs which involved a multitude of cooperating institu-

tions and interdisciplinary cooperation of scientists from various geoscientific fields continued to dominate the scene in the early 2000s. However, with the increasing number of recording devices and the capacity to deploy instruments over large areas, tomographic methodologies such as teleseismic tomography became viable. Many earth scientists started to prefer long-term deployments of instruments using natural events as energy sources instead of short-term projects using expensive controlled sources.

Figure 10.7.5-02 contains a summary of projects carried out since 1950. The map shows all locations for which data, recognized as marine projects, are available in the database at the U.S. Geological Survey in Menlo Park, California. Some of the projects which were discussed in the individual "Oceans" subchapters for the Oceans do not show up in Figure 10.7.5-02 or on the maps shown for each decade at the end of the corresponding chapter. The reasons for these discrepancies are the same as those discussed above for Figure 10.7.5-01.

10.8. THE STATE OF THE ART AROUND 2006 AND OUTLOOK

We have terminated our historical review of controlled-source seismic experiments with projects planned and carried out until the end of 2005. The large number of recording devices available nowadays as well as the ability to record continuously over long time periods has allowed researchers to extend seismic surveys into three dimensions. This also has enabled researchers to avoid problems which may arise if using man-made sources such as quarry blasts, borehole or underwater explosions, but rather, at least for projects on land, to plan for seismic tomography surveys with teleseismic and/or local events as energy sources. Nevertheless, the interest in studying details of crustal and upper-mantle structure has remained until present, and many new projects were performed just recently or are in the planning stage, because many details of crustal structure can only be achieved by controlled-source seismic near-vertical incidence

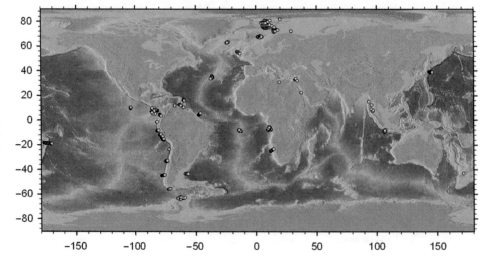

Figure 10.7.5-01. Seismic refraction measurements in the Atlantic, Pacific, and Indian Oceans performed between 2000 and 2008 (data points from papers published until 2010 in easily accessible journals and books).

1950–2010: worldwide oceanic crust with bathymetry < – 250 m

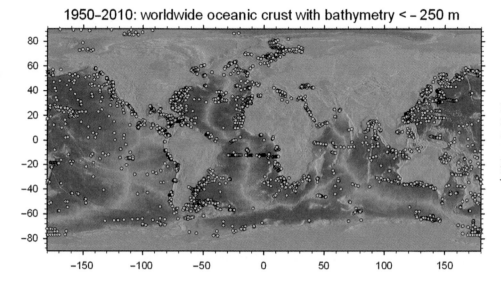

Figure 10.7.5-02. Seismic refraction measurements in the Atlantic, Pacific and Indian Oceans performed between 1950 and 2008 (data points from papers published until 2010 in easily accessible journals and books).

reflections seismics and/or seismic-refraction/wide-angle reflection experiments with densely spaced controlled sources and a multitude of seismic recorders.

In 2009, controlled-source seismology could have celebrated its 160th birthday, since Robert Mallet in 1849 for the first time used dynamite explosions to measure the speed of elastic waves in surface rocks (Mintrop, 1947; Dewey and Byerly, 1969). In 1947 Mintrop wrote the first history on the first 100 years of explosion seismology (Mintrop, 1947; Appendix 3-1 and 3-2). Since then an enormous amount of explosion-seismic data around the world has been compiled, and data and results were published in thousands of papers distributed in numerous journals and books, making a complete collection of explosion-seismic projects and their data and interpretations almost impossible.

Though not the scope of this publication, dealing mainly with academic non-profit research activities, it is of interest to note, that two companies, GSI in the United States and SEISMOS in Germany, which were founded in 1925 and mentioned in some detail in Chapter 3.3, both celebrated their 80th anniversary in 2005. As of 2005, after several mergers and acquisitions, the heritages of GSI and Western Geophysical still existed, along with several pioneering European companies such as GECO, SEISMOS, and PRAKLA, as part of the seismic contracting company Western-Geco. Many other companies using reflection seismology in hydrocarbon exploration, hydrology, engineering studies, and other applications have been formed since the method was first invented. Major service companies today include Compagnie Générale de Géophysique, Veritas DGC, and Petroleum Geo-Services. Most major oil companies also have actively conducted research into seismic methods as well as collected and processed seismic data using their own personnel and technology. For more details of the worldwide history of exploration geophysics, the reader is referred to Lawyer et al. (2001), who have compiled a personalized history on the commercial application of controlled-source seismology starting at the time of World War I.

Reflection seismology has also found widespread applications in non-commercial research by academic and government scientists around the world. Many of those projects were described in the foregoing individual chapters, according to the decade when they were accomplished.

Textbooks on seismic theory, published in large numbers over the decades, have accompanied and guided the development of interpretation methods to deal with the proper interpretation of active source seismic data obtained on land and at sea (e.g., Nettleton, 1940; Bullen, 1947; Worzel and Ewing, 1948; Grant and West, 1965; Musgrave, 1967; Maxwell, 1970; Kosminskaya, 1971; Officer, 1974; Červený et al., 1977; Kennett, 1983; Bullen and Bolt, 1985; Yılmaz, 1987; Lay and Wallace, 1995; Aki and Richards, 1980, 2002; Jones, 1999; Kennett, 2001; Chapman, 2004; Borcherdt, 2009). There are in addition numerous articles that address various aspects concerning the theory for interpretations of controlled-source seismology data (e.g., Willmore and Bancroft, 1960; Steinhart et al., 1961c; Ewing, 1963a; Ludwig et al., 1970; Fuchs and Müller, 1971; Bessonova et al., 1974; Braile and Smith, 1975; Giese, 1976b; McMechan and Mooney, 1980; Červený and Horn, 1980; Spence et al., 1984; Červený, 1979, 1985; Zelt and Smith, 1992; Hole, 1992; Hole and Zelt, 1995; Zelt, 1999). Similarly, compilations of interpretation methods commonly used were also summarized from time to time (e.g., James and Steinhart, 1966; Mooney, 1989; Braile et al., 1995; Mechie, 2000). Levander et al. (2007) provided a convenient review of theory and application of controlled-source seismic data.

The idea to compile all data available worldwide was already born, shortly after Mintrop's first review, in the early 1950s when, e.g., Macelwane (1951) and Reinhardt (1954) compiled all hitherto published explosion seismic data in tables, and when Closs and Behnke (1961, 1963) constructed worldwide crustal cross sections, based on the knowledge obtained until the end of the 1950s. A tabular collection of then known facts of physics in general was published by the Springer Publishing Company

(Heidelberg–New York–Tokyo) in the early 1950s in the series of handbooks ("Landolt-Börnstein"): Numbers and Functions from Natural Sciences and Techniques, of which Volume III (Bartels and ten Bruggencate, 1953) was dedicated to astronomy (P. ten Bruggencate, editor) and geophysics (J. Bartels, editor) and appeared in 1952.

Reviews on seismic data and results were repeatedly published, mainly on regional scales, as summarized in Tables 2.1 and 2.2 of Chapter 2 including Moho and other contour maps (e.g., Steinhart and Meyer, 1961; Morelli et al., 1967; Kosminskaya, 1969; Healy and Warren, 1969; Belyaevsky et al., 1973; Warren and Healy, 1973; Woollard, 1975; Giese et al., 1976a; Meissner et al., 1987b; Braile et al., 1989; Freeman and Mueller, 1992; Pavlenkova, 1996; Li and Mooney, 1998; Collins et al., 2003; Grad et al., 2009). Finlayson (2010) compiled the complete history of deep seismic profiling from 1946 to 2006 across the Australian continent and its margins.

In the early 1980s, the Springer Publishing Company (Heidelberg–New York–Tokyo) edited a new series of the handbooks ("Landolt-Börnstein") presenting a worldwide overview on Natural Sciences and Techniques, available around the end of the seventies in numerical form and functions. Group V was dedicated to Geophysics and Space Research, with subvolume V/2 in particular to Geophysics of the Solid Earth, the Moon and the Planets. In this volume V/2, a worldwide compilation of explosion seismic data on crust and uppermost mantle structure was published in tabular form as representative velocity-depth functions for the crust together with the corresponding location maps for Europe, North America, and the rest of the world (Prodehl, 1984).

At about the same time the earliest 3-D seismic velocity model of the Earth's crust was published by Soller et al. (1981), who assembled one of the first compilations of global Moho depths and upper mantle velocity. An up-to-date summary of our knowledge on seismic structure of the continental lithosphere, based on controlled-source seismic data, and a summary of the seismic structure of the oceanic crust and passive continental margins were compiled by Mooney et al. (2002) and Minshull (2002), respectively, in the framework of an *International Handbook of Earthquake and Engineering Seismology*, edited by Lee et al. (2002).

More than a decade after Prodehl's and Soller's global compilations, a $2° \times 2°$ cell model called 3SMAC was constructed (Nataf and Ricard, 1996). It was derived using both seismological data and non-seismological constraints such as chemical composition, heat flow, and hotspot distribution, from which estimates of seismic velocities and the density in each layer were made. Two years later, CRUST 5.1 was introduced (Mooney et al., 1998) incorporating twice the amount of active source seismic data than 3SMAC. At this time, regional compilations of depth-to-Moho values for the Middle East and North Africa were also published (Seber et al., 2001). The $5° \times 5°$ resolution of CRUST 5.1, however, was still too coarse for regional studies. In 2000, CRUST 2.0 updated the ice and sediment thickness information

of CRUST 5.1 at $1° \times 1°$ resolution, while basically redistributing the crustal thickness data onto a $2° \times 2°$ grid.

Without the addition of a significant amount of new data, however, this redistribution did little to clarify the crustal structure at higher resolution. Therefore, under the leadership of Walter D. Mooney and Shane Detweiler, since the 1990s a database, the Global Seismic profiles Catalog (GSC), is being assembled at the Office for Earthquake Studies, U.S. Geological Survey, in Menlo Park, California.

As was discussed at the end of Chapter 9, this continuously growing database, having assembled more than 50 years of data on seismic crustal and uppermost-mantle structure studies, had already throughout the 1990s enabled the construction of various worldwide syntheses of crustal parameters or the compilation of syntheses on selected continental and oceanic areas (e.g., Christensen and Mooney, 1995; Mooney et al., 1998; Mooney, 2002, 2007; Chulick and Mooney, 2002).

The GSC is a computer database and master catalog of 1-D global crustal and mantle structures profiles constructed from the vast amount of velocity and layer thickness data derived from various types of seismic surveys. This includes data from refraction and reflection seismology, receiver functions, tomographic inversion, and earthquake modeling. Thus, it contains material derived from all types of seismic survey techniques used to determine seismic wave velocities and the depth and thickness of layers within the Earth's crust.

An extensive literature search has been undertaken to track down as many of the seismic survey publications as possible. A large number of these publications have been carefully scrutinized, and the appropriate data for the GSC have been extracted and put into a standardized format for inclusion in the database. At present, the database contains over 10,000 one-dimensional P- and S-wave profiles, making it the largest such catalog in existence. The database is continually being updated and enlarged—there currently exist enough data to create several thousand additional profiles, with enough publications each year to create hundreds more.

The database is in ASCII text format, and is designed for easy manipulation using spreadsheet applications (e.g., Microsoft Excel), or by application of advanced programming languages, especially on either personal computers or workstations. It is easily expandable, correctable and modifiable using standard word processors or by specially designed FORTRAN or other language code. New types of data can be and have been added to the GSC (e.g., tectonic age, velocity gradient, seismic coda, gravimetric and electrical resistivity data), which increases the sophistication of the database. In addition, portions of the GSC have been available via the Internet (http://earthquake.usgs.gov/research/structure/crust/NAm_data.txt)

A database like the GSC has a wide variety of potential use in a wide range of geophysical and seismological applications. It has been used to construct several 3-D global and regional seismic structure models of the Earth's crust, including the above mentioned $5° \times 5°$ average global block model (CRUST 5.1,

Mooney et al., 1998), and structural models for North America (Chulick and Mooney, 2002) and South America (G. Chulick, 2010, written commun.).

Second, these models, in turn, can serve as input for mathematically modeling other properties of the crust and the Earth as a whole, including crustal density variations, regional gravity anomalies, and seismic wave traveltime (Chulick, 1997). For example, by applying an appropriate seismic velocity-density relationship (Christensen and Mooney, 1995), one can create a regional 3-D density model for the crust.

The full data set available up until the end of 2008 and collected in the Global Seismic profiles Catalog (GSC) has served as a basis for the construction of the following contour maps.

Figures 10.8-01 to 10.8-03 present the data points which served as basis to draw the Moho contour maps shown in the following figures (Figs. 10.8-04 to 10.8-13). For a better view, the map is shown in different scales, for the whole world (Fig. 10.8-01), as well as enlarged for both the western and the eastern hemisphere (Figs. 10.8-02A and 10.8-03B).

Furthermore, the map was converted into polar views to show the available data points around the South Pole (Antarctica) (Fig. 10.8-02B) and around the North Pole (Fig. 10.8-03A). The location maps of available data points clearly show where gaps in data still exist in spite of 60 years of concerted efforts of scientists around the world to unravel the structure of the Earth's crust. The majority of data gaps are naturally located in the hostile Arctic and Antarctic regions. But also around the equator, particularly

in Africa and South America large areas exist where no seismic measurements have taken place. In part it may be due to hostile environments prohibiting extensive seismic research, but also political circumstances in many cases are contra productive for pure scientific work.

Figures 10.8-04 and 10.8-05 present crustal thickness maps of the world, Figure 10.8-04 showing the topography with data points and Moho depth contour lines overlain, Figure 10.8-05 showing the basement age (sediments removed) with contour lines of Moho depths overlain.

The average crustal thickness under continents is around 40 km. Exceptions are western, central, and southern Europe as well as the Basin and Range province of western North America, where the average crustal thickness is around 30 km. Areas with particular thick crust (up to 70 km) coincide with the topographic highest mountain ranges—the Himalayas in central Asia and the Andes in South America. The central parts of the Atlantic, Pacific, and Indian Oceans are enclosed by the 10-km depth contour, indicating average crustal thickness being less than 10 km.

More details of the crustal structure underneath the continents can be viewed in the individual maps for the different continents. We have ordered them in the same sequence, as the individual continents were discussed for each decade in the previous chapters.

The map for Europe (Fig. 10.8-06) shows crustal thicknesses of 40–45 km under the East European platform including the

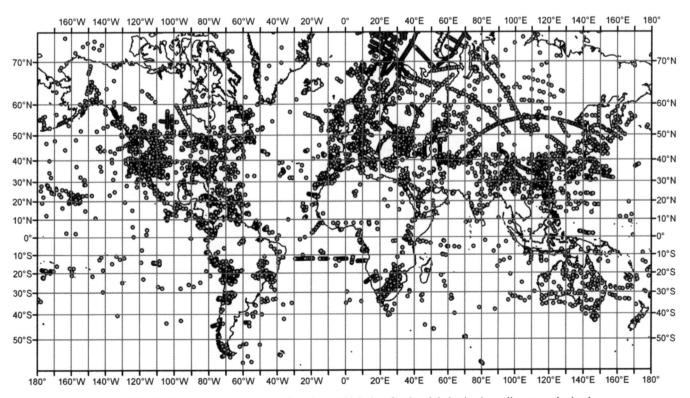

Figure 10.8-01. World map showing locations from which data for the global seismic studies were obtained.

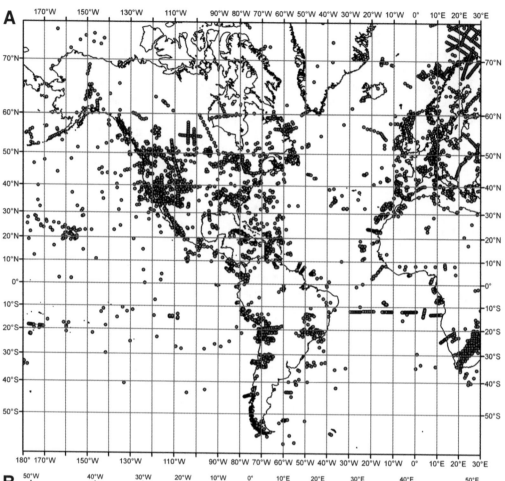

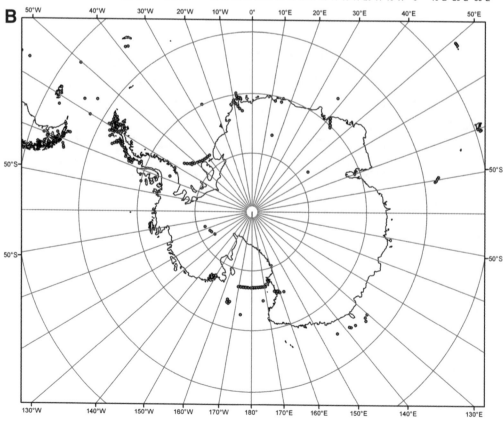

Figure 10.8-02. (A) Western hemisphere showing locations from which data for the global seismic studies were obtained. (B) Map of the South Pole region showing locations from which data for the global seismic studies were obtained.

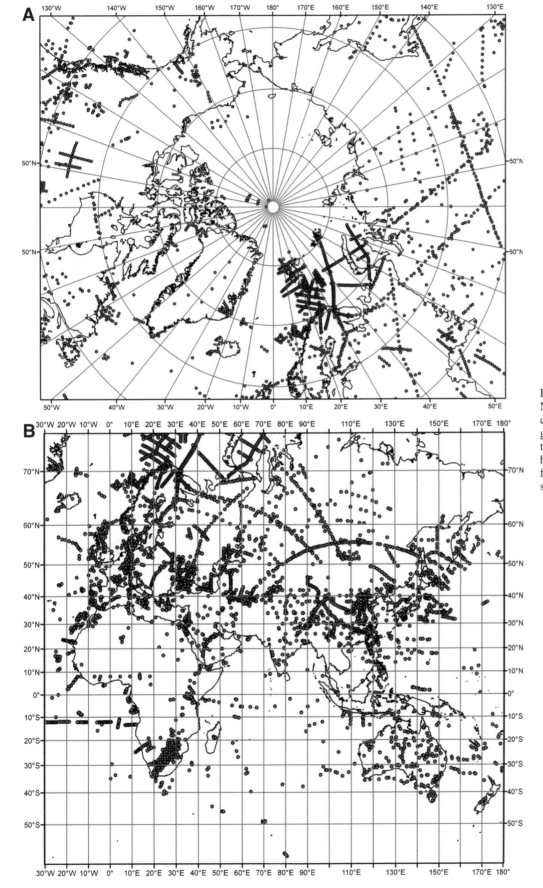

Figure 10.8-03. (A) Map of the North Pole region showing locations from which data for the global seismic studies were obtained. (B) Map of the eastern hemisphere showing locations from which data for the global seismic studies were obtained.

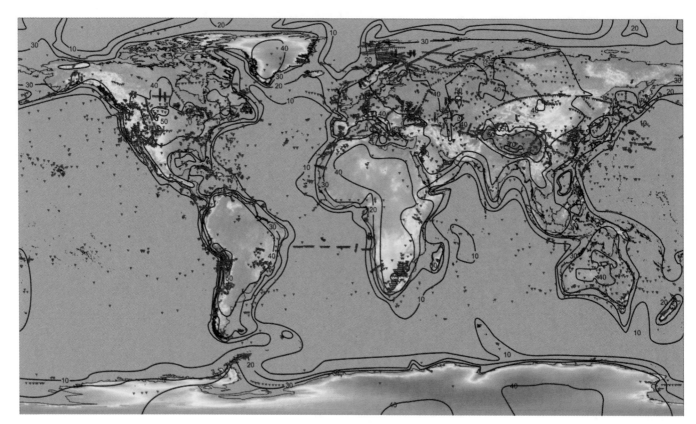

Figure 10.8-04. Topographic map showing data points and crustal thickness around the world as derived from global seismic studies, based on the Global Seismic Profiles Catalog.

Urals, under the Fennoscandian Shield, and—barely visible—under the Alps. For western, central, and southern Europe, 30 km thickness is the average. The increase in crustal thickness in the southeastern corner of the map indicates already the crustal thickening toward the Himalayas, shown in Figure 10.8-07, where up to 70 km Moho depth have been measured. The whole of northern Asia is characterized by 40–45 km average thickness. Under Japan and adjacent island arcs to the north and the south, crustal thickness is around 30 km

Crustal thickness under most of North America (Fig. 10.8-08) is 40–45 km, except for the relatively narrow Atlantic coastal plain and a wider area along the west coast including the Basin and Range province—not clearly seen on this map—and Alaska.

The Australian continent (Fig. 10.8-09) has a thick crust of 40–45 km in its center, but most of the continent has a slightly thinner crust with Moho depths between 30 and 40 km. The offshore areas to the northwest and north as well as to the southeast toward New Zealand apparently are underlain by a shallow continental crust near 20 km thickness in average. Also the southern shelf area is rather wide.

Both for Africa (Fig. 10.8-10) and for South America (Fig. 10.8-11), there is little known on crustal structure of the interior of both continents In Africa, the Afro-Arabian rift region, South Africa with Namibia, and Morocco were partly covered by seismic-refraction campaigns; in South America, most seismic surveys dealt with the Andean region. The continental shelves of both continents are evidently quite narrow. It is evident that the scale chosen for the Moho maps is too rough to recognize details such as narrow graben features with elevated crust-mantle boundary. Narrow mountain ranges with root structures or oceanic features like mid-ocean ridges do not appear as outstanding anomalies on these maps.

Finally, crustal thickness maps were compiled for the South Pole region (Fig. 10.8-12) and for the North Pole region (Fig. 10.8-13). While it is not surprising that the crust under Antarctica is around 40 km thick, the Arctic Ocean is in wide areas underlain by an oceanic crust of less than 10 km thickness.

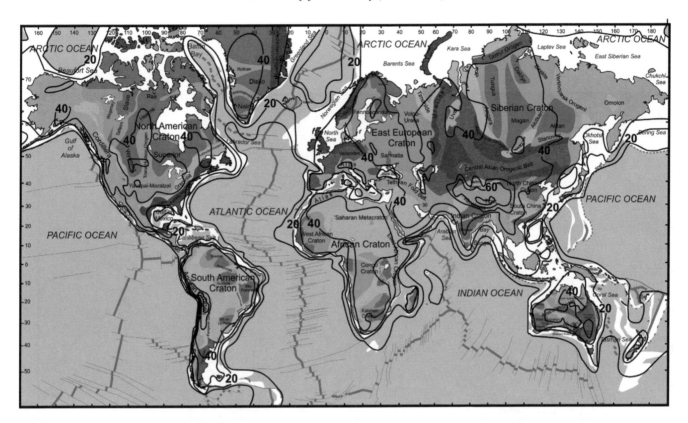

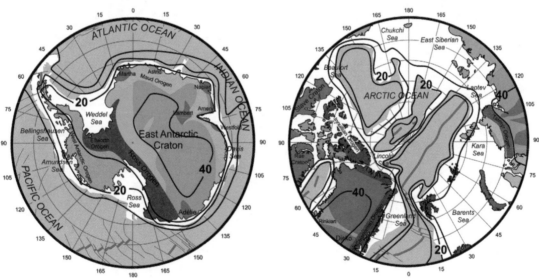

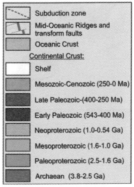

Figure 10.8-05. Basement age map showing crustal thickness around the world as derived from global seismic studies, based on the Global Seismic Profiles Catalog.

Figure 10.8-06. Basement age map of Europe showing Moho depth contour lines as derived from global seismic studies, based on the Global Seismic Profiles Catalog.

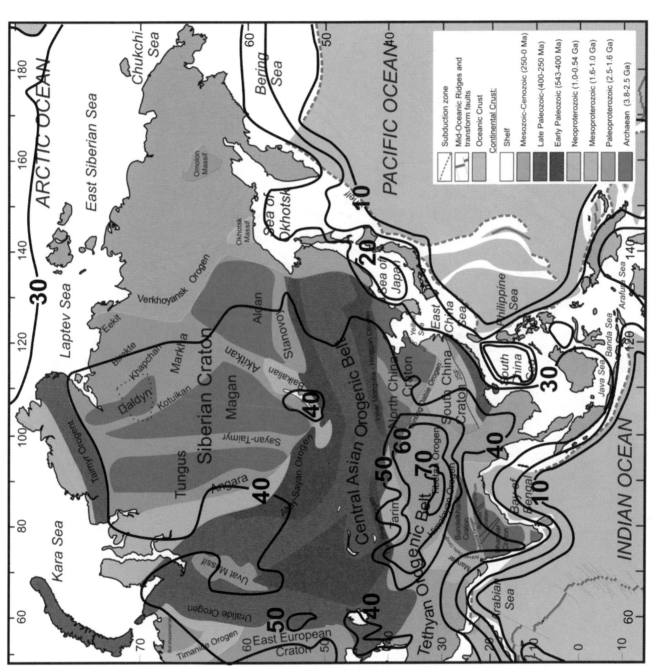

Figure 10.8-07. Basement age map of Asia showing Moho depth contour lines as derived from global seismic studies, based on the Global Seismic Profiles Catalog.

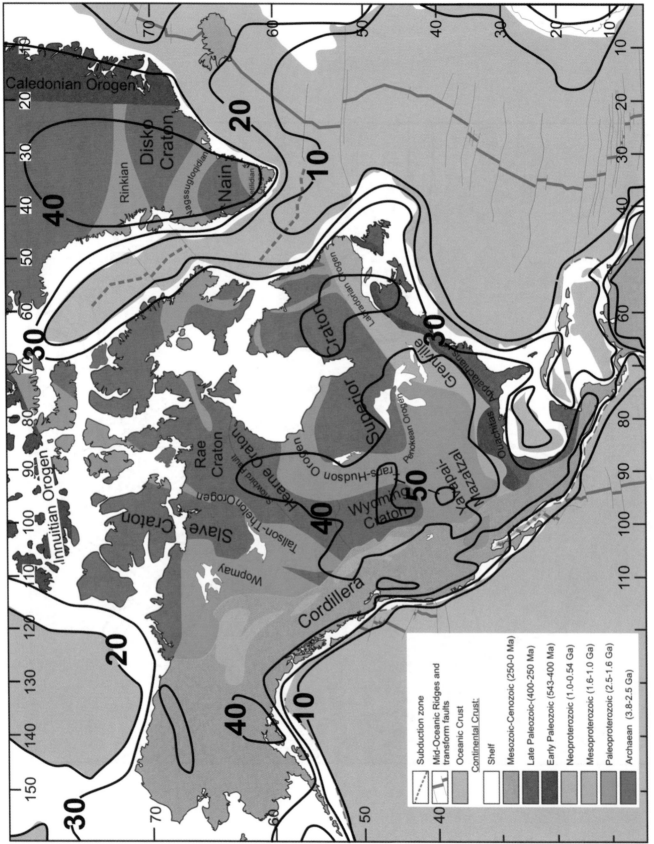

Figure 10.8-08. Basement age map of North America showing Moho depth contour lines as derived from global seismic studies, based on the Global Seismic Profiles Catalog.

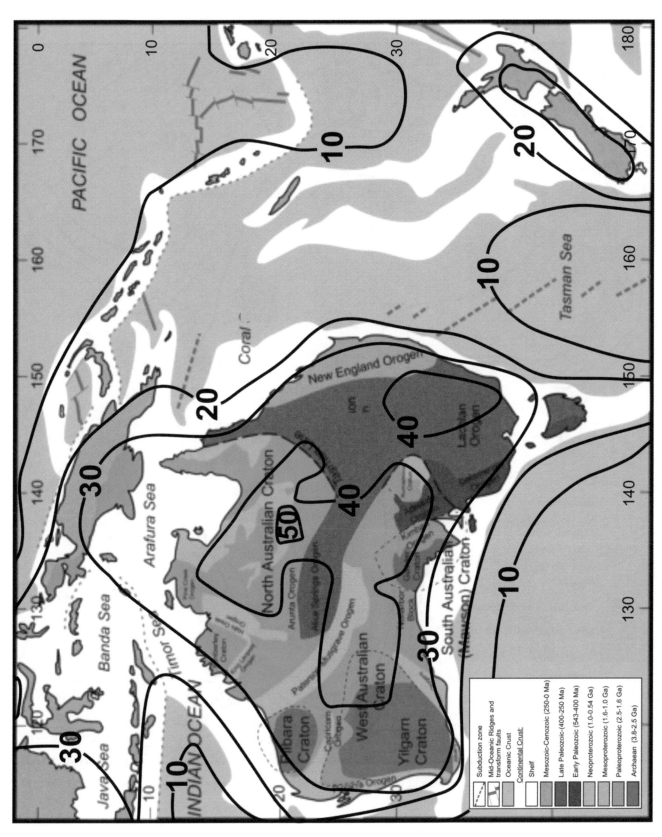

Figure 10.8-09. Basement age map of Australia and New Zealand showing Moho depth contour lines as derived from global seismic studies, based on the Global Seismic Profiles Catalog.

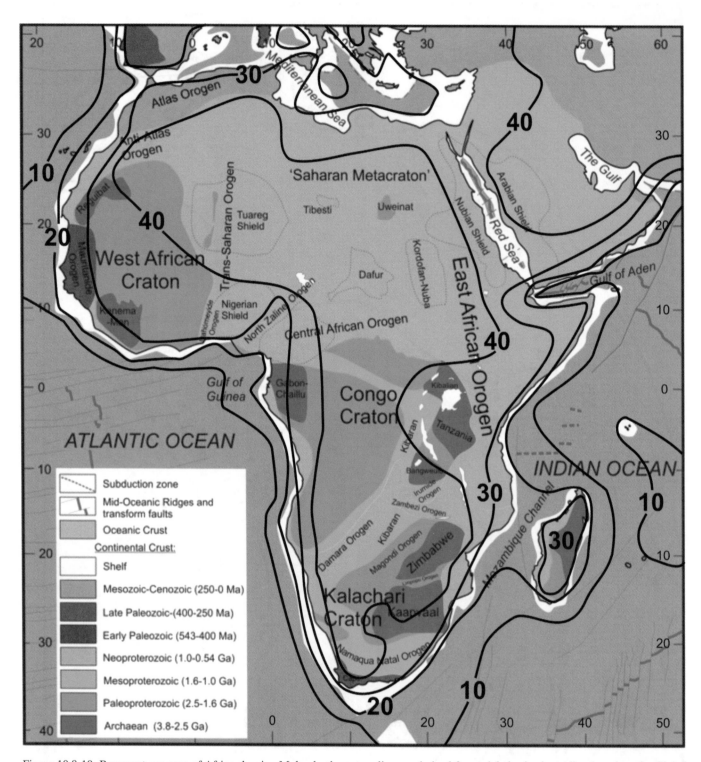

Figure 10.8-10. Basement age map of Africa showing Moho depth contour lines as derived from global seismic studies, based on the Global Seismic Profiles Catalog.

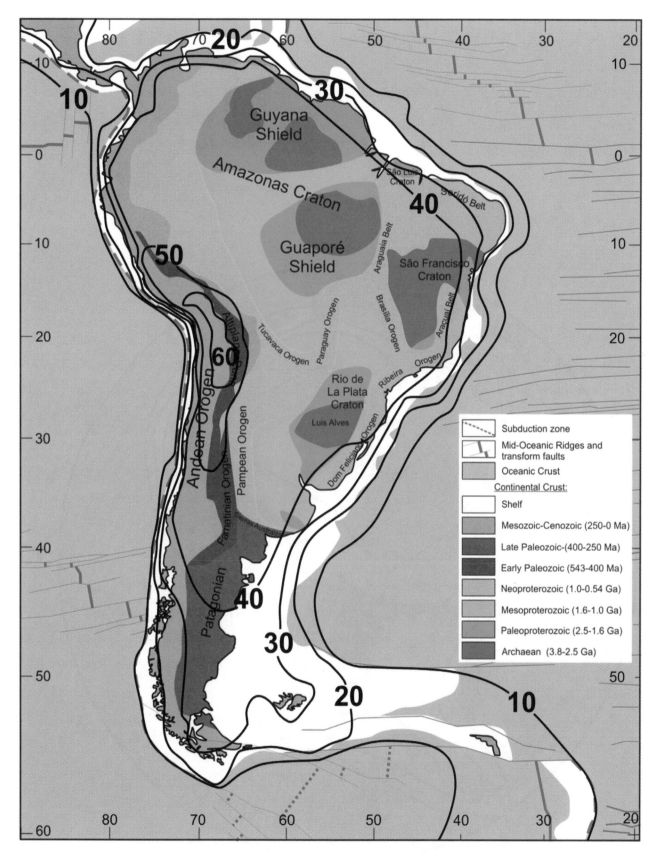

Figure 10.8-11. Basement age map of South America showing Moho depth contour lines as derived from global seismic studies, based on the Global Seismic Profiles Catalog.

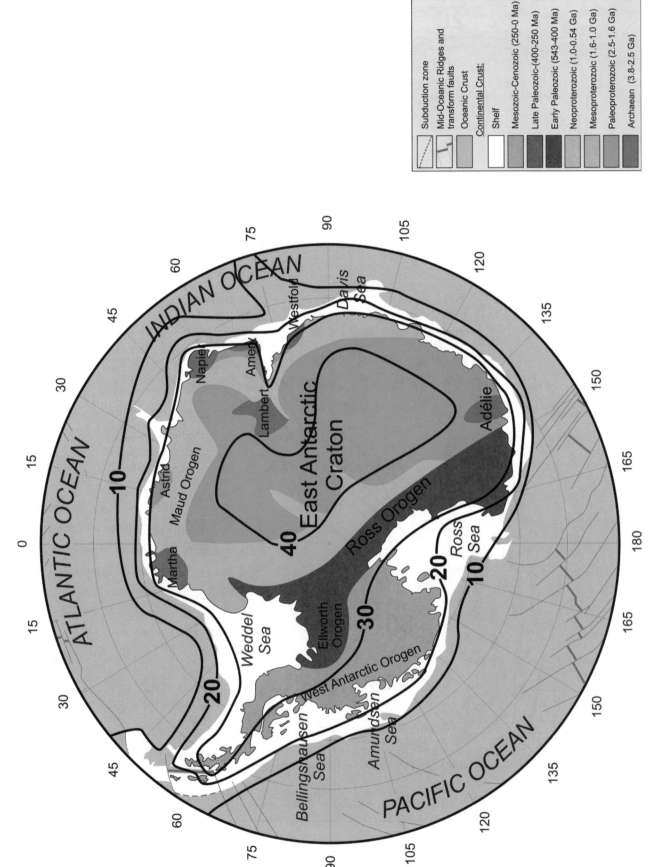

Figure 10.8-12. Basement age map of Antarctica and surrounding oceans showing Moho depth contour lines as derived from global seismic studies, based on the Global Seismic Profiles Catalog.

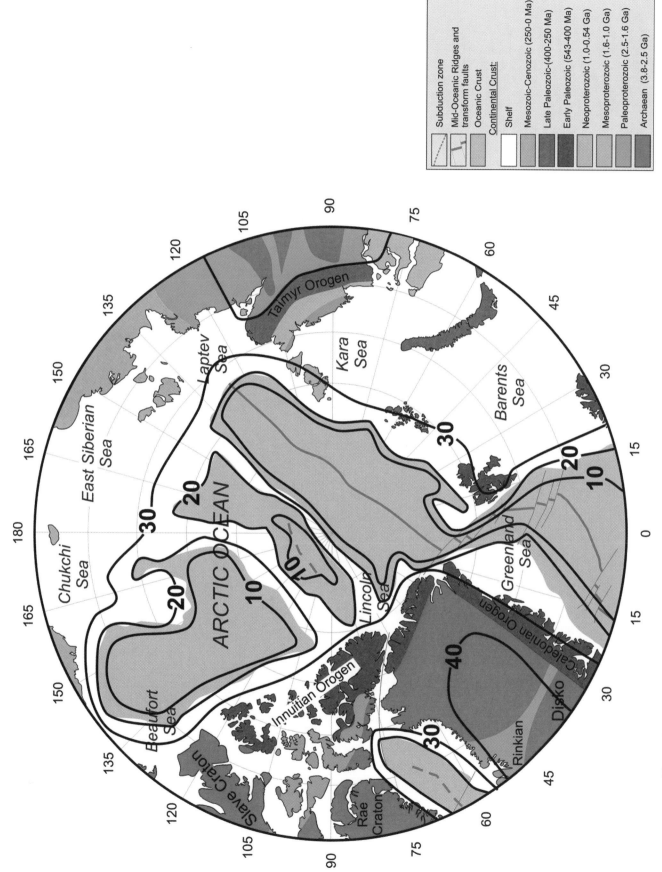

Figure 10.8-13. Basement age map of the North Pole Region (Arctic Ocean and surroundings) showing Moho depth contour lines as derived from global seismic studies, based on the Global Seismic Profiles Catalog.

REFERENCES CITED
[2] to [10] = Number of chapter in which reference is cited.

Abbot, H.L., 1878, On the velocity of transmission of earth waves: New Haven, Connecticut, American Journal of Science and the Arts, v. 15, p. 178–184. [3]

Abramovitz, T., Landes, M., Thybo, H., Jacob, B., and Prodehl, C., 1999, Crustal velocity structure across the Tornquist and Iapetus suture zones—A comparison based on MONA LISA and VARNET data: Tectonophysics, v. 314, p. 69–82, doi:10.1016/S0040-1951(99)00237-1. [9]

Abramovitz, T., Thybo, H., and MONA LISA Working Group, 1998, Seismic structure across the Caledonian Deformation Front along MONA LISA profile 1 in the southeastern North Sea: Tectonophysics, v. 288, p. 153–176, doi:10.1016/S0040-1951(97)00290-4. [9]

Achauer, U., Evans, J.R., and Stauber, D.A., 1988, High-resolution seismic tomography of compressional wave velocity structure at Newberry volcano, Oregon Cascade Range: Journal of Geophysical Research, v. 93, p. 10,135–10,147. [8]

Achauer, U., Maguire, P.K.H., Mechie, J., Green, W.V., and the KRISP Working Group, 1992, Some remarks on the structure and geodynamics of the Kenya rift, in Ziegler, P.A., ed., Geodynamics of rifting, volume II: case history studies on rifts: North and South America and Africa: Tectonophysics, v. 213, p. 257–268. [9]

Achauer, U., and the KRISP Working Group, 1994, New ideas on the Kenya rift based on the inversion of the combined dataset of the 1985 and 1989/1990 seismic tomography experiments, in Prodehl, C., Keller, G.R., and Khan, M.A., eds., Crustal and upper mantle structure of the Kenya rift: Tectonophysics, v. 236, p. 305–329, doi:10.1029/JB093iB09p10135. [9]

Agamemnone, G., 1937, I risultati scientifici d'una grande mina: Renili'eontri ilell'Accademia Nazionale dei Lincei, Cl. Sci. Fis., Mat. e Nat., (6) 25, p. 601–607. [3]

Agger, H.E., and Carpenter, E.W., 1965, A crustal study in the vicinity of the Eskdalemuir seismological array station: Geophysical Journal of the Royal Astronomical Society, v. 9, p. 69–83. [6]

Aichroth, B., Ye, S., Feddersen, J., Maistrello, M., and Pedone, R., 1990, A compilation of data from the 1986 European Geotraverse experiment (main line) from Genova to Kiel—Tables. Open-File Report 90-1, Geophysical Institute, University of Karlsruhe, 115 p. [8]

Aichroth, B., Prodehl, C., and Thybo, H., 1992, Crustal structure along the central segment of the EGT, in Freeman, R., and Mueller, St., eds., The European Geotraverse, Part 8: Tectonophysics, v. 207, p. 43–64. [2] [8]

Aki, K., and Richards, P.G., 1980, Quantitative seismology; theory and methods (2 volumes): San Francisco, W.H. Freeman, 932 p. [2] [10]

Aki, K., and Richards, P.G., 2002, Quantitative seismology, Second Edition: Sausalito, California, University Science Books. [2] [10]

Alden, A., 2010, The Greatest Science Program in History: http://geology.about.com/cs/escibasics/a/aa101203a.htm. [7] [8]

Aldrich, L.T., Asada, T., Bass, M.N., Hales, A.L., Tuve, M.A., and Wetherill, G.W., 1960, The Earth's Crust: Washington, D.C., Carnegie Institution, Year Book, v. 59, p. 202–208. [6]

Al-Kindi, S., White, N., Sinha, M., England, R., and Tiley, R., 2003, Crustal trace of a hot convective sheet: Geology, v. 31, p. 207–210, doi:10.1130/0091-7613(2003)031<0207:CTOAHC>2.0.CO;2. [8]

Allen, A., 1966, Seismic refraction investigations in the Scotia Sea: Scientific report of British Antarctica Survey no. 55. [6]

Allis, R.G., 1986, Mode of crustal shortening adjacent to the Alpine fault, New Zealand: Tectonics, v. 5, p. 15–32, doi:10.1029/TC005i001p00015. [9]

Allmendinger, R.W., Farmer, H., Hauser, E., Sharp., J., Von Tish, D., Oliver, J., and Kaufman, S., 1986, Phanerozoic tectonics of the Basin and Range—Colorado Plateau transition from COCORP data and geologic data: a review, in Barazangi, M., and Brown, L., eds., Reflection seismology: the continental crust: American Geophysical Union, Geodynamics Series, v. 14, p. 257–267. [2] [8]

Allmendinger, R.W., Hauge, T.A., Hauser, E.C., Potter, C.J., Klemperer, S.L., Nelson, K.D., Knuepfer, P., and Oliver, J., 1987, Overview of the COCORP 40°N transect, western United States: the fabric of an orogenic belt: Geological Society of America Bulletin, v. 98, p. 308–319, doi:10.1130/0016-7606(1987)98<308:OOTCNT>2.0.CO;2. [2] [8] [9]

ALP 2002 Working Group, 2003, The ALP 2002 refraction experiment and its relation to TRANSALP, in Conference on Seismic exploration of the Alpine lithosphere, Trieste Italy, 10–12 February 2003, extended abstract, www.alp2002.info/bibliography/Transalp03_abstr.htm. [10]

Alpine Explosion Seismology Group, 1975, A lithospheric seismic profile along the axis of the Alps, 1975—I. First results: PAGEOPH, v. 114, p. 1109–1130, doi:10.1007/BF00876205. [2] [7]

Alsdorf, D., Makovsky, Y., Zhao, W, Brown, L.D., Nelson, K.D., Klemperer, S.L., Hauck, M., Ross, A., Cogan, M., Clark, M., Che, J., and Kuo, J., 1998, INDEPTH (International Deep Profiling of Tibet and the Himalya) multichannel seismic reflection data: description and availability: Journal of Geophysical Research, v. 103, p. 26,993–26,999. [9]

Alvarez, L.W., Alvarez, W., Asaro, F., and Michel, H.V., 1980, Extraterrestrial cause for the Cretaceous–Tertiary extinction: Science, v. 208, p. 1095–1108, doi:10.1126/science.208.4448.1095. [9]

Alvarez-Marron, J., Perez-Estaun, A., Danobeitia, J.J., Pulgar, J.A, Martinez Catalan, J.R., Marcos, A., Bastida, F., Ayarza Arribas, P., Aller, J., Gallart, J., Gonzales-Lodeiro, F., Banda, E., Comas, M.C., and Cordoba, D., 1996, Seismic structure of the northern continental margin of Spain from ESCIN deep seismic profiles: Tectonophysics, v. 264, p. 355–363. [2] [9]

Alvarez-Marron, J., Rubio, E., and Torne, M., 1997, Subduction-related structures in the North Iberian Margin: Journal of Geophysical Research, v. 102, p. 22,497–22,511, doi:10.1029/97JB01425. [9]

ANCORP Working Group, 1999, Seismic reflection image revealing offset of Andean subduction–zone earthquake locations into oceanic mantle: Nature, v. 397, p. 341–344, doi:10.1038/16909. [9]

ANCORP Working Group, 2003, Seismic imaging of a convergent continental margin and plateau in the central Andes (Andean Continental Research Project 1996 (ANCORP'96): Journal of Geophysical Research, v. 108, B7, 2328, doi:10.1029/2002JB001771. [2] [9]

Anderson, D.L., 1971, The plastic layer of the earth's mantle, in Wilson, T., ed., Continents adrift; Readings from Scientific American: San Francisco, Freeman and Co., p. 28–35. [7]

Anderson, R.N., and Noltimier, H.C., 1973, A model for the horst and graben structure of mid-ocean ridge crests based upon spreading velocity and basalt delivery to the oceanic crust: Geophysical Journal of the Royal Astronomical Society, v. 34, p. 137–147. [7]

Angenheister, G., Bögel, H., Gebrande, H., Giese, P., Schmidt-Thomé, P., and Zeil, W., 1972, Recent investigations of surficial and deeper crustal structure of the eastern and southern Alps: Geologische Rundschau, v. 61, p. 349–395, doi:10.1007/BF01896323. [6]

Angenheister, G.H., 1927, Beobachtungen bei Sprengungen: Zeitschrift für Geophysik, v. 3, p. 28–33. [2] [3]

Angenheister, G.H., 1928, Seismik, in Geiger, G., and Scheel, K., eds., Handbuch d. Physik: Berlin, Springer, v. VI, p. 566–622. [2] [3]

Angenheister, G.H., 1935, Geophysical exploration in the uppermost layer of the Earth's crust and their practical importance: Research Progress 1, p. 126–131. [3]

Angenheister, G.H., 1942, Ausbreitung der Bodenschwingungen bei grossen Sprengungen und oberflächennahem Erdstoß: Göttingische Gelehrte Anzeigen 204, p. 7. [3]

Angenheister, G.H., 1950, Fortschreitende elastische Wellen in planparallelen Platten: Post mortem veröffentlicht durch O. Förtsch: Gerlands Beiträge zur Geophysik, v. 61, no. 4, p. 296–308. [3]

Ankeny, L.A., Braile, L.W., and Olsen, K.H., 1986, Upper crustal structure beneath the Jemez Mountains volcanic field, New Mexico, determined by three-dimensional inversion of seismic refraction and earthquake data: Journal of Geophysical Research, v. 91, p. 6188–6198. [2] [7] [8] [9]

Ansorge, J., 1968, Die Struktur der Erdkruste an der Westflanke der Zone von Ivrea: Schweizerische Mineralogische und Petrographische Mitteilungen, v. 48, p. 247–254. [6]

Ansorge, J., 1975, Die Feinstruktur des obersten Erdmantels unter Europa und dem mittleren Nordamerika [Ph.D. thesis]: Univeity of Karlsruhe, 111 p. [2] [6] [7]

Ansorge, J., and Mueller, S., 1971, The fine structure of the upper mantle in Europe and in North America: Proceedings of the 12th General Assembly of the European Seismological Commission (Luxembourg 1970), Comm. A-13 de l'Observatoire Royal de Belgique, Sér. Géophys. no. 101, p. 196–197. [7]

Ansorge, J., and Mueller, S., 1979, The structure of the earth's crust and upper mantle from controlled source observations: Schweizerische Mineralogische und Petrographische Mitteilungen, v. 59, p. 133–140. [2] [7]

Ansorge, J., Emter, D., Fuchs, K., Lauer, J.P., Mueller, St., and Peterschmitt, E., 1970, Structure of the crust and upper mantle in the rift system around the Rhinegraben, *in* Illies, H., and Mueller, St., eds., Graben problems: Stuttgart, Schweizerbart, p. 191–197. [7]

Ansorge, J., Bonjer, K.-P., and Emter, D., 1979a, Structure of the uppermost mantle from long-range seismic observations in southern Germany and the Rhinegraben area: Tectonophysics, v. 56, p. 31–48, doi:10.1016/0040-1951(79)90005-2. [4]

Ansorge, J., Mueller, St., Kissling, E., Guerra, I., Morelli, C., and Scarascia, S., 1979b, Crustal cross section across the zone of Ivrea-Verbano from the Valais to the Lago Maggiore: Bollettino di Geofisica Teorica e Applicata, v. 21, p. 149–157. [2] [7]

Ansorge, J., Prodehl, C., and Bamford, D., 1982, Comparative interpretation of explosion seismic data: Journal of Geophysics, v. 51, p. 69–84. [7] [8] [10]

Ansorge, J., Blundell, D., and Mueller, S., 1992, Europe's lithosphere—seismic structure, *in* Blundell, D., Freeman, R., and Mueller, S., eds., A continent revealed—the European Geotraverse: Cambridge University Press, p. 33–69. [2] [7] [8]

Antoine, J.W., and Ewing, J.I., 1963, Seismic refraction measurements on the margins of the Gulf of Mexico: Journal of Geophysical Research, v. 68, p. 1975–1996, doi:10.1029/JZ068i007p01975. [5] [6]

Aoki, H., Tada, T., Sasaki, Y., Ooida, T., Muramatu, I., Shimamura, H., and Furuya, I., 1972, Crustal structure in the profile across central Japan as derived from explosion seismic observations: Journal of Physics of the Earth, v. 20, p. 197–223. [2] [6] [9] [10]

Aric, K., Hirschleber, H.B., Menzel, H., and Weigel, W., 1970, Über die Struktur der Großen Meteor-Bank nach seismischen Ergebnissen: Meteor Forschungsergebnisse, Deutsche Forschungsgemeinschaft: Borntraeger, Berlin-Stuttgart, Reihe C Geologie und Geophysik, C3, p. 48–64. [6]

Arita, K., Ikawa, T., Yamamoto, A., Saito, M., Nishida, Y., Satoh, H., Kimura, G., Watanabe, T., Ikawa, T., and Kuroda, T., 1998, Crustal structure and tectonics of the Hikada collision zone, Hokkaido (Japan), revealed by vibroseis seismic reflection and gravity surveys: Tectonophysics, v. 290, p. 197–210, doi:10.1016/S0040-1951(98)00018-3. [2] [9]

Arlitt, R., Patzwahl, R., and Zeyen, H., 1993, 1-D modelling of the lithospheric structure beneath the western Iberian peninsula using the the ILIHA–DSS data, *in* Mezcua, J., and Carreno, E., eds., Iberian lithosphere, heterogeneity and anisotropy (ILIHA): Monografia no. 10, Instituto Geografico Nacional, Madrid, Spain, p. 229–249. [8]

Asada, T., and Shimamura, H., 1976, Observation of earthquakes and explosions at the bottom of the western Pacific: Structure of oceanic lithosphere revealed by Longshot experiment, *in* Sutton, G.H., Manghnani, M.H., Moberly, R., and McAffe, E.U., eds., The geophysics of the Pacific Ocean basin and its margin: American Geophysical Union, Geophysical Monograph 19, p. 135–154. [2] [7]

Asada, T., and Shimamura, H., 1979, Long-range refraction experiments in deep ocean: Tectonophysics, v. 56, p. 67–82, doi:10.1016/0040-1951(79)90009-X. [2] [7]

Asada, T., Steinhart, J.S., Rodriguez, A., Tuve, M.A., and Aldrich, L.T., 1961, The earth's crust: Washington, Carnegie Institution, Year Book, v. 60, p. 244–250. [6]

Asada, T., Shimamura, H., Asano, S., Kobayashi, K., and Tomoda, Y., 1983, Explosion experiments on long-range profiles in the northwestern Pacific and the Mariana Sea, *in* Hilde, T.W.C., and Ueda, S., eds., Geodynamics of the Western Pacific–Indonesian Region, American Geophysical Union, Geodynamics Series, v. 11, p. 105–120. [2] [7]

Asano, S., Okada, H., Yoshii, T., Yamamoto, K., Hasegawa, T., Ito, K., Suzuki, S., Ikami, A., and Hamada, K., 1979, Crust and upper mantle structure beneath northeastern Honshu, Japan, as derived from explosion seismic observations: Journal of the Physics of the Earth, v. 27 Suppl., p. 1–13. [7]

Asano, S., Yamada, T., Suyehiro, K., Yoshii, T., Misawa, Y., and Iizuka, S., 1981, Crustal structure in a profile off the Pacific coast of northeastern Japan by the refraction method with ocean bottom seismometers: Journal of the Physics of the Earth, v. 29, p. 267–281. [7] [9]

Asano S., Yoshii, T., Kubota, S., Sasaki, Y., Okada, H., Suzuki, S., Masuda, T., Mutakami, H., Nishide, N., and Inatani, H., 1982, Crustal structure of Izu Peninsula, central Japan, as derived fro explosion seismic observations, 1. Mishima–Shimoda Profile: Journal of the Physics of the Earth, v. 30, p. 367–387. [2] [7]

Ashcrost, W.A., 1970, Crustal structure of the South Shetland Island and Bransfield Strait: Scientific report of British Antarctica Survey no. 66. [6]

Asudeh, I., Green, A.G., and Forsyth, D.A., 1988, Canadian expedition to study the Alpha Ridge complex: results of the seismic refraction survey: Geophysical Journal, v. 92, p. 283–301, doi:10.1111/j.1365-246X.1988.tb01140.x. [8]

Asudeh, I., Anderson, F., Parmelee, J., Vishnubhatla, S., Munro, P., and Thomas, J., 1992, A portable refraction seismograph PRS1: Geological Survey of Canada Open-File Report 2478, p. 34 p. [8]

Atwater, T., 1970, Implications of plate tectonics for the Cenozoic tectonic evolution of western North America: Geological Society of America Bulletin, v. 81, p. 3513–3536, doi:10.1130/0016-7606(1970)81[3513:IOPTFT]2.0.CO;2. [9]

Avedik, F., and Hieke, W., 1981, Reflection seismic profiles from the Central Ionian Sea (Mediterranean) and their geodynamic interpretation. Meteor Forschungsergebnisse, Deutsche Forschungsgemeinschaft: Berlin-Stuttgart, Borntraeger, Reihe C Geologie und Geophysik, C34, p. 49–64. [2] [7]

Avedik, F., Renard, V., Buisine, D., and Cornic, J.-Y., 1978, Ocean bottom refraction seismograph (OBRS): Marine Geophysical Researches, v. 3, p. 357–379, doi:10.1007/BF00347673. [2] [7]

Avedik, F., Geli, L., Gaulier, J.M., and Le Formal, J.P., 1988, Results from three refraction profiles in the northern Red Sea (above 25°N) recorded with an ocean bottom vertical seismic array, *in* LePichon, X., and Cochran, J.R., eds., The Gulf of Suez and Red Sea rifting: Tectonophysics, v. 153, p. 89–101. [8]

Avedik, F., Renard, V., Allenou, J.P., and Morvan, B., 1993, "Single bubble" airgun array for deep exploration: Geophysics, v. 58, p. 366–382, doi:10.1190/1.1443420. [8]

Avedik, F., Hirn, A., Renard, V., Nicolich, R., Olivet, J.L., Sachpazi, M.,1996, "Single-bubble" marine source offers new perspectives for lithospheric exploration: Tectonophysics, v. 267, p. 57–71, doi:10.1016/S0040-1951(96)00112-6. [10]

Averill, M.G., 2007, A lithospheric investigation of the southern Rio Grande rift [Ph.D. thesis]: Univeristy of Texas at El Paso, 141 p. [2] [10]

Ayarza, P., Martínez-Catalán, J.R., Gallart, J., Pulgar, J.A., and Danobeitia, J.J., 1998, ESCIN 3.3: A seismic image of the Variscan crust in the hinterland of NW Iberian Massif: Tectonics, v. 17, p. 171–186, doi:10.1029/97TC03411. [9]

Azbel, I.Ya., Buyanov, A.F., Ionkis, V.T., Sharov, N.V., and Sharova, V.P., 1989, Crustal structure of the Kola peninsula from inversion of deep seismic sounding data, *in* Freeman, R., Knorring, M. von, Korhonen, H., Lund, C., and Muller, St., eds., The European Geotraverse, Part 5: The POLAR Profile: Tectonophysics, v. 162, p. 87–99. [2] [8]

Baba, T., and Cummins, R.P., 2005, Contiguous rupture areas of two Nankai Trough earthquakes revealed by high-resolution tsunami waveform inversion: Geophysical Research Letters, v. 32, doi:10.1029/2004GL022320. [10]

BABEL Working Group, 1991, Reflectivity of a Proterozoic shield: examples from BABEL seismic profiles across Fennoscandia, *in* Meissner, R., Brown, L., Dürbaum, H.-J., Franke, W., Fuchs, K., and Seifert, F., eds., Continental lithosphere: deep seismic reflections: American Geophysical Union, Geodynamics Series, v. 22, p. 77–86. [2] [8] [9]

BABEL Working Group, 1993a, Deep seismic reflection/refraction interpretation of crustal structure along BABEL profiles A and B in the southern Baltic Sea: Geophysical Journal International, v. 112, p. 325–343, doi:10.1111/j.1365-246X.1993.tb01173.x. [8]

BABEL Working Group, 1993b, Integrated seismic studies of the Baltic shield using data in the Gulf of Bothnia region: Geophysical Journal International, v. 112, p. 305–324, doi:10.1111/j.1365-246X.1993.tb01172.x. [8]

Babuska, V., Plomerova, J., and the Bohemian Working Group, 2003, Seismic experiment searches for active magmatic source in deep lithosphere, central Europe: Eos (Transactions, American Geophysical Union), v. 84, p. 409, 416–417. [10]

Baier, B., Berckhemer, H., Gajewski, D., Green, R., Grimsel, C., Prodehl, C., and Vees, R., 1983, Deep seismic sounding in the area of the Damara Orogen (Namibia), *in* Martin, H., and Eder, F.W., eds., Intracontinental Fold Belts: Berlin-Heidelberg, Springer, p. 885–900. [2] [7] [9]

Bakun, W.H., and Lindt, A.G., 1985, The Parkfield, California, earthquake prediction experiment: Science, v. 229, p. 619–624, doi:10.1126/science.229.4714.619. [9] [10]

Baldridge, W.S., Braile, L.W., Fehler, M.C., and Moreno, F.A., 1997, Science and sociology butt heads in tomography experiment in sacred mountains. Eos (Transactions, American Geophysical Union), v. 78, p. 417, 422–423. [2] [9]

Bamford, S.A.D., 1971, An interpretation of first-arrival data from the Continental Margin refraction experiment: Geophysical Journal of the Royal Astronomical Society, v. 24, p. 213–229. [2] [6] [8]

Bamford, S.A.D., 1972, Evidence for a low-velocity zone in the crust beneath the western British Isles: Geophysical Journal of the Royal Astronomical Society, v. 30, p. 101–105. [6]

Bamford, D., 1973, Refraction data in western Germany—a time-term interpretation: Zeitschrift für Geophysik, v. 39, p. 907–927. [2] [7]

Bamford, D., Faber, S., Fuchs, K., Jacob, B., Kaminski, W., King, R., Nunn, K., Prodehl, C., and Willmore, P., 1976, A lithospheric seismic profile in Britain—I. Preliminary results: Geophysical Journal of the Royal Astronomical Society, v. 44, p. 145–160. [2] [7]

Bamford, D., Nunn, K., Prodehl, C., and Jacob, B., 1978, LISPB IV. Crustal structure of northern Britain: Geophysical Journal of the Royal Astronomical Society, v. 54, p. 43–60. [2] [7] [8]

Bamford, D., Jentsch, M., and Prodehl, C., 1979, P$_n$ anisotropy studies in northern Britain and the eastern and western United States: Geophysical Journal of the Royal Astronomical Society, v. 57, p. 397–430. [2] [7]

Banda, E., and Ansorge, J., 1980, Crustal structure under the central and eastern part of the Betic cordillera: Geophysical Journal of the Royal Astronomical Society, v. 63, p. 515–532. [2] [7]

Banda, E., Ansorge, J., Boloix, M., and Cordoba, D, 1980, Structure of the crust and upper mantle beneath the Balearic islands (western Mediterranean): Earth and Planetary Science Letters, v. 49, p. 219–230, doi:10.1016/0012-821X(80)90066-7. [2] [7]

Banda, E., Danobeitia, J.J., Surinach, E., and Ansorge, J., 1981a, Features of crustal structure under the Canary islands: Earth and Planetary Science Letters, v. 55, p. 11–24, doi:10.1016/0012-821X(81)90082-0. [2] [7] [9]

Banda, E., Surinach, E., Aparicio, A., Sierra, J., and Ruiz de la Parte, E., 1981b, Crust and upper mantle structure of the central Iberian Meseta: Geophysical Journal of the Royal Astronomical Society, v. 67, p. 779–789. [7]

Banda, E., Ranero, C.R., Danobeitia, J.J., and Rivero, A., 1992, Seismic boundaries of the eastern Central Atlantic Mesozoic crust from multichannel seismic data: Geological Society of America Bulletin, v. 104, p. 1340–1349, doi:10.1130/0016-7606(1992)104<1340:SBOTEC>2.3.CO;2. [2] [8]

Banda, E., Gallart, J., Garcia-Duenas, V., Danobeitia, J.J., and Makris, J., 1993, Lateral variation of the crust in the Iberian peninsula: new evidence from the Betic Cordillera: Tectonophysics, v. 221, p. 53–66, doi:10.1016/0040-1951(93)90027-H. [8]

Banda, E., Torné, M., and the Iberian Atlantic Margins Group, 1995, Iberian Atlantic Margins Group investigates deep structure of ocean margins: Eos (Transactions, American Geophysical Union), v. 76, no. 3, p. 25, 28–29. [2] [9]

Barazangi, M., and Brown, L., eds., 1986a, Reflection seismology: a global perspective: American Geophysical Union, Geodynamics Series, v. 13, p. 311 p. [2] [8]

Barazangi, M., and Brown, L., eds., 1986b, Reflection seismology: The continental crust: American Geophysical Union, Geodynamics Series, v. 14, p. 339 p. [2] [8]

Barchi, M., Minelli, G., Magnani, B., and Mazzotti, A., 2003, Line CROP 03: northern Apennines, in Scrocca, D., Doglioni, C., Innocenti, F., Manetti, P., Mazzotti, A., Bertelli, L., Burbi, L., and D'Offizi, S., eds., 2003 CROP Atlas: seismic reflection profiles of the Italian crust: Memorie Descrittive della Carta Geologica d'Italia, v. 62, p. 127–136 [9]

Barker, D.H.N., Sutherland, R., Henrys, S., and Bannister, S., 2009, Geometry of the Hikurangi subduction thrust and upper plate, North Island, New Zealand: Geochemistry Geophysics Geosystems, v. 10, Q02007, doi:10.1029/2008GC002153. [2] [10]

Barranco, L.M., Ansorge, J., and Banda, E., 1990, Seismic refraction constraints on the geometry of the Ronda peridotitc assif (Betic Cordillera, Spain): Tectonophysics, v. 184, p. 379–392, doi:10.1016/0040-1951(90)90450-M. [2] [7]

Barrett, D.L., Berry, M., Blanchard, J.E., Keen, M.J., and McAllister, R.E., 1964, Seismic studies on the eastern seabord of Canada: the Atlantic coast of Nova Scotia: Canadian Journal of Earth Science, v. 1, p. 10–22. [6]

Barsch, O., and Reich, H., 1930, Seismische Arbeiten in Norddeutschland, Beiträge zur physikalischen Erforschung der Erdrinde, Preuss: Geol. Landesanstalt, Berlin, Heft 3, 58 p. [3]

Bartels, J., and ten Bruggencate,P., 1953, Astronomie und Geophysik, Landolt Börnstein, 6th ed., vol III: Berlin-Göttingen-Heidelberg, Springer. [10]

Bartelsen, H., Lueschen, E., Krey, Th., Meissner, R., Schmoll, H., and Walther, Ch., 1982, The combined seismic reflection-refraction investigation of the Urach geothermal anomaly, in Haenel, R., ed., The Urach geothermal project (Swabian Alb, Germany): Stuttgart, Schweizerbart, p. 247–262. [2] [7] [8]

Barton, D.C., 1929, The seismic method of mapping geologic structure: Geophysical Prospecting: Transactions of the American Institute of Mining and Metallurgical Engineering, v. 81, p. 572–624. [3]

Barton, P.J., 1986, Comparison of deep reflection and refraction structures in the North Sea, in Barazangi, M., and Brown, L., eds., Reflection seismology: a global perspective: American Geophysical Union, Geodynamics Series, v. 13, p. 297–300. [2] [8]

Barton, P.J, and Wood, R., 1984, Tectonic evolution of the North Sea basin: crustal stretching and subsidence: Geophysical Journal of the Royal Astronomical Society, v. 79, p. 987–1022, doi:10.1111/j.1365-246X.1984.tb02880.x. [6] [8] [9]

Barton, P.J., Owen, T.R.E., and White, R.S., 1990, The deep structure of the east Oman continental margin: preliminary results and interpretation, in Leven, J.H., Finlayson, D.M., Wright, C., Dooley, J.C., and Kennett, B.L.N., eds., Seismic probing of continents and their margins: Tectonophysics, v. 173, p. 319–331. [2] [8]

Bates, A., and Hall, D.H., 1975, Upper mantle structure in southern Saskatchewan and western Manitoba from Project Edzoe: Canadian Journal of Earth Science, v. 12, p. 2134–2144. [6]

Båth, M., 1960, Crustal structure of Iceland: Journal of Geophysical Research, v. 65, p. 1793–1807, doi:10.1029/JZ065i006p01793. [5]

Batini, F., Cameli, G.M., Lazzarotto, A., and Liotta, D., 2003, Line CROP 18: Southern Tuscany, in Scrocca, D., Doglioni, C., Innocenti, F., Manetti, P., Mazzotti, A., Bertelli, L., Burbi, L., and D'Offizi, S., eds., CROP Atlas: seismic reflection profiles of the Italian crust: Memorie Descrittive della Carta Geologica d'Italia, Roma, v. 62, p. 137–144. [9]

Bauer, K., Neben, S., Schreckenberger, B., Emmermann, R., Hinz, K., Fechner, N., Gohl, K., Schulze, A., Trumbull, R.B., and Weber, K., 2000, Deep structure of the Namibia continental margin as derived from integrated geophysical studies: Journal of Geophysical Research, v. 105, p. 25,829–25,853, doi:10.1029/2000JB900227. [2] [9] [10]

Bayer, R., et Équipe de Profil Alpes ECORS-CROP, 1987, First results of a deep seismic profile through the western Alps (ECORS-CROP Program): Comptes rendus de l'Académie de sciences, Paris, 105, série II, p. 1461–1470. [8]

Bayer, U., Scheck, M., Rabbel, W., Krawczyk, C.M., Götze, H.-J., Stiller, M., Beilecke, T., Marotta, A.-M., Barrio-Alvers, L., and Kuder, J., 1999, An integrated study of the NE-German Basin: Tectonophysics, v. 314, p. 285–307, doi:10.1016/S0040-1951(99)00249-8. [9]

Bean, C.J., and Jacob, A.W.B., 1990, P-wave anisotropy in the lower lithosphere: Earth and Planetary Science Letters, v. 99, p. 58–65, doi:10.1016/0012-821X(90)90070-E. [2] [8]

Beaudoin, B.C., Perkins, G., Fuis, G.S., and Luetgert, J.H., 1989, Data report for the 1987 seismic-refraction survey, Alaska Range and Fairbanks South deployments: Menlo Park, California, U.S. Geological Survey Open-File Report 89-321, 114 p. [8]

Beaudoin, B.C., Fuis, G.S., Nockleberg, W.J., and Christensen, N.I., 1992, Thin, low-velocity crust beneath the southern Yukon-Tanana terrane, east-central Alaska: Journal of Geophysical Research, v. 97, p. 1921–1942, doi:10.1029/91JB02881. [8]

Beaudoin, B.C., Fuis, G.S., Lutter, W.J., Mooney, W.D., and Moore, T.E., 1994, Crustal velocity structure of the northern Yukon-Tanana upland, central Alaska: results from TACT refraction/wide-angle reflection data: Geological Society of America Bulletin, v. 106, p. 981–1001, doi:10.1130/0016-7606(1994)106<0981:CVSOTN>2.3.CO;2. [8]

Behnke, C., 1971, Explosion seismology data generalization—examples from the Eastern Alps: Proceedings of the 12th General Assembly of the European Seismological Commission (Luxembourg 1970), Obs. Royal de Belgique, Comm. Sér. A, no 13, Sér. Geophys. 101, p. 177–181. [6]

Behnke, C., and Giese, P., 1970, Refraction-seismic investigations 1956–1969 to explore the earth's crust in the Alps (in German). Unpublished Open-File Report, Niedersächs. Landesamt f. Bodenforschung, Hannover, and Free University Berlin, p. D1–D43 and p. 1–43. [6]

Behnke, C., Giese, P., Prodehl, C., de Visintini, G., 1962, Seismic refraction investigations in the Dolomites for the exploration of the earth's crust in the Eastern Alpine area 1961: Bollettino di Geofisica Teorica e Applicata, v. 4, p. 110–132. [6]

Behrendt, J.C., 1986, Structural interpretation of multichannel seismic reflection profiles crossing the southeastern United States and the adjacent continental margin—décollements, faults, Triassic (?) basins and Moho reflections, *in* Barazangi, M., and Brown, L., eds., Reflection seismology: the continental crust: American Geophysical Union, Geodynamics Series, v. 14, p. 201–213. [2] [7] [8]

Behrendt, J.C., Green, A.G., Lee, M.W., Hutchinson, D.R., Cannon, W.F., Milkereit, B., Agena, W.F., and Spencer, C., 1989, Crustal extension in the Midcontinent Rift System—results from GLIMPCE deep seismic reflection profiles over Lakes Superior and Michigan, *in* Mereu, R.F., Mueller, St., and Fountain, D.M., eds., Properties and processes of earth's lower crust: American Geophysical Union, Geophysical Monograph 51, p. 81–89. [8]

Behrendt, J.C., Hutchinson, D.R., Lee, M., Thornber, C.R., Trehu,A., Cannon, W., and Green, A., 1990, GLIMPCE seismic reflection evidence of deep-crustal and upper-mantle intrusions and magmatic underplating associated with the Midcontinent Rift System of North America, *in* Leven, J.H., Finlayson, D.M., Wright, C., Dooley, J.C., and Kennett, B.L.N., eds., Seismic probing of continents and their margins: Tectonophysics, v. 173, p. 595–615. [8]

Behrens, K., Hansen, J., Flüh, E.R., Goldflam, S., and Hirschleber, H., 1986, Seismic investigations in the Skagerrak and Kattegat, *in* Galson, D.A., and Mueller, St., eds., The European Geotraverse, Part 2: Tectonophysics, v. 128, p. 209–228. [2] [8]

Behrens, K., Goldflam, S., Heikkinen, P., Hirschleber, H., Lindquist, G., and Lund, C.E., 1989, Reflection seismic measurements across the Granulite belt of the POLAR profile in the northern Baltic Shield, northern Finland, *in* Freeman, R., Knorring, M. von, Korhonen, H., Lund, C., and Muller, St., eds., The European Geotraverse, Part 5: The POLAR Profile: Tectonophysics, v. 162, p. 101–111. [8]

Beloussov, V.G., Vol'vovski, B.S., Vol'vovski, I.S., and Ryaboi, V.Z., 1962, Experimental recording of deep reflected waves: Bull. (Izvestiya) Acad. Sci., U.S.S.R., Geophys. Ser., English translation, no. 8 (1962), p. 662–669. [2] [6]

Belyaevsky, N.A., Borisov, A.A., Vol'vovsky, I.S., and Schukin, Yu.K., 1968, Transcontinental crustal sections of the U.S.S.R. and adjacent areas: Canadian Journal of Earth Science, v. 5, p. 1067–1078. [6]

Belyaevsky, N.A., Borisov, A.A., Fedinsky, V.V., Fotiadi, E.E., Subbotin, S.I., and Volvovsky, I.S., 1973, Structure of the earth's crust on the territory of the U.S.S.R., *in* Mueller, S., ed., The structure of the earth's crust, based on seismic data: Tectonophysics, v. 20, p. 35–45. [2] [6] [10]

Bentley, C.R., 1973, Crustal structure of Antarctica, *in* Mueller, S., ed., The structure of the earth's crust, based on seismic data: Tectonophysics, v. 20, p. 229–240. [2] [5] [6]

Bentley, C.R., and Worzel, J.L., 1956, Geophysical investigations in the emerged and submerged Atlantic coastal plain, Part X: Continental slope and continental rise south of the Grand Banks: Bulletin of the Geological Society of America, v. 67, p. 1–18, doi:10.1130/0016-7606 (1956)67[1:GIITEA]2.0.CO;2. [2] [5]

Benz, H.M., Unger, J.D., Leith, W.S., Mooney, W.D., Solodilov, L., Egorkin, A.V., and Ryaboy, V.Z., 1992, Deep seismic sounding in northern Eurasia: Eos (Transactions, American Geophysical Union), v. 73, no. 28, p. 297, 300. [8]

Beránek, B., Dudek, A., Feifar, M., Hrdlicka, A., Suk M., Zoukova, M., and Weiss, J., 1972a, Czechoslovakia, *in* Sollogub, V.B., Prosen, D., and Militzer, H., Crustal structure of central and southeastern Europe based on the results of explosion seismology (publ. in Russian, 1971). English translation edited by Szénás, Gy., 1972, Geophysical Transactions, spec. ed., Müszaki Könyvkiado, Budapest, chapter 24, p. 87–98. [2] [6]

Beránek, B., Weiss, J., Hrdlicka, A., Dudek, A., Zoukova, M., Suk, M., Feifar, M., Militzer, H., Knothe, H., Mituch, E., Posgay, K., Uchman, J., Sollogub, V.B., Chekunov, A.V., Prosen, D., Milovanovic, B., and Roksandic, M., 1972b, The results of the measurements along the international profiles, *in* Sollogub, V.B., Prosen, D., and Militzer, H., eds., Crustal structure of central and southeastern Europe based on the results of explosion seismology (publ. in Russian 1971). English translation edited by Szénás, Gy., 1972, Geophysical Transactions, spec. ed., Müszaki Könyvkiado, Budapest, chapter 3, p. 133–139. [2] [6]

Beránek, B., Mayerova, M., Zounková, M., Guterch, A., Materzok, R., and Pajchel, J., 1973, Results of deep seismic sounding along International profile in Czechoslovakia and Poland: Studia in Geophysica et Geodaetica, Academy of Science, CSSR, Prague, v. 17, p. 205-217. [7]

Berckhemer, H., 1968, Topographie des "Ivrea-Körpers" abgeleitet aus seismischen und gravimetrischen Daten: Schweizerische Mineralogische und Petrographische Mitteilungen, 48, p. 235–246. [6]

Berckhemer, H., 1970, MARS 66—a magnetic tape recording equipment for deep seismic sounding: Zeitschrift für Geophysik, v. 36, p. 501–518. [6]

Berckhemer, H., Baier, B., Bartelsen, H.,Behle, A., Burckhardt, H., Gebrande, H., Makris, J., Menzel, H., Miller, H., and Vees, R., 1975, Deep seismic soundings in the Afar region and on the highland of Ethiopia, *in* Pilger, A., and Rösler, A., eds., Afar Depression of Ethiopia: Stuttgart, Schweizerbart, p. 89–107. [2] [7]

Berg, E., 1973, Crustal structure in Alaska, *in* Mueller, S., ed., The structure of the earth's crust, based on seismic data: Tectonophysics, v. 20, p. 165–182. [5] [6]

Berg, J.W., Cook, K.L., and Narans, H.D., 1960, Seismic investigation of crustal structure in the eastern part of the Basin and Range Province: Bulletin of the Seismological Society of America, v. 50, p. 511–535. [2] [5]

Berg, Jr., J.W., Trembley. L., Emelia, D.W., Hutt, J.R., King, J.M., Long, L.T., McKnight, W.R., Sarmah, S.K., Souders, R., Thiruvathukal, J.V., and Vossler, D.A., 1966, Crustal refraction profile, Oregon coast range. Bulletin of the Seismological Society of America, v. 56, p. 1357–1362. [6]

Bergerat, F., Mugnier, J.-L., Guellec, S., Truffert, C., Cazes, M., Damotte, B., and Roure, F., 1990, Extensional tectonics and subsidence of the Bresse basin: an interpretation from ECORS data: Mémoires Société Géologique France, N.S., v. 156, p. 145–156. [8]

Bernabini, M., and Manetti, P., 2003, CROP project: goals and organization, *in* Scrocca, D., Doglioni, C., Innocenti, F., Manetti, P., Mazzotti, A., Bertelli, L., Burbi, L., and D'Offizi, S., eds., CROP Atlas: seismic reflection profiles of the Italian crust: Memorie Descrittive della Carta Geologica d'Italia, Roma, 62, p. 9–14. [8] [9]

Bernabini, M., Nicolich, R., and Polino, R., 2003, Seismic lines CROP-ECORS across the Western Alps, *in* Scrocca, D., Doglioni, C., Innocenti, F., Manetti, P., Mazzotti, A., Bertelli, L., Burbi, L., and D'Offizi, S., eds., CROP Atlas: seismic reflection profiles of the Italian crust: Memorie Descrittive della Carta Geologica d'Italia, Roma, 62, p. 89–96. [8]

Berrocal, J., Marangoni, Y., de Sa, N.C., Fuck, R., Soares, J.E.P., Dantas, E., Perosi, F., and Fernandes, C., 2004, Deep seismic refraction and gravity crustal model and tectonic deformation in Tocantins Province, Central Brazil: Tectonophysics, v. 388, p. 187–199, doi:10.1016/j.tecto.2004 .04.033. [2] [9]

Berry, M.J., 1973, Structure of the crust and upper mantle in Canada, *in* Mueller, S., ed., The structure of the earth's crust, based on seismic data: Tectonophysics, v. 20, p. 183–201. [2] [6]

Berry, M.J., and Forsyth, D.A., 1975, Structure of the Canadian Cordillera from seismic refraction and other data: Canadian Journal of Earth Science, v. 12, p. 182–208. [2] [6]

Berry, M.J., and Fuchs, K., 1973, Crustal structure of the Superior and Grenville provinces of the northeastern Canadian Shield: Bulletin of the Seismological Society of America, v. 63, p. 1393–1432. [2] [6]

Berry, M.J., and Mair, J.A., 1977, The nature of the earth's crust in Canada, *in* Heacock, J.G. ed., The earth's crust: American Geophysical Union Geophysical Monograph 20, p. 319–348. [2] [7]

Berry, M.J., and West, G.F., 1966, A time–term interpretation of the first-arrival data of the 1963 Lake Superior experiment, *in* Steinhart, J.S., and Smith, T.J., eds., The earth beneath the continents: Washington, D.C., American Geophysical Union, Geophysical Monograph 10, p. 166–180. [6]

Berry, M.J., Jacoby, W.R., Niblett, E.R., and Stacey, R.A., 1971, A review of geophysical studies in the Canadian Cordillera: Canadian Journal of Earth Science, v. 8, p. 788–801. [6]

Berthelsen, A., 1983, The early (800–300 Ma) crustal evolution of the off-shield regions of Europe, *in* Galson, D.A., and Mueller, S., eds., Proceedings of the First Workshop on the EGT project—the northern segment: Strasbourg, European Science Foundation, p. 125–142. [8]

Berzin, R., Oncken, O., Knapp, J.H., Perez-Estaun, A., Hismatulin, T., Yunusov, N., and Lipilin, A., 1996, Orogenic evolution of the Ural Mountains: Results from an integrated seismic experiment: Science, v. 274, p. 220–221, doi:10.1126/science.274.5285.220. [9]

Berzin, R.G., Yurov, Yu.G., and Pavlenkova, N.I., 2002, CDP and DSS data along the Uchta-Kem profile (the Baltic Shield): Tectonophysics, v. 355, p. 187–200, doi:10.1016/S0040-1951(02)00141-5. [2] [9]

Bessonova, E.N., Fishmen, V.M., Ryaboi, V.Z., and Sitnikova, G.A., 1974, The tau method for inversion and traveltime. I. Deep seismic sounding data: Geophysical Journal of the Royal Astronomical Society, v. 36, p. 377–398. [2] [7] [10]

Betlej, K., Gadomska, B., Gorczynski, L., Guterch, A., Milolajzak, A., and Uchman, J., 1967, Deep seismic sounding on the profile Starachowice-

Radzyn Podlaski: Publications of the Institute of Geophysics, Polish Academy of Sciences, v. 14, p. 41–46. [6]

Bezdan, S., and Hajnal, Z., 1998, Expanding spread profiles across the Trans-Hudson orogen, *in* Klemperer, S.L., and Mooney, W.D., eds., Deep seismic probing of the continents, II: a global survey: Tectonophysics, v. 288, p. 83–91. [9]

Bibee, L.D., and Shor, G.G., 1976, Compressional wave anisotropy in the crust and upper mantle: Geophysical Research Letters, v. 3, p. 639–642, doi:10.1029/GL003i011p00639. [2] [7]

Biella, G.C., Gelati, R., Maistrello, M., Mancuso, M., Massiotta, P., and Scarascia, S., 1987, The structure of the upper crust in the Alps-Apennines region deduced from seismic refraction data, *in* Freeman, R., and Mueller, St., eds., The European Geotraverse, Part 3: Tectonophysics, v. 142, p. 71–85. [2] [8]

Birch, F., Schairer, J.F., and Spicer, H.C., eds., 1942, Handbook of physical constants: Geological Society of America Special Paper 36, 323 p. [4] [5]

Birkenmajer, K., Guterch A., Grad M., Janik T., and Perchuc, E., 1990, Lithospheric transect Antarctic Peninsula–South Shetland Islands, West Antarctica: Polish Polar Research, v. 11, p. 241–258. [8]

BIRPS and ECORS, 1986, Deep seismic reflection profiling between England, France and Ireland: Journal of the Geological Society London, 143, p. 45–52, doi:10.1144/gsjgs.143.1.0045. [8]

Birt, C.S., Maguire, P.K.H., Khan, M.A., Thybo, H., Keller, G.R., and Patel, J., 1997, The influence of pre–existing structures on the evolution of the southern Kenya rift valley—evidence from seismic and gravity studies, *in* Fuchs, K., Altherr, R., Müller, B., and Prodehl, C., eds., Structure and dynamic processes in the lithosphere of the Afro-Arabian rift system: Tectonophysics, v. 278, p. 211–242. [9]

Bleibinhaus, F., and Gebrande, H., 2006, Crustal structure of the Eastern Alps along the TRANSALP profile from wide-angle seismic tomography, *in* Gebrande, H., Castellarin, A., Lueschen, E., Millahn, K., Neubauer, F., and Nicolich, R., eds., TRANSALP—a transect through a young collisional orogen: Tectonophysics, v. 414, p. 51–69. [9]

Bleibinhaus, F., Beilecke, T., Bram, K., and Gebrande, H., 1999, A seismic velocity model for the SW Baltic Sea derived from BASIN'96 refraction seismic data: Tectonophysics, v. 314, p. 269–283, doi:10.1016/S0040 -1951(99)00248-6. [2] [9]

Bleibinhaus, F., Brueckl, E., and ALP 2002 Working Group, 2006, Wide-angle observations of ALP 2002 shots on the TRANSALP profile: linking the two DSS projects, *in* Gebrande, H., Castellarin, A., Lueschen, E., Millahn, K., Neubauer, F., and Nicolich, R., eds., TRANSALP—a transect through a young collisional orogen: Tectonophysics, v. 414, p. 71–78. [10]

Bleibinhaus, F., Hole, J.A., Ryberg, T., and Fuis, G.S., 2007, Structure of the California Coast Ranges and San Andreas Fault at SAFOD from seismic waveform inversion and reflection imaging: Journal of Geophysical Research, v. 112, B06315, doi:10.1029/2006JB004611. [10]

Blümling, P., and Prodehl, C., 1983, Crustal structure beneath the eastern part of the Coast Ranges (Diablo Range) of central California from explosion-seismic and near-earthquake data, *in* Ansorge, J., and Mereu, R.F., eds., Probing the earth's lithosphere by controlled source seismology: Physics of the Earth and Planetary Interiors, v. 31, p. 313–326. [6] [7]

Blundell, D.J., and Docherty, J.I.C., 1987, A programme of deep seismic reflection profiling across Europe. Final report, EC contract no. EN3G 0016 UK (H), unpublished. [2] [8]

Blundell, D.J., and Parks, R., 1969, A study of the crustal structure beneath the Irish Sea: Geophysical Journal of the Royal Astronomical Society, v. 17, p. 45–62. [6]

Blundell, D., Freeman, R., and Mueller, S., eds., 1992, A continent revealed—the European Geotraverse: Cambridge University Press, 275 p. [2] [8]

Boatwright, J., Blair, L., Catchings, R.D., Goldman, M.R., Perosi, F., and Steedman, C.E., 2004, Using twelve years of USGS refraction lines to calibrate the Brocher and others (1997) 3D velocity model of the Bay Area: U.S. Geological Survey Open-File Report 2004-1282, 34 p. [9]

Bocin, A., Stephenson, R., Tryggvason, A., Panea, I., Mocanu, V., Hauser, F., and Matenco, L., 2005, 2.5D seismic velocity modeling in the southeastern Romanian Carpathians Orogen and its foreland: Tectonophysics, v. 410, p. 273–291, doi:10.1016/j.tecto.2005.05.045. [10]

Bock, G., Achauer, U., Alinaghi, A., Ansorge, J., Bruneton, M., Friederich, W., Grad, M., Guterch, A., Hjelt, S.-E., Hyvönen, T., Ikonen, J.-P., Kissling, E., Komminaho, K., Korja, A., Heikkinen, P., Kozlovskaya, E., Neusky, M.V., Pavlenkova, N., Pedersen, H., Plomerova, J., Raita, T., Riznichenko,

O., Roberts, R.G., Sandoval, S., Sanina, I.A., Sharov, N., Tiikkainen, J., Volosov, S.G., Wielandt, E., Wylegalla, K., Yliniemi, J., and Yurov, Y., 2001, Seismic Probing of Fennoscandian Lithosphere: Eos (Transactions, American Geophysical Union), v. 82, no. 50, p. 621, 628–629, doi:10.1029/01EO00356. [9]

Bogdanova, S.V., Gorbatschev, R., and Stephenson, R.A., 2001, EUROBRIDGE: Paleoproterozoic accretion of Fennoscandia and Sarmatia: Tectonophysics, v. 339, p. vii–x, doi:10.1016/S0040-1951(01)00030-0. [9]

Bogdanova, S., Gorbatschev, R., Grad, M., Janik, T., Guterch, A., Kozlovskaya, E., Motuza, G., Skridlaite, G., Starostenko, V., Taran, L., and EUROBRIDGE and POLONAISE Working Groups, 2006, EUROBRIDGE: new insight into the geodynamic evolution of the East European Craton, *in* Gee, D.G., and Stephenson, R.A., eds., European lithosphere dynamics: Geological Society of London Memoir 32, p. 599–625. [9]

Bohm, M., Lüth, S., Echtler, H., Asch, G., Bataille, K., Bruhn, C., Rietbrock, A., and Wigger, P., 2002, The Southern Andes between 36° and 40°S latitude: seismicity and average seismic velocities: Tectonophysics, v. 356, p. 275–289, doi:10.1016/S0040-1951(02)00399-2. [9] [10]

Bois, C., and ECORS Scientific Parties, 1991, Post-orogenic evolution of the European crust studied from ECORS deep seismic profiles, *in* Meissner, R., Brown, L., Dürbaum, H.-J., Franke, W., Fuchs, K., and Seifert, F., eds., Continental lithosphere: deep seismic reflections: American Geophysical Union, Geodynamics Series, v. 22, p. 69–76. [2] [8] [9]

Bois, C., Cazes, M., Damotte, B., Galdeano, A., Hirn, A., Mascle, A., Matte, P., Raoult, J.F., and Torreilles, G., 1986, Deep seismic profiling of the crust in northern France: the ECORS project, *in* Barazangi, M., and Brown, L., eds., Reflection seismology: a global perspective: American Geophysical Union, Geodynamics Series, v. 13, p. 21–29. [2] [8]

Bois, C., Damotte, B., Mascle, A., Cazes, M., Hirn, A., and Biju-Duval, B., 1987, Operations and main results of the ECORS project in France: Geophysical Journal of the Royal Astronomical Society, v. 89, p. 279–286. [8]

Bois, C., Pinet, B., and Roure, F., 1989, Dating lower crustal features in France and adjacent areas from deep seismic profiles, *in* Mereu, R.F., Mueller, St., and Fountain, D.M., eds., Properties and processes of earth's lower crust: American Geophysical Union, Geophysical Monograph 51, p. 17–31. [8]

Bolt, B.A., 1962, A seismic experiment using quarry blasts near Sydney: Australian Journal of Physics, v. 15, p. 293–300. [2] [5]

Bolt, B.A., Doyle, H.A., and Sutton, D.J., 1958, Seismic observations from the 1956 atomic explosions in Australia: Geophysical Journal of the Royal Astronomical Society, v. 1, p. 135–145. [2] [5]

Bonjer, K.-P., 1997, Seismicity pattern and style of seismic faulting at the eastern border of the southern Rhinegraben, *in* Fuchs, K., Altherr, R., Müller, B., and Prodehl, C., eds., Stress and stress release in the lithosphere—structure and dynamic processes in the rifts of western Europe: Tectonophysics, v. 275, p. 41–69. [9]

Bonjer, K.-P., Kaminski, W., and Kind, R., 1974, Seismic observations in Germany of a 10 t explosion off Scotland: Journal of Geophysics, v. 40, p. 259–264. [7]

Booth-Rea, G., Ranero, C.R., Martinez-Martinez, J.M., and Grevemeyer, I., 2007, Crustal types and Tertiary tectonic evolution of the Alboran Sea, western Mediteranean: Geochemistry Geophysics Geosystems, v. 8, no. 10, doi:10.1029/2007GC001639. [9]

Borcherdt, R.D., 2009, Viscoelastic waves in layered media: Cambridge, U.K., Cambridge University Press, 305 p. [2] [10]

Borus, H., Flüh, E., Grevemeyer, I., Kopp, C., and Lelgemann, H., 2001, Crustal structure, *in* Spiess, V., Flüh, E., and Schott, F., eds., Trans Atlantic 2000, Meteor-Berichte M47, no. 01-2, part 2. Leitstelle METEOR, Institute Meereskunde, University of Hamburg, p. 2-24–2-35. [2] [9]

Bosshard, E., and MacFarlane, D.-J., 1970, Crustal structure of the western Canary islands from seismic refraction and gravity data: Journal of Geophysical Research, v. 75, no.26, p. 4901–4918, doi: 10.1029 /JB075i026p04901. [9]

Bott, M.H.P., and Gunnarson, K., 1980, Crustal structure of the Iceland-Faeroe Ridge: Journal of Geophysics, v. 47, p. 221–227. [2] [6] [7]

Bott, M.H.P., Holder, A.P., Long, R.E., and Lucas, A.L., 1970, Crustal structure beneath the granites of South-West England, *in* Newall, G., and Rast, N., eds., Mechanism of igneous intrusion: Geology Journal Special Issue, 2, p. 93–102. [6]

Bott, M.H.P., Browitt, C.W.A., and Stacey, A.P., 1971, The deep structure of the Iceland-Faeroe Ridge: Marine Geophysical Research, v. 1, p. 328–351. [2] [6] [7]

Bott, M.H.P., Sunderland, J., Smith, P.J., Casten, U., and Saxov, S., 1974, Evidence for continental crust beneath the Faeroe islands: Nature, v. 248, p. 202–204, doi:10.1038/248202a0. [7]

Bott, M.H.P., Nielsen, P.H., and Sunderland, J., 1976, Converted P-waves originating at the continental margin between the Iceland-Faeroe Ridge and the Faeroe block: Geophysical Journal of the Royal Astronomical Society, v. 44, p. 229–238. [7]

Bott, M.H.P., Armour, A.R., Himsworth, E.M., Murphy, T., and Wylie, G., 1979, An explosion seismology investigation of the continental margin west of the Hebrides, Scotland, at 58°N, *in* Keen, C.E. ed., Crustal properties along passive margins: Tectonophysics, v. 59, p. 217–231, doi:10.1016/0040-1951(79)90046-5. [2] [7]

Bott, M.H.P., Green, A.S.P., Long, R.E., and Stevenson, D.L., 1983, Preliminary deep crustal structure beneath northern England from CSSP: Geophysical Journal of the Royal Astronomical Society, v. 73, p. 285. [2] [8]

Bott, M.H.P., Long, R.E., Green, A.S.P., Lewis, A.H.J., Sinha, M.C., and Stevenson, D.L., 1985, Crustal structure south of the Iapetus suture beneath northern England: Nature, v. 314, p. 724–727, doi:10.1038/314724a0. [8]

Bottinga, Y., Hirn, A., and Steinmetz, L., 1973, Implications de l'existenced'un canal à moindre vitesse sous Moho : Bulletin de la Société Géologique de France, (7), 15, p. 500–505. [7]

Braile, L.W., and Smith, R.B., 1975, Guide to the interpretation of crustal refraction profiles: Geophysical Journal of the Royal Astronomical Society, v. 40, p. 145–176. [2] [7] [10]

Braile, L.W., Smith, R.B., Keller, G.R., Welch, R.M., and Meyer, R.P., 1974, Crustal structure across the Wasatch Front from detailed seismic refraction studies: Journal of Geophysical Research, v. 79, p. 2669–2676, doi:10.1029/JB079i017p02669. [2] [7]

Braile, L.W., Smith, R.B., Ansorge, J., Baker, M.R., Sparlin, M.A., Prodehl, C., Schilly, M.M., Healy, J.H., Mueller, St., and Olsen, K.H., 1982, The Yellowstone–Snake River Plain seismic profiling experiment: Crustal structure of the eastern Snake River Plain: Journal of Geophysical Research, v. 87, p. 2597–2609, doi:10.1029/JB087iB04p02597. [7]

Braile, L.W., Hinze, W.J., von Frese, R.R.B., and Keller, G.R., 1989, Seismic properties of the crust and uppermost mantle of the conterminous United States and adjacent Canada, *in* Pakiser, L.C., and Mooney, W.D., eds., Geophysical framework of the continental United States: Geological Society of America Memoir 172, p. 655–680. [2] [8] [10]

Braile, L.W., Wang, B., Daudt, C.R., Keller, G.R., and Patel, J.P., 1994, Modelling the 2-D seismic velocity structure across the Kenya rift, *in* Prodehl, C., Keller, G.R., and Khan, M.A., eds., Crustal and upper mantle structure of the Kenya rift: Tectonophysics, v. 236, p. 251–269. [9]

Braile, L.W., Keller, G.R., Mueller, S., and Prodehl, C., 1995, Seismic techniques, *in* Olsen, K.H. ed., Continental rifts: Evolution, structure, tectonics: Amsterdam, Elsevier, p. 61–92. [2] [10]

Braile, L.W., Keller, G.R., Wendlandt, R.F., Morgan, P., and Khan, M.A., 1995, The East African rift system, *in* Olsen, K.H., ed., Continental rifts: evolution, structure, tectonics: Elsevier, p. 213–231. [2] [9]

Branson, J.C., Moss, F.J., and Taylor, F.J., 1976, Deep crustal reflection seismic test survey, Mildura, Victoria, and Broken Hill, NSW, 1968, Bureau of Mineral Resources, Australia, report 183. [6] [7]

Breivik, A.J., Mjelde, R., Grogan, P., Shimamura, H., Murai, Y., Nishimura, Y., and Kuwano, A., 2002, A possible Caledonide arm through the Barents Sea imaged by OBS data: Tectonophysics, v. 355, p. 67–97, doi:10.1016/S0040-1951(02)00135-X. [2] [9]

Brewer, J.A., and Oliver, J.E., 1980, Seismic reflection studies of deep crustal structure: Annual Reviews of Earth and Planetary Science, v. 8, p. 205–230, doi:10.1146/annurev.ea.08.050180.001225. [2] [6] [7]

Brewer, J.A., Smithson, S.B., Oliver, J.E., Kaufman, S., and Brown, L.D., 1980, The Laramide orogeny: evidence from COCORP seismic reflection profiles in the Wind River mountains, Wyoming: Tectonophysics, v. 62, p. 165–189, doi:10.1016/0040-1951(80)90191-2. [7]

Brewer, J.A., Allmendinger, L.D., Brown, L.D., Oliver, J.E., and Kaufman, S., 1982, COCORP profiling across the northern Rocky Mountain front, Part I: Laramide structure of the front in Wyoming: Geological Society of America Bulletin, v. 93, p. 1242–1252, doi:10.1130/0016-7606 (1982)93<1242:CPATRM>2.0.CO;2. [7]

British National Committee for Geodesy and Geophysics, 1948, Report of seismic work on the North German explosions, 1946–1947, Oslo. [4]

Brocher, T.M., 1994, Data report for piggyback wide-angle recordings of the 1993 San Francisco Bay Area, California, seismic refraction experiment: U.S. Geological Survey Open-File Report 94-448, 19 p. [9]

Brocher, T.M., and Moses, M.J., 1990, Wide-angle seismic recordings obtained during the TACT multichannel reflection profiling in the northern Gulf of Alaska: Menlo Park, California, U.S. Geological Survey Open-File Report 90-663, 40 p. [8]

Brocher, T.M., and Moses, M.J., 1993, Onshore-offshore wide-angle seismic recordings of the San Francisco Bay Area seismic imaging experiment (BASIX): the five-day recorder data: U.S. Geological Survey Open-File Report 93-276, 89 p. [9]

Brocher, T.M., Hart, P.E., and Carle, S.F., 1990, Feasibilty study of the seismic reflection method in Amaragosa Desert, Nye County, Nevada: U.S. Geological Survey Open-File Report 89-133, 150 p. [2] [8] [9]

Brocher, T.M., Moses, M.J., Fisher, M.A., Stephens, C.D., and Geist, E.L., 1991a, Images of the plate boundary beneath southern Alaska, *in* Meissner, R., Brown, L., Dürbaum, H.-J., Franke, W., Fuchs, K., and Seifert, F., eds., Continental lithosphere: deep seismic reflections: American Geophysical Union, Geodynamics Series, 22, p. 241–246. [2] [8]

Brocher, T.M., Klemperer, S.L., ten Brink, U., and Holbrook, W.S., 1991b, Wide-angle seismic profiling of San Francisco Bay Area faults: preliminary results from BASIX: Eos (Transactions, American Geophysical Union), v. 72, p. 446. [2] [9]

Brocher, T.M., Moses, M.J., and Lewis, S.D., 1992, Wide-angle seismic recordings obtained during seismic reflection profiling by the S.P. Lee offshore the Loma Prieta epicenter: U.S. Geological Survey Open-File Report 92-245, 63 p. [9]

Brocher, T.M., Moses, M.J., and Trehu, A.M., 1993a, Onshore-offshore wide-angle seismic recordings from central Oregon: the five-day recorder data: U.S. Geological Survey Open-File Report 93-318, 24 p. [8] [9]

Brocher, T.M., Carr, M.D., Fox, K.F., and Hart, P.E., 1993b, Seismic reflection profiling across Tertiary extensional structures in the eastern Amaragosa Desert, Basin and Range, USA: Geological Society of America Bulletin, v. 105, p. 30–46, doi:10.1130/0016-7606(1993)105 <0030:SRPATE>2.3.CO;2. [8] [9]

Brocher, T.M., McCarthy, J., Hart, P., Holbrook, W.S., Furlong, K.P., McEvilly, T.V., Hole, J.A., and Klemperer, S.L., 1994, Seismic evidence for a lower–crustal detachment beneath San Francisco Bay, California: Science, v. 265, p. 1436–1439, doi:10.1126/science.265.5177.1436. [9]

Brocher, T.M., Allen, R.M., Stone, D.B., Wolf, L.W., and Galloway, B.K., 1995a, Data report for onshore-offshore wide-angle seismic recordings in the Bering–Chukchi Sea, western Alaska and eastern Siberia: U.S. Geological Survey Open-File Report 95-650, 57 p. [2] [9]

Brocher, T.M., Davis, M.J., Clarke, S.H., Jr., and Geist, E.L., 1995b, Onshore-offshore wide-angle seismic recordings in October 1994 near Cape Blanco, Oregon. U.S. Geological Survey Open-File Report 95-819, 69 p. [9]

Brocher, T.M., Clayton, R.W., Klitgord, K.D., Bohannon, R.G., Sliter, R., McRaney, J.K.ardner, J.V., and Keene, J.B., 1995c, Multi-channel seismic-reflection profiling on the R/V *Maurice Ewing* during the Los Angeles region seismic experiment (LARSE), California: U.S. Geological Survey Open-File Report 95-228, 71 p. [9]

Brocher, T.M., Hart, P.E., Hunter, W.C., and Langenheim, V.E., 1996, Hybrid-source seismic reflection profiling across Yucca Mountain, Nevada—regional lines 2 and 3: U.S. Geological Survey Open-File Report 96-28, 110 p. [9]

Brocher, T.M., Parsons, T., Creager, K.C., Crosson, R.S., Symons, N.P., Spence, G.D., Zelt, B.C., Hammer, P.T.C., Hyndman, R.D., Mosher, D.C., Trehu, A.M., Miller, K.C., ten Brink, U.S., Fisher, M.A., Pratt, T.L., Alvarez, M.G., Beaudoin, B.C., Louden, K.E., and Weaver, C.S., 1999, Wide-angle seismic recordings from the 1998 seismic hazards investigations of Puget Sound (SHIPS), western Washington and British Columbia: U.S. Geological Survey Open-File Report 99-314, 123 p. [9]

Brocher, T.M., Pratt, T.L., Miller, K.C., Trehu, A.M., Snelson, C.M., Weaver, C.S., Creager, K.C., Crosson, R.S., ten Brink, U.S., Alvarez, M.G., Harder, S.H., and Isudeh, I., 2000, Report for explosion and earthquake data acquired in the 1999 seismic hazards investigation of Puget Sound (SHIPS), Washington: U.S. Geological Survey Open-File Report 00-318, 81 p. [9]

Brocher, T.M., Parsons, T., Blakely, M.I., Christensen, N.I., Fisher, M.A., Wells, R.E., and the SHIPS Working Group, 2001, Upper crustal structure in Puget Lowland, Washington: results from 1998 Seismic Hazards Investigation of Puget Sound: Journal of Geophysical Research, v. 106, p. 13,541–13,564, doi:10.1029/2001JB000154. [2] [9]

Brocher, T.M., Pratt, T.L., Spence, G.D., Riedel, M., and Hyndman, R.D., 2003, Wide-angle seismic recordings from the 2002 Georgia Basin geohazards initiative, northwestern Washington and British Columbia: Menlo Park, California, U.S. Geological Survey Open-File Report 03-160, 34 p. [10]

Brocher, T.M., Fuis, G.S., Lutter, W.J., Christensen, N.I., and Ratchkovski, N.A., 2004a, Seismic velocity models for the Denali fault zone along the Richardson highway, Alaska: Bulletin of the Seismological Society of America, v. 94, p. S85–S106, doi: 10.1785/0120040615. [2] [8]

Brocher, T.M., Moses, M.J., and Lewis, S.D., 2004b, Evidence for oceanic crust beneath the continental margin west of the San Andreas fault from onshore-offshore wide-angle seismic recordings, in Wells, R.E. ed., The Loma Prieta, California, Earthquake of October 17, 1989—Geologic Setting and Crustal Structure: U.S. Geological Survey Professional Paper, 1550-E, p. 107–125. [9]

Brockamp, B., 1931a, Die Mächtigkeit des grönländischen Inlandeises. Zeitschrift für Gletscherkunde, Bd. XXIII, Heft 4/5, p. 277–284. [3]

Brockamp, B., 1931b, Seismische Beobachtungen von Steinbruchsprengungen: Zeitschrift für Geophysik, v. 7, p. 295–317. [3] [4]

Bröckel, K. von, ed., 1994, Meteor-Berichte M0, no. 94-5. Leitstelle METEOR, Institute Meereskunde, University of Hamburg, 126 p. [6] [8]

Brown, A.R., 1970, Deep crustal reflection studies, Amadeus and Ngalia Basins, Northern Territory, 1969, Bureau of Mineral Resources, Australia, report 1962/61. [6]

Brown, L.D., 2006, A review of the seismic evidence for partial melting of the Tibetan crust from Project INDEPTH: Eos (Transactions, American Geophysical Union), v. 87, no. 36, Western Pacific Geophysical Meetong Supplement, Abstract S44A–02. [9]

Brown, L.D., Krumhanst, P.K., Chapin, C.E., Sanford, A.R., Cook, F.A., Kaufman, S., Oliver, J.E., and Schilt, F.S., 1979, COCORP seismic reflection studies of the Rio Grande rift, in Riecker, R.E., ed., Rio Grande rift: tectonics and magmatism: Washington, D.C., American Geophysical Union, p. 169–184. [2] [7]

Brown, L.D., Chapin, C.E., Sanford, A.R., Kaufman, S., and Oliver, J.E., 1980, Deep structure of the Rio Grande rift from seismic reflection profiling: Journal of Geophysical Research, v. 85, p. 4773–4800. [7]

Brown, L.D., Ando, C., Klemperer, S., Oliver, J., Kaufman, S., Czuchra, Walsh, T., and Isachsen, Y.W., 1983a, Adirondack-Appalachian crustal structure: the COCORP northeast traverse: Geological Society of America Bulletin, v. 94, p. 1173–1184. [2] [7] [8]

Brown, L.D., Serpa, L., Setzer, T., Oliver, J., Kaufman, S., Lillie, R., Steiner, D., and Steeples, D.W., 1983b, Intracrustal complexity in the U.S. mid-continent: preliminaey results from COCORP surveys in NE Kansas: Geology, v. 11, p. 25–30. [2] [8]

Brown, L.D., Barazangi, M., Kaufman, S., and Oliver, J.E., 1986, The first decade of COCORP: 1974–1984, in Barazangi, M., and Brown, L., eds., Reflection seismology: a global perspective: American Geophysical Union, Geodynamics Series, v. 13, p. 107–120. [2] [7] [8]

Brown, L.D., Zhao, W., Nelson, K.D., Hauck, M., Alsdorf, D., Ross, A., Cogan, M., Clark, M., Liu, X., and Che, J., 1996, Bright spots, structure, and magmatism in southern Tibet from INDEPTH seismic reflection profiling: Science, v. 274, p. 1688–1690. [2] [9]

Brown, R.D., Vedder, J.G., Wallace, R.E., Roth, E.F., Yerkes, R.F. Castle, R.O., Waananen, A.O., Page, R.E., and Eaton, J.P., 1967, The Parkfield-Cholame, California, earthquakes of June–August 1966—surface geologic effects, water-resources aspects, and preliminary seismic data: U.S. Geological Survey Professional Paper 579, 66 p. [9] [10]

Brueckl, E., Bodoky, T., Hegedus, E., Hrubcova, P., Gosar, A., Grad, M., Guterch, A., Hajnal, Z., Keller, G.R., Sicak, A., Sumanovac, F., Thybo, H., Weber, F., and ALP 2002 Working Group, 2003, ALP 2002 seismic experiment: Czech Academy of Sciences, Prague, Czech Republic, Studia in Geophysica et Geodaetica, v. 47, p. 671–679, doi:10.1023/A:1024780022139. [2] [10]

Brueckl, E.P, Bleibinhaus, F., Gosar, A., Grad, M., Guterch, A., Hrubcova, P., Keller, G.R., Majdanski, M., Sumanovac, F., Tiira, T., Yliniemi, J., Hegedus, E., and Thybo, H., 2007, Crustal structure due to collisional and escape tectonics in the Eastern Alps region based on profiles Alp01 and Alp02 from the ALP 2002 seismic experiment: Journal of Geophysical Research, v. 112, B06308, doi:10.1029/2006JB004687. [10]

Brueckl, E., Behm, M., Decker, K., Grad, M., Guterch, A., Keller, G.R., and Thybo, H., 2010, Crustal structure and active tectonics in the Eastern Alps: Tectonics, v. 29, TC2011, doi:10.1029/2009TC002491. [10]

Brun, J.-P., Wenzel, F., and the ECORS-DEKORP team, 1991, Crustal scale structure of the southern Rhine Graben from ECORS-DEKORP seismic reflection data: Geology, v. 19, p. 758–762, doi:10.1130/0091-7613(1991)019<0758:CSSOTS>2.3.CO;2. [2] [8]

Brunet, M.-F., Volozh, Y.A., Antipov, M.E., and Lobkovsky, L.I., 1999, The geodynamic evolution of the Precaspian Basin (Kazakhstan) along a north-south section, in Stephenson, R.A., Wilson, M., and Starostenko, V.I., eds., EUROPROBE GeoRift, volume 2: Intraplate tectonics and basin dynamics of the East European Craton and its margins: Tectonophysics, v. 313, p. 85–106. [9]

Bullard, E.C., and Gaskell, T.F., 1941. Submarine seismic investigations: Royal Society of London Proceedings, ser. A, no. 971, v. 177, p. 476–499. [2] [3] [5]

Bullard, E.C., Gaskell, T.F., Harland, W.B., and Kerr-Grant, C., 1940, Seismic investigations on the Palaeozoic floor of east England: Royal Society of London Philosophical Transactions, ser. A, v. 239, p. 29–94. [2] [3] [5]

Bullen, K.E., 1939, The crustal structure of the New Zealand Region as inferred from Studies of Seismic Waves, Proceedings of the 6th Pacific Scientific Congress, p. 103. [3]

Bullen, K.E., 1947, Introduction to the theory of seismology: Cambridge, U.K., Cambridge University Press. [2] [10]

Bullen, K.E., and Bolt, B.A., 1985, An introduction to the theory of seismology, 4th ed.: Cambridge, U.K., Cambridge University Press. [2] [10]

Bunce, E.T, Crampin, S., Hersey, J.B., and Hill, M.N., 1964, Seismic refraction observations on the continental boundary west of Britain: Journal of Geophysical Research, v. 69, no. 18, p. 3853–3864, doi:10.1029/JZ069i018p03853. [2] [6]

Bunch, A.W.H., 1980, Crustal development of the Reykjanes Ridge from seismic refraction, in Jacoby, W., Björnsson, A., and Möller, D., eds., Iceland: Evolution, active tectonic, and structure: Journal of Geophysics, v. 47, p. 261–264. [2] [7]

Buness, H., 1990, A compilation of data from the 1983 European Geotraverse experiment from the Ligurian Sea to the Southern Alps: Open-File Report, Institut für Geophysik, FU Berlin, Germany, 75 p. [8]

Bungenstock, H., Closs, H., and Hinz, K., 1966, Seismische Untersuchungen im nördlichen Teil des Arabischen Meeres (Golf von Oman): Erdöl und Kohle, Erdgas, Petrochemie, 19 (4), p. 237–243. [6]

Burek, B., and Dueker, K., 2005, Lithospheric stratigraphy beneath the Southern Rocky Mountains, USA, in Karlstrom, K.E., and Keller, G.R., eds., The Rocky Mountain Region: An evolving lithosphere—tectonics, geochemistry, and geophysics: Washington, D.C., American Geophysical Union Geophysical Monograph 154, p. 317–328. [9]

Burkhardt, H., and Vees, R., 1975, The Afar 1972 seismic source signals: Generation, control and comparison with profile records, in Pilger, A., and Rösler, A., eds., Afar Depression of Ethiopia, Schweizerbart, Stuttgart, 108–113. [7]

Byerly, P., 1946, The seismic waves from the Prot Chicago explosion: Bulletin of the Seismological Society of America, v. 36, p. 331–348. [2] [4]

Byerly, P., and Wilson, T.J., 1935a, The central California earthquakes of May 16, 1933, and June 7, 1934: Bulletin of the Seismological Society of America, v. 25, p. 223–246. [3]

Byerly, P., and Wilson, T.J., 1935b, The Richmond quarry blast of Aug. 16, 1934: Bulletin of the Seismological Society of America, v. 25, p. 259–268. [2] [3]

Byrne, G.F., Jacob, A.W.B., Mechie, J., and Dindi, E., 1997, Seismic structure of the upper mantle beneath the southern Kenya rift from wide-angle data, in Fuchs, K., Altherr, R., Müller, B., and Prodehl, C., eds., Structure and dynamic processes in the lithosphere of the Afro-Arabian rift system: Tectonophysics, v. 278, p. 243–260. [9]

Caielli, G., Capizzi, P., Corsi, A., de Franco, R., Luzio, D., de Luca, L., and Vitale, M., 2003, Wide-angle sea-land connections as an integration of the CROP MARE 2 project, in Scrocca, D., Doglioni, C., Innocenti, F., Manetti, P., Mazzotti, A., Bertelli, L., Burbi, L., and D'Offizi, S., eds., CROP Atlas: seismic reflection profiles of the Italian crust: Memorie Descrittive della Carta Geologica d'Italia, Roma, 62, p. 55–74. [9]

Calvert, A.J., and Potts, C.G., 1985, Seismic evidence for hydrothermally altered mantle beneath old crust in the Tydeman fracture zone: Earth and Planetary Science Letters, v. 75, p. 439–449, doi:10.1016/0012-821X(85)90187-6. [2] [8]

Camelbeeck, T., and Iranga, M.D., 1996, Deep crustal earthquakes and active faults along the Rukwa trough, eastern Africa: Geophysical Journal International, v. 124, p. 612–630, doi:10.1111/j.1365-246X.1996.tb07040.x. [9]

Canales, J.P., Detrick, R.S., Bazin, S., Harding, A.J., and Orcutt, J.A., 1998, Off-axis crustal thickness across and along the East Pacific Rise within the MELT area: Science, v. 280, p. 1218–1221, doi:10.1126/science.280.5367.1218. [2] [9]

Canales, J.P., Detrick, R.S., Lin, J., Collins, J.A., and Toomey, D.R., 2000, Crustal and upper mantle seismic structure beneath the rift mountains and across a non–transform offset at the Mid-Atlantic Ridge (35°N): Journal of Geophysical Research, v. 105, p. 2699–2719, doi:10.1029/1999JB900379. [2] [9]

Canales, J.P., Ito, G., Detrick, R.S., and Sinton, J., 2002, Crustal thickness along the western Galapagos Spreading Center and compensation of the Galapagos Swell: Earth and Planetary Science Letters, v. 203, p. 311–327, doi:10.1016/S0012-821X(02)00843-9. [9]

Cann, J.R., 1970, New model of the structure of the oceanic crust: Naturre, v. 226, p. 928–930, doi:10.1038/226928a0. [6]

Carbonell, R., Perez-Estaun, A., Gallart, I., Díaz, J., Kashubin, S., Mechie, J., Stadtlander, R., Schulze, A., Knapp, J., and Morozov, A., 1996, Crustal root beneath the Urals: wide-angle seismic evidence: Science, v. 274, p. 222–224, doi:10.1126/science.274.5285.222. [9]

Carbonell,R., Sallares, V., Pous, J., Danobeitia, J.J., Queralt, P., Ledo, J,.J., and Garcia Duenas, V., 1998a, A multidisciplinary geophysical study in the Betic chain (southern Iberian Peninsula), in Klemperer, S.L., and Mooney, W.D., eds., Deep seismic probing of the continents, II: a global survey: Tectonophysics, v. 288, p. 137–152. [2] [9]

Carbonell, R., Lecerf, D., Itzin, M., Gallart, J., and Brown, D., 1998b, Mapping the Moho beneath the southern Urals with wide-angle reflections: Geophysical Research Letters, v. 25, p. 4229–4232, doi:10.1029/1998GL900107. [9]

Carbonell, R., Gallart, I., and Torne, J., 2000a, Deep seismic profiling of the continents and their margins: Tectonophysics, v. 329, 359 p. [2] [9]

Carbonell, R., Gallart, I., Perez-Estaun, A., Díaz, J.., Kashubin, S., Mechie, J., Wenzel F., and Knapp, J., 2000b, Seismic wide-angle constraints on the crust of the southern Urals: Journal of Geophysical Research, v. 105, p. 13,755–13,777, doi:10.1029/2000JB900048. [2] [9]

Carbonell, R., Simancas, F., Juhlin, C., Pous, J., Perez-Estaun, A., Gonzales-Lodeiro, F., Munoz, G., Heise, W., and Ayarza, P., 2004, Geophysical evidence of a mantle derived intrusion in SW Iberia: Geophysical Research Letters, v. 31(11), L11601, p. 1–4, doi:10.1029/2004GL019684. [2] [10]

Carbotte, S.M., 2001, Mid-ocean ridge seismic structure, in Steele, J., Thorpe, S., and Turekian, K., eds., Encyclopedia of Ocean Sciences: Amsterdam, Academic Press, Elsevier, p. 1788–1798. [10]

Carder, D.S., Tocher, D., Bufe, C., Stewart, S.W., Eisler, J., and Berg, E., 1967, Seismic wave arrivals from LONGSHOT, 0°–27°: Bulletin of the Seismological Society of America, v. 57, p. 573–590. [6]

Carder, D.S., Qamar, A., and McEvilly, T.V., 1970, Trans-California seismic profile—Pahute Mesa to San Francisco Bay: Bulletin of the Seismological Society of America, v. 60, p. 1829–1846. [6]

Carlson, R.L., 1998, Seismic velocities in the uppermost oceanic crust: age dependence and the fate of layer 2A: Journal of Geophysical Research, v. 103, p. 7069–7077, doi:10.1029/97JB03577. [7]

Carmichael, D., Carpenter, G., Hubbard, A., McCamy, K., and McDonald, W., 1973, A recording ocean bottom seismograph: Journal of Geophysical Research, v. 78, p. 8748–8750, doi: 10.1029/JB078i035p08748. [7]

Carton, H., Singh, S.C., Hirn, A., Bazin, S., de Voogd, B., Vigner, A., RIcolleau, A., Cetin, S., Oçakoglu, N., Karakoç, F., Sevilgen, V., 2007, Seismic imaging of the three-dimensional architecture of the Cinarcik Basin along the North Anatolian Fault: Journal of Geophysical Research, v. 112, B06101, p. 1–17. doi:10.1029/2006JB004548. [10]

Cassinis, R., 1986, The geophysical exploration of the upper crust from the Ligurian coast to the northern margin of the Po valley: problems and results, in Galson, D.A., and Mueller, St., eds., The European Geotraverse, Part 2: Tectonophysics, v. 128, p. 381–394. [8]

Cassinis, R., Finetti, I., Morelli, C., Steinmetz, L., and Vecchia, O., 1969, Deep seismic refraction research on Sicily: Bollettino di Geofisica Teorica e Applicata, v. 11, p. 140–160. [6] [8]

Catchings, R.D., and Kohler, W.M., 1996, Reflected seismic waves and their effect on strong shaking during the 1989 Loma Prieta, California, earthquake: Bulletin of the Seismological Society of America, v. 86, p. 1401–1416. [9]

Catchings, R.D., and Mooney, W.D., 1988a, Crustal structure of the Columbia Plateau; evidence for continental rifting: Journal of Geophysical Research, v. 93, p. 459–471, doi:10.1029/JB01p00459. [2] [8] [9]

Catchings, R.D., and Mooney, W.D., 1988b, Crustal structure of East-Central Orogon: relation between Newberry volcano and regional crustal structure: Journal of Geophysical Research, v. 93, p. 10,081–10,094, doi:10.1029 /JB093iB09p10081. [8] [9]

Catchings, R.D., and Mooney, W.D., 1991, Basin and Range crustal and upper mantle structure, northwest to central Nevada: Journal of Geophysical Research, v. 96, p. 6247–6267, doi:10.1029/91JB00194. [8]

Catchings, R., and PASSCAL Working Group, 1988, The 1986 PASSCAL Basin and Range lithospheric seismic experiment: Eos (Transactions, American Geophysical Union), v. 69, p. 593, 596–598, doi:10.1029/88EO00174. [2] [8]

Catchings, R.D., M.J., Rymer, M.R. Goldman, J.A. Hole, R. Huggins, and C. Lippus, 2002, High-resolution seismic velocities and shallow structure of the San Andreas Fault Zone at Middle Mountain, Parkfield, California: Bulletin of the Seismological Society of America, v. 92, p. 2493–2503, doi:10.1785/0120010263. [10]

Catchings, R.D., Goldman, M.R., Rymer, M.J., Gandhok, G., and Fuis, G.S., 2003, Data report for the main line of PSINE seismic survey across the San Andreas fault and SAFOD site near Parkfield, California: U.S. Geological Survey Open-File Report 03-84, 42 p. [10]

Cernobori, L., Hirn, A., McBride, J.H., Nicolich, R., Petronio, L., Romanelli, M., and STREAMERS/PROFILES Working Group, 1996, Crustal image of the Ionian basin and its Calabrian margins: Tectonophysics, v. 264, p. 175–189, doi:10.1016/S0040-1951(96)00125-4. [9]

Červený, V., 1979, Ray theoretical seismograms for laterally inhomogeneous structures: Journal of Geophysics, v. 46, p. 335–342. [2] [7] [8] [10]

Červený, V., 1985, Gaussian-beam synthetic seismograms: Journal of Geophysics, v. 58, p. 44–72. [2] [9] [10]

Červený, V., and Horn, F., 1980, The ray seismic method and dynamic ray tracing system for three-dimensional inhomogeneous media: Bulletin of the Seismological Society of America, v. 70, p. 47–77. [2] [7] [8] [10]

Červený, V., and Pšenčik, I., 1984a, Data set 1: Synthetic record sections for a 2-D laterally inhomogeneous structure and explanatory notes, in Finlayson, D.M., and Ansorge, J., eds., Workshop proceedings: interpretation of seismic wave propagation in laterally heterogeneous structures: Bureau of Mineral Resources, Geology and Geophysics, report 258, Canberra, Australia, p. 3–14. [8]

Červený, V., and Pšenčik, I., 1984b, Data set 1: Model Zurich, computation of synthetic record sections, in Finlayson, D.M., and Ansorge, J., eds., Workshop proceedings: interpretation of seismic wave propagation in laterally heterogeneous structures: Bureau of Mineral Resources, Geology and Geophysics, report 258, Canberra, Australia, p. 15–39. [8]

Červený, V., and Pšenčik, I., 1984c, Numerical modelling of seismic wave fields in 2-D laterally varying layered structures by the ray method, in Engdahl, E.R., ed., Documentation of earthquake algorithms: Boulder, Colorado, World Data Center (A) for Solid Earth Geophysics, Rep. S-35, p. 36. [7] [9]

Červený, V., Molotkov, I.A., and Pšenčik, I., 1977, Ray method in seismology: Univerzita Karlova, Praha, 214 p. [2] [7] [8] [10]

Chabert, A., Ravaut, C., Readman, P.W., O'Reilly, B.M., and Shannon, P.M., 2006, Crustal structure of the Hatton Basin (North Atlantic) from a wide-angle seismic experiment: European Geophysical Union, 2006 Annual Meeting, Geophysical Research Abstracts, v. 8, 07929. [2] [10]

Chandra, N.N., and Cumming, G.L., 1972, Seismic refraction studies in western Canada: Canadian Journal of Earth Science, v. 9, p. 1099–1109. [6]

Chapman, C.H. 2004, Fundamentals of seismic wave propagation: Cambridge, U.K. Cambridge University Press, 608 p. [2] [10]

Chapman, C.H., and Orcutt, J.A., 1980, Inversion of seismic refraction data: Eos (Transactions, American Geophysical Union) v. 61, p. 304. [8]

Chapman, C.H., and Orcutt, J.A., 1985, Least-squares fitting of marine seismic refraction data: Geophysical Journal of the Royal Astronomical Society, v. 82, p. 339–374. [8]

Charlier, Ch., 1947, Deuxième rapport sur l'explosion d'Heligoland: Service Séismologique et gravimétrique, Série S, no. 3, Observatoire Royal de Belgique a Uccle, 27 p. [4]

Charvis, P., and Operto, S., 1999, Structure of the Cretaceous Kerguelen volcanic province (southern Indian Ocean) from wide-angle seismic data: Journal of Geodynamics, v. 28, p. 51–71, doi:10.1016/S0264-3707 (98)00029-5. [9]

Chen, B., and Gao, W., 1988, Crusatl structure along the Tiajiwen-Shayan DSS profile, in Developments in the Research of Deep Structures of China's Continent: Beijing, Geological Publishing House, p. 152–168. [8]

Chen, X., Wu, Y., Du, P., Li, J., Wu, Y., Jiang, G., and Zhao, J., 1988, Crustal velocity structure at the two sides of Longmenshan Tectonic Belt, in Developments in the Research of Deep Structures of China's Continent: Beijing, Geological Publishing House, p. 97–113. [8]

Cheung, H.P.Y., and Clowes, R.M., 1981, Crustal structure from P- and S-wave analyses: ocean bottom seismometer results in the northeast Pacific: Geophysical Journal of the Royal Astronomical Society, v. 65, p. 47–73. [2] [7]

Chian, D., Louden, K.E., Minshull, T.A., and Whitmarsh, R.B., 1999, Deep structure of the ocean-continent transition in the southern Iberia Abyssal Plain from seismic refraction profiles: Ocean Drilling Programme (Legs 149 and 197) transect: Journal of Geophysical Research, v. 104, p. 7443–7462, doi:10.1029/1999JB900004. [9]

Choudhury, M., Giese, P., and de Visintini, G., 1971, Crustal structure of the Alps—Some general features from explosion seismology: Bollettino di Geofisica Teorica e Applicata, v. 13, p. 211–240. [6]

Christensen, N.I., and Mooney, W.D., 1995, Seismic velocity structure and composition of the continental crust: Journal of Geophysical Research, v. 100, p. 9761–9788, doi:10.1029/95JB00259. [2] [8] [9] [10]

Christensen, N.I., and Salisbury, M.H., 1975, Structure and constitution of the lower oceanic crust: Reviews of Geophysics, v. 13, p. 57–86, doi:10.1029 /RG013i001p00057. [2]

Christensen, N.I., and Salisbury, M.H., 1982, Lateral heterogeneity in the seismic structure of the oceanic crust inferred from velocity studies in the Bay of Islands ophiolite, Newfoundland: Geophysical Journal of the Royal Astronomical Society, v. 68, p. 675–688. [7] [8]

Christensen, N.I., and Smewing, J.D., 1981, Geology and seismic structure of the northern section of the Oman ophiolite: Journal of Geophysical Research, v. 86, p. 2545–2555, doi:10.1029/JB086iB04p02545. [7]

Christeson, G.L., Shaw, P.R., and Garmany, J.D., 1997, Shear and compressional wave structure of the East Pacific Rise, 9°–10°N: Journal of Geophysical Research, v. 102, p. 7821–7835, doi:10.1029/96JB03901. [2] [9]

Christeson, G.L., McIntosh, K.D., Shipley, T.H., Flueh, E.R., and Goedde, H., 1999, Structure of the Costa Rica convergent margin, offshore Nicoya Peninsula: Journal of Geophysical Research, v. 104, p. 25,443–25,468, doi:10.1029/1999JB900251. [2] [9]

Christeson, G., Nakamura, Y., Buffler, R., Morgan, J., and Warner, M., 2001, Deep crustal structure of the Chicxulub impact crater: Journal of Geophysical Research, v. 106, p. 21,751–21,769, doi:10.1029/2001JB000337. [9]

Chuaqui, L., and McEvilly, T.V., 1968, Detailed crustal structure within the central California large-scale seismic array, *in* Annual Report, Seismographic Station, University of California, Berkeley. [6]

Chulick, G.S., 1997, Comprehensive seismic survey database for developing three-dimensional models of the Earth's crust: Seismological Research Letters, v. 68, p. 734–742. [10]

Chulick, G.S., and Mooney, W.D., 2002, Seismic structure of the crust and uppermost mantle of North America and adjacent oceanic basins: a synthesis: Bulletin of the Seismological Society of America, v. 92, p. 2478–2492, doi:10.1785/0120010188. [2] [9] [10]

Clark, S.A., Zelt, C.A., Magnani, M.B., and Levander, A., 2008, Characterizing the Caribbean–South American plate boundary at 64°W using wide-angle data: Journal of Geophysical Research, v. 113, B07401, doi:10.1029/2007JB005329. [10]

Cleary, J., 1967, P times to Australian stations from nuclear explosions: Bulletin of the Seismological Society of America, v. 57, p. 773–781. [5]

Cleary, J., 1973, Australian crustal structure, *in* Mueller, S., ed., The structure of the earth's crust, based on seismic data: Tectonophysics, v. 20, p. 241–248. [2] [5] [6]

Clee, T.E., Barr, KG., and Berry, M.J., 1974, Fine structure of the crust near Yellowknife: Canadian Journal of Earth Science, v. 11, p. 1534–1549. [2] [6]

Clément, C., Hirn, A., Charvis, P., Sachpazi, M., and Marnelis, F., 2000, Seismic structure and the active Hellenic subduction in the Ionian islands: Tectonophysics, v. 329, p. 141–156, doi:10.1016/S0040-1951(00)00193-1. [2] [9]

Clément, C., Sachpazi, M., Charvis, P. Graindorge, M. Laigle, M., A. Hirn, A., and G. Zafiropoulos, G., 2004, Reflection-refraction seismics in the Gulf of Corinth: hints at deep structure and control of the deep marine basin: Tectonophysics, v. 391, p. 85–95. [9]

Clement, W.P., Carbonell, R., and Smithson, S.B., 1994, Shear-wave splitting in the lower crust beneath the Archean crust of southwest Greenland: Tectonophysics, v. 232, p. 195–210, doi:10.1016/0040-1951(94)90084-1. [2] [8]

Clitheroe, G., Gudmundsson, O., and Kennett, B.L.N., 2000, The crustal thickness of Australia: Journal of Geophysical Research, v. 105, p. 13,697–13,713, doi:10.1029/1999JB900317. [2] [9]

Closs, H., 1969, Explosion seismic studies in western Europe, *in* Hart, P.J., ed., The earth's crust and upper mantle: American Geophysical Union, Geophysical Monograph 13, p. 178–188. [2] [5] [6]

Closs, H, 1972, Cruise to the Norwegian Sea with F.S. Planet, August 4–21, 1969, Meteor Forschungsergebnisse, Deutsche Forschungsgemeinschaft,

Borntraeger, Berlin-Stuttgart, Reihe C Geologie und Geophysik, C8, p. 1–9. [2] [6]

Closs, H., 1974, Die geophysikalische Reichsaufnahme und ihre Vorgeschichte, *in* Birett, H., Helbig, K., Kertz, W., and Schmucker, U., eds., Zur Geschichte der Geophysik: Berlin-Heidelberg-New York, Springer, p. 115–130. [3]

Closs, H., and Behnke, C., 1961, Fortschritte der Anwendung seismischer Methoden in der Erforschung der Erdkruste: Geologische Rundschau, v. 51, p. 315–330, doi:10.1007/BF01820003. [2] [5] [10]

Closs, H., and Behnke, C., 1963, Progress in the use of seismic methods in the exploration of the Earth's crust: International Geology Review, v. 5, no. 8, p. 945–956, doi:10.1080/00206816309474688. [2] [5] [10]

Closs, H., and Labrouste, Y., 1963, Séismologie: Recherches séismologiques dans les Alpes occidentals au moyen de grandes explosions een 1956, 1958 et 1960: Centre National de la Recherche Scientifique, Mémoir Collectif, Année Géophysique Internationale, Série XII, Fasc. 2, 241 p. [2] [5]

Closs, H., Dietrich, G., Hempel, G., Schott, W., and Seibold, E., 1968, "Atlantische Kuppelfahrt" 1967 mit dem Forschungsschiff METEOR: Meteor Forsch. Erg. 5, p. 1–71 Reihe A. [2] [6] [9]

Closs, H., Bungenstock, H, and Hinz, K, 1969a, Ergebnisse seismischer Untersuchungen in nördlichen Arabischen Meer, ein Beitrag zur Internationalen Indischen Ozean-Expedition: Meteor Forschungsergebnisse, Deutsche Forschungsgemeinschaft, Borntraeger, Berlin-Stuttgart, Reihe C Geologie und Geophysik, C2, p. 1–28. [2] [6]

Closs, H, Dietrich, G, Hempel, G, Schott, W, and Seibold, E, 1969b, Atlantische Kuppenfahrten 1967 mit dem Forschungsschiff "Meteor": Meteor Forschungsergebnisse, Deutsche Forschungsgemeinschaft, Borntraeger, Berlin-Stuttgart, Reihe A Allgemeines, Physik und Chemie des Meeres, A5, p. 1–71. [6]

Closs, H., Hinz, K., Maucher, A, 1972, Mittelmeerfahrten 1969c (Nr. 17) und 1971 (Nr. 22) des Forschungsschiffes "Meteor": Meteor Forschungsergebnisse, Deutsche Forschungsgemeinschaft, Borntraeger, Berlin-Stuttgart, Reihe A Allgemeines, Physik und Chemie des Meeres, A10, p. 31–50. [2] [6]

Clowes, R.M., 1993, Variations in continental crustal structure in Canada from LITHOPROBE seismic reflection and other data: Tectonophysics, v. 219, p. 1–27, doi:10.1016/0040-1951(93)90284-Q. [2] [8] [9]

Clowes, R.M., editor, 1997, LITHOPROBE Phase V proposal—evolution of a continent revealed. LITHOPROBE Secretariat: University of British Columbia, Vancouver, B.C., 292 p. [8]

Clowes, R.M., and Green, A.G., editors, 1994, Seismic reflection profiling of the continents and their margins: Tectonophysics, v. 232, 450 p. [2] [9]

Clowes, R.M., and Kanasewich, E.R., 1970, Seismic attenuation and the nature of reflecting horizons within the crust: Journal of Geophysical Research, v. 75, p. 6693–6705, doi:10.1029/JB075i032p06693. [6]

Clowes, R.M., and Kanasewich, E.R., 1972, Digital filtering of deep crustal seismic reflections: Canadian Journal of Earth Science, v. 9, p. 434–451. [6]

Clowes, R.M., Kanasewich, E.R., and Cumming, G.L., 1968, Deep crustal seismic reflections at near-vertical incidence: Geophysics, v. 33, p. 441–451, doi:10.1190/1.1439942.[2] [6]

Clowes, R.M., Green, A.G., Yorath, C.J., Kanasewich, E.R., West, G.F., and Garland, G.D., 1984, LITHOPROBE—a national program for studying the third dimension of geology: Journal of the Canadian Society of Exploration Geophysicists, v. 20, p. 23–39. [8]

Clowes, R.M., Spence, G.D., Ellis, R.M., and Waldron, D.A., 1986, Structure of the lithosphere in a young subduction zone: results from reflection and refraction studies, *in* Barazangi, M., and Brown, L., eds., Reflection seismology: the continental crust: American Geophysical Union, Geodynamics Series, v. 14, p. 313–321. [8] [9]

Clowes, R.M., Brandon, M.T., Green, A.G., Yorath, C.J., Sutherland Brown, A., Kanasewich, E.R., and Spencer, C., 1987a, LITHOPROBE—southern Vancouver Island: Cenozoic subduction complex imaged by deep seismic reflections: Canadian Journal of Earth Science, v. 24, p. 31–51. [8] [9]

Clowes, R.M., Yorath, C.J., and Hyndman, R.D., 1987b, Reflection mapping across the convergent margin of western Canada: Geophysical Journal of the Royal Astronomical Society, v. 89, p. 79–84. [8]

Clowes, R.M., Cook, F.A., Green, A.G., Keen, C.E., Ludden, J.N., Percival, J.A., Quinlan, G.M., and West, G.F., 1992, LITHOPROBE—new prespectives on crustal evolution: Canadian Journal of Earth Science, v. 29, p. 1813–1864. [2] [8] [9]

Clowes, R.M., Zelt, C.A., Amor, J.R. and Ellis, R.M., 1995, Lithospheric structure in the southern Canadian Cordillera from a network of seismic refraction lines: Canadian Journal of Earth Science v. 32, p. 1485–1513, doi:10.1139/e95-122. [2] [8] [9]

Clowes, R.M., Calvert, A.J., Eaton, D.W., Hajnal, Z., Hall, J., and Ross, G.M., 1996, LITHOPROBE reflection studies of Archean and Proterozoic crust in Canada: Tectonophysics, v. 264, p. 65–88, doi:10.1016/S0040-1951 (96)00118-7. [9]

Clowes, R.M., Cook, F.A., Hajnal, Z., Hall, J., Lewry, J., Lucas, S., and Wardle, R., 1999, Canada's LITHOPROBE project—collaborative multidisciplinary geoscience research leads to new understanding of continental evolution: Episodes, v. 22, p. 3–20. [2] [9]

Clowes, R.M., Burianyk, M.J.A., Gorman, A.R., and Kanasewich, E.R., 2002, Crustal velocity structure from SAREX, the Southern Alberta Refraction Experiment: Canadian Journal of Earth Science, v. 39, p. 351–373, doi:10.1139/e01-070. [2] [9]

Clowes, R.M., Hammer, P.T.C., Fernandez-Viejo, G., and Welford, J.K., 2005, Lithospheric structure in northwestern Canada from Lithoprobe seismic refraction and related studies: a synthesis: Canadian Journal of Earth Science, v. 42, p. 1277–1293, doi:10.1139/e04-069. [2] [9]

Cohen, T.J., and Meyer, R.P., 1966, The mid-continent gravity high: gross crustal structure, in Steinhart, J.S., and Smith, T.J., eds., The earth beneath the continents: American Geophysical Union, Geophysical Monograph 10, p. 141–165. [6]

Colburn, R.H., and Walter, A.W., 1984, Data report for two seismic-refraction profiles crossing the epicentral region of the 1983 Coalinga, California, earthquakes: Menlo Park, California, U.S. Geological Survey Open-File Report 84-643, 58 p. [8]

Collette, B.J., Lagaay, R.A., Ritsema, A.R., and Schouten, J.A., 1970, Seismic investigations in the North Sea, 3 to 7: Geophysical Journal of the Royal Astronomical Society, v. 19, p. 183–199. [6]

Collier, J.S., and Singh, S.C., 1997, Detailed structure of the top of the melt body beneath the East Pacific Rise at 9°40′N from waveform inversion of seismic reflection data: Journal of Geophysical Research, v. 102, p. 20,287–20,304, doi:10.1029/97JB01514. [8]

Collier, J.S., and Singh, S.C., 1998, Poisson's ratio structure of young oceanic crust: Journal of Geophysical Research, v. 103, p. 20,981–20,996, doi:10.1029 /98JB01980. [8]

Collier, J.S., and Sinha, M.C., 1992, Seismic mapping of a magma chamber beneath the Valu Fa Ridge, Lau Basin: Journal of Geophysical Research, v. 97, p. 14,031–14,053, doi:10.1029/91JB02751. [8]

Collier, J.S., Minshull, T.A., Kendall, J.-M., Whitmarsh, R.B., Rümpger, G., Joseph, P., Samson, P., Lane, C.I., Sansom, V., Vermeesch, P.M., Hammond, J., Wookey, J., Teanby, N., Ryberg, T.M., and Dean, S.M., 2004, Rapid continental breakup and microcontinent formation in the western Indian Ocean: Eos (Transactions, American Geophysical Union), v. 85, no. 46, p. 481, 487. [2] [10]

Collins, C.D.N., 1978, Crustal structure of the central Bowen basin, Queensland: BMR Journal of Australian Geology and Geophysics, v. 3, p. 203–209. [2] [7]

Collins, C.D.N., 1991, The nature of the crust-mantle boundary under Australia from seismic evidence, in Drummond, B.J., ed., The Australian Lithosphere: Geological Society of Australia Special Publication 17, p. 67–80. [2] [9]

Collins, C.D.N., Drummond, B.J., and Nicoll, M.G., 2003, Crustal thickness pattern in the Australian continent: Geological Society of Australia Special Paper 273, p. 121–128. [2] [9] [10]

Collins, J.A., Brocher, T.M., and Karson, J.A., 1986, Two-dimensional seismic reflection modeling of the inferred fossil oceanic crust/mantle transition in the Bay of Islands ophiolite: Journal of Geophysical Research, v. 91, p. 12,520–12,538, doi:10.1029/JB091iB12p12520. [7] [8]

Colombi, B., Giese, P., Luongo, G., Morelli, C., Riuscetti, M., Scarascia, S., Schütte, K.-G., Strowald, J., and de Vissintini, G., 1973, Preliminary report on the seismic refraction profile Gargano-Salerno-Palermo-Pantelleria (1971): Bollettino di Geofisica Teorica e Applicata, v. 15, p. 225–254. [2] [7] [9]

Comas, M.C., Danobeitia, J.J., Alvarez-Marron, J., and Soto, J.I., 1995, Crustal reflections and the structure in the Alboran Basin: preliminary results of the ESCI-Alboran survey: Revista de la Sociedad Geológica de España, v 8, p. 529–542. [9]

Conrad, V., 1925, Laufzeitkurven des Tauernbebens vom 28.11.1923. Mitt. Erdb. Komm., Wien, Akad. Wiss., Neue Folge, no.59. [3]

Conrad, V., 1926, Laufzeitkurven eines alpinen Bebens: Zeitschrift für Geophysik, v. 2, p. 34–35. [3]

Conrad, V., 1928, Das Schwadorfer Beben vom 8. Oktober 1927: Gerlands Beiträge zur Geophysik, v. 20, p. 240–277. [3]

Constantinescu, P., and Cornea, I., 1972, Roumania, in Sollogub, V.B., Prosen, D., and Militzer, H., 1972, Crustal structure of central and southeastern Europe based on the results of explosion seismology (publ. in Russian 1971). English translation edited by Szénás, Gy., 1972: Geophysical Transactions, spec. ed., Müszaki Könyvkiado, Budapest, chapter 27, p. 113–117. [2] [6]

Conway, A., Dentith, M.C., Doody, J.J., and Hall, J., 1987, Preliminary interpretation of upper crustal structure across the Midland valley of Scotland from two east-west seismic-refraction profiles: Journal of the Geological Society of London, v. 144, p. 865–870, doi:10.1144/gsjgs.144.6.0865. [8]

Cook, F.A., ed., 1995, The Southern Canadian Cordillera transect of Lithoprobe: Canadian Journal of Earth Science, v. 32, p. 1483–1824, doi:10.1139 /e95-121. [2]

Cook, F.A., and Erdmer, P., eds., 2005, The Lithoprobe Slave–Northern Cordillera Lithospheric Evolution (SNORCLE) transect: Canadian Journal of Earth Science (special issue), v. 42, p. 865–1311, doi:10.1139/e05-067. [2] [9]

Cook, F.A., Albaugh, D.S., Brown, L.D., Kaufman, S., Oliver, J.E., and Hatcher, R.D., Jr., 1979a, Thin-skinned tectonics in the crystalline southern Appalachians: COCORP seismic reflection profiling of the Blue Ridge and Piedmont: Geology, v. 7, p. 563–567, doi:10.1130/0091-7613(1979)7 <563:TTITCS>2.0.CO;2. [2] [7]

Cook, F.A., McCullar, D.B., Decker, E.R., and Smithson, S.B., 1979b, Crustal structure and evolution of the southern Rio Grande Rift, in Riecker, R.E., ed., Rio Grande rift: tectonics and magmatism: Washington, D.C., American Geophysical Union, p. 195–208. [7]

Cook, F.A., Brown, L.D., Kaufman, S., Oliver, J.E., and Peterson, T.A., 1981, COCORP seismic profiling of the Appalachian orogen beneath the coastal plain of Georgia: Geological Society of America Bulletin, part I, v. 92, p. 738–748, doi:10.1130/0016-7606(1981)92<738:CSPOTA>2.0.CO;2. [7]

Cook, F.A., Coffin, K.C., Lane, L.S., Dietrich, J.R., and Dixon, J., 1987, Structure of the southeast margin of the Beaufort–Mackenzie basin, Arctic Canada, from crustal seismic-reflection data: Geology, v. 15, p. 931–935, doi:10.1130/0091-7613(1987)15<931:SOTSMO>2.0.CO;2. [8]

Cook, F.A., van der Velden, A.J., Hall, K.W., and Roberts, B.J., 1998, Tectonic delamination and subcrustal imbrication of the Precambrian lithosphere in northwestern Canada, mapped by LITHOPROBE: Geology, v. 26, p. 839–842, doi:10.1130/0091-7613(1998)026<0839:TDASIO>2.3.CO;2. [9]

Cook, F.A., van der Velden, A.J., Hall, K.W., and Roberts, B.J., 1999, Frozen subduction in Canada's Northwest Territories: LITHOPROBE deep lithospheric reflection profiling in the western Canadian Shield: Tectonics, v. 18, p. 1–24, doi:10.1029/1998TC900016. [9]

Cordoba, D., and Banda, E., 1980, Gradientes de velocidad en la corteza y manto superior de las islas baleares (Mar Mediterraneo): Geofisica Internacional, Mexico, v. 19, p. 285–303. [7]

Cordoba, D., Banda, E., and Ansorge, J., 1987, The Hercynian crust in northwestern Spain: a seismic survey: Tectonophysics, v. 132, p. 321–333, doi:10.1016/0040-1951(87)90351-9. [2] [8]

Cormier, M.H., Detrick, R.S., and Purdy, G.M., 1984, Anomalously thin crust in oceanic fracture zones: new seismic constraints from the Kane fracture zone: Journal of Geophysical Research, v. 89, p. 10,249–10,266, doi:10.1029/JB089iB12p10249. [2] [7] [8]

Cornea, I., Radulescu, F., Pompilian, A., and Sova, A., 1981, Deep seismic soundings in Romania: PAGEOPH, v. 119, p. 1144–1156, doi:10.1007 /BF00876693. [7]

Coruh, C., Costain, J.K., Hatcher, R.D., Jr., Pratt, T.L., Williams, R.T., and Phinney, R.A., 1987, Results from regional vibroseis profiling: Appalachian ultradeep core hole site study: Geophysical Journal of the Royal Astronomical Society, v. 89, p. 147–156. [2] [8]

Costain, J.K., Hatcher, R.D., Jr., and Coruh, C., 1989a, Appalachian ultradeep core hole (ADCOH) site investigation: regional seismic lines and geologic interpretation, in Bally, A.W., and Palmer, A.R., eds., The Geology of North America—An overview: Boulder, Colorado, Geological Society of America, Geology of North America, vol. A, Plate 8. [8]

Costain, J.K., Hatcher, R.D., Jr., and Coruh, C., 1989b, Appalachian ultradeep core hole (ADCOH) site investigation: regional seismic lines and geologic interpretation, in Hatcher, R.D., Jr., Thomas, W.A., and Viele, G.W., eds., The Appalachian-Ouachita orogen in the United States: Boulder, Colorado, Geological Society of America, Geology of North America, vol. F-2, Plate 12. [8]

Cotton, J.A., and Catchings, R.D., 1988, Data report for the 1984 U.S. Geological Survey central Columbia Plateau seismic-refraction experiment, Washington-Oregon: Menlo Park, California, U.S. Geological Survey Open-File Report 88-226, 56 p. [8]

Cotton, J.A., and Catchings, R.D., 1989, Data report for the 1983 U.S. Geological Survey east-central Oregon seismic-refraction experiment, Washington-Oregon: Menlo Park, California, U.S. Geological Survey Open-File Report 89-124, 30 p. [8]

Cram, Jr., I.H., 1961, A crustal structure refraction survey in south Texas: Geophysics, v. 26, p. 560–573, doi:10.1190/1.1438915. [6]

Crawford, W.C., Hildebrand, J.A., Dorman, L.M., Webb, S.C., and Wiens, D.A., 2003, Tonga ridge and Lau Basin crustal structure from seismic refraction data: Journal of Geophysical Research, v. 108, no. B4, doi:10.1029/2001JB001435. [2] [10]

Creaser, B., and Spence, G., 2005, Crustal structure across the northern Cordillera, Yukon Territory, from seismic wide-angle studies: Omineca Belt to Intermontane Belt: Canadian Journal of Earth Science, v. 42, p. 1187–1203, doi:10.1139/e04-093. [9]

Criley, E., and Eaton, J., 1978, Five-day recorder seismic system: Menlo Park, California, U.S. Geological Survey Open-File Report 78-266, 85 p. [8]

Csontos, L., Nagymarosy, A., Horvath, F., and Kovac, M., 1992, Tertiary evolution of the Intra-Carpathian area: a model: Tectonophysics, v. 208, p. 221–241, doi:10.1016/0040-1951(92)90346-8. [9]

Czuba, W., Grad, M., and Guterch, A., 1999, Crustal structure of northwestern Spitsbergen from DSS measurements: Polish Polar Research, v. 20, p. 131–148. [2] [8] [9]

Czuba, W., Grad, M., Luosto, U., Motuza, G., Nasedkin, V., and POLONAISE P5 Working Group, 2001, Crustal structure of the East European Craton along the POLONAISE'97 P5 profile: Acta Geophysica Polonica, v. 49, no. 2, p. 145–168. [9]

Czuba, W., Grad, M., Luosto, U., Motuza, G., Nasedkin, V., and POLONAISE P5 Working Group, 2002, Upper crustal seismic structure of the Mazury complex and Mazowsze massif within East European Craton in NE Poland: Tectonophysics, v. 360, p. 115–128, , doi:10.1016/S0040-1951(02)00352-9. [9]

Czuba, W., Ritzmann, O., Nishimura, Y., Grad, M., Mjelde, R., Guterch, A., and Jokat, W., 2004, Crustal structure of the continent-ocean transition zone along two deep seismic transects in northwestern Spitsbergen: Polish Polar Research, v. 25, p. 205–221. [2] [9]

Dachev, K., Petkov, I., Velchev, Ts., Andronova, E., Mikhailov, S., 1972, Bulgaria, *in* Sollogub, V.B., Prosen, D., and Militzer, H., eds., Crustal structure of central and southeastern Europe based on the results of explosion seismology (publ. in Russian 1971). English translation edited by Szénás, Gy., 1972: Geophysical Transactions, spec. ed., Müszaki Könyvkiado, Budapest, chapter 26, p. 106–112. [2] [6]

Dadlez, R., Grad, M., and Guterch, A., 2005, Crustal structure below the Polish Basin: is it composed of proximal terranes derived from Baltica?: Tectonophysics, v. 411, p. 111–128, doi:10.1016/j.tecto.2005.09.004. [9]

Dahl-Jensen, T., Thybo, H., Hopper, J., and Rosing, M., 1998, Crustal structure at the SE Greenland margin from wide-angle and normal-incidence seismic data, *in* Klemperer, S.L., and Mooney, W.D., eds., Deep seismic probing of the continents, II: a global survey: Tectonophysics, v. 288, p. 191–198. [2] [9]

Daignières, M., Gallart, J., Banda, E., and Hirn, A., 1982, Implications of the seismic structure for the orogenic evolution of the Pyrenean range: Earth and Planetary Science Letters, v. 57, p. 88–100, doi:10.1016/0012-821X(82)90175-3. [7]

Dainty, A.M., Keen, C.E., Keen, M.J., and Blanchard, J.E., 1966, Review of geophysical evidence on crust and upper mantle structure on the eastern seabord of Canada, *in* Steinhart, J.S., and Smith, T.J., eds., The earth beneath the continents: Washington, D.C., American Geophysical Union Geophysical Monograph 10, p. 349–368. [6] [7]

Daley, M.A., Ambos, E.L., and Fuis, G.S., 1985, Seismic refraction data collected in the Chugach Mountains and along the Glenn Highway in southern Alaska in 1984: Menlo Park, California, U.S. Geological Survey Open-File Report 78-266, 44 p. [8]

Daly, R.A., 1914, Igneous rocks and their origin: New York, McGraw-Hill, 561 p. [3]

Danobeitia, J.J., Arguedas, M., Gallart, J., Banda, E., and Makris, J., 1992, Deep seismic configuration of the Valencia Trough and its Iberian and Balearic borders from extensive refraction-wide angle reflection seismic profiling, *in* Banda, E., and Santanach, P., eds., Geology and Geophysics

of the Valencia Trough, Western Mediterranean: Tectonophysics, v. 302, p. 37–55. [2] [8]

Darbyshire, F.A., Bjarnason, I.Th., White, R.S., and Flóvenz, O.G., 1998, Crustal structure above the Iceland mantle plume imaged by the ICEMELT refraction profile: Geophysical Journal International, v. 135, p. 1131–1149, doi:10.1046/j.1365-246X.1998.00701.x. [2] [9]

Dash, B.P., Ball, M.M., King, G.A., Butler, L.W., and Rona, P.A., 1976, Geophysical investigation of the Cape Verde archipelago: Journal of Geophysical Research, v. 81, p. 5249–5259, doi:10.1029/JB081i029p05249. [7]

Dauphin, J.P., and Simoneit, B.R.T., 1991, The Gulf and peninsula province of the Californias: American Association of Petroleum Geologists Memoir 47, 834 p. [8]

Davey, F.J., 1987, Seismic reflection measurements behind the Hikurangi convergent margin, southern North Island, New Zealand: Geophysical Journal of the Royal Astronomical Society, v. 89, p. 443–448. [2] [8]

Davey, F.J., 2005, A Mesozoic crustal suture on the Gondwana margin in the New Zealand region: Tectonics, v. 24, no. 4, TC4006, doi:1029/2004TC001719. [2] [9]

Davey, F.J., and Broadbent, M., 1980, Seismic refraction measurements in Fiordland, southwest New Zealand: New Zealand Journal of Geology and Geophysics, v. 23, no. 4, p. 395–406. [2] [6] [7]

Davey, F.J., and Jones, L.E.A., eds., 2004, Continental lithosphere: Tectonophysics, v. 388, 297 p. [2] [9] [10]

Davey, F.J., and Lodolo, E., 1995, Crustal seismic reflection measurements across an ensialic back-arc basin (Bay of Plenty, New Zealand): Bollettino di Geofisica Teorica e Applicata, v. 37, p. 25–37. [2] [9]

Davey, F.J., and Smith, E.G.C., 1983, A crustal seismic reflection-refraction experiment across the subducted Pacific Plate under Wellington, New Zealand: Physics of the Earth and Planetary Interiors, v. 31, p. 327–33, doi:10.1016/0031-9201(83)90092-4. [2] [7]

Davey, F.J., and Stern, T.A., 1990, Crustal seismic observations across the convergent plate boundary, North Island, New Zealand: Tectonophysics, v. 173, p. 283–296, doi:10.1016/0040-1951(90)90224-V. [2] [8]

Davey, F.J., Hampton, M., Childs, J., Fisher, M., Lewis, K., and Pettinga, J.R., 1986, Structure of a growing accretionary prism, Hikurangi margin, New Zealand: Geology, v. 14, p. 663–666, doi:10.1130/0091-7613 (1986)14<663:SOAGAP>2.0.CO;2. [2] [8]

Davey, F.J., Henrys, S.A., and Lodolo, E., 1995, Asymmetric rifting in a continental back-arc environment, North Island, New Zealand: Journal Volcanology and Geothermal Research, v. 68, p. 209–238, doi:10.1016/0377-0273(95)00014-L. [9]

Davey, F.J., Henyey, T., Holbrook, W.S., Okaya, D., Stern, T.A., Melhuish, A., Henrys, S., Anderson, H., Eberhart-Phillips, D., McEvilly, T., Uhrhammer, R., Wu, F., Jirazek, G.R., Wannamaker, P.E., Caldwell, G., and Christensen, N., 1998, Preliminary results from a geophysical study across a modern, continent-continent collisional plate boundary—the Southern Alps, New Zealand, *in* Klemperer, S.L., and Mooney, W.D., eds., Deep seismic probing of the continents, II: a global survey: Tectonophysics, v. 288, p. 221–235. [2] [9]

Davies, D., and Francis, T.J.G., 1964, The crustal structure of the Seychelles Bank: Deep-Sea Research, v. 11, p. 921–927. [6]

Davy, B.R., Hoernle, K., and Werner, R. 2008, Hikurangi Plateau: Crustal structure, rifted formation, and Gondwana subduction history: Geochemistry Geophysics Geosystems, v. 9, Q07004, doi:10.1029/2007GC001855. [2] [10]

Davydova, N.I., Pavlenkova, N.I., Tulina, Yu.V., and Zverev, S.M., 1985, Crustal structure of the Barents Sea from seismic data, *in* Husebye, E.S., Johnson, G.L., and Kristoffersen, Y., eds., Geophysics of the Polar Region: Tectonophysics, v. 114, p. 213–231. [7] [8]

Dawson, P.B., and Stauber, D.A., 1986, Data report for a three-dimensional high-resolution P-velocity structural investigation of Newberry Volcano, Oregon, using seismic tomography: Menlo Park, California, U.S. Geological Survey Open-File Report 86-352, 45 p. [8]

de Quervain, A., 1931, Beitrag zur experimentellen Bestimmung der Geschwindigkeit der Erdbebenwellen in den obersten Schichten. (a) Sprengung bei Alpnach am 25. März 1922. (b) Die Registrierung der Explosion des Forts Falconara bei Spezia am 28. September 1922. Jahresber. Schweizer. Erdbebendienstes, 1931, p. 7–9. [3]

de Voogd, B., Serpa, L., and Brown, L., 1988, Crustal extension and magmatic processes: COCORP profiles from Death Valley and the Rio Grande rift: Geological Society of America Bulletin, v. 100, p. 1550–1567, doi:10.1130/0016-7606(1988)100<1550:CEAMPC>2.3.CO;2. [2] [8]

de Voogd, B., Keen, C., and Kay, W.A., 1990, Fault reactivation during Mesozoic extension in eastern offshore Canada, *in* Leven, J.H., Finlayson, D.M., Wright, C., Dooley, J.C., and Kennett, B.L.N., eds., Seismic probing of continents and their margins: Tectonophysics: v. 173, p. 567–580. [8]

de Voogd, B., Nicolich, R., Olivet, J.I., Fanucci, F., Burrus, J., Maufret, A., Pascal, G., Argnani, A., Auzende, J.M., Bernabini, M., Bois, C., Carmignani, L., Fabbri, A., Finetti, I., Galdeano, A., Gorini, C.Y., Labaume, P., Laiat, D., Patriat, P., Pinet, B., Ravat, J., Ricci Lucci, F., and Vernassa, S., 1991, First deep seismic reflection transect from the Gulf of Lions to Sardinia (ECORS-CROP profiles in Western Mediterranean): Geodynamics, v. 22, p. 265–274. [8]

de Voogd, B., Truffert, C., Chamot-Rooke, N., Huchon, P., Lallemant, S., and Le Pichon, X., 1992, Two-ship deep seismic soundings in the basin of the Eastern Mediterraean Sea (Pasiphae Cruise): Geophysical Journal International, v. 109, p. 536–552, doi:10.1111/j.1365-246X.1992.tb00116.x. [9]

Dean, S.M., Minshull, T.A., Whitmarsh, R.B., and Louden, K.E., 2000, Deep structure of the ocean-continent transition in the southern Iberia Abyssal Plain from seismic refraction profiles: II. The IAM-9 transect at 40° 20′ N: Journal of Geophysical Research, v. 105, p. 5859–5886, doi:10.1029/1999JB900301. [9]

Deep Probe Working Group, 1998, Probing the Archean and Proterozoic Lithosphere of Western North America: GSA Today, v. 8, no. 7, p. 1–17. [9]

DEKORP and OROGENIC PROCESSES Working Groups, 1999, Structure of the Saxonian Granulites—geological and geophysical constraints on the exhumation of HP/HT-rocks: Tectonics, v. 18, p. 756–773, doi:10.1029/1999TC900030. [9]

DEKORP Research Group, 1985, First results and preliminary interpretation of deep reflection seismic recordings along DEKORP 2 South: Journal of Geophysics, v. 57, p. 137–163. [2] [8]

DEKORP Research Group, 1988, Results of the DEKORP 4/KTB Oberpfalz deep seismic reflection investigations: Journal of Geophysics, v. 62, p. 69–101. [2] [8]

DEKORP Research Group, 1991, Results of the DEKORP 1 (BELCORP-DEKORP) deep seismic reflection studies in the western part of the Rhenish Massif: Geophysical Journal International, v. 106, p. 203–227, doi:10.1111/j.1365-246X.1991.tb04612.x. [2] [8]

DEKORP Research Group, 1994, The deep reflection seismic profiles DEKORP 3/MVE-90: Zeitschrift für Geologische Wissenschaften, v. 22, p. 623–825. [2] [8] [9]

DEKORP Research Group (A), 1994a, Profile DEKORP 3/MVE-90: Reflection seismic field measurements and data processing: Zeitschrift für Geologische Wissenschaften, v. 22, no. 6, p. 627–646. [2] [9]

DEKORP Research Group (B), 1994b, Crustal structure of the Saxothuringian Zone: Results of the deep seismic profile MVE-90(East): Zeitschrift für Geologische Wissenschaften, v. 22, no. 6, p. 647–769. [9]

DEKORP/BASIN Research Group, 1998, Survey providdes seismic insights into an old suture zone: Eos (Transactions, American Geophysical Union), v. 79, no. 12, p. 151–159. [9]

DEKORP/BASIN Research Group, 1999, The deep structure of the NE German Basin—Constraints on the controlling mechanisms of intracontinental basin development: Geology, v. 27, p. 55–58. [9]

Demachi, T., Hasemi, A., Iwasaki, T., and Okudera, M., 2004, Three-dimensional P wave velocity structure of shallow crust beneath the northern Awaji Island derived from refraction seismic explosion: Earth, Planets, and Space, v. 56, p. 473–477. [9]

Den, N., Ludwig, W.J., Ewing, J.I., Murauchi, S., Hotta, H., Edgar, N.T., Yoshii, T., Asanuma, T., Hagiwara, K., Sato, T., and Ando, S., 1969, Seismic refraction measurements in the Northwest Pacific Basin: Journal of Geophysical Research, v. 74, no. 6, p. 1421–1434, doi:10.1029/JB074i006p01421. [2] [6]

Denham, D., Simpson, D.W., Gregson, P.J., and Sutton, D.J., 1972, Travel times and amplitudes from explosions in northern Australia: Geophysical Journal of the Royal Astronomical Society, v. 28, p. 225–235. [2] [6] [7]

Dentith, M.C., and Hall, J., 1990, MAVIS: Geophysical constraints on the structure of the Carboniferous basin of West Lothian: Transactions of the Royal Society of Edinburgh, v. 81, no. 2, p. 117–126. [2] [8]

Dentith, M.C., Dent, V.F., and Drummond, B.J., 2000, Deep crustal structure in the southwestern Yilgarn Craton, Western Australia: Tectonophysics, v. 325, p. 227–255, doi:10.1016/S0040-1951(00)00119-0. [7] [8]

DESERT Group, 2004, The crustal structure of the Dead Sea transform: Geophysical Journal International, v. 156, p. 655–681, doi:10.1111/j.1365-246X.2004.02143.x. [2] [9]

DESERT Team, 2000, Multinational geoscientific research effort kicks off in the Middle East: Eos (Transactions, American Geophysical Union), v. 81, p. 609, 616–617. [9]

Detrick, R.S., and Purdy, G.M., 1980, The crustal structure of the Kane frscture zone from seismic refraction studies: Journal of Geophysical Research, v. 85, p. 3759–3778, doi:10.1029/JB085iB07p03759. [2] [7] [8]

Detrick, R.S., Buhl, P., Vera, E., Mutter, J., Orcutt, J., Madsen, J., and Brocher, T., 1987, Multichannel seismic imaging of a crustal magma chamber along the East Pacific Rise: Nature, v. 326, p. 35–41, doi:10.1038/326035a0. [2] [8] [9]

Detrick, R.S., Harding, A.J., Kent, G.M., Orcutt, J.A., Mutter, J.C., and Buhl, P., 1993a, Seismic structure of the southern East Pacific Rise: Science, v. 259, p. 499–503, doi:10.1126/science.259.5094.499. [9]

Detrick, R.S., White, R.S., and Purdy, G.M., 1993b, Crustal structure of North Atlantic fracture zones: Reviews of Geophysics, v. 31, p. 439–458, doi:10.1029/93RG01952. [8]

Detrick, R., Collins, J., Stephen R. and Swift, S., 1994, In situ evidence for the nature of the seismic layer 2/3 boundary in oceanic crust: Nature, v. 370, p. 288–290, doi:10.1038/370288a0. [2] [8]

Dewey, J., And Byerly, P., 1969, The early history of seismometry: Bulletin of the Seismological Society of America, v. 59, no. 1, p. 183–227. [2] [3] [10]

Díaz, J., and Gallart, J., 2009, Crustal structure beneath the Iberian peninsula and surrounding waters: a new compilation of deep seismic sounding results: Physics of the Earth and Planetary Interiors, v. 173, p. 181–190, doi:10.1016/j.pepi.2008.11.008. [2] [8] [9] [10]

Diebold, J.B., Stoffa, P.L., and Study Group LASE, 1988, A large aperture seismic experiment in the Baltimore Canyon trough, *in* Sheridan, R.E., and Grow, J.A., The Atlantic Continental Margin: Boulder, Colorado, Geological Society of America, Geology of North America, vol. I-2, p. 387–398. [8]

Dieterle, G., and Peterschmitt, E., 1964, Expedition Antarctique Belge 1959: sondages seismiques en Terre de la Reine Maud: Memoires de l'Academie Royale des Sciences d'Outre-Mer, Classe des Sciences techniques, N.S., Tome XIII, fasc. 4, p. 101 p. [2] [5]

Diment, W.H., Stewart, S.W., and Roller, J.C., 1961, Crustal structure from the Nevada Test Site to Kingman, Arizona, from seismic and gravity observations: Journal of Geophysical Research, v. 66, p. 201–214, doi:10.1029/JZ066i001p00201. [2] [5] [6]

Ding, W., Huang, C., Cao, J., and Jiang, G., 1988, Crustal structures along Suixian-Xian DSS profile, *in* Developments in the Research of Deep Structures of China's Continent. Geological Publishing House, Beijing, p. 38–47. [8]

Dix, C.H., 1965, Reflection seismic crustal studies: Geophysics, v. 30, p. 1068–1084. [6]

DOBREflection-2000 and DOBREfraction '99 Working Groups, 2002, DOBRE Studies Evolution of Inverted Intra-cratonic Rifts in Ukraine: Eos (Transactions, American Geophysical Union), v. 83, p. 323, 326–327. [9]

DOBREfraction '99 Working Group, 2003, "DOBREfraction '99"—velocity model of the crust and upper mantle beneath the Donbass Foldbelt (East Ukraine): Tectonophysics, v. 371, p. 81–110. [2] [9]

Dobrin, M.B., 1976, Introduction to geophysical prospecting, 3rd. ed.: McGraw-Hill, New York, 630 p. [7]

Dohr, G., 1957, Ein Beitrag der Reflexionsseismik zur Erforschung des tieferen Untergrundes: Geologische Rundschau, v. 46, p. 17–26, doi:10.1007/BF01802879. [5] [6] [8]

Dohr, G., 1959, Über die Beobachtungen von Reflexionen aus dem tieferen Untergrund im Rahmen routinemässiger reflexionsseismischer Messungen: Zeitschrift für Geophysik, v. 25, p. 280–300. [2] [5] [6]

Dohr, G., 1972, Reflexionsseismische Tiefensondierung: Zeitschrift für Geophysik, v. 38, p. 193–220. [8]

Dohr, G., 1983, Ergebnisse geophysikalischer Untersuchungen über den Bau des Nordwestdeutschen Beckens: Erdoel-Erdgas, v. 99, p. 252–267. [6]

Dohr, G., and Fuchs, K., 1967, Statistical evaluation of deep crustal reflections in Germany: Geophysics, v. 32, p. 951–967, doi:10.1190/1.1439908. [2] [6]

Dohr, G., Bachmann, G.H., and Grosse, S., 1989, Das Norddeutsche Becken. Ergebnisse geophysikalischer Arbeiten zur Untersuchung des tieferen Untergrundes: Niedersächsische Akademie der Geowissenschaften, 2. Hannover, Veroefft, p. 4–47. [7] [8]

Doody, J.J., and Brooks, M.,1982. Recent seismic refraction studies in SW England: Geophysical Journal of the Royal Astronomical Society, v. 69, p. 278. [2] [7]

Dooley, J.C., 1979, A geophysical profile across Australia at 29°S: BMR Journal of Australian Geology and Geophysics, v. 4, p. 353–359. [7]

Dooley, J.C., and Moss, F.J., 1988, Deep crustal reflections in Australia 1957–1973—II. Crustal models: Geophysical Journal of the Royal Astronomical Society, v. 93, p. 239–249. [2] [6] [7]

Dorman, L.M., 2001, Seismology sensors, *in* Steele, J., Thorpe, S., and Turekian, K., eds., Encyclopedia of Ocean Sciences: Academic Press, Elsevier, Amsterdam, p. 2737–2744. [10]

Doyle, H.A.,1957, Seismic recordings of atomic explosions in Australia: Nature, v. 180, p. 132–134, doi:10.1038/180132a0. [5]

Doyle., H.A., Everingham, I.B., and Hogan, T.K., 1959, Seismic recordings of large explosions in southeastern Australia: Australian Journal of Physics, v. 12, p. 222. [2] [5]

Drake, C.L., and Girdler, R.W., 1964, A geophysical study of the Red Sea: Geophysical Journal, v. 8, p. 473–495, doi:10.1111/j.1365-246X.1964.tb06303.x. [5] [6]

Drake, C.L., Worzel, J.L., and Beckmann, W., 1952, Seismic refraction measurements in the Gulf of Maine: Bulletin of the Seismological Society of America, v. 42 (12.2), p. 1244–1245. [2] [4] [5]

Drummond, B.J.,1979, A crustal profile across the Archaean Pilbara and northern Yilgarn cratons, northwest Australia: BMR Journal of Australian Geology and Geophysics, v. 4, p. 171–180. [2] [7]

Drummond, B.J., 1981, Crustal structure of the Precambrian terrains of northwest Australia from seismic refraction data: BMR Journal of Australian Geology and Geophysics, v. 6, p. 123–135. [2] [7]

Drummond, B.J., 1988, A review of crust–upper mantle structure in the Precambrian areas of Australia and implications for Precambrian crustal evolution: Precambrian Research, v. 40-41, p. 101–116, doi:10.1016/0301-9268(88)90063-0. [2] [7] [8]

Drummond, B.J., ed., 1991, The Australian lithosphere: Geological Society of Australia Special Publication 17, 208 p. [2] [8]

Drummond, B.J., Collins, C.D.N., and Gibson, G., 1979, The crustal structure of Papua and northwest Coral Sea: BMR Journal of Australian Geology and Geophysics, v. 4, p. 341–351. [7]

Drummond, B.J., Goleby, B.R., Goncharov, A.G., Wyborn, L.A.I., Collins, C.D.N., and MacCready, T., 1998, Crustal-scale structures in the Proterozoic Mount Isa Inlier of north Australia, *in* Klemperer, S.L., and Mooney, W.D., eds., Deep seismic probing of the continents, II: a global survey: Tectonophysics, v. 288, p. 43–56. [2] [9]

Drummond, B.J., Goleby, B.R., and Swager, C.P., 2000a, Crustal signature of late Archaean tectonic episodes in the Yilgarn craton, western Australia: evidence from deep seismic sounding: Tectonophysics, v. 329, p. 193–221, doi:10.1016/S0040-1951(00)00196-7. [2] [9]

Drummond, B.J., Barton, T.J., Korsch, R.J., Rawlinson, N., Yeates, A.N., Collins, C.D.N., and Brown, A.V., 2000b, Evidence for crustal extensiuon and inversion in eastern Tasamania, Australia, during Neoproterozoic and Early Palaeozoic: Tectonophysics, v. 329, p. 1–21, doi:10.1016/S0040-1951(00)00185-2. [2] [9]

Drummond, B.J., Lyons, P., Goleby, B.R., and Jones, L., 2006, Constraining models of the tectonic setting of the giant Olympic Dam iron oxide-copper-gold deposit, South Australia, using deep seismic reflection data: Tectonophysics, v. 420, p. 91–103, doi:10.1016/j.tecto.2006.01.010. [2] [10]

Dürbaum, H.-J., Fritsch, J., and Nickel, H., 1971, Untersuchungen über den Charakter der Reflexionen aus grossen Tiefen: Beih. Geologisches Jahrbuch, v. 90, p. 99–133. [8]

Dürbaum, H.-J., Hinz, K., and Makris, J., 1977, Seismic studies in the Cretan Sea: "Meteor" Forschungsergebnisse, Reihe C, no. 27, Borntraeger, Berlin-Stuttgart, 45 p. [7]

Dürbaum, H.-J., Reichert, C., Sadowiak, P., and Bram, K., eds., 1992, Integrated Seismics Oberpfalz 1989, data evaluation and interpretation as of October 1992, KTB report 92-5/DEKORP report, NLFB Hannover: 373 p. [2] [8]

Dürbaum, H.-J., Dohr, G., and Meissner, R., 1997, Das Deutsche Kontinentale Reflexionsseismische Programm (DEKORP), *in* Neunhöfer, H., Börngen, M., Junge, A., and Schweitzer, J., eds., Zur Geschichte der Geophysik in Deutschland: Jubiläumsschrift zur 75jährigen Wiederkehr der Gründung der Deutschen Geophysikalischen Gesellschaft, Hamburg, p. 149–155. [8]

Durrheim, R.J., 1986, Recent reflection seismic developments in the Witwatersrand basin, *in* Barazangi, M., and Brown, L., eds., Reflection seismology: a global perspective: American Geophysical Union, Geodynamics Series, v. 13, p. 77–83. [2] [8]

Durrheim, R.J., 1987, Seismic reflection and refraction studies of the deep structure of the Agulhas Bank: Geophysical Journal of the Royal Astronomical Society, v. 89, p. 395–398. [2] [8]

Durrheim, R.J., Nicolaysen, L.O., and Corner, B., 1991, A deep seismic reflection profile across the Archean-Proterozoic Witwatersrand basin, South Africa, *in* Meissner, R., Brown, L., Dürbaum, H.-J., Franke, W., Fuchs, K., and Seifert, F., eds., Continental lithosphere: deep seismic reflections: American Geophysical Union, Geodynamics Series, v. 22, p. 213–224. [2] [8]

Duschenes, J., Potts, C., and Rayner, M.P.W., 1985, Cambridge deep ocean geophone: Marine Geophysical Researches, v. 7, p. 455–466, doi:10.1007/BF00368950. [8]

Eaton, J.P., 1963, Crustal structure between Eureka, Nevada, and San Francisco, California, from seismic-refraction measurements: Journal of Geophysical Research, v. 68, p. 5789–5806. [6]

Eaton, J.P., 1966, Crustal structure in northern and central California from seismic evidence, *in* Bailey, E.H., ed., Geology of Northern California: California Division of Mines Geological Bulletin, v. 190, p. 419–426. [6]

Ebeniro, J.O., Nakamura, Y., Sawyer, D.S., and O'Brien, W.P., Jr., 1988, Sedimentary and crustal structure of the northwestern Gulf of Mexico: Journal of Geophysical Research, v. 93, p. 9075–9092, doi:10.1029/JB093iB08p09075. [8]

ECORS Pyrenees Team, 1988, The ECORS deep reflection seismic survey across the Pyrenees: Nature, v. 331, p. 508–511, doi:10.1038/331508a0. [2] [8]

Edel, J.B., Fuchs, K., Gelbke, C., Prodehl, C., 1975, Deep structure of the southern Rhinegraben area from seismic-refraction investigations: Zeitschrift für Geophysik, v. 41, p. 333–356. [2] [7] [8]

Edgar, N.T., Ewing, J.I., and Hennion; J., 1971, Seismic refraction and reflection in Caribbean Sea: American Association of Petroleum Geologists Bulletin, v. 55 (6), p. 833–870. [2] [5]

Edwards, R.A., Whitmarsh, R.B., and Scrutton, R.A., 1997, The crustal structure across the transform continental margin off Ghana, eastern equatorial Atlantic: Journal of Geophysical Research, v. 102, p. 747–772, doi:10.1029/96JB02098. [2] [9]

Egger, A., 1990, A comprehensive compilation of seismic refraction data along the southern segment of the European Geotraverse from the northern Appenines to the Sardinia Channel (1979–1985). Open-File Report, Institut für Geophysik, ETH-Hönggerberg, Zurich, Switzerland, 202 p. [8]

Egger, A., 1992, Lithospheric structure along a transect from the northen Apennines to Tunisia derived from seismic refraction data [Ph.D. thesis no. 9675]: ETH Zurich, 207 p. [8]

Egger, A., Demartin, M., Ansorge, J., Banda, E., and Maistrello, M., 1988, The gross structure of the crust under Corsica and Sardinia, *in* Freeman, R., Berthelsen, A., and Mueller, St., eds., The European Geotraverse, Part 4: Tectonophysics, v. 150, p. 363–389. [2] [8]

Egloff, F., Rihm, R., Makris, J., Izzeldyn, Y.A., Bobsien, M., Meier, K., Junge, P., Noman, T., and Warsi, W., 1991, Contrasting styles of the eastern and western margins of the southern Red Sea: the 1988 SONNE experiment, *in* Makris, J., Mohr, P., and Rihm, R., eds., Red Sea: birth and early history of a new oceanic basin: Tectonophysics, v. 198, p. 329–353. [2] [8]

Egorkin, A.V., 1999, Deep seismic studies based on three-component recording of ground motions: Izvestija, Physics of the Solid Earth, v. 35, p. 566–585 (translated from Fizika Zemli, no. 7-8, 1999, p. 44–64). [8]

Egorkin, A.V., and Chernyshov, N.M., 1983, Pecularities of mantle waves from long-range profiles: Journal of Geophysics, v. 54, p. 30–34. [8]

Egorkin, A.V., and Pavlenkova, N.I., 1981, Studies of mantle structure of U.S.S.R. territory on long-range seismic profiles: Physics of the Earth and Planetary Interiors, v. 25, p. 12–26, doi:10.1016/0031-9201(81)90125-4. [2] [7]

Egorkin, A.V., Zuganov, S.K., Pavlenkova, N.I., and Chernyshev, N.M., 1987, Results of lithosphere studies from long-range profiles in Siberia: Tectonophysics, v. 140, p. 29–47, doi:10.1016/0040-1951(87)90138-7. [8]

Egorkin, A.V., Kosminskaya, I.P., and Pavlenkova, N.I., 1991, Multi-wave studies of the continental lithospheres: Physics of the Earth, v. 9, p. 82–96. [2] [8]

Eiby, G.A., 1955, New Zealand crustal structure: Nature, v. 176, p. 32, doi:10.1038/176032a0. [5]

Eiby GA, 1957, Crustal Structure project: the Wellington Profile: New Zealand Department of Science and Industrial Research, Geophysical Memoir, v. 5, p. 6–28. [5]

El-Isa, Z., Mechie, J., Prodehl, C., Makris, J., and Rihm, R., 1987a, A crustal structure study of Jordan derived from seismic refraction data:

Tectonophysics, v. 138, p. 235–253, doi:10.1016/0040-1951(87)90042-4. [2] [8] [9]

El-Isa, Z., Mechie, J., and Prodehl, C., 1987b, Shear velocity structure of Jordan from explosion seismic data: Geophysical Journal of the Royal Astronomical Society, v. 90, p. 265–281. [8]

Ellis, R.M., Spence, G.D., Clowes, R.M., Waldron, A., Jones, I.F., Green, A.G., Forsyth, D.A., Mair, J.A., Berry, M.J., Mereu, R.F., Kanasewich, E.R., Cumming, G.L., Hajnal, Z., Hyndman, R.D., McMechan, G.A., and Loncarevic, B.D., 1983, The Vancouver Island seismic project: A COCRUST onshore-offshore study of a covergent margin: Canadian Journal of Earth Science, v. 20, p. 719–741. [8]

Ellsworth, W.L., 1990, Earthquake history, 1769–1989, in Wallace, G.E., ed., The San Andreas fault system, California: U.S. Geological Survey Professional Paper 1515, p. 153–187. [9]

Emmermann, R., and Wohlenberg, J., eds., 1989, The German continental deep drilling program (KTB)—site-selection studies in the Oberpfalz and Schwarzwald: Springer, Berlin-Heidelberg, 553 p. [2] [8]

Enciu, D.M., Knapp, C.C., and Knapp, J.H., 2009, Revised crustal architecture of the southeastern Carpathian foreland from active and passive seismic data: Tectonics, v. 28, TC4013, 20 p., doi:10.1029/2008TC002381. [10]

Enderle, U., 1998, Signaturen in refraktionsseismischen Daten als Abbild geodynamischer Prozesse [Ph.D. thesis]: University of Karlsruhe, 220 p. [9]

Enderle, U., Tittgemeyer, M., Itzin, M., Prodehl, C., and Fuchs, K., 1997, Scales of structure in the lithosphere—images of processes, in Fuchs, K., Altherr, R., Müller, B., and Prodehl, C., eds., Stress and stress release in the lithosphere—structure and dynamic processes in the rifts of western Europe: Tectonophysics, v. 275, p. 165–198. [9]

Enderle, U., Schuster, K., Prodehl, C., Schulze, A., and Bribach, J., 1998a, The refraction seismic experiment GRANU95 in the Saxothuringian belt, SE-Germany: Geophysical Journal International, v. 133, p. 245–259, doi:10.1046/j.1365-246X.1998.00462.x. [2] [9]

Enderle, U., Spindler, S., Goertz, A., Prodehl, C., Schulze, A., and Schuster, K., 1998b, A compilation of data from the GRANU95 seismic-refraction experiment in SW-Germany. Open-File Report 98-1, Geophysical Institute, Univeristy of Karlsruhe, 208 p. [9]

Engelhard, L., 1998, 50 Jahre Helgolandsprengung: Mitteilungen Deutsche Geophys. Ges., ISSN 0934-6554, Nr. 2/1998, p. 2–32. [3] [4]

England, R.W., 2000, Deep structure of North West Europe from deep seismic profiling: the link between basement tectonics and basin development, in Mohriak, W., and Talwani, M., eds., Atlantic Rifts and Continental Margins: American Geophysical Union, Geophysical Monograph 115, p. 57–83. [8]

England, R.W., and Hobbs, R.W., 1997, The structure of the Rockall Trough imaged by deep seismic reflection profiling: Journal of the Geological Society of London, v. 154, p. 497–502, doi:10.1144/gsjgs.154.3.0497. [2] [8]

England, R.W., McBride, J.H., and Hobbs, R.W., 2005, The role of Mesozoic rifting in the opening of the NE Atlantic: evidence from deep seismic profiling across the Faroe-Shetland Trough: Journal of the Geological Society of London, v. 162, p. 661–673, doi:10.1144/0016-764904-076. [2] [8] [9]

English, W.A., ed., 1939, Seismograph prospecting for oil: American Institute Mining and Metallurgical Engineers, technical publication no. 1059, 29 p. [3]

Epili, D., and Mereu, R.F., 1989, The GLIMPCE seismic experiment: Onshore refraction and wide-angle reflection observations from a fan line over the Lake Superior Midcontinent Rift System, in Mereu, R.F., Mueller, St., and Fountain, D.M., eds., Properties and processes of earth's lower crust: American Geophysical Union, Geophysical Monograph 51, p. 93–101. [8]

Erslev, E.A., 2005, 2D Laramide geometries and kinematics of the Rocky Mountains, western U.S.A, in Karlstrom, K.E., and Keller, G.R., eds., The Rocky Mountain Region: An evolving lithosphere—tectonics, geochemistry, and geophysics: Washington, D.C., American Geophysical Union Geophysical Monograph 154, p. 7–20. [9]

EUGEMI Working Group, 1990, The European Geotraverse seismic refraction experiment of 1986 from Genova, Italy to Kiel, Germany, in Freeman, R., and Mueller, St., eds., The European Geotraverse, Part 6: Tectonophysics, v. 176, p. 43–57. [6] [8]

EUGENO-S Working Group, 1988, Crustal structure and tectonic evolution of the transition between the Baltic Shield and the North German Calidonides (the EUGENO-S Project), in Freeman, R., Berthelsen, A., and Mueller, St., eds., The European Geotraverse, Part 4: Tectonophysics, v. 150, p. 253–348. [2] [8]

EUROBRIDGE Seismic Working Group, 1999, Seismic velocity structure across the Fennoscandia-Sarmatia suture of the East European Craton beneath the EURO BRIDGE profile through Lithuania and Belarus: Tectonophysics, v. 314, p. 193–217, doi:10.1016/S0040-1951(99)00244-9. [2] [9]

EUROBRIDGE'95 Seismic Working Group, 2001, EUROBRIDGE'95: deep seismic profiling within the East European Craton: Tectonophysics, v. 339, p. 153–175. [9]

European Science Foundation, 1990, European Geotraverse Project (EGT) 1983–1990, final report: European Science Foundation, Strasbourg, 67 p. [2] [8]

Evenchick, C.A., Gabrielse, H., and Snyder, D., 2005, Crustal structure and lithology of the northern Canadian Cordillera: alternative interpretations of SNORCLE seismic reflection lines 2a and 2b: Canadian Journal of Earth Science, v. 42, p. 1149–1161, doi:10.1139/e05-009. [9]

Everingham, I.B., 1965, The crustal structure of the south-west of Western Australia: Bureau of Mineral Resources Australia, Report, 1965/1967. [2] [6]

Evison, F.F., Ingham, C.E., and Orr, R.H., 1959, Thickness of the earth's crust in Antarctica: Nature, v. 183, 4657, p. 306–308, doi:10.1038/183306a0. [5]

Ewing, G.N., Dainty, A.M., Blanchard, J.E., and Keen, M.J., 1966, Seismic studies on the eastern seabord of Canada: the Appalachian system, I: Canadian Journal of Earth Science, v. 3, p. 89–109. [6]

Ewing, J.I., 1963a, Elementary theory of seismic refraction and reflection measurements, in Hill, M.N., ed., The Sea vol. 3, The earth beneath the Sea: New York–London, Interscience Publishers, p. 3–19. [2] [5] [6] [10]

Ewing, J.I., 1963b, The mantle rocks, in Hill, M.N., ed., The Sea vol. 3, The earth beneath the Sea: New York–London, Interscience Publishers, p. 103–109. [5] [6]

Ewing, J.I., 1969, Seismic model of the Atlantic Ocean, in Hart, P.J., ed., The earth's crust and upper mantle: American Geophysical Union, Geophysical Monograph 13, p. 220–225. [6] [10]

Ewing, J.I., and Ewing, M., 1959, Seismic refraction measurements in the Atlantic Ocean basins, in the Mediterranean Sea, on the Mid Atlantic Ridge and in the Norwegian Sea: Bulletin of the Geological Society of America, v. 70, p. 291–318, doi:10.1130/0016-7606(1959)70 [291:SMITAO]2.0.CO;2. [5]

Ewing, J.I., and Ewing, M., 1961, A telemetering ocean-bottom seismograph: Journal of Geophysical Research, v. 66, p. 3863–3878, doi:10.1029 /JZ066i011p03863. [3] [5]

Ewing, J.I., and Ewing, M., 1966, Marine seismic studies: Transactions, American Geophysical Union, v. 47, no. 1, p. 276–279. [6]

Ewing, J.I., and Ewing, M., 1970, General observations—seismic reflection, in Maxwell, A.E., ed., The Sea, new concepts of ocean floor evolution: New York, Wiley-Interscience, p. 53–84. [2] [5] 6] [10]

Ewing, J.I., and Houtz, R., 1979, Acoustic stratigraphy and structure of the oceanic crust, in Talwani, M., Harrison, C.G., and Hayes, D.E., eds., Deep drilling results in the Atlantic Ocean: Washington, D.C., American Geophysical Union, p. 1–14. [7]

Ewing, J.I., and Meyer, R.P., 1982, Rivera Ocean seismic experiment (ROSE) overview: Journal of Geophysical Research, v. 87, p. 8345–8357, doi:10.1029/JB087iB10p08345. [2] [7]

Ewing, J.I., and Nafe, J.E., 1963, The unconsolidated sediments, in Hill, M.N., ed., The Sea vol. 3, The earth beneath the Sea: New York–London, Interscience Publishers,p. 73–84. [5]

Ewing, J.I., and Purdy, G.M., 1982, Upper crustal velocity structure in the ROSE area of the East Pacific Rise: Journal of Geophysical Research, v. 87, p. 8397–8402, doi:10.1029/JB087iB10p08397. [7]

Ewing, J.I., and Zaunere, R., 1964, Seismic profiling with a pneumatic sound source: Journal of Geophysical Research, v. 69, p. 4913–4915, doi:10.1029/JZ069i022p04913. [5] [6]

Ewing, J.I., Antoine, J.W., and Ewing, M., 1960, Geophysical measurements in the western Caribbean Sea and in the Gulf of Mexico: Journal of Geophysical Research, v. 65, p. 4087–4126, doi:10.1029/JZ065i012p04087. [5] [6]

Ewing, J.I., Ewing, M., Aitken, T., and Ludwig, W.J., 1968, North Pacific sediment layers measured by seismic profiling, in Knopoff, L., Drake, C.L., Hart, P.J., eds., The crust and upper mantle of the Pacific area: American Geophysical Union, Geophysical Monograph 12, p. 147–173. [6]

Ewing, M., and Press, F., 1952, Crustal structure and surface wave dispersion, Part II: Bulletin of the Seismological Society of America, v. 42, p. 315–325. [5]

Ewing, M., and Press, F., 1955, Geophysical contrasts between continents and oceanic basins, *in* Poldervaart, A., ed., Crust of the earth (a symposium): Geological Society of America Special Paper 62, p. 1–6. [4]

Ewing, M., and Vine, 1938, Deap-sea measurements without wires and cables: Transactions, American Geophysical Union, part 1, p. 248–251. [3] [5]

Ewing, M., Crary, A.P., and Lohse, J.M., 1934, Seismological observations on quarry blasting: Transactions, American Geophysical Union, 15 (Annual Meeting I), p. 91–94. [3]

Ewing, M., Crary, A.P., and Rutherford, H.M., 1937, Geophysical investigations in the emerged and submerged Atlantic coastal plain, Part I: Bulletin of the Geological Society of America, v. 48, p. 753–802. [2] [3]

Ewing, M., Woollard, G.P., and Vine, A.C., 1939, Geophysical investigations in the emerged and submerged Atlantic coastal plain, Part III: Bulletin of the Geological Society of America, v. 50, p. 257–296. [3] [5]

Ewing, M., Woollard, G.P., and Vine, A.C., 1940, Geophysical investigations in the emerged and submerged Atlantic coastal plain, Part IV: Bulletin of the Geological Society of America, v. 51, p. 1821–1840. [3] [5]

Ewing, M., Woollard, G.P., Vine, A.C., and Worzel, J.L., 1946, Recent results in submarine geophysics: Bulletin of the Geological Society of America, v. 57, p. 909–934. [3] [5]

Ewing, M., Worzel, J.L., Hersey, J.B., Press, F., and Hamilton, G.R., 1950, Seismic refraction measurements in the Atlantic Ocean basin, Part I: Bulletin of the Seismological Society of America, v. 40, p. 233–242. [2] [4] [5]

Ewing, M., Sutton, G.H., and Officer, C.B., 1952, Seismic refraction measurements in the Atlantic Ocean basin, Part IV: Bulletin of the Seismological Society of America, v. 42, p. 1353. [4] [5]

Ewing, M., Sutton, G.H., and Officer, C.B., 1954, Seismic refraction measurements in the Atlantic Ocean, Part. VI: Typical deep stations, North America Basin: Bulletin of the Seismological Society of America, v. 44, p. 21–38. [2] [4] [5]

Ewing, M., Ludwig, W.J., and Ewing, J.I., 1964, Sediment distribution in the oceans: the Argentine basin: Journal of Geophysical Research, v. 69 (10), p. 2003–2032, doi:10.1029/JZ069i010p02003. [6]

Ewing, M., Eittreim, S., Truchan, M., and Ewing, J.I., 1969, Sediment distribution in the Indian Ocean: Deep-Sea Research, v. 16, p. 231–248. [6]

Ewing, M., Hawkins, L.V., and Ludwig, W.J., 1970, Crustal structure of the Coral Sea: Journal of Geophysical Research, v. 75, no. 11: p. 1953–1962, doi:10.1029/JB075i011p01953. [2] [6]

Explosion Seismology Group Pyrenees, 1980, Seismic reconnaissance of the structure of the Pyrenees: Ann. Geophys., v. 36, p. 135–140. [7]

Faber, S., 1978, Refraktionsseismische Untersuchung der Lithosphäre unter den Britischen Inseln [Ph.D. thesis]: University of Karlsruhe, 132 p. [7]

Faber, S., and Bamford, D., 1979, Lithospheric structural contrasts across the Caledonides of northern Britain, *in* Fuchs, K., and Bott, M.H.P., eds., Structure and compositional variations of the lithosphere and the asthenosphere: Tectonophysics, v. 56, p. 17–30. [2] [7] [8]

Faber, S., and Bamford, D., 1981, Moho offset beneath northern Scotland: Geophysical Journal of the Royal Astronomical Society, v. 67, p. 661–672. [7]

Fagot, M.G., 1986, Development of a deep-towed seismic system: a new capability for deep-ocean acoustic measurements, *in* Akal, T., and Berkson, J.M., eds., Ocean Seismo-Acoustics: New York, Plenum Press, p. 853–862. [8]

Fahlquist, D.A., and Hersey, J.B., 1969, Seismic refraction measurements in the western Mediterranean Sea: Bulletin de l'Institut Oceanographique Monaco, v. 67, p. 1368. [6] [7]

Faleide, J.I., Gudlaugsson, S.T., Eldholm, O., Myhre, A.M., and Jackson, H.R., 1991, Deep seismic transects across the sheared western Barents Sea–Svalbard continental margin: Tectonophysics, v. 189, p. 73–89, doi:10.1016/0040-1951(91)90488-E. [7] [8]

Fanucci, F., and Morelli, D., 2003. The CROP profiles across the Western Mediterranean basins, *in* Scrocca, D., Doglioni, C., Innocenti, F., Manetti, P., Mazzotti, A., Bertelli, L., Burbi, L., and D'Offizi, S., eds., CROP Atlas: seismic reflection profiles of the Italian crust: Memorie Descrittive della Carta Geologica d'Italia, Roma, v. 62, p. 166–170. [8]

Fernández Viejo, G., and Clowes, R.M., 2003, Lithospheric structure beneath the Archaean Slave Province and Proterozoic Wopmay orogen, northwestern Canada, from a lithoprobe refraction/wide-angle reflection survey: Geophysical Journal International, v. 153, p. 1–19, doi:10.1046/j.1365-246X .2003.01807.x. [9]

Fernández-Viejo, G., Gallart, J., Javier, J., Pulgar, A., Cordoba, D., and Danobeitia, J.J., 2000, Seismic signature of Variscan and Alpine tectonics in NW Iberia: Crustal structure of the Cantabrian Mountains and Duero basin: Journal of Geophysical Research, v. 105, p. 3001–3018, doi:10.1029/1999JB900321. [2] [9]

Ferrucci, F., Gaudiosi, G., Hirn, A., and Nicolich, R., 1991, Ionian Basin and Calabrian Arc: new elements by DSS data: Tectonophysics, v. 195, p. 411–419, doi:10.1016/0040-1951(91)90223-F. [9]

Fessenden, R.A., 1917, Methods and apparatus for locating ore bodies. U.S. Patent 1,240,328; September 18, 1917. [3]

Finckh, P., and Frei, W., 1991, Seismic reflection profiling in the Swiss Rhône valley. Part 1: Seismic reflection field work, seismic processing and seismic results of the Roche-Voury and Turtmann and Agarn lines: Eclogae Geologicae Helvetiae, v. 84, p. 345–357. [8]

Finckh, P., Ansorge, J., Mueller, St., and Sprecher, Chr., 1984, Deep crustal reflections from a Vibroseis survey in northern Switzerland: Tectonophysics, v. 109, p. 1–14, doi:10.1016/0040-1951(84)90167-7. [8]

Finckh, P., Frei, W., Fuller, B., Johnson, R., Mueller, St., Smithson, S., and Sprecher, Chr., 1986, Detailed crustal structure from a seismic reflection survey in northern Switzerland, *in* Barazangi, M., and Brown, L., eds., Reflection seismology: a global perspective: American Geophysical Union, Geodynamics Series, v. 13, p. 43–54. [2] [8]

Finetti, I.R., 2003, The CROP profiles across the Mediterranean Sea (CROP MARE 1 and 2), *in* Scrocca, D., Doglioni, C., Innocenti, F., Manetti, P., Mazzotti, A., Bertelli, L., Burbi, L., and D'Offizi, S., eds., CROP Atlas: seismic reflection profiles of the Italian crust: Memorie Descrittive della Carta Geologica d'Italia, Roma, v. 62, p. 171–184. [9]

Finetti, I., and Morelli, C., 1973, Geophysical exploration of the Mediterranean Sea: Bollettino di Geofisica Teorica e Applicata, v. 15, p. 263–271. [7]

Finlayson, D.M., 1977, Seismic travel-times to east Papua from USSR nuclear explosions: BMR Journal of Australian Geology and Geophysics, v. 2, p. 209–216. [7]

Finlayson, D.M., 1982, Seismic crustal structure of the Proterozoic North Australian craton between Tennant Creek and Mount Isa: Journal of Geophysical Research, v. 87, p. 10,569–10,578, doi:10.1029/JB087iB13p10569. [2] [7]

Finlayson, D.M., 2010, A chronicle of deep seismic sounding profiling across the Australian continent and its margins, 1946–2006: D.M. Finlayson, Canberra, 255 p. [2] [5] [6] [7] [8] [9] [10]

Finlayson, D.M., and Ansorge, J., 1984, Workshop proceedings: interpretation of seismic wave propagation in laterally heterogeneous structures: Bureau of Mineral Resources, Geology and Geophysics, report 258, Canberra, Australia, 207 p. [8] [10]

Finlayson, D.M., and Collins, C.D.N., 1993, Lithospheric velocity structures under the southern New England Orogen: evidence for underplating at the Tasman Sea margin: Australian Journal of Earth Sciences, v. 40, p. 141–153, doi:10.1080/08120099308728071. [2] [8]

Finlayson, D.M., and Cull, J.P., 1973, Time-term analysis of New Britain–New Ireland island arc structures: Geophysical Journal of the Royal Astronomical Society, v. 33, p. 265–280. [6]

Finlayson, D.M., and Mathur, S.P., 1984, Seismic refraction and reflection features of the lithosphere in northern and eastern Australia, and continental growth: Annales Geophysiae, v. 2, no. 6, p. 711–722. [8]

Finlayson, D.M., and McCracken, HM., 1981, Crustal structure under the Sydney basin and Lachlan fold belt, determined from explosion seismic studies: Journal of the Geological Society of Australia, v. 28, p. 177–190. [7]

Finlayson, D.M., Cull, J.P., Wiebenga, W.A., Furumoto, A.S., and Webb, J.P., 1972, New Britain–New Ireland crustal seismic refraction investigations 1967 and 1969: Geophysical Journal of the Royal Astronomical Society, v. 29, p. 245–253. [2] [6]

Finlayson, D.M., Cull, J.P., and Drummond, B.J., 1974, Upper mantle structure from the Trans-Australia seismic refraction data: Journal of Geological Society Australia, v. 21, p. 447–458. [2] [7]

Finlayson, D.M., Muirhead, K.J., Webb, J.P., Gibson, G., Furumoto, A.S., Cooke, R.J.S., and Russell, A.S., 1976, Seismic investigation of the Papuan Ultramfic belt: Geophysical Journal of the Royal Astronomical Society, v. 44, p. 45–60. [7]

Finlayson, D.M., Drummond, B.J., Collins, C.D.N., and Connelly, J.B., 1977, Crustal structures in the region of the Papuan Ultramafic belt: Physics of the Earth and Planetary Interiors, v. 14, p. 13–29, doi:10.1016/0031 -9201(77)90043-7. [2] [7]

Finlayson, D.M., Prodehl, C., Collins, C.D.N., 1979, Explosion seismic profiles, and implications for crustal evolution, *in* southeastern Australia: BMR Journal of Australian Geology and Geophysics, v. 4, p. 243–252. [2] [7]

Finlayson, D.M., Leven, J.H., and Wake-Dyster, K.D., 1989, Large-scale lenticles in the lower crust under an intra-continental basin in eastern Australia, *in* Mereu, R.F., Mueller, St., and Fountain, D.M., eds., Properties and processes of earth's lower crust: American Geophysical Union, Geophysical Monograph 51, p. 3–16. [2] [8]

Finlayson, D.M., Johnstone, D.W., Owen, A.J., and Wake-Dyster, K.D., 1996, Deep seismic images and the tectonic framework of early rifting in the Otway Basin, Australian southern margin: Tectonophysics, v. 264, p. 137–152, doi:10.1016/S0040-1951(96)00123-0. [2] [9]

Finlayson, D.M., Collins, C.D.N., Lukaszyk, I., and Chudyk, E.C., 1998, A transect across Australia's southern margin in the Otway Basin region: crustal architecture and the nature of rifting from wide-angle seismic profiling, *in* Klemperer, S.L., and Mooney, W.D., eds., Deep seismic probing of the continents, II: a global survey: Tectonophysics, v. 288, p. 177–189. [2] [9]

Finlayson, D.M., Korsch, R.J., Glen, R.A., Leven, J.H. and Johnstone, D.W., 2002, Seismic imaging and crustal architecture across the Lachlan Transverse Zone, a possible early cross-cutting feature of eastern Australia: Australian Journal of Earth Sciences, v. 49, p. 311–321, doi:10.1046/j.1440-0952.2002.00917.x. [2] [9]

FIRE consortium, 2006, FIRE (Finnish reflection experiment), *in* Grad, M., Booth, D., and Tiira, T., eds., European Seismological Commission (ESC), Subcommission D—Crust and Upper Mantle Structure, Activity Report 2004–2006, [2] [10]

Fisher, M.A., Brocher, T.M., Hyndman, R.D., Trehu, A.M., Weaver, C.S., Creager, K.C., Crosson, R.S., Parsons, T., Cooper, A.K., Mosher, D.C., Spence, G.D., Zelt, B.C., Hammer, P.T.C., ten Brink, U.S., Pratt, T.L., Miller, K.C., Childs, J.R., Cochrane, G.R., Chopra, S., and Walia, R., 1999, Seismic survey probes urban earthquake hazards in Pacific Northwest: Eos (Transactions, American Geophysical Union), v. 80, no. 2, p. 13–17, doi:10.1029/99EO00011. [9]

Flecha, I., Palomeras, I., Carbonell, R., Simancas, F., Ayarza, P., Matas, J., Gonzales Lodeiro, F., and Perez-Estaun, A., 2009, Seismic imaging and modelling of the lithosphere of SW Iberia: Tectonophysics, v. 472, p. 148–157, doi:10.1016/j.tecto.2008.05.033. [10]

Fliedner, M.M., and Klemperer, S.L., 1999, Structure of an island-arc: Wide-angle seismic studies in the eastern Aleutian Islands, Alaska: Journal of Geophysical Research, v. 104, p. 10,667–10,694, doi:10.1029/98JB01499. [2] [9]

Fliedner, M.M., Klemperer, S.L., and Christensen, N.I., 2000, Three-dimensional seismic model of the Sierra Nevada arc, California, and its implications for crustal and upper mantle composition: Journal of Geophysical Research, v. 105, p. 10,899–10,921, doi:10.1029/2000JB900029. [9]

Flueh, E.R., Milkereit, B., Meissner, R., Meyer, R.P., Ramirez, J.E., Quintero, J.d.C., and Udias, A., 1981, Seismic refraction observations in northwestern Colombia at latitude 5.5°N: Zentralblatt für Geologie und Paläontologie, Teil I: 1981, p. 231–242. [2] [7]

Flueh, E.R., Mooney, W.D., Fuis, G.S., and Ambos, E.L., 1989, Crustal structure of the Chugach Mountains, southern Alaska: a study of peg-leg multiples from a low-velocity zone: Journal of Geophysical Research, v. 94, p. 16,023–16,035, doi:10.1029/JB094iB11p16023. [8]

Flueh, E.R., Fisher, M., Scholl, D., Parsons, T., ten Brink, U., Klaeschen, D., Kukowshi, N., Trehu, A., Childs, J., Bialas, J., and Vidal, N., 1997, Scientific teams analyze earthquake hazards of the Cascadia subduction zone: Eos (Transactions, American Geophysical Union), v. 78, p. 153–157, doi:10.1029/97EO00097. [9]

Flueh, E.R., Vidal, N., Ranero, C.R., Hojka, A., von Huene, R., Bialas, J., Hinz, K., Cordoba, D., Danobeitia, J.J., and Zelt, C., 1998a, Seismic investigation of the continental margin off- and onshore Valparaiso, Chile, *in* Klemperer, S.L., and Mooney, W.D., eds., Deep seismic probing of the continents, II: a global survey: Tectonophysics, v. 288, p. 251–263. [2] [9]

Flueh, E.R., Klaeschen, D., Weinrebe, W., and Reichert, C., 1998b, Investigation of small submarine volcanic cones at Ninetyeast Ridge by high resolution seismic investigations and bathymetry: Eos (Transactions, American Geophysical Union), v. 79, p. 872. [9]

Flueh, E.R., Grevemeyer, I., and Reichert, C., 1999, Ocean site survey reveals anatomy of a hotspot track: Eos (Transactions, American Geophysical Union), v. 80, p. 77, doi:10.1029/99EO00052. [9]

Forsyth, D.A., Berry, M.J., and Ellis, R.M., 1974, A refraction survey across the Canadian Cordillera at 54°N: Canadian Journal of Earth Sciences, v. 12, p. 539–557. [6]

Forsyth, D.A., Morel-a-l'Huissier, Asudeh, I., and Green, A.G., 1986, Alpha Ridge and Iceland—products of the same plume?: Journal of Geodynamics, v. 6, p. 197–214, doi:10.1016/0264-3707(86)90039-6. [2] [8]

Forsyth, D.A., Overton, A., Stephenson, R.A., Embry, A.F., Ricketts, B.D., and Asudeh, I., 1990, Delineation of sedimentary basins using seismic techniques on Canada's arctic continental margin, *in* Pinet, B., and Bois, C., eds., The potential of deep seismic profiling for hydrocarbon exploration: Editions Technip, Paris, p. 225–236. [2] [8]

Förtsch, O., 1951, Analyse der seismischen Registrierungen der Grosssprengung bei Haslach im Schwarzwald am 28 April 1948: Jahrb. f. Bodenforschung, 66, p. 65–80. [4]

Förtsch, O., and Schulze, G.A., 1948, Seismik der Fernsprengungen: Naturforschung und Medizin in Deutschland 1939–1946, 18, p. 44–51. [5]

Fouqué, F., and Lévy, M., 1889, Expériences sur la vitesse de propagation des secousses: Mémoires de l'Académie des Sciences, Paris, v. 30, no. 2, p. 57–77. [3]

Fowler, C.M.R., 1976, Crustal structure of the Mid-Atlantic ridge crest at 37°N: Geophysical Journal of the Royal Astronomical Society, v. 47, p. 459–491. [2] [7] [8]

Fowler, C.M.R., 1978, The Mid-Atlantic Ridge: structure at 45°N: Geophysical Journal of the Royal Astronomical Society, v. 54, p. 167–183. [2] [7] [8]

Fowler, C.M.R., 2005, The solid earth—an introduction to global geophysics: Cambridge University Press, 685 p. [2]

Fowler, C.M.R., and Keen, C.E., 1979, Oceanic crustal structure—Mid-Atlantic Ridge at 45°N: Geophysical Journal of the Royal Astronomical Society, v. 56, p. 219–226. [7]

Francis, T.J.G., and Raitt, R.W., 1967, Seismic refraction measurements in the southern Indian Ocean: Journal of Geophysical Research, v. 72, p. 3015–3041, doi:10.1029/JZ072i012p03015. [2] [6]

Francis, T.J.G., and Shor, G.G., 1966, Seismic refraction measurements in the northwest Indian Ocean: Journal of Geophysical Research, v. 71, p. 427–449. [2] [6]

Francis, T.J.G., Davies, D., and Hill, M.N., 1966, Crustal structure between Kenya and the Seychelles: Royal Society of London Philosophical Transactions, A259, p. 240–261. [2] [6]

Franke, D., Neben, S., Schreckenberger, B., Schulze, A., Stiller, M., and Krawczyk, C.M., 2006, Crustal structure across the Colorado Basin, offshore Argentina: Geophysical Journal International, v. 165, p. 850–864, doi:10.1111/j.1365-246X.2006.02907.x. [2] [9]

Franke, W., Bortfeld, R.K., Brix, M., Drozdzewski, G., Dürbaum, H.J., Giese, P., Janoth, W., Jödicke, H., Reichert, C., Scherp, R., Schmoll, J., Thomas, R., Thünker, M., Weber, K., Wiesner, M.G., and Wong, H.K., 1990, Crustal structure of the Rhenish Massif: results of the deep seismic reflection lines DEKORP 2-North and 2-North-Q: Geologische Rundschau, v. 79, p. 523–566, doi:10.1007/BF01879201. [2] [8]

Franke, W., Haak, V., Oncken, O., and Tanner, D., eds., 2000, Orogenic processes: quantification and modelling in the Variscan belt: Geological Society London Special Publication 179, 459 p. [2]

Freeman, R., and Mueller, St., eds., 1992, A continent revealed—the European Geotraverse—Atlas of compiled data: Cambridge University Press, 13 maps and CD (EGT database). [2] [8] [10]

Freeman, B., Klemperer, S.L., and Hobbs, R.W., 1988, The deep structure of northern England and the Iapetus suture zone from BIRPS deep seismic reflection profiles: Journal of the Geological Society of London, v. 145, p. 727–740, doi:10.1144/gsjgs.145.5.0727. [2] [8]

Freeman, R., Knorring, M. von, Korhonen, H., Lund, C., and Muller, St., eds., 1989, The European Geotraverse, Part 5: The POLAR Profile: Tectonophysics, v. 162, p. 1–171. [8]

Fuchs, K., 1968, The reflection of spherical waves from transition zones with arbitrary depth-dependent elastic moduli and density: Journal of Physics of the Earth, v. 16, Special Issue, p. 27–41. [6]

Fuchs, K., 1969, On the properties of deep crustal reflectors: Zeitschrift für Geophysik, v.35, p. 133–149. [6]

Fuchs, K., 1975, Seismische Anisotropie des oberen Erdmantels und Intraplatten-Tektonik: Geologische Rundschau, v. 64, p. 700–716, doi:10.1007/BF01820691. [6] [7]

Fuchs, K., 1977, Seismic anisotropy of the subcrustal lithosphere as evidence for dynamical processes in the upper mantle: Geophysical Journal of the Royal Astronomical Society, v. 49, p. 167–179. [6] [7]

Fuchs, K., 1983, Recently formed elastic anisotropy and petrological models of the continental subcrustal lithosphere in southern Germany: Physics of the Earth and Planetary Interiors, v. 31, p. 93–118, doi:10.1016/0031-9201(83)90103-6. [7]

Fuchs, K., and Müller, G., 1971, Computation of synthetic seismograms with the reflectivity method and comparison with observations: Geophysical

Journal of the Royal Astronomical Society, v. 23, p. 417–433. [2] [6] [7] [8] [10]

Fuchs, K., and Vinnik, L.P., 1982, Investigation of the subcrustal lithosphere and asthenosphere by controlled source seismic experiments on long-range profiles, *in* Palmason, G., ed., Continental and oceanic rifts: American Geophysical Union, Geodynamics Series, v. 8, p. 81–98. [7]

Fuchs, K., Mueller, S., Peterschmitt, E., Rothé, J.P., Stein, A., Strobach, K., 1963a, Krustenstruktur der Westalpen nach refraktionsseismischen Messungen: Gerlands Beiträge zur Geophysik, v. 72, p. 149–169. [5] [6]

Fuchs, K., Mueller, S., Peterschmitt, E., Rothé, J.P., Stein, A., Strobach, K., 1963b, Essais d'interprétation séismique. VI.A. Essai d'interprétation no. 1, *in* Closs, H., and Labrouste, Y., eds., Séismologie: Recherches séismologiques dans les Alpes occidentals au moyen de grandes explosions een 1956, 1958 et 1960: Mémoir Collectif, Année Géophysique Internationale, Centre National de la Recherche Scientifique, Série XII, Fasc. 2, p. 118–176. [5] [6]

Fuchs, K., Vinnik, L.P., and Prodehl, C., 1987, Exploring heterogeneities of the continental mantle by high resolution seismic experiments, *in* Fuchs, K., and Froidevaux, C., eds., Composition, structure and dynamics of the lithosphere-asthenosphere system: American Geophysical Union, Geodynamics Series, v. 16, p. 137–154. [2] [7]

Fuchs, K., Altherr, R., Müller, B., and Prodehl, C., eds., 1997, Structure and dynamic processes in the lithosphere of the Afro-Arabian rift system: Tectonophysics, v. 278, no. 1-4, 352 p. [2] [6] [9]

Fuis, G.S., and Mooney, W.D., 1990, Lithospheric structure and tectonics from seismic-refraction and other data, *in* Wallace, G.E., ed., The San Andreas fault system, California: U.S. Geological Survey Professional Paper 1515, p. 207–236. [9] [10]

Fuis, G.S., Mooney, W.D., Healy, J.H., McMechan, G.A., and Lutter, W.J., 1984, A seismic refraction survey of the Imperial Valley region, California: Journal of Geophysical Research, v. 89, p. 1165–1189, doi:10.1029/JB089iB02p01165. [2] [7] [9]

Fuis, G.S., Zucca, J.J., Mooney, W.D., and Milkereit, B., 1987, A geologic interpretation of seismic-refraction results in northeastern California: Geological Society of America Bulletin, v. 98, p. 53–65, doi:10.1130/0016-7606(1987)98<53:AGIOSR>2.0.CO;2. [8]

Fuis, G.S., Ambos, E.L., Mooney, W.D., Christensen, N.I., and Geist, E., 1991, Crustal structure of accreted terranes in southern Alaska, Chugach Mountains and Copper River basin, from seismic refraction results: Journal of Geophysical Research, v. 96, p. 4187–4227, doi:10.1029/90JB02316. [2] [8]

Fuis, G.S., Okaya, D.A., Clayton, R.W., Lutter, W.J., Ryberg, T., Brocher, T.M., Henyey, T.M., Benthien, M.L., Davis, P.M., Mori, J., Catchings, R.D., ten Brink, U.S., Kohler, M.D., Klitgord, K.D., and Bohannon, R.G., 1996, Images of crust beneath southern California will aid study of earthquakes and their effects: Eos (Transactions, American Geophysical Union), v. 77, p. 173, 176. [2] [9]

Fuis, G.S., Murphy, J.M., Lutter, W.J., Moore, T.E., and Bird, K.J., 1997, Deep seismic structure and tectonics of northern Alaska: crustal-scale duplexing with deformation extending into the upper mantle: Journal of Geophysical Research, v. 102, p. 20,873–20,896, doi:10.1029/96JB03959. [2] [8] [9]

Fuis, G.S., Murphy, J.M., Okaya, D.A., Clayton, R.W., Davis, P.M., Thygesen, K., Baher, S.A., Ryberg, T., Benthien, M.L., Simila, G., Perron J.T., Yong, A.K., Reusser, L., Lutter, W.J., Kaip, G., Fort, M.D., Asudeh, I., Sell, R., Vanschaak, J.R., Criley, E.E., Kaderabek, R., Kohler, W.M., and Magnuski, N.H., 2001a, Report for borehole explosion data acquired in the 1999 Los Angeles Region Seismic Experiment (LARSE II), southern California: Part 1, description of the survey: U.S. Geological Survey Open-File Report 01-408, 74 p. [9]

Fuis, G.S., Ryberg, T., Godfrey, N.J., Okaya, D.A., and Murphy, J.M., 2001b, Crustal structure and tectonics from the Los Angeles basin to the Mojave Desert, southern California: Geology, v. 29, p. 15–18, doi:10.1130/0091-7613(2001)029<0015:CSATFT>2.0.CO;2. [9]

Fuis, G.S., Clayton, R.W., Davis, P.M., Ryberg, T., Lutter, W.J., Okaya, D.A., Hauksson, E., Prodehl, C., Murphy, J.M., Benthien, M.L., Baher, S.A., Kohler, M.D., Thygesen, K., Simila, G., and Keller, G.R., 2003, Fault systems of the 1971 San Fernando and 1994 Northridge earthquakes, southern California: Relocated aftershocks and seismic images from LARSE II: Geology, v. 31, p. 171–174, doi:10.1130/0091-7613(2003)031<0171:FSOTSF>2.0.CO;2. [9]

Fuis, G.S., Moore, T.E., Plafker, G., Brocher, T.M., Fisher, M.A., Mooney, W.D., Nokleberg, W.J., Page, R.A., Beaudoin, B.C., Christensen, N.I.,

Levander, A.R., Lutter, W.J., Saltus, R.W., and Ruppert, N.A., 2008, Trans-Alaska Crustal Transect and continental evolution involving subduction underplating and synchronous foreland thrusting: Geology, v. 36, p. 267–270, doi:10.1130/G24257A.1. [9]

Funck, T., Hopper, J.R., Larsen, H.C., Louden, K.E., Tucholke, B.E., and Holbrook, W.S., 2003, Crustal structure of the ocean-continent transition at Flemish Cap: seismic refraction results: Journal of Geophysical Research, v. 108 (B11), 2531, doi:10.1029/2003JB002434. [2] [8] [9]

Furumoto, A.S., Wiebenga, W.A., Webb, J.P., and Sutton, G.H., 1973, Crustal structure of the Hawaiian archipelago, northern Melanesia, and the central Pacific basin by seismic refraction methods, *in* Mueller, S., ed., The structure of the earth's crust, based on seismic data: Tectonophysics, v. 20, p. 153–164. [2] [6]

Gajewski, D., 1989, Compressional and shear-wave velocity models of the Schwarzwald derived from seismic refraction data, *in* Emmermann, R., and Wohlenberg, J., eds., The continental deep drilling program (KTB): Berlin, Heidelberg, New York, Springer Verlag, p. 363–383. [8]

Gajewski, D., and Prodehl, C., 1985, Crustal structure beneath the Swabian Jura, SW Germany, from seismic refraction investigations: Journal of Geophysics, v. 56, p. 69–80. [7] [8]

Gajewski, D., and Prodehl, C., 1987, Seismic refraction investigation of the Black Forest, *in* Freeman, R., and Mueller, St., eds., The European Geotraverse, Part 3: Tectonophysics, v. 142, p. 27–48. [2] [8]

Gajewski, D., and Pšenčik, I., 1987, Computation of high-frequency seismic wavefields in 3-D laterally inhomogeneous anisotropic media: Geophysical Journal of the Royal Astronomical Society,v. 91, p. 383–411. [8]

Gajewski, D., and Pšenčik, I., 1988, Ray synthetic seismograms for a 3-D anisotropic lithospheric structure: Physics of the Earth and Planetary Interiors, v. 51, p. 1–23, doi:10.1016/0031-9201(88)90017-9. [8]

Gajewski, D., Holbrook, W.S., and Prodehl, C., 1987, A three-dimensional crustal model of southwest Germany derived from seismic refraction data, *in* Freeman, R., and Mueller, St., eds., The European Geotraverse, Part 3: Tectonophysics, v. 142, p. 49–70. [2] [8]

Gajewski, D., Prodehl, C., Ritter, J., and Feddersen, J., 1988, A compilation of data from the 1984 seismic-refraction experiment in SW-Germany: Open-File Report 88-1, Geophysics Institute, University of Karlsruhe, 110 p. [8]

Gajewski, D., Prodehl, C., and Zeis, S., 1989, A compilation of data from the Wildflecken-1982 seismic-refraction experiment in southern Germany: Open-File Report 89-1, Geophysical Institute, University of Karlsruhe, 41 p. [8]

Gajewski, D., Schulte, A., Riaroh, D., and Thybo, H., 1994, Deep seismic sounding in the Turkana depression, northern Kenya rift, *in* Prodehl, C., Keller, G.R., and Khan, M.A., eds., Crustal and upper mantle structure of the Kenya rift: Tectonophysics, v. 236, p. 165–178. [9]

Galitzin, Fürst B., 1914, Vorlesungen über Seismometrie (ed. by O. Hecker): Teubner, Leipzig-Berlin, 538 p. [3]

Gallart, J., Daignieres, M., Banda, E., Surinach, E., and Hirn, A., 1980, The eastern Pyrenean domain: lateral variations at crust-mantle level: Annales de Géophysique, v. 36, p. 141–158. [2] [7]

Gallart, J., Rojas, H., Díaz, J., and Danobeitia, J.J., 1990, Features of deep crustal structure and transition offshore-onshore at the Iberian flank of the Valencia Trough (Western Mediterranean): Journal of Geodynamics, v. 12, p. 233–252, doi:10.1016/0264-3707(90)90009-J. [8]

Gallart, J., Vidal, N., Danobeitia, J.J., and the ESCI-Valencia Trough Working Group, 1994, Lateral variations in the deep crustal structure at the Iberian margin of the Valencia trough imaged from seismic reflection methods: Tectonophysics, v. 232, p. 59–75, doi:10.1016/0040-1951(94)90076-0. [2] [9]

Gallart, J., Vidal, N., and Danobeitia, J.J., 1995a, Multichannel seismic image of the crustal thinning at the NE Iberian margin combing normal and wide-angle reflection data: Geophysics Research Letters, v. 22, no. 4, p. 489–492, doi:10.1029/94GL03272. [9]

Gallart, J., Díaz, J., Vidal, N., and Danobeitia, J.J., 1995b, The base of the crust in the Betics–Alboran Sea transition: evidences of an abrupt structural variation from wide-angle ESCI data: Acta Geológica Hispanica, v. 8, p. 519–528. [9]

Gallart, J., Díaz, J., Nercessian, A., Mauffret, A., and Dos Reis, T., 2001, The eastern end of the Pyrenees: seismic features at the transition to the NW Mediterranean: Geophysical Research Letters, v. 28, no. 11, p. 2277–2280, doi:10.1029/2000GL012581. [9]

Galperin, E., and Kosminskaya, I.P., 1958, Characteristics of the methods of deep seismic sounding on the sea: Bulletin of the Academy of Science, USSR, Geophysics Series 7, p. 475–483. [4] [5]

Galperin, E., and Kosminskaya, I.P., eds., 1964, Structure of the earth's crust in the transition zone between Asia and the Pacific: Moscow, Nauka (in Russian). [2] [5] [6]

Galvé, A., Hirn, A., Mei, J., Gallart, J., de Voogd, B., Lepine, J.-C., Díaz, J., Youxue, W., and Hui, Q., 2002, Modes of raising northeastern Tibet probed by explosion seismology: Earth and Planetary Science Letters, v. 203, p. 35–43, doi:10.1016/S0012-821X(02)00863-4. [2] [9]

Galvé, A., Jiang, M., Hirn, A., Sapin, M., Laigle, M., de Voogd, B., Gallart, J., and Qian, H., 2006, Explosion seismic P and S velocity and attenuation constraints on the lower crust of the North-Central Tibetan Plateau, and comparison with South Tibet–Himalayas: implications on composition, mineralogy, temperatures and tectonic evolution: Tectonophysics, v. 412 (3–4), p. 141–157, doi:10.1016/j.tecto.2005.09.010. [9]

Gamburtsev, 1952, The seismic deep sounding of the earth's crust (in Russian): Doklady Nauk SSSR, v. 87 (6), p. 943–946.[2] [4]

Ganchin, Y.V., Smithson, S.B., Morozov, I.B., Smythe, D.K., Garipov, V.Z., Karaev, N.A., and Kristofferson, Y., 1998, Seismic studies around the Kola Superdeep Borehole, Russia, in Klemperer, S.L., and Mooney, W.D., eds., Deep seismic probing of the continents, II: a global survey: Tectonophysics, v. 288, p. 1–16. [2] [9]

Gane, P.G., Hales, A.L., and Oliver, H.O., 1946, A seismic investigation of the Witwatersrand earth tremors: Bulletin of the Seismological Society of America, v. 36, p. 49–80. [4]

Gao, R., Cheng, X.-Z., and Ding, Q., 1995, Preliminary geodynamic model of Golmud-Ejin Qi geoscience transect: Acta Geophysica Sinica, v. 38, p. 3–14. [9]

Gao, R., Huang, D., Lu, D., et al., 2000, Deep seismic reflection profile across the juncture zone between the Tarim basin and the West Kunlun Mountains: Chinese Science Bulletin, v. 45, p. 2281–2286, doi:10.1007/BF02886369. [9]

Garcia-Duenas, V., Banda, E., Torne, M., Cordoba, D., and ESCI-Beticas Working Group, 1994, A deep seismic reflection survey across the Betic Chain (southern Spain): first results: Tectonophysics, v. 232, p. 77–89, doi:10.1016/0040-1951(94)90077-9. [9]

Garrick, R.A., 1968, A reinterpretation of the Wellington crustal refraction profile: New Zealand Journal of Geology and Geophysics, v. 11, p. 1280–1294. [5]

Gaskell, T.F., and Swallow, J.C., 1951, Seismic refraction experiments in the North Atlantic: Nature, v. 167, 4253, p. 723–724, doi:10.1038/167723a0. [2] [4] [5]

Gaskell, T.F., and Swallow, J.C., 1952, Seismic refraction experiments in the Pacific: Nature, v. 170, 4337, p. 1010–1012, doi:10.1038/1701010a0. [2] [5]

Gaskell, T.F., and Swallow, J.C., 1953, Seismic refraction experiments in the Indian Ocean and the Mediterranean Sea: Nature, v. 172, p. 535, doi:10.1038/172535b0. [2] [5]

Gaskell, T.F., Hill, M.N., and Swallow, J.C., 1958, Seismic measurements made by H.M.S. *Challenger* in the Atlantic, Pacific and Indian Oceans and in the Mediterranean Sea, 1950–1953: Philosophical Transactions of the Royal Society of London, A251, p. 23–83, doi:10.1098/rsta.1958.0008. [5] [6]

Gaulier, J.M., LePichon, X., Lyberis, N., Avedik, F., Geli, L., Moretti, I., Deschamps, A. and Hafez, S., 1988, Seismic study of the crust of the northern Red Sea and Gulf of Suez, in LePichon, X., and Cochran, J.R., eds., The Gulf of Suez and Red Sea rifting: Tectonophysics, v. 153, p. 55–88. [2] [8]

Gebrande, H., 1976, A seismic ray–tracing method for two-dimensional inhomogeneous media, in Giese, P., Prodehl, C., and Stein, A., eds., Explosion seismology in central Europe—Data and Results: Berlin-Heidelberg-New York, Springer, p. 162–167. [7] [8]

Gebrande, H., Miller, H., and Einarson, P., 1980, Seismic structure along RRISP–profile I, in Jacoby, W., Björnsson, A., and Möller, D., eds., Iceland: Evolution, active tectonic, and structure: Journal of Geophysics, v. 47, p. 239–249. [7]

Gebrande, H., Bopp, M., Neurieder, P., and Schmidt, T., 1989, Crustal structure in the surroundings of the KTB drill site as derived from refraction and wide-angle seismic observations, in Emmermann, R., and Wohlenberg, J., eds., The German continental deep drilling program (KTB)—Site-selection studies in the Oberpfalz and Schwarzwald. Springer, Berlin-Heidelberg, p. 151–176. [2] [8]

Gebrande, H., Bopp, M., Meichelböck, M., and Neurieder, P., 1991, 3-D wide-angle investigations in the KTB surroundings as part of the "Integrated Seismics Oberpfalz 1989 (ISO89)," first results, in Meissner, R., Brown,

L., Dürbaum, H.-J., Franke, W., Fuchs, K., and Seifert, F., eds., Continental lithosphere: deep seismic reflections: American Geophysical Union, Geodynamics Series, v. 22, p. 147–160. [8]

Gebrande, H., Castellarin, A., Lueschen, E., Millahn, K., Neubauer, F., and Nicolich, R., 2006, TRANSALP—a transect through a young collisional orogen: Tectonophysics, v. 414, p. 1–282, doi:10.1016/j.tecto.2005.10.030. [2] [9] [10]

Gee, D.G., and Stephenson, R.A., editors, 2006, European lithosphere dynamics: Geological Society of London, Memoirs, v. 32, 662 p. [2] [9] [10]

Gee, D.G., and Zeyen, H.J., eds., 1996, EUROPROBE—Lithospheric Dynamics: origin and evolution of continents: EUROPROBE Secretariat, Upssala University, 138 p. [2]

Gerdom, M., Trehu, A.M., Flueh, E.R., and Klaeschen, D., 2000, The continental margin off Oregon from seismic investigations: Tectonophysics, v. 329, p. 79–97, doi:10.1016/S0040-1951(00)00190-6. [9]

Gerecke, F., 1933, Wellentypen, Strahlengang und Tiefenberechnung bei seismischen Eisdickenmessungen auf dem Rhonegletscher [Ph.D. thesis]: Univeristy of Goettingen, Germany, 28 p. [3]

German Research Group for Explosion Seismology, 1964, Crustal structure in Western Germany: Zeitschrift für Geophysik, v. 30, p. 209–234. [6]

Gettings, M.E., Blank, H.R., Mooney, W.D., and Healy, J.H., 1986, Crustal structure of southwestern Saudi Arabia: Journal of Geophysical Research, v. 91, p. 6491–6512, doi:10.1029/JB091iB06p06491. [7] [8]

Gettrust, J.F., Furukawa, K., and Kempner, W.B., 1982, Variation in young oceanic crust and upper mantle: Journal of Geophysical Research, v. 87, p. 8435–8445, doi:10.1029/JB087iB10p08435. [7]

Gibbs, A.K., Payne, B., Setzer, T., Brown, L.D., Oliver, J.E., and Kaufman, S., 1984, Seismic reflection study of the Precambrian crust of central Minnesota: Geological Society of America Bulletin, v. 95, p. 280–294, doi:10.1130/0016-7606(1984)95<280:SSOTPC>2.0.CO;2. [2] [7]

Gibbs, J.F., 1972, unpublished interpretation (see Warren et al., 1972). [6]

Gibbs, J.F., and Roller,J.C., 1966, Crustal structure determined by seismic refraction measurements between the Nevada test site and Ludlow, California, in Geological Survey Research 1966, U.S. Geological Survey Professional Paper, 550-D, p. D125–D131. [6]

Giese, P., 1968a, Versuch einer Gliederung der Erdkruste im nördlichen Alpenvorland, in den Ostalpen und in Teilen der Westalpen mit Hilfe charakteristischer Refraktions–Laufzeit–Kurven sowie einer geologischen Deutung: Geophys. Abh. Inst. Meteorol. u. Geophys., FU Berlin 1 (2), p. 202 p. [4] [6]

Giese, P., 1968b, Die Struktur der Erdkruste im Bereich der Ivrea-Zone: Schweizerische Mineralogische und Petrographische Mitteilungen, v. 48, p. 261–284. [6]

Giese, P., 1976a, Problems and tasks of data generalization, in Giese, P., Prodehl, C., and Stein, A., eds., Explosion seismology in central Europe—data and results: Berlin-Heidelberg-New York, Springer, p. 137–145. [2] [6]

Giese, P., 1976b, Depth calculation, in Giese, P., Prodehl, C., and Stein, A., eds., Explosion seismology in central Europe—data and results: Berlin-Heidelberg-New York, Springer, p. 146–161. [6] [10]

Giese, P., 1976c, Results of the generalized interpretation of the deep-seismic sounding data, in Giese, P., Prodehl, C., and Stein, A., eds., Explosion seismology in central Europe—data and results: Berlin-Heidelberg-New York, Springer, p. 201–214. [6]

Giese, P., 1985, The structure of the upper lithosphere between the Liguran Sea and the Southern Alps, in Galson, D.A., and Mueller, S., eds., Proceedings of the Second Workshop on the Geotraverse Project, the southern segment: European Science Foundation, Strasbourg, p. 143–153. [8]

Giese, P., and Morelli, C., 1973, Crustal structure of Italy—Some general features from explosion seismology: Bulletin of the Geological Society of Greece, v. 10, p. 94–98. [6]

Giese, P., and Prodehl, C., 1976, Main features of crustal structure in the Alps, in Giese, P., Prodehl, C., and Stein, A., eds., Explosion seismology in central Europe—data and results: Berlin-Heidelberg-New York, Springer, p. 347–375. [2] [6]

Giese, P., and Schütte, K.-G., 1980, Resultados das medias de sismica de refracao a l'este da Serra do Espinhaco, M.G., Brazil, in Zeil, W., ed., Proyectos de la Deutsche Forschungsgemeinschaft, Grupo de Trabajo: Investigacion Geoscientifica en Latinoamerica: Deutsche Forschungsgemeinschaft, Bonn, Germany, p. 44–50. [7]

Giese, P., and Stein, A., 1971, An attempt of a generalized interpretation of deep-sounding measurements in the area between the North Sea and the Alps: Zeitschrift für Geophysik, v. 37, p. 237–272. [6]

Giese, P., Prodehl, C., and Behnke, C., 1967, Ergebnisse refraktionsseismischer Messungen 1965 zwischen dem Französischen Zentralmassiv und den Westalpen: Zeitschrift für Geophysik, v. 33, p. 215–261. [2] [6]

Giese, P., Morelli, C., Prodehl, C., and Vecchia, O., 1971, Crust and upper mantle beneath the southern zone of Ivrea: Proceedings of the 12th General Assembly of the European Seismological Commission (Luxembourg 1970), Obs. Royal de Belgique, Comm. Sér. A–no 13, Séries Geophysics 101, p. 182–183. [6]

Giese, P., Morelli, C., and Steinmetz, L., 1973, Main features of crustal structure in western and central Europe based on data of explosion seismology, *in* Mueller, St., ed., The structure of the earth's crust based on seismic data: Tectonophysics, v. 20, p. 367–379. [2] [6] [7] [9]

Giese, P., Prodehl, C., and Stein, A., eds., 1976a, Explosion seismology in central Europe—data and results: Berlin-Heidelberg-New York, Springer, 429 p. [2] [4] [5] [6] [10]

Giese, P., Hinz, E., Prodehl, C., Schröder, H., and Stein, A., 1976b, Description of profiles, *in* Giese, P., Prodehl, C., and Stein, A., eds., Explosion seismology in central Europe—data and results: Berlin-Heidelberg-New York, Springer, p. 73–112. [6]

Giese, P., Makris, J., Akashe, B., Röwer, P., Letz, H., and Mostaanpour, M., 1984, The crustal structure in southern Iran derived from seismic explosion data: Neues Jahrbuch für Geologie und Paläontologie Abhandlungen, v. 168, p. 230–243. [2] [7]

Giese, P., Reutter, K.-J., Jacobshagen, V., and Nicolich, R., 1982, Explosion seismic crustal studies in the Alpine-Mediterranean region and their implications to tectonic processes, *in* Berckhemer, H., and Hsü, K., eds., Alpine–Mediterranean geodynamics: Geodynamics Series, American Geophysical Union, 7, p. 39–73. [8]

Giese, P., Scheuber, E., Schilling, F., Schmitz, M., and Wigger, P., 1999, Crustal thickening processes in the Central Andes and the different natures of the Moho–discontinuity, *in* Reutter, K.-J., ed., Central Andean deformation: Journal of South American Earth Science, v. 12, p. 201–220. [2] [9]

Ginzburg, A., Makris, J., Fuchs, K., Prodehl, C., Kaminski, W., and Amitai, U., 1979a, A seismic study of the crust and upper mantle of the Jordan–Dead Sea rift and their transition towards the Mediterranean Sea: Journal of Geophysical Research, v. 84, p. 1569–1582, doi:10.1029/JB084iB04p01569. [2] [7] [9]

Ginzburg, A., Makris, J., Fuchs, K., Perathoner, B., and Prodehl, C., 1979b, Detailed structure of the crust and upper mantle along the Jordan–Dead Sea rift: Journal of Geophysical Research, v. 84, p. 5605–5612, doi:10.1029/JB084iB10p05605. [7] [8]

Ginzburg, A., Mooney, W.D., Walter, A.W., Lutter, W.J., and Helay, J.H., 1983, Deep structure of northern Mississippi embayment: American Association of Petroleum Geologists Bulletin, v. 67, p. 2031–2046. [7]

Ginzburg, A., Makris, J., and Nicolich, R., 1986, European Geotraverse: a seismic refraction profile across the Ligurian Sea, *in* Galson, D.A., and Mueller, St., eds., The European Geotraverse, Part 1: Tectonophysics, v. 126, p. 85–97. [8]

Gish, D.M., Keller, G.R., and Sbar, M.L., 1981, A refraction study of deep crustal structure in the Basin and Range; Colorado Plateau of eastern Arizona: Journal of Geophysical Research, v. 86, p. 6029–6038, doi:10.1029/JB086iB07p06029. [7]

Given, J.W., and Helmberger, D.V., 1983, Inversion of SH body wave seismograms for the shear velocity of the western U.S.: Eos (Transactions, American Geophysical Union), v. 64, p. 763. [8]

Glen, R.A., Korsch, R.J., Direen, N.G., Jones, L.E.A., Johnstone, D.W., Lawrie, K.C., Finlayson, D.M., and Shaw, R.D., 2002, Crustal structure of the Ordovician Macquarie Arc, Eastern Lachlan Orogen, based on seismic reflection profiling: Australian Journal of Earth Sciences, v. 49, p. 323–348, doi:10.1046/j.1440-0952.2002.00925.x. [2] [9]

Glocke, A., and Meissner, R., 1976, Near-vertical reflections recorded at the wide-angle profile in the Rhenish Massif, *in* Giese, P., Prodehl, C., and Stein, A., eds., Explosion seismology in central Europe—data and results: Berlin-Heidelberg-New York, Springer, p. 252–256. [6]

Gobert, B., Hirn, A., and Steinmetz, L., 1972, Shots of profile II, recordd on land, north of the Pyrenees, *in* Leenhardt, O., Gobert, B., Hinz, K., Hirn, A., Hirschleber, H., Hsü, H.J., Refubatti, A., Rudant, J.-P., Rudloff, R., Ryan, W.B.F., Snoek, M., and Steinmetz, L., eds., Results of the Anna cruise—three north-south seismic profiles through the western Mediterranean Sea: Bulletin du Centre Recherches Pau–SNPA, v. 6, p. 373–381. [7]

Godfrey, N.J., Beaudoin, B.C., Lendl, C., Meltzer, A., and Luetgert, J.H., 1995, Data report for the 1993 Mendocino Triple Junction seismic experiment: U.S. Geological Survey Open-File Report, 95-275, 83 p. [9]

Godfrey, N.J., Meltzer, A.S., Klemperer, S.L., Trehu, A.M., Leitner, B., Clarke, Jr., S.H., and Ondrus, A., 1998, Evolution of the Gorda Escarpment, San Andreas fault and Mendocino triple junction from multichannel seismic data collected across the northern Vizcaino block, offshore northern California: Journal of Geophysical Research, v. 103, p. 23,813–23,825, doi:10.1029/98JB02138. [9]

Godfrey, N.J., Davey, F., Stern, T.A., and Okaya, D., 2001, Crustal structure and thermal anomalies of the Dunedin Region, South Island, New Zealand: Journal of Geophysical Research, v. 106, p. 30,835–30,848. [9]

Gohl, K., and Smithson, S.B., 1993, Structure of the Archean crust and passive margin of southwest Greenland from seismic wide-angle data: Journal of Geophysical Research, v. 98, p. 6623–6638, doi:10.1029/93JB00016. [8]

Gohl, K., Smithson, S.B., and Kristoffersen, Y., 1991, The structure of the Archean crust in SW Greenland from seismic wide-angle data: a preliminary anaysis, *in* Meissner, R., Brown, L., Dürbaum, H.-J., Franke, W., Fuchs, K., and Seifert, F., eds., Continental lithosphere: deep seismic reflections: American Geophysical Union, Geodynamics Series, v. 22, p. 53–57. [8]

Gohn, G.S., ed., 1983, Studies related to the Charleston, South Carolina, earthquake of 1886—tectonics and seismicity: U.S. Geological Survey Professional Paper 1313. [8]

Goldflam, P., Weigel, W., and Loncarevic, B.D., 1980, Seismic structure along RRISP–profile I on the southeast flank of the Reykjanes Ridge, *in* Jacoby, W., Björnsson, A., and Möller, D., eds., Iceland: Evolution, active tectonic, and structure: Journal of Geophysics, v. 47, p. 250–260. [7]

Goldman, M.R., Fuis, G.S., Luetgert, J.H., and Geddes, D.J., 1992, Data report for the TACT 1987 seismic refraction survey: Fairbanks North and Olnes deployments: Menlo Park, California, U.S. Geological Survey Open-File Report 92-196, 98 p. [8]

Goldschmidt-Rokita, A., Hansch, K.F., Hirschleber, H.B., Iwasaki, T., Kanazawa, T., Shimamura, H., and Sellevoll, M.A., 1994, The ocean/continent transition along a profile through the Lofoten basin, Northern Norway: Marine Geophysical Research, v. 16, p. 201–224, doi:10.1007/BF01237514. [8]

Goleby, B.R., Drummond, B.J., Korsch, R.J., Willcox, J.B., O'Brien, G.W., and Wake-Dyster, K.D., 1994, Review of recent results from continental deep seismic profiling in Australia: Tectonophysics, v. 232, p. 1–12, doi:10.1016/0040-1951(94)90072-8. [2] [9]

Goleby, B.R., Korsch, R.J., Fomin, T., Bell, B., Nicoll, M.G., Drummond, B.J. and Owen, A.J., 2002, A preliminary 3D geological model of the Kalgoorlie region, Yilgarn Craton, Western Australia based on deep seismic reflection and potential field data: Australian Journal of Earth Sciences, v. 49, p. 917–933, doi:10.1046/j.1440-0952.2002.00967.x. [2] [9]

Goleby, B.R., Blewett, R.S., Korsch, R.J., Champion, D.C., Cassidy, K.F., Jones, L.E.A., Groenewald, P.B., and Henson, P., 2004, Deep seismic reflection profiling in the Archaean northeastern Yilgarn Craton, Western Australia: implications for crustal architecture and mineral potential: Tectonophysics, v. 388, p. 119–133, doi:10.1016/j.tecto.2004.04.032. [2] [9] [10]

Goleby, B.R., Blewett, R.S., Fomin, T., Fishwick, S., Reading, A.M., Henson, P.A., Kennett, B.L.N., Champion, D.C., Jones, L., Drummond, B.J., and Nicoll, M., 2006, An integrated multi-scale 3D seismic model of the Archaean Yilgarn Craton, Australia: Tectonophysics, v. 420, p. 75–90, doi:10.1016/j.tecto.2006.01.028. [9] [10]

Gomes, P.O., Gomes, B.S., Palma, J.J.C., Jinno, K., and de Souza, J.M., 2000, Ocean-continent transition and tectonic framework of the oceanic crust at the continental margin off NE Brazil: Results of LEPLAC project, *in* Mohriak, W., and Talwani, M., eds., Atlantic Rifts and Continental Margins: American Geophysical Union, Geophysical Monograph 115, p. 261–291. [2] [9]

Goncharov, A.G., Lizinsky, M.D., Collins, C.D.N., Kalnin, K.A., Fomin, T.N., Drummond, B.J., Goleby, B.R., Platonenkova, L.N., 1998, Intracrustal "Seismic isostasy" in the Baltic Shield and Australian Precambrian cratons from deep seismic profiles and the Kola Superdeep bore hole data, *in* Braun, J., Dooley, J.C., Goleby, B.R., van der Hilst, R.D., and Klootwijk,C.T., eds., Structure and evolution of the Australian continent: American Geophysical Union, Geodynamics Series, v. 26, p. 119–138. [9]

Gorini, C., Le Marrec, A., and Mauffret, A., 1993, Contribution to the structural and sedimentary history of the Gulf of Lions (Western Mediterranean), from the ECORS profiles, industrial seismic profiles and well data: Bulletin de la Société Géologique de France, v. 164, p. 353–363. [8]

Gorman, A.R., Clowes, R.M., Ellis, R.M., Henstock, T.J., Spence, G.D., Keller, G.R., Levander, A., Snelson, C.M., Burianyk, M.J.A., Kanasewich, E.R., Asudeh, I., Hajnal, Z., and Miller, K.C., 2002, Deep Probe: imaging the roots of western North America: Canadian Journal of Earth Science, v. 39, p. 375–398, doi:10.1139/e01-064. [2] [9]

Goslin, J., Recq, M., and Schlich, R., 1981, Structure profonde du plateau de Madagascar: relations avec le plateau de Crozet: Tectonophysics, v. 76, p. 75–97, doi:10.1016/0040-1951(81)90254-7. [2] [7]

Grad M., and Tripolsky A.A., 1995, Crustal structure from P and S waves and petrological models of the Ukrainian shield: Tectonophysics, v. 250, 89–112, doi:10.1016/0040-1951(95)00045-X. [7]

Grad, M., Shiobara, H., Janik, T., Guterch, A., and Shimamura, H., 1997, Crustal model of the Bransfield Rift, West Antarctica, from detailed OBS refraction experiments: Geophysical Journal International, v. 130, p. 506–518, doi:10.1111/j.1365-246X.1997.tb05665.x. [2] [9]

Grad M., Keller G.R., Thybo, H., Guterch A., and POLONAISE Working Group, 2002, Lower lithosphere structure beneath the Trans-European Suture Zone from POLONAISE'97 seismic profiles: Tectonophysics, v. 360, p. 153–168, doi:10.1016/S0040-1951(02)00350-5. [9]

Grad M., Jensen, S.L., Keller G.R., Guterch A., Thybo, H., Janik, T., Tiira, T., Yliniemi, J., Luosto, U., Motuza, G., Nasedkin, V., Czuba, W., Gaczynski, E., Sroda, P., Miller, K.C., Wilde-Piorko, M., Komminaho, K., Jacyna, J., and Korabliova, L., 2003, Crustal structure of the Trans–European Suture Zone region along POLONAISE'97 seismic profile P4: Journal of Geophysical Research, v. 108, 2541, doi:10.1029/2003JB002426. [9]

Grad M., Guterch A., Keller G.R., Janik, T., Hegedüs E., Vozar, J., Slaczka, A., Tiira, T., and Yliniemi, J., 2006, Lithospheric structure beneath trans-Carpathian transect from Precambrian platform to Pannonian basin: CELEBRATION 2000 seismic profile CEL05: Journal of Geophysical Research, v. 111, B03301, doi:1029/2005JB003647. [9]

Grad M., Guterch A., Mazur, S., Keller, G.R., Špicak, A., Hrubcová, P., Geissler, W.H., and SUDETES 2003 Working Group, 2008, Lithospheric structure of the Bohemian Massif and adjacent Variscan belt in central Europe based on Profile S01 from the SUDETES 2003 experiment: Journal of Geophysical Research, v. 113, B10304, doi:10.1029/2007JB005497. [10]

Grad M., Tiira T., and ESC Working Group, 2009, The Moho depth map of the European Plate: Geophysical Journal International, v. 176, p. 279–292, doi: 10.1111/j.1365-246X.2008.03919.x. [10]

Graeber, F.M., and Asch, G., 1999, Three-dimensional models of P wave velocity and P- to S-velocity ratio in the southern central Andes by simultaneous inversion of local earthquake data: Journal of Geophysical Research, v. 104, p. 20,237–20,256, doi:10.1029/1999JB900037. [9]

Granet, M., Stoll, G., Dorel, G., Achauer, U., Poupinet, G., and Fuchs, K., 1995, Massif Central (France): new constraints on the geodynamical evolution from teleseismic tomography: Geophysical Journal International, v. 121, p. 33–48, doi:10.1111/j.1365-246X.1995.tb03509.x. [9]

GRANIT Transect, 1992, in Rybalka, V., and Kashubin, S., eds., Methods and results of investigations: Ekaterinburg, URGK I UTP VNTGeo, 113 p. (in Russian). [9]

Grant, A.C., 1975, Structural modes of the western margin of the Labrador Sea, in van der Linden, W.J.M., and Walde, J.A., eds., Offshore Geology of Eastern Canada, vol. II, Geological Survey of Canada Paper 74-30, p. 217–231. [7]

Grant, F.S., and West, G.F., 1965, Interpretation theory in applied geophysics: New York, McGraw-Hill, 584 p. [2] [10]

Green, A.G., Berry, M.J., Spencer, C.P., Kanasewich, E.R., Chiu, S., Clowes, R.M., Yorath, C.J., Stewart, D.B., Unger, J.D., and Poole,W.H., 1986, Recent seismic reflection studies in Canada, in Barazangi, M., and Brown, L., eds., Reflection seismology: a global perspective: American Geophysical Union, Geodynamics Series, v. 13, p. 85–97. [2] [7] [8]

Green, A.G., Cannon, W.F., Milkereit, B., Hutchinson, D.R., Davidson, A., Behrendt, J.C., Spencer, C., Lee, M.W., Moerl-a-l'Hussier, P.,and Agena, W.F., 1989, A "GLIMPCE" of the deep crust beneath the Great Lakes, in Mereu, R.F., Mueller, St., and Fountain, D.M., eds., Properties and processes of earth's lower crust: American Geophysical Union, Geophysical Monograph 51, p. 65–80. [2] [8]

Green, A.G., Milkereit, B., Morel-a-l'Huissier, P., Spencer, C., Ellis, R.M., Clowes, R.M., and Ansorge, J., 1990a, Studies of laterally heterogeneous structures using seismic refraction and reflection data: Geological Survey of Canada Paper 89-13: 224 p. [2] [8] [10]

Green, A.G., Clowes, R.M., and Ellis, R.M., 1990b, Crustal studies across Vancouver Island and adjacent offshore margin: Geological Survey of Canada paper 89-13, p. 3–25. [2] [8]

Green, R.W.E., 1973, A portable multi-channel recorder and a data processing system: Bulletin of the Seismological Society of America, v. 63, p. 423–431. [7]

Green, R.W.E., and Hales, A.L., 1966, Seismic refraction measurements in the southwestern Indian Ocean: Journal of Geophysical Research, v. 71, p. 1637–1648. [6]

Green, W.V., Achauer, U., and Meyer, R.P., 1991, A three-dimensional seismic image of the crust and upper mantle beneath the Kenya rift: Nature, v. 354, p. 199–203, doi:10.1038/354199a0. [9]

Greenroyd, C.J., Peirce, C., Rodger, M., Watts, A.B., and Hobbs, R.W., 2006, Crustal structure of the French Guiana continental margin—implications for rifting and the early spreading of the eqatorial Atlantic: European Geophysical Union, Annual Meeting 2006, Geophysical Research Abstracts, 8, 07788. [2] [10]

Gregersen, S., Flüh, E.R., Moeller, C., and Hirschleber, H., 1987, Seismic data of the EUGENO-S project (European Geotraverse Northern segment Southern part): Open-File Report, Department of Seismology, Danish Geodetic Institute, Charlottenlund, Denmark. [8]

Gregersen, S., Voss, P., and TOR Working Group, 2002, Summary of project TOR delineation of a stepwise, sharp, deep lithosphere transition across Germany-Denmark-Sweden: Tectonophysics, v. 360, p. 61–73, doi:10.1016/S0040-1951(02)00347-5. [9]

Grevemeyer, I., Weigel, W., Whitmarsh, R.B., Avedik, F., and Deghani, G.A., 1997, The Aegir Rift: crustal structure of an extinct spreading axis: Marine Geophysical Research, v. 19, p. 1–23. [8]

Grevemeyer, I., Weigel, W., and Jennrich, C., 1998, Seismic structure and crustal ageing at 14°S on the East Pacific Rise: Geophysical Journal International, v. 135, p. 573–584, doi:10.1046/j.1365-246X.1998.00673.x. [2] [9]

Grevemeyer, I., Flueh, E.R., Reichert, C., Bialas, J., Kläschen, D., and Kopp, C., 2001a, Crustal architecture and deep structure of the Ninetyeast Ridge hotspot trail from active-source ocean bottom seismology: Geophysical Journal International, v. 144, p. 414–431, doi:10.1046/j.0956-540X.2000.01334.x. [2] [9]

Grevemeyer, I., Weigel, W., Schüssler, S., and Avedik, F., 2001b, Crustal and upper mantle seismic structure and lithospheric flexure along the Society Island hotspot chain: Geophysical Journal International, v. 147, p. 123–140, doi:10.1046/j.0956-540x.2001.01521.x. [2] [8] [9]

Grevemeyer, I., Aric, K., Booth-Rea, G., Böhme, G., Faria, B., Fröhlich, J., Greenroyd, G., Haase, C., Hahn, J., Kahl, G, Krabbenhöft, A., Neiss, H., Nöske, M., Osenhirt, W.-T., Peirce, C., Rodger, M., Schnabel, M., Schlesinger, A., Schwenk, A., Wäbs, S., and Weinrebe., W., 2009, Cape Verde Hotspot: A seismic refraction study of isostasy and magmatic underplating, in Steinfeld, R., Rhein, M., Brandt, P., Grevemeyer, I., Reston, T., Devey, C., and Lackschewitz, K., eds., Oceanography, geology and geophysics of the South Equatorial Atlantic. Meteor-Berichte M62, no. 09-1. Leitstelle METEOR, Institute Meereskunde, University of Hamburg, p. 3-1–3-40. [2] [10]

Griffiths, D.J., and Westbrook, G.K., 1992, Deep Geology, in Duff, P.McL.D., and Smith, A.J., eds., Geology of England and Wales: Geological Society of London. [7]

Griffiths, D.H., King, R.F., Khan, M.A., and Blundell, D.J., 1971, Seismic refraction line in the Gregory rift: Nature, v. 229, p. 69–71. [2] [6]

Grobys, J.W.G., Gohl, K., Davy, B., Uenzelmann-Neben, G., Deen, T., and Barker, D., 2007, Is the Bounty Trough off eastern New Zealand an aborted rift?: Journal of Geophysical Research, v. 112: B03103, doi:10.1029/2005JB004229. [2] [10]

Grobys, J.W.G., Gohl, K., Uenzelmann-Neben, G., Davy, B., and Barker, D., 2009, Extensional and magmatic nature of the Campbell Plateau and Great South Basin from deep crustal studies: Tectonophysics, v. 472, no. 1-4, p. 213–225, doi:10.1029/2005JB004229. [2] [10]

Groß, K., Micksch, U., and TIPTEQ Research Group, Seismics Team, 2008, The reflection seismic survey of project TIPTEQ—the inventory of the Chilean subduction zone at 38.2° S: Geophysical Journal International, v. 172, p. 565–571, doi:10.1111/j.1365-246X.2007.03680.x. [10]

Groupe d'Etudes des Explosions Alpines (Closs, H., and Labrouste, Y., editors), 1963, Recherches seismologiques dans les Alpes Occidentales au moyen de grandes explosions en 1956, 1958 et 1960: Ventre National de la Recherche Scientifique, Serie XII, Fasc. 2: 241 p. [5]

Groupe Grands Profils Sismiques and German Research Group for Explosions Seismology, 1972, A long-range seismic profile in France from the Bretagne to the Provence: Annales de Géophysique, v. 28, p. 247–256. [7]

Guellec, S., Mugnier, J.-L., Tardy, M., and Roure, F., 1990, Neogene evolution of the western Alpine foreland in the light of ECORS data and balanced cross section: Mémoires de la Société Géologíque de France, N.S., 156, p. 165–184. [2] [8]

Guggisberg, B., and Berthelsen, A., 1987, A two-dimensional velocity-model for the lithosphere beneath the Baltic Shield and its possible tectonic significance: Terra Cognita, v. 7, p. 631–638. [2] [7] [8] [9]

Guggisberg, B., Kaminski, W., and Prodehl, C., 1991, Crustal structure of the Fennoscandian Shield: A traveltime interpretation of the long-range FENNOLORA seismic refraction profile, in Freeman, R., Huch, M., and Mueller, St., eds., The European Geotraverse, Part 7: Tectonophysics, v. 195, p. 105–137. [2] [7] [8]

Gutenberg, B., 1914, Über Erdbebenwellen. VIIA. Beobachtungen an Registrierungen von Fernbeben in Göttingen und Folgerungen über die Konstitution des Erdkörpers: Nachrichten der Gesellschaft der Wissenschaften zu Göttingen, Mathematisch-physikalische Klasse, p. 1–52. [3]

Gutenberg, B., 1924, Dispersion und Extinktion von seismischen Oberflächenwellen und der Aufbau der obersten Erdschichten: Physikalische Zeitschrift, v. 25, p. 377–381. [3]

Gutenberg, B., 1926, Die Geschwindigkeit der Erdbebenwellen in den obersten Erdschichten und ihr Einfluss auf die Ergebnisse einiger Probleme der Seismometrie: Gerlands Beiträge zur Geophysik, v. 15, p. 51–63. [3]

Gutenberg, B., 1932a, Beobachtungen von Erdbebenwellen, in Handbuch der Geophysik, vol. 4, ed.: Berlin: Gebr. Bornträger, p. 151–263. [3]

Gutenberg, B., 1932b, Travel-time curves at small distances and wave velocities in southern California: Gerlands Beiträge zur Geophysik, v. 35, p. 6–45. [3]

Gutenberg, B., 1934, The propagation of the longitudinal waves produced by the Long Beach earthquake: Gerlands Beiträge zur Geophysik, v. 41, p. 114–120. [3]

Gutenberg, B., 1936, The structure of the Earth's crust and the spreading of the continents: Bulletin of the Geological Society of America, v. 47, p. 1587–1610. [3]

Gutenberg, B., 1943a, Seismological evidence for the roots of mountains: Bulletin of the Geological Society of America, v. 54, p. 473–498. [4]

Gutenberg, B., 1943b, Eathquakes and structure in southern California: Bulletin of the Geological Society of America, v. 54, p. 499–526. [4]

Gutenberg, B., 1946, Interpretation of records obtained from the New Mexico atomic bomb test, July 16, 1945: Bulletin of the Seismological Society of America, v. 36, p. 327–330. [2] [4]

Gutenberg, B., 1951a, Travel times from blasts in southern California: Bulletin of the Seismological Society of America, v. 41. [4]

Gutenberg, B., 1951b, Crustal layers of continents and oceans: Geological Society of America Bulletin, v. 62, p. 427–440, doi:10.1130/0016-7606 (1951)62[427:CLOTCA]2.0.CO;2. [2] [4]

Gutenberg, B., 1952, Waves from blasts recorded in southern California: American Geophysical Union Transactions, v. 33, p. 427–431. [5]

Gutenberg, B., 1955, Wave velocities in the earth's crust, in Poldervaart, A., ed., Crust of the earth (a symposium): Geological Society of America Special Paper 62, p. 19–34. [2] [4]

Gutenberg, B., and Richter, C.F., 1946, Seismic waves from atomic bomb tests: Transactions, American Geophysical Union, v. 27, p. 776. [2] [4]

Gutenberg, B., Wood, H.O., and Buwalda, J.P., 1932, Experiments testing seismographic methods for determining crustal structure: Bulletin of the Seismological Society of America, v. 22, p. 185–246. [3]

Guterch, A., 1974, Refraction studies of structure of the earth's crust and upper mantle with deep seismic sounding method on the territory of Poland: Acta Geophysica Polonica, v. 22, p. 225–246. [7]

Guterch, A., 1977, Structure and physical properties of the earth's crust in Poland in the light of new data of DSS: Publications of the Institute of Geophysics, Polish Academy of Sciences, A14 (115), p. 347–357. [2] [7]

Guterch, A., and Sellevoll, M.A., 1981, Seismic studies of the earth's crust structure in the West Spitsbergen and Greenland Sea carried out in 1976 and 1978 [abst.]: Eos, v. 62, p. 217. [7]

Guterch, A., Uchman, J., and Wojtczak-Gadomska, B., 1967, Investigations on the earth's crustal structure in Poland by means of deep seismic sounding: Publications of the Institute of Geophysics, Polish Academy of Sciences, v. 14, p. 115–127. [2] [6]

Guterch, A., Materzok, R., and Pajchel, J., 1973, Preliminary results of deep seismic soundings on the southeastern part of international profile VII: Publications of the Institute of Geophysics, Polish Academy of Sciences, v. 60, p. 53–62. [6]

Guterch, A., Pajchel, J., Perchuc, E., Kowalski, J., Duda, S., Komber, J., Bojdys, G., and Sellevoll, M.A., 1978, Seismic reconnaissance measurement on the crustal structure in the Spitsbergen region 1976, University of Bergen, Seismological Observatory, Bergen: 61 p. [7]

Guterch, A., Grad, M., Materzok, R., and Toporkiewicz, S., 1983, Structure of the earth's crust of the Permian basin in Poland: Acta Geophysica Polonica, v. 31, p. 121–138. [2] [7]

Guterch, A., Grad, M., Janik, T., Perchuc, E., and Pajchel, J., 1985, Seismic studies of the crustal structure in West Antarctica 1979–1980—preliminary results, in Husebye, E.S., Johnson, G.L., and Kristoffersen, Y., eds., Geophysics of the polar regions: Tectonophysics, v. 114, p. 411–429. [2] [7]

Guterch, A., Grad, M., Materzok, R., Perchuc, E., and Toporkiewicz, 1986, Results of seismic crustal studies in Poland 1969–1985 (in Polish): Publications of the Institute of Geophysics, Polish Academy of Sciences, A-17 (192), p. 3–83. [8]

Guterch, A., Grad, M., Materzok, R., Perchuc, E., Janik, T., Gaczynski, E., Doan, T.T., Bialek, T., Gadomski, D., Meynarski, S., and Toporkiewicz, 1991a, Structure of the lower crust of the Paleozoic platform in Poland from seismic wide-angle and near-vertical reflection surveys: Publications of the Institute of Geophysics, Polish Academy of Sciences, A-19 (236), p. 41–61. [2] [6] [7] [8]

Guterch, A., Luosto, U., Grad, M., Yliniemi, J., Gaczynski, E., Korhonen, H., Janik, T., Lindblom, P., Materzok, R., and Perchuc, E., 1991b, Seismic studies of crustal structure in the Tornquist-Tesseyre-Zone in northwestern Poland: Publications of the Institute of Geophysics, Polish Academy of Sciences, A-19 (236), p. 147–156. [2] [8]

Guterch, A., Grad M., Janik T., Materzok R., Luosto U., Yliniemi J., Lück E., Schulze A., Förste K., 1994, Crustal structure of the transition zone between Precambrian and Variscan Europe from new seismic data along LT-7 profile (NW Poland and eastern Germany). Comptes rendus de l'Académie de sciences, Paris, 319, ser.II, 1489–1496. [8]

Guterch, A., Grad M., Janik T., and Sroda, P., 1998, Polish geodynamic expeditions—seismic structure of West Antarctica: Polish Polar Research, v. 19, p. 113–123. [8] [9]

Guterch, A., Grad, M., Thybo, H., Keller, G.R., and the POLONAISE Working Group, 1999, POLONAISE '97—an international seismic experiment between Precambrian and Variscan Europe in Poland: Tectonophysics, v. 314, p. 101–121, doi:10.1016/S0040-1951(99)00239-5. [2] [9]

Guterch, A., Grad, M., and Keller, G.R., 2001, Seismologists celebrate the new millennium with an experiment in central Europe: Eos (Transactions, American Geophysical Union), v. 82, p. 529, 533–534, doi:10.1029/01EO00313. [9]

Guterch, A., Grad M., Spicak A., Brückl E., Hegedüs E., Keller G.R., Thybo H. and CELEBRATION 2000, ALP 2002, SUDETES 2003 Working Groups, 2003a, An Overview of Recent Seismic Refraction Experiments in Central Europe: Studia in Geophysica et Geodaetica, Academy of Science, Czech Republic, Prague, v. 47, 651–658. [2] [9] [10]

Guterch, A., Grad, M., Keller, G.R., Posgay, K., Vozar, J., Spicak, A., Brueckl, E., Hajnal, Z., Thybo, H., Selvi, O., and CELEBRATION 2000 Experiment Team, 2003b, CELEBRATION 2000 seismic experiment: Studia in Geophysica et Geodaetica, Academy of Science, Czech Republic, Prague, v. 47, p. 659–669. [2] [9]

Guterch, A., Grad, M., and Keller, G.R., 2007, Crust and lithospheric structure—long range controlled source seismic experiments in Europe, in Romanowicz, B., and Dziewonski, A., eds., Seismology and structure of the earth: Amsterdam, Elsevier, Treatise on Geophysics, vol. 1, p. 533–558. [2] [10]

Haalk, H., 1934, Lehrbuch der angewandten Geophysik (p. 300–305): Verlag Gebrüder Bornträger. [3]

Hajnal, Z., Fowler, M.R., Mereu, R.F., Kanasewich, E.R., Cumming, G.L., Green, A.G., and Mair, J.A., 1984, An initial analysis of the earth's crust under the Williston Basin: 1979 COCRUST experiment: Journal of Geophysical Research, v. 89, p. 9381–9400, doi:10.1029/JB089iB11p09381. [2] [7] [8]

Hajnal, Z., Reilkoff, B., Posgay, K., Hegedus, E., Takacs, E., Asudeh, I., Mueller, St., Ansorge, J., and DeIaco, R., 1996, Crustal-scale extension in the central Pannonian basin: Tectonophysics, v. 264, p. 191–204, doi:10.1016/S0040-1951(96)00126-6. [2] [9]

Hajnal, Z., Ansdell, K.M., and Ashton, K.E., eds., 2005a, The Trans-Hudson Orogen Transect of Lithoprobe: Canadian Journal of Earth Science (special issue), v. 42, p. 379–761, doi:10.1139/e05-053. [2] [9]

Hajnal, Z., Lewry, J., White, D.J., Ashton, K., Clowes, R., Stauffer, M., Gyorfi, I., and Takacs, E., 2005b, The Sask Craton and Hearne Province margin: seismic-reflection studies in the western Trans-Hudson Orogen: Canadian Journal of Earth Science, v. 42, p. 403–419, doi:10.1139/e05-026. [2] [9]

Hales, A.L., 1973, The crust of the Gulf of Mexico: a discussion, *in* Mueller, S., ed., The structure of the earth's crust, based on seismic data: Tectonophysics, v. 20, p. 217–225. [2] [6]

Hales, A.L., and Asada, T., 1966, Crustal structure in coastal Alaska, *in* Steinhart, J.S., and Smith, T.J., eds., The earth beneath the continents: American Geophysical Union Geophysical Monograph 10, p. 420–432. [5]

Hales, A.L., and Nation, J.B., 1972, A crustal structure profile on the Agulhas bank: Bulletin of the Seismological Society of America, v. 62, p. 1029–1051. [6]

Hales, A.L., and Nation, J.B., 1973a, A seismic refraction survey in the Northern Rocky Mountains: More evidence for an intermediate crustal layer: Geophysical Journal of the Royal Astronomical Society, v. 35, p. 381–399. [6]

Hales, A.L., and Nation, J.B., 1973b, A seismic refraction study in the Indian Ocean: Bulletin of the Seismological Society of America, v. 63, p. 1951–1966. [2] [6] [7]

Hales, A.L., and Rynn, J.M.W., 1978, A long-range, controlled source seismic profile in northern Australia: Geophysical Journal of the Royal Astronomical Society, 55, p. 633–644. [7]

Hales, A.L., Helsley, C.E., Dowling, J.J., and Nation, J.B., 1967, The east coast on-shore off-shore experiment, 1. The first arrivals: Air Force Office of Scientific Research Final Report AF0SR 67-0852, 121 p. [6]

Hales, A.L., Helsley, C.E., Dowling, J.J., and Nation, J.B., 1968, The east coast onshore-offshore experiment: 1. The first arrival phases: Bulletin of the Seismological Society of America, v. 58, p. 757–819. [2] [6]

Hales, A.L., Helsley, C.E., and Nation, J.B., 1970, Crustal study on the Gulf coast of Texas: American Association of Petroleum Geologists Bulletin, v. 54, p. 2040–2057. [6]

Hales, A.L., Muirhead, K.J., and Rynn, J.M.W., 1980, A compressional velocity distribution for the upper mantle: Tectonophysics, v. 63, p. 309–348, doi:10.1016/0040-1951(80)90119-5. [2] [7]

Hall, J., 1978, LUST—a seismic refraction survey of the Lewisian basement complex in NW Scotland: Journal of the Geological Society London, v. 135, p. 555–563, doi:10.1144/gsjgs.135.5.0555. [2] [7]

Hall, J., and Quinian, G., 1994, A collisional crustal fabric pattern recognized from seismic reflection profiles of the Appalachian/Caledonide orogen: Tectonophysics, v. 232, p. 31–42, doi:10.1016/0040-1951(94)90074-4. [8]

Hall, J., Quinian, G., Marillier, F., and Keen, C., 1990, Diiping shear zones and the base of the crust in the Appalachians, offshore Canada, *in* Leven, J.H., Finlayson, D.M., Wright, C., Dooley, J.C., and Kennett, B.L.N., eds., Seismic probing of continents and their margins: Tectonophysics, v. 173, p. 581–593. [8]

Hall, J., Louden, K.E., Funck, T., and Deemer, S., 2002, Geophysical characteristics of the continental crust along the Lithoprobe Eastern Canadian Shield Onshore–Offshore Transect (ESCOOT): a review: Canadian Journal of Earth Science, v. 39, p. 569–587, doi:10.1139/e02-005. [9]

Hamilton, R.M., 1986, Seismic reflection studies by the U.S. Geological Survey, *in* Barazangi, M., and Brown, L., eds., Reflection seismology: a global perspective: American Geophysical Union, Geodynamics Series, v. 13, p. 99–106. [2] [7] [8]

Hamilton, R.M., Ryall, A., and Berg, E., 1964, Crustal structure southwest of the San Andreas fault from quarry blasts: Bulletin of the Seismological Society of America, v. 54, p. 67–77. [6]

Hammer, P.T.C., and Clowes, R.M., 2004, The accreted terranes of northwestern British Columbia, Canada: lithospheric velocity structure and tectonics: Journal of Geophysical Research, v. 109, B06305, doi:1029/2003JB002749, 19p. [9]

Hammer, P.T.C., Dorman, L., Hildebrand, J.A., and Cornuelle, B.D., 1994, Jasper Seamount structure: seafloor seismic refraction tomography: Journal of Geophysical Research, v. 99, p. 6731–6752. [9]

Hammer, P.T.C., Clowes, R.M., and Ellis, R.M., 2000, Crustal structure of NW British Columbia and SE Alaska from seismic wide-angle studies: Coast Plutonic Complex to Stikinia: Journal of Geophysical Research, v. 105, p. 7961–7981, doi:10.1029/1999JB900378. [2] [9]

Hampel, A., Kukowski, N., Bialas, J., and Huebscher, C., 2004, Ridge subduction at an erosive margin: the collision zone of the Nazca ridge in southern Peru: Journal of Geophysical Research, v. 109 (2), doi:10.1029/2003JB002593. [9]

Hanson, K., Berg, E., and Gedney, L., 1968, A seismic refraction profile and crustal structure in central interior Alaska: Bulletin of the Seismological Society of America, v. 58, p. 1657–1665. [2] [6]

Harding, A., 2001, Seismic structure, *in* Steele, J., Thorpe, S., and Turekian, K., eds., Encyclopedia of Ocean Sciences: Amsterdam, Academic Press, Elsevier, p. 2731–2737. [10]

Harrington, P.K., Barker, P.F., and Griffiths, D.H., 1972, Crustal structure of the South Orkney Islands area from seismic refraction and magnetic measurements, *in* Adie, R.J., ed., Antarctic Geology and Geophysics, Universitets Forlaget, Oslo (UUGS series B number1), p. 27–32. [6]

Harris, L.D., Harris, A.G., de Witt, W., Jr., and Bayer, K.C., 1981, Evaluation of southern eastern overthrust belt beneath Blue Ridge–Piedmont thrust: American Association of Petroleum Geologists Bulletin, v. 65, p. 2497–2505. [2] [7]

Harris, R.N., Walter, A.W., and Fuis, G.S., 1988, Data report for 1980–1981 seismic refraction profiles in the western Mojave desert, California: U.S. Geological Survey Open-File Report 88-580, Menlo Park, California: 65 p. [2] [8]

Harrison, A., and White, R.S., 2006, Lithospheric structure of an active backarc basin: the Taupo Volcanic Zone, New Zealand: Geophysical Journal International, v. 167, p. 968–990, doi:10.1111/j.1365-246X.2006.03166.x. [2] [10]

Hart, P.J., ed., 1969, The Earth's crust and upper mantle: American Geophysical Union, Geophysical Monograph, v. 13, 735 p. [2] [6]

Hart, P.J., 1954, Variation of velocity near the Mohorovičić discontinuity under Maryland and northeastern Virginia [thesis]: Cambridge, Massachsetts, Harvard University. [6]

Hart, P.J., 1964, Upper Mantle Project: American Geophysical Union Transactions, v. 45, p. 423–428. [6]

Hasbrouck, W.P., 1964, A seismic reflection crustal study in central eastern Colorado [Ph.D. thesis]: Colorado School of Mines, 133 p. [6]

Hashizume, M., Oike, K., Asano, S., Hamaguchi, H., Okada, H., Murauchi, S., Shima, E., and Nogoshi, M., 1968, Crustal structure in the profile across the northeastern part of Honshu, Japan, as derived from explosion seismology observations, part 2: Bulletin of the Earthquake Research Institute, University of Tokyo, v. 46, p. 607–630. [8]

Hatcher, R.D., Jr., Costain, J.K., Coruh, C., Phinney, R.A., and Williams, R.T., 1987, Tectonic implications of new Appalachian Ultradeep Core Hole (ADCOH) seismc reflection data from the crystalline southern Appalachians: Geophysical Journal of the Royal Astronomical Society, v. 89, p. 157–162. [8]

Hauksson, E., and Haase, J.S., 1997, Three-dimensional Vp and Vp/Vs velocity models of the Los Angeles basin and central Transverse Ranges, California: Journal of Geophysical Research, v. 102, p. 5423–5453, doi:10.1029/96JB03219. [9]

Hauser, E.C., and Lundy, J., 1989, COCORP deep reflections: Moho at 50 km (16s) beneath the Colorado Plateau: Journal of Geophysical Research, v. 94, p. 7071–7081, doi:10.1029/JB094iB06p07071. [8]

Hauser, E.C., Gephart, J., Latham, L., Brown, L.D., Kaufman, S., Oliver, J.E., and Lucchitta, I., 1987a, COCORP Arizona transect: strong crustal reflections and offset Moho beneath the transition zone: Geology, v. 15, p. 1103–1106, doi:10.1130/0091-7613(1987)15<1103:CATSCR>2.0.CO;2. [2] [8]

Hauser, E.C., Potter, C., Hauge, T., Burgess, S., Burtch, S., Mutschler, J., Allmendinger, R., Brown, L., Kaufman, S., and Oliver, L., 1987b, Crustal structure of eastern Nevada from COCORP deep seismic reflection data: Geological Society of America Bulletin, v. 99, p. 833–844, doi:10.1130/0016-7606(1987)99<833:CSOENF>2.0.CO;2. [9]

Hauser, F., O'Reilly, B.M., Jacob, A.W.B., Shannon, P.M., Makris, J., and Vogt, U., 1995, The crustal structure of the Rockall Trough: differential stretching without underplating: Journal of Geophysical Research, v. 100, p. 4097–4116, doi:10.1029/94JB02879. [2] [8] [10]

Hauser, F., Raileanu, V., Prodehl, C., Bala, A., Schulze, A., and Denton, P., 2000, The Seismic-Refraction Project VRANCEA-99: Geophysical Institute, University of Karlsruhe, Open-File Report. [9]

Hauser, F., Raileanu, F., Fielitz, W., Bala, A., Prodehl, C., and Polonic, G., 2001, The crustal structure between the southeastern Carpathians and the Moesian platform from a refraction seismic experiment in Romania: Tectonophysics, v. 340, p. 233–256, doi:10.1016/S0040-1951(01)00195-0. [2] [9]

Hauser, F., Prodehl, C., Landes, M, and the VRANCEA Working Group, 2002, Seismic experiments target earthquake–prone region in Romania: Eos (Transactions, American Geophysical Union), v. 83, p. 457, 462–463. [9] [10]

Hauser, F.,Raileanu, V., Fielitz, W., Dinu, C., Landes, M., Bala, A., and Prodehl, C., 2007a, Seismic crustal structure between the Transylvanian Basin and the Black Sea, Romania: Tectonophysics, v. 430, p. 1–25, doi:10.1016 /j.tecto.2006.10.005. [2] [9] [10]

Hauser, F., O'Reilly, B.M., and Readman, P.W., 2007b, The Porcupine Irish Margins Project: first data examples from an onshore/offshore seismic experiment in SW Ireland [abst.]: 50th Annual Irish Geological Research Meeting, School of Environmental Sciences, University of Ulster, Coleraine, Northern Ireland, 23–25 February. [2] [10]

Hawkins, L.V., Hennion, J.F., Nafe, J.E., and Doyle, H.A., 1965a, Marine seismic refraction studies on the continental margin to the South of Australia: Deep Sea Research, v. 12, p. 479–495. [2] [6]

Hawkins, L.V., Hennion, J.F., Nafe, J.E., and Thyer, R.F., 1965b, Geophysical investigations in the area of the Perth Basin, Western Australia: Geophysics, v. 30, p. 1026–1052, doi:10.1190/1.1439686. [2] [6]

Hawman, R,B., Chapman, M.C., Powell, C.A., Clippard, J.E., and Ahmed, H.O., 2001, Wide-angle reflection profiling with quarry blasts in the Eastern Tennessee Seismic Zone: Seismological Research Letters, v. 72, p. 108–122. [2] [9]

Hayes, R.C., 1936, Seismic waves and crustal structure in the New Zealand Region: New Zealand Journal of Science and Technology, v. 17, 1; Dom. Obs. 101 (S.26). [3]

Hayes, D.E., Houtz, R.E., Jarrard, R.D., Mrozowski, C.L., and Watanabe, T., 1978, A geophysical Atlas, East and Southeast Asian Seas, crustal structure; Map scale at equator 1:6,442,194: Chart compilation by the Office for the International Decade of Ocean Exploration (I.D.O.E.) of the National Science Foundation: Geological Society of America Map and Chart Series MC-25. [7]

Heacock, J.G., ed., 1971, The structure and physical properties of the Earth's crust: American Geophysical Union Geophysical Monograph 14, 348 p. [2] [6]

Heacock, J.G., ed., 1977, The Earth's crust—its nature and properties: American Geophysical Union Geophysical Monograph 20, 754 p. [2] [7]

Healy, J.H., 1963, Crustal structure along the coast of California from seismic-refraction measurements: Journal of Geophysical Research, v. 68, p. 5789–5806. [6]

Healy, J.H., and Peake, L.G., 1975, Seismic velocity structure along a section of the San Andreas fault near Bear Valley, California: Bulletin of the Seismological Society of America, v. 65, p. 1177–1197. [6]

Healy, J.H., and Warren, D.H., 1969, Explosion seismic studies in North America, in Hart, P.J., ed., The earth's crust and upper mantle: American Geophysical Union Geophysical Monograph 13, p. 208–220. [2] [6] [10]

Healy, J.H., Mooney, W.D., Blank, H.R., Gettings, M.E., Kohler, W.M., Lamson, R.J., and Leone, L.E., 1982, Saudi Arabian seismic deep-refraction profile: Final Project Report, Saudi Arabian Deputy Ministry of Mineral Resources, Open-File Report, USGS OFR-02-37, 429 p., including appendices; also U.S. Geological Survey Open-File Report 83-390. [2] [7] [8]

Hecker, O., 1900, Ergebnisse der Messung von Bodenbewegungen bei einer Sprengung: Gerlands Beiträge zur Geophysik, v. 4, p. 98–104. [3]

Hecker, O., 1922, Die Explosionskatastrophe von Oppau am 21 September 1921 nach den Aufzeichnungen der Erdbebenwarten. Veröff. D. Hauptstation Erdbebenforsch. Jena, Heft 2, p. 3–18. [3]

Heitzmann, P., Frei, W., Lehner, P., and Valasek, P., 1991, Crustal indentation in the Alps—an overview of reflection seismic profiling in Switzerland, in Meissner, R., Brown, L., Dürbaum, H.-J., Franke, W., Fuchs, K., and Seifert, F., eds., Continental lithosphere: deep seismic reflections: American Geophysical Union, Geodynamics Series, v. 22, p. 161–176. [2] [8]

Helmberger, D.V., 1968, The crust-mantle transition in the Bering Sea: Bulletin of the Seismological Society of America, v. 58, p. 179–214. [6] [7] [8]

Henry, W.J., Mechie, J., Maguire, P.K.H., Khan, M.A., Prodehl, C., Keller, G.R., and Patel, J., 1990, A seismic investigation of the Kenya rift valley: Geophysical Journal International, v. 100, p. 107–130, doi:10.1111 /j.1365-246X.1990.tb04572.x. [2] [8]

Henrys, S.A., Davey, F.J., and OBrien, B., 1995, Crustal structure of the Southern Hikurangi subduction zone, New Zealand (abst.): Eos (Transactions, American Geophysical Union), v. 76, no. 46, Fall Meeting Supplement, F550. [2] [9]

Henrys, S., Reyners, M., and Bibby, H., 2003a, Exploring the plate boundary structure of the North Island, New Zealand: Eos (Transactions, American Geophysical Union), v. 84, no. 31, p. 289, 294–295, doi:10.1029 /2003EO310002. [10]

Henrys, S.A., Bannister, S., Pecher, I.A., Davey, F., Stern, T., Stratford, W., White, R., Harrison, T., Nishimura, Y., and Yamada, A., 2003b, New Zealand North Island Geophysical Transect (NIGHT): Field Acquisition Report: Institute of Geological and Nuclear Sciences Science Report 2003/19, Institute of Geological and Nuclear Sciences Science, Lower Hutt, New Zealand. [2] [10]

Henrys, S., Reyners, M., Pecher, I., Bannister, S., Nishimura, Y., and Maslen, G., 2006, Kinking of the subducting slab by escalator normal faulting beneath the North Island of New Zealand: Geology, v, 34, p. 777–780. [10].

Henstock, T.J., and Levander, A., 2000, Lithospheric evolution in the wake of the Mendocino Triple Junction: structure of the San Andreas fault system at 2 Ma: Geophysical Journal International, v. 140, p. 233–247, doi:10.1046/j.1365-246x.2000.00010.x. [9]

Henstock, T.J., Levander, A., and Hole, J.A., 1997, Deformation in the lower crust of the San Andreas fault system in northern California: Science, v. 278, p. 650–653, doi:10.1126/science.278.5338.650. [9]

Henstock, T.J., Levander, A., Snelson, C.M., Keller, G.R., Miller, K.C., Harder, S.H., Gorman, A.R., Clowes, R.M., Burianyk, M.J.A., and Humphreys, E.D., 1998, Probing the Archean and Proterozoic lithosphere of western North America: GSA Today, v. 8, p. 1–5, 16–17. [9]

Henyey, T.E., Okaya, D.A., Frost, E.G., and McEvilly, T.V., 1987, CALCRUST (1985) seismic reflection survey, Whipple Mountains detachment terrane, California: an overview: Geophysical Journal of the Royal Astronomical Society, v. 89, p. 111–118. [8]

Herglotz, G., 1907, Über das Benndorfsche Problem der Fortpflanzungsgeschwindigkeit der Erdbebenstrahlen: Phys. Zeitschrift, Band 8, 145–147. [3]

Herquel, G., Wittlinger, G., and Guilbert, J., 1995, Anisotropy and crustal thickness of northern Tibet. New constraints for tectonic modelling: Geophysical Research Letters, v. 22, p. 1925–1928, doi:10.1029/95GL01789. [9]

Herrin, E., 1969, Regional variations of P-wave velocity in the upper mantle beneath North America, in Hart, P.J., ed., The earth's crust and upper mantle: American Geophysical Union, Geophysical Monograph 13, p. 242–246. [6]

Herron, T.J., Ludwig, W.J., Stoffa, P.L., Kan, T.K., and Bühl, P., 1978, Structure of the East Pacific Rise crest from multichannel seismic reflection data: Journal of Geophysical Research, v. 83, p. 798–804, doi:10.1029 /JB083iB02p00798. [2] [7] [8]

Hersey, J.B., 1963, Continuous reflection profiling, in Hill, M.N., ed., The Sea vol. 3. The earth beneath the Sea: Interscience Publ., New York-London, p. 47–72. [4] [5]

Hersey, J.B., and Ewing, M, 1949, Seismic reflections from beneath the ocean floor: Transactions, American Geophysical Union, v. 30, p. 5–14. [4]

Hersey, J.B., Officer, C.B., Johnson, H.R., and Bergstrom, S., 1952, Seismic refraction observations north of the Brownson Deep: Bulletin of the Seismological Society of America, v. 42, p. 291–306. [4] [5]

Hersey, J.B., Bunce, E.T., Wyrick, R.F., and Dietz, F.T., 1959, Geophysical investigations of the continental margin between Cape Henry, Virginia, and Jacksonville, Florida: Bulletin of the Geological Society of America, v. 70, p. 437–466, doi:10.1130/0016-7606(1959)70[437:GIOTCM]2.0.CO;2. [2] [5]

Hess, H.H., 1954, Geological hypotheses and the Earth's crust under the oceans: Royal Society London Proceedings, ser. A, v. 222, p. 341–348. [5]

Hess, H., 1964, Seismic anisotropy of the uppermost mantle under the oceans: Nature, v. 203, p. 629–631, doi:10.1038/203629a0. [6]

Hickman, S., Zoback, M., Younker, L., and Ellsworth, W., 1994, Deep scientific drilling in the San Andreas fault zone: Eos (Transactions, American Geophysical Union), v. 75, p. 137, 140, 142, doi:10.1029/94EO00830. [9] [10]

Hicks, N.O., 2001, Lithospheric structure of the Basin and Range province, southwestern Colorado Plateau: Southeastern California, southern Nevada, and western Arizona [M.Sc.thesis]: University of Texas at El Paso, United States, 162 p. [2] [9]

Hildenbrand, T.G., Schweig, E.S., Catchings, R.D., Langenheim, V.E., Mooney, W.D., Pratt, T.L., and Stanley, W.D., 1995, Crustal geophysics gives insight into New Madrid seismic zone: Eos (Transactions, American Geophysical Union), v. 76, no. 7, p. 65, 68–69. [9]

Hill, D.P., 1969, Crustal structure of the island of Hawaii from seismic-refraction measurements: Bulletin of the Seismological Society of America, v. 59, p. 101–130. [2] [6]

Hill, D.P., 1972, Crustal and upper mantle structure of the Columbia plateau from long range seismic-refraction measurements: Geological Society of America Bulletin, v. 83, p. 1639–1648, doi:10.1130/0016-7606(1972)83 [1639:CAUMSO]2.0.CO;2. [6]

Hill, D.P., 1976, Structure of Long Valley caldera, California, from a seismic refraction experiment: Journal of Geophysical Research, v. 81, p. 745–753, doi:10.1029/JB081i005p00745. [2] [7] [8]

Hill, D.P., and Pakiser, 1966, Crustal structure between the Nevada Test Site and Boise, Idaho, from seismic refraction measurements, *in* Steinhart, J.S., and Smith, T.J., eds., The earth beneath the continents: American Geophysical Union Geophysical Monograph 10, p. 391–420. [6]

Hill, D.P., Kissling, E., Luetgert, H., and Kradolfer, U., 1985, Constraints on the upper crustal structure of the Long Valley–Mono Craters volcanic complex, eastern California, from seismic refraction measurements: Journal of Geophysical Research, v. 90, p. 11,135–11,150, doi:10.1029 /JB090iB13p11135. [2] [8]

Hill, M.N., 1952, Seismic refraction shooting in an area of the East Atlantic: Philosophical Transactions of the Royal Society of London, ser. A, v. 244 (890), p. 561–594. [2] [4] [5]

Hill, M.N., ed., 1963a, The Sea vol. 3, The earth beneath the Sea: New York-London, Interscience Publ., 963 p. [2] [5] [6]

Hill, M.N., ed., 1963b, Single ship seismic refraction shooting, *in* Hill, M.N., ed., The Sea vol. 3, The earth beneath the Sea: New York-London, Interscience Publ., p. 39–46. [5]

Hill, N.M., and King, W.B.R., 1953, Seismic prospecting in the English Channel and its geological interpretation: Geological Society Quarterly Journal, v. 109 (pt. 1), p. 1–18. [2] [5]

Hill, N.M., and Laughton, A.S., 1954, Seismic observations in the eastern Atlantic: Royal Society of London Proceedings, ser. A, v. 222, p. 348–356. [5]

Hill, N.M., and Swallow, J.C., 1950, Seismic experiments in the Atlantic: Nature, v. 165, p. 193–194, doi:10.1038/165193c0. [2] [4]

Hinz, K, 1969, The Great Meteor Seamount: Results of seismic reflection measurements with a pneumatic sound sorce, and their geological interpretation: Meteor Forschungsergebnisse, Deutsche Forschungsgemeinschaft, Borntraeger, Berlin-Stuttgart, Reihe C Geologie und Geophysik, C2, 63–77. [6]

Hinz, K., 1972, Results of seismic refraction investigations (project ANNA) in the western Mediterranean Sea, south and north of the island of Mallorca, *in* Leenhardt, O., Gobert, B., Hinz, K., Hirn, A., Hirschleber, H., Hsü, H.J., Refubatti, A., Rudant, J.-P., Rudloff, R., Ryan, W.B.F., Snoek, M., and Steinmetz, L., Results of the Anna cruise—three north-south seismic profiles through the western Mediterranean Sea: Bulletin du Centre Recherches Pau–SNPA, v. 6, p. 405–426. [7]

Hinz, K., Seibold, E., and Wissmann, G., 1974, Continental slope anticline and unconformities off West Africa: Meteor Forschungsergebnisse, Deutsche Forschungsgemeinschaft, Borntraeger, Berlin-Stuttgart, Reihe C Geologie und Geophysik, C17, p. 67–73. [2] [7]

Hinz, K., Makris, J., Weigel, W., and Wissmann, G., 1977, Seismic studies in the Cretan sea, 4. Synoptic considerations and their geotectonic implications: Meteor Forschungsergebnisse, Deutsche Forschungsgemeinschaft, Borntraeger, Berlin-Stuttgart, Reihe C Geologie und Geophysik, C27, p. 44–45. [7]

Hinz, K., Neben, S., Schreckenberger, B., Roeser, H.A., Block, M., Goncalves de Sousa, K., and Meyer, H., 1999, The Argentine continental margin north of 48°S: sedimentary successions, volcanic activity during breakup: Marine Petrology and Geology, v. 16, p. 1–25, doi:10.1016/S0264-8172 (98)00060-9. [9]

Hirn, A., and Perrier, G., 1974, Deep seismic sounding in the Limagnegraben, *in* Illies,J.H., and Fuchs,K., eds., Approaches to taphrogenesis: Stuttgart, Schweizerbart, p. 329–340. [7]

Hirn, A., and Sapin, M., 1976, La croute terrestre sous la Corse: données sismiques: Bulletin de la Société Géologíque de France, v. 18, p. 1195–1199. [7]

Hirn, A., and Sapin, M., 1984, The Himalayan zone of crustal interaction: suggestions from explosion seismology: Ann. Geophys., v. 2, p. 123–130. [2] [8]

Hirn, A., Kind, R., Steinmetz, L., and Fuchs, K., 1973, Long-range profiles in Western Europe: II. Fine structure of the lower lithosphere in France (southern Bretagne): Zeitschrift für Geophysik, v. 39, p. 363–384. [2] [7]

Hirn, A., Prodehl, C., and Steinmetz, L., 1975, An experimental test of models of the lower lithosphere in Bretagne (France): Ann. Géophys., v. 31, p. 517–530. [2] [7]

Hirn, A., Steinmetz, L., and Sapin, M., 1977, A long range seismic profile in the western Mediterranean basin: structure of the upper mantle: Ann. Géophys., v. 33, p. 373–384. [2] [7]

Hirn, A., Daignières, M., Gallart, J., and Vadell, M., 1980, Explosion seismic sounding of throws and dips in the continental Moho: Geophysical Research Letters, v. 7, p. 263–266, doi:10.1029/GL007i004p00263. [7]

Hirn, A., Jobert, G., Wittlinger, G., Xin, X-Z, and Yuan, G-E, 1984a, Main features of the upper lithosphere in the unit between the High Himalayas and the Yarlung Zangbo Jiang suture: Ann. Geophys., v. 2, p. 113–118. [8]

Hirn, A., Lepine, J-C., Jobert, G., Sapin, M., Wittlinger, G., Xin, X-Z, Yuan, G-E, Jing, W-X, Wen, T-J, Bai, X-S, Pandey, M.R., and Tater, J.M., 1984b, Crustal structure and variability of the Himalayan border of Tibet: Nature, v. 307 (no. 5946), p. 23–25, doi:10.1038/307023a0. [8] [9]

Hirn, A., Nercessian, A., Sapin, M., Jobert, G., Xin, X-Z, Yuan, W-X, and Wen, T-J, 1984c, Lhasa block and bordering sutures—a continuation of a 500-km Moho traverse through Tibet: Nature, v. 307 (no. 5946), p. 25–27, doi:10.1038/307025a0. [8] [9]

Hirn, A., Jiang, M., Sapin, M., et al., 1995, Seismic anisotropy as an indicator of mantle flow beneath the Himalayas and Tibet: Nature, v. 375, p. 571–574, doi:10.1038/375571a0. [9]

Hirn, A., Sachpazi, M., Sliqi, R., McBride, J.H., Marnelis, F., Cernobori, L., and STREAMERS-PROFILES Group, 1996, A traverse of the Ionian islands front with coincident normal incidence and wide-angle seismics: Tectonophysics, v. 264, p. 35–49, doi:10.1016/S0040-1951(96)00116-3. [2] [9]

Hirsch, K.K., Bauer, K., and Scheck-Wenderoth, M., 2009, Deep structure of the western South African passive margin—Results of a combined approach of seismic, gravity and isostatic investigations: Tectonophysics, v. 470, no. 1-2, p. 57–70, doi:10.1016/j.tecto.2008.04.028. [2] [10]

Hirschleber, H., Rudloff, R., and Snoek, M., 1972, Preliminary results of seismic measurements in the Gulf of Lion, *in* Leenhardt, O., Gobert, B., Hinz, K., Hirn, A., Hirschleber, H., Hsü, H.J., Refubatti, A., Rudant, J.-P., Rudloff, R., Ryan, W.B.F., Snoek, M., and Steinmetz, L., eds., Results of the Anna cruise—three north-south seismic profiles through the western Mediterranean Sea: Bulletin du Centre Recherches Pau–SNPA, v. 6, p. 373–381. [7]

Hirschleber, H.B., Lund, C.-E., Meissner, R., Vogel, A., and Weinrebe, W., 1975, Seismic investigations along the "Blue Road" traverse: Journal of Geophysics, v. 41, p. 135–148. [2] [7]

Hobson, G.D., 1967, Hudson Bay crustal seismic experiment: time and distance data: Canadian Journal of Earth Science, v. 4, p. 879–899. [2] [6]

Hochstein, M.P., 1968, Seismic measurements in the Cook Islands, Southwest Pacific Ocean: New Zealand Journal of Geology and Geophysics, v. 10 (6), p. 1499–1526. [2] [6]

Hodgson, E.A., 1942, Velocity of elastic waves and structure of the crust in the vicinity of Ottawa, Canada: Bulletin of the Seismological Society of America, v. 32, p. 249–255. [4]

Hodgson, J., 2001, A seismic and gravity study of the Leinster Granite: SE Ireland [unpublished Ph.D. thesis]: Dublin Institute for Advanced Studies, Ireland. [8] [9]

Hodgson, J.A., Readman, P.W., O'Reilly, B.M., Kennan, P., Harder, S., Keller, R., and Thybo, H., 2000, Leinster Granite Seismic Project (LEGS): Preliminary Results of a Geophysical Study, *in* Jacob, A.W.B., Bean, C.J., and Jacob, S.T.F., eds., Proceedings of the 1999 CCSS Workshop, Dublin 1999, Communications of the Dublin Institute for Advanced Studied, Series D, Geophysics Bulletin, v. 49, p. 80–81. [2] [9]

Hodgson, J.H., 1947, Analysis of traveltimes from rockbursts at Kirkland Lake, Ontario: Bulletin of the Seismological Society of America, v. 37, p. 5–17. [2] [4]

Hodgson, J.H., 1953, A seismic survey in the Canadian Shield, I, II, Dominion Observ. Pub., Ottawa, 16, p. 113–163; 169–181. [2] [4] [5] [6]

Hoffman, L.R., and Mooney, W.D., 1983, A seismic study of Yucca Mountain and vicinity, southern Nevada; data report and preliminary results: U.S. Geological Survey Open-File Report 83-588, 50 p. [2] [8] [9]

Holbrook, W.S., 1990, The crustal structure of the northwestern Basin and Range province, Nevada, from wide-angle seismic data: Journal of Geophysical Research, v. 95, p. 21,843–21,869, doi:10.1029/JB095iB13p21843. [8]

Holbrook, W.S., and Mooney, W.D., 1987, The crustal structure of the the the axis of the Great Valley, California, from seismic refraction measurements, *in* Asano, S., and Mooney, W.D., eds., Seismic studies of the continental lithosphere: Tectonophysics, v. 140, p. 49–63. [8]

Holbrook, W.S., Gajewski, D., Krammer, A., and Prodehl, C., 1988, An interpretation of wide-angle shear-wave data in Southwest Germany: Poisson's ratio and petrological implications: Journal of Geophysical Research, v. 93, p. 12,081–12,106, doi:10.1029/JB093iB10p12081. [8]

Holbrook, W.S., Mooney, W.D., and Christensen, N.I., 1992a, The seismic velocity structure of the deep continental crust, *in* Fountain, D.M., Arculus, R.,and Kay, R.W., eds., Continental lower crust: Development in Geotectonics: Amsterdam, Elsevier, v. 23, p. 1–43. [2] [8]

Holbrook, W.S., Purdy, G.M., Collins, J.A., Sheridan, R.E., Musser, D.L., Glover, I.L., Talwani, M., Ewing, J.I., Hawman, R., and Smithson, S., 1992b, Deep velocity structure of rifted continental crust, U.S. Mid-Atlantic margin, from EDGE wide-angle reflection/refraction data: Geophysical Research Letters, v. 19, p. 1699–1702, doi:10.1029/92GL01799. [2] [8] [9]

Holbrook, W.S., Reiter, E.C., Purdy, G.M., and Toksöz, M.N., 1992c, Image of the Moho across the continent-ocean transition, U.S. east coast: Geology, v. 20, p. 203–206, doi:10.1130/0091-7613(1992)020<0203:IOTMAT >2.3.CO;2. [8]

Holbrook, W.S., Lizarralde, D., McGeary, S., Diebold, J., Bangs, N., and Klemperer, S.L., 1994a, Wide-angle ocean-bottom seismie data from the Aleutian Arc: Preliminary results of the 1994 Survey (abst.): Eos (Transactions, American Geophysical Union), v. 75, no. 44, p. 643. [9]

Holbrook, W.S., Purdy, G.M., Sheridan, R.E., Glover, L., Talwani, M., Ewing, J., and Hutchinson, D., 1994b, Seismic structure of the U.S. Mid-Atlantic continental margin: Journal of Geophysical Research, v. 99, p. 17,871–17,891, doi:10.1029/94JB00729. [9]

Holbrook, W.S., Reiter, E.C., Purdy, G.M., Sawyer, D., Stoffa, P.L., Austin, J.A., Oh, J., and Makris, J., 1994c, Deep structure of the U.S. Atlantic continental margin, offshore South Carolina, from coincident ocean bottom and multichannel seismic data: Journal of Geophysical Research, v. 99, p. 9155–9178, doi:10.1029/93JB01821. [8]

Holbrook, W.S., Brocher, T.M., ten Brink, U.S. and Hole, J.A., 1996, Crustal structure of a transform plate boundary: San Francisco Bay and the central California continental margin: Journal of Geophysical Research, v. 101, p. 22,311–22,334, doi:10.1029/96JB01642. [9]

Holder, A.P., and Bott, M.H.P., 1971, Crustal structure in the vicinity of Southwest England. Geophysical Journal of the Royal Astronomical Society, v. 23, p. 465–489. [6]

Hole, J.A., 1992, Nonlinear high-resolution three-dimensional seismic travel time tomography: Journal of Geophysical Research, v. 97, p. 6553–6562, doi:10.1029/92JB00235. [2] [9] [10]

Hole, J.A., and Zelt, B.C., 1995, 3-D finite-difference reflection traveltimes: Geophysical Journal International, v. 121, p. 427–434, doi:10.1111 /j.1365-246X.1995.tb05723.x. [2] [9] [10]

Hole, J.A., Brocher, T.M., Klemperer, S.L., Parsons, T., Benz, H.M., and Furlong, K.P., 2000, Three-dimensional seismic velocity structure of the San Francisco Bay area: Journal of Geophysical Research, v. 105, p. 13,859–13,874, doi:10.1029/2000JB900083. [9]

Hole, J.A., Catchings, R.D., Clair, K.C.St. Rymer, M.J., Okaya, D.A., and Carney, B.J., 2001, Steep-dip seismic imaging of the shallow San Andreas fault near Parkfield: Science, v. 294, p. 1513–1515, doi:10.1126/science .1065100. [10]

Hole, J.A., Zelt, C.A., and Pratt, R.G., 2005, Advances in controlled-source seismic imaging: Eos (Transactions, American Geophysical Union), v. 86, p. 177 and 181. [9] [10]

Hole, J.A., Ryberg, T., Fuis, G.S., Bleibinhaus, F., and Sharma, A.K., 2006, Structure of the San Andreas fault zone at SAFOD from a seismic refraction survey: Geophysical Research Letters, v. 33, L07312, doi:10.1029 /2005GL025194. [2][10]

Holliger, K., and Klemperer, S.L., 1990, Gravity and deep seismic reflection profiles across the North Sea rifts, *in* Blundell, D.J., and Gibbs, A., eds., Tectonic evolution of the North Sea rifts: Oxford University Press, p. 78–96. [2] [8]

Hooft, E.E.E., Detrick, R.S., and Kent, G.M., 1997, Seismic structure and indicators of magma budget along the southern East Pacific Rise: Journal of Geophysical Research, v. 102, p. 27319–27340, doi:10.1029/97JB02349. [9]

Hooft, E.E.E., Detrick, R.S., Toomey, D.R., Collins, J.A., and Lin, J., 2000, Crustal thickness and structure along three contrasting spreading segments of the Mid-Atlantic Ridge, 33.5°–35°N: Journal of Geophysical Research, v. 105, p. 8205–8226, doi:10.1029/1999JB900442. [2] [9]

Hopkins, W., 1848, Report of the British Association for the Advancement of Science, Oxford, June 1847, London (1848), part II, sec. 2, p. 33. [3]

Hopper, J.R., Funck, T., Tucholke, B.E., Louden, K.E., Holbrook, W.S., and Larsen, H.C., 2006, A deep seismic investigation of the Flemish Cap margin: implications for the origin of deep reflectivity and evidence for asymmetric break-up between Newfoundland and Iberia: Geophysical Journal International, v. 164, p. 501–515, doi:10.1111/j.1365-246X .2006.02800.x. [9]

Horsefield, S.J., Whitmarsh, R.B., White, R.S., and Sibuet, J.-C., 1994, Crustal structure of the Goban Spur rifted continental margin, NE Atlantic: Geophysical Journal International, v. 119, p. 1–19, doi:10.1111/j.1365-246X .1994.tb00909.x. [2] [8]

Horvath, F., Bada, G., Szafian, P., Tari, G., Adam, A., and Cloetingh, S., 2006, Formation and deformation of the Pannonian basin: constraints from observational data, *in* Gee, D.G., and Stephenson, R.A., eds., European lithosphere dynamics: Geological Society of London Memoir 32, p. 191–206. [9]

Hosford, A., Lin, J., and Detrick, R.S., 2001, Crustal evolution over the last 2 m.y. at the Mid-Atlantic Ridge OH-1 segment, 35°N: Journal of Geophysical Research, v. 106, p. 13,269–13,285, doi:10.1029/2001JB000235. [9]

Houtz, R., 1976, Seismic properties of layer 2A in the Pacific: Journal of Geophysical Research, v. 81, p. 6321–6331, doi:10.1029/JB081i035p06321. [7]

Houtz, R., and Ewing, J., 1976, Upper crustal structure as a function of plate age: Journal of Geophysical Research, v. 81, p. 2490–2498, doi:10.1029 /JB081i014p02490. [7]

Houtz, R., Ewing, J., and Buhl, P., 1970, Seismic data from sonobuoy stations in the northern and equatorial Pacific: Journal of Geophysical Research, v. 75, no. 26, p. 5093–5111, doi:10.1029/JB075i026p05093. [2] [6]

Howie, J.M., Miller, K.C., and Savage, W.U., 1993, Integrated crustal structure across the south central California margin: Santa Lucia escarpment to the San Andreas fault: Journal of Geophysical Research, v. 98, p. 8173–8196, doi:10.1029/93JB00025. [2] [8]

Hrubcova, P., Sroda, P., Spicak, A., Guterch, A., Grad, M., Keller, G.R., Brueckl, E., and Thybo, H., 2005, Crustal and uppermost mantle structure of the Bohemian Massif based on CELEBRATION 2000 data: Journal of Geophysical Research, v. 110, B11305, doi:10.1029/ 2004JB003080. [9]

Hu, H., Chen, X., Zhang, B., Song, W., Xiao, Z., and He, Z., 1988, Crustal structures along Suixian-Anyang DSS profile, *in* Developments in the Research of Deep Structures of China's Continental Geological Publishing House, Beijing, p. 48–60. [8]

Hubbard, S.S., Coruh, C., and Costain, J.K., 1991, Paleozoic and Grenvillan structures in the southern Appalachians: extended interpretation of seismic reflection data: Tectonics, v. 10, p. 141–170, doi:10.1029/90TC01854. [8]

Hübscher, C., Gohl, K., and Thorwart, M., 2003, Refraction seismics, *in* Pätzold, J., Bohrmann, G., and Hübscher, C., eds., Black Sea–Mediterranean–Red Sea: Berichte M52, no. 03-2, part 2. Leitstelle METEOR, Institute Meereskunde, University of Hamburg, p. 2-6–2-8. [2] [10]

Hughes, S., and Luetgert, J.H., 1991, Crustal structure of the western New England Appalachians and the Adirondack Mountains: Journal of Geophysical Research, v. 96, p. 16,471–16,494, doi:10.1029/91JB01657. [2] [8]

Hughes, S., and Luetgert, J.H., 1992, Crustal structure of the southeastern Grenville province, northern New York state and eastern Ontario: Journal of Geophysical Research, v. 97, p. 17,455–17,479, doi:10.1029/92JB01793. [9]

Hughes, S., Hall, J., and Luetgert, J.H., 1994, The seismic velocity structure of the Newfoundland Appalachian orogen: Journal of Geophysics Research, v. 99, p. 13,663–13,653, doi:10.1029/94JB00653. [8]

Husebye, E.S., Ro, H.E., Kinck, J.J., and Larsson, F.R., 1988, Tectonic studies in the Skagerrak province: the "MOBIL Search" cruise: Norges Geologisk Undersøkelse Special Publication 3, p. 14–20. [9]

Hussenoeder, S.A., Collins, J.A., Kent, G.M., Detrick, R.S., and the TERA Group, 1996, Seismic analysis of the axial magma chamber reflector along the southern East Pacific Rise from conventional reflection profiling: Journal of Geophysical Research, v. 101, p. 22087–22105, doi:10.1029/96JB01907. [9]

Hussong, D.M., Edwards, P.B., Johnson, S.H., Campbell, J.F., and Sutton, G.H., 1976, Crustal structure of the Peru-Chile trench between 8° and 12°S latitude, *in* Sutton, G.H., Manghnani, M.H., Moberly, R., and McAffe, E.U., eds., The geophysics of the Pacific Ocean basin and its margin: American Geophysical Union, Geophysical Monograph 19, p. 71–85. [2] [7]

Hutchins, K., 1969, Arctic geophysics: Arctic, v. 22, no. 3, p. 225–232. [6]

Hutchinson, D.R., Grow, J.A., Klitgord, K.D., and Detrick, R.S., 1986, Moho reflections from the Long Island Plateau, eastern United States, *in* Barazangi, M., and Brown, L., eds., Reflection seismology: the continental crust: American Geophysical Union, Geodynamics Series, v. 14, p. 173–187. [2] [7] [8]

Hutchinson, D.R., Trehu, A.M., and Klitgord, K.D., 1987, Structure of the lower crust beneath the Gulf of Maine: Geophysical Journal of the Royal Astronomical Society, v. 89, p. 189–194. [2] [8]

Hyndman, R.D., Wang, K., and Yamano, M., 1995, Thermal constraints on the seismogenic portion of the southwestern Japan subduction thrust: Journal of Geophysical Research, v. 100, p. 15,373–15,392, doi:10.1029/95JB00153. [9]

Hynes, A., and Ludden, J.N., eds., 2000, The Lithoprobe Abitibi-Grenville Transect: Canadian Journal of Earth Science, v. 37, p. 115–516. [9]

Ibs-von Seht, M., Blumenstein, S., Wagner, R., Hollnack, D., and Wohlenberg, J., 2001, Seismicity, seismotectonics and crustal structure of the southern Kenya rift—new data from the Lake Magadi area: Geophysical Journal International, v. 146, p. 439–453, doi:10.1046/j.0956-540x.2001.01464.x. [9]

Iidaka, T., Takeda, T., Kurashimo, E., Kawamura, T., Kaneda, Y., and Iwasaki, T., 2004, Configuration of subducting Philippine Sea plate and crustal structure in the central Japan region: Tectonophysics, v. 388, p. 7–20, doi:10.1016/j.tecto.2004.07.002. [2] [10]

Iidaka, T., Kato, A., Kurashimo, E., Iwasaki, T., Hirata, N., Katao, H., Hirose, I., and Miyamachi, H., 2009, Fine structure of P-wave velocity distribution along the Atotsugawa fault, central Japan: Tectonophysics, v. 472, p. 95–104, doi:10.1016/j.tecto.2008.06.016. [10]

Ikami, A., 1978, Crustal structure in the Shizuoka district, central Japan as derived from explosion seismic observations: Journal of the Physics of the Earth, v. 26, p. 299–331. [2] [7]

Ikami, A., Ito, K., Sasaki, Y., and Asano, S., 1982, Crustal structure in the profile across Shikoku, Japan, as derived from the off Sakaide explosions (in Japanese with English abstract): Journal of the Seismological Society of Japan, v. 35, p. 367–375. [2] [7] [9]

Ikami, A., Yoshii, T., Kubota, S., Sasaki, Y., Hasemi, A., Moriya, T., Miyamachi, H., Matsu'ura, R.S., and Wada, K., 1986, A seismic refraction profile in and around Nagano Prefecture, central Japan: Journal of the Physics of the Earth, v. 34, p. 457–474. [2] [8]

Ikeda, Y., Iwasaki, T., Sato, H., Matsuta, N., and Kozawa, T., 2004, Seismic reflection profiling across the Itoigawa-Shizuoka Tectonic Line at Matsumoto, central Japan: Earth, Planets, and Space, v. 56, p. 1315–1321. [10]

Ikeda, Y., Iwasaki, T., Kano, K., Ito, T., Sato, H., Tajikara, M., Kikuchi, S., Higashinaka, M., Kozawa, T., and Kawanaka, T., 2009, Active nappe with a high slip late: Seismic and gravity profiling across the southern part of the Itoigawa-Shizuoka Tectonic Lie, central Japan: Tectonophysics, v. 472, p. 72–85, doi:10.1016/j.tecto.2008.04.008. [10]

ILIHA DSS Group, 1993a, A deep seismic sounding investigation on lithospheric heterogeneity and anisotropy beneath the Iberian Peninsula, *in* Badal, J., Gallart, J., and Paulssen, H., eds., Seismic studies of the Iberian Peninsula: Tectonophysics, v. 221, p. 35–51. [2] [8]

ILIHA DSS Group, 1993b, A deep seismic sounding investigation of lithospheric heterogeneity and anisotropy beneath the Iberian Peninsula, *in* Mezcua,J., and Carreno, E., eds., Iberian lithosphere, heterogeneity and anisotropy (ILIHA): Monografia no. 10, Instituto Geografico Nacional, Madrid, Spain, p. 105–127. [8]

Illies, J.H., and Mueller, S., eds., 1970, Graben problems: Stuttgart, Schweizerbart, 316 p. [7]

Italian Explosion Seismology Group and Institute of Geophysics, ETH Zurich, 1981, Crust and upper mantle structures in the Southern Alps from deep seismic sounding profiles (1977, 1978) and surface wave dispersion analysis: Bollettino di Geofisica Teorica e Applicata, v. 92, p. 297–330. [7]

Ito, K., Umeda, Y., Sato, H., Hirata, N., Kawanaka, T., and Ikawa, T., 2006, Deep seismic surveys in the Kinki district: Shingu-Maizuru line: Bulletin of the Earthquake Research Institute, University of Tokyo, v. 81, p. 239–245. [10]

Ito, T., 2000, Crustal structure of the Hidaka collision zone and its foreland fold-and-thrust belt, Hokkaido, Japan: Journal of the Japanese Association of Petroleum Technology, v. 65, p. 103–119 (in Japanese with English abstract). [2] [9]

Ito, T., 2002, Active faulting, lower crustal delamination and ongoing Hidaka arc-arc collision, Hokkaido, Japan, *in* Fujinawa, Y., and Yoshida, A., eds., Seismotectonics in convergent plate boundary: Tokyo, Terrapub, p. 219–224. [2] [9]

Ito, T., Iwasaki, T., and Thybo, H., eds., 2009a, Deep seismic profiling of the continents and their margins: v. 472, p. 1–341. [2] [10]

Ito, T., Kojima, Y., Kodaira, S., Sato, H., Kaneda, Y., Iwasaki, T., Kurashimo, E., Tsumura, N., Fujiwara, A., Miyauchi, T., Hirata, N., Harder, S., Miller, K., Murata, A., Yamakita, S., Onishi, M., Abe, S., Sato, T., and Ikawa, T., 2009b, Crustal structure of southwest Japan, revealed by the integrated seismic experiment southwest Japan 2002: Tectonophysics, v. 472, p. 124–134, doi:10.1016/j.tecto.2008.05.013. [10]

Ivandic, M., Grevemeyer, I., Bialas, J., and Petersen, C.J., 2010, Serpentinization in the trench–outer rise region offshore of Nicaragua: constraints from seismic refraction and wide-angle data: Geophysical Journal International, v. 180, p. 1253–1264, doi:10.1111/j.1365-246X.2009.04474.x. [2] [10]

Iwasaki, T., Shiobara, H., Nishizawa, A., Kanazawa, T., Suyehiro, K., Hirata, N., Urabe, T., and Shimamura, H., 1989, A detailed subduction structure in the Kuril trench deduced from ocean bottom seismographic refraction studies: Tectonophysics, v. 165, p. 315–336, doi:10.1016/0040-1951(89)90056-5. [8]

Iwasaki, T., Hirata, N., Kanazawa, T., Melles, J., Suyehiro, K., Urabe, T., Moller, L., Makris, J., and Shimamura, H., 1990, Crustal and upper mantle structure in the Ryukyu Island Arc deduced from deep seismic sounding: Geophysical Journal International, v. 102, p. 631–651, doi:10.1111/j.1365-246X.1990.tb04587.x. [8]

Iwasaki, T., Yoshii, T., Moriya, T., Kobayashi, A., Nishiwaki, M., Tsutsui, T., Iidaka, T., Ikami, A., and Masuda, T., 1993, Seismic refraction study in the Kitakami region, northern Honshu, Japan: Journal of the Physics of the Earth, v. 41, p. 165–188. [9]

Iwasaki, T., Yoshii, T., Moriya, T., Kobayashi, A., Nishiwaki, M., Tsutsui, T., Iidaka, T., Ikami, A., and Masuda, T., 1994, Precise P and S wave velocity structures in the Kitakami massif, northern Honshu, Japan, from a seismic refraction experiment: Journal of Geophysical Research, v. 99, p. 22187–22204. [2] [8] [9]

Iwasaki, T., Ozel, O., Moriya, T., Sakai, S., Suzuki, S., Aoki, G., Maaeda, T., Iidaka, T., 1998, Lateral structural variation across a collision zone in central Hokkaido, Japan, as revealed by wide-angle seismic experiments: Geophysical Journal International, v. 132, p. 435–457, doi:10.1046/j.1365-246x.1998.00454.x. [2] [8] [9]

Iwasaki, T., Kato, W., Abe, S., Ichinose, Y., Umino, N., Okada, T., Koshiya, S., Kosuga, M., Saka, M., Sato, H., Shimizu, N., Takeda, T., Tsumura, N., Noda, K., Hasegawa, A., Hirata, N., Watanabe, K., Ikawa, T., and Ohguchi, T., 1999, Seismic refraction observations at the Sen'ya fault zone, northern Honshu, Japan: Bulletin of the Earthquake Research Institute, University of Tokyo, v. 74, p. 49–62 (in Japanese). [9]

Iwasaki, T., Kato, W., Moriya, T., Hasemi, A., Umino, N., Okada, T., Miyashita, T., Mizogami, T., Takeda, T., Sekine, S., Matsushima, T., Tashiro, K., and Miyamachi, H., 2001a, Extensional structure in northern Honshu arc as inferred from seismic refraction/wide-angle reflection profiling: Geophysical Research Letters, v. 28, no. 12, p. 2329–2332, doi:10.1029/2000GL012783. [2] [9]

Iwasaki, T., Sato, H., Hirata, N., Ito, T., Moriya, T., Kurashimo, E., Kawanaka, T., Kozawa, T., Ichinose, Y., Saka, M., Takeda, T., Kato, W., Yoshikawa, T., Arita, K., Takanami, T., Yamamoto, A., Yoshii, T., and Ikawa, T., 2001b, Seismic reflection experiment in the northern part of the Hikada collision zone, Hokkaido, Japan: Bulletin of the Earthquake Research Institute, University of Tokyo, v. 76, p. 115–127 (in Japanese). [9]

Iwasaki, T., Yoshii, T., Ito, T., Sato, H., and Hirata, N., 2002, Seismological features of island arc crust as inferred from recent seismic expeditions in Japan: Tectonophysics, v. 355, p. 53–66, doi:10.1016/S0040-1951(02)00134-8. [2] [9]

Iwasaki, T., Adachi, K., Moriya, T., Miyamachi, H., Matsushima, T., Miyashita, K., Takeda, T., Taira, T., Yamada, T., and Ohtake, K., 2004, Upper and middle crustal deformation of an arc-arc collision across Hokkaido, Japan, *inferred from seismic refraction/wide-angle reflection experiments: Tectonophysics, v. 388, p. 59–73, doi:10.1016/j.tecto.2004.03.025. [2] [9]

Iyer, H.M., Pakiser, L.C., Stuart, D.J., and Warren, D.H., 1969, Project Early Rise: seismic probing of the upper mantle: Journal of Geophysical Research, v. 74, p. 4409–4441, doi:10.1029/JB074i017p04409-02. [2] [6] [7]

Jackson, H.R., Forsyth, D.A., and Johnson, G.L., 1986, Oceanic affinities of the Alpha Ridge, Arctic Ocean: Marine Geology, v. 73, p. 237–261, doi:10.1016/0025-3227(86)90017-4. [8]

Jackson, W.H., and Pakiser, L.C., 1965. Seismic study of crustal structure in the southern Rocky Mountains: U.S. Geological Survey Professional Paper 525-D, p. 85–92. [6]

Jackson, W.H., Stewart, S.W., and Pakiser, L.C., 1963, Crustal structure in eastern Colorado from seismic refraction measurements: Journal of Geophysical Research, v. 68, p. 5777–5787. [2] [6]

Jacob, A.W.B., 1975, Dispersed shots at optimum depth—an efficient seismic source for lithospheric studies: Journal of Geophysics, v. 41, p. 63–70. [7]

Jacob, A.W.B., and Willmore, P.L., 1972, Teleseismic P waves from a 10 ton explosion: Nature, v. 236, p. 305, doi:10.1038/236305a0. [7]

Jacob, A.W.B., Kaminski, W., Murphy, T., Phillips, W.E.A., and Prodehl, C., 1985, A crustal model for a northeast-southwest profile through Ireland: Tectonophysics, v. 113, p. 75–103, doi:10.1016/0040-1951(85)90111-8. [2] [8]

Jacob, A.W.B., Bean, C.J., Nolte, B., and Prodehl, C., 1991, P-wave sections in a realistic anisotropic lithosphere: Geophysical Journal International, v. 107, p. 709–714, doi:10.1111/j.1365-246X.1991.tb01430.x. [8]

Jacob, A.W.B., Vees, R., Braile, L.W., and Criley, E., 1994, Optimization of wide-angle seismic signal-to-noise ratios and P-wave transmission in Kenya, *in* Prodehl, C., Keller, G.R., and Khan, M.A., eds., Crustal and upper mantle structure of the Kenya rift: Tectonophysics, v. 236, p. 61–79. [9]

Jacob, A.W.B., Delvaux, D., and Khan, M.A., eds., 1997, Lithospheric structure, evolution and sedimentation in continental rifts. Proceedings of the IGCP 400 Meeting, Dublin, Ireland: Communications of the Dublin Institute for Advanced Studies, Series D, Geophysical Bulletin, v. 48, p. 172 p. [2] [9] [10]

Jacob, A.W.B., Bean, C.J., and Jacob, S.T.F., eds., 2000, Active and passive seismic techniques reviewed. Proceedings of the 1999 CCSS Workshop, Dublin, Ireland: Communications of the Dublin Institute for Advanced Studies, Series D, Geophysical Bulletin, v. 49: 117 p. [2] [3] [9] [10]

Jacobson, R.S., Shor, G.G. Jr., Kieckhefer, R.M., and Pudy, G.M., 1979, Seismic refraction and reflection studies in the Timor-Aru Trough system and Australian continental shelf, *in* Watkins, J.S., Montadert, L., and Dickerson, P., eds., Geological and Geophysical Investigations of Continental Margins: American Association of Petroleum Geologists Memoir 29, p. 209–222. [2] [7]

Jacoby, W., Björnsson, A., and Möller, D., eds., 1980, Iceland: Evolution, active tectonic, and structure: Journal of Geophysics, v. 47, p. 1–277. [2]

Jaiswal, P., Zelt, C.A., Bally, A.W., and Dasgupta, R., 2008, 2-D traveltime and waveform inversion for improved seismic imaging: Naga thrust and fold belt, India: Geophysical Journal International, v. 173, p. 642–658, doi:10.1111/j.1365-246X.2007.03691.x. [2] [10]

James, D.E., and Steinhart, J.S., 1966, Structure beneath continents: a critical review of explosion studies from 1960–1965, *in* Steinhart, J.S., and Smith, T.J., eds., The earth beneath the continents: American Geophysical Union, Washington, D.C., Geophysical Monograph 10, p. 293–333. [2] [6] [10]

James, D.E., Smith, T.J., and Steinhart, J.S., 1968, Crustal structure of the middle Atlantic states: Journal of Geophysical Research, v. 73, p. 1983–2007, doi:10.1029/JB073i006p01983. [6] [8]

Janik, T., 1997, Seismic crustal structure of the Bransfield Strait, West Antarctica: Polish Polar Research, v. 18, p. 171–225. [2] [8] [9]

Janik, T., Yliniemi, J., Grad M., Thybo, H., Tiira, T., and POLONAISE'97 Working Group, 2002, Crustal structure across the TESZ along POLONAISE'97 seismic profile P2 in NW Poland: Tectonophysics, v. 360, p. 129–152, doi:10.1016/S0040-1951(02)00353-0. [9]

Janik, T., Grad M., Guterch A., Dadlez, R., Yliniemi, J., Tiira, T., Keller G.R., Gaczynski, E., and CELEBRATION 2000 Working Group, 2005, Lithospheric structure of the Trans-European Suture Zone along the TTZ–CEL03 seismic transect (from NW to SE Poland): Tectonophysics, v. 411, p. 129–156, doi:10.1016/j.tecto.2005.09.005. [9]

Janik, T., Sroda, P., Grad M., and Guterch A., 2006, Moho depth along the Antarctic Peninsula and crustal structure across the landward projection of the Hero Fracture Zone, *in* Fütterer, D.K., Damaschke, D., Kleinschmidt, G., Miller, H., and Tessensohn, F., eds., Antarctica: contributions to global earth sciences: Berlin-Heidelberg-New York, Springer, p. 229–236. [9]

Jarchow, C.M., 1991, Investigations of magmatic underplating beneath the northwestern Basin and Range province, Nevada, seismic data acquisition and tectonic problems of the Columbia plateau, Washington, and the nature of the Mohorovičić discontinuity worldwide [Ph.D. thesis]: Department of Geophysics, Stanford University, Stanford, California, Chapter 2: The nature of the Mohorovičić discontinuity, p. 2,01–2,53. [3]

Jeffreys, H., 1926, On near earthquakes: Monthly Notices of the Royal Astronomical Society, Geophysics Supplement, v. 1, p. 385–402. [3]

Jeffreys, H., 1929, The Earth: Its origin, history and physical constitution: Cambridge University Press, England, 346 p. [3]

Jeffreys, H., 1947, On the Burton–on–Trent explosion of 1944 November 27: Monthly Notices of the Royal Astronomical Society, Geophysics Supplement, v. 5, p. 99–104. [4]

Jensen, L.W., Steiner, D., Brown, L.D., Kaufman, S., and Oliver, J.E., 1979, COCORP deep crustal seismic reflection studies in the Michigan basin: Michigan Basin Geological Society Symposium, 16. [2] [7]

Jentsch, M., 1979, Reinterpretation of a Deep-Seismic-Sounding profile on the Ukrainian Shield: Journal of Geophysics, v. 45, p. 355–372. [6]

Jentsch, M., Bamford, D., Emter, D., and Prodehl, C., 1982, A seismic-refraction investigation of the basement structure in the Urach geothermal anomaly, southern Germany, *in* Haenel, R. ed., The Urach geothermal project (Swabian Alb, Germany): Stuttgart, Schweizerbart, p. 231–245. [2] [7]

Jiang, M., Galvé, A., Hirn, A., de Voogd, B., Laigle, M., Su, H.P., Díaz, J., Lépine, J.C., Wang, Y.X., 2006, Crustal thickening and variations in architecture from the Qaidam basin to the Qang Tang (North-Central Tibetan Plateau), from wide angle reflection seismology: Tectonophysics, v. 412, no. 3-4, p. 121–140, doi:10.1016/j.tecto.2005.09.011. [9]

Jödicke, H., Untiedt, J., Olgemann, W., Schulte, L., and Wagenitz, V., 1983, Electrical conductivity structure of the crust and upper mantle beneath the Rhensh Massif, *in* Fuchs, K., von Gehlen, K., Mälzer, H., Murawski, H., and Semmel, A., eds., Plateau uplift: the Rhenish Massif—a case history: Berlin-Heidelberg, Springer, p. 288–302. [8]

Johnson, L.R., 1965. Crustal structure between Lake Mead, Nevada, and Mono Lake, California: Journal of Geophysical Research, v. 70, p. 2863–2872, doi:10.1029/JZ070i012p02863. [6]

Johnson, S.H., and Couch, R.W., 1970, Crustal structure in the North Cascade Mountains of Washington and British Columbia from seismic refraction measurements: Bulletin of the Seismological Society of America, v. 60, p. 1259–1269. [6]

Jones, E.J.W., 1999, Marine Geophysics: Chichester-New York, John Wiley and Sons Ltd., 466 p. [2] [5] [6] [8] [10]

Joppen, M., and White, R.S., 1990, The structure and subsidence of Rockall Trough from two-ship seismic experiments: Journal of Geophysical Research, v. 95, p. 19821–19837, doi:10.1029/JB095iB12p19821. [2] [8]

Joset, A., and Holtzscherer, J.J., 1953, Etudes des vitesses de propagation des ondes seismiques sur l'Inlandsis du Groenland: Annals of Geophysics, v. 9, p. 330–344. [5]

Juhlin, C., Kashubin, S., Knapp, J.H., Makovsky, V., and Ryberg, T., 1995, Project conducts seismic reflection profiling in the Urals Mountains: Eos (Transactions, American Geophysical Union), v. 76, p. 193–196, doi:10.1029/95EO00110. [9]

Juhlin, C., Knapp, J.H., Kashubin, S., and Bliznetov, M., 1996, Crustal evolution of the Middle Urals based on seismic reflection and refraction data: Tectonophysics, v. 264, p. 21–34, doi:10.1016/S0040-1951(96)00115-1. [2] [9]

Junger, A., 1951, Deep basement reflections in Big Horn County, Montana: Geophysics, v. 16, p. 499–505, doi:10.1190/1.1437698. [2] [4]

Kaila, K.L., 1986, Tectonic framework of Narmada-Son lineament—a continental rift system in central India from deep seismic soundings, *in* Barazangi, M., and Brown, L., eds., Reflection seismology: a global perspective: American Geophysical Union, Geodynamics Series, 13, p. 133–150. [8]

Kaila, K.L., Krishna, V.G., Chowdhury, K.R., and Narain, H., 1978, Structure of the Kashmir Himalaya from deep seismic soundings: Journal of the Geological Society of India, 19, p. 1–20. [2] [7]

Kaila, K.L., Chowdhury, K.R., Reddy, P.R., Krishna, V.G., Hainarain, Subbotin, S.I., Sollogub, V.B., Chekunov, A.V., Kharetchko, G.E., Lazarenko, M.A., and Ilchenko, T.V., 1979, Crustal structure along Kavali-Udipi profile in the Indian Peninsular shield from deep seismic soundings: Journal of the Geological Society of India, v. 20, p. 307–333. [2] [7]

Kaila, K.L., Krishna, V.G., and Mall, D.M., 1981a, Crustal structure along Mehmadabad-Billimora profile in the Cambay basin, India, from deep seismic soundings: Tectonophysics, v. 76, p. 99–130, doi:10.1016/0040-1951(81)90255-9. [2] [7] [8]

Kaila, K.L., Murty, P.R.K., Rao, V.K., and Kharetchko, G.E., 1981b, Crustal structure from deep seismic soundings along the Koyna II (Kelsi-Loni) profile in the Deccan trap area, India: Tectonophysics, v. 73, p. 365–384, doi:10.1016/0040-1951(81)90223-7. [2] [7]

Kaila, K.L., Reddy, P.R., Dixit, M.M., and Koteswara, R.P., 1985, Crustal structure across the Narmada-Son lineament, central India, from deep seismic soundings: Journal of the Geological Society of India, v. 26, p. 465–480. [8]

Kaila, K.L., Tewari, H.C., Chowdhury, K.R., Rao, V.K., Sridhar, A.R., and Mall, D.M., 1987a, Crustal structure of the northern part of the Proterozoic Cuddapah basin of India from deep seismic soundings and gravity data: Tectonophysics, v. 140, p. 1–12, doi:10.1016/0040-1951(87)90136-3. [8]

Kaila, K.L., Tewari, H.C., and Mall, D.M., 1987b, Crustal structure and delineation of Gondwana basin in the Mahanadi delta area, India, from deep seismic soundings: Journal of the Geological Society of India, v. 29, p. 293–308. [8]

Kaila, K.L., Murthy, P.R.K., Mall, D.M., Dixit, M.M., and Sarkar, D., 1987c, Deep seismic soundings along Hirapur-Mandla profile, central India: Geophysical Journal of the Royal Astronomical Society, v. 89, p. 399–404. [8]

Kaila, K.L., Murthy, P.R.K., Rao, V.K., and Venkateswaralu, N., 1990a, Deep seismic sounding in the Godavari graben and Godavari (coastal) basin, India: Tectonophysics, v. 173, p. 307–317, doi:10.1016/0040-1951(90)90226-X. [8]

Kaila, K.L., Tewari, H.C., Krishna, V.G., Dixit, M.M., Sarkar, D., and Reddy, M.S., 1990b, Deep seismic sounding studies in the north Cambay and Sanchor basins, India: Geophysical Journal International, v. 103, p. 621–637, doi:10.1111/j.1365-246X.1990.tb05676.x. [8]

Kaila, K.L., Reddy, P.R., Mall, D.M., Venkateswaralu, N., Krishna, V.G., and Prasad, A.S.S.S.R.S., 1992, Crustal structure of the west Bengal basin, India, from deep seismic sounding investigations: Geophysical Journal International, v. 111, p. 45–66, doi:10.1111/j.1365-246X.1992.tb00554.x. [8]

Kaila, K.L., Murty, P.R.K., Rao, N.M., Rao, I.B.P., Rao, P.K., Sridhar, A.R., Murthy, A.S.N., Rao, V.V., and Prasad, B.R., 1996, Structure of the crystalline basement in the West Bengal basin, India, as determined from DSS studies: Geophysical Journal International, v. 124, p. 175–188, doi:10.1111/j.1365-246X.1996.tb06362.x. [8]

Kaminski, W. Bamford, D., Faber, S., Jacob, B., Nunn, K., Prodehl, C., 1976, A lithospheric seismic profile in Britain–II. Recording of a local earthquake: Zeitschrift für Geophysik, v. 42, p. 103–110. [7]

Kan, R-J, Hu, H X, Zeng, R-S, Mooney, W.D., and McEvilly, T.V., 1986, Crustal structure of Yunnan province, China, from seismic refraction profiles: Science, v. 234, p. 433–437, doi:10.1126/science.234.4775.433. [2] [8]

Kan, R., Hu, H., Zeng, R., Mooney, W.D., and McEvilly, T.V., 1988, Crustal structure and evolution of the Yunnan province, China, from seismic refraction profile, in Developments in the Research of Deep Structures of China's Continent: Geological Publishing House, Beijing, p. 267–276. [8]

Kanasewich, E.R., 1966, Deep crustal structure under the plains and Rocky Mountains: Canadian Journal of Earth Sciences, v. 3, p. 937–946. [6]

Kanasewich, E.R., and Cumming, G.L., 1965, Near-vertical incidence seismic reflections from the "Conrad" discontinuity: Journal of Geophysical Research, v. 70, p. 3441–3446, doi:10.1029/JZ070i014p03441. [2] [6]

Kanasewich, E.R., Clowes, R.M., and McCloughan, C.H., 1969, A buried Precambrian rift in western Canada: Tectonophysics, v. 8, p. 513–527, doi:10.1016/0040-1951(69)90051-1. [6]

Kanasewich, E.R., Hajnal, Z., Green, A.G., Cumming, G.L., Mereu, R.F., Clowes, R.M., Morel-a-l'Hussier, P., Chiu, S., Congram, A.M., Macrides, C.G., and Shariar, M., 1987, Seismic studies of the crust under the Williston Basin: Canadian Journal of Earth Science, v. 24, p. 2160–2171, doi:10.1139/e87-205. [7] [8]

Kanasewich, E.R., Burianyk, M.J.A., Ellis, R.M., Clowes, R.M., White, D.J., Coté, T., Forsyth,D.A., Luetgert, J.H., and Spence, G.D., 1994, Crustal velocity structure of the Omineca belt, southeastern Canadian Cordillera: Journal of Geophysical Research, v. 99, p. 2653–2670, doi:10.1029/93JB03108. [8]

Kanasewich, E.R., Burianyk, M.J.A., Dubuc, G.P., Lemieux, J.F., and Kalantzis, F., 1995, Three-dimensional seismic reflection studies of the Alberta Basement: Canadian Journal of Exploration Geophysics, v. 31, p. 1–10. [2] [9]

Kane, M.F., and Pakiser, L.C., 1961, Geophysical study of subsurface structure in southern Owens Valley, California: Geophysics, v. 26, p. 12–26, doi:10.1190/1.1438835. [5]

Kaneda, Y., Nishide, N., Sasaki, Y., Asano, S., Yoshii, T., Ichinose, Y., and Saka, M., 1979, Explosion seismic observaation of reflected waves from the Mohorovičić discontinuity and crustal structure in western Kanto district: Journal of the Physics of the Earth, v. 27, p. 511–526. [2] [7]

Kanestrom, R., and Haugland, K., 1971, Part II. Profile section 3-4 (Trans-Scandinavian Deep Sounding Project 1969), in Deep seismic sounding

in Northern Europe: Swedish Natural Science Research Council (NFR), Stockholm, p. 76–91. [6]

Kao, H., Gao, R., Rau, R.-J., Shi, D., Chen, R.-Y., Guan, T., and Wu, F.T., 2001, Seismic image of the Tarim basin and its collision with Tibet: Geology, v. 29, p. 575–578, doi:10.1130/0091-7613(2001)029<0575:SIOTTB>2.0.CO;2. [9]

Karlstrom, K.E., and Keller, G.R., eds., 2005, The Rocky Mountain Region: An evolving lithosphere—tectonics, geochemistry, and geophysics: Washington, D.C., American Geophysical Union Geophysical Monograph 154, 441 p. [2] [6] [9]

Karlstrom, K.E., Whitmeyer, S.J., Dueker, K., Williams, M.L., Bowring, S.A., Levander, A., Humphreys, E.D., Keller, G.R., and the CD-ROM Working Group, 2005, Synthesis of results from the CD-ROM experiment: 4-D image of the lithosphere beneath the Rocky Mountains and implications for understanding the evolution of continental lithosphere, in Karlstrom, K.E., and Keller, G.R., eds., The Rocky Mountain Region: An evolving lithosphere—tectonics, geochemistry, and geophysics: Washington, D.C., American Geophysical Union Geophysical Monograph 154, p. 421–441. [9]

Kashubin, S., Juhlin, C., Friberg, M., Rybalka, A., Petrov, G., Kashubin, A., Bliznetsov, M., and Steer, D., 2006, Crustal structure of the Middle Urals based on seismic reflection data, in Gee, D.G., and Stephenson, R.A., eds., European lithosphere dynamics: Geological Society, London, Memoir 32, p. 427–442. [2] [9] [10]

Kashubin, S., Tryggvason, A., Juhlin, C., Rybalka, A.V., Kashubina, T.V., and Shkred, I.G., 2009, The Krasnouralsky profile in the Middle Urals, Russia: a tomographic approach to vintage DSS data: Tectonophysics, v. 472, p. 249–263, doi:10.1016/j.tecto.2008.08.026. [9]

Katz, S., 1955, Seismic study of crustal structure in Pennsylvania and New York: Bulletin of the Seismological Society of America, v. 45, p. 303–325. [2] [5] [6] [8]

Katz, S., and Ewing, M., 1956, Seismic refraction measurements in the Atlantic Ocean, Part VII: Atlantic Ocean basin, west of Bermuda: Bulletin of the Seismological Society of America, v. 67, p. 475–510.[2] [5]

Katz, S., Edwards, R.S., and Press, F., 1953, Seismic-refraction profile across the Gulf of Maine: Bulletin of the Seismological Society of America, v. 64, p. 249–251. [2] [4]

Keen, C.E., 1979, Thermal history and subsidence of rifted continental margins—evidence for wells on the Nova Scotia and Labrador shelves: Canadian Journal of Earth Science, v. 16, p. 505–522, doi:10.1139/e79-046. [7]

Keen, C.E., and Barrett, D.L., 1971, A measurement of seismic anisotropy in the northeast Pacific: Canadian Journal of Earth Science, v. 8, p. 1056–1064. [2] [6] [7]

Keen, C.E., and Barrett, D.L., 1981, Thinned and subsided continental crust on the rifted margin of Eastern Canada: crustal structure, thermal evolution and subsidence history: Geophysical Journal of the Royal Astronomical Society, v. 65, p. 443–465. [2] [7]

Keen, C.E., and Tramontini, C., 1970, A seismic refraction survey on the mid-Atlantic ridge: Geophysical Journal of the Royal Astronomical Society, v. 20, p. 473–491. [2] [6] [7]

Keen, C.E., Barrett, D.L., Manchester, K.S., and Ross, D.I., 1972, Geophysical studies in Baffin Bay and some tectonic implications: Canadian Journal of Earth Science, v. 9, p. 239–256. [2] [7]

Keen, C.E., Kay, W.A., and Roest, W.R., 1990, Crustal anatomy of a transform continental margin, in Leven, J.H., Finlayson, D.M., Wright, C., Dooley, J.C., and Kennett, B.L.N., eds., Seismic probing of continents and their margins: Tectonophysics, v. 173, p. 527–544. [2] [8]

Keller, G.R., and Baldridge, W.S., 1999, The Rio Grande rift: a geological and geophysical overview: Rocky Mountain Geology, v. 34, p. 121–130, doi:10.2113/34.1.121. [9]

Keller, G.R., and Hatcher, R.D., Jr., 1999, Some comparisons of the structure and evolution of the southern Appalachian-Ouachita orogen and portions of the Trans-European Suture Zone region: Tectonophysics, v. 314, p. 43–68, doi:10.1016/S0040-1951(99)00236-X. [9]

Keller, G.R., Smith, R.B., and Braile, L.W., 1975, Crustal structure along the Great Basin–Colorado Plateau transition from seismic refraction studies: Journal of Geophysical Research, v. 80, p. 1093–1098, doi:10.1029/JB080i008p01093. [7]

Keller, G.R., Braile, L.W., and Schlue, J.W., 1979, Regional crustal structure of the Rio Grande rift from surface wave dispersion measurements, in Riecker, R.E., ed., Rio Grande rift; tectonics and magmatism: Washington, D.C., American Geophysical Union, Special Publication, p. 115–126. [7] [8]

Keller, G.R., Khan, M.A., Morgan, P., Wendlandt, R.F., Baldridge, W.S., Olsen, K.H., Prodehl, C., and Braile, L.W., 1991, A comparative study of the Rio Grande and Kenya rifts, *in* Gangi, A.F., ed., World rift systems: Tectonophysics, v. 197, p. 355–372. [8]

Keller, G.R., Braile, L.W., Davis, P.M., Meyer, R.P., Mooney, W.D., and the KRISP Working Group, 1992, Kenya Rift International Seismic Project, 1989–1990 experiment: Eos (Transactions, American Geophysical Union), v. 73, p. 345, 349–350, doi:10.1029/91EO00265. [9]

Keller, G.R., Mechie, J., Braile, L.W., Mooney, W.D., and Prodehl, C., 1994a, Seismic structure of the uppermost mantle beneath the Kenya rift, *in* C. Prodehl, G.R. Keller, and M.A. Khan, eds., Crustal and upper mantle structure of the Kenya rift: Tectonophysics, v. 236, p. 201–216. [9]

Keller, G.R., Prodehl, C., Mechie, J., Fuchs, K., Khan, M.A., Maguire, P.K.H., Mooney, W.D., Achauer, U., Davis, P.M., Meyer, R.P., Braile, L.W., Nyambok, I.O., and Thompson, G.A., 1994b, The East African rift system in the light of KRISP 1989–1990, *in* Prodehl, C., Keller, G.R., and Khan, M.A., eds., Crustal and upper mantle structure of the Kenya rift: Tectonophysics, v. 236, p. 465–483. [9]

Keller, G.R., Snelson, C.M., Sheehan, A.F., and Dueker, K.G., 1998, Geophysical studies of crustal structure in the Rocky Mountain region: a review. Rocky Mountain Geology, v. 33, p. 217–228. [6] [9]

Keller, G.R., Karlstrom, K.E., and Farmer, G.L., 1999, Tectonic evolution in the Rocky Mountain region: 4-D imaging of the continental lithosphere: Eos (Transactions, American Geophysical Union), v. 80, p. 493, 495, 498. [9]

Keller, G.R., Karlstrom, K.E., Williams, M.L., Miller, K.C, Andronicos, C., Levander, A., Snelson, C., and Prodehl, C, 2005a, The dynamic nature of the continental crust-mantle boundary: crustal evolution in the Southern Rocky Mountain region as an example, *in* Karlstrom, K.E., and Keller, G.R., eds., The Rocky Mountain Region: An evolving lithosphere—tectonics, geochemistry, and geophysics: Washington, D.C., American Geophysical Union Geophysical Monograph 154, p. 403–420. [9]

Keller, G.R., Klemperer, S.L., and Mooney, W.D., 2005b, PASSCAL instrumentation for controlled source seismology—a brief history: unpublished report. [2] [9] [10]

Kelly, K.R., Ward, R.W., Treitel, S., and Alford, R.M., 1976, Synthetic seismograms: a finite difference approach: Geophysics, v. 41, p. 2–27, doi:10.1190/1.1440605. [9]

Kempner, W.C., and Gettrust, J.F., 1982a, Ophiolites, synthetic seismograms, and oceanic crustal structure, 1: Comparison of ocean bottom seismometer data and synthetic seismograms from the Bay of Islands ophiolite: Journal of Geophysical Research, v. 87, p. 8447–8462, doi:10.1029/JB087iB10p08447. [7] [8]

Kempner, W.C., and Gettrust, J.F., 1982b, Ophiolites, synthetic seismograms, and oceanic crustal structure, 2: A comparison of synthetic seismograms of the Samail Ophiolite, Oman, and the ROSE refraction data from the East Pacific Rise: Journal of Geophysical Research, v. 87, p. 8462–8476, doi:10.1029/JB087iB10p08463. [7] [8]

Kennedy, G.C., 1959, The origin of continents, mountain ranges, and ocean basins: American Scientist, v. 47, p. 491–504. [5]

Kennett, B.L.N., 1974, Reflections, rays, and reverberations: Bulletin of the Seismological Society of America, v. 64, p. 1685–1696. [7]

Kennett, B.L.N., 1977, Towards a more detailed seismic picture of the oceanic crust and mantle: Marine Geophysics Research, v. 3, p. 7–42, doi:10.1007/BF00309792. [7]

Kennett, B.L.N., 1983, Seismic wave propagation in stratified media: Cambridge, U.K., Cambridge University Press, 342 p. [2] [7] [10]

Kennett, B.L.N., 2001, The seismic wavefield—Volume I: Introduction and theoretical development: Cambridge, U.K., Cambridge University Press. [2] [10]

Kent, G.M., Harding, A.J., and Orcutt, J.A. 1993, Distribution of magma beneath the East Pacific Rise between the Clipperton transform and the 9°17′N deval from forward modelling of common depth point data: Journal of Geophysical Research, v. 98, p. 13945–13969, doi:10.1029/93JB00705. [8]

Kent, G.M., Harding, A.J., and Orcutt, J.A., 1994, Uniform accretion of oceanic crust south of the Garrett transform at 14°15′S on the East Pacific Rise: Journal of Geophysical Research, v. 99 (B5), p. 9097–9116, doi:10.1029/93JB02872. [9]

Keranen, K., Klemperer, S.L., Gloagen, R., and the EAGLE Working Group, 2004, Three-dimensional seismic imaging of a protoridge axis in the Main Ethiopian rift: Geology, v. 32, p. 949–952, doi:10.1130/G20737.1. [10]

Khan, M.A., Maguire, P.K.H., Henry, W., Higham, M., Prodehl, C., Mechie, J., Keller, G.R., and Patel, J., 1989, A crustal seismic refraction line along the axis of the S. Kenya rift: Journal of African Earth Science, v. 8, p. 455–460, doi:10.1016/S0899-5362(89)80038-7. [8]

Khan, M.A., Mechie, J., Birt, C., Byrne, G., Gaciri, S., Jacob, B., Keller, G.R., Maguire, P.K.H., Novak, O., Nyambok, I.O., Patel, J.P., Prodehl, C., Riaroh, D. Simiyu, S., and Thybo, H., 1999, The lithospheric structure of the Kenya Rift as revealed by wide-angle seismic measurements, *in* Mac Niocaill, C., and Ryan, P.D., eds., Continental tectonics: Geological Society, London, Special Publication 164, p. 257–269. [9]

Kieckhefer, R.M., Shor, G.G.Jr., Curray, J.R., Sugiarta, W., and Hehuwat, F., 1980, Seismic refraction studies of the Sunda trench and forearc basin: Journal of Geophysical Research, v. 85, p. 863–889, doi:10.1029/JB085iB02p00863. [2] [7]

Kim, K.Y., Lee,J.M., Moon, W., Baag, C-E., Jung, H., and Hong, M.H., 2007, Crustal structure of the southern Korean pensinsula from seismic waves generated by large explosions in 2002 and 2004: Pure and Applied Geophysics, v. 164, p. 97–113, doi:10.1007/s00024-006-0149-4. [2] [10]

Kind, R., 1974, Long-range propagation of seismic energy in the lower lithosphere: Journal of Geophysics, v. 40, p. 189–202. [7]

Kind, R., Ni, J., Zhao, W., Wu, J., Yuan, X., Zhao, L., Sandvol, E., Reese, C., Nabelek, J., and Hearn, T., 1996, Mid-crustal low velocity zone beneath the southern Lhasa block: results from the INDEPTH-II earthquake recording program: Science, v. 274, p. 1692–1694, doi:10.1126/science.274.5293.1692. [9]

Kirk, R.E., Whitmarsh, R.B., and Langford, J.J., 1982, A three-component ocean bottom seismograph for controlled source and earthquake seismology: Marine Geophysical Researches, v. 5, p. 327–341, doi:10.1007/BF00305568. [8]

Kirk, R.E., Robertson, K., Whitmarsh, R.B., and Miles, P.R., 1991, A technique for conducting seismic refraction experiments on the ocean bed using bottom shots: Marine Geophysical Researches, v. 13, p. 153–160. [8]

Kleffman, S., Davey, F., Melhuish, A., Okaya, D., Stern, T., 1998, Crustal structure in the central South Island, New Zealand, from the Lake Pukaki seismic experiment: New Zealand Journal of Geology and Geophysics, v. 41, p. 39–49, doi:10.1080/00288306.1998.9514789. [2] [9]

Klemperer, S.L., 2006, Crustal flow in Tibet: geophysical evidence for the physical state of the Tibetan lithosphere, and inferred patterns of active flow, *in* Law, R.D., Searle, M.P., and Godin, L., eds., Channel flow, ductile extrusion and exhumation in continental collision zones: Geological Society London, spec. publ., 268, p. 39–70. [9]

Klemperer, S.L., and Hobbs, R.W., 1991, The BIRPS Atlas: Deep seismic reflection profiles around the British Isles: Cambridge University Press, 124 p. and 100 seismic sections. [2] [8] [9]

Klemperer, S.L., and Hurich, C.A., 1990, Lithospheric structure of the North Sea from deep seismic reflection profiling, *in* Blundell, D.J., and Gibbs, A., eds., Tectonic evolution of the North Sea rifts: Oxford University Press, p. 37–63. [2] [8]

Klemperer, S.L., and Mooney, W.D., eds., 1998a, Deep seismic probing of the continents, I: a global survey: Tectonophysics, v. 288, p. 298 p. [2] [9]

Klemperer, S.L., and Mooney, W.D., eds., 1998b, Deep seismic probing of the continents, II: general results and new methods: Tectonophysics, v. 286: 292 p. [2] [9]

Klemperer, S.L., Hauge, T.A., Hauser, E.C., Oliver, J., and Potter, C.J., 1986, The Moho in the northern Basin and Range province, Nevada, along the COCORP 40°N seismic-reflection transect: Geological Society of America Bulletin, v. 97, p. 603–618, doi:10.1130/0016-7606(1986)97<603:TMITNB>2.0.CO;2. [8]

Klemperer, S.L., Miller, E.L., Grantz, A., Scholl, D.W., Childs, J.R., Bogdanov, N.A., Belykh, I.N., Gnibidenko, H., Galloway, B., Hicks, B., Cole, F., and Toro, J., 2002, Crustal structure of the Bering and Chukchi shelves; deep seismic reflection profiles across the North American continent between Alaska and Russia, *in* Miller, E.L., Grantz, A., and Klemperer, S.L., eds., Tectonic Evolution of the Bering Shelf-Chukchi Sea-Artic Margin and Adjacent Landmasses: Geological Society of America Special Paper 360, p. 1–24. [9]

Kluth, C.F., and Coney, P.J., 1981, Plate tectonics of the Ancestral Rocky Mountains: Geology, v. 9, p. 10–15, doi:10.1130/0091-7613(1981)9<10:PTOTAR>2.0.CO;2. [9]

Knapp, C.C., Knapp, J.H., and Connor, J.A., 2004, Crustal-scale structure of the South Caspian basin revealed by deep seismic reflection profiling: Marine and Petroleum Geology, v. 21, p. 1073–1081, doi:10.1016/j.marpetgeo.2003.04.002. [10]

Knapp, J.H., Diaconescu, C.C., Bader, M.A., Sokolov, V.B., Kashubin, S.N., and Rybalka, A.V., 1998, Seismic reflection fabrics of continental collision and post–orogenic extension in the Middle Urals, central Russia, *in* Klemperer, S.L., and Mooney, W.D., eds., Deep seismic probing of the continents, II: a global survey: Tectonophysics, v. 288, p. 115–126. [2] [9]

Knothe, H., and Schröder, E., 1972, The German Democratic Republic, *in* Sollogub, V.B., Prosen, D., and Militzer, H., 1972, Crustal structure of central and southeastern Europe based on the results of explosion seismology (published in Russian 1971). English translation edited by Szénás, Gy., 1972: Geophysical Transactions, special edition, Müszaki Könyvkiado, Budapest, chapter 23, p. 80–86. [2] [6]

Knothe, H., and Walther, K.F., 1968, Vorbereitung und Durchführung einer seismischen Tiefensondierung im Grenzgebiet DDR-CSSR: Freiberger Forschungshefte, C239. [6]

Knott, C.G., 1919, The propagation of earthquake waves through the earth and connected problems: Royal Society of Edinburgh Proceedings, v. 39, p. 157–208. [3]

Kodaira, S., Goldschmidt-Rokita, A., Hartmann, J.M., Hirschleber, H.B., Iwasaki, T., Kanazawa, T., Krahan, H., Tomita, S., and Shimamura, H., 1995, Crustal structure of the Lofoten continental margin, off northern Norway, from ocean-bottom seismographic studies: Geophysical Journal International, v. 121, p. 907–924, doi:10.1111/j.1365-246X.1995.tb06447.x. [8]

Kodaira, S., Iwasaki, T., Urabe, T., Kanazawa, T., Egloff, F., Makris, J., and Shimamura, H., 1996, Crustal structure across the middle Ryukyu trench obtained from ocean bottom seismographic data: Tectonophysics, v. 263, p. 39–60, doi:10.1016/S0040-1951(96)00025-X. [8]

Kodaira S., Takahashi, N., Park, J.-O., Mochizuki, K., Shinohara, M., and Kimura, S., 2000, Western Nanakai Trough seismogenic zone: results from wide-angle ocean bottom seismic survey: Journal of Geophysical Research, v. 105, p. 5887–5905, doi:10.1029/1999JB900394. [2] [9]

Kodaira S., Kurashimo, E., Park, J.-O., Takahash, N., Nakanishi, A., Miura, S., Iwasaki, T., Hirata, N., Ito, K., and Kaneda, Y., 2002, Structural factors controlling the rupture process of as megathrust earthquake at the Nankai trough seismogenic zone: Geophysical Journal International, v. 149, p. 815–835, doi:10.1046/j.1365-246X.2002.01691.x. [2] [9] [10]

Kodaira, S., Iidaka, T., Kato, A., Park, J.-O., Iwasaki, T., and Kaneda, Y., 2004, High pore fluid pressure may cause silent slip in the Nankai Trough: Science, v. 304, p. 1295–1298, doi:10.1126/science.1096535. [2] [10]

Kodaira, S., Iidaka, T., Nakanishi, A., Park, J.-O., Iwasaki, T., Kaneda, Y., 2005, Onshore-offshore seismic transect from the eastern Nankai Trough to central Japan crossing a zone of the Tokai slow slip event: Earth, Planets and Space, 57, p. 943–959. [10]

Kodaira, S., Hori, T., Ito, A., Miura, S., Fujie, G., Park, J.-O., Baba, T., Sakaguchi, H., Kaneda, Y., 2006, A cause of rupture segmentation and synchronization in the Nankai trough revealed by seismic imaging and numerical simulation: Journal of Geophysical Research, v. 111, B09301, doi:10.1029/2005JB004030. [10]

Kodaira, S., Sato, T., Takahashi, N., Ito, A., Tamura, Y., Tatsumi, Y., and Kaneda, Y., 2007a, Seismological evidence for variable growth of crust along the Izu intraoceanic arc: Journal of Geophysical Research, v. 112: doi:10.1029/2006BJ004593. [2] [10]

Kodaira, S., Sato, T., Takahashi, N., Miura, S., Tamura, Y., Tatsumi, Y., and Kaneda, Y., 2007b, New seismological constraints on growth of continental crust in the Izu-Bonin intra-oceanic arc: Geology, v. 35, p. 1031–1034, doi:10.1130/G23901A.1. [10]

Kogan, A.L., 1972, Results of deep seismic soundings of the earth's crust in East Antarctica, *in* Adie, R.J., ed., Antarctic Geology and Geophysics: Universitets Forlaget, Oslo (UUGS series B number1), p. 485–489. [6]

Kohler, M.D., and Davis, P.M., 1997, Crustal thickness variations in southern California from Los Angeles Region Seismic Experiment (LARSE) passive phase teleseismic travel times: Bulletin of the Seismological Society of America, v. 87, p. 1330–1334. [9]

Köhler, R., 1974, Anfänge der Reflexionsseismik in Deutschland, *in* Birett, H., Helbig, K., Kertz, W., and Schmucker, U., eds., Zur Geschichte der Geophysik: Berlin-Heidelberg-New York, Springer, p. 99–113. [3]

Kohler, W.M., and Catchings, R.D., 1994, Data report for the 1993 seismic refraction experiment in the San Francisco Bay Area, California: U.S. Geological Survey Open-File Report 94-241, 71 p. [9]

Kohler, W.M., and Fuis, G.S., 1988, Data report for the 1979 seismic-refraction experiment in the Imperial Valley, California: U.S. Geological Survey Open-File Report 88-255, 96 p. [7]

Kohler, W.M., Fuis, G.S., and Berge, P.A., 1987, Data report for the 1979–1985 seismic-refraction surveys in northeastern California: U.S. Geological Survey Open-File Report 87-625, Menlo Park, California, 99 p. [8]

Kong, X., Wang, Q., and Xiong, S., 1996 . Comprehensive geophysics and lithosphere structure in the western Xiang (Tibet) plateau: Science in China (Series D), v. 39, p. 348–358. [9]

Kopp, C., Fruehn, J., Flueh, E.R., Reichert, C., Kukowski, N., Bialas, J., and Klaeschen, D., 2000, Structure of the Makran subduction zone from wide-angle and reflection seismic data: Tectonophysics, v. 329, p. 171–191, doi:10.1016/S0040-1951(00)00195-5. [2] [9]

Korenaga, J., Holbrook, W.S., Kent, G.M., Kelemen, P.B., Detrick, R.S., Larsen, H.-C., Hopper, J.R., and Dahl-Jensen, T., 2000, Crustal structure of the southeast Greenland margin from joint refraction and reflection seismic tomography: Journal of Geophysical Research, v. 105, p. 21,591–21,614, doi:10.1029/2000JB900188. [2] [9]

Korsch, R.J., Wake-Dyster, K.D., O'Brien, P.E., Finlayson, D.M. and Johnstone, D.W., 1992, Geometry of Permian to Mesozoic sedimentary basins in eastern Australia and their relationship to the New England Orogen, *in* Rickard, M.J., Harrington, H.J., and Williams, P.R., eds., Basement Tectonics 9: Dordrecht, Kluwer Academic Publishers, p. 85–108. [2] [8]

Korsch, R.J., Johnstone, D.W., and Wake-Dyster, K.D., 1997, Crustal architecture of the New England Orogen based on deep seismic reflection profiling: Geological Society of Australia Special Publication 19, p. 29–51. [2] [9]

Korsch, R.J., Goleby, B.R., Leven, J.H., and Drummond, B.J., 1998, Crustal architecture of central Australia based on seismic reflection profiling, *in* Klemperer, S.L., and Mooney, W.D., Deep seismic profiling of the continents, II: a global survey: Tectonophysics, v. 288, p. 57–69, doi:10.1016/S0040-1951(97)00283-7. [2] [8] [9]

Korsch, R.J., Barton, T.J., Gray, D.R., Owen, A.J. and Foster, D.A., 2002, Geological interpretation of a deep seismic reflection transect across the boundary between the Delamerian and Lachlan orogens, in the vicinity of The Grampians, Western Victoria: Australian Journal of Earth Sciences, v. 49, p. 1057–1075, doi:10.1046/j.1440-0952.2002.00963.x. [2] [9]

Kosminskaya, I.P., 1969, Explosion seismology: introduction, *in* Hart, P.J., ed., The earth's crust and upper mantle: American Geophysical Union Geophysical Monograph 13, p. 177. [6] [10]

Kosminskaya, I.P., 1971, Deep seismic sounding of the earth's crust and upper mantle: Translated by G.V. Keller: New York-London, Consultants Bureau, 184 p. [2] [7] [10]

Kosminskaya, I.P., and Pavlenkova, N.I., 1979, Seismic models of inner parts of the Euro-Asian continent and its margins: Tectonophysics, v. 59, p. 307–320, doi:10.1016/0040-1951(79)90052-0. [7]

Kosminskaya, I.P., and Riznichenko, Y.V., 1964, Seismic studies of the earth's crust in Eurasia, *in* Odishaw, H., ed., Research in Geophysics, Vol. 2, Solid Earth and Interface Phenomena: Cambridge, MIT Press, p. 81–122. [2] [6]

Kosminskaya, I.P., Belyaevsky, N.A., and Volvovsky, I.S., 1969, Explosion seismology in the USSR, *in* Hart, P.J., ed., The earth's crust and upper mantle: American Geophysical Union Geophysical Monograph 13, p. 195–208. [2] [6]

Kosminskaya, I.P., Zverev, S.M., and Udintsev, G.B., 1973, Soviet seismic studies of the Earth's crust in the Pacific Ocean during the International Upper Mantle Project—a summary, *in* Mueller, S., ed., The structure of the earth's crust, based on seismic data: Tectonophysics, v. 20, p. 147–151. [2] [6] [7]

Kostyuchenko, S.L., Egorkin, A.V., and Solodilov, L., 1999, Structure and genetic mechanisms of the Precambrian rifts of the East-European Platform in Russia by integrated seismic, gravity and magnetic data, *in* Stephenson, R.A., Wilson, M., and Starostenko, V.I., eds., EUROPROBE GeoRift, volume 2: Intraplate tectonics and basin dynamics of the East European Craton and its margins: Tectonophysics, v. 313, p. 9–28. [9]

Kostyuchenko, S.L., Sapozhnikov, R., Egorkin, A., Gee, D.G., Berzin, R., and Solodilov, L., 2006, Crustal structure and tectonic model of northeastern Baltica, based on deep potential data, *in* Gee, D.G., and Stephenson, R.A., eds., European lithosphere dynamics: Geological Society of London, Memoir 32, p. 521–539. [2] [9] [10]

Kozlovsky, Y., ed., 1988, The superdeep well of the Kola peninsula: Berlin-Heidelberg-New York, Springer, 558 p. [8] [9]

Krabbenhoeft, A., Bialas, J., Kopp, H., Kukowski, N., and Huebscher, C., 2004, Crustal structure of the Peruvian continental margin from wide-angle seismic studies: Geophysical Journal International, v. 159, p. 749–764, doi:10.1111/j.1365-246X.2004.02425.x. [2] [9]

Krawczyk, C.M., and the SPOC Team, 2003, Amphibious seismic survey images plate interface at 1960 Chile earthquake: Eos (Transactions, American Geophysical Union), v. 84, p. 301, 304–305. [2] [9] [10]

Krawczyk, C.M., Stiller, M., and DEKORP/BASIN Research Group, 1999, Reflection seismic constraints and Moho beneath on Paleozoic crustal structure the NE German Basin: Tectonophysics, v. 314, p. 241–253, doi:10.1016/S0040-1951(99)00246-2. [2] [9]

Krawczyk, C.M., Stein, E., Choi, S., Oettinger, G., Schuster, K., Götze, H.-J., Haak, V., Oncken, O., Prodehl, C., and Schulze, A., 2000, Constraints on distribution and exhumation mechanisms of high-pressure rocks from geophysical studies—the Saxothuringian case between the Bray Fault and Elbe Line, *in* Franke, W., Haak, V., Oncken, O., and Tanner, D., eds., Orogenic processes: quantification and modelling in the Variscan belt: Geological Society of London Special Publication 179, p. 303–322, 2000. [2] [9]

Krawczyk, C.M., Eilts, F., Lassen, A., and Thybo, H., 2002, Seismic evidence of Caledonian deformed crust and uppermost-mantle structure in the northern part of the Trans-European Suture Zone, SW Baltic Sea: Tectonophysics, v. 360, p. 215–244, doi:10.1016/S0040-1951(02)00355-4. [9]

Krawczyk, C.M., Mechie, J., Lueth, S., Tasarova, Z., Wigger, P., Stiller, M., Brasse, H., Echtler, H.P., Araneda, M., and Bataille, K., 2006, Geophysical signatures and active tectonics at the south-central Chilean margin, *in* Oncken, O., Chong, G., Franz, G., Giese, P., Goetze, H.-J., Ramos, V.A., Strecker, M.R., and Wigger, P., eds., The Andes—active subduction orogeny. Frontiers in Earth Sciences Series: Berlin-Heidelberg-New York, Springer, p. 171–192. [2] [9] [10]

Krey, T., Schmidt, G., and Seelis, K.-H., 1961, Über die Möglichkeit, den reflexionsseismisch erfassbaren Tiefenbereich zu erweitern: Erdöl und Kohle, v. 14, p. 521–526. [6]

Krishna V.G., Rao C.V.R.K., Gupta H.K., Sarkar D., and Baumbach M., 1999, Crustal seismic velocity structure in the epicentral region of the Latur earthquake (September 29, 1993), southern India: inferences from modelling of the aftershock seismograms: Tectonophysics, v. 304, p. 241–255, doi:10.1016/S0040-1951(99)00028-1. [2] [9]

KRISP Working Group, 1987, Structure of the Kenya rift from seismic refraction: Nature, v. 325, no. 6101, p. 239–242, doi:10.1038/325239a0. [8]

KRISP Working Group, 1991, A compilation of data from the 1990 Kenya Rift International Seismic Project KRISP 90 seismic refraction–wide-angle reflection experiment: Geophysical Institute, University of Karlsruhe, Open-File Report 91-1, 80 p. [9]

KRISP Working Group, 1995a, Group takes a fresh look at the lithosphere underneath southern Kenya: Eos (Transactions, American Geophysical Union), v. 76, p. 73, 81–82. [9]

KRISP Working Group, 1995b, A compilation of data from the 1994 Kenya Rift International Seismic Project KRISP 94 seismic refraction–wide-angle reflection experiment and accompanying gravity measurements: Geophysical Institute, University of Karlsruhe, Open-File Report 95-1, 65 p. [9]

KRISP Working Party, 1991, Large-scale variation in lithospheric structure along and across the Kenya Rift: Nature, v. 354, p. 223–227, doi:10.1038/354223a0. [9]

Kurashimo, E., Tokunaga, M., Hirata, N., Iwasaki, T., Kodaira, S., Kaneda, Y., Ito, K., Nishida, R., Kimura, S., and Ikawa, T., 2002, Geometry of the subducting Philippine Sea Plate and the crustal and upper mantle structure beneath the eastern Shikoku Island revealed by seismic refraction/wide-angle reflection profiling: Zisin, v. 54, p. 489–505. [9]

Kurashimo, E., Iwasaki, T., Hirata, N., Ikawa, T., Kaneda, Y., and Onishi, M., 2007, Crustal structure of the southwestern Kuril Arc sited in the eastern part of Hokkaido, Japan, inferred from seismic refraction/reflection experiments: Earth, Planets, and Space, v. 59, p. 375–380. [9]

Kutschale, H., 1966, Arctic Ocean geophysical studies: the southern half of the Siberia basin: Geophysics, v. 31, p. 683–710, doi:10.1190/1.1439804. [2] [6]

Labrouste, Y., Choudhury, M., and Perrier, G., 1963, Essais d?interprétation séismique. VI.B. Essai d'interprétation no. 2, *in* Closs, H., and Labrouste, Y., eds., Séismologie: Recherches séismologiques dans les Alpes occidentals au moyen de grandes explosions een 1956, 1958 et 1960: Mémoir

Collectif, Année Géophysique Internationale, Centre National de la Recherche Scientifique, Série XII, Fasc. 2, p. 176–201. [5] [6]

Labrouste, Y., Baltenberger, P., Perrier, G., and Recq, M., 1968, Courbes dégale profondeur de la discontinuité de Mohorovičić dans le sud-est de la France: Comptes rendus de l'Académie de sciences, France, v. 266, p. 663–665. [6]

Lafond, C.C., and Levander, A., 1995, Migration of wide-aperture onshore-offshore seismic data, central California: seismic images of late stage subduction: Journal of Geophysical Research, v. 100, p. 22,231–22,243, doi:10.1029/95JB01968. [2] [9]

Lafoy, Y., Geli, L., Klingelhoefer, K., Vially, R., Sichler, B., and Nouze, H., 2005, Discovery of continental stretching and oceanic spreading in the Tasman Sea: Eos (Transactions, American Geophysical Union), v. 86, no. 10, p. 101, 104–105, doi:10.1029/2005EO100001. [2] [10]

Laigle, M., Hirn, A., Sachpazi, A., and Roussos, N., 2000, North Aegean crustal deformation: a new active fault imaged to 10 km depth by marine reflection seismic: Geology, v. 28, p. 71–74, doi:10.1130/0091-7613(2000)28<71:NACDAA>2.0.CO;2. [9]

Laigle, M., Bécel, A., de Voogd, B., Hirn, A., Taymaz, T., Ozalabey, S., and members of the SEISMARMARA Leg1, 2007, A first deep seismic survey in the Sea of Marmara: deep basins and whole crust architecture and evolution: Earth and Planetary Science Letters, v. 270, no. 3-4, p. 168–179, doi:10.1016/j.epsl.2008.02.031. [2] [10]

Lampshire, L.D., Coruh, C., and Costain, J.K., 1994, Crustal structures and the eastern extent of lower Paleozoic shelf strata within the central Appalachians: a seismic reflection interpretation: Geological Society of America Bulletin, v. 106, p. 1–18. [8]

Landes, M., 2001, VARNET-96: data processing and seismic sections: Geophysical Institute, University of Karlsruhe, Open-File Report, 340 p., doi:10.1046/j.1365-246X.2000.00035.x. [9]

Landes, M., Prodehl, C., Hauser, F., Jacob, A.W.B., Vermeulen, N.J., and Mechie, J., 2000, VARNET-96: Influence of the Variscan and Caledonian orogenies on crustal structure in SW Ireland: Geophysical Journal International, v. 140, p. 660–676, doi:10.1046/j.1365-246X.2000.00035.x. [8] [9]

Landes, M., O'Reilly, B.M., Readman, P.W., Shannon, P.M., and Prodehl, C., 2003, VARNET-96: three-dimensional upper crustal velocity structure of SW Ireland: Geophysical Journal International, v. 153, p. 424–442, doi:10.1046/j.1365-246X.2003.01911.x. [2] [9]

Landes, M., Fielitz, W., Hauser, F., Popa, M., and the CALIXTO Group, 2004a, 3-D upper-crustal tomographic structure across the Vrancea seismic zone, Romania: Tectonophysics, v. 382, p. 85–102, doi:10.1016/j.tecto.2003.11.013. [9]

Landes, M., Hauser, F., Raileanu, V., Bala, A., Prodehl, C., Bribach, J., Harder, S., Hegedues, E., Keller, R.G., Stephenson, R.A., Mocanu, V., Dinu, C., and Diaconescu, C., 2004b, VRANCEA 2001 data processing and seismic sections: Geophysical Institute, University of Karlsruhe, Open-File Report, 57 p. and seismic sections. [9] [10]

Landes, M., Ritter, J.R.R., Do, V.C., Readman, P.W. and O Reilly, B.M., 2004c, Passive teleseismic experiment explores the deep subsurface of southern Ireland: Eos (Transactions, American Geophysical Union), v. 85, p. 337, 340–341, doi:10.1029/2004EO360002. [9] [10]

Landes, M., Ritter, J.R.R., Readman, P.W. and O Reilly, B.M, 2005, The Irish crustal structure and its signatures from the Caledonian and Variscan orogenies: Terra Nova, v. 17, p. 111–120. [2] [8] [9] [10]

Landisman, M., and Mueller, S., 1966, Seismic studies of the earth's crust in continents; part II: Analysis of wave propagation in continents and adjacent shelf areas: Geophysical Journal of the Royal Astronomical Society, v. 10, p. 539–554. [4] [6]

Landisman, M., Mueller, S., and Mitchell, B.J., 1971, Review of evidence for velocity inversions in the continental crust, *in* Heacock, J.G., ed., The structure and physical properties of the earth's crust: American Geophysical Union Geophysical Monograph 14, p. 11–34. [6]

Langinen, A.E., Lebedeva-Ivanova, N.N., Gee, D.G., and Zamansky, Yu.Ya., 2009, Correlations between the Lomonossov Ridge, Marvin Spur and adjacent basins of the Arctic Ocean based on seismic data: Tectonophysics, v. 472, p. 309–322, doi:10.1016/j.tecto.2008.05.029. [2] [8]

Larkin, S.P., McCarthy, J., and Fuis, G.S., 1988, Data report for the PACE 1987 seismic refraction survey, west-central Arizona: U.S. Geological Survey Open-File Report 88-694, Menlo Park, California: 95 p. [8]

LASE Study Group, 1986, Deep structure of the U.S. east coast passive margin from large aperture seismic experiments (LASE): Marine and Petroleum Geology, v. 3, p. 234–242, doi:10.1016/0264-8172(86)90047-4. [2] [8]

Lassen, N.A., Thybo, H., and Berthelsen, A., 2001, Reflection seismic evidence for Caledonian deformed sediments above Sveconorwegian basement in the southwestern Baltic Sea: Tectonics, v. 20, no. 2, p. 268–276, doi:10.1029/2000TC900028. [9]

Latham, T.S., Best, J., Chaimov, T., Oliver, J., Brown, L., and Kaufman, S., 1988, COCORP profiles from the Montana plains: the Archean cratonic crust and a lower crustal anomaly beneath the Williston basin: Geology, v. 16, p. 1073–1076, doi:10.1130/0091-7613(1988)016<1073:CPFTMP >2.3.CO;2. [2] [8]

Lau, K.W.H., Louden, K.E., Funck, T., Tucholke, B.E., Holbrook, W.S., Hopper, J.R., and Larsen, H.C., 2006a, Crustal structure across the Grand Banks–Newfoundland Basin Continental Margin—I. Results from a seismic refraction profile: Geophysical Journal International, v. 167, p. 127–156, doi:10.1111/j.1365-246X.2006.02988.x. [8] [9]

Lau, K.W.H., Louden, K.E., Deemer, S., Hall, J., Hopper, J.R., Tucholke, B.E., Holbrook, W.S., and Larsen, H.C., 2006b, Crustal structure across the Grand Banks–Newfoundland Basin Continental Margin—II. Results from a seismic reflection profile: Geophysical Journal International, v. 167, p. 157–170, doi:10.1111/j.1365-246X.2006.02989.x. [9]

Laughton, A.S., and Tramontini, C., 1970, Recent studies of the crustal structure in the Gulf of Aden: Tectonophysics, v. 8, p. 359–375, doi:10.1016/0040 -1951(69)90043-2. [6]

Laughton, A.S., Matthews, D.H., and Fisher, R.L., 1970, The structure of the Indian Ocean, *in* Maxwell, A.E., ed., 1970, The Sea, new concepts of ocean floor evolution: New York, Wiley-Interscience, v. 4, p. 543–586. [6]

Lawyer, L.C., Bates, C.C., and Rice, R.B., 2001, Geophysics in the affairs of mankind—a personalized history of exploration geophysics: Society of Exploration Geophysicists, Tulsa, Oklahoma, U.S.A., 205 p. [3] [10]

Lay, T., and Wallace, T.C., 1995, Modern global seismology: Acad. Press, San Diego–New York, 517 p. [2] [10]

Leaver, D.S., Mooney, W.D., and Kohler, W.M., 1984, A seismic refraction study of the Oregon Cascades: Journal of Geophysical Research, v. 89, p. 3121–3134, doi:10.1029/JB089iB05p03121. [8] [9]

Lebedeva-Ivanova, N.N., Zamansky, Yu.Ya., Langinen, A.E., and Sorokin, M.Yu., 2006, Seismic profiling across the Mendeleev Ridge at 82°N: evidence of continental crust: Geophysical Journal International, v. 165, p. 527–544, doi:10.1111/j.1365-246X.2006.02859.x. [2] [8] [9]

Leclair, A.D., Percival, J.A., Green, A.G., Wu, J., West, G.F., and Wang, W., 1994, Seismic reflection profiles across the central kapuskasing uplift: Canadian Journal of Earth Science, v. 31, p. 1016–1026, doi:10.1139/e94-092. [8]

Lee, W., Kanamori, H., Jennings, P.C., and Kisslinger, C., eds., 2002, International Handbook of Earthquake and Engineering Seismology. Academic Press, Amsterdam. Part A: p. 1–933, Part B: p. 934–1945. [2] [10]

Leenhardt, O., 1972, Organization, *in* Leenhardt, O., Gobert, B., Hinz, K., Hirn, A., Hirschleber, H., Hsü, H.J., Refubatti, A., Rudant, J.-P., Rudloff, R., Ryan, W.B.F., Snoek, M., and Steinmetz, L., Results of the Anna cruise—three north-south seismic profiles through the western Mediterranean Sea: Bulletin du Centre Recherches Pau–SNPA, v. 6, p. 373–381. [7]

Leenhardt, O., Gobert, B., Hinz, K., Hirn, A., Hirschleber, H., Hsü, H.J., Refubatti, A., Rudant, J.-P., Rudloff, R., Ryan, W.B.F., Snoek, M., and Steinmetz, L., 1972, Results of the Anna cruise—three north-south seismic profiles through the western Mediterranean Sea: Bulletin du Centre Recherches Pau–SNPA, v. 6, p. 365–452. [2] [7]

Leet, L.D., 1936, Seismological data on surface layers in New England: Bulletin of the Seismological Society of America, v. 26, p. 129–145. [3] [4]

Leet, L.D., 1938, Travel times for New England: Bulletin of the Seismological Society of America, v. 28, p. 45–48. [2] [3]

Leet, L.D., 1941, Trial travel times for northeastern America: Bulletin of the Seismological Society of America, v. 31, p. 325–334. [3]

Leet, L.D., 1946, The velocity of P in the granitic layer: Transactions, American Geophysical Union, v. 27, p. 631–635. [4]

Leet, L.D., and Birch, F., 1942, Seismic velocities, *in* Birch, F., Schairer, J.F., and Spicer, H.C., eds., 1942, Handbook of physical constants: Geological Society of America Special Paper 36, p. 93–101. [4] [5]

Lehmann, I., 1948, On two explosions in Danish waters in the autumn of 1946: Geofisica Pura ed Applicata, v. 12, p. 145–161. [4]

Leitner, B., Trehu, A.M., and Godfrey, N.J., 1998, Crustal structure of the northwestern Vizcaino block and Gorda Escarpment, offshore northern California, and implications for postsubduction deformation of a paleo-accretionary margin: Journal of Geophysical Research, v. 103, p. 23,795–23,812, doi:10.1029/98JB02050. [9]

Le Pichon, X., and Cochran, J.R., eds., 1988, The Gulf of Suez and Red Sea rifting: Tectonophysics, v. 153, p. 1–320. [2]

Le Pichon, X., Houtz, R.E., Drake, C.L., and Nafe, J.E., 1965, Crustal structure of the mid-ocean ridges: Journal of Geophysical Research, v. 70, p. 319–329, doi:10.1029/JZ070i002p00319. [5]

Lerch, D.W., Klemperer, S.L., Glen, J.M.G., Ponce, D.A., Miller, E.L., and Colgan, J.P., 2007, Crustal structure of the northwestern Basin and Range province and its transition to unextended volcanic plateaus: Geochemistry, Geophysics, Geosystems, 8Q02011, doi:10.1029/2006GC001429. [2] [10]

Lessel, K., 1998, Die Krustenstruktur der Zentralen Anden in Nordchile (21–24°S), abgeleitet aus 3D-Modellierungen refraktionsseismischer Daten: Berliner Geowissenschaftliche Abhandlungen, Reihe B, Band 31: 185 p. [9]

Levander, A., Fuis, G.S., Wissinger, E.S., Lutter, W.J., Oldow, J.S., and Moore, T.E., 1994, Seismic images of the Brooks Range fold and thrust belt, Arctic Alaska, from an integrated seismic reflection/refraction experiment: Tectonophysics, v. 232, p. 13–30, doi:10.1016/0040-1951(94)90073-6. [8]

Levander, A., Zelt, C., and Magnani, M.B., 2005, Crust and upper mantle velocity structure of the Southern Rocky Mountains from the Jemez Lineament to the Cheyenne Belt, *in* Karlstrom, K.E., and Keller, G.R., eds., The Rocky Mountain Region: An evolving lithosphere—tectonics, geochemistry, and geophysics: Washington, D.C., American Geophysical Union Geophysical Monograph 154, p. 293–308. [9]

Levander, A., Schmitz, M., Ave Lallemant, H., Zelt, C.A., Sawyer, D.S., Magnani, M.B., Mann, P., Christeson, G., Wright, J., Pavlis, G., and Pindell, J., 2006, The BOLIVAR and GEODINOS projects: investigating continental growth and deformation along the SE Caribbean plate boundary: Eos (Transactions, American Geophysical Union), v. 87, p. 97–100. [10]

Levander, A., Zelt, C., and Symes, W.W., 2007, Crust and lithospheric structure—active source studies of crust and lithospheric structure, *in* Romanowicz, B., and Dziewonski, A., eds., Treatise on Geophysics, vol 1, Seismology and structure of the earth: Amsterdam, Elsevier, p. 247–288. [2] [10]

Leven, J.H., 1980, The application of synthetic seismograms to the interpretation of crustal and upper mantle structure [Ph.D. thesis]: Canberra, School of Earth Sciences, ANU, 235 p. [7]

Leven, J.H., and Lindsay, J.F., 1995, A geophysical investigation of the southern margin of the Musgrave Block, South Australia: AGSO Journal of Australian Geology and Geophysics, v. 16, no. 1-2, p. 155–161. [9]

Leven, J.H., Finlayson, D.M., Wright, C., Dooley, J.C., and Kennett, B.L.N., eds., 1990, 3rd international symposium on deep seismic profiling of the continents and their margins: Tectonophysics, v. 173, p. 1–645. [2] [8]

Lewis B.T.R., and Garmany, J.D., 1982, Constraints on the structure of the East Pacific Rise from seismic refraction data: Journal of Geophysical Research, v. 87, p. 8417–8425. [7]

Lewis, B.T.R., and Snydsman, W.E., 1979, Fine structure of the lower crust on the Cocos plate: Tectonophysics, v. 55, p. 87–105, doi:10.1029 /JB087iB10p08417. [2] [7]

Lhiuzou Explosion Research Group, 1988, Observation of Lhiuzou explosion and the crustal structure in eastern Guangxi, *in* Developments in the Research of Deep Structures of China's Continent. Geological Publishing House, Beijing, p. 246–252. [8]

Li, S., and Mooney, W.D., 1998, Crustal structure of China from deep seismic sounding profiles, *in* Klemperer, S.L., and Mooney, W.D., eds., Deep seismic probing of the continents, II: a global survey: Tectonophysics, v. 288, p. 105–113. [2] [8] [9] [10]

Li, S., Mooney, W.D., and Fan, J., 2006, Crustal structure of mainland China from deep seismic sounding data: Tectonophysics, v. 420, p. 239–252, doi:10.1016/j.tecto.2006.01.026. [2] [10]

Liao, Q., Wang, Z., Zhu, Z., and Wu, X., 1988, Crust and upper mantle structure in the Quanzhou-Shantou region of China, *in* Developments in the Research of Deep Structures of China's Continent: Geological Publishing House, Beijing, p. 227–235. [8]

Liebscher, H.J., 1962, Reflexionshorizonte der tieferen Erdkruste im bayerischen Alpenvorland, abgeleitet aus Ergebnissen der Reflexionsseismik: Z. für Geophysik, v. 28, p. 162–184. [2] [4] [5] [6]

Liebscher, H.J., 1964, Deutungsversuche für die Struktur der tieferen Erdkruste nach reflexionsseismischen und gravimetrischen Messungen im deutschen Alpenvorland: Zeitschrift für Geophysik, v. 30, p. 51–96, 115–126. [2] [4] [5] [6]

Light, M.P.R., Maslanyi, M.P., Greenwood, R.J., and Banks, N.L., 1993, Seismic sequence stratigraphy and tectonics offshore Namibia, *in* Williams, G., and Dobb, A., eds., Tectonics and seismic sequence stratigraphy : Geological Society of London Special Publication 71, p. 163–191. [9]

Lindeque, A.S., Ryberg, T., Stankiewicz, J., Weber, M.H., and de Wit, M.J., 2007, Deep crustal seismic reflection experiment across the southern Karoo Basin, South Africa: South African Journal of Geology, v. 110, no. 2-3, p. 419–438, doi:10.2113/gssajg.110.2-3.419. [2] [10]

Lindwall, D.A., 1988, A two-dimensional seismic investigation of crustal structure under the Hawaiian islands near Oahu and Kauai: Journal of Geophysical Research, v. 93, p. 12,107–12,122, doi:10.1029 /JB093iB10p12107. [8]

Linke, F. and Wiechert, E., 1903, Monatsberichte über seismische Registrierungen in Göttingen: Königliches Geophysikalisches Institut zu Göttingen (Eigenverlag), 4 p. [3]

Liu, C.-S., Reed, D.L., Lundberg, N., Moore, G.F., McIntosh, K.D., Nakamura, Y., Wang, T.K., Chen, T.H., Lallemand, S., 1995, Deep seismic imaging of the Taiwan arc-continent collision zone: Eos (Transactions, American Geophysical Union), v. 76, no. 46, p. F635. [9]

Liu, M., Mooney, W.D., Li, S., Okaya, N., and Detweiler, S., 2006, Crustal Structure of the Northeastern Margin of the Tibetan Plateau from the Songpan-Ganzi Terrane to the Ordos Basin: Tectonophysics, v. 420, p. 253–266, doi:10.1016/j.tecto.2006.01.025. [2] [9] [10]

Lizarralde, D., and Holbrook, W.S., 1997, U.S. Mid-Atlantic margin structure and early thermal evolution: Journal of Geophysical Research, v. 102, p. 22,855–22,875, doi:10.1029/96JB03805. [8] [9]

Lizarralde, D., Holbrook, W. and Oh, J., 1994, Crustal structure across the Brunswick magnetic anomaly, offshore Georgia, from coincident ocean bottom And multi-channel seismic data: Journal of Geophysical Research, v. 99, p. 21741–21,757. [8]

Long, D.T., Cox, S.C., Bannister, S., Gerstenberger, M.C., and Okaya, D., 2003, Upper crustal structure beneath the eastern Southern Alps and the Mackenzie Basin, New Zealand, derived from seismic reflection data: New Zealand Journal of Geology and Geophysics, v. 46, p. 21–39, doi:10.1080 /00288306.2003.9514993. [2] [9]

Long, G.H., Brown, L.D., and Kaufman, S., 1978, A deep seismic reflection survey across the San Andreas fault near Parkfield, California: Eos (Transactions, American Geophysical Union), v. 59, p. 385. [2] [7]

Long, R.E., Maguire, P.K.H., and Sundarlingham, K., 1973, Crustal structure of the East African rift zone: Tectonophysics, v. 20, p. 269–281, doi:10.1016/0040-1951(73)90116-9. [6] [7]

Lorenzo, J.M., Mutter, J.C., Larson, R.L., and the Northwest Australia Study Group, 1991, Development of the continent-ocean transform boundary of the southern Exmouth Plateau: Geology, v. 19, p. 843–846. [8]

Louie, J.N., Thelen, W., Smith, S.B., Scott, J.B., Clark, M., and Pullammanappallil, S., 2004, The northern Walker Lane refraction experiment: Pn arrivals and the northern Sierra Nevada root: Tectonophysics, v. 388, p. 253–269, doi:10.1016/j.tecto.2004.07.042. [2] [10]

Lowe, C., and Jacob, A.W.B., 1989, A north-south seismic profile across the Caledonian suture zone in Ireland: Tectonophysics, v. 168, p. 297–318, doi:10.1016/0040-1951(89)90224-2. [2] [8]

Lu, D., and Wang, X., 1990, The crustal structure and deep internal processes in the Tuotuohe-Golnud area of north Qinghai-Xizang plateau: Bulletin of the Chinese Academy of Geological Science, v. 21, p. 227–237. [9]

Lucas, S.B., White, D.J., Hajnal, Z., Lewry, J., Green, A., Clowes, R., Zwanzig, H., Ashton, K., Schledewitz, D., Stauffer, M., Norman, AS., Williams, P.F., and Spence, G., 1994, Three-dimensional collisional structure of the Trans-Hudson Orogen, Canada: Tectonophysics, v. 232, p. 161–178, doi:10.1016/0040-1951(94)90082-5. [9]

Ludden, J.N., ed., 1994, The Abitibi-Grenville Lithoprobe transect seismic reflection results, part I: the western Grenville Province and Pontiac Subprovince: Canadian Journal of Earth Science, v. 31, p. 227–446. [2] [9]

Ludden, J.N., ed., 1995, Results from the Abitibi-Grenville Lithoprobe transect, part II: the Abitibi Greenstone Belt: Canadian Journal of Earth Science, v. 32 (2). [2] [9]

Ludden, J.N., and Hynes, A., 2000a, The Lithoprobe Abitibi-Grenville transect: Canadian Journal of Earth Science, v. 37, p. 115–476, doi:10.1139 /cjes-37-2-3-115. [2] [9]

Ludden, J.N., and Hynes, A., 2000b, The Lithoprobe Abitibi–Grenville transect: two billion years of crust formation and recycling in the Precambrian Shield: Canadian Journal of Earth Science, v. 37, p. 459–476, doi:10.1139 /cjes-37-2-3-459. [9]

Ludwig, W.J., Ewing, J.I., Ewing, M., Murauchi, S., Den, N., Asano, S., Hotta, H., Hayakawa, M., Asanuma, T., Ichikawa, K., and Noguchi, I., 1966, Sediments and structure of the Japan trench: Journal of Geophysical Research, v. 71, p. 2121–2137. [2] [5] [6]

Ludwig, W.J., Nafe, J.E., Simpson, E.S.W., and Sacks, S., 1968, Seismic refraction measurements on the southeast African continental margin: Journal of Geophysical Research, v. 73, p. 3707–3719, doi:10.1029 /JB073i012p03707. [2] [6]

Ludwig, W.J., Nafe, J.E., and Drake, L.E., 1970, General observations—seismic refraction, *in* Maxwell, A.E., ed., The Sea, new concepts of ocean floor evolution: Wiley-Interscience, New York, p. 53–84. [2] [6] [10]

Lueschen, E., 1992, ASTRA 1991: First seismic field experiment in German-Russian cooperation (in German): DGG-Mitteilungen (News of the German Research Society), v. 3, p. 24–39. [2] [9]

Lueschen, E., Wenzel, F., Sandmeier, K.-J., Menges, D., Rühl, Th., Stiller, M., Janoth, W., Keller, F., Söllner, W., Thomas, R., Krohe, A., Stenger, R., Fuchs, K., Wilhelm, H., and Eisbacher, G., 1987, Near-vertical and wide-angle seismic surveys in the Black Forest, SW Germany: Journal of Geophysics, v. 62, p. 1–30. [2] [8]

Lueschen, E., Wenzel, F., Sandmeier, K.-J., Menges, D., Rühl, Th., Stiller, M., Janoth, W., Keller, F., Söllner, W., Thomas, R., Krohe, A., Stenger, R., Fuchs, K., Wilhelm, H., and Eisbacher, G., 1989, Near-vertical and wide-angle seismic surveys in the Schwarzwald, *in* Emmermann, R., and Wohlenberg, J., eds., The German continental deep drilling program (KTB): Springer, Berlin-Heidelberg, p. 297–362. [8]

Lueschen, E., Nicolich, R., Cernobori, L., Fuchs, K., Kern, H., Kruhl, J.H., Persoglia, S., Romanelli, M., Schenk, V., Siegesmund, S., and Tortorici, L. (Italian-German Lower Crust Group), 1992, A seismic reflection-refraction experiment across the exposed lower crust in Calabria (southern Italy): first results: Terra Nova, v. 4, p. 77–86. [2] [9]

Lueschen, E., Lammerer, B., Gebrande, H., Millahn, K., Nicolich, R., and TRANSALP Working Group, 2004, Orogenic structure of the Eastern Alps, Europe, from TRANSALP deep seismic reflection profiling: Tectonophysics, v. 388, p. 85–102, doi:10.1016/j.tecto.2004.07.024. [2] [9]

Lueschen, E., Borrini, D., Gebrande, H., Lammerer, B., Millahn, K., Neubauer, F., Nicolich, R., and TRANSALP Working Group, 2006, TRANSALP—deep crustal Vibroseis and explosive seismic profiling in the Eastern Alps, *in* Gebrande, H., Castellarin, A., Lueschen, E., Millahn, K., Neubauer, F., and Nicolich, R., eds., TRANSALP—a transect through a young collisional orogen: Tectonophysics, v. 414, p. 9–38. [9] [10]

Luetgert, J.H., Mann, C.E., and Klemperer, S.L., 1987, Wide-angle deep crustal reflections in the northern Appalachians: Geophysical Journal of the Royal Astronomical Society, v. 89, p. 183–189. [2] [8]

Luetgert, J.H., Hughes, S., Cipar, J., Mangino, S., Forsyth, D., and Asudeh, I., 1990, Data report for O-NYNEX—the 1988 Grenville-Appalachian seismic refraction experiment in Ontario, New York and New England: Menlo Park, California, U.S. Geological Survey Open-File Report 90-426, 51 p. [2] [8]

Luetgert, J., Benz, H., Criley, E., and Song-Lin, L., 1992, Data report for a seismic-refraction/wide-angle reflection investigation of the Atlantic coastal plain in South Carolina: Menlo Park, California, U.S. Geological Survey Open-File Report 92-723, 35 p. [9]

Luetgert, J.H., Mooney, W., Trehu, A., Nabelek, J., Keller, G.R., Miller, K., Asudeh, I., and Isbell, B., 1993, Data report for a seismic-refraction/wide-angle reflection investigation of the Puget Basin and Willamette Valley in western Washington and Oregon: U.S. Geological Survey Open-File Report 93-347, 71 p. [9]

Luetgert, J., Benz, H., and Madabhushi, S., 1994, Crustal structure beneath the Atlantic coastal plain of South Carolina: Seismological Research Letters, v. 65, p. 180–191 [9]

Lueth, S., Wigger, P., and ISSA Research Group, 2003, A crustal model along 39°S from a seismic refraction profile—ISSA 2000, Revista Geolgica de Chile, v. 30, no. 1, p. 83–101. [2] [9] [10]

Lund, C.-E., 1979, The fine structure of the lower lithosphere underneath the Blue Road Profile in northern Scandinavia: Tectonophysics, v. 56, p. 111–122, doi:10.1016/0040-1951(79)90017-9. [7]

Lund, C.-E., Gorbatschev, R., and Smirnov, A., 2001, A seismic model of the Precambrian crust along the coast of southeastern Sweden: the Coast Profile wide-angle airgun experiment and the southern part of FEN-NOLORA revisited: Tectonophysics, v. 339, p. 93–111, doi:10.1016 /S0040-1951(01)00135-4. [2] [9]

Luosto, U., and Korhonen, H., 1986, Crustal structure of the Baltic Shield based on off-Fennolora refraction data, *in* Galson, D.A., and Mueller, St., eds., The European Geotraverse, Part 2: Tectonophysics, v. 128, p. 183–208. [2] [8]

Luosto, U., Lanne, E., Korhonen, H., Guterch, A., Grad, M., Materzok, R., and Prechuc, E., 1984, Deep structure of the earths crust on the SVEKA profile in central Finland: Ann. Geophys., v. 2, p. 559–570. [2] [8]

Luosto, U., Flueh, E.R., Lund, C.E., and Working Group, 1989, The crustal structure along the POLAR profile from seismic refraction investigations, *in* Freeman, R., Knorring, M. von, Korhonen, H., Lund, C., and Muller, St., eds., The European Geotraverse, Part 5: The POLAR Profile: Tectonophysics, v. 162, p. 51–85. [2] [8]

Luosto, U., Tiira, T., Korhonen, H., Azbel, I., Burmin, V., Buyanov, A., Kosminskaya, I., Ionkis, V., and Sharov, N., 1990, Crust and upper mantle structure along the DSS Baltic profile in SE Finland: Geophysical Journal International, v. 101, p. 89–110, doi:10.1111/j.1365-246X.1990.tb00760.x. [8]

Lutter, W.J., Nowack, R.L., and Braile, L.W., 1990, Seismic imaging of upper crustal structure using traveltimes fom the PASSCAL Ouachita experiment. Journal of Geophysical Research, v. 95, p. 4633–4646. [9]

Lutter, W.J., Roberts, P.M., Thurber, C.H., Steck, L., Fehler, M.C., Stafford, D.G., Baldridge, W.S., and Zeichert, T.A., 1995, Teleseismic P-wave image of crust and upper mantle structure beneath the Valles caldera, New Mexico: initial results from the 1993 JTEX passive array: Geophysical Research Letters, v. 22, p. 505–508, doi:10.1029/94GL03220. [9]

Lutter, W.J., Fuis, G.S., Thurber, C.H., and Murphy, J., 1999, Tomographic images of the upper crust from the Los Angeles basin to the Mojave Desert, California: results from the Los Angeles Region Seismic Experiment: Journal of Geophysical Research, v. 104, p. 25,543–25,565, doi:10.1029/1999JB900188. [9]

Lutter, W.J., Fuis, G.S., Ryberg, T., Okaya, D.A., Clayton, R.W., Davis, P.M., Prodehl, C., Murphy, J.M., Langenheim, V.E., Benthien, M.L., Godfrey, N.J., Christensen, N.I., Thygesen, K., Thurber, C.H., Simila, G., and Keller, G.R., 2004, Upper crustal structure from the Santa Monica Mountains to the Sierra Nevada, southern California: Tomographic results from the Los Angeles Regional Seismic Experiment, Phase II (LARSE II): Bulletin of the Seismological Society of America, v. 94, p. 619–632, doi:10.1785/0120030058. [9]

Mabey, D.R., 1960a, Gravity survey of the western Mojave Desert, California: U.S. Geological Survey Professional Paper 316-D, p. 51–73. [5]

Mabey, D.R., 1960b, Regional gravity survey of part of the Basin and Range province: U.S. Geological Survey Professional Paper 400-B, p. 283–285. [5]

Macelwane, J.B., 1951, Evidence on the interior of the earth derived from seismic studies, *in* Gutenberg, B., ed., Internal constitution of the earth, 2nd ed.: Dover Publ., Inc., Chapter X, p. 227–304. [2] [3] [4] [10]

Macgregor-Scott, N., and Walter, A.W., 1985, Data report, two earthquake–source refraction profiles crossing epicentral region of the 1983 Coalinga, California, earthquakes: U.S. Geological Survey Open-File Report 85-435, Menlo Park, California: 64 p. [8]

Macgregor-Scott, N., and Walter, A., 1988, Crustal velocities near Coalinga, California, modelled from a combined earthquake/explosion refraction profile: Bulletin of the Seismological Society of America, v. 78, p. 1475–1490. [8]

MacKenzie, D.P., 1978, Some remarks on the development of sedimentary basins: Earth and Planetary Science Letters, v. 40: 25–32, doi:10.1016/0012-821X(78)90071-7. [8]

Mackenzie, G.D., Shannon, P.M., Jacob, A.W.B., Morewood, N.C., Makris, J., Gaye, M., Egloff, F., 2002, The velocity structure of the sediments in the southern Rockall Basin: results from new wide-angle seismic modelling: Marine and Petroleum Geology, v. 19, p. 989–1003, doi:10.1016/S0264-8172(02)00133-2. [2] [9] [10]

Mackenzie, G.D., Thybo, H., and Maguire, P. K.H., 2005, Crustal velocity structure across the Main Ethiopian Rift: results from two-dimensional wide-angle seismic modelling: Geophysical Journal International, v. 162 (3), p. 994–1006, doi:10.1111/j.1365-246X.2005.02710.x. [10]

Magnani, M.B., Levander, A., Miller, K.C., Eshete, T., and Karlstrom, K.E., 2005, Seismic investigation of the Yavapai–Mazatzal transition zone and the Jemez lineament in northern New Mexico, *in* Karlstrom, K.E., and Keller, G.R., eds., The Rocky Mountain Region: An evolving lithosphere—tectonics, geochemistry, and geophysics: Washington, D.C., American Geophysical Union Geophysical Monograph 154, p. 227–238. [9]

Magnani, M.B., Zelt, C.A., Levander, A., and Schmitz, M., 2009, Crustal structure of the South American-Caribbean plate boundary at 67°W from controlled source seismic data: Journal of Geophysical Research, v. 114, p. B02312, doi:10.1029/2008JB005817. [2] [10]

Maguire, P.K.H., and SEIS–UK, 2002, A new dimension for UK seismology. Astronomy and Geophysics, v. 42, p. 23–25. [2] [9] [10]

Maguire, P.K.H., Swain, C.J., Masotti, R., and Khan, M.A., 1994, A crustal and uppermost mantle cross-sectional model of the Kenya rift derived from seismic and gravity data, *in* Prodehl, C., Keller, G.R., and Khan, M.A., eds., Crustal and upper mantle structure of the Kenya rift: Tectonophysics, v. 236, p. 217–249. [9]

Maguire, P.K.H., Ebinger, C.J., Stuart, G.W., Mackenzie, G.D., Whaler, K.A., Kendall, J.-M., Khan, M.A., Fowler, C.M.R., Klemperer, S.L., Keller, G.R., Harder, S,., Furman, T., Mickus, K., Asfaw, L., and Abebe, B., 2003, Geophysical Project in Ethiopia Studies Continental Breakup: Eos (Transactions, American Geophysical Union), v. 84, no. 35, p. 337, 342–343, doi:10.1029/2003EO350002. [2] [9] [10]

Maguire, P.K.H., Keller, G.R., Klemperer, S.L., Mackenzie, G.D., Keranen, K., Harder, S., O'Reilly, B.M., Thybo, H., Asfaw, L., and Amha, M., 2006, Crustal structure of the northern Main Ethiopian Rift from the EAGLE controlled survey; a snapshot of incipient lithospheric breakup, *in* Yigu, G., Ebinger, C.J., and Maguire, P.K.H., eds., The Afar volcanic province within the East African rift system: Geological Society of London, Special Publication 29, p. 269–291. [10]

Maguire, P.K.H., England, R.W., and Hardwick, A.J., 2011, LISPB DELTA, a lithospheric seismic profile in Britain: analysis and interpretation of the Wales and southern England section: Journal of the Geological Society of London, v. 168, p. 61–82, doi:10.1144/0016-76492010-030. [7]

Mahadevan, T.M., 1994, Deep continental structure of India: a review: Bangladore, Geological Society India Memoir 28, 569 p., doi:10.1190/1.1440679. [2] [7] [8]

Mair, J.A., and Lyons, J.A., 1976, Seismic reflection techniques for crustal structure studies: Geophysics, v. 41, p. 1272–1290. [7]

Maistrello, M., Scarascia, S., Corsi, A., Egger, A., and Thouvenot, F., 1990, EGT 1985 southern segment. Compilation of data from the seismic refraction experiments in Tunisia and Pelagian Sea: Open-File Report, CNR—Istituto per la Geofisica della Litosfera, Milano, Italy, Part I: 115 p., Part II: 79 plates. [8]

Maistrello, M., Scarascia, S., Ye, S., and Hirn, A., 1991, EGT 1986 central segment. Compilation of seismic data (additional profiles and fans) in Northern Apennines, Po Plain, Western and Southern Alps. Open-File Report, IGL—Istituto per la Geofisica della Litosfera, Milano, Italy, Part A: 24 p., Part B: 56 plates. [8]

Majdanski, M., Grad, M., Guterch, A., and SUDETES Working Group, 2006, 2-D seismic tomographic and ray tracing modelling of the crustal structure across the Sudetes Mountains based on SUDETES 2003 experiment data: Tectonophysics, v. 413, p. 249–269. [10]

Makovsky, Y., and Klemperer, S.L., 1999, Measuring the seismic properties of Tibetan hot spots: evidence for free aqueous fluids in the Tibetan middle crust: Journal of Geophysical Research, v. 104 (B5), p. 10,795–10,825, doi:10.1029/1998JB900074. [9]

Makovsky, Y., Klemperer, S.L., Ratschbacher, L., Brown, L.D., Li, M., Zhao, W, and Men, M., 1996, INDEPTH Wide-angle reflection observation of P-wave to S-wave conversion from crustal bright spots in Tibet: Science, v. 274, p. 1690–1691, doi:10.1126/science.274.5293.1690. [9]

Makovsky, Y., Klemperer, S.L., Ratschbacher, L., and Alsdorf, D., 1999, Midcrustal reflector on INDEPTH wide-angle profiles: an ophiolitic slab beneath the India-Asia suture in southern Tibet? Tectonics, v. 18, p. 793–808, doi:10.1029/1999TC900022. [9]

Makris, J., 1976, A dynamic model of the Hellenic arc deduced from geophysical data: Tectonophysics, v. 36, p. 339–346, doi:10.1016/0040-1951(76)90108-6. [7]

Makris, J., 1977, Geophysical investigations of the Hellenides: Hamburger Geophysikalische Einzelschriften, University of Hamburg, v. 34, p. 1–124. [2] [7]

Makris, J., 1985, Geophysics and geodynamic implications for the evolution of the Hellenides, *in* Stanley, J., and Wezel, F.-C., eds., Geological evolution of the Mediterranean Basin: New York–Berlin, Springer, p. 231–248. [7]

Makris, J., and Stobbe, C., 1984, Physical properties and state of the crust and upper mantle of the Eastern Mediterranean Sea deduced from geophysical data: Marine Geology, v. 55, p. 347–363, doi:10.1016/0025-3227(84)90076-8. [7]

Makris, J., and Vees, R., 1977, Crustal structure of the Aegean Sea and the islands of Crete of Evia and Crete, Greece, obtained by refractional seismic experiments: Journal of Geophysics, v. 42, p. 329–341. [2] [7]

Makris, J., Weigel, W., and Koschyk, K., 1977, Seismic studies in the Cretan Sea—3. crustal models of the Cretan Sea deduced from refraction seismic measurements and and gravity data: Berlin-Stuttgart, Meteor Forschungsergebnisse, Borntraeger, C 27, p. 31–43. [2] [7]

Makris, J., Ben Avraham, Z., Behle, A., Ginzburg, A., Giese, P., Steinmetz, L., Whitmarsh, R.B., and Eleftheriou, S., 1983a, Seismic refraction profiles between Cyprus and Israel and their interpretation: Geophysical Journal of the Royal Astronomical Society, v. 75, p. 575–591. [7]

Makris, J., Allam, A., Mokhtar, T., Basahel, A., Dehghani, G.A., and Bazari, M., 1983b, Crustal structure in the northwestern region of the Arabian shield and its transition to the Red Sea: Bulletin Faculty of Earth Science, King Abdulaziz University, v. 6, p. 435–447. [7]

Makris, J., Demnati, A., and Klussmann, J., 1985, Deep seismic soundings in Morocco and a crust and upper mantle model deduced from seismic and gravity data: Annales Geophysicae, v. 3, p. 369–380. [2] [7] [8]

Makris, J., Nicolich, R., and Weigel, W., 1986, A seismic study in the Western Ionian Sea: Annales Geophysicae, v. 6, B, p. 665–678. [9]

Makris, J., Egloff, R., Jacob, A.W.B., Mohr, P., Murphy, T., and Ryan, P., 1988, Continental crust under the southern Porcupine Seabight west of Ireland: Earth and Planetary Science Letters, v. 89, p. 387–397, doi:10.1016/0012 -821X(88)90125-2. [8]

Makris, J., Mohr, P., and Rihm, R., eds., 1991, Red Sea: birth and early history of a new oceanic basin: Tectonophysics, v. 198, p. 129–466. [2]

Malinowski, M., Zelazniewicz, A., Grad, M., Guterch, A., Janik, T., and CELEBRATION Working Group, 2005, Seismic and geological structure of the crust in the transition from Baltica to Palaeozoic Europe in SE Poland—CELEBRATION 2000 experiment, profile CEL02: Tectonophysics, v. 401, p. 55–77, doi:10.1016/j.tecto.2005.03.011. [9]

Malinowski, M., Grad, M., Guterch, A., Takács, E., Śliwiński, Z., Antonowicz, L., Iwanowska, E., Keller, G.R., and Hegedüs, E., 2007, Effective sub-Zechstein salt imaging using low-frequency seismic—results of the GRUNDY 2003 experiment across the Variscan front in the Polish Basin: Tectonophysics, v. 439, p. 89–106, doi:10.1016/j.tecto.2007.03.006. [2] [10]

Mallet, R., 1852, Report to the British Association for the Advancement of Science, Part 1, p. 272–320. [2] [3]

Mandler, H.A.F., and Clowes, R.M., 1998, The HIS bright reflector: further evidence for extensive magmatism in the Precambrian of western Canada, *in* Klemperer, S.L., and Mooney, W.D., eds., Deep seismic probing of the continents, II: a global survey: Tectonophysics, v. 288, p. 71–81. [2] [9]

Mandler, H.A.F., and Jokat, W., 1998, The crustal structure of central East Greenland: results from combined land-sea seismic refraction experiments: Geophysical Journal International, v. 135, p. 63–76, doi:10.1046 /j.1365-246X.1998.00586.x. [2] [8] [9]

Mansfield, R.H., and Evernden, J.F., 1966, Long-range seismic data from the Lake Superior seismic experiment, 1963–1964, *in* Steinhart, J.S., and Smith, T.J., eds., The earth beneath the continents: Washington, D.C., American Geophysical Union Geophysical Monograph 10, p. 249–269. [2] [6]

Marillier, F., Tomassino, A., Patriat, Ph., and Pinet, B., 1988, Deep structure of the Aquitaine Shelf: constraints from expanding spread profiles on the ECORS Bay of Biscay transect: Marine and Petroleum Geology 5, p. 65–74, doi:10.1016/0264-8172(88)90040-2. [2] [8]

Marillier, F., Hall, J., Hughes, S., Louden, K., Reid, I., Roberts, B., Clowes, R., Coté, T., Fowler, J., Guest, S., Lu, H., Luetgert, J., Quinlan, G., Spencer, C., and Wright, J., 1994, LITHOPROBE East onshore-offshore seismic refraction survey—constraints on interpretation of reflection data in the Newfoundland Appalachians: Tectonophysics, v. 232, p. 43–58, doi:10.1016/0040-1951(94)90075-2. [2] [8] [9]

Martin, M., Ritter, J.R.R., and the CALIXTO Working Group, 2005, High-resolution teleseismic body-wave tomography beneath SE-Romania—I. implications for three-dimensional crustal correction strategies with a new crustal velocity model: Geophysical Journal International, v. 162, p. 448–460, doi:10.1111/j.1365-246X.2005.02661.x. [9] [10]

Martin, M., Wenzel, F., and the CALIXTO Working Group, 2006, High-resolution teleseismic body–wave tomography beneath SE-Romania—II. Imaging of a slab detachment scenario: Geophysical Journal International, v. 164, p. 579–595, doi:10.1111/j.1365-246X.2006.02884.x. [9] [10]

Martínez-Martínez, J.M.; Soto, J.I. and Balanyá, J.C., 1997, Crustal decoupling and intracrustal flow beneath domal exhumed core complexes, Betics (SE Spain): Terra Nova, v. 9, p. 223–227, doi:10.1111/j.1365-3121.1997. tb00017.x. [9]

Massé, R.P., 1973, Compressional velocity distribution beneath central and eastern North America: Bulletin of the Seismological Society of America, v. 63, p. 911–935. [2] [6] [7]

Masson, F., Jacob, A.W.B., Prodehl, C., Readman, P.W., Shannon, P.M., Schulze, A. and Enderle, U., 1998, A wide-angle seismic traverse through the Variscan of southwest Ireland: Geophysical Journal International, v. 134, p. 689–705, doi:10.1046/j.1365-246x.1998.00572.x. [8] [9]

Masson, F., Hauser, F., and Jacob, A.W.B., 1999, The lithospheric trace of the Iapetus Suture in SW Ireland from teleseismic data: Tectonophysics, v. 302, p. 83–89, doi:10.1016/S0040-1951(98)00279-0. [9] [10]

Mathur, S.P., 1974, Crustal structure in southwestern Australia from seismic and gravity data: Tectonophysics, v. 24, p. 151–182, doi:10.1016/0040 -1951(74)90135-8. [2] [6]

Mathur, S.P., 1983, Deep reflection experiments in northeastern Australia, 1976–1978. Geophysics, v. 48, p. 1588–1597, doi:10.1190/1.1441441. [2] [7]

Mathur, S.P., 1984, Improvements in seismic reflection techniques for studying the lithosphere in Australia: Tectonophysics, v. 105, p. 373–381, doi:10.1016/0040-1951(84)90214-2. [7]

Matsu'ura, R.S., Yoshii, T., Moriya, T., Miyamachi, H., Sasaki, Y., Ikami, A., and Ishida, M., 1991, Crustal structure of a seismic-refraction profile across the Melian and Akaishi tectonic lines, central Japan: Bulletin of the Earthquake Research Institute, v. 66, p. 497–516. [2] [8] [10]

Matthews, D.H., and Cheadle, M.J., 1986, Deep reflections from the Caledonides and Variscides west of Britain and comparison with the Himalayas, *in* Barazangi, M., and Brown, L., eds., Reflection seismology: a global perspective: American Geophysical Union, Geodynamics Series, v. 13, p. 5–19. [2] [8]

Matthews, D.H., and Smith, C., eds., 1987, Deep seismic reflection profiling of the continental lithosphere: Geophysical Journal of the Royal Astronomical Society, v. 89, p. 1–448. [2] [8]

Matuzawa, T., Matumoto, T., and Asano, S., 1959, On the crustal structure derived from observations of the second Hokoda explosion: Bulletin of the Earthquake Research Institute, v. 37, p. 509–524. [2] [5] [8]

Maurain, C., Eblé, L., and Labrouste, H., 1925, Sur les ondes séismiques des explosions de la Courtine: Journal de Physique et le Radium, v. 6, p. 65–78, doi:10.1051/jphysrad:019250060306500. [3]

Maurer, H., and Ansorge, J., 1992, Crustal structure beneath the northern margin of the Swiss Alps, *in* Freeman, R., and Mueller, St., eds., The European Geotraverse, Part 8: Tectonophysics, v. 207, p. 165–181. [8]

Maxwell, A.E., ed., 1970, The Sea, new concepts of ocean floor evolution: New York, Wiley-Interscience, v.4, I: 791 p., II: 664 p. [2] [5] [6] [10]

Mayer, G., May, P.M., Plenefisch, T., Echtler, H., Lüschen, E., Wehrle, V., Müller, B., Bonjer, K.-P., Prodehl, C., Wilhelm, H., and Fuchs, K., 1997, The deep crust of the southern Rhinegraben: reflectivity and seismicity as images of dynamic processes, *in* Fuchs, K., Altherr, R., Müller, B., and Prodehl, C., eds., Stress and stress release in the lithosphere—structure and dynamic processes in the rifts of western Europe: Tectonophysics, v. 275, p. 15–40. [2] [9]

Mayerova, M., Novotny, M., and Fejfar, M., 1994, Deep seismic sounding in Czechoslovakia, *in* Bucha, V., and Blizkovsky, M., eds., Crustal stucture of the Bohemian Massif and the West Carpathians: Berlin-Heidelberg-New York, Springer, p. 13–20. [7]

McBride, J.H., White, R.S., Smallwood, J.R., and England, R.W., 2004, Must magmatic intrusion in the lower crust produce reflectivity? Tectonophysics, v. 388, p. 271–297, doi:10.1016/j.tecto.2004.07.055. [9]

McCamy, K. and Meyer, R.P., 1964, A correlation method of apparent velocity measurement: Journal of Geophysical Research, v. 69, p. 691–699. [5] [6]

McCamy, K., and Meyer, R.P., 1966, Crustal results of fixed multiple shots in the Mississippi embayment, *in* Steinhart, J.S., and Smith, T.J., eds., The earth beneath the continents: American Geophysical Union Geophysical Monograph 10, p. 370–381. [6]

McCarthy, J., and Hart, J., 1993, Data report for the 1991 bay area seismic imaging experiment (BASIX): U.S. Geological Survey Open-File Report 93-301, 26 p. [9]

McCarthy, J., Mutter, J.C., Morton, J.L., Sleep, N.H., and Thompson, G.A., 1988, Relic magma chamber structures preserved within the Mesozoic North Atlantic crust?: Geological Society of America Bulletin, v. 100, p. 1423–1436, doi:10.1130/0016-7606(1988)100<1423:RMCSPW>2.3.CO;2. [2] [8]

McCarthy, J., Larkin, S.P., Fuis, G., Simpson, R.W., and Howard, K.A., 1991, Anatomy of a metamorphic core complex: seismic refraction/wide-angle reflection profiling in southeastern California and western Arizona: Journal of Geophysical Research, v. 96, p. 12,259–12,291, doi:10.1029 /91JB01004. [2] [8]

McCarthy, J., Kohler, W., and Criley, E., 1994, Data report for the PACE 1989 seismic refraction survey, northern Arizona: U.S. Geological Survey Open-File Report 94-138, 88 p. [2] [8]

McDonald, M.A., Webb, S.C., Hildebrand, J.A., Cornuelle, B.D., and Fox, C.G., 1994, Seismic structure and anisotropy of the Juan de Fuca Ridge at 45°N: Journal of Geophysical Research, v. 99, p. 4857–4873, doi:10.1029 /93JB02801. [2] [8]

McEvilly, T.V., 1966, Crustal structure estimation within a large-scale array: Geophysical Journal, v. 11, p. 13–17, doi:10.1111/j.1365-246X.1966 .tb03489.x. [6]

McGeary, S., 1987, Nontypical BIRPS on the margin of the northern North Sea: the SHET survey: Geophysical Journal of the Royal Astronomical Society, v. 89, p. 231–238. [2] [8]

McGinnis, I.D., Bowen, R.H., Erickson, J.M., Allred, B.J., and Kreamer, J.L., 1985, East-west boundary in McMurdo Sound: Tectonophysics, v. 114, p. 341–356, doi:10.1016/0040-1951(85)90020-4. [2] [8]

McMechan, G.A., and Mooney, W.D., 1980, Asymptotic ray theory and synthetic seismograms for laterally varying structures: theory and application to the Imperial Valley, California: Bulletin of the Seismological Society of America, v. 70, p. 2021–2035. [2] [7] [8] [10]

McNamara, D.E., Walter, W.R., Owens, T.J., and Ammon, C.J., 1996, Upper mantle velocity structure beneath the Tibetan plateau from Pn travel time tomography: Journal of Geophysical Research, v. 102, p. 493–505, doi:10.1029/96JB02112. [9]

Meador, P.J., and Hill, D.P., 1983, Data report for the July–August 1982 seismic-refraction experiment in the Mono craters–Long Valley region, California: Menlo Park, California, U.S. Geological Survey Open-File Report 83-708, 53 p. [2] [8]

Meador, P.J., Hill, D.P., and Luetgert, J.H., 1985, Data report for the July–August 1983 seismic-refraction experiment in the Mono craters–Long Valley region, California, axial seismic refraction profiles: Menlo Park, California, U.S. Geological Survey Open-File Report 85-708, 70 p. [2] [8]

Meador, P.J., Ambos, E.L., and Fuis, G.S., 1986, Data report for the North and South Richardson Highway profiles, TACT, seismic-refraction survey, southern Alaska: Menlo Park, California, U.S. Geological Survey Open-File Report 86-274, 51 p. [8]

Mechie, J., 2000, Popular tools for interpreting seismic refraction/wide-angle reflection data, in Jacob, A.W.B., Bean, C.J., and Jacob, S.T.F., eds., 2000, Proceedings of the 1999 CCSS Workshop held in Dublin, Ireland: Communications of the Dublin Institute for Advanced Studies, Series D: Geophysical Bulletin, v. 49, p. 25–27. [9] [10]

Mechie, J., and Prodehl, C., 1988, Crustal and uppermost mantle structure beneath the Afro-Arabian rift system, in Le Pichon, X., and Cochran, J.R., eds., The Gulf of Suez and Red Sea rifting: Tectonophysics, v. 153, p. 103–121. [2] [7] [8] [9]

Mechie, J., Prodehl, C., and Fuchs, K., 1983, The long-range seismic refraction experiment in the Rhenish Massif, in Fuchs, K., von Gehlen, K., Mälzer, H., Murawski, H., and Semmel, A., eds., Plateau uplift: The Rhenish Massif—a case history: Berlin-Heidelberg, Springer, p. 260–275. [2] [7]

Mechie, J., Prodehl, C., and Koptschalitsch, G., 1986, Ray path interpretation of the crustal structure beneath Saudi Arabia: Tectonophysics, v. 131, p. 333–352, doi:10.1016/0040-1951(86)90181-2. [7]

Mechie, J., Egorkin, A.V., Fuchs, K., Ryberg, T., Solodilov, L. and Wenzel, F., 1993, P-wave mantle velocity structure beneath northern Eurasia from long-range recordings along the profile Quartz: Physics of the Earth and Planetary Interiors, v. 79, p. 269–286. [2] [7[[8]

Mechie, J., Keller, G.R., Prodehl, C., Gaciri, S., Braile, L.W., Mooney, W.D., Gajewski, D., and Sandmeier, K.-J.,1994, Crustal structure beneath the Kenya rift from axial profile data, in Prodehl, C., Keller, G.A., and Khan, M.A., eds., Crustal and upper mantle structure of the Kenya rift: Tectonophysics, v. 236, p. 179–200. [9]

Mechie, J., Prodehl, C., Keller, G.R., Khan, M.A., and Gaciri, S.J., 1997, A model for structure, composition and evolution of the Kenya rift, in Fuchs, K., Altherr, R., Müller, B., and Prodehl, C., eds., Structure and dynamic processes in the lithosphere of the Afro-Arabian rift system: Tectonophysics, v. 278, p. 95–119. [9]

Mechie, J., Abu-Ayyash, K., Ben-Avraham, Z., El-Kelani, R., Mohsen, A., Rümpker, G., Saul, J., and Weber, M., 2005, Crustal shear velocity structure across the Dead Sea transform from two-dimensional modelling of DESERT project explosion seismic data: Geophysical Journal International, v. 160, p. 910–924, doi:10.1111/j.1365-246X.2005.02526.x. [9]

Mechie, J., Abu-Ayyash, K., Ben-Avraham, Z, El-Kelani, R, Qabbani, L., Weber, M., and DESIRE Group, 2009, Crustal structure of the southern Dead Sea basin derived from project DESIRE wide-angle seismic data: Geophysical Journal International, doi:10.1111/j.1365-246X.2009.04161.x. [2] [10]

Meissner, R., 1966, An interpretation of the wide-angle measurements in the Bavarian Molasse Basin: Geophysical Prospecting, v. 14, p. 7–16, doi:10.1111/j.1365-2478.1966.tb01740.x. [2] [6] [8]

Meissner, R., 1967a, Zum Aufbau der Erdkruste, Ergebnisse der Weitwinkelmessungen im bayerischen Molassebecken, Teil 1 und 2: Gerlands Beiträge zur Geophysik, v. 76, p. 211–254, 295–314. [6]

Meissner, R., 1967b, Exploring deep interfaces by seismic wide angle measurements: Geophysical Prospecting, v. 15, p. 598–617, doi:10.1111/j.1365-2478 .1967.tb01806.x. [6]

Meissner, R., 1973, The "Moho" as a transition zone: Geophysical Surveys, v. 1, p. 195–216, doi:10.1007/BF01449763. [6]

Meissner, R., 1976, Comparison of wide-angle measurements in the USSR and the Federal Republic of Germany, in Giese, P., Prodehl, C., and Stein, A., eds., Explosion seismology in central Europe—data and results: Berlin-Heidelberg-New York, Springer, p. 380–384. [6]

Meissner, R., 1986, The continental crust—a geophysical approach: London, Academy Press, 426 p. [2]

Meissner, R., 2000, The mosaic of terranes in central Europe as seen by deep reflection studies, in Mohriak, W., and Talwani, M., eds., Atlantic Rifts and Continental Margins, American Geophysical Union Geophysical Monograph 115, p. 33–55. [8]

Meissner, R., and Bortfeld, R.K., 1990, DEKORP–Atlas. Results of Deutsches Kontinentales Reflexionsseismisches Programm: Berlin-Heidelberg, Springer. [2] [8] [9]

Meissner, R., and Sadowiak, P., 1992, The terrane concept and its manifestation by deep reflection studies in the Variscides: Terra Nova, v. 4, p. 598–607. [8]

Meissner, R., Bartelsen, H., Glocke, A., and Kaminski, W., 1976a, An interpretation of wide-angle measurements in the Rhenish Massif, in Giese, P., Prodehl, C., and Stein, A., eds., Explosion seismology in central Europe—data and results: Berlin-Heidelberg-New York, Springer, p. 245–251. [2] [6]

Meissner, R., Flüh, E.R., Stibane, F., and Berg, E., 1976b, Dynamics of the active plate boundary in southwest Colombia according to recent geophysical measurements: Tectonophysics, v. 35, p. 115–136, doi:10.1016/0040 -1951(76)90032-9. [7]

Meissner, R., Bartelsen, H., and Murawski, H., 1980, Seismic reflection and refraction studies for investigating fault zones along the Geotraverse Rhenoherzynikum: Tectonophysics, v. 64, p. 59–84. [2] [7]

Meissner, R., Bartelsen, H., and Murawski, H., 1981, Thin-skinned tectonics in the northern Rhenish Massif, Germany: Nature, v. 290, p. 399–401, doi:10.1038/290399a0. [8]

Meissner, R., Springer, M., Murawski, H., Bartelsen, H., Flüh, E.R., and Dürschner, H., 1983, Combined seismic reflection-refraction investigations in the Rhenish Massif and their relation to recent tectonic movements, in Fuchs, K., von Gehlen, K., Mälzer, H., Murawski, H., and Semmel, A., eds., Plateau uplift: the Rhenish Massif—a case history: Berlin-Heidelberg, Springer, p. 276–287. [2] [7]

Meissner, R., Wever, T., and Bittner, R., 1987a, Results of DEKORP 2–S and other reflection profiles through the Variscides: Geophysical Journal of the Royal Astronomical Society, 89, p. 319–324. [8]

Meissner, R., Wever, T., and Flueh, E.R., 1987b, The Moho in Europe—implications for crustal development: Annales Geophysicae, v. 5B, p. 357–364. [2] [8] [10]

Meissner, R., Brown, L., Dürbaum, H.-J., Franke, W., Fuchs, K., and Seifert, F., eds., 1991a, Continental lithosphere: deep seismic reflections: American Geophysical Union, Geodynamics Series, v. 22, p. 450 p. [2] [8]

Meissner, R., and the DEKORP Research Group, 1991b, The DEKORP surveys: Major results in tectonic and reflective styles, in Meissner, R., Brown, L., Dürbaum, H.-J., Franke, W., Fuchs, K., and Seifert, F., eds., Continental lithosphere: deep seismic reflections: American Geophysical Union, Geodynamics Series, v. 22, p. 69–76. [2] [8]

Meissner, R., Tilmann, F., and Haines, S., 2004, About the lithospheric structure of central Tibet, based on seismic data from the INDEPTH III profile: Tectonophysics, v. 380, p. 1–25, doi:10.1016/j.tecto.2003.11.007. [9]

Melhuish, A., Sutherland, R., Davey, F.J., and Larmarche, G., 1999, Crustal structure and neotectonics of the Puysegur oblique subduction zone, New Zealand: Tectonophysics, v. 313, p. 335–362, doi:10.1016/S0040-1951(99)00212-7. [2] [9]

Mellman, G.R., 1980, A method of body-wave waveform inversion for the determination of the earth structure: Geophysical Journal of the Royal Astronomical Society, v. 62, p. 481–504. [8]

MELT Seismic Team, 1998, Imaging the deep seismic structure beneath a mid-ocean ridge: the MELT experiment: Science, v. 280, p. 1215–1217, doi:10.1126/science.280.5367.1215. [9]

Meltzer, A.S., Levander, A.R., and Mooney, W.D., 1987, Upper crustal structure, Livermore Valley and vicinity, California coast ranges: Bulletin of the Seismological Society of America, v. 77, p. 1655–1673. [8]

Meltzer, A., Sarker, G., Beaudoin, B., Seeber, L., and Armbruster, J., 2001, Seismic characterization of an active metaporphic massif, Nanga Parbat, Pakistan Himalaya: Geology, v. 29, p. 651–654, doi:10.1130/0091-7613(2001)029<0651:SCOAAM>2.0.CO;2. [9]

MEMSAC Working Group, 1993, Uniform seismic data recording format explored: Eos (Transactions, American Geophysical Union), v. 74, no. 37, 421, doi:10.1029/93EO00472. [9]

Menard, H.W., 1964, Marine geology of the Pacific: New York, McGraw Hill. [6]

Menke, W., West, M., Brandsdottir, B., and Sparks, D., 1998, Compressional and shear velocity structure of the lithosphere in northern Iceland: Bulletin of the Seismological Society of America, v. 88, p. 1561–1571. [2] [9]

Mereu, R.F., 2000a, The complexity of the crust and Moho under the southeastern Superior and Grenville provinces of the Canadian Shield from seismic refraction–wide-angle reflection data: Canadian Journal of Earth Science, v. 37, p. 439–458, doi:10.1139/cjes-37-2-3-439. [8] [9]

Mereu, R.F., 2000b, The effect of small random crustal reflectors on the complexity of Pg and PmP coda: Physics of the Earth and Planetary Interiors, v. 120, p. 183–199, doi:10.1016/S0031-9201(99)00164-8. [9]

Mereu, R.F., and Hunter, J.A., 1969, Crustal and upper mantle structure under the Canadian Shield from Project Early Rise data: Bulletin of the Seismological Society of America, v. 59, p. 147–155. [6]

Mereu, R.F., and Jobidon, G., 1971, A seismic investigation of the crust and Moho on a line perpendicular to the Grenville Front: Canadian Journal of Earth Science, v. 8, p. 1553–1583. [6]

Mereu, R.F., Majumdar, S.C., and White, R.E., 1977, The structure of the crust and upper mantle under the highst ranges of the Canadian Rockies from a seismic refraction survey: Canadian Journal of Earth Science, v. 14, p. 196–208. [6]

Mereu, R.F., Wang, D., Kuhn, O., Forsyth, D.A., Green, A.G., Morel, P., Buchbinder, G.R., Crossley, D., Schwarz, E., duBerger, R., Brooks, C., and Clowes, R., 1986, The 1982 COCRUST experiment across the Ottawa-Bonnechere graben and Grenville front in Ontario and Quebec: Geophysical Journal of the Royal Astronomical Society, v. 84, p. 491–514. [2] [8]

Mereu, R.F., Baerg, J., and Wu, J., 1989, The complexity of the continental lower crust and Moho from PmP data: results from COCRUST experiments, in Mereu, R.F., Mueller, St., and Fountain, D.M., eds., Properties and processes of earth's lower crust: American Geophysical Union Geophysical Monograph 51, p. 103–119. [2] [7] [8]

Merkel, R.H., and Alexander, S.S., 1969, Use of correlation analysis to interpret continental margin ECOOE refraction data: Journal of Geophysical Research, v. 74, p. 2683–2697, doi:10.1029/JB074i010p02683. [6]

Meyer, R.P., Steinhart, J.S., and Woollard, G.P., 1958, Seismic determination of crstal structure in the central plateau of Mexico (abst.): Transactions, American Geophysical Union, v. 39, p. 525. [5] [9]

Meyer, R.P., Steinhart, J.S., and Woollard, G.P., 1961a, Central plateau, Mexico, 1957, in Steinhart, J.S., and Meyer, R.P., eds., 1961, Explosion Studies of Continental Structure: Carnegie Institution of Washington, Publication no. 622, p. 199–225. [2] [5] [9]

Meyer, R.P., Steinhart, J.S., and Bonini, W.E., 1961b, Montana, 1959, in Steinhart, J.S., and Meyer, R.P., eds., 1961, Explosion Studies of Continental Structure: Carnegie Institution of Washington, Publication no. 622, p. 305–343. [2] [5]

Meyer, R.P., Mooney, W.D., Hales, A.L., Helsley, C.E., Woollard, G.P., Hussong, D.M., Kroenke, L.W., and Ramirez, J.E., 1976, Project Narino III: Refraction observation across a leading edge, Malpelo island to the Colombian Cordillera Occidental, in Sutton, G.H., Manghnani, M.H.,

Moberly, R., and McAffe, E.U., eds., The geophysics of the Pacific Ocean basin and its margin: American Geophysical Union Geophysical Monograph 19, p. 105–132. [2] [7]

Mezcua, J., and Carreno, E., eds., 1993, Iberian Lithosphere Heterogeeity and Anisotropy—ILIHA: Monografia no. 10, Istituto Geografico Nacional, Madrid, Spain, 329 p. [8]

Michel, B., 1978, La croûte entre vallée du Rhin et vallée du Rhône: interprétation de profils sismiques et résultats structureaux [thèse]: University of Paris VII, 133 p. [7]

Milkereit, B., and Flueh, E.R., 1985, Saudi Arabian refraction profile: Crustal structure of the Red Sea–Arabian Shield transition: Tectonophysics, v. 111, p. 283–298, doi:10.1016/0040-1951(85)90289-6. [7]

Millahn, K., Lueschen, E., Gebrande, H., Lammerer, and TRANSALP Working Group, 2006, TRANSALP—cross-line recording during the seismic reflection transect in the Eastern Alps, in Gebrande, H., Castellarin, A., Lueschen, E., Millahn, K., Neubauer, F., and Nicolich, R., eds., TRANSALP—a transect through a young collisional orogen, Tectonophysics, v. 414, p. 39–49. [9]

Miller, H., and Gebrande, H., 1976, Crustal structure in southeastern Bavaria derived from seismic-refraction measurements by ray-tracing methods, in Giese, P., Prodehl, C., and Stein, A., eds., Explosion seismology in central Europe—data and results: Berlin-Heidelberg-New York, Springer, p. 339–346. [7]

Miller, H., Ansorge, J., Aric, K., and Perrier, G., 1978, Preliminary results of the lithospheric seismic Alpine longitudinal profile, 1975, from France to Hungary, in Closs, H., Roeder, D., and Schmidt, K., eds., Alps, Appenines, Hellenides: Stuttgart, Schweizerbart, p. 33–39. [2] [7]

Miller, H., Mueller, St., and Perrier, G., 1982, Structure and dynamics of the Alps: a geophysical inventory, in Berckhemer, H., and Hsü, K., eds., Alpine-Mediterranean geodynamics: American Geophysical Union Geodynamics Series, v. 7, p. 175–203. [7]

Miller, K.C., Keller, G.R., Grindley, J.M., Luetgert, J.H., Mooney, W.D., and Thybo, H., 1997, Crustal structure along the west flank of the Cascades; western Washington: Journal of Geophysical Research, v. 102, p. 17,857–17,873, doi:10.1029/97JB00882. [9]

Milne, J., 1885, Seismic experiments: Transactions of the Seismological Society of Japan, v. 8, p. 1–82. [3]

Minshull, T.A., 2002, Seismic structure of the oceanic crust and rifted continental margins, in Lee, W., Kanamori, H., Jennings, P.C., and Kisslinger, C., eds., International Handbook of Earthquake and Engineering Seismology, Part A: Amsterdam, Academic Press, p. 911–924. [2] [5] [6] [7] [8] [9] [10]

Mintrop, L., 1930, On the history of the seismic method for the investigation of underground formations and mineral deposits. Part II. Seismos: Hannover, GmbH, 128 p. [3]

Mintrop, L., 1947, 100 Jahre physikalische Erdbebenforschung und Sprengseismik. Die Naturwissenschaften, v. 34, p. 257–262, 289–295, doi:10.1007/BF00589855. [2] [3] [10]

Mintrop, L., 1949a, Wirtschaftliche und wissenschaftliche Bedeutung geophysikalischer Verfahren zur Erforschung von Gebirgsschichten und nutzbaren Lagerstätten. Berg—und Hüttenmännische Monatshefte, 94 (8/9): Wien, Springer, p. 198–211. [3]

Mintrop, L., 1949b, On the stratification of the earth's crust according to seismic studies of a large explosion and of earhquakes: Geophysics, v. 14, p. 321–336, doi:10.1190/1.1437540. [4]

Mitchell, B.J., and Landisman, M., 1971, Geophysical measurements in the southern Great Plains, in Heacock, J.G., ed., The structure and physical properties of the earth's crust: American Geophysical Union Geophysical Monograph 14, p. 77–93. [6]

Mithal, R., and Mutter, J.C., 1989, A low velocity zone within the layer 3 region of 118 Myr old oceanic crust in the western North Atlantic: Geophysical Journal, v. 97, p. 275–294, doi:10.1111/j.1365-246X.1989.tb00501.x. [8]

Mitra, S., Priestley, K., Bhattacharyya, A.K., and Gaur, V., 2005, Crustal structure and earthquake focal depths beneath northeastern India and southern Tibet: Geophysical Journal International, v. 160, p. 227–248, doi:10.1111/j.1365-246X.2004.02470.x. [9]

Mituch, E., and Posgay, K., 1972, Hungary, in Sollogub, V.B., Prosen, D., and Militzer, H., 1972, Crustal structure of central and southeastern Europe based on the results of explosion seismology (publ. in Russian 1971). English translation edited by Szénás, Gy., 1972: Geophysical Transactions, spec. ed., Müszaki Könyvkiado, Budapest, chapter 28, p. 118–129. [2] [6]

Miura, S., Kodaira, S., Nakanishi, A., Tsuru, T., Takahashi, N., Hirata, N., and Kaneda, Y., 2003, Structural characteristics controlling the seismicity of southern Japan Trench fore-arc region, revealed by ocean bottom seismographic data: Tectonophysics, v. 363, p. 79–102, doi:10.1016/S0040-1951 (02)00655-8.[2] [9] [10]

Miura, S., Takahashi, N., Nakanishi, A., Tsuru, T., Kodaira, S., and Kaneda, Y., 2005, Structural characteristics off Miyagi forearc region, the Japan trench seismogenic zone, deduced from a wide-angle reflection and refraction study: Tectonophysics, v. 407, p. 165–188, doi:10.1016/j.tecto.2005 .08.001. [2] [9] [10]

Mjelde, R., Sellevoll, M.A., Shimamura, H., Iwasaki, T., and Kanazawa, T., 1992, A crustal study off Lofoten, N. Norway, by use of 3-component ocean bottom seismographs: Tectonophysics, v. 212, p. 269–288, doi:10.1016 /0040-1951(92)90295-H. [2] [8] [9] [10]

Mjelde, R., Kodeira, S., Shimamura, H., Kanazawa, T., Shiobara, H., Berg, E.W., and Riise, O., 1997, Crustal structure of the central part of the Voring basin, mid-Norway, from ocean-bottom seismographs: Tectonophysics, v. 277, p. 235–257, doi:10.1016/S0040-1951(97)00028-0. [2] [9]

Mjelde, R., Digranes, P., Van Schaak, M., Shimamura, H., Shiobara, H., Kodeira, S., Naess, O., Sorenes, N., and Thorbjoernsen, T., 1998, Crustal structure of the northern part of the Voring Basin, mid-Norway margin, from ocean-bottom seismic and gravity data: Journal of Geophysical Research, v. 106, p. 6769–6791, doi:10.1029/2000JB900415. [2] [9] [10]

Mjelde, R., Digranes, P., Shimamura, H., Shiobara, H., Kodeira, S., Brekke, H., Egebjerg, T., Sorenes, N., and Vaagnes, E., 2001, Crustal structure of the outer Voring Plateau, offshore Norway, from wide-angle seismic and gravity data: Tectonophysics, v. 293, p. 175–205, doi:10.1016/S0040 -1951(98)00090-0. [9] [10]

Moeller, L., and Makris, J., 1989, An Ocean Bottom Seismic Station for general use—technical requirements and applications, *in* Hoefeld, J., Mitzlaff, A., and Polomsky, S., eds., Europe and the Sea, Marine Sciences and Technology in the 1990s: German Committee for Marine Sciences and Technology, Hamburg, p. 196–211. [8]

Mohorovičić, A., 1910, Das Beben vom 8.X.1909: Jahrbuch des meteorologischen Observatoriums in Zagreb für 1909, IX, IV. Teil, Abschn. 1, Zagreb. [3]

Mohriak, W.U., Bassetto, M., and Santos Vieira, I., 1998, Crustal architecture and tectonic evolution of the Sergipe-Alagoas and Jacoipe basins, offshore northeastern Brazil, *in* Klemperer, S.L., and Mooney, W.D., eds., Deep seismic probing of the continents, II: a global survey: Tectonophysics, v. 288, p. 199–220. [2] [9]

MONA LISA Working Group, 1997, Deep seismic investigations of the lithosphere in the southeastern North Sea: Tectonophysics, v. 269, p. 1–19, doi:10.1016/S0040-1951(96)00111-4. [2] [9]

Montrasio, A., Nicolich, R., and Sciesa, E., 2003, The profile CROP Alpi Centrali, *in* Scrocca, D., Doglioni, C., Innocenti, F., Manetti, P., Mazzotti, A., Bertelli, L., Burbi, L., and D'Offizi, S., eds., CROP Atlas: seismic reflection profiles of the Italian crust: Memorie Descrittive della Carta Geologica d'Italia, Roma, v. 62, p. 96–106. [8]

Mooney, W.D., 1987, Seismology of the continental crust and upper mantle: Reviews in Geophysics, v. 25, p. 1168–1176, doi:10.1029/ RG025i006p01168. [8]

Mooney, W.D., 1989, Seismic methods to determine earthquake source parameters and lithospheric structure, *in* Pakiser, L.C., and Mooney, W.D., eds., Geophysical framework of the continental United States: Geological Society of America Memoir 172, p. 249–284. [2] [10]

Mooney, W.D., 2002, Continental crust, *in* Encyclopedia of Physical Science and Technology, 3rd ed., v. 3, p. 635–657. [2] [9] [10]

Mooney, W.D., 2007, Crust and lithospheric structure-global crustal structure, *in* Romanowicz, B., and Dziewonski, A., eds., Seismology and structure of the earth: Treatise on Geophysics, v. 1: Amsterdam, Elsevier, p. 361–417. [10]

Mooney, W.D., and Braile, L.W., 1989, The seismic structure of the continental crust and upper mantle of North America, *in* Bally, A.W., and Palmer, A.R., eds., The geology of North America—an overview: Boulder, Colorado, Geological Society of America, Geology of North America, v. A, p. 39–52. [2] [8]

Mooney, W.D., and Brocher, T.M., 1987, Coincident seismic reflection/refraction studies of the continental lithosphere: a global review: Reviews in Geophysics, v. 25, p. 723–742, doi:10.1029/RG025i004p00723. [2] [8]

Mooney, W.D., and Meissner, R., 1992, Multi-generic origin of crustal reflectivity: a review of seismic reflection profiling of the continental lower

crust and Moho, *in* Fountain, D.M., Arculus, R., and Kay, R.W., eds., Continental lower crust: Amsterdam, Elsevier, p. 45–79. [2] [8]

Mooney, W.D., and Prodehl, C., 1978, Crustal structure of the Rhenish Massif and adjacent areas; a reinterpretation of existing seismic-refraction data: Journal of Geophysics, v. 44, p. 573–601. [7]

Mooney, W.D., and Prodehl, C., 1984, A comparison of crustal sections: Arabian Shield to the Red Sea, *in* Mooney, W.D., and Prodehl, C., eds., Proceedings of the 1980 Workshop of the IASPEI on the seismic modeling of laterally varying structures: Contributions based on data from the 1978 Saudi Arabian refraction profile: U.S. Geological Survey Circular 937, p. 140–153. [7] [8] [10]

Mooney, W.D., and Weaver, C.S., 1989, Regional crustal structure and tectonics of the Pacific coastal states; California, Oregon, and Washington, *in* Pakiser, L.C., and Mooney, W.D., eds., Geophysical framework of the continental United States: Geological Society of America Memoir 172, p. 129–161. [8]

Mooney, W.D., Meyer, R.P., Laurence, J.P., Meyer, H., and Ramirez, J.E., 1979, Seismic refraction studies of the western cordillera, Colombia: Bulletin of the Seismological Society of America, v. 69, p. 1745–1761. [2] [7]

Mooney, W.D., Andrews, M.C., Ginzburg, A., Peters, D.A., and Hamilton, R.M., 1983, Crustal structure of the northern Mississippi embayment and a comparison with other continental rift zones: Tectonophysics, v. 94, p. 327–348, doi:10.1016/0040-1951(83)90023-9. [2] [7]

Mooney, W.D., Gettings, M.E., Blank, H.R., and Healy, J.H., 1985, Saudi Arabian seismic-refraction profile: a traveltime interpretation of crust and upper mantle structure: Tectonophysics, v. 111, p. 173–246, doi:10.1016 /0040-1951(85)90287-2. [2] [7]

Mooney, W.D., Laske, G., and Masters, T.G., 1998, CRUST 5.1: A global crustal model at 5°× 5°: Journal of Geophysical Research, v. 103, p. 727–747, doi:10.1029/97JB02122. [2] [9] [10]

Mooney, W.D., Prodehl, C., and Pavlenkova, N.I., 2002, Seismic velocity structure of the continental lithosphere from controlled source data, *in* Lee, W., Kanamori, H., Jennings, P.C., and Kisslinger, C., eds., International Handbook of Earthquake and Engineering Seismology, Part A: Amsterdam, Academic Press, p. 887–910. [2] [8] [9] [10]

Moore, J.C., Diebold, J., Fisher, M.A., Sample, J., Brocher, T., Talwani, M., Ewing, J., von Huene, R., Rowe, C., Stone, D., Stevens, C., and Sawyer, D., 1991, EDGE deep seismic reflection transect of the eastern Aleutian arch-trench layered lower crust reveals underplating and continental growth: Geology, v. 19, p. 420–424, doi:10.1130/0091-7613(1991)019 <0420:EDSRTO>2.3.CO;2. [9]

Moores, E.M., 1973, Geotectonic significance of ultramafic rocks: Earth-Science Reviews, v. 9, p. 241–258, doi:10.1016/0012-8252(73)90093-7. [7]

Moreira, V.S., Mueller, St., Mendes, A.S., Prodehl, C., 1977, Crustal structure of southern Portugal, Proceedings of the 15th General Assembly of European Seismological Commission (Krakov, 1976): Publications of the Institute of Geophysics, Polish Academy of Sciences, A-4(115), part I, p. 413–426. [2] [7]

Morelli, C., 1993, Risultati di 31 anni (1956–1986) di DSS e 7 anni (1986–1992) di CROP in Italia: Atti XII° Convegno GNGTS, CNR Roma, p. 3–30. [7] [8]

Morelli, C., 2003, A historical perspective to the CROP project, *in* Scrocca, D., Doglioni, C., Innocenti, F., Manetti, P., Mazzotti, A., Bertelli, L., Burbi, L., and D'Offizi, S., eds., CROP Atlas: seismic reflection profiles of the Italian crust: Memorie Descrittive della Carta Geologica d'Italia, Roma, v. 62, p. 1–7. [7]

Morelli, C., and Nicolich, R., 1980, Evolution and tectonics of the western Mediterranean and surrounding areas: Instituto Geográfico Nacional Madrid, publ. no. 201, p. 1–65. [2] [7]

Morelli, C., and Nicolich, R., 1990, A cross section of the lithosphere along the European Geotraverse southern segment (from the Alps to Tunisia), *in* Freeman, R., and Mueller, St., eds., The European Geotraverse, Part 6: Tectonophysics, v. 176, p. 229–243. [8]

Morelli, C., Bellemo, S., Finetti, I., and de Visintini, G., 1967, Preliminary depth contour maps for the Conrad and Moho discontinuities in Europe: Bollettino di Geofisica Teorica e Applicata, v. 9, p. 142–157. [2] [6] [10]

Morelli, C., Giese, P., Carrozzo, M.T., Colombi, B., Gurra, I., Hirn, A., Letz, H., Nicolich, R., Prodehl, C., Reichert, C., Röwer, P., Sapin, M., Scarascia, S., and Wipper, P., 1977, Crustal and upper mantle structure of the northern Appennines, the Ligurian Sea, and Corsica, derived from seismic and gravimetric data: Bollettino di Geofisica Teorica e Applicata, v. 19, p. 199–260. [2] [6] [7]

Morewood, N.C., Shannon, P.M., Mackenzie, G.D., O'Reilly, B.M., Jacob, A.W.B., and Makris, J., 2003, Deep seismic experiments investigate crustal development at continental margins: Eos (Transactions, American Geophysical Union), v. 84, p. 225, 228. [9] [10]

Morgan, J.V., and Barton, P.J., 1990, A geophysical study of the Hatton bank volcanic margin: a summary of the results of a combined seismic, gravity and magnetic experiment, *in* Leven, J.H., Finlayson, D.M., Wright, C., Dooley, J.C., and Kennett, B.L.N., eds., Seismic probing of continents and their margins: Tectonophysics, v. 173, p. 517–526. [2] [8]

Morgan, J.V., Christeson, G.L., and Zelt, C.A., 2002, Testing the resolution of a 3D velocity tomogram across the Chicxulub crater: Tectonophysics, v. 355, p. 215–226, doi:10.1016/S0040-1951(02)00143-9. [9]

Morgan, J.V., Warner, M., Urrutia-Fuccugauchi, J., Gulick, S., Christeson, G., Barton, P., Rebolledo-Vieyra, M., and Melosh, J., 2005, Chicxulub crater seismic survey prepares way for future drilling: Eos (Transactions, American Geophysical Union), v. 86, no. 36, p. 325, 328. [2] [9]

Moriya T., Okada, H., Matsuhima, T., Asano, S., Yoshii, T., and Ikami, A., 1998, Collision structure in the upper crust beneath the southwestern foot of the Hidaka Mountains Hokkaido, Japan, as derived from explosion seismic observations: Tectonophysics, v. 290, p. 181–196, doi:10.1016/S0040-1951(98)00011-0. [8]

Morozov, I.B., Smithson, S.B., Hollister, L.S., and Diebold, J.B., 1998, Wide-angle seismic imaging across accreted terranes, southeastern Alaska and western British Columbia: Tectonophysics, v. 299, p. 281–296, doi:10.1016/S0040-1951(98)00208-X. [9]

Morozov, I.B., Smithson, S.B., Chen, J., and Hollister, L.S., 2001, Generation of new continental crust and terrane accretion in southeastern Alaska and western British Columbia: Tectonophysics, v. 341, p. 49–67, doi:10.1016/S0040-1951(01)00190-1. [9]

Morozova, E.A., Morozov, I.B., and Smithson, S.B., 1999, Heterogeneity of the uppermost mantle beneath Russian Eurasia from the ultra-long-range profile QUARTZ: Journal of Geophysical Research, v. 104, p. 20,329–20,348, doi:10.1029/1999JB900142. [8]

Morozova, E.A., Wan, X., Chamberlain, K.R., Smithson, S.B., Johnson, R., and Karlstrom, K.E., 2005, Inter–wedging nature of the Cheyenne belt—Archean-Proterozoic suture defined by seismic reflection data, *in* Karlstrom, K.E., and Keller, G.R., eds., The Rocky Mountain Region: An evolving lithosphere—tectonics, geochemistry, and geophysics: Washington, D.C., American Geophysical Union Geophysical Monograph 154, p. 217–226. [9]

Morris, G.B., Raitt, R.W., and Shor, G.G., 1969, Velocity anisotropy and delay time maps of the mantle near Hawaii: Journal of Geophysical Research, v. 74, p. 4300–4316, doi:10.1029/JB074i017p04300. [6] [7]

Morton, J.L., and Sleep, N.H., 1985, Seismic reflection from a Lau Basin magma chamber, *in* Scholl, D.W., and Vallier, T.L., eds., Geology and offshore resources of Pacific island arcs—Tonga region. Earth Science Series, vol. 2, p. 441–453, Circum-Pacific Council for Energy and Mineral Resources, Houston, Texas. [2] [8]

Morton, J.L., Sleep, N.H., Normark, W.R., and Tompkins, D.H., 1987, Structure of the southern Juan de Fuca Ridge from seismic reflection records: Journal of Geophysical Research, v. 92, p. 11315–11326, doi:10.1029/JB092iB11p11315. [2] [8]

Moss, F.J., and Dooley, J.C., 1988, Deep crustal reflectionrecordings in Australia 1957–1973, I. Data acquisition and presentation: Geophysical Journal of the Royal Astronomical Society, v. 93, p. 229–237. [2] [5] [6] [7]

Moss, F.J., and Mathur, S.P., 1986, A review of continental reflection profiling in Australia, *in* Barazangi, M., and Brown, L., eds., Reflection seismology: a global perspective: American Geophysical Union, Geodynamics Series, v. 13, p. 67–76. [2] [7] [8]

Mostaanpour, M.M., 1984, Einheitliche Auswertung krustenseismischer Daten in Westeuropa: Darstellung von Krustenparametern und Laufzeitanomalien: Berliner Geowissenschaftliche Abhandlungen, Reihe B, Band 10: 90 p. [7]

Mothes, H, 1929, Neue Ergebnisse der Eisseismik: Zeitschrift für Geophysik, v. 5, p. 120. [3]

Mueller, S., ed., 1973, The structure of the earth's crust based on seismic data: Tectonophysics, v. 20, 391 p. [2]

Mueller, S., and Landisman, M., 1966, Seismic studies of the earth's crust in continents; part I: Evidence for a low-velocity zone in the upper part of the lithosphere: Geophysical Journal of the Royal Astronomical Society, v. 10, p. 525–538. [4] [6]

Mueller, St., Peterschmitt, E., Fuchs, K., and Ansorge, J., 1967, The rift structure of the crust and upper mantle beneath the Rhinegraben. Rhinegraben Progress Report 1967, Abh. geol. Landesamt Baden–Württemberg, 6, p. 108–113. [7]

Mueller, S., Prodehl, C., Mendes, A.S. Moreira, V.S., 1973, Crustal structure in the southwestern part of the Iberian peninsula, *in* Mueller, St., ed., The structure of the earth's crust based on seismic data: Tectonophysics, v. 20, p. 307–318. [7]

Mueller, S., Prodehl, C. Mendes, A.S. Moreira, V.S., 1974, Deep–seismic sounding experiments in Portugal: Proceedings of the 13th General Assembly of European Seismological Commission (Brasov 1972), Techn. Econ. Studies, Bukarest, ser. D, no. 10, part II, 339–349. [7]

Mueller, S., Ansorge, J., Egloff, R., and Kissling, E., 1980, A crustal cross section along the Swiss Geotraverse from the Rhinegraben to the Po Plain: Eclogae Geologicae Helvetiae, v. 73, p. 463–483. [7]

Muirhead, K.J., and Simpson, D.W., 1972, A three-quarter watt seismic station: Bulletin of the Seismological Society of America, v. 62, p. 985–990. [7]

Muirhead, K.J., Cleary, J.R., and Finlayson, D.M., 1977, A long-range seismic profile in southeastern Australia: Geophysical Journal of the Royal Astronomical Society, v. 48, p. 509–520. [7]

Müller, G., 1970, Exact ray theory and its application to the reflection of elastic waves from vertically inhomogeneous media: Geophysical Journal of the Royal Astronomical Society, v. 21, p. 261–283. [6]

Müller, G., and Fuchs, K., 1976, Inversion of seismic records with the aid of synthetic seismograms, *in* Giese, P., Prodehl, C., and Stein, A., eds., Explosion seismology in central Europe—data and results: Berlin-Heidelberg-New York, Springer, p. 178–188. [6]

Müller, H.K., 1934, Beobachtung der Bodenbewegung in drei Komponenten bei Sprengungen: Zeitschrift für Geophysik, v. 10, p. 40–58. [3]

Muller, M.R., Robinson, C.J., Minshull, T.A., White, R.S., and Bickle, M.J., 1997, Thin crust beneath Ocean Drilling Program borehole 735B at the Southwest Indian Ridge?: Earth and Planetary Science Letters, v. 148, p. 93–107, doi:10.1016/S0012-821X(97)00030-7. [2] [9]

Muller, M.R., Minshull, T.A., and White, R.S., 1999, Segmentation and melt supply at the Southwest Indian Ridge: Geology, v. 27, p. 867–870, doi:10.1130/0091-7613(1999)027<0867:SAMSAT>2.3.CO;2. [2] [9]

Murauchi, S., Den, N., Asano, S., Hotta, H., Yoshii, T., Asanuma, T., Hagiwara, K., Ichikawa, K., Sato, T., Ludwig, W.J., Ewing, J.I., Edgar, N.T., and Houtz, R.E., 1968, Crustal structure of the Philippine Sea: Journal of Geophysical Research, v. 73, no. 10, p. 3143–3171, doi:10.1029/JB073i010p03143. [2] [6]

Murphy, J.M., 1988, USGS FM cassette seismic-refraction recording system: Menlo Park, California, U.S. Geological Survey Open-File Report 88-570, 43 p. [2] [7] [8]

Murphy, J.M., 1990, Data report for the Great Valley, California, axial seismic refraction profiles: Menlo Park, California, U.S. Geological Survey Open-File Report 89-494, 36 p. [8]

Murphy, J.M., and Luetgert, J.H., 1986, Data report for the Maine-Quebec cross-strike seismic-refraction profile: Menlo Park, California, U.S. Geological Survey Open-File Report 86-47, 71 p. [8]

Murphy, J.M., and Luetgert, J.H., 1987, Data report for the 1984 Maine along-strike seismic-refraction profiles: Menlo Park, California, U.S. Geological Survey Open-File Report 87-133, 134 p. [8]

Murphy, J.M., and Walter, A.W., 1984, Data report for a seismic refraction investigation: Morro Bay to the Sierra Nevada, California: Menlo Park, California, U.S. Geological Survey Open-File Report 84-642, 37 p. and 12 plates.[2] [8]

Murphy, J.M., Catchings, R.D., Kohler, W.M., Fuis, G.S., and Eberhart-Philliips, D., 1992, Data report for 1991 active–source seismic profiles in the San Francisco Bay Area, California: U.S. Geological Survey Open-File Report 92-570, 45 p. [9]

Murphy, J.M., Fuis, G.S., Levander, A.R., Lutter, W.J., Criley, E.E., Hentys, S.A., Asudeh, I., and Fowler, J.C., 1993, Data report for a 1990 seismic reflection/refraction experiment in the Brooks Range, Arctic Alaska: Menlo Park, California, U.S. Geological Survey Open-File Report 93-265, 128 p. [8] [9]

Murphy, J.M., Fuis, G.S., Okaya, D.A., Thygesen, K., Baher, S.A., Ryberg, T., Kaip, G., Fort, M.D., Asudeh, I., and Sell, R., 2002, Report for borehole explosion data acquired in the 1999 Los Angeles Region Seismic Experiment (LARSE II), southern California: Part 2, data tables and plots. U.S. Geological Survey Open-File Report 02-179, 252 p. [9]

Musacchio, G., Mooney, W.D., Luetgert, J.H., and Christensen, N.I., 1997, Composition of the crust in the Grenville and Appalachian provinces of North America inferred from V_p/V_s ratios: Journal of Geophysical Research, v. 102, p. 15,225–15,241, doi:10.1029/96JB03737. [8]

Musacchio, G., White, D.J., Asudeh, I., and Thomson, C.J., 2004, Lithospheric structure and composition of the Archean western Superior Province from seismic refraction/wide-angle reflection and gravity modelling: Journal of Geophysical Research, v. 109, B3, B03304 10.1029/2003JB002427. [2] [9]

Musgrave, A.W., 1967, Seismic refraction prospecting: Tulsa, Oklahoma, The Society of Exploration Geophysicists, 604 p. (Section 1: History of early refraction work, p. 1–11). [2] [3] [10]

Mutter, C.Z., and Mutter, J.C., 1993, Variations in thickness of layer 3 dominate oceanic crustal structure: Earth and Planetary Science Letters, v. 117, p. 295–317, doi:10.1016/0012-821X(93)90134-U. [8]

Mutter, J.C., and Jongsma, D., 1978, The pattern of the pre-Tasman Sea rift system and the geometry of breakup: Bulletin of the Australian Society of Exploration Geophysicists, v. 9, p. 70–75, doi:10.1071/EG978070. [7]

Mutter, J.C., and Karner, G.D., 1979, The continental margin off northeast Australia, in Stephenson, P.J., and Henderson, R.A., eds., Geophysics and Geology of Northeast Australia: Geological Society of Australia (Queensland Division), p. 47–69. [6]

Mutter, J.C., and NATO Study Group, 1985, Multichannel seismic images of the oceanic crust's internal structure: evidence for a magma chamber beneath the Mesozoic Mid-Atlantic Ridge: Geology, v. 9, p. 629–632, doi:10.1130/0091-7613(1985)13<629:MSIOTO>2.0.CO;2. [8]

Mutter, J.C., Larson, R.L., and Northwest Australia Sudy Group, 1989, Extension of the Exmouth Plateau, offshore northwestern Australia: Deep seismic reflection/refraction evidence for simple and pure shear mechanism: Geology, v. 17, p. 15–18, doi:10.1130/0091-7613(1989)017<0015:EOTEPO>2.3.CO;2. [2] [8]

Nafe, J.E., and Drake, C.L., 1957, Variation with depth in shallow and deep water marine sediments of porosity, density and the velocities of compressional and shear waves: Geophysics, v. 22, p. 523–552, doi:10.1190/1.1438386. [5]

Nafe, J.E., and Drake, C.L., 1963, Physical properties of marine sediments, in Hill, M.N., ed., The Sea v. 3., The earth beneath the Sea: New York-London, Interscience Publ., p. 794–815. [5]

Nagumo, S., Ouchi, T., Kasahara, J., Koresawa, S., Tomoda, Y., Kobayashi, K., Furumoto, A.S., Odegard, M.E., and Sutton, G.H., 1981, Sub-Moho seismic profile in the Mariana Basin–ocean bottom seismograph long-range explosion experiment: Earth and Planetary Science Letters, v. 53, p. 93–102, doi:10.1016/0012-821X(81)90030-3. [2] [7]

Nakamura, Y., Donoho, P.L., Roper, P.H., and McPherson, P.M., 1987, Large-offset seismic surveying using ocean bottom seismographs and airguns: instrumentation and field techniques: Geophysics, v. 52, p. 1601–1611, doi:10.1190/1.1442277. [8]

Nakanishi, A., Kurashimo, E., Tatsumi, Y., Yamaguchi, H., Miura, S., Kodaira, S., Obana, K., Takahashi, N., Tsuru, T., Kaneda, Y., Iwasaki, T., and Hirata, N., 2009, Crustal evolution of the southwestern Kuril Arc, Hokkaido Japan, deduced from seismic velocity and geochemical structure: Tectonophysics, v. 472, p. 105–123, doi:10.1016/j.tecto.2008.03.003. [2] [9]

Narans, H.D. Jr., Berg, J.W., Cook, K.L., 1961, Sub-basement seismic reflections in northern Utah: Journal of Geophysical Research, v. 66, p. 599–603. [6]

Nataf, H.-C., and Ricard, Y., 1996, 3SMAC: An a priory tomographic model of the upper mantle based on geophysical modelling: Physics of the Earth and Planetary Interiors, v. 95, p. 101–122, doi:10.1016/0031-9201 (95)03105-7. [10]

NAT Study Group, 1985, North Atlantic Transect: a wide-aperture, two-ship multichannel seismic investigation of the oceanic crust: Journal of Geophysical Research, v. 90, p. 10,321–10,341, doi:10.1029 /JB090iB12p10321. [2] [8]

Navin, D., Peirce, C., and Sinha, M.C., 1998, The RAMESSES experiment—II. Evidence for accumulated melt beneath a slow spreading ridge from wide-angle refraction and multichannel reflection seismic profiles: Geophysical Journal International, v. 135, p. 746–772, doi:10.1046/j.1365-246X.1998.00709.x. [2] [9]

Nelson, K.D., Lillie, R., de Voogd, B., Brewer, J., Oliver, J., Kaufman, S., Brown, L., and Viele, G.W., 1982, COCORP seismic reflection profiling in the Ouachita Mountains of western Arkansas:geometry and geologic interpretation: Tectonics, v. 1, p. 413–430, doi:10.1029/TC001i005p00413. [2] [8]

Nelson, K.D., Arnow, J.A., McBride, J.H., Willemin, R.J., Oliver, J., Brown, L., and Kaufman, S., 1985, New COCORP profiling on the southeastern U.S. coastal plains, part I: Late Paleozoic suture and Mesozoic rift basin: Geology, v. 13, p. 714–717, doi:10.1130/0091-7613(1985)13<714:NCPITS>2.0.CO;2. [2] [8]

Nelson, K.D., McBride, J.H., Arnow, J.A., Wille, D.M., Brown, L., Oliver, J., and Kaufman, S., 1987, Rsults of recent COCORP profiling in the southeastern United States: Geophysical Journal of the Royal Astronomical Society, v. 89, p. 141–146. [8]

Nelson, K.D., Zhao, W., Brown, L.D., Kuo, J., Che, J., Liu, X., Klemperer, S.L., Makovsky, Y., Meissner, R., Mechie, J., Kind, R., Wenzel, F., Ni, J., Nabelek, J., Leshou, C., Tan, H., Wei, W., Jones, A.G., Booker, J., Unsworth, M., Kidd, W.S.F., Hauck, M., Alsdorf, D., Ross, A., Cogan, M., Wu, C., Sandvol, E., and Edwards, M., 1996, Partially molten middle crust beneath southern Tibet: synthesis of project INDEPTH results: Science, v. 274, p. 1684–1688, doi:10.1126/science.274.5293.1684. [2] [9]

Nemeth, B., and Hajnal, Z., 1998, Structure of the lithospheric mantle beneath the Trans-Hudson orogen, Canada, in Klemperer, S.L., and Mooney, W.D., eds., Deep seismic probing of the continents, II: a global survey: Tectonophysics, v. 288, p. 93–104. [9]

Nemeth, B., Hajnal, Z., and Lucas, S.B., 1996, Moho signature from wide-angle reflections: preliminary results of the 1993 Trans-Hudson Orogen refraction experiment: Tectonophysics, v. 264, p. 111–121, doi:10.1016 /S0040-1951(96)00121-7. [9]

Nemeth, B., Clowes, R.M., and Hajnal, Z., 2005, Lithospheric structure of the Trans-Hudson Orogen from seismic refraction-reflection studies: Canadian Journal of Earth Science, v. 42, p. 435–456, doi:10.1139/e05-032. [2] [9]

Neprochnov, Yu.P., 1960, On the Choice of Optimal Explosion Conditions during Marine Seismic Refraction Studies: Razvedochnaya i Promyslovaya Geofizika, No. 35, 12–15. [5]

Neprochnov, Yu.P., 1989, Study of the lower crust and upper mantle using ocean bottom seismographs, in Mereu, R.F., Mueller, St., and Fountain, D.M., eds., Properties and processes of earth's lower crust: American Geophysical Union, Geophysical Monograph 51, p. 159–168. [2] [7] [8]

Neprochnov, Yu.P., Elnikov, I.N., and Kholopov, B.V., 1967, The structure of the earth's crust in the Indian Ocean according to results of seismic investigations carried out during the 36th voyage of the survey vessel "Vityaz": Dokl. Acad. Nauk SSSR, v. 174, no. 2, p. 429–431. [6]

Neprochnov, Yu.P., Rykunov. L.N., and Sedov, V.V., 1968, Experience of the Application of Automated Bottom Seismographs during Deep Seismic Sounding Studies: Izv. AN SSSR, Fiz. Zem., no. 11, 25–35. [6]

Neprochnov, Yu.P., Kosminskaya, I.P., and Malovitsky, Ya.P., 1970, Structure of the crust and upper mantle of the Black and Caspian Seas: Tectonophysics, v. 10, p. 517–525, 531–538. [6]

Nercessian, A., Mauffret, A., Dos Reis, T., Vidal, N., Gallart, J., and Díaz, J., 2001, Deep reflection seismic images of the crustal thinning in the eastern Pyrenees and western Gulf of Lion: Journal of Geodynamics, v. 31, p. 211–225, doi:10.1016/S0264-3707(00)00029-6. [2] [9]

Nettleton, L.L., 1940,. Geophysical prospecting for oil: New York, McGraw-Hill, 444 p. (History of seismic prospecting, p. 232–234). [2] [3] [10]

Neves, F., and Singh, S.C., 1996, Sensitivity study of seismic reflection/refraction data: Geophysical Journal International, v. 126, p. 470–476, doi:10.1111 /j.1365-246X.1996.tb05303.x. [9]

Nicolas, A., Hirn, A., Nicolich, R., and Poolini, R., 1990, Lithospheric wedging in the Western Alps inferred from the ECORS-CROP traverse: Geology, v. 18, p. 587–590, doi:10.1130/0091-7613(1990)018<0587:LWITWA>2.3.CO;2. [8]

Nicolich, R., Laigle, M., Hirn, A., Cernobori, L., and Gallart, J., 2000, Crustal structure of the Ionian margin of Sicily: Etna volcano in the frame of regional evolution: Tectonophysics, v. 329, p. 121–139, doi:10.1016/S0040-1951(00)00192-X. [9]

Nielsen, C., Sandrin, A., Nielsen, L., and Thybo, H., 2006, Reflection seismic image of a large mafic batholith—the ESTRID 2005 profile: European Geophysical Union, Annual Meeting 2006, Geophysical Research Abstracts, v. 8, 05963. [10]

Nishizawa, A., and Suyehiro, K., 1986, Crustal structure across the Kuril Trench off southeastern Hokkaido by airgun-OBS profiling: Geophysical Journal of the Royal Astronomical Society, v. 86, p. 371–397. [8]

Nolet, G., Dost, B., and Paulssen, H., 1986, Intermediate wavelength seismology and the NARS experiment: Annales. Geophys., v. 4, p. 305–314. [8]

Novak, O., Prodehl, C., Jacob, B., and Okoth, W., 1997a, Crustal structure of the Chyulu Hills, southeastern Kenya, in Fuchs, K., Altherr, R., Müller, B., and Prodehl, C., eds., Structure and dynamic processes in the lithosphere of the Afro-Arabian rift system: Tectonophysics, v. 278, p. 171–186. [9]

Novak, O., Ritter, J.R.R., Altherr, R., Garasic, V., Volker, F., Kluge, C., Kaspar, T., Byrne, G.F., Sobolev, S.V., and Fuchs, K., 1997b, An integrated model for the deep structure of the Chyulu Hills volcanic field, Kenya, in Fuchs,

K., Altherr, R., Müller, B., and Prodehl, C., eds., Structure and dynamic processes in the lithosphere of the Afro-Arabian rift system: Tectonophysics, v. 278, p. 187–209. [9]

Nyblade, A.A., and Langston, C.A., 2002, Broadband seismic experiments probe the East African rift: Eos (Transactions, American Geophysical Union), v. 83, no. 37, p. 405, 408–409, doi:10.1029/2002EO000296. [9]

Nyblade, A.A., Birt, C., Langston, C.A., Owens, T.J., and Last, R.J., 1996, Seismic experiment reveals rifting of craton in Tanzania: Eos (Transactions, American Geophysical Union), v. 77, p. 517, 521. [9]

O'Brien, P.N.S., 1968, Lake Superior crustal structure—a reinterpretation of the 1963 seismic experiment: Journal of Geophysical Research, v. 73, p. 2669–2689, doi:10.1029/JB073i008p02669. [6]

Ocola, L.C., and Meyer, R.P., 1972, Crustal low-velocity zones under the Peru-Bolivia Altiplano: Geophysical Journal of the Royal Astronomical Society, v. 30, p. 199–209. [2] [6] [9]

Officer, C.B., Jr., 1955a, A deep-sea seismic reflection profile: Geophysics, v. 20, p. 270–282, doi:10.1190/1.1438139. [5]

Officer, C.B., Jr., 1955b, Southwest Pacific crustal structure: American Geophysical Union Transactions, v. 36, p. 449–459. [2] [5]

Officer, C.B., 1974, Introduction to theoretical geophysics: New York-Heidelberg-Berlin, Springer, 385 p. [2] [10]

Officer, C.B., and Ewing, M., 1954, Geophysical investigations in the emerged and submerged Atlantic coastal plain, Part VII: continental shelf, continental slope, and continental rise south of Nova Scotia: Bulletin of the Geological Society of America, v. 65, p. 653–670, doi:10.1130/0016-7606(1954)65[653:GIITEA]2.0.CO;2. [2] [5]

Officer, C.B., Ewing, M., and Wuenschel, P.C., 1952, Seismic refraction measurements in the Atlantic Ocean, Part IV: Bermuda, Bermuda Rise and Nares Basin: Bulletin of the Geological Society of America, v. 63, p. 777–808, doi:10.1130/0016-7606(1952)63[777:SRMITA]2.0.CO;2. [2] [3] [4] [5]

Officer, C.B., Ewing, J.I., Hennion, J.F., Harkrider, D.G., and Miller, D.E., 1959, Geophysical investigations in the eastern Caribbean: Summary of 1955 and 1956 cruises: London, Pergamon Press, Physics and Chemistry of the Earth, v. 3, p. 17–109. [5]

Oh, J., Phillips, J.D., Austin, J.A.Jr., and Stoffa, P.L., 1991, Deep penetration seismic reflection images across the United States continental margin, *in* Meissner, R., Brown, L., Dürbaum, H.-J., Franke, W., Fuchs, K., and Seifert, F., eds., Continental lithosphere: deep seismic reflections: American Geophysical Union, Geodynamics Series, v. 22, p. 225–240. [2] [8]

Ohmura, T., Moriya, T., Piao, C., Iwasaki, T., Yoshii, T., Sakai, S., Takeda, T., Miyashita, K., Yamazaki, F., Ito, K., Yamazaki, A., Shimada, Y., Tashiro, K., and Miyamachi, H., 2001, Crustal stricture in and around the region of the 1995 Kobe Earthquake deduced from a wide-angle and refraction seismic exploration: Island Arc, v. 10, p. 215–227, doi:10.1046/j.1440-1738.2001.00319.x. [2] [9]

Okada, H., Suzuki, S., Moriya, T., and Asano, S., 1973, Crustal structure in the profile across the southern part of Hokkaido, Japan, as derived from explsion seismic observations: Journal of the Physics of the Earth, v. 21, p. 329–354. [2] [6] [9]

Okada, H., Moriya, T., Masuda, T., Hasegawa, T., Asano, S., Kasahara, K., Ikami, A., Aoki, H., Sasaki, Y., Hurukawa, N., and Matsumura, K., 1978, Velocity anisotropy in the Sea of Japan as revealed by big explosions: Journal of the Physics of the Earth, v. 26 Suppl., p. S491–S502. [2] [7]

Okada, H., Asano, S., Yoshii, T., Ikami, A., Suzuki, S., Hasegawa, T., Yamamoto, K., Ito, K., and Hamada, K., 1979, Regionality of the upper mantle around northeastern Japan as revealed by big explsions at sea: I. SEIHA–I explosion experiment: Journal of the Physics of the Earth, v. 27, p. S15–S32 ppl. [2] [7]

Okaya, D., Henrys, S., and Stern, T., 2002, Double-sided onshore-offshore seismic imaging of a plate boundary: "super-gathers" across South Island, New Zealand: Tectonophysics, v. 355, p. 247–263, doi:10.1016/S0040-1951(02)00145-2. [9]

Okaya, D., Stern T., Davey F., Henrys, S., and Cox, S., 2007, Continent-continent collision at the Pacific/Indo-Australian plate boundary: background, motivation, and principal results, *in* Okaya, D., Stern T., and Davey F., eds., A continental plate boundary: tectonics at South Island, New Zealand: Washington, D.C., American Geophysical Union, Geophysical Monograph 175, p. 1–18. [9]

O'Leary, D.M., Clowes, R.M. and Ellis, R.M., 1993, Crustal velocity structure in the southern Coast Belt, British Columbia: Canadian Journal of Earth Science, v. 30, p. 2389–2403. [8]

Oliver, H.W., Pakiser, L.C., and Kane, M.F., 1961, Gravity anomalies in the central Sierra Nevada, California: Journal of Geophysical Research, v. 66, p. 4265–4271, doi:10.1029/JZ066i012p04265. [5]

Oliver, J.E., 1986, A global perspective on seismic reflection profiling of the contninental crust, *in* Barazangi, M., and Brown, L., eds., Reflection seismology: a global perspective: American Geophysical Union, Geodynamics Series, v. 13, p. 5–19. [2] [7]

Oliver, J.E., 1996, Shocks and rocks—seismology in the plate tectonics revolution: Washington, D.C., American Geophysical Union, History of Geophysics, v. 6, 139 p. [7]

Oliver, J.E., and Kaufman, S., 1976, Profiling the Rio Grande rift: Geotimes, v. 21, p. 20–23. [7]

Oliver, J.E., Dobrin, M., Kaufman, S., Meyer, R., and Phinney, R., 1976, Continuous seismic reflection profiling of the deep basement, Hardeman county, Texas: Geological Society of America Bulletin, v. 87, p. 1537–1546, doi:10.1130/0016-7606(1976)87<1537:CSRPOT>2.0.CO;2. [2] [7] [8]

Olsen, K.H., 1983, The role of seismic refraction data for studies of the origin and evolution of continental rifts: Tectonophysics, v. 94, p. 349–370, doi:10.1016/0040-1951(83)90024-0. [7]

Olsen, K.H., ed., 1995, Continental rifts: Evolution, structure, tectonics: Amsterdam, Elsevier, 466 p. [2] [8] [9]

Olsen, K.H., Keller, G.R., and Stewart, J.N., 1979, Crustal structure along the Rio Grande rift from seismic refraction profiles, *in* Riecker, R.E., ed., Rio Grande rift: tectonics and magmatism: Washington, D.C., American Geophysical Union, p. 127–144. [2] [7]

Olsen, K.H., Cash, D.J., and Stewart, J.N., 1982, Mapping the northern and eastern extent of the Socorro midcrustal magma body by wide-angle seismic reflections: New Mexico Geological Society Guidebook, 33rd field conference, Albuquerque country II, p. 179–186. [7]

Olsen, K.H., Braile, L.W., Stewart, J.N., Daudt, C.R., Keller, G.R., Ankeny, L.A., and Wolff, J.J., 1986, Jemez Mountains volcanic field, New Mexico: time-term interpretation of the CARDEX seismic experiment and comparison with Bouguer gravity: Journal of Geophysical Research, v. 91, p. 6175–6187. [8] [9]

Operto, S., and Charvis, P., 1996, Deep structure of the southern Kerguelen plateau (southern Indian Ocean) from ocean bottom seismometer wide angle data: Journal of Geophysical Research, v. 101, p. 25,077–25,103, doi:10.1029/96JB01758. [2] [9]

Orcutt, J.A., 1987, Structure of the earth: oceanic crust and uppermost mantle: Reviews of Geophysics, v. 25, p. 1177–1196, doi:10.1029 /RG025i006p01177. [2] [8]

Orcutt, J.A., and Dorman, L.M., 1977, An oceanic long range explosion experiment: Journal of Geophysics, v. 43, p. 257–263. [2] [7]

Orcutt, J.A., Kennett, B.L.N., Dorman, L.M., and Prothero, W.A., 1975, Evidence for a low-velocity zone underlying a fast-spreading rise crest: Nature, v. 256, p. 475–476, doi:10.1038/256475a0. [7] [8]

Orcutt, J.A., Kennett, B.L.N., and Dorman, L.M., 1976, Structure of the East Pacific Rise from an ocean bottom seismometer survey: Geophysical Journal of the Royal Astronomical Society, v. 45, p. 305–320. [2] [6] [7]

Orcutt, J.A., Jordan, T.H., Menard, H.W., and Natland, J., 1983, Ngendei seismic experiment: refraction and reflection studies: Eos (Transactions, American Geophysical Union), v. 64, p. 269. [8]

Orcutt, J.A., McClai, J.S., and Burnett, M., 1984, Seismic constraints on the generation, evolution and structure of the oceanic crust, *in* Gass, G.I., Lippard, S.J., and Shelton, A.W., eds., Ophiolites and oceanic lithosphere. Special Publication Geological Society London, Blackwell Science Publ., Oxford, v. 13, p. 7–16. [2] [8]

O'Reilly, B.M., Shannon, P.M., and Vogt, U., 1991, Seismic studies in the North Celtic Sea basin: implications for basin development: Journal of the Geological Society London, v. 148, p. 191–195, doi:10.1144/gsjgs .148.1.0191. [2] [8]

O'Reilly, B.M., Hauser, F., Jacob, A.W.B., Shannon, P.M., Makris, J., and Vogt, U., 1995, The transition between the Erris and the Rockall basins: new evidence from wide-angle seismic data: Tectonophysics, v. 241, p. 143–163, doi:10.1016/0040-1951(94)00166-7. [8] [10]

O'Reilly, B.M., Hauser, F., Jacob, A.W.B., and Shannon, P.M., 1996, The lithosphere below the Rockall Trough: wide-angle seismic evidence for extensive serpentization: Tectonophysics, v. 255, p. 1–23, doi:10.1016/0040-1951 (95)00149-2. [8] [10]

O'Reilly, B.M., Hauser, F., Ravaut, C., Shannon, P.M., and Readman, P.W., 2006, Crustal thinning, mantle exhumation and serpentinization in the Porcupine Basin, offshore Ireland: evidence from wide-angle seismic data:

Journal of the Geological Society, v. 163, p. 775–787, doi:10.1144/0016 -76492005-079. [2] [10]

Osler, J.C., and Louden, K.E., 1992, Crustal structure of an extinct rift axis in the Labrador Sea: preliminary results from a seismic refraction survey: Earth and Planetary Science Letters, v. 108, p. 243–258, doi:10.1016/0012-821X (92)90026-R. [2] [8]

Ostrovsky, A.A., 1993, Study of the earth's crust deep seismic structure along the Baltic Sea profile: Institute Oceanology Russian Acadademy of Science and GEOMAR, Kiel, unpublished report. [2] [8]

Owen, T.R.E., and Barton, P.J., 1990, The Cambridge digital seismic recorder for land and marine use, *in* Leven, J.H., Finlayson, D.M., Wright, C., Dooley, J.C., and Kennett, B.L.N., eds., Seismic probing of continents and their margins: Tectonophysics, v. 173, p. 145–154. [8]

Owens, T.J., and Zandt, G., 1997, Implications of crustal property variations for models of Tibetan Plateau evolution: Nature, v. 387, p. 37–43, doi:10.1038 /387037a0. [9]

Ozel, O., Iwasaki, T., Moriya, T., Sakai, S., Maeda, T., Piao, C., Yoshii, T., Tsukada, S., Ito, A., Suzuki, M., Yamazaki, A., and Miyamachi, H., 1999, Crustal structure of central Japan and its petrological implications: Geophysical Journal International, v. 138, p. 257–274, doi:10.1046/j.1365-246x .1999.00859.x. [9]

Page, B.M., and Brocher, T.M., 1993, Thrusting of the central California margin over the edge of the Pacific plate during the transform regime: Geology, v. 21, p. 635–638, doi:10.1130/0091-7613(1993)021<0635:TOTCCM >2.3.CO;2. [9]

Pakiser, L.C., 1961, Gravity, volcanism, and crustal deformation in Long Valley, California: U.S. Geological Survey Professional Paper 424-B, p. 250–253. [5]

Pakiser, L.C., 1963, Structure of the crust and upper mantle in the western United States: Journal of Geophysical Research, v. 68, p. 5747–5756. [2] [6]

Pakiser, L.C., and Hill, D.P., 1963, Crustal structure in Nevada, and southern Idaho from nuclear explosions: Journal of Geophysical Research, v. 68, p. 5757–5766. [6]

Pakiser, L.C., and Mooney, W.D., eds., 1989, Geophysical framework of the continental United States: Geological Society of America Memoir 172, 826 p. [2] [8]

Pakiser, L.C., and Steinhart, J.S., 1964, Explosion seismology in the western hemisphere, *in* Odishaw, H., ed., Research in Geophysics, Vol. 2, Solid Earth and Interface Phenomena, MIT Press,Cambridge, p. 123–147. [6]

Pakiser, L.C., and Zietz, I., 1965. Transcontinental crustal and upper-mantle structure. Reviews in Geophysics, v. 3, p. 505–520, doi:10.1029 /RG003i004p00505. [6]

Pakiser, L.C., Press, F., and Kane, M.F., 1960, Geophysical investigation of Mono Basin, California: Geological Society of America Bulletin, v. 71, p. 415– 448, doi:10.1130/0016-7606(1960)71[415:GIOMBC]2.0.CO;2. [5]

Palmason, G., 1971, Crustal structure of Iceland from explosion seismology: Societas Scientiarum Islandica, Reykjavik, Iceland, 187 p. [7]

Palomeras, I., Carbonell, R., Flecha, I., Simancas, F., Ayarza, P., Matas, J., Martinez Poyatos, D., Azor, A., Gonzales Lodeiro, F., and Perez-Estaun, A., 2009, Nature of the lithosphere across the Variscan orogen of SW Iberia: Dense wide-angle seismic reflection data: Journal of Geophysical Research, v. 114: B02302, doi:10.1029/2007/JB005050. [2] [10]

Panea, I., Stephenson, R., Knapp, C., Mocanu, V., Drijkoningen, G., Matenco, V., Knapp, J., and Prodehl, C., 2005, Near-vertical seismic reflection image using a novel acquisition technique across the Vrancea Zone and Foscani Basin, southeastern Carpathians (Romania), *in* Cloetingh, S., Matenco, L. Bada, G. Dinu, C. and Mocanu, V., eds., The Carpathians-Pannonian Basin System: Natural Laboratory for Coupled Lithospheric-Surface Processes: Tectonophysics, v. 410, p. 293–310. [9] [10]

Park, J.-O., Tsuru, T., Kodaira, S., Cummins, P.R., and Kaneda, Y., 2002, Splay fault branching along the Nankai subduction zone: Science, v. 297, p. 1157–1160, doi:10.1126/science.1074111. [2] [9] [10]

Parotto, M., Cavinato, G.P., Miccadei, E., and Tozzi, M., 2003, Line CROP 11: Central Apennines, *in* Scrocca, D., Doglioni, C., Innocenti, F., Manetti, P., Mazzotti, A., Bertelli, L., Burbi, L., and D'Offizi, S., eds., CROP Atlas: seismic reflection profiles of the Italian crust: Memorie Descrittive della Carta Geologica d'Italia, Roma, 62, p. 145–154. [9]

Parsiegla, N., Gohl, K., and Uenzelmann-Neben, G., 2007, Deep crustal structure of the sheared South African continental margin: first results of the Agulhas-Karoo Geoscience Transect: South African Journal of Geology, v. 110, no. 2-3, p. 393–406, doi:10.2113/gssajg.110.2-3.393. [2] [10]

Parsons, E., ed., 2002, Crustal structure of the coastal and marine San Francisco Bay region, California: U.S. Geological Survey Professional Paper 1658, 145 p. [9]

Parsons, T., 1998, Seismic reflection evidence that the Hayward fault extends into the lower crust of the San Francisco Bay area, California: Bulletin of the Seismological Society of America, v. 88, p. 1212–1223. [9]

Parsons, T., and Zoback, M.L., 1997, Three-dimensional upper crustal velocity structure beneath San Francisco Peninsula, California: Journal of Geophysical Research, v. 102, p. 5473–5490, doi:10.1029/96JB03222. [9]

Parsons, T., McCarthy, J., Kohler, W.M., Ammon, C.J., Benz, H.M., Hole, J.A., and Criley, E.E., 1996, Crustal structure of the Colorado Plateau, Arizona: Application of new long-offset seismic data analysis techniques: Journal of Geophysical Research, v. 101, p. 11,173–11,194, doi:10.1029 /95JB03742. [8]

Parsons, T., Trehu, A.M., Luetgert, J.H., Miller, K.C., Kilbride, F., Wells, R.E., Fisher, M.A., Flueh, E., ten Brink, U.S., and Christensen, N.I., 1998, A new view into the Cascade subduction zone and volcanic arc: implications for earthquake hazards along the Washington margin: Geology, v. 26, p. 199–202, doi:10.1130/0091-7613(1998)026<0199:ANVITC >2.3.CO;2. [9]

Parsons, T., Wells, R.E., Trehu, A.M., Fisher, M.A., Flueh, E., and ten Brink, U.S., 1999, Three-dimensional velocity structure of Siletzia and other accreted terranes in the Cascadia forearc of Washington: Journal of Geophysical Research, v. 104, p. 18,015–18,039, doi:10.1029/1999JB900106. [9]

Pascal, G., Torné, M., Buhl, P., Watts, A.B., and Mauffret, A., 1992, Crustal and velocity structure of the Valencia trough (western Mediterranean). Part I. Detailed interpretation of five Exapnding Spread Profiles, *in* Banda, E., and Santanach,P., eds., Geology and Geophysics of the Valencia Trough, Western Mediterranean: Tectonophysics, v. 203, no. 1-4, p. 21–35. [2] [8]

PASSCAL Working Group, 1988, The 1986 PASSCAL Basin and Range lithospheric seismic experiment: Eos (Transactions, American Geophysical Union), v. 69, p. 593, 596–598, doi:10.1029/88EO00174. [8]

Patzwahl, R., 1998, Plattengeometrie und Krustenstruktur am Kontinentalrand Nord–Chiles aus weitwinkelseismischen Messungen: Berliner Geowissenschaftliche Abhandlungen, Reihe B30, 150 p. [8] [9]

Patzwahl, R., Mechie, J., Schulze, A., and Giese, P., 1999, Two-dimensional velocity models of the Nazca plate subduction zone between 19.5°S and 25°S from wide-angle seismic measurements during the CINCA95 project: Journal of Geophysical Research, v. 104, p. 7293–7317, doi:10.1029 /1999JB900008. [2] [9]

Pavlenkova, N.I., 1996, Crust and upper mantle structure in northern Eurasia from seismic data: Advances in Geophysics, v. 37, p. 1–133, doi:10.1016 /S0065-2687(08)60269-1. [2] [5] [6] [7] [8] [10]

Pavlenkova, N.I., Solodilov, L.N., and Mooney, W.D., 1993, 1993 CCSS Workshop proceedings volume. [9] [10]

Pavlenkova, N.I., Pilipenko, V.N., Verpakhovskaya, A.O., Pavlenkova, G.A., and Filonenko, V.P., 2009, Crustal structure in Chile and Okhotsk Sea regions: Tectonophysics, v. 472, p. 28–38, doi:10.1016/j.tecto.2008.08.018. [5]

Peddy, C.P., Pinet, B., Masson, D., Scrutton, R., Sibuet, J.C., Warner, M.R., Lefort, J.P., and Shroeder, I.J., (BIRPS and ECORS), 1989, Crustal structure of the Goban Spur continental margin, Northeast Atlantic, from deep reflection profiling: Journal of the Geological Society of London, v. 146, p. 427–437. [2] [8]

Peirce, C., and Barton, P. J., 1991, Crustal structure of the Madeira–Tore Rise, eastern North Atlantic—results of a DOBS wide-angle and normal incidence seismic experiment in the Josephine Seamount region: Geophysical Journal International, v. 106, p. 357–378, doi:10.1111/j.1365-246X.1991 .tb03898.x. [2] [8]

Percival, J.A., ed., 1994, The Kapuskasing Transect of LITHOPROBE: Canadian Journal of Earth Science, v. 31, p. 1013–1286, doi:10.1139/e94-091. [2] [8]

Percival, J.A., and Helmstaedt, H., eds., 2006, The Western Superior Province Lithoprobe and NATMAP transects: Canadian Journal of Earth Science, v. 43, p. 743–1117, doi:10.1139/E06-063. [2]

Percival, J.A., Shaw, D.M., Milkereit, B., et al., 1991, A closer look at deeper crustal reflections: Eos (Transactions, American Geophysical Union), v. 72, p. 337–340, doi:10.1029/90EO10264. [8]

Perez-Estaun, A., Pulgar, J.A, Banda, E., Alvarez-Manon, J., and ESCI-N Research Group, 1994, Crustal structure of the external Variscides in northwest Spain from deep seismic reflection profiling: Tectonophysics, v. 232, p. 91–118, doi:10.1016/0040-1951(94)90078-7. [9]

Perkins, W.E., and Phinney, R.A., 1971, A reflection study in the Wind River uplift, Wyoming: American Geophysical Union Monograph 14, p. 41–50. [2] [6]

Perrier, G., and Ruegg, J.C., 1973, Structure profonde du Massif Central français: Annales de Géophysique, v. 29, p. 435–502. [2] [6] [7]

Peterschmitt, E., Menzel, H., and Fuchs, K., 1965. Seismische Messungen in den Alpen. Die Beobachtungen auf dem NE-Profil Lago Lagorai 1962 und ihre vorläufige Auswertung: Zeitschrift für Geophysik, v. 31, p. 41–49. [6]

Petkovic, P., Collins, C.D.N., and Finlayson, D.M., 2000, A crustal transect between Precambrian Australia and the Timor Trough across the Vulcan Sub-basin: Tectonophysics, v. 329, p. 23–38, doi:10.1016/S0040-1951(00)00186-4. [2] [9]

Phinney, R.A., 1986, A seismic cross section of the New England Appalachians: the orogen exposed, in Barazangi, M., and Brown, L., eds., Reflection seismology: the continental crust: American Geophysical Union, Geodynamics Series, v. 14, p. 157–172. [7]

Phinney, R.A., and Roy-Chowdhury, K., 1989, Reflection seismic studies of crustal structure in the eastern United States, in Pakiser, L.C., and Mooney, W.D., eds., Geophysical framework of the continental United States: Geological Society of America Memoir 172, p. 613–653. [2] [7] [8] [9]

Pilger, A., and Rösler, A., eds., 1975, Afar Depression of Ethiopia: Stuttgart, Schweizerbart, 410 p. [7]

Pilger, A., and Rösler, A., eds., 1976, Afar between continental and oceanic rifting: Stuttgart, Schweizerbart, 216 p. [7]

Pinet, B., Montadert, L., and ECORS Scientific Party, 1987, Deep seismic reflection and refraction profiling along the Aquitaine shelf (Gulf of Biscay): Geophysical Journal of the Royal Astronomical Society, v. 89, p. 305–312. [2] [8]

Pinet, B., Sibuet, J.-C., Lefort, J.-P., Shroeder, I.J., and Montadert, L., 1991, Structure profonde de la marge des entrés de la Manche et du plateau continental celtique: le profil WAM: Mémoires de la Société Géologique de France, v. 159, p. 167–183. [8]

Plafker, G., and Mooney, W.D., 1997, Introduction to the special section: The Trans-Alaska Crustal Transect (TACT) across Arctic Alaska: Journal of Geophysical Research, v. 102, p. 20,639–20,643, doi:10.1029/97JB01048. [2] [8]

Podvin, P., and Lecomte, I., 1991, Finite difference computation of traveltimes in very contrasted velocity models: a massively parallel approach and its associated tools: Geophysical Journal International, v. 105, p. 271–284, doi:10.1111/j.1365-246X.1991.tb03461.x. [9]

Poehls, K.A., 1974, Seismic refraction on the Mid-Atlantic Ridge at 37°N: Journal of Geophysical Research, v. 79, p. 3370–3373, doi:10.1029/JB079i023p03370. [7]

Poldervaart, A., ed., 1955, Crust of the earth (a symposium): Geological Society of America Special Paper 62, p. 762. [2][4]

Pontoise, B., and Latham, G., 1982, Etude par refraction de la structure interne de l'arc des Tonga, in Contribution a l'Etude Geodynamique du Sud-Ouest Pacifique, ORSTOM Institute, Paris, p. 237–254. [10]

Pope, K.O., Ocampo, A.C., Kinsland, G.L., and Smith, R., 1996, Surface expression of the Chicxulub crater, Mexico: Geology, v. 24, p. 527–530, doi:10.1130/0091-7613(1996)024<0527:SEOTCC>2.3.CO;2. [9]

Potter, C.J., Sanford, W.E., Yoos, T.R., Prussen, E.I., Keach, R.W., II, Oliver, J.E., Kaufman, S., and Brown, L.D., 1986, COCORP deep seismic reflection traverse of the interioir of the North American cordillera, Washington and Idaho: implications for orogenic evolution: Tectonics, v. 5, p. 1007–1026, doi:10.1029/TC005i007p01007. [2] [8]

Potter, C.J., Allmendinger, R.W., Hauser, E.C., and Oliver, J.E., 1987, COCORP deep seismic reflection traverses of the U.S. Cordillera: Geophysical Journal of the Royal Astronomical Society, v. 89, p. 99–104. [8]

Powell, C.M.R., Sinha, M.C., Carter, P.W., and Leonard, J.R., 1986, A sea-bottom multichannel hydrophone array: Marine Geophysical Researches, v. 8, p. 277–292, doi:10.1007/BF00305487. [8]

Pratt, T.L., Coruh, C., and Costain, J.K., 1987, Lower crustal reflections in central Virginia: Geophysical Journal of the Royal Astronomical Society, v. 89, p. 163–170. [8]

Pratt, T.L., Coruh, C., and Costain, J.K., 1988, A geophysical study of the Earth's crust in central Virginia: implications for Appalachian crustal structure: Journal of Geophysical Research, v. 93, p. 6649–6667, doi:10.1029/JB093iB06p06649. [2] [8]

Pratt, T.L., Meagher, K.L., Brocher, T.M., Yelin, T., Norris, R., Hultgrien, L., Barnett, E., and Weaver, C.S., 2003, Earthquake recordings from the 2002 Seattle seismic hazard investigation of Puget Sound (SHIPS), Washington State: Menlo Park, California, U.S. Geological Survey Open-File Report 03-361, 72 p. [10]

Press, F., 1960, Crustal structure in the California-Nevada region: Journal of Geophysical Research, v. 65, p. 1039–1051, doi:10.1029/JZ065i003p01039. [5]

Press, F., and Beckmann, W., 1954, Geophysical investigations in the emerged and submerged Atlantic coastal plain, Part VIII: Grand Banks and adjacent shelves: Bulletin of the Geological Society of America, v. 65, p. 299–314, doi:10.1130/0016-7606(1954)65[299:GIITEA]2.0.CO;2. [2] [5]

Priestley, K., Davey, F.J., 1983, Crustal structure of Fiordland, south western New Zealand from seismic refraction measurements: Geology, v. 11, p. 660–663, doi:10.1130/0091-7613(1983)11<660:CSOFSN>2.0.CO;2. [7]

Prodehl, C., 1964, Auswertung von Refraktionsbeobachtungen im bayerischen Alpenvorland (Steinbruchsprengungen bei Eschenlohe 1958-1961) in Hinblick auf die Tiefenlage des Grundgebirges: Zeitschrift für Geophysik, v. 30, p. 161–181. [6]

Prodehl, C., 1965, Struktur der tieferen Erdkruste in Südbayern und längs eines Querprofiles durch die Ostalpen, abgeleitet aus refraktionsseismischen Messungen bis 1964: Bollettino di Geofisica Teorica e Applicata, v. 7, p. 35–88. [2] [6]

Prodehl, C., 1970, Seismic refraction study of crustal structure in the western United States. Geological Society of America Bulletin, v. 81, p. 2629–2646, doi:10.1130/0016-7606(1970)81[2629:SRSOCS]2.0.CO;2. [6] [7]

Prodehl, C., 1976, Record sections for selected local earthquakes recorded by the Central California Microearthquake Network: Menlo Park, California, U.S. Geological Survey Open-File Report 77-37, 310 p. [7]

Prodehl, C., 1977, The structure of the crust–mantle boundary beneath different tectonic areas of North America and Europe as derived from explosion seismology, in Heacock, J.G., ed.: The earth's crust: American Geophysical Union Geophysical Monograph, v. 20, p. 349–369. [6] [7]

Prodehl, C., 1979, Crustal structure of the western United States—a reinterpretation of seismic-refraction measurements from 1961 to 1963 in comparison with the crustal structure of central Europe: U.S. Geological Survey Professional Paper 1034, 74 p. [2] [6] [7] [8]

Prodehl, C., 1984, Structure of the earth's crust and upper mantle, in Hellwege, K.-H., editor in chief, Landolt Börnstein New Series: Numerical data and functional relationships in science and technology. Group V, Volume 2a: K. Fuchs and H. Soffel, eds., Physical properties of the interior of the earth, the moon and the planets: Berlin-Heidelberg, Springer, p. 97–206. [2] [7] [10]

Prodehl, C., 1985, Interpretation of a seismic refraction survey across the Arabian Shield in western Saudi Arabia: Tectonophysics, v. 111, p. 247–282, doi:10.1016/0040-1951(85)90288-4. [7]

Prodehl, C., 1998, Crustal record sections—a guide to pattern recognition: Open-File Report, Geophysical Institute, University of Karlsruhe, 131 p. [2]

Prodehl, C., and Aichroth, B., 1992, Seismic investigations along the European Geotraverse in Central Europe, in Kern, H., and Gueguen, Y., eds., Structure and composition of the lower continental crust: Terra Nova, v. , p. 14–24. [8]

Prodehl, C., and Lipman, P.W., 1989, Crustal structure of the Rocky Mountain region, in Pakiser, L.C., and Mooney, W.D., eds., Geophysical framework of the continental United States: Geological Society of America Memoir 172, p. 249–284. [5] [6] [7] [8] [9]

Prodehl, C., and Mechie, J., 1991, Crustal thinning in relationship to the evolution of the Afro–Arabian rift system—a review of seismic-refraction data, in Makris, J., Mohr, P., and Rihm, R., eds., Red Sea: Birth and early history of a new oceanic basin: Tectonophysics, v. 198, p. 311–327. [8]

Prodehl, C., and Pakiser, L.C., 1980, Crustal structure of the southern Rocky Mountains from seismic measurements: Geological Society of America Bulletin, v. 91(I), p. 147–155, doi:10.1130/0016-7606(1980)91<147:CSOTSR>2.0.CO;2. [2] [6] [7]

Prodehl, C., Moreira, V.S., Mueller, S., and Mendes, A.S., 1975, Deep-seismic sounding experiments in central and southern Portugal: Proceedings of the 14th General Assembly of the European Seismological Commission (Trieste 1974), Akad. Wiss. DDR, Berlin, p. 261–266. [7]

Prodehl, C., Ansorge, J., Edel, J.B., Emter, D., Fuchs, K., Mueller, St., and Peterschmitt, E., 1976a, Explosion-seismology research in the central and southern Rhinegraben—a case history, in Giese, P., Prodehl, C., and Stein, A., eds., Explosion seismology in central Europe—data and results: Berlin-Heidelberg-New York, Springer, p. 313–328. [4] [6] [7]

Prodehl, C., Hirn, A., Kind, R., Steinmetz, L., and Fuchs, K., 1976b, Elastic properties of the lower lithosphere obtained by large-scale seismic experiments in France, *in* Strens, R.G.J., ed., The physics and chemistry of rocks and minerals: New York-London-Sydney, John Wiley and Sons Ltd., p. 239–243. [7]

Prodehl, C., Schlittenhardt, J., Stewart, S.W., 1984, Crustal structure of the Appalachian Highlands in Tennessee, *in* Zwart, H.J., Behr, H.J., Oliver, J.E., eds., Appalachian and Hercynian fold-belts: Tectonophysics, v. 109, p. 61–76. [2] [6] [7] [9]

Prodehl, C., Keller, G.R., Khan, M.A., eds., 1994a, Crustal and upper mantle structure of the Kenya rift: Tectonophysics, v. 236, no. 1-4, 483 p. [2] [6] [9]

Prodehl, C., Mechie, J., Achauer, U., Keller, G.R., Khan, M.A., Mooney, W.D., Gaciri, S.J., and Obel, J.D., 1994b, The KRISP 90 seismic experiment—a technical review, *in* Prodehl, C., Keller, G.R., and Khan, M.A., eds., Crustal and upper mantle structure of the Kenya rift: Tectonophysics, v. 236, p. 33–60. [9]

Prodehl, C., Jacob, B., Thybo H., Dindi, E., and Stangl, R., 1994c, Crustal structure of the northeastern flank of the Kenya rift, *in* C. Prodehl, G.R. Keller, and M.A. Khan, eds., Crustal and upper mantle structure of the Kenya rift: Tectonophysics, v. 236, p. 271–290. [9]

Prodehl, C., Mueller, St., and Haak, V., 1995, The European Cenozoic rift system, *in* Olsen, K.H., ed., Continental rifts: evolution, structure, tectonics: Amsterdam, Elsevier, p. 133–212. [2] [7] [9]

Prodehl, C., Fuchs, K., and Mechie, J., 1997a, Seismic-refraction studies of the Afro-Arabian rift system—a brief review, *in* Fuchs, K., Altherr, R., Müller, B., and Prodehl, C., eds., Structure and dynamic processes in the lithosphere of the Afro-Arabian rift system: Tectonophysics, v. 278, p. 1–13. [2] [6] [9]

Prodehl, C., Ritter, J.R.R., Mechie, J., Keller, G.R., Khan, M.A., Jacob, B. Fuchs, K., Nyambok, I.O., Obel, J.D., and Riaroh, D., 1997b, The KRISP 94 lithospheric investigation of southern Kenya—the experiments and their main results, *in* Fuchs, K., Altherr, R., Müller, B., and Prodehl, C., eds., Structure and dynamic processes in the lithosphere of the Afro-Arabian rift system: Tectonophysics, v. 278, p. 121–147. [9]

Prodehl, C., Johnson, R.A., Keller, G.R., Snelson, C. and Rumpel, H.-M., 2005, Background and overview of previous controlled source seismic studies, *in* Karlstrom, K.E., and Keller, G.R., eds., The Rocky Mountain Region: An evolving lithosphere—tectonics, geochemistry, and geophysics: Washington, D.C., American Geophysical Union Geophysical Monograph 154, p. 201–216. [6] [9]

Prosen, D., Milovanovic, B., and Roksandic, M., 1972, Yugoslavia, *in* Sollogub, V.B., Prosen, D., and Militzer, H., eds., Crustal structure of central and southeastern Europe based on the results of explosion seismology (publ. in Russian 1971). English translation edited by Szénás, Gy., 1972: Geophysical Transactions, spec. ed., Müszaki Könyvkiado, Budapest, chapter 25, p. 99–105. [2] [6]

Pulgar, J.A, Gallart, J., Fernandez-Viejo, G., Perez-Estaun, A., Alvarez-Manon, J., and ESCIN Group, 1996, Seismic image of the Cantabrian Mountains in the western extension of the Pyrenean belt from integrated reflection and refraction data: Tectonophysics, v. 264, p. 1–19, doi:10.1016/S0040-1951(96)00114-X. [9]

Purdy, G.M., 1982, The variability in seismic structure of layer 2 nearthe East Pacific Rise at 12°N: Journal of Geophysical Research, v. 87, p. 8403–8416, doi:10.1029/JB087iB10p08403 [7]

Purdy, G.M., 1986, A determination of the seismic velocity structure of sediments using both sources and receivers near the ocean floor: Marine Geophysical Researches, v. 8, p. 75–91, doi:10.1007/BF02424829. [8]

Purdy, G.M., and Detrick, R.S., 1986, Crustal structure of the Mid-Atlantic Ridge at 23°N from seismic refraction studies: Journal of Geophysical Research, v. 91, p. 3739–3762, doi:10.1029/JB091iB03p03739. [8]

Puzyrev, N.N., Mandelbaum, N.N., Krylov, S.V., Mishenkin, B.P., Krupskaya, G.V., and Petrick, G.V., 1973, Deep seismic investigations in the Baikal rift zone: Tectonophysics, v. 20, p. 85–95, doi:10.1016/0040-1951(73)90098-X. [6]

Puzyrev, N.N., Mandelbaum, N.N., Krylov, S.V., Mishenkin, B.P., Petrick, G.V., and Krupskaya, G.V., 1978, Deep structure of the Baikal and other continental rift zones from seismic data: Tectonophysics, v. 45, p. 15–22, doi:10.1016/0040-1951(78)90219-6. [6]

Quinlan, G., ed., 1998, Lithoprobe East transect: Canadian Journal of Earth Science, v. 35, p. 1203–1346, doi:10.1139/cjes-35-11-1203. [2]

Radulescu, D.P., Cornea, I., Sandulescu, M., Constantinescu, P., Radulescu, F., and Pompilian, A., 1976, Structure de la croûte terrestre en Roumanie. Essai d'interprétation des études sismiques profondes: Anuarul Institutului de Geologie şi Geofizică, v. 50, p. 5–36. [2] [7]

Radulescu, F., and Pompilian, A., 1991, Twenty-five years of deep seismic soundings in Romania (1966–1990): Rev. Roum. Géophysique, Bucharest, v. 35, p. 89–97. [6]

Rai, S.S., Priestley, K., Gaur, V.K., Singh, M.P., and Searle, M., 2006, Configuration of the Indian Moho beneath the NW Himalaya and Ladakh: Geophysical Research Letters, v. 33, L15308, doi:10.1029/2006GL026076. [2] [9]

Raileanu, V., Bala, A., Hauser, F., Prodehl, C. and Fielitz, W., 2005, Crustal properties from S-wave and gravity data along a seismic refraction profile in Romania, *in* Cloetingh, S., Matenco, L., Bada, G., Dinu, C., and Mocanu, V., eds., The Carpathians-Pannonian Basin System: Natural Laboratory for Coupled Lithospheric-Surface Processes: Tectonophysics, v. 410, p. 251–272. [9]

Raitt, R.W., 1949, Studies of ocean-bottom structure off southern California with explosive waves: Bulletin of the Geological Society of America, v. 60 (12.2): 1915. [2] [4]

Raitt, R.W., 1956, Seismic refraction studies of the Pacific Ocean basin, Part I: Crustal thickness of the central equitorial Pacific: Bulletin of the Geological Society of America, v. 67, p. 1623–1640, doi:10.1130/0016-7606 (1956)67[1623:SSOTPO]2.0.CO;2. [2] [4] [5]

Raitt, R.W., 1963, The crustal rocks, *in* Hill, M.N., ed., The Sea, vol.3. Interscience Publ., New York–London, p. 85–102. [2] [5] [6] [10]

Raitt, R.W., 1964, Geophysics of the South Pacific, *in* Odishaw, H., ed., Research in Geophysics, Vol. 2, Solid Earth and Interface Phenomena, MIT Press, Cambridge, p. 223–241. [2] [5] [6]

Raitt, R.W., Fisher, R.L., and Mason, R.G., 1955, Tonga trench, *in* Poldervaart, A., ed., Crust of the earth (a symposium): Geological Society of America Special Paper 62, p. 237–254. [5] [10]

Raitt, R.W., Shor, G.G., Francis, T.J., and Morris, G.B., 1969, Anisotropy of the Pacific upper mantle: Journal of Geophysical Research, v. 74, p. 3095–3109, doi:10.1029/JB074i012p03095. [6] [7]

Raitt, R.W., Shor, Jr., G.G., Morris, G.B., and Kirk, H.K., 1971, Mantle anisotropy in the Pacific Ocean: Tectonophysics, v. 12, p. 173–186, doi:10.1016/0040-1951(71)90002-3. [2] [6] [7]

Rajendra Prasat, B., Tewari, H.C., Vijaya Rao, V., Dixit, M.M., and Reddy, P.R., 1998, Structure and tectonics of the Proterozoic Aravalli-Delhi fold belt in northwestern India from deep seismic reflection studies, *in* Klemperer, S.L., and Mooney, W.D., eds., Deep seismic probing of the continents, II: a global survey: Tectonophysics, v. 288, p. 31–41. [2] [9]

Ramsay, D.C., Colwell, J.B., Coffin, M.F., Davies, H.L., Hill, P.J., Pigram, C.J., and Stagg, H.M. J., 1986, New findings from the Kerguelen Plateau: Geology, v. 14, p. 589–593, doi:10.1130/0091-7613(1986)14<589:NFFTKP>2.0.CO;2. [2] [8]

Rankin, D.S., Ravindra, R., and Zwicker, D., 1969, Preliminary interpretation of the first refraction arrivals in Gaspe from shots in Labrador and Quebec: Canadian Journal of Earth Science, v. 6, p. 771–774. [6]

Ransome, C.R., 1979, A crustal model for Cardigan Bay: Geophysical Journal of the Royal Astronomical Society, v. 57, p. 254. [2] [7]

Rao, G.S.P., and Tewari, H.C., 2005, The seismic structure of the Saurashtra crust in northwest India and its relationship with the Reunion plume: Geophysical Journal International, v. 160, p. 318–330. [7]

Ratcliffe, N.M., Burton, W.D., D'Angelo, R.M., and Costain, J.K., 1986, Low-angle extensional faulting, reactivated mylonites, and seismic reflection geometry of the Newark basin margin in eastern Pennsylvania: Geology, v. 14, p. 766–770, doi:10.1130/0091-7613(1986)14<766:LEFRMA>2.0.CO;2. [2] [8]

Raum, T., Mjelde, R., Digranes, P., Shimamura, H., Shiobara, H., Kodaira, S., Haatveldt, G., Sorenes, N., and Thorbjornsen, S., 2002, Crustal structure of the southern part of the Voring Basin, mid-Norway margin, from wide-angle seismic and gravity data: Tectonophysics, v. 355, p. 99–126, doi:10.1016/S0040-1951(02)00136-1. [2] [9] [10]

Reddy, P.R., Rajendra Prasat, B., Vijaya Rao, V., Kare, P., Kesava Rao, G., Murty, A.S.N., Sarkar, D., Raju, S., Rao, G.S.P., and Sridhar, V., 1995, Deep seismic reflection profiling along the Nandsi-Kunjer section of Nagaur-Jhalawar transect: preliminary results, *in* Sinha-Roy, S., and Gupta, K.R., eds., Continental crust of northwestern and central India: Geological Society India, Memoir 31, p. 353–372. [9]

Reich, H., 1937, Erfahrungen mit dem seismischen Refraktionsverfahren bei der geophysikalischen Reichsaufnahme: Beiträge zur angewandten Geophysik, 7, p. 1–16. [3]

Reich, H., 1953, Über seismische Beobachtungen der Prakla an Reflexionen aus grosser Tiefe bei den Steinbruchsprengungen in Blaubeuren am 4.3. und 10.5. 1952, Geologisches Jahrbuch, v. 68, p. 225–240. [2] [5] [6]

Reich, H., 1958, Seismische und geologische Ergebnisse der 2 to–Sprengung im Tiefbohrloch Tölz I am 11.12.54: Geologisches Jahrbuch, v. 75, p. 1–46. [5]

Reich, H., 1960, Zur Frage der geologischen Deutung seismischer Grenzflächen in den Alpen: Geologische Rundschau, v. 50, p. 465–473, doi:10.1007/BF01786862. [5]

Reich, H., Schulze, G.A., and Förtsch, O., 1948, Das geophysikalische Ergebnis der Sprengung von Haslach im südlichen Schwarzwald: Geologische Rundschau, v. 36, p. 85–96, doi:10.1007/BF01791919. [2] [4]

Reich, H., Förtsch, O., and Schulze, G.A., 1951, Results of seismic observations in Germany on the Heligoland explosion of April 18, 1947: Journal of Geophysics Research, v. 56, p. 147–156, doi:10.1029/JZ056i002p00147. [2] [4]

Reichert, J.C., 1993, Ein geophysikalischer Beitrag zur Erkundung der Tiefenstruktur des Nordwestdeutschen Beckens längs des refraktionsseismischen Profils NORDDEUTSCHLAND 1975/76: Geologisches Jahrbuch, E, v. 50, 87 p. [2] [7]

Reid, I.D., 1988, Crustal structure beneath the southern Grand Banks: Seismic refraction results and their implications: Canadian Journal of Earth Science, v. 25, p. 760–772. [8]

Reid, I.D., 1994, Crustal structure of a non-volcanic rifted margin east of Newfoundland: Journal of Geophysical Research, v. 99, p. 15,161–15,180, doi:10.1029/94JB00935. [8]

Reinhardt, H.G., 1954, Steinbruchsprengungen zur Erforschung des tieferen Untergrundes. Freiberger Forschungsh., C15, p. 9–91. [2] [3] [4] [5] [10]

Research Group for Explosion Seismology, 1951, Explosion-seismic observations: Bulletin of the Earthquake Research Institute, Tokyo, v. 29, p. 97–105. [2] [5]

Research Group for Explosion Seismology, 1952, 1953, 1954, Explosion-seismic observations: Bulletin of the Earthquake Research Institute, Tokyo, v. 30, p. 279–292; 31, p. 281–289; 32, p. 79–86. [5]

Research Group for Explosion Seismology, 1958, Crustal Structure in Northern Kwanto District by Explosion Seismic Observations. Part I Description of Explosions and Observations: Bulletin of the Earthquake Research Institute, Tokyo, v. 36, p. 329–348. [5]

Research Group for Explosion Seismology, 1959, Observations of seismic Waves from the Second Hokoda Explosion: Bulletin of the Earthquake Research Institute, Tokyo, v. 37, p. 495–508. [5]

Research Group for Explosion Seismology, 1966a, Explosion seismological research in Japan, in Steinhart, J.S., and Smith, T.J., eds., The earth beneath the continents: American Geophysical Union, Washington, D.C., Geophysical Monograph 10, p. 334–348. [2] [6]

Research Group for Explosion Seismology, 1966b, Crustal structure in the western part of Japan derived from the observation of the first and second Kurayoshi and Hanabusa explosions, Zisin, v. 19, p. 107–124. [2] [6]

Research Group for Explosion Seismology, 1973, Crustal structure of Japan as derived from explosion seismic data, in Mueller, S., ed., The structure of the earth's crust, based on seismic data: Tectonophysics, v. 20, p. 129–135. [2] [6] [7]

Research Group for Explosion Seismology, 1975, Regionality of the upper mantle in northeastern Japan from experiments of large shots at sea: Marine Science Month, v. 7, p. 672–677 (in Japanese with English abstract). [7]

Research Group for Explosion Seismology, 1977, Regionality of the upper mantle around northeastern Japan as derived from explosion seismic observations and its seismological implications: Tectonophysics, v. 37, p. 117–130, doi:10.1016/0040-1951(77)90042-7. [7]

Research group for Explosion Seismology, 1992a, Explosion seismic observation on the Kii peninsula, southwestern Japan (Kawachinagano-Kiwa profile): Bulletin of the Earthquake Research Institute, University of Tokyo, v. 67, p. 37–56. [2] [8]

Research Group for Explosion Seismology, 1992b, Explosion seismic observervations in the Kitakami region: Bulletin of the Earthquake Research Institute, University of Tokyo, v. 67, p. 437–461. [8] [9]

Research Group for Explosion Seismology, 1993, Explosion seismic observervations in the Hokkaido region, Japan (Tsubetsu and Monbetsu profile): Bulletin of the Earthquake Research Institute, University of Tokyo, v. 68, p. 209–229 (in Japanese). [9]

Research Group for Explosion Seismology, 1994, Explosion seismic observervations in the center of Honshu, Japan (Agatsuma-Kanazawa profile):

Bulletin of the Earthquake Research Institute, University of Tokyo, v. 69 (in Japanese). [2] [9]

Research Group for Explosion Seismology, 1995, Explosion seismic observation in the central to western part of Honshu, Japan. Fujihashi-Kamigori profile: Bulletin of the Earthquake Research Institute, University of Tokyo, v. 70, p. 9 ff (in Japanese). [2] [8] [9]

Research Group for Explosion Seismology, 1996, Seismic refraction experiment in the northern Kanto district (Shimogo-Kiryu profile): Bulletin of the Earthquake Research Institute, University of Tokyo, v. 71, p. 73–101 (in Japanese). [2] [9]

Research Group for Explosion Seismology, 1997, Seismic refraction experiment in and around a source region of the 1995 Kobe earthquake (Keihoku-Seidan profile): Bulletin of the Earthquake Research Institute, University of Tokyo, v. 72, p. 69–117 (in Japanese). [2] [9]

Research Group for Explosion Seismology, 1999, Seismic refraction/wide-angle experiment across the northern Honshu arc: Bulletin of the Earthquake Research Institute, University of Tokyo, v. 74, p. 63–122 (in Japanese). [9]

Research Group for Explosion Seismology, 2002a, Seismic refraction/wide-angle experiment across the foreland area of the Hikada collision zone, Hokkaido, Japan (Ohtaki-Urahoro profile): Bulletin of the Earthquake Research Institute, University of Tokyo, v. 77, p. 139–172 (in Japanese). [9]

Research Group for Explosion Seismology, 2002b, Seismic refraction/wide-angle experiment across the foreland area of the Hikada collision zone, Hokkaido, Japan (Ohtaki-Biratori profile): Bulletin of the Earthquake Research Institute, University of Tokyo, v. 77, p. 173–198 (in Japanese). [9]

Research Group for Explosion Seismology, 2008, Dense seismic refraction/wide-angle reflection experiment in the eastern flank of the Ou backbone range, northern Honshu, Japan: Bulletin of the Earthquake Research Institute, University of Tokyo, v. 83, p. 43–75. [2] [9]

Research Group for Lithospheric Structure in Tunisia, 1992, The EGT'85 seismic experiment in Tunisia: a reconnaissance of the deep structures, in Freeman, R., and Mueller, St., eds., The European Geotraverse, Part 8: Tectonophysics, v. 207, p. 245–267. [2] [8]

Research Group on Underground Structure in the Kobe-Hanshin Area, 1997, Seismic refraction experiment in the Kobe-Hanshin area: Bulletin of the Earthquake Research Institute, University of Tokyo, v. 72, p. 1–18 (in Japanese). [2] [9]

Reston, T.J., Gaw, V., Pennell, J., Klaeschen, D., Stubenrauch, A., and Walker, I., 2004, Extreme crustal thinning in the south Porcupine Basin and the nature of the Porcupine Median High: implications for the formation of non-volcanic rifted margins: Journal of the Geological Society, London, v. 161, p. 1–16, doi:10.1144/0016-764903-036. [2] [9] [10]

Reston, T., Baxmann, M., Brunn, W., Chabert, A., Fekete, N., Gernigon, L., Hasenclever, J., Ingenfeld, R., Ivanova, A., Jones, S., Klein, G., Kriwanek, S., Liersch, P., Neiss, H., Ochsenhirt, W.T., Rogers, E., Steffen, K.P., Thierer, P., Thurow, U., Wagner, G., and Weiland, H., 2006, Changes in structure of the Earth's crust associated with progressive extension of the Porcupine Rift Basin, in Hebbeln, D., Pfannkuche, O., Reston, T., and Ratmeyer, V., eds., Northeast Atlantic 2004, Meteor-Berichte M61, no. 06-2. Leitstelle METEOR, Institute Meereskunde, University of Hamburg, p. 2-1-2-32. [2] [10]

Reston, T., Bialas, J.,DaCosta, J., Fekete, N., Hacker, S., Hagen, C., Hopper, J., Jacobsen, I., Jones, S., Klaeschen, D., Maczassek, K., Netzeband, G., Oestmann, F., Orth, S., Planert, L., Schwenk, A., Thurow, U., Vinding Fallesen, M., Winkler, V., Zamora, N., and Zillmer, M., 2009, Geophysical studies near the Ascension Transform: Evolution of ridge segmentation and crustal structure, in Steinfeld, R., Rhein, M., Brandt, P., Grevemeyer, I., Reston, T., Devey, C., and Lackschewitz, K., eds., Oceanography, geology and geophysics of the South Equatorial Atlantic. Meteor-Berichte M62, no. 09-1. Leitstelle METEOR, Institute Meereskunde, University of Hamburg, p. 4-1-4-8. [2] [10]

Reutter, K.-J., ed., 1999, Central Andean deformation: Journal of South American Earth Science, v. 12, p. 99–235. [2]

Reyners, M.E., Eberhart-Phillips, D., Stuart, G., and Nishimura, Y., 2006, Imaging subduction from the trench to 300 km depth beneath the central North Island, New Zealand, with V_p and V_p/V_s: Geophysical Journal International, v. 165, no. 2, p. 565–583, doi:10.1111/j.1365-246X.2006.02897.x. [10]

Reynolds, C.A., 1978, Boundary conditions for the numerical solution of wave propagation problems: Geophysics, v. 43, p. 1099–1110, doi:10.1190/1.1440881. [9]

Rhinegraben Research Group for Explosion Seismology, 1974, The 1972 seismic refraction experiment in the Rhinegraben—first results, *in* Illies J.H., and Fuchs, K., eds., Approaches to taphrogenesis: Stuttgart, Schweizerbart, p. 122–137. [7]

Richardson, K.R., Smallwood, J.R., White, R.S., Snyder, D., and Maguire, P.K.H., 1998, Crustal structure beneath the Faeroe islands and the Faeroe-Iceland Ridge: Tectonophysics, v. 300, p. 159–180, doi:10.1016/S0040-1951(98)00239-X. [9]

Rietbrock, A., Haberland, C., Bataille, K., Dahm, T., and Oncken, O., 2005, Studying the seismogenic coupling zone with a passive seismic array: Eos (Transactions, American Geophysical Union), v. 86, no. 32, p. 293, 297. [2] [9] [10]

Rihm, R., Makris, J., and Möller, L., 1991, Seismic surveys in the northern Red Sea: asymmetric crustal structure, *in* Makris, J., Mohr, P., and Rihm, R., eds., Red Sea: birth and early history of a new oceanic basin: Tectonophysics, v. 198, p. 279–295. [2] [7] [8]

Ritter, J.R.R., 2001, Students explore history of the Göttingen institute of geophysics: Eos (Transactions, American Geophysical Union), v. 82, p. 224, doi:10.1029/01EO00123. [3]

Ritter, J.R.R., and Achauer, U., 1994, Crustal tomography of the central Kenya rift, *in* Prodehl, C., Keller, G.R., and Khan, M.A., eds., Crustal and upper mantle structure of the Kenya rift: Tectonophysics, v. 236, p. 291–304. [9]

Ritter, J.R.R., and Kaspar, T., 1997, A tomography study of the Chyulu Hills, Kenya, *in* Fuchs, K., Altherr, R., Müller, B., and Prodehl, C., eds., Structure and dynamic processes in the lithosphere of the Afro-Arabian rift system: Tectonophysics, v. 278, p. 149–169. [9]

Ritter, J.R.R., Fuchs, K., Kaspar, T., Lange, F.E.I., Nyambok, I.O., and Stangl, R.L., 1995, Seismic images illustrate the deep roots of the Chyulu Hills volcanic area, Kenya: Transactions, American Geophysical Union, v. 76, no. 28, p. 273, 278. [9]

Ritter, J.R.R., Meyer, R., Keyser, M., Olejniczak, R., and Barth, A., eds., 2000, History of seismology in Goettingen: Unpubl. Mskr., Geophysical Institute, University of Karlsruhe [3]

Ritzmann, O., Jokat, W., Czuba, W., Guterch, A., Mjelde, R., and Nishimura, Y., 2004, A deep transect from Hovgard Ridge to northwestern Svalbard across the continental-ocean transition: a sheared margin study: Geophysical Journal International, v. 157, p. 683–702, doi:10.1111/j.1365-246X.2004.02204.x. [9]

Robertsson, J.O.A., Blanch, J.O., and Symes, D.D., 1994, Viscoelastic finite-difference modeling. Geophysics, v. 59, p. 1444–1456, doi:10.1190/1.1443701. [10]

Röwer, P., Prodehl, C., and Giese, P., 1977, The seismic refraction profile Lac Nègre–Genova–La Spezia, *in* Morelli, C., et al., eds., Crustal and upper mantle structure of the northern Appennines, the Ligurian Sea, and Corsica, derived from seismic and gravimetric data: Bollettino di Geofisica Teorica e Applicata, v. 75-76, p. 249–252. [2] [6]

Rohr, K.M.M., Milkereit, B., and Yorath, C.J., 1988, Asymmetric deep crustal structure across the Juan de Fuca Ridge: Geology, v. 16, p. 533–537, doi:10.1130/0091-7613(1988)016<0533:ADCSAT>2.3.CO;2. [2] [8]

Roller, J.C., 1964, Crustal structure in the vicinity of Las Vegas, Nevada, from seismic and gravity observations, *in* Geological Survey Research 1964: U.S. Geological Survey Professional Paper 475-D, p. D108–D111. [6]

Roller, J.C., 1965. Crustal structure in the eastern Colorado Plateau province from seismic-refraction measurements: Bulletin of the Seismological Society of America, v. 55, p. 107–119. [6] [8]

Roller, J.C., and Healy, J.H., 1963, Crustal structure between Lake Mead, Nevada, and Santa Monica Bay, California from seismic-refraction measurements: Journal of Geophysical Research, v. 68, p. 5837–5849. [6]

Roller, J.C., and Jackson,W.H., 1966, Seismic-wave propagation in the upper mantle: Lake Superior, Wisconsin, to Denver, Colorado, *in* Steinhart, J.S., and Smith, T.J., eds., The earth beneath the continents: Washington, D.C., American Geophysical Union, Geophysical Monograph 10, p. 270–275. [6]

Roller, J.C., Borcherdt, R.D., and Borcherdt, C., 1967, Unpubl. interpretation (see Prodehl et al., 1984). [6]

Romanowicz, B., and Dziewonski, A., eds., 2007, Seismology and structure of the earth: Amsterdam, Elsevier, Treatise on Geophysics, vol. 1, 858 p. [2]

Rondenay, S., Bostock, M.G., Hearn, T.M., White, D.J., and Ellis, R.M, 2000, Lithospheric assembly and modification of the SE Canadian Shield: Abitibi-Grenville teleseismic experiment: Journal of Geophysical Research, v. 105, p. 13,735–13,754, doi:10.1029/2000JB900022. [9]

Rosaire, E.E., and Lester, O.C., Jr., 1932, Seismological discovery and partial detail of Vermillion Bay salt dome, Louisiana: American Association of Petroleum Geologists Bulletin, v. 16, p. 1221–1229. [2] [3]

Rosendahl, B.R., and Groschel-Becker, H., 2000, Architecture of the continental margin in the Gulf of Guinea as revealed by reprocessed deep-imaging seismic data, *in* Mohriak, W., and Talwani, M., eds., Atlantic Rifts and Continental Margins: Washington, D.C., American Geophysical Union, Geophysical Monograph 115, p. 85–103. [8]

Rosendahl, B.R., Raitt, R.W., Dorman, L.M., Bibee, L.D., Hussong, D.M., and Sutton, G.H., 1976, Evolution of oceanic crust, 1, a physical model of the East Pacific Rise rest derived from seismic refraction data: Journal of Geophysical Research, v. 81, p. 5294–5304, doi:10.1029/JB081i029p05294. [7]

Rosendahl, B.R., Rogers, J.J.W., and Rach, N.M., eds., 1989, African rifting. Journal of African Earth Science, v. 8 (special issue), p. 135–629. [9]

Rosendahl, B.R., Groschel-Becker, H., Meyers, J., and Kaczmarick, K., 1991, Deep seismic reflection studies of a passive margin, southeastern Gulf of Guinea: Geology, 19, p. 291–295, doi:10.1130/0091-7613(1991)019<0291:DSRSOA>2.3.CO;2. [2] [8]

Roslov, Yu.V., Sakoulina, T.S., and Pavlenkova, N.I., 2009, Deep seismic investigations in the Barents and Kara Seas: Tectonophysics, v. 472, p. 301–308, doi:10.1016/j.tecto.2008.05.025. [2] [9]

Ross, A.R., Brown, L.D., Pananont, P., Nelson, K.D., Klemperer, S., Haines, S., Wenjin, Z., and Jingru, G., 2004, Deep reflection surveying in central Tibet: lower-crustal layering and crustal flow: Geophysical Journal International, v. 156, p. 15–128, doi:10.1111/j.1365-246X.2004.02119.x. [9]

Ross, G.M., ed., 1999, The LITHOPROBE Alberta Basement Transect: Bulletin of Canadian Petroleum Geology, v. 47, p. 331–594. [9]

Ross, G.M., ed., 2000, The Lithoprobe Alberta Basement Transect: Canadian Journal of Earth Science, v. 37, p. 1447–1650, doi:10.1139/cjes-37-11-1447. [2] [9]

Ross, G.M., ed., 2002, The Lithoprobe Alberta Basement Transect: Canadian Journal of Earth Science, v. 39, p. 287–437, doi:10.1139/e02-011. [2] [9]

Ross, G.M., Eaton, D.W., Boerner, D.E., and Clowes, R.M., 1997, Geologists probe buried Craton in western Canada: Eos (Transactions, American Geophysical Union), v. 78, p. 493–494, 497, doi:10.1029/97EO00302 [9]

Rothé, J.P., and Peterschmitt, E., 1950, Etude séismique des explosions d'Haslach: Ann. Institute Physique du Globe, University of Strasbourg, Nouvelle Ser. 5, part 3: Géophysique. [4]

Rothé, J.P., and Sauer, K., eds., 1967, The Rhinegraben progress report 1967, Abh. Geol. Landesamt Baden–Württemberg 6: 146 p. [7]

Rothé, J.P., Lacoste, J., Bois, C., Dammann, Y., and Hée, A., 1924, Etude de la propagation de l'ébranlement des explosions de la Courtine, Comparaison avec l'explosion d'Oppau: Bureau Centre Séismologique Internationale Pub., série A, travaux sci., fasc. 1, p. 82–98. [3]

Rothé, J.P., Peterschmitt, E., and Stahl, P., 1948, Les ondes séismique des explosions d'Haslach: Comptes rendus de l'Académie de sciences, v. 227, p. 354–356. [4]

Rotstein, Y., Yuval, Z., and Trachtman, P., 1987, Deep seismic reflection studies in Israel—an update: Geophysical Journal of the Royal Astronomical Society, 89, p. 389–394. [2] [8]

Roure, F., Bergerat, F., Damotte, B., Mugnier, J.L., and Polino, R., eds., 1996, The ECORS-CROP Alpine seismic traverse: Bulletin de la Société Géologique de France, v. 170, p. 1–113. [8]

RRISP Working Group, 1980, Reykjanes Ridge Iceland seismic experiment (RRISP), *in* Jacoby, W., Björnsson, A., and Möller, D., eds., Iceland: Evolution, active tectonic, and structure: Journal of Geophysics, v. 47, p. 228–238. [2] [7]

Ruegg, J.C., 1975, Main results about the crustal and upper mantle structure of the Djibouti region (T.F.A.I.), *in* Pilger, A., and Rösler, A., eds., Afar Depression of Ethiopia: Stuttgart, Schweizerbart, p. 120–134. [2] [7]

Ruffman, A., and van Hinte, J.E., 1973, Orphan Knoll, a 'chip' off the North American 'Plate', *in* Hood, P.J., ed., Earth Science Symposium on Offshore Eastern Canada: Geological Survey of Canada Paper 72-23, p. 407–429. [7]

Rumpel, H-M., 2003, Crustal structure and shear zones in the Southern Rocky Mountains from refraction/wide-angle reflection data [Ph.D. thesis]: University of Karlsruhe, 150 p. [9]

Rumpel, H-M, Prodehl, C, Snelson, C M, and Keller, G R, 2005, Results of the CD-ROM project seismic refraction/wide-angle reflection experiment: the upper and middle crust, *in* Karlstrom, K.E., and Keller, G.R., eds., The Rocky Mountain Region: An evolving lithosphere—tectonics, geochemistry, and geophysics: Washington, D.C., American Geophysical Union Geophysical Monograph 154, p. 257–269. [9]

Rumpfhuber, E., Keller, G.R., Sandvol, E., Velasco, A.A., and Wison, D.C., 2009, Rocky Mountain evolution: tying continental dynamics of the Rocky Mountains and Deep Probe seismic experiments with receiver functions: Journal of Geophysical Research, v. 114, B08301, doi:10.1029/2008JB005726. [9]

Rumpfhuber, E.M., and Keller, G.R., 2009, An integrated analysis of controlled- and passive- source seismic data across an Archean-Proterozoic suture zone in the Rocky Mountains, USA: Journal of Geophysical Research, v. 114, B08305, doi:10.1029/2008JB005886. [9]

Ruppert, S., Fliedner, M.M., and Zandt, G., 1998, Thin crust and active upper mantle beneath the southern Sierra Nevada in the western United States, *in* Klemperer, S.L., and Mooney, W.D., eds., Deep seismic probing of the continents, I: general results and new methods: Tectonophysics, v. 286, p. 237–252. [9]

Ryaboi, V.Z., 1966, Kinematic and dynamic characteristics of deep waves associated with boundaries in the crust and upper mantle (from a DSS study on the profile Kopet Dagl–Aral Sea): Izv. Erath Physics 3, p. 4–82. (DSS in 1962–1963). [7]

Ryall, A., and Stuart, D.J., 1963, Traveltimes and amplitudes from nuclear explosions: Nevada Test Site to Ordway, Colorado: Journal of Geophysical Research, v. 68, p. 5821–5835. [6]

Ryberg, T., and Fuis, G.S., 1998, The San Gabriel Mountains bright reflective zone: possible evidence of young mid-crustal thrust faulting in southern California, *in* Klemperer, S.L., and Mooney, W.D., eds., Deep seismic probing of the continents, I: general results and new methods: Tectonophysics, v. 286, p. 31–46. [9]

Ryberg, T., Wenzel, F., Egorkin, A.V., and Solodilov, L., 1998, Properties of the mantle transition zone in northern Eurasia: Journal of Geophysical Research, v. 103, p. 811–822, doi:10.1029/97JB02837. [2] [8]

Ryberg, T., Fuis, G.S., Bauer, K., Hole, J.A., Bleibinhaus, F., and Rymer, M., 2005, Upper crustal reflectivity of the central California Coast Range near the San Andreas Observatory at Depth (SAFOD) [poster]: Eos (Transactions, American Geophysical Union) December 2005. [10]

Rykunov, L.N., and Sedov, V.V., 1967, An ocean-bottom seismograph: Isvestija: Physics of Solid Earth, v. 8, p. 537–541. [7]

Rynn, J.M., and Reid, I.D., 1983, Crustal structure of the western Arafura Sea from ocean bottom seismograph data: Journal of the Geological Society of Australia, v. 30, p. 59–74. [2] [7]

Sachpazi, M., Hirn, A., Clement, Ch., Laigle, M., and Roussos, N., 2003, Evolution to fastest opening of the Gulf of Corinth continental rift from deep seismic profiles: Earth and Planetary Science Letters, v. 216, p. 243–257, doi:10.1016/S0012-821X(03)00503-X. [2] [9]

Sadowiak, P., Wever, T., and Meissner, R., 1991, Deep seismic reflectivity patterns in specific tectonic units of western and central Europe: Geophysical Journal International, v. 105, p. 45–54, doi:10.1111/j.1365-246X.1991.tb03443.x. [2] [8]

Sain, K., Bruguier, N., Murty, A.S.N., and Reddy, P.R., 2000, Shallow velocity structure along the Hirapur-Mandla profile using traveltime inversion of wide-angle seismic data, and its tectonic implications: Geophysical Journal International, v. 142, p. 505–515, doi:10.1046/j.1365-246x.2000.00176.x. [8]

Sakoulina, T.S., Telegin, A.N., Tikhonova, I.M., Verba, M.L., Matveev, Y.I., Vinnick, A.A., Kopylova, A.V., and Dvornikov, L.G., 2000, The results of deep seismic investigations on geotraverse in the Barents Sea from Kola peninsula to Franz-Joseph Land: Tectonophysics, v. 329, p. 319–331, doi:10.1016/S0040-1951(00)00201-8. [2] [9]

Salisbury, M.H. and Christensen, N.I., 1978, The seismic velocity structure of a traverse through the Bay of Islands ophiolite complex, Newfoundland, an exposure of ancient oceanic crust and upper mantle: Journal of Geophysical Research, v. 83, p. 805–817, doi:10.1029/JB083iB02p00805. [7] [8]

Sallares, V., Danobeitia, J.J., Flueh, E.R., and Leandro, G., 1999, Wide angle seismic velocity structure across the Middle American Trench in northern Costa Rica: Journal of Geodynamics, v. 27, p. 327–344, doi:10.1016/S0264-3707(98)00007-6. [2] [9]

Sallares, V., Charvis, P., Flueh, E.R., and Bialas, J., 2003, Seismic structure of Cocos and Malpelo ridges and implications for hotspot-ridge interaction: Journal of Geophysical Research, v. 108, 2564, doi:10.1029/2003JB002431. [2] [9] [10]

Sallares, V., Charvis, P., Flueh, E.R., Bialas, J., and the SALIERI Scientific Party, 2005, Seismic structure of the Carnegie ridge and the nature of the Galápagos hotspot: Geophysical Journal International, v. 161, p. 763–788, doi:10.1111/j.1365-246X.2005.02592.x. [2] [9] [10]

Sandmeier, K.-J., and Wenzel,F., 1986, Synthetic seismograms for a complex model: Geophysics Research Letters, v. 13, p. 22–25, doi:10.1029/GL013i001p00022. [8] [9]

Sandrin, A., and Thybo, H., 2008, Deep seismic investigations of crustal extensional structures in the Danish Basin along the ESTRID-2 profile: Geophysical Journal International, v. 173, p. 623–641, doi:10.1111/j.1365-246X.2008.03759.x. [10]

Sapin, M., and Hirn, A., 1974, Results of explosion seismology in the southern Rhône valley: Annales de Géophysique, v. 30, p. 181–202. [2] [7]

Sapin, M., and Prodehl, C., 1973, Long-profiles in western Europe—I. Crustal structure between the Bretagne and the Central Massif of France: Annales de Géophysique, v. 29, p. 127–145. [2] [7]

Sapin, M., Wang, X.-J., Hirn, A., and Xu, Z.X., 1985, A deep seismic sounding in the crust of the Lhasa block: Annales de Géophysique, v. 3, p. 637–646. [8] [9]

Sarkar, D., Chandrakala, K., Padmavathi Devi, P., Sridhar, A.R., Sain, K., and Reddy, P.R., 2001, Crustal velocity structure of western Dharwar Craton, South India: Journal of Geodynamics, v. 31, p. 227–241, doi:10.1016/S0264-3707(00)00021-1. [7]

Sarnthein, M., Seibold, E., Grobe, H., and Schumacher, S., 2008, Data Compilation of the Research Vessel METEOR (1964). WDC-MARE Reports 0006 (2008), ISSN 1862-4022. Alfred Wegener Institute for Polar and Marine Research, Bremerhaven, Germany [2] [6] [7] [8]

Sasatani, T., Yoshii, T., Ikami, A., Tanada, T., Nishiki, T., and Kato, S. 1990, Upper crustal structure under the central part of Japan: Miyota-Shikishima profile: Bulletin of the Earthquake Research Institute, University of Tokyo, v. 65, p. 33–48. [2] [8]

Satarugsa, P., and Johnson, R.A., 1998, Crustal velocity structure beneath the flank of the Ruby Mountains metamorphic core complex: results from normal-incidence to wide-angle seismic data: Tectonophysics, v. 295, p. 369–395, doi:10.1016/S0040-1951(98)00015-8. [9]

Satarugsa, P., and Johnson, R.A., 2000, Constraints on crustal composition beneath a metamorphic core complex: results from 3-component wide-angle seismic data along the eastern flank of the Ruby Mountains, Nevada: Tectonophysics, v. 329, p. 223–250, doi:10.1016/S0040-1951(00)00197-9. [9]

Sato, H., Hirata, H., Ito, T., Tsumura, N., and Ikawa, T., 1998, Seismic reflection profiling across the seismogenic fault of the 1995 Kobe earthquake, southwestern Japan, *in* Klemperer, S.L., and Mooney, W.D., eds., Deep seismic probing of the continents, I: general results and new methods: Tectonophysics, v. 286, p. 19–30. [2] [9]

Sato, H., Hirata, N., Iwasaki, T., Matsubara, M., and Ikawa, T., 2002, Deep seismic profiling across Ou backbone range Northern Honshu Island, Japan: Tectonophysics, v. 355, p. 41–52, doi:10.1016/S0040-1951(02)00133-6. [2] [9]

Sato, H., Iwasaki, T., Kawasaki, S., Ikeda, Y., Matsuta, N., Takeda, T., Hirata, N., and Kawanaka, T., 2004, Formation and shortening deformation of a back-arc rift basin revealed by deep seismic profiling, central Japan: Tectonophysics, v. 388, p. 47–58, doi:10.1016/j.tecto.2004.07.004. [2] [10]

Sato, H., Hirata, N., Koketsu, K., Okaya, D., Abe, S., Kobayashi, R., Matsubara, M., Iwasaki, T., Ito, T., Ikawa, T., Kawanaka, T., and Harder, S., 2005, Earthquake Source Fault beneath Tokyo: Science, v. 309, p. 462–464, doi:10.1126/science.1110489. [2] [10]

Sato, H., Ito, K., Abe, S., Kato, N., Iwasaki, T., Hirata, N., Ikawa, T. and Kawanaka, T., 2009, Deep seismic reflection profiling across active reverse faults in the Kinki Triangle central Japan: Tectonophysics, v. 472, p. 86–94, doi:10.1016/j.tecto.2008.06.014. [2] [10]

Sato, T., Sato, T., Shinohara, M., Hino, R., Nishino, M., and Kanazawa, T., 2006, P-wave velocity structure of the margin of the southeastern Tsushima Basin in the Japan Sea using ocean bottom seismometers and airguns: Tectonophysics, v. 412, p. 159–171, doi:10.1016/j.tecto.2005.09.001. [2] [9] [10]

Scandone, P., Mazzotti, A., Fradelizio, G.I., Patacca, E., Stucchi, E., Tozzi, M., and Zanzi, L., 2003, Line CROP 04: southern Apennines, *in* Scrocca, D., Doglioni, C., Innocenti, F., Manetti, P., Mazzotti, A., Bertelli, L., Burbi, L., and D'Offizi, S., eds., CROP Atlas: seismic reflection profiles of the Italian crust: Memorie Descrittive della Carta Geologica d'Italia, Roma, v. 62, p. 155–166 [9]

Scarascia, S., and Cassinis, R., 1997, Crustal structures in the central-eastern Alpine sector: a revision of the available DSS data: Tectonophysics, v. 271, p. 157–188, doi:10.1016/S0040-1951(96)00206-5. [8]

Scarascia, S., Lozej, A., and Cassinis, R., 1994, Crustal structures of the Ligurian, Tyrrhenian and Inian seas and adjaacent onshore areas interpreted from wide-angle seismic profiles: Bollettino di Geofisica Teorica e Applicata, v. 36, p. 5–19. [7]

Scheidegger, A.E., and Willmore, P.L., 1957, The use of a least-squares method for the interpretation of data from seismic surveys: Geophysics, v. 22, p. 9–22, doi:10.1190/1.1438348. [6]

Schenk, V., 1990, The exposed crustal cross section of southern Calabria, Italy: structure and evolution of a segment of Hercynian crust, *in* Salisbury, M.H., and Fountain, D.M., eds., Exposed cross section of the continental crust: Dordrecht, Kluwer, p. 21–42. [9]

Scherwath, M., Stern, T., Davey, F., Okaya, D., Holbrook, W.S., Davies, R., and Kleffmann, S., 2003, Lithospheric structure across oblique continental collision in New Zealand from wide-angle P wave modeling: Journal of Geophysical Research, v. 108 (B12), 2566, doi: 10.1029/2002JB002286. [9]

Scherwath, M., Flueh, E., Grevemeyer, I., Tilmann, F., Contreras-Reyes, E., and Weinrebe, W., 2006, Investigating subduction zone processes in Chile: Eos (Transactions, American Geophysical Union), v. 87, no. 27, p. 265, 270. [10]

Scheuber, E., and Giese, P., 1999, Architecture of the Central Andes—a compilation of geoscientific data along a transect at 21°S, *in* Reutter, K.-J., ed., Central Andean deformation: Journal of South American Earth Science, v. 12, p. 103–107, doi:10.1029/2002JB002286. [9]

Schmidt, A., 1888, Ein Beitrag zur Dynamik der Erdbeben: Jahreshefte des Vereins für Vaterländische Naturkunde in Württemberg, Stuttgart, v. 44, p. 248–270. [3]

Schmidt-Aursch, M.C., and Jokat, W., 2005, The crustal structure of central East Greenland—I: from the Caledonian orogen to the Tertiary igneous province: Geophysical Journal International, v. 160, p. 736–752, doi:10.1111/j.1365-246X.2005.02514.x. [2] [9] [10]

Schmidt-Aursch, M.C., Jokat, W., and Miller, H., 2006, Deutscher Geräte-Pool für amphibische Seismologie-DEPAS-Mariner Teil: German Geophysical Society Annual Meeting 6–9 March 2006, A18 SO, Abstracts, p. 87. [2] [10]

Schmincke, H.-U., and Rihm, R., eds., 1994, Ozeanvulkan 1993, Meteor–Berichte M24, no. 94-2. Leitstelle METEOR, Institute Meereskunde, University of Hamburg, 88 p. (Seismic refraction, p. 14, 56–62). [9]

Schmitz, M., 1993, Kollisionsstrukturen in den Zentralen Anden: Berliner Geowissenschaftliche Abhandlungen, B20, 127 p. [8]

Schmitz, M., Lessel, K., Giese, P., Wigger, P., Araneda, M., Bribach, J., Graeber, F., Grunewald, S., Haberland, C., Lüth, S., Röwer, P., Ryberg, T., and Schulze, A., 1999, The crustal structure beneath the Central Andean forearc and magmatic arc as derived from seismic studies—the PISCO 94 experiment in northern Chile: Journal of South American Earth Science, v. 12, p. 237–260, doi:10.1016/S0895-9811(99)00017-6. [2] [9]

Schmitz, M., Chalbaud, D., Castillo, J., and Izarra, C., 2002, The crustal structure of the Guayana Shield, Venezuela, from seismic refraction and gravity data: Tectonophysics, v. 345, p. 103–118, doi:10.1016/S0040-1951(01)00208-6. [2] [9] [10]

Schmitz, M., Martins, A., Izarra, C., Jacome, M.I., Sanchez, J., and Rocabado, V., 2005, The major features of the crustal structure in northeastern Venezuela from deep wide-angle seismic observations and gravity modeling: Tectonophysics, v. 399, p. 109–124, doi:10.1016/j.tecto.2004.12.018. [2] [9] [10]

Schmitz, M., Bezada, M., Avila, J., Vieira, E., Yanez, M., Levander, A., Zelt, C.A., Magnani, M.B., Jacome, M.I., and the BOLIVAR Active Seismic Working Group, 2008, Crustal thickness variations in Venezuela from deep seismic observations: Tectonophysics, v. 459, p. 14–26, doi:10.1016/j.tecto.2007.11.072. [10]

Schmoll, J., Bittner, R., Dürbaum, H.-J., Heinrichs, T., Meissner, R., Reichert, C., Rühl, T., and Wiederhold, H., 1989, Oberpfalz deep seismic reflection survey and velocity studies, *in* Emmermann, R., and Wohlenberg, J., eds., The German continental deep drilling program (KTB)—site-selection studies in the Oberpfalz and Schwarzwald: Berlin-Heidelberg, Springer, p. 99–149. [8]

Schneider, W.A., Ranzinger, K.A., Balch, A.H., and Kruse, C., 1992, A dynamic programming approach to first arrival traveltime computation in media with arbitrarily distributed velocities: Geophysics, v. 57, p. 39–50, doi:10.1190/1.1443187. [9]

Schnuerle, P., Liu, C.-S., Lallemand, S.E., and Reed, D.L., 1998, Structural insight into the south Ryukyu margin: effects of the subducting Gagua

Ridge, *in* Klemperer, S.L., and Mooney, W.D., eds., Deep seismic probing of the continents, II: a global survey: Tectonophysics, v. 288, p. 237–250. [2] [9]

Schulte-Pelkum, V., Gaspar, M., Sheehan, A., Pandey, M.R., Sapkota, S., Bilham, R., and Wu,. F., 2005, Imaging the Indian subcontinent beneath the Himalaya: Nature, v. 435, p. 1222–1225, doi:10.1038/nature03678. [9]

Schultz, A.P., and Crosson, R.S., 1996, Seismic velocity structure across the central Washington Cascade Range from refraction interpretation with earthquake sources: Journal of Geophysical Research, v. 101, p. 27,899–27,915, doi:10.1029/96JB02289. [9]

Schulz, G., 1957, Reflexionen aus dem kristallinen Untergrund des Pfälzer Berglandes: Zeitschrift für Geophysik, v. 23, p. 225–235. [5]

Schulze, A., and Weber, M., 2006, Deutscher Pool für amphibische Seismologie—DEPAS-Landteil: German Geophysical Society Annual Meeting 6–9 March 2006, A17 SO, Abstracts, p.86. [2] [10]

Schulze, G.A., 1947, Seismische Ergebnisse der Helgoland-Sprengung: Die Naturwissenschaften, v. 34, p. 288, doi:10.1007/BF00589869. [2] [4]

Schulze, G.A., 1974, Anfänge der Krustenseismik, *in* Birett, H., Helbig, K., Kertz, W., and Schmucker, U., eds., Zur Geschichte der Geophysik: Berlin-Heidelberg-New York, Springer, p. 89–98. [2] [3] [4] [5]

Schulze, G.A., and Förtsch, O., 1950, Die seismischen Beobachtungen bei der Sprengung auf Helgoland am 18.4.47 zur Erforschung des tieferen Untergrundes: Geologisches Jahrbuch, v. 64, p. 204–242. [4]

Schumann, G., and Oesberg, R., 1966, Beiträge zur Erkundung des tieferen Untergrundes des Thüringer Beckens durch Auswertung tiefer Reflexionen: Geophysik und Geologie—Beiträge zur Synthese zweier Wissenschaften, v. 9, Leipzig. [6]

Schweitzer, J., and Lee, W.H.K., 2003, Old seismic bulletins to 1920: a collective heritage from early seismologists, *in* Lee, W., Kanamori, H., Jennings, P.C., and Kisslinger, C., eds., International Handbook of Earthquake and Engineering Seismology, Part B: Amsterdam, Academic Press, p. 1718–1723. [3]

Scott, C., Shillington, D., Minshull, T., Edwards, R., and White, N., 2006, Structure of the eastern Black Sea basin inferred from wide-angle seismic data: Europ. Geophys. Union, Ann. Mtg 2006, Geophys. Res. Abstracts, 8: 07319. [2] [10]

Scrocca, D., Doglioni, C., Innocenti, F., Manetti, P., Mazzotti, A., Bertelli, L., Burbi, L., and D'Offizi, S., eds., 2003, CROP Atlas: seismic reflection profiles of the Italian crust: Memorie Descrittive della Carta Geologica d'Italia, Roma, v. 62, 194 p. [2] [8] [9]

Scrutton, R.A., 1972, The crustal structure of Rockall Plateau Microcontinent: Geophysical Journal of the Royal Astronomical Society, v. 27, p. 259–275. [2] [6]

Seber, D., Sandvol, E., Sandvol, C., Brindisi, C., and Barazangi, M., 2001, Crustal model for the Middle East and North Africa region: implications for the isostatic compensation mechanism: Geophysical Journal International, v. 147, p. 630–638, doi:10.1046/j.0956-540x.2001.01572.x. [10]

Sellevoll, M.A., 1973, Mohorovičić discontinuity beneath Fennoscandia and adjacent parts of the Norwegian Sea and the North Sea: Tectonophysics, v. 20, p. 359–366, doi:10.1016/0040-1951(73)90123-6. [2] [6]

Sellevoll, M.A., ed., 1982, Seismic crustal studies on Spitsbergen 1978: University of Bergen, Seismological Observatory, Bergen, 62 p. [2] [7]

Sellevoll, M.A., and Warrick, R.E., 1971, A refraction study of crustal structure in southern Norway: Bulletin of the Seismological Society of America, v. 61, p. 457–471. [6]

Sellevoll, M.A., Duda, S.J., Guterch, A., Pajchel, J., Perchuc, E., and Thyssen, F., 1991, Crustal stucture in the Svalbard region from seismic measurements: Tectonophysics, v. 189, p. 55–71, doi:10.1016/0040-1951(91)90487-D. [7] [8]

Shackleford, P.R.J., and Sutton, D.J., 1979, Crustal structure in South Australia using quarry blasts, *in* Denham, D., ed., Crust and upper mantle of Southeast Australia: Bureau of Mineral Resources, Australia, report 1979/2, abstr. [7]

Shannon, P.M., Jacob, A.W.B., Makris, J., O'Reilly, B.M., Hauser, F., and Vogt, U., 1994, Basin evolution in the Rockall region, North Atlantic: First Break, v. 12, p. 515–522. [2] [8] [10]

Shannon, P.M., Jacob, A.W.B., O'Reilly, B.M., Hauser, F., Readman, P.W., and Makris, J., 1999, Structural setting, geological development and basin modelling in the Rockall Trough, *in* Fleet, A.J., and Boldy, S.A.R., eds., Petroleum Geology of Northwest Europe, Proceedings, 5th Conference: Geological Society London, Petroleum Geology, v. 86, p. 421–431. [8] [10]

Sharpless, S.W., and Walter, A.W., 1988, Data report for the 1986 San Luis Obispo, California, seismic-refraction survey: Menlo Park, California, U.S. Geological Survey Open-File Report 88-35, 48 p. [2] [8]

Shaw, P.R., and Orcutt, J.A., 1985, Waveform inversion of seismic refraction data and application to young Pacific crust: Geophysical Journal of the Royal Astronomical Society, v. 82, p. 375–414. [8]

Shearer, P., and Orcutt, J.A., 1985, Anisotropy in the oceanic lithosphere—theory and observations from the Ngendei seismic refraction experiment in the southwest Pacific: Geophysical Journal of the Royal Astronomical Society, v. 80, p. 493–526. [2] [8]

Sheehan, A., Schulte-Pelkum, V., Boyd, O., and Wilson, C., 2005, Passive source seismology of the Rocky Mountain region, in Karlstrom, K.E., and Keller, G.R., eds., The Rocky Mountain Region: An evolving lithosphere—tectonics, geochemistry, and geophysics: Washington, D.C., American Geophysical Union Geophysical Monograph 154, p. 309–315. [9]

Sheridan, R.E., Musser, D.L., Glover, M.I., Talwani, M., Ewing, J., Holbrook, S., Purdy, G.M., Hawman, R., and Smithson, S., 1991, EDGE deep seismic reflection study of U.S. mid-Atlantic continental margin: Eos (Transactions, American Geophysical Union), v. 72, p. 273. [8]

Shimamura, H., Asada, T., Suyehiro, K., Yamada, T., and Inatani, H., 1983, Longshot experiments to study velocity anisotropy in the oceanic lithosphere of the northwestern Pacific: Physics of the Earth and Planetary Interiors, v. 31, p. 348–362, doi:10.1016/0031-9201(83)90094-8. [2] [7]

Shinohara, M., Suyehiro, K., Matsuda, S., and Ozawa, K., 1993, Digital recording ocean bottom seismometer using portable digital audio tape recorder: Journal of the Japanese Society Marine Surveys Technol., v. 5, p. 21–31 (in Japanese with English abstract). [9]

Shor, G.G., 1955, Deep reflections from southern California blasts: Transactions, American Geophysical Union, v. 36, p. 133–138. [6]

Shor, G.G., 1963, Refraction and reflection techniques and procedure, in Hill, M.N., ed., The Sea vol. 3, The earth beneath the Sea: New York-London, Interscience Publ., p. 20–38. [4] [5]

Shor, G.G., and Pollard, D.D., 1963, Seismic investigations of Seychelles and Saya de Malha banks, northwest Indian Ocean: Science, v. 142, p. 48–49, doi:10.1126/science.142.3588.48. [2] [6]

Shor, G.G., and Raitt, R.W., 1958, Seismic studies in the southern California continental borderland, in Seccion IX, Geofisica Aplicada, Congreso Geologia Internacional XX Sesion, Mexico, D.F., p. 243–259. [6]

Shor, G.G., and Raitt, R.W., 1969, Explosion seismic refraction studies of the crust and upper mantle in the Pacific and Indian Oceans, in Hart, P.J., ed., The earth's crust and upper mantle: American Geophysical Union, Geophysical Monograph 13, p. 225–230. [2] [6]

Shor, G.G., Dehlinger, P., Kirk, H.K., and French, W.S., 1968, Seismic refraction studies off Oregon and northern California: Journal of Geophysical Research, v. 73, p. 2175–2194, doi:10.1029/JB073i006p02175. [2] [6] [8]

Shor, G.G., Menard, H.W., and Raitt, R.W., 1970, Structure of the Pacific basin, in Maxwell, A.E., ed., The Sea, new concepts of ocean floor evolution: New York, Wiley-Interscience, 4, II, p. 3–27. [2] [6]

Shor, G.G., Kirk, H., and Menard, H., 1971, Crustal structure of the Melanesian area: Journal of Geophysical Research, v. 76 (11), p. 2562–2586, doi:10.1029/JB076i011p02562 [6]

Sick, C., Yoon, M.-K., Rauch, K., Buske, S., Lueth, S., Araneda, N., Bataille, K., Chong, G., Giese, P., Krawczyk, C.M., Mechie, J., Meyer, H., Oncken, O., Reichert, C., Schmitz, M., Shapiro, S., Stiller, M., and Wigger, P., 2006, Seismic images of accretive and erosive subduction zones from the Chilean margin, in Oncken, O., Chong, G., Franz, G., Giese, P., Goetze, H.-J., Ramos, V.A., Strecker, M.R., and Wigger, P., eds., The Andes—active subduction orogeny: Frontiers in Earth Sciences Series: Berlin-Heidelberg-New York, Springer, p. 147–169. [2] [9]

Simpson, F.L., Haak, V., Khan, M.A., Sakkas, V., and Meju, M., 1997, The KRISP-94 magnetotelluric survey of early 1995: first results, in Fuchs, K., Altherr, R., Müller, B., and Prodehl, C., eds., Structure and dynamic processes in the lithosphere of the Afro-Arabian rift system: Tectonophysics, v. 278, p. 261–271. [9]

Singh, S.C., Hague, P.J., and McCaughey, M., 1998a, Study of the crystalline crust from a two-ship normal-incidence and wide-angle experiment, in Klemperer, S.L., and Mooney, W.D., eds., Deep seismic probing of the continents, I: general results and new methods: Tectonophysics, v. 286, p. 79–91. [2] [9]

Singh, S.C., Kent, G.M., Collier, J.S., Harding, A.J., and Orcutt, J.A., 1998b, Melt to mush variations in crustal magma properties along the ridge crest at the southern East Pacific Rise: Nature, v. 394, p. 874–878, doi:10.1038/29740.[9]

Singh, S.C., Harding, A.J., Kent, G.M., Sinha, M.C., Combier, V., Bazin, S., Tong, C.H., Pye, J.W., Barton, P.J., Hobbs, R.W., White, R.S., and Orcutt, J.A., 2006a, Seismic reflection images of the Moho underlying melt sills at the East Pacific Rise: Nature, v. 442, p. 287–290, doi:10.1038/nature04939. [2] [9]

Singh, S.C., Crawford, W.C., Carton, H., Seher, T., Combier, V., Cannat, M., Canales, J.P., Düsünür, D., Escartin, J., and Miranda, M., 2006b, Discovery of a magma chamber and faults beneath a Mid-Atlantic Ridge hydrothermal field: Nature, v. 442, p. 1029–1032, doi:10.1038/nature05105. [2] [10]

Sinha, M.C., and Louden, K.E., 1983, The Oceanographer fracture zone, I. crustal structure from seismic refraction studies: Geophysical Journal of the Royal Astronomical Society, v. 75, p. 713–736. [9]

Sinha, M.C., Owen, T.R.E., and Masoon, M., 1981a, An ocean–bottom hydrophone recorder for seismic refraction experiments: Marine Geophysical Research, v. 5, p. 173–187. [7]

Sinha, M.C., Louden, K.E., and Parsons, B., 1981b, The crustal structure of the Madagaskar Ridge: Geophysical Journal of the Royal Astronomical Society, v. 66, p. 351–377. [7]

Sinha, M.C., Navin, D.A., MacGregor, L.M., Constable, S., Peirce, C., White, A., Heinson, G., and Inglis, M.A., 1999, Evidence for accumulated melt beneath the slow-spreading Mid-Atlantic Ridge, in Cann, J.R., Elderfield, H., and Laughton, A., eds., Mid-ocean ridges—dynamics of processes associated with creation of new oceanic crust: Cambridge, U.K., Cambridge Universtity Press, p. 17–37. [2] [9]

Sinno, Y.A., Keller, G.R., and Sbar, M.L., 1981, A crustal seismic refraction study in west-central Arizona: Journal of Geophysical Research, v. 86, p. 5032–5038, doi:10.1029/JB086iB06p05023. [2] [7]

Sinno, Y.A., Daggett, P.H., Keller, G.R., Morgan, P., and Harder, S.H., 1986, Crustal structure of the southern Rio Grande rift determined from seismic refraction profiling: Journal of Geophysical Research, v. 91, p. 6143–6156, doi:10.1029/JB091iB06p06143 [2] [7] [8]

Skoko, D., and Mokrović, J., 1982, Andrija Mohorovičić: Skolska Knjiga, Zagreb. [3]

Slack, P.D., Davis, P.M., and the Kenya Rift International Seismic Project (KRISP) Working Group, 1994, Attenuation and velocity of P-waves in the mantle beneath the East African rift, Kenya, in Prodehl, C., Keller, G.R., and Khan, M.A., eds., Crustal and upper mantle structure of the Kenya rift: Tectonophysics, v. 236, p. 331–358. [9]

Slichter, L.B., 1951, Crustal structure in the Wisconsin area: Office of Naval Research Report N9ONR-86200: Institute of Geophysics, Los Angeles, California, University of California. [6]

Smallwood, J.R., Staples, R.K., Richardson, K.R., and White, R.S., 1999, Crust generated above the Iceland mantle plume: from continental rift to oceanic spreading center: Journal of Geophysical Research, v. 104, p. 22,885–22,902, doi:10.1029/1999JB900176. [2] [9]

Smith, E.G.C., Stern, T., O'Brien, B., 1995, A seismic velocity profile across central South Island, New Zealand, from explosion data: New Zealand Journal of Geology and Geophysics, v. 38, p. 565–570, doi:10.1080/00288306.1995.9514684. [2] [8]

Smith, L.K., White, R.S., Kusznir, N.J., the iSIMM team, 2005, Structure of the Hatton Basin and adjacent continental margin: Geological Society of London, Petroleum Geology Conference Series, v. 6, p. 947–956, doi:10.1144/0060947. [10]

Smith, R.B., and Braile, L.W., 1994, The Yellowstone hotspot: Journal of Volcanology and Geothermal Research, v. 61, p. 121–187, doi:10.1016/0377-0273(94)90002-7. [7]

Smith, R.B., and Christiansen, R.L., 1980, Yellowstone Park as a window on the Earth's interior: Scientific American, v. 242, p. 104–117, doi:10.1038/scientificamerican0280-104. [7]

Smith, R.B., Schilly, M.M., Braile, L.W., Ansorge, J., Lehmann, J.L., Baker, M.R., Prodehl, C., Healy, J.H., Mueller, St., and Greensfelder, R.W., 1982, The 1978 Yellowstone–eastern Snake River Plain seismic profiling experiment: Crustal structure of the Yellowstone region and experiment design: Journal of Geophysical Research, v. 87, p. 2583–2596, doi:10.1029/JB087iB04p02583. [2] [7]

Smith, T.J., Steinhart, J.S., and Aldrich, L.T., 1966, Crustal structure under Lake Superior, in Steinhart, J.S., and Smith, T.J., eds., The earth beneath the continents: Washington, D.C., American Geophysical Union, Geophysical Monograph 10, p. 181–197. [6]

Smithson, S.B., and Johnson, R.A., 1989, Crustal structure of the western United States based on reflection seismology, *in* Pakiser, L.C., and Mooney, W.D., eds., Geophysical framework of the continental United States: Geological Society of America Memoir 172, p. 577–612. [2] [8] [9]

Smithson, S.B., Brewer, J.A., Kaufman, S., Oliver, J.E., and Hurich, C.A., 1979, Structure of the Laramide Wind River uplift, Wyoming, from COCORP deep reflection data and from gravity data: Journal of Geophysical Research, v. 84, p. 5955–5972, doi:10.1029/JB084iB11p05955. [2] [7]

Smythe, D.K., Smithson, S.B., Gillen, C., Humphreys, C., Kristoffersen, Y., Karaev, N.A., Garipov, V.Z., Pavlenkova, N.I., and the KOLA–92 Working Group, 1994, Project images crust, collects seismic data in world's largest borehole: Eos (Transactions, American Geophysical Union), v. 75, p. 473–476, doi:10.1029/94EO01089. [9]

Snelson, C.M., 2001, Investigating crustal structure in western Washington and in the Rocky Mountains: implications for seismic hazards and crustal growth [Ph.D. thesis]: University of Texas at El Paso, 234 p. [2] [9]

Snelson, C.M., Henstock, T.J., Keller, G.R., Miller, K.C., Levander, A., 1998, Crustal and uppermost mantle structure along the DEEP PROBE seismic profile: Rocky Mountain Geology, v. 33, p. 181–198. [2] [9]

Snelson, C.M., Keller, G.R., Miller, K.C., Rumpel, H.-M., and Prodehl, C., 2005, Regional crustal structure derived from the CD-ROM 99 seismic refraction/wide-angle reflection profile: the lower crust and upper mantle, *in* Karlstrom, K.E., and Keller, G.R., eds., The Rocky Mountain Region: An evolving lithosphere—tectonics, geochemistry, and geophysics: Washington, D.C., American Geophysical Union Geophysical Monograph 154, p. 271–291. [9]

Snelson, C.M., Brocher, T.M., Miller, K.C., Pratt, T.L., and Trehu, A.M., 2007, Seismic amplification within the Seattle basin, Washington state: insights from SHIPS seismic tomography experiments: Bulletin of the Seismological Society of America, v. 97, p. 1432–1448, doi:10.1785/0120050204. [9]

Snyder, D., and Hobbs, R., 1999, The BIRPS Atlas II. A second decade of deep seismic reflection profiling (3 CD-ROM pack): Geological Society of London. [8]

Snyder, D.B., Spencer, C., and Warner, M. (convenors), 1997, Workshop meeting on controlled source seismology (CCSS), 4–8 June 1996, Cambridge, UK: Journal of Conference Proceedings, v. 1, p. 1–96. [2] [9] [10]

Snyder, D.B., Hobbs, R.W., and Chicxulub Working Group, 1999, Ringed structural zones with deep roots formed by the Chicxulub impact: Journal of Geophysical Research, v. 104, p. 10,743–10,755, doi:10.1029/1999JB900001. [2] [9]

Snyder, D.B., Eaton, D.W., Hurich, C.A., eds., 2006, Seismic probing of continents and their margins: Tectonophysics, v. 420, p. 1–520, doi:10.1016/j.tecto.2006.01.008. [2] [10]

Soller, D.R., Ray, R.D., and Brown, R.D., 1981, A global crustal thickness map: Open-File Report, Phoenix Corporation, McLean, Virginia, 51 p. [2] [7] [10]

Sollogub, V.B., 1969, Seismic crustal studies in southeastern Europe, *in* Hart, P.J., ed., The earth's crust and upper mantle: American Geophysical Union, Geophysical Monograph 13, p. 189–195. [6] [7]

Sollogub, V.B., and Chekunov, A.V., 1972, The Ukrainian Soviet Socialistic Republic, *in* Sollogub, V.B., Prosen, D., and Militzer, H., 1972, Crustal structure of central and southeastern Europe based on the results of explosion seismology (publ. in Russian 1971). English translation edited by György, S., 1972: Geophysical Transactions, special edition, Müszaki Könyvkiado, Budapest, chapter 21, p. 44–68. [2] [6]

Sollogub, V.B., Prosen, D., and Militzer, H., 1972, Crustal structure of central and southeastern Europe based on the results of explosion seismology (publ. in Russian 1971). English translation edited by Szénás, Gy., 1972: Geophysical Transactions, special edition, Müszaki Könyvkiado, Budapest, 172 p. [2] [6]

Sollogub, V.B., Litvinenko, I.V., Chekunov, A.V., Ankudinov, S.A., Ivanov, A.A., Kalyuzhnaya, L.T., Kokorina, L.K., and Tripolsky, A.A., 1973a, New D.S.S.-data on the crustal structure of the Baltic and Ukraiian shields: Tectonophysics, v. 20, p. 67–84, doi:10.1016/0040-1951(73)90097-8. [2] [6] [7]

Sollogub, V.B., Prosen, D., and Co-Workers, 1973b, Crustal structure of central and southeastern Europe by data of explosion seismology: Tectonophysics, v. 20, p. 1–33, doi:10.1016/0040-1951(73)90093-0. [6]

Sornes, A., 1968, Pn time-term survey Norway-Scotland 1967: ARPA Supplementary scientific report No 612-1. [2] [6]

Sorokin, Yu.M., Ya, Zamansky, Yu.Ya., Langinen, A.Ye., Jackson, H.R., and Macnab,R., 1999, Crustal structure of the Makarov Basin, Arctic Ocean determined by seismic refraction: Earth and Planetary Science Letters, v. 168, p. 187–199. [2] [8]

Sparlin, M.A., Braile, L.W., and Smith, R.B., 1982, Crustal structure of the eastern Snake River Plain determined from ray trace modelling of seismic refraction data: Journal of Geophysical Research, v. 87, p. 2619–2633, doi:10.1029/JB087iB04p02619. [7]

Spence, G.D., and McLean, N.A.,1998, Crustal seismic velocity and density structure of the Intermontane and Coast belts, southwestern Cordillera: Canadian Journal of Earth Science, v. 35, p. 1362–1379, doi:10.1139/cjes-35-12-1362. [8]

Spence, G.D., Whittall, K.P., and Clowes, R.M., 1984, Practical synthetic seismograms for laterally varying media calculated by asymptotic ray theory: Bulletin of the Seismological Society of America, v. 74, p. 1209–1223. [2] [7] [8] [10]

Spencer, C., Green, A., and Luetgert, J.H., 1987, More seismic evidence on the location of Grenville basement beneath the Appalachians of Quebec—Maine: Geophysical Journal of the Royal Astronomical Society, v. 89, p. 183–189. [8]

Spencer, C., Green, A., Morel-a-l'Huissier, P., Milkereit, B., Luetgert, J.H., Steward, D.B., Unger, J.D., and Phillips, J.D., 1989, Allochthonous units I the northern Appalachians: results from the Quebec—Maine seismic reflection and refraction surveys: Tectonics, v. 8, p. 667–696. [2] [8]

Spieth, M.A., Hill, D.P., and Geller, R.J., 1981, Crustal structure in the northwestern foothills of the Sierra Nevada from seismic refraction experiments: Bulletin of the Seismological Society of America, v. 71, p. 1075–1087. [7]

Sponheuer, W., and Gerecke, F., 1949, Die Sprengung in Grosseutersdorf bei Kahla (Thueringen) am 1. February 1947, Veroeff. Zentralinst. Erdbebenforsch. Jena, Heft 51, p. 48–56. [4]

Spudich, P., and Orcutt, J., 1980, A new look at the seismic velocity structure of the oceanic crust: Reviews of Geophysics and Space Physics, v. 18, p. 627–645, doi:10.1029/RG018i003p00627. [2] [6] [7] [8]

Sroda, P., Grad, M., and Guterch, A., 1997, Seismic models of the earth's crustal structure between the South Pacific and the Antarctic Peninsula, *in* Ricci, C.A., ed., The Antarctic Region: Geological evolution and processes: Terra Antarctica Publication, Siena, p. 685–689. [9]

Sroda, P., Czuba, W., Grad M., Guterch, A., Gaczynski, E., and POLONAISE Working Group, 2002, Three-dimensional seismic modelling of crustal structure in the TESZ region based on POLONAISE'97 data: Tectonophysics, v. 360, p. 169–185, doi:10.1016/S0040-1951(02)00351-7. [9]

Stadtlander, R., Mechie, J., and Schulze, A., 1999, Deep structure of the southern Ural mountains as derived from wide-angle seismic data: Geophysical Journal International, v. 137, p. 501–515, doi:10.1046/j.1365-246X.1999.00794.x. [9]

Stangl, R., 1990, Die Struktur der Lithosphäre in Schweden, abgeleitet aus einer gemeinsamen Interpretation de P- und S-Wellen Registrierungen auf dem FENNOLORA–Profil. [Ph.D. thesis]: University of Karlsruhe, 187 p. [7] [8]

Stankiewicz, J., Ryberg, T., Schulze, A., Lindeque, A., Weber, M.H., and de Wit, M.J., 2007, Initial results from wide-angle seismic refraction lines in the southern Cape: South African Journal of Geology, v. 110, no. 2-3, p. 407–418, doi:10.2113/gssajg.110.2-3.407. [2] [10]

Stankiewicz, J., Parsiegla, N., Ryberg, T., Gohl, K, Weckmann, U., Trumbull, R., and Weber, M., 2008, Crustal structure of the southern margin of the African continent: Results from geophysical experiments: Journal of Geophysical Research, v. 113, B10313, doi:10.1029/2008JB005612. [10]

Staples, R.K., White, R.S., Brandsdottir, B., Menke, W., Maguire, P.K.H., and McBride, J.H., 1997, Faeroe–Iceland Ridge experiment, 1, Crustal structure of northeastern Iceland: Journal of Geophysical Research, v. 102, p. 7849–7866, doi:10.1029/96JB03911. [9]

Starostenko, V., Omelchenko, V., and Tolkunov, A., 2006, Refraction/wide-angle reflection seismic profile "DOBRE-2" project Near-Azov massif and Crimea, shelves of Azov and Black Seas, Ukraine (DSS and CDP), *in* Grad, M., Booth, D., and Tiira, T., eds., European Seismological Commission (ESC), Subcommission D—Crust and Upper Mantle Structure, Activity Report 2004–2006. [2] [10]

Stäuble, M., and Pfiffner, O.A., 1991, Processing, interpretation and modeling of seismic reflection data in the Molasse basin of eatern Switzerland: Eclogae Geologicae Helvetiae, v. 84, p. 151–175. [8]

Steele, J., Thorpe, S., and Turekian, K., eds., 2001, Encyclopedia of Ocean Sciences: Elsevier, Amsterdam, Academic Press. [2] [10]

Steeples, D.W., and Miller, R.D., 1989, Kansas reflection profiles: Kansas Geological Survey Bulletin, v. 226, p. 129–164. [2] [6]

Steer, D.N., Knapp, J.H., Brown, L.D., Rybalka, A.V., and Sokolov, V.B., 1995, Crustal structure of the Middle Urals based on reprocessing of Russina seismic reflection data: Geophysical Journal International, v. 123, p. 673–682, doi:10.1111/j.1365-246X.1995.tb06883.x. [9]

Stein, A., and Schröder, H., 1976, Shotpoint data of quarry blasts 1958–1972, *in* Giese, P., Prodehl, C., and Stein, A., eds., Explosion seismology in central Europe—data and results: Springer, Berlin-Heidelberg-New York, p. 54–61. [6]

Stein, C.A., and von Herzen, R.P., 2001, Geophysical heat flow, *in* Steele, J., Thorpe, S., and Turekian, K., eds., Encyclopedia of Ocean Sciences: Amsterdam, Academic Press, Elsevier, p. 1149–1157. [10]

Steinhart, J.S., 1961, The continental crust from explosions: a review, *in* Steinhart, J.S., and Meyer, R.P., eds., Explosion Studies of Continental Structure: Carnegie Institution of Washington, Publication no. 622, p. 7–73. [3]

Steinhart, J.S., and Meyer, R.P., eds., 1961, Explosion Studies of Continental Structure: Carnegie Institution of Washington, Publication no. 622, 409 p. [2] [5] [6] [10]

Steinhart, J.S., and Smith, T.J., eds., 1966, The earth beneath the continents: Washington, D.C., American Geophysical Union, Geophysical Monograph 10, 663 p. [2] [6]

Steinhart, J.S., Meyer, R.P., and Woollard, G.P., 1961a, Arkansas-Missouri, 1958, *in* Steinhart, J.S., and Meyer, R.P., eds., Explosion Studies of Continental Structure: Carnegie Institution of Washington, Publication no. 622, p. 226–247. [2] [5]

Steinhart, J.S., Meyer, R.P., and Woollard, G.P., 1961b, Wisconsin–Upper Michigan, 1958–1959, *in* Steinhart, J.S., and Meyer, R.P., eds., Explosion Studies of Continental Structure: Carnegie Institution of Washington, Publication no. 622, p. 248–304. [2] [5]

Steinhart, J.S., Smith, T.J., and Meyer, R.P., 1961c, Theoretical background, *in* Steinhart, J.S., and Meyer, R.P., eds., Explosion Studies of Continental Structure: Carnegie Institution of Washington, Publication no. 622, p. 74–125. [5] [10]

Steinhart, J.S., Smith, T.J., Sacks, I.S., Summer, R., Suzuki, Z., Rodriguez, A., Lomnitz, C., Tuve, M.A., and Aldrich, L.T., 1963, Seismic studies: Carnegie Institution of Washington, Year Book, v. 62, p. 280–289. [6]

Steinmetz, L., Hirn, A., and Perrier, G., 1974, Réflexions sismiques à la base de l'asthénosphère: Annales de Géophysique, v. 30, p. 173–180. [2] [7]

Steinmetz, L., Whitmarsh,R.B., Moreira, V.S., 1977, Upper mantle structure beneath the Mid-Atlantic Ridge north of the Azores based on observations of compressional waves: Geophysical Journal of the Royal Astronomical Society, v. 50, p. 353–380. [2] [7]

Steinmetz, L., Ferrucci, F., Hirn, A., Morelli, C., and Nicolich, R., 1983, A 550-km-long Moho traverse in the Tyrrhenian Sea from O.B.S. recorded P_n waves: Geophysical Research Letters, v. 10, p. 428–431, doi:10.1029/GL010i006p00428. [2] [7]

Stephen, R.A., 1981, Seismic anisotropy observed in upper oceanic crust: Geophysical Research Letters, v. 8, p. 865–868, doi:10.1029/GL008i008p00865. [7]

Stephen, R.A., 1985, Seismic anisotropy in the upper oceanic crust: Journal of Geophysical Research, v. 90, p. 11,383–11,396, doi:10.1029/JB090iB13p11383. [7]

Stephen, R.A., Louden, K.E., and Matthews, D.H., 1980, The oblique seismic experiment on DSDP Leg 52: Geophysical Journal of the Royal Astronomical Society, v. 60, p. 289–300. [7]

Stephenson, R.A., Morel-a-l'Huissier, P., Zelt, C.A., Mereu, R.F., Kanasewich, E.R., Hajnal, Z., Northey, D.J., and West, G.F., 1989, Crust and upper mantle structure and the origin of the Peace River arch: Bulletin of Canadian Petroleum Geology, v. 37, p. 224–235. [2] [8]

Stephenson, R.A., Yegorova, T., Brunet, M.-F., Stovba, S., Wilson, M., Starostenko, V., Saintot, A., and Kusznir, N., 2006, Late Paleozoic intra- and pericratonic basins in the East European Craton and its margins., *in* Gee, D.G., and Stephenson, R.A., eds., European lithosphere dynamics: Geological Society of London Memoir 32, p. 463–479. [9]

Steppe, J.A., and Crosson, R.S., 1978, P-velocity models of the southern Diablo Range, California, from inversion of earthquake and explosion arrival times: Bulletin of the Seismological Society of America, v. 68, p. 357–367. [6]

Stern, T.A., 1985, A back-arc basin formed within continental lithosphere: the Central Volcanic Region of New Zealand: Tectonophysics, v. 112, p. 385–409, doi:10.1016/0040-1951(85)90187-8. [2] [8]

Stern, T.A., and Davey, F.J., 1987, A seismic investigation of crustal and upper mantle structure within the Central Volcanic Region of New Zealand: New Zealand Journal of Geology and Geophysics, v. 30, p. 217–231. [8] [9]

Stern, T.A., and McBride, J., 1998, Seismic exploration of continental strike-slip zones, *in* Klemperer, S.L., and Mooney, W.D., eds., Deep seismic probing of the continents, I: General results and new methods: Tectonophysics, v. 286, p. 63–78. [9]

Stern, T.A., Davey, F.J., and Smith, E.G.C., 1986, Crustal structure studies in New Zealand, *in* Barazangi, M., and Brown, L., eds., Reflection seismology: a global perspective: American Geophysical Union, Geodynamics Series, v. 13, p. 121–132. [2] [5] [8]

Stern, T.A., Smith, E.G.C., Davey, F.J., and Muirhead, K.J., 1987, Crustal and upper mantle structure of the northwestern North Island, New Zealand, from seismic refraction data: Geophysical Journal of the Royal Astronomical Society, v. 91, p. 913–916. [8]

Stern, T.A., Okaya, D, Kleffman, S, Scherwath, M, Henrys, S, and Davey, F, 2007, Geophysical exploration and Dynamics of the Alpine Fault Zone, In: Okaya, D., Stern T., and Davey F., eds., A continental plate boundary: tectonics at South Island, New Zealand: Washington, D.C., American Geophysical Union, Geophysical Monograph 175, p. 207–233. [9]

Stewart, D.B., Unger, J.D., Phillips, J.D., Goldsmith,R., Poole,W.H., Spencer, C.P., Green, A.G., Loiselle, M.C., and St-Julien, P., 1986, The Quebec–western Maine seismic reflection profile: setting and first year results, *in* Barazangi, M., and Brown, L., eds., Reflection seismology: the continental crust: American Geophysical Union, Geodynamics Series, v. 14, p. 189–199. [2] [8]

Stewart, S.W., 1968a, Preliminary comparison of seismic travel-times and inferred crustal structure adjacent to the San Andreas fault in the Diablo and Gabilan Ranges of central California, *in* Dickinson, W.R., and Grantz, A., eds., Proceedings of Conference on Geologic Problems of the San Andreas Fault System: Stanford University, School of Earth Science, p. 218–230. [6] [7]

Stewart, S.W., 1968b, Crustal structure in Missouri by seismic refraction methods: Bulletin of the Seismological Society of America, v. 58, p. 291–323. [6]

Stewart, S.W., and Pakiser, L.C., 1962, Crustal structure in eastern New Mexico interpreted from the Gnome explosion: Bulletin of the Seismological Society of America, v. 52, p. 1017–1030. [6]

Stiller, M., 1991, 3-D vertical incidence seismic reflection survey at the KTB location, Oberfalz, *in* Meissner, R., Brown, L., Dürbaum, H.-J., Franke, W., Fuchs, K., and Seifert, F., eds., Continental lithosphere: deep seismic reflections: American Geophysical Union, Geodynamics Series, v. 22, p. 101–113. [8]

Stoffa, P. L., and Buhl, P., 1979, Two ship multichannel seismic experiments for deep crustal studies: Expanded spread and constant offset profiles: Journal of Geophysical Research, v. 84, p. 7645–7660. [7] [8]

Stratford, W.R., and Stern, T.A., 2006, Crust and upper mantle structure of a continental backarc: central North Island, New Zealand: Geophysical Journal International, v. 166, p. 469–484, doi:10.1111/j.1365-246X.2006.02967.x. [10]

Subbotin, S., Sollogub, V.B., Prosen, D., Dragasevic, T., Mituch, E., and Posgay, K., 1968, Crustal structure of southeastern Europe according to the data of deep seismic soundings: Bollettino di Geofisica Teorica e Applicata, v. 10, p. 241–264. [6]

Sumanovac, F., Oreskovic, J., Grad, M., and ALP 2002 Working Group, 2009, Crustal structure at the contact of the Dinarides and Pannonian basin based on 2-D seismic and gravity interpretation of the Alp07 profile in the ALP 2002 experiment: Geophysical Journal International, v. 179. P. 615–633. [10]

Summers, T.P., Westbrook, G.K., and Hall, J., 1982, The Western Isles Seismic Experiment: Geophysical Journal of the Royal Astronomical Society, v. 69, p. 279. [2] [7] [8]

Sun, W., Zhu, Z., Zhang, L., Song, S., Zhang, C., and Zheng, Y., 1988, Exploration of the crust and upper mantle in North China, *in* Developments in the Research of Deep Structures of China's Continent: Beijing, Geological Publishing House, p. 19–37. [8]

Sutherland, R., 1999, Cenozoic bending of New Zealand basement terranes and Alpine Fault displacement: a brief review: New Zealand Journal of

Geology and Geophysics, v. 42, p. 295–301, doi:10.1080/00288306.1999 .9514846. [9]

Sutton, G., Maynard, G., and Hussong, D., 1971, Widespread occurrence of a high-velocity basal layer in the Pacific crust found with repetitive sources and sonobuoys, *in* Heacock, J.G., ed., The structure and physical properties of the earth's crust: American Geophysical Union Geophysical Monograph 14, p. 193–209. [2] [6]

Sutton, G.H., Manghnani, M.H., Moberly, R., and McAffe, E.U., eds., 1976, The geophysics of the Pacific Ocean basin and its margin: American Geophysical Union, Geophysical Monograph 19, 480 p. [7]

Suvorov, V.D., Melnik, E.A., Thybo, H., Perchuc, E., and Parasotka, B.S., 2006, Seismic velocity model of the crust and uppermost mantle around the Mirnyi kimberlite field in Siberia: Tectonophysics, v. 420, p. 49–73, doi:10.1016/j.tecto.2006.01.009. [2] [8]

Suyehiro, K., and Nishizawa, A., 1994, Crustal structure and seismicity beneath the forearc off northeastern Japan: Journal of Geophysical Research, v. 99, p. 22,331–22,347, doi:10.1029/94JB01337. [2] [8]

Suyehiro, K., Takahashi, N., Ariie, Y. Yokoi, Y., Hino, R., Shinohara, M., Kanazawa, T., Hirata, N., Tokuyama, H., and Taira, A., 1996, Continental crust, crustal underplating and low-Q upper mantle beneath an oceanic island arc: Science, v. 272, p. 390–392, doi:10.1126/science.272.5260.390. [2] [9]

Swain, C.J., Maguire, P.K.H., and Khan, M.A., 1994, Geophysical experiments and models of the Kenya rift before 1989, *in* C. Prodehl, G.R. Keller, and M.A. Khan, eds., Crustal and upper mantle structure of the Kenya rift: Tectonophysics, v. 236, p. 23–32. [8]

Szénás, Gy., ed., 1972, English translation of: Sollogub, V.B., Prosen, D., and Militzer, H., 1971, Crustal structure of central and southeastern Europe based on the results of explosion seismology (publ. in Russian): Geophysical Transactions, special edition, Müszaki Könyvkiado, Budapest, 172 p. [2] [6]

Taber, J.J., and Lewis, B.T.R., 1986, Crustal structure of the Washington continental margin from refraction data: Bulletin of the Seismological Society of America, v. 76, p. 1011–1024. [8]

Taira, T., Moriya, T., Miyamachi, H., Wada, N., Hirano, S., Otsuka, K., Matsubura, W., and Maruyama, Y., 2002, Seismic refraction experiment in the northeastern part of the Hokkaido, Japan: Bulletin of the Earthquake Research Institute, University of Tokyo, v. 77, p. 225–230 (in Japanese). [2] [9]

Takahashi, N., Kodaira, S., Tsuru, T., Park, J-O., Kaneda, Y., and Kinoshita, H., 2000, Detailed plate boundary structure off northeast Japan: Geophysical Research Letters, v. 27 (13), p. 1977–1980, doi:10.1029/2000GL008501. [9] [10]

Takahashi, N., Kodaira, S., Tsuru, T., Park, J-O., Kaneda, Y., Suyehiro, K., Kinoshita, H., Abe, S., Nishino, M., and Hino, R., 2004, Seismic structure and seismogenesis off Sanriku region, northeastern Japan: Geophysical Journal International, v. 159, no. 1, p. 129–145, doi:10.1111/j.1365-246X.2004 .02350.x. [2] [9] [10]

Takahashi, N., Kodaira, S., Tatsumi, Y., Kaneda, Y., and Suyehiro, K., 2008, Structure and growth of the Izu-Bonin-Mariana arc crust: 1. Seismic constraint on crust and mantle structure of the Mariana arc–back-arc system: Journal of Geophysical Research, v. 113, B01104, doi:10.1029 /2007JB005120. [10]

Takeda, T., 1997, Reanalysis from seismic refraction data in Nagano basin area, Japan—crustal structure in central Japan [M.S. thesis]: University of Tokyo, 26 p. [10]

Talwani, M., and Abreu, V., 2000, Inferences regarding initiation of oceanic crust formation from the U.S. coast margin and conjugate South Atlantic margins, *in* Mohriak, W., and Talwani, M., eds., Atlantic Rifts and Continental Margins: American Geophysical Union, Geophysical Monograph, v. 115, p. 211–233. [9]

Talwani, M., Sutton, G.H., and Worzel, J.L., 1959, A crustal section across the Puerto Rico Trench: Journal of Geophysical Research, v. 64, p. 1545–1555. [5]

Talwani, M., Le Pichon, X., and Ewing, M., 1965, Crustal structure of the mid-ocean ridges: Journal of Geophysical Research, v. 70, p. 341–352, doi:10.1029/JZ070i002p00341. [5]

Talwani, M., Windisch, C.C., and Langseth, M.G., Jr., 1971, Reykjanes Ridge crest: a detailed geophysical study: Journal of Geophysical Research, v. 76, p. 473–491, doi:10.1029/JB076i002p00473. [7]

Talwani, M., Mutter, J., Houtz, R., and Konig, M., 1979, The crustal structure and evolution of the area underlying the magnetic quiet zone on the margin south of Australia, *in* Watkins, J.S., Montadert, L., and Dickerson, P., eds.,

Geological and Geophysical Investigations of Continental Margins: American Association of Petroleum Geologists Memoir 29, p. 151–175. [2] [7]

Tandon, K., Brown, L., and Hearn, T., 1999, Deep structure of the northern Rio Grande rift beneath the San Luis basin (Colorado) from a seismic reflection survey: implications for rift evolution: Tectonophysics, v. 302, p. 41–56, doi:10.1016/S0040-1951(98)00224-8. [9]

Tarkov, A.P., and Basula, I.V., 1983, Inhomogeneous structure of the Voronezh shield lithosphere from explosion seismology data: Physics of the Earth and Planetary Interiors, v. 31, p. 281–292. [2] [7]

Tatel, H.E., and Tuve, M.A., 1955, Seismic exploration of a continental crust, *in* Poldervaart, A., ed., Crust of the earth (a symposium): Geological Society of America Special Paper 62, p. 35–50. [2] [5]

Tatel, H.E., and Tuve, M.A., 1956, Seismic crustal measurements in Alaska (abst.): Transactions, American Geophysical Union, v. 37, p. 360. [2] [5]

Tatel, H.E., and Tuve, M.A., 1958, Carnegie seismic expedition to the Andes, 1957 (abst.): Transactions, American Geophysical Union, v. 39, p. 580. [5] [9]

Taylor, S.R., 1989, Geophysical framework of the Appalachians and adjacent Grenville Province, *in* Pakiser, L.C., and Mooney, W.D., eds., Geophysical framework of the continental United States: Geological Society of America Memoir 172, p. 317–348. [8] [9]

Temmler, T., Franke, D., Heyde, I., and Neben, S., 2006, Marine seismische Untersuchung und refraktionsseismische Modellierung im Pelotas Becken—offshore Uruguay: German Geophysical Society, Annual Meeting, Bremen, 6–9 March 2006, A17 SO, Abstracts p. 170–171. [2] [10]

ten Brink, U.S., Bannister, S., Beaudoin, B.C., and T.A. Stern. 1993, Geophysical investigations of the tectonic boundary between East and West Antarctica: Science, v. 261, p. 45–50, doi:10.1126/science.261.5117.45. [2] [9]

ten Brink, U.S., Drury, R.M., Miller, G.K., 1996, Los Angeles region seismic experiment (LARSE), California—off-shore seismic refraction data: Open-File Report 96-27, 9 p. and plates. [9]

ten Brink, U.S., Al-Zoubi, A.S., Flores, C.H., Rotstein, Y., Qabbani, I., Harder, S.H., and Keller, G.R., 2006, Seismic imaging of deep low-velocity zone beneath the Dead Sea basin and transform fault: implications for strain localization and crustal rigidity: Geophysical Research Letters, v. 33, L24314, doi:10.1029/2006GL027890. [2] [9] [10]

Teng, J.W., 1979, Geophysical investigations of the earth's crust and upper mantle in China: Acta Geophysica Sinica, v. 22, p. 346–350. [2] [7]

Teng, J.-W., 1987, Explosion study of the structure and seismic velocity distribution of the crust and upper mantle under the Xizang (Tibet) plateau: Geophysical Journal of the Royal Astronomical Society, v. 89, p. 405–414. [2] [7] [8]

Teng, J.W., et al., 1974, The crustal structure of the eastern Qaidam basin from deep reflected waves: Acta Geophysica Sinica, v. 17, p. 122–135. [7]

Teng, J.W., Xiong, S., Yin, Z., Xu, Z., Wang, X., and Lu, D., 1983a, Structures of the crust and upper mantle pattern and velocity distributional characteristics at the northern region of the Himalaya Mountains: Acta Geophysica Sinica, v. 26, p. 525–540. [8]

Teng, J.W., Sun, K-Z, Xiong, S-B, Yin, Z-X.Y.H., and Chen, L-F, 1983b, Deep seismic reflection waves and structure of the crust from Dangxung to Yadong on the Xizang Plateau (Tibet): Physics of the Earth and Planetary Interiors, v. 31, p. 293–306, doi:10.1016/0031-9201(83)90089-4. [8]

Tewari, H.C., Dixit, M.M., Rajendra Prasat, B., Rao, N.M., Venkateshwarlu, N., Vijaya Rao, V., Kare, P., Kesava Rao, G., Raju, S., and Kaila, K.L., 1995, Deep crustal reflection studies across the Aravalli-Delhi fold belt: results from the northwestern part, *in* Sinha-Roy, S., and Gupta, K.R., eds., Continental crust of northwestern and central India: Geological Society of India Memoir 31, p. 383–402. [9]

Tewari, H.C., Dixit, M.M., Rao, N.M., Venkateshwarlu, N., and Vijaya Rao, V., 1997, Crustal thickening under the Paleo/Mesoproterozoic Delhi fold belt in northwestern India: evidence from deep reflection profiling: Geophysical Journal International, v. 129, p. 657–668, doi:10.1111/j.1365-246X.1997.tb04501.x. [9]

Teysseire, R., and Teysseire, B., 2003, Wawrzyniec Karol de Teysseire: a pioneer in the study of "cryptotectonics": Eos (Transactions, American Geophysical Union), v. 83, p. 541, 546. [9]

The research group for the 2003 Hinagu fault seismic expedition, 2008, Seismic expedition in the Hinagu fault area, Kyushu island, Japan: Bulletin of the Earthquake Research Institute, University of Tokyo, v. 83, p. 103–130. [9]

Thouvenot, F., Senechal, G., Hirn, A., and Nicolich, R., 1990, The ECORS-CROP wide-angle reflection seismics: constraints on deep interfaces beneath the Alps, *in* Roure, F., Heitzmann, P., and Polino, R., eds.,

Deep structure of the Alps: Mémoires de la Société Géologíque Suisse, Zürich, 1; Vol. spec. Society Geol. It., Roma, 1, p. 97–106. [8]

Thouvenot, F., Poupinet, G., Kashubin, S.N., Matte, Ph., Egorkin, A.V., and Solodilov, L.N., 1994, Deep structure of the Urals from wide-angle reflection seismic and teleseismic data: Abstracts from the XIX EGS Meeting, Genoble, 25–29 April 1994, C33. [9]

Thurber, C., Roecker, S., Lutter, W., and Ellsworth, W., 1996, Imaging the San Andreas fault with explosion and earthquake sources: Eos (Transactions, American Geophysical Union), v. 77, p. 45, 57–58, doi:10.1029/96EO00032. [9]

Thybo, H., ed., 2002, Deep seismic probing of the continents and their margins: Tectonophysics, v. 355, 263 p. [2] [9] [10]

Thybo, H., Pharaoh, T., and Guterch, A., 1999, Introduction (Symposium SE19 on Geophysical Investigations of the TESZ, annual meeting of the EGS in Nice 1998): Tectonophysics, v. 314, p. 1–5, doi:10.1016/S0040-1951 (99)00233-4. [9]

Thybo, H., Janik, T., Omelchenko, V.D., Grad, M., Garetsky, R.G., Belinsky, A.A., Karatayev, G.I., Zlotski, G., Knudsen, M.E., Sand, R., Yliniemi, J., Tiira, T., Luosto, U., Komminaho, K., Giese, R., Guterch, A., Lund, C.-E., Kharitonov, O.M., Ilchenko, T., Lysynchuk, D.V., Skobolev, V.M., and Doody, J.J., 2003, Upper lithospheric seismic velocity structure across the Pripyat Trough and the Ukrainian Shield along the EUROBRIDGE '97 profile: Tectonophysics, v. 371, p. 41–79, doi:10.1016/S0040-1951 (03)00200-2. [2] [9]

Thybo, H., Sandrin, A., Nielsen, L., Lykke-Andersen, H., and Keller, G.R., 2006, Seismic velocity structure of a large mafic intrusion in the crust of central Denmark from project ESTRID: Tectonophysics, v. 420, p. 105–122, doi:10.1016/j.tecto.2006.01.029. [2] [10]

Tittgemeyer, M., Wenzel, F., Fuchs, K., and Ryberg, T., 1996, Wave propagation in a multiple-scattering upper mantle—observations and modelling: Geophysical Journal International, v. 127, p. 492–502, doi:10.1111 /j.1365-246X.1996.tb04735.x. [9]

Todd, B.J., Reid, I., and Keen, C.E., 1988, Crustal structure across the Southwest Newfoundland Tansform Margin: Canadian Journal of Earth Science, v. 25, p. 744–759. [2] [8]

Tolstoy, I., Edwards, R.S., and Ewing, M., 1953, Seismic refraction measurements in the Atlantic Ocean, Part III: Bulletin of the Seismological Society of America, v. 43, p. 35–48. [5]

Tomek, C., 1993, Deep crustal structure beneath the central and inner Carpathians: Tectonophysics, v. 226, p. 417–431, doi:10.1016/0040-1951 (93)90130-C. [8]

Tomek, C., Dvorakova, L., Ibrmajer, I., Jiricek, R., and Korab, T., 1987, Crustal profiles of active continental collisional belt: Czechoslovak deep seismic reflection profiling in the West Carpathians: Geophysical Journal of the Royal Astronomical Society, v. 89, p. 383–388. [2] [8]

Tomek, C., Dvorakova, V., and Vrana, S., 1997, Geological interpretation of the 9HR and 503 M seismic profiles in western Bohemia: Journal of Geological Sciences Prague, v. 47, p. 43–51. [9]

Toomey, D.R., Purdy, G.M., Solomon, S.C., and Wilcock, W.S.D., 1990, The three-dimensional seismic velocity structure of the East Pacific Rise near latitude 9°30′N: Nature, v. 347, p. 639–645, doi:10.1038/347639a0. [2] [8]

Toomey, D.R., Hooft, E.E., and Detrick, R.S., 2001, Crustal thickness variations and internal structure of the Galapagos Archipelago: Eos (Transactions, American Geophysical Union), v. 82, no. 47, Fall Meeting Supplement, Abstract T42B-0939. [9]

Toporkiewicz, 1986, The structure of the consolidated basement: Publications of the Institute of Geophysics, Polish Academy of Sciences, A-13 (160), p. 101–117. [2] [6] [7]

Torné, M., Pascal, G., Buhl, P., Watts, A.B., and Mauffret, A., 1992, Crustal and velocity structure of the Valencia trough (western Mediterranean). Part II. A combined refraction/wide-angle reflection and near-vertical reflection study, in Banda, E., and Santanach,P., eds., Geology and Geophysics of the Valencia Trough, Western Mediterranean: Tectonophysics, v. 203, no. 1-4, p. 1–20. [2] [8]

Tramontini, C.B., and Davies, D., 1969, A seismic refraction survey in the Red Sea: Geophysical Journal, v. 17, p. 225–241, doi:10.1111/j.1365-246X .1969.tb02323.. [6]

TRANSALP Working Group, 2001, European orogenic processes research transects the Eastern Alps: Eos (Transactions, American geophysical Union), v. 82, p. 453, 460–461, doi:10.1029/01EO00269. [9]

TRANSALP Working Group, 2002, First deep seismic reflection images of the Eastern Alps reveal giant crustal wedges and transcrustal ramps: Geophysical Research Letters, v. 29, no. 10, p. 92.1–92.4, doi:10.1029 /2002GL014911. [9]

Trehu, A.M., 1991, Tracing the subducted oceanic crust beneath central California continental margin: results from ocean bottom seismometers deployed during the 1986 Pacific Gas and Electric EDGE experiment: Journal of Geophysical Research, v. 96, p. 6493–6506, doi:10.1029/90JB00494. [8]

Trehu, A.M., and Nakamura, Y., 1993, Onshore-offshore wide-angle seismic recordings from central Oregon: the ocean bottom seismometer data: U.S. Geological Survey Open-File Report 93-317, 30 p. [2] [8] [9]

Trehu, A.M., and the Mendocino Working Group, 1995, Pulling the rug out from under California: seismic images of the Mendocino Triple Junction region: Eos (Transactions, American Geophysical Union), v. 76, no. 38, p. 369, 380–381, doi:10.1029/95EO00229. [9]

Trehu, A.M., and Wheeler, W.H.,IV, 1987, Possible evidence in the seismic data of profile SJ-6 for subducted sediments beneath the Coast Ranges of California, USA, in Walter, A.W., and Mooney, W.D., 1987, Commission on Controlled Source Seismology (CCSS) proceedings of the 1985 workshop interpretation of seismic wave propagation in laterally heterogeneous terranes, Susono, Shizouka, Japan August 15–18, 1985: Interpretations of the SJ-6 seismic reflection/refraction profile, south central California, USA: U.S. Geological Survey Open-File Report 87-73, p. 91–104. [8]

Trehu, A.M., Ballard, A., Dorman, L.N., Gettrust, J.F., Klitgord, K.D., and Schreiner, A., 1989a, Structure of the lower crust beneath the Carolina trough, U.S. Atlantic continental margin: Journal of Geophysical Research, v. 94, p. 10,585–10,600, doi:10.1029/JB094iB08p10585. [8]

Trehu, A.M., Klitgord, K.D., Sawyer, D.S., and Buffler, R.T., 1989b, Atlantic and Gulf of Mexico continental margins, in Pakiser, L.C., and Mooney, W.D., eds., Geophysical framework of the continental United States: Geological Society of America Memoir 172, p. 349–382. [2] [7] [8]

Trehu, A.M., Holt, T., Shi, J., Nakamura, Y., Brocher, T.M., and Moses, M., 1990, Preliminary results from the 1989 Oregon onshore-offshore seismic imaging experiment: Eos (Transactions, American Geophysical Union), v. 71, p. 1588. [9]

Trehu, A.M., Azevedo, S., Nabelek, J., Luetgert, J., Mooney, W., Asudeh, I., and Isbell, B., 1993, Seismic-refraction data across the Coast Range and Willamette Basin in central Oregon: the 1991 Pacific Northwest experiment: U.S. Geological Survey Open-File Report, 93-319, 31 p. [8] [9]

Trehu, A.M., Asudeh, I., Brocher, T.M., Luetgert, J., Mooney, W., Nabelek, J., and Nakamura, Y., 1994, Crustal architecture of the Cascade forearc: Science, v. 266, p. 237–243, doi:10.1126/science.266.5183.237. [9]

Trey, H., Cooper, A.K., Pellis, G., DellaVedova, B., Cochrane, G., Brancolini, G., and Makris, J., 1999, Transect across the West Antarctic rift system in the Ross Sea, Antarctica: Tectonophysics, v. 301, p. 61–74, doi:10.1016 /S0040-1951(98)00155-3. [2] [8]

Truffert, C., Burg, J-P., Cazes, M., Bayer, R., Damotte, B., and Rey, D., 1990, Structures crustales sous le Jura et la Bresse: constraints sismiques et gravimetriques le long des profiles ECORS Bresse–Jura et Alpes II: Mémoires de la Société Géologíque de France, N.S., 156, p. 157–164. [8]

Tryggvason, E., and Qualls, B.R., 1967, Seismic refraction measurements of crustal structure in Oklahoma: Journal of Geophysical Research, v. 72, p. 3738–3740, doi:10.1029/JZ072i014p03738 [6]

Tsumura, N., Ikawa, H., Ikawa, T., Shinohara, M., Ito, T., Arita, K., Moriya, T., Kimura, G., and Ikawa, T., 1999, Delamination-wedge structure beneath the Hidaka collision zone, central Hokkaido, Japan inferred from seismic reflection profiling: Geophysical Research Letters, v. 26, p. 1057–1060, doi:10.1029/1999GL900192. [2] [9]

Tsuru, T., Park, J-O., Miura, S., Kodaira, S., Kido, Y., and Hayashi, T., 2002, Along-arc structural variation of the plate boundary at the Japan Trench margin: Implication of interpolate coupling: Journal of Geophysical Research, v. 107, p. 2375, doi:10.1029/2001JB001664. [2] [9] 10]

Tucholke, B.E., Houtz, R.E., and Barrett, D.M., 1981, Continental crust beneath the Agulhas Plateau, southwest Indian Ocean: Journal of Geophysical Research, v. 86, p. 3791–3806, doi:10.1029/JB086iB05p03791. [2] [7]

Turekian, K.K., 2001, Origin of the oceans, in Steele, J., Thorpe, S., and Turekian, K., eds., Encyclopedia of Ocean Sciences: Amsterdam, Academic Press, Elsevier, p. 2055–2058. [10]

Tuve, M.A., 1948, Seismic sounding: Carnegie Institution of Washington, Year Book, v. 47, p. 60–65. [4]

Tuve, M.A., 1953, The earth's crust: Carnegie Institution of Washington, Year Book, v. 52, p. 103–108. [6]

Tuve, M.A., and Tatel, H.E., 1950a, Coherent seismic wave patterns: Science, v. 112, p. 452–543. [4]

Tuve, M.A., and Tatel, H.E., 1950b, Seismic observations, Corona (California) blast, 1949: Transactions, American Geophysical Union, v. 31, p. 324. [4]

Tuve, M.A., Goranson, R.W., Greig, J.W., Rooney, W.J., Doak, J.B., and England, J.L., 1948, Studies of deep crustal layers by explosive shots: American Geophysical Union Transactions, v. 29, p. 772. [2] [4] [5]

Tuve, M.A., Tatel, H.E., and Hart, P.J., 1954, Crustal structure from seismic exploration: Journal of Geophysical Research, v. 59, p. 415–422, doi:10.1029/JZ059i003p00415. [5]

Twaltwadzse, G., 1950, The structure of the earth's crust in the upper Kartli (in Russian): Soobshchenija Akad. Nauk Grusinskoi SSR 11 (8), p. 479–482. [2] [4]

Uchman, J., 1972, Poland, *in* Sollogub, V.B., Prosen, D., and Militzer, H., 1972, Crustal structure of central and southeastern Europe based on the results of explosion seismology (publ. in Russian 1971). English translation edited by Szénás, Gy., 1972: Geophysical Transactions, special edition, Müszaki Könyvkiado, Budapest, chapter 22, p. 75–79. [2] [6]

Uchman, J., 1973, Results of deep seismic soundings along international profile V: Publications of the Institute of Geophysics, Polish Academy of Sciences, v. 60, p. 47–52. [6]

United Observing Group of the Yongping Explosion, SSB, 1988, Yongping explosion and deep structure in southeast China, *in* Developments in the Research of Deep Structures of China's Continent: Beijing, Geological Publishing House, p. 140–153. [7] [8]

Usami, T., Mikumo, T., Shima, E., Tamaki, I., Asano, S., Asada, T., and Matuzawa, T., 1958, Crustal Structure in Northern Kwanto District by Explosion Seismic Observations. Part II. Models of crustal structure: Bulletin of the Earthquake Research Institute, v. 36, p. 349–357. [2] [5]

Valasek, P., Mueller, St., Frei, W., and Holliger, K., 1991, Results of NFP 20 seismic reflection profiling along the Alpine section of the European Geotraverse (EGT): Geophysical Journal International, v. 105, p. 85–102, doi:10.1111/j.1365-246X.1991.tb03446.x. [2] [8]

Van Avendonk, H.J.A., Holbrook, W.S., Okaya, D., Austin, J.K., Davey, F.J., Stern, T., 2004, Continental crust under compression: a seismic refraction study of South Island Geophysical Transect I, South Island, New Zealand: Journal of geophysical research, Solid earth, v. 109, no. B6, B06302, doi:10.1029/2003JB002790. [9]

Van der Hilst, R.D., Kennett, B., Christie, D., and Grant, J., 1994, Project SKIPPY explores the lithosphere and mantle beneath Australia: Eos (Transactions, American Geophysical Union), v. 75, p. 177–181. [2] [9]

Van der Hilst, R.D., Kennett, B.L.N., and Shibutani, T., 1998, Upper mantle structure beneath Australia from portable array deployments, *in* Braun, J., Goleby, B., and Van der Hilst, R., eds., Structure and evolutin of the Australian continent: American Geophysical Union Geodynamics Series, v. 26, p. 39–57, doi:10.1029/94EO00857. [2] [9]

Van der Velden, A.J., van Staal, C.R., and Cook, F.A., 2004, Crustal structure, fossil subduction, and the tectonic evolution of the Newfoundland Appalachians: Geological Society of America Bulletin, v. 116, p. 1485–1498, doi:10.1130/B25518.1. [2] [8]

VanWagoner, T.M., Crosson, R.S., Creager, K.C., Medema, G., Preston, L., Symons, N.P., and Brocher, T.M., 2002, Crustal structure and relocated earthquakes in the Puget Lowland, Washington, from high-resolution seismic tomography: Journal of Geophysical Research, v. 107, no. 12, p. 22-1–22-23. [10]

Veevers, J.J., and Cotterill, D., 1978, Western margin of Australia: evolution of a rifted arch system: Geological Society of America Bulletin, v. 89, p. 337–355, doi:10.1130/0016-7606(1978)89<337:WMOAEO>2.0.CO;2. [8]

Vera, E.E., and Mutter, J.C., 1988, Crustal structure in the ROSE area of the East Pacific Rise: one-dimensional travel time inversion of sonobuoys and expanding spread profiles: Journal of Geophysical Research, v. 93, p. 6635–6648, doi:10.1029/JB093iB06p06635. [7]

Vera, E.E., and Diebold, J.B., 1994, Seismic imaging of oceanic layer 2A between 9°30'N and 10°N on the East Pacific Rise from two-ship wide-aperture profiles: Journal of Geophysical Research, v. 99, p. 3031–3041, doi:10.1029/93JB02107. [8]

Vera, E.E., Mutter, J.C., Buhl, P., Orcutt, J.A., Harding, A.J., Kappus, M.E., Detrick, R.S., and Brocher, T.M., 1990, The structure of 0- to 0.2-m.y.-old oceanic crust at 9°N on the East Pacific Rise from expanded spread profiles: Journal of Geophysical Research, v. 95, p. 15529–15556, doi:10.1029/JB095iB10p15529. [8]

Vergne, J., Wittlinger, G., Farra, V., and Su, H., 2003, Evidence for upper crustal anisotropy in the Songpan–Ganze (northeastern Tibet) terrane: Geophysical Research Letters, v. 30, doi:10.1029/2002GL016847. [9]

Vidal, N., Klaeschen, D., Kopf, A., Docherty, C., von Huene, R., and Krasheninnikov, V.A., 2000, Seismic images at the convergence zone from south of Cyprus to the Syrian coast, eastern Mediterranean: Tectonophysics, v. 329, p. 157–170, doi:10.1016/S0040-1951(00)00194-3. [9]

Vidale, J., 1988, Finite-difference calculaton of travel times: Bulletin of the Seismological Society of America, v. 78, p. 2062–2076. [9]

Vine, F.J., 2001, Magnetics, *in* Steele, J., Thorpe, S., and Turekian, K., eds., Encyclopedia of Ocean Sciences: Amsterdam, Academic Press, Elsevier, p. 1515–1525. [10]

Vine, F.J., and Moores, E.M., 1972, A model for the gross structure, petrology, and magnetic properties of oceanic crust, *in* Shagam, R., Hargraves, R.B., Morgan, F.J., Van Houten, F.B., Burk, C.A., Holland, H.D., and Hollister, L.C., eds., Studies in earth and space sciences: a memoir in honour of Harry Hammond Hess: Geological Society of America Memoir 132, p. 195–205. [6]

Vinnik, L.P., and Ryaboi, V.Z., 1981, Deep structure of the East-European platform according to seismic data: Physics of the Earth and Planetary Interiors, v. 25, p. 27–37, doi:10.1016/0031-9201(81)90126-6. [7]

Vogel, A., ed., 1971, Deep seismic sounding in Northern Europe: Swedish Natural Science Research Council (NFR), Stockholm, 98 p. [2] [6]

Vogel, A., and Lund, C-E., 1971, Profile section 2-3, *in* Deep seismic sounding in Northern Europe: Swedish Natural Science Research Council (NFR), Stockholm, p. 62–75. [2] [6]

Vogt, U., Makris, J., O'Reilly, B.M., Hauser, F., Readman, P.W., and Jacob, A.W.B., 1998, The Hatton basin and continental margin: crustal structure from wide-angle seismic and gravity data: Journal of Geophysical Research, v. 103, p. 12,545–12,566, doi:10.1029/98JB0060. [8] [10]

Von Huene, R., and Flueh, E.R., 1995, A review of marine geophysical studies along the Middle America Trench off Costa Rica and the problematic seaward terminus of continental crust, *in* Seyfried, H., and Hellmann, W., eds., Profile 7: Stuttgart, Germany, University of Stuttgart, p. 143–159. [9]

Von Huene, R., and Scholl, D.W., 1991, Observations at convergent margins concerning sediment subduction, subduction erosion, and the growth of continental crust: Reviews of Geophysics, v. 29, no, 3, p. 279–316, doi:10.1029/91RG00969. [8]

Von Knorring, M., and Lund, C-E., 1989, Description of the POLAR profile transect display: Tectonophysics, v. 162, p. 165–171, doi:10.1016/0040-1951(89)90362-4. [8]

Voss, M., Jokat, W., and Schmidt-Aursch, M., 2006, Crustal structure of the East Greenland margin north of Jan Mayen Fracture Zone, *in* Grad, M., Booth, D., and Tiira, T., eds., European Seismological Commission (ESC), Subcommission D—Crust and Upper Mantle Structure, Activity Report 2004–2006. [10]

Voss, M., Schmidt-Aursch, M., and Jokat, W., 2009, Variations in magmatic processes along the East Greenland volcanic margin: Geophysical Journal International, v. 177, p. 755–782, doi:10.1111/j.1365-246X.2009.04077.x. [10]

Waldhauser, F., Kissling, E., Ansorge, J., and Mueller, St., 1999, Three-dimensional interface modelling with two-dimensional seismic data: the Alpine crust-mantle boundary: Geophysical Journal International, v. 135, p. 264–278, doi:10.1046/j.1365-246X.1998.00647.x. [9]

Walter, A.W., 1990, Upper-crustal velocity structure near Coalinga, as determined from seismic-refraction data, *in* Rymer, M.J., and Ellsworth, W.L., eds., The Coalinga, California, earthquake of May 2, 1983: U.S. Geological Survey Professional Paper 1487, p. 23–39. [2] [8]

Walter, A.W., and Mooney, W.D., 1982, Crustal structure of the Diablo and Gabilan Ranges, central California: a reinterpretation of existing data: Bulletin of the Seismological Society of America, v. 72, p. 1567–1590. [2] [6] [7] [9]

Walter, A.W., and Mooney, W.D., 1987, Commission on Controlled Source Seismology (CCSS) proceedings of the 1985 workshop interpretation of seismic wave propagation in laterally heterogeneous terranes, Susono, Shizouka, Japan, August 15–18, 1985: Interpretations of the SJ-6 seismic reflection/refraction profile, south central California, USA: U.S. Geological Survey Open-File Report 87-73, 132 p. [2] [8] [9] [10]

Walther, C., and Flueh, E.R., 1993, The POLAR profile revisited: combined P- and S-wave interpretation: Precambrian Research, v. 64, p. 153–168, doi:10.1016/0301-9268(93)90073-B. [8]

Wang, C-Y, Zeng, R-S, Mooney, W.D., and Hacker, B.R., 2000, A crustal model of the ultrahigh-pressure Dabie Shan orogenic belt, China, derived from deep seismic refraction profiling: Journal of Geophysical Research, v. 105, p. 10,857–10,869, doi:10.1029/1999JB900415. [9]

Wang, Y., Mooney, W.D., Yuan, X., and Coleman, R.G., 2003, The crustal structure from the Altai Mountains to the Altyn Tagh fault, northwest China: Journal of Geophysical Research, v. 108, B6, 2322, doi: 10.1029 /2001JB000552: ESE 7-1-7-15. [2] [8] [9]

Ward, G., and Warner, M., 1991, Lower crustal lithology from shear wave seismic reflection data, *in* Meissner, R., Brown, L., Dürbaum, H.-J., Franke, W., Fuchs, K., and Seifert, F., eds., Continental lithosphere: deep seismic reflections: American Geophysical Union, Geodynamics Series, v. 22, p. 343–349. [2] [8]

Ward, W.C., Keller, G., Stinnesbeck, W., and Adatte, T., 1995, Yucatan subsurface stratigraphy: implications and constraints for the Chicxulub impact: Geology, v. 23, p. 873–876, doi:10.1130/0091-7613(1995)023<0873:YNSSIA >2.3.CO;2. [9]

Wardle, R.J., and Hall, J., eds., 2002a, Proterozoic evolution of the northeastern Canadian Shield: Lithoprobe Eastern Canadian Shield Onshore-Offshore Transect (ECSOOT): Canadian Journal of Earth Science, v. 39, p. 563–897, doi:10.1139/e02-029. [2] [9]

Wardle, R.J., and Hall, J., 2002b, Proterozoic evolution of the northeastern Canadian Shield. Onshore-offshore transect (ECSOOT), introduction and summary: Canadian Journal of Earth Science, v. 39, p. 563–567, doi:10.1139 /e02-029. [9]

Warren, D.H., 1968a, Transcontinental geophysical survey (35°–39°N) seismic refraction profiles of the crust and upper mantle from 113° to 125°W longitude: U.S. Geological Survey Miscellaneous Investigations Map I-532-D. [6]

Warren, D.H., 1968b, Transcontinental geophysical survey (35°–39°N) seismic refraction profiles of the crust and upper mantle from 100° to 113°W longitude: U.S. Geological Survey Miscellaneous Investigations Map I-533-D. [6]

Warren, D.H., 1968c, Transcontinental geophysical survey (35°–39°N) seismic refraction profiles of the crust and upper mantle from 87° to 100° W longitude: U.S. Geological Survey Miscellaneous Investigations Map I-534-D. [6]

Warren, D.H., 1968d, Transcontinental geophysical survey (35°–39°N) seismic refraction profiles of the crust and upper mantle from 74° to 87°W longitude: U.S. Geological Survey Miscellaneous Investigations Map I-535-D. [6]

Warren, D.H., 1969, A seismic survey of crustal structure in central Arizona: Geological Society of America Bulletin, v. 80, p. 257–282, doi:10.1130 /0016-7606(1969)80[257:ASSOCS]2.0.CO;2. [2] [6] [7] [8]

Warren, D.H., 1981, Seismic-refraction measurements of crustal structure near Santa Rosa and Ukiah, California, *in* McLaughlin, R.J., and Donnelly-Nolan, I.M., eds., Research in The Geysers–Clear Lake Geothermal Area, Northern California: U.S. Geological Survey Professional Paper 1141, p. 167–181. [7] [9]

Warren, D.H., and Healy, J.H.,1973, The crust in the conterminous United States, *in* Mueller, S., ed., The structure of the earth's crust, based on seismic data: Tectonophysics, v. 20, p. 203–213. [2] [6] [10]

Warren, D.H., and Jackson, W.H., 1968, Surface seismic measurements of the Project Gasbuggy explosion at intermediate distance ranges: U.S. Geological Survey Open-File Report, 45 p. [6]

Warren, D.H., Healy, J.H., and Jackson, W.H., 1966, Crustal seismic measurements in southern Mississippi: Journal of Geophysical Research, v. 71, p. 3437–3458. [6]

Warren, D.H., Healy, J.H., Bohn, J., and Marshall, P.A., 1972, Crustal calibration of the Large Aperture Seismic Array (LASA), Montana: U.S. Geological Survey Open-File Report, p. 1–58 and A1–A105. [6]

Warren, D.H., Healy, J.H., Bohn, J., and Marshall, P.A., 1973, Crustal structure under Lasa from seismic refraction measurements: Journal of Geophysical Research, v. 78, p. 8721–8734, doi:10.1029/JB078i035p08721. [2 [6]

Warrick, R.E., Hoover, D.B., Jackson, W.H., Pakiser, L.C., and Roller, J.C., 1961, The specification and testing of a seismic-refraction system: Geophysics, v. 26, p. 820–824, doi:10.1190/1.1438963. [6]

Watts, A.B., ten Brink, U.S., Buhl, P., and Brocher, T.M., 1985, A multichannel seismic study of lithospheric flexure across the Hawaiian-Emporer seamount chain: Nature, v. 315, p. 105–111, doi:10.1038/315105a0. [2] [8]

Weatherby, B.B., 1940, The history and development of seismic prospecting: Geophysics, v. 5, p. 215–230, doi:10.1190/1.1441805. [3]

Weber, J.R., 1986, The Alpha Ridge: Gravity, seismic and magnetic evidence for a homogenous, mafic crust: Journal of Geodynamics, v. 6, p. 117–136, doi:10.1016/0264-3707(86)90036-0. [8]

Weber, J.R., 1990, The structures of the Alpha Ridge, Arctic Ocean and Iceland–Faeroe Ridge, North Atlantic: Comparisons and implications for the evolution of the Canada Basin: Marine Geology, v. 93, p. 43–68, doi:10.1016/0025-3227(90)90077-W. [8]

Weber, M., Mechie, J., Abu-Ayyash, K., Ben-Avraham, Z., El-Kelani, R., Qabbani, I., and DESIRE Group, 2007, Seismic wide-angle reflection/refraction profiling from the DESIRE project reveals the deep structure across the Southern Dead Sea Basin: Eos (Transactions, American Geophysical Union), v. 88, no. 52, Fall Meeting Suppl., Abstract T43A-1077. [9] [10]

Wefer, G., Weigel, W., Pfannkuche, O., eds., 1991, Ostatlantik–Expedition. Meteor-Berichte M5, no. 91-1. Leitstelle METEOR, Institute Meereskunde, University of Hamburg, 166 p. (Chapters 5.31–5.3.2, Reflexions und Refraktionsseismik, p. 18–24, 84–104) [9]

Wegler, U., and Lühr, B.-G., 2001, Scattering behaviour at Merapi volcano (Java) revealed from an active seismic experiment: Geophysical Journal International, v. 145, p. 579–592, doi:10.1046/j.1365-246x.2001.01390.x. [2] [9]

Weichert, D.H., and Whitham, K., 1969, Calibration of the Yellowknife Seismic Array with first zone explosions: Geophysical Journal of the Royal Astronomical Society, v. 18, p. 461–476. [2] [6]

Weigel, W., 1974, Crustal structure under the Ionian Sea: Journal of Geophysics, v. 40, p. 137–140. [2] [6] [7]

Weigel, W., and Grevemeyer, I., 1999, The Great Meteor seamount: seismic structure of a submerged intraplate volcano: Journal of Geodynamics, v. 28, p. 27–40, doi:10.1016/S0264-3707(98)00030-1. [2] [6] [9]

Weigel, W., and Wissmann, G., 1977, A first crustal section from seismic observations west of Mauretania, *in* Proceedings, XV General Assembly of the European Seismological Commission (Krakov 1976): Publications of the Institute of Geophysics, Polish Academy of Sciences, A-4 (115), p. 369–380. [7]

Weigel, W., Flüh, E., Miller, H., Butzke, A., Deghani, A., Gebbhardt, V., Harder, I., Hepper, J., Jokat, W., Kläschen, D., Schübler, S., and Zhao-Groekart, Z., 1995, Investigations of the East Greenland continental margin between 70° and 72°N by deep seismic sounding and gravity studies: Marine Geophysics Research, v. 17, p. 167–199, doi:10.1007/BF01203425. [2] [8]

Welford, J.K., and Clowes, R.M., 2004, Deep 3-D seismic reflection imaging of Precambrian sills in southwestern Alberta, Canada: Tectonophysics, v. 388, p. 161–172, doi:10.1016/j.tecto.2003.11.015. [9]

Welford, J.K., Clowes, R.M., Ellis, R.M., Spence, G.D., Asudeh, I., and Hajnal, Z., 2001, Lithospheric structure across the craton—Cordilleran transition of northeastern British Columbia: Canadian Journal of Earth Science, v. 38, p. 1169–1189, doi:10.1139/cjes-38-8-1169. [9]

Wentworth, C.M., and Zoback, M.D., 1990, Structure of the Coalinga area and thrust origin of the earthquake, *in* Rymer, M.J., and Ellsworth, W.L., eds., The Coalinga, California, earthquake of May 2, 1983: U.S. Geological Survey Professional Paper 1487, p. 41–68. [8]

Wentworth, C.M., Walter, A.W., Bartow, J.A., and Zoback, M.D., 1983, Evidence on the tectonic setting of the 1983 Coalinga earthquakes from deep reflection and refraction profiles across the southeastern end of Kettleman Hills, *in* Bennett, J.H., and Sherburne, R.W., The 1983 Coalinga, California, earthquakes: California Division of Mines and Geology Special Publication, v. 66, p. 113–126. [8]

Wentworth, C.M., Zoback, M.D., and Bartow, J.A., 1984, Tectonic setting of the 1983 Coalinga earthquakes from seismic reflection profiles: a progress report, *in* Rymer, M.J., and Ellsworth, W.L., eds., Proceedings of Workshop XXVII: Mechanics of the May 2, 1983 Coalinga earthquake: U.S. Geological Survey Open-File Report 85-44, p. 19–30. [8]

Wentworth, C.M., Zoback, M.D., Griscom, A., Jachow, R.C., and Walter, D., 1987, A transect across the Mesozoic accretionary margin of central California: Geophysical Journal of the Royal Astronomical Society, v. 89, p. 105-110. [8]

Wenzel, F., 1997, Strong Earthquakes: A challenge for geosciences and civil engineering—a new Collaborative Research Center in Germany: Seismological Research Letters, v. 68, p. 438–443. [9]

Wenzel, F., Sandmeier, K.-J., and Waelde, W., 1987, Properties of the lower crust from modeling refraction and reflection data: Journal of Geophysical Research, v. 92, p. 11,575–11,583, doi:10.1029/JB092iB11p11575. [8]

Wenzel, F., Brun, J.-P., and the ECORS-DEKORP Working Group, 1991, A deep reflection seismic line across the northern Rhine Graben from ECORS-DEKORP deep seismic reflection data: Earth and Planetary Science Letters, v. 104, p. 140–150, doi:10.1016/0012-821X(91)90200-2. [2] [8]

Wenzel, F., Lungu, D. and Novak, O., eds., 1998a, Vrancea Earthquakes: Tectonics, Hazard and Risk Mitigation. Selected papers of the First International Workshop on Vrancea Earthquakes, Bucharest, November 1–4, 1997, Kluwer Academic Publishers, Dordrecht, Netherlands, 374 p. [9]

Wenzel, F., Achauer, U., Enescu, D., Kissling, E. Russo, R., Mocanu, V. and Mussachio, G., 1998b, The final stage of plate detachment; International tomographic experiment in Romania aims to a high-resolution snapshot of this process: Eos (Transactions, American Geophysical Union), v. 79, p. 589, 592–594, doi:10.1029/98EO00427. [9]

Wernicke, B., Clayton, R., Ducea, M., Jones, C.H., Park, S., Ruppert, S., Saleeby, J., Snow, J.K., Sqires, L., Fliedner, M., Jiracek, G., Keller, R., Klemperer, S., Luetgert, J., Malin, P., Miller, K., Mooney, W., Oliver, H., and Phinney, R., 1996, Origin of high mountains in the continents: the southern Sierra Nevada: Science, v. 271, p. 190–193, doi:10.1126/science.271.5246.190. [9]

White, D.J., Ansorge, J., Bodoky, T.J., and Hajnal, Z., eds., 1996, Seismic reflection profiling of the continents and their margins: Tectonophysics, v. 264: 392 p. [2] [9]

White, D.J., Forsyth, D., Asudeh, I., Carr, S.D., Wu, H., Easton, R.M., and Mereu, R.F., 2000, A seismic-based cross-section of the Grenville Orogen in southern Ontario and western Quebec: Canadian Journal of Earth Science, v. 37, p. 183–192, doi:10.1139/cjes-37-2-3-183. [8] [9]

White, D.J., Thomas, M.D., Jones, A.G., Hope, J., Nemeth, B., and Hajnal, Z., 2005, Geophysical transect across a Paleoproterozoic continent-continent collision zone: The Trans-Hudson Orogen: Canadian Journal of Earth Science, v. 42, p. 385–402, doi:10.1139/e05-002. [9]

White, R.E., 1969, Seismic phases recorded in South Australia and their relation to crustal structure: Geophysical Journal of the Royal Astronomical Society, v. 17, p. 249–261. [2] [6]

White, R.S., and Whitmarsh, R.B., 1984, An investigation of seismic anisotropy due to cracks in the upper oceanic crust at 45°N, Mid-Atlantic Ridge: Geophysical Journal of the Royal Astronomical Society, v. 79, p. 439–467. [2] [7] [8]

White, R.S., Detrick, R.S., Sinha, M.C., and Cormier, M.-H., 1984, Anomalous seismic crustal structure of oceanic fracture zones: Geophysical Journal of the Royal Astronomical Society, v. 79, p. 779–798. [9]

White, R.S., Detrick, R.S., Mutter, J.C., Buhl, P., Minshull, T.A., and Morris, E., 1990, New seismic images of the oceanic crustal structure: Geology, v. 18, p. 462–465, doi:10.1130/0091-7613(1990)018<0462:NSIOOC>2.3.CO;2. [2] [8]

White, R.S., McKenzie, D., and O'Nions, R.K., 1992, Oceanic crustal thickness from seismic measurements and rare earth element inversions: Journal of Geophysical Research, v. 97, p. 19,683–19,715, doi:10.1029/92JB01749. [7]

White, W.R.H., and Savage, J.C., 1965, A seismic refraction and gravity study of the earth's crust in British Columbia: Bulletin of the Seismological Society of America, v. 55, p. 463–486. [6]

White, W.R.H., Bone, M.N., and Milne, W.G., 1968, Seismic refraction surveys in British Columbia—a preliminary interpretation, in Knopoff, L., Drake, C.L., and Pembroke, J.H., eds., The crust and upper mantle of the Pacific area: Washington, D.C., American Geophysical Union, Geophysical Monograph 12, p. 81–93. [6]

Whitman, D., and Catchings, R.D., 1987, Data report for the vertical-component seismic refraction data obtained during the 1986 PASSCAL Basin and Range lithospheric seismic experiment, northern Nevada: Menlo Park, California, U.S. Geological Survey Open-File Report 87-415, 72 p. [2] [8]

Whitmarsh,R.B., 1975, Axial intrusion zone beneath the Median Valley of the Mid-Atlantic Ridge at 37°N detected by explosion seismology: Geophysical Journal of the Royal Astronomical Society, v. 42, p. 189–215, doi:10.1111/j.1365-246X.1975.tb05857.x. [2] [7]

Whitmarsh, R.B., and Calvert, A.J., 1986, Crustal structure of Atlantic fracture zones—I. The Charlie-Gibbs fracture zone: Geophysical Journal of the Royal Astronomical Society, v. 85, p. 107–138. [2] [8]

Whitmarsh, R.B., and Lilwall, R.C., 1983, Ocean-bottom seismographs, in Bott, M.H.P., Saxov, X., Talwani, M., and Thiede, J., eds., Structure and development of the Greenland-Scotland Ridge: Plenum, New York, p. 257–286. [7] [8]

Whitmarsh,R.B., Ginzburg, A., and Searle, R.C., 1982, The structure and origin of the Azores-Biscay Rise, northeast Atlantic Ocean: Geophysical Journal of the Royal Astronomical Society, v. 70, p. 79–107. [2] [7]

Whittaker, A., and Chadwick, R.A., 1983, Deep seismic reflection profiling onshore United Kingdom: First Break, 1/8, p. 9–13. [7] [8]

Wiechert, E., 1903, Theorie der automatischen Seismographen: Abhandlungen der Königliche Gesellschaft der Wissenschaften zu Göttingen: Mathematisch-physikalische Klasse, p. 3–128. [3]

Wiechert, E., 1907, Was wissen wir von der Erde unter uns?: Deutsche Rundschau, v. 33, p. 376–394. [3]

Wiechert, E., 1910, Bestimmung des Weges der Erdbebenwellen im Erdinnern: Physikalische Zeitschrift, v. 1, p. 294–311. [3]

Wiechert, E., 1923, Untersuchungen der Erdrinde mit dem Seismometer unter Benutzung künstlicher Erdbeben: Gesellschaft der Wissenschaften zu Göttingen Nachrichten, Mathematisch-hysikalische Klasse, p. 57–70. [2] [3]

Wiechert, E., 1926, Untersuchungen der Erdrinde mit Hilfe von Sprengungen: Geologsiche Rundschau, v. 17, p. 339–346, doi:10.1007/BF01802787. [2] [3] [4]

Wiechert, E., 1929, Seismische Beobachtungen bei Steinbruchssprengungen: Zeitschrift für Geophysik, v. 5, p. 159–162, [2] [3]

Wiechert, E., and Zoeppritz, K., 1907, Über Erdbebenwellen. 1.Teil Theoretisches über die Ausbreitung der Erdbebenwellen (E. Wiechert), 2.Teil Laufzeitkurven (K. Zoeppritz). Königliche Gesellschaft der Wissenschaften zu Göttingen Nachrichten, Mathematisch-physikalische Klasse, p. 415–459. [3]

Wielandt, E., 1972, Anregung seismischer Wellen durch Unterwasserexplosionen [Ph.D. thesis]: University of Karlsruhe, 120 p. [7]

Wielandt, E., 1975, Generation of seismic waves by underwater explosions: Geophysical Journal of the Royal Astronomical Society, v. 40, p. 421–439. [7]

Wigger, P., 1986, Krustenseismische Untersuchungen in Nord-Chile und Süd-Bolivien: Berliner Geowissenschaftliche Abhandlungen, v.A66, p. 31–48. [2] [8]

Wigger, P., and Harder, S., 1986, Seismologische und krustenseismische Untersuchungen im Atlas-System, Marokko: Berliner Geowissenschaftliche Abhandlungen, v. A66, p. 273–288. [8]

Wigger, P., Araneda, M., Giese, P., Heinsohn, W.-D., Röwer, P., Schmitz, M., and Viramonte, J., 1991, The crustal structure along the Central Andean Transect derived from seismic refraction investigations, in Omarini, R., and Götze, H.-J., eds., Central Andean Transect, Nazca Plate to Chaco Plains, southwestern Pacific, northern Chile and northern Argentina: Global Geoscience Transect 8, co-published by Inter Union Commission on the Lithosphere and American Geophysical Union, p. 13–19. [8]

Wigger, P., Asch, G., Giese, P., Heinsohn, El Alami, S.O., and Ramdani, F., 1992, Crustal structure along a traverse across the Middle and High Atlas mountains derived from seismic refraction studies: Geologishe Rundschau, v. 81, p. 237–248. [2] [8]

Wigger, P., Schmitz, M., Araneda, M., Asch, G., Baldzuhn, S., Giese, P., Heinsohn, W.-D., Martinez, E., Ricaldi, E., Röwer, P., and Viramonte, J., 1994, Variation of the crustal structure of the southern Central Andes deduced from seismic refraction investigations, in Reutter, K.-J., Scheuber, E., Wigger, P., eds., Tectonics of the Southern Central Andes: New York, Springer, p. 23–48. [2] [8]

Wiggins, S.M., Dorman, L.M., Cornuelle, B.D., and Hildebrand, J.A., 1996, Hess Deep rift valley structure from seismic tomography: Journal of Geophysical Research, v. 101, p. 22,335–22,353, doi:10.1029/96JB01230. [2] [9]

Wilde-Piorko, M., Grad, M., and TOR Working Group, 2002, Crustal structure variation from the Precambrian to Palaeozoic platforms in Europe imaged by the inversion of teleseismic receiver functions—procect TOR: Geophysical Journal International, v. 150, p. 261–270, doi:10.1046/j.1365-246X.2002.01699.x. [9]

Will, M., 1976, Calculation of travel times and ray paths for lateral inhomogeneous media, in Giese, P., Prodehl, C., and Stein, A., eds., Explosion seismology in central Europe—data and results: Berlin-Heidelberg-New York, Springer, p. 168–177. [7] [8]

Willden, R., 1965, Seismic refraction measurements of crustal structure between American Falls Reservoir, Idaho, and Flaming Gorge Reservoir, Utah, in Geological Survey Research, U.S. Geological Survey Professional Paper, 525-C, p. C44–C50. [6]

Williams, A.J., Brocher, T.M., Mooney, W.D., and Boken, A., 1999, Data report for seismic refraction survey conducted from 1980 to 1982 in the Livermore Valley and the Santa Cruz Mountains, California: Menlo Park, California, U.S. Geological Survey Open-File Report 99-146, 78 p. [2] [8]

Williamson, P.E., Collins, C.D.N., and Falvey, D.A., 1989, Crustal structure and formation of a magnetic quiet zone in the Australian southern margin: Marine and Petroleum Geology, v. 6, p. 221–231, doi:10.1016/0264-8172(89)90002-0. [8]

Willmore, P.L., 1949, Seismic experiments on the North German explosions, 1946–1947: Philosophical Transactions, A, no. 843, p. 123–151. [4]

Willmore, P.L., 1973, Crustal structure in the region of the British Isles: Tectonophysics, v. 20, p. 341–357, doi:10.1016/0040-1951(73)90122-4. [2] [6] [8]

Willmore, P.L., and Bancroft, A.M., 1960, The time-term approach to refraction seismology: Geophysical Journal of the Royal Astronomical Society, v. 3, p. 419–432. [6] [10]

Willmore, P.L., and Scheidegger, A.E., 1956, Seismic observations in the Gulf of Lawrence: Transactions, Royal Society of Canada, v. 111, no. 50, p. 21–38. [6] [10]

Willmore, P.L., Hales, A.L., and Gane, P.G., 1952, A seismic investigation of crustal structure in the western Transvaal: Bulletin of the Seismological Society of America, v. 42, p. 53–80. [2] [4] [5]

Wilson, J.M., and Fuis, G.S., 1987, Data report for the Chemehuevi, Vidal, and Dutch Flat lines, PACE, seismic-refraction survey, southeastern California and western Arizona: Menlo Park, California, U.S. Geological Survey Open-File Report 87-86, 75 p. [8]

Wilson, J.M., Meador, P., and Fuis, G.S., 1987, Data report for the 1985 TACT seismic refraction survey, south-central Alaska: Menlo Park, California, U.S. Geological Survey Open-File Report 87-440, 78 p. [8]

Wilson, J.M., McCarthy, J., Johnson, R.A., and Howard, K.A., 1991, An axial view of a metamorphic core complex: crustal structure of the Whipple and Chemehuevi Mountains, southeastern California and western Arizona: Journal of Geophysical Research, v. 96, p. 12,293–12,311, doi:10.1029/91JB01003. [8]

Winardhi, S., and Mereu, R.F., 1997, Crustal velocity structure of the Superior and Grenville provinces of the southeastern Canadian Shield: Canadian Journal of Earth Science, v. 34, p. 1167–1184, doi:10.1139/e17-094. [9]

Wissinger, E.S., Levander, A., and Christensen, N.I., 1997, Seismic images of crustal duplexing and continental subduction in the Brooks Range: Journal of Geophysical Research, v. 102, p. 20,847–20,871, doi:10.1029/96JB03662. [8]

Wittlinger, G., Tapponnier, P., Poupinet, G., Mei, J., Shi, D., Herquel, G., and Masson, F., 1998, Tomographic evidence for localized lithosphere shear along the Altyn Tagh fault: Science, v. 281, p. 74–76, doi:10.1126/science.282.5386.74. [9]

Wittlinger, G., Vergne, J., Tapponnier, P., et al., 2004, Teleseismic imaging of subducting lithosphere and Moho offsets beneath western Tibet: Earth and Planetary Science Letters, v. 221, p. 117–130, doi:10.1016/S0012-821X(03)00723-4. [9]

Wojtczak-Gadomska, B., Guterch, A., and Uchman, J., 1964, Preliminary results of deep seismic soundings in Poland: Bulletin de l'Acadamie Polonaise des Sciences, Série des Sci. Géol. et Géogr., v. XII, no. 4, p. 205–211. [6]

Wolf, L.W., and Cipar, J.J., 1993, Through thick and thin: a new model for the Colorado Plateau from seismic refraction data from Pacific to Arizona crustal experiment: Journal of Geophysical Research, v. 98, p. 19,881–19,894, doi:10.1029/93JB02163. [8]

Wood, H.O., and Richter, C.F., 1931, A study of blasting recorded in southern California: Bulletin of the Seismological Society of America, v. 21, p. 28–46. [2] [3]

Wood, H.O., and Richter, C.F., 1933, A second study of blasting recorded in southern California: Bulletin of the Seismological Society of America, v. 23, p. 95–110. [2] [3]

Woollard, G.P., 1975, The interrelationship of crustal and upper mantle parameter values in the Pacific: Reviews of Geophysics and Space Physics, v. 13, p. 87–137, doi:10.1029/RG013i001p00087. [2] [7] [10]

Woollard, G.P., and Ewing, M., 1939, Structural geology of the Bermuda islands: Nature, v. 143, p. 898, doi:10.1038/143898a0. [2] [3]

Working Group for Deep Seismic Sounding in Spain 1974–1975, 1977, Deep seismic soundings in southern Spain: PAGEOPH, v. 115, p. 721–735. [7]

Worzel, J.L., and Ewing, M., 1948, Propagation of sound in the ocean: Geological Society of America Memoir 27, 35 p. [2] [10]

Wright, C., Goleby, B.R., Collins, C.D.N., Korsch, R.J., Barton, T., Sugiharto, S., and Greenhalgh, S.A., 1990, Deep seismic reflection profiling in central Australia: Tectonophysics, v. 173, p. 247–256, doi:10.1016/0040-1951(90)90221-S. [2] [8]

Wright, J.A., and Hall, J., 1990, Deep seismic probing in the Nosop basin, Botswana: cratons, mobile belts and sedimentary basins, *in* Leven, J.H., Finlayson, D.M., Wright, C., Dooley, J.C., and Kennett, B.L.N., eds., Seismic probing of continents and their margins: Tectonophysics, v. 173, p. 333–343. [2] [8]

Wrinch, D., and Jeffreys, H., 1923, On the seismic waves from the Oppau explosion of 1921, Sept. 21, Monthly Notices: Geophysical Journal of the Royal Astronomical Society, Suppl., v. 1, p. 15–22. [3]

Yamanaka, Y., and Kikuchi, M., 2004, Asperity map along the subduction zone in northeastern Japan inferred from regional seismic data: Journal of Geophysical Research, v. 109, B07307, doi:10.1029/2003JB002683. [10]

Yan, Q.Z., and Mechie, J., 1989, A fine structural section through the crust and lower lithosphere along the axial region of the Alps: Geophysical Journal International, v. 98, p. 465–488, doi:10.1111/j.1365-246X.1989.tb02284.x. [2] [7]

Ye, S., 1991, Crustal structure beneath the central Swiss Alps derived from seismic refraction data [Ph.D. thesis]: Swiss Federal Institute of Technology Zürich (ETH), 114 p. [2] [8]

Ye, S., Flueh, E.R., Klaeschen, D., and von Huene, R., 1997, Crustal structure along the EDGE transect beneath the Kodiak shelf off Alaska derived from OBH seismic refraction data: Geophysical Journal International, v. 130, p. 283–302, doi:10.1111/j.1365-246X.1997.tb05648.x. [2] [9]

Ye, S., Canales, J.P., Rihm, R., Danobeitia, J.J., and Gallart, J., 1999, A crustal transect through the northern and northeastern part of the volcanic edifice of Gran Canaria, Canary islands: Journal of Geodynamics, v. 28, p. 3–26, doi:10.1016/S0264-3707(98)00028-3. [2] [9]

Yilmaz, Ö.,1987, Seismic data processing, *in* Neitzel, E.B., ed., Investigations in Geophysics: Tulsa, Oklahoma, Society of Exploration Geophysics. [2] [10]

Yin, Z., Teng, J., and Liu, H., 1990, The 2-D crustal structure study in the Yadong-Damxung region of the Xizang plateau: Bulletin of the Chinese Academy of Geological Sciences, v. 21, p. 239–245. [9]

Yoon, M.-K., Baykulov, M., Dümmong, S., Brink, H.-J., and Gajewski, D., 2008, Reprocessing of deep seismic reflection data from the North German Basin with the common reflection surface stack: Tectonophysics, doi:10.1016/J.tecto.2008.05.010. [2] [7] [8]

Yoon, M., Buske, S., Lüth, S., Schulze, A., Shapiro, S.A., Stiller, M., and Wigger, P., 2003, Along-strike variations of crustal reflectivity relaeted to the Andean subduction process: Geophysical Research Letters, v. 30, no. 4, 1160, doi:10.1029/2002GL015848. [2] [9]

Yoshii. T., 1994, Crustal structure of the Japanese Islands revealed by explosion seismic observations: Zisin, v. 46, p. 479–491 (in Japanese with English abstract). [2] [7]

Yoshii, T., and Asano, S., 1972, Time-term analyses for explosio seismic data: Journal of the Physics of the Earth, v. 20, p. 47–57. [2] [6] [7]

Yoshii, T., Sasaki, Y., Tada, T., Okada, H., Asano, S., Muramatsu, I., Hashizume, M., and Moriya, T., 1974, The third Kurayoshi explosion and the crustal structure in the western part of Japan: Journal of the Physics of the Earth, v. 22, p. 109–121. [2] [6]

Yoshii, T., Okada, H., Asano, S., Ito, K., Hasegawa, T., Ikami, A., Moriya, T., Suzuki, S., and Hamada, K., 1981, Regionality of the upper mantle around northeastern Japan as revealed by big explosions at sea II. SEIHA-2 and SEHA-3 experiment: Journal of Physics of the Earth, v. 29, p. 201–220. [7]

Yoshii, T., Asano, S., Kubota, S., Sasaki, Y., Okada, T., Moriya, T., and Murakami, H., 1985, Crustal Structure of Izu Peninsula, central Japan, as derived from explosion seismic observations, 2. Ito-Matsuzaki profile: Journal of the Physics of the Earth, v. 33, p. 435–451. [2] [7] [8]

Yuan, X., Wang, S., Li, L., and Zhu, J., 1986, A geophysical investigation of deep structure in China, *in* Barazangi, M., and Brown, L., eds., Reflection seismology: a global perspective: American Geophysical Union, Geodynamics Series, v. 13, p. 151–160. [2] [7] [8]

Yuan, X., Ni, J., Kind, R., Mechie, J., and Sandvol, E., 1997, Lithospheric and upper mantle structure of southern Tibet from a seismological passive source experiment: Journal of Geophysical Research, v. 102, p. 27,491–27,500, doi:10.1029/97JB02379. [9]

Zehnder, C.M., Mutter, J.C., and Buhl, P., 1990, Deep seismic and geochemical constraints on the nature of rift-induced magmatism during breakup of the North Atlantic, *in* Leven, J.H., Finlayson, D.M., Wright, C., Dooley, J.C., and Kennett, B.L.N., eds., Seismic probing of continents and their margins: Tectonophysics, v. 173, p. 545–565. [2] [8]

Zeilhuis, A., and van der Hilst, R.D., 1996, Upper-mantle shear velocity beneath eastern Australia from inversion of waveforms from SKIPPY portable arrays: Geophysical Journal International, v. 127, p. 1–16, doi:10.1111/j.1365-246X.1996.tb01530.x. [2] [9]

Zeis, St., Gajewski, D., and Prodehl, C., 1990, Crustal structure of Southern Germany from seismic-refraction data, *in* Freeman, R., and Mueller, St., eds., The European Geotraverse, Part 6: Tectonophysics, v. 176, p. 59–86. [2] [8]

Zelt, B.C., Ellis, R.M., Clowes, R.M., Kanasewich, E.R., Asudeh, I., Hajnal, Z., Luetgert, J.H., Ikami, A., Spence, G.D., and Hyndman, R.D., 1992, Crust and upper mantle velocity structure of the Intermontane belt, southern Canadian Cordillera: Canadian Journal of Earth Science, v. 29, p. 1530–1548, doi:10.1139/e92-121. [8]

Zelt, B.C., Ellis, R.M., and Clowes, R.M., 1993, Crustal velocity structure in the eastern Insular and southernmost coast belts, Canadian Cordillera: Canadian Journal of Earth Science, v. 30, p. 1014–1027, doi:10.1139/e93-085. [8]

Zelt, B.C., Ellis, R.M., Clowes, R.M., and Hole, J.A., 1996, Inversion of three-dimensional wide-angle seismic data from the southwestern Canadian Cordillera: Journal of Geophysical Research, v. 101, p. 8503–8529, doi:10.1029/95JB02807. [9]

Zelt, C.A., 1998, Lateral velocity resolution from three-dimensional seismic refraction data: Geophysical Journal International, v. 135, p. 1101–1112, doi:10.1046/j.1365-246X.1998.00695.x. [2] [9]

Zelt, C.A., 1999, Modelling strategies and model assessment for wide-angle seismic traveltime data: Geophysical Journal International, v. 139, 183–204, doi:10.1046/j.1365-246X.1999.00934.x. [2] [9] [10]

Zelt, C.A., and Barton, P.J., 1998, Three-dimensional seismic refraction tomography: a comparison of two methods applied to data from the Faeroe Basin: Journal of Geophysical Research, v. 103, p. 7187–7210, doi:10.1029/97JB03536 [9]

Zelt, C.A., and Smith, R.B., 1992, Seismic traveltime inversion for 2-D crustal velocity structure: Geophysical Journal International, v. 108, p. 16–34, doi:10.1111/j.1365-246X.1992.tb00836.x. [2] [9] [10]

Zelt, C.A., and White, D.J. 1995, Crustal structure and tectonics of the southeastern Canadian Cordillera: Journal of Geophysical Research, v. 100, p. 24,255–24,273, doi:10.1029/95JB02632. [9]

Zelt, C.A., Hojka, A.M., Flueh, E.R., and McIntosh, K.D., 1999, 3D simultaneous seismic refraction and reflection tomography of wide-angle data from the central Chilean margin: Geophysical Research Letters, v. 26, p. 2577–2580, doi:10.1029/1999GL900545. [9]

Zelt, C.A., Sain, K., Naumenko, J.V., and Sawyer, D.C., 2003, Assessment of crustal velocity models using seismic refraction and reflection tomography: Geophysical Journal International, v. 153, p. 609–626, doi:10.1046/j.1365-246X.2003.01919.x. [9] [10]

Zeyen, H.J., Banda, E., Gallart, J., and Ansorge, J., 1985, A wide angle seismic reconnaissance survey of the crust and upper mantle in the Celtiberian chain of eastern Spain: Earth and Planetary Science Letters, v. 75, p. 393–402. [2] [7]

Zeyen, H.J., Novak, O., Landes, M., Prodehl, C., Driad, L., and Hirn, A., 1997, Refraction-seismic investigations of the northern Massif Central (France), *in* Fuchs, K., Altherr, R., Müller, B., and Prodehl, C., eds., Stress and stress release in the lithosphere—structure and dynamic processes in the rifts of western Europe: Tectonophysics, v. 275, p. 99–118, doi:10.1016/0012-821X(85)90182-7. [2] [9]

Zhang, P.-Z., and Klemperer, S.L., 2005, West-east variation in crustal thickness in northern Lhasa block, central Tibet, from deep seismic sounding data: Journal of Geophysical Research, v. 110, B09403, doi:10.1029/2004JB003139. [9]

Zhang, S., et al., 1988, Interpretation of the Menyuan-Pingliang-Weinan DSS profile in West China, *in* Developments in the Research of Deep Structures of China's Continent: Beijing, Geological Publishing House, p. 61–88. [8]

Zhang, Z., Yang, L., Teng, J., Badal, J., and Harris, J.M., 2005, Mirror-image symmetry between sedimentary basins and consolidated crust in China: Menlo Park, California, U.S. Geological Survey, Office for Earthquake Studies, unpublished report, 30 p. [8]

Zhao, J., Mooney, W.D., Zhang, X., Li, Z., Jin, Z., and Okaya, N., 2006, Crustal structure across the Altyn Tagh Range at the northern margin of the Tibetan plateau and tectonic implications: Earth and Planetary Science Letters, v. 241, p. 804–814, doi:10.1016/j.epsl.2005.11.003. [2] [9] [10]

Zhao, L.-S., Sen, M.K., Stoffa, P., and Frohlich, C., 1996, Application of very fast simulated annealing to the determination of the crustal structure neneath Tibet: Geophysical Journal International, v. 125, p. 355–370, doi:10.1111/j.1365-246X.1996.tb00004.x. [2] [9]

Zhao, W., Nelson, K.D., and Project INDEPTH Team, 1993, Deep seismic reflection evidence for continental underthrusting beneath Tibet: Nature, v. 366, p. 557–559, doi:10.1038/366557a0. [2] [9]

Zhao, W., Mechie, J., Guo, J., Wenzel, F., Meissner, R., Ratschbacher, L., Steentoft, H., Husen, S., Brauner, H.-J., Jiang, D., Frisch, W., and Hauff, S.-F., 1997, Seismic mapping of crustal structures beneath the Indus-Yarling suture, Tibet: Terra Nova, v. 9, p. 42–46, doi:10.1046/j.1365-3121.1997.d01-7.x. [2] [9]

Zhao, W., Mechie, J., Brown, L.D., Guo, J., Haines, S., Hearn, T., Klemperer, S.L., Ma, Y.S., Meissner, R., Nelson, K.D., Ni, J.F., Pananont, P., Rapine, R., Ross, A., and Saul, J., 2001, Crustal structure of central Tibet as derived from project INDEPTH wide angle seismic data: Geophysical Journal International, v. 145, p. 486–498, doi:10.1046/j.0956-540x.2001.01402.x. [2] [9]

Ziegler, P.A., ed., 1992a, Geodynamics of rifting, volume I. Case history studies on rifts: Europe and Asia: Tectonophysics, v. 215, p. 1–363. [2]

Ziegler, P.A., ed., 1992b, Geodynamics of rifting, volume II. Case history studies on rifts: North and South America and Africa: Tectonophysics, v. 213, p. 1–284. [2]

Ziegler, P.A., ed., 1992c, Geodynamics of rifting, volume III. Thematic discussions: Tectonophysics, v. 215, p. 1–253. [2]

Zoback, M.D., and Wentworth, C.M., 1986, Crustal studies in central California using an 800-channel seismic reflection recording system, *in* Barazangi, M., and Brown, L., eds., Reflection seismology: a global perspective: American Geophysical Union, Geodynamics Series, v. 13, p. 183–196. [8]

Zucca, J.J., and Hill, D.P., 1980, Crustal structure of the southeast flank of Kilauea volcano, Hawaii, from seismic refraction measurements: Bulletin of the Seismological Society of America, v. 70, p. 1149–1159. [7]

Zucca, J.J., Hill, D.P., and Kovach, R.L., 1982, Crustal structure of the Mauna Loa volcano, Hawaii, from seismic refraction and gravity data: Bulletin of the Seismological Society of America, v. 72, p. 1535–1550. [2] [7]

Zucca, J.J., Fuis, G.S., Milkereit, B., Mooney, W.D., and Catchings, R.D., 1986, Crustal structure of northern California from seismic-refraction data: Journal of Geophysical Research, v. 91, p. 7359–7382, doi:10.1029/JB091iB07p07359 [2] [8]

Zurek, B., and Dueker, K., 2005, Lithospheric stratigraphy beneath the Southern Rocky Mountains, USA, *in* Karlstrom, K.E., and Keller, G.R., eds., The Rocky Mountain Region: An evolving lithosphere—tectonics, geochemistry, and geophysics: Washington, D.C., American Geophysical Union Geophysical Monograph 154, p. 317–328. [9]

Zverev, S.M., ed., 1967, Problems in deep seismic sounding: New York-London, Consultants Bureau, 166 p. [6]

Zverev, S.M., and Kosminskaya, I.P., eds., 1980, Seismic models of the lithosphere for the major geostructures on the territory of the USSR: Moscow, Publishing House Nauka, 180 p. [2] [7]

Zverev, S.M., and Tulina, Yu.V., eds., 1971, Deep seismic sounding of the Earth's crust in the Sakhalin-Hokkaido-Primorye zone: Moscow, Nauka, 286 p. (in Russian). [6]

Zverev, S.M., and Tulina, Yu.V., 1973, Pecularities in the deep structure of the Sakhalin-Hokkaido-Primorye zone, *in* Mueller, S., ed., The structure of the earth's crust, based on seismic data: Tectonophysics, v. 20, p. 115–127. [6]

Zverev, S.M., Litvinenko, I.V., Palmason, G., Yaroshevskaya, G.A., Osokin, N.N., and Akhmetjev, M.A., 1980a, A seismic study of the rift zone in northern Iceland, *in* Jacoby, W., Björnsson, A., and Möller, D., eds., Iceland: Evolution, active tectonic, and structure: Journal of Geophysics, v. 47, p. 191–201. [7]

Zverev, S.M., Litvinenko, I.V., Palmason, G., Yaroshevskaya, G.A., and Osokin, N.N., 1980b, A seismic study of the axial rift in southwest Iceland, *in* Jacoby, W., Björnsson, A., and Möller, D., eds., Iceland: Evolution, active tectonic, and structure: Journal of Geophysics, v. 47, p. 202–210. [7]

MANUSCRIPT ACCEPTED BY THE SOCIETY 8 DECEMBER 2010